Diphyllobothrium sp. in water. He noted that *Taenia* eggs do not hatch, thus establishing the principle that helminth eggs may be physiologically different.

1838
–1839
Schleiden and Schwann formulated the cell theory.

1841   G. G. Valentin in Switzerland discovered the first known trypanosome from the blood of the fish *Salmo fario*.

1843   A. Dubini, an Italian physician, described the hookworm *Ancylostoma duodenale* from man. This worm was first known in 1838.

1845   F. Dujardin demonstrated the relationship between cysticerci and *Taenia* adults.

1848   Josiah Nott of New Orleans postulated that mosquitoes transmitted both malaria and yellow fever, thus advancing the vector concept.

1849   G. Gros discovered the first parasitic amoeba of humans, *Entamoeba gingivalis*.

1851   M. Bilharz discovered *Schistosoma haematobium*, *Hymenolepis nana*, and *Heterophyes heterophyes*, which were later described by von Siebold and Bilharz.

1851
–1852
F. Küchenmeister recovered cysticerci from suitable intermediate hosts fed *Taenia* eggs, and recovered adult worms from hosts fed cysticerci. Thus, the life cycle pattern of a cyclophyllidean cestode became established.

1853   C. T. E. von Siebold recovered adult *Echinococcus granulosus* from dogs that had been fed hydatid cysts.

1857   P. H. Malmstem in Stockholm described the first parasitic ciliate of humans, *Balantidium coli*.

1857
–1859
F. R. Leuckart and R. Virchow independently unravelled the life cycle of *Trichinella spiralis*.

1859   **Charles Darwin put forth the theory of organic evolution based on natural selection in "*The Origin of the Species.*"** Claude Bernard demonstrated that parasitic worms contain polysaccharides, thus initiating studies on the biochemistry of parasites.

1865
–1866
Gregor Mendel formulated the basic laws of genetics.

1869   N. M. Melnikov, a Russian zoologist, demonstrated that the dog louse, *Trichodectes canis*, is the intermediate host for the tapeworm *Dipylidium caninum*.

1869   H. Krabbe, a Danish physician, discovered that each order of birds possesses its particular tapeworms, thus suggesting host specificity.

1875   F. A. Lösch discovered *Entamoeba histolytica* in St. Petersburg, Russia.

1878   Patrick Manson, working in China, observed the development of *Wuchereria bancrofti* in the mosquito *Culex quiquefasciatus* and established the essential role of this vector. A year earlier he confirmed the hypothesis that this nematode was the cause of elephantiasis.

1880   C. L. A. Laveran discovered the malaria-causing organism *Plasmodium malariae* (or possible *P. falciparum*) in human erythrocytes, and first saw *P. vivax*, which Grassi and Felletti (1890) named *Haemamoeba vivax*.

1881
–1883
A. P. W. Thomas and K. G. F. R. Leuckart independently worked out the first trematode life cycle experimentally, that of *Fasciola hepatica*.

1882   **E. Metchinkoff, a Russian then working at Messina, recognized the phenomenon of phagocytosis, thus initiating the concept of cellular immunity so important in parasitism. He was awarded the Nobel Prize in 1908.**

1883   G. Bunge, a Swiss chemist, first demonstrated acid production by a parasitic worm, thus was a pioneer in helminth physiology.

1886   Alfred Giard introduced the concept of parasitic castration and its implications in the area of parasite-caused biologic effects.

1892   Johannes Müller of Berlin discovered and named *Entoconcha*, the first known parasitic snail, thus establishing that parasitism occurs among molluscs.

1888   **Charles Ricket discovered humoral immunity when he found that rabbits injected with the blood of another infected with staphylococci became immune. He was awarded the Nobel prize.** (Ricket's finding was expounded upon by K. Landsteiner and P. Ehrlich.)

GENERAL PARASITOLOGY

Second Edition

GENERAL PARASITOLOGY

Second Edition

THOMAS C. CHENG
Medical University of South Carolina
Charleston

Academic Press College Division
Harcourt Brace Jovanovich, Publishers
Orlando   San Diego   San Francisco   New York
London   Toronto   Montreal   Sydney   Tokyo   São Paulo

To Anne

Academic Press, Inc.
Orlando, Florida 32887

United Kingdom Edition Published by Academic Press, Inc.
(London) Ltd., 24/28 Oval Road, London NW1 7DX

ISBN: 0-12-170755-5
Library of Congress Catalog Card Number: 86-70105

Printed in the United States of America

CONTENTS

v

PREFACE

Over a dozen years have passed since the first edition of this textbook was published. As is to be expected, tremendous progress has been made in the study of zooparasites and the nature of parasitism. This is especially true in the case of the protozoans and helminths of medical and economic importance. Although it was recognized that a new edition was sorely needed, the decisions as to what information should be added and what should be deleted were monumental. Continuing the original intent, this book is meant to be a teaching tool rather than a reference volume for seasoned investigators. It is meant to supplement formal lectures, but at the same time to provide students with sufficient information as to where more detailed review articles and primary research reports can be located. For this reason, a number of recent review articles are included in the reference sections at the ends of the chapters.

As a teaching tool, the book had to include certain basic information, although considered to be "classical," to provide background information for students being introduced to parasites and their ways of life. Although the frontiers of modern parasitology are primarily involved in biochemical and immunologic probes on the one hand and quantitative ecologic (including epidemiologic) investigations on the other, we decided that it is essential that certain anatomic, life history, and basic physiologic information be retained, although in many instances in an abbreviated form.

The classification of parasites is integral to understanding these organisms; however, at the request of many colleagues who teach parasitology, the taxonomy of each group considered in this volume has been relegated to the end of each chapter. The material incorporated in the text is presented in a broad taxonomic order. My many years of classroom experience have taught me that this is the most effective approach to introducing students to the large array of parasites. Some, of course, will disagree.

Again, at the advice of colleagues, a concerted effort has been made to provide more information pertaining to species of medical importance. Although the sections devoted to amoebiasis, malaria, schistosomiasis, hookworm disease, and other parasite-caused human diseases have been expanded, this volume is still intended for the biologist rather than the medical or veterinary student. Consequently, detailed discussions of symptomology, pathology, chemotherapy, and control have been omitted. Nevertheless, sufficient information pertaining to those parasites of medical, veterinary, and economic importance (especially in the marine environment) has been presented to permit greater appreciation of their pathobiology (in the broad sense).

The world continues to be a shrinking arena, and a course in parasitology, among other objectives, will permit the student to appreciate why the World Health Organization has included five parasitic diseases—filariasis, schistosomiasis, leishmaniasis, trypanosomiasis, and malaria—among the world's six major, unconquered human diseases. Although it is recognized that research aimed at reducing or eradicating the major parasitic diseases of humans and food-producing animals is essential from the standpoint of human welfare, the study of parasites as unique organisms and the probing of the basic mechanisms underlying the phenomenon of parasitism must be perpetuated as intellectual thrusts if parasitology is to continue as an integral part of the modern study of life and its manifestations. It is with this view in mind that this book was written.

Many years ago when the idea of a general parasitology textbook first came to mind, two of the objectives were to present information that would permit the reader to gain appreciation of the rich history of the discipline, and to gain some feeling as to how widespread parasitism as a way of life is in the animal kingdom. With the continuous pressure to keep the number of pages within reason, the objective of a historical sketch of parasitology has gone by the board, although a chronological tabulation of major discoveries is presented on the front and back end sheets. Even the second objective has suffered to some extent. The less frequently encountered groups of zooparasites have been relegated to abbreviated presentations in the final chapter. However, every attempt has been made to present the remarkable diversity in morphology, ecology, life cycle patterns, physiology, and in instances where known, the biochemistry of the parasitic fauna.

Parasitology, like any aspect of human endeavor, needs the infusion of new ideas and experimental approaches to resolving the numerous unanswered questions. Hopefully, the reading of this textbook coupled with the formal lectures will excite some budding biologists to accept the challenge of entering the field as professionals. History has taught us that many of the major advances in science have been achieved as a result of serendipity. However, to quote Pasteur, ''Chance favors the prepared mind.'' Hopefully, this contribution will assist in the preparation of a few minds.

As stated, the interlude between the first edition and this edition has been over 12 years. This does not mean that there existed a dozen years of freedom and idleness. Considerable time has been spent reading, collecting notes, deciding what changes were warranted in view of the intent of this book, writing, and rewriting, among other duties.

Many persons were kind enough to criticize the several versions of the typescript and make recommendations. Among these, special appreciation is due Dr. Frank J. Etges of the University of Cincinnati, the late Dr. Marietta Voge of the University of California at Los Angeles, Dr. Burton J. Bogitsh of Vanderbilt University, and Dr. Hisao P. Harai of the University of Calgary. If errors exist in this volume, I alone am responsible.

Finally, I wish to acknowledge the hospitality provided by Anne Whitelaw at ''Mount Hope'' in Virginia, where the original versions of several chapters were written under serene conditions.

THOMAS C. CHENG

Charleston, South Carolina

ACKNOWLEDGMENTS

I wish to acknowledge the numerous authors who contributed a large number of the illustrations incorporated in this volume. Although the majority of these, especially the line drawings, have been redrawn from the originals, listed below are the references to the primary sources. In addition, the sources for the tabular material included in this book are also listed. I am grateful to the many publishers, journal editors, and authors who have given me their permission to use figures and tables from their publications.

Achmerov, A. K. (1941). *C. R. Acad. Sci.* URSS **30.**

Aikawa, M., Hepler, P. K., Huff, C. G., and Sprinz, H. (1966). *J. Cell Biol.* **28.**

Aikawa, M., Miller, L. H., Johnson, G., and Rabbege, G. (1978). *J. Cell. Biol.* **77.**

Alexeieff, A. G. (1912). *Arch. Zool. Exp. Gen.* **9.**

Alicata, J. E. (1962). *Canad. J. Zool.* **40.**

Arnold, J. G. (1929). *Zoologica* **23.**

Artigas, P. (1926). *Bol. Biol.* **1.**

Askew, R. R. (1971). "Parasitic Insects." American Elsevier, New York.

Atkins, D. (1933). *J. Mar. Biol. Ass.* U.K. **19.**

Atkinson, G. F. (1889). *Bull. Ala. Agr. Exp. Sta.* (N.S.) **9.**

Augener, H. (1930). *Zool. Anz.* **62.**

Avers, C. J. (1976). "Cell Biology." Van Nostrand, New York.

Bacot, A. W., and Martin, C. J. (1914). *J. Hyg.* **13.**

Badanin, N. V. (1929). *Russ. Zh. Trop. Med.* **7.**

Baer, J. (1952). "Ecology of Animal Parasites." Univ. Illinois Press, Urbana, Illinois.

Baker, E. W., and Strandtmann, R. W. (1948). *J. Parasitol.* **34.**

Baker, E. W., and Wharton, G. W. (1952). "An Introduction to Acarology." Macmillan, New York.

Barrett, J. T. (1983). "Textbook of Immunology." Mosby, St. Louis, Missouri.

Ball, S. J. (1916). *J. Morphol.* **27.**

Balozet, L. (1937). *Arch. Inst. Pasteur Tunis* **26.**

Bashikirova, E. I. (1941). *Tr. Bashkir. Vet. Stants.* **3.**

Bate, C. S., and Westwood, J. O. (1868). "A History of the British Sessile-Eyed Crustacea." Vol. 2. Van Voorst, London.

Baur, A. (1864). *Nova Acta Leopold. Carol.* **31.**

Baylis, H. A. (1929). "A Manual of Helminthology, Medical and Veterinary." Baillière, London.

Baylis, H. A. (1944). *Ann. Mag. Nat. Hist.* **11.**

Baylis, H. A., and Lane. C. (1920). *Proc. Zool. Soc. London,* 1920.

Beachley, R. G., and Bishopp, F. C. (1942). *Va. Med. Month.* **69.**

Beaver, P. C. (1952). *Amer. J. Clin. Pathol.* **22.**

Beaver, P. C., and Danaraj, T. G. (1958). *Amer. J. Trop. Med. Hyg.* **7.**

Beck, D. E., and Braithwaite, L. F. (1960). "Invertebrate Zoology Laboratory Workbook." Burgess, Minneapolis, Minnesota.

Beck, J. W., and Davies, J. E. (1981). "Medical Parasitology," 3rd ed. Mosby, St. Louis, Missouri.

Becker, E. R., and Talbott, M. (1927). *Iowa St. Coll. J. Sci.* **1.**

Bellanti, J. A. (1971). "Immunology." Saunders, Philadelphia, Pennsylvania.

Benham, W. B. (1901). *In* "Treatise on Zoology" (E. R. Lankester, ed.). A. C. Black, London.

Benjamini, E., and Feingold, B. F. (1970). *In* "Immunity to Parasitic Animals" (G. J. Jackson, R. Herman, and I. Singer, eds.), Vol. 2. Appleton-Century-Crofts, New York.

Bennett, H. J. (1936). *Ill. Biol. Monogr.* **15.**

Berge, T. O. (ed.) (1975). U.S. Dept. HEW Publ. No. (CDE) 75-8301.

Berlese, A. (1910). *Redia* **6.**

Bernier, G. M. (1970). *Progr. Allergy* **14.**

Bettendorf, H. (1897). *Zool. Jahrb. Abt. Anat.* **10.**

Bigelow, H. B. (1909). *Mus. Comp. Zool. Mem. Harvard* **37.**

Bird, A. F. (1971). "The Structure of Nematodes." Academic Press, New York.

Bishop, A. (1938). *Parasitology* **30.**

Bishopp, F. C. (1915), *U.S. Dept. Agr. Bull.* **248.**

Bishopp, F. C. (1935). *Annu. Rep. Smithsonian Inst.* 1933.

Blanchard, R. (1888). "Traité du Zoologie Médicale." 2 vols. J.-B. Baillière & fils, Paris.

Blanchard, R. (1891). *Mem. Soc. Zool. Fr.* **4.**

Blauvelt, W. E. (1945) Ph.D. Thesis, Cornell University, Ithaca, New York.

Bocquet, C., and Stock, J. H. (1957). *Proc. Kon. Ned. Akad. Wetensch.* **60.**

Bogitsh, B. J. (1961). *Proc. Helminthol. Soc. Wash.* **28.**

Boldt, M. (1910). *Zool. Anz.* **36.**

Borrer, D. J., and DeLong, D. M. (1954). "An Introduction to the Study of Insects." Holt, Rinehart and Winston, New York.

Bosma, N. J. (1934). *Trans. Amer. Microsc. Soc.* **53.**

Boyd, E. (1948). *Proc. Entomol. Soc. Wash.* **50.**

Brandes, G. P. H. (1899). *Abh. Naturf. Ges. Halle* **21.**

Braun, M. (1900). *Zentr. Bakteriol. Parasitenk. Infektionskr.* **27.**

Braune, R. (1913). *Arch. Protistenk.* **32.**

Brinkman, A. (1927). *Nyt. Mag. Naturvid.* **65.**

Brown, F. (ed.) (1950). "Selected Invertebrate Types." Wiley, New York.

Brumpt, E. J. A. (1900). *Bull. Soc. Zool. Fr.* **25.**

Brumpt, E.J.A. (1910). "Précis de Parasitologie." Masson et Cie, Paris.

Brumpt, E.J.A. (1936). "Précis de Parasitologie," 5th ed. Masson et Cie, Paris.

Brumpt, E.J.A. (1949). "Précis de Parasitologie," 6th ed. Masson et Cie, Paris

Brumpt, E.J.A., and Joyeux, C. (1912). *Bull. Soc. Pathol. Etol.* **5.**

Bryam, J. E., and Fisher, F. M. Jr. (1974). *Tissue and Cell.* **6.**

Burgdorfer, W., and Brinton, L. P. (1975). *Ann. N.Y. Acad. Sci.* **266.**

Bürger, O. (1891). *Zool. Jahrb. Abt. Anat.* **4.**

Burt, D. R. R. (1936). *Ceylon J. Sci.* **19.**

Burton, P. R. (1964). *J. Morphol.* **115.**

Butschli, O. (1800–1889) *In* "Bronns Klassen und Ordnungen des Tierreiches." Vols. 1–3. Akademische Verlagagesellschaft, Leipzig.

Byam, W., and Archibald, R. G. (1921–1923). "The Practice of Medicine in the Tropics," Vols. 1–3. Hodder and Stoughton, London.

Bychowsky, B. E. (1957). "Monogenetic Trematodes, their Systematics and Phylogeny." Amer. Inst. Biol. Sci., Washington, D.C. (English transl. 1961).

Bychowsky, B. J. (1933). *Zool. Anz.* **105.**

Cable, R. M. (1958). "An Illustrated Laboratory Manual of Parasitology." Burgess, Minneapolis, Minnesota.

Camerano, L. (1897). *Mem. Reale Accad. Sci. Torino Ser.* (2) **47.**

Cameron, A. E. (1937). *Highl. Agr. Soc. Scot. Trans. Ser.* (5) **49.**

Cameron, T. W. M. (1923). *J. Helminthol.* **1.**

Cameron, T. W. M. (1956). "Parasites and Parasitism." Wiley, New York.

Carvalho, J. (1942). *J. Parasitol.* **28.**

Caullery, M., and Mesnil, F. (1901). *Arch. Anat. Microsc.* **4.**

Caullery, M., and Mesnil, F. (1903). *Ann. Fac. Sci. Marseilles* **13.**

Caullery, M., and Mesnil, F. (1920). *Bull. Biol. Fr. Belg.* **54.**

Cerfontaine, P. (1896). *Arch. Biol.* **14.**

Chagas, C. (1910). *Mem. Inst. Oswaldo Cruz* **1.**

Chandler, A. C., and Read, C. P. (1961). "Introduction to Parasitology." Wiley, New York.

Chatton, E. (1952). *In* "Traité de Zoologie." (P.P. Grassé, ed.). Masson et Cie, Paris.

Chen. T. T. (1948). *J. Morphol.* **83.**

Cheng, T. C. (1960). *J. Tenn. Acad. Sci.*. **35.**

Cheng, T. C. (1963). *Ann. N.Y. Acad. Sci.* **113.**

Cheng, T. C. (1967). *Advan. Mar. Biol.* **5.**

Cheng, T. C. (1968). *Pacific Sci.* **22.**

Cheng, T. C. (1970). "Symbiosis: Organisms Living Together." Pegasus: Bobbs-Merrill, Indianapolis, Indiana.

Cheng, T. C. (1971). *In* "Aspects of the Biology of Symbiosis" (T. C. Cheng, ed.). University Park Press, Baltimore, Maryland.

Cheng, T. C. (1975). *J. Invert. Pathol.* **26.**

Cheng, T. C., and Bier, J. W. (1972). *Parasitology* **64.**

Cheng, T. C., and James, H. A. (1960). *Trans. Amer. Microsc. Soc.* **79.**

Cheng, T. C., and Thakur, A. (1967). *J. Parasitol.* **53.**

Chitwood, B. G. (1930). *J. Morph. Physiol.* **49.**

Chitwood, B. G. (1931). *Z. Morphol. Ockol. Tiere* **23.**

Chitwood, B. G. (1932). *Z. Parasitenk.* **5.**

Chitwood, B. G., and Chitwood, M. B. (1933). *Z. Zellforsch, Mikrosk. Anat.* **22.**

Chitwood, B. G., and Wehr, E. E. (1933). *Z. Parasitenk.* **7.**

Chitwood, M. B., Valesquez, C., and Salazar, N. G. (1968). *J. Parasitol.* **54.**

Christensen, R. O. (1929). *J. Parasitol.* **16.**

Christie, J. R. (1931). *J. Agr. Res.* **42.**

Christie, J. R. (1936). *J. Agric. Res.* **52.**

Ciurea, I. (1924). *Parasitology* **16.**

Clausen, E. (1915). Thesis, Univ. Neuchâtel.

Cleveland, L. R. (1950). *J. Morphol.* **86.**

Cleveland, L. R. (1950). *J. Morphol.* **87.**

Cleveland, L. R., Hall, S. R., Sanders, E. P., and Collier, J. (1934). *Mem. Amer. Acad. Arts Sci.* **17.**

Coe, W. R. (1902). *Zool. Anz.* **25.**

Cohn, E. (1912). Inaug. Diss., Königsberg.

Comstock, J. H. (1949). "An Introduction to Entomology," 9th ed. Comstock, Ithaca, New York.

Connell, F. H. (1930). *Univ. Calif. Publ. Zool.* **36.**

Cooley, R. A. (1938). *Nat. Inst. Health Bull.* **171.**

Cooley, R. A. (1946). *Nat. Inst. Health Bull.* **187.**

Cooley, R. A., and Kohls, G. M. (1944). *Amer. Midl. Natur. Monogr. Ser.* **1.**

Cooper, A. R. (1914). *Trans. Roy. Can. Inst.* **10.**

Cooper, A. R. (1918) *Ill. Biol. Monogr.* **4.**

Cort, W. W., Hussey, K. L., and Ameel, D. J. (1960). *J. Parasitol.* **46.**

Craig, C. F., and Faust, E. C. (1951). "Clinical Parasitology." Lea & Febiger, Philadelphia, Pennsylvania.

Cram, E. B. (1931). *U.S. Dept. Agr. Tech. Bull.* **227.**

Crowell, R. M. (1960). *Bull. Ohio Biol. Surv.* **1.**

Crusz, H. (1957). *J. Parasitol.* **43.**

Daengsvang, S. (1980). "A Monograph on the Genus *Gnathostoma* & Gnathostomiasis in Thailand." Southeast Asian Medical Information Center, Tokyo, Japan.

Datta, M. N. (1936). *Rec. Ind. Mus.* **38.**
Datta, M. N. (1940). *Rec. Ind. Mus.* **42.**
Daugherty, J. W. (1954). *Proc. Soc. Exp. Biol. Med.* **85.**
Davis, B.D., Dulbecco, R., Eisen, H.N., and Ginsberg, H.S. (1980). "Microbiology," 3rd ed. Harper & Row, New York.
Davis, H. S. (1917). *Bull. U.S. Bur. Fish.* **35.**
Davis, H. S. (1924). *Rep. U.S. Comm. Fish.* 1923, Append. 8.
Dawes, B. (1940). *Parasitology* **32.**
Dawes, B. (1946). "The Trematoda." Cambridge Univ. Press, London.
Dawes, B. (1947). "The Trematodes of British Fishes." Roy. Soc., London.
Delage, Y. (1884). *Arch. Zool. Exp. Gen.* **2.**
de Man, J. G. (1888). *Mem. Soc. Zool. Fr.* **1.**
de Man, J. G. (1907). *Mem. Soc. Zool. Fr.* **20.**
Dennis, E. W. (1932). *Univ. Calif. Publ. Zool.* **36.**
Dietschy, J. M. (1967). *Fed. Proc.* **26.**
Dietz, E. (1909). *Zool. Anz.* **34.**
Dietz, E. (1910). *Zool. Jahrb. Suppl.* **12.**
Dikmans, G., and Mapes, C. R. (1950). *Proc. Helminthol. Soc. Wash.* **17.**
Diller, W. F., Jr. (1928). *J. Morphol. Physiol.* **46.**
Dixon, K. E. (1965). *Parasitology* **55.**
Dixon, K. E. (1966). *Parasitology* **56.**
Dixon, K. E., and Mercer, E. H. (1964). *Quart. J. Microsc. Sci.* **105.**
Dobrovolny, C. G. (1939). *Trans. Amer. Microsc. Soc.* **58.**
Dobrovolny, C. G., and Ackert, J. (1934). *Parasitology* **26.**
Dönges, J. (1964). *Z. Parasitenk.* **24.**
Dogiel, V. (1906). *Mitt. Zool. Sta. Neapel* **18.**
Dogiel, V. (1927). *Arch. Protistenk.* **59.**
Dogiel, V. (1929). *Arch. Protistenk.* **68.**
Dogiel, V. A. (1964). "General Parasitology." Oliver and Boyd, Aberdeen, Scotland. (Transl. by Z. Kabata.)
Dollfus, R. Ph. (1942). *Arch. Mus. Hist. Nat. (Paris)* **19.**
Doucet, J. (1965). *Mem. Off. Rech. Sci. Tech. Outre-Mer.* **14.**
Dougherty, E. C. (1945). *Proc. Helminthol. Soc. Wash.* **12.**
Douthitt, H. (1915). *Ill. Biol. Monogr.* **1.**
Dove, W. E., Hall, D. G., and Hull, J. B. (1932). *Ann. Entomol. Soc. Amer.* **25.**
Dove, W. E. and Shelmire, B. (1932). *J. Parasitol.* **18.**
Dubois, G. (1938). *Mém. Soc. Neuch. Sci. Nat.* **6.**
Duboscq, O., and Collin, B. (1910). *C. R. Acad. Sci. (Paris)* **151.**

Edmondson, W. T. (ed.) (1959). "Freshwater Biology." Wiley, New York.
Ejsmont, L. (1926). *Bull. Int. Acad. Cracovie*, 1926.
Ejsmont, L. (1929). *Bull. Int. Acad. Polon. Sci. Lett. Cracovie C1. Sci. Math. Natur.* **2.**
Ejsmont, L. (1932). *C. R. Soc. Biol.* **110.**
Ejsmont, L. (1936). *Roczn. Prac. Nauk. Zrzesz. Asyst. Univ. J. Pilsudsk. Warszawie* **1.**
Ekbaum, E. K. (1933). *Contr. Can. Biol. Fish.* **8.**
Elliott, A. M. (1968). "Zoology," 4th ed. Appleton-Century-Crofts, New York.
Ellis, M. M. (1912). *Zool. Anz.* **39.**
El Mofty, M. M. (1961). Ph.D. Thesis, Univ. Dublin, Ireland.
El Mofty, M. M., and Smyth, J. D. (1960). *Nature (London)* **186.**

Elton, C., Ford, E., and Backer, I. (1931). *Proc. Zool. Soc. London* 1931.
Essex, H. E. (1927). *Ill. Biol. Monogr.* **11.**
Ewing, H. E. (1921). *U.S. Dept. Agr. Bull.* **986.**
Ewing, H. E. (1944). *J. Parasitol.* **30.**
Eysenhardt, C. G. (1829). *Verh. Berlin Ges. Naturf. Fr.* **1.**

Fain, A. (1961). *Sci. Zool.* **92.**
Fairbairn, D. (1958). *Can. J. Zool.* **36.**
Fairbairn, D. (1960). *In* "Host Influence in Parasite Physiology" (L. A. Stauber, ed.). Rutgers Univ. Press, New Brunswick, New Jersey.
Faust, E. C. (1930). *Parasitology* **22.**
Faust, E. C. (1932). *Quart. Rev. Biol.* **7.**
Faust, E. C. (1937). "Human Helminthology." Williams & Wilkins, Baltimore, Maryland.
Faust, E. C., Meleny, H. E., and Annandale, N. (1924). *J. Parasitol.* **11.**
Faust, E. C., Campbell, H. E., and Kellogg, C. R. (1929). *Amer. J. Hyg.* **9.**
Fawcett, D. W. (1981). "The Cell," 2nd ed. Saunders, Philadelphia, Pennsylvania.
Ferris, G. F. (1916). *Entomol. News* **27.**
Ferris, G. F. (1918). *Can. Entomol.* **50.**
Feyel, T. (1936). *Arch. Anat. Microsc.* **32.**
Finnegan, S. (1931). *Proc. Zool. Soc. London* 1931.
Fischthal, J. H., and Allison, L. N. (1942). *Trans. Amer. Microsc. Soc.* **61.**
Fouquet, D. (1876). *Arch. Zool. Exp. Gen.* **5.**
Fox, I., and García-Moll, I. (1962). *J. Parasitol.* **48.**
Francis, E. (1919). *Pub. Health Rep.* **34.**
Frederickson, D. W. (1978). *J. Parasitol.* **64.**
Fuhrmann, O. (1928). *Bull. Soc. Neuch. Sci. Nat.* **52.**
Fuhrmann, O. (1928). *In* "Kükenthal und Krumbachs Handbuch der Zoologie," Vol. 2. G. Fischer, Jena.
Fuhrmann, O. (1931). *In* "Kükenthal's Handbuch der Zoologie," Vol. 2. G.Fischer, Jena.
Fuhrmann, O. (1932). *Mem. Univ. Neuschâtel* **8.**
Fuhrmann, O. (1933). *Rev. Suisse Zool.* **40.**
Furgason, W. H. (1940). *Arch. Protistenk.* **94.**

Gage, S. H. (1893). "The Lake and Brook Lampreys of New York." Wilder Quarter-Century Book, Ithaca, New York.
Galliard, H. (1934). *Ann. Parasitol.* **12.**
Gallien, L. (1937). *Ann. Parasitol. Hum. Comp.* **15.**
Geiman, Q. M. and Ratcliff, H. L. (1936). *Parasitology* **28.**
Gerstaeker, A. (1881). *In* "Bronns Klassen und Ordnungen des Tierreiches," Vol. 5. Akademische Verlagsgesellschaft, Leipzig.
Giard, A. (1911-1913). "Oeuvres Diverses," Lab d'Evol. Êtres Organ., Paris.
Giard, A., and Bonnier, J. (1893). *Bull. Sci. Fr. Beig.* **25.**
Gibbons, I. R., and Grimstone, A. V. (1960). *J. Biophys. Biochem. Cytol.* **7.**
Goto, S. (1894). *J. Coll. Sci. Tokyo Univ.* **8.**

Gould, S. E. *et al.* (1955). *Amer. J. Pathol.* **31.**

Grabiec, S., Guttowa, A., and Michajlow, W. (1964). *Bull. Acad. Pol. Sci. Cl. II* **12.**

Granata, L. (1925). *Arch. Protistenk.* **50.**

Grandjean, F. (1938). *Bull. Soc. Zool. Fr.* **63.**

Grassi, P.P., and Foà, A. (1911). *Atti Reale Accad. Lincei Ser.* **5.**

Guberlet, J. E. (1928). *J. Helminthol.* **6.**

Guberlet, J. E. (1933). *Pubbl. Sta. Zool. Napoli* **12.**

Haldane, J. B. S., and Huxley, J. (1927). "Animal Biology." Oxford Univ. Press, London.

Haley, A. J., and Winn, H. E. (1959). *Trans. Amer. Fish. Soc.* **88.**

Halkin, H. (1901). *Arch. Biol.* **18.**

Hall, D. G. (1932). *Amer. J. Hyg.* **16.**

Hall, M. C. (1916). *Proc. U.S. Nat. Mus.* **50.**

Hall, M. C. (1919). *Proc. U.S. Nat. Mus.* **55.**

Hall, M. C. (1921). *Proc. U. S. Nat. Mus.* **59.**

Halton, D. W. (1964). Ph.D. Thesis, Univ. of Leeds, England.

Harada, I. (1935). *Mem. Fac. Sci. Agr. Taihoku Imp. Univ.* **14.**

Hargis, W. (1956). *Proc. Helminthol. Soc. Wash.* **28.**

Harper, W. F. (1929). *Parasitology* **21.**

Harris, J. E., and Crofton, H. D. (1957). *J. Exp. Biol.* **34.**

Hart, J. F. (1936). *Trans. Amer. Microsc. Soc.* **55.**

Harwood, R. F., and James, M. T. (1979). "Entomology in Human and Animal Health." MacMillan, New York.

Hasselmann, G. E. (1926). *Bol. Inst. Brasil. Sci.* **2.**

Haswell, W. A. (1893). *Macleay Mem. Vol. Linn. Soc. N.S. Wales.*

Hatt. P. (1931). *Arch. Zool. Exp. Gen.* **72.**

Hauschka, T. S. (1943). *J. Morphol.* **73.**

Hawking, F. (1962). *Ann. N.Y. Acad. Sci.* **98.**

Hein, W. (1904). *Z. Wiss. Zool.* **76.**

Hegner, R. W., and Taliaferro, W. H. (1924). "Human Protozoology," Macmillan, New York.

Hegner, R. W., Root, F. M., and Augustine, D. L. (1929). "Animal Parasitology." Appleton-Century, New York.

Heinze, K. (1937). *Z. Parasitenk.* **9.**

Helfer, H., and Schottke, E. (1935). *In* "Bronns Klassen und Ordnungen des Tierreiches," Vol. 5. Akademische Verlagsgesellschaft, Leipzig.

Herms, W. B. (1950). "Medical Entomology," 4th ed. Macmillan, New York.

Hesse, E. (1909). *Arch. Zool. Exp. Gen.* **3.**

Hesse, R. (1892). *Z. Wiss. Zool.* **54.**

Heymons, R. (1935). *In* "Bronns Klassen und Ordnungen des Tierreiches," Vol. 5. Akademische Verlagsgesellschaft, Leipzig.

Hirst, S. (1922). *Econ. Ser. Brit. Mus. (Nat. Hist.)* **13.**

Hogue, M. J. (1921). *Amer. J. Hyg.* **1.**

Holt, P. C. (1963). *J. Tenn. Acad. Sci.* **38.**

Hortsmann, H. J. (1962). *Z. Parasitenk.* **21.**

Hoskin, G. P., and Cheng, T. C. (1970). *Proc. Symp. Mollusca Part III.* Marine Biol. Ass. India.

Hsiung, T. S. (1930). *Iowa St. Coll. J. Sci.* **4.**

Hughes, R. C. (1927). *Trans. Amer. Microsc. Soc.* **46.**

Hughes, R. C. (1929). *Trans. Amer. Microsc. Soc.* **48.**

Humes, A. G. (1942). *Ill. Biol. Monogr.* **18.**

Humphries, D. A. (1967). *Anim. Behav.* **15.**

Hungerford, H. B. (1919). *J. Parasitol.* **5.**

Hunter, G. W. III. (1927). *Ill. Biol. Monogr.* **11.**

Hunter, G. W. III, Swartzwelder, J. C., and Clyde, D. F. (1976). "Tropical Medicine." 5th ed. Saunders, Philadelphia, Pennsylvania.

Hunter, W. S., and Vernberg, W. B. (1955). *Exp. Parasitol.* **4.**

Hyman, L. H. (1940). "The Invertebrates: Protozoa through Ctenophora." McGraw-Hill, New York.

Hyman, L. H. (1951). "The Invertebrates: Acanthocephala, Aschelminthes, and Entoprocta." McGraw-Hill, New York.

Hyman, L. H. (1951). "The Invertebrates: Platyhelminthes and Rhynchocoela." McGraw-Hill, New York.

Ingles, L. G. (1936). *Trans. Amer. Microsc. Soc.* **55.**

Ito, S., Vinson, J. W., and McGuire, T. J. Jr. (1975). *Ann. N.Y. Acad. Sci.* **266.**

Iwanow, A. W. (1937). *Acta Zool.* **18.**

Jahn, T.L., and Kuhn, L. (1932). *Biol. Bull.* **62.**

James, B. L. Bowers, E. A., and Richards, J. G. (1966). *Parasitology* **56.**

James, M.T. (1947). *U.S. Dept. Agr. Misc. Pub. No. 631.*

James, M.T., and Harwood, R.F. (1969). "Herms's Medical Entomology," 6th ed. Macmillan, New York.

Janicki, C. (1928). *Rabot. Volzhsk. Biol. Stantsii* **10.**

Janicki, C., and Rasin, K. (1930). *Ztschr. Wissensch. Zool.* **136.**

Jacques, H. E. (1947). "How to Know the Insects." Brown, Dubuque, Iowa.

Jobling, B. (1939). *Parasitology* **31.**

Johnson, P. T., and Hertig, M. (1970). *Exp. Parasitol.* **27**

Johnston, T. H. (1931). *Aust. J. Exp. Biol. Med. Sci.* **8.**

Johnstone, J. (1911). *Rep. Liverpool Sea Fish, Invest. Year* 1911

Jones, F. G. W. (1959). *In* "Plant Pathology: Problems and Progress 1908–1958." Univ. Wisconsin Press, Madison.

Joyeux, C.E. (1927). *Bull. Inst. Clin. Quir.* **3.**

Joyeux, C. E., and Baer, J. G. (1936). *Bull Soc. Pathol. Exot.* **29.**

Kalantaryan, E. V. (1928). *Tr. Gos. Inst. Eksper. Vet. Moskva* **5.**

Kasschau, M. R., and Mansour, T. E. (1982). *Nature* **296.**

Kates, K. C. (1943). *Amer. J. Vet. Res.* **4.**

Katz, M., Despommier, D.D., and Gwadz, R. (1982). "Parasitic Diseases." Springer-Verlag, New York.

Kearn, G. C. (1962). *J. Mar. Biol. Ass. U.K.* **42**

Kearn, G. C. (1963). *J. Mar. Biol. Ass. U.K.* **43**

Kearn, G. C. (1971). *In* "Ecology and Physiology of Parasites" (A.M. Fallis, ed.). Univ. Toronto Press, Toronto.

Keilin, D., and Nuttall, G.H.F. (1930). *Parasitology* **22.**

Keilen, D., and Robinson, V. (1933). *Parasitology* **25.**

Kenney, M. (1973). "Scope Monograph on Pathoparasitology." Upjohn, Kalamazoo, Michigan.

Kessell, Q. C. (1924). *Parasitology* **16.**

Kirby, H. (1926). *Univ. Calif. Publ. Zool.* **29.**

Kirby, H. (1932). *Univ. Calif. Publ. Zool.* **37.**

Kirby, H., and Honigberg, B. (1950), *Univ. Calif. Publ. Zool.* **55**

Klein, B. M. (1932). *Ergeb. Biol.* **8.**

Koehler, R., and Vaney, C. (1912). *Bull. Sci. Fr. Belg.* **46.**

Kofoid, C. A., and MacLennan, R. F. (1932). *Univ. Calif. Publ. Zool.* **39.**

Kolbe, H. J. (1889). "Einführung in die Kenntnis der Insekten." F. Dümmler, Berlin.

Komai, T. (1922). "Studies on Two Aberrant Ctenophores—*Coeloplana* and *Gastrodes.*" Publ. by author, Kyoto.

Komiya, Y. (1939). *Z. Parasitenk.* **10.**

Kozloff, E. N. (1948). *J. Morphol.* **83.**

Kozloff, E. N. (1969). *J. Parasitol.* **55.**

Kreis, H. A. (1929). *Capita Zool.* **2.**

Kröyer, H. N. (1837). *Naturh. Tidsskr.* **1.**

Kruidenier, F. J. (1953). *J. Morphol.* **92.**

Kudo, R. R. (1913). *Zool. Anz.* **41**

Kudo, R. R. (1921). *J. Morphol.* **35.**

Kudo, R. R. (1921). *J. Parasitol.* **7.**

Kudo, R. R. (1922). *Parasitology* **14.**

Kudo, R. R. (1926). *Arch. Protistenk.* **53.**

Kudo, R. R. (1934). *Ill. Biol. Monogr.* **13.**

Kudo, R. R. (1936). *Arch. Protistenk.* **87.**

Kudo, R. R. (1943). *J. Morphol* **72.**

Kudo, R. R. (1966). "Protozoology," 5th ed. Thomas, Springfield, Illinois.

Kudo, R. R., and Daniels, E. W. (1963). *J. Protozool.* **10.**

Kühn, A. (1921). Morphologie der Tiere in Bildern. I. Flagellaten." G. Borntraeger, Berlin.

Kükenthal, W. (1931). "Handbuch der Zoologie. Eine Naturgeschichte der Stamme des Tierreiches," Vols. 1–3. G. Fischer, Jena.

Laake, E.W., Cushing, E.C. and Parish, H.E., (1936). *U.S. Dept. Agr. Tech. Bull.* **500.**

Ladda, R., Aikawa, M., and Sprinz, H. (1969). *J. Parasitol.* **55.**

Laird, M. (1958). *Can J. Zool.* **36.**

Lameere, A. (1916). *Bull. Biol. Fr. Belg.* **50.**

Lameere, A. (1936). *Rec. Inst. Zool. Torley-Rousseau* **6.**

Lapage, G. (1935). *Parasitology* **27.**

Lapage, G. (1958). "Parasitic Animals." W. Heffer, Cambridge, England.

Lebour, M. V. (1908). *Rep. Sci. Invest. Northumberland Sea Fish. Comm.* 1907.

Lebour, M. V. (1911). *Parasitology* **4.**

Lebour, M. V. (1916). *J. Mar. Biol. Ass. U.K.* **11.**

Lee, D. L. (1965). "The Physiology of Nematodes." Oliver and Boyd, Edinburgh.

Lee, D. L. (1965). *Parasitology* **55.**

Lee, D. L. (1966). *Advan. Parasitol.* **4.**

Leeson, H. S. (1941). *Parasitology* **33.**

Léger, L (1892). *Tablettes Zool.* **3.**

Léger, L. (1926). *Trav. Lab. Hydro. Pisc.* **18.**

Léger, L., and Duboscq, O. (1902). *Arch. Parasitol.* **6.**

Léger, L., and Duboscq, O. (1925). *Trav. Sta. Zool. Wimereux* **9.**

Léger, L., and Hesse, E. (1916). *C.R. Soc. Biol.* **79**

Leidy, J. (1853). *Trans. Amer. Phil. Soc. [N.S.]* **10.**

Leiper, R. T. (1913). *Trans. Roy. Soc. Trop. Med. Hyg.* **6.**

Lennon, R. E. (1954). *Fish. Bull.* **98**

Leuckart, R. (1860). "Bau und Entwick lungsgeschichte der Pentastomen nach Untersuchungen besonders von *Pent. taenioides* und *P. denticulatum.*" C.F. Winter'sche Verlagshandlung, Leipzig.

Linton, E. (1942). *Proc. U.S. Nat. Mus.* **64**

Lipin, A. (1911). *Zool. Jahrb., Abt. Anat. Ont. Tiere* **42.**

Llewellyn, J. (1963). *Advan. Parasitol.* **1.**

Lloyd, J. H. (1920). *Proc. Zool. Soc. London,* 1920.

Lönneberg, E. (1889). *Bihang, K. Sv. Vet. Akad. Handl.* **14.**

Löser, E. (1965). *Z. Parasitenk.* **25.**

Long, L. H., and Wiggins, N. E. (1939). *J. Parasitol.* **25.**

Looss, A. (1894). *Bibl. Zool.* **6.**

Looss, A. (1896). *Mem. Inst. Egypt.* **3.**

Looss, A. (1899). *Zool. Jahrb. Syst.* **12.**

Looss, A. (1902). *Zentralbl. Bakteriol. Parasitenk. Infektionskr. Abt. 1 Orig.* **31.**

Lühe, M. F. L. (1900). *Zentralbl. Bakteriol. Parasitenk. Infektionskr. Abt. 1* **28.**

Lühe, M. F. L. (1909). *In* "Die Susswasserfauna Deutschlands" (A. Brauer, ed.), Vol. 17. G. Fischer, Jena.

Lumsden, R. D. (1966). *Z. Parasitenk.* **27.**

Lumsden, R. D., Oaks, J. A., and Alworth, W. L. (1970). *J. Parasitol.* **56.**

Lynch, J. E. (1933). *Quart J. Microsc. Sci. [N.S.]* **76.**

Lynch, J. E. (1945). *J. Parasitol.* **31.**

Lynch, J. E., and Noble, A. E. (1931). *Univ. Calif. Publ. Zool.* **36.**

Lyons, K. M. (1972). *In* "Ecology and Physiology of Parasites." (A.M. Fallis, ed.). Univ. Toronto Press, Toronto, Canada.

McCaig, M. L. O., and Hopkins, G. A. (1965). *Parasitology* **55.**

McConnaughey, B. H. (1951). *Univ. Calif. Publ. Zool.* **55.**

McIntosh, A. (1940). *Proc. Helminthol. Soc. Wash.* **7.**

Mackin, J. G., Owen, H. M., and Collier, A. (1950). *Science* **111.**

MacLannan, R. F. (1944). *Trans. Amer. Microsc. Soc.* **63.**

MacLaren, N. (1904). *Jena Z. Med. Naturwiss.* **38.**

Maggenti, A. (1981). "Nematology." Springer-Verlag, New York.

Malek, E. A., and Cheng, T. C. (1974). "Medical and Economic Malacology." Academic Press, New York.

Manter, H. W. (1947). *Amer. Midl. Natur.* **38.**

Maplestone, P. A. (1930). *Rec. Ind. Mus.* **32.**

Marchal, P. (1904). *Arch. Zool. Exp. Gen.* **2.**

Marcus, E. (1929). *In* "Bronns Klassen und Ordnungen des Tierreiches," Vol. 5, Akademische Verlagsgesellschaft, Leipzig.

Markewitsch, A. P. (1957). "Parasitic Copepoda of Fish of USSR." Moscow. (In Russian).

Marlatt, C. L. (1934). *Farmer's Bull.* **754.**

Martiis, L. (1911). *Arch. Protistenk.* **23.**

Martin, W. E. (1969). *Biol. Bull.* **137.**

Matheson, R. (1950). "Medical Entomology." Comstock, Ithaca, New York.

Mavor, J. W., (1915). *J. Parasitol.* **2.**

May, H. (1919). *Ill. Biol. Monogr.* **5.**

Mehra, H. R. (1934). *Bull. Acad. Sci. United Prov Allahabad* **3.**

Meinert, F. (1892). *Ent. Medd. Kjøbenhavn* **3.**

Merton, H. (1913). *Abh. Senckenberg Naturforsch. Ges.* **35.**

Metcalf, C. L., and Flint, W. P. (1962). "Destructive and Useful Insects." McGraw-Hill, New York.

Metcalf, M. M. (1932). *Bull. U.S. Nat. Mus.* **120.**

Meves, F. (1920). *Arch. Mikrosk. Anat. Entwicklungsmech.* **94.**

Meyer, A. (1931). *Arch. Zellforsch. Mikro. Anat.* **14.**

Michael, A. D. (1903). "British Tyroglyphidae," Vol. 2. Roy. Soc., London.

Millemann, R. E., and Knapp, S. E. (1970). *In* "A Symposium on Diseases of Fishes and Shellfishes" (S. F. Snieszko, ed.). Amer. Fish. Soc., Washington, D.C.

Minchin, E. A. (1912). "An Introduction to the Study of the Protozoa, with Special Reference to the Parasitic Forms." Arnold, London.

Miyazaki, I. (1960). *Exp. Parasitol.* **9.**

Mönnig, H.O. (1934). "Veterinary Helminthology and Entomology." Baillière, London.

Mönnig, H. O. (1949). *Wolboer* **21.**

Montgomery, T. (1898). *Bull. Mus. Comp. Zool. Harvard* **32.**

Montgomery, T. (1903). *Zool. Jahrb. Abt. Anat.* **18.**

Monticelli, F. S. (1892). Festschrift 70 Geburtstag R. Leuckart. Engelmann, Leipzig.

Moore, D. V. (1946). *J. Parasitol.* **32.**

Moore, J. P. (1959). *In* "Freshwater Biology" (W. T. Edmondson, ed.). Wiley, New York.

Moorthy, V. N. (1937). *J. Parasitol.* **23.**

Morgan, B. B., Hawkins, P. A. (1949). "Veterinary Helminthology." Burgess, Minneapolis, Minnesota.

Mori, T. (1935). *Dobutsugaku Zasshi* **47.**

Morris, G. P., and Halton, D. W. (1971). *J. Parasitol* **57.**

Morseth, D. J. (1967). *J. Parasitol.* **53.**

Mosgovey, A. A. (1951). *Tr. Gelmintol. Lab. Akad. Nauk SSSR* **5.**

Müller, G. (1927). *Z. Morphol. Okol. Tiere* **7.**

Mueller, J. F. (1963). *Ann. N.Y. Acad. Sci.* **113.**

Mueller, J. F. (1965). *J. Parasitol.* **51.**

Mueller, J. F., and Van Cleave, H. J. (1932). *Roosevelt Wildl. Ann. 3, No.* **2.**

Müller-Eberhard, H.J. (1969). *Ann. Rev. Biochem.* **38.**

Murray, J. (1905). *Trans. Roy. Soc. Edinburgh* **41.**

Nagakura, K. (1930). *Jap. J. Zool.* **3.**

Nagaty, H. F. (1937). *Egypt. Univ. Fac. Med. Publ.* **12.**

Neal, T. J., and Barnett, H. C. (1961). *Ann. Entomol. Soc. Amer.* **52.**

Nelson, G. S. (1962). *J. Helminthol.* **36.**

Netter, F. H. (1962). "Digestive System." Ciba Pharm. Co.

Neveu-Lemaire, M. (1936). "Traité d'Helminthologies Médicale et Vétérinaire." Vigot, Paris.

Newstead, R., and Potts, W. H. (1925). *Ann. Trop. Med. Parasitol.* **19.**

Nicoll, W. (1911). *Proc. Zool. Soc. London* **2.**

Nigrelli, R. F. (1936). *Zoologica* **21.**

Noble, A. E. (1936). *J. Parasitol.* **22.**

Noble, E. R., and Noble, G. A. (1971). "Parasitology: The Biology of Animal Parasites." 3rd ed. Lea & Febiger, Philadelphia, Pennsylvania.

Noble, G. A. (1975). *J. Parasitol.* **61.**

Nouvel, H. (1933). *Ann. Inst. Oceanogr. (Monaco)* [N.S.] **13.**

Nuttall, G. H. F., and Warburton, C. (1908). *Proc. Cambridge Phil. Soc.* **14.**

Nybelin, O. (1917). *Kgl. Sv. Vetenskapsakad. Handl.* **53.**

Nybelin, O. (1922). *Goteborgs Kgl. Vetenskapsakad. Handl.* **26.**

Nybelin, O. (1931). *In* "Natural History of Juan Ferandez and Easter Island" (C. Skottsberg, ed.), Vol. 3, Uppsala.

Nyberg, P. A., Bauer, D. H., and Knapp, S. E. (1968). *J. Protozool.* **15.**

Odhner, T. (1902). *Zentrbl. Bakteriol. Parasitenk, Infetionskr.* **1.**

Odhner, T. (1910). *Zool. Anz.* **35.**

Olsen, O. W. (1974). "Animal Parasites: Their Biology and Life Cycles," 2nd ed. University Park Press, Baltimore, Maryland.

Onji, Y., and Nishio, T. (1915). *Iji Shinbun. No.* **949.**

Ortlepp, R. J. (1922). *Rec. Zool. Soc. London,* 1922.

Owen, Nemanic, Stevens (1979).

Ozaki, Y. (1934). *Proc. Imp. Acad. Tokyo* **10.**

Palombi, A. (1934). *Pubbl. Sta. Zool. Napoli* **14.**

Pan, C. T. (1965). *Amer. J. Trop. Med. Hyg.* **14.**

Paterson, N. F. (1958). *Parasitology* **48.**

Paul, A. A. (1938). *J. Parasitol.* **24.**

Pence, D. B. (1967). *J. Parasitol.* **53.**

Penn, G. H. Jr. (1942). *J. Parasitol.* **28.**

Pennak, R. W. (1953). "Freshwater Invertebrates of the United States." Ronald Press, New York.

Perkins, F. O. (1969). *J. Invert. Pathol.* **13.**

Perkins, F. O., and Castagna, M. (1971). *J. Invert. Pathol.* **17.**

Perkins, F. O., Zwerner, D. E., and Dias, R. K. (1975). *J. Parasitol.* **61.**

Perkins, R. C. L. (1905). *Hawaii Sugar Planters' Ass. Exp. Sta. Bull.* **1.**

Petrov, A. (1930). *Zool. Anz.* **86.**

Phifer, K. (1960). *J. Parasitol.* **46.**

Pierce, W. D. (1909). *U.S. Nat. Mus. Bull.* **66.**

Pierce, W. D. (1918). *Proc. U.S. Nat. Mus.* **54.**

Pitelka, D. R. (1970). *J. Protozool.* **17.**

Plate, L. (1886). *Z. Wiss. Zool.* **43.**

Plate, L. (1914). *Z. Naturwiss.* **51.**

Plahn, M. (1905). *Zool. Anz.* **29.**

Poinar, G. O. Jr. (1975). "Entomogenous Nematodes." E. J. Brill, Leiden, Netherlands.

Poluszynski, G. (1930). *Tieralrztl. Rundsch.* **36.**

Porchet-Hennère, E., and Richard, A. (1971). *J. Protozool.* **18.**

Porter,A. (1942). *Proc. Zool. Soc. London,* 1941.

Potter, C. C. (1937). *Proc. Helminthol. Soc. Wash.* **4.**

Pratt, H. S. (1903). *Mark Ann. Vol. Art.* **2.**

Price, E. W. (1929). *Proc. U.S. Nat. Mus.* **75.**

Price, E. W. (1931). *J. Parasitol.* **18.**

Price, E. W. (1932). *J. Parasitol.* **19.**

Price, E. W. (1934). *J. Wash. Acad. Sci.* **24.**

Price, E. W. (1935). *Proc. Helminthol. Soc. Wash.* **2.**

Price, E. W. (1938). *J. Wash. Acad. Sci.* **28.**
Price, E. W. (1940). *Proc. Helminthol. Soc. Wash.* **7.**

Raevskaya, Z. A. (1931). *Ztschr. Infektionskr. Haustiere* **40.**
Railliet, A. (1884). *Bull. Mem. Soc. Central. Med. Vet.* **2.**
Railliet, A. (1895). "Traité de Zoologie Médicale et Agricole." Asselin and Houzeau, Paris.
Ransom, B. H. (1909). *Bull. U.S. Nat. Mus.* **69.**
Ransom, B. H. (1911). *U.S. Dept. Agr. Bur. Animal Ind. Circ.* **127.**
Rauther, M. (1914). *Zool. Anz.* **43.**
Ray, S. M., and Chandler, A. C. (1955). *Exp. Parasitol.* **4.**
Read, C. P. (1956). *Exp. Parasitol.* **5.**
Read, C. P., and Simmons, J. E., Jr. (1963). *Physiol. Rev.* **43.**
Read, C. P., and Voge, M. (1954). *J. Parasitol.* **40.**
Rees, B. (1960). *Parasitology* **50.**
Rees, F. G. (1940). *Parasitology* **32.**
Rees, F. G. (1943). *Parasitology* **35.**
Reich, P. R. (1976). "Scope Manual of Hematology," 5th ed. Upjohn, Kalamazoo, Michigan.
Reichenow, E. (1919). *Sitzungsber. Naturf. Fr. Berlin* **2.**
Reichenow, E. (1923). *Arch. Schiffs. Trop. Hyg.* **27.**
Remane, A. (1929). *In* "Die Tierwelt der Nord- und Ostsee" (J. G. Grimpe and E. Wagler, eds.), Lief 16. Akademische Verlagsgesellschaft, Leipzig.
Ridley, R. K. (1968). *J. Parasitol.* **54.**
Ridley, R. K. (1969). *J. Parasitol.* **55.**
Riehm, G. (1881). *Z. Ges. Naturwiss.* **54.**
Riepen, O. (1933). *Z. Wiss. Zool.* **143.**
Ringo, D. L. (1963). *J. Protozool.* **10.**
Robinson, D. L. H. (1960). *Ann. Trop. Med. Parasitol.* **54.**
Römer, F. (1896). *Abh. Senckenberg Naturforsch. Ges.* **23.**
Rogers, T. D., Scholes, V. E., and Schlichting, H. E. (1972). *J. Protozool.* **19.**
Rogers, W. P. (1960). *Proc. Roy. Soc.* **B152.**
Rogers, W. P. (1966). *In* "Biology of Parasites" (E. J. L. Soulsby, ed.). Academic Press, New York.
Rohde, K. (1971). *Zool. Jb. Anat. Bd.* **88.**
Rohde, K. (1973). *Int. J. Parasit.* **3.**
Rose, J. H. (1960). *Res. Vet. Sci.* **1.**
Ross, I. C., and McKay, A. C. (1929). *Bull. Counc. Sci. Ind. Res. Comm. Aust.* **43.**
Rothschild, M., and Clay, T. (1952). "Fleas, Flukes and Cuckoos. A Study of Bird Parasites." Collins, London.
Rudzinska, M. A., and Trager, W. (1977). *Canad. J. Zool.* **55.**
Ruszkowski, J. S. (1932). *Bull. Int. Akad. Polon. Sci. Lett. Cl. Sci. Math. Nat. Ser.* **B7.**

Sambon, L. W. (1922). *J. Trop. Med. Hyg.* **25.**
Sanders, E. P., and Cleveland, L. R. (1930). *Arch. Protistenk.* **70.**
Sarasin, P., and Sarasin, F. (1887). *Ergeb. Forschung. Ceylon* **1.**
Sars, G. O. (1885). *Rep. Sci. Res. Vuy. H.M.S. Challenger* **13.**
Savel, J. (1955). *Rev. Pathol. Comp.* **55.**
Schauinsland, H. (1886). *Sitzbungsber Ges. Morphol. Physiol. Munch.* **2.**
Scheer, B. (1953). "Comparative Physiology." Wiley, New York.
Scheltema, R. S. (1962). *J. Parasitol.* **48.**
Schmidt, G. D. (1970). "How to Know the Tapeworms." Wm. C. Brown, Dubuque, Iowa.

Schmidt, G. D., and Roberts, L. S. (1981). "Foundations of Parasitology," 2nd ed. Mosby, St. Louis, Missouri.
Scholtyseck, E. O. (1979). "Fine Structure of Parasitic Protozoa." Springer-Verlag, New York.
Schrader, F. (1921). *J. Parasitol.* **7.**
Schuberg, A. (1910). *Arb. Kaiserl. Biol. Land-Forstwert.* **33.**
Schuster, F. L. (1968). *J. Parasitol.* **54.**
Scott, A. (1904). *Trans. Liverpool Biol. Soc.* **18.**
Scott, D. B., Nylen, M. U., von Brand, T., and Pugh, M. H. (1962). *Exp. Parasitol.* **12.**
Sedar, A. W., and Porter, K. R. (1955). *J. Biophys. Biochem. Cytol.* **1.**
Self, J. T. (1971). *Trans. Am. Microsc. Soc.* **90.**
Self, J. T., and Kuntz, R. E. (1957). *J. Parasitol.* **43.**
Senaud, J., and Černá, Z. (1969). *J. Protozool.* **16.**
Seurat, L. G. (1916). *C.R. Soc. Biol.* **79.**
Shakhtakhtinskaia, Z. (1949). *Tr. Gelmintol. Lab. Akad. Nauk SSSR* **2.**
Shipley, A. E., and Hornell, J. (1906). *In* "Herdman's Report to the Government of Ceylon on the Pearl Oyster Fisheries of the Gulf of Manaar," Vol. 11. Roy. Soc. London.
Short, R. B. (1961). *J. Parasitol.* **47.**
Shumway, W. (1924). *J. Parasitol.* **11.**
Siedlecki, M. (1902). *Bull. Int. Acad. Sci. Cracovie* **8.**
Silvestri, F. (1937). *Mus. Comp. Zool. Bull. Harvard* **81.**
Smart, J. (1943). "A Handbook for the Identification of Insects of Medical Importance." British Museum (Natural History), London.
Smit, B. (1931). *Onderstepoort J. Vet. Sci. Anim. Ind.* **17.**
Smith, G. W. (1910). *Quart. J. Microsc. Sci.* **55.**
Smith, G. W. (1911). *Quart. J. Microsc. Sci.* **57.**
Smith, T., and Johnson, H. P. (1902). *J. Exp. Med.* **6.**
Smyth, J. D. (1976). "Introduction to Animal Parasitology," 2nd ed. Hodder & Stoughton, Kent, England.
Smyth, J. D. (1963). Tech. Comm. No. 34, Commonwealth Bureau of Helminthol. St. Albans, England.
Smyth, J. D., and Halton, D. W. (1983). "The Physiology of Trematodes." Cambridge Univ. Press, London.
Smyth, J. D. (1969). "The Physiology of Cestodes." Oliver and Boyd, Edinburgh.
Smyth, J. D. (1969). *Parasitology* **59.**
Smyth, J. D., Miller, H. J., and Hawkins, A. B. (1967). *Exp. Parasitol.* **21.**
Snodgrass, R. E. (1935). "Principles of Insect Morphology." McGraw-Hill, New York.
Southwell, T. (1925). *Ann. Trop. Med. Parasitol.* **19.**
Southwell, T. (1928). *Ann. Trop. Med. Parasitol.* **22.**
Southwell, T. (1930). *In* "Fauna of British India," Vol. I.
Sponholtz, G. M., and Short, R. B. (1976). *J. Parasitol.* **62.**
Sprague, V., Beckett, R. L., and Sawyer, T. K. (1969). *J. Invert. Pathol.* **14.**
Stabler, R. M. (1941). *J. Morphol. Physiol.* **69.**
Stabler, R. M., and Chen, T. T. (1936). *Biol. Bull.* **70.**
Stafford, J. (1905). *Zool. Anz.* **28.**
Stauber, L. A. (1945). *Biol. Bull.* **88.**
Stekhoven, J. H. (1927). *Proc. Sect. Sci. K. Akad. Wetensch. Amsterdam* **30.**

Stekhoven, J. H., and Teuniseen, R. (1938). *Explor. Parc Nat. Albert Mission de Witte* **22.**
Stekhoven, J. H., Adam, W., and de Coninck, L. A. (1933). *Mem. Mus. Roy. Hist. Nat. Belg.* **58.**
Stephens, W. M. (1968). *Oceans* **1.**
Steuer, A. (1902). *Zool. Anz.* **25.**
Stiles, C. W. (1896). *Proc. U.S. Nat. Mus.* **19.**
Stiles, C. W. (1906). *U.S. Publ. Health Serv. Hyg. Lab. Bull.* **25.**
Stiles, C. W. (1906). *U.S. Publ. Health Serv. Hyg. Lab. Bull.* **40.**
Stiles, C. W., and Goldberger, J. (1910). *U.S. Publ. Health Serv. Hyg. Lab. Bull.* **60.**
Stirewalt, M. A., and Evans, A. S. (1955). Proj. Rep. NM 005 048-02-32. Naval Med. Res. Inst., Bethesda, Maryland.
Stirewalt, M. A., and Kruidenier, F. J. (1961). *Exp. Parasitol.* **11.**
Stiven, A. E. (1965). *J. Invert. Pathol.* **7.**
Storer, T. I. (1972). "General Zoology," 5th ed. McGraw-Hill, New York.
Storer, T. I., and Usinger, R. L. (1968). "Elements of Zoology," 3rd ed. McGraw-Hill, New York.
Strandtmann, R. W. (1948). *J. Parasitol.* **34.**
Strauss, P. R. (1971). *J. Protozool.* **18.**
Strickland, G. T., (ed.) (1984). "Hunter's Tropical Medicine." 6th ed. Saunders, Philadelphia, Pennsylvania.
Stunkard, H. W. (1922). *Amer. Mus. Nov.* **39.**
Stunkard, H. W. (1926). *Anat. Rec.* **165.**
Stunkard, H. W. (1940). *Amer. J. Trop. Med.* **20.**
Sutherland, J. L. (1933). *Quart. J. Microsc. Sci.* **76.**
Swezy, O. (1923). *Univ. Calif. Publ. Zool.* **20.**
Szidat, L. (1955). *Arch. Hydrobiol.* **50.**

Tanabe, H. (1922). *Okayama Igakkai Zasshi* **385.**
Taylor, C. V. (1920). *Univ. Calif. Publ. Zool.* **19.**
Thakur, A. S., and Cheng, T. C. (1968). *Parasitology* **58.**
Thapar, G. S. (1925). *J. Helminthol.* **3.**
Thapar, G. S. (1930). *Annu. Mag. Nat. Hist.* **6.**
Theodor, O. (1967). "An Illustrated Catalogue of the Rothschild Collection of Nycteribiidae." British Museum (Natural History), London.
Thomas, A. P. (1883). *Quart. J. Microsc. Soc.* **23.**
Thomas, L. J. (1929). *J. Parasitol.* **15.**
Thomas, L. J. (1931). *J. Parasitol.* **17.**
Törnquist, N. (1931). *Goeteborgs. Kugl. Vetensk.-Och. Vett.-Samh. Handl. Ser.* **B2.**
Toryu, Y. (1933). *Cloq. Sci. Rep. Tohoku Imp. Univ. 4s(Biol.)* **8.**
Trägårdh, I. (1946). *Lunds. Univ. Arsskr. Afd.* **42.**
Trägårdh, I. (1950). *Ark. Zool. (Ser. 2)* **25.**
Trager, W. (1953). *Ann. N.Y. Acad. Sci.* **56.**
Travassos, L. (1937). *Mem. Inst. Oswaldo Cruz.* **32.**
Triffitt, M., and Oldham, J. (1927). *J. Helminthol.* **5.**
Troisi, R. L. (1933). *Trans. Amer. Microsc. Soc.* **52.**
Tubangui, M. A. (1935). *Philipp. J. Sci.* **50.**
Tubangui, M. A., and Masilungan, V. A. (1938). *Philipp. J. Sci.* **66.**
Tyzzer, E. E. (1929). *Amer. J. Hyg.* **10.**

Ubelaker, J. E., Allison, V. F., and Specian, R. D. (1973). *J. Parasitol.* **59.**

Ubelaker, J. E., Cooper, N. B., and Allison, V. F. (1970). *J. Invert. Pathol.* **16.**
Ulmer, M. J. (1951). *Trans. Amer. Microsc. Soc.* **70.**

Van Beneden, P. J. (1850). *Bull. Acad. Roy. Sci. Belg.* **17.**
Van Cleave, H. J. (1919). *Bull. Ill. Nat. Hist. Surv.* **13.**
Van Cleave, H. J. (1935). *J. Parasitol.* **21.**
Van Cleave, H. J. (1937). *Goeteborgs Kugl. Vetensk. Och. Vett.-Samh. Handl., Ser.* **B5.**
Van Cleave, H. J. (1939). *J. Parasitol.* **25.**
Van Cleave, H. J. (1940). *Rep. Hancock Pac. Exped.* **2.**
Van Cleave, H. J. (1941). *Quart. Rev. Biol.* **16.**
Van Cleave, H. J., and Lincicome, D. R. (1939). *Parasitology* **31.**
Van Cleave, H. J., and Pratt, E. M. (1940). *J. Parasitol.* **26.**
Vaney, C. (1914). *C.R. 9th Congr. Int. Zool. Monaco, 1913.*
Vávra, J. (1976). *Comp. Pathobiol.* **1.**
Vergeer, T. (1928). *J. Amer. Med. Ass.* **90.**
Vernberg, W. B. (1963). *Ann. N.Y. Acad. Sci.* **113.**
Vernberg, W. B., and Hunter, W. S. (1956). *Exp. Parasitol.* **5.**
Vernberg, W. B., and Hunter, W. S. (1963). *Exp. Parasitol.* **14.**
Vickerman, K. (1969). *J. Protozool.* **16.**
Vincent, M. (1927). *Parasitology* **19.**
Vitzthum, H. G. (1925). *Duet. Ent. Z.* **4.**
Vitzthum, H. G. (1926). *Zool. Jahrb. Abt. Syst.* **52.**
Vitzthum, H. G. (1929). *Tierwelt Mitteleurop.* **3.**
Vitzthum, H. G. (1930). *Zool. Jahrb.* **60.**
Vitzthum, H. G. (1931). *Mem. Mus. Hist. Nat. Belg. (Hors. Ser.)* **5.**
Vitzthum, H. G. (1935). *Z. Parasitenk.* **7.**
Vitzthum, H. G. (1940). *In* "Bronns Klassen und Ordnungen des Tierreiches," Vol. 5. Akademische Verlagsgesellschaft, Leipzig.
Voeltzow, A. (1888). *Arb. Zool. Inst. Würzburg* **8.**
Voge, M. (1969). *In* "Problems in Systematics of Parasites." (G. D. Schmidt, ed.). University Park Press, Baltimore, Maryland.
Voge, M., and Heyneman, D. (1957). *Univ. Calif. Publ. Zool.* **59.**
Vogel, H., and Minning, W. (1940). *Arch. Schiffs-Tropen-Hyg.* **44.**
Voltzenlogel, S. (1902). *Zool. Jahrb. Abt. Anat.* **16.**
von Brand, T. (1939). *J. Parasitol.* **25.**
von Brand, T. (1952). "Chemical Physiology of Endoparasitic Animals." Academic Press, New York.
von Brand, T. (1966). "Biochemistry of Parasites." Academic Press, New York.
von Brand, T. (1973). "Biochemistry of Parasites," 2nd ed. Academic Press, New York.
von Brand, T., and Weinbach, E. C. (1965). *Comp. Biochem. Physiol.* **14.**
von Brand, T., Scott, D. B., Nylen, M. U., and Pugh, M. H. (1965). *Exp. Parasitol.* **16.**

Wagner, O. (1917). *Jena Z. Naturwiss.* **55.**
Ward, H. B. (1916). *J. Parasitol.* **2.**
Ward, H. B. (1921). *J. Parasitol.* **7.**
Ward, H. B., and Whipple, G. C. (1918). "Freshwater Biology." Wiley, New York.
Wardle, R. A. (1932). *Contr. Can. Biol. Fish. [N.S.]* **7.**
Wardle, R. A. (1935). *Biol. Bd. Can. Bull.* **45.**
Watson, E. E., (1911). *Univ. Calif. Publ. Zool.* **6.**

Weidner, E. (1976). *J. Cell Biol.* **71.**

Weiner, D. (1960). *Vet. Med.* **55.**

Weinstein, P. P. (1955). *Am. J. Trop. Med. Hyg.* **4.**

Weinstain, P. P. (1960). *In* "Host Influence on Parasite Physiology" (L. A. Stauber, ed.). Rutgers Univ. Press, New Brunswick, New Jersey.

Wenrich, D. H. (1943). *J. Morphol.* **72.**

Wenrich, D. H. (1944). *Amer. J. Trop. Med.* **24.**

Wenrich, D. H., and Emmerson, M. A. (1933). *J. Morphol.* **55.**

Wenyon, C. M. (1926). "Protozoology." Baillière, London.

Wenyon, C. M. (1932). *Trans. Roy. Soc. Trop. Med. Hyg.* **25.**

Wesenberg-Lund, C. J. (1931). *Kgl. Dan. Vidensk. Selsk. Skr. Naturv. Math. Afd. 9,* **4.**

Wharton, G. W. (1941). *Smithsonian Misc. Coll.* **99.**

Wharton, G. W. (1946). *Ecol. Monogr.* **16.**

Whitfield, P. J. (1979). "The Biology of Parasitism." University Park Press, Baltimore, Maryland.

Whitman, C. O. (1882). *Mitt. Zool. Sta. Neapel* **4.**

Whittick, R. J. (1943). *In* Smart's "A Handbook for the Identification of Insects of Medical Importance." British Museum (Natural History) London.

Wickler, W. (1968). "Mimicry in Plants and Animals." World Univ. Library, Hampshire, England.

Willey, R. L. Bowen, W. R., and Durban, E. (1970). *Science* **170.**

Williams, A. B. (1965). *Fish. Bull. U.S. Fish Wildl. Serv.* **65.**

Williams, H. H. (1968). *Parasitology* **58.**

Willmann, C. (1941). *Stud. Oboru. Vseob. Kras. Nauk.* **8.**

Wilson, C. B. (1905). *Proc. U.S. Nat. Mus.* **25.**

Wilson, C. B. (1911). *Proc. U.S. Nat. Mus.* **39.**

Wilson, C. B. (1915). *Proc. U.S. Nat. Mus.* **47.**

Wilson, C. B. (1917). *Proc. U.S. Nat. Mus.* **53.**

Wilson, C. B. (1944). *Proc. U.S. Nat. Mus.* **94.**

Wilson, C. B. (1959). *In* "Freshwater Biology" (W. T. Edmondson, ed.). Wiley, New York.

Wisniewski, L. W. (1930). *Mem. Acad. Polon. Sci. Cl. Sci. Math. Nat.* **B2.**

Witenberg, G. (1934). *Arch. Zool. Ital.* **20.**

Witenberg, G. C. (1938). Liv. Jub. Prot. Lauro Travassos. Brazil.

Wolffhügel, K. (1900). Inaug. Dissert. Basel.

Wolffhügel, K. (1938). *Hyg. Infektionsk. Haustiere* 53.

Woodhead, A. E. (1929). *Trans. Amer. Microsc. Soc.* **48.**

Woodland, W. N. F. (1923). *Quart. J. Microsc. Sci.* **67.**

Woodland, W. N. F. (1927). *Proc. Zool. Soc. London,* 1927.

Woodland, W. N. F. (1937). *Proc. Zool. Soc. London,* 1937.

Yamaguti, S. (1934). *Jap. J. Zool.* **5.**

Yamaguti, S. (1935). *Jap. J. Zool.* **6.**

Yamaguti, S. (1938). *Jap. J. Zool.* **8.**

Yamaguti, S. (1939). *Jap. J. Zool.* **8.**

Yamaguti, S. (1940). *Jap. J. Zool.* **9.**

Yamaguti, S. (1941). *Jap. J. Zool.* **9.**

Yeh, L. S. (1959). *J. Helminthol.* **33.**

Yokogawa, M., and Yoshimura, H. (1965). *Amer. J. Trop. Med. Hyg.* **14.**

Yorke, W., and Maplestone, P. A. (1926). "The Nematode Parasites of Vertebrates." J. & A. Churchill, London.

Yosii, N. (1931). *J. Fac. Sci. Imp. Univ. Tokyo* **2.**

Zaman, V. (1979). "Atlas of Medical Parasitology." ADIS Press, Balgowlah, NSW, Australia.

Zeder, J. G. H. (1800). "Erster Nachtrag zur Naturgeschichte der Eingeweidewürmer von J. A. C. Goetze." Leipzig.

Zelinka, C. (1888). *Z. Wiss. Zool.* **47.**

Zeller, E. (1892). *Z. Wiss. Zool.* **22.**

Zschokke, F. (1907). *Zentralbl. Bakteriol, Parasitenk. Infektionskr.* **44.**

Zwetkow, W. N. (1926). *Arch. Russ. Protistenk.* **5.**

1

PARASITISM AND SYMBIOSIS

Parasitology, the study of parasites and their relationship to their hosts, is one of the most fascinating phases of biology. This discipline actually encompasses several approaches to the study of parasitic organisms. Through the years the field has enjoyed contributions from those who have studied parasites and parasitism from the phylogenetic, ecologic, morphologic, physiologic, chemotherapeutic, immunologic, and nutritional standpoints, and with great strides being made in cellular and molecular biology and biochemistry, spectacular advances have been made from the chemical viewpoint. Irrespective of the approach, the point to bear in mind is that parasitism is a function of the whole organism. In other words, although manifestations of the parasitic way of life may be appreciated at the molecular, cellular, tissue, and populational levels of organization, it is the whole organism that practices parasitism. However, one cannot overemphasize the interdisciplinary role of parasitology. As is the case in increasingly more of the branches of the biological sciences, parasitologists are examining parasites and parasitism at all levels of organization, ranging from the populational and macroecologic to the microecologic and biochemical levels. It should be noted also that in the case of public health or medical parasitology, a sociogeographic or a socioeconomic approach is also important (Weisbrod *et al.*, 1973; Heyneman, 1977, 1984).

In recent years, the World Health Organization has proclaimed that of the six major unconquered human diseases complexes in the world, five, schistosomiasis (p. 332), malaria (p. 194), filariasis (p. 536), African trypanosomiasis (p. 130), and leishmaniasis (p. 121), are parasitic in the traditional sense. The sixth, leprosy, is caused by a bacterium. As a consequence, interest in medical parasitology has been rekindled among biomedical scientists. Also, because of the realization of the economic implications of parasitic infections among domestic animals, the importance of veterinary parasitology has long been recognized.

With significant advances being made throughout the world in the area of aquaculture, including mariculture, with the goal of providing additional sources of protein for the world's starving peoples, the need to understand and control destructive parasites of fish and certain edible invertebrates is becoming vividly clear. Thus, it is apparent that understanding parasites and parasitism is important not only from the viewpoint of the advancement of knowledge but also from the practical viewpoint.

DEFINITIONS

The concept of parasitism is often misunderstood because of the complexity of the relationship between the parasite and its host; consequently, this section is devoted to brief definitions of the general terminology often used in the description of organisms living together.

Any animal, plant, or protist that spends a portion or all of its life intimately associated with another organism of a different species is considered to be a **symbiont** (or symbiote), and the relationship is designated as **symbiosis**. The term symbiosis as used here does not imply mutual or unilateral metabolic dependency.

Four subordinate categories of symbiotic relationships are commonly distinguished, although the lines of demarcation between these are often tenuous. Specifically, under the broad heading of symbiosis, types of associations known as phoresis, commensalism, parasitism, and mutualism have been identified.

Phoresis. The term phoresis means "to carry." During this type of relationship, that which is usually the smaller of the two species—the **phoront**—is mechanically carried in or on the larger species—the **host**. No metabolic interaction or dependency occurs. An example of phoresis is the transport of bacteria on the legs of flies. Although the fly host provides involuntary transportation for the bacteria, the life processes and welfare of both partners are not obligatorily dependent on what is essentially an accidental relationship.

Commensalism. During this type of relationship, both the host and the commensal "eat at the same table." In other words, the spatial proximity of the two partners permits the commensal to feed on substances captured or ingested by the host.

An example of commensalism in the marine environment is the relationship between *Amphiprion percula* (the clownfish) or *Kentrocapros aculeatus* (the trunkfish) and large, tropical sea anemones (Fig. 1.1). Living within the tentacular zone of the anemone, the fish are not injured by the host's stinging nematocysts. In fact, it has been reported that these commensal fish are capable of protecting themselves from the sting of their coelenterate hosts' nematocysts by secreting a mucous envelope around themselves. This protective coat is laid down only after an initial acclimation period during which the fish are stung, though not severely. After acclimation, the fish are afforded protection while living between the anemone's tentacles in that their predators are attacked by the host's nematocysts. In addition to shelter and protection, these fish also share their host's food.

The protection afforded by the mucous coat, however, is not permanent. If *Amphiprion* should leave the tentacular zone of the sea anemone, as it commonly does, it has to become reacclimated, although, as a rule, the time required for reacclimation is shorter than the initial acclimation period.

Although the sharing of food and the providing of shelter are the main features of commensalism, a number of unique behavioral patterns are commonly associated with certain commensals. An example can be found in a number of fish that live commensalistically with the common, tropical, hat-pin urchin *Diadema* (Fig. 1.2). The sharp, needlelike spines of this sea urchin are continuously moving, and if a shadow, such as that cast by a potential predator, falls on an urchin, the tips of the spines quickly converge toward the shadow. This is a protective mechanism. When such fish as *Aeoliscus strigatus* (the shrimpfish) and *Diademichthys deversor* (the clingfish) are associated with *Diadema*, they invariably hover perpendicularly, with their tubular snouts pointed toward the body of the sea urchin and their elongate bodies held parallel to the spines (Fig. 1.2). In this position, these fish are least apt to be pierced by the urchin's spines. Furthermore, as the spines move, the fish sway with the movement. The ability of these commensals to orient themselves with respect to the urchin and to respond to the motion of the spines is governed by their sense of sight.

In summary, during a commensalitic relationship, the commensal shares the host's food and, in many

Fig. 1.1. Commensalism. A clownfish, *Amphiprion*, living among the tentacles of an anemone. The nematocysts of the anemone would kill other small fish, but *Amphiprion* secretes a mucus that prevents the anemone's tentacles from discharging their nematocysts. (Photograph by W. M. Stephens, 1968.)

instances, is also provided with shelter; however, as a rule, the commensal is not directly dependent on the host metabolically and, hence, can be separated from it and still survive, carrying on its biologic functions without the necessity of experimentally providing some factor or factors of host origin.

Parasitism. Parasitism is defined as an intimate and obligatory relationship between two heterospecific organisms during which the **parasite**, usually the smaller of the two partners, is metabolically dependent on the host. The relationship may be permanent, as in the case of tapeworms found in the intestines of mammals, or temporary, as during the feeding of mosquitoes, leeches, and ticks on their hosts' blood. Parasitism is said to be obligatory because the parasite cannot normally survive if it is prevented from making contact with its host. Of course, it should be borne in mind that in some instances methods are now available for maintaining or culturing certain species of parasites *in vitro*, but these techniques generally involve highly complex chemical media and apparatuses to ensure the maintenance of rather exact physiochemical conditions.

Parasites are said to be metabolically dependent on their hosts, and it is dependency that makes the relationship an obligatory one. Not all of the metabolic requirements of parasites are yet known or clearly defined, but sufficient information is available to give some clues of the categories of dependencies involved. These are discussed in detail in Chapter 2. However, it should be pointed out that one of the most commonly encountered dependencies is nutrition. Unlike commensals, parasites derive their nutrient requirements directly from their host, generally from the latter's stored nutrients. The ingestion of host cells by the common liver fluke, *Fasciola hepatica*, while within its bovine or sheep host is an example. The uptake of certain nutrients, primarily hemoglobin, from the enveloping host red blood cell by the malaria-causing organisms, *Plasmodium* spp., is another example. In the case of blood-sucking parasites, such as ticks, leeches, and mosquitoes, it is obvious that nutrients come directly from the host.

Because the intimate relationship between a parasite and its host generally involves the exposure of the host to antigenic substances of parasite origin, whether these molecules comprise the body of the parasite (**somatic antigens**) or molecules secreted or excreted by the parasite (**metabolic antigens**), antibodies are usually synthesized by the host in response to the parasite (see Chapter 3). Thus, unlike phoresis and commensalism, parasitism, in addition to involving metabolic dependency on the part of the parasite, also generally involves immunologic response on the part of the host.

Several types of parasitism are recognized. An organism that does not absolutely depend on the parasitic way of life, but is capable of adapting to it if

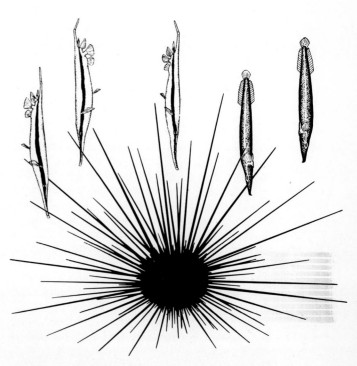

Fig. 1.2. Commensalism. Three shrimpfish, *Aeoliscus*, vertically oriented among the spines of a hat-pin urchin, *Diadema*. The two fish on the right are clingfishes, *Diademichthys*, which are also found similarly associated with *Diadema*. (Modified after Davenport, 1966.)

placed in such a relationship, is known as a **facultative parasite**. If an organism is completely dependent on the host during a segment or all of its life cycle, the parasite is known as an **obligatory parasite**. If an organism accidentally acquires an unnatural host and survives, it is known as an **incidental parasite**.

An **erratic parasite** is one that wanders into an organ in which it is not usually found; a **periodic** or **sporadic parasite** is one that visits its host intermittently to obtain some metabolic requirement; a **pathogenic parasite** is one that is the causative agent of a disease state in the host. The disease may be chronic or acute.

Parasites that live within the body of their host in locations such as the alimentary tract, liver, lungs, and urinary bladder are known as **endoparasites**. Those attached to the outer surfaces of their host, or superficially embedded in the host's body surface, are known as **ectoparasites**.

The host is commonly the larger of the two species in the symbiotic relationship. It may be classified as:

4

Parasitism and Symbiosis

(1) a **definitive** or **final host**, if the parasite attains sexual maturity in it; (2) an **intermediate host**, if it serves as a temporary but essential environment for the completion of the parasite's life cycle (for example, molluscs and arthropods commonly serve as first and second intermediate hosts in which digenetic trematodes complete a part of their development); and (3) a **transfer** or **paratenic host**, if it is not necessary for the completion of the parasite's life cycle, but is utilized as a temporary refuge and vehicle for reaching an obligatory host, usually the definitive host. Arthropods and other invertebrates that serve as hosts, as well as carriers, for protozoan and other smaller parasites are referred to as **vectors**; for example, various species of mosquitoes serve as vectors for the protozoan malarial parasites, *Plasmodium* spp. Although arthropods comprise the most common group of vectors, this role is not restricted to them. As Lee (1971) has pointed out, various groups of worms, including trematodes, cestodes, and nematodes, are also known to serve as vectors for various microorganisms. From the evolutionary standpoint, some intermediate hosts may have been definitive hosts at one time. On the other hand, other intermediate hosts may have been transfer or paratenic hosts.

Animals that become infected and serve as a source from which other animals can be infected are known as **reservoir hosts**. For example, it is now known that in Malaysia and elsewhere in Southeast Asia, the filarial worm *Brugia malayi* causes a type of filariasis in humans. This parasite is transmitted to humans through the bite of mosquitoes belonging to the genera *Mansonia* and *Anopheles*. Although infections are established from person to person via the mosquito vector, *B. malayi* can also be transmitted from cats and monkeys to humans via the mosquito. Thus, infected cats and monkeys serve as reservoir hosts from which human infections can be established.

One other major parasitologic term should be defined at this point. As our knowledge of parasitic diseases advances, it has become increasingly evident that a number of parasites can cause diseases in both humans and animals. The term **zoonosis** has been coined to designate such diseases. Although the term zoonotic diseases is commonly applied to those diseases transmissible to humans from other vertebrates, it is artificial to employ this term in this restricted manner, for it is also correct to refer to diseases transmissible from invertebrates to humans as being zoonotic.

As indicated in many of the following chapters, zoonoses form a significant part of human parasitic diseases.

Finally, the concept of **hyperparasitism** is being introduced. A **hyperparasite** is an organism which parasitizes another parasite. Many instances of this are known. For example, Sprague (1964) has reported a protozoan, *Nosema dollfusi*, as a hyperparasite of the larval stage of a flatworm (trematode), *Bucephalus cuculus*, which, in turn, is a parasite of the American oyster (Fig. 1.3), and Perkins *et al.* (1975) have reported the occurrence of another protozoan, *Urosporidium spisuli*, in the body cells of the nematode *Sulcascaris sulcata*, which in turn, is a parasite of the surf clam, *Spisula solidissima*, and other marine molluscs (Fig. 1.4).

In recent years, as a result of interest in finding biologic control agents to substitute for chemical agents traditionally used to control intermediate hosts or vectors of pathogenic parasites, hyperparasites have been considered as candidate agents. For example, Lai and Canning (1980) have reported that the protozoan *Nosema algerae*, normally a parasite of mosquitoes, will become hyperparasites of the larval stages of the human-infecting blood fluke, *Schistosoma mansoni*, if the infective stage of *N. algerae* is fed to the snail *Biomphalaria glabrata*, a normal intermediate host of *S. mansoni*. Hyperparasitized blood fluke larvae in snails do not produce as many cercariae. Cercariae represent the stage of *S. mansoni* ineffective to humans (p. 336). The use of protozoan hyperparasites as bio-

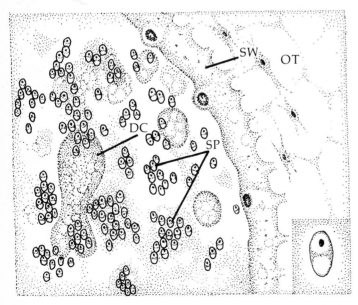

Fig. 1.3. Hyperparasitism. Drawing of a portion of a sporocyst of the trematode *Bucephalus* sp. showing brood chamber filled with *Nosema dollfusi* spores and cellular debris containing spores. *Bucephalus*, in turn, is a parasite of the American oyster, *Crassostrea virginica*. Insert, a single spore stained with Heidenhain's hematoxylin. DC, developing cercaria of *Bucephalus*; OT, oyster tissue; SP, spores of *N. dollfusi*; SW, sporocyst wall of *Bucephalus*.

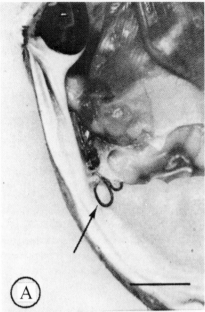

Fig. 1.4. Hyperparasitism. A. Photograph of the nematode *Sulcascaris sulcata* embedded between adductor and foot retractor muscles of the surf clam, *Spisula solidissima*. The dark coloration of the nematode is due to presence of the hyperparasite *Urosporidium spisuli*. Bar = 10 mm. **B.** Hyperparasitized *S. sulcata* containing vermiform sporocysts of *U. spisuli*. Bar = 0.5 mm. (After Perkins *et al.*, 1975; with permission of Journal of Parasitology.)

logic control agents has been reviewed by Canning (1977).

Mutualism. The fourth subcategory of symbiosis is mutualism. In this instance the **mutualist** and the host are metabolically dependent on each other. Classic examples of this type of relationship include the lichens, which represent composites of certain species

of algae and fungi. During such a relationship the alga synthesizes an excess of certain organic compounds, and these molecules are utilized by the fungus. In return, the fungus provides the alga with water, minerals, and protection from desiccation and high intensities of light.

It is of interest to note that the term symbiosis was first coined in 1879 by a German botanist, Heinrich Anton De Bary, to describe the relationship between certain species of algae and fungi living together to form lichens. Subsequently, as the nature of this association became apparent, the designation of symbiosis was adopted by some to mean mutualism. However, De Bary did not imply mutual dependence when he introduced the term, and in following the original definition, symbiosis is now widely accepted in the broad sense, and mutualism, as defined above, is considered to be one subordinate type of heterospecific relationship.

Another example of mutualism is the relationship between certain species of flagellated protozoans living in the gut of woodroaches and termites. As a result of the lifelong investigations by the late L. R. Cleveland, a great deal is known about the mutual metabolic dependencies of some of these flagellates and the woodroach *Cryptocercus punctulatus*. It is known, for example, that these flagellates, which depend almost entirely on a carbohydrate diet, acquire their nutrients in the form of wood chips ingested by their host (Fig. 1.5). In return, they are capable of synthesizing and

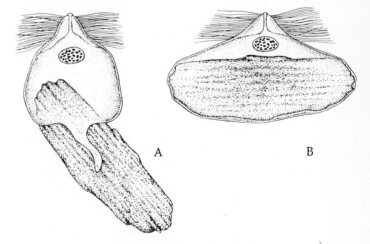

Fig. 1.5. Ingestion of wood chips by the flagellate *Leidyopsis* **sp.** The flagellate on the left (**A**) is beginning to ingest a wood chip by phagocytosis, while the specimen on the right (**B**) contains an unusually large chip. (Redrawn after Swezy, 1923.)

secreting cellulases, the cellulose-digesting enzymes, which are utilized by the roaches in their digestion. Amazingly, the woodroaches, although active wood ingesters, are incapable of synthesizing their own cellulases and are thus dependent on the mutualists. This concept is readily tested experimentally by defaunating the roaches, i.e., removing their flagellate mutualists, by placing them in a chamber containing a relatively high concentration of oxygen. Since oxygen is considerably more toxic to the flagellates than to the woodroaches, these protozoans die while their hosts are left unharmed. Roaches treated in this manner soon die of starvation despite continuing to actively ingest wood.

The relationship between *Cryptocercus* and its mutualistic flagellates involves more than the dietary story described. Another intriguing aspect involves the influence of the woodroach's molting hormone, ecdysone, on the sexual cycle of the flagellates. Specifically, it is only when the host's ecdysone titer reaches a certain level that sexual reproduction among the flagellates occurs. Thus, sexuality among the mutualists is dependent on the host's hormone.

Trager (1970) has written an interesting book, *Symbiosis*, in which he discusses numerous examples of mutualism, including bacteroids in roaches, legumes and rhizobia, algae in invertebrates, and microorganisms in the alimentary tract of vertebrates.

Now that symbiosis and the subordinate categories of heterospecific relationships have been defined, it is important to point out that these are man-made definitions which are, in most instances, extremely useful in categorizing natural symbiotic associations. Certain associations, however, tend to overlap between two of the subcategories. As symbolized in Figure 1.6, it is possible to find some that would qualify as both phoretic and commensalistic, or as both commensalistic and parasitic. Such overlapping

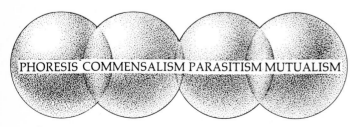

Fig. 1.6. Schematic drawing illustrating overlapping between the four categories of symbiosis. Notice that there is less overlapping between commensalism and parasitism than between phoresis and commensalism and etween parasitism and mutualism. (After Cheng, 1970.)

relationships may be thought of as being transitional and reflect the evolutionary shifts from one to another. In fact, it has been suggested that one should be able to find a complete gradation between, for example, parasitism and mutualism, in which during the initial shift the parasite gives off some metabolic by-product which is not required, but can be utilized, by the host. Eventually, the host not only becomes metabolically dependent on this by-product of parasite origin but also may become dependent on other factors, and the relationship thus evolves into a mutualistic one.

Cleaning symbiosis. In addition to the categories of symbiosis discussed, another unique type of heterospecific relationship occurs in nature which can be considered symbiotic. This type of association has been designated **cleaning symbiosis** and is a relationship during which certain animals, known as cleaners, remove ectoparasites, bacteria, diseased and injured tissues, and unwanted food particles from cooperating hosts (Fig. 1.7). For example, in the marine environment certain species of fish and crustaceans are engaged in cleaning symbiosis with other animals, primarily larger species of sublittoral fish. The opinion is often expressed that this type of relationship is mutually beneficial in that it results in the removal of deleterious and unwanted materials from the host, and during the process the cleaners are furnished with food. Cleaning symbiosis, however, should not be confused with true mutualism, in which the metabolic exchange is of a considerably more intimate physiologic and chemical nature and the relationship is obligatory for the two species involved.

Cleaning symbiosis has been studied almost exclusively by naturalists and ethologists. Several interesting features of this type of relationship have been recognized. As an example, ethologists have reported that cleaners almost always perform their function at specific sites known as **cleaning stations.** Furthermore, hosts are apparently able to recognize cleaners and do not react aggressively when approached by one. Interested readers should consult the review by Losey (1971).

It is being pointed out that cleaners are not limited to fish and crustaceans. It is known, for example, that various species of remoras remove ectoparasites and debris from their pelagic hosts, and in Monterey Bay, California, the ocean sunfish, *Mola mola*, is cleaned by sea gulls. Also, cleaning symbiosis is not limited to marine organisms. Comparable terrestrial associations also occur. The crocodile, for example, is commonly cleaned by the Egyptian plover; domestic cattle are cleaned by egrets; and the rhinoceros by tick birds.

PATHOLOGY OF PARASITISM

In 1675, Wepfer, a German pathologist, reported that the staggering condition commonly referred to as "gid" in sheep and cattle is caused by a bladder (the larval stage or cysticercus) of the tapeworm *Multiceps multiceps* in their brains. This was the first objective observation on the path- ologic effect of a parasite on its host. In time, as it became increasingly more evident that certain parasites are pathogenic to humans and domestic animals, studies aimed at describing the pathology of parasitism became more common.

Hoeppli (1959) has contributed a scholarly volume on the history of parasitology among primitive and ancient civilizations, and Foster (1965) has authored a history of great discoveries in parasitology in the western world. More recently, Kean and his colleagues (1978) have edited a definitive two-volume treatise on the history of medical parasitology, Harrison (1978) has contributed a fascinating history of the war against malaria, and Warren and Bowers (1983) have edited a volume on the impact of parasitic diseases worldwide.

Because of the medical and economic implications of certain parasitic diseases, it is not surprising that the pathology of parasitic infections in humans and domestic animals has received more than its share of attention. In fact, in many ways, the study of parasitism can be thought of as a pathological discipline since alterations in hosts due to the presence of parasites is, in fact, the study of deviations from the normal state in hosts. Such alterations, however, need not result in clinical manifestations of a disease state.

It is noted that parasitic infections also may regulate host populations. For example, Perrin and Powers (1980) have reported that 11 to 14% of deaths among spotted dolphins (*Stenella* spp.) are caused by nematode infections in the brain. For a theoretical discussion of the regulation of animal populations by pathogenic parasites, see May (1983).

As modern pathology advances, increasingly more investigators have come to realize that describable deviations from the normal state have a physiologic and/or biochemical basis. Thus, the subdiscipline of molecular pathology has come into being. Where known, the biochemical bases for certain parasitic diseases are presented in subsequent chapters. At this point, some fairly commonly encountered alterations in parasitized hosts are being considered.

EFFECTS OF PARASITES ON HOSTS

Parasitism is still being defined by some, especially medical and veterinary parasitologists, as a relationship during which one of the partners (the parasite) inflicts some degree of injury to the other (the host). Among biologists, this defi- nition has been essentially discarded since, in many instances, interpretation of the "infliction of injury" is extremely difficult. Furthermore, anthropomorphism, which is scientifically unacceptable, is often introduced into the picture.

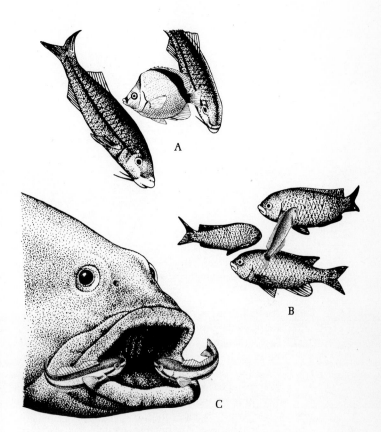

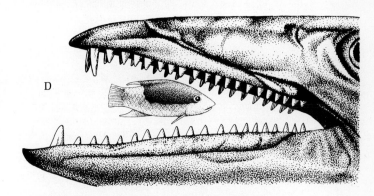

Fig. 1.7. Four examples of cleaning symbiosis as found among marine fish. A. A butterfly fish, *Chaetodon nigrirostris*, cleaning two Mexican goatfish, *Pseudopeneus dentatus*. **B.** A senorita, *Oxyjulis californica*, cleaning a group of blacksmiths, *Chromis punctipinnis*. **C.** Two neon gobies, *Elecatinus oceanops*, cleaning inside the mouth of a grouper, *Epinephelus* sp. **D.** A Spanish hogfish, *Bodianus rufus*, cleaning inside the mouth of a barracuda, *Sphyraena barracuda*. (After Cheng, 1970.)

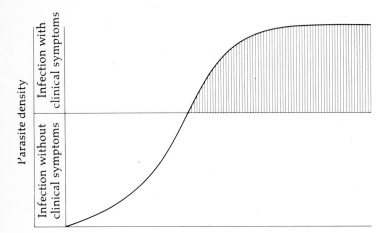

Fig. 1.8. Graph showing the correlation of disease with clinical symptoms and greater parasite density.

Nevertheless, as stated, many species of parasites do cause some degree of alteration within their hosts and, although some of these may result in disease, it need not be the case. It should be borne in mind that parasitic diseases, especially those caused by metazoan parasites, are usually a function of parasite density. In other words, since many species of zooparasites, unlike bacteria and viruses, do not multiply within their hosts, the number of established parasites is at best equal to the number initially introduced.

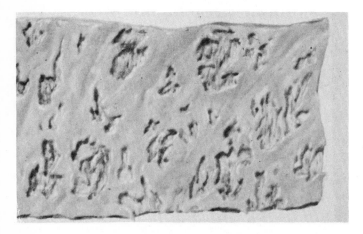

Fig. 1.9. **Amoebiasis.** Large ulcerated areas on mucosal surface of human colon resulting from confluence of smaller ulcers caused by *Entamoeba histolytica*. (After Netter, 1962 © 1962 CIBA Pharmaceutical Co., Division of CIBA-GEIGY Corporation. Reproduced, with permission, from the CIBA Collection of Medical Illustrations by Frank H. Netter, M.D. All rights reserved.)

Furthermore, the onset of recognizable disease symptoms is dependent on the presence of a sufficient number of parasites as well as the physiologic state of the host. Thus, in many instances no clinical symptoms are apparent when small numbers of parasites are present (Fig. 1.8). With this in mind, some of the better known causes of parasitic diseases are discussed below.

Destruction of Host's Tissues

Not all parasites are capable of destroying the host's tissues, and even among those that do, the degree of damage varies greatly. Some parasites injure the host's tissues during the process of entering; others inflict tissue damage after they have entered; still others induce histopathologic changes by eliciting cellular immunologic response to their presence. A combination of these three types of injury may also occur. The hookworms, *Necator americanus* and *Ancylostoma duodenale*, exemplify the first instance, for the infective larvae of these nematodes inflict extensive damage to cells and un- derlying connective tissues during penetration of the host's skin. The cercariae of certain schistosomes that cause "swimmer's itch," while penetrating the host's skin, cause inflammation and damage to the surrounding tissues. This represents histopathologic alterations of the first and third categories. Although cercariae-caused dermatitis is extremely irritating, fortunately, because of host incompatibility, these worms usually do not become established in the host's blood.

Various helminths, such as acanthocephalans and certain flukes and tapeworms, when armed with attachment hooks and spines, often irritate the cells lining the lumen of their host's intestine while they are holding on. In most cases the damage is minute, but repeated irritations over long periods can result in appreciable damage. Moreover, the microscopic lesions resulting from such irritation can become sites for secondary infections by bacteria and other microorganisms. The amoebic dysentery-causing protozoan, *Entamoeba histolytica*, actively lyses the epithelial cells lining the host's large intestine, causing large ulcerations that are not only damaging in themselves, but are also sites for secondary bacterial infections (Fig. 1.9). Such ulcerations result from a combination of mechanical destruction and the action of secreted enzymes. This amoeba also is known to cause large abcesses in the host's liver (Fig. 1.10).

During migration within its human host, larvae of the large nematode *Ascaris lumbricoides* pass through the lungs, causing physical damage to lung tissue (Fig. 1.11) and inducing a profound immunologic reaction of an asthmalike nature, commonly termed "ascaris pneumonia." This is an example of histopathologic alteration reflecting the second and third categories of damage.

Ancylostoma duodenale, one of the human-infecting

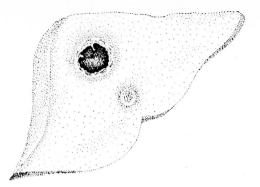

Fig. 1.10. Amoebiasis. Drawing showing large lesion in human liver due to invasion by *Entamoeba histolytica.*

hookworms, is a good example of a parasite that causes both internal and external tissue damage. The external damage phase has been discussed. Once established within the host's intestine, this roundworm produces lesions in the gut wall as well as anemia when present in large numbers by engulfing small pieces of tissue and large volumes of blood.

Histopathologic studies of parasite-damaged tissues have revealed that cell damage other than removal by ingestion or from mechanical disruption is of three major types. (1) **Parenchymatous** or **albuminous degeneration** occurs when the cells become swollen and packed with albuminous or fatty granules, the nuclei become indistinct, and the cytoplasm appears pale. This type of damage is characteristic of liver, cardiac muscle, and kidney cells. (2) **Fatty degeneration** means that the cells become filled with an abnormal amount of fat deposits, giving them a yellowish appearance. Liver cells commonly display this type of degeneration when in contact with parasites.

(3) **Necrosis** occurs when any type of cell degeneration persists. The cells finally die, giving the tissue an opaque appearance. As the result of encystment of *Trichinella spiralis* in mammalian skeletal muscle cells, necrosis of the surrounding tissues is followed by calcification.

Tissue Changes

One of the possible consequences of parasitism associated with cell and tissue parasites is a change in the growth pattern of the affected tissue. Some of these changes may be serious, whereas others are structural and have no serious systemic importance to the whole organism. Such changes, for our purposes, can be divided into four main types.

Hyperplasia. Hyperplasia is an accelerated rate of cell division resulting from an increased level of cell metabolism. This leads to a greater total number of cells, but not necessarily an increase in their absolute size. Hyperplasia, as associated with parasitism, commonly follows inflammation and is the consequence of an excessive level of tissue repair. When liver flukes, *Fasciola* spp., occur in the bile duct of their host, there is a thickening of this duct. This exemplifies the hyperplastic condition resulting from excessive division of the epithelial lining of the duct stimulated by the presence of the parasite.

Hypertrophy. Hypertrophy is an increase in cell size. This condition is commonly associated with intracellular parasites. For example, during the erythrocytic phase of *Plasmodium vivax* (p. 196), the parasitized red blood cells are commonly enlarged.

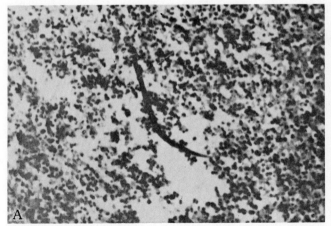

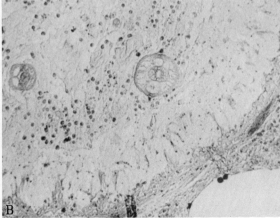

Fig. 1.11. Larva migrans. A. Larva of *Ascaris lumbricoides* during migration through lung of human host. Notice displacement of host cells. (Courtesy of Armed Forces Institute of Pathology, negative No. 80930.) **B.** Section of human lung showing bronchiole filled with mucopurulent material containing a larva of *A. lumbricoides* cut transversely through the esophagus and midintestine. (After Beaver and Danaraj, 1958. Courtesy of Dr. P. C. Beaver, Tulane University, Louisiana.)

Another interesting example of hyperplasia lies in a parasitized invertebrate. When the spermatogonial cells of the annelid *Polymnia nebulosa* are parasitized by the protozoan *Caryotropha mesnili*, hypertrophy of the host cells occurs, involving both the nucleus and cytoplasm (Fig. 1.12). Some of the surrounding cells undergo similar changes and eventually fuse with the infected cell to form a giant multinucleated cell.

Metaplasia. Metaplasia describes the changing of one type of tissue into another without the intervention of embryonic tissue. When the fluke *Paragonimus westermani* occurs in human lungs, it is surrounded by a wall of host tissue composed of epithelial cells and elongate fibroblasts. Since it is known that epithelial cells and large quantities of fibroblasts do not normally occur in lungs, it can be inferred that these encapsulating cells have resulted from the transformation of certain other types of cells in the lungs, hence metaplasia.

Neoplasia. Neoplasia is the growth of cells in a tissue to form a new structure, for example, a tumor. The neoplastic tumor (1) is not inflammatory, (2) is not required for the repair of organs, and (3) does not conform to a normal growth pattern. Neoplasms may be **benign**, i.e., remain localized and do not invade adjacent tissues, or **malignant**, i.e., invade adjacent tissues or move (metastasize) to other parts of the body through the blood or lymph. Cancers are malignant neoplasms. Several species of parasites have been associated with tumors, including cancers, in mammals (Table 1.1).

Competition for Host's Nutrients

Competition for the host's nutrients by parasites to a detrimental point is probably the first type of damage that comes to one's mind. Although some have doubted the importance of parasites in this regard, since the amount of food a microscopic parasite can utilize seems negligible, more recent studies of nutritional requirements of parasites, especially endoparasites, have indicated that depletion of a host's nutrients by parasites may have serious consequences, especially when the parasite density is sufficiently great. The so-called broad fish tapeworm, *Diphyllobothrium latum*, in humans is known to cause an anemia similar to pernicious anemia because of the affinity this parasite has for vitamin B_{12}. This worm can absorb 10 to 50 times as much B_{12} as other tapeworms. Since B_{12} plays an

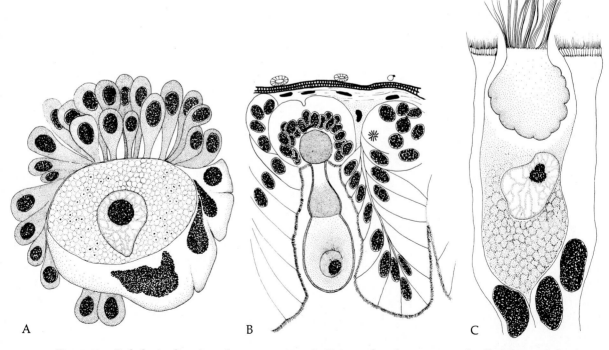

A   B   C

Fig. 1.12. Pathologic alterations due to parasites. A. Hypertrophy of spermatogonial cell of the annelid *Polymnia nebulosa* parasitized by the protozoan *Carytropha mesnili*. (Redrawn after Siedlecki, 1902.) **B.** Invasion of the intestinal epithelium of *Gryllomorpha* by the gregarine *Clepsidrina davini* causing host cells to fuse into a syncytium. (Redrawn after Léger and Duboscq, 1902.) **C.** Hypertrophy of an intestinal epithelial cell of *Blaps* parasitized by the gregarine *Stylorhynchus longicollis*. Only the epimerite of the gregarine is visible in this drawing. (Redrawn after Léger and Duboscq, 1902.)

important role in blood formation, its uptake by *D. latum* results in anemia (von Bonsdorff, 1977).

Studies on nutritional requirements have shown that tapeworms absorb not only simple sugars from their hosts but also certain amino acids, some constituents of yeast in the host's diet, and other nutritional essentials. These substances are essentially derived from two sources: those comprising the intestinal contents of the host and those derived from the host's tissues. Thus, it is not only possible but probable in cases involving large numbers of parasites that an appreciable amount of such materials is drained from the host, and in instances of undernourished hosts—poor sanitary conditions conducive to parasitic infections often go hand in hand with undernourishment—this drainage has considerable effect.

Utilization of Host's Nonnutritional Materials

In some cases parasites also feed on host substances other than stored or recently acquired nutrients. The endo- and ectoparasites that feed on the host's blood are examples. It is extremely difficult to estimate the amount of blood any organism can rob from its host. Table 1.2 lists some estimated amounts taken in by a few blood-feeding species. From the data presented, it should be obvious that the blood lost through parasitic infections can constitute an appreciable amount over a period of time. It has been estimated that 500 human hookworms can cause the loss of 250 ml of blood, or 1/24 of the total volume of blood, each day. This estimate may be too high since others have estimated that no more than 50 ml of blood are removed per day. Nevertheless, the loss of even 50 ml of blood per day constitutes a serious drainage of blood cells, hemoglobin, and serum. For a review of anemias caused by parasites, see Jennings (1976).

Table 1.1. Parasites Associated with Tumor Formation (Neoplasia) in Mammals[a]

Group	Parasite	Host	Site of Tumor
Protozoa	*Eimeria stiedae*	Rabbit	Liver
Trematoda	*Schistosoma mansoni*	Human	Intestine and liver
	Schistosoma haematobium	Human	Bladder
	Schistosoma japonicum	Human	Intestine
	Paragonimus westermani	Tiger	Lung
	Clonorchis sinensis	Human	Liver
Cestoda	*Cysticercus fasciolaris*[*]	Rats	Liver
	Echinococcus granulosus	Human	Lung
Nematoda	*Gongylonema neoplasticum*	Rat	Tongue
	Spirocerca lupi[*]	Dog	Esophagus

[a] Only those marked with an asterisk have been unequivocally established. (Data from Schwabe, 1955, and Smyth, 1962.)

Mechanical Interference

Relatively little is known about injuries to hosts resulting from mechanical interferences by parasites. Probably the best-known case of this type of damage is elephantiasis. In humans infected with the filarial nematode *Wuchereria bancrofti* (p. 537), the adult worms are lodged in the lymphatic ducts. The continuous increase of the number of worms, coupled with the aggregation of connective tissue may result in complete blockage of the lymph flow; excess fluid behind the blockage then seeps through the walls of the lymph ducts into the surrounding tissues, causing edema. In time, the buildup of scar tissue and fluid leads to enlargement of the limbs, breasts, or scrotum, a condition known as elephantiasis (Fig. 1.13).

Table 1.2. Blood Intake of Some Blood-feeding Parasites[a]

Species of Parasite	Host	Number of Parasites	Amount of Blood Lost from Host
Ticks			
Ixodes ricinus (larvae)	Sheep	1000	5 ml
Ixodes ricinus (adult female)	Sheep	1	1 ml[b]
Leeches			
Haemadipsa zeylanica	Humans and animals	—	Sufficient in heavy infections to cause anemia
Limnatis nuotica	Humans and animals	—	Sufficient in heavy infections to cause anemia
Nematodes			
Ancylostoma caninum	Dogs and humans	1	0.5 ml each day
Ancylostoma duodenale	Humans	500	250 ml each day
Necator americanus	Humans	500	250 ml each day
Haemonchus contortus	Sheep	4000	60 ml each day

[a] Tabulated from data collected and presented by Lapage (1958).
[b] Heavily infected sheep may lose 250 ml of blood per week.

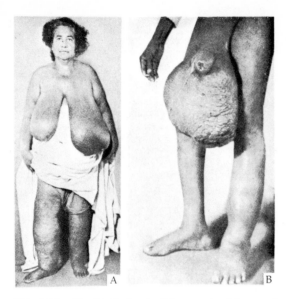

Fig. 1.13. **Elephantiasis.** **A.** Enlargement of legs and breasts of woman in the Cook Islands. (After Hunter *et al.*, 1976; with permission of W. B. Saunders). **B.** Enlargement of scrotom and left leg. (After A. Fisher in Hunter *et al.*, 1976; with permission of W. B. Sanders).

The sheer occupancy of a large portion of the liver and other organs of humans and many herbivores by hydatid cysts of the tapeworm *Echinococcus granulosus* constitutes another type of mechanical interference (Fig. 1.14). These fluid-filled cysts can attain a diameter of several centimeters; one cyst removed from a

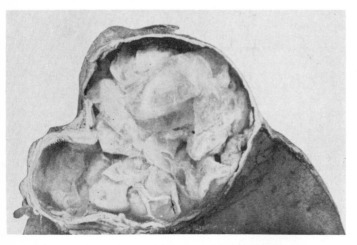

Fig. 1.14. **Hydatid disease.** Hydatid cyst of *Echinococcus granulosus* containing daughter cysts in liver. (After Hunter *et al.*, 1960.)

woman in Australia contained 50 quarts of fluid! *Coenurus cerebralis*, the cysticercus larva of the dog tapeworm *Multiceps multiceps*, is known to exert extreme pressure on the brain and spinal cord of sheep. Infected sheep, which serve as the second intermediate host, are said to suffer from "staggers" or "gid" because of their staggering movements resulting from neurological damage to the brain.

It is known that chicken erythrocytes infected with the avian malaria organism, *Plasmodium gallinaceum*, tend to stick together, thus clogging the fine capillaries. This also occurs in cases of primate malaria. Blood vessels blocked by infected blood cells often rupture. Those rupturing in the area of the brain permit blood to leak into the brain tissue, thus causing the death of the host.

Suffocation of fish whose gills are parasitized by monogenetic trematodes and dinoflagellates is another good example of mechanical damage to hosts.

Effects of Toxins, Poisons, and Secretions

Specific poisons or toxins, egested, secreted, or excreted by parasites, have been cited in many cases as the cause of irritation and damage to hosts. This aspect of parasitology requires a great deal of research since toxins are often cited as the causative factors in these irritations when no definite proof is at hand. Isolation followed by characterization and assay of toxic substances is the only reliable means of verifying their existence, and this involves extremely painstaking procedures.

A good example of an irritating parasite secretion that elicits an allergic reaction in the host is that which causes schistosome cercarial dermatitis. The severe inflammatory reaction of the host's tissue strongly suggests that the fluke secretes some substance that causes the inflammation, and, indeed, such a secretion is known to exist. In the case of bloodsucking insects such as mosquitoes, the swelling resulting from the bites represents the host's response to the irritating salivary secretions of the insect.

A known parasite toxin (or allergin) is the body fluid of the nematodes *Parascaris equorum* and other ascarids. The ability of this fluid to irritate the host's cornea and mucous membranes of the nasopharyngeal cavity is well known among parasitologists who work with ascarid worms for any extended period. Such persons commonly develop severe allergic reactions to the parasites' body fluids. Some experiments have had to be terminated because of this phenomenon.

Another recently discovered parasite toxin is that produced by the intestinal amoeba *Entamoeba histolytica*. As discussed in detail later (p. 160), there are pathogenic and nonpathogenic strains of this amoeba. It is now known that pathogenic strains contain a potent toxin that will not only kill cells maintained in culture but also produce toxic symptoms in parasitized mammalian hosts (Lushbaugh *et al.*, 1978a, 1979; Bos, 1979).

The specific actions of toxins of parasite origin have been studied in a number of instances. For example, Daugherty and Herrick (1952) have reported that in chickens parasitized by the protozoan *Eimeria tenella* there is a depletion of stored glycogen and marked hyperglycemia or increase in the blood sugar level. Further study has revealed that homogenates of the parasite inhibit glycolysis in tissues *in vitro*, but do not interfere with the degradation of fructose-1, 6-phosphate. The *in vitro* experiments have also revealed that the homogenates interfere with the esterification of inorganic phosphate. From this information, it seems possible that some phases of intermediary carbohydrate metabolism, especially the phosphorylative steps of the Embden-Meyerhof glycolytic pathway (p. 52), are disrupted during infection with *E. tenella*. In this instance, the specific activity of the toxin in the parasite extract has been fairly well defined.

OTHER PARASITE-INDUCED ALTERATIONS

Besides the types of pathologic alterations discussed above, there are other interesting ones somewhat peculiar to parasitism. These are discussed briefly at this point.

Sex Reversals

Among the most interesting alterations in parasitized animals are the secondary manifestations resulting from damage to some specific organ. Giard (1911–1913) and Smith (1910, 1911) have reported that in crabs parasitized by the crustacean *Sacculina*, the gonadal tissue is drastically changed in males, but not females. Seventy percent of the parasitized male crabs acquired secondary female characteristics—the abdomen broadens, appendages are modified to grasp eggs, and chelae become smaller (Fig. 1.15). Histologic examinations of the testes revealed that the testicular cells were at various stages of degeneration. This has been found to be true also of ovarian tissue in parasitized females. However, if the parasites are removed from the male, the remaining testicular cells regenerate to form a hermaphroditic gonad that is responsible for the changes in secondary sex characteristics. Since ovarian cells cannot regenerate, the removal of the parasites from the females is not followed by changes in secondary sex characteristics. Similarly, another parasitic crustacean, *Peltogaster carvatus*, has the same effect on the male and female hermit crabs, *Eupagurus excavatus*.

The destruction of gonadal tissues by a parasite is known as **parasitic castration**, and this subject has been reviewed by Reinhard (1956). Changes in sex-related characteristics are known as **sex reversal**,

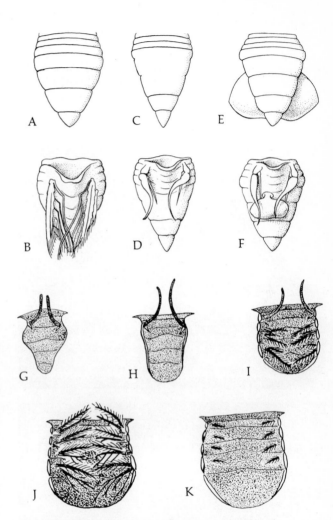

Fig. 1.15. Modifications of external sex characteristics as a result of parasitism.

A–F. Modification of abdomen of *Carcinus maenas* due to parasitism by *Sacculina*. **A, B,** abdomen of normal female (A = dorsal view, B = ventral view); **C, D,** abdomen of normal male (C = dorsal view, D = ventral view); **E, F,** abdomen of parasitized male (E = dorsal view, F = ventral view). (Redrawn after Giard, 1911– 1913.)

G–K. Abdomen of *Inachus mauritanicus* showing modifications due to parasitism by *Sacculina*. **G,** normal male; **H, I,** parasitized males; **J,** normal female; **K,** parasitized female. (Redrawn after Smith, 1910–1911.)

although in most instances these are merely modifications in secondary sex characteristics rather than true change in sex.

Many studies focus on *Sacculina* because this unique crustacean parasite is known to affect the metabolism and sex life of its hosts. In one crab host, *Carcinus maenas*, the females and males are metabolically different, as manifested in the chemical composition of the blood. In males, the blood is generally colorless except just prior to molting, when, because of the presence of tetronerythrin, it becomes pink. In sexually

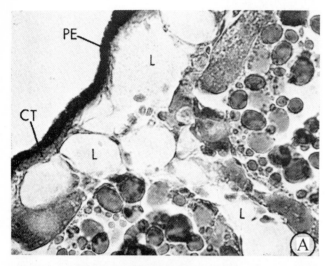

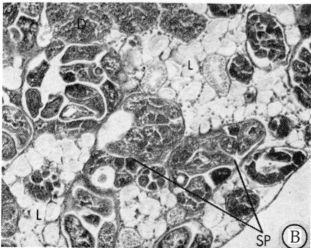

Fig. 1.16. Parasitic castration. A. Photomicrograph showing single layer of Leydig cells (connective tissue cells) in nonparasitized mudflat snail, *Ilyanassa obsoleta*, situated between healthy gonadal cells and body surface and between eggs (×476). **B.** Photomicrograph showing sporocysts of *Zoogonus lasius* occupying space usually filled with gonadal cells. Note that the intersporocyst spaces are filled with Leydig cells (×119). CT, connective tissue; L, Leydig cell; PE, pigmented surface epithelium; SP, sporocysts of *I. obsoleta*. (After Cheng *et al.*, 1973; with permission of Academic Press.)

mature females the blood is yellowish because of the presence of lutein. The fat content varies, being 0.198% in mature females, 0.086% in "pink-blooded" males, and 0.059% in "white-blooded" males. Smith (1910, 1911) has shown that in *Sacculina*-infected males the fat content in the blood is increased considerably, approaching the level found in females. In short, *Sacculina* causes males to alter their metabolism to be more femalelike.

Parasitic Castration

This interesting phenomenon need not lead to sex reversal; in fact, it is exceptional when it occurs. Furthermore, castration due to parasites need not be complete, i.e., it can be manifested as no more than a reduction in sperm or egg production.

The mechanisms underlying parasitic castration are still not understood. Before we can discuss the phenomenon, it is essential that definitions of types of castration be provided. In the **direct** type, the parasite develops in the host's gonad and eventually replaces the reproductive cells. In such instances the castration is usually complete, i.e., the host becomes sterile. In the second type, referred to as the **indirect** type, the parasite is not physically situated in the host's gonad; rather, total or partial atrophy of reproductive cells commences when the parasite, although within the host's body, is situated at some distance from the gonad.

An example of direct parasitic castration occurs in the mudflat snail, *Ilyanassa obsoleta*, which commonly occurs on intertidal mudflats along the Atlantic and Pacific coasts of the United States. Cheng *et al.* (1973) have reported that about 2% of these snails along the New Jersey shore are castrated by the trematode *Zoogonus lasius*. A larval stage of this parasite, known as sporocysts, develops in the snail's gonad (Fig. 1.16) and, as a result, the host's reproductive cells are almost totally absent. In their place are large, vacuolated connective tissue cells. Since trematode sporocysts do not possess a mouth, the destruction of the host's gonad could not be due to mechanical ingestion and/or destruction. Therefore, this type of direct parasitic castration is called **direct chemical castration**. There is evidence that the sporocysts of *Z. lasius* secrete a molecule that causes the destruction of the host's reproductive cells, as well as inhibits game to genesis (Pearson and Cheng, 1985).

Another example of direct castration also occurs in a marine snail. Cooley (1962) has reported complete castration of the southern oyster drill, *Thais haemastoma*, when another larval stage of certain trematodes, known as rediae, occurs in its gonad. In this case the rediae are those of *Parorchis acanthus*. The destruction of the oyster drill's gonads in this instance has resulted from direct ingestion by the rediae, each of which possesses a mouth. This type of parasitic castration is called **direct mechanical castration**.

Since the parasites are not in direct contact with the host's gonad in indirect parasitic castration, the basis for it must be chemical, i.e., the parasite must secrete some molecule which directly or indirectly causes the atrophy of the host's reproductive cells, or competitively inhibit gonadal tissue by consuming large amounts of host nutrients.

Two examples of indirect (chemical) castration can also be found in molluscs. Hosier and Goodchild (1970) reported that when the freshwater snail *Menetus dilatatus* is parasitized by the larvae of the trematode *Spirorchis scripta*, the host's gonad ceases to produce reproductive cells before the parasite reaches it. Also, McClelland and Bourns (1969) reported that the suppression of reproductive cell production in the freshwater snail *Lymnaea stagnalis* occurs before larvae (sporocysts) of the trematode *Trichobilharzia ocellata* reach the gonads.

Application of parasitic castration as a biologic control method for reducing the number of invertebrate intermediate hosts of pathogenic parasites, (e.g., snails that transmit schistosomiasis) has been advocated but has yet to be employed in the field. A great deal needs to be learned about the basis of parasitic castration, especially direct and indirect chemical castration.

Enhanced Growth

Another interesting aspect of parasite-induced change in hosts is enhanced growth. Although parasites are generally considered to be detrimental to their hosts and to cause loss of energy and poor health, instances are known in which the occurrence of parasites actually induces enhanced growth of the host. Among invertebrate hosts, workers of the ant *Pheidole commutata* become much larger when parasitized by the nematode *Mermis*. The abdomen sometimes enlarges to eight times the normal size. The explanation given for this unusual condition is that the parasitized ants engorge themselves and consequently undergo extensive growth. Whether it is true remains to be shown.

Enhanced growth in molluscs parasitized by larval trematodes has been suspected since 1934, when Wesenberg-Lund reported that the freshwater snail *Lymnaea auriculata* infected with trematode larvae are larger than uninfected ones. This Danish investigator has also postulated that the alleged increase in growth, which he termed **gigantism**, was the result of excessive consumption of food by infected snails to meet the demands of their parasites. This initial report was followed by several contributed by M. Rothschild in England, who reported that in nature the estuarine gastropod *Hydrobia ulvae* is larger when parasitized by trematode larvae. Furthermore, Rothschild suggested that the alleged enhanced growth of parasitized snails was the result of parasitic castration. Subsequently, it was widely accepted that parasitized molluscs grow faster; however, controlled experi-

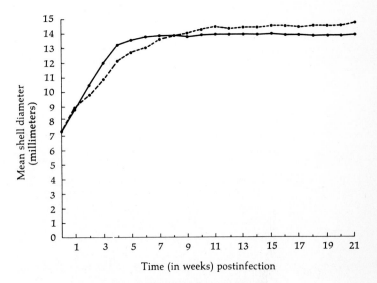

Fig. 1.17. Enhanced growth of snail due to parasitism. Comparative growth rates (shell diameter) of *Biomphalaria glabrata* in- fected (solid line) and noninfected (dashed line) with *Schistosoma mansoni*. The differences are statistically significant. (Data from Pan, 1965.)

mental data were not available. Finally, Chernin (1960) and Pan (1965) subjected the hypothesis of enhanced growth in parasitized molluscs to carefully controlled experimental testing. They traced the growth rates as reflected in measurements of shell diameter of the freshwater snail *Biomphalaria glabrata* infected with the larval stages of the human blood fluke, *Schistosoma mansoni*. These investigators have reported that infected snails have greater shell diameters during the initial 36–42 days postinfection, but the growth of the uninfected control snails caught up with the experimentals 43–56 days postinfection (Fig. 1.17). Thus, it would appear that if true enhanced growth does occur in parasitized molluscs, it is limited to the early growth phase. It should be noted, however, that Cheng (1971), as the result of determining the weights of whole snails, shells, and soft tissues, using the snail *Nitocris dilatatus* naturally infected with larvae of the trematode *Acanthatrium anaplocami*, and another snail, *Physa sayii*, experimentally infected with the larvae of another trematode, *Echinostoma revolutum*, has found that although the mean weights of parasitized snails of both species were significantly greater, the increased weight was attributable not to increased soft tissue mass but to heavier shells (Table 1.3). Concurrent quantitative analysis of the calcium concentrations in the tissues of parasitized and nonparasitized snails has revealed that there are greater Ca^{++} concentrations in parasitized ones (Table 1.4). This led to the examination of snail tissues and, as a result, the conclusion has been drawn that the so-called enhanced growth in parasitized *Nitocris dilatatus* and *Physa sayii* is not true growth but thicker and larger shells resulting from the increased deposition of calcium. In these instances invasion by parasites causes certain calcium-storing cells in the mollusc's digestive gland to release their stored products, and the calcium, carried in blood, becomes deposited in the shell-secreting mantle and eventually is deposited in the shell.

There are validated instances of enhanced growth in invertebrate hosts. For example, West (1960) and Fisher and Sanborn (1962) have shown that larvae of the beetle *Tribolium* attain larger sizes when parasitized by the protozoan *Nosema*. Furthermore, infected beetle larvae undergo as many as six supernumerary molts and commonly die as giant larvae that weigh twice as much as nonparasitized controls. The enhanced growth

in parasitized *Tribolium*, however, is somewhat different from that in parasitized molluscs in that the weights of both parasitized and nonparasitized beetles increase at the same rate until the 12th day. It is only after that day that a more rapid increase in body weight among members of the infected group occurs. An explanation for the enhanced growth may be extrapolated from the studies of Fisher and Sanborn (1962, 1964). In brief, they have demonstrated that *Nosema* implanted into allatectomized nymphs (nymphs in which the corpora allata are removed) of the roach *Blaberus* replaces the corpora allata in function. If the situation in *Tribolium* is similar to that in *Blaberus*, then the enhanced growth in parasitized *Tribolium* may be attributed, directly or indirectly, to the production of a growth-stimulating hormone by *Nosema*.

Evidences for enhanced growth in vertebrates due to parasitism are somewhat better substantiated than in invertebrates. For example, Mueller (see Mueller, 1966, for review) has examined the growth of mice in which larvae (spargana) of the tapeworm *Spirometra mansonoides* had been implanted. He determined that mice harboring six to eight parasites grew faster than nonparasitized ones (Fig. 1.18). It is now known that,

Table 1.3. Weights of Whole Snails, Soft Tissues, and Dried Shells of Nonparasitized *Nitocris dilatatus* and Those Parasitized by *Acanthatrium anaplocami* and Sporocysts Removed from the Latter[a]

	Parasitized Snails[b](35)	Nonparasitized Snails[b] (45)	Sporocysts	Statistical Significance at 5% Level
Mean wt. of whole snails (gm ± SD)	0.396 ± 0.144	0.329 ± 0.137		+
Mean wt. of soft tissues (gm ± SD)	0.126 ± 0.055	0.114 ± 0.045		−
Mean dry wt. of shells (gm ± SD)	0.221 ± 0.084	0.162 ± 0.071		+
Mean wt. in each parasitized snail (gm ± SD)			0.024 ± 0.011	

[a] After Cheng (1971).
[b] Number of specimens in parentheses.

Table 1.4. Calcium Ion Concentrations in 0.5 ml of Extract, Soft Tissues per Snail, and per Gram of Soft Tissues of Nonparasitized *Nitocris dilatatus* and Those Parasitized by *Acanthatrium anoplocami*[a]

	No. of Snails	Ca^{2+} Conc./0.5 ml Aqueous Extract (mg/0.5 ml Extract) Mean ± SD	Ca^{2+} Conc. in Soft Tissues/ Snail (mg) Mean ± SD	Ca^{2+} Conc./gm of Soft Tissues (mg) Mean ± SD
Parasitized	35	0.046 ± 0.020	0.092 ± 0.042	0.730 ± 0.170
Nonparasitized	45	0.027 ± 0.014	0.054 ± 0.029	0.473 ± 0.052
Statistical significance at 5% level		+	−	+

[a] After Cheng (1971).

again, in this instance, the parasites produce a growth hormonelike substance (Mueller and Reed, 1968; Steelman *et al.*, 1970). Similarly, it is also known that the weights of rats parasitized by the blood protozoan *Trypanosoma lewisi* increase more rapidly than those of unparasitized rats.

In summary, it is now known that the presence of certain species of parasites in limited numbers will cause the enhanced growth of their hosts, vertebrates and invertebrates, and, in two instances, it has been demonstrated that the enhanced growth is due to the stimulation of growth-promoting molecules secreted by the parasites.

PHYSIOLOGY AND BIOCHEMISTRY OF PARASITISM

An integral part of modern parasitology is the study of the relationship between the host and the parasite. This intimate relationship invariably involves physiologic and biochemical as well as morphologic adaptations, and hence investigations of parasite physiology and biochemistry have become important aspects of the discipline. Questions such as "how are parasites metabolically dependent on their hosts?" "how do parasites affect their hosts?" and "how do hosts affect their parasites?" are basic to the discipline. Also, since the antigenicity of parasites is due to somatic as well as metabolic antigens, the chemical compositions of their soma and metabolites have received increased attention. Consequently, it is imperative that persons interested in parasitism and parasites should become familiar with the physiology and biochemistry of these animals. Knowledge of their enzymatic activities, the pathways resulting in the synthesis of energy, protective mechanisms, secretions and excretions, composition, respiration, and metabolism of parasites in general are of great importance to our understanding of the parasitic way of life.

From the practical viewpoint, as Cheng (1977) has summarized, the scientific principle underlying modern chemotherapeutic research is to identify some metabolic process within a pathogenic parasite that does not occur or is less sensitive in the host. When such a process is discovered, then potential therapeutic agents could be developed to inhibit the process in the parasite and by so doing kill the pathogen, but not affect the host deleteriously.

Although biologists have long been interested in the physiology and chemistry of parasites, most of our knowledge in this area has come to light during the 20th century with the availability of new techniques and the advent of basic concepts derived from molecular and cell biology.

Research in such indispensable areas as morphology, taxonomy, development, phylogenetic relationships, chemotherapy, and pathology is still

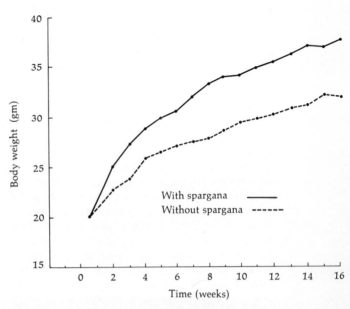

Fig. 1.18. Enhanced growth due to parasitism. Comparative body weight increases in laboratory mice implanted with *Spirometra mansonoides* spargana. Each of the infected mice received seven worms. Each point on the graph represents the mean weight of eight animals. (Redrawn after Mueller, 1963.)

progressing, but in recent years the trend has been to advance our understanding of the immunology, physiology, and biochemistry of parasitism. In 1952, von Brand first published a monograph on the chemical physiology of endoparasites, a thorough revision of which appeared in 1973 entitled *Biochemistry of Parasites*. Van den Bossche (1972, 1976) has edited two advanced treatises, *Comparative Biochemistry of Parasites* and *Biochemistry of Parasites and Host-Parasite Relationships*, which should be read by interested students. These books, along with the establishment of the journals *Experimental Parasitology* and *Molecular and Biochemical Parasitology* as well as the increasing number of biochemical, immunological, and physiological papers in other parasitological and physiology journals, reflect the upsurge of interest in these areas of parasitology.

ECOLOGY OF PARASITISM

Just as the study of the physiology and biochemistry of parasitism has launched the discipline into new frontiers, so have modern studies on the ecology of parasitism. The "state of the art" in this area is reflected by the volumes edited by Fallis (1971), Canning and Wright (1972), Kennedy (1976), and Esch (1977). In brief, the study of relationships between hosts and parasites may be considered an ecological subject. Conceptually, the host is the environment in which the parasite lives. Furthermore, the host's environment, be it a tropical jungle, a freshwater pond, or the open ocean, affects the parasite through its host.

Investigations of ecologic relationships have revealed that free-living and parasitic organisms do not represent two distinct groups. A gradient occurs between the two extremes. For example, among the platyhelminths, the free-living planarians are capable of synthesizing their own digestive enzymes, and the chemical constituents of their bodies are derived from materials ingested or absorbed from the environment. On the other hand, the ectoparasitic monogeneids (p. 273), also capable of synthesizing their own digestive enzymes, depend mainly on the blood of their hosts (primarily fishes and amphibians) for nutrients. Although most of the chemical building blocks used in their bodies are derived from the blood of the host, this is not the only source, since oxygen and possibly other chemicals, can be derived from the aquatic environment. The endoparasitic digenetic trematodes (p. 299) can synthesize their own enzymes, but all the constituents of their body tissues are synthesized from those obtained within their hosts, some even from the host's tissues. Finally, the tapeworms are largely dependent on their hosts for the digestion of food, for present information indicates these worms cannot synthesize many of the essential digestive enzymes. Food predigested by the host is absorbed through ultramicroscopic microvilli and other microstructures situated on or immediately beneath the body surface of the tapeworm and used within its body. Thus, it is apparent that there is a continuum between free-living planarians and obligate endoparasitic tapeworms. Quantitative studies of enzymatic activities, source of body chemicals, etc., could lead to quantitative analyses of the dependency of parasites on their hosts.

Of course, the ecology of parasites is not limited to that discussed in the previous paragraph. Equally interesting are the ecologic factors that influence the distribution of parasites and how parasites select their niches. Some salient aspects of this broad topic are considered below.

Population Dynamics (r- and K-selections)

In recent years, borrowing a page from quantitative ecologists interested in the regulation of populations of free-living organisms, some parasite ecologists (Force, 1975; Jennings and Calow, 1975; Esch *et al.*, 1977) have proposed that what have been designated r- and K-selection could be conceptually useful in understanding the evolution of parasites. By definition, **r-selection** occurs where the factors comprising the selective forces on the organisms (r-strategists) are unstable and the environmental conditions are variable. On the other hand, **K-selection** occurs where the factors acting on the organisms (K-strategists) are relatively stable over a period of time. Species that are r-stategists are characterized by high fecundity rates, high mortality, short life spans, and population sizes that are variable in time, usually below the carrying capacity of the environment (Pianka, 1970). Species that are K-strategists are characterized by relatively low fecundity and mortality, longer life spans, and relatively stable population sizes. Thus, populations of r-strategist are controlled by density-independent factors, while K-strategists are controlled by density-dependent factors.

In the case of digenetic trematodes (p. 299), it may be concluded that they are r-strategists since they have high biotic potential and high mortality as a result of the selective pressures present in their unstable environments (their habitats are different at practically every phase of their complex life cycles).

It should be remembered that r- and K-strategies are relative. In other words, species B may be an r-strategist compared to species C, but a K-strategist when compared to species A. A continuum exists as depicted below:

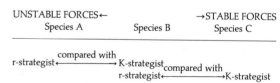

UNSTABLE FORCES←   →STABLE FORCES
  Species A   Species B   Species C

r-strategist ←———compared with———→ K-strategist
                     r-strategist ←——compared with——→ K-strategist

Fig. 1.19

Distribution of Parasites

The zoogeography, ecology, and population dynamics of parasites continue to be intriguing areas of research. The ecology of parasites can be divided into two categories—that concerned with the relationship between parasite and the exterior environment, directly or indirectly, is designated **macroecology**, while that concerned with the relationship between parasite and its immediate environment provided by the host is designated **microecology**. Some of the more important principles of macroecology and microecology are considered below.

Macroecology. The far-flung distribution of a definitive host need not mean that its parasites are also widely distributed, especially if specific intermediate hosts and vectors are involved. If an intermediate host is absent in a given geographic area, even if the definitive host is abundantly present, the parasite population will eventually die out since the reproduction of the parasite, which is dependent on completion of its life cycle, is not possible. Therefore, maintenance of a specific species of parasite in an area depends on the availability of all of its hosts. Because of this dependency, factors governing the survival of the hosts indirectly govern the presence of parasites.

By means of normal evolutionary processes—i.e., random mutations acted on by forces of natural selection—species of parasites have become acclimated to certain types of hosts. For example, the trematode *Calicotyle kroyeri* is specific for a limited number of marine skates and rays, and the liver lancet fluke *Dicrocoelium dendriticum* is found only in pastoral areas where the small land snail *Cionella lubrica* and the ant *Formica fusca*—the first and second intermediate hosts—occur. The human malarial parasite, *Plasmodium vivax*, is only endemic to areas where suitable mosquito vectors can survive. Thus, the geographic limits of the definitive host do not necessarily represent the same limits for the parasite, as is most obviously exemplified in the case of human parasites. The human population is widely distributed, but certain species of human parasites are not as widely distributed, particularly if intermediate hosts are involved. For example, human trypanosomiasis and filariasis are not endemic to North America. If no intermediate host is involved, as in the case of the protozoan *Entamoeba coli*, the geographic distribution of the parasite usually coincides with that of humans.

The geographic distributions of certain wellknown species of parasites, especially those parasitizing humans and domestic animals, have been fairly well determined. Thus, the hemoflagellate *Trypanosoma brucei rhodesiense*—the causative agent of African sleeping sickness—is limited to Zimbabwe, Kenya, Malawi, Mozambique, Tanzania, and areas of eastern Uganda, and the "salmon poisoning" fluke, *Nanophyetus salmincola*, is limited to North America and eastern Siberia. In most cases of parasites of wild animals, however, geographic distributions remain hazy.

Factors That Influence Parasite Density and Distribution. The presence or absence of a number of biologic, chemical, and physical factors in the environment directly or indirectly affects the densities and distributions of parasites. The more important of these are considered below.

Flora. Vegetation that serves as food and shelter for hosts, both intermediate and definitive, greatly influences the parasite population. This is particularly evident in the case of helminth parasites. For example, various aquatic molluscan hosts of digenetic trematodes survive only where plants in the water and deciduous trees on the banks are abundant. Some of the aquatic plants provide not only food but also oxygen for the molluscs. Leaves dropping from the trees very often serve as food for aquatic snails. If such flora are sparse or absent, the molluscan population declines, and the likelihood that the trematodes will complete their life cycles is proportionally diminished.

It is noted that the presence of certain plants associated with the aquatic environment may retard, rather than enhance, the survival of certain species of parasites. For example, occurrence of the berry-bearing plant known as endod, *Phytolacca dodecandra*, in parts of east Africa causes reduction in the number of freshwater snails, *Biomphalaria*, which are the intermediate hosts for the human-infecting blood fluke, *Schistosoma mansoni*. This is because the endod berries, dropping into water or placed therein by women who use them as a laundry detergent, are poisonous to the snails. Thus, the reduction in number of *Biomphalaria* results in reduction of the schistosome parasite. A potentially useful molluscicide against *Biomphalaria* has been developed by Lemma *et al.* (1972) from endod berries, but it has been of only limited practical use.

Fauna. Since parasitism can occur only if two different species of organisms enter the symbiotic relationship, the presence and abundance of the host species are of critical importance. In nature, the presence of prey is absolutely necessary to carnivores and indirectly influences the parasite density. The importance of the faunistic population forming the food web, therefore, is obvious. For example, larvae of the tapeworm *Taenia pisiformis* are found in the liver and mesenteries of rabbits and will develop into adults only when the viscera of infected rabbits are ingested by a carnivorous mammal—commonly a wild cat. Thus, the presence and abundance of wild rabbits and wild cats are vital to the maintenance of this tapeworm.

Since most species of animals are migratory to some extent, this feature of hosts has important implications relative to parasite distribution. Hosts carry their parasites with them as they migrate, whether it is rapid or over a long period of time; however, in time they tend to lose most of their original parasites and pick up new ones that are endemic to the new habitats. For example, the freshwater fish *Lota lota* is a member of a family, Gadidae, which otherwise includes exclusively marine species, and therefore it is believed to have migrated to fresh water from its original habitat. An examination of the endoparasites of *L. lota* will reveal that it carries a few marine parasites from its past but harbors many more recently acquired freshwater parasites.

Another example of loss of original parasites rests with migratory whales and porpoises. These marine mammals lose their original endoparasitic helminths when they reach different environments. This observation is due to a large degree on what has been mentioned earlier. Specifically, the absence of appropriate intermediate hosts for some of their original parasites in new habitats has caused the interruption in the survival and transmission of the parasites, and thus they are gradually lost. On the other hand, the presence of intermediate hosts which transmit new parasites permits the acquisition of the latter if all requisites for their establishment exist.

Water. Water plays a major role in the maintenance of many types of parasitic fauna. Many sporadic parasites, such as mosquitoes, can complete their development only when bodies of water are present. Furthermore, the absence of water would impede the development of a large number of helminths that use aquatic invertebrates as intermediate hosts. Moreover, the infective form of many parasites, particularly flatworms, is free-swimming and requires water in which to migrate and reach its host. Such is the case with the cercariae of certain digenetic trematodes and the ciliated larvae of certain monogeneids and tapeworms.

Not only is water of prime importance in the maintenance of certain parasites, but its physical state may also be influential. There is evidence that the velocity of flow influences the parasite population, as shown by Rowan and Gram (1959) who demonstrated that when the number of *Schistosoma mansoni* cercariae per unit volume of water is constant, experimentally exposed mice acquire heavier worm infections in fast-flowing than in slow-flowing water. The explanation is that during a given period, more cercariae come in contact with each mouse in fast water than in slow water.

Not only do physical factors such as temperature, oxygen and carbon dioxide concentrations, pH, salinity, and mineral content influence the number and survival rates of intermediate hosts, but these factors also influence the longevity of free-swimming stages of certain helminths, such as the coracidia of certain tapeworms and the miracidia and cercariae of certain trematodes. In addition, these factors obviously affect the aquatic stages of mosquitoes and other arthropod parasites.

Host Population Density and Behavior. Population densities of transport, and intermediate and definitive hosts affect the parasite population density, for the latter is directly dependent on the former. In addition to population densities, feeding and other behavior patterns of hosts affect the parasite density.

For example, the predatory feeding habits of many definitive hosts make possible the active intake of larval parasites than can only complete their development after being ingested by the host.

Although experimental evidence is scanty, there is reason to believe that the parasite population, particularly metazoan parasites, is an indicator of the biotic productivity of the environment. Thus, if the fish population in a lake harbors a large number of individuals and varieties of parasites, a concerted survey, in most instances, will reveal a rich assortment and a large population of free-living fauna, primarily invertebrates. When one considers that metazoan parasites generally require intermediate hosts—commonly invertebrates—the use of the number and variety of parasites as an indicator of biotic wealth in an area is understandable.

From the foregoing it should be clear that a multitude of factors, such as the flora, fauna, presence and temperature of water, pH, salinity, and mineral content, as well as the terrain and host population density and behavior, are all important in governing the population and distribution of the parasitic fauna.

Influence of Seasons. Anyone who has collected host animals from one type of habitat throughout the year will appreciate that fluctuations in both the number and kind of parasites occur throughout the seasons. This is especially true in temperate climes where the more marked seasonal changes are sharply reflected in the biotic organic life. For example, Elton *et al.* (1931) have reported that in the approximately 700 specimens of the rodent *Apodemus sylvatus* they studied in England, the nematode *Heligmosomum dubium* was the most common parasite. As indicated in Figure 1.20, there were seasonal differences during the 3 years studied. The cause for these differences, however, is uncertain.

Another example of seasonal fluctuation of parasite density in a population of hosts is provided by the snail *Helisoma trivolvis* in a lake in eastern Pennsylvania. A year-round study I conducted many years ago indicated that the percentage of snails parasitized by the larval stages (rediae) of the trematode *Echinostoma revolutum* increases significantly, reaching the 45% level during the fall, drops off during the winter, hits its lowest point, 3%, during the early spring, and then gradually increases to about 10% during the summer. One major reason for this seasonal fluctuation is that *E. revolutum* adults in this lake lives in migratory ducks, which are most abundant during the fall. Consequently, with the visitation of large numbers of infected ducks and their passing eggs in feces, relatively large numbers of miracidia hatching from eggs are present in the water, and these actively infect *H. trivolvis*. Another reason is because *H. trivolvis* usually burrows into the bottom mud during the colder months, where it is less apt to be infected by miracidia.

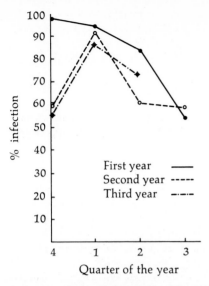

Fig. 1.20. Seasonal fluctuations. The incidence of parasitization of *Apodemus sylvaticus* by the nematode *Heligmosomum dubium* in different quarters of the year in 3 consecutive years, beginning with the last quarter, i.e., October–December. (Data from Elton *et al.*, 1931.)

It is noted that instances are known in which the biology of endoparasites is influenced by seasonal changes. This is especially true in ectothermic hosts. One of the most spectacular examples of this has been reported by Sukhanova (1959, 1962), who found that temperature tolerance of the protozoan *Opalina ranarum* in the frog *Rana temporaria* varies with the time of year. She found that the protozoan would survive for 6.32–11.49 minutes at 38°C during winter (January), 7.83–15.59 minutes at the same temperature during spring (April), 17.74–22.0 minutes during summer (July), and 11.35–15.62 minutes during the fall (September). It is thus evident that *Opalina* can tolerate its upper lethal temperature limit (38°C) better during the summer. Sukhanova's data serve as an example of how an endoparasite can be influenced by its macroenvironment, i.e., the environment in which its host is found. To demonstrate that the tolerance of *Opalina* in frogs to 38°C is indeed influenced by the temperature of the macroenvironment, Sukhanova reported that she could alter the survival time of the protozoan if the hosts were experimentally placed in different temperatures.

It is noted that in addition to seasonal alterations in the quantity and quality of parasitic fauna, year-to-year changes are also known to occur.

Microecology. As stated, microecology deals with relationships between parasites and their intra-host habitats. It is evident from earlier discussions that a host provides its parasites with shelter, nutrition, and other life-sustaining factors. Many aspects of

microecology are discussed in subsequent chapters. At this point, some major principles will be considered.

Host Specificity. One of the most fascinating aspects of parasitism is the phenomenon of **host specificity.** Host specificity is defined as the adaptability of a species of parasite to a certain species or group of hosts. The mechanisms responsible for host specificity are not completely known but are undoubtedly complex and varied, for the degree of specificity differs from species to species.

Wenrich (1935) has postulated that host specificity among parasites has evolved along two main lines:

1. Some parasites have adapted themselves to many varieties of hosts. This is particularly true among protozoan parasites, such as members of the genus *Trypanosoma*, which parasitize hundreds of vertebrate hosts of all major classes. Species of the protozoan parasite *Eimeria* are again good examples of parasites having a wide range of hosts. *Eimeria* spp. are known to parasitize annelids, arthropods, and vertebrates. Although the genus is well represented by species parasitizing a variety of hosts, individual species are often very host specific.

2. The second group of parasites has been limited to a small category of hosts. This type of limited host specificity is convincingly demonstrated by the monogenetic trematodes. For example, monogeneids of the family Hexabothriidae have been reported only as ectoparasites of the Elasmobranchii and are further restricted to members of the class Chondrichthyes. In addition, monogenetic trematodes of the genus *Gyrodactylus* appear to be restricted to teleost fish.

Obviously, there are intermediates between these two main streams of host-specific adaptations. The digenetic flukes in their adult stages are generally limited to one or two groups of vertebrate hosts and hence display some degree of host specificity, although, as striking exceptions, flukes of the family Lecithodendriidae parasitize fish, amphibians, reptiles, birds, and mammals. On the other hand, flukes of the family Bucephalidae are found only in fish. This indicates that there are varying degrees of host specificity among this group of parasites. The same may be said of tapeworms.

As our knowledge of host specificity increases, it is now generally agreed that the number of currently recognized species will be reduced because species reported from different hosts and established on that basis may be found to be infective to additional hosts and thus shown to be identical. Also, morphologic differences that presently constitute the criteria for the definition of new species have been proved in some instances to be simply the result of the influence of the particular host and hence merely intraspecific variations.

Another interesting problem in the study of host specificity is that of geographically linked specificities. For example, the Brazilian strain of the human fluke *Schistosoma mansoni* cannot utilize the Puerto Rican strain of the snail *Biomphalaria glabrata* as the intermediate host, although it can utilize native Brazilian snails of the same species. Similarly, Hsu and Hsu (1968) have reported that two strains of *Schistosoma japonicum*, designated the Changhua or CH strain and the Ilan or IL strain, both native to Formosa, are essentially noninfective to the snails *Oncomelania hupensis* from the People's Republic of China and *O. quadrasi* from the Philippines, but can utilize *O. formosana* from Formosa as the intermediate host. Both *O. hupensis* and *O. quadrasi* are compatible snail hosts of *Schistosoma japonicum* in the People's Republic of China and the Philippines, respectively. It is also of interest to note that both the CH and IL strains of *Schistosoma japonicum* are noninfective to humans, although they will infect other mammals, especially rats, small insectivores, and certain infrahuman primates, and consequently have been designated **zoophilic** strains.

The biologic bases for host specificity are far from completely understood, although in the case of the schistosomes mentioned the phenomenon has a genetic basis. Newton (1953), who crossed a Brazilian strain of *Biomphalaria glabrata* with a Puerto Rican strain, found that the hybrids have a susceptibility index between those of the two parental strains. This information indicates that the compatibility-incompatibility phenomenon is genetically controlled. It is known that if a hybrid strain of *B. glabrata* is continuously back-crossed for several generations, the progeny become refractory to infection by the Puerto Rican strain of *S. mansoni*. This suggests that the genes responsible for incompatibility are recessive to those governing compatibility and, as the result of back-crossing, occur in the progeny in the homozygous condition, and hence the phenotypic characteristic, of being incompatible to the Puerto Rican strain of *S. mansoni*.

Host specificity need not always have a genetic or physiologic basis; the governing factor could be ecologic or structural; of course, the structural or morphologic basis for host specificity has a genetic basis.

An example of ecological specificity was reported by Schiller (1959). He tested the ability of the tapeworm *Vampirolepis nana** to infect a variety of mammals, including some which are not known to harbor this parasite in nature. It was found that some of the latter are susceptible, thus indicating that the reason that they do not harbor *V. nana* in nature is the lack of opportunity. The prevention of infection in nature due to ecologic barriers is known as **ecological nonspecificity**, and the ability of a parasite to infect a compatible host as the result of ecologic opportunity is designated **ecological specificity**.

The basis for ecological specificity need not always be spatial, i.e., the accessibility of the infective form of the parasite to the host. It could have food selection as a basis. For example, Schiller has shown that the grey squirrel is readily infected with *V. nana* in the laboratory if it is maintained on a diet of dog biscuits, but if it is fed its natural food consisting of acorns or mushrooms, it cannot be infected. Thus, it would appear that one of the reasons why the grey squirrel is not naturally infected is because, in nature, its diet lacks a component supplied by dog biscuits.

Some known cases also suggest that some native biochemical component of the host may govern host specificity. For example, Smyth and Haslewood (1963) have proposed that the biochemical composition of a mammalian host's bile could act as one possible selective agent in determining host specificity. The composition of bile varies greatly relative to bile salts, conjugation of bile salts, fatty acids, pigments, and rate of secretion. If a parasite, such as the hydatid tapeworm *Echinococcus granulosus*, enters an unfavorable host, the host's bile can rapidly destroy the invader by lysing its body tegument, or in some way proving toxic to the parasite, or interfering with its metabolism. If the helminth enters a favorable host, the host's bile can provide the stimulus necessary for the inauguration of the next growth phase in the parasite's life cycle, hatching if an egg is introduced, scolex evagination if a larval tapeworm is introduced, or excystation if an encysted trematode is introduced. Furthermore, the host's bile may provide a surface-activating agent that may be essential for the parasite's metabolic activities at the ribosomal or mitochondrial level.

Bile is not the only agent that can serve as the stimulant for further development once the parasite enters a compatible host. Physical factors, such as pH, temperature, and salt concentration, and biochemical factors, such as enzymes and essential amino acids, may influence compatibility. As an example, Cheng and Thakur (1967) have demonstrated that the excystation of the metacercaria of the trematode *Philophthalmus gralli* is stimulated by a specific temperature. In nature, this parasite encysts on aquatic vegetation, and when ingested by a bird it eventually reaches the

---

* Also known as *Hymenolepis nana*.

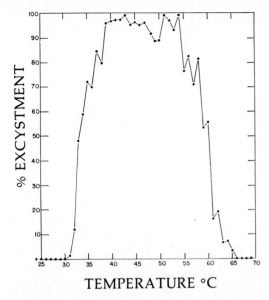

Fig. 1.21. Excystment of metacercariae of *Philophthalmus gralli* in relation to ambient temperature. (After Cheng and Thakur, 1967.)

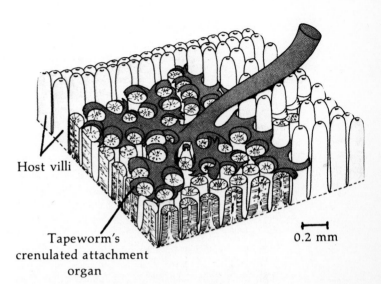

Host villi

Tapeworm's crenulated attachment organ

0.2 mm

Fig. 1.22. Compatibility based on structure. Attachment of the tapeworm *Phyllobothrium piriei* to the intestinal villi of its elasmobranch host. Notice how the crenulated attachment organ of the worm fits exactly around the host's villi. (After Williams, 1968; with permission of *Parasitology.*)

eyes, where it develops to maturity. As indicated in Figure 1.21, 90% or more of the encysted metacercariae shed their envelopes between 39° and 54°C. The body temperatures of birds range from 41.2° to 43.5°C. Thus, the excystation of metacercariae, which is essential for further migration and development, is stimulated by the avian host's body temperature. If a parasite is eaten by a nonavian animal, the incompatible body temperature will either kill the parasite or fail to stimulate excystation. In this instance the host's body temperature regulates compatibility.

An example of a structural basis of compatibility is that between the tapeworm *Phyllobothrium piriei* and its elasmobranch ray host. Williams (1968) has reported that the holdfast organ of this tapeworm is of such a shape that each crenulation of the lobed edge fits snugly around a short villus projecting from the lining of the spiral valve region of the host (Fig. 1.22). Thus, the ability of *P. piriei* to cling on to its ray host is based largely in a structural fit between parasite and host.

Finally, as will be explained in greater detail in Chapter 3, the host's internal defense mechanisms, including immunity, are known to play an important role in governing host specificity. In other words, the manifestation of a minimal amount of host response in the form of cellular response (encapsulation, phagocytosis, etc.) and/or production of antibodies is an important mechanism governing host specificity.

Parasitic Niches　There is a principle in ecology known as **Gause's rule**, which states that two species having essentially the same niche cannot coexist in the same habitat. An organism's niche is its special

place in its habitat, including how it interacts with other organisms associated with it. An organism's habitat is where it lives. Thus, two species of intestinal parasites may occupy the same habitat, for example, the host's small intestine, but if the two species are to remain distinct, that is, maintain separate gene pools, they must interact differently with their place of abode and other associated biota. An example of Gause's rule as applied to parasites lies again with the tapeworm *P. piriei* and another tapeworm, *Pseudanthobothrium hanseni*. Both of these worms are anchored to the villi lining the ray host's spiral valve region; however, there are marked differences. If the spiral valve of the ray, *Raja radiata*, is examined closely, it can be separated into seven tiers. The anterior three tiers bear longer villi while tiers 5, 6, and 7 bear short villi. The holdfast organ of *P. hanseni* is of such a shape that it fits precisely around the longer villi of tiers 1, 2, and 3, whereas the holdfast organ of *P. piriei* fits the shorter villi of tiers 5, 6, and 7 snugly. Thus, although *P. hanseni* and *P. piriei* occupy essentially the same habitat, i.e., the host's spiral valve region, they do not share the same niche, and this difference prevents interbreeding between the two tapeworm species.

Biotic Potential among Parasites.　As a general rule, animals that have adopted the parasitic way of life possess greater biotic potential—i.e., they produce more progeny.

In the case of parasitic protozoans, very little work has been done in actually measuring their rates of reproduction. We do know that one mature, eight-nucleate cyst of *Entamoeba coli*, when taken into the alimentary tract of the host, excysts as a multinucleated metacystic amoeba which gives rise to eight meta-cystic trophozoites by binary fission, and a four-nucleate cyst of *Entamoeba histolytica* after excystation through rapid divisions, also gives rise to eight young trophozoites. Free-living species of amoebae under normal conditions have little or no occasion to encyst and undergo intracystic nuclear division, which is followed by cytoplasmic division (cytokinesis) upon excystation, thus increasing the number of individuals. In addition to this mode of reproduction, free-living and parasitic species both are capable of binary fission in the trophozoite form, thus increasing the number of daughter amoebae. Therefore, it is apparent that parasitic amoebae can produce more individuals in a given period, because they encyst more often (Fig. 1.23).

Among parasitic helminths, the biotic potential is indeed great. A single human pinworm, *Enterobius vermicularis*, is capable of producing 4,672 to 16,888 (mean 11,105) eggs during its life span. A single female *Ascaris lumbricoides*, the common human intestinal roundworm, is capable of producing 200,000 eggs daily and has a total capacity of 27,000 eggs in its uterus. *Raillietina demerariensis*, the Celebes tapeworm of humans, is estimated to bear 200 to 250 eggs per gravid proglottid and is comprised of approximately 5000 such proglottids. These tremendously large numbers of eggs probably represent the highest reproductive rates among living animals. Indirectly, the fecundity of parasites is manifested by the magnitude of parasitic infections that exist despite the mortality rates. In 1947 Stoll announced the astounding fact that there were an estimated 2,220,000,000 cases of human helminthic infections in the world. This is a challenging public health problem, not to mention the millions of infections in domestic and wild animals, as well as those in lower vertebrates and invertebrates.

The reproductive rate of insects also is extremely great. It has been stated that, excluding the protozoans and nematodes, there are more insects on the face of the earth than any other group of animals.

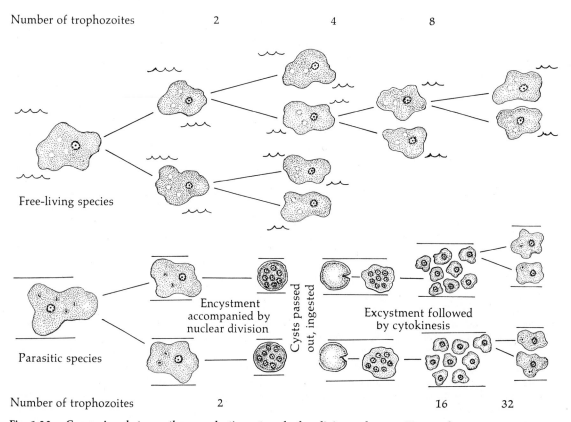

Fig. 1.23. Comparison between the reproductive rates of a free-living and a parasitic amoeba.

Certainly the biotic potential of some parasitic insects must rank among the highest in the animal kingdom.

The striking exception to the rule of greater biotic potential among parasites is the parasitic molluscs, which give rise to fewer progeny than their free-living relatives. This undoubtedly is due to evolutionary changes associated with the more efficient means these animals have for parasitizing their hosts.

It is not surprising that the biotic potential is so high among parasites if one considers the tremendous odds parasites must surmount to perpetuate their species. Protozoan parasites that require an arthropod vector to carry them from one vertebrate host to another must await the appropriate vector, and civilization with all its modern insecticides and biologic control techniques is not making life any easier for them. Digenetic trematodes exhibit what is considered one of the most complex types of life histories among animals. Most of them require one, two, three, or even four intermediate hosts to complete their life cycles. Think of the difficulties confronting these minute animals. Often the metacercaria (the last larval stage in most species of trematodes) remains encysted in an intermediate host and must be ingested by a compatible definitive host before it can complete its development. The parasite itself is helpless. The same obstacles in varying degree confront the cestodes, the spiny-headed worms, and other groups. Little wonder the sexually mature individuals lay so many eggs; it is the parasite's way of overcoming these odds in preserving the species. Only when more efficient methods of completing the life cycle have evolved, as among the parasitic molluscs, does the number of eggs produced decrease.

EVOLUTION OF PARASITES

Where and when did parasites arise? There is no clear-cut answer. In postulating the origin of parasitism, all agree that parasites arose from free-living progenitors. It follows that endoparasites have resulted from free-living forms that were accidentally introduced into the host. In this new environ- ment, spontaneous mutations occurred, and the more proficient mutants thrived, exemplifying the concept of "survival of the fittest." Among the tapeworms, the continuous appearance of mutant forms has resulted in the establishment of species that lack an alimentary tract. Such mutants have obviously become highly successful, for the environment consists of a matrix of digested and partially digested nutrients that can be absorbed. Thus, the lack of an alimentary tract, though seemingly "degenerate," is actually a more advanced and efficient condition. Furthermore, tapeworm enzymatic systems, modes of locomotion, digestion, and sensation have also become more efficiently adapted to parasitism. Similar modifications have taken place among other groups of parasites.

Preadaptation

In considering the evolution of parasites from free-living ancestors, the concept of **preadaptation** is a useful one. The term does not imply any type of predestination on the part of the potential parasite to become parasitic. It merely means that the organism, while free living, has potentialities other than its normal adaptive characteristics, but the former may never play an adaptive role while the organism is free living. However, if for some reason its environment becomes modified, for example, if it should accidentally enter a potential host, these previously nonmanifest potentialities become of critical importance for survival in the new environment. Preadaptations can be structural, physiologic, or both.

An example of preadaptation to parasitism is found among certain free-living nematodes. These worms are commonly found associated with various insects, primarily beetles, in decaying organic material upon which they feed. It is conceivable that those nematodes, which are present-day intestinal parasites of beetles, were accidentally introduced into the insect's gut through ingestion and were able to survive because of a previously unexpressed potentiality of being able to survive in environments with little oxygen.

Another example lies with such cave-dwelling arthropods as mites. Ancestors of the extant species of mites that are ectoparasitic on bats most probably were preadapted to parasitism because, being guanophilic, they were accustomed to a high nitrogen content in their diet.

Numerous other examples of preadaptation to parasitism may be cited. Most of them are of a physiologic nature. In fact, because of the more frequent occurrence of physiologic preadaptations, it is generally agreed that this category of preadaptive features is more important during the transition to parasitism than are morphologic types.

Some Possible Evolutionary Pathways

When the body architecture, development, and composition of the major groups of parasitic flatworms, the trematodes or flukes, and the cestodes or tapeworms are compared with those of the free-living flatworms, the turbellarians, sufficient similarities exist to warrant their being considered to be members of the same phylum. This implies that the parasitic and free-living flatworms have originated from a common ancestral stock. Indeed, as Heyneman (1960) has pointed out, all of the flatworms most probably have originated from a primitive ancestor that gave rise to a modern-day group of comparatively primitive free-living flatworms known as the acoels (members of the turbellarian order Acoela). From the acoel stock have risen all of the other groups of turbellarians, including the rhabdocoels (members of the order Rhabdocoela) (Fig. 1.24). The natural habitats of the modern-day rhabdocoels include marine, freshwater, and terrestrial environments. Even more significant is the fact that a number of rhabdocoels are consistently found associated with other organisms; for example, they occur on a number of marine plants, especially algae. Even more dramatic is the fact that a number of species

have been found to be endosymbiotic in other turbellarians, molluscs, echinoderms, sipunculids, crustaceans, and annelids. Although the exact nature of their relationship with their hosts has not been examined to the point at which definite statements can be made of the existence or nonexistence of metabolic dependency, superficial observations, coupled with the rather consistent association of these worms with other marine invertebrates, suggest that these associations are more than accidental phoretic ones. As an example, in Hawaiian waters, a small, white rhabdocoel is commonly found attached to the red and white hydroid *Pennaria*. The coloration of this worm is a perfect example of color camouflage. It blends in so remarkably with the white area on the surface of *Pennaria* that unless subjected to critical examination,

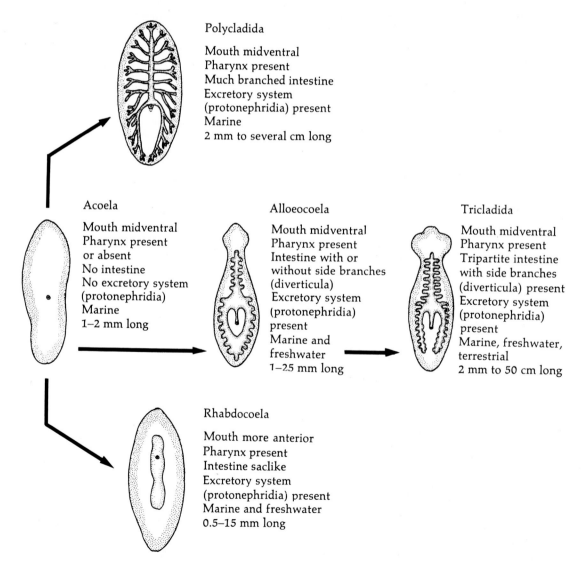

Polycladida

Mouth midventral
Pharynx present
Much branched intestine
Excretory system
(protonephridia) present
Marine
2 mm to several cm long

Acoela

Mouth midventral
Pharynx present
or absent
No intestine
No excretory system
(protonephridia)
Marine
1–2 mm long

Alloeocoela

Mouth midventral
Pharynx present
Intestine with or
without side branches
(diverticula)
Excretory system
(protonephridia)
present
Marine and
freshwater
1–25 mm long

Tricladida

Mouth midventral
Pharynx present
Tripartite intestine
with side branches
(diverticula) present
Excretory system
(protonephridia)
present
Marine, freshwater,
terrestrial
2 mm to 50 cm long

Rhabdocoela

Mouth more anterior
Pharynx present
Intestine saclike
Excretory system
(protonephridia) present
Marine and freshwater
0.5–15 mm long

Fig. 1.24. Turbellarian phylogeny. Hypothetical evolutionary relationships between the five orders of turbellarians. (After Cheng, 1970.)

it usually escapes detection. Although this rhabdocoel is known also to occur on algae situated in the proximity of *Pennaria*, its common occurrence on *Pennaria*, plus its color adaptation, suggest the beginning of a commensalistic relationship, at least in the area where phoresis and commensalism overlap. The reason for pointing out the fairly common occurrence of rhabdocoels in association with other organisms is that one group of rhabdocoels, the dalyellioids (members of the section Dalyellioida of the suborder Lecithophora), shows remarkable structural similarities to the trematodes. These structural similarities, coupled with the relatively common occurrence of symbiotic species, some of which are found in molluscs, have led to the postulation that the parasitic trematodes have evolved from the dalyellioid stock. The fact that some dalyellioids are symbionts of molluscs is of particular significance since, except for a few exotic exceptions, all of the digenetic trematodes, the largest order of the Trematoda, utilize a molluscan host during their life cycles. Relative to this theory, Heyneman has postulated that the evolution of present-day parasitic digenetic trematodes followed the course depicted in Figure 1.25 and outlined below.

1. An ancestral dalyellioid rhabdocoel entered a commensalistic relationship with a mollusc. Specifically, its larval life was spent within the mollusc's mantle cavity, and the larval generations derived their nutritional requirements from the host's secreted mucus. The adult was free living.

2. In time, the entering or initial larval form penetrated the mollusc's tissues. The ability to synthesize and secrete the enzymes required to facilitate penetration presumably had been acquired as an adaptation to feeding on the host's mucus. Adults developing from intramolluscan larvae escaped and remained free living.

3. The greater biotic potential of most parasitic animals has been discussed. Trematodes are no exception to this rule. In fact, among the Digenea, not only are literally hundreds, if not thousands, of eggs produced by each adult individual as the result of sexual reproduction, but the number of individuals is also greatly increased as the result of asexual reproduction during the intramolluscan larval stages. It is believed that during the third phase of adaptation to parasitism, the parasite developed the ability to reproduce asexually.

4. Finally, a mechanism was evolved by which the ancestral adultlike organism, escaping from the mollusc, could be introduced into another host, which in the case of trematodes was some vertebrate. Such a mechanism may be in the form of encystment on vegetation, the encysted form later eaten by the vertebrate;

penetration by the form escaping from the mollusc into a second host within which it encysts, and the second host is later ingested by the definitive vertebrate host; or direct penetration by the postmolluscan form into the definitive vertebrate host.

The four hypothetical steps involved during the evolution of a modern-day digenetic trematode represent an example of the evolution of a parasitic relationship from the phoresis-commensalism line. With minor modifications, a similar pathway can be postulated for a number of other groups of parasites.

Is there a hypothetical pathway for the direct origin of parasitism from a free-living organism? An example is the ciliated protozoans of the genus *Tetrahymena* (Fig. 1.26). These protozoans are usually free living in fresh water, but they are preadapted to parasitism, some being highly pathogenic to their hosts. Among their known hosts are mosquito larvae, amphipods, midges, rotifers, gastropods, breams, salamanders, chicks, trout, various amphibians, oligochaetes, guppies, the African lungfish, the kissing gourami, the platyfish, tardigrades, clams, catfish, and even humans. It is noted that at least two species of *Tetrahymena*, *T. limacis*, and *T. rostrata*, are true pathogenic endoparasites of slugs (Brooks, 1968).

Adaptation to Multiple Hosts

Where more than one host is involved in completion of the parasite's life cycle, the multihost species usually may be considered to be more evolutionarily advanced because presumably the parasite has become adapted, in time, to several species of hosts. The modern intermediate hosts may well have been definitive hosts at one time. This concept is quite popular, but in studying the evolution of parasites within the same phylum, and in light of biotic potentials and life-cycle patterns, perhaps the advancement in parasitism is best represented by a bell-shaped curve (Fig. 1.27). That is, the increasing number of intermediate hosts signifies evolutionary advancement only up to a certain point, whereafter the elimination of certain intermediate hosts should be considered to be a more advanced condition because the parasite enhances its chances of reaching the definitive host by eliminating one or more intermediate hosts.

This concept is borne out by the liver fluke *Fasciola*. Here the parasite requires only one intermediate host; cercariae emerging from the snail host encyst as metacercariae on aquatic vegetation, and when these are ingested by the definitive host, they eventually mature into adults. It is hypothesized that a second intermediate host was at one time involved in the

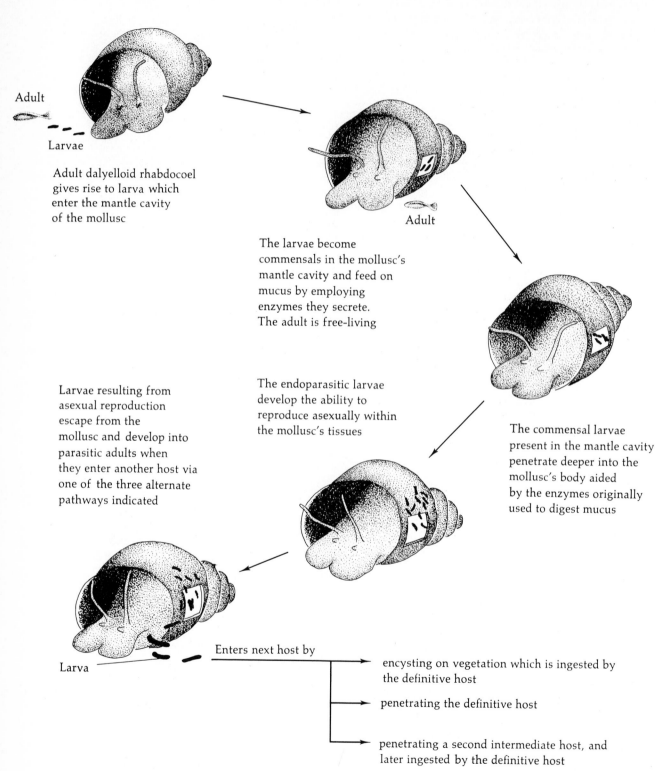

Adult

Larvae

Adult dalyelloid rhabdocoel
gives rise to larva which
enter the mantle cavity
of the mollusc

Adult

The larvae become
commensals in the mollusc's
mantle cavity and feed on
mucus by employing
enzymes they secrete.
The adult is free-living

Larvae resulting from
asexual reproduction
escape from the
mollusc and develop into
parasitic adults when
they enter another host via
one of the three alternate
pathways indicated

The endoparasitic larvae
develop the ability to
reproduce asexually within
the mollusc's tissues

The commensal larvae
present in the mantle cavity
penetrate deeper into the
mollusc's body aided
by the enzymes originally
used to digest mucus

Larva

Enters next host by

encysting on vegetation which is ingested by
the definitive host

penetrating the definitive host

penetrating a second intermediate host, and
later ingested by the definitive host

Fig. 1.25. Adaptation to endoparasitism. Hypothetical pathway by which a free-living dalyellioid turbellarian could have gradually adapted to endoparasitism.

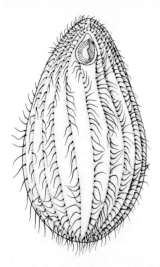

Fig. 1.26. *Tetrahymena pyriformis,* **a ciliated protozoan capable of facultative parasitism.** The body of a living specimen is pliable and measures about 50×30 μm. (Redrawn after Furgason, 1940.)

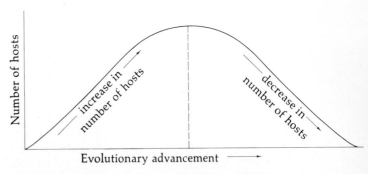

Fig. 1.27. Hypothetical scheme showing correlation of the evolution of parasitic animals with the number of obligatory hosts.

Fasciola life cycle; however, as an adaption for facilitating entry into their terrestrial, herbivorous hosts (cattle, sheep), this parasite discarded the requirement for the second intermediate host, and the present species encyst on vegetation. The highly developed branched in- testinal tract of these liver flukes suggests their more advanced state, compared to that of other digeneans, in which two, three, or more intermediate hosts are required.

Another example is the rodent tapeworm *Vampirolepis nana.* This worm was, until recently, considered to be a member of the genus *Hymenolepis.* The species of *Hymenolepis* require two hosts, a vertebrate definitive host and an invertebrate intermediate host, to complete their life cycles. In the case of *V. nana,* however, the necessity of an intermediate host is eliminated although, if presented with one, it will undergo larval development within it. This evidence strongly suggests that the life cycle of *V. nana* originally involved an intermediate host, but during its relatively recent evolution, it has discarded its dependency on such a host. The loss of its dependency on an intermediate host, coupled with certain structural differences, has caused this tapeworm to be transferred to the genus *Vampirolepis* from *Hymenolepis.*

Coevolution of Host and Parasite

Observations exist which suggest that the coevolution of both host and parasite occurs. For example, in regions of Africa where trypanosomiasis is endemic, indigenous ruminants are mildly infected with insignificant morbidity while recently imported ruminants suffer virulent infections which are commonly fatal (Allison, 1982). Such observations suggest that the

concurrent evolution of indigenous hosts and the trypanosomes has resulted in the lack of virulence. It also follows theoretically that the parasite evolves toward avirulence, provided that transmissibility and the duration of infectiousness are independent of virulence. It is noted, however, that this assumption does not always hold true. For detailed discussions of coevolution of host and parasite, see Anderson and May (1982) and Levin (1982).

Some Evolutionary Patterns

As a result of theoretical speculations about the evolution of parasites and parasitism, the following three generalizations have been proposed. It must be borne in mind, however, that these are broad concepts to which there are numerous exceptions.

Szidat's Hypothesis. The more evolutionarily specialized the host group, the more specialized are the parasites; conversely, the more primitive the host group, the less specialized are the parasites. Based on this hypothesis, the degree of specialization of the parasitic fauna may serve as a clue to the phylogenetic position of the host.

Fahrenholz's Hypothesis. The common ancestors of modern parasites were themselves parasites of the common ancestors of present-day hosts. Based on this hypothesis, the degrees of relationships between modern parasites may provide clues as to the parentage of modern hosts. This hypothesis finds its strongest supporting evidence in parasites of wild animals. The advent of domestic animals and the close living conditions between humans and such animals have brought about many exceptions.

Eichler's Hypothesis. Hosts belonging to a large taxonomic group, for example, a family consisting of numerous species, will harbor a greater diversity of parasites than hosts belonging to a restricted taxonomic group.

Interrelationship Between Types of Symbiosis

The question may now be raised as to what the evolutionary relationship is between the various categories of symbiosis. The concept that parasitism has evolved from commensalism, and commensalism, in turn, has evolved from phoresis is a popular one. Such evolutionary pathways undoubtedly have occurred and are still occurring in nature. However, there is reason to believe that not all parasitic relationships have arisen from commensalistic ones or vice versa. Although the concept that parasitism is a more primitive type of relationship was very popular at one time, this belief is no longer accepted as an irrefutable dogma. One of the reasons for this is as follows.

If one examines an array of closely related parasites, particularly tissue parasites, in a large number of species of hosts in nature, one would expect to find the severest host tissue reaction in those hosts which are believed to be relatively newly acquired ones. This assumption is based on the concept that the host is able to "recognize self from nonself" and the foreignness, or "nonselfness," of the parasite is most evident when the relationship is relatively new. Furthermore, this hypothesis can be tested in a number of instances by experimentally introducing a parasite into the tissues of a host in which it is not normally found. In such a case, the implanted parasite often does cause more severe reactions. Based on such studies, it has been concluded that since parasitism commonly elicits host response, it must represent the earliest stage of a heterospecific relationship which in time, as the result of changes resulting from spontaneous mutations, becomes less foreign and thus has evolved toward commensalism. On the other hand, two lines of reasoning tend to refute this assumption, at least in part.

First, experimental evidences are now available that indicate that pathogenicity need not always be associated with relatively new relationships. To cite an example, when the malaria-causing protozoan *Plasmodium* is experimentally injected into a variety of hosts, it will cause death in some, moderate malaria in others, and no detectable symptoms in still others. Since these are all "new" hosts, it must be concluded that, at least in the case of *Plasmodium*, not all new hosts respond so drastically as to result in a disease. Similarly, when *Trypanosoma brucei brucei*, a pathogenic, flagellated protozoan parasite of donkeys, horses, mules, camels, cattle, and dogs, is experimentally inoculated into several species and subspecies of the American deer mouse,

Peromyscus, some species will contract acute infections and die, others will exhibit pronounced resistance to the disease, and still others will develop subacute to chronic disease. That such variations should occur in closely related "new" hosts also suggests that not all new hosts react drastically to a new parasite and that innate or native resistance associated with the host species must be taken into consideration.

Turning now to the possible origin of mutualism, it is entirely conceivable that most instances of this type of relationship had their origin in parasitism, in which one of the partners (the parasite) is metabolically dependent on the other (the host). In time, the host becomes obligately dependent on one or more metabolic by-products or end products produced by the parasite and the relationship becomes a mutually dependent one, hence, mutualism. Exceptions to this hypothetical pathway, however, exist. For example, it is known that the relationship between certain species of algae (zooxanthellae) and corals represents mutualism, with the algae supplying the coral with oxygen and removing carbon dioxide and, in addition, increasing the metabolic efficiency of the host by secreting trace amounts of vitaminlike or hormonelike factors and possibly other diffusable organic substances. In return, the coral contributes inorganic by-products that are utilized by the algae. Various investigators have shown, however, that despite the mutual exchange of molecules, the relationship is not obligatory since the coral can survive, although not as effectively, if the algae are removed by either chemical treatment or maintenance in the dark. Similarly, algae released from corals can be readily cultured as independent organisms. Thus, the coral-zooxanthellae mutualism appears to be a facultative one rather than one evolving from a unilateral, metabolically dependent relationship, i.e., parasitism.

In summary, we must conclude that the various categories of symbiosis could have arisen in two ways. Along one path, one should consider phoresis, commensalism, parasitism, and mutualism as independent forms of relationships, each having arisen separately from the other and dependent on the particular partners entering into the relationship. Along the second path, the subtle gradient between these categories of symbiotic relationships suggests that, at least in certain instances, one type of symbiosis may have evolved from another.

The Model Approach to Parasitology

During the past three decades there has been an increase in popularity of the so-called model approach for the elucidation of basic biologic processes. This approach involves the choice of an organism, irrespective of its economic or medical importance, which appears to be ideally suited for the experimental evaluation of some process. For example, the use of the fruitfly *Drosophila melanogaster*, the fungus *Neurospora crassa*, the bacterium *Escherichia coli* and the T phages (viruses) in genetic studies is well known to all students of biology. Similarly, the use of the giant axon of the squid for neurophysiologic studies, and the eggs of sea urchins, chicks, and amphibians for embryologic investigations are well known. Although the use of parasitic animals as models has been practised for many years, Smyth (1969) has brought the value of this approach into sharp focus. He has pointed out that certain helminth parasites, such as the tapeworm *Echinococcus granulosus*, have all the requisites of being advantageous model organisms since (1) their life cycles are well known, (2) they are relatively easily maintained in the laboratory, and (3) the more important morphologic and biochemical characteristics of their phenotypes have been identified.

Smyth has proposed that *E. granulosus* could be an ideal model for understanding the genetic basis of differentiation, which is a fundamental problem of interest to all developmental biologists. The complete life cycle of this tapeworm is presented in a later chapter (p. 410). At this point it is sufficient to mention that *E. granulosus* utilizes two hosts during its life cycle. The larval hydatid cyst (Fig. 1.28) may occur in a variety of mammals, and when this cyst, which encloses numerous protoscolices, is ingested by the definitive host, usually a dog feeding on carcasses, each protoscolex eventually develops into an adult tapeworm.

A fundamental feature of the protoscolex is its ability to differentiate in two directions. If for some reason the hydatid cyst is ruptured and protoscolices leak out into an adjacent tissue in the intermediate host, each develops into a vesicular mass which, in turn, differentiates into another cyst (secondary hydatidosis) and the cyst gives rise to additional protoscolices. On the other hand, if protoscolices are ingested by a dog, they will evaginate, become attached to the host's intestinal mucosa, and differentiate into segmented or strobilate adult worms. This unique

heterogeneous morphogenetic condition provides an ideal model for studying the genetic basis of differentiation. In other words, what factors determine whether a protoscolex will differentiate into a cyst or a strobilate adult? Smyth has advanced a system of hypothetical control circuits based on the Jacob-Monod model of gene control which would provide an explanation of how the switch from "larval" to "adult" differentiation could occur. This system (Fig. 1.29) assumes the occurrence of a regulator gene, RG_1, which controls development into another larva and another regulator gene, RG_2, which controls differentiation into an adult. Now, under the influence of the larval environment (in the intermediate host), an inducer is produced which blocks the action of RG_1 so that the larval operator genes become functional and proceed to direct the synthesis of "larval" mRNA (messenger RNA) and, in turn, "larval" proteins, so that the protoscolex develops into another larval cyst. Similarly, if the protoscolex should reach the dog's

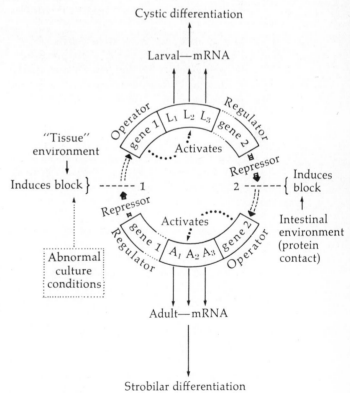

Fig. 1.29. **Jacob–Monod model.** Hypothetical control circuit based on the Jacob–Monod model of gene action as applied to the differentiation of *Echinococcus granulosus* into cystic or strobilar forms. (Modified after Smyth, 1969.)

Fig. 1.28. **Hydatid cysts of** *Echinococcus granulosus*. Several unilocular hydatid cysts in lung of a sheep. Each cyst contains several protoscolices. (After J. Jensen in Schmidt and Roberts, 1981; with permission of C. V. Mosby).

gut, this "adult" environment produces an inducer, which blocks the action of RG_2, and consequently the adult operator genes become operative, resulting in the synthesis of "adult" mRNA and subsequently "adult" proteins. As the consequence, the protoscolex develops into an adult worm.

An alternative hypothesis is that the adult regulator gene RG_2 is normally inactive but becomes active when the "larval" operator gene or operon is functional, perhaps activated by a product of larval synthesis. Consequently, as long as larval synthesis is occurring, adult development is suppressed. However, if the environment becomes unsuitable for larval development, for example, when the protoscolex is introduced into the intestine of a dog, the adult operon is automatically activated when RG_2 returns to its inactive state and strobilization results.

The working model provided by *Echinococcus granulosus* appears to be a promising tool for understanding the genetic basis of developmental biology. The same hypothesis could be applied to the understanding of larval (asexual) reproduction and adult (sexual) reproduction in the life cycle of digenetic trematodes (p. 302).

It is most likely that many were initiated to the study of meiosis and the fertilization process by studying the cytology of the gametic cells of the horse nematode, *Ascaris megalocephala*. This model, first described in detail by van Beneden in 1883, is an example of a parasite which, because of its convenient chromosomal complement ($2n = 4$), expedited our understanding of gametogenesis and fertilization.

For those whose interests focus on medical or veterinary parasitology, the so-called model approach has a slightly different connotation. Since humans and valuable domestic animals are usually not suitable experimental animals, understanding of their diseases, including parasitic ones, must depend on finding a model from among more readily available model animals. A good example of this approach is associated with understanding hepatic amoebiasis. As will be discussed elsewhere (p. 159), the pathogenic amoeba *Entamoeba histolytica* may invade the liver and cause the development of abscesses. Since human cases of hepatic amoebiasis do not lend themselves to experimentation leading to a better understanding of the mechanisms responsible for this aspect of the disease, a model was searched for. Earlier studies have revealed that, although liver lesions will develop in hamsters and rabbits, these are granulomatous, i.e., filled with granular blood cells. This is not the usual pathology associated with human cases and, therefore, the models were not ideal. Now, Lushbaugh *et al.* (1978b) have

shown that if the amoebae are subjected to multiple passage through the livers of hamsters with intervening recovery in culture, their virulence is increased. Furthermore, these amoebae will produce large, fluid-filled abscesses characteristic of human liver amoebiasis in the hamster liver four to six weeks after inoculation. This important discovery has now provided a model for studying the pathophysiologic basis of human hepatic amoebiasis.

It should be apparent from the information presented in this chapter that parasitology is a many-headed dragon or, more aesthetically pleasing, a many-blossomed flowering bush (Cheng, 1973). The discipline welcomes individuals of all persuasions, be they molecular biologists or ecologists. The frontiers are wide open for individuals with keen minds, intellectual inquisitiveness, and ambition. In the following chapter, I have attempted to point out how parasitism can be analyzed and, I hope, the reader will readily appreciate that much remains to be done in the elucidation of parasitism. Parasitology remains one aspect of modern biology in which the discoveries of molecular and cell biology and quantitative ecology can be adapted with profit.

REFERENCES

Allison, A. C. (1982). Coevolution between hosts and infectious disease agents, and its effects on virulences. *In* "Population Biology of Infectious Diseases." (R. M. Anderson and R. M. May, eds.), pp. 245–267. Springer-Verlag, New York.

Anderson, R. M., and May, R. M. (1982). Coevolution of hosts and parasites. *Parasitology* **85**, 411–426.

Bos, H. J. (1979). *Entamoeba histolytica*: Cytopathogenicity of intact amoebae and cell-free extracts; isolation and characterization of an intracellular toxin. *Exp. Parasitol.* **47**, 369–377.

Brooks, W. M. (1968). Tetrahymenid ciliates as parasites of the gray garden slug. *Hilgardia* **39**, 205–276.

Canning, E. U. (1977). New concepts of Microsporida and their potential in biological control. *In* "Parasites, Their World and Ours." (A. M. Fallis, ed.), pp. 101–140. Roy. Soc. Canada, Ottawa, Canada.

Canning, E. U., and Wright, C. A. (eds.). (1972). "Behavioural Aspects of Parasite Transmission." Academic Press, London.

Cheng, T. C. (1970). "Symbiosis." Bobbs-Merrill, Indianapolis, Indiana.

Cheng, T. C. (1971). Enhanced growth as a manifestation of parasitism and shell deposition in parasitized mollusks. *In* "Aspects of the Biology of Symbiosis." (T. C. Cheng, ed.), pp. 103–137. University Park Press, Baltimore, Maryland.

Cheng, T. C. (1973). The future of parasitology: One person's view. *Bios.* **44**, 163–171.

Cheng, T. C. (1977). The control of parasites: the role of the parasite uptake mechanisms and metabolic interference in parasites as related to chemotherapy. *Proc. Helminth. Soc. Wash.* **44**, 2–17.

Cheng, T. C., and Thakur, A. S. (1967). Thermal activation and inactivation of *Philophthalmus gralli* metacercariae. *J. Parasitol.* **53**, 212–213.

Cheng, T. C., Sullivan, J. T., and Harris, K. R. (1973). Parasitic castration of the marine prosobranch gastropod *Nassarius obsoletus* by sporocysts of *Zoogonus rubellus* (Trematoda): histopathology. *J. Invertebr. Pathol.* **21**, 183–190.

Chernin, E. (1960). Infection of *Australorbis glabratus* with *Schistosoma mansoni* under bacteriologically sterile conditions. *Proc. Soc. Exp. Biol. Med.* **105**, 292–296.

Cooley, N. R. (1962). Studies on *Parorchis acanthus* (Trematoda: Digenea) as a biological control for the southern oyster drill, *Thais haemastoma*. *Fish. Bull. Fish Wildl. Serv.* **62**, 77–91.

Daugherty, J. W., and Herrick, C. A. (1952). Cecal coccidiosis and carbohydrate metabolism in chickens. *J. Parasitol.* **38**, 298–304.

Elton, C., Ford, E., and Backer, I. (1931). The health and parasites of a wild mouse population. *Proc. Zool. Soc. (London)* pp. 657–721.

Esch, G. W. (ed.). (1977). "Regulations of Parasite Populations." Academic Press, New York.

Esch, G. W., Hazen, T. C., and Aho, J. M. (1977). Parasitism and r- and K-selection. *In* "Regulation of Parasite Populations." (G. W. Esch, ed.), pp. 9–62. Academic Press, New York.

Fallis, A. M. (ed.). (1971). "Ecology and Physiology of Parasites." University of Toronto Press, Toronto, Canada.

Fisher, F. M., Jr., and Sanborn, R. C. (1962). Production of insect juvenile hormone by the microsporidian parasite *Nosema*. *Nature (London)* 194, 1193.

Fisher, F. M. Jr., and Sanborn, R. C. (1964). *Nosema* as a source of juvenile hormone in parasitized insects. *Biol. Bull.* **126**, 235–252.

Force, D. C. (1975). Succession of r and k strategists in parasitoids. *In* "Evolutionary Strategies of Parasitic Insects and Mites." (P. W. Price, ed.), pp. 112–129. Plenum, New York.

Foster, W. D. (1965). "A History of Parasitology." E. & S. Livingstone, Edinburgh, Scotland.

Giard, A. (1911–1913). "Oeuvres Diverses." Paris.

Harrison, G. (1978). "Mosquitoes, Malaria & Man: A History of the Hostilities Since 1880." E. P. Dutton, New York.

Heyneman, D. (1960). On the origin of complex life cycles of digenetic flukes. In "Libro Homenaje al Dr. Eduardo Caballero y Caballero." Sec. Educ. Publ., Inst. Politec. Nac., Esc. Nac. Cienc. Biol., Mexico, D.F. pp. 133–152.

Heyneman, D. (1977). Parasitic diseases in relation to environment, customs, and geography. *In* "Parasites, Their World and Ours. (A. M. Fallis, ed.), pp. 1–24. Royal Soc. Canada, Ottawa, Canada.

Heyneman, D. (1984). Development and disease: a dual dilemma. *J. Parasitol.* **70**, 3–17.

Hoeppli, R. (1959). "Parasites and parasitic Infections in Early Medicine and Science." University of Malaya Press, Singapore.

Hosier, D. W., and Goodchild, C. G. (1970). Suppressed egg-laying by snails infected with *Spirorchis scripta* (Trematoda: Spirorchiidae). *J. Parasitol.* **56**, 302–304.

Hsu, S. Y. L., and Hsu, H. F. (1968). The strain complex of *Schistosoma japonicum* in Taiwan, China. *Z. Tropenmed. Parasitol.* **19**, 43–59.

Jennings, F. W. (1976). The anaemias of parasitic infections. *In* "Pathophysiology of Parasitic Infection." (E. J. L. Soulsby, ed.), pp. 41–67. Academic Press, New York.

Jennings, J. B., and Calow, P. (1975). The relationship between high fecundity and the evolution of entoparasitism. *Oecologia* **21**, 109–115.

Kean, B. H., Mott, K. E., and Russell, A. J. (eds.). (1978). "Tropical Medicine and Parasitology: Classical Investigations." Vols. I and II. Cornell University Press, Ithaca, New York.

Kennedy, C. R. (ed.). (1976). "Ecological Aspects of Parasitology." North Holland, Amsterdam, Holland.

Lai, P. F., and Canning, E. U. (1980). Infectivity of a microsporidium of mosquitoes (*Nosema algerae*) to larval stages of *Schistosoma mansoni* in *Biomphalaria glabrata*. *Intern. J. Parasit.* **10**, 293–301.

Lee, D. L. (1971). Helminths as vectors of micro-organisms. *In* "Ecology and Physiology of Parasites." (A. M. Fallis, ed.). pp. 104–122. University of Toronto Press, Toronto, Canada.

Lemma, A., Brody, G., Newell, G. W., Parkhurst, R. M., and Skinner, W. A. (1972). Studies on the molluscicidal properties of endod (*Phytolacca dodecandra*). I. Increased potency with butanol extraction. *J. Parasitol.* **58**, 104–107.

Levin, B. R. (1983). Evolution of parasites and hosts (group report). *In* "Population Biology of Infectious Diseases." (R. M. Anderson and R. M. May, eds.), pp. 213–243. Springer-Verlag, New York.

Losey, G. S., Jr. (1971). Communication between fishes in cleaning symbiosis. *In* "Aspects of the Biology of Symbiosis" (T. C. Cheng, ed.), pp. 45–76. University Park Press, Baltimore, Maryland.

Lushbaugh, W. B., Kairalla, A. B., Cantey, J. R., Hofbauer, A. F., Pittman, J. C., and Pittman, F. E. (1978a). Citotoxicity (sic) of a cell free extract of *Entamoeba histolytica*. *Arch. Invest. Med. (Mexico)* **9**, 233–236.

Lushbaugh, W. B., Kairalla, A. B., Loadholt, C. B., and Pittman, F. E. (1978b). Effect of hamster liver passage on the virulence of axenically cultivated *Entamoeba histolytica*. *Am. J. Trop. Med. Hyg.* **27**, 248–254.

Lushbaugh, W. B., Kairella, A. B., Hofbauer, A. F., Cantey, J. R., and Pittman, F. E. (1979). Isolation of cytotoxin-enterotoxin from *Entamoeba histolytica*. *J. Infect. Dis.* **139**, 9–17.

May, R. M. (1983). Parasitic infections as regulators of animal populations. *Am. Sci.* **71**, 36–45.

McClelland, G., and Bourns, T. K. R. (1969). Effects of *Trichobilharzia ocellata* on growth, reproduction, and survival of *Lymnaea stagnalis*. *Exp. Parasitol.* **24**, 137–146.

Mueller, J. F. (1966). Host-parasite relationships as illustrated by the cestode *Spirometra mansonoides*. In "Host-Parasite Relationships" (J. E. McCauley, ed.), pp. 15–58. Oregon State University Press, Corvallis, Oregon.

Mueller, J. F., and Reed, P. (1968). Growth stimulation induced by infection with *Spirometra mansonoides* sparga in propylthiouracil-treated rats. *J. Parasitol.* **54**, 51–54.

Newton, W. L. (1953). The inheritance of suceptibility to infection with *Schistosoma mansoni* in *Australorbis glab-*

ratus. Exp. Parasitol. **2**, 242–257.

Pan, C. T. (1965). Studies on the host-parasite relationship between *Schistosoma mansoni* and the snail *Australorbis glabratus. Am. J. Trop. Med. Hyg.* **14**, 931–976.

Pearson, E. J., and Cheng, T. C. (1985). Studies on parasitic castration: Occurrence of a gametogenesis-inhibiting factor in extract of *Zoogonus lasius* (Trematoda). *J. Invertebr. Pathol.* **46**, 239–246.

Perkins, F. O., Zwerner, D. E., and Dias, R. K. (1975). The hyperparasite, *Urosporidium spisuli* sp.n. (Haplosporea), and its effects on the surf clam industry. *J. Parasitol.* **61**, 944–949.

Perrin, W. F., and Powers, J. E. (1980). Role of a nematode in natural mortality of spotted dolphins. *J. Wildl. Man.* **44**, 960–963.

Pianka, E. R. (1970). On r- and K-selection. *Am. Natural.* **104**, 592–597.

Reinhard, E. G. (1956). Parasitic castration of Crustacea. *Exp. Parasitol.* **5**, 79–107.

Rowan, W. B., and Gram, A. L. (1959). Relation of water velocity of *Schistosoma mansoni* infection in mice. *Am. J. Trop. Med. Hyg.* **8**, 630–634.

Schiller, E. L. (1959). Experimental studies on morphological variation in the cestode genus *Hymenolepis*. IV. Influence of the host on variation in *H. nana. Exp. Parasitol.* **8**, 581–590.

Smith, G. W. (1910). Studies in the experimental analysis of sex. *Quart. J. Microsc. Sci.* **55**, 225–240.

Smith, G. W. (1911). Studies in the experimental analysis of sex. Part 7. Sexual changes in the blood and liver of *Carcinus maenus. Quart. J. Microsc. Sci.* **57**, 251–256.

Smyth, J. D. (1969). Parasites as biological models. *Parasitology* **59**, 73–91.

Smyth, J. D., and Haslewood, G. A. D. (1963). The bio-chemistry of bile as a factor in determining host specificity in intestinal parasites, with particular reference to *Echinococcus granulosus. Ann. N.Y. Acad. Sci.* **113**, 234–260.

Sprague, V. (1964). *Nosema dollfusi* n.sp. (Microsporidia, Nosematidae), a hyperparasite of *Bucephalus cuculus* in *Crassostrea virginica. J. Protozool.* **11**, 381–385.

Steelman, S. L., Morgan, E. R., Cuccaro, A. J., and Glifzer, M. S. (1970). Growth hormone-like activity in hypophysectomized rats implanted with *Spirometra mansonoides* spargana. *Proc. Soc. Exp. Biol. Med.* **133**, 269–273.

Stoll, N. R. (1947). This wormy world. *J. Parasitol.* **33**, 1–18.

Sukhanova, K. M. (1959). Temperature adaptations in the parasitic protozoa of amphibian hosts. *Tsitologia* **1**, 587–600. (In Russian).

Sukhanova, K. M. (1962). Temperature adaptations of *Opalina ranarum* in the course of one annual cycle. *Tsitologia* **4**, 250–275. (In Russian).

Trager, W. (1970). "Symbiosis." Van Nostrand Reinhold, New York.

Warren, K. S., and Bowers, J. Z. (eds.). (1983). "Parasitology: A Global Perspective." Springer-Verlag, New York.

Williams, H. H. (1968). *Phyllobothrium piriei* sp.nov. (Cestoda: Tetraphyllidea) from *Raja naevus* with a comment on its habitat and mode of attachment. *Parasitology* **58**, 929–937.

Wenrich, D. H. (1935). Host-parasite relations between parasitic Protozoa and their hosts. *Proc. Am. Phil. Soc.* **75**, 605–650.

West, A. F. (1960). The biology of a species of *Nosema* (Sporozoa: Microsporidia) parasitic in the flour beetle *Tribolium confusum. J. Parasitol.* **46**, 745–754.

Weisbrod, B. A., Andreano, R. L., Baldwin, R. E., Epstein, E. H., and Kelley, A. C. (1973). "Disease and Economic Development. The Impact of Parasitic Diseases in St. Lucia." University of Wisconsin Press, Madison, Wisconsin.

Zekhnov, M. I. (1949). Dynamics of the parasite fauna of *Colaeus monedula. Uch. Zap. Vologod. Gos. Pedagog. Inst.* (*Biol.*) **5**, 29–116. (In Russian).

2

THE MICROENVIRONMENT AND THE PHASES OF PARASITISM

As exemplified in Chapters 4 through 21, parasitic animals are represented in practically every phylum. It should thus be evident to anyone with even a passing acquaintance with the biotic world that we can expect to find parasitism in practically every habitat. Consequently, it would be futile to render accounts of the characteristics of the numerous types of habitats in which parasites are found, especially in the case of ectoparasites. The habitats of these, obviously, approximate those of their hosts, and their physiologic processes are influenced by their environments, as are those of their hosts. In considering the life processes of endoparasites, however, one should be somewhat familiar with the physicochemical nature of the sites in which they live. Hence, it is the intent of this chapter to give some general accounts of selected habitats within animal hosts where endoparasites are most commonly found.

When one considers that during their life spans, numerous parasites pass from one type of habitat within one host to another within another host, one cannot help but be awed at the remarkable adaptability of these organisms. The principle that has evolved from such observations is that the well-adapted parasite must be able to exploit the advantages and to withstand the hazards of its niche. During the course of the evolution of both host and parasite, numerous ingenious mechanisms have evolved as the result of natural selection. These enable endoparasites not only to survive under rather unique conditions, but also to take advantage of such conditions in order to survive and perpetuate their species. Some of these unique habitats and examples of how some parasites have adapted to them are discussed below.

Within vertebrates, the most common sites in which endoparasites are found are the alimentary canal and its associated organs, the circulatory system, the respiratory system, the coelom, and within certain cells. The volume entitled *Ecological Aspects of Parasitology*, edited by C. R. Kennedy (1976), should be consulted by those interested in the various types of microhabitats within the vertebrate body in which parasites occur. Only the alimentary tract and blood are briefly considered below, but the principles exemplified are equally applicable to the remaining sites.

THE VERTEBRATE ALIMENTARY CANAL

One cannot help but wonder at first glance why the vertebrate alimentary canal is one of the favored habitats of endoparasites. It is dark; a battery of carbohydrate-, protein-, and fat-digesting enzymes are present; its pH ranges from 1.5 to 8.4; it undergoes rapid and continuous chemical, physiologic, and physical changes associated with entry, digestion, absorption, and passage of waste; and it is practically devoid of oxygen. All of these represent obstacles the parasite must cope with in order to become established. Nevertheless, certain parasites have not only adapted to these but have also capitalized on certain seemingly hazardous obstacles.

pH. In mammals, the pH of the alimentary tract varies greatly from region to region (Fig. 2.1). For example, the pH of the mouth is about 6.7, but may range from 5.6 to 7.6. The stomach, on the other hand, is usually strongly acidic resulting from the secretion of hydrochloric acid by certain lining cells, known as parietal cells, but it may vary between 1.49 and 8.38; that of a mouse may vary between 3.26 and 6.24; that of a cow between 2.0 and 4.1; and that of a sheep between 1.05 and 3.60. In the duodenum the pH is usually slightly acidic in humans, being about 6.7, although it is alkaline in cats and goats (pH 8.2–8.9). The alkalinity is due to the secretion of bile, pancreatic juices, and other alkaline substances. In the case of humans, these secretions are usually insufficiently alkaline to neutralize the strongly acidic

stomach contents passing into the duodenum. It is noted, however, that fluctuations in pH occur in the duodenum. For example, Mettrick (1971) has found that in rats the pH in the duodenum falls (i.e., becomes more acidic) after feeding and remains low for four hours. Interestingly, when rats are infected with the tapeworm *Hymenolepis diminuta*, the pH in the small intestine drops significantly below that in uninfected rats, with a mean pH of 6.29. Mettrick has attributed this to acidic secretions from the parasites, and also observed that as the small intestine becomes more acidic, the tapeworms migrate anteriorly. The review by Mettrick and Podesta (1974) on this and related phenomena is recommended.

Cysts of protozoan parasites, eggs or larvae of parasitic worms, and other categories of parasites which must pass through the mouth and stomach before they can become established in the small or large intestine obviously must be able to withstand these drastic alterations in pH. Furthermore, certain parasites are able to take advantage of the situation to enhance their chances of being established. Examples of how certain parasites have taken advantage of specific pHs in the host's digestive tract are provided by several species of tapeworms. For example, excystation of the cysticercoid larvae of *Hymenolepis diminuta* is initiated by acid pepsin, and complete excystation occurs under the influence of the rat host's body temperature (37°C). Similarly, *Taenia taeniaeformis* cysticerci excyst in the small intestine of cats after initial action by acid pepsin. Although the enzymatic action of pepsin undoubtedly plays a role in the excystation processes of both species, the pHs occurring in the gut activates the enzyme, hence, the acidity is of critical importance to the establishment of both *H. diminuta* and *T. taeniaeformis*. The examples provided by these two species of tapeworms indicate that these parasites are not only able to survive under a variety of pHs but are also dependent on the acidity of the stomach to stimulate further development.

Enzymes. Digestive enzymes are secreted along certain portions of the vertebrate alimentary tract (Table 2.1). These are highly efficient at degrading ingested foodstuffs. Yet, it is well known that all successful intestinal parasites are not harmed by their hosts' digestive enzymes for reasons not yet completely understood. The general explanation often given is that not only are the surfaces of intestinal parasites highly resistant to digestion, but also these organisms are capable of secreting substances that block the activity of the enzymes (for review, see von Brand, 1973).

For example, a polypeptide that is antitryptic is produced by the nematode *Ascaris*; it is similar to a trypsin inhibitor that occurs in beef pancreas. Also, Green (1957) has shown that the body wall of *Ascaris* contains two antienzymes, a trypsin and a chymotrypsin inhibitor. The antichymotrypsin activity is

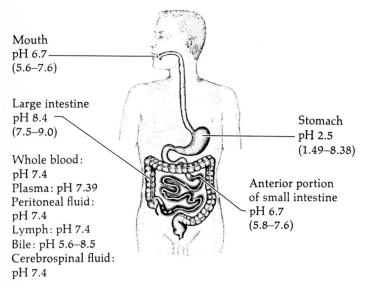

Mouth
pH 6.7
(5.6–7.6)

Large intestine
pH 8.4
(7.5–9.0)

Whole blood:
pH 7.4
Plasma: pH 7.39
Peritoneal fluid:
pH 7.4
Lymph: pH 7.4
Bile: pH 5.6–8.5
Cerebrospinal fluid:
pH 7.4

Stomach
pH 2.5
(1.49–8.38)

Anterior portion
of small intestine
pH 6.7
(5.8–7.6)

Fig. 2.1. Hydrogen ion concentrations (pH) at various levels of the human alimentary tract. (After Cheng, 1970.)

Table 2.1. Digestive Enzymes Occurring in Different Regions of the Human Digestive Tract

Region of Digestive Tract	Enzymes Present
Salivary glands	Amylase, erepsin, lipase, phosphatase
Esophagus	Lipase
Stomach	Lipase, pepsin, phosphatase, rennin, urease
Pancreas	Amylase, erepsin, lipase, maltase, trypsin
Small intestine	Amylase, enterokinase, erepsin, sucrase, lipase, phosphatase
Large intestine	Phosphatase

inhibited by heat (80°C) and trichloroacetic acid, whereas the antitrypsin activity is not. Peanasky and Laskowski (1960) have since isolated and purified the chymotrypsin inhibitor.

In addition to the role of antienzymes in protecting parasites from being digested in their hosts, two other factors may also serve this function: (1) the chemical incompatibility of the body surface in the case of certain endoparasitic helminths, and (2) the selective impermeability of individual cells. As an example of the first, Bird (1955) suggested that larvae of the nematode *Haemonchus contortus* are resistant to pepsin digestion because of the absence of the aromatic amino acids lysine and arginine in the worm's cuticle. It is known that pepsin readily attacks proteins that include a high content of these aromatic amino acids.

Besides being able to resist digestion, most parasites have developed mechanisms to take advantage of their hosts' digestive capabilities by utilizing the simple sugars, fatty acids and glycerol, and amino acids resulting from the breakdown of carbohydrates, fats, and proteins, respectively, comprising the hosts' diets. Many parasites actively compete with their hosts' mucosal cells for these nutrients, and in some instances of heavy infection , the hosts may suffer from nutrient deficiency as the result of the competition. In the case of vitamin B_{12}, for example, the broad fish tapeworm *Diphyllobothrium latum* is known to absorb such large quantities that a type of anemia due to vitamin B_{12} deficiency develops in some human hosts (von Bonsdorff, 1977).

Although many intestinal parasites utilize the breakdown products resulting from their hosts' digestion, some are capable of taking in larger molecules or particulate food either by phagocytosis, as in the case of intestinal amoebae, or by ingestion through the mouth, as in the case of flukes. Hence, it is not surprising that a number of digestive enzymes have been found in the food vacuoles and intestines of many parasites. For example, a protease, acid and alkaline phosphatases, and an esterase have been detected in *Haplometra cylindracea*, a lung fluke of frogs, which feeds primarily on their blood. Similarly, both alkaline and acid phosphatases, as well as an esterase, have been found in *Opisthioglyphe ranae*, an intestinal trematode of frogs. This fluke feeds on its host's tissues, mucus, and blood. Among the endosymbiotic amoebae, carbohydrases have been detected in a number of species, including *Entamoeba histolytica*, an intestinal parasite of humans and other mammals, *Entamoeba gingivalis*, a parasite found in human mouths, *Balantidium coli*, an intestinal ciliate of pigs and humans, and in many species of mutualistic protozoans found in the rumen of cows.

In searching for enzymes associated with endoparasites, several have been found associated with body surfaces. These may be of parasite origin, designated **intrinsic** enzymes, or they may be derived from the host, designated **extrinsic** enzymes.

Intrinsic Enzymes. These enzymes are involved in digestion of nutrients at the body surface as well as being involved in penetration and migration of molecules through host tissues. Surface digestion is best known among tapeworms. For example, intrinsic phosphohydrolase activity is associated with the surface of the tapeworm *Hymenolepis diminuta*. This enzyme mediates the hydrolytic breakdown of otherwise impermeable phosphate esters such as fructose 1,6-diphosphate. The hydrolysis of this compound results in the release of inorganic phosphate that can be absorbed. Fructose, however, can be only minimally taken up by *H. diminuta*. Similarly, ribonuclease (RNase) activity associated with the body surface of *H. diminuta* mediates the hydrolysis of ribonucleic acid (RNA) to yield nucleosides to be taken up by the tapeworm.

Extrinsic Enzymes. As will be discussed in detail later (p. 391), it has been demonstrated that certain host digestive enzymes become more efficient when adsorbed to the parasite surface, a phenomenon known as **membrane** or **contact digestion.** For example, the activity of pancreatic α-amylase of host origin is enhanced when it is adsorbed onto the body surfaces of the tapeworms *Hymenolepis diminuta, H. microstoma,* and *Moniezia expansa.* Since this enzyme is not produced by these worms, it must be an extrinsic enzyme which becomes bound to a fuzzy layer on the parasites' surface known as the **glycocalyx.** This arrangement results in a spatial advantage of proximity between the breakdown products resulting from enzymatic activity and the parasite's surface which absorbs them. The binding of extrinsic enzymes is selective, as seen in the acanthocephalan *Moniliformis dubius* which can bind host amylase but not trypsin, chymotrypsin, or lipase.

Table 2.2. Oxygen Tensions in Some Selected Habitats of Parasites[a]

Habitat	Host Species	O₂ Tension
Arterial blood	Human, dog, fish	70–100
Venous blood (heart)	Human, horse, duck	37–40
Venous blood (portal vein)	Dog, cat, etc.	49–66
Peritoneal cavity	Rabbit, rat, cat	28–40
Pleural cavity	Human, monkey	12–39
Bile	Cattle, sheep, dog	0–30
Abomasum (near mucosa)	Sheep	4–13
Rumen (gases)	Cattle, sheep, goat	0–2
Stomach (gases)	Humans	0–70
Small intestine (near mucosa)	Sheep, rat	4–30
Small intestine (gases)	Horse, cattle, dog	0–6
Small intestine (gases)	Pig	8–65
Large intestine (gases)	Horse, cattle, rabbit	0–5

[a] Figures given as millimeters of mercury (mm Hg).
(Data from von Brand, 1952.)

Parasitism

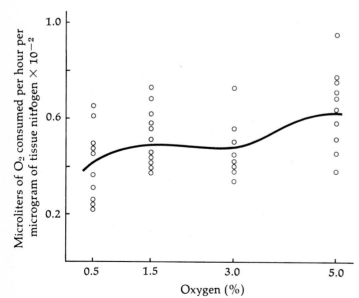

Fig. 2.2. Respiration pattern. Oxygen consumption of *Himasthla quissetensis* rediae at different oxygen tensions at 30°C. (Redrawn after Vernberg, 1963.)

Oxygen Tension. While some data are available on the oxygen tensions occurring in the main regions along the alimentary tracts of certain vertebrates, and these are usually lower than those found at other sites within the hosts' bodies (Table 2.2), one must be wary of accepting such measurements as being representative of O₂ tensions of the parasites' habitats. Rogers (1949) has shown, by employing microelectrodes, that the oxygen tension in the lumen adjacent to the mucosa of the small intestine of the rat is three times greater than that in the bulk of the intestinal content. This, however, does not negate the fact that intestinal parasites live in environments with low oxygen tensions. It is noted that just because there is some oxygen present in the parasite's microenvironment does not mean that the parasite is aerobic, i.e., requires oxygen to carry on its energy producing metabolic processes (p. 55). The classic example of this was reported by von Brand (1952) who pointed out that *Trichomonas tenax*, a parasitic flagellate found in the human mouth and pharynx, in the gingival crevices between teeth, and tonsillar crypts, respectively, is an obligate anaerobe, i.e., it cannot easily survive in the presence of oxygen, although there is obviously an ample supply in the mouth. How, then, can this paradox be resolved? Von Brand has explained that the bacteria living in association with *T. tenax* are aerobic, and as the result of their continuous utilization of the oxygen present in the gingival crevices, there is essentially no free oxygen at these sites.

Because of the low oxygen tensions along the intestinal tract of vertebrates, most parasitic flatworms and roundworms living at these sites are capable of anaerobic metabolism (p. 54). Some even have very special metabolic mechanisms that enhance energy production in environments low in, or devoid of, oxygen (p. 40). As a rule, however, these endoparasites are not obligatory anaerobes since, if oxygen is made available, they can also carry on aerobic metabolism.

Although most helminth parasites are capable of aerobic metabolism, there are some evidences that indicate that an atmosphere of high oxygen tension is toxic. For example, it is known that *Ascaris lumbricoides*, although capable of aerobic respiration, will die in about an hour when placed in an atmosphere of pure oxygen.

Conformers and Regulators. Aerobic animals, both free living and parasitic, fall into two categories depending on their response to varying ambient oxygen tensions. Among members of the first group, known as **conformers**, oxygen consumption rate decreases or increases proportionately with the amount of available oxygen, whereas among members of the second group, known as **regulators**, oxygen consumption is relatively independent of the ambient oxygen tension down to a low critical value (known as the critical oxygen tension or Pc) below which, their oxygen consumption declines rapidly.

During the life cycles of certain parasites, especially those that alternate between habitats with different oxygen tensions, the organism may be a conformer at one stage and a regulator at another. For example, Vernberg (1963) has shown that the rediae of the trematode *Himasthla quissetensis*, which live in a low-oxygen-tension microhabitat within their snail host, is a regulator. Figure 2.2 depicts the oxygen consumption pattern of these rediae at four different oxygen tensions. It is evident from these data that the rediae regulate their respiration rate somewhat independently down to 1.5% oxygen, and even when the oxygen level drops to 0.5%, the respiration rate is only 39% less than at the 5% level. On the other hand, the free living cercariae of *H. quissetensis* are typical conformers (Fig. 2.3). When the partial pressure of oxygen in their environment (seawater) is lowered, the respiration rate decreases in direct proportion to the decrease in oxygen tension. When these cercariae are placed in water containing only 0.5% oxygen, they become inactive after 12 hours and moribund or dead after 18 hours. A modification of the typical conformer pattern is illustrated by the cercariae of another marine trematode, *Zoogonus lasius*, which also conform to decreasing oxygen tensions down to the 3% level (Fig. 2.4), but maintain relatively constant oxygen uptake below this level.

At one time it was thought that the respiratory pattern of conformers was due to difficulties in the diffusion of oxygen to the various tissues of the body when the ambient oxygen tensions are low. Although this may partially explain the case of bulky organisms, it appears unlikely in such small and delicate animals as most trematode cercariae. It is now believed that rather than the problem solely involving diffusion, the proportional decrease of oxygen consumption may be the result of less efficiency in some key process of an enzymatic nature at low oxygen tensions.

Oxygen Debt. Another interesting feature associated with oxygen utilization is the phenomenon known as **oxygen debt**, a process not exclusive to parasites. Oxygen debt is manifested as a temporary increase in oxygen consumption by organisms previously exposed to anaerobic conditions. For example, when the larvae of *Eustrongylides ignotus*, a nematode parasite of ducks and chicks that can also infect cats and rabbits, and *Ascaris lumbricoides* adults are initially maintained under anaerobic conditions and then supplied with oxygen, their oxygen consumption is elevated by 30 and 60 to 100%, respectively. The explanation given for oxygen debt is that when oxygen is not available, incompletely oxidized metabolic products such as pyruvic acid or lactic acid (p. 54) are formed, and these are either excreted, accumulated in tissues, or both. When oxygen becomes available once again, a greater than normal amount is utilized

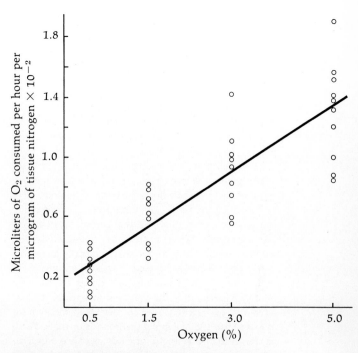

Fig. 2.3. Respiration pattern. Oxygen consumption by *Himasthla quissetensis* cercariae at different oxygen tensions at 30°C. (Redrawn after Vernberg, 1963.)

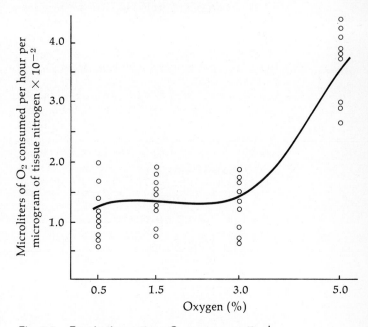

Fig. 2.4. Respiration pattern. Oxygen consumption by *Zoogonus lasius* cercariae at various oxygen tensions at 30° C. (Redrawn after Vernberg, 1963.)

in conjunction with the metabolism of these stored metabolites.

Oxygen Tension in Invertebrates. Less is known about the oxygen content in invertebrate digestive tracts. Among known instances is that in the gut of the common woodroach, *Cryptocercus punctulatus*. A number of mutualistic flagellates occur in the gut of this insect (p. 138), and it is well known that these mutualists are extremely sensitive to oxygen, being readily killed when exposed to it. Thus, the gut of *Cryptocercus* must be essentially free of oxygen.

Although molluscs represent the most common group of intermediate hosts for trematodes, very little is known about the oxygen tensions in their digestive tracts. It is known, however, that certain bivalve molluscs, especially oysters (*Crassostrea*), mussels (*Mytilus*), and mud clams (*Rangia*), become essentially anaerobic when their valves are closed. The oxygen tension in the pallial fluid of these bivalves drops significantly within minutes after the valves shut and the tissues commence to switch over to anaerobic metabolism (Hochachka, 1980). Among gastropods or snails, there is reason to believe that in the tissues of the digestive gland, an outpocket of the intestinal tract, oxygen tension is extremely low. In fact, larval trematodes that inhabit the gastropod digestive gland depend almost exclusively on carbohydrate metabolism for energy production (p. 51), suggesting that there is insufficient oxygen to permit the oxidation-reduction processes involved in lipid metabolism. Interestingly, cercariae which usually escape from the snail and become free swimming in well-oxygenated water, switch to aerobic metabolism and can utilize stored lipids as an energy source.

Aerobic respiration falls into two types: the so-called mammalian type, which is inhitbited by antimycin A, and the alternative type, which is antimycin A-insensitive. Fry and Jenkins (1984) compared respiration in the adults of ten species of nematodes and found that both types occur in these parasites. Furthermore, the alternative type is of comparable activity in all of these parasites; however, the mammalian type show variations. The extent of this latter type is correlated with both the body diameter and habitat of the worm, being greater in thinner worms and in those whose habitats are more aerobic.

Other Gases. Not only oxygen, but nitrogen, hydrogen, carbon dioxide, and methane usually occur in varying quantities in the vertebrate gut (Table 2.3). Among these, the occurrence of relatively large quantities of CO_2 is of particular interest because not only does this gas play an important role in regulating the intracellular pH of the parasites but also may stimulate certain physiologic processes within parasites which are essential for their establishment. The best-known example of the latter role is that demonstrated by Rogers (1966), who has shown that dissolved CO_2 at low redox potentials (reduction-oxidation potentials or the reducing, i.e., removal of oxygen or positive charges, and oxidizing, i.e., addition of oxygen or positive charges, powers of a solution) can stimulate the larvae of *Ascaris lumbricoides* within eggs to release a hatching fluid containing enzymes (an esterase, a chitinase, and possibly a protease), which are responsible for hatching. The pH of the dissolved CO_2 is also important (Fig. 2.5). Similarly, evidences suggest that dissolved CO_2 at 38°C is also an essential component of the hatching stimulus in the case of *Toxocara mystax*, an intestinal nematode of cats, and *Ascaridia galli*, the common intestinal nematode of chickens. In the case of these three species of roundworms, the parasites have, through evolutionary adaptations, or perhaps as the result of pre-adaptation, become dependent on the CO_2 in their hosts' digestive tracts to stimulate hatching and consequently allow them to become established.

CO_2 Fixation. As briefly stated earlier, endoparasitic helminths of mammals have evolved novel ways to cope with their essentially anaerobic habitats. One of these is **CO_2 fixation**, that is, in these organisms, CO_2 is incorporated into other molecules. The notion that CO_2 fixation occurs is based on the results of studies that show (1) the dependence of growth and metabolism on CO_2, (2) the disappearance of CO_2 from anaerobic bicarbonate-containing culture media, and (3) the dependence of glycogen utilization on CO_2. Fairbairn (1954), in a pioneering study,

Table 2.3. Percent Composition of Gases in the Small Intestine of Several Vertebrates

Gas	Cattle, Sheep, and Goat	Horse	Dog	Pig	Rabbit	Goose
CO_2	62–92	15–43	15.92	2.16–79.89	13.56–75	2.04–87.83
O_2	0	0.57–0.76	0.29	0.08–8.2	0–0.19	0–3.62
CH_4	0.04–6.6	0	—	0–28.29	2–2.83	0–13.51
H_2	0–37	20–24	26.48	0–39.56	7.72–18	0.72–20.06
N_2	1	37–60	57.28	2.02–92	6–75.71	67.92–85.28

(Data modified after Read, 1950.)

showed that in *Heterakis gallinae*, an intestinal parasite of chickens, CO_2 is fixed under anaerobic conditions and appears in proprionate and probably succinate, i.e., CO_2 taken up by worms becomes incorporated in these two metabolites. The probable pathways by which this occurs are discussed later (p. 366). It is now known that CO_2 is fixed by trematodes such as *Fasciola*, nematodes such as *Ascaris*, and tapeworms such as *Hymenolepis*.

In view of the above, it is now apparent that CO_2 not only stimulates further development on the part of endoparasites but also becomes incorporated into their metabolism. The review by Bryant (1975) on this topic gives additional details.

Physiologic and Structural Changes. In addition to the digestive enzymes present in the vertebrate gut, other secretions necessary for normal digestion also occur. No doubt such secretions as hydrochloric acid and bile have profound influences on intestinal parasites. The influence of the ambient pH in the stomach and small intestine due to hydrochloric acid has been mentioned. At this point some aspects of the influence of bile are discussed.

Bile. The importance of bile on parasites living in the small intestines and bile ducts of mammals should be obvious since at these sites the parasites are in direct contact with this substance. Bile is a yellow to olive-green fluid (because of the presence of bilirubin and biliverdin, both being degradation products of hemoglobin from destroyed red blood cells) secreted by the liver and carried to the small intestine by the hepatic duct. This duct gives off a branch, the cystic duct, by which bile can pass into and out of the gallbladder, where about 50 ml are stored in humans (Fig. 2.6). Between the junction of the hepatic and cystic ducts and the small intestine, bile flows through the common bile duct. Bile is largely a liver excretion, and in humans 500 to 1000 ml are formed daily. It contains inorganic salts, mucin, nucleoproteins, cholesterol, pigments, lipids, CO_2, ammonia, urea, and purine derivatives. The major constituents of mammalian bile are listed in Table 2.4, and some properties of human and animal biles are listed in Table 2.5. As in the intestine, the oxygen tension of bile is low, ranging from 0 to 30 mm Hg. Thus, it is essentially an anaerobic habitat.

The functions of bile in digestion are to lower the surface tension and increase the emulsification of fats, thus aiding in their digestion and absorption. Its ability to perform these functions is almost exclusively due to the presence of bile salts, which are in the form of sodium salts of glycocholate and taurocholate conjugated with the amino acids taurine and glycine. Salts other than glycocholate and taurocholate are found in mammals other than humans. For example, the lithocholate, chenodeoxycholate, and deoxycholate are other common mammalian bile salts. It is of interest to point out that glucose is also present in bile and may

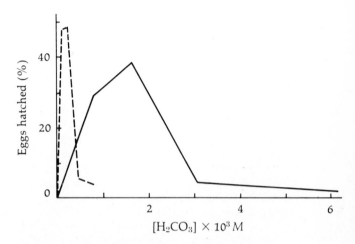

Fig. 2.5. **Effects of physical factors on hatching.** The hatching of *Ascaris lumbricoides* eggs at different concentrations of dissolved carbon dioxide and hydrogen ions. The medium was bicarbonate–carbon dioxide buffer at pH 7.3 (broken line) and pH 6.0 (solid line) containing 0.02 *M* sodium dithionite under mixtures of nitrogen and carbon dioxide. (Redrawn after Rogers, 1966.)

be of nutritional value to both host and parasites.

Bile is known to play at least three important roles in assisting the successful establishment of parasites. (1) Because of the presence of $NaHCO_3$ (sodium bicarbonate), bile is commonly alkaline (Table 2.5). Its alkalinity acts as a buffer against the acidity of the

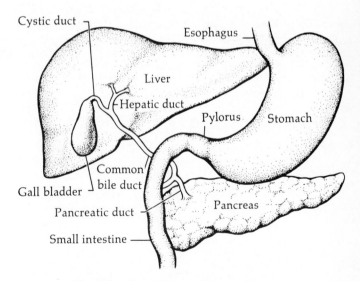

Fig. 2.6. **Mammalian digestive tract.** Drawing showing anatomic relationship between digestive tract, liver, gallbladder, and pancreas in mammals. (After Cheng, 1970.)

gastric hydrochloric acid, thus enabling parasites to survive. (2) Bile is also known to stimulate certain parasites to undergo further development. As examples, bile salts have been shown to stimulate the activation or excystment of a number of tapeworm larvae (Table 2.6), without which these worms could not undergo further development. It has been suggested that the specificity of the bile-tapeworm stimulation phenomenon may be a controlling factor in host specificity, i.e., if the bile salts of a certain host fail to stimulate further development of a parasite, that parasite would not become established in that host; if it does, the parasite is specifically adapted to that host. (3) Certain parasites, such as *Fasciola hepatica* and *F. gigantica*, liver flukes of cattle and sheep, live within the common bile duct. In this confining habitat, the acidic metabolic wastes produced by the parasites no doubt would be deleterious to the parasites themselves were it not for the fact that the alkalinity of the bile counteracts the acidic metabolites.

The function of bile in parasitism serves as another example of the influence of the host's digestive physiology on the establishment of parasites. The interaction between the digestion-associated organs, including the liver and gallbladder, and parasites has been reviewed by Cheng (1976).

Since bile salts appear to be specifically absorbed in the distal portion of the small intestine, it might be expected that a gradient in bile salt concentration exists in the small intestine. Indeed, Dietschy (1967) has demonstrated that gradients of conjugated and nonconjugated bile salts occur in the rat intestine, and these gradients are not linear (Fig. 2.7). Hepatic bile includes a concentration of bile salts at about 33 mM when delivered to the duodenum, at which point it is rapidly diluted to about 8 mM. Subsequently, as bile passes down the alimentary tract, the absorption of water and solutes occurs more rapidly than the diffusion of bile salts into the mucosa. Consequently, there is an increase in the bile salt concentration until the region of specific absorption is reached (about segment 8). In the distal part of the small intestine there is a rapid decline in the intraluminal concentration of bile salts.

Since, as indicated, bile salts influence the establishment and further development of certain parasites, the

Table 2.4. Quantitative Measurements of the Major Constituents of Bile in Several Animals.[a]

	Human	Ox	Pig	Rabbit	Guinea Pig	Dog	Rat[b]
Total lipids	—	100–160	—	—	140	—	—
Bilirubin	1,000	—	32–62	87–131	—	92–170	8–9
Cholesterol	630	37	130–180	10–120	—	80–100	12.7
Fatty acids	970	370	820–2,000	—	—	1,600–5,000	—
Total bile salts	5,180	7,200	8,500–12,000	1,100–2,600	780	7,900–15,000	—

[a] The amounts are presented in mg/100 ml of bile.
[b] No gallbladder.
(Data modified after Dittmer, 1961.)

Table 2.5. Some Properties of Animal Biles[a]

Species	O₂ Tension (mm Hg)	Daily Secretion Per Kilo Body Wt. (ml)	pH Hepatic Bile	pH Gallbladder Bile	Δ°C	Cholesterol Content (mg/%)	Glucose Content (mg/%)
Crow	—	13–110	—	—	—	—	—
Goose	—	6–18	—	—	—	—	—
Rat	—	40–60	—	—	—	—	—
Guinea pig	—	160–200	7.7–7.8	7.2–9.1	—	—	—
Rabbit	—	74–219	—	6.4–6.7	—	100–120	20
Sheep	0–30	21–30	5.9–6.7	6.0–6.7	−0.59 to −0.6	—	—
Ox	—	—	—	6.7–7.5	−0.53 to −0.6	30–70	—
Cat	—	14	—	5.3	—	—	—
Dog	0–30	8–76	7.1–8.2	5.2–6.9	—	110–140	55–88
Human	—	15	6.2–8.5	5.6–8.0	−0.56	—	—

[a] The oxygen tension of bile is low (0–30 mm Hg) so that, like the gut, the bile duct is essentially an anaerobic habitat. It is noted that the pH range of bile from the liver (hepatic bile) is generally higher than that stored in the gallbladder. (After Smyth, 1962.)

fluctuation in bile salt concentrations along the host's digestive tract undoubtedly influences the site of parasite establishment.

It is of interest to note that the bile of carnivores includes mostly salts of cholic acids whereas that of herbivores contains high levels of salts of deoxycholic acid. Hence, the hydatid tapeworm, *Echinococcus granulosus*, can only develop and live in carnivores because it is rapidly lysed by sodium deoxycholate but is unaffected by cholate (Smyth, 1962). This is another example of how bile can govern compatibility or incompatibility between host and parasite.

Peristalsis. When food reaches a certain part of the esophagus, the circular muscles behind the food contract, resulting in narrowing of the esophagus and exerting pressure on the food, pushing it posteriorly. A sequence of these contractions, known as peristalsis, occurs until the food is pushed into the stomach. In the stomach, vigorous muscular contractions of the gastric wall cause continuous churning of food. When the food, now known as chyme, reaches the small and large intestines, peristalsis continues, pushing both chyme and feces toward the anus. These peristaltic motions represent one type of physical obstacle intestinal parasites must successfully confront and overcome, or else they would be swept out of the host along with the undigested material. Indeed, in some instances this does happen when the parasite for some reason can no longer cope with this physical pressure. In other instances, such as with intestinal bacteria and protozoans, many are continuously passed out in their hosts' feces but their rapid reproductive rates enable these microorganisms to maintain a dynamic equilibrium between those being voided and those present in the intestine. Multicellular parasites, on the other hand, have much more complex life cycles, usually do not reproduce in their vertebrate hosts' intestines; consequently they cannot maintain this type of dy-

namic equilibrium. However, they have acquired other means of overcoming this handicap. The topic of mechanical anchorage of symbionts to hosts has been reviewed by Nachtigall (1974). Reviewed below are some of the more salient mechanisms employed

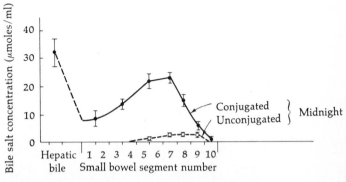

Fig. 2.7. Bile salts. Bile salt concentrations down the length of the small intestine of the rat. Rats (200–225 gm) were fed food and water and then killed at midnight when their stomachs and intes- tines were filled with food. The small bowel was divided into 10 segments of equal length, numbered for purposes of identification from 1 to 10, anterior to posterior. The intestinal contents of each segment were centrifuged, and the concentration of conjugated and unconjugated bile salts in the supernatant fraction was quantified by use of thin-layer chromatography. In another group of animals, the concentration of bile salt in hepatic bile was determined in samples of bile obtained within 10 minutes after cannulation of the common hepatic duct. (From Dietschy, 1967.)

Table 2.6. Some Factors Contributing to the Excystation of Larval Cestodes[a]

Cestode Species	Acid Pepsin (Effect on Cyst Digestion)	Trypsin (Effect on Excystment)	Bile Salts (Effect on Excystment)	Temperature (Effect on Excystment) 18–26°C	37°C
Hymenolepis citelli	Initiates	None[b]	Activation only	None	Excyst
H. diminuta	Initiates	None[b]	Some excystment	None	Excyst
Vampirolepis nana	Initiates	None[b]	Activation only	None	Excyst
Taenia taeniaeformis	Essential	None	Some excystment	Excyst	Excyst
T. solium	Essential	None[b]	Excyst		Excyst
T. pisiformis	Unessential		Excyst	Excyst	Excyst
T. tenuicollis			Excyst	Excyst	Excyst
Taeniarhynchus saginatus			Excyst	Excyst	Excyst
Oochoristica symmetrica	Unessential	None	Excyst	Excyst	Excyst
Raillietina kashiwariensis	Unessential	Excyst[c]	Some excystment	None	Excyst[d]

[a] After Read and Simmons, 1963.

[b] Produces excystment if bile salts are present.

[c] Pancreatin active, lipase relatively inactive, amylase not tested.

[d] Temperature at 40° to 42°C.

by metazoan parasites, as well as some of those employed by protozoan parasites.

Because of their microscopic sizes, many species of intestinal protozoans, unless they are entangled in luminal chyme as are most intestinal bacteria, are not greatly affected by peristalsis. Furthermore, many

protozoans, such as *Entamoeba histolytica*, are generally lodged in the minute crypts of the intestinal wall. This is not to say that this and other species of intestinal protozoans are not passed out in the hosts' feces; many are. In fact, one of the diagnostic techniques used to determine whether an individual is suffering from amoebiasis is to search for trophic and cystic amoebae in the patient's diarrheic stools. Another human parasite, the flagellate *Giardia lamblia*, has a ventral surface shaped like a sucking disc by which it holds on to the host's mucosa. Still other groups of intestinal protozoans, such as the gregarine apicomplexans parasitic in invertebrates, have specialized regions of the head by which they are attached to their hosts' intestinal cells (Fig. 2.8).

In the case of the intestinal helminths, which represent a large group of gut-dwelling parasites, specialized body structures have evolved to help cope with the problem. Among tapeworms, not only are their long and flattened bodies intimately held against their hosts' intestinal mucosa and bend with each peristaltic motion, but each worm possesses a holdfast

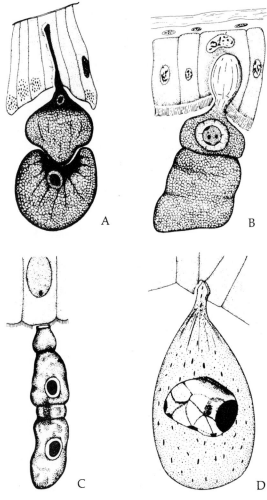

Fig. 2.8. Examples of some apicomplexans attached to the epithelial lining of their hosts' digestive tracts. A. The gregarine *Garnhamia aciculata* attached to the midgut of the Ceylon silverfish, *Peliolepisma calva*. (Redrawn after Crusz, 1957.) **B.** The gregarine *Pileocephalus striatus* attached to the gut of the larva of the insect *Ptychoptera contaminata*. (Redrawn after Léger and Duboscq, 1925.) **C.** Linearly arranged trophozoites (sporadins) of the gregarine *Nematopsis legeri* attached to the ciliated gut epithelium of the crustacean *Eriphia spinifrons*. (Redrawn after Hatt, 1931.) **D.** Trophozoite of the gregarine *Merogregarina macrospora* attached to the gut of the ascidian *Amaroucium* sp. (Redrawn after Porter, 1940.)

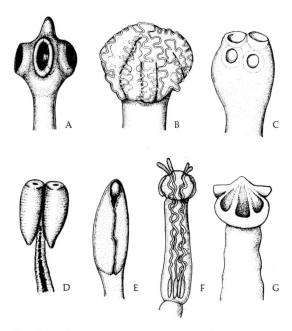

Fig. 2.9. Several types of tapeworm scolices. A. Scolex of *Baerietta baeri* (Cyclophyllidea), a parasite of the toad *Bufo asiaticus* in China. **B.** Scolex of *Phyllobothrium lactuca* (Tetraphyllidea), a parasite of elasmobranch fish in Europe and the Orient. **C.** Scolex of *Proteocephalus coregoni* (Proteocephala), a parasite of the fish *Coregonus atikameg* from the Hudson Bay. **D.** Scolex of *Bothridium pithonis* (Pseudophyllidea), a parasite of the python, *Python molurus*. **E.** Scolex of *Adenocephalus pacificus* (Pseudophyllidea), a parasite of sea lions. **F.** Scolex of *Diplootobothrium springeri* (Trypanorhyncha), a parasite of the dogfish *Platysqualus tudes* from Florida. **G.** Scolex of *Glaridacris hexacotyle* (Caryophyllidea), a parasite of the sucker (fish) *Catostomus* sp. from Arizona. (Redrawn after various authors.)

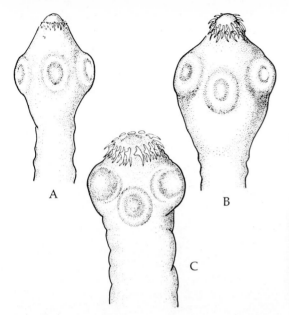

Fig. 2.10. Examples of some tapeworms with armed scolices. A. *Scolex of Echinococcus granulosus.* **B.** Scolex of *Multiceps packi.* **C.** Scolex of *Taenia solium.*

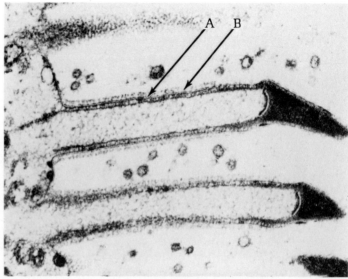

Fig. 2.11. Electron micrograph of portion of outer surface of tegument of the cestode *Schistocephalus solidus* **showing two adjacent microtriches each consisting of a proximal shaft and an electron-dense apex. A.** Plasma membrane covering tegument. **B.** Glycocalyx adhering to surface membrane. (After McCaig and Hopkins, 1965.)

organ, known as the **scolex**, at its anterior end. The shapes, sizes, and accessory structures of tapeworm scolices vary considerably among the thousands of known species (Fig. 2.9). These differences comprise one of the common criteria used in classifying tapeworms.

In addition to the scolical suckers, some species possess an apical snout known as a **rostellum**. The rostellum of many species, including *Echinococcus granulosus*, the hydatid tapeworm (Fig. 2.10), *Multiceps packii*, a canine tapeworm (Fig. 2.10), and *Taenia solium*, the pork tapeworm of humans (Fig. 2.10), is armed with one or more circular rows of hooks which are embedded in the host's intestinal wall to further aid in attachment.

In spite of the efficiency of the scolex, with its suckers and rostellar hooks, doubt may still be expressed as to how such a relatively small structure as the scolex can possibly prevent an entire tapeworm from being swept away with the flow of luminal contents. After all, as in the case of *Hymenolepis diminuta*, a rat tapeworm, the unarmed (hookless) scolex is only 300 μm wide and not much longer, and each sucker is only 100 to 200 μm in diameter. Studies with the electron microscope have revealed surface projections known as **microtriches** extending from the entire surface of the body of this worm (Fig. 2.11). It has been proposed that these microtriches, through contact with the microvilli of the striated border in the host's small intestine, may serve as another means of resisting the intestinal current. Microtriches have been

demonstrated on all tapeworms that have been studied with the electron microscope.

Most intestinal flukes possess two powerful muscular suckers strategically situated on the body, one at the front end and one at or near the middle of the ventral body surface. These suckers actually hold on to the host by grasping a piece of the mucosa (Fig. 2.12). In addition, tegumental spines on their body surfaces also play an adhesive role.

Among nematodes, although the structure of the mouth and the anterior portion of the alimentary tract is not significantly different from those of free-living species, these structures are highly efficient at holding on to the host's intestinal wall. In addition, certain species, such as the hookworms, possess cuticular teethlike structures, known as buccal teeth or plates, with which they can grasp and tear into their hosts' mucosal tissues (Fig. 2.13).

One of the most detailed studies on the holdfast mechanism of an endoparasitic helminth is that by Hammond (1966, 1967) on the spinyheaded worm (acanthocephalan) *Acanthocephalus ranae* (Fig. 2.14). This parasite attaches itself to the intestinal wall of its anuran host by embedding its eversible proboscis into its host's gut wall. Furthermore, the proboscal hooks, which are directed backward, become firmly anchored

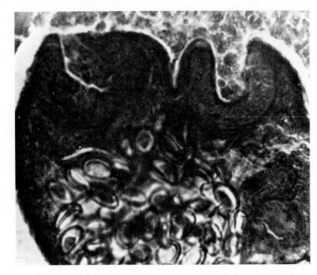

Fig. 2.12. **Function of digenean suckers.** Photomicrograph showing both oral sucker and acetabulum of *Hasstilesia tricolor* holding onto intestinal mucosa of rabbit host.

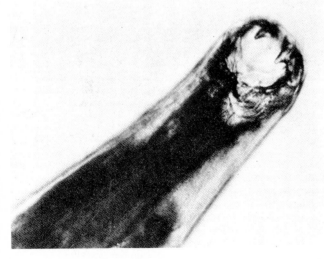

Fig. 2.13. **Hookworm teeth.** Dorsal view of the Old World hookworm, *Ancylostoma duodenale*, showing powerful ventral teeth. (Armed Forces Institute of Pathology Neg. No. N−41730−2.)

in the gut wall when the neck retractor muscles retract, and by so doing pull the worm's body tightly against the host. Also, as the muscles contract, fluid from a pair of internal pouches, known as **lemnisci**, is squeezed into spaces in the proboscis wall. This causes the diameter of the proboscis to expand, thus locking the parasite firmly to the host's intestinal wall. This

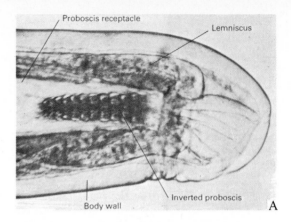

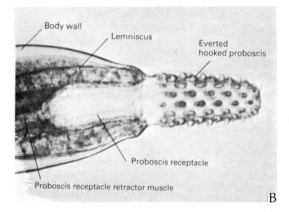

Fig. 2.14. **Proboscis and associated structures of *Acanthocephalus ranae*. A.** Inverted proboscis. **B.** Everted proboscis. (After R. A. Hammond in Whitfield, 1979; with permission of University Park Press.)

intricate holdfast mechanism involving hydraulic pressure and hooks provides extremely firm anchorage.

Mucosa. It has been stated that many of the intestinal parasites are at least partially intimately associated with the intestinal mucosa. Hence, the biochemical and structural composition of the mucosa may profoundly influence the nature of substances available to parasites.

The vertebrate mucosa has a structurally elaborate brush border of microvilli facing the lumen of the gut. Recent evidences suggest that the brush border functions in digestion as well as the separation of functional digestion and absorption. For example, Miller and Crane (1961) have found the digestive disaccharidases (maltase, isomaltase, sucrase, lactase, and trehalase) in the microvilli. Furthermore, Overton *et al.* (1965) and Johnson (1967) have demonstrated that these enzymes are associated with surface particles about 50 Å in diameter attached by salts to the luminal surface of the brush border. Crane (1967) has suggested that the monosaccharides resulting from the action of the hydrolytic enzymes associated with these particles are more efficiently absorbed than free monosaccharides added to the intestinal lumen. This kinetic advantage of utilizing disaccharides intimately associated with the brush border undoubtedly has important effects on the availability of sugars to parasites in contact with the mucosa, especially those that appear to be sharply limited in the quality of carbohydrates that can be absorbed and metabolized.

It is apparent from this discussion that the many features of the alimentary tract greatly influence the suitability or unsuitability of this environment for various parasites. For those interested in comparable information pertaining to the avian alimentary tract, the detailed review by Crompton and Nesheim (1976) is recommended.

MAMMALIAN BLOOD

The circulatory system, especially blood cells and plasma, represents another favored habitat for certain parasites, protozoans and worms. Like the alimentary tract, this system is also extremely rich in nutrients, as shown by Dittmer (1961) in a review of the chemical composition of blood. For such parasites as the fluke *Schistosoma mansoni*, blood is an ideal habitat because this fluke possesses a mouth and all necessary digestive enzymes to utilize the plasma proteins and blood cells as nutrients. In addition, these parasites can utilize the fats (triglycerides, lecithin, cholesterol), carbohydrates, and amino acids present in blood.

The gaseous content of blood varies with the host species and whether it is arterial or venous blood. In humans, for example, there is a 27% difference in the O_2 tension between arterial and venous blood, the former being higher. In birds, which have a higher metabolic rate, the difference may be as great as 60%. The CO_2 tension is remarkably constant in blood, remaining at about 40 mm Hg. The pH is also relatively constant, varying only slightly above or below pH 7.0, the average pH value being 7.4 in most birds and mammals.

Although also present to some extent in the lower digestive tract, the major deterrent of blood as a suitable habitat for many parasites is the occurrence of antibodies (immunoglobulins). These, as will be described in detail in Chapter 3, are complex molecules synthesized in response to antigenic challenge, including invading parasites. It is noted, however, that parasites that live in the blood, either in plasma or blood cells, have evolved intricate mechanisms to overcome the deleterious effects of antibodies. These adaptations are discussed in subsequent chapters.

THE PHASES OF PARASITISM

In recent years impressive advances have been made in the study of parasites and parasitism. This has been due, in part, to advances in scientific instrumentation, but even more important, parasitologists have come to realize that the phenomenon of parasitism can be understood only if it is conceptually subdivided into its component phases and each of these subjected to analysis. Although parasitism may be subcategorized in a variety of ways, Cheng's (1967) proposal that three major phases be recognized is being adopted with minor modifications. The phases proposed are (1) host-symbiont contact, (2) establishment of the symbiont, and (3) escape of the symbiont. Each of the major phases can be further subdivided as outlined in Table 2.7. In the following sections, examples of each of the phases of parasitism are presented.

HOST-SYMBIONT CONTACT

Since, by definition, parasites are metabolically dependent on their hosts, for a parasite to be successful in not only maintaining its own life but also reproducing and thus perpetuating its species, it must successfully become attached to or enter a suitable host. As indicated in subsequent chapters, many parasites have one or more free-living stages during their life cycles. The free-living stages may either actively or passively make contact with the host. In the first instance, the infective form approaches the host and makes contact, while in the latter, it is the host that makes contact with the infective stage of the parasite. Gradients between these two approaches exist. For example, the

Table 2.7. The Phases of Parasitism and the Factors Influencing Each

---

Host-symbiont contact
  Active contact
    Contact influenced by chemotaxis
    Contact influenced by other taxes
    Contact influenced by parasite selectivity
    Contact influenced by nature of substrate
  Passive contact

Preparation for entry

Establishment of the symbiont
  Niche selection and attachment
  Overcoming the host's defense mechanisms
  Acquisition of adequate nourishment
  Dependency on host's enzymes
  Developmental and growth stimuli of host origin
  Pathogenicity of parasite

Escape of the symbiont
  Active escape
  Passive escape

infective form of the parasite, such as hookworm larvae (p. 506), may be attracted to the skin of the host, but the behavior of the host, i.e., being in the proximity of the parasite and exposing bare skin, is also necessary.

ACTIVE CONTACT

As indicated in Table 2.7, contact between an active, infective organism, and its host may be (1) accidental, (2) influenced by chemotaxis, (3) influenced by other taxes, (4) influenced by the visual or chemosensitivity of the parasites, or (5) influenced by the nature of the substrate. Naturally, combinations of two or more of these factors may be involved.

Accidental Contact. This type of contact is limited primarily to phoresis. For example, the settling of hydroids on the shells of marine bivalves.

Contact Influenced by Chemotaxis. Many symbionts, ranging from ectocommensals to endoparasites, in varying degrees, seek and contact their hosts. It is primarily with such active symbionts that specific attraction to the host is suspected.

Among parasites, the favored model for studying chemotactic attraction involves trematode miracidia and their molluscan intermediate hosts. This topic has been reviewed by Cheng (1967), Ulmer (1971),

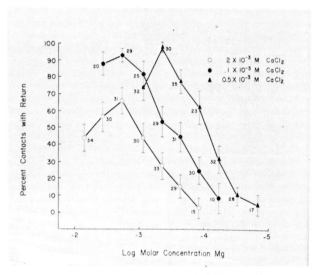

Fig. 2.15. Activity of *Schistosoma mansoni* miracidia. Stimulatory effects of various concentrations of MgCl$_2$ and CaCl$_2$. The assays were conducted at pH 7.5. Numbers at plot points represent the numbers of miracidia tested. The vertical lines represent standard deviations. Notice that the lower Ca/Mg ratios are more attractive to miracidia. (After Aponholtz and Short, 1976; with permission of *Journal of Parasitology*).

Chernin (1974), and MacInnis (1976). In brief, ample evidences now exist which indicate that chemotaxis is involved in miracidium-mollusc contact (Wright, 1959; von Plempel, 1964; MacInnis, 1965; Chernin, 1970). In fact, Chernin has proposed the term **miraxone** to designate the chemoattractant(s) emitted by snails to attract miracidia.

The chemical nature of miraxone still remains elusive, although it is known that the material emitted by the snail *Biomphalaria glabrata* is thermostable and water soluble. Based on more recent evidence, the possibility exists that a variety of materials can serve as the chemoattractant. These include certain amino acids (glycine, tyrosine, proline, histidine, phenylalanine, methionine, leucine, asparagine, cysteine, valine, aspartic acid), butyric acid, the sugar galactose, and weak dilutions of hydrochloric and acetic acids (MacInnis, 1965). On the other hand, based on the reasoning that snails must remove calcium from their environment for shell formation, the levels of calcium and related ions in water conditioned by snails may affect miracidial behavior, Sponholtz and Short (1976) have shown that water containing less calcium and more magnesium, i.e., lower Ca/Mg ratios, are attractive to miracidia of *Schistosoma mansoni* (Fig. 2.15). Also, Stibbs *et al.* (1976) have reported that a major component of water conditioned by the snail *Biomphalaria glabrata*, an intermediate host for *S. mansoni*, is MgCl$_2$ (0.24 mM), which elicits an excellent response from *S. mansoni* miracidia. The conditioned water also included calcium. They concluded that miraxonal activity in snail-conditioned water can be attributed to the magnesium content.

Although the examples given of chemotactic response by parasites have concentrated on miracidia and molluscs, there is ample evidence that infective stages of some other categories of parasites are also chemotactically attracted to their hosts. With the establishment of the occurrence of this phenomenon by examining the miracidium-mollusc relationship in some detail, there is no need to cite additional examples; however, other examples are found in subsequent chapters. It should be pointed out, however, that chemotactic response on the part of parasites is not in a direct, straightline path toward the host, but commonly follows a circuitous route.

Contact Influenced by Other Taxes. Since chemotaxis is operative only within short distances, it would appear possible that other mechanisms exist that tend to bring the symbiont and its host into the proximity of one another from greater distances.

Not all of the mechanisms involved in host-habitat location are yet known; however, the natural taxes of both symbiont and host undoubtedly serve as major forces in bringing the partners into each other's proximity. Thus, for example, the negative geotaxis and positive phototaxis of *Schistosoma mansoni* miracidia serve to bring these ciliated larvae to the habitat of their molluscan host which, as a rule, is found in the subsurficial region of bodies of water, clinging to the underside of vegetation. Similarly, phototaxis among certain larval parasitic nematodes has been reported (see Croll, 1975, for review).

Alterations in Taxes. It is of interest to mention the studies by Welsh (1930, 1931) on tactic responses, which revealed that once host-symbiont contact is made, the original response may become altered and serve to further advantage in permitting successful establishment of the symbiont.

Working with the mite *Unionicola ypsilophorus* var. *haldemani*, which inhabits the mantle cavity and gills of the freshwater mussel *Anodonta cataracta*, Welsh found that the tactic responses of the mite could be reversed by a host factor. Specifically, he demonstrated that if *U. ypsilophorus*, removed from its host, is washed free of a factor of host origin, it is positively phototactic. If the host factor is added to an aquarium containing positively phototactic mites, these immediately reverse to negative phototaxis. Welsh has proposed that the host factor may be a decomposition product of mucus or some other protein from the gills. He has demonstrated that the taxis-reversing factor is specific. By testing materials from eight species of bivalves on three species of mites, he found that only materials from natural hosts cause reversal of taxis.

These interesting findings by Welsh indicate that the natural taxes of a symbiont may be altered once it makes contact with the host. In the case of the mites studied, Welsh has said, "This reversal may be considered adaptive, for, aided by a positive chemotropism and stereotropism, it enables the mites to enter and remain within the host."

Contact Influenced by Parasite Selectivity. It is known that true predators among arthropods and vertebrates with specialized food habits may locate their prey by highly precise behavior involving, at least in part, olfactory recognition of the prey. Although this type of precise recognition during predation has been demonstrated among certain snails and nudibranchs, its importance in symbiotic relationships has been little studied except in the case of a few parasitic insects. It is noted, however, that considerable work has been done on the structure of sensory organs of parasitic worms, especially the nematodes (see McLaren, 1976, for review), but owing to the lack of satisfactory electrophysiological techniques at this time, the exact functions of these body surface receptors remain uncertain. Nevertheless, there is no doubt that at least some are involved in host recognition.

Among parasitic insects two classical examples may be cited. Laing (1937) has demonstrated that the chalcid insect *Alysia manducator* may be attracted by olfaction to the environment of its host, which is the blowfly larva. The parasité is attracted, as the result of a powerful chemotaxis, to some factor(s) in decom-

posing meat. Thus, although *A. manducator* is not directly attracted to its host, its selective preference for rotting meat, which is its host's habitat, indirectly aids in its making contact with the preferred host.

Another example of host preference among parasitic insects concerns the parasitic wasp *Nemeritis*. Thorpe and Jones (1937) have demonstrated that if the eggs of *Nemeritis* are artificially introduced into an unnatural host, the wax moth *Meliphora*, instead of its natural host, *Ephestia*, the resulting adults give a strong olfactometric response to *Meliphora* when given a choice between it and a blank. These workers, however, were not successful in attempts to demonstrate a preference of conditioned *Nemeritis* for *Meliphora* over *Ephestia*. Nevertheless, it is of interest to point out that wasps that develop in the unnatural host become sufficiently conditioned so as to recognize the new host. This model is useful in considering the origin of new host–symbiont associations. It is conceivable that many such associations arose when during a brief free developmental period in the life of the symbiont it happened to encounter a potentially new host in some habitat differing from that of its natural host. If the new host produces some factor(s) similar to, but not identical with, that with which the symbiont has been associated in the natural host, it is conceivable that an effective conditioning to the new host factor(s) may bring about a change in host preference, aided perhaps by a shortening or loss of the free stage, and resulting in subsequent generations becoming genetically isolated in or on their new host. From then on, the course of evolution of the symbiont would be controlled by that of its new environment—the new host (Davenport, 1955).

PASSIVE CONTACT

If the infective organism is passive, in the majority of instances its mode of contacting the host is dependent on the latter's feeding habits. This is exemplified by a number of parasites of humans. For example, the intestinal amoeba *Entamoeba histolytica* infects humans when its nonmotile cyst is ingested as a contaminant of food or drink. Similarly, the infective eggs of such intestinal round worms as *Ascaris lumbricoides*, *Enterobius vermicularis*, and *Trichuris trichiura* enter humans by ingestion. In these instances, poor sanitation is an important factor in host-parasite contact.

In some cases, the host's feeding habits also play a major role in host-parasite contact despite the fact that the infective parasite is motile. This is especially true of aquatic invertebrate hosts, such as bivalve molluscs. For example, in the case of sedentary, filter-feeding molluscs, smaller parasitic organisms are drawn into the host by its currents, and if they successfully pass through the selective process of the host's gill apparatus, can become established as endoparasites or, if they do not, certain species could become attached to the exterior of the soft tissues as ectosymbionts. The thigmotrichous ciliates, often found in the mantle cavity or on the gills and palps of estuarine pelecypods, belong to the second group, whereas the ciliate *Trichodina myicola*, found in the

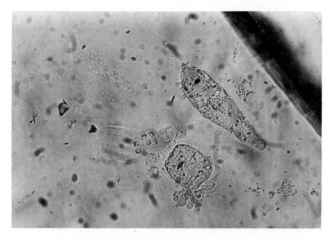

Fig. 2.16. Miracidial response. Photomicrograph of ciliated miracidium of *Fasciola gigantica* and one that has shed its epidermal plates after 1 minute of contact with 1:10 dilution of the hemolymph from its normal molluscan host *Galba ollula*. (After Cheng, 1968.)

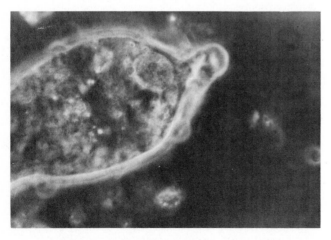

Fig. 2.17. Miracidial secretion. Phase-contrast photomicrograph showing invagination of apical papilla of *Fasciola gigantica* miracidium and secretion of lytic enzyme after exposure to undiluted *Galba ollula* hemolymph. (After Cheng, 1968.)

alimentary canal of the soft-shell clam, *Mya arenaria*, belongs to the first.

Commonly, cannibalistic hosts become parasitized while feeding on other animals that are parasitized. Cheng and Alicata (1965) have reported that the transfer of the third-stage larvae of the nematode *Angiostrongylus cantonensis* from one land snail, *Achatina fulica*, to another can be effected by this method.

PREPARATION FOR ENTRY

An interesting aspect of host-parasite contact that deserves increased emphasis is preparation for entry. This does not occur with all endoparasites; rather, it is limited to certain helminths. The phenomenon is best illustrated by the miracidia of certain species of trematodes. Dawes (1960) has expressed the view that the shedding of the ciliated epidermis of these larvae is a prerequisite for successful penetration of the snail intermediate host (p. 312). This may be true for the fasciolid trematodes (p. 347), but not for the schistosomes, under natural conditions. Schistosome miracidia usually retain their ciliated epidermis until they enter the molluscan intermediate host (Lengy, 1962). Cheng (1968) has shown that a "host factor" present in the hemolymph of the snail *Galba ollula* not only stimulates the miracidium of *Fasciola gigantica* to shed its epidermis (Fig. 2.16), but also causes it to invert its apical papilla to form a cuplike structure in preparation to hold on to the snail host's epidermis (Fig. 2.17). The question of how a miracidium outside of the mollusc can come in contact with hemolymph is answered by the fact that molluscs, except for cephalopods, have open circulatory systems and hemolymph in sinuses are sporadically leaked to the exterior by a process known as diapedesis.

Lytic Secretions. Most parasites that gain entry by penetrating the surface of their hosts secrete lytic enzymes to facilitate this process.

It should be emphasized that invasive parasites enter their hosts not only from external surfaces but also through the intestinal mucosa,. Although the actual invasion processes of most species of parasites appear similar superficially, the chemical mechanisms involved may be different. Specifically, the secretions may alter the host tissues in a number of ways, including affecting the ground substances and basement membranes of integuments and intestinal linings, increasing the amount of free solvent water, depolymerizing carbohydrate-protein complexes, and lysing or digesting the surface cells. In all cases that have been studied, proteolytic enzymes have been identified as components of lytic secretions. In the case of *Schistosoma mansoni* cercariae, hyaluronidase and collagenaselike enzymes also are involved. Hyaluronidase hydrolyzes hyaluronic acid, a constituent of the intercellular ground substance which cements cells together, thus rendering the intercellular spaces amenable to passage. Collagenase mediates the breakdown of collagen, a constituent of basement membranes.

ESTABLISHMENT OF THE SYMBIONT

Even if the parasite makes contact with a potential host, there is no assurance that it will become established and continue normal growth and development. As outlined in Table 2.7, a number of factors are known to influence successful establishment. The principal ones are (1) successful habitat selection and attachment, (2) overcoming the host's defense mechanisms, (3) deriving adequate nourishment, (4) dependency on the host's enzyme systems, (5) provision of the appropriate developmental and growth stimuli, and (6) pathologic changes induced by the parasite. These factors are discussed below.

Habitat Selection and Attachment

Parasites, both endo- and ectoparasitic species, are very specific as to habitat requirements. Adults of the liver flukes, *Fasciola* spp., usually only occur in the bile ducts of their definitive host, as is the tapeworm *Hymenolepis microstoma*. Similarly, certain species of lice are found only at certain sites on the hosts' body surfaces (p. 639). Even the intestinal parasites are quite specific as to which segment of the intestinal tract they occupy. There is little doubt that appropriate habitat selection based on biochemical, physiologic, and physical needs is extremely important if a parasite is to become established. Numerous examples of this principle are given in subsequent chapters.

In addition to occupying a compatible site, each parasite must become mechanically attached. This attachment not only must prevent the involuntary shifting of the organism from its habitat but in some instances is essential for its physiologic well-being. Such is the case with the tapeworm *Echinococcus granulosus* with its scolex embedded in its definitive host's mucosal crypts (p. 410). As stated earlier, various species of helminths are well adapted for holding on (p. 44), and this is also true for ectoparasitic arthropods and practically every other category of zooparasite.

Host's Defense Mechanisms. There is little doubt that one of the major factors affecting the successful establishment of parasites, especially endoparasites, is their ability to overcome the host's internal defense mechanisms. This aspect of parasitology has been investigated with such intensity that the next chapter is entirely devoted to it. At this point it suffices to state that all parasites elicit host response upon exposure and for establishment the parasite must be capable of overcoming this.

Adequate Nourishment. Although significant progress has been made in understanding the nutritional requirements of parasites, especially in those species which are of medical and economic importance, progress has been slow because very often it is only with the establishment of chemically defined *in vitro* culture media that such requirements can be critically analyzed. Current methods for culturing protozoan and helminth parasites have been described in a volume edited by Taylor and Baker (1978). It is noted, however, that many methods do not involve chemically defined media and hence are not useful tools to the nutrition biochemist.

It must also be borne in mind that a thorough understanding of the nourishment of parasites involves more than the study of the metabolic fate of molecules taken into the body. It is equally important to realize how specific parasites take up nutrients and in what form. These aspects of parasite biology are discussed in detail in the subsequent chapters dealing with specific groups. At this point, the general metabolic schemes employed in the synthesis of energy as they occur in living organisms, including parasites, will be considered. In other words, the more common metabolic pathways that occur after the molecules are taken in by means of phagotrophy, pinocytosis, or ingestion followed by absorption and/or pincytosis through the intestinal wall will be discussed. In addition to serving as fuel for energy production, lipids, carbohydrates, and especially proteins are utilized in the construction of cells, tissues, and a variety of essential molecules, such as enzymes and hormones.

For the sake of convenience, nutrients can be classified as carbohydrates, lipids or fats, and proteins, and their metabolism is being considered under these subheadings. Some parasites, or certain stages during the life cycles of certain species, utilize modifications of these common metabolic pathways.

Carbohydrate Metabolism. During digestion following the ingestion of carbohydrates, polysaccharides are broken down to simple sugars, commonly glucose. This process occurs either extracellularly, i.e., within the lumen of digestive tract, or intracellularly, i.e., within cells, or both, depending on the species. In those species in which nutrients are taken up through their body surfaces, intracellular digestion is the rule. The resulting monosaccharides are absorbed through the intestinal wall if luminal digestion occurs. In either case, the sugars are circulated in the blood or some other body fluid if a circulatory system is absent. As blood glucose is transported to the various cells of the body, much of it permeates the cell membranes and becomes available for intracellular energy production by means of carbohydrate metabolism. The remaining sugars, if not required immediately, are polymerized and stored as glycogen (Fig. 2.18) in certain types of cells. The biochemical sequence of events leading to the synthesis of glycogen from monosaccharides is known as **glycogenesis**. It is from this source that the glycogen found in the parenchymal cells and muscles of parasitic worms comes.

The amount of stored glycogen in parasites varies with the species, sex, and where it lives. Only 1% of the dry body weight of monogenetic trematodes,

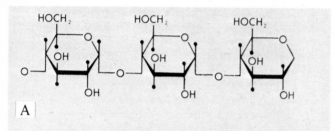

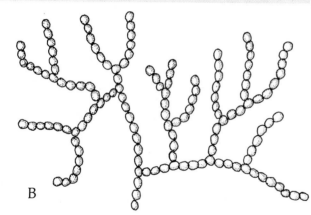

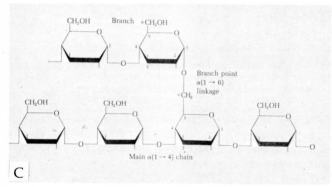

Fig. 2.18. Glycogen and amylopectin. A. Structure of a segment of a glycogen molecule showing that this polysaccharide is a polymer of glucose. **B.** Schematic representation of a glycogen molecule showing that the glucose chains are branched. **C.** Structure of a portion of a molecule of amylopectin showing an $\alpha(1 \rightarrow 6)$ branch point as well as a main $\alpha(1 \rightarrow 4)$ chain.

which are primarily ectoparasitic on the gills of fish, is glycogen, while among digenetic trematodes, which are endoparasites, stored glycogen makes up 2 to 30% of the dry weight. Tapeworms, which are also endoparasites, contain even greater quantities of glycogen, ranging from 20 to 60% of the dry weight. Similarly, trematodes and spinyheaded worms (acanthocephalans), which are endoparasites, store relatively large quantities of glycogen, ranging from 10 to 60% of their dry weights.

Besides species differences, there are differences in the amounts of stored glycogen between the sexes in dioecious species. For example, among schistosomes, male worms store significantly more glycogen (14–30% of dry weight) than females (3.5% of dry weight).

In the case of schistosomes in mammalian hosts, Cornford *et al.* (1983) have demonstrated that if the hosts (mouse and hamster) are starved, glycogen is reduced in male worms within hours but not in female worms. This establishes the principle that there are differences between the sexes of the parasite relative to the rate of glycogen utilization.

It is noted at this point that stored glycogen in parasites, especially worms, can be utilized as an energy source. This is readily demonstrated by depriving parasites maintained in culture of carbohydrates or all nourishment. Under such conditions, rapid depletion of stored glycogen occurs. Similarly, as has been demonstrated by Cornford *et al.* (1983), if a host is starved, its parasites (male schistosomes in their case) rapidly utilize stored glycogen as an energy source. Another example rests with chickens harboring the tapeworm *Raillietina cesticillus*. If the birds are starved, stored glycogen in the parasite's body is reduced from about 25% to 1.5% of its dry weight in 24 hours.

Although some parasitic protozoans store glycogen, another highly branched polysaccharide, amylopectin (Fig. 2.18), occurs in many species.

Now the question can be asked: How does glycogen serve as an energy source?

If glycogen is the starting material in carbohydrate metabolism, it may be first degraded to glucose. The subsequent steps involved, referred to as **glycolysis** or the **Embden-Meyerhof pathway**, occur in the cytoplasm. These can be divided into two phases, the preparatory and the oxidative phases.

The Preparatory Phase. The initial step of this phase is the phosphorylation of glucose to form glucose phosphate. During this process, an ATP (adenosine triphosphate) molecule, or, in some instances, phosphoric acid, gives up a phosphate molecule, which becomes bonded to the glucose. If ATP is the phosphorylative agent, as is commonly the case, it loses one phosphate molecule and becomes ADP (adenosine diphosphate).

The second step involves a series of rearrangements of the glucose molecule so that glucose phos-

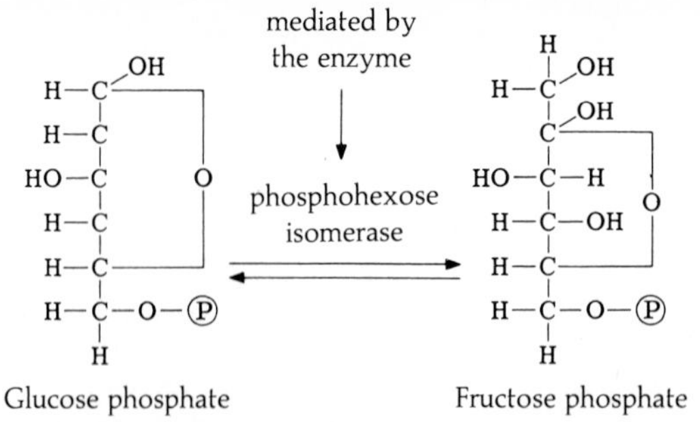

Glucose phosphate   Fructose phosphate

Fig. 2.19

phate becomes fructose phosphate. This process is illustrated by the reaction depicted in Figure 2.19. If fructose is the carbohydrate initially used as the fuel, it enters the metabolic pathway at this point by phosphorylation.

The third step involves the bonding of another phosphate molecule to fructose phosphate to form fructose diphosphate. The additional phosphate is also donated by a molecule of ATP which, after giving up the phosphate, becomes ADP.

Next, fructose, a 6-carbon sugar, splits between its third and fourth carbons, and with slight rearrangements, the original fructose diphosphate becomes two identical 3-carbon molecules known as phosphoglyceraldehyde (PGAL). This process, which represents the fourth step, is depicted in Figure 2.20.

The fifth and final step of the preparative phase occurs simultaneously with the initial step of the oxidative phase. The preparative phase is represented by another phosphorylation during which another phosphate is attached to that terminal of each molecule of PGAL, which is without a phosphate, thus giving rise

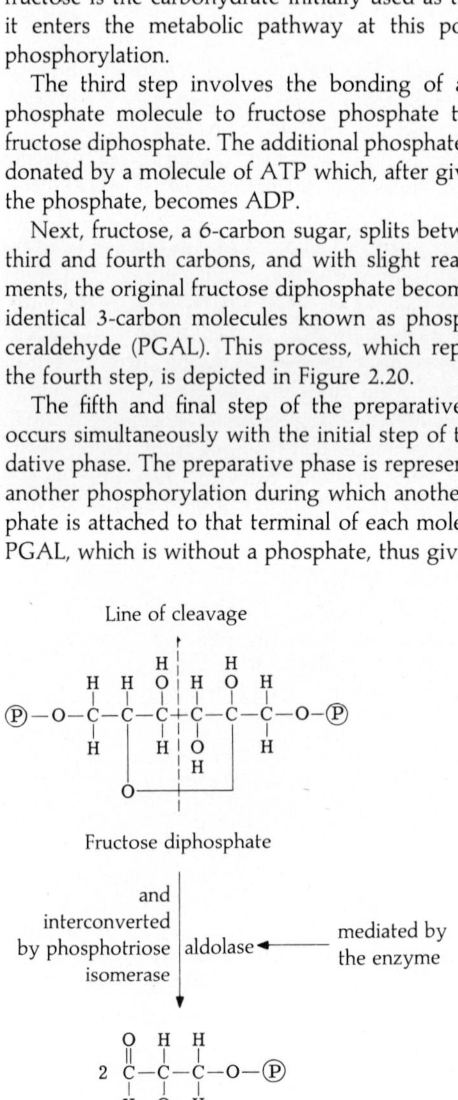

Line of cleavage

Fructose diphosphate

and interconverted by phosphotriose isomerase   aldolase ◀—— mediated by the enzyme

Phosphoglyceraldehyde (PGAL)

Fig. 2.20

to two molecules of diphosphoglyceric acid. This time, the phosphorylating agent (or phosphate donor) is phosphoric acid drawn from the mineral supply within the cell.

The Oxidative Phase. At this point the first step of the oxidative phase, known as oxidative dehydrogenation, occurs. This involves the acquisition of a high energy bond by the last phosphate added to each diphosphoglyceric acid molecule and the removal of two hydrogen atoms, which are accepted by two molecules of nicotinamide adenine dinucleotide (NAD), one by each. Diagrammatically, the processes described in this paragraph are shown in Figure 2.21.

The second step of the oxidative phase involves the transfer of the high-energy phosphate from each of the two molecules of diphosphoglyceric acid to two molecules of ADP, one to each. As the result, two molecules of ATP are now formed. It is recalled that two molecules of ATP were converted to ADP during the preparatory phase of glycolysis. Now, two ADPs are converted to ATPs. The net result up to this point is no gain or loss of energy- containing phosphate bonds, or for our purpose, energy.

Upon the transfer of the high energy phosphates, the two molecules of diphosphoglyceric acid now become two molecules of phosphoglyceric acid (PGA). Now an oxidative dehydrogenation, the third step, occurs during which PGA loses two hydrogens that combine with one oxygen from the PGA molecule to form water and, as the result, a high-energy bond ($\sim$) is formed. The resulting molecule is phosphoenolpyruvic acid. This step is illustrated by Figure 2.22. It is noted that since the last oxidative dehydrogenation did not result in the release of hydrogen per se but water, no NAD was necessary as a hydrogen acceptor.

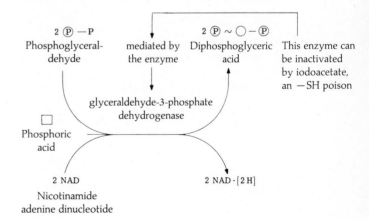

Fig. 2.21

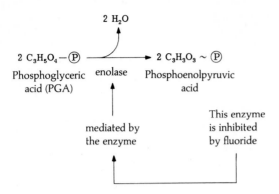

Fig. 2.22

The fourth step involves the transfer of the high energy phosphate from each of the two molecules of phosphoenolpyruvic acid to two ADPs, one to each, thus forming two ATPs. At the last evaluation, the net result was no gain or loss of energy. At this point, however, there are two ATPs, or two high-energy phosphate bonds gained.

Upon losing its high-energy phosphate, and incidentally gaining a hydrogen from ADP, phosphoenolpyruvic acid now becomes pyruvic acid or pyruvate.

If the cell is respiring under anaerobic conditions, pyruvate serves as a hydrogen acceptor from NAD and forms either lactic acid or alcohol and carbon dioxide. The process is known as **fermentation** or **anaerobic metabolism** and is illustrated in Figure 2.23. Under these conditions, the two ATPs gained represent the total energy produced.

On the other hand, if cellular respiration occurs under aerobic conditions, pyruvate need not serve as a hydrogen acceptor and can be further utilized for the formation of energy via the pathway known as the citric acid cycle to be discussed later. Furthermore, the two [2H] attached to each of the two molecules of NAD may pass on to oxygen and in so doing yield three ATPs per [2H], hence, an additional gain of six ATPs.

In most animal cells degradation of glucose to

pyruvate via glycolysis is more rapid anaerobically than aerobically. This inhibition of glycolysis by oxygen is known as the **Pasteur effect.** Conversely, when oxygen consumption is inhibited by glucose, the process is known as the **Crabtree effect.** Since glycolysis in certain species of helminths persists in the presence of high oxygen tensions, i.e., no or limited Pasteur effect, this lends support to the view that anaerobiosis is fundamental to the energy metabolism of helminths. Some species, however, do exhibit a Pasteur effect. These include larval *Schistosoma mansoni,* and the larvae of the nematodes *Ascaris lumbricoides, Eustrongyloides ignotus, Nippostrongylus brasiliensis,* and *Litomosoides carinii.* The Crabtree effect most probably occurs in these parasites with a very active glycolytic pathway.

Lipid Metabolism. Lipids or fats can also be transformed to pyruvic acid in the cytoplasm via several steps. The pyruvic acid resulting from this pathway, like that formed via the glycolytic pathway, can be further metabolized to produce energy.

During the breakdown of fats, each lipid molecule is reduced to its major component parts—three molecules of fatty acids and one molecule of glycerol. This process is shown in Figure 2.24. Glycerol, a 3-carbon molecule, is transformed into phosphoglyceraldehyde (PGAL) which, as we have already seen, is the intermediary metabolite also formed when the fructose diphosphate molecule cleaves during the fifth step of glycolysis. From this step on, the PGAL resulting from lipid metabolism goes on to become pyruvate in the same manner as that resulting from glycolysis.

Fatty acids resulting from the digestion of fats undergo further degradation until a number of small

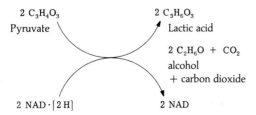

Fig. 2.23

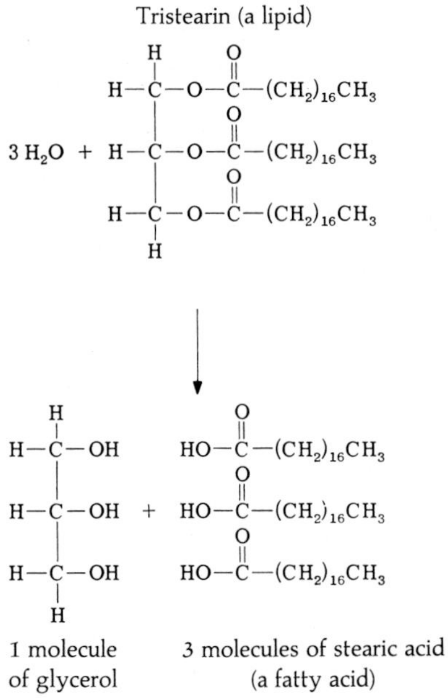

Tristearin (a lipid)

1 molecule of glycerol   3 molecules of stearic acid (a fatty acid)

Fig. 2.24

2-carbon units are formed. These are converted into a compound known as acetyl coenzyme A or acetyl-CoA. Acetyl, a 2-carbon molecule, is a derivative of acetic acid which does not normally exist by itself but is combined with a carrier molecule known as coenzyme A, hence the acetyl-CoA referred to above. The fate of this compound will be traced later.

Protein Metabolism. Proteins employed as fuel in the synthesis of energy are first broken down to their constituent amino acids. Each amino acid is then deaminated, i.e., the $-NH_2$, or amine radical, of each amino acid is removed. The remaining portion of each amino acid undergoes various reactions, which will not be considered in detail, to form either pyruvate or acetyl-CoA. These pathways, which occur in cytoplasm, are summarized in Figure 2.25.

The Citric Acid Cycle. It is apparent from the above discussion that the end product of carbohydrate metabolism via glycolysis is pyruvate. The end products of lipid metabolism are pyruvate and acetyl-CoA, and those of protein metabolism are also pyruvate and acetyl-CoA. Although not considered in detail here, the 3-carbon pyruvate is transformed into the 2-carbon acetyl through the loss of a molecule of CO_2, the addition of hydrogen, and rearrangement of the molecule. Acetyl, in turn, becomes attached to coenzyme A by a sulfur bond (-S-). In an abbreviated form, this conversion of pyruvate to acetyl-CoA may be expressed as in Figure 2.26. The conversion of pyruvate to acetyl-CoA is also an energy-producing process. Specifically, it produces six ATPs.

It is appropriate to mention at this point that other 2- and 3-carbon molecules present in the cell cytoplasm can also be utilized in energy production. The 3-carbon molecules, by one pathway or another, are converted to pyruvate, which, in turn is converted to acetyl-CoA by the pathway given. The 2-carbon molecules are converted to acetyl which becomes linked to CoA form acetyl-CoA. It should thus be apparent that acetyl-CoA is the product of all organic molecules to be used for energy production. It, in turn, is the starting point for further breakdown leading to the liberation of usable energy via a cyclic pathway interchangeably known as the **citric acid cycle**, the **tricarboxylic acid cycle**, or the **Krebs' cycle** (Fig. 2.27). The sites at which the steps involved in this cycle are on mitochondrial membranes.

The citric acid cycle commences when oxaloacetic acid, a 4-carbon compound normally found in cells, combines with the 2-carbon acetyl-CoA to form the 6-carbon citric acid. This process, mediated by the citrate condensing enzyme, is illustrated by Figure 2.28. The next step involves the formation of another 6-carbon acid, *cis*-aconitic acid, from citric acid, mediated by the enzyme aconitase. A molecule of water is given off during the process. Figure 2.29 expresses this reaction.

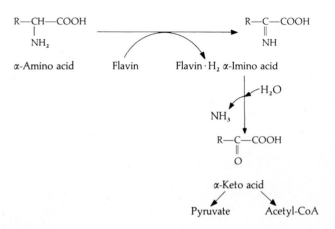

Fig. 2.25

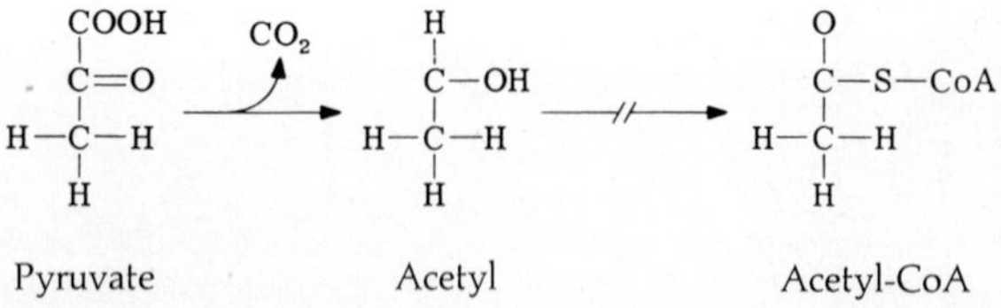

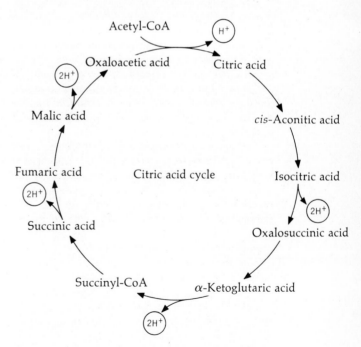

Fig. 2.26

Fig. 2.27. The citric acid cycle showing steps at which hydrogen is given off.

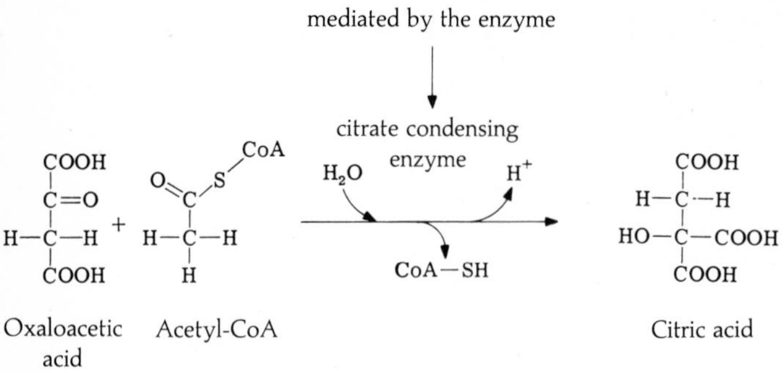

mediated by the enzyme

↓

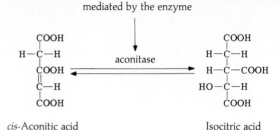

Fig. 2.28

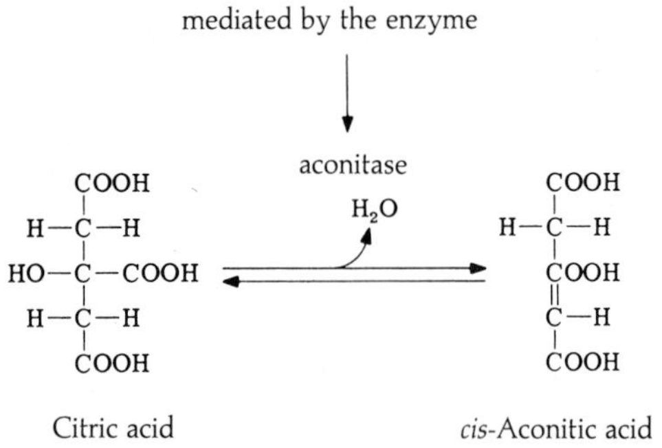

mediated by the enzyme

↓

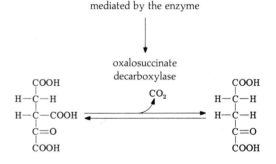

Fig. 2.29

mediated by the enzyme

↓

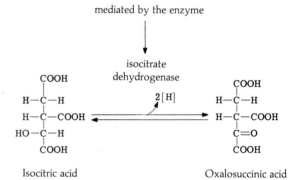

cis-Aconitic acid   Isocitric acid

Fig. 2.30

mediated by the enzyme

↓

Isocitric acid   Oxalosuccinic acid

Fig. 2.31

Now, *cis*-aconitic acid, mediated by the enzyme aconitase, becomes another 6-carbon molecule, isocitric acid. This is indicated in Figure 2.30.

Now, isocitric acid, mediated by the enzyme isocitrate dehydrogenase, becomes another 6-carbon molecule known as oxalosuccinic acid. During this process, as indicated in Figure 2.31, two atoms of hydrogen are given off.

The next step involves the 6-carbon oxalosuccinic acid, which becomes a 5-carbon molecule known as α-ketoglutaric acid. This process, mediated by an enzyme known as oxalosuccinate decarboxylase, involves the giving off of a molecule of CO_2. The carbon incorporated in the CO_2 accounts for the reduction from a 6-carbon compound to a 5-carbon compound. This step is illustrated in Figure 2.32.

The next step involves the transformation of the 5-carbon α-ketoglutaric acid to the 4-carbon succinyl-CoA. The process is mediated by the enzyme α-ketoglutarate dehydrogenase. During this step two atoms of hydrogen are given off, the carbon lost is given off as a molecule of CO_2, and the succinyl becomes linked to CoA present in the cell by an -S-bond. These are indicated in Figure 2.33.

mediated by the enzyme

↓

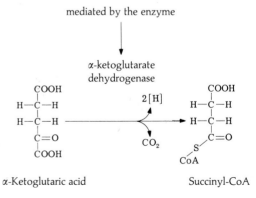

Oxalosuccinic acid   α-Ketoglutaric acid

Fig. 2.32

mediated by the enzyme

↓

α-Ketoglutaric acid   Succinyl-CoA

Fig. 2.33

Following this reaction, succinyl-CoA becomes the 4-carbon succinic acid. The enzyme involved is succinate thiokinase. During this step, CoA and the sulfur bond are given off as shown in Figure 2.34.

The next step (Fig. 2.35) involves alteration of succinic acid to fumaric acid, another 4-carbon molecule. The enzyme involved is succinate dehydrogenase and, as the designation of this enzyme suggests, hydrogen is given off, in this case, 2[H].

During the next step (Fig. 2.36) fumaric acid becomes malic acid, a 4-carbon molecule. The enzyme involved is fumarase.

Finally, malic acid becomes oxaloacetic acid (Fig. 2.37), also a 4-carbon molecule. The enzyme involved is malate dehydrogenase and during the process, 2[H] are given off. The oxaloacetic acid thus formed can now recombine with acetyl-CoA to repeat the cycle by forming citric acid.

Electron Transport (Hydrogen Transfer). It is recalled at this point that hydrogen has been given off at five steps involved in the citric acid cycle (Fig. 2.27). In each instance the dehydrogenation is mediated by a specific enzyme known as a **dehydro-**

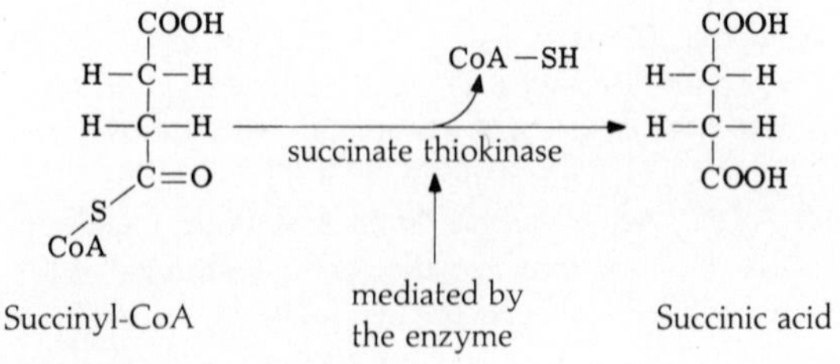

Fig. 2.34

Fig. 2.35

Fig. 2.36

Fig. 2.37

genase. Ultimately, these hydrogens must be accepted by oxygen within the cell. In order to provide oxygen, aerobic organisms must breathe and the oxygen taken in eventually reaches the cells. It is of interest to note, however, that the hydrogen given off during the citric acid cycle does not combine with oxygen immediately to form water. Rather, the hydrogen is first passed along a series of intermediate hydrogen (or electron) carriers and it is the last of these carriers that passes the hydrogen to oxygen. This series of electron transfers occurs within mitochondria. The advantage of such a system is that additional usable energy is formed during these intermediate transfer processes. The eventual acceptance of hydrogen by oxygen is also an energy producing process.

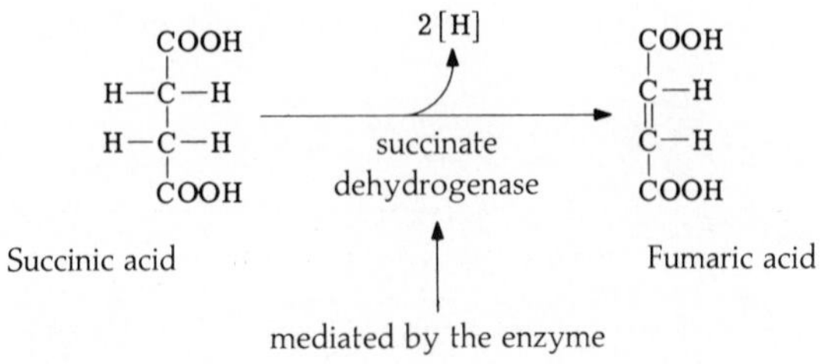

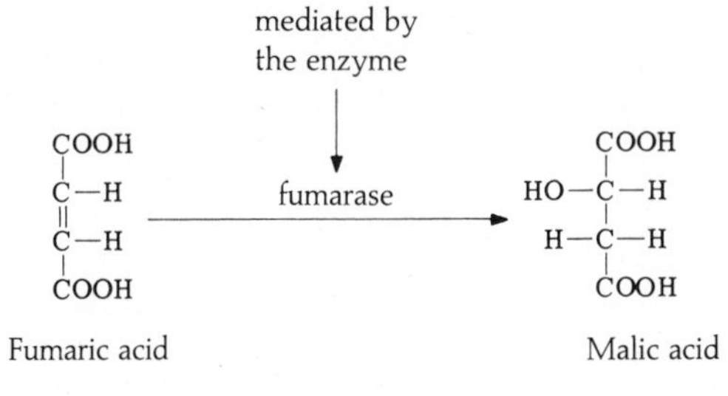

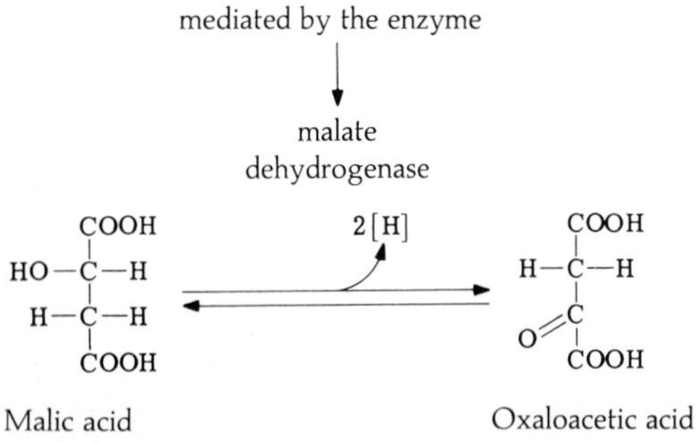

What are these intermediate hydrogen or electron transferring carriers? We have already seen the NAD is a hydrogen acceptor. In some cases NADP (nicotinamide adenine dinucleotide phosphate) can also serve as a hydrogen acceptor, but NAD is the usual acceptor within mitochondria. In addition to NAD and NADP, another type of carrier coenzymes, known as **flavin nucleotides**, performs the same function. Two specific types of flavin nucleotides are involved—flavin adenine dinucleotide (FAD) and flavin mononucleotide (FMN). Both the nicotinamide-containing coenzymes, i.e., NAD and NAPD, and the flavin nucleo- tides, i.e., FAD and FMN, are known as primary acceptors since they accept hydrogen or electrons resulting from the dissociation of an atom of hydrogen as those given off during the citric acid cycle. The primary acceptors then pass on the hydrogen (electron) to what is known as the **cytochrome system.** This system may be considered the third category of electron carriers. It consists of several coenzymes, namely, cytochrome b, ubiquinone or carrier Q, cytochrome c_1, cyctochrome c, cytochrome a, and cytochrome a_3 or cytochrome oxidase. These carriers do not carry H as such but the e^- resulting

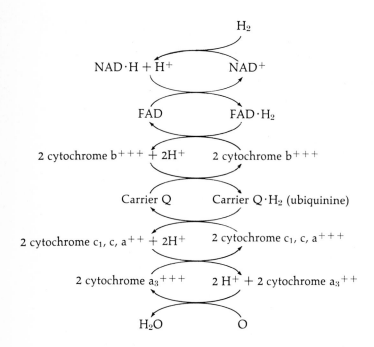

Fig. 2.38. Schematic drawing of the cytochrome or "electron cascade" system by which the e^- **resulting from the dissociation of H is transported.** NAD, nicotinamide adenine dinucleo- tide (formerly known as DPN or diphosphopyridine nucleotide); FAD, flavin adenine dinucleotide. The cytochromes c_1, c, and a function independently and in succession, although they are shown as operating in a group in this figure to save space. Notice that at the end the hydrogen combines with oxygen to form water.

from the dissociation of H to form H^+ and e^-. The cytochromes either transfer the e^- directly to the next coenzyme or give up the e^-, which recombines with a H^+ to form atomic H. If the latter occurs, the atomic H redissociates and the resulting e^- is picked up by the next coenzyme carrier. Since the system described represents, in essence, the passing of electrons from one coenzyme carrier to the next (Fig. 2.38), the term **electron cascade** is commonly used to describe it.

The last cytochrome coenzyme, namely cytochrome a_3 or cytochrome oxidase, gives up the e^- to H^+ to form H which, in turn, combines with oxygen to form water (Fig. 2.38). Quantitatively, a total of 12 ATPs are formed as the result of the transfer of hydrogen (or electrons) given off with each turn of the citric acid cycle. Since actually two molecules of oxaloacetic acid combine with two molecules of acetyl-CoA each time, two turns of the citric acid cycle occur concurrently. Consequently, 2×4 [2H] = 8 [2H] are given off during the citric acid cycle and hence $2 \times 12 = 24$ ATPs are formed as the result of the citric acid cycle combined with the hydrogen transfer ("electron cascade") process.

In totaling the number of ATPs formed, it is recalled that the conversion of pyruvic acid to acetyl-CoA contributed six ATPs and each of the two turns of the citric acid cycle, combined with the electron cascade, yielded 12, hence a total of $6 + 12 + 12 = 30$ ATPs. When carbohydrate metabolism via glycolysis is taken into account, eight additional ATPs may be added. Hence the net gain of 38 ATPs is realized when one molecule of glucose is metabolized aerobically via the glycolytic and citric acid cycle pathways (Fig. 2.39).

It is of historic interest that the earliest observations on the cytochromes (electron transport system)

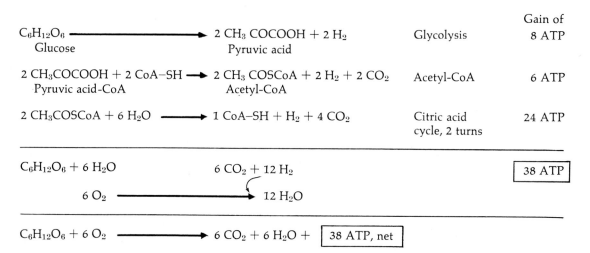

Fig. 2.39. Energy production. Quantitative summary of the net number of ATPs gained as the result of glucose metabolism (respiration). The last equation represents the net input and output.

were made by Keilin (1925) on *Ascaris* among other animals. The excellent review of this system in parasitic protozoans and helminths by Bryant (1970) is recommended.

The question to be asked at this point is: How much energy is associated with ATP? The answer is that each mole (molecular weight of the compound in grams) of ATP stores 7 to 8 $\times 10^3$ calories of energy. Hence one mole of glucose, if completely metabolized via the glycolytic and citric acid cycle pathways, provides 38×7 (or 8) $\times 10^3$ calories of energy.

It is known that glycolysis is not involved during protein and lipid metabolism. Hence, it is apparent that carbohydrate metabolism via glycolysis followed by the aerobic citric acid cycle is a much more efficient pathway of energy production. Since many endoparasites live in environments where free oxygen is lacking or at a minimum, and since the utilization of lipids and proteins as fuel for energy production via the citric acid cycle is dependent on the availability of oxygen, many parasites depend, at least in part, on glycolysis and fermentation, both being anaerobic processes, for the production of energy. Thus, as a general rule it may be stated that endoparasites depend heavily on carbohydrate metabolism as their energy-generating mechanism. Exceptions, however, are known. These include the protozoans *Trypanosoma lewisi* (p. 128) and *Leishmania braziliensis* (p. 123), the tapeworms *Taenia hydatigera* and *Echinococcus granulosus* (p. 408), and the nematodes *Ascaris lumbricoides* (p. 519) and *Litomosoides carinii* (p. 543).

Monophosphate Shunt. The pathways given for carbohydrate, lipid, and protein metabolism during the production of energy are the common ones usually encountered in both free-living and parasitic or-

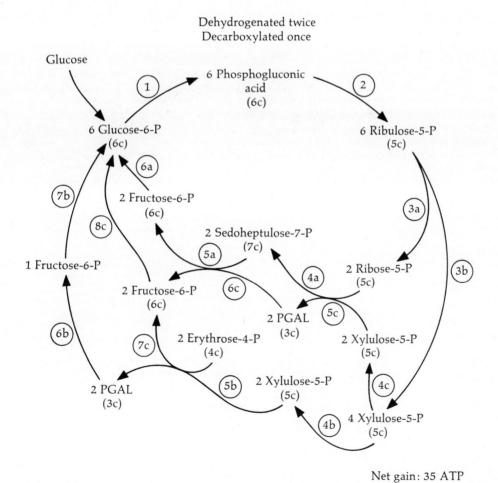

Fig. 2.40. The hexose monophosphate shunt. Phosphates are indicated by P and glyceraldehyde phosphate is represented by PGAL. The number of carbons in each molecule is indicated in brackets. Notice that three alternate pathways, a, b, and c, exist.

ganisms. It is known, however, that several alternate pathways may occur in different species or tissues. These alternate pathways are not correlated in any way with the ecological or phylogenetic position of the organisms and thus appear to be characteristic of a given species or tissue only. One of these alternates is the **hexose monophosphate shunt.** It has been reported to occur in certain helminth parasites. In brief, this shunt involves the utilization of glucose as the carbohydrate fuel. The steps involved (Fig. 2.40) do not occur within mitochondria but in the cytoplasm. These involve the phosphorylation of glucose to form glucose-6-phosphate after which six molecules of glucose-6-phosphate are dehydrogenated twice in succession and decarboxylated once, with NADP molecules serving as the hydrogen acceptors. During this process, six molecules of CO_2 are given off and six molecules of ribulose phosphate remain. Next, the ribulose phosphate molecules undergo a complex series of rearrangements during which 4-, 5-, and 7-carbon sugars are intermediates at certain points. The ultimate result is the appearance of five molecules of glucose-6-phosphate, one less than at the start. The net result of the cycle is the conversion of one out to six glucose molecules to 6 CO_2. The net energy gain is 35 ATPs, which compares favorably with the 38 ATPs resulting from the metabolism of one molecule of glucose via glycolysis and the citric acid cycle.

Alternative End Products. In the previous discussion of carbohydrate metabolism from glycogen to pyruvate via glycolysis, it was pointed out that the typical end product is lactic acid. Although this is the general scheme, many species of parasites produce a wide variety of alternative end products. Some of these are tabulated in Table 2.8. From such information it can be stated that fermenters such as *Schistosoma mansoni* and the filarial nematodes *Brugia pahangi* and *Dipetalonema viteae* excrete lactic acid almost exclusively under both anaerobic and aerobic conditions. Many others excrete volatile fatty acids; for example, under aerobic conditions, the blood-dwelling trypanosomes excrete acetate and some ciliated protozoans excrete acetate, proprionate, and butyrate.

The end products of carbohydrate metabolism may vary with the life cycle stage or the environment. For example, Rew and Douvres (1983) have demonstrated that in the case of *Oesophagostomum radiatum*, the nodular worm of cattle, when third-stage larvae are incubated in a simple salt solution containing radioactive glucose, they excrete CO_2 and acetic, propionic, and lactic acids. Larvae in third molt, fourth-

Table 2.8. End Products of Carbohydrate Catabolism in Some Helminth Parasites

Species	Anaerobic	Aerobic
Trematodes		
Dicrocoelium dendriticum	Lactate, acetate, propionate, succinate	Lactate (?)
Fasciola hepatica	Acetate, propionate, lactate (trace), isobutyrate, isovalerate, 2-methylbutyrate, succinate	Similar to anaerobic
Schistosoma mansoni	Lactate	Lactate
Echinostoma liei	n-Valerate, n-hexanoate, propionate, butyrate, acetate, 2-methylbutyrate, lactate, succinate	Similar to anaerobic
Cestodes		
Hymenolepis diminuta	Lactate, acetate, succinate	Similar to anaerobic but less succinate
Echinococcus granulosus		
(adult)	Succinate, lactate, acetate	Similar to anaerobic
(larva)	Lactate, acetate, succinate, ethanol	Lactate, pyruvate, acetate, less succinate, ethanol
Moniezia expansa	Lactate, succinate	Similar to anaerobic but less succinate
Nematodes		
Ancylostoma caninum	Acetate, propionate, isobutyrate (trace), 2-methylbutyrate	Similar to anaerobic
Ascaridia galli	Lactate, propionate, acetate (?)	Similar to anaerobic
Ascaris lumbricoides	2-Methylvalerate, 2-methylbutyrate, acetate (trace), n-valerate, propionate, isobutyrate, tiglate, n-caproate, acetoin	Similar to anaerobic
Angiostrongylus cantonensis	Lactate	Similar to anaerobic
Haemonchus contortus	Acetate, proprionate, propanol, lactate (trace), succinate, ethanol	Similar to anaerobic
Heterakis gallinae	Acetate, propionate, lactate (trace), pyruvate	Similar to anaerobic
Litomosoides carinii	Lactate, acetate	Less lactate, more acetate
Nippostrongylus brasiliensis	Lactate, succinate	Lactate, pyruvate (trace)
Brugia pahangi	Lactate	Lactate, acetate
Acanthocephalans		
Moniliformis dubius	Lactate (trace), succinate, acetate, butyrate, ethanol, possibly formate	Similar to anaerobic
Echinorhynchus gadi	Lactate, succinate	Lactate

stage larvae, and adults all excrete CO_2, acetic, propionic, and lactic acids at twice the rate of third-stage larvae plus an additional product, methylbutyric acid. If the environment is altered, in this case if the worms are incubated in a complex medium, the major excretory products of the various stages are isobutyric and 3-methylbutyric acids.

Among helminths that have been studied, the nematodes *Ascaris lumbricoides* and *Trichinella spiralis*, the liver fluke *Fasciola hepatica*, and the acanthocephalan *Moniliformis dubius* excrete a variety of volatile fatty acids, including acetate, proprionate, butyrate, and to lesser amounts, valerate, isovalerate, isobutyrate, and caproate. Also, it is known that the proportions of fatty acids produced varies with ambient conditions. For example, *Fasciola* excretes less proprionate and proportionally more acetate with increasing O_2 tension. Ethyl alcohol is an end product of glycolysis in some parasites such as the larval holdfast organs of the tapeworm *Echinococcus granulosus*, the larvae and adults of the tapeworm *Taenia taeniaeformis*, and the acanthocephalan *Moniliformis dubius*.

Regulation of carbohydrate metabolism. The question that has been asked is: What regulates the types of metabolic end products excreted? It is known that the ratio of the activities of two key enzymes in carbohydrate catabolism is responsible, at least in part. Specifically, the ratio of the activities of pyruvate kinase (PK) and phosphoenolpyruvate carboxykinase (PEPCK) is responsible. PK mediates the pathway shown in Figure 2.41. PEPCK mediates the pathway shown in Figure 2.42.

High PK activity will lead to the production of lactate from PEP and high PEPCK activity will lead to the production of succinate or its metabolites. Thus, it is apparent that PEP represents a major branch point in carbohydrate metabolism. Hence, parasites like the schistosomes, which are homolactic fermenters, have a high PK:PEPCK ratio, while succinate producers have a much lower PK:PEPCK ratio (Table 2.9). It has been proposed that the control of the direction of metabolism from PEP to either lactate or succinate in intestinal parasites is mediated by CO_2 partial pressure in the host's intestinal tissue (Chappell, 1980).

Dependency on Host's Enzymes. It is known that a number of species of tapeworms and trematodes are only capable of taking up and utilizing the

phosphoenolpyruvate + ADP (Adenosine diphosphate)
$\rightleftarrows$ pyruvate + ATP (Adenosine triphosphate)

Fig. 2.41

oxaloacetate + GTP (Guanosine triphosphate)
$\rightleftarrows$ phosphoenolpyruvate + CO_2 + GDP (Guanosine diphosphate)

Fig. 2.42

Table 2.9. Pyruvate Kinase (PK)/Phosphoenolpyruvate Carboxykinase (PEPCK) Ratio in Some Parasitic Helminths

Species	Ratio	Major End Products
Trematodes		
Dicrocoelium dendriticum	0.05	Propionate, acetate
Fasciola hepatica	0.25–0.4	Propionate, acetate
Schistosoma mansoni	5–10	Lactate
Cestodes		
Moniezia expansa	0.1	Succinate, lactate
Hymenolepis diminuta	0.2	Succinate, lactate
Schistocephalus solidus (plerocercoid)	1.7	Acetate, propionate
Nematodes		
Ascaris lumbricoides	0.04	2-Methylbutyrate, 2-methylvalerate
Trichinella spiralis (larva)	0.3	*n*-Valerate
Setaria cervi	0.4	Lactate
Dictyocaulus viviparus	1.0	Lactate
Nippostrongylus brasiliensis	2.0	Lactate
Litomosoides carinii	2.8	Lactate, acetate
Dirofilaria immitis	10.6	Lactate
Acanthocephalan		
Moniliformis dubius	0.34	Ethanol

breakdown products of larger molecules. For example, a number of intestinal trematodes feed on partially digested foods of their hosts as well as on mucus and mucosal cells derived from the hosts' intestines. Thus, these parasites are nutritionally dependent upon their hosts' digestive enzymes. Specific examples are cited in subsequent chapters.

Another phase of dependency on hosts' enzymes is presented in greater detail on pp. 22, 43. This involves the enzymatic digestion of the cysts of certain species of cestodes and trematodes so that further development can occur once these parasites reach their hosts' alimentary tracts.

A third phase of dependency on enzymes of host origin has been hypothesized for various species of microsporidan tissue parasites of vertebrates and invertebrates (p. 214). These intracellular, protozoan parasites do not include mitochondria in their cytoplasm; however, when examined with the electron microscope, large numbers of host mitochondria can be seen clustered around each parasite (Fig. 2.43). It has been suggested that these parasites are dependent upon the mitochondrial respiratory activities of their hosts for energy production and hence are dependent on their hosts' mitochondrial enzymes.

Developmental and Growth Stimuli of Host Origin. As mentioned in the previous section, certain larval helminths require the action of certain digestive enzymes of their hosts to cause them to excyst so that

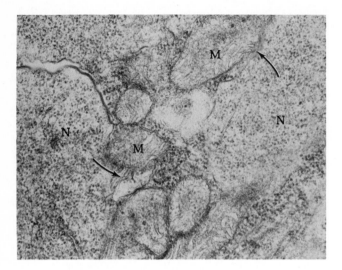

Figure 2.43. *Nosema*-**host relationship.** Electron micrograph showing microtubular connections between the protozoan parasite *Nosema bombycis* and mitochondria of its honeybee host (arrows). N, *Nosema*; M, host mitochondria. (Courtesy of A. Cali.)

they can continue their development. In these instances, the hosts' enzymes may be thought of as developmental stimuli. Such stimuli need not be confined to enzymes. They may be in the form of bile salts and other materials (p. 41).

In the case of bile salts, it is now known that these stimulate further development of parasites by (1) affecting membrane permeability, (2) initiating activity on the part of encapsulated larvae, (3) acting synergistically with the host's digestive enzymes to digest away the restricting capsule of the parasite, and (4) metabolically affecting the establishing parasite so that it becomes more suited to the new environment.

Because bile salts are amphipathic, i.e., each molecule includes a hydrophilic and a hydrophobic terminal, they act as efficient surface-active agents or **surfactants**. Consequently, bile salts can exert a profound effect on biologic membranes, affecting both the protein and lipid moieties. As a result, permeability of the surface membranes of parasite eggs and cysts is increased and the entry of water and digestive enzymes through what was previously an essentially impermeable covering is now possible.

As stated, bile salts can stimulate movement of previously immobile, encysted parasites, such as the encysted stage of the protozoan *Eimeria* (p. 186), the encysted metacercariae (p. 326) of several species of digenetic trematodes including *Cyathocotyle bushiensis*, an intestinal parasite of birds, and the liver fluke, *Fas-*

ciola hepatica. Similarly, bile salts are known to activate the encysted larvae of certain tapeworms and acanthocephalans.

Bile may act synergistically with host digestive enzymes to cause breakdown of cyst walls thus releasing the enclosed parasites. For example, the lytic activities of mammalian trypsin and lipase are increased in the presence of bile salts. Additional examples of host stimulation of parasite growth and development are scattered throughout subsequent chapters.

As new information is being contributed, it is becoming increasingly apparent that compatible hosts do in many ways influence the growth and differentiation processes of their parasites and thus enhance their normal sequence of development, or, conversely, that incompatible hosts in some manner inhibit the normal developmental sequences of their parasites.

Pathologic Changes Induced by Parasites. From the standpoint of survival value and perpetuation of the species, it would be deleterious to a parasite to cause the death of its host, at least prior to its escape or that of its germ cell-bearing progeny. Consequently, the degree and lethality of the pathologic alterations induced by the parasite influences their establishment.

It is not the author's intent to present a synopsis of pathologic changes associated with various types of parasitism. Many examples of such are presented in subsequent chapters, and a general discussion of cytopathology is given on p. 9. The intent is to emphasize some principles.

When one views and learns to appreciate the variety and prevalence of parasitism, it will also become apparent that the majority of zooparasites are avirulent. Parasitic diseases are commonly chronic. This is not to say that such diseases are unimportant from the standpoint of the host's health.

When examined at the cellular level, commonly, although not uniformly, parasitism by animals results in dramatic tissue and cellular reactions, for example, intense granuloma formation occurs only when the parasite is a comparatively recent associate in terms of evolutionary time. This principle is vividly exemplified in parasites of invertebrates where cellular internal defense mechanisms accompanied by little or no humoral reaction occurs. For example, larval trematodes in their natural molluscan hosts may cause varying degrees of damage with little or no cellular reaction. However, if such larvae enter a less compatible or incompatible mollusc, or if they should migrate into an organ where they are not usually found, the cellular reactions can be profound (Cheng, 1967).

Unlike instances involving invertebrate hosts, some degree of tissue reaction usually occurs when parasites invade the tissues of vertebrates. These vary from inconsequential to severe. The review by

Poynter (1966) of tissue reactions to selected nematodes is recommended as is the volume edited by Vinken *et al.* (1978) on the histopathology of parasitic infections in the mammalian nervous system.

ESCAPE OF THE SYMBIONT

Although the details of escape from hosts have not been investigated extensively, there is little doubt that from the viewpoint of perpetuation of the species and the symbiosis, it is equally as important that symbionts, including parasites, escape from their hosts as it is to make contact and/or enter. From what is known, the escape process can be divided into two major categories, **active** and **passive**, although in many instances a combination of the two occurs.

Active escape is exemplified by trematode cercariae leaving molluscs, although some evidence suggests that the process is enhanced by mechanical pressure on the part of the host. It is obvious that the departure of such temporary ectoparasites as mosquitoes and certain acarines (p. 574) involve active motility on their part.

Passive escape is best exemplified by the passage of the eggs of most species of intestinal helminths out of their hosts in feces. In some instances, however, even this involves activity on the part of the parasites. For example, although the eggs of the human pinworm, *Enterobius vermicularis* (p. 528), are passed out through the host's anus, the female worm does migrate posteriorly to deposit her eggs in the perianal zone.

As stated, a combination between active and passive escape does occur in some instances. For example, microfilariae of certain species of nematodes are removed from the definitive host by an insect vector (intermediate host), but this can only occur as the result of the nematodes being in the peripheral circulation as the result of periodic migration to that site (p. 547). It could be argued that microfilarial periodicity is affected by physiologic pressure of host origin; nevertheless, the motile parasites, responding to such pressures, do accumulate in the peripheral circulation.

In addition to the more conventional methods of escape mentioned, some dramatic ones, involving spectacular participation on the part of the host or parent, are known. The method by which certain species of clams belonging to the genus *Lampsilis* aid their parasitic young to escape and become attached to their hosts is described at this point not only to serve as an example of a spectacular escape mechanism, but also as one of mimicry.

Many of the rivers in North America are inhabitated by clams belonging to the family Unionidae (p. 742). The fertilized eggs of these clams are transferred to a special area of the gill chamber known as the **marsupium** or brood chamber where the eggs develop

into larvae known as **glochidia**. These are released from the parent in bursts, and as many as 300,000 glochidia may be released at one time. The glochidia of several species of clams, including *Lampsilis*, are parasitic. These must clamp onto the gills of a fish where they feed on the superficial tissues and possibly blood. After a period of parasitic existence, the glochidia develop into young adults, drop off, and mature into free-living adults.

Anyone familiar with the flow and nature of rivers can readily appreciate that these extremely small glochidia cannot reach the gills of fishes unaided. They must be actively taken up by the fish. In the case of those species found in relatively stagnant waters, these larval clams could fall to the bottom and later be scooped up by bottom-feeding fish. If the river is swift and the bottom is sandy, this mechanism for contacting the fish host is unsatisfactory. Species of *Lampsilis* living in such waters have developed a unique method of passing on their parasitic glochidia. The female adults of many develop an outgrowth of the mantle margin when carrying larvae. This outgrowth may be shaped like a papilla, a lobe, a flap, or a ribbon, and lies near the exhalant aperture. In some species this outgrowth is even more spectacular. In *Lampsilis nasuta*, for example, a white spot moves conspicuously up and down in a gap between the two mantle margins. The outgrowth of the mantle margin of *Lampsilis radiata* is in the form of a bifurcated appendage, which is capable of twitching movements and gives the appearance of a fish's tail. Still more

Figure 2.44. Mimicry. Gravid female of the clam *Lampsilis ovata ventricosa* with highly modified mantle edge that resembles a fish. (After Wickler, 1968.)

elaborate outgrowths are known. That of *Lampsilis fasciola* resembles a fish and that of *Lampsilis ovata ventricosa* is amazingly fishlike (Fig. 2.44). It bears an anterior "head" with an eye-spot and a posterior "tail." Furthermore, it is partially brightly colored, being gray on the outside and orange on the inside with a longitudinal black stripe. With such elaborate outgrowths as baits, fishes are attracted, and as soon as the unsuspecting fish casts a shadow over the clam, the latter expels thousands of glochidia that become attached to the fish's gills. This mechanism for ensuring contact between glochidia and their fish hosts has been confirmed in the laboratory.

The review by Ulmer (1971) on site-finding behavior of helminths includes some information on the mechanisms involved in escape of parasites as does the monograph by Chappell (1980).

REFERENCES

Bird, A. F. (1955). Importance of proteases as factors involved in the exsheathing mechanism of infective nematode larvae of sheep. *Science* **121**, 107.

Bryant, C. (1970). Electron transport in parasitic helminths and protozoa. *Adv. Parasitol.* **8**, 139–171.

Bryant, C. (1975). Carbon dioxide utilisation, and the regulation of respiratory metabolic pathways in parasitic helminths. *Adv. Parasitol.* **13**, 35–69.

Campbell, W. C., and Todd, A. C. (1955). *In vitro* metamorphosis of the miracidium of *Fascioloides magna* (Bassi, 1875) Ward, 1917. *Trans. Amer. Microsc. Soc.* **74**, 225–228.

Chappell, L. H. (1980). "Physiology of Parasites." John Wiley, New York.

Cheng, T. C. (1967). Marine molluscs as hosts for symbioses. *Adv. Marine Biol.* **5**, 1–424.

Cheng, T. C. (1968). The compatibility and incompatibility concept as related to trematodes and molluscs. *Pac. Sci.* **22**, 141–160.

Cheng, T. C. (1976). Liver and other digestive organs. *In* "Ecological Aspects of Parasitology" (C. R. Kennedy, ed.), pp. 287–302. North-Holland, Amsterdam.

Cheng, T. C., and Alicata, J. E. (1965). On the modes of infection of *Achatina fulica* by the larvae of *Angiostrongylus cantonensis*. *Malacologia* **2**, 267–274.

Chernin, E. (1970). Behavioral responses of miracidia of *Schistosoma mansoni* and other trematodes to substances emitted by snails. *J. Parasitol.* **56**, 287–296.

Chernin, E. (1974). Some host-finding attributes of *Schistosoma mansoni* miracidia. *Am. J. Trop. Med. Hyg.* **23**, 320–327.

Cornford, E. M., Diep, C. P., and Rowley, G. A. (1983). *Schistosoma mansoni*, *S. japonicum*, *S. haematobium*; glycogen content and glucose uptake in parasites from fasted and control hosts. *Exp. Parasitol.* **56**, 397–408.

Crane, R. K. (1967). Structural and functional organization of an epithelial cell brush border. *In* "Intracellular Transport" (K. B. Warren, ed.). Academic Press, New York.

Croll, N. A. (1975). Behavioral analysis of nematode movement. *Adv. Parasitol.* **13**, 71–122.

Crompton, D. W. T., and Nesheim, M. C. (1976). Host-parasite relationships in the alimentary tract of domestic birds. *Adv. Parasitol.* **14**, 95–194.

Davenport, D. (1955). Specificity and behavior in symbioses. *Quart. Rev. Biol.* **30**, 29–46.

Dawes, B. (1960). A study of the *Fasciola hepatica* and an account of the mode of penetration of the sporocyst into *Lymnaea trunculata*. "Libro Homen-je al Dr. Eduardo Cabellero y Caballero," Jubileo 1940–1960, pp. 95–111. Escuela Nacional de Ciencias Biologicas, Mexico.

Dietschy, J. M. (1967). Effects of bile salts on the intermediate metabolism of the intestinal mucosa. *Fed. Proc. Fed. Am. Soc. Exp. Biol.* **26**, 1589–1598.

Dittmer, D. S. (ed.) (1961). "Biological Handbooks: Blood and Other Body Fluids." *Fed. Am. Soc. Exp. Biol.*, Washington, D.C.

Fairbairn, D. (1954). The metabolism of *Heterakis gallinae*. II. Carbon dioxide fixation. *Exp. Parasitol.* **3**, 52–63.

Fry, M. and Jenkins, D. C. (1984). Nematoda, aerobic respiratory pathways of adult parasitic species. *Exp. Parasitol.* **57**, 86–92.

Green, N. M. (1957). Protease inhibitors from *Ascaris lumbricoides*. *Biochem. J.* **66**, 416–419.

Hammond, R. A. (1966). The proboscis mechanism of *Acanthocephalus ranae*. *J. Exp. Biol.* **45**, 203–213.

Hammond, R. A. (1967). The mode of attachment within the host of *Acanthocephalus ranae* (Schrank, 1788), Lühe, 1911. *J. Helminth.* **41**, 321–328.

Hochachka, P. W. (1980). "Living without Oxygen." Harvard Univ. Press, Cambridge, Massachusetts.

Johnson, C. F. (1967). Disaccharidase localization in hamster intestine brush borders. *Science* **155**, 1670.

Keilin, D. (1925). On cytochromes, a respiratory pigment, common to animals, yeast, and higher plants. *Proc. R. Soc. Lond.*, 312–339.

Kennedy, C. R. (ed.) (1976). "Ecological Aspects of Parasitology." North-Holland, Amsterdam, Holland.

Laing, J. (1937). Host-finding by insect parasites. I. Observations on the finding of hosts by *Alysia manducator*, *Mormoniella vitripennis* and *Trichogramma evencescens*. *J. Anim. Ecol.* **6**, 298–317.

Lengy, L. (1962). Studies on *Schistosoma bovis*. *Isr. Bull. Res.* **10E**, 57–58.

MacInnis, A. J. (1965). Responses of *Schistosoma mansoni* miracidia to chemical attractants. *J. Parasitol.* **51**, 731–746.

MacInnis, A. J. (1976). How parasites find hosts: some thoughts on the inception of host-parasite integration. *In* "Ecological Aspects of Parasitology" (C. R. Kennedy, ed.). pp. 3–20. North-Holland, Amsterdam.

McLaren, D. J. (1976). Nematode sense organs. *Adv. Parasitol.* **14**, 195–265.

Mettrick, D. F. (1971). *Hymenolepis diminuta*: pH changes in rat intestinal contents and worm migration. *Exp. Parasitol.* **29**, 386–401.

Mettrick, D. F., and Podesta, R. B. (1974). Ecological and physiological aspects of helminth-host interactions in the mammalian gastrointestinal canal. *Adv. Parasitol.* **12**, 183–278.

Miller, D., and Crane, R. K. (1961). The digestive function of the epithelium of the small intestine. II. Localization of disaccharide hydrolysis in the isolated brush border portion of intestinal epithelial cells. *Biochim. Biophys. Acta* **52**, 293–298.

Nachtigall, W. (1974). "Biological Mechanisms of Attachment: the Comparative Morphology and Bioengineering or Organs for Linkage, Suction and Adhesive." Springer-Verlag, Berlin.

Overton, J., Eicholz, A., and Crane, R. K. (1965). Studies on the organization of brush border in intestinal epithelial cells. II. Fine structure of fractions of tris-disrupted hamster brush border. *J. Cell Biol.* **26**, 693–706.

Peanasky, R. J., and Laskowski, M. (1960). Chymotrypsin inhibitor from *Ascaris. Biochim. Biophys. Acta* **37**, 167–169.

Poynter, D. (1966). Some tissue reactions to the nematode parasites of animals. *Adv. Parasitol.* **4**, 321–383.

Rew, R. S., and Douvres, F. W. (1983). *Oesophagostomum radiatum*: Glucose metabolism of larvae grown *in vitro* and adults grown *in vivo. Exp. Parasitol.* **55**, 179–187.

Rogers, W. P. (1949). On the relative importance of aerobic metabolism in small nematode parasites of the alimentary tract. I. Oxygen tensions in the normal environment of the parasites. *Aust. J. Sci. Res.* **28**, 166–174.

Rogers, W. P. (1966). Exsheathing and hatching mechanisms in helminths. *In* "Biology of Parasites" (E. J. L. Soulsby, ed.) pp. 33–40. Academic Press, New York.

Smyth, J. D. (1962). Lysis of *Echinococcus granulosus* by surface-active agents in bile and the role of this phenomenon in determining host specificity in helminths. *Proc. R. Soc. Lond.* B, **156**, 553–572.

Sponholtz, G. M., and Short, R. B. (1976). *Schistosoma mansoni* miracidia: stimulation by calcium and magnesium. *J. Parasitol.* **62**, 155–157.

Stibbs, H. H., Chernin, E., Ward, S., and Karnovsky, M. L. (1976). Magnesium emitted by snails alters swimming behavior of *Schistosoma mansoni* miracidia. *Nature* **260**, 702–703.

Taylor, A. E. R., and Baker, J. R. (eds.). (1978). "Methods of Cultivating Parasites *in Vitro*." Academic Press, London.

Thorpe, W. H., and Jones, F. G. W. (1937). Olfactory conditioning in a parasitic insect and its relation to the problem of host selection. *Proc. R. Soc. Lond.* B, **124**, 56–81.

Ulmer, M. J. (1971). Site-finding behaviour in helminths in intermediate and definitive hosts. *In* "Ecology and Physiology of Parasites." (A. M. Fallis, ed.), pp. 123–159. University of Toronto Press, Toronto.

Vernberg, W. B. (1963). Respiration of digenetic trematodes. *Ann. N.Y. Acad. Sci.* **113**, 261–271.

Vinken, P. J., Bruyn, G. W., and Klawans, H. L. (eds.) (1978). "Infections of the Nervous System" Part III. North-Holland, Amsterdam.

von Bonsdorff, B. (1977). "Diphyllobothriasis in Man." Academic Press, London.

von Brand, T. (1952). "Chemical Physiology of Endoparasitic Animals." Academic Press, New York.

von Brand, T. (1973). "Biochemistry of Parasites." 2nd ed. Academic Press, New York.

von Plempel, M. (1964). Chemotaktische Anlockung der Miracidien von *Schistosoma mansoni* durch *Australorbis glabratus. Z. Naturforsch.* **196**, 268–269.

Welsh, J. H. (1930). Reversal of phototropism in a parasitic water mite. *Biol. Bull.* **59**, 165–169.

Welsh, J. H. (1931). Specific influence of the host on the light responses of parasitic water mites. *Biol. Bull.* **61**, 497–499.

Wright, C. A. (1959). Host-localization by trematode miracidia. *Ann. Trop. Med. Parasitol.* **53**, 288–292.

3

IMMUNITY TO PARASITES

As stated in the previous chapter, to become successfully established in a host, a parasite must be able to overcome the potential host's internal defense mechanisms, whether in the form of true immunity or some other mechanism. A great deal of attention has been paid to this aspect of parasitism so that immunoparasitology has evolved into a popular and important aspect of the discipline. It is not my intent to present a comprehensive review of immunology as related to animal parasites. Because of the introductory nature of this volume, certain principles of invertebrate and vertebrate internal defense mechanisms are presented as background information. Interested readers are referred to the comprehensive, two-volume treatise edited by Jackson *et al.* (1969, 1970), which covers practically all aspects of immunoparasitology, although it is somewhat outdated. For more brief, specialized reviews, see Wakelin (1978) and Barriga (1981). Also, the volume edited by Soulsby (1972) and the review by Sinclair (1970) should be consulted.

SELF VS NONSELF

The central theme in immunology is the recognition of self from nonself by an animal. In immunoparasito-

logy, the animal is the host and the parasite is either "self" or "nonself." In other words, the parasite is recognized as foreign (nonself) by the host and is reacted against, or the parasite in some way camouflages itself and consequently is recognized by the host as part of self and therefore is not reacted against. The successful immunologic camouflaging on the part of a parasite by producing hostlike molecules on its body surface is referred to as **molecular mimicry** (Damian, 1964). Examples of this are cited later.

HOST REACTIONS

When a host recognizes the parasite as nonself, it generally reacts against the invader in two ways: **cellular** (or **cell mediated**) **reactions** and **humoral reactions**. In the first, specialized cells become mobilized to arrest and usually eventually destroy the parasite, whereas in the second, specialized molecules in the circulatory system interact with the parasite, usually resulting in its immobilization and destruction. In the case of vertebrate hosts, humoral reactions primarily involve glycoprotein molecules known as **antibodies** or **immunoglobulins**. A more detailed discussion of these molecules is presented later. At this point, a general consideration of the internal defense mechanisms of invertebrate hosts is presented.

INVERTEBRATE INTERNAL DEFENSE MECHANISMS

It is now generally agreed that invertebrates do not synthesize immunoglobulins against foreign substances, including zooparasites. Consequently, strictly speaking, the reactions in invertebrates to nonself materials should not be considered to be true immunologic responses; however, the fact that certain types of reactions are elicited by the invasion of parasites and other foreign materials has caused them to be commonly referred to as "immune" responses.

As shown in Figure 3.1, the internal defense mechanisms of animals, both invertebrates and vertebrates, are of two types: **innate** and **acquired**. Innate (or natural) mechanisms are those that are assumed to be genetically mediated rather than expressions of previous experience with some nonself material. In other words, innate mechanisms represent the native capabilities of the host to act against an invader. Acquired mechanisms are those that develop in the host in response to previous exposure to nonself material. Theoretically, both innate and acquired internal defense mechanisms can be of two types: **cellular** and **humoral**.

INNATE INTERNAL DEFENSE MECHANISMS

Although it has only been in recent years that concerted attention has been directed toward invertebrate internal defense mechanisms, it has become evident that innate mechanisms are of prime impor-

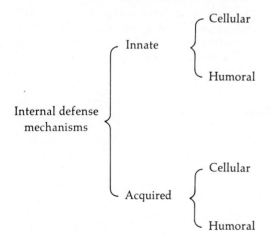

Fig. 3.1. Classification of internal defense mechanisms in animals.

tance in host-parasite relationships in these hosts. When a parasite contacts an invertebrate host for the first time, it is confronted with host reactions that may or may not deter it from successful establishment. The nature of such reactions has not been thoroughly investigated in all cases; nevertheless, differences in host susceptibility have been widely recognized. For example, Ward (1963) has shown that when the mosquito *Culex pipiens* ingests malarial parasites (*Plasmodium cathemerium* and *P. relictum*), various degrees of susceptibility occur. Although it is known that these variations are due to a single genetic factor that lacks dominance, the genetic expression of the factor is not known. Similarly, Kartman (1953), through selection and breeding experiments, has established that strains of the mosquito *Aedes aegypti* are either susceptible or refractory to the dog heartworm, *Dirofilaria immitis*. Other evidence for the genetic control of compatibility or incompatibility has been contributed by MacDonald (1962) who has found that the susceptibility of *Aedes aegypti* to the filarial worm *Brugia malayi* is controlled by a sex-linked recessive gene in the mosquito.

There is also evidence that the susceptibility of naive (i.e., previously uninfected) molluscs to helminths, especially larval trematodes, is genetically controlled, and in these instances, the phenotypic manifestations are better understood since cellular reaction in the form of encapsulation (see below) occurs and commonly results in the death of the parasites (Cheng, 1968; Cheng and Rifkin, 1970). Although innate defense mechanisms have commonly been considered to be nonspecific in the past, recent evidence suggests that specific recognition exists at the molecular level (Cheng, 1985).

Discussed below are the known types of phenotypic manifestations of innate internal defense mechanisms in invertebrates, especially molluscs and insects, which comprise the most extensively studied groups. Briefly, such mechanisms can be categorized as: (1) phagocytosis, (2) encapsulation, (3) nacrezation, (4) melanization, and (5) humoral factors.

Phagocytosis

When a foreign parasite small enough to be phagocytosed invades an invertebrate, it is usually phagocytosed by the host's leucocytes. This is true of most invertebrates but it has been most extensively studied in molluscs and insects. Furthermore, it is primarily granulocytes, i.e., blood cells with numerous cytoplasmic granules (which are lysosomes), that are the most active phagocytic cells (Foley and Cheng, 1975; Ratcliffe and Rowley, 1979).

Immediately prior to phagocytosis, there is an increase in the number of phagocytes within the host. This phenomenon, known as **leucocytosis**, is stimulated by the presence of the parasites; however, the responsible mechanism(s) remains undetermined. It is also not completely known how invertebrate phagocytes "recognize" the nonself nature of the invading parasites. The currently accepted explanation is that phagocytosis actually consists of three phases: (1) attraction of phagocytes to the nonself material, commonly by chemotaxis; (2) attachment of the foreign material to the surface of the phagocyte; usually involving a specific chemical binding site; and (3) internalization of the foreign substance, i.e., engulfment by the phagocyte (Fig. 3.2).

The fate of phagocytosed parasites has been traced by a number of investigators (see review by Cheng, 1967). Briefly, the foreign material may (1) be degraded intracellularly; (2) be transported by phagocytes across epithelial borders to the exterior; or (3) remain undamaged within phagocytes and some, such as certain bacteria, may even multiply within these host cells. Which of these paths is taken is dependent on the nature of the nonself material.

It should be pointed out that not all small parasites are phagocytosed. Some, such as the haplosporidan *Haplosporidium nelsoni* in oysters (p. 217), are rarely phagocytosed. The reason why this should happen remains undetermined; however, based on evidences derived from other similar phenomena, it is hypothesized that the oyster phagocytes recognize *H. nelsoni* as "self" and consequently do not attack it. On the other hand, if for some reason the surface of the parasite becomes altered so that it is now recognized as "nonself," it is phagocytosed (Fig. 3.3).

Encapsulation

All invertebrates that have been investigated are capable of encapsulating foreign materials, including parasites, that are too large to be phagocytosed. Encapsulation involves the enveloping of an invading nonself mass by cells and/or fibers of host origin. Again, by far the majority of studies on encapsulation in invertebrates have been on insect and molluscan hosts. Studies directed at arthropods, primarily insects, have been reviewed by Salt (1970) and Ratcliffe and Rowley (1979), whereas those directed at molluscs have been reviewed most recently by Cheng and Rifkin (1970).

In brief, when the parasite enters the host and is "recognized" as nonself, usually the first sign of encapsulation is an increase in the number of phagocytic leucocytes, hence leucocytosis. Many of these cells migrate toward the parasite and form a capsule of discrete cells around it. There is reason to believe that the host cells are chemically attracted to the invading parasite. Evidences suggest that, at least in some instances, the chemotaxis of leucocytes is to certain molecules incorporated in the parasite's body surface and in other instances to secreted molecules. In the case of trematodes and cestodes, the body surface molecules are usually mucopolysaccharides. The chemoattractant secreted by protozoans and bacteria could be a relatively small peptide (Howland and Cheng, 1981). In the case of nematodes, secreted exsheathing (molting) fluid usually serves as the chemoattractant. Thus, the initial stages of encapsulation resemble phagocytosis; in fact, it has been suggested that the initial phase of cellular encapsulation may represent aborted attempts by host cells to phagocytose the parasite. However, because of the latter's size, successful phagocytosis is not possible.

With time, the number of host cells surrounding the parasite increases. Pan (1965) has suggested that some of the leucocytes of snails (*Biomphalaria glabrata*) transform into epithelioid cells which become intimately abutted to form a wall surrounding the parasite. In insect hosts, the host's leucocytes may become fused to form a syncytial tunic.

It is of interest to note that, as is the case with phagocytosis, not all parasites too large to be phagocytosed become encapsulated. Furthermore, in the case of the blood fluke *Schistosoma mansoni* in its nor-

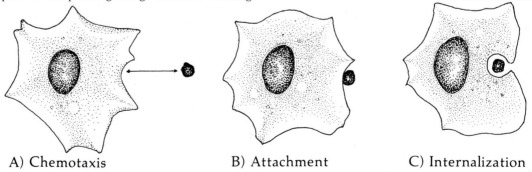

A) Chemotaxis   B) Attachment   C) Internalization

Fig. 3.2. Three phases of phagocytosis. A. Chemotactic attraction between phagocyte and foreign particle. **B.** Attachment of foreign particle to surface of phagocyte. **C.** Internalization of foreign particle.

mal snail host, *Biomphalaria glabrata*, healthy cercariae and sporocysts elicit little or no encapsulation; however, if these stages in the trematode's life cycle should become moribund, conspicuous encapsulation results. This suggests that the healthy parasite, as the result of its intimate association with the snail through evolutionary time, has become so adapted that it is recognized as self and hence no host reaction occurs; but if it becomes chemically altered, it is immediately recognized as nonself and is encapsulated.

Although encapsulation in insects is almost exclusively of the cellular type, a second major type occurs in molluscs. Here, the capsule is comprised primarily of fibers. Rifkin and Cheng (1968) have studied an example of this type of encapsulation in the American oyster, *Crassostrea virginica*, directed toward the larva of the tapeworm *Tetragonocephalum* (= *Tylocephalum*). The process is initiated when the invading parasite compresses the surrounding connective tissue cells (Leydig cells). This triggers the host cells to synthesize the precursor of the fibrous material which becomes deposited intercellularly. The fibers, which resemble reticulin rather than true collagen, gradually become concentrically deposited in layers around the parasite (Rifkin *et al.*, 1969).

After the parasite is encapsulated by fibers, two types of host cells migrate into the matrix of the capsule. One is the molluscan granulocyte and the other is the so-called brown cell. Although the chemical mechanisms still remain a mystery, the migration of these cells into the capsule is followed by the death of the parasite. Eventually the parasite's tissues are completely disintegrated and the fragments become phagocytosed by host granulocytes. Other types of encapsulation are known in molluscs. Those interested in a more detailed account are referred to Cheng and Rifkin (1970).

The fates of encapsulated parasites differ. As described above, when larvae of *Tetragonocephalum* invade an incompatible mollusc such as the American oyster or the clam *Tapes semidecussata*, it is encapsulated and destroyed. On the other hand, if a larva of the nematode *Angiostrongylus cantonensis* is encapsulated in the giant African snail, *Achatina fulica*, it continues to develop normally. The reason for this difference remains unresolved.

Nacrezation

Nacrezation, or pearl formation, is another type of cellular defense mechanism known in molluscs. When certain helminth parasites, especially the metacercariae of the trematode *Meiogymnophallus minutus*, occur between the inner surface of the shell (nacreous layer) and the mantle of marine bivalves, the mantle is stimulated to secrete nacre that becomes deposited around the parasites. In so doing, a pearl is formed and the enclosed parasite is killed.

The mechanism of nacrezation has been reviewed

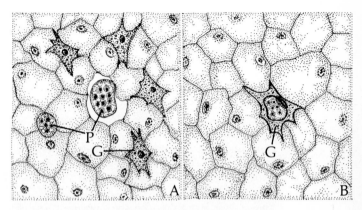

Fig. 3.3. *Haplosporidium nelsoni* **in oyster Leydig (connective) tissue. A.** Nonphagocytosed multinucleated plasmodia. **B.** Plasmodium that had become phagocytosed after the molecular structure of its surface had been altered. G, Oyster granulocyte. P, Plasmodia of *H. nelsoni*.

by Tsujii (1960). It is of interest to note that nacrezation need not be associated with parasites since a grain of sand on the mantle also can initiate the process. The Japanese have perfected this technique to such a state that the cultivation of pearls is a major industry in Japan.

Melanization

Another type of internal defense mechanism in insects against parasites, especially helminths and arthropods, is melanization. This process is characterized by the deposition of the black-brown pigment melanin around the invading parasite. Melanization is detrimental to the parasite and may lead to its death by interfering with such vital activities as hatching, molting, or feeding (Fig. 3.4). Although many are of the

Figure 3.4. Melanization. Photomicrograph showing the nematode *Heterotylenchus autumnalis* encapsulated and covered with melanin in the hemocoel of a larval house fly, *Musca domestica*. (Courtesy of A. J. Nappi.)

opinion that melanization is associated with certain types of host blood cells (Salt, 1963), there is some evidence that the process is direct, without the intervention of blood cells (Bronskill, 1962; Esslinger, 1962).

The principal evidences supporting the theory that cells are involved in melanization are (1) that melanin deposition almost always occurs after cellular encapsulation and (2) the finding by Taylor (1969) of premelanosomes (organelles intimately involved in melanin synthesis) in hemocytes of the cockroach *Leucophaea maderae*. It is possible that the deposition of melanin may be the result of the degeneration of host hemocytes and the subsequent release of premelanosomes.

Chemically, melanization is the result of enzymatic oxidation of polyphenols, primarily by the enzyme tyrosinase. The speed of the melanin reaction varies, depending on the host insect. For example, Salt (1955) has observed melanization around the parasite *Nemeritis* in the beetle *Tenebrio* within 24 hours after parasitization, after 48 hours in the fly *Calliphora*, and after 24 to 96 hours in the lepidopteran *Diataraxia*.

If the parasite is a nematode, melanization commonly first appears in the areas of the mouth and anus of the invader. If the parasite is another insect, melanin deposition also usually first appears in the parasite's mouth and anus, but, in addition, dark stripes around the body coinciding with its intersegmental membranes may also occur. Gradually, however, the entire body surface of the parasite is covered with melanin.

Humoral Factors

The role of innate humoral factors in invertebrates as defense mechanisms against parasites has not been extensively studied, although there are some evidences for their existence, especially in molluscs.

Innate humoral factors in invertebrates fall into two functional categories: (1) those that are directly parasitocidal, and (2) those that enhance cellular reactions. As an example of the first, it is known that the tissue extracts of several species of marine molluscs contain a constituent that is lethal to the cercariae of the trematode *Himasthla quissetensis*. In this case, the reason the cercariae are able to penetrate and survive in the molluscs is because the hosts' hemolymph contains a substance which induces the secretion of a protective cyst wall around the parasite (Cheng *et al.*, 1966).

The occurrence of innate humoral factors in arthropods is suggested by Hynes and Nicholas (1958), who found that many penetrating acanthors (larvae) of the acanthocephalan *Polymorphus minutus* attached to the gut of the crustacean *Gammarus* were moribund or dead. The parasites were not encapsulated, suggesting that their death was caused by contact with the host's hemolymph.

Examples of invertebrate humoral factors that enhance cellular reactions lie with the naturally occurring agglutinins or lectins. These glycoprotein molecules enhance phagocytosis of the nonself material and consequently, although chemically different, represent the invertebrate version of vertebrate opsonins.

In recent years, increased attention paid to invertebrate humoral factors has revealed two other categories of such factors: (1) secreted lysosomal enzymes, and (2) synthesized antimicrobial molecules. Both of these, unlike the innate humoral factors mentioned, are acquired. Brief descriptions of these acquired humoral factors follow.

Lysosomal Enzymes.   Lysosomes, which constitute one type of membrane-bound organelle in cells, occur abundantly in certain types of invertebrate hemocytes, the granulocytes. Lysosomes include a variety of acid hydrolases, collectively known as lysosomal enzymes (Table 3.1). It has been demonstrated (see Cheng, 1979, for review) that when molluscs and insects are challenged with foreign material, including parasites, there is hypersynthesis of at least certain lysosomal enzymes and these are released into the serum in which the hemocytes are bathed. Consequently, when parasites make contact with the elevated levels of lysosomal enzymes, at least some are killed. The killing mechanism may be direct, i.e., one or more of the enzymes attack and destroy the foreign invader, or indirect, i.e., one or more of the enzymes cause chemical alteration of the parasite's body surface so that it now becomes recognized as foreign and is attacked by the host's hemocytes. The concept of lysosomal enzymes acting as one type of acquired humoral factor is schematically illustrated in Figure 3.5.

Antimicrobial Molecules.   As stated earlier, invertebrates do not synthesize immunoglobulins. Recently, however, Boman (1981) and Steiner *et al.* (1981) have demonstrated that when certain insects are challenged with microorganisms, they respond by synthesizing antimicrobial molecules which are quite different from vertebrate immunoglobulins. For example, if the cecropia moth, *Hyalophora cecropia*, is challenged with *Escherichia coli* and several other species of Gram-negative bacteria, the insect synthesizes two small basic proteins (P9A and P9B) which will kill these bacteria.

VERTEBRATE IMMUNITY

Immunology, as one would expect, began with studies on vertebrates, especially mammals. The credit has been generally given to Charles Ricket, who initiated the discipline in 1888, although Eli Metchnikoff had

Table 3.1. Some representative lysosomal enzymes, their pH optima, and reactions mediated.

Lysosomal enzyme	pH optimum	Reaction
Oxidoreductase		
Peroxidase	5–6	Donor + H_2O_2 —oxidized donor + $2H_2O$
Hydolases acting on carboxylic esters		
Arylesterase	5	Cleaves carboxylic esters of 2-naphthol and other aromatic alcohols
Phospholipase A_2	4.5	Cleaves the fatty acyl ester linkage to carbon 2 of the glyceryl moiety in phosphatidylcholine and other phospholipids
Hydrolases acting on phosphoric monoesters		
Acid phosphatase	3–6	Inorganic phosphate is released from glycero-2-phosphate, AMP, 4-nitrophenyl phosphate, 1-naphthol phosphate
Phosphoprotein phosphatase	5.5–6	Releases phosphate ion from serine phosphate residues of phosphoproteins, e.g., casein, phosvitin
Hydrolases acting on phosphoric diesters		
Deoxyribonuclease II	3.8–5.5	Endo-cleavage of double-stranded DNA, often both strands at same point
Sphingomyelin phosphodiesterase	4.8–5	Cleaves sphingomyelin to yield phosphocholine and acylsphingosine
Ribonuclease II	5.4–6.7	Endo-cleavage of RNA, leaving 3′-phosphate terminals, 2′, 3′-cyclic phosphates are intermediates
Hydolases acting on sulfuric esters		
Sulfatase A	5–5.6	Liberates sulfate ion from such substrates as nitrocatechol sulfate, cerebroside 3-sulfate, and ascorbic acid 2-sulfate
Chondroitin-6-sulfatase	4.8	Liberates sulfate ion from 6-sulfogalactosaminyl residues in chondroidin-6-sulfate
Hydrolases acting on glycosides		
Lysozyme	6.2	Cleaves linkage of N-acetylmuramic acid to N-acetylglucosamine in polysaccharide component of some bacterial cell walls and N-acetylglucosamine linkage in chitin
Neuraminidase	4–4.5	Cleaves nonreducing terminal α-glycosidic linkages of N-acetylneuraminic acid in glycoproteins and glycolipids
α-Glucosidase	4–5	Cleaves nonreducing terminal α-glucosyl residues from glycogen, maltose, and other oligosaccharises
β-Glucuronidase	4.3–5	Cleaves nonreducing terminal β-glucuronosyl residues from glycoaminoglycans and conjugated steroids
Hyaluronidase	3.5–4.1	Cleaves N-acetylglucosaminide linkages in hyaluronate, chondroitin-4-sulfate, and chondroitin-6-sulfate
Hydrolases cleaving peptide bonds near ends of polypeptides: exopeptidases		
Lysosomal carboxypeptidase A	5	Cleaves C-terminal residues from peptides with broad specificity (excluding arginine and lysine)
Lysosomal carboxypeptidase B	6.2	Cleaves C-terminal residues from peptides with broad specificity (including arginine and lysine)
Lysosomal dipeptidase	5.5	Cleaves dipeptides with broad specificity
Hydrolases cleaving peptide bonds away from ends of polypeptides: endopeptidases		
Cathepsin B	3.5–6	Demonstrates papainlike specificity as endopeptidase
Cathepsin E	2.5	Acts on proteins similarly to pepsin
Hydrolase acting on amide bonds other than peptides		
Acylsphingosine deacylase	4–4.8	Cleaves fatty acyl amide linkages in ceramides
Hydrolase acting on acid anhydrides		
Nucleoside triphosphatase	4–5.2	Cleaves ATP and other nucleoside triphosphates to yield diphosphate
Hydrolase acting on nitrogen-sulfur bonds		
Heparin sulfamatase	4.5–5.1	Liberates sulfate ion from N-sulfoglucosaminyl residues in heparin and heparan sulfate

studied invertebrate internal defense mechanisms in 1882. Because of the rapid development of vertebrate immunology and its obvious practical implications, almost all of the fundamental principles of the discipline have been based on findings derived from the vertebrate system. Consequently, before delving into the immunity of vertebrates to animal parasites, it appears appropriate to review briefly some of the theories and principles of immune mechanisms.

A parasite, whether protozoan or metazoan, is comprised of a number of different molecules some of which are antigenic to the host. An **antigen** is any substance that is capable, under appropriate conditions, of inducing the synthesis of antibodies and of reacting specifically in some detectable manner with such antibodies. Thus, theoretically all zooparasites contain multiple antigens.

The term **antibody** refers to proteins synthesized in response to the administration of an antigen and which react specifically with that antigen and, to a variable extent, with molecules of similar structure. Although an antibody is said to be formed in response to an antigen, there are some naturally occurring serum proteins with the structural properties of immunoglobulins that react specifically with certain antigens although the organisms in which they occur have not been previously exposed to these antigens. These naturally occurring immunoglobulins are known as **innate** or **natural antibodies**.

A large variety of macromolecules are antigenic. Virtually all proteins, many polysaccharides, nucleoproteins, lipoproteins, numerous synthetic polypeptides, and even some small molecules, if they are suitably linked to proteins or polypeptides, can serve as antigens. In order for a molecule to be antigenic, certain operational parameters must be met; for example, the quantity of the potentially antigenic material introduced and the route and frequency of the introductions all influence antigenicity. Immunogenicity is not an inherent property of a macromolecule, e.g., unlike its molecular weight or absorption spectrum, but is operationally dependent on the biologic system and conditions employed.

Antigenic Determinants

The reaction between an antigen and the corresponding antibody involves an actual combination of the two. The nature of this combination is discussed below; however, it is useful at this point to distinguish between the entire antigenic molecule and its **antigenic determinants**, i.e., those restricted portions of the antigen that determine the specificity of the antigen-antibody reaction.

Antigenic determinants are limited in size, being equivalent in volume to perhaps four or five amino acid residues. For example, if NO_2 groups are attached to rabbit serum proteins and injected into rabbits, the antisera (antibody-containing sera) produced will react not only with the nitrated proteins of the rabbit, but also with those of the horse, chicken, or other animals. However, the antisera will not react with

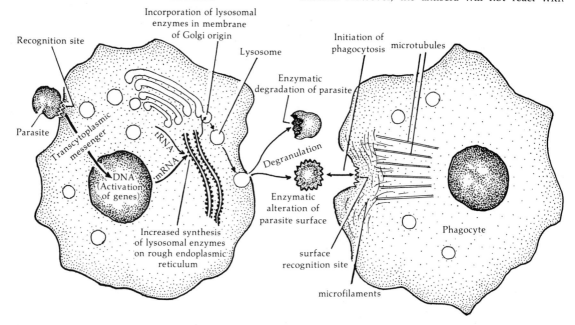

Fig. 3.5. Cellular events leading to hypersynthesis of lysosomal enzymes and their subsequent discharge into serum (degranulation).

nonnitrated serum proteins. Thus, the antibodies are capable of recognizing the nitro groups or some other uniquely altered structures in the nitrated proteins used as antigens. In this case, the NO_2 group is the antigenic determinant. The classic example of a determinant was contributed by Landsteiner and Lampl in 1920. They found that rabbits injected with *p*-azobenzene arsonate-globulin form antibodies that react with this protein and other proteins containing *p*-azobenzene arsonate, but *p*-azobenzene arsonate itself does not elicit antibody production in the host. Substances such as *p*-azobenzene arsonate are defined as **haptens**, i.e., they are not immunogenic by themselves but will react selectively with antibodies of the appropriate specificity. The antigenic determinant and hapten concept is illustrated in Figure 3.6.

Antigens of Zooparasites

Parasites, including protozoans, are comprised of a variety of molecules. Some of these are temporary constituents of their soma and others are more permanent. Consequently, a single parasite may be considered a conglomerate of antigens, each with its antigenic determinants. Thus, it is not surprising that the more refined studies on immunity to parasites have revealed the multiple antigenicity of these organisms. Furthermore, antibodies produced as the result of antigenic stimulation by one species or one stage in the life cycle of a parasite may react with certain antigens of another species or another developmental stage. Such cross-reactions indicate that these multiantigenic parasites share common antigens.

When considering antigens of parasites, it is important to remember that two categories of antigens exist: **somatic** antigens and **metabolic** antigens. Somatic antigens are those associated with molecules comprising the soma of the parasite, whereas metabolic antigens are those associated with secretions and excretions. For example, during the molting process of nematodes, an exsheathing or molting fluid is involved (p. 482). This fluid is highly antigenic and has been employed with considerable success in the vaccination of hosts as in the case of immunizing sheep against the nematode *Haemonchus contortus* by the use of its exsheathing fluid antigens.

ANTIBODIES

Antibodies synthesized by vertebrates are immunoglobulins. Immunoglobulins are proteins which share many antigenic, structural, and biologic similarities, but in which the various types differ in their primary amino acid sequence, thus accounting for their highly specific functions.

In mammals, including humans, five different classes of immunoglobulins are known, each with a distinct chemical structure and, in most cases, a speci-

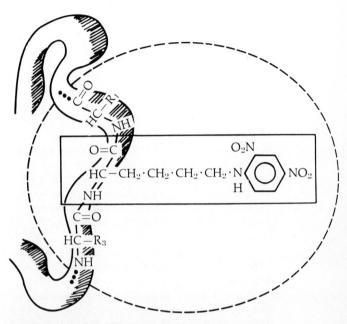

Fig. 3.6.   A representative haptenic group. A 2,4-dinitrophenyl (DNP) group substituted in the ε-NH_2 group of a lysine residue. The haptenic group is outlined by the solid line and the antigenic determinant is visualized as the area outlined by the broken line. Amino acid residues contributing to the antigenic determinant need not be the nearest covalently linked neighbors of the ε-DNP-lysine residue as shown; they could be parts of distant segments of the polypeptide chain hooked back to become contiguous with the DNP-lysyl residue. (After Davis *et al.*, 1967.)

fic biologic role. These classes are listed in Table 3.2, along with some of their properties.

Of the five classes of immunoglobulins, γG (gamma globulin, IgG) is the most abundant. In immunized animals it achieves significant concentrations in the vascular system, as well as extravascularly in tissues. It has a relatively long half-life of about 23 days, can cross the placenta, and is able to fix complement. γG provides the bulk of immunity against invading organisms, including zooparasites.

γA (alpha globulin, IgA) is the second most abundant type of immunoglobulin. It is produced in high concentrations by lymphoid tissues lining the gastrointestinal, respiratory, and urogenital tracts. When secreted, each γA molecule is combined with a protein designated the **secretory component** which is believed to protect the globulin against proteolytic enzymes. γA does not fix complement.

γM (mu globulin, IgM) is the largest of the immunoglobulin molecules in terms of size, and because

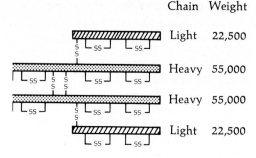

Light   22,500

Heavy   55,000

Heavy   55,000

Light   22,500

H   NH₂

Fig. 3.7.   Schematic diagram of γ_G showing the two light and two heavy chains. Notice the interchain disulfide bonds. The amino terminal end is at the right and the carboxyl terminal end is at the left. (After Bellanti, 1971.)

of this characteristic, is essentially limited to the vascular system. This class of globulins is a very efficient agglutinator of particulate antigens (bacteria, blood protozoans, etc.) and fixes complement with a high degree of efficiency. γM plays its most important role as a protective molecule during the initial few days of the primary immune response. When a parasite is introduced into a host for the first time, the synthesis of γM and γG begins almost simultaneously; however, the amount of γM peaks within a few days and then declines more rapidly than the level of γG.

The importance of γD (delta globulin, IgD) relative to parasitism has not yet been ascertained; however, a definite antibody function has been associated with it, including cases of penicillin hypersensitivity in humans.

γE (epsilon globulin, IgE) occurs only in trace amounts in serum. It has the ability to attach to skin (hence it is sometimes designated the **cytotropic antibody**; it is also referred to as the **reaginic antibody**) and to initiate aspects of allergic reactions. Like γA, γE is synthesized primarily in the linings of the respiratory and intestinal tracts. γE is commonly produced in helminthic infections.

Many studies have been made of the structures and chemical compositions of immunoglobulins. Interested individuals are referred to the volumes by Nossal (1969) and Hood *et al.* (1978). In brief, each immunoglobulin is made up of four polypeptide chains held together by disulfide bonds (Fig. 3.7). Two of the chains are smaller, each with a molecular weight of 22,000, and are designated **light chains**. The other two, each with a molecular weight of 55,000, are called **heavy chains**. A chemically different type of heavy chain exists for each of the five classes of im-

munoglobulins and is responsible for the antigenic differences that occur between them. It is also the heavy chains that are responsible for the observed biologic differences between the classes of immunoglobulins. The five types of heavy chains are known as γ, α, μ, ζ, and ε chains (Table 3.2).

There are two different types of light chains in each of the immunoglobulin classes. These are designated the kappa (κ) and lambda (λ) light chains. Thus, there are ten possible combinations of heavy and light chains, and all ten are normally found in any individual.

Any immunoglobulin may be designated by its heavy and light chain composition. For example, a γG molecule could be $\gamma 2\kappa 2$, $\alpha 2\lambda 2$, etc.

In the higher molecular weight immunoglobulins, the four basic chains, two heavy and two light, are repeated. For example, γM generally exists as a pentomer, with each of the five units held together by weak disulfide bonds (Fig. 3.8).

Table 3.2.   Some Physical and Biologic Properties of Human Immunoglobulin Classes[a]

Class	Mean Serum Concentration (mg/100 ml)	Molecular Weight	$s_{20 \cdot w}$	Mean Survival T2 (Days)	Biological Function	Heavy Chain Designation	No. of Subclasses
γ_G or IgG	1240	150,000	7	23	1. Fix complement 2. Cross placenta 3. Heterocytotropic antibody	γ	4
γ_A or IgA	280	170,000	7, 10, 14	6	1. Secretory antibody	α	2
γ_M or IgM	120	890,000	19	5	1. Fix complement 2. Efficient agglutination	μ	2
γ_D or IgD	3	150,000	7	2.8	?	δ	—
γ_E or IgE	0.03	196,000	8	1.5	1. Reaginic antibody 2. Homocytotropic antibody	ε	—

[a] After Bellanti, 1971.

As a result of studies directed at determining the amino acid sequence of both light and heavy chains of immunoglobulins, it is known that light chains possess a region of **constancy** and a region of **variability** which are approximately equal (Fig. 3.9). In the region of constancy, the carboxyl half, the molecular sequence is almost absolutely constant. On the other hand, there is extensive variability in the amino terminal half, or region of variability, of the molecule.

The primary sequence of the heavy chains is different for each of the immunoglobulin classes, but the amino acid sequence variation of the heavy chains is very similar to that of light chains. They also contain a region of variability and one of constancy. The first is represented by the amino terminal end comprised of the first 110 or so amino acids, and the constancy region is comprised of the remaining 330 residues. It is noted, however, that variations due to genetic markers and to heavy chain subclass markers do exist in the constant region.

As the result of finding regions of variability and constancy in both light and heavy chains, the concept of "two genes—one polypeptide chain" has been proposed to account for the constant and variable portions of the immunoglobulin peptide chain. According to this theory, one gene, designated C, encodes the constant half of the peptide chain, and another gene, designated V, encodes the variable half. In some way, either the two genes, the two messengers, or the two half-chains link up to form a complete polypeptide chain.

Antibody Synthesis

When a host is immunologically challenged, a functionally specialized type of lymphocyte, known interchangeably as the **antigen-sensitive cell**, the **recognizing cell**, or the **receptor cell**, is the first to receive the signal. Specifically, the immune response-stimulating antigen, or **immunogen**, becomes bonded to antibodies or antibodylike molecules on the cell surfaces of receptor cells.

At this point it needs to be pointed out that there are two major populations of **lymphocytes**: B and T cells. B cells, formed in bone marrow and subsequently processed through the bursa of Fabricius of birds (hence its designation), a lymphoid organ attached to the intestine near the cloaca, and through the liver and spleen of mammals, possess receptors comprised of antibodies at a concentration of 10^5 molecules per cell. T cells, formed in the thymus, also have cell surface receptors with properties similar to those of B cells, but their molecular nature is still uncertain. Binding of an antigen to a receptor initiates either a humoral or a cellular immune response in vertebrates, depending on whether a B cell or a T cell receptor is stimulated. Each lymphocyte carries only one type of specific receptor and consequently will respond to only a few closely related antigenic determinants. A mammal,

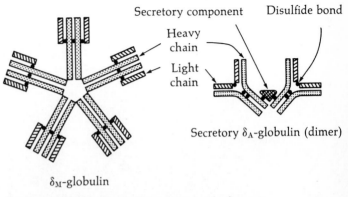

δ_M-globulin

Fig. 3.8. Models of γ_M and secretory γ_A. The former is shown in its usual pentameric form while the latter is shown attached to a secretory component. Notice the absence of the light-heavy interchain bonds in γ_A which is the predominant subclass (γ_{A_2}). (After Bellanti, 1971.)

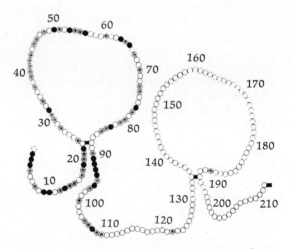

Fig. 3.9. Composite drawing of light chain (κ chain) illustrating variation in amino acid sequence. The amino terminal end of the peptide chain is at the left and the carboxyl terminal is at the right. Each circle represents one of the 214 amino acids of the chain. The open circles (O) indicate residues where only one amino acid has been found; the shaded circles (⊙) designate positions where two alternate amino acids have been detected; and the solid circles (●) indicate where three or more different amino acids have occurred. Disulfide bonds are indicated by bars. The region of variability is confined almost entirely to the amino terminal half of the molecule while the region of constancy is limited to the carboxyl terminal half, except for position 191. (After Bernier, 1970.)

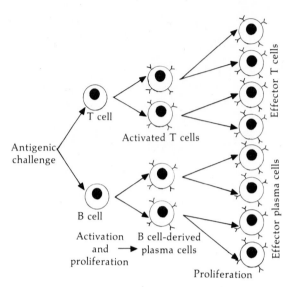

Fig. 3.10. Schematic diagram showing activation and proliferation of T and B lymphocytes of vertebrate immune system upon antigenic challenge.

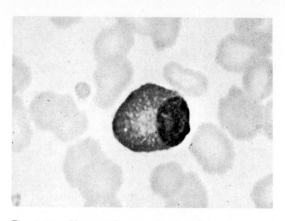

Fig. 3.11. Human plasma cell. The cell is characterized by an eccentric nucleus, perinuclear clear zone (or pale-staining area adjacent to the nucleus), and dark purple cytoplasm.

however, contains 10^8 to 10^{12} lymphocytes and possesses the capacity to respond to an enormous variety of immunogens.

Once the immunogen is bound to the receptor site of a lymphocyte (B or T cell), the host cell is stimulated to proliferate and differentiate. As a consequence, clones of progeny lymphocytes are formed. Each of

the cells of the clone displays surface receptors of the same **idiotype** as the original stimulated cell. In the process of proliferation, some progeny differentiate into **effector cells**, the functional end products of the immune response (Fig. 3.10). The B lymphocyte effector cells are known as **plasma cells** (Fig. 3.11). These secrete antibodies of the same idiotype and antigen-recognition specificity as their cell-surface receptors.

T lymphocytes are antigenically stimulated to differentiate into several types of effector cells with different functions. One of these is known as the **cytotoxic** or **killer T cells** (T_c cells). These eliminate foreign cells directly. Killer cells accomplish their function either through direct contact with surface membranes of target cells, for example, tumor cells, or through the secretion of nonspecific, nonantibody

Table 3.3. Cell Types and Effector Mechanisms Triggered by or Involved in Immune Reactions[a]

	Humoral Factors	
Cell Type	*Agents Responsible for Mobilization of Cells*	*Cell Product*
Nonspecific		
Macrophages (monocytes)	Chemotactic factors, migration inhibitory factor (MIF)	Processed immunogen
Granulocytes		
Neutrophils	Chemotactic factors (complement-associated and bacterial factors)	Kallikreins (producing kinins), SRS-A, basic peptides
Eosinophils	Identical with neutrophils; specific chemotactic factors	?
Basophils	?	Vasoactive amines
Platelets	Factors producing platelet aggregates (thrombin, collagen)	Vasoactive amines
Specific		
Plasma cells	?	Antibody
Lymphocytes	?	Antibody, MIF, interferon, cytotoxin, transforming factor, "transfer factor" and others

[a] After Bellanti, 1971.

mediator molecules known as **lymphokines**. Lymphokines can poison the foreign cells or tissues (lymphotoxins), stimulate phagocytosis by macrophages (macrophage activating factor), or attract inflammatory cells to the site of injury (chemotaxis).

Other types of effector T cells are responsible for delayed hypersensitivity (T_D cells), for amplifying T_c cell differentiation and proliferation (T_A cells), for helping B cell differentiation and proliferation (T_H cells), and for suppressing immune responses (T_S cells).

The currently accepted notion of the synthesis and secretion of immunoglobulins by B lymphocytes with specific idiotypes is known as the **selection theory**; i.e., B cells already committed to the production of a particular antibody (immunoglobulin) are stimulated by the introduction of the immunogen to proliferate

and to elaborate the immunoglobulin.

Cells of the reticuloendothelial system of vertebrates are the sites of elaborating cell products associated with the immune mechanism, including antibodies. These cells are strategically distributed throughout the body, including the lining of the lymphatic and vascular systems. The specific cell types, the mechanisms which trigger them, and their cell products are tabulated in Table 3.3.

Macrophages. Macrophages occur in various tissues including the blood, where they are called **monocytes** (Fig. 3.12). Unlike granulocytes, macrophages

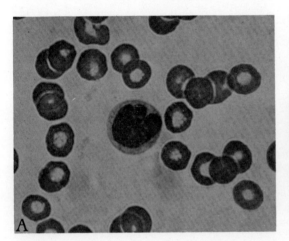 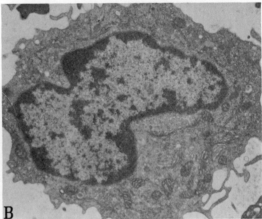

Fig. 3.12. Human macrophages (monocytes). A. Macrophage from peripheral blood, Wright-Giemsa stain, × 1400. (After Malinin in Bellanti, 1971.) **B.** Electron microphage of macrophage, × 11,000. (After McFarland in Bellanti, 1971.)

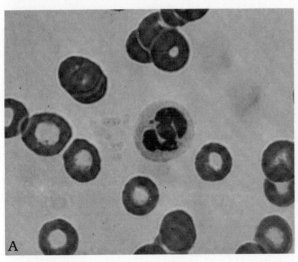 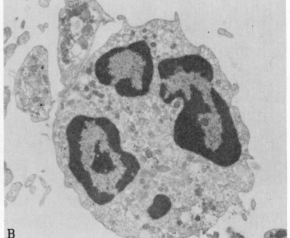

Fig. 3.13. Human neutrophils or polymorphonuclear leucocytes from peripheral blood. A. Neutrophil showing segmented nucleus, Wright-Giemsa stain, × 1350.(After Malinin in Bellanti, 1971.) **B.** Electron micrograph of neutrophil showing cytoplasmic granules, × 10,500. (After McFarland in Bellanti, 1971.)

are capable of dividing in tissues. Some macrophages are held in place by reticular fibers. All macrophages are capable of phagocytosis and, hence, are important as cellular defense mechanisms in the arrest and/or removal of foreign materials, including certain parasites. This process is known as **clearance**. The phagocytosis of foreign materials is sometimes facilitated by

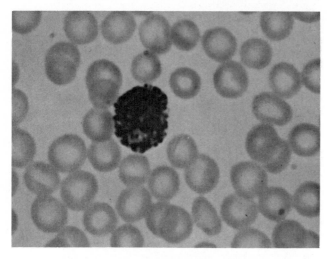

Fig. 3.14. **Basophil.** Human basophilic leucocyte (basophil) from peripheral blood showing relatively large basophilic granules in the cytoplasm, Wright-Giemsa stain, × 1400 (After Bellanti, 1971.)

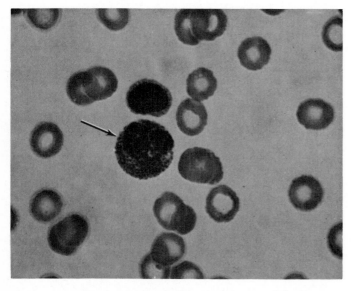

Fig. 3.15. **Eosinophil.** Human eosinophilic leucocyte (eosinophil) (arrow) and a small lymphocyte from peripheral blood, Wright-Giemsa stain, × 1400. (After Malinin in Bellanti, 1971.)

antibodies. Specifically, the foreign particle becomes coated with an antibody known as an **opsonin** and as a consequence is phagocytosed more efficiently and rapidly. In addition to true opsonins, **complement**, a series of sequentially reacting serum proteins, may also be involved as an amplifier of phagocytosis (p. 80).

Macrophages are also involved in delayed hypersensitivity reactions (p. 82) and are attracted to an area of injury by a number of chemotactic factors, some derived from the complement system. For a review of what is known about phagocytosis, especially in mammals, consult the volume edited by Karnovsky and Bolis (1982).

Granulocytes. Granulocytes have their origin in bone marrow and are released into the circulatory system at a sufficient rate to replace dying cells. There are three types of morphologically distinguishable granulocytes in mammals and other vertebrates. The first type, known as the **neutrophil** or **neutrophilic polymorphonuclear leucocyte** (Fig. 3.13), comprises from 60 to 70% of the total number of leucocytes in the peripheral circulation of the adult human. The neutrophil, being a terminal cell type of myeloid differentiation, does not divide. Its primary function is to phagocytose and digest particulate foreign material, especially certain virulent bacteria. Furthermore, during inflammation, the body's reaction to injury, neutrophils migrate to and congregate at the site of injury, such as the bite of a tick or mosquito, through the mediation of certain components of complement (p. 80).

Neutrophils are also known to be directly or indirectly involved in the production of a substance known as '"slow reactive substance" (SRS-A), which is a humoral factor containing a fatty acid.

The second type of granulocyte is the **basophil** (Fig. 3.14). It constitutes about 0.5% of the peripherally circulating leucocytes. Basophils, along with platelets, contain vasoactive amines such as histamine and serotonin. The secretion of these amines is believed to be triggered by contact with antigen-antibody complexes through complement-dependent or complement-independent mechanisms.

The third type of granulocyte is the **eosinophil** (Fig. 3.15). Eosinophils make up from 1 to 3% of circulating leucocytes. These cells are of particular interest to parasitologists because eosinophilia, i.e., an increase in the number of eosinophils, is characteristic of almost all helminth infections in mammals. Although some early evidence suggested that they are attracted by antigen-antibody complexes, more recent studies suggest that they respond to complement-derived chemotactic factors.

Plasma Cells. Plasma cells are of primary interest in immunology because, upon antigenic stimulation, they elaborate antibodies. These cells (Fig. 3.16) are characterized by their RNA-rich cytoplasm, thus suggesting active protein (antibody) synthesis. Since the enclosed RNA is readily stained with pyronin, these cells are sometimes referred to as **pyroninophils**.

ANTIGEN-ANTIBODY INTERACTIONS

Antigen-antibody interactions can be divided into three categories: (1) primary, (2) secondary, and (3) tertiary (Table 3.4).

PRIMARY INTERACTION

The initial or primary antigen-antibody interaction is the basic event during which the antigen is bound to two or more available sites on the antibody molecule

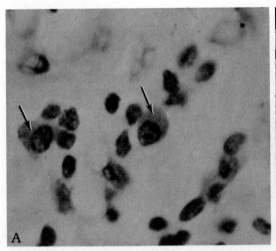 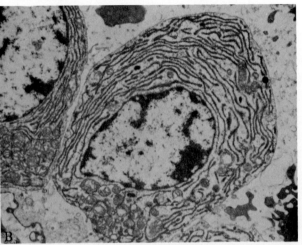

Fig. 3.16. Human plasma cells. A. Plasma cells (arrows) from section of inflamed aorta, hematoxylin and eosin stain, × 1000. (After Malinin in Bellanti, 1971.) **B.** Electron micrograph of plasma cell showing well-developed endoplasmic reticulum, × 11,000. (After McFarland in Bellanti, 1971.)

Table 3.4.   Schematic Representation of Primary, Secondary, and Tertiary Antigen-Antibody Reactions[a]

Primary Ag + Ab reaction

Ammonium sulfate method (Farr technique)
Equilibrium dialysis
Immunofluorescence, radiolabeling, ferritin labeling

Secondary manifestations *in vitro*	Tertiary manifestations *in vitro*	
	Beneficial Effects	**Deleterious Effects**
Precipitation	Precipitation *in vivo* (?)	Lupus erythematosus
Agglutination	Hemagglutination, leucoagglutination, bacterial agglutination	Nonhemolytic transfusion reactions
Complement-dependent reactions	Bacterial lysis, hemolysis	Severe cytolytic reactions
Cytolysis (cytotoxicity)	Phagocytosis	Deficiency may lead to enhanced susceptibility to infection
Phagocytosis-promoting activity (opsonization)		
Chemotaxis	Chemotaxis	Deficiency may lead to enhanced susceptibility to infection
Permeability	Altered permeability	Edema
Neutralization	Antimicrobial immunity Toxin neutralization Viral neutralization	
Cytotropic	?	Allergic diseases of humans (anaphylaxis)

· [a] After Bellanti, 1971.

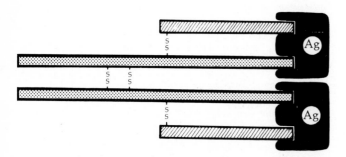

Fig. 3.17. Schematic representation of the binding of antigen to antibody sites. (After Bellanti, 1971.)

(Fig. 3.17). This type of interaction is rarely directly visible but, as indicated in Table 3.4, can be measured by such techniques as the ammonium sulfate precipitation (Farr's) technique and labeling antibody or antigen with fluorescent, electron-dense, radioactive, or enzymatic markers. More commonly, it is through the occurrence of secondary and tertiary events that the primary interaction is ascertained.

SECONDARY INTERACTIONS

Secondary manifestations of the antigen-antibody inter- action include agglutination, precipitation, complement-dependent reactions, neutralization, immobilization, and some other specialized types of reactions.

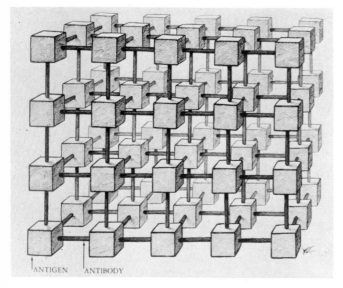

Fig. 3.18. Schematic representation of antigen–antibody lattice formation. (After Bellanti, 1971.)

Agglutination

This type of assay system used to detect the occurrence of antibodies depends not so much on the antibody produced as on the physical form of the antigen. If the antigen is naturally or experimentally attached to particulate matter, for example, bacterial cells, blood cells, latex, or bentonite particles, the antigen-antibody interaction results in the clumping of the antigenic particles. This phenomenon is known as agglutination.

Precipitation

Again, this type of assay system depends on the form of the antigen. In this case, the antigen is in soluble form and the antigen-antibody complex forms large aggregates which are insoluble.

In the case of both agglutination and precipitation, a two-stage, reversible chemical interaction occurs. In the first stage, serum antibodies in the host react with specific antigens on the foreign molecule. The ease of this combination is dependent on several factors, especially pH, ionic strength, and temperature, and is made possible by means of van der Waal's forces. When this binding has reached equilibrium, the second stage, known as **lattice formation**, occurs. During lattice formation, the unbound receptor sites on the antibody molecules become attached to suitable receptors on additional antigen molecules, forming a lattice (Fig. 3.18).

Lattice formation is also specific. Since the two antigenic receptor sites of a divalent antibody are identical, an antibody with one specificity can link only identical antigens together.

In precipitation, since the antigens are soluble, a fairly large lattice must be formed before it can be visually appreciated. Therefore, a large number of antibody molecules are required. On the other hand, in agglutination the antigen is a part of a large, insoluble particle, and comparatively few antibody molecules are required to form a visible lattice. Consequently, agglutination is a more sensitive assay method for antibody detection. It is also of interest to note that on a molar basis, γM is a more efficient agglutinator than γG, but γG is a better precipitin than γM.

Complement. Although this term was originally coined to designate an auxiliary factor in serum which, acting on an antibody-coated cell, would lead to lysis of the cell, complement is now known to be comprised of a complex set of interacting proteins. Specifically, it is known that the complement sequence consists of nine functional entities or eleven discrete proteins identified as C1q, C1r, C1a, C2 ... C9 (Table 3.5).

The coating of foreign particles or cells with the first four functional components of complement (C1q, C1r, C1a, and C2, C3, C4) renders the foreign material immediately susceptible to phagocytosis. Furthermore, the interactions of these four functional components can also generate a portion of C3 (known as C3a) or a portion of C5 (known as C5a) which, in turn, will mediate an acute inflammatory response and attract leucocytes to the area of the invading parasite. It is convenient to regard complement as an array of substrates from which mediators of the acute inflammatory response can be generated. For detailed discussions of the complement system, the review by Mayer (1973) is recommended.

Complement-dependent reactions. Complement may be involved in a number of types of antigen-antibody interactions. In the case of agglutination, for example, complement may participate. In addition, lysis, phagocytosis, chemotaxis, and altered permeability are commonly complement dependent (Table 3.4).

Lysis is the destruction of the cell membrane of the foreign cell as the result of the action of a specific antibody on a surface antigen mediated through the activation of the complete complement sequence. In parasitology, lysis is usually manifested when the foreign cell is a protozoan parasite and is ruptured and killed as the result.

Phagocytosis, as explained earlier, is the engulfment of the nonself material by host cells. This activity in vertebrates is enhanced through the action of antigen, antibody, and complement. This occurs in two ways. (1) Accumulation of leucocytes, or leucocytosis, in vertebrates is commonly brought about through the activation of the complement sequence in which several chemotactic principles are initiated, (2) Certain antibodies, known as **opsonins**, become coated onto the foreign material and this enhances phagocytosis. These opsonins are mediated by the activation of the complement sequence.

Altered permeability of the foreign cell is also associated with the combination of antigen and antibody attributable to specific mediators generated by the activation of the complement cascade by the immune complex.

TERTIARY INTERACTIONS

Tertiary interactions are defined as *in vivo* expressions of antigen-antibody interactions. At times these may be of survival value to the host, but at other times they may lead to disease through immunologic injury. In the first instance, the antibodies produced serve a protective function, forming the basis for total or partial antiparasitic resistance. On the other hand, deleterious effects can occur. A brief discussion of the more salient aspects of immunologic injury is presented below.

The pathologic effects of immunologic processes, or hypersensitivity, are classified into four types (Table 3.6). **Type I**, or **immediate hypersensitivity**, results in a reaction within minutes after the introduction of a soluble antigen into a previously sensitized host. This type of immediate hypersensitivity is known as **anaphylaxis**. Generalized anaphylaxis is manifested in a variety of ways, depending on the shock organs of the host species. For example, in guinea pigs, minutes after challenge with antigen, the animal will scratch, sneeze, and cough and may convulse, collapse, and die. This is primarily due to respiratory impairment resulting from constriction of the smooth muscles in the bronchioles and to bronchial edema. Consequently, a rapid drop in blood pressure and a generalized increase in vascular permeability may also be part of the shock reaction. In rabbits, the shock organ is the heart, and cardiac failure is the usual cause of death during anaphylaxis. In humans, the signs of anaphylactic shock are itching, erythema, vomiting, abdominal cramps, diarrhea, and respiratory distress. In severe cases, laryngeal edema and vascular collapse may lead to death.

Table 3.6. Immunologic Mechanisms of Tissue Injury[a]

Type	Manifestation	Mechanism
I	Immediate hypersensitivity reactions	γ_E and other immunoglobulins
II	Cytotoxic antibody	γ_G and γ_M
III	Antigen-antibody complexes	γ_G mainly
IV	Delayed hypersensitivity (cell-mediated)	Sensitized lymphocytes

[a] After Bellanti, 1971.

Table 3.5. Properties of Human Complement Proteins[a]

Properties	Clq	Clr	Cla	C2	C3	C4	C5	C6	C7	C8	C9
Serum concentration (μg/ml)	190	—	22	20–40	1,200	430	75	—	—	<10	<10
Sedimentation coefficient (S)	11.1	7.0	4.0	5.5	9.5	10.0	8.7	5–6	5–6	8.0	4.5
Approximate molecular weight	400,000	—	79,000	117,000	185,000	240,000	—	—	—	150,000	79,000
Relative electrophoretic mobility	γ_2	β	α_2	β_2	β_1	β_1	β_1	β_2	β_2	γ_1	α
Carbohydrate (%)	15	—	—	—	2.7	14	19	—	—	—	—
Reactive SH	—	—	—	2(?)	1–2	—	—	—	—	—	—

[a] From Müller-Eberhard, 1969.

Type II hypersensitivity, or **cytotoxic antibody response**, occurs when the antibodies are directed to tissue antigens. There are two main mechanisms for such injury. (1) The antibody may react with the host's own tissue cells, causing cytolysis and killing by activation of all nine components of complement (p. 80), (2) When the antitissue antibody reacts with its antigen, the complex can interact with the host's phagocytic cells, which adhere to the immunoglobulin, sometimes enhanced by the fixation of complement through the third component. The host's own erythrocytes, for example, can not only be lysed but also phagocytosed by other cells (cells of the reticuloendothelial system). Also, the fixation of antibody and complement can cause accumulation of such host cells as neutrophils and eosinophils, which release injurious constituents. Because the host's antibodies affect its own tissues in type II reactions, the term **autoimmunity** is commonly used to describe the condition.

Type III reaction is manifested as tissue injury produced by antigen-antibody complexes. In brief, after an antigen has combined with an antibody within the host, certain mechanisms, such as phagocytosis, eventually eliminate the complex. However, during the process, a variety of inflammatory reactions occur which result in tissue injury.

Type IV reaction is known as **delayed hypersensitivity**. This condition is a manifestation of cell or cell-mediated immunity. Delayed hypersensitivity is defined as an increased reactivity to specific antigens mediated not by antibodies but by cells. It is termed "delayed" because of its slow onset, taking 24 hours to reach maximal intensity. The fact that it is cell-mediated was first unequivocally demonstrated by Landsteiner and Chase in the 1940s when they demonstrated that the delayed response can be transferred to an unreactive recipient by cells but not by serum.

Morphologically, delayed hypersensitivity can be appreciated as an accumulation of host cells (neutrophils, macrophages, and lymphocytes) at the site of antigen concentration. It should be emphasized that delayed hypersensitivity results on secondary, rather than initial, contact with the antigen.

Primary and Challenge Infections and Anamnesis

In animals with true immunologic capabilities, i.e., those that have the capacity to synthesize immunoglobulins (e.g., mammals), there is an interesting feature known as **anamnesis** or **memory response**. Prior to discussing this feature, two terms need to be defined: **primary** and **challenge** infections. A primary infection is the one during which the host experiences its initial exposure to the antigen while a challenge infection refers to all subsequent exposures.

A host exposed to a primary infection rapidly eliminates the antigen, and this must precede immune elimination. Immune elimination is the result of the combination of antigen with antibody. This is known as the **primary response**. As depicted in Figure 3.19, there is a latent phase associated with the primary response. No identifiable antibody occurs in the serum during the primary response; however, antibody secreted by single cells can be readily detected. As the latent period ends, the primary antibody response becomes demonstrable throughout the entire animal. The blood antibody titer increases during the next few days, but does not reach a high level. Subsequently, the antibody level increases for a few weeks, plateaus, and then begins to drop. As depicted in Figure 3.19, the initial shape of the primary response curve is sigmoidal with an extended decay period.

If the host is subjected to a challenge infection, any remaining antibody is rapidly removed by combining with antigen; subsequently, there is a fall in the detectable antibody level in the blood (Fig. 3.19). Almost immediately thereafter, within days, there is a spectacular rise in the antibody level. Actually, the second rise in antibody level can be 10 to 50 times higher than the primary response. This accelerated enhanced response to the challenge infection is known as the

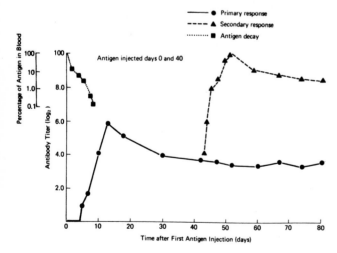

Fig. 3.19. Antigen decay (elimination) and primary and secondary immunoglobulin synthesis curves. The antigen elimination curve shows the three phases of equilibration, metabolic elimination, and immune elimination, the latter beginning at about the fifth day. Circulating antibodies are not detectable until about the fifth day. Notice how the secondary immunoglobulin response following the readministration of antigen at the fortieth day reaches a very high titer compared to the primary response (examplifying anamnesis). (After Barrett, 1983; with permission of C. V. Mosby.)

anamnestic or memory response. In other words, the host's immune system "remembers" that it has been confronted with the same antigen previously and remains primed for a second encounter.

One can induce anamnesis at any time after the primary response, sometimes even after several years when the primary response antibody titer has dropped to zero; however, the secondary response will not be as spectacular as the one induced by an earlier challenge infection.

Certain generalities can be stated relative to certain classes of immunoglobulins produced during primary and challenge infections. γM is rich in the antibody response to primary infections. It has a half-life of about 8 to 10 days. The antibody formed in response to challenge infection consists of more γG, which has a half-life of 25 to 40 days.

Anamnesis is characteristic of vertebrate hosts and has yet to be conclusively demonstrated in invertebrates.

Molecular Mimicry. Parasitologists have long been amazed that endoparasites, both protozoans and helminths, can survive for such long durations in "immunologically hostile" environments, i.e., in hosts where there are relatively high levels of immunoglobulins. The process has been described by which certain parasites are recognized as self and consequently do not invoke immunologic reactions in their hosts, a phenomenon adapted through thousands or millions of years of evolution. In relatively recent years, considerable research has focused on the mechanisms underlying how such parasites can refrain from stimulating their hosts immunologically. There is now sufficient information to conclude that such parasites are immunologically inert because they present their hosts with surfaces and/or metabolic immunogens that are antigenically similar to those of the hosts so that they are recognized as self. This phenomenon of antigen sharing between parasite and host has been termed **molecular mimicry** (Damian, 1964). The topic has been reviewed by Damian (1979).

The sharing of antigens between host and parasite was first demonstrated by Sprent (1962). Subsequently, Damian (1967), Capron *et al.* (1968), and others have demonstrated that helminth parasites not only share antigens with their hosts but also with other species within their class, and even with helminths belonging to a different class. Examples of parasites sharing antigens with their hosts are listed in Tables 3.7 and 3.8; examples of trematodes sharing antigens with other species of trematodes are listed in Table 3.9. Antigen sharing between species of helminths represents one of the major difficulties in the development of species-specific antigens for immunologic diagnostic tests. As an example, as indicated in Table 3.7, the human lung fluke *Paragonimus westermani* shares five antigens in common with the Chinese liver

Table 3.7. Number of Antigens Shared Between Selected Species of Digenetic Trematodes and Their Intermediate and Definitive Hosts

	No. of Antigens Shared	
Species	*With Intermediate Host*	*With Definitive Host*
Fasciola hepatica	4 (*Lymnaea truncatula*)	6 (cattle)
Dicrocoelium dentriticum	?	6 (sheep)
Paragonimus westermani	3 (*Melanoides tuberculata*)	4 (dog)
Clonorchis sinensis	?	3 (rabbit)
Opisthorchis felineus	?	3 (cat)
Schistosoma mansoni	6 (*Biomphalaria glabrata*)	5 (hamster)
Schistosoma haematobium	4 (*Bulinus truncatus*)	4 (hamster)
Schistosoma japonicum	4 (*Oncomelania nosophora*)	4 (rabbit)

(Data from Capron *et al.*, 1968.)

Table 3.8. Shared Antigens Between Selected Species of Cestodes and Their Intermediate and Definitive Hosts

	No. of Antigens Shared	
Species	*With Intermediate Host*	*With Definitive Host*
Taeniarhynchus saginatus	6 (cattle)	1 (human)
Taenia solium (*Cysticercus cellulosae*)	4 (pig)	1 (human)
Taenia pisiformis	7 (rodent)	0 (dog)
Taenia hydatigera	6 (rodent)	0 (dog)
Moniezia expansa	? (mites)	4 (sheep)
Anoplocephala magna	? (mites)	3 (horse)
Dipylidium caninum	? (fleas)	4 (dog)
Vampirolepis nana	? (fleas)	2 (rodents)
Echinococcus granulosus		
Hydatid fluid	3–15 (sheep)	?
Protoscolices	4 (sheep)	0 (dog)
Hydatid membrane	8 (sheep)	?

(Data from Capron *et al.*, 1968.)

fluke *Clonorchis sinensis*. Therefore, if a skin-test antigen comprised of whole, homogenized *P. westermani* is used for diagnosis, individuals harboring *C. sinensis* would also present a positive reaction.

With regard to the sharing of antigens between host and parasite resulting in the inability of the host to recognize the parasite as nonself, some controversy exists as to how the phenomenon originated. Three

Table 3.9. Sharing of Antigens Between Trematodes: Number of Antigens Shared Between Different Species of Digeneans as Revealed by Immunoelectrophoresis

	Fasciola hepatica	*Dicrocoelium dendriticum*	*Clonorchis sinensis*	*Paragonimus westermani*	*Schistosoma mansoni*	*Schistosoma haematobium*	*Schistosoma japonicum*
Fasciola hepatica	25	6	5	5	5	5	4
Dicrocoelium dendriticum	6	19	5	5	3	3	2
Clonorchis sinensis	5	5	21		5	5	4
Paragonimus westermani	5	5	5	20	4	4	2
Opisthorchis felineus	4	3	11	4	2	2	1
Schistosoma mansoni	5	3	5	4	21	19	11
Schistosoma haematobium	5	3	5	4	19	21	11
Schistosoma japonicum	4	2	4	2	11	11	23

(Data from Capron *et al.*, 1968.)

possibilities exist, and it remains to be conclusively demonstrated which one (or perhaps more) is operative: (1) mimicry by natural selection, (2) mimicry by host induction, and (3) mimicry resulting from incorporation of host antigens.

Mimicry by Natural Selection. According to this hypothesis (Fig. 3.20), parasite antigens, through natural selection, have evolved to more nearly approach those of its host. This concept implies that the more ancient a particular host-parasite association is, the more likely it is that the parasite antigens would resemble those of its host.

Mimicry by Host Induction. According to this hypothesis (Fig. 3.20), the host in some manner is capable of inducing the parasite to produce hostlike antigens. Acceptance of this hypothesis depends on the belief that the parasite possesses a series of genes which code for host antigens; furthermore, these genes would be activated only when induced by factors of host origin.

Mimicry Resulting from Incorporation of Host Antigens. According to this hypothesis (Fig. 3.20), the parasite is capable of adsorbing or absorbing host antigens onto its body surface, and such molecules serve to mask the exposed parasite antigens.

Of the three hypotheses presented, experimental data on schistosomes appear to support host-induced (Damian *et al.*, 1973) and incorporation of host-antigens mimicry (Clegg *et al.*, 1971; Clegg, 1972).

Now that some of the principles of immunity have been described, additional examples of immunity in vertebrates to zooparasites will be presented. In considering immunity in vertebrates to parasites, one should bear in mind that host immunity is influenced by a number of factors including the genetics of the host, nutrition, number of parasites invading, and age and physiological state of the host. Furthermore, it must be remembered that total immunity, i.e., complete refractiveness after immunization, is rare when the invading organism is an animal parasite. Rather, the host is often only partially immune to further infection. Also, the partial immunity may be reflected, not only in a lesser parasite density, but merely as stunted parasites or some other manifestation.

Both protozoan and metazoan endoparasites can be categorized as either blood or tissue parasites. Even the so-called lumen-dwelling forms, such as those occurring in the intestine, are essentially tissue parasites since usually at least a portion of their bodies are embedded in host tissues. In either case, one can expect antibody production only if the parasite's body or its excretions and secretions are in direct contact with immunologically competent host cells. The following review of immune reactions in vertebrates to parasites is presented under the subheadings of protozoan and metazoan blood and tissue parasites.

Protozoan Blood Parasites

As is presented in subsequent chapters, a number of protozoan parasites, especially the hemoflagellates, are found primarily within the circulatory systems of their vertebrate hosts. As the result of their presence, the host almost always develops some degree of acquired immunity. This immunity, either complete or partial, may be maintained only while the parasites are present—the phenomenon is known as **premunition** (or concomitant immunity)—or it may be a **sterile** immunity, i.e., it persists after the complete disappearance of the parasites.

An example of premunition occurs in animals infected with certain species of *Babesia*. Riek (1963) has reported that infected cattle retain acquired immunity as long as the parasitemia (parasites present in blood) persists, usually for only a few months. Once the parasite is eliminated immunity is lost and the hosts become susceptible to reinfection. On the other hand, Joyner and Davies (1967) have reported that in the case of cattle infected with *Babesia divergens* and *B. bigemina*, a persistent immunity lingers after the parasites are removed. Similarly, a protective immunity persists in cattle long after *Theileria parva* has disappeared, even enduring for the life of the host. These are examples of sterile immunity.

A great deal of attention has been paid to the malaria-causing parasites from the standpoint of immunity. In the case of the avian-infecting species of *Plasmodium*, premunition usually occurs; however, a sterile immunity can be produced with the rodent malaria-causing agents *Plasmodium berghei* and *P. vinckei* (Corradetti, 1963; Cox, 1966). In the case of *P. berghei*, the sterile immunity is gradually lost. In the case of human malarias, both premunition and sterile immunity can be demonstrated, although neither is usually complete. Furthermore, because of the complex life cycles of these species of *Plasmodium*, antibodies produced against one stage may not be effective against another. For example, it is known that immune serum from West Africans is effective against East African strains of *Plasmodium falciparum*, but only against the asexual stages in blood (p. 198) and hence is not very effective in preventing secondary infections. Protective antibodies against malaria are primarily of the γG class, an immunoglobulin that can be transmitted from mother to fetus, endowing the newborn with antimalarial immunity for approximately 6 months. In addition to γG, γM, γA, and possibly γD

are also produced against malaria (Tobie *et al.*, 1966; McFarlane and Voller, 1966).

Untreated African trypanosomiasis usually kills the host, but this does not mean that some degree of immunity does not develop. It has been demonstrated that certain breeds of cattle develop a partial immunity if repeatedly infected with trypanosomes, but this immunity wanes without continued exposure. Similarly, it has been shown that a breed of N'dama cattle, when maintained in a tsetse fly belt, shows only transient parasitemia after a challenge infection. However, if the cattle are reared outside of a tsetse fly belt, they show a much more persistent parasitemia, thus indicating that partial immunity probably exists. The antibodies produced against certain trypanosomes kill the para-

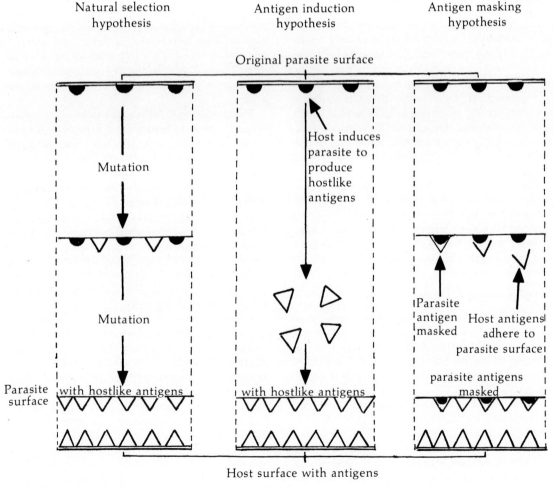

Fig. 3.20.   Molecular mimicry in parasites. Three hypotheses to explain occurrence of host-like antigens on parasite surface. (Modified after Smyth, 1976.)

sites by interfering with their oxygen consumption (Desowitz, 1959).

In African human trypanosomiasis, the immunoglobulins produced are primarily of the γM class, as indicated by marked elevations of γM in serum and cerebrospinal fluid.

Specificity of Immune Response.   Because blood protozoans consist of a number of antigens, it is not surprising that, owing to the existence of common antigens, cross-immunity has been reported, as is the case among helminths (p. 83). For example, there is cross-immunity between *Babesia rodhaini* and *B. microti*. On the other hand, immunity to malaria is generally more specific (Cohen and McGregor, 1963; Voller *et al.*, 1966), but not absolute.

Immunization.   Many attempts have been made to develop active immunization against blood protozoans by employing killed or attenuated parasites as antigens. The results have been mixed and, even where protection has been conferred, the immunity is often incomplete or has been effective only against small challenge infections. Nevertheless, although mass immunization programs are not yet feasible, it is known that improved techniques hold promise in the killing or attenuating of whole parasites to use as antigens: freezing and thawing (Johnson *et al.*, 1963), formalin treatment (Lapierre and Rousset, 1961a,b; Soltys, 1964), β-propiolactone treatment (Soltys, 1965), and X-ray irradiation (Sanders and Wallace, 1966).

Another technique with potential value for immunization is the use of avirulent strains of parasites. Weiss and DeGuisti (1966), for example, found that a strain of *Plasmodium berghei* which had lost its infectivity for mice after serial passage through tissue culture, maintained its ability to induce a sterile immunity in the host if injected.

Protozoan Tissue Parasites

Many protozoans are primarily tissue parasites in their vertebrate hosts, while others, such as *Plasmodium* spp., are tissue parasites during one phase of their life cycles and blood parasites during another. This brief introduction to immunity to protozoan tissue parasites is limited to a few examples to establish the principles. The reviews by Stauber (1970) and Kozar (1970) are recommended to those desiring greater detail.

As stated earlier, it is only when parasites make contact with their hosts' tissues in such a manner that their antigenic properties are recognized that antibodies are produced. Thus, when the amoeba *Entamoeba histolytica* resides in its hosts' intestinal lumina,

no detectable antibodies occur. However, when this amoeba invades the mucosa, circulatory system, and other tissues, antibodies become evident, although these are not protective. Nevertheless, these antibodies are significant from the standpoint of immunodiagnosis (Kessel *et al.*, 1965; Powell *et al.*, 1966). As in the case of the other tissue-invading protozoans, the antibodies produced are of the γG and γM classes. The production and functions of the other immunoglobulin classes as related to tissue protozoans remain uncertain.

In the case of *Leishmania tropica* (p. 123), the parasites invade the macrophages of the host's skin and multiply therein. If the sores produced heal spontaneously, the host becomes totally immune to further infection for life. However, if the sore is surgically removed before spontaneous recovery, protective immunity does not occur, and the host remains susceptible to further infection. It is of interest to note that circulating antibodies cannot be demonstrated in cutaneous *L. tropica* infections but are readily demonstrated in visceral *L. donovani* infections. In general, human immunologic response to leishmaniasis is primarily of the cell-mediated type, with circulating antibodies playing only a minor protective role. Cell-mediated immunity is manifested by a population of T cells that develops when stimulated by parasite antigens. These T cells destroy the host's macrophages in which the parasites occur and hence also kill the parasite.

The immunologic pattern associated with *L. tropica* does not appear to hold true for *L. braziliensis*, a mucocutaneous parasite. In the latter case, although complement-fixing antibodies are formed, the host is not protected from further infection after the primary lesions heal. Thus, although premunition occurs, protective sterile immunity does not.

In visceral leishmaniasis caused by *L. donovani*, the host usually dies unless treated and cured; but, in rare instances of spontaneous cure, the host is conferred complete immunity for life. This also occurs if the host is cured by drugs. It appears that if premunition occurs, it is ineffective in destroying the parasite, but sterile immunity is developed.

Immunity to the Mediterranean form of kala azar caused by *Leishmania infantum* (or *L. tropica infantum*) provides another interesting feature. The disease affects children under 5 years old but is rare in adults. This cannot be explained by the development of immunity during childhood since adults visiting in endemic areas who have never been exposed to the parasite are also unaffected. According to Taub (1956), the phenomenon can be explained by the presence of an adult serum component, in which the parasite cannot live.

Trypanosoma cruzi, unlike the other human-infecting trypanosomes, spends only a short period of time in the blood. During the rest of the time, it occurs within the vertebrate host's cells, particularly those of the reticuloendothielal system and cardiac muscle. Available evidence indicates that immunity exists in endemic areas (Muniz, 1962) but it remains uncertain whether this immunity is initiated by the blood or tissue stages. In experimentally infected mice, Norman and Kagan (1960) and Nussenzweig *et al.* (1963) have shown that the blood stages are important in eliciting antibody production, but it is uncertain whether these data can be extrapolated to humans since the blood stages in mice are prolonged. As in the case of *Plasmodium berghei*, an initial challenge with an avirulent strain of *Trypanosoma cruzi* in mice can induce complete resistance to a subsequent challenge with a virulent strain.

In the case of *Histomonas meleagridis* (p. 117) infection in turkeys, young birds up to 1 year old are the most susceptible, but older ones can become infected, and recovery sometimes occurs. Recovered hosts are immune to reinfection, and protection can be induced by immunization with avirulent strains.

Immunity to *Toxoplasma gondii*, another tissue parasite of endotherms (p. 189), has been reviewed by Beattie (1963). In adult humans the acute phase of the disease is rare (p. 191) but antibodies are common, suggesting protective immunity. The antibodies are, at least in part, of the γM class. It has been suggested that γG occurs during the chronic phase of toxoplasmosis and γM occurs during the acute phase (Remington and Miller, 1966).

In congenital infections with *Toxoplasma gondii*, death usually occurs. If the infant survives, hydrocephalus and other pathologic effects may occur (Fig. 3.21). In such children, the liver, spleen, lungs, heart, and other organs may become involved. As in adults, antibodies are produced. First, γM is produced which acts against the parasites by preventing their spread, and later, γG is produced which destroys the parasites. This explains the acquisition of immunity.

Special Manifestations of Antigen-Antibody Reactions as Related to Protozoan Parasites. Studies on immunity to protozoan parasites in the blood of mammals, especially the trypanosomes, have revealed an interesting aspect of immunoparasitology that is unique. Specifically, it has been known for many years that humans infected with the African sleeping sickness-causing protozoans *Trypanosoma brucei gambiense* and *Trypanosoma brucei rhodesiense* portray fluctuations in the number of parasites in the blood, a phenomenon known as **cyclic parasitemia**. Investigations directed at understanding the responsible mechanism have revealed that these parasites portray **antigenic variation**. In other words, they are referred to as **variable antigenic types**. Each species has a highly complicated antigenic structure, which gives rise to strain and variant populations. As a population of these trypanosomes increases in number within a host, its antigenicity stimulates the host to produce

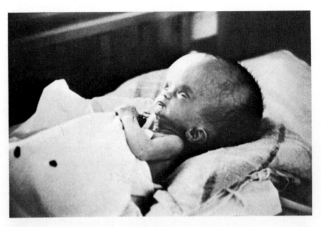

Fig. 3.21. Photograph of a child with hydrocephalus due to congenital toxoplasmosis. (Courtesy of Dr. H. Zaiman.)

antibodies against it, and, as a result, kills most of the parasites (Fig. 3.22). However, a small number of individuals that are antigenically different survive. These divide to produce a second population of trypanosomes (Fig. 3.22). The members of the second population stimulate the host's immune mechanisms to produce antibody$_2$ by exposing antigen$_2$. As a result, most of the members of the second population of trypanosomes are killed. The few survivors are antigenic variants that produce antigen$_3$ (Fig. 3.22). This sequence of antigenic variations results in the above mentioned fluctuations of parasitemia and clinical relapses which continue until the host dies or the parasites are eliminated.

Experimental studies have revealed the frequency with which new antigenic variants can arise. For example, in the case of *Trypanosoma congolense* in sheep, a new variant arises every ninth day, and in the case of *Trypanosoma vivax* in sheep and calves, a new variant arises every seventh day.

Two major subclasses of antigens have been identified from trypanosomes. The first subclass is comprised of the **bound antigens**, which remain stable during a relapsing infection. These antigens include enzymes and nucleoproteins and are only weakly antigenic. The second subclass is comprised of the unstable **surface-coat antigens**. The surface-coat antigens are the glycoprotein **exoantigens**, i.e., antigens secreted by the trypanosomes, which are present in the serum of infected animals and induce the production of protective antibodies.

Because of the occurrence of antigenic variants among trypanosomes, the development of an immunizing vaccine for humans and domestic animals against trypanosomiasis is most difficult.

Blood Helminths

Two major groups of helminth parasites occur in their vertebrate hosts' blood: adult schistosomes (p. 332) and certain microfilariae (p. 537).

Although much attention has been paid to the antigens of schistosomes, our knowledge about those antigens that stimulate acquired resistance and the mechanism on which this resistance is based is still incomplete. Since it is possible to induce marginal protective immunity only by vaccinating hosts with dead worms, extracts, and secretions, it is apparent that the stimulation of resistance is associated with the living parasite.

Live *Schistosoma mansoni* cercariae can be employed to induce immunity in the rhesus monkey. Specifically, Smithers and Terry (1965, 1967) have reported that if 100 live cercariae are introduced into a monkey, a high level of immunity develops against a subsequent challenge 16 weeks later. If the number of cercariae used is less than 100, or if the subsequent challenge

occurs sooner than 16 weeks, the results are more variable. The relatively small number of normal cercariae required to induce a significant level of immunity when compared with the results involving irradiated cercariae is of interest. In the case of irradiated cercariae, Sadun *et al.* (1964) have reported that even 25,000 cercariae irradiated at 400 rep (rad equivalent, physical) and given as a single exposure does not induce protective immunity. Since it is known that the irradiation of cercariae at this level does not prevent their invasion and migration to the liver, but does prevent their development to fully mature adults, it may be inferred that effective immunity requires contact with living adult worms. This hypothesis has been strengthened by the studies of Smithers and Terry (1967), who transplanted living adult schistosomes into the hepatic portal vessels of monkeys. They have found that this procedure confers immunity to the host against cercarial penetration, although the transplantation of dead adult worms does not.

As a result of experimental studies on *S. mansoni* infections in rodents and rhesus monkeys, it is known that a premune condition develops in which a primary infection renders the host immune to subsequent reinfection. In schistosomiasis, this condition is known as **concomitant immunity**, because the primary infection is not killed by the immune response. This phenomenon has led to the suggestion that it may be

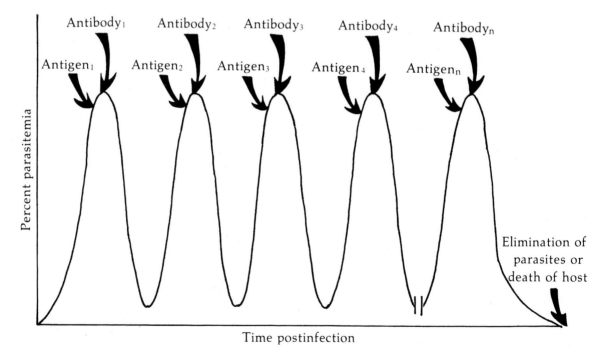

Fig. 3.22. Antigenic variation and relapsing infections with trypanosomes. Antigens 1, 2, 3, 4, ... *n* are variants produced by a population of trypanosomes in a host. Each variant induces the production of a variant-specific antibody, which eliminates the majority of the trypanosomes. A small number if trypanosomes, however, survive because they possess the new variant antigen against which the host has not yet synthesized antibody. The survivors multiply, resulting in a rising parasitemia until there is sufficient variant-specific antibody circulation to largely eliminate the subpopulation of parasites. This phenomenon continues until the parasites are completely eliminated or the host dies.

possible to immunize against schistosomiasis by using attenuated cercariae, i.e., cercariae so treated (usually by radiation) that they will not undergo further development.

The basis for concomitant immunity has been sought by many investigators. As a result, it is generally accepted that it arises through the sharing of common antigens by host and parasite. Furthermore, evidences suggest that the parasites bind antigenic materials of host origin to their body surfaces. For example, schistosomes grown to maturity in mice and then surgically transferred to the hepatic portal system of rhesus monkeys that had been immunized against mouse red blood cells die within 24 hours. On the other hand, worms maturing in mice and then transferred to nonimmunized monkeys survive. These results support the idea that schistosomes bind host antigens to their surfaces, and by so doing become camouflaged, rendering them immunologically indistinguishable from host tissues. Worms of the secondary or **challenge** infections appear not to be able to bind host antigen at a rate sufficient to elude the antibodies produced against the primary infection. For a detailed review of immunity in mice to schistosomes, see Dean (1983).

Immunologic studies pertaining to filarial worms have revealed some interesting findings. It is well known that the microfilariae of such filarial worms as *Wuchereria* (p. 537), *Brugia* (p. 539), and *Dirofilaria* (p. 544) can survive for long periods in blood and seem unaffected by circulating antibodies. For example, *Dirofilaria repens* microfilariae transfused into uninfected dogs survive for 2 months to 3 years. Similarly, *Litomosoides carinii*, a parasite of the cotton rat (p. 543), can persist for months, and *Mansonella ozzardi* (p. 545) and *Mansonella perstans* (p. 545) can survive in humans for 2 to 3 years. Exceptions do occur; for example, *Loa loa* microfilariae (p. 544) disappear by the fourth day after transfusion (Mazzotti and Palomo, 1957).

From longevity data on microfilariae, it would appear that the majority of species in their normal hosts invoke very little immune response (Smithers, 1968). This does not mean that no antibodies are formed; rather, the antibody titers are so low that they are essentially ineffective as protective mechanisms. Under abnormal conditions, however, suppression of microfilaremia (microfilariae in blood) can be achieved. For example, Wong (1964) has demonstrated that the repeated injection of large numbers of *Dirofilaria* microfilariae into uninfected dogs will induce the formation of antibodies that can destroy the microfilariae. It is believed that these antibodies do not occur in naturally infected dogs.

If microfilariae are introduced into an unnatural but at least partially compatible host, a more striking immune response occurs. For example, Singh and Raghavan (1962) have reported that the duration of patent microfilaremia of *Litomosoides carinii* in its normal host, the cotton rat, is a year of more; however, when this nematode is introduced into the albino rat, the average duration of patent microfilaremia is only about 18 weeks.

It is important to note that antibodies lethal to microfilariae do not affect adult worms (Ramakrishnan *et al.*, 1962), thus suggesting that these antibodies are stage specific. For more thorough reviews of immunity to nematodes, see Thorson (1970) and Sinclair (1970).

Tissue Helminths

A number of helminth parasites occur in their host's tissues other than blood. *Trichinella spiralis* larva is an example (p. 489). Furthermore, it bears repeating that many intestinal parasites are at least partially embedded in their hosts' intestinal mucosa at some stage in their development and hence may be thought of as tissue parasites. The tapeworm *Echinococcus granulosus* is an example (p. 408).

Among nematode tissue parasites, antibody production in the host is manifested in a variety of ways.

1. The establishment of the parasite can be totally or partially inhibited. For example, Michel (1962) has shown that a calf given a single dose of 2000 *Dictyocaulus viviparus* larvae can resist the establishment of over 90% of a second infection given 11 days later.

2. The normal development of the parasite within the immunized host may be inhibited. For example, Sommerville (1960) has found that the development of *Cooperia curticei* in sheep is inhibited in the late fourth stage. The stage at which this inhibition occurs varies from one nematode species to another, but is consistent with each species. In many cases the worms are inhibited while migrating through tissues.

3. The parasite does not grow as rapidly or reach its normal size. This type of reaction has been demonstrated in *Litomosoides carinii* (MacDonald and Scott, 1953), *Trichinella spiralis* (Semrad and Coors, 1951), and others.

4. The parasite develops anatomic abnormalities. Michel (1967), for example, has reported that an increased proportion of female *Ostertagia ostertagi* fail to develop a vulvar flap when introduced into calves that have been previously infected with the same parasite.

5. The parasite's life cycle is altered. For example, the eggs laid by *Strongyloides* spp. in naive pigs hatch into infective larvae; however, if the worms occur in pigs that had been previously infected with a homologous parasite, the eggs

give rise to larvae of the free-living sexual generation (Varjú, 1966).

6. Adult worms are eliminated from the host sooner than in nonimmunized hosts. In this type of immune response, the worms may be gradually eliminated at a more or less constant rate, or the elimination may be abrupt.

In addition to the manifestations of immunity listed, almost all tissue nematodes provoke a cellular response. This occurs as cellular infiltration accompanied by edema and is suggestive of hypersensitivity.

It bears remembering that most nematode infections and, indeed, helminth infections in general, stimulate the mammalian host to produce γE (reaginic antibodies), but this immunoglobulin does not appear to be exclusively responsible for limiting an infection. The parasite antigens that stimulate γE production are known as **allergens**.

There is a relationship between γE production, eosinophilia, and immunopathology in many nematode infections, and it may be mediated by anaphylaxis, involving the synthesis and secretion of histamine from mast cells at the site of invasion or the habitat of the parasite.

Immunity to tapeworm infections is primarily the result of the antigenic properties of the preadult stages of the parasite. These stages, unlike the adults, are tissue dwelling and consequently are more antigenically intimate. This does not mean that adult tapeworms are nonimmunogenic. Indeed, it has been demonstrated that immunoglobulins (γA, γE, and γG) are synthesized in response to adults of *Hymenolepis diminuta* (p. 417), *H. microstoma* (p. 418), *Raillietina cesticillus* (p. 416), and *Echinococcus granulosus* (p. 408). Hopkin's (1980) comprehensive review of immunity to *H. diminuta* should be consulted for details.

Although antibodies do not generally cause the death of adult tapeworms, affected worms undergo retarded growth plus destrobilation, i.e., their body segments slough off. If a host is treated with immunosuppressant drugs such as cortisone, and methotrexate, or antilymphocyte serum, destrobilation generally does not occur. Antibodies of various classes are bound to the tapeworm surface, and it has been postulated that this binding may interfere with the absorption of essential nutrients.

Tissue Trematodes

Most studies of immunity to tissue trematodes have been carried out on *Clonorchis sinensis*, the oriental liver fluke (p. 355); *Fasciola hepatica*, the cattle and sheep liver fluke (p. 347); and *Paragonimus westermani*, the oriental lung fluke (p. 358). In the case of *C.*

sinensis, circulating antibodies can be detected 4 weeks after the experimental infection of rabbits. These antibodies decrease 16 to 25 weeks after infection. Sun and Gibson (1969) have reported precipitating antibodies in rabbits against the metabolic products of *C. sinensis* after 5 weeks of infection; the antibodies persisted for 21 weeks. In spite of these positive *in vitro* tests, there is an absence of protective immunity in animals infected with this liver fluke. This emphasizes the principle that positive *in vitro* serologic reactions need not mean that protective immunity exists.

Data available on *Fasciola hepatica* are similar to those on *Clonorchis sinensis* (see Platzer, 1970 for review). Although the presence of antibodies has been detected with various *in vitro* techniques, field studies of *F. hepatica* infection in goats and sheep have given no indication of acquired immunity (Pantalouris, 1965). In cattle, nonspecific resistance develops because of bile duct calcification (Sinclair, 1967). On the other hand, Lang (1967) has found evidence for acquired immunity in mice experimentally infected with *F. hepatica*. The condition is appreciated as fewer worms reaching the host's bile duct, and those that do, reach this site earlier. Lang has suggested that this may be due to delayed sensitivity.

Studies on *Paragonimus* have resulted in essentially the same type of results as with *Fasciola* and *Clonorchis*. *In vitro* serologic tests on parasitized hosts have revealed low-titer complement-fixing and precipitating antibodies, and there may be a positive anaphylactic test; however, protective immunity has not been observed.

In the case of *Vampirolepis nana* (p. 416), a primary egg infection in mice results in a lowered but consistent reinfection rate (Bailenger *et al.*, 1961; Heyneman, 1962). This reinfection immunity develops rapidly following primary infection with eggs, in 1 or 2 days, and varies with the number of eggs administered. This immunity to *V. nana* is somewhat unusual for platyhelminth parasites since it persists for almost the life span of the mouse host after the parasites are removed or have died. It should be remembered, however, that in the case of *V. nana*, there is a prolonged extraintestinal (tissue) development phase and the antibodies produced are believed to be directed at larval antigens rather than at the adult worms.

Under certain conditions, adult tapeworms induce protective immunity. Specifically, Weinmann (1964, 1966) has shown that *Hymenolepis microstoma*, *H. diminuta*, and *H. citelli* can stimulate protective immunity in mice if given enough time and if the hosts are repeatedly infected. Tan and Jones (1967, 1968) have demonstrated a protective immunity against *H. microstoma* in rodents when the primary challenge consists of irradiated worms.

Special Manifestations of Antigen-Antibody Reactions as Related to Helminth Parasites. Several types of immunologic reactions to whole-animal helminth parasites will be mentioned briefly. These pertain to trematode and nematode parasites and actually represent specialized forms of antigen-antibody reactions.

Cercarienhüllenreaktion or CHR. Vogel and Minning (1949) were the first to demonstrate that when living cercariae of *Schistosoma mansoni* are placed in serum from an infected mammalian host, a translucent, irregular envelope is formed around each cercaria (Fig. 3.23). The envelope usually binds the cercaria to the substrate. There is little doubt that the CHR is a specialized form of precipitin reaction.

Cercarial Agglutination. When *Schistosoma mansoni* cercariae are placed in Kahn tubes and mixed with sera from mice or hamsters previously infected with this parasite, they will cohere in clumps (Fig. 3.24). This technique for demonstrating antibodies against *S. mansoni* antigens was first reported by Liu and Bang (1950). Stirewalt and Evans (1955) have reported that the cercariae must be suspended in physiologic salt solution at 25°C before the antiserum is added in order to achieve impaired motility and agglutination of the cercariae. This diagnostic technique, obviously a form of classical agglutination, is not as reproducible as the CHR.

Cercarial Precipitation. Papirmeister and Bang (1948) have reported that if *Schistosoma mansoni* cercariae are placed in sera from humans and monkeys previously infected with this parasite, a finely granular or globular deposit develops around the parasites. This reaction, however, is inconsistent and of doubtful value as a diagnostic tool.

Miracidial Immobilization. When living miracidia of certain species of trematodes, for example, *Schistosoma mansoni*, are placed in antisera of hosts infected with the adult stage, a reaction between the parasite antigens and host antibodies occurs and can be visualized as miracidial immobilization.

Excretory-Secretory (ES) Precipitation. When nematodes are placed in specific antisera, an opaque, whitish precipitate forms at the oral and anal orifices. This precipitate represents the interaction of the antibody with antigens emitted from these body apertures.

Immunity to Arthropod Parasites

As indicated in subsequent chapters (Chapters 16–20), the majority of parasitic arthropods are ectoparasitic, and many of these are hematophagous. Immune reactions directed to these parasites, therefore, must be in response to antigens introduced as the result of the parasites' feeding habits.

Cellular reactions to the insertion of the parasite's mouthparts vary in magnitude, depending on both the arthropod's mode of feeding and on the host. Generally, the trauma resulting from species that lacerate the host's skin with their bladelike mandibles, e.g., deerflies, blackflies, tsetse flies, is more severe

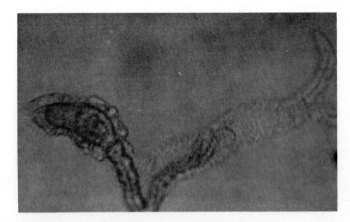

Fig. 3.23.   The CHR reaction. *Schistosoma mansoni* cercaria showing pericercarial reaction envelope after being suspended in infected mouse serum for 60 minutes. (After Stirewalt and Evans, 1955.)

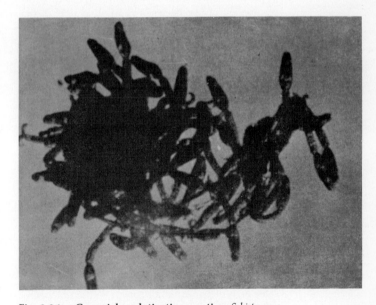

Fig. 3.24.   Cercarial agglutination reaction. *Schistosoma mansoni* cercariae in saline-diluted infected mouse serum after incubation at 25°C. (After Stirewalt and Evans, 1955.)

than that caused by parasites with piercing-sucking mouthparts, e.g., mosquitoes and fleas. Of course, the duration of feeding, the size of the arthropod, and the physiologic state of the host all contribute to the degree of trauma. Furthermore, it must be noted that evidences suggest that the mouthparts of arthropods, coated with a cuticular exoskeleton, are essentially nonantigenic in terms of directly eliciting cellular response. However, these mouthparts are commonly contaminated with microbes and molecules that do elicit response.

Undoubtedly the most antigenic of substances associated with hematophagous arthropods are the various secretions. Evidence for the existence of salivary gland secretions in mosquitoes has been provided by Hudson *et al.* (1960) working with *Aedes stimulans*. They have found that bites by mosquitoes with their salivary glands removed are more painful than those of normal mosquitoes. This suggests the presence of an anesthetic component in the mosquito's saliva. In the case of the tick *Dermacentor andersoni* (p. 582), an oral secretion originating in the salivary glands hardens into a clear material molded around the mouthparts and the host's skin, thus enabling the tick to remain attached for a considerable period of time. Similarly, it is known that the cat flea, *Ctenocephalides felis* (p. 627), secretes a powerful spreading factor which causes the softening of the host's dermal tissue, thus facilitating feeding. Another type of oral secretion by hematophagous arthropods are the anticoagulants (Metcalf, 1945). In addition, numerous other compounds, including hemagglutinins, toxins, sugars, amino acids, peptides, proteins, phenolic compounds, etc., have been detected associated with the mouthparts of blood-sucking insects. Some of these are antigenic. For a review of the role of tick salivary glands, see Binnington and Kemp (1980).

Reactions to antigens introduced by biting arthropods are primarily in the form of immediate and delayed hypersensitivities (pp. 81, 82). The immediate type normally consists of erythema and edema at the site of the bite, and is usually accompanied by intense pruritus. In some hosts, the skin reaction is delayed until 15 to 23 hours after the bite. This delayed reaction usually consists of erythema accompanied by papulation, and the site eventually becomes necrotic.

In Table 3.10 are tabulated the identities of the cellular constituents of the host response to an insect bite.

McLaren *et al.* (1983) have reported that small clusters of basophils appear at the primary bite sites within 24 hours after the tick *Ornithodoruos tatakovskyi* feeds on a guinea pig, and by 72 hours these cells constitute about 11% of the total leucocytes. If the tick feeds on the host a second time, an augmented cellular infiltrate dominated by basophils (48–56% of total cells) occurs. Despite the occurrence of a strong cutaneous basophil response of the kind that mediates immune rejection of prolonged-feeding ixodid ticks (p. 582), guinea pigs showed no resistance to fast-feeding argasid ticks (p. 580). It has been suggested that argasid ticks probably complete their blood meals prior to the arrival of basophils at the bite sites. It is also noted that guinea pig basophils at secondary *O. tartakovskyi* bite sites exhibit three types of structural alterations: (1) **piecemeal alterations** involving a vesicular degranulation mechanism; (2) an **anaphylactic type of alteration** involving single or compound exocytosis of whole granules; and (3) **cytotoxic alterations** culminating in complete disintegration. These cellular responses are attributed to salivary secretions of the tick.

Antibodies have been demonstrated in mammals subjected to repeated bites by hematophagous insects. Furthermore, these antibodies can be transferred to another host via cells of lymphoid origin (Allen, 1964) and hence are cell-associated immunoglobulins. In addition to this type of antibody, circulating antibodies have also been found in sera from humans sensitized to bites of a particular arthropod. When intradermally injected into nonsensitized humans, the antisera cause the sensitization of the injection sites so that challenge bites by that arthropod elicit immediate skin reaction.

The topic of immunity to arthropods has been authoritatively reviewed by Benjamini and Feingold (1970).

The role of immunology in the study of parasitism is an important one. Successful immunization against parasitic diseases remains one of the major areas of

Table 3.10. The Sequence of Events in Cellular Responses in the Skin Reactivity of Guinea Pigs to Flea Bites[a]

Stage	Skin Reactivity	Cellular Response at Bite Site	
		20 Minutes after Bite	*24 Hours after Bite*
I	No reactivity	No cellular abnormalities	No cellular abnormalities
II	Delayed	No cellular abnormalities	Intense monocytic and lymphocytic dermal infiltration
III	Immediate and delayed	Eosinophilic infiltration	Intense monocytic and lymphocytic dermal infiltration
IV	Immediate	Eosinophilic infiltration	Very mild, if any, monocytic and lymphocytic infiltration
V	No reactivity	No cellular abnormalities	No cellular abnormalities

[a] After Benjamini and Feingold, 1970 in Jackson *et al.* (1970).

emphasis in research. Although some progress has been achieved, this method of preventing the major parasitic diseases of humans and domestic animals has yet to be realistic and effective. Attenuated parasites, homogenized parasite tissue, dead parasites, soluble parasite antigens, controlled live parasites, and heterologous protection have been attempted; yet, except in a few instances, effective vaccines are still in the future. A part of the difficulty has been mentioned earlier, i.e., antigenic variation in trypanosomiasis and concomitant immunity in schistosomiasis. The malaria-causing *Plasmodium* spp. also bring about concomitant immunity.

Success in developing vaccines has been achieved by using attenuated larvae to protect cattle against the nematode lungworm, *Dictyocaulus viviparus*. Similarly, success has been achieved in protecting sheep against *Dictyocaulus filaria* and dogs against the hookworm *Ancylostoma caninum*. A live vaccine has been developed against *Babesia bovis*, a blood protozoan parasite of calves, in Australia. The parasite used is obtained after rapid passage of *B. bovis*-infected blood through splenectomized calves. This procedure causes the parasite to become less virulent and does not infect the tick vector (Callow, 1977). Consequently, the life cycle of this parasite is interrupted.

A promising recent approach to developing a broad-spectrum vaccine for parasitic diseases has arisen from studies on tumor immunology. This is based on the concept of shared antigens. Specifically, following administration of BCG (Bacille Calmette Guérin) vaccine, nonspecific immunity is developed for a wide range of parasites, including rodent babesias (p. 208), malarias (p. 194), *Leishmania tropica* (p. 123), *Trypanosoma cruzi* (p. 131), and the tapeworms *Echinococcus granulosus* (p. 408) and *E. multilocularis* (p. 412). Such successes in the laboratory have lent credence to the idea that to take advantage of shared antigens among parasites and microorganisms may prove to be the most profitable approach in vaccine development.

Recently, Enea *et al.* (1984) and McCutchan *et al.* (1984) have devised methods of isolating and cloning the genes responsible for the synthesis of the major surface protein, known as the **circumsporozoite protein**, of the sporozoites of the malarial parasite *Plasmodium falciparum*. Thus, it is now possible to mass produce the circumsporozoite protein and test its effectiveness as a vaccine against one type of human malaria.

It needs to be mentioned that a number of relatively recently developed immunologic techniques, including the production of monoclonal antibodies (Köhler and Milstein, 1975) and the enzyme-linked immunosorbent assay (ELISA) (Engvall and Perlmann, 1971, 1972; Van Weeman and Schurs, 1971), have come into use in immunoparasitology. The landmark application of monoclonal antibodies in parasitology is that by Potocnjak *et al.* (1982). These authors had previously reported the production of a monoclonal mouse antibody directed against the circumsporozoite protein of the murine malaria parasite, *Plasmodium berghei*. This antibody protects mice against *P. berghei* infection. By injecting mice with this antibody, they produced a second monoclonal antibody against the idiotype of the anticircumsporozoite protein immunoglobulin used as the immunogen. This anti-idiotypic antibody reacted with a structure closely associated with the antigen-binding site of the anticircumsporozoite protein antibody. The inhibition of the interaction of these two antibodies by *P. berghei* sporozoites constitutes the basis of a radioimmunoassay which permits quantitative determination of small amounts of the relevant circumsporozoite antigen, even when present in a crude extract.

The ELISA immunoassay involves an antigen or antibody linked to an enzyme. Degradation of the enzyme substrate by the enzyme-linked antigen-antibody complex can be measured photometrically and is proportional to the concentration of the antigen or antibody in the test material. This technique has been applied with success for the detection of parasites or antibodies to parasites (Voller *et al.*, 1976).

It should be apparent from this discussion that great strides are being made in immunoparasitology. The essay by Perlmann (1983) on advances is recommended.

REFERENCES

Allen, J. R. (1964). Some properties of oral secretion of mosquitoes. Ph.D. Thesis, Queen's University, Kingston, Ontario, Canada.

Bailenger, J., Baudoin, M., and Pautrizel, R. (1961). Etude de l'immunité des rongeurs à l'égard d'*Hymenolepis nana* (von Siebold, 1852, Blanchard, 1891). Ann. Parasitol. **36**, 595–611.

Barriga, O. O. (1981). "The Immunology of Parasitic Infections." University Park Press, Baltimore.

Beattie, C. P. (1963). Immunity to *Toxoplasma*. In "Immunity to Protozoa" (P. C. C. Garnham, A. E. Pierce, and I. Roitt, eds.), pp. 253–258. Blackwell, Oxford.

Benjamini, E., and Feingold, B. F. (1970). Immunity to arthropods. In "Immunity to Parasitic Animals" (G. J. Jackson, R. Herman, and I. Singer, eds.), Vol. 2, pp. 1061–1134. Appleton-Century-Crofts, New York.

Binnington, K. C., and Kemp, D. H. (1980). Role of tick salivary glands in feeding and disease transmission. *Adv. Parasit.* **18**, 315–339.

Boman, H. G. (1981). Insect responses to microbial infections. In "Microbial Control of Insects, Mites and Plant Diseases, 1970–1980," pp. 769–784. Academic Press, New York.

Bronskill, J. F. (1962). Encapsulation of rhabditoid nematodes in mosquitoes. *Can. J. Zool.* **40**, 1269–1275.

Callow, L. L. (1977). Vaccination against bovine babesiosis. *Adv. Exp. Med. Biol.* **93**, 121–149.

Capron, A., Bignet, J., Vernes, A., and Afchain, D. (1968). Structure antigeniques des helminthes. Aspects immunologiques des relations host-parasite. *Pathol. Biol.*, **16**, 121–138.

Cheng, T. C. (1967). Marine molluscs as hosts for symbioses. *Adv. Mar. Biol.* **5**, 1–424.

Cheng, T. C. (1968). The compatibility and incompatibility concept as related to trematodes and molluscs. *Pac. Sci.* **22**, 141–160.

Cheng, T. C. (1979). The role of hemocytic hydrolases in the defense of molluscs against invading parasites. *Haliotis* **8**, 193–209.

Cheng, T. C. (1985). Evidences for molecular specificities involved in molluscan inflammation. *Comp. Pathobiol,* **8**, 129–142.

Cheng, T. C., and Rifkin, E. (1970). Cellular reactions in marine molluscs in response to helminth parasitism. *In* ''Diseases of Fishes and Shellfishes'' (S. F. Snieszko, ed.), p. 443. Am. Fish. Soc. Symp. No. 5, Washington, D.C.

Cheng, T. C., Shuster, C. N., Jr., and Anderson, A. H. (1966). Effects of plasma and tissue extracts of marine pelecypods on the cercaria of *Himasthla quissetensis*. *Exp. Parasitol.* **19**, 9–14.

Clegg, J. A. (1972). The schistosome surface in relation to parasitism. *In* ''Functional Aspects of Parasite Surfaces.'' (A. E. R. Taylor, ed.). *Symp. Br. Soc. Parasitol.* **10**, 23–40. Blackwell, Oxford.

Clegg, J. A., Smithers, S. R., and Terry, R. J. (1971). Acquisition of host antigens by *Schistosoma mansoni* during cultivation *in vitro. Nature* **232**, 653–654.

Cohen, S., and McGregor, I. A. (1963). Gamma globulin and acquired immunity to malaria. *In* ''Immunity to Protozoa'' (P. C. Garnham, A. E. Pierce, and I. Roitt, eds.), pp. 123–159. Blackwell, Oxford.

Corradetti, A. (1963). Acquired sterile immunity in experimental protozoal infections. *In* '' Immunity to Protozoa'' (P. C. Garnham, A. E. Pierce, and I. Roitt, eds.), pp. 69–77. Blackwell, Oxford.

Cox, H. W. (1966). A factor associated with anemia and immunity in *Plasmodium knowlesi* infections. *Mil. Med.* (Suppl.) **131**, 1195–1200.

Damian, R. T. (1964). Molecular mimicry: antigen sharing by parasite and host and its consequences. *Am. Nat.* **98**, 129–149.

Damian, R. T. (1967). Common antigens between *Schistosoma mansoni* and the laboratory mouse. *J. Parasitol.* **53**, 60–64.

Damian, R. T. (1979). Molecular mimicry in biological adaptation. *In* ''Host-Parasite Interfaces'' (B. B. Nickol, ed.). pp. 103–126. Academic Press, New York.

Damian, R. T., Greene, N. D., and Hubbard, W. J. (1973). Occurrence of mouse a_2-macroglobulin antigenic determinants on *Schistosoma mansoni* adults, with evidence on their nature. *J. Parasitol.* **59**, 64–73.

Dean, D. A. (1983). *Schistosoma* and related genera: acquired resistance in mice. *Exp. Parasitol.* **55**, 1–104.

Desowitz, R. S. (1959). Studies on immunity and host-parasite relationships. I. The immunological response of resistant and susceptible breeds of cattle to trypanosomal challenge. *Ann. Trop. Med. Parasitol.* **53**, 292–313.

Enea, V., Ellis, J., Zavala, F., Arnot, D. E., Asavanich, A., Masuda, A., Quakyi, I., and Nussenzweig, R. S. (1984). DNA cloning of *Plasmodium falciparum* circumsporozoite gene: amino acid sequence of repetitive epitope. *Science* **225**, 628–630.

Engvall, E., and Perlmann, P. (1971). Enzyme-linked immunosorbent assay (ELISA). Quantitative assay of IgG. *Immunochemistry*, **8**, 871–874.

Engvall, E., and Perlmann, P. (1972). Enzyme-linked immunosorbent assay (ELISA). III. Quantitation of specific antibodies by enzyme labelled anti0immunoglobulin in antigen coated tubes. *J. Immunol.* **109**, 129–135.

Esslinger, J. H. (1962). Behavior of microfilariae of *Brugia pahangi* in *Anopheles quadrimaculatus*. *Am. J. Trop. Med. Hyg.* **11**, 749–758.

Foley, D. A., and Cheng, T. C. (1975). A quantitative study of phagocytosis by hemolymph cells of the pelecypods *Crassostrea virginica* and *Mercenaria mercenaria. J. Invert. Pathol.* **25**, 189–197.

Heyneman, D. (1962). Studies on helminth immunity. I. Comparison between luminal and tissue phases of infection in the white mouse by *Hymenolepis nana* (Cestoda: Hymenolepididae). *Am. J. Trop. Hyg.* **11**, 46–63.

Hood, L. E., Weissman, I. L., and Wood, W. B. (1978). ''Immunology.'' Benjamin/Cummings, Menlo Park, California.

Hopkins, C. A. (1980). Immunity and *Hymenolepis diminuta*. *In* ''Biology of the Tapeworm *Hymenolepis diminuta*.'' (H. P. Arai, ed.). pp. 551–614. Academic Press, New York.

Howland, K. H., and Cheng, T. C. (1982). Identification of bacterial chemoattractants for oyster (*Crassostrea virginica*) hemocytes. *J. Invert. Pathol.* **39**, 123–132.

Hudson, A., Bowman, L., and Orr, C. W. M. (1960). Effect of absence of saliva on blood feeding by mosquitoes. *Science* **131**, 1730–1731.

Hynes, H. B. N., and Nicholas, W. L. (1958). The resistance of *Gammarus* sp. to infection by *Polymorphus minutus* (Goeze, 1782) (Acanthocephala). *Ann. Trop. Med. Parasitol.* **52**, 376–383.

Jackson, G. J., Herman, R., and Singer, I. (eds.). (1969). ''Immunity to Parasitic Animals,'' Vol. 1. Appleton-Century-Crofts, New York.

Jackson, G. J., Herman, R., and Singer, I. (eds.) (1970). ''Immunity to Parasitic Animals,'' Vol. 2. Appleton-Century-Crofts, New York.

Johnson, P., Neal, R. A., and Gall, D. (1963). Protective effect of killed trypanosome vaccines with incorporated adjuvants. *Nature (London)* **200**, 83.

Joyner, L. P., and Davies, S. F. M. (1967). Acquired resistance to *Babesia divergens* in experimental calves. *J. Protozool.* **14**, 260.

Karnovsky, J. L., and Bolis, L. (eds.) (1982). ''Phagocytosis—Past and Future.'' Academic Press, New York.

Köhler, G., and Milstein, C. (1975). Continuous cultures of fused cells secreting antibody of predefined specificity. *Nature,* **256**, 495–497.

Kartman, L. (1953). Factors influencing infection of the mosquito with *Dirofilaria immitis* (Leidy, 1856). *Exp. Parasitol.* **2**, 27–78.

Kessel, J. F., Lewis, W. P., Pasquel, C. M., and Turner, J. A. (1956). Indirect haemagglutination and complement fixation tests in amebiasis. *Am. J. Trop. Med. Hyg.* **14**, 540–550.

Kozar, A. (1970). Toxoplasmosis and coccidiosis in mammalian hosts. *In* "Immunity to Parasitic Animals" (G. J. Jackson, R. Herman, and I. Singer, eds.), Vol. 2, pp. 871–912. Academic Press, New York.

Lang, B. Z. (1967). Host-parasite relationships of *Fasciola hepatica* in the white mouse. II. Studies on acquired immunity. *J. Parasitol.* **53**, 21–30.

Lapierre, J., and Rousset, J. J. (1961a). Etude de l'immunite dans les infections à *Trypanosoma gambiense* chez le souris blanche. Variations antigeniques au cours des crises trypanolytiques. *Bull. Soc. Pathol. Exot.* **54**, 332–335.

Lapierre, J., and Rousset, J. J. (1961b). Caractères biologiques d'une souche virulente de *Trypanosoma gambiense*. Immunization par vaccin tués. *Bull. Soc. Pathol. Exot.* **54**, 336–345.

Liu, C., and Bang, F. B. (1950). Agglutination of cercariae of *Schistosoma mansoni* by immune sera. *Proc. Soc. Exp. Biol. Med.* **74**, 68–72.

MacDonald, E. M., and Scott, J. A. (1953). Experiments on immunity in the cotton rat to the filarial worm, *Litomosoides carinii. Exp. Parasitol.* **2**, 174.

MacDonald, W. W. (1962). The genetic basis of susceptibility to infection with semiperiodic *Brugia malayi* in *Aedes aegypti. Ann. Trop. Med. Parasitol.* **56**, 373–382.

Mayer, M. M. (1973). The complement system. *Sci. Am.* **229**, 54–66.

Mazzotti, L., and Palomo, E. (1957). A note on the survival of the microfilariae of *Mansonella ozzardi. Bull. W.H.O.* **16**, 696–699.

McCutchan, T. F., Hansen, J. L., Dame, J. B., and Mullins, J. A. (1984). Mung bean nuclease cleaves *Plasmodium* genomic DNA at sites before and after genes. *Science* **225**, 625–628.

McFarlane, H., and Voller, A. (1966). Studies on immunoglobulins of Nigerians. Part II. Immunoglobulins and malarial infection in Nigerians. *J. Trop. Med. Hyg.* **69**, 104–107.

McLaren, D. J., Worms, M. J., Brown, S. J., and Askenase, P. W. (1983). *Ornithodorus tartakovki*: Quantitation and ultrastructure of cutaneous basophil responses in the guinea pig. *Exp. Parasit.* **56**, 153–168.

Metcalf, R. L. (1945). The physiology of the salivary glands of *Anopheles quadrimaculatus. J. Nat. Malar. Soc.* **4**, 271–278.

Michel, J. F. (1962). Studies on resistance to *Dictyocaulus* infection. IV. The rate of acquisition of protective immunity in infection of *D. viviparus. J. Comp. Pathol.* **72**, 281–285.

Michel, J. F. (1967). Morphological changes in a parasitic nematode due to acquired resistance of the host. *Nature (London)* **215**, 520.

Muniz, J. (1962). Immunidade na doenca de Chagas (trypanosomiasis americana). *Mem. Inst. Oswaldo Cruz* **60**, 103–147.

Nossal, G. J. V. (1969). "Antibodies and Immunity." Basic Books, New York.

Norman, L., and Kagan, I. G. (1960). Immunologic studies in

Trypanosoma cruzi. II. Acquired immunity in mice infected with avirulent American strains of *T. cruzi. J. Infec. Dis.* **107**, 168–174.

Nussenzweig, V., Kloetzel, J., and Deane, L. M. (1963). Acquired immunity in mice infected with strains of immunological types A and B of *Trypanosoma cruzi. Exp. Parasitol.* **14**, 233–239.

Pan, C. T. (1965). Studies on the host-parasite relationship between *Schistosoma mansoni* and the snail *Australorbis glabratus. Am. J. Trop. Med. Hyg.* **14**, 931.

Pantalouris, E. M. (1965). "The Common Liver Fluke, *Fasciola hepatica* L." Pergamon Press, Oxford.

Papirmeister, B., and Bang, F. B. (1948). The *in vitro* action of immune sera on cercariae of *Schistosoma mansoni. Am. J. Hyg.* **48**, 74–80.

Perlmann, P. (1983). Immunology and parasitic diseases. *In* "Parasitology: A Global Perspective." (K. S. Warren and J. Z. Bavers, eds.) p. 228–235. Springer-Verlag, New York.

Platzer, E. G. (1970). Trematodes of the liver and lung. *In* "Immunity to Parasitic Animals" (G. J. Jackson, R. Herman, and I. Singer, eds.), Vol. 2, pp. 1009–1019. Appleton-Century-Crofts, New York.

Potocnjak, P., Zavala, F., Nussenzweig, R., and Nussenzweig, V. (1982). Inhibition of idiotype–anti-idiotype interaction for detection of a parasite antigen: a new immunoassay. *Science* **215**, 1637–1639.

Powell, S. J., Maddison, S. E., Hodgson, R. G., and Elsdon-Dew, A. (1966). Amoebic gel-diffusion precipitin test. Clinical evaluation in acute amoebic dysentery. *Lancet* **12**, 566–567.

Ramakrishnan, S. P., Singh, D., and Krishnaswami, A. K. (1962). Evidence of acquired immunity against microfilariae of *Litomosoides carinii* in albino rats with mite-induced infection. *Indian J. Malariol.* **16**, 263–268.

Ratcliffe, N. A., and Rowley, A. F. (1979). Role of hemocytes in defense against biological agents. *In* "Insect Hemocytes: Development, Forms, Functions, and Techniques" (A. P. Gupta, ed.), pp. 331–414. Cambridge University Press, Cambridge, England.

Remington, J. S., and Miller, M. J. (1966). 19 S and 7 S anti-*Toxoplasma* antibodies in diagnosis of acute congenital and acquired toxoplasmosis. *Proc. Soc. Exp. Biol. Med.* **121**, 357–363.

Riek, R. F. (1963). Immunity to babesiosis. *In* "Immunity to Protozoa" (P. C. C. Garnham, A. E. Pierce, and I. Roitt, eds.), pp. 160–179. Blackwell, Oxford.

Rifkin, E., and Cheng, T. C. (1968). The origin, structure, and histochemical characterization of encapsulating cysts in the oyster *Crassostrea virginica* parasitized by the cestode *Tylocephalum* sp. *J. Invert. Pathol.* **10**, 54–64.

Rifkin, E., Cheng, T. C., and Hohl, H. R. (1969). An electron-microscope study of the constituents of encapsulating cysts in the American oyster, *Crassostrea virginica*, formed in response to *Tylocephalum* metacestodes. *J. Invert. Pathol.* **14**, 211–226.

Sadun, E. H., Bruce, J. I., and Macomber, P. B. (1964). Parasitologic, pathologic and serologic reactions to *Schistosoma mansoni* in monkeys exposed to irradiated cercariae. *Am. J. Trop. Med. Hyg.* **13**, 548–557.

Salt, G. (1955). Experimental studies in insect parasitism. VIII. Host reactions following artificial parasitization. *Proc. Roy. Soc. Edin.* B144, 380–398.

Salt, G. (1970). "The Cellular Defense Reactions of Insects." Cambridge University Press, London.

Sanders, A., and Wallace, F. G. (1966). Immunization of rats with irradiated *Trypanosoma lewisi. Exp. Parasitol.* **18**, 301–304.

Semrad, J. E., and Coors, M. J. (1951). The rise and fall of immunity to *Trichinella spiralis* in the albino rat and its effect on the growth and reproduction of the parasite. *Trans. Ill. State Acad. Sci.* **44**, 205–208.

Sinclair, K. B. (1967). Pathogenesis of *Fasciola* and other liver-flukes. *Helminth. Abstr.* **36**, 115–134.

Sinclair, I. J. (1970). The relationship between circulating antibodies and immunity to helminthic infections. *Adv. Parasitol.* **8**, 97–138.

Singh, D., and Raghaven, N. G. S. (1962). The duration of patent infection of *Litomosoides carinii* in the albino rat. *Indian J. Malariol.* **16**, 193–201.

Smithers, S. R. (1968). Immunity to blood helminths. *In* "Immunity to Parasites" (A. E. R. Taylor, ed.), pp. 55–56. Blackwell, Oxford.

Smithers, S. R., and Terry, R. J. (1965). Naturally acquired resistance to experimental infections of *Schistosoma mansoni* in the rhesus monkey (*Macaca mulatta*). *Parasitology* **55**, 701–710.

Smithers, S. R., and Terry, R. J. (1967). Resistance to experimental infection with *Schistosoma mansoni* in rhesus monkeys induced by the transfer of adult worms. *Trans. Roy. Soc. Trop. Med. Hyg.* **61**, 517–533.

Soltys, M. A. (1964). Immunity to trypanosomiasis. V. Immunization of animals with dead trypanosomes. *Parasitology* **54**, 585–591.

Soltys, M. A. (1965). Immunologic properties of *Trypanosoma brucei* inactivated by β-propiolactone. *Proc. Sec. Intern. Congr. Protozool., London*, p. 138.

Sommerville, R. I. (1960). The growth of *Cooperia curticei* (Giles, 1892), a nematode parasite of sheep. *Parasitology* **50**, 261–267.

Soulsby, E. J. L. (ed.) (1972). "Immunity to Animal Parasites." Academic Press, New York.

Sprent, J. F. A. (1962). Parasitism, immunity and evolution. *In* "The Evolution of Living Organisms" (G. W. Leeper, ed.), pp. 149–165. Melbourne University Press, Melbourne, Australia.

Stauber, L. A. (1970). Leishmanias. *In* "Immunity to Parasitic Animals" (G. J. Jackson, R. Herman, and I. Singer, eds.), Vol. 2, pp. 739–765. Appleton-Century-Crofts, New York.

Steiner, H. Hultmark, D., Entström, Å., Bennich, H., and Boman, H. G. (1981). Sequence and specificity of two antibacterial proteins involved in insect immunity. *Nature* **292**, 246–248.

Stirewalt, M. A., and Evans, A. S. (1955). Serologic reactions in *Schistosoma mansoni* infections. I. Cercaricidal,

precipitation, agglutination, and CHR phenomena. Project Report NM 005 048–02–32, Naval Medical Research Institute, Bethesda, Maryland.

Sun, T., and Gibson, J. B. (1969). Antigens of *Clonorchis sinensis* in experimental and human infections. An analysis by gel-diffusion technique. *Am. J. Trop. Med. Hyg.* **18**, 241–252.

Tan, B. D., and Jones, A. W. (1967). Autoelimination by means of X-rays: distinguishing the crowding factor from others in premunition caused by the mouse bile duct cestode, *Hymenolepis microstoma. Exp. Parasitol.* **20**, 250–255.

Tan, B. D., and Jones, A. W. (1968). Resistance of mice to reinfection with bile-duct cestode, *Hymenolepis microstoma. Exp. Parasitol.* **22**, 250–255.

Taub, J. (1956). The effect of normal human serum on *Leishmania. Bull. Res. Counc. Isr.* **6E**, 55–57.

Taylor, R. L. (1969). A suggested role for the polyphenolphenoloxidase system in invertebrate immunity. *J. Invert. Pathol.* **14**, 427–428.

Thorson, R. E. (1970). Direct-infection nematodes. *In* "Immunity to Parasitic Animals" (G. J. Jackson, R. Herman, and I. Singer, eds.), Vol. 2, pp. 913–961. Appleton-Century-Crofts, New York.

Tobie, J. E., Abele, D. C., Wolff, S. M., Contacos, P. G., and Evens, C. B. (1966). Serum immunoglobulin levels in human malaria and their relationship to antibody production. *J. Immunol.* **97**, 498–505.

Tsujii, T. (1960). Studies on the mechanism of shell- and pearl-formation in Mollusca. *J. Fac. Fish. Prefect. Univ. Mie* **5**, 1–70.

Van Weeman, B. K., and Schurs, A. H. W. M. (1971). Immunoassay using antigen-enzyme conjugates. FEBS Letters **15**, 232.

Varjú, L. (1966). Studies on strongyloidosis. VII. The nature of changes in the developmental course of swinestronglyloides. *Z. Parasitenk.* **28**, 175–192.

Vogel, H., and Minning, W. (1949). Weitere Beobachtungen über die Cercarienhüllenreaktion, eine Seropräzipitation mit lebenden Bilharzia-Cercarien. *Z. Tropenmed. Parasitol.* **1**, 378–386.

Voller, A., Garnham, P. C. C., and Targett, G. A. T. (1966). Cross immunity in monkey malaria. *J. Trop. Med. Hyg.* **69**, 121–123.

Wakelin, D. (1978). Genetic control of susceptibility and resistance to parasitic infection. *Adv. Parasit.* **16**, 219–308.

Walker, I. (1959). Die Abwehrreaktion des Wirtes *Drosophila melanogaster* gegen die zoophage Cynipide *Pseudeucoila bochei* Weld. *Rev. Suisse Zool.* **66**, 569–632.

Ward, P. A. (1963). Neutrophil chemotactic factors and related clinical diseases. *Arthritis Rheum.* **13**, 181–186.

Weinmann, C. J. (1964). Host resistance to *Hymenolepis nana.* II. Specificity of resistance to reinfection in the direct cycle. *Exp. Parasitol.* **15**, 514–526.

Weinmann, C. J. (1966). Immunity mechanisms in cestode infections. *In* "Biology of Parasites" (E. J. L. Soulsby, ed.), pp. 301–320. Academic Press, New York.

Weiss, M. L., and DeGuisti, D. L. (1966). Active immunization against *Plasmodium berghei* malaria in mice. *Am. J. Trop. Med. Hyg.* **15**, 472–482.

Wong, M. M. (1964). Studies on microfilaremia in dogs. II. Levels of microfilaremia in relation to immunologic responses of the host, *Am. J. Trop. Med. Hyg.* **13**, 66–77.

INTRODUCTION TO THE PARASITIC PROTOZOA

Mastigophora—The Flagellates

Protozoans are commonly referred to as unicellular organisms because their bodies resemble a single cell that includes all the basic characteristics of the cells of metazoans. They are, however, considered acellular by some (Hyman, 1940). While the majority of protozoans are free living in various aquatic and moist environments, many species are mutualists, commensals, and true parasites. Some of the parasitic species, such as *Trypanosoma brucei brucei* and *Entamoeba histolytica*, are highly pathogenic to their vertebrate hosts, and hence are of concern to veterinary and human medicine.

The sizes of protozoans vary greatly. If the individual masses of all known species are compared with the masses of other types of cells, most protozoans would fall within the lower half of the range scale. Table 4.1 gives some idea of the comparative masses of various types of cells.

GENERAL MORPHOLOGY OF PROTOZOA

As is the case with all organisms, the classification of the Protozoa is in continuous flux. However, a special committee of the Society of Protozoologists has pre-

sented a revised classification (Levine *et al.*, 1980). According to this report, the Protozoa is considered a subkingdom divided into seven phyla. The first of these, Sarcomastigophora, includes the subphyla Mastigophora (the flagellates), Opalinata (the opalinids), and Sarcodina (the amoebae). The other phyla are Apicomplexa (the gregarines, coccidia, piroplasms, and related protozoans), Myxozoa (the myxosporans and actinosporans), Microspora (the microsporans), Labyrinthomorpha (the labyrinthomorphs), Ascetospora (the ascetosporans), and the Ciliophora (the ciliates and suctorians).

CYTOPLASMIC ZONES

The cytoplasm of the protozoan trophozoite (vegetative form) is surrounded by some form of cell membrane or a specialized rigid or semirigid covering. The cytoplasm is commonly divided into two areas—the peripheral **ectoplasm** and the medullary **endoplasm**. The consistency and appearance of the ecto- and endoplasms differ among species, and the delineation between the two areas may also vary markedly. In

Table 4.1. Comparative Masses of Various Types of Cells[a]

Cell or Organism	Mass (gm)
Dinosaur egg, ostrich egg, cycad ovule	10^2 to 10^3
Valonia macrophysa (mature)	10^1
Valonia ventricola (mature)	10^0
Nitella (large internode)	10^{-1}
Frog egg	10^{-2} to 10^{-3}
Human striated muscle	10^{-4}
Human ovum	10^{-5}
Large *Paramecium*	10^{-6}
Large-sensory neuron of dog	10^{-6}
Average *Vorticella*	10^{-7}
Human smooth muscle fiber	10^{-7}
Human liver cell	10^{-7}
Entamoeba histolytica	10^{-8}
Frog erythrocyte	10^{-9}
Plasmodium spp.	10^{-9}
Human sperm	10^{-9}
Small protozoa (*Monas*)	10^{-9}
Anthrax bacillus	10^{-11}
Tubercle and pus bacteria	10^{-12}
Smallest bacteria	10^{-14}
Filterable virus[b]	10^{-15}

[a] Partially after Haldane and Huxley (1927).

[b] Limit of light microscope.

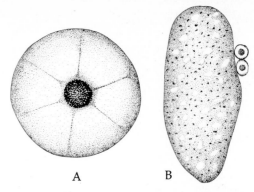

Fig. 4.1. Types of protozoan nuclei. A. Vesicular nucleus of the amoeba *Entamoeba invadens.* **B.** Compact nucleus of the ciliate *Paramecium aurelia* with two adjacent vesicular micronuclei. (Redrawn after Kudo, 1966 Protozoology, 5th ed. Charles C Thomas.)

certain species the separation between the two cytoplasmic zones appears hypothetical.

NUCLEI

Protozoans are eukaryotes and therefore possess well-defined nuclei. Some have one nucleus, others have two or more essentially identical nuclei, and still others have two distinct types of nuclei—a **macronucleus** and one or more **micronuclei**. In members of the phyla Sarcomastigophora, Labyrinthomorpha, Apicomplexa, Ascetospora, Myxozoa, and Microspora there are one or more nuclei of the same type, whereas in the Ciliophora there are generally two types of nuclei.

The macronucleus, when present, is typically larger and is associated with trophic activities. The micronucleus is concerned with reproductive activities.

In addition to characterizing nuclei as either macro- or micronuclei, two morphologically distinct types of nuclei have been described—the **vesicular nucleus** and the **compact nucleus**. These two types are often referred to in the differentiation of species.

VESICULAR NUCLEUS

The nuclear membrane of a vesicular nucleus, although delicate, is visible (Fig. 4.1). Throughout the nucleo-

plasm, the chromatin material may be lightly diffused, but one or more prominent bodies, the **endosomes**, are apparent. These bodies do not stain with the Feulgen reaction, a cytochemical test for DNA. Endosomes are believed by some to be analogous to nucleoli although, unlike the latter, they typically maintain their identity through the various phases of mitosis. The vesicular nucleus is usually found in species of Sarcodina and Mastigophora.

COMPACT NUCLEUS

The compact nucleus contains a seemingly larger amount of chromatin scattered as minute granules or clumps throughout the nucleus, and the nuclear membrane is less conspicuous (Fig. 4.1). Nuclei of this type are generally larger and assume varying shapes, ranging from rounded to ovate. They may also be club- or rod-shaped, filamentous, or dendritic. The compact nucleus of the common parasite of cockroaches, *Nyctotherus ovalis*, is 20 μm or more in diameter. Compact nuclei are usually found among the Ciliphora.

VACUOLES

The cytoplasm of most protozoans contains one or more vacuoles. These appear as light, rounded "vesicles" floating among the cytoplasmic granules. Vacuoles are differentiated physiologically into **contractile** and **food** vacuoles. Contractile vacuoles function as osmoregulatory organelles, and food vacuoles—conspicuous in holozoic and parasitic species—as sites of digestion of food inclusions. Other types of vacuoles are known; for example, **concrement vacuoles** are characteristic of ciliates belonging to the family Butschliidae, which inhabit the alimentary canal of mammals, and those belonging to the family Paraisotrichidae, which are endoparasitic in the caecum and colon of horses (Fig. 4.2). These highly specialized vacuoles, situated singly in the

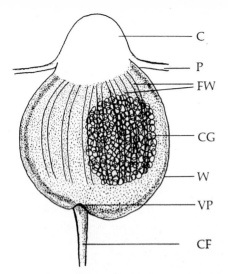

Fig. 4.2. Concrement vacuole of *Blepharoprosthium* sp. C, cap; P, pellicle; FW, fibrils of wall; CG, concrement grains; W, wall; VP, vacuolar pore; CF, centripetal fibril. (Redrawn after Dogiel, 1929.)

anterior third of the organism's body, are composed of a pellicular cap, a permanent vacuolar wall, concrement granules, and two fibrillar systems. During division by transverse fission, the anterior daughter retains the vacuole, while a new one is formed in the posterior daughter from the pellicle into which concrement granules flow. No surface pores are present, and the vacuoles are believed to be sensory in function.

CYTOPLASMIC INCLUSIONS

RIBOSOMES

Ribosomes (or polyribosomes) are the sites of protein synthesis and have regularly been recognized in the cytoplasm of protozoan and metazoan cells. When examined with the electron microscope they are found to be located on the endoplasmic reticulum and are derived from it (Fig. 4.3). Some, however, do become detached and remain free in the cytoplasm. These organelles appear as minute, RNA-rich bodies.

MITOCHONDRIA

Chemically, mitochondria are made up of proteins, phospholipids, glycerides, cholesterols, and a small quantity of RNA and DNA. More than half the solids present are proteins, mostly in the form of enzymes. These minute bodies can be demonstrated in protozoans by supravital staining with Janus green B, Janus red, or osmium tetroxide and may be spherical, ovoid, cylindrical, or filamentous.

In the parasite *Monocystis*, found in the coelom and seminal vesicles of oligochaetes, mitochondria can be demonstrated in the form of minute rods throughout

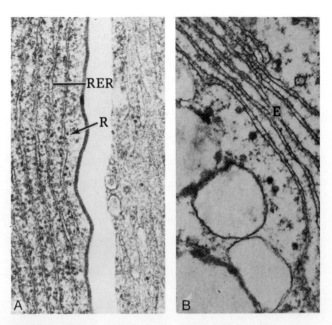

Fig. 4.3. Electron micrographs of endoplasmic reticulum (ER) and ribosomes. A. Portion of cyst of the microsporidan parasite *Nosema* showing rough endoplasmic reticulum; the dense bodies associated with the parallel ER are ribosomes. (Courtesy of Dr. A. Cali.) **B.** Portion of the cytoplasm of an oocyst of the protozoan *Aggregata* showing smooth (without associated ribosomes) ER. (After Porchet-Hennère and Richard, 1971.) E, smooth endoplasmic reticulum; R, ribosomes; RER, rough endoplasmic reticulum.

the asexual phase of the parasite's life cycle, but, with the commencement of sporulation, the mitochondria decrease both in size and number and are not visible in the definitive spore.

Mitochondria are the sites of intracellular aerobic metabolism (respiration). Lindberg and Ernster (1954) arrived at this conclusion when it was discovered that the principal enzymes involved in the Krebs, or citric acid, cycle (p. 55) are associated with mitochondria. The structure of mitochondria has been successfully defined by electron microscopy (Fig. 4.4). Each mitochondrion is delimited by two membranes. The inner membrane forms an undulating surface, the folds of which are known as **cristae** if they are membranous, and as **tubuli** if they are tubular. Cristae (or tubuli) provide a substantial increase in the amount of inner membrane that can be accommodated within the mitochondrial body, and thus serve as a device to allow the insertion of many more of the respiratory enzymes that are associated with this membrane.

The mitochondrial **matrix** is a semisolid system enveloped by the inner membrane. Except for ribosomes and DNA fibers that can be observed in cells prepared with special care, no other regularly structural components occur in this matrix. The matrix also contains salts and other solutes, as well as a variety of enzymes that catalyze oxidations of nutrients in energy-yielding reactions. About 50% of the matrix material is protein, which often appears as fine fibrils.

GOLGI APPARATUS (BODY)

Functionally, Golgi bodies are associated with the concentrating and channeling, including secreting, of certain intracellular materials, especially proteins. This encompasses the packaging of lysosomal enzymes and the formation of secretory vesicles; the synthesis of mucus; and the storage of lipids and proteins. The Golgi apparatus was originally described by Hirschler (1914) in the parasitic protozoan *Monocystis ascidiae.* The arrangement and number of these organelles is inconsistent, even in a given species. Furthermore, the number of Golgi complexes varies at various stages in the life cycle. For example, in *Ichthyophthirius*, a ciliate ectoparasite of fishes, and in *Protoopalina*, an intestinal parasite of amphibians, the Golgi apparatus disappears when the protozoans are encysted, but arises *de novo* upon excystment. Even the distribution of these organelles varies at different stages in the life cycle of gregarines. Similar variations undoubtedly also occur in other groups of protozoans.

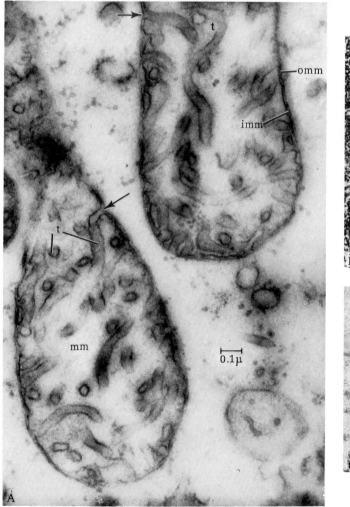

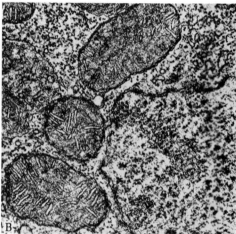

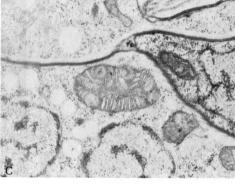

Fig. 4.4.   Electron micrographs of protozoan mitochondria. A. Two mitochondria in the cortical cytoplasm of the ciliate *Paramecium,* showing profiles of tubuli projecting into lumina. (After Sedar and Porter, 1955.) **B.** Portion of the cytoplasm of the flagellate *Astasia longa* showing mitochondria with cristae projecting into lumina (arrow). (After Ringo, 1963.) **C.** Portion of cytoplasm of the flagellate *Leishmania tarentolae* showing mitochondrion. (After Strauss, 1971.) imm, Inner mitochondrial membrane; mm, mitochondrial matrix; omm, outer mitochondrial membrane; t, tubular cristae.

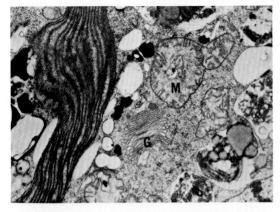

Fig. 4.5.   Golgi and mitochondrion. Portion of cytoplasm of the flagellate *Euglena gracilis* showing Golgi body (G) and mitochondrion (M). (After Rogers, Scholes, and Schlichting, 1972.)

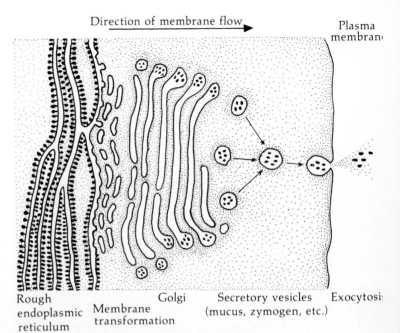

Direction of membrane flow

Plasma membrane

Rough endoplasmic reticulum

Membrane transformation

Golgi

Secretory vesicles (mucus, zymogen, etc.)

Exocytosis

Fig. 4.6.   Membrane transformation. Schematic diagram showing sequence of events involved in membrane transformation and travel from sites of synthesis at the rough endoplasmic reticulum, through the Golgi apparatus, and final fusion with the plasma membrane during exocytosis.

Within the cytoplasm, the Golgi apparatus appears as osmiophilic and argentophilic granules which, when examined with the electron microscope, appear as stacks of flattened sacs (**cisternae**) associated with small vesicles and vacuoles of various sizes (Fig. 4.5). Chemically, they are composed of phospholipids, lipoproteins, and proteins.

In addition to being involved in the packaging, transport, and secretion of various proteins, among other functions, the Golgi apparatus also appears to be involved in membrane transformation. Specifically, there is evidence that new membrane is synthesized at the endoplasmic reticulum and transferred to the Golgi apparatus, where modifications occur. Fully modified membrane is subsequently added to the cell membrane when Golgi vesicles are fused with it during exocytosis (Fig. 4.6).

LYSOSOMES

Another type of cytoplasmic inclusion found in protozoan as well as metazoan cells is the lysosome. Its centrifugal properties are between those of mitochondria and ribosomes. Each lysosome is bound by a

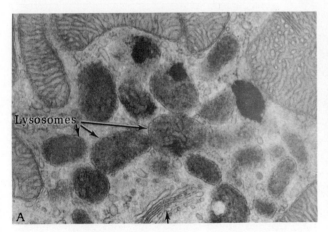

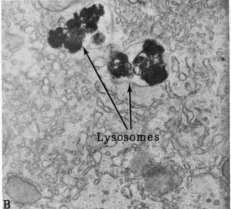

Fig. 4.7.   Electron micrographs of lysosomes. A. A cluster of lysosomes each with a delimiting membrane. × 29,000. **B.** Two lysosomes enclosing dense inclusions. × 25,000. (After Fawcett, 1966.)

unit membrane (Fig. 4.7) within which occur acid hydrolases such as acid phosphatase, aryl sulfatase, cathepsin, aminopeptidase, and others (Table 3.1).

Lysosomes play a role in the turnover of normal metazoan cells and in the autolysis of protozoans. Also, they are involved in the intracellular degradation of phagocytosed or pinocytosed materials. For example, when an amoeba engulfs foreign material, including host cells in the case of certain parasitic species, the foreign material is enveloped in vacuoles known as **phagosomes**. Shortly thereafter, phagosomes fuse with lysosomes, and the enzymes enclosed within the latter are released into the phagosomes and effect the degradation (or intracellular digestion) of the ingested material.

In addition to the above, it has been mentioned (p. 70) that lysosomal enzymes originating in lysosomes may play an important role in influencing compatibility or incompatibility between host and parasite.

MICROBODIES

Although first discovered in 1954 in electron micrographs of mouse kidney cells, microbodies have since been found to be ubiquitous in all eukaryotic cells, including protozoans. Each microbody ranges in length from about 0.2 to 1.7 μm and may be spherical, ovoid,

$$RH_2 + O_2 \xrightarrow{\text{oxidase}} R + H_2O_2$$
$$\text{hydrogen peroxide}$$

$$H_2O_2 \xrightarrow{\text{catalase}} H_2O + 1/2O_2$$

Fig. 4.9. Function of catalase in microbody. Reactions demonstrating the production of hydrogen peroxide (above) and subsequent conversion of the toxic hydrogen peroxide to water and oxygen (below).

or dumbbell-shaped (Fig. 4.8). They usually occur as clusters adjacent to the endoplasmic reticulum. Although exceptions occur, the major characteristic of microbodies is the occurrence of the enzyme catalase, which removes hydrogen peroxide. In fact, as originally demonstrated by C. deDuve, the major function of microbodies is the removal of harmful hydrogen peroxide that has been generated in a prior reaction by catalase (Fig. 4.9). It is also known that microbodies are involved in gluconeogenesis and biosynthesis of the amino acids glycine and serine.

Microbodies originate as buds pinched off the endoplasmic reticulum. Before this process is completed, enzymes synthesized along rough endoplasmic reticulum accumulate in the endoplasmic reticulum lumen and become sequestered in each microbody as it is pinched off.

HYDROGENOSOMES

Parasitic trichomonad flagellates (p. 108) are predominantly anaerobic but can tolerate oxygen. They carry on respiration in the presence of oxygen but have no cytochromes (p. 57) and no functional Krebs cycle (p. 55). Furthermore, *Tritrichomonas* and other trichomonads, like a few other known anaerobic protozoans, have no mitochondria. However, they do include microbodylike organelles whose single membrane envelopes a granular, unstructured matrix (Fig. 4.10). These organelles have been designated **hydrogenosomes**.

Unlike microbodies, hydrogenosomes include no catalase and flavin oxidase activity. These enzymes occur only in the cytosol of flagellates with hydrogenosomes. Furthermore, uniquely, an electron transfer pathway (p. 57) exists in which a hydrogenase transfers electrons to protons and molecular hydrogen is formed. Thus, hydrogenosomes are entirely unlike either microbodies or mitochondria since protons, rather than molecular oxygen, act as the terminal electron acceptors.

CYTOPLASMIC FOOD RESERVES

These inclusions have been described from various species of parasitic protozoans. The nature and amount of such stored nutrients vary with the environment. Glycogen and/or amylopectin have been reported in the cysts of *Ichthyophthirius*, in the cysts of certain

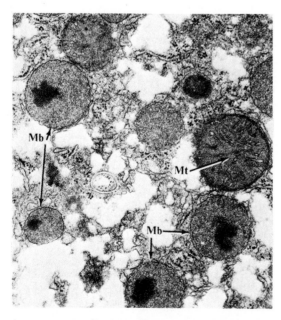

Fig. 4.8. Microbodies. Electron micrograph showing microbodies each with a core inclusion. Mb, microbodies; Mt, mitochondrion. (After E. H. Newcomb and S. E. Frederick in Avers, 1976; with permission of Van Nostrand.)

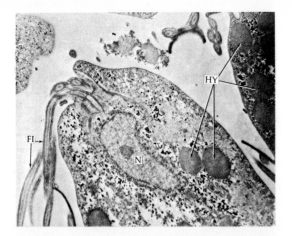

Fig. 4.10. Hydrogenosome. Electron micrograph of a portion of two specimens of *Pentatrichomonas* showing hydrogenosomes. Notice that mitochondria are absent. × 25,800. (After Mattern and Honigberg in Avers, 1976; with permission of Van Nostrand.) HY, hydrogenosome; N, nucleus.

intestinal amoebae of mammals, and in other parasitic protozoans. This stored food material is formed shortly before encystation and is completely utilized prior to excystation. In certain gregarines, amylopectin is formed within a mitochondrial sphere, which disappears after the amylopectin mass reaches a certain size. The same has been reported in *Ichthyophthirius*. In addition to glycogen or amylopectin, lipid droplets may also be present.

OTHER CYTOPLASMIC INCLUSIONS

Other types of inclusions, such as smooth endoplasmic reticulum, also are present in protozoans. The reader is referred to any recent text on cell biology for discussions of these. The monograph by Scholtyseck (1979) is especially recommended.

OTHER STRUCTURES

Detailed discussions of such structures as pigment granules, pyrenoids, and chromatophores—all found among the plantlike Phytomastigophorea—and

Table 4.2. Protein Content of Some Parasitic Protozoa

Species	Protein in Percent of Dry Weight
Flagellate	
Trypanosoma cruzi, culture form	43–53
Apicomplexan	
Eimeria acervulina, oocysts	41
Ciliate	
Entodinium caudatum	25

photoreceptors—found among euglenae and dinoflagellates—are omitted from this text because they are found primarily in free-living species, although they occur in symbiotic species of *Euglena*. In addition, certain specialized organelles are present in some specific groups of protozoans; these are discussed in the following sections devoted to each taxon.

CHEMICAL COMPOSITION OF PROTOZOA

PROTEINS

Since proteins constitute most of the soma and all of the enzymes of protozoans, it is not surprising that those species that have been studied have revealed a relatively high content (Table 4.2).

In addition to pure proteins, conjugated forms such as mucopolysaccharides and lipoproteins also occur. For example, it is known that in *Trypanosoma cruzi* and *Tritrichomonas foetus* the antigenic fractions are in the form of conjugates of proteins and polysaccharides.

Special nitrogen-including molecules that represent stored materials rather than integral parts of the soma and enzyme systems also have been reported in certain protozoans. For example, **volutin**, composed of free nucleic acids, is stored as a reserve for the nucleus and is accumulated in coccidians (p. 183) and trypanosomes, which undergo rapid nuclear divisions. Volutin does not occur in metazoans. The volume by Gutteridge and Coombs (1977) on protozoan bio-

Table 4.3. Lipid Content of Some Parasitic Protozoa[a]

Species	Lipids in Percent of	
	Fresh Tissues	Dry Tissues
Flagellates		
Crithidia fasciculata	2.99	
Trypanosoma cruzi, culture form	3.12	20.1
Apicomplexan		
Goussia gadi	3.6	22.0
Eimeria acervulina (oocyst)		14.4
Plasmodium knowlesi		28.8
Ciliates		
Entodinium simplex		6.3
Isotricha intestinalis		9.1
Isotricha prostoma		7.7

[a] After von Brand, 1973.

chemistry is recommended to those desiring further information.

LIPIDS

Various investigators have attempted to determine the presence and amount of lipids in parasitic protozoans. Such studies have employed two sets of techniques; one involving cytochemical procedures for the localization of the lipids, and the other involving bio-chemical methods, including thin-layer and gas chromatography and infrared or ultraviolet spectroscopy for the quantitative and qualitative analysis of lipids.

As indicated in Table 4.3, there is considerable variation in the amount of lipids in parasitic protozoans, nor is there any apparent correlation between the parasites' habitats and the amount of lipids present.

Fractionation of stored lipids in *Plasmodium knowlesi* carried out by Morrison and Jeskey (1947) has revealed that unsaturated fatty acids account for a relatively high percentage of the total lipids (Table 4.4). This differs from the situation in the parasitic helminths, in which phospholipids and unsaponifiable matter constitute the larger fractions. Von Brand (1962) has demonstrated that in the culture form of *Trypanosoma cruzi* the fatty acids of the acetone-insoluble lipid fraction contains 12 members ranging from C_{13} to C_{18}, including branched, unbranched, saturated, monounsaturated, and doubly unsaturated fatty acids. Of these, the occurrence of relatively large amounts of a C_{15} fatty acid is unusual since fatty acids with an odd number of carbons usually occur in small amounts in animals.

Some evidence indicates that the quantity of lipids varies between the sexes. For example, among gregarines (p. 178), the female of an encysted pair includes more deposited lipids than the male. Similarly, lipids occur in the macrogametocytes and spores of *Eimeria* spp., but not in the microgametocytes, schizonts, and merozoites.

Finally, seasonal variations in lipid content also occur. In the case of *Opalina ranarum*, for example, a large accumulation of fat occurs in the fall, but it is almost completely depleted toward the end of winter.

CARBOHYDRATES

Carbohydrates stored in the bodies of parasitic protozoans are generally in the form of glycogen and/or amylopectin. **Amylopectin**, formerly known as para-glycogen, is a type of animal starch similar to glycogen that can be dissolved in water, but with difficulty; glycogen is not water soluble. Amylopectin was first discovered in gregarines and later found in coccidians and endoparasitic ciliates. It is not known to occur in metazoans. Within the bodies of protozoans, depending on the species, glycogen and amylopectin are distributed either in definite locations, randomly, or in, but not restricted to, definite organelles. To exemplify the first instance, the large glycogen mass in the cysts of *Iodamoeba* is generally found at the pole opposite the nucleus (Fig. 4.11). In young binucleated cysts of *Entamoeba*, the large glycogen vacuole is always located between the two nuclei. The amylopectin found in gregarines is located primarily in the deutomerite (p. 179); that found in trophozoites of

Table 4.4. Fatty Acids Identified in Some Parasitic Protozoa[a]

Columns under **Flagellates**: Crithidia sp. (Phospholipids)[b], Crithidia sp. (Neutral Lipids)[b], Leishmania enriettii[b,c], Leishmania tarentolae[b,c], Trypanosoma cruzi (Acetone-soluble Lipids)[b,c], Trypanosoma cruzi (Acetone-insoluble Lipids)[b,c], Trypanosoma lewisi[b,d], Trypanosoma lewisi[b,c]; under **Ciliates**: Isotricha intestinalis[e], Entodinium simplex[e]; under **Apicomplexan**: Plasmodium knowlesi[f].

Chain Length	Unsaturation	Crithidia sp. (Phospholipids)[b]	Crithidia sp. (Neutral Lipids)[b]	Leishmania enriettii[b,c]	Leishmania tarentolae[b,c]	Trypanosoma cruzi (Acetone-soluble)[b,c]	Trypanosoma cruzi (Acetone-insoluble)[b,c]	Trypanosoma lewisi[b,d]	Trypanosoma lewisi[b,c]	Isotricha intestinalis[e]	Entodinium simplex[e]	Plasmodium knowlesi[f]
8	0					+						
9	0					+						
10	0			+		+						
10	1					+						
12	0			+	+	+						
13	0						+					
14	0	+	+	+	+	+	+	+	+			
14	1			+		+	+			+	+	
15	0					+	+					
15	1					+						
16	0	+	+	+	+	+	+	+	+	+	+	
16	1			+		+	+	+	+			
16	2				+	+						
17	0					+	+					
18	0	+	+	+	+	+	+	+	+	+	+	+
18	1	+	+	+	+	+	+	+	+	+		+
18	2	+	+	+	+	+	+	+	+	+		
18	3	+	+	+				+	+			
18	4				+							
20	2				+				+			
20	3	+	+		+				+			
20	4	+			+				+	+		
20	5								+	+		
22	4	+	+		+				+			
22	5	+	+		+				+	+		
22	6				+				+			

[a] Data compiled by von Brand, 1966.
[b] Method used, gas chromatography.
[c] Culture form.
[d] Bloodstream form.
[e] Method used, paper chromatography.
[f] Method used, chemical fractionation.

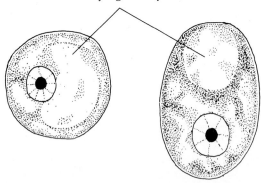

Glycogen body

Fig. 4.11. Cysts of *Iodamoeba butschlii*. Size is about 6 to 15 μm in greatest diameter; large glycogen body in each.

Nyctotherus lies between the anterior end of the body and the macronucleus. Randomly scattered glycogen granules occur in *Tritrichomonas foetus* and in the trophozoites of various intestinal amoebae. Polysaccharides found in various rumen-dwelling ciliates are commonly found in specific organelles but are not limited to these. In these ciliates, glycogen is stored primarily in the complex skeletal plates.

Little is known about the distribution of simple sugars in parasitic protozoans. Among flagellates, Fairbairn (1958) has reported very small amounts of glucose (0.06% of dry weight) and trehalose (0.05%) in the culture form of *Trypanosoma cruzi*, and Williamson and Desowitz (1961) have reported 0.1–1.0 μg of glucosamine per milligram of protein in hydrolyzed trypanosomes. On the other hand, Loran *et al.* (1956) have reported considerably greater amounts of glucosamine in the amoeba *Entamoeba histolytica*.

Stored polysaccharides in parasites are primarily in the form of polymers of D-glucopyranose. These are either glycogen, which is by far the most common, or amylopectin. The glycogens isolated from such flagellates as *Tritrichomonas foetus* and *Trichomonas gallinae* have comparable molecular weights, $3–4 \times 10^6$, although that from *T. foetus* is longer chained (15) than that of *T. gallinae* (9) (Manners and Ryley, 1955).

Ciliates, especially those in the rumens of ungulates, include relatively large quantities of polysaccharides. Masson and Oxford (1951), for example, have reported that in *Isotricha* 70% of the dry substance is amylopectin, and Abou Akkada and Howard (1960) have reported 64% in *Entodinium*. Hemoflagellates, on the other hand, are apparently rather poor in stored polysaccharides; for example, Gerzeli (1955) has reported no cytochemically demonstrable polysaccharide in the bloodstream form of trypanosomes.

As is the case with lipids, amounts of polysaccharides vary with the stage and sex of the protozoan. For example, relatively large amounts occur in the macrogametocytes, oocysts, and mature asexual stages of *Eimeria* spp., but little or none occur in microgametocytes, young schizonts, and other stages.

Seasonal fluctuations in the amount of stored glycogen also is known to occur as in the flagellate *Cryptobia helicis*, a parasite in the reproductive organs of various snails. Such fluctuations do not occur in endoparasitic helminths. In *C. helicis*, glycogen is not present during the summer but is present during the winter. This is true probably because snails are ectothermic hosts whose metabolism fluctuates greatly with the temperature. Seasonal variations in the amount of glycogen in the host are known to occur, and therefore it is not surprising that the metabolism of the parasite, which is totally dependent on the host, should also fluctuate.

INORGANIC SUBSTANCES

In addition to proteins, lipids, carbohydrates, and combinations of these, several inorganic substances have been demonstrated that are important in the metabolism of protozoan parasites. From microincineration of *Opalina*, a parasite in the large intestine of frogs and toads, Scott and Horning (1932) have demonstrated ash deposits that indicate the presence of inorganic substances. These substances are present in the myonemes, cilia, and basal granules. Analyses of the ash have revealed that calcium oxide is deposited in the cytoplasm. Similar ash deposits are found in the same sites and in the nucleus of *Nyctotherus*, a parasitic protozoan from the colon of amphibians. Analyses have revealed that the cytoplasm is rich in sodium, while silica is present in the cytopharynx. Both calcium and iron are found lining the walls of food vacuoles in the cytoplasm. In the woodroach mutualist *Trichonympha*, most of the inorganic materials are present in the neuromotor system, cytoplasm, and nucleus, and the body surface is rich in calcium compounds.

Because phosphorus plays an important role in several metabolic processes (phosphorylations) it is not surprising that this element has been demonstrated in various parasitic protozoans, such as the trypanosomes. The synthesis of volutin, for example, is apparently correlated to the occurrence of phosphorus, since it has been demonstrated that when the free-living flagellate *Haematococcus pluvialis* is cultured in a phosphorus-rich medium, volutin is accumulated in large quantities, and the volutin disappears when the culture is changed to a phosphorus-free one. This is of interest to parasitologists since some parasitic flagellates also accumulate volutin.

In the avian malaria-causing *Plasmodium gallinaceum*, Kruszynski (1951) has reported the presence of potassium, sodium, calcium, and phosphorus; however, in *Plasmodium berghei*, the same investigator (Kruszynski, 1952) has reported that calcium does not occur, although it is present in the erythrocytes of the rodent host, thus implying that the parasite can derive its required calcium from the host.

Those interested in a much more extensive discussion of the chemical constituents of parasitic protozoans are referred to von Brand's (1973) monograph.

NUTRIENT ACQUISITION

The topic of how protozoans obtain their food has been adequately summarized by Grell (1973). Briefly, the mechanisms for the uptake of nutrients by these organisms can be categorized as being of three types: (1) phagocytosis (or phagotrophy), (2) saprozoic, and (3) pinocytosis.

PHAGOCYTOSIS

Phagocytosis, or the ingestion of particulate foods, is considered the most common mechanism among protozoans. It occurs among amoebae, apicomplexans, myxozoans, ascetosporans, microsporans, flagellates, and ciliates. Among amoebae, phagocytosis generally is accomplished by pseudopodial engulfment (Fig. 4.12). In the case of the flagellate *Dientamoeba fragilis*, a parasite of humans (p. 118), a gulletlike structure is formed which receives the particulate food (Fig. 4.13).

Phagotrophy by the malarial parasites, *Plasmodium* spp., was first reported by Rudzinska and Trager (1957, 1959) who demonstrated by electron microscopy that these protozoans are capable of engulfing the hemoglobin present in the surrounding host erythrocyte by a mechanism comparable to certain

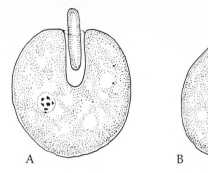

Fig. 4.13. Gulletlike structure in *Dientamoeba fragilis*. **A.** Tubular gullet containing a partially ingested bacterium. **B.** Opening of gullet at apex of cytoplasmic extension. (Redrawn after Wenrich, 1944.)

types of amoeboid action. Specifically, a food vacuole is formed by invagination of one portion of the parasite's body surface, and the hemoglobin entering the vacuole is completely engulfed when the aperture to the vacuole is closed off (Fig. 4.14). In addition to *Plasmodium*, other species, such as *Monocystis* (p. 180), also are capable of ingesting host cells by phagotrophy.

Among flagellates, phagocytosis also occurs. For example, in the case of *Histomonas meleagridis*, a parasite of birds (p. 117), a temporary gullet is formed which permits the intake of particulate material (Fig. 4.15). In the case of the flagellates occurring in the guts of woodroaches and termites, phagotrophy is accomplished by a process comparable to amoeboid engulfment (Fig. 4.15).

SAPROZOIC NUTRITION

Saprozoic feeding involves the intake of nutrients through the protozoan's body surface, be it a pellicle or a plasma membrane; hence, the process is commonly referred to as **permeation**. Relatively little is known about the mechanisms involved, although there are evidences that suggest that sometimes more than simple diffusion occurs. This is true even for certain metal ions; for example, Klein (1961) has shown that the uptake of potassium by the amoeba *Acanthamoeba* occurs against a concentration gradient, i.e., potassium

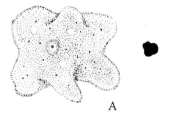
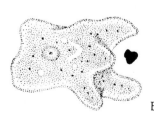
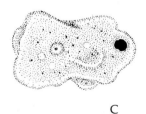

Fig. 4.12. Phagocytosis. Stages during the engulfment of particulate matter by amoeboid phagocytosis.

moves from an area of lesser concentration on the exterior to an area of greater concentration inside the amoeba. Similarly, Min and Cosgrove (1963) have shown that monosaccharides enter the flagellate *Crithidia luciliae* by active transport against a gradient.

PINOCYTOSIS

Pinocytosis, or "cell drinking," has been known since the 1930s to occur in protozoans. This process involves the formation of delicate invaginations of the body surface through which the exogenous material is permitted to enter (Fig. 4.17). Subsequently, vacuoles containing the material are pinched off as **pinocytotic vesicles** at the inner terminal of each pinocytotic canal. With the advent of the electron microscope, increasingly more evidence is now available which suggests that pinocytosis is a fairly common uptake mechanism among parasitic protozoans. For example, it has been shown that trypanosomes engage in this mechanism for nutrient uptake.

The difference between phagocytosis (phagotrophy) and pinocytosis appears to be one of degree, the former involving the uptake of "particulate" material (including macromolecules) and the latter involving fluids, and also molecules.

The review article by Conner (1967) about transport phenomena and those by Muller (1967) and

Nilsson (1979) concerned with digestion in protozoans are recommended to those desiring greater detail.

PROTOZOAN ENCYSTATION

Many parasitic protozoans are capable of encystation. A typical protozoan cyst is composed of a rounded mass of protoplasm enveloped within a rigid or semirigid wall secreted by the trophozoite during the process of encystation. The cyst wall may be single layered or multilayered.

Cysts of parasitic protozoans serve four primary functions: (1) they protect against unfavorable environmental conditions; (2) they function as sites for reorganization and nuclear division, followed by multiplication upon excystation; (3) they function in attachment; and (4) they serve as a means of transmission from one host to another. The first function is exemplified by the cysts of the human pathogen *Entamoeba histolytica*, which are formed when the environment within the host becomes unfavorable to the

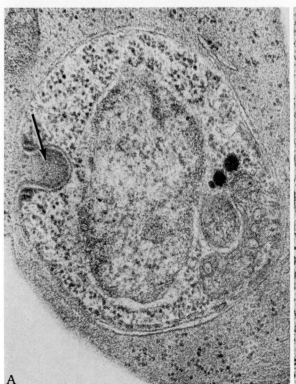
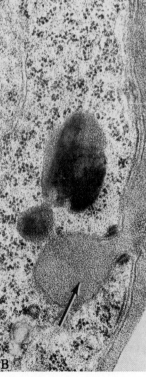

Fig. 4.14. Electron micrographs of erythrocytic stages of *Plasmodium cathemerium,* **a causative agent of avian malaria. A.** Young schizont within host's erythrocyte taking in hemoglobin and cytoplasm by phagotrophy (arrow) ($\times 56,000$). **B.** Gametocyte within host's erythrocyte taking in hemoglobin and cytoplasm by phagotrophy (arrow) ($\times 43,000$). (After Aikawa *et al.,* 1966.)

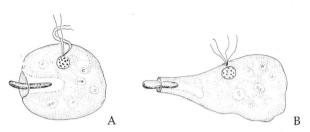

Fig. 4.15. Temporary gullet in *Histomonas meleagridis.*
A. Partly ingested bacterium in gullet opening on body
surface. **B.** Elongate gullet in snoutlike projection of body.
(Redrawn after Wenrich, 1943.)

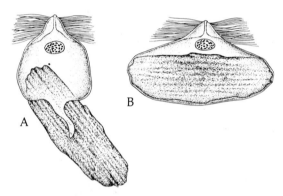

Fig. 4.16. Phagotrophy. Drawings showing ingestion of
a wood chip by the flagellate *Leidyopsis* sp. **A.** Flagellate
beginning to ingest a wood chip. **B.** Flagellate containing
large chip. (After Cheng, 1970.)

trophozoites. Such cysts, after being passed out in
feces, remain viable for many weeks under normal
moist conditions and for days at higher and lower
temperatures and during periods of desiccation.

Encysted *Entamoeba histolytica* also undergo reorganization and nuclear division. Shortly after the
trophozoite encysts, cytoplasmic reorganization occurs
followed by nuclear divisions, so that the mature cyst
encloses four nuclei. If mature cysts are reintroduced
into a suitable host, excystation occurs, and the escaping metacystic trophozoite with four nuclei undergo
both nuclear and cytoplasmic divisions so that eight
small trophozoites are produced from the protoplasm
in each *E. histolytica* cyst (p. 160).

Some cysts, such as those of the fish epithelium-
infecting ciliate *Ichthyophthirius*, serve an additional
function—attachment. When the encysted ciliate falls
to the substratum, the cyst wall adheres to it, holding
the parasite in place until excystation.

Finally, intestinal protozoans, such as *Entamoeba*

histolytica, Iodamoeba butschlii, and *Giardia lamblia,*
usually are transmitted to new hosts or become re-
established in the same host through the swallowing
of cysts. Thus, the cysts serve as the transmission
stage.

Protozoan cyst walls vary in chemical composition.
Cysts of *Entamoeba histolytica* and *Endolimax nana,*
both human parasitic amoebae, and that of the human
intestinal flagellate *Giardia lamblia* contain proteins of
keratinlike or elastinlike albuminoids that on acid hydrolysis are found to be composed of lysine, histidine,
arginine, tyrosine, glutamic acid, and glycine. The
cysts of ciliates also are known to be proteinaceous.

CONDITIONS FAVORING ENCYSTATION

Encysted protozoans can withstand adverse environmental conditions that would kill the unprotected
trophozoite. The resistance of *Entamoeba histolytica*
cysts, for example, is well known. They can withstand
extremes in such physical conditions as heat, drying,
and deleterious chemical irritants (p. 159). The longevity of protozoan cysts varies greatly, depending
on the species and environmental factors. Laboratory
experiments suggest, however, that the conditions
prevailing at the time of encystation determine longevity to a great extent. For example, cysts formed in
cultures that have not dried do not live as long as
those formed after the culture has dried.

Factors that favor encystation are not yet all
known. In many species the process appears to be a
response to nutrient deficiency. In addition, desiccation, increased concentration of dissolved salts,
changes in temperature, low pH, decreased oxygen
supply, accumulation of waste products, and crowding all appear to contribute to encystation. Among
intestinal protozoans, absorption of water from the
host's intestine, with a resulting increase in the concentration of certain substances in the intestine, also
favors encystation. This may be why encysted
amoebae are seldom found in diarrheic stools where
trophozoites abound. What the relative importance of
these factors may be to encystation *in vivo* is uncertain. Again, the various influencing factors appear to
have different effects on different species.

CLASSIFICATION OF PROTOZOA

In following the classification recommended by the
Society of Protozoologists (Levine *et al.,* 1980), the
subkingdom Protozoa is divided into seven phyla:
Sarcomastigophora, Apicomplexa, Myxozoa, Microspora, Labyrinthomorpha, Ascetospora, and Ciliophora
(for characteristics of each phylum, see the end of this
chapter and the following three chapters). All members of the Sarcomastigophora utilize either pseudopodia or flagella as locomotor organelles during some

stage in their life cycles. Among other characteristics, the members of Apicomplexa, Microspora, Ascetospora, Myxozoa, and Labyrinthomorpha possess no flagella or pseudopodia in their definitive stage; however, such organelles occur at some alternative stage in the life cycles of certain species.

The illustrated guide to the Protozoa edited by Lee *et al.* (1985) is highly recommended to those interested in classification.

PHYLUM SARCOMASTIGOPHORA

The Sarcomastigophora includes those protozoans commonly referred to as amoebae and flagellates, which comprise the subphyla Sarcodina and Mastigophora, respectively. In addition, the so-called opalinids, which constitute the subphylum Opalinata, are also assigned to this phylum.

SUBPHYLUM MASTIGOPHORA

The Mastigophora, commonly known as flagellates, includes all the protozoans that possess one or more flagella in their trophozoite form. The majority of the flagellates are free living and are found in various habitats, but a large number are symbiotic in or on both invertebrates and vertebrates. Most endosymbiotic species inhabit their hosts' digestive tracts, circulatory systems, and tissues. That flagellates are capable of swimming has aided in their adaptation to different habitats within their hosts. Unlike amoebae, which require a solid surface on which to glide, flagellates can survive in a liquid medium and thus are well adapted for living in their host's blood, lymph, and cerebrospinal fluid. The flagellated trypanosomes exhibit a further adaptation for life in a liquid medium that is evident in their body forms. They are elongate torpedo-shaped and thus can swim in their host's body fluids with little resistance.

The relationships between flagellates and their hosts need not result in disease. Some flagellates apparently do little damage and are maintained in their

Table 4.5. Some Important Parasitic Flagellates of Humans and Domestic Animals

Flagellate	Principal Hosts	Habitat	Main Characteristics	Disease
Leishmania				
L. donovani	Humans, dogs, cats, horses, sheep, cattle	Spleen, bone marrow, liver, monocytes	2–4 μm in diameter, eccentric rounded nucleus	Kala-azar
L. tropica	Humans, dogs	Dermal sores	Indistinguishable from *L. donovani*	Oriental sore
L. braziliensis	Humans, dogs, monkeys	Dermal sores, especially mucous membranes	Indistinguishable from *L. donovani*	Mucocutaneous leishmaniasis
Trypanosoma				
T. equiperdum	Horses, cattle, donkeys	Genitalia and internal reproductive organs	25–28 μm long, 1–2 μm wide	Dourine
T. theileri	Cattle	Blood	60–70 μm long, 4–5 μm wide, myonemes	Nonpathogenic
T. melophagium	Sheep	Blood	50–60 μm long	Nonpathogenic
T. evansi	Horses, mules, donkeys, cattle, dogs, camels, elephants	Blood	25 μm long	Surra
T. equinum	Horses (S. Amer.)	Blood	20–25 μm long, no blepharoplast	Mal de Caderas
T. hippicum	Horses (Panama), mules	Blood	16–18 μm long	Murrina
T. brucei	Horses, donkeys, mules, camels, cattle, swine, dogs	Blood	Pleomorphic, 15–30 μm long	Fatal nagana
T. simiae	Pigs, monkeys, sheep, goats	Blood	Seldom with free flagellum	Virulent
T. congolense	Cows, other domestic animals	Blood	Monomorphic, 8–19 μm long, 3 μm wide; no free flagella	Bovine trypanosomiasis
T. vivax	Ruminants, equines	Blood	15.5–30.5 μm long, free flagellum	Virulent
T. brucei gambiense	Humans, monkeys, antelopes, dogs	Blood, lymph	15–30 μm long, 1–3 μm wide, spiral undulating membrane	Gambian trypanosomiasis (sleeping sickness)
T. brucei rhodesiense	Humans, wild game, domestic animals	Blood, lymph	Usually indistinguishable from *T. gambiense*	Rhodesian trypanosomiasis (sleeping sickness)
T. cruzi	Humans, cats, dogs, monkeys, squirrels	Blood	C- or U-shaped, 20 μm long	Chagas' disease
T. americanum	Cattle	Blood	17–25 μm long or longer	Nonpathogenic?
T. lewisi	Rats		Approximately 30 μm long	Nonpathogenic

Table 4.5. Some Important Parasitic Flagellates of Humans and Domestic Animals (continued)

Flagellate	Principal Hosts	Habitat	Main Characteristics	Disease
Trichomonas				
T. canistomae	Dogs	Mouth	Four anterior flagella, one trailing; 9 μm long; 3.4 μm wide	Nonpathogenic
T. felistomae	Cats	Mouth	8.3 μm long, 3.3 μm wide	Nonpathogenic
T. gallinae	Turkeys, chickens, pigeons	Upper digestive tract, liver	6.2–18.9 μm long, 2.3–8.5 μm wide	Avian trichomoniasis
T. gallinarum	Turkeys, chickens, pigeons	Caecum	Pear-shaped, 9–12 μm long, 6–8 μm wide	Nonpathogenic?
T. anseri	Geese	Caecum	Oval body, 7.9 μm long, 4.7 μm wide, large cytosome	Nonpathogenic
T. vaginalis	Women	Vagina	10–30 μm long, 10–12 μm wide, cytostome inconspicuous	Vaginitis
T. hominis	Humans	Intestine	5–20 μm long	Nonpathogenic
T. tenax	Humans	Mouth	10–30 μm long	Nonpathogenic
Tritrichomonas				
T. equi	Horses	Colon, caecum	Three anterior flagella, undulating membrane, slender axostyle, 4–6.5 μm long	Nonpathogenic
T. foetus	Cattle	Genital tract	Pear-shaped, three anterior flagella, undulating membrane, axostyle, 10–25 μm long, 3–5 μm wide	Tritrichomonas abortion
T. suis	Pigs	Intestine	Three anterior flagella, undulating membrane, axostyle, 8–10 μm long	Nonpathogenic
T. eberthi	Chickens	Caecum	9 μm long, 4–6 μm wide	Nonpathogenic
Retortamonas				
R. ovis	Sheep	Intestine	Pear-shaped, two flagella of length of body, 5.2 μm long, 3–3.7 μm wide	Nonpathogenic
R. cuniculi	Rabbits	Caecum	Posterior flagellum thick, one-half as long as body; 7.5–13 μm long; 5.5–9.5 μm wide	Nonpathogenic
R. intestinalis	Humans	Intestine	Pleomorphic, 4–9 μm long, 3–4 μm wide, cytostome one-third length of body	Diarrhea?
Pleuromonas jaculans	Chickens	Caecum	One short anterior flagellum, one long trailing flagellum, 5–12 μm long, 5 μm wide	Nonpathogenic
Pentatrichomonas				
P. hominis	Humans, other primates, dogs, cats	Intestine	Three to five anterior flagella	Nonpathogenic
P. gallinarum	Turkeys, chickens, guinea-fowls	Caecum	Pear-shaped, five anterior flagella, one trailing flagellum, axostyle, 6–8 μm long	Caecal lesions
Histomonas meleagridis	Chickens, turkeys, ducks, geese	Caecum, liver, other tissues	Pleomorphic, four flagella	Blackhead histomoniasis
Monocercomonas				
M. ruminantium	Sheep, cattle	Rumen, prepuce	Three anterior flagella, 12–14 μm long, 8–10 μm wide	Nonpathogenic
M. cuniculi	Rabbits	Caecum	Nucleus ellipsoidal, axostyle protrudes from posterior end, 5–14 μm long	Nonpathogenic
M. gallinarum	Chickens	Caecum	5–8 μm long, 3–4 μm wide, pear-shaped, three anterior flagella, one longer trailing flagellum	Nonpathogenic
Chilomastix mesnili	Humans	Large intestine	Pear-shaped, three free anterior flagella, one flagellum in cytostome	Nonpathogenic
Embadomonas intestinalis	Humans	Large intestine	Slipper-shaped, two unequal anteriorly directed flagella, large cytostome	Nonpathogenic
Tricercomonas intestinalis	Humans	Large intestine	Pear-shaped, three anterior flagella, one trailing flagellum	Nonpathogenic
Giardia lamblia	Humans	Intestine	Pear-shaped, bilaterally symmetrical, two nuclei, two axostyles, four pairs of flagella	Diarrhea
Hexamita spp.	Turkeys, pigeons, chickens	Small intestine	Elongate, two anterior nuclei, two pairs anterior flagella, two trailing flagella	Severe enteritis
Callimastix				
C. equi	Horses	Colon, caecum	Kidney-shaped, 12–15 flagella at hilus, 12–18 μm long, 7–10 μm wide	Nonpathogenic
C. frontalis	Cows	Rumen	Anterior end disc-shaped, 12 flagella, 30 μm long	Nonpathogenic

hosts for long periods, while others are apparently beneficial to their hosts. Other flagellates are known to be highly pathogenic. A list of flagellates of significance in human and veterinary medicine is given in Table 4.5.

The Mastigophora includes a large number of heterogeneous forms undoubtedly divergent in their immediate lines of descent. Although all bear flagella, they are, in other respects, quite different.

MORPHOLOGY

The Flagellum

The specific organelle common to all mastigophorans is the flagellum. This locomotor structure is also found on intermediate forms of certain amoebae and other groups of protozoans but is characteristic of the definitive form of all flagellates. The number of flagella ranges from one to many, depending on the species.

The single flagellum is a filamentous cytoplasmic projection. When seen under the light microscope, the cytoplasm can be observed to form a sheath around the axial filament or the **axonema** (Fig. 4.18). The axonema arises from a **kinetosome** (basal body) (Fig. 4.19) situated within the body proper.

Among members of the family Trypanosomatidae, a unique organelle known as the **kinetoplast** is closely associated with and is usually posterior to the kinetosome. The kinetoplast consists of an electron-dense rod of spirally coiled DNA within an envelope and is continuous with an adjacent mitochondrion (Fig. 4.20). Under the light microscope, the kinetosome and kinetoplast usually appear as a single structure. The cytoplasmic sheath is contractile and tapers toward the free end. When studied with the electron microscope, the axial filament is composed of two central

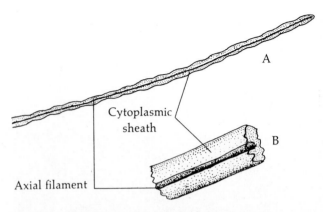

Fig. 4.18. Flagellum as seen with the light microscope. A. Single flagellum showing cytoplasmic sheath surrounding axial filament. **B.** Enlarged drawing of one segment.

and nine peripheral microtubules (Fig. 4.21). Interested readers are referred to Lewis (1975) and Scholtyseck (1979) for greater detail.

Studies on the kinetosome by use of the electron microscope have revealed that it consists of a short cylinder comprised of nine groups of three microtubules. The microtubules in each triplet are so close together that they share a common wall where they touch (Fig. 4.19). If a kinetosome is viewed from the base, distally, it is clear that the triplets are rotated inward in a clockwise fashion, with fine filaments projecting in a cartwheel-like manner toward a central hub from each triplet (Fig. 4.22). The "lumen" of each

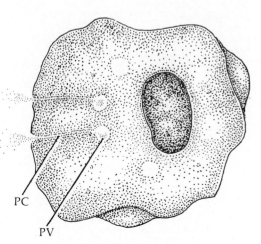

Fig. 4.17. Pinocytosis. Cell undergoing pinocytosis. Notice occurrence of pinocytotic canal (PC) and pinocytotic vesicle (PV).

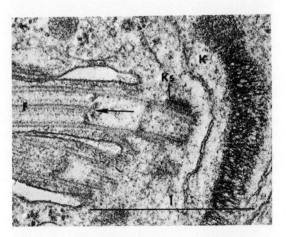

Fig. 4.19. Kinetosome. Electron micrograph showing kinetosome at base of flagellum of *Trypanosoma cruzi* epimastigote. (After Aikawa and Sterling, 1974; with permission of Academic Press.) F, flagellum; K, kinetoplast; KS, kinetosome; arrow pointing to basal plate.

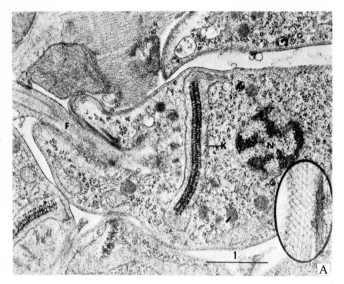

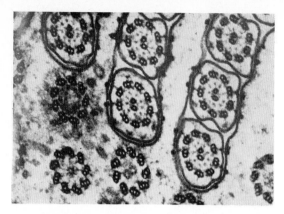

Fig. 4.21. **Electron micrograph of cross-sections of flagella of** *Pseudotrichonympha* **from the gut of a termite.** Note that the median fibrils are single and the marginal ones are double. Sections through kinetosomes are seen below the flagella. (After Gibbons and Grimstone, 1960.)

Fig. 4.20. **Kinetoplast. A.** Electron micrograph showing kinetoplast of *Trypanosoma cruzi* epimastigote. Insert: para-flagellar rod showing latticelike appearance. (After Aikawa and Sterling, 1974; with permission of Academic Press.) F, flagellum; K, kinetoplast; N, nucleus. **B.** Electron micrograph of promastigote of *Leishmania* sp. showing kinetoplast. **C.** Drawing of B. CEM, cell membrane; DB, dark bodies; ER, endoplasmic reticulum; F, flagellum; FP, flagellar pocket; KI, kinetoplast; MI, mitochondrion; MT, microtubules; MU, multilamellar structure; N, nucleus; V, vesicle. (After Scholty-seck, 1979; with permission of Springer-Verlag.)

kinetosome is open to and appears to be continuous with the cytoplasm of the body. The skewing of the kinetosomal triplets becomes less pronounced distally, and the cartwheel disappears. At the proximal end, one microtubule in each triplet tapers to an end, and the kinetosome is closed by an electron-dense discoid structure known as the **terminal plate**. This plate is often situated very close to the cell surface where the shaft of the flagellum protrudes.

The kinetosome is of great importance in the formation and function of the flagellum. Not only are the nine peripheral pairs of microtubules in each axoneme direct extensions of two microtubules in each triplet in the kinetosome, but the kinetosome is also respon-

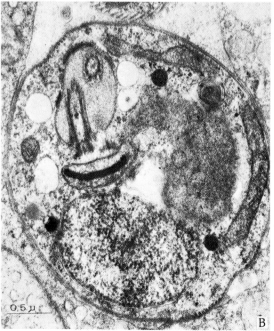

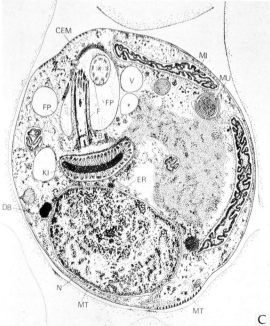

sible for the formation of the axoneme. When the flagellate divides, the kinetosomes replicate themselves (or serve as organizing centers for such replication) prior to formation of the new flagellum. Thus, the kinetosome closely resembles the centriole of other animal cells.

In some flagellates, a Golgi apparatus is closely associated with the kinetosome. If a periodic fibril, the **parabasal filament**, connects the Golgi to the kinetosome, the Golgi apparatus is referred to as a **parabasal body**. Its function is believed to be similar to that of Golgi apparatuses in other cells. More than one parabasal body may be present in each flagellate, and their sizes fluctuate with the availability of nourishment to the organism (Grimstone, 1959). During certain periods, such as when the flagellate divides, the parabasal body may become invisible under the light microscope.

Motility of Flagella. Flagellar movement aids in the propelling and directing of the organism's movement, assists in procuring food in some instances, and may perhaps even serves tactile and secretory functions, i.e., emit mating substances (**gamones**).

The mechanics of flagellar action has long been of interest to cell biologists. Flagellar undulations commonly originate at the base and move toward the free end. Although various types of flagellar movements are known, including movement at only the tip, the source of the required energy and the mechanism(s) that controls the movements remain incompletely understood. Some evidences suggest that the two central microtubules transmit excitation, while the nine peripheral pairs of microtubules are the sites of adenosine triphosphate (ATP) splitting (Hayashi, 1961), mediated by ATPase found in the protein constituents of the flagellum (Gibbons and Rowe, 1965). Probably the best hypothesis for flagellar movement is that the basic mechanism is essentially the same as that of muscle contraction—that is, there exists a special arrangement of protein chains whose interactions cause contraction and extension by telescoping. The energy is supplied by ATP. The volume edited by Hatano *et al.* (1979) should be consulted by those desiring further details.

Other Organelles

Certain parasitic flagellates possess various specialized organelles that serve specific functions. These organelles, considered below, are utilized also as diagnostic features.

The Axostyle. The axostyle is a supporting structure embedded along the longitudinal axis in the cytoplasm (Fig. 4.23). It may appear as a fine filament in some species and as a broad hyaline rod in others. It may extend partially along, totally along, or beyond the length of the body, and may possess a broad, rounded anterior end known as the **capitulum**. In stained specimens, the axostyle may appear homo-

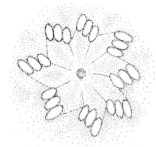

Fig. 4.22. **Kinetosome structure.** View of a kinetosome in cross-section from the base distally, showing skewed triplets.

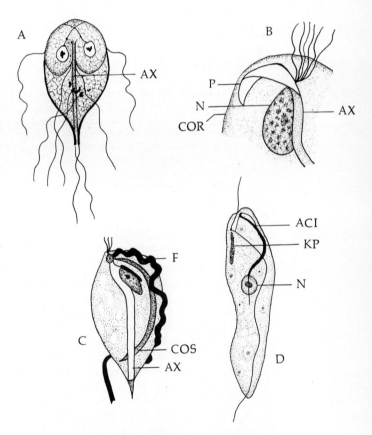

Fig. 4.23. **Specialized organelles of flagellates. A.** *Giardia lamblia*, an intestinal parasite of humans, showing axostyle. **B.** *Hexamastix citelli*, showing pelta and axostyle. (Modified after Kirby and Honigberg, 1950.) **C.** *Tritrichomonas muris*, an intestinal parasite of mice, showing costa and axostyle. (Modified after Kirby and Honigberg, 1950.) **D.** *Cryptobia helicis* from the gonads of various pulmonate snails, showing aciculum. (Modified after Kozloff, 1948.) ACI, aciculum; AX, axostyle; COR, cortex; COS, costa; F, flagellum; KS, kinetosome; N, nucleus; P, pelta.

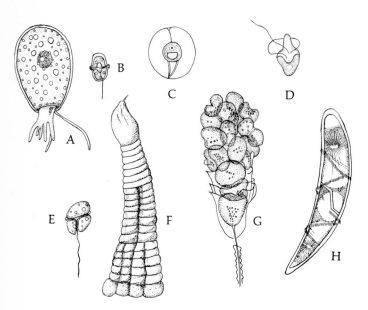

Fig. 4.24. Parasitic dinoflagellates. A. *Oodinium ocellatum,* recently detached from host's gill. **B.** Free-living flagellated stage of *O. ocellatum.* (Redrawn after Nigrelli, 1936.) **C.** *Chytriodinium parasiticum,* found in copepod eggs. (Redrawn after Dogiel, 1906.) **D.** *Duboscquella tintinnicola* swarm cell. (Redrawn after Duboscq and Collin, 1910.) **E.** Free-living swarm cell of *Haplozoon clymellae.* **F.** *Haplozoon clymellae,* mature colony. (Redrawn after Shumway, 1924.) **G.** *Apodinium mycetoides,* swarm cell formation. (Redrawn after Chatton, 1952.) **H.** *Blastodinium spinulosum.* (Redrawn after Chatton, 1952.)

geneous, granular, naked, or sheathed. When stained with iron hematoxylin, only some axostyles are deeply stained, thus suggesting differences in chemical composition. In certain species of woodroach gut-inhabiting flagellates, the numerous slender axostyles are known as **axial filaments**.

When an axostyle-bearing flagellate divides by longitudinal fission, new axostyles are formed *de novo* in the daughter cells.

Studies with the electron microscope have revealed that the tubelike axostyle is comprised of a sheet of microtubules. This complex may extend from the region of the kinetosomes to the posterior end, where it may protrude through the plasma membrane.

The Pelta. The pelta is a crescent-shaped membrane that can be demonstrated in certain species of flagellates, such as *Trichomonas,* by staining with Bodian's silver or some comparable stain (Fig. 4.23). Again, electron microscopy has revealed that the pelta is comprised of microtubules. This is a long and wide organelle that supports the anterior end of cer-

tain flagellates. It is connected with the anterior terminal of the axostylar capitulum.

The Costa. The costa is a thin, rodlike structure running along the base of the undulating membrane of certain flagellates, such as *Trypanosoma* spp. and *Tritrichomonas* spp. (Fig. 4.23). It too is probably a supporting structure. Its fine structure has been reported by Inoki *et al.* (1959).

The Aciculum. The aciculum, a bent, finlike organelle, has been demonstrated in *Cryptobia helicis,* a flagellate parasitic in the reproductive organ of pulmonate snails (Fig. 4.23). The function of this structure is uncertain; but it may be a supporting organelle.

Myonemes. In several parasitic flagellates, contractile myonemes have been demonstrated, which may be thought of as a primitive type of muscle fibril that enables the protozoans to contract and extend their bodies. More commonly, myonemes are found in various free-living ciliates such as *Stentor* and *Zoothamnium,* and in free-living flagellates such as *Leptodiscus* and *Craspedotella.*

Other organelles and structures found in parasitic flagellates are common to all protozoans and have been discussed previously.

SYSTEMATICS AND BIOLOGY

The subphylum Mastigophora is subdivided into two classes—Phytomastigophorea (the "plantlike" flagellates) and Zoomastigophorea (the "animal-like" flagellates). The diagnostic characteristics of both classes and the subordinate taxa are given at the end of the chapter.

CLASS PHYTOMASTIGOPHOREA

Most of these plant-like flagellates are free-living, being capable of holophytic, mixotrophic, holozoic, or saprozoic existences in various aquatic habitats. The essay by Dodge (1979) is recommended to those interested in this group.

All the symbiotic members of the Phytomastigophorea belong to the order Dinoflagellida, with a few exotic exceptions. As examples, species of *Astasia* of the order Euglenida have been reported in larger protozoans such as *Spirostomum, Amoeba proteus,* and *Stentor* and *Chrysidella* of the order Prymnesiida has been found in foraminiferans and radiolarians (see Ball, 1968, for complete review).

ORDER DINOFLAGELLIDA

The parasitic dinoflagellates all belong to the family Blastodidiidae and are ecto- and endoparasites of both plants and animals. Representative genera include *Oodinium, Chytriodinium, Duboscquella, Haplozoon, Apodinium,* and *Blastodinium.* Discussion of the biology of a few representative species follows.

Oodinium ocellatum. This flagellate is ectoparasitic on the gill filaments of marine fish, especially coral fish (Fig. 4.24), causing the so-called velvet disease which may reach epizootic proportions in aquariums (Laird, 1956). This small (average 60 by 50 μm), ovoid organism is encased within a shell or test that completely surrounds the body cytoplasm except for a terminal aperture through which a broad flagellum and cytoplasmic processes protrude. It is by these processes that the organism is attached to its host. Within the cytoplasm is a large spherical nucleus along with many chromatophores, starch granules, and a stigma. The smaller organisms remain attached to the fish while growing and finally drop off. They ultimately grow as large as 150 μm in diameter. The unattached organism withdraws its flagellum and processes, and the aperture is closed off by a cellulose secretion. Within the now completely sealed test, the organism undergoes division and eventually gives rise to as many as 128 cells. Each one of these daughter cells becomes flagellated and divides once more. The resulting flagellated **swarm cells** (also known as swarmers) rupture out of the test and actively seek the gills of a fish host, to which they become attached. Once situated, each swarm cell develops a test, extends cytoplasmic processes and a flagellum, and begins to increase in size. The closely related *O. poucheti*, an ectoparasite of tunicates, and *O. limneticum*, a parasite of freshwater fishes, have similar life cycles. This last species bears green chromatophores but no flagellum or stigma. The nature of the relationship between the various species of *Oodinium* and their hosts remains uncertain; however, it appears to be an obligatory one.

The "velvet disease" may lead to death resulting from suffocation, since the parasites are commonly attached in large number to the surface of the gills.

Haplozoon clymellae. *H. clymellae* is an intestinal parasite of the polychaete *Clymella torquata*. The mature forms are gathered in colonies of 250 or more individuals arranged in a pyramidal or linear fashion (Fig. 4.24). The swarm cells are unicellular with two flagella.

Chytriodinium parasiticum. This parasite was originally reported from the eggs of copepods (Fig. 4.24). The young grow at the expense of the eggs' contents, and on reaching maximal size, divide into many daughter cells. Each daughter cell gives rise to swarmers, each of which bears four flagella. These swarmers escape from the egg to seek new copepod eggs.

Apodinium spp. These dinoflagellates are ectoparasitic on tunicate gill slits and are binucleate and colorless (Fig. 4.24).

Blastodinium. Endoparasitic in copepods, *Blastodinium* adults are spindle shaped and arched, and the noncellulose envelope often bears two spiral rows of bristles (Fig. 4.24).

ORDER EUGLENIDA

Included in the order Euglenida is the genus *Colacium*. At least one member of this genus has been reported as a symbiont of aquatic arthropods. Most of the other members of the Euglenida are free-living holophytes, that is, they can synthesize their nutrient requirements, usually by photosynthesis.

Willey *et al.* (1970) reported nonflagellated *Colacium* in the rectums of damselflies (Fig. 4.25) in interdunal ponds in Indiana during the winter months, but these hosts are devoid of *Colacium* during the other months of the year. It is known that damselflies complete one life cycle per year. Eggs are laid during the summer; they then hatch, and the young nymphs grow rapidly until their habitats are frozen over. The overwintering nymphs grow very little. In the spring they resume growth and the adults emerge in June. The damselflies and *Colacium* live separately during

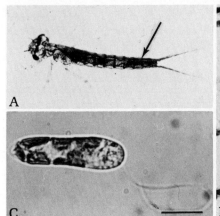

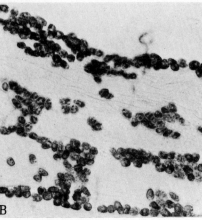

Fig. 4.25. *Colacium* **in damselfly nymphs. A.** Nymphal damselfly, *Anomalagrion hastatum,* carrying euglenoids in its rectum (arrow). **B.** Rectum removed from damselfly showing arrangement of euglenoids. **C.** Flagellated *Colacium* (Scale represents 10 μm). (After Willey *et al.,* 1970. *Science* **170**, 80–81. © 1970 by the American Association for the Advancement of Science.

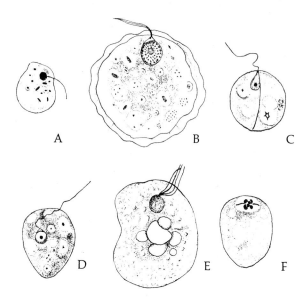

Fig. 4.26. *Histomonas meleagridis.* **A.** Specimen from caecum of chicken. **B.** Large specimen from pheasant. **C.** Rounded form. **D.** Amoeboid form. **E.** Specimen from caecum of chicken, showing food vacuoles. **F.** Stage with dividing nucleus. (Redrawn after Wenrich, 1943, and Bishop, 1938.)

the spring, summer, and fall. Sometime toward the end of fall, the *Colacium* is attracted to the insect and enters through the anus. The flagellate loses its flagellum in preparation of entering the host and lacks a functional flagellum while living as an endosymbiont. It escapes during the spring through the anus, the evacuation apparently being triggered by the rising temperature. A flagellum develops on each escaping flagellate.

The nature of the relationship between this species of *Colacium* and its host remains uncertain; it is possible that the host derives some of its required oxygen from its photosynthetic euglenids. Since the oxygen content in frozen pond water is low compared with that during the warmer months, this seasonal symbiosis may be beneficial, if not essential.

Other species of *Colacium* are ectoparasitic on aquatic invertebrates, especially arthropods.

CLASS ZOOMASTIGOPHOREA

A large number of zoomastigophoreans are parasitic. The class is divided into eight orders, seven of which include parasites (see classification at the end of this chapter).

ORDER TRICHOMONADIDA

Included in the order Trichomonadida are the families Monocercomonadidae and Trichomonadidae. Mem-

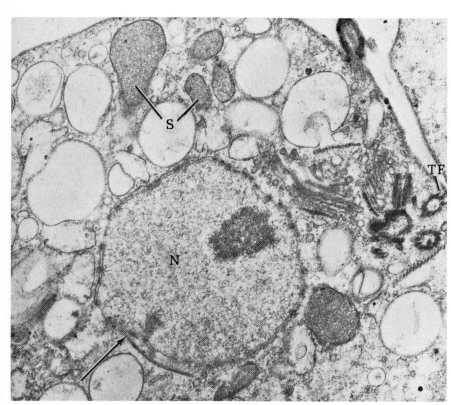

Fig. 4.27. *Histomonas meleagridis.* Electron micrograph of a portion of a trophozoite showing cytoplasm-filled sacs that resemble mitochondria superficially. × 24,500. (After Schuster, 1968.) S, Cytoplasm-filled sacs; TF, transitional fibers; N, nucleus.

bers of the first are parasites of insects; however, *Histomonas* and *Dientamoeba* are parasites of vertebrates.

Histomonas. *Histomonas meleagridis* is one of the better known parasitic trichomonads (Fig. 4.26). Not only is this species interesting in itself, but it is also of great economic importance, because it is the etiologic agent of "blackhead" enterohepatitis, also known as histomoniasis, which is lethal to chickens, turkeys, and other fowl. It is worldwide in distribution. In the United States, histomoniasis was responsible for a loss of $9.3 million to the chicken and turkey industries during 1951–1960.

This parasite is interesting because it has both amoeboid and flagellated stages. The organism as found in the caecum of birds (which is rare) or in culture is a flagellated amoeboid form. These small amoeboid forms measure 5–30 μm in diameter and always possess one flagellum; however, there are usually four kinetosomes, the basic number in trichomonads. Food vacuoles present in the amoeboid form include blood cells, bacteria, or starch granules. Studies with the electron microscope have revealed the presence of a pelta, a V-shaped parabasal body, and a structure resembling an axostyle, but the absence of mitochondria (Fig. 4.27).

The form of *H. meleagridis* within tissues (Fig. 4.28) has no flagellum, although kinetosomes occur near the nucleus. Three extracellular, intertissue stages have

been described from the bird host: the **invasive**, **vegetative**, and **resistant** stages.

The invasive stage (or form) measures 8–17 μm in diameter and resembles a small amoeba. It feeds by phagocytosis, and the food vacuoles contain particles but not bacteria. This stage is found in new lesions in the bird's caecal wall and liver.

The vegetative stage (or form) occurs in older lesions and is somewhat larger, measuring 12–21 μm wide. This stage, less active than the invasive stage, is commonly tightly packed in liver or caecal lesions. After dissolving host tissues with secreted proteolytic enzymes, the vegetative form feeds on predigested tissues.

The so-called resistant stage (or form) is no more resistant than the other stages. It is compact, 4–11 μm in diameter, and the ectoplasm forms a dense layer. This stage, found scattered or packed together in hepatic lesions, also feed on predigested host tissues.

In turkeys, infections are acquired either through the ingestion of the parasite from the fecal droppings of infected birds or through the ingestion of eggs of the coparasite *Heterakis gallinae*, a nematode. These eggs are very resistant and may survive for several months in soil. Hence, a contaminated poultry range may remain dangerous to turkeys for many months.

The role of *Heterakis gallinae* in the transmission of *Histomonas* is interesting. The protozoan, ingested by the worm, undergoes development and multiplication in the nematode; thus, the worm should be considered a true intermediate host, After initially invading the nematode's intestinal cells and multiplying, the protozoans break out into the pseudocoel and invade the worm's ovary. Here they feed and multiply extracellularly and move down the ovary with the developing oogonia, then penetrate the oocytes. Feeding and multiplication continue in the oocytes and newly formed eggs. When infected nematode eggs are passed out of a bird host, the protozoan eventually invades the tissues of the developing larval nematode, especially those of the digestive and reproductive systems. When the larvae-containing nematode eggs are ingested by a bird, the protozoan is carried into a new definitive host. This method of transmission is known as **transovarian passage**.

In addition to the methods of transmission already mentioned, Lund *et al.* (1966) have reported that earthworms, *Lumbricus*, *Allolobophora*, and *Eisenia*, can serve as vectors for transmission to birds as the result of ingesting infected *Heterakis gallinae* eggs.

Young birds are more often infected than older ones. Turkeys, especially young ones, are readily

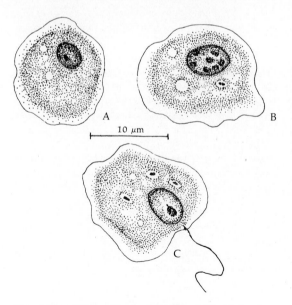

Fig. 4.28. *Histomonas meleagridis.* **A.** Specimen from tissue (liver). Note the absence of flagellum. **B.** Specimen in transitional stage from lumen of caecum. Note the presence of pseudopodia and the distribution pattern of chromatin, which suggest the initiation of division. No flagellum, however, is present. **C.** Specimen that has adapted to the caecal environment. Note the presence of flagellum.

killed by this parasite, although infected chickens, especially older ones, usually survive and become carriers. Therefore, it is not advisable to raise turkeys on a range where chickens have been kept. Birds that recover from histomoniasis are immune to reinfection. In addition, as suggested above, susceptibility decreases with age. The review by McDougald and Reid (1978) on *H. meleagridis* and related flagellates should be consulted by those desiring more details.

The pathogenicity of *H. meleagridis* appears to be a synergistic effect; that is, when the protozoan is present alone in the avian host, no tissue destruction occurs (Doll and Franker, 1963), but if the protozoan is accompanied by certain species of bacteria, the classical symptoms of histomoniasis develop (Franker and Doll, 1964; Bradley and Reid, 1966). Thus, avian histomoniasis, like in the case of mammalian amoebiasis (p. 161), is apparently the product of the combined effort on the part of a protozoan and certain bacteria.

Dientamoeba. *Dientamoeba fragilis* (Fig. 4.29), a fairly common, essentially nonpathogenic parasite of humans, has also been reported from the intestine of macaque monkeys in the Philippines. Originally thought to be a true amoeba, it has been shown by Camp *et al.* (1974) to be more closely related to *Histomonas* of the family Monocercomonadidae. Its geographic range is probably worldwide, infecting about 4% of the human population. The trophozoite, which is 4–18 μm in diameter, is actively amoeboid. Both one and two nuclei occur. The nuclei contain a prominent endosome surrounded by four or five minute chromatin granules. The binucleate forms are actually dividing individuals arrested at telophase. The cystic form is unknown, and transmission from one host to another presumably occurs via the trophozoite, which is extremely viable and capable of motility up to 48 hours after leaving the host in feces. Dobell (1940) had suggested that transmission may

be brought about with trophozoites becoming lodged in the eggs of nematode coparasites that are ingested by the host. Indeed, the pinworm (*Enterobius vermicularis*) has been suspected. This hypothesis is very reasonable in light of the role played by *Heterakis gallinae* in the life history of *Histomonas meleagridis* in fowls.

D. fragilis is pathogenic, causing episodes of flatulence with mild diarrhea.

Trichomonas and *Pentatrichomonas.* As stated, the second family of the order Trichomonadida is Trichomonadidae. It includes a large number of flagellates, most, if not all, are parasitic.

Three species of trichomonads are known to parasitize humans, although only one, *Trichomonas vaginalis*, is pathogenic. The other two human-infecting species are *Trichomonas tenax* from the mouth and *Pentatrichomonas hominis* from the colon. A rather high proportion of the world's population harbors one or more of these three species. Although no exact figures are available, various conservative samplings suggest that 25% of women and 10% of men are infected with *T. vaginalis*; 10% of the population harbor *T. tenax*; and 2% harbor *P. hominis*. In addition to humans, various simians are infected with these three species, with *P. hominis* being the most common and *T. vaginalis* being relatively rare. It is likely that both human and simian hosts have inherited these parasites, so to speak, from their common ancestral stock.

The three human-infecting species of trichomonads are morphologically similar, but not identical, and show some physiologic differences as well. For example, *T. vaginalis* requires a higher pH, ranging from 4.0 to 5.0, for optimal growth. It has been suggested that *T. tenax* was the first of the three to infect hominoids and that invasion of the colon by *P. hominis* was established when some of the more adaptable specimens of *T. tenax* were swallowed. The modern *T. vaginalis* may well have originated as a wanderer from the colon.

Trichomonas tenax (= *T. buccalis*) is a fairly common species found in the tartar and gum of the mouth (Fig. 4.30). It measures 5–16 μm long by 2–15 μm wide

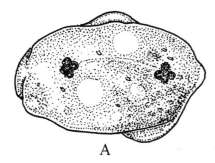
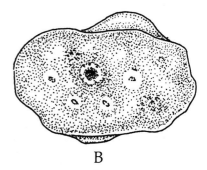

A   B

Fig. 4.29. *Dientamoeba fragilis.* **A.** Binucleate trophozoite showing four chromatin bodies within each telophase nucleus. **B.** Uninucleate trophozoite.

and bears four anteriorly directed flagella and a fifth associated with an undulating membrane. It is generally nonpathogenic.

Trichomonas vaginalis, measuring 7–32 μm long by 3–12 μm wide, also with four anterior flagella and another associated with an undulating membrane, is often found in the vaginal tract of women, where it commonly causes inflammation and whitish to yellowish secretion (Fig. 4.30). This parasite is capable of secreting a toxic substance that injures cells grown in tissue culture. Studies have revealed that different strains of *T. vaginalis* exist, most being of such low pathogenicity that the infected person is essentially asymptomatic. Other strains, however, cause acute inflammation, itching, and a copious white discharge swarming with flagellates. *T. vaginalis* feeds on bacteria, leucocytes, and cell exudates and is itself phagocytosed by host monocytes. Although generally thought of as a parasite of women, *T. vaginalis* is also fairly commonly found in the male urethra and in the prostate gland, suggesting that males serve as vectors of trichomoniasis by means of sexual intercourse.

Pentatrichomonas hominis is a highly motile and flexible species that measures 5–20 μm long and 7–14 μm wide, with a cytostome near the anterior end and with three to five, usually five, anteriorly project-

ing flagella (Fig. 4.30). As is the case with other trichomonads, no encysted form is known. Transmission from the intestinal tract of one human to another is thought to occur directly via the flagellated form. To validate this hypothesis, the eminent protozoologist Dobell swallowed a culture of a strain of *P. hominis* from a monkey, and became infected. The apparent lack of cysts does not appear to be a handicap, because the trophozoites can endure considerable changes in environmental conditions although they cannot survive very well in highly diluted sewage. Specimens can survive for at least 24 hours in feces-contaminated milk, suggesting that transmission could occur through contaminated food and drink. In addition to being a parasite of humans and simians, *P. hominis* can infect dogs, cats, mice, and other rodents, such hosts serving as natural reservoirs for human infections.

The review by Honigberg (1978) of trichomonads important in human medicine is recommended to those desiring greater details.

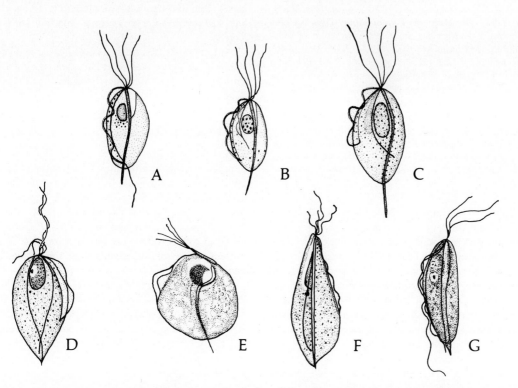

Fig. 4.30.   Some representative flagellates. A. *Pentatrichomonas hominis* from human, 5–20 μm long. **B.** *Trichomonas tenax* from human mouth, 5–16 μm long. **C.** *Trichomonas vaginalis* from human, 7–32 μm long by 10–20 μm wide. (**A–C** redrawn after Wenrich, 1944.) **D.** *Trichomonas gallinae* from a hawk, 6–19 μm long by 2–9 μm wide. **E.** *Trichomonas gallinae* from a turkey. **F.** *Trichomonas gallinae* from a domestic pigeon. (**D–F** redrawn after Stabler, 1941.) **G.** *Tritrichomonas foetus* from cattle, 10–15 μm long. (Redrawn after Wenrich and Emmerson, 1933.)

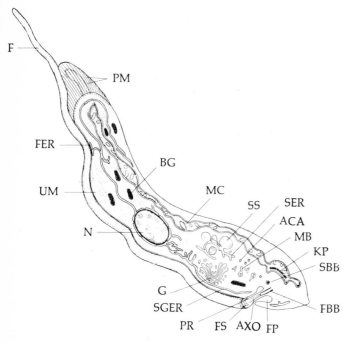

F
PM
FER
BG
UM
MC
SS
SER
ACA
N
MB
KP
SBB
G
SGER
FBB
PR
FS
AXO
FP

Fig. 4.31. Structure of *Trypanosoma*. Drawing of a specimen of *Trypanosoma* showing fine structure as determined by electron microscopy. ACA, acanthosome; AXO, axonome; BG, bacilliform dense granular body; F, flagellum; FBB, first basal body; FER, flagellum-associated endoplasmic reticulum; FP, flagellar pocket; FS, flagellar swelling; G, Golgi apparatus; KP, kinetoplast; MB, multivesicular body; MC, mitochondrial canal; N, nucleus; PM, pellicular microtubules; PR, paraxial rod; SBB, second (barren) basal body); SER, smooth endoplasmic reticulum; SGER, subtending granular endoplasmic reticulum; SS, sac of secretion; UM, undulating membrane. (Modified after Vickerman, 1969.)

Avian Trichomoniasis. In fowls, such as pigeons, turkeys, and chickens, *Trichomonas gallinae* is a fairly common parasite found in the upper digestive tract. This pear-shaped flagellate, which measures 6.2–8.9 μm long, is the causative agent of avian trichomoniasis (Fig. 4.30). This disease is not fatal in some cases, and yet extremely lethal in others, leading some to believe that two strains exist. On the other hand, some suspect that this parasite has different effects on different hosts (Levine and Brandley, 1940), since new hosts (canaries, hawks, quails, and bobwhites) experimentally infected become ill. Stabler and Engley (1946), through a series of experiments involving bacteria-free (axenic) and bacteria-bearing (monoxenic) cultures of *T. gallinae*, have demonstrated that some of the squabs infected with the bacteria-free cultures died. The other type of culture had no drastic effect.

These experiments suggest that *T. gallinae* is pathogenic only in the absence of bacteria.

Some immunity is conferred on birds that have been previously exposed to *T. gallinae*. For example, pigeons that have survived infections as squabs are symptomless carriers. There also appears to be cross-immunity between the various strains. Infection with a relatively harmless strain produces immunity against virulent strains.

Various other species of trichomonads are known in an array of hosts, including amphibians, rodents, and termites. The review by Honigberg (1978) of species of veterinary importance is recommended.

Tritrichomonas. In veterinary protozoology, *Tritrichomonas foetus* (Fig. 4.30) is an important pathogen because infections in cows result in abortions by destruction of the placental attachments and the removal of the aborted fetus intact, and by destruction of the fetal membranes that are retained within the uterus. In the first type of abortion, the cow may reconceive after the loss of the fetus; however, in the second type, the cow becomes permanently sterile as the result of chronic endometritis. Transmission of *T. foetus* infections occurs during copulation, and bulls are known to be susceptible. The preputial cavity is the preferred site of infection. Once a bull becomes infected, the infection becomes permanent and can be transmitted to other cows during mating. Sheep, deer, and hamsters can also become infected with this flagellate.

As in the case of *Trichomonas gallinae*, a number of strains of *Tritrichomonas foetus* have been distinguished serologically. Also, immunity is conferred although it is generally incomplete, i.e., reinfections are possible but these are usually less severe.

ORDER KINETOPLASTIDA

This order includes many species parasitic in plants, invertebrates, and vertebrates. The kinetoplastids are generally small flagellates with one to four flagella. The body is plastic, but not amoeboid. Reproduction occurs by longitudinal fission, and during the life cycle, individuals can assume more than one form (the phenomenon is known as **polymorphism**). This is especially true among members of the family Trypanosomatidae.

Parasitic members of this order are included in three families—the monoflagellar Trypanosomatidae (suborder Trypanosomatina); the biflagellar Cryptobiidae (suborder Bodonina), members of which possess an undulating membrane; and the biflagellar Bodonidae (suborder Bodonina), members of which do not possess an undulating membrane.

Family Trypanosomatidae

Members of this family are all parasitic and are found in their hosts' blood and other tissues. Their bodies are typically elongate and more or less flattened, and a single flagellum arises from a kinetosome near which is located a kinetoplast, The kinetosome varies in location from the anterior, to the middle, to the posterior end of the body. When it is located posteriorly, the basal portion of the flagellum forms the outer margin of the undulating membrane (Fig. 4.31).

Developmental and Definitive Forms. Life cycles of members of the Trypanosomatidae include more than one stage. The designations of these stages are presented below.

Amastigote. The **amastigote** (Fig. 4.32) is rounded or oval, with a nucleus, kinetosome, and kinetoplast. The flagellum is reduced to a tiny fibril that is totally embedded in the cytoplasm, and hence no free flagellum occurs.

Promastigote. The **promastigote** (Fig. 4.32) is elongate and has a comparatively large nucleus. A free flagellum arises from a kinetosome located near the anterior end of the body.

Opisthomastigote. The **opisthomastigote** (Fig. 4.32) is limited to members of the genus *Herpetomonas.* It is characterized by a postnuclear kinetosome and kinetoplast, with a flagellum arising from the former. The flagellum passes through the body and emerges from the anterior terminal of the body as a free structure.

Epimastigote. The **epimastigote** (Fig. 4.32) is characterized by a juxtanuclear kinetosome and kinetoplast and has a flagellum that arises from the former and emerges from the side of the body to run along a short undulating membrane.

Trypomastigote. The **trypomastigote** (Fig. 4.32) is characterized by its postnuclear kinetosome and kinetoplast. The flagellum arising from the kinetosome emerges from the side of the body, runs along a long undulating membrane, and is free anteriorly.

Choanomastigote. The **choanomastigote** (Fig. 4.32) is the formal designation given to the so-called barleycorn stage of members of the genus *Crithidia.* It is characterized by an antenuclear kinetoplast, and the flagellum arising from it emerges to the exterior through a wide funnel-shaped reservoir.

As illustrated in Fig. 4.32, members of the Trypanosomatidae have at least two forms in their life cycles. For example, members of the genus *Leishmania* exhibit both the promastigote form (in insect hosts) and the amastigote form (as intracellular parasite of the vertebrate reticuloendothelial system).

Representative Trypanosomatids

Genus *Leishmania*. *Leishmania donovani, L. tropica*, and *L. braziliensis* are the three human-infecting species. These are indistinguishable morphologically. In the vertebrate host (humans, dogs, etc.), the parasites are ovoid, measure 1.5–3 μm in diameter, and possess a distinct nucleus, kinetosome, kinetoplast, flagellar fibril, and cytoplasmic vacuoles. All are tissue parasites of the reticuloendothelial system in their vertebrate hosts.

Leishmania donovani*. This organism (Fig. 4.33) is the causative agent of visceral leishmaniasis, Dumdum fever, or kala-azar, an often fatal disease of humans (Fig. 4.33). In the mammalian body, the parasite is found in macrophages, certain leucocytes, spleen,

---

*For a detailed discussion of subspecies of *Leishmania donovani* that occur in the New World, see Lainson (1983).

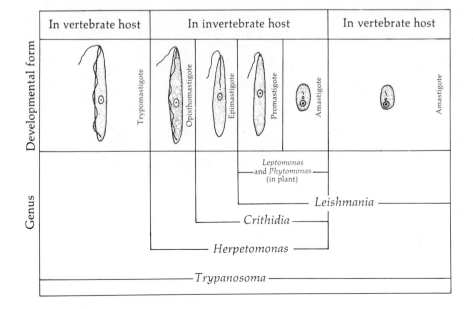

Fig. 4.32. Morphologic differences between members of the family Trypanosomatidae and forms attained by members of various genera during their development.

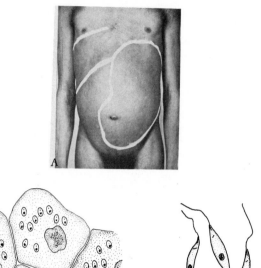

Fig. 4.33. *Leishmania donovani.* **A.** Chronic kala-azar due to *L. donovani* depicting extreme splenomegaly and hepatomegaly. (After Hunter, Frye, and Swartzwelder, 1960.) **B.** Amastigotes (Leishman bodies) in endothelial cell of spleen. **C.** Flagellated promastigotes from culture.

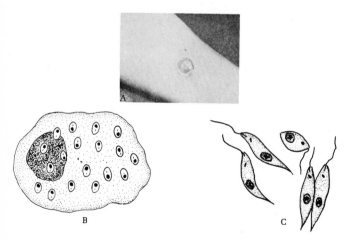

Fig. 4.34. *Leishmania tropica.* **A.** Cutaneous sore due to *L. tropica* infection. (After Ash and Spitz in Hunter, Swartzwelder, and Clyde, 1976.) **B.** Amastigotes (Leishman bodies) in mononuclear host cell. **C.** Flagellated promastigotes from culture.

liver, bone marrow, lymph glands, intestinal mucosa, and certain other cells. According to Stauber (1966), the varied distribution of amastigotes in man and animals is largely dependent on the abundance and distribution of phagocytic cells of the lymphoid macrophage system. In other words, the variety of locations where amastigotes occur is increased with prolonged infection and as inflammation in various parts of the body results in these sites being populated by infected phagocytes migrating from the blood as exudate cells. Typically the spleen and liver are enlarged because of an increase in the fibrous elements and the number of macrophages (Fig. 4.33).

Recent epidemiologic and clinical studies have revealed the occurrence of several varieties (or strains) of *L. donovani.* The Mediterranean-Middle Asian variety occurs throughout the Mediterranean Basin and extends through southern Russia to China. The common sandfly vectors are *Phlebotomus major, P. chinensis, P. perniciosus,* and *P. longicuspis.* This parasite utilizes dogs, jackals, and foxes as reservoir hosts, and it primarily infects young children. A second variety occurs in northeast India and Bangladesh. Its most important vector is *P. argentipes.* It apparently has no natural reservoir hosts, and it infects mainly human adults and adolescents. A more virulent but clinically similar variety occurs in East Africa. Its vectors are *P. martini* and *P. orientalis,* and it may employ wild rodents as reservoirs.

The New World variety is widespread in Central and South America. It causes primarily a zoonotic infection among foxes and dogs, and its vector is *Phlebotomus longipalpis.* It needs to be noted that the patterns and distribution of the varieties of human leishmaniasis are known to change (Hutt *et al.,* 1973).

During feeding on an infected mammal, the sandfly ingests amastigotes, which transform into large numbers of small promastigotes in the fly's gut in about 3 days. These result from the longitudinal division of the original promastigotes that develop from the amastigotes once the latter reach the insect's gut. The promastigotes migrate anteriorly to the fly's pharynx and mouth cavity, where most of them are found by the fifth day. From these anterior regions the promastigotes can easily migrate to the proboscis, ready to be introduced into the new human or animal host. Upon being inoculated by the sandfly, they actively establish themselves via the circulatory system as intracellular amastigotes in the mammal.

A number of small mammals, including the hamster and mouse, can serve as experimental hosts, and these can be infected by bypassing the sandfly vector. Spleen hemogenate containing amastigotes from an infected mammal is inoculated directly into the recipient.

Stauber *et al.* (1954) have shown that infected hamsters exposed to high environmental temperatures of 34° to 35°C are sometimes cured. As in the case of numerous other pathogens, prolonged cultivation tends to reduce virulence.

The hamster is a favored experimental host because it usually does not develop age resistance to leishmanial infections. Furthermore, it does not show much natural immunity. The course of the infection is dependent on the virulence of the strain of *Leishmania*, the susceptibility of the host, and the number of parasites introduced. If a virulent strain is introduced, the infected hamster generally becomes gravely ill in less than 1 month. On the other hand, if a less virulent strain is used, it may take 6 months before death occurs.

Since the natural method of transmission of the human-infecting species of *Leishmania* is by sandfly vectors, there is a certain degree of seasonal fluctuation in the number of new infections, correlated with the abundance and activity of the sandflies.

Leishmania tropica. *L. tropica* causes cutaneous leishmaniasis, a relatively mild skin disease commonly known as the Oriental sore (Fig. 4.34). This species is endemic to those countries of Europe and North Africa bordering the Mediterranean Sea. In Asia it is endemic to Syria, Armenia, Israel, southern Russia, China, Vietnam, and India. *L. tropica* has also been reported from Peru, Bolivia, Brazil, the Guianas, and Mexico (Fig. 4.35). Unlike *L. donovani*, *L. tropica* is primarily an intracellular parasite in endothelial cells around cutaneous sores. Beginning as a papule, the sore erupts and spreads, forming cutaneous eruptions most commonly on the hands, feet, legs, and face.

A variant clinical form of cutaneous leishmaniasis

occurs in South and Central America and Ethiopia. It is referred to as **diffuse cutaneous leishmaniasis**. Several months or years after the initial infection by the sandfly, the parasites disseminate widely and cause nodules and plaquelike lesions under the skin. The face, arms, and legs are most commonly affected, and the lesions resemble lepromatous leprosy. Patients with this form of cutaneous leishmaniasis have a specific type of immune deficiency (Hutt *et al.*, 1973). The vectors for *L. tropica* are *Phlebotomus sergenti*, *P. major*, *P. papatasii*, *P. caucasicus*, and a few less known species of sandflies. The life history of *L. tropica* parallels that of *L. donovani*. In addition to utilizing humans as the mammalian host, this species also infects dogs and cats in China and a few Mideastern countries. The monkey, bullock, and brown bear are known to be naturally infected in the Mideast, as are horses and gerbils.

*Leishmania braziliensis.** The third human-infecting species, *L. braziliensis* (Fig. 4.36), is primarily neotropical in distribution being found in Brazil, Paraguay, Peru, Argentina, Bolivia, Uruguay, Venezuela, Ecuador, Colombia, Panama, Costa Rica, and Mexico. Human infections also have been reported from the

---

*For a detailed discussion of the subspecies of *Leishmania braziliensis*, see Lainson (1983).

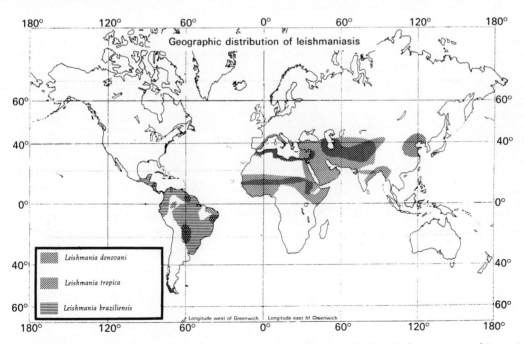

Fig. 4.35.   Geographic distribution of the three major species of human-infecting *Leishmania*. (Armed Forces Inst. Pathol., Neg. No. 68-1805-2.)

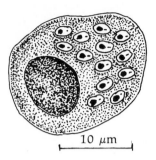

10 μm

Fig. 4.36. *Leishmania braziliensis.* Amastigotes within host's macrophage.

Sudan, Somali Republic, Kenya, Italy, India, and China (Fig. 4.35). It is commonly referred to as the American species since it is by far more common in South and Central America. The parasite appears to prefer the mucous membranes of the nose, tongue, and mouth (Fig. 4.37).

This disease caused by *L. braziliensis* is known as mucocutaneous leishmaniasis, **espundia**, **uta**, or chiclero ulcer. Interestingly, it does not occur in the high Andes mountains in its distribution range. Furthermore, the clinical manifestations vary along this range, resulting in considerable confusion as to the identity of the parasite. Typically, inoculation of promastigotes by the bite of the sandfly results in a small, red papule on the skin. This itchy primary lesion ulcerates in 1 to 4 weeks and heals within 6 to 15 months. This is followed by the appearance of secondary lesions on

the body. In Mexico and Central America the secondary lesion usually appears on the ear, resulting in its destruction. This is common among chicleros, forest-dwelling natives who harvest the gum of chicle trees. Recent investigators tend to recognize the parasite as a separate species, *L. mexicana*. Spiny rats, other forest rodents, kinkajous, dogs, and cats are known to serve as reservoir hosts in this region.

In Venezuela and Paraguay, the lesions often appear as flat, ulcerated plaques that remain open and oozing. The disease syndrome is called **pian bois** in that area.

In the more southerly range of *L. braziliensis*, the lesions tend to metastasize and spread from the primary lesion to mucocutaneous regions. The secondary lesion may appear before the primary has healed, or it may be many years (up to 24) before the secondary symptoms appear. The secondary lesion often involves the nasal and buccal tissues, causing degeneration of cartilaginous and soft tissues (Fig. 3.37). Furthermore, secondary bacterial infection is common. The disease is referred to as **espundia** or **uta** in this area; it may last for many years, and death may result from secondary infection or respiratory complications. This condition is also known to occur in Ethiopia (Cheng, 1983).

The life history of *L. braziliensis*, parallels that of the two other human-infecting species; however, in this case *Phlebotomus intermedius*, *P. squamipes*, *P. panamensis*, and several other species of sandflies of the genus *Lutzomyia* are the primary insect vectors. It has been shown that direct infections from sore to sore are possible.

Effect of Diets on Transmission of Leishmania. A fascinating aspect of the transmission of *Leishmania* spp. by sandflies has to do with the diet of the vector. It was known that kala-azar and *Phlebotomus argentipes* in India have a similar geographic distribution. Furthermore, it is not difficult to demonstrate that the insects do become heavily infected with *Leishmania* promastigotes when fed on leishmaniasis victims. However, experimental trials at infecting humans through the bite of the insect were mostly unsuccessful until it was discovered that the substitution of a diet of raisin juice or glucose solution for blood, after the initial blood meal, made the vectors highly infective. Apparently *Leishmania* does not survive in *Phlebotomus* in nature, or at least not in sufficient numbers to produce infections, unless the fly follows a blood meal with the ingestion of some fruit juice rich in glucose. It still remains undetermined what the preferred fruit juice is.

Not only does the diet of the invertebrate host influence the course of leishmaniasis, but that of the mammalian host does also. Actor (1960) has shown that the administration of desoxypyridoxine (a pyridoxine analog) to mice fed either a basic or a pyridoxine-free diet prior to infection, results in an increase in the rate of accumulation of parasites. Daily adminis-

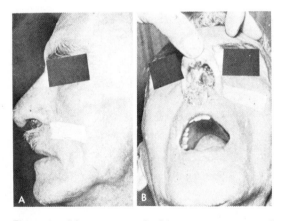

Fig. 4.37. **Mucocutaneous leishmaniasis.** Lesions caused by *Leishmania braziliensis* on patient with the disease for more than 15 years. **A.** Lateral view showing tapir nose. **B.** Frontal view showing destruction of nasal septum. (After Chavarria in Hunter, Swartzwelder, and Clyde, 1976; with permission of W. B. Saunders.)

tration of pyridoxine to these deficient mice reverses this effect. Pantothenic acid deficiency also influences the parasite density. The injection of pantoyl taurine (a pantothenic acid analog) and/or the deletion of pantothenic acid from the diet of infected mice results in a suppression of parasite numbers during the first 15 days of infection. After this time, an increase in the rate of accumulation of parasites occurs in pantothenate-deficient mice. Finally, Actor has also demonstrated that if mice are placed on a protein-free diet prior to or following infection, they harbor significantly higher parasite burdens than those of pair-fed or *ad libitum* controls. In the latter case, it is possible that the protein deficiency results in a reduction of the host's ability to synthesize antibodies.

Behavior in Sandflies. Johnson and Hertig (1970) have reported that two distinct species of *Leishmania* behave differently in their sandfly hosts, and these differences may be useful in the identification of different species. Specifically, these investigators have found that all Panamanian strains of *L. braziliensis*, whether isolated from humans, other mammals, or wild sandflies, when introduced into the sandflies

Lutzomyia sanguinaria or *L. gomezi* after passage through experimentally infected guinea pigs, produce infections characterized by the growth of promastigotes in the hindgut, especially in the so-called hind triangle (Fig. 4.38), with or without growth in the midgut. Over 90% of the flies infected with Panamanian *L. braziliensis* had promastigotes in their hindguts, and almost half of these flagellates were found only in the hind triangle. A Peruvian strain of *L. braziliensis* behaved similarly in *Lutzomyia sanguinaria* and *L. gomezi*. On the other hand, two varieties of *Leishmania mexicana* (considered by some as a strain of *L. braziliensis**), from Guatemala and British Honduras, are typically found (as promastigotes) only in the midgut of the two species of sandflies. These studies

---

*For a detailed discussion of *Leishmania mexicana* and subspecies, see Lainson (1983).

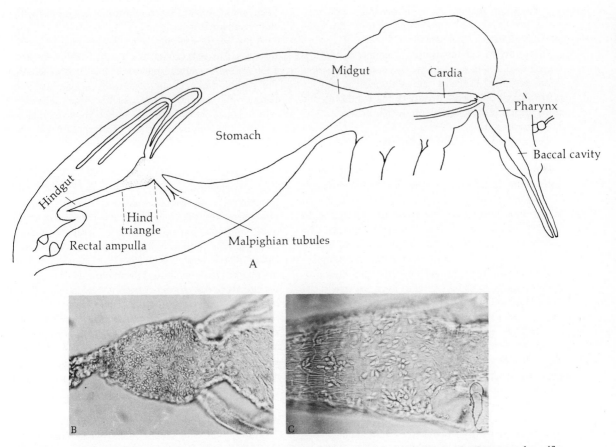

Fig. 4.38. *Leishmania braziliensis* **in the sandflies** *Lutzomyia gomezi* **and** *L. sanguinaria.* **A.** Drawing of sandfly showing position of the hind triangle. **B.** Photomicrograph showing *L. braziliensis* attached to wall of hind triangle of *Lutzomyia gomezi* 3 days after infection. **C.** Photomicrograph showing *L. braziliensis* attached to wall of hind triangle of *Lutzomyia sanguinaria* 7 days after infection. (After Johnson and Hertig, 1970.)

indicate that both *Leishmania braziliensis* and *L. mexicana* have preferred habitats within the sandfly host.

Other Leishmania *Species.* *Leishmania caninum* of dogs and children from the Mediterranean Basin, *L. infantum* of children in China and the Mediterranean, and *L. mysxi* in the dormouse are related species the biology of which is not well known. These species are all morphologically similar to the better known species. *L. chamaeleonis* was described in cloaca of the lizard, *Chamaeleon vulgaris.* It is doubtful whether *L. chamaeleonis* is similar to the mammal-infecting species because promastigotes appear to be the only form occurring in its life cycle. Furthermore, this species lives in its host's intestine. Other species of saurian leishmanias may occasionally leave the intestine and migrate into the blood. Still other species apparently have found the blood an ideal habitat and only occur there. It may be postulated that the intestine- and blood-dwelling species are comparatively primitive and have not adapted to an intracellular existence. Little is known about the transmission of these presumably primitive leishmanias.

Immunity to Leishmaniasis. Stauber (1963) has contributed a review of immunity to *Leishmania* spp. More recently, Williams and Vasconcellos Coelho (1978) have published a review of the taxonomy and transmission of *Leishmania* spp. which includes a section on immunity.

Leishmanias, especially *L. donovani*, are unusual in that they are entirely intracellular parasites in their host's reticuloendothelial system. In fact, it occurs within the lysosomes of the host's macrophages (Chang and Dwyer, 1978). Proteolytic enzymes that attack other invaders of the blood do not appear to destroy leishmanias. Within the macrophages, the parasites multiply by binary fission at an estimated rate of once every 24 hours. Stauber has shown that in experimentally infected hamsters, the parasites proliferate rapidly in the spleen and liver, especially the liver. This rapid reproduction rate decreases sharply in the liver after 7 to 8 days. It has been suggested that cessation of reproductive activity is most probably due to the destruction of the originally infected cells. The parasites are then released into the blood, where they are confronted with destructive forces (antibodies, etc.) for the first time.

If rats or rabbits are exposed to infection they are resistant. On the other hand, the golden hamster and cotton rat are highly susceptible. These findings suggest that different host species vary in their ability to resist infection by this parasite. Hence, immunity, at least in part, may be responsible for host specificity. Then again, strain differences among both host and parasite also contribute to host specificity. Humans occasionally recover spontaneously from visceral leishmaniasis, whereas hamsters almost always die. Thus, humans appear to be more resistant than hamsters.

A certain degree of acquired resistance to reinfection does occur. Successful vaccination of humans has been achieved by immunizing the host with living, avirulent promastigotes of *Leishmania* of a rodent-infecting strain against subsequent infection by promastigotes of a human-infecting strain.

It is noted that Mukkada *et al.* (1985) have found that the metabolism (respiration, catabolism of energy substrates, incorporation of precursors into macromolecules) of amastigotes of *Leishmania donovani*, which, as stated, are endoparasitic in their host's macrophages, operates optimally at pH 4.0 to 5.5, and with all activities decreasing sharply above this pH range. On the other hand, promastigotes (culture forms) carry out the same metabolic activities optimally at or near pH 7.0. This adaptation by *L. donovani* amastigotes to an acidic environment may account in part for their unusual ability to survive and multiply within the acidic milieu of their host's phagolysosomes (formed from the fusion of lysosomes with the phagosome).

Although *L. donovani* often causes visceral leishmaniasis, *L. tropica* causes cutaneous leishmaniasis, and *L. braziliensis* causes mucocutaneous leishmaniasis, the pathologies are not completely species specific because the extent of the infection does influence which tissues are infected. It is noted, however, that *L. donovani* tends to be more host tissue-specific than the other two species.

That immunity to *Leishmania* spp. occurs is amply documented by the more classical studies cited above. Research aimed at understanding immunity to these protozoans in recent years has been directed toward two objectives: (1) deciphering the constituent antigens and metabolites of leishmanias with the goal of devising more accurate diagnostic tests, and (2) deciphering the operative mechanisms underlying immunity to leishmaniasis. Information pertaining to both of these objectives has been reviewed by Zuckerman and Lainson (1977). In brief, studies directed toward the first objective, not surprisingly, have revealed that antigenic overlapping and group-specific cross-reactivity among various species of *Leishmania* and between *Leishmania* spp. and *Trypanosoma* spp. occur. This has caused the major problem in the development of specific immunodiagnostic tests. On the other hand, Dwyer (1973) has demonstrated that stage-specific antigens exist among *Leishmania* spp.

Studies with the second objective in mind have revealed that, unlike many other protozoan diseases, immunity to leishmaniasis is dominated by cell-mediated reactions although antibodies are formed (Garnham and Humphrey, 1969; Dumonde, 1973;

Zuckerman, 1975). It is noted, however, that the antibodies are not thought to be of critical importance in protection.

Finally, it is noted that most leishmaniasis patients who have overcome an initial bout of cutaneous or visceral leishmaniasis are radically cured and do not harbor latent infection (Neal *et al.*, 1969). Only a small minority of individuals are unable to mount an effective sterilizing response and continue to harbor parasites indefinitely. Such persons are said to suffer from **leishmaniasis recidiva**.

For a succinct account of all aspects of human leishmaniasis, see the World Health Organization (WHO) report (1984).

Genus *Leptomonas*. Among the genera belonging to the Trypanosomatidae, *Leptomonas* is probably the most primitive. During the life cycle of the members of this genus, these flagellates (promastigotes) lose their free flagella, round up as amastigotes, and become cysts that are the infective forms. When cysts are ingested by a prospective host, excystation occurs and each escaping organism develops a new flagellum, and takes on the typical elongate shape of the adult (or promastigote). Under certain conditions, the flagellum of the adult may be lost and the body becomes attached to the epithelial cells lining the host's gut.

Leptomonas ctenocephali. This organism (Fig. 4.39) is a parasite occurring in the hindgut of the dog flea, *Ctenocephalides canis.* The promastigote usually is found in the foregut. When the parasite reaches the hindgut, it becomes an amastigote which is passed out in the host's feces and later infects a new host when it is ingested, in the encysted form, by another larval flea. Other species of *Leptomonas* include *L. butschlii*, the first species known, in the gut of the nematode *Trilobus gracilis*; *L. patellae* in various insects; and *L. pyraustae* in corn borers. Most *Leptomonas* species

appear to be nonpathogenic; however, Paillot (1927) has reported a pathologic condition in the European corn borer, *Pyrausta nubilalis*, caused by *L. pyraustae.* The gut apparently is the primary site of infection, but cases are known in which the flagellate has invaded the surrounding tissues.

Genus *Phytomonas*. Members of the genus *Phytomonas* are morphologically similar to those of *Leptomonas*. These flagellates are parasitic in plants and transmitted by insects.

Genus *Crithidia*. The genus *Crithidia* includes those species parasitic in arthropods, primarily insects, and various other invertebrates. In the adult stage, members of this genus differ from promastigotes in the point of flagellar origin. As with *Leptomonas*, there is an amastigote stage during the life cycle, which eventually becomes situated within a spherical cyst that is passed out in the host's feces. Ingestion of cysts by another susceptible host is the mechanism of infection. Upon entering the new host, excystation occurs, and the emerging parasite develops a new free flagellum and develops into a choanomastigote measuring 4–10 μm long. Instances are known in which attached forms occur, as in the genus *Leptomonas*.

Crithidia fasciculata (Fig. 4.39) is found in the intestine of various species of mosquitoes, including *Anopheles maculipennis* and *Culiseta incidens*. The amastigote is found in the host's hindgut. A cyst wall is formed prior to its passage out of the host. *Crithidia gerridis* is a parasite of the common water bug *Gerris fossarum*.

In the case of *Crithidia fasciculata*, Clark *et al.* (1964) have reported that the relatively high percentage of infection in the mosquito *Culiseta incidens* in nature can be accounted for by larva-to-larva and adult-to-larva transmission. Specifically, they have shown that choanoflagellates discharged from one infected larva in feces can enter another via ingestion. Similarly, choanoflagellates discharged from adult mosquitoes can infect larvae.

Although members of the genus *Crithidia* are generally considered nonpathogenic, occasionally deaths of their arthropod hosts have been reported. For example, Langridge (1966) has reported mass mortalities of honey bees in Australia allegedly due to an unidentified species of *Crithidia*.

Carbohydrate Uptake. Min (1965) has demonstrated that the uptake of monosaccharides by *Crithidia luciliae*, a parasite of insects, occurs by two mechanisms: (1) there is an active transport mechanism operative when the external concentration of the sugars is low (0.5 mM) and which is dependent on a supply of metabolic energy; and (2) facilitated diffusion

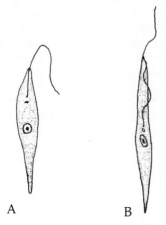

Fig. 4.39.   Trypanosomatids. A. *Leptomonas ctenocephali.* (Redrawn after Wenyon, 1926.) **B.** *Crithidia fasciculata.*

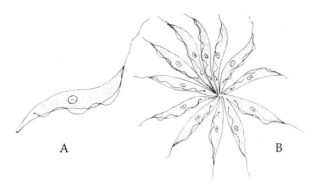

A   B

Fig. 4.40. ***Trypanosoma lewisi* from rat blood. A.** Trypomastigote. **B.** Rosette formation.

occurs when the external sugar concentration is high (20 mM). Earlier, Cosgrove (1963) had reported that *Crithidia* spp. can utilize ribose, fructose, galactose, glucose, mannose, sucrose, and raffinose. In addition, one strain of *C. fasciculata* was found to be able to utilize xylose, cellobiose, and maltose.

Genus *Trypanosoma*. The genus *Trypanosoma* includes literally hundreds of species, all blood and lymph parasites of vertebrates and invertebrates. Determination of the various species is a frustrating and difficult task, because morphologically many different "species" are identical but are found naturally in different hosts and hence are generally considered host-specific species. In addition to species similarities, trypanosomes tend to be pleomorphic (possessing many forms).

During the life cycle of *Trypanosoma* spp., the organisms attain the amastigote, promastigote, epimastigote, and trypomastigote forms. The first three are found in the vector, if one exists, and the trypomastigote form occurs in the definitive host.

The biology of *T. lewisi** (Fig. 4.40), a widely distributed parasite of rats, is elucidated here because we know more about this species, which has been studied critically by a number of investigators.

Life Cycle of Trypanosoma lewisi. The rat becomes infected by this organism while licking its fur. During this process, freshly deposited rat flea feces including parasites or infected fleas are ingested. After several hours of incubation, the trypanosomes (trypomastigotes) appear in the circulating blood as extracellular parasites. These multiply rapidly by longitudinal fission for a week or more, often resulting in rosette formation (Fig. 4.40). When the flea, *Cerato-*

phyllus fasciatus, feeds on the rat's blood, it also takes in the circulating trypomastigotes. Approximately 25% of the ingested trypanosomes persist in the flea, and each one penetrates an epithelial cell lining the host's stomach. Within the cell, the flagellate curls up and eventually becomes rounded. Sporulation (asexual reproduction) occurs within its enlarged and rounded body, resulting in many small trypomastigotes. These enlarge, rupture out of the cell, and become active within the flea's gut, seeking another cell to penetrate and thus repeat the sporogonic cycle. Eventually, those trypomastigotes that are free in the intestinal lumen metamorphose into epimastigotes and migrate posteriorly to the rectal region, some simultaneously undergoing longitudinal fission. In the rectum the epimastigotes become attached to the lining and become pleomorphic while undergoing longitudinal fission. Finally, some transform into trypomastigotes and are passed out in the host's feces. Infection of another rat is initiated when contaminated feces of an infected flea is ingested by that rat or introduced into the wound caused by a flea bite. The second type of transmission is known as **mechanical transmission**.

Within the rat host the introduced metacyclic trypomastigotes transform into epimastigotes and commence to reproduce by longitudinal fission while situated in the visceral blood capillaries. After about 6 days, the parasites appear in the peripheral blood as rather "fat" trypomastigotes. A rat, once infected, becomes immune to reinfections even long after the parasite has disappeared from its blood.

The development of *Trypanosoma lewisi* in the posterior portion of the gut of the insect host is characteristic of a group of trypanosomes, including *T. melophagium* of sheep in sheep ticks, *T. theileri* of cattle in tabanids, and *T. cruzi* of humans and other animals in triatomid bugs. This type of development, known as **stercorarian** (or posterior station) development, is believed to be more primitive than the **salivarian** (or anterior station) type of development of African sleeping sickness trypanosomes. The salivarian type of development is discussed on p. 130.

In the laboratory, experimental infection of rats with *T. lewisi* can be readily achieved through intraperitoneal injection of two or three drops of blood from an infected rat suspended in citrated physiologic saline to prevent clotting. This technique for the transfer of infection, which does not involve development in an insect host, has its counterpart in nature. Vectors with contaminated mouthparts can transmit trypanosomes from one host to another without the flagellates passing through the gut-developmental stages in the vector.

The relative ease with which *T. lewisi* can be maintained in the laboratory in albino rats, although periodic transfer to new hosts is necessary, has provided an ideal experimental situation. As a result, much has been learned about the physiology of this trypanosome.

---

* *Trypanosoma lewisi* belongs to the subgenus *Herpetosoma* and hence is commonly referred to as *Trypanosoma (Herpetosoma) lewisi.*

Antibodies in Infected Rats. Multiplication of trypomastigotes of *Trypanosoma lewisi* in the blood of a rat decreases in approximately 6 days after the infection because of the appearance of a reproduction-inhibiting antibody, **ablastin** (Taliaferro, 1924, 1932). Some of the parasites are killed at this time, but a small population of slender trypomastigotes remains. These are infective to fleas but do not undergo further reproduction within the rat host. Furthermore, surviving antigenic variants (p. 87) cannot repopulate the blood owing to the presence of ablastin, which prevents reproduction. At the time when ablastin is exerting its maximal effect on the trypanosome population, another antibody response kills the remaining epimastigotes and most trypomastigotes, expressing the predominant somatic antigens (D'Alesandro, 1970, 1972; Yasuda and Dusanic, 1971). This immune response precipitates the trypanocidal crisis at the peak of infection (Fig. 4.41). Only antigenically variant trypomastigotes survive this crisis. Ultimately, another trypanocidal antibody response directed against the variant antigens terminates the infection.

Ablastin, a unique antibody, is a globulin with many characteristics of a typical antibody, but its action differs in that it inhibits reproduction. Nucleic acid and protein synthesis by the trypanosome are inhibited, as is the uptake of nucleic acid precursors (D'Alesandro, 1970; Lumsden, 1972). Although the site and mode of action of ablastin is still uncertain, assumptions can be made concerning its activity. The specificity of the antibody is probably directed against sites for metabolite transport or recognition on the trypanosome's cell membrane. With cellular uptake of key metabolites blocked or reduced, the trypanosomes are indirectly forced to assume a maintenance state (nonreproducing trypomastigote form) from one of assimilation and growth (reproducing epimastigote form). Ablastin appears to be a γG molecule (Eisen, 1974). For a detailed review of the possible action of ablastin, see Mansfield (1977).

Pathogenicity. *Trypanosoma lewisi* has no visible deleterious effect on the rat host and, as a result of the appearance of ablastin and the trypanocidal antibodies, the infection soon dies out. However, if *T. equiperdum*, a hemoflagellate of horses, is introduced into the rat, it is able to reproduce without being inhibited and generally kills the host within a week. The cause for this striking difference in the pathogenicity of the two species remains uncertain, although it is at least in part due to genetic differences that are manifested as metabolic differences. One such difference has to do with carbohydrate metabolism. Pathogenic trypanosomes are generally greater consumers of carbohydrates than nonpathogenic species, but they fail to oxidize the carbohydrates completely. A heavy burden is thus placed on the carbohydrate reserves of the host, and possibly because of toxic byproducts, the host's carbohydrate metabolism is compromised. In horses, *T. equiperdum* produces a chronic disease with eventual

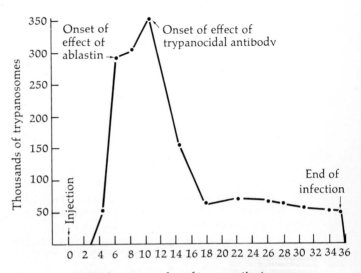

Fig. 4.41. Graph showing number of trypomastigotes of *Trypanosoma lewisi* per cubic millimeter of blood during the course of experimental infection in rats. (Redrawn after Taliaferro, 1941.)

involvement of the nervous system and subsequent death.

Vitamin deficiences in the host may increase the pathogenicity of trypanosomes. For example, *T. lewisi* can become pathogenic in rats deficient in pantothenic acid.

It is noted that natural infections of humans with *T. lewisi* have been reported (Shrivastava and Shrivastava, 1974). The transient parasitemias observed in peripheral blood were accompanied by fever, which disappeared when the parasitemia was eliminated.

Trypanosomes in Homologous and Heterologous Hosts. An animal in which a parasite is naturally found is referred to as a **homologous** host. For instance, the rat is the homologous host for *Trypanosoma lewisi*, whereas mice and other animals are **heterologous** hosts. Certain heterologous hosts, such as mice for *T. lewisi*, can sustain the parasite, although less efficiently. Other heterologous hosts are completely refractile—that is, they do not sustain the parasite.

T. lewisi can be experimentally adapted to other mammalian hosts. For example, *T. lewisi* can be grown in mice when the parasite inoculum is mixed with normal rat serum. Serum globulins can be shown to coat the trypanosomes and to promote the infection in the mouse by either masking dominant antigenic groups or providing some nutritive function (Greenblatt *et al.*, 1969; D'Alesandro, 1970, 1972).

Once trypanosomes become adapted to an initially refractile or semirefractile heterologous host, drastic changes in the parasite's biology often result. There may be an alteration of virulence for the normal host as well as loss of the parasite's capability to undergo cyclical transmission in the arthropod vector. Changes in the trypanosome's antigenic constitution, metabolism, drug sensitivity, and even morphology may also result from the parasite's adaptation to the abnormal host.

Metabolism. Trypanosomes vary greatly in their terminal oxidation mechanisms. The respiration of *T. lewisi*, *T. duttoni* of rats, *T. theileri* of cattle, *T. cruzi* of humans, and other related species is markedly inhibited by cyanide but is relatively insensitive to known inhibitors of sulfhydryl enzymes. Thus, it appears that this group of flagellates includes a cytochrome system. Indeed, in *T. lewisi* components of the cytochrome system have been demonstrated, as has the light-reversible carbon monoxide inhibition of respiration. As further proof, succinate oxidation in *T. lewisi* has been shown to be sensitive to antimycin A. For a detailed discussion of the cytochromes of flagellates and other protozoans, see Pettigrew (1979).

African Human Trypanosomiasis

The manifestation of African human trypanosomiasis is the sleeping sickness. Although trypanosomes were first discovered in 1841, human and animal trypanosomiasis are ancient diseases that have scourged mankind in Africa for centuries. Africans lying prostrate, drooling from the mouth, insensitive to pain, and later dying must have been a common sight to early slave traders. Without doubt trypanosomes have been introduced into North America and other parts of the world by infected slaves. Fortunately, the absence of compatible vectors has prevented the spread of this disease.

Although somewhat on the decline, primarily because of the introduction of modern insecticides and control techniques that have reduced the number of insect vectors and as a result of drugs, trypanosomiasis is by no means eradicated. Not only does human trypanosomiasis hinder human endeavors, but animal trypanosomiasis tends to reduce the number of both domesticated animals and game, which indirectly affects cultures dependent on such animals for food and as beasts of burden. It is known, for example, that in parts of Nigeria, cattle that have been imported to help relieve the protein-deficient condition in the local population rapidly succumb to trypanosomiasis, and consequently the dietary problem has not been resolved. One can justifiably claim that trypanosomiasis is one of the primary reasons for the lack of advancement in endemic areas of Africa.

On the other hand, ironically, these flagellates have played important roles in advancing science. Obser-

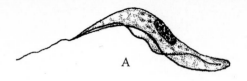

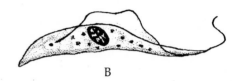

Fig. 4.42. African human-infecting trypanosomes. A. *Trypanosoma brucei rhodesiense.* **B.** *Trypanosoma brucei gambiense.*

vations on trypanosomes gave Ehrlich the idea that specific drugs could be found that would attack specific pathogens. His pursuance of this concept led to his discovery of "606" to combat syphilis. It was also through the study of trypanosomes that the concept of drug resistance by microorganisms became known.

The two human-infecting subspecies of African trypanosomes are *Trypanosoma brucei rhodesiense* and *T. brucei gambiense. Trypanosoma brucei rhodesiense*, the causative agent of the virulent East African or Rhodesian sleeping sickness, is endemic to southeastern coastal Africa (Fig. 4.42). It is a typical trypomastigote, tapering at both ends and measuring 15–30 by 1–3 μm. The kinetosome is situated near the posterior end of the body. The common insect vector is the tsetse fly, *Glossina morsitans.* In addition to parasitizing humans, this species is found in wild game and domestic animals in endemic areas. The reservoir hosts serve as sources of new human infections.

Trypanosoma brucei gambiense causes Gambian or Central African sleeping sickness (Fig. 4.42). The disease is confined to central Africa, on each side of the equator. Domestic and wild animals, such as pigs, antelopes, buffaloes, and reed bucks, serve as reservoir hosts. Experimentally, *T. brucei gambiense* has been introduced into rats, and amazingly, organisms have been found in the placental blood of infected rats as well as in the livers of embryos. Transplacental infections, however, are not common, although such cases are known even in humans. The introduction of *T. brucei gambiense* into the circulation of the vertebrate host usually occurs through the bite of the tsetse flies *Glossina palpalis* and *G. tachinoides* (p. 663). The mechanism of transmission differs from that of *T. lewisi* but is identical to that of *T. brucei rhodesiense* in that the flagellate is introduced through the bite, for the infective form is in the salivary glands (salivarian) and not in the feces (stercorarian) (although infection through the feces has also been reported). In addition to being a blood parasite, *T. brucei gambiense* also invades the lymphatic glands and the central nervous system. For an authoritative review of African sleeping sickness caused by *T. brucei rhodesiense* and *T. brucei gambiense*, see de Raadt and Seed (1977).

Salivarian Development

Development of *Trypanosoma lewisi* and related species in the posterior portion of the insect vector's gut is referred to as **stercorarian development** (p. 128), characteristic of the more primitive species. African human trypanosomes, on the other hand, undergo **salivarian development** within tsetse flies. Actually, three groups of salivarian trypanosomes can be recognized as forming an evolutionary series. (1) The **vivax group** consists of *T. vivax* of cattle, sheep, goats and horses, from tsetse flies, and *T. uniforme* of cattle, sheep, and goats, from tsetse flies. This group develops only in the proboscis of the vector. (2) The **congolense group** consists of *T. congolense* of horses, cattle, and sheep, from tsetse flies; *T. dimorphon* of horses, from tsetse flies; and *T. simiae* of monkeys, from tsetse flies. Trypanosomes of this group first enter the stomach of the vector upon ingestion. Later they migrate anteriorly to the pharynx. (3) The **brucei group** includes *T. brucei brucei* of horses, mules, etc., from tsetse flies; the two African human-infecting subspecies, *T. brucei rhodesiense* and *T. brucei gambiense*, from tsetse flies; *T. suis* of pigs from tsetse flies; and *T. evansi* of horses, mules, donkeys, cattle, camels, and elephants, from tabanids and other blood-sucking flies. Trypanosomes of this group first enter the vector's stomach, where some development occurs. They then move forward and invade the salivary glands, where the infective metacyclic forms develop.

Salivarian trypanosomes infect the vertebrate host via the vector's bites.

For a review of host susceptibility to African trypanosomiasis, see Murray *et al.* (1982).

American Human Trypanosomiasis

In 1909, Chagas found that the thatched-roof huts in the villages of the State of Minas Gerais, Brazil, were infested with large blood-sucking bugs, *Panstrongylus megistus*, which were infected with flagellates. When he inoculated the flagellates into guinea pigs and monkeys, an acute infection developed. Later surveys revealed that among infants and young children inhabiting infested huts, an acute disease characterized by fever, swollen glands, anemia, and nervous disturbances prevailed. Today we recognize this disease to be the American form of trypanosomiasis, caused by *Trypanosoma cruzi*.

Trypanosoma cruzi* is a smaller species, being C- or U-shaped and averaging 20 μm in length (Fig. 4.43). Unlike other trypanosomes, *T. cruzi* is never found

Fig. 4.43. *Trypanosoma cruzi* trypomastigote from experimentally infected rat.

dividing in the blood of its mammalian host. Furthermore, it can become established as an intracellular parasite in greatly swollen cells in various human tissues, especially in heart cells, striated muscles, the central nervous system, and glands. As an intracellular parasite, *T. cruzi* assumes the amastigote form and undergoes rapid division, resulting in numerous individuals. These individuals later develop into trypomastigotes and escape into the blood when the host cells rupture. In chronic cases of *T. cruzi* infections, trypomastigotes are not seen in the blood, because the circulating antibodies readily kill them. However, large numbers of parasites can be demonstrated in tissue cells, where they are apparently protected from antibodies. In order to survive, trypomastigotes rupturing out of cells must invade another cell almost immediately, before they are affected by antibodies.

The insect vector is usually the cone-nosed or kissing bug, *Panstrongylus megistus*, or a few closely related species. However, *T. cruzi* is not very host specific as far as the insect vector is concerned. The parasite can develop not only in almost all species of reduviid bugs, but also in bedbugs, ticks, and even in the body cavities of caterpillars. For a detailed review of *Trypanosoma cruzi* and the disease caused by it, see Fife (1977).

The disease caused by *T. cruzi* is known as South American trypanosomiasis or Chagas' disease. It is commonly, but not exclusively, a disease of infants and children and is widely distributed in Central and South America. Since 1968, *T. cruzi* is known to occur in southern Asia (Malaysia, India, and most probably Thailand, Vietnam, Laos, Indonesia, Taiwan, and perhaps Japan) (Weinman, 1977). In nature, cats, dogs, bats, armadillos, rodents, and other mammals have been found to serve as reservoir hosts. In southern Asia, macaque monkeys are suspected to be reservoir hosts for human infections. Animal infections are known to exist in the southern and western United States and have been reported in raccoons in Maryland. This flagellate can be introduced into a large variety of mammals under experimental conditions.

Within the invertebrate host, which becomes infected by ingesting infected blood, the trypomasti-

---

*Referred to by some as *Schizotrypanum cruzi*.

gotes undergo longitudinal fission in the stomach and intestine and assume the epimastigote form while continuing to multiply. After 8 to 10 days, the infective forms (known as **metacyclic** forms) appear in the vector's rectum and are, therefore, of the stercorarian type. The metacyclic forms are passed out in feces. Infection of the vertebrate host is accomplished in the same fashion as in *Trypanosoma lewisi*—that is, the feces are rubbed into the wound caused by the bite or into the eyes or mucous membranes in human infections, or the feces of infected vectors are ingested in animal infections.

Trypanosoma brucei brucei is extremely pleomorphic (Fig. 4.44). This parasite averages 20 μm in length and is transmitted to donkeys, horses, mules, camels, cattle, sheep, cats, and dogs by various species of tsetse flies, primarily *Glossina morsitans*, although *G. pallidipes* and *G. swynnertoni* also are common vectors. In addition to tsetse flies, other blood-sucking arthropods may transmit *T. brucei brucei* mechanically. Trypanosomes can enter through intact mucous membranes of the alimentary canal, conjunctiva, or abraded skin (Soltys *et al.*, 1973). This trypanosome is extremely virulent to domestic animals, especially horses and camels, and the disease, known as **nagana**, is generally lethal. Various wild animals are infected in nature, particularly antelopes, in which the parasite

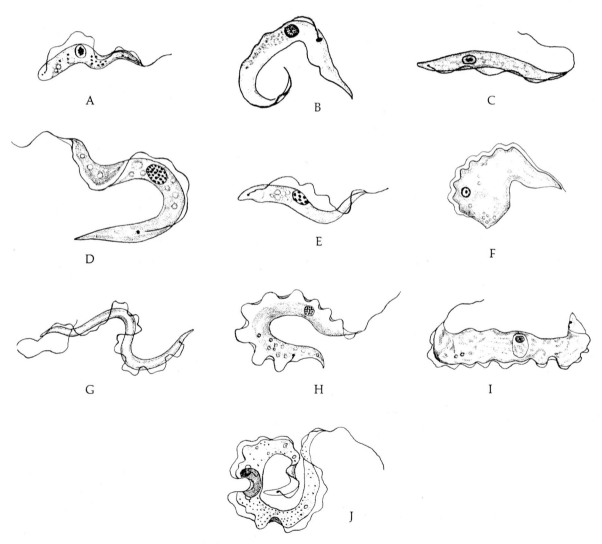

Fig. 4.44.   Some representative species of *Trypanosoma*. A. *T. brucei brucei* from mule. **B.** *T. melophagium* from sheep. **C.** *T. duttoni* from mice. **D.** *T. theileri* from cattle. **E.** *T. evansi* from horse. **F.** *T. rotatorium* from frogs. **G.** *T. diemyctyli* from the newt *Triturus*. **H.** *T. giganteum* from *Raja oxyrhynchus*. **I.** *T. percae* from the fish *Perca fluviatilis*. **J.** *T. granulosum* from the eel *Anguilla vulgaris*.

produces a benign infection. Most mammals can be infected except for baboons and humans. It is endemic to areas of Africa where *Glossina* spp. thrive.

Trypanosoma theileri (=*T. americanum*) is a North American parasite of cattle (Fig. 4.44). It measures 17–25 μm in length and, when maintained *in vitro*, assumes only the epimastigote form. Myonemes have been demonstrated within its cytoplasm. It is apparently nonpathogenic and is transmitted by biting tabanid flies. *Trypanosoma melophagium* is a parasite of sheep (Fig. 4.44). It is 50–60 μm long with delicate drawn-out ends. Infection is established when sheep ingest the sheep ked, *Melophagus ovinus*, or its feces. The infective trypomastigote is not transmitted by the bite of the insect. The organism disappears in sheep 1 to 3 months after the keds have been eradicated. *Trypanosoma evansi*,* the causative agent of the fatal disease **surra** in camels in North Africa, is known also to parasitize horses, dogs, cattle, donkeys, and mules (Fig. 4.44). The disease is transient in cattle and is known to occur in the Sudan, Panama, Venezuela, Brazil, Bolivia, Paraguay, Argentina, Burma, and India. Although a number of specific names have been given to trypanosomes similar or identical to *T. evansi* in various parts of the world, Hoare (1972) has advanced the opinion that these are all synonyms of *T. evansi*. The review by Woo (1977) is recommended to those desiring greater detail.

Other important species of *Trypanosoma* include *T. equiperdum*,* the causative agent of the chronic disease of horses and donkeys, **dourine**, in various parts of the world, including the United States. Unlike the other species of *Trypanosoma*, *T. equiperdum* does not require a vector but is transmitted directly during coitus. Infections can be transmitted in the laboratory through injection of infected blood. Furthermore, it has been demonstrated that it can be mechanically transmitted by stable- and horseflies.

Two other species, *Trypanosoma duttoni* and *T. rotatorium*, are commonly encountered in the laboratory (Fig. 4.44). The first is a blood parasite of mice and is often used in experimental work. As in *T. lewisi* infections, the reproduction-inhibiting ablastin and trypanocidal antibodies have been shown to be present in mice infected with *T. duttoni*. *Trypanosoma rotatorium* (Fig. 4.44) occurs in the blood of tadpoles and frogs. The leech, *Placobdella marginata*, is the vector in some areas.

Pathology in Invertebrate Hosts

Trypanosomes sometimes injure their invertebrate hosts. Grewal (1957) has reported that when *T. rangeli*, a parasite of humans and dogs in South America, is introduced into the bug *Rhodnius prolixus*, severe

damage results. During the first nymphal instars of this bug, the trypanosomes inhibit molting. When this species is introduced into bedbugs, the pathogenicity is even greater, causing significant numbers of these invertebrate hosts to die.

According to Watkins (1971), who has made a comprehensive study of the pathogenicity of *T. rangeli* in *R. prolixus*, a variety of morphologic, physiologic, and biochemical abnormalities occur and the death of the bug may be due to autointoxication caused by damaged muscular, nervous, and tracheal systems, or to *T. rangeli*-induced inhibition of *Nocardia rhodnii*, an essential mutualist normally occurring in the esophagus and midgut of *R. prolixus*.

Family Cryptobiidae

The family Cryptobiidae (of the suborder Bodonina), which includes the genus *Cryptobia* (Fig. 4.45), is characterized by two flagella, one free and the other forming an undulating membrane. Most species of this family are parasites in the reproductive organs of invertebrates, primarily molluscs, although *Cryptobia borreli*, *C. salmositica*, and *C. cyprini* occur in the blood of fish.

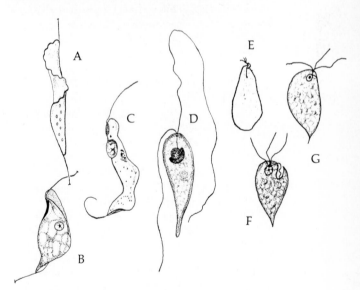

Fig. 4.45. Some parasitic flagellates. A. *Cryptobia borreli* from the blood of various fishes. (Redrawn after Mavor, 1915.) **B.** *Cryptobia helicis* from reproductive organs of pulmonate snails. (Redrawn after Kozloff, 1948.) **C.** *Cryptobia borreli*; fixed and stained specimen. (Redrawn after Mavor, 1915.) **D.** *Proteromonas lacertae* from the gut of lizards. (Redrawn after Kuhn, 1921.) **E.** *Mixotricha paradoxa* from the gut of termites. (Redrawn after Sutherland, 1933.) **F.** *Enteromonas hominis* from human intestine. **G.** *Chilomastix mesnili* from human intestine.

---

*Both *Trypanosoma evansi* and *T. equiperdum* are assigned to the subgenus *Trypanozoon* and hence are commonly referred to as *Trypanosoma (Trypanozoon) evansi* and *Trypanosoma (Trypanozoon) equiperdum*, respectively.

In the case of *Cryptobia salmositica*, a blood parasite of the coho salmon (*Oncorhynchus kisutch*), the torrent sculpin (*Cottus rhotheus*), and other teleosts in Washington State, the salmonid leech, *Piscicola salmositica*, serves as the vector. For a review of cryptobiid flagellates in the blood of freshwater and marine fish, see Becker (1977).

Family Bodonidae

Members of the family Bodonidae (of the suborder Bondonina) are characterized by two flagella originating at the anterior end of the body, one directed anteriorly and the other trailing. There is no undulating membrane. The parasitic members of this family include *Proteromonas*, found in the digestive tract of various lizards (Fig. 4.45), and *Embadomonas*, found primarily in the gut of arthropods. Those interested in the taxonomy of the Bodonidae should consult the paper by Hollande (1952).

ORDER RETORTAMONADIDA

The major diagnostic characteristics of the members of the Order Retortamonadida are listed at the end of this chapter.

Genus *Chilomastix*. Probably the best known genus belonging to this order is *Chilomastix*. All of the species occur in the intestine of vertebrates. *Chilomastix mesnili* (Fig. 4.45) occurs in the large intestine of humans, chimpanzees, orangutans, monkeys, and pigs. About 3.5% of the human population of the United States and 6% of the world population harbor this flagellate. This pear-shaped protozoan, measuring 5–24 μm long, is sometimes found in diarrheic feces. *Chilomastix intestinalis* occurs in guinea pigs, *C. cuniculi* in rabbits, and *C. gallinarum* in chickens. Transmission of the various species occurs via cysts.

The trophozoite of *C. mesnili* possesses a surficial, longitudinal, spiral groove along the middle of the body. Also, there is a sunken cytostomal groove near the anterior end. Along each side of the cytostome is a cytoplasmic cytostomal fibril. The cytostome leads into the cytopharynx, where endocytosis occurs. Four flagella, one longer than the others, emerge from kinetosomes situated at the anterior end of the body. The kinetosomes are interconnected by microfibrils. One of the flagella is very short; it curves back into the cytostome, where it undulates.

The nonpathogenic *C. gallinarum*, 10–13 μm long, is commonly encountered in poultry.

Genus *Retortamonas*. Many species belonging to this genus are parasitic in the intestines of invertebrates; for example, *R. gryllotalpae* (7–14 μm long) is found in the mole cricket, *Gryllotalpa gryllotalpa* (Fig.

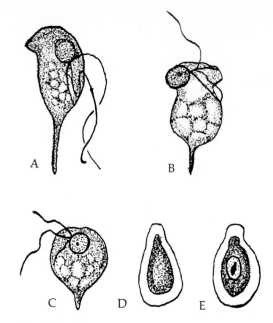

Fig. 4.46.   Some species of *Retortamonas*. A. *R. gryllotalpae* from intestine of the mole cricket, *Gryllotalpa gryllotalpa*. **B.** *R. blattae* from colon of cockroaches. **C.** *R. intestinalis* from human intestine. **D.** Unstained cyst of *R. intestinalis*. **E.** Stained cyst of *R. intestinalis*. (All redrawn after Wenyon, 1926, 1932.)

4.46), and *R. blattae* (6–9 μm long) in cockroaches (Fig. 4.46). Some, however, are nonpathogenic parasites of vertebrates, for example, *R. caviae* (4–7 by 2.4–4.3 μm) in the caecum of guinea pigs, and *R. intestinalis* (Fig. 4.46) in the human intestine. This latter species is pleomorphic but is often pyriform or oval with a drawn-out posterior end. It measures 4–9 by 3–4 μm and has a large cytostome. It possesses two flagella, one of which extends anteriad, whereas the other emerges from the cytostomal groove and trails posteriad. Its cysts (Fig. 4.46), each measuring 4.5–7 μm long and including a single nucleus, has been found in diarrheic feces, but it is doubtful whether this flagellate is responsible for the diarrhea. This species has also been reported from monkeys and chimpanzees. It has been suggested that *R. intestinalis* may be an insect symbiont that has been accidentallly introduced into the human through ingestion of an infected insect. The type species of the genus is a parasite of crane flies.

ORDER DIPLOMONADIDA

Genus *Giardia*. Members of the genus *Giardia* are representative of the Diplomonadida. These flagellates are characterized by a blunt anterior terminal and a tapering posterior end (Figs. 4.47 and 4.48). The dorsal surface of the body is convex, and the ventral surface is concave, forming an adhesive disc. There are two nuclei, two axostyles, four pairs of flagella

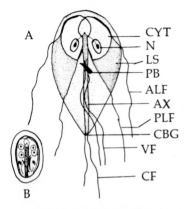

Fig. 4.47. *Giardia lamblia.* A. Trophozoite. B. Cyst. ALF, anterior lateral flagellum; AX, axostyle; CBG, caudal basal granule; CF, caudal flagellum; CYT, cytostome; LS, lateral shield; N, nucleus; PB, Parabasal body; PLF, posterior lateral flagellum; VF, ventral flagellum.

Fig. 4.48. *Giardia muris.* Scanning electron micrograph showing dorsal (upper) and ventral (lower) aspects of *G. muris.* The upper trophozoite is attached to the microvillous surface, and the lower trophozoite reveals the adhesive disc surrounded by the ventrolateral flange and paired ventral flagella. Prior attachment sites appear as circular indentations formed by edges of adhesive disc. (After Owen, Nemanic, and Stevens, 1979.)

present, and no cytostome. Cysts are the transmission stage. Numerous species of *Giardia* have been described from many vertebrates, and one has even been found in a nematode. The hosts include tadpoles, frogs, many species of rodents, rabbits, humans, other primates, and other vertebrates. Unfortunately, the descriptions of most of the species have been based on the concept that a different host species must involve a different parasite species. Filice (1952) has pointed out that the morphology of *Giardia* varies considerably with such factors as host diet, and he considers many of the described species "races."

Giardia lamblia. Of the numerous species of *Giardia*, *G. lamblia* of humans holds a prominent place (Fig. 4.47). This flagellate, measuring 8–16 μm in length by 5–12 μm in width, is found in the small intestine, particularly in the duodenum and occasionally invades the bile ducts, causing cholecystitis (Iwata and Araki, 1960). In life, each individual holds onto the host's intestinal mucosa by employing its concave ventral body surface as a suction cup. Its distribution is cosmopolitan. Giardiasis, the disease caused by this parasite, is highly contagious. A summary of surveys of 134,966 persons throughout the world has revealed that the infection rates range from 2.4 to 67.5%. In recent years, water-borne giardiasis has reached epidemic proportions in several areas of the United States (Aspen, Colorado, 123 cases; Rome, New York, 4800 cases; Camas, Washington, 600 cases; Berlin, New Hampshire, 750 cases) (Craun, 1979). Widespread infections among children commonly occur as a result of unsanitary conditions in day-care centers. Also, Americans traveling in the Soviet Union have frequently become infected with *G. lamblia*.

Clinical manifestations of *G. lamblia* infection can range from asymptomatic cyst passage to severe malabsorption. Prominent symptoms include diarrhea, abdominal cramps, fatigue, weight loss, flatulence, anorexia, and nausea. An incubation period of 1 to 8 weeks is typical, and the mean duration of acute illness is often 2 to 3 months (Wolfe, 1979). Some destruction of the host's mucosal cell membranes by the physical action of the adhesive disc of *Giardia* may also interfere with absorption of nutrients. The presence of large numbers of *Giardia* adhering to the intestinal wall is known to cause mechanical interference with absorption of fats. Thus, there is the possibility of vitamin deficiencies, especially fat-soluble vitamins, in hosts. For a detailed review of the species of *Giardia* and giardiasis, see Meyer and Radulescu (1979).

Encystment. Although many of the factors favoring the encystation of *Giardia* remain in doubt, the lowering of the intestinal pH to about 6.7 in rats causes encystation. On the other hand, dilution of the feces and a short exposure to normal body temperature apparently are sufficient to induce encystation.

Because giardiasis has become recognized as a common human parasitic disease in the United States, a special symposium was sponsored by the U.S. Environmental Protection Agency (Jakubowski and Hoff, 1979). The proceedings from this meeting is recommended to these desiring further details.

Other Species of Giardia. *Giardia muris,* measuring 7–13 µm long, is a fairly common parasite of rodents (Fig. 4.48). I recall seeing thousands of these flagellates lining the intestinal wall of a white-footed deer mouse in Virginia. Other species of *Giardia* have been described, but after critical study, Filice (1952) considers these invalid.

Genus *Hexamita.* Members of *Hexamita* are closely related to those of *Giardia.* Their bodies are pyriform and include two vesicular nuclei located near the anterior end, one on each side of the midsagittal plane. There are eight flagella present; six originating anteriorly and two trailing. Two axostyles, which may be tubular rather than rodlike, are present. *Hexamita* includes free-living as well as parasitic species.

Hexamita meleagridis. Parasitic species of *Hexamita* include *H. meleagridis* found in the small intestine and sometimes in the caecum and bursa of Fabricius of turkeys, quails, pheasants, and partridges. It causes an infectious catarrhal enteritis known as hexamitiasis in young birds. It has been reported from the United States, Great Britain, and South America. It probably occurs elsewhere also. In the United States, hexamitiasis causes losses amounting to millions of dollars to the turkey industry annually.

H. meleagridis, measuring 6–12 µm long by 2–5 µm wide, multiplies by longitudinal fission. Transmission from one avian host to another occurs via feed and water contaminated with the fecal droppings of infected birds. Cysts of *H. meleagridis* have been reported but not confirmed, although certain other species of *Hexamita*—for example, *H. salmonis* in the intestine of various species of trout and salmon—are known to encyst. Interested readers are referred to Becker (1977) for a comprehensive review of the species of *Hexamita* parasitic in fish.

Other Parasitic Hexamita *Species.* Other *Hexamita* species that are obligatory parasites include *H. columbae,* measuring 5–9 µm in length by 2.5–7 µm in width, found in the intestine of pigeons; *H. muris,* 7–9 by 2–3 µm, found in the intestine of rats, mice, hamsters, and various wild rodents; and *H. intestinalis,* 10–16 µm long, found in the intestine of frogs. A few

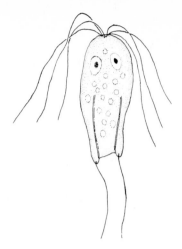

Fig. 4.49. *Hexamita nelsoni,* **a facultative parasite of the digestive tract of the American oyster,** *Crassostrea virginica.*

species are parasites of invertebrates. For example, *H. periplanetae,* 5–8 µm long, is found in cockroaches.

Hexamita nelsoni. Among the free-living species, which are generally found in stagnant water, *H. nelsoni* is of considerable interest to those interested in marine parasitology (Fig. 4.49). This flagellate, originally thought to be *H. inflata,* another free-living species, is broadly oval with a truncate posterior end. It measures 13–25 by 9–15 µm. It is a fairly common saprobic member of the community of microorganisms found at the mud-water interface in coastal areas. It can become a facultative parasite of oysters, primarily found in the stomach, especially when anaerobic conditions and other environmental factors in oyster bed sediments reach such a level that the oysters are placed under stress.

Laird (1961) suggested that hexamitiasis in oysters may be a major cause of mass oyster mortalities, especially if conditions continue to render the environment of oysters physiologically unfavorable, thus causing them to become progressively less resistant to facultative parasites. This concept has gained favor in recent years relative to protozoan facultative parasites. At any rate, *H. nelsoni* (or some related species) has been implicated as the cause of oyster mortalities in Prince Edward Island, Canada, the Pacific Northwest of the United States, and Holland.

Seasonal fluctuations in the number of *H. nelsoni* parasitizing the American oyster, *Crassostrea virginica,* have been studied by Scheltema (1962). *Hexamita nelsoni* is most abundant in oysters (54.5%) in the Delaware Bay during winter and early spring (Fig. 4.50). By culturing *H. nelsoni* in the laboratory at different temperatures, Scheltema has found that this flagellate multiplies at low temperatures, but its rate of reproduction increases with increasing water temperature up to approximately 20°C. Temperatures above 25°C are lethal. Because the metabolic rate of oysters is decreased at low temperatures, Scheltema has postulated that during the winter and early spring when the temperature of water in Delaware Bay is low, the metabolism of the host is decreased, and *H. nelsoni* is apparently capable of reproducing rapidly enough to be found in large numbers. During later spring and summer, however, when the host's metabolism is increased, its internal defense mechanisms (leucocytosis, phagocytosis; see p. 67) remove the parasite at a greater rate than the parasite's reproduction rate. Thus, fewer parasites are present during warmer periods. If the metabolic rate of the oyster is reduced during the warm months, resulting from stress for example, large numbers of *Hexamita* are present because its reproductive rate is enhanced by the higher temperature, and at the same time the host is not as efficient at removing it. These physiologic findings appear to corroborate Laird's hypothesis.

Feng and Stauber (1968) have studied experimental hexamitiasis in the oyster *Crassostrea virginica*. They have reported that the *in vivo* growth of *H. nelsoni* orally and intracardially introduced into oysters is also temperature dependent. Specifically, when 3.2×10^6 *Hexamita* were injected intracardially into each oyster and maintained at 6°C, the flagellates increased in number after a lag phase of 8 days and all of the infected oysters died within 18 days. When 4.0×10^5 *Hexamita*, a smaller dosage, were introduced into each oyster and maintained at 12°C, nine of ten died within 20 days. When the dosage injected was reduced to 1.0×10^3 *Hexamita* per oyster and maintained at 12°C, half survived. Furthermore, even if 8.8×10^5 *Hexamita* were inoculated and the oysters maintained at 18°C, not even a patent infection was observed in the 2-week period. These data indicate

that both the number of parasites and the ambient temperature are crucial factors in hexamitiasis in oysters. As might be expected, the larger the number of parasites introduced, the more deaths result, and the higher the ambient temperature (within reason), the fewer deaths result, since the internal defense mechanisms of the host, especially phagocytosis (p. 67), are more efficient, for example, when maintained at 18°C as compared to 12°C.

ORDER OXYMONADIDA

All the members of this relatively small order are symbiotic, primarily in invertebrates. Representative genera include *Oxymonas, Pyrsonympha, Dinenympha, Notila*, and *Saccinobaculus* (Fig. 4.51). All of these occur in the intestines of termites and woodroaches and are mutualists. The major characteristics by which the members of Oxymonadida can be recognized are listed in the section devoted to classification of flagellates at the end of this chapter.

ORDER HYPERMASTIGIDA

The members of this order include all the multiflagellated species found in the alimentary tract of termites, woodroaches, and cockroaches. The major identification characteristics of the members of this order and the two subordinate suborders, Lophomonadina and Trichonymphina, are presented in the section devoted to the classification of flagellates at the end of this chapter.

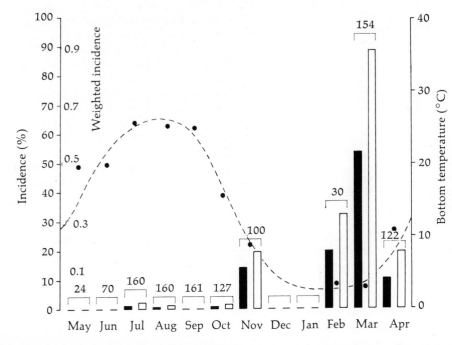

Fig. 4.50. Incidence of *Hexamita nelsoni* in the stomach of dredge samples of living oysters, *Crassostrea virginica*, taken from Delaware Bay between May, 1959, and April, 1960. Solid bars denote percentage incidence, open bars denote index of weighted incidence. Numbers above bars are the sample sizes on which the data from each month are based. Points on the temperature curve indicate single measurements of bottom temperature taken at the time oyster collections were made. (After Scheltema, 1962.)

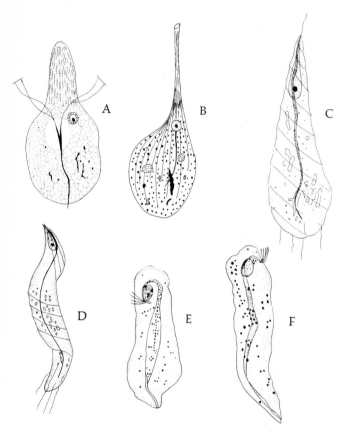

Fig. 4.51. Some symbiotic flagellates of insects. A. *Oxymonas dimorpha* parasitic in the gut of the termite *Neotermes simplicicornis* in California and Arizona; 17–195 by 14–165 μm. **B.** Attached nonflagellated form of *O. dimorpha*. (Redrawn after Connell, 1930.) **C.** *Pyrsonympha vertens* in the gut of the termite *Reticulitermes flavipes*, 100–150 μm long. **D.** *Dinenympha gracilis* in the gut of *Reticulitermis flavipes*, 25–50 by 6–12 μm. (**C** and **D**, redrawn after Kudo, 1966. "Protozoology," 5th ed. Charles C Thomas.) **E.** *Notila proteus* in the gut of the woodroach *Cryptocercus punctulatus*. (Redrawn after Cleveland, 1950.) **F.** *Saccinobaculus ambloaxostylus* in the gut of the woodroach; 65–110 by 18–26 μm. (Redrawn after Cleveland, 1950.)

Generally, the number of hypermastigids per insect host is large. Hungate (1939) has estimated the mass of these flagellates to be one-seventh to one-fourth of the total weight of termite workers, Cleveland (1925a) has stated that it may be high as one-half, and Katzin and Kirby (1939) have estimated it to be one-third of the total body weight. In any case, the number of flagellates is large, and in the instances of the cellulose-digesting species found in termites of the families Kalotermitidae, Mastotermitidae, and Rhinotermitidae,

the mutualistic relationship is essential to both host and flagellate.

Rigid host specificity occurs among the Hypermastigida. Flagellates of one host, when transferred to another (a process known as **transfaunation**), do not survive or survive only briefly. The relationship between the host and the flagellates is so firmly entrenched that definite correlations of the metabolic processes of both organisms exist as discussed below.

For those interested in the taxonomy of the Hypermastigida, a key to the families of the order and a listing of representative species are given by Kudo (1966). This order includes two suborders: Lophomonadina and Trichonymphina.

Lophomonadina

The suborder Lophomonadina includes *Lophomonas* (Fig. 4.52) characterized by being ovoid or elongate, with a small, anteriorly situated vesicular nucleus, and an axostyle composed of many filaments. The known species, including *L. blattarum*, occur in the colon of cockroaches. These flagellates include a cyst stage in their life cycles. Also included in the Lophomonadina are the genera *Joenia* and *Staurojoenina* in the gut of termites (Fig. 4.52).

Trichonymphina

The suborder Trichonymphina includes the genera *Barbulanympha*, *Rhynchonympha*, *Trichonympha*, *Urinympha* (Fig. 4.52), among others. All members of these genera are found in the woodroach *Cryptocercus punctulatus*.

Mutualistic Relationship

The greatly enlarged hindgut of the woodroach *Cryptocercus punctulatus* harbors numerous mutualistic flagellates of different species. The late L. R. Cleveland and his associates conducted many studies on the effects of certain hormones of the host on these mutualists (Cleveland, 1960). Details of these elaborate studies have been summarized by Cheng (1970). In brief, a hormone of the roach host, known as ecdysone, is known to stimulate the initiation of gametogenesis in the flagellates, i.e., their sexual cycles, which terminates when a male and a female gamete fuse.

The *Cryptocercus* ecdysone-flagellate story serves as an example of how an invertebrate host's hormone not only affects, but is necessary for, the initiation of one indispensable aspect of the biology of a symbiont.

Some striking differences exist between the termite-flagellate relationship and that between *Cryptocercus* and its flagellates. In *Cryptocercus*, the dense flagellate population is maintained throughout life, starting 2 to 3 days after hatching. In all termites except members of the family Termitidae, large numbers of flagellates also occur in the hindgut, but these die at each of the four initial molts, and the subsequent instar has to be reinfected by feeding on the feces of an infected termite. The flagellates do not die at the time of the final molt, during which the fifth instar develops into the adult. This departure from the *Cryptocercus*-flagellate pattern is interesting, and no one has yet satisfactorily demonstrated what causes the death of the flagellates. It should also be pointed out that in sexually mature adult termites of most genera, the protozoan population declines with age, increase in size, and greater reproductive activity of the hosts. In time, all the flagellates disappear. Sexuality usually does not occur among termite-infecting flagellates. Only rarely does it occur in certain genera, strongly suggesting that it is of no consequence in their reproduction. For additional examples of mutualism involving protozoans and other microorganisms, see Trager (1970) and Whitfield (1979).

PHYSIOLOGY AND BIOCHEMISTRY OF PARASITIC FLAGELLATES

OXYGEN REQUIREMENTS

All parasitic protozoans can be grouped into two categories according to their oxygen requirements—aerobic types require some amount of molecular oxygen, and anaerobic types require none. Some anaerobic species are **obligatory anaerobes**, that is, they cannot utilize oxygen, and, furthermore, the presence of the

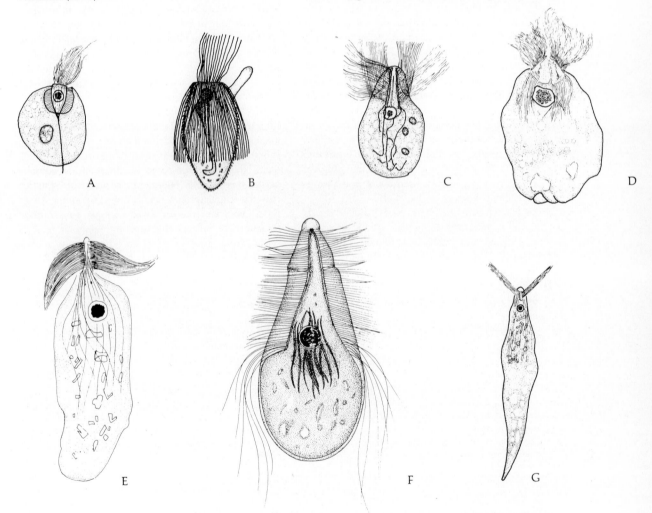

Fig. 4.52. Some hypermastigidans. A. *Lophomonas blattarum* from the colon of the cockroach *Blatta orientalis*, 25–30 μm long. (Redrawn after Kudo, 1926.) **B.** *Joenia annectens* from gut of termite. (Redrawn after Grassi and Foà, 1911.) **C.** *Staurojoenina assimilis* from gut of termite, 105–190 μm long. (Redrawn after Kirby, 1926.) **D.** *Barbulanympha ufalula* from the gut of the woodroach *Cryptocercus punctulatus*, 250–340 μm by 175–275 μm. (Redrawn after Cleveland *et al.*, 1934.) **E.** *Rhynchonympha tarda* from the gut of *Cryptocercus punctulatus*, 130–215 μm by 30–70 μm. (Redrawn after Cleveland *et al.*, 1934.) **F.** *Trichonympha collaris* from the gut of the termite *Zootermopsis angusticollis*, about 150 μm long. (Redrawn after Kirby, 1932.) **G.** *Urinympha talea* from the gut of *Cryptocercus punctulatus*, 75–300 μm by 15–50 μm. (Redrawn after Cleveland *et al.*, 1934.)

slightest amount of oxygen is toxic. Other anaerobic species are **facultative anaerobes**—they can survive without molecular oxygen if it is not available. Krogh (1941) has postulated that true anaerobic flagellates do not exist, because they require some amount of oxygen. This hypothesis is probably true, although many flagellates, such as those inhabiting the gut of termites, are extremely sensitive to molecular oxygen and are readily killed when exposed to small amounts. Many such flagellates have been considered anaerobes. Very little oxygen exists in the gut of termites, for the oxygen tension under the chitinous exoskeleton of insects is only 2 to 18% of that found in air, and these subskeletal areas have much more contact with the external environment than the gut. Nevertheless, even the slightest amount of oxygen is sufficient for these microorganisms. Whether true obligatory anaerobes exist among the parasitic flagellates remains in doubt, although *Trichomonas tenax* is apparently one. At first this report appears questionable, since *T. tenax* is known to live in the human mouth, where oxygen abounds. However, von Brand (1946) has suggested that what little oxygen is present in the gingival cervices surrounding teeth is used up by the accompanying bacterial flora.

It has been shown that some of the trichomonads can be cultured *in vitro* in serum with little or no oxygen. Willems *et al.* (1942) have demonstrated that in cultures of *Trichomonas hepatica*, optimum respiration occurs when oxygen forms 5 to 10% of the gas phase. Forty percent of the organisms died when cultured in the presence of air, and 75% died when maintained in pure oxygen. These results indicate that this species should be considered a **microaerobe**—that is, one which requires only trace quantities of oxygen. In the light of experiments such as this, Krogh's hypothesis may be correct, since many flagellates considered to be anaerobes may well be microaerobes or facultative anaerobes at best. Such is the case among certain trypanosomes that thrive best at low molecular oxygen tensions (Moulder, 1948).

Although oxygen is utilized by hemoflagellates, the rate of consumption varies between those found in the blood and those in tissue or culture forms. For example, 10^8 organisms of the blood form of *Trypanosoma brucei gambiense* utilize 170 mm³/hour of oxygen, whereas 10^8 organisms of the culture form utilize only 14 mm³/hour. Tissue forms consume even less. In addition, various physical and chemical factors influence oxygen consumption. Although oxygen tension and ionic concentrations appear to have little effect, temperature and the availability of sugar have pronounced effects on oxygen consumption. Elevated temperatures are accompanied by greater consumption, up to a point. The abundance of glucose in the medium is correlated with a high rate of oxygen consumption. With the depletion of glucose in the medium, the respiratory rate declines and eventually becomes negligible. This can be explained by the fact that the flagellates can metabolize glucose in energy production, and the increased oxygen consumption undoubtedly reflects the greater metabolic rate resulting from glucose utilization. The rates of oxygen consumption by some flagellates in the presence and absence of sugar are tabulated in Table 4.6.

Table 4.6. Rates of Oxygen Consumption of Some Parasitic Protozoa at an Oxygen Tension of Approximately 160 mm Hg[a]

| Species | Stage[b] | Temp. (°C) | Absence of Sugar | | Presence of Sugar | | RQ in the | |
			mm³/10⁸/hour	mm³/mg n/hour	mm³/10⁸/hour	mm³/mg n/hour	Absence of Sugar	Presence of Sugar
Endotrypanum schaudinni	C	30	4	29	47	309		
Leishmania enriettii	C	30	3	36	24	251		
Strigomonas oncopelti	C	30		146		350–600	0.9	
Trypanosoma congolense	B	37			136			1.0
Trypanosoma congolense	C	30			38			0.9
Trypanosoma cruzi	C	30	9	71	23	178		
Trypanosoma equiperdum	B	37			149			
Trypanosoma brucei gambiense	B	37			161			0.09
Trypanosoma brucei gambiense	C	29			18			0.97
Trypanosoma lewisi	B	37		40		600		0.97
Trypanosoma vivax	B	37			167			
Trichomonas batracharum	C	37	82		127		1.07	
Trichomonas vaginalis	C	37	96		269			1.02
Tritrichomonas foetus	C	37	120	176	350	350		

[a] Data compiled by von Brand, 1973.
[b] C, culture forms; B, bloodstream forms.

The age of trypanosomes appears to influence the respiratory rate. For example, young dividing *Trypanosoma lewisi* show a lower oxygen consumption than older nondividing individuals. This phenomenon is related to the presence of ablastin, which interferes with the oxidative glucose metabolism of young individuals, since this antibody is produced early in the infection.

A correlation between respiratory intensity and age of the organism is demonstrated also by a number of other species of parasitic flagellates including *Trichomonas vaginalis*, *Tritrichomonas foetus*, *Leishmania donovani*, and *Trypanosoma cruzi*. In each case, there is a decline in oxygen consumption as the organism becomes older (von Brand, 1973).

Undoubtedly ectoparasitic flagellates, such as *Oodinium ocellatum*, and the swarm cells of parasitic Phytomastigophorea are true aerobes, for they are continuously bathed in oxygenated water in their habitats.

The sensitivities of various flagellates to molecular oxygen are not only interesting areas of study but also have provided parasitologists with an ideal experimental tool. For example, Cleveland (1925a, b) has demonstrated that when various hosts, such as termites, cockroaches, earthworms, frogs, goldfish, and salamanders, are placed in more than a normal abundance of oxygen, their endosymbionts are killed without the hosts themselves suffering any injury. This technique has provided workers with symbiont-free hosts for experimental work. Cleveland has also pointed out that the symbionts of different hosts vary in their response to various oxygen tensions. For example, the flagellates of the termite *Leucotermes* are eliminated by an oxygen tension of 1 atm within 24 hours, but those of *Reticulitermes* and *Cryptotermes* are still alive after 10 days when subjected to the same tension. The flagellates *Leptomonas* and *Polymastix* of cockroaches succumbed within 40 minutes. On the average, flagellate parasites are 67.5 times more susceptible to oxygen than are their hosts.

GROWTH REQUIREMENTS

Because of the relative ease with which flagellates can be cultured *in vitro*, more is known concerning the nutritional requirements of these protozoans. For many, a simple 1.5% nutrient agar slant covered with sterile Ringer's solution containing 1 to 20 parts horse serum is an effective culture medium. Unlike amoebae, many flagellates can be cultured axenically (i.e., without accompanying bacteria), thus eliminating the problem of determining what is actually being utilized by the protozoan and what by the bacteria.

The addition of other materials to basic culture media is beneficial in the culture of a number of flagellates. For example, among earlier studies not involving chemically defined culture media, Cailleau (1938 and earlier papers) has demonstrated that *Tritrichomonas foetus* can be grown in human serum and *T. columbae*

in pigeon serum only when cholesterol and ascorbic acid are added. The same investigator has shown that in culturing *T. batrachorum*, cholesterol and linoleic acid are needed. Guthrie (1946), knowing that ascorbic acid is beneficial when added to *T. foetus* cultures, replaced this vitamin with thioglycolate and found that the growth was just as rich, suggesting that ascorbic acid *per se* is not the immediate requirement but its reducing ability is the beneficial property.

As the culture of flagellates progressed, attempts have been made to devise axenic media and chemically defined ones, that is, media in which each ingredient is quantitatively known, and hence cannot contain sera, yeast extract, or any other biologic entity. Trager (1974) has contributed an authoritative review of the nutrition and biosynthetic capabilities of flagellates as related to *in vitro* cultivation, and Taylor and Baker (1978) have edited a volume in which are listed the available methods for culturing flagellates, including *Giardia*, *Trichomonas*, *Leishmania*, and *Trypanosoma*.

As progress is being made in *in vitro* culture, many investigators have directed their research toward the mechanisms involved in the uptake and utilization of nutrients. The volume by Gutteridge and Coombs (1977) as well as the review by Goodwin (1979) and the volumes edited by Kreier (1977, 1978) should be consulted for greater detail. The following represents a brief summary of what is known.

STORED ENERGY

No trypanosomatid has been shown to contain large quantities of stored polysaccharides; however, varying amounts of lipids have been reported from these flagellates. Energy stores must exist in such species as *Trypanosoma cruzi*, *Crithidia fasciculata*, and *Leishmania* spp. since they have high endogenous respiratory rates. The identity of stored nutrients has yet to be clarified, although Rogerson and Gutteridge (1980) have reported that triglyceride is the main energy reserve in *Trypanosoma cruzi*. One possibility is that polyphosphate, a polymer composed of phosphate monomers, functions as an energy store. Such molecules have been isolated from *C. fasciculata* and other trypanosomatids. A number of functions have been attributed to polyphosphates, including energy reserves (as phosphagens), phosphorus stores, and a role in controlling orthophosphate levels. Relative to the role of these molecules as stored energy, it has been suggested that the more primitive flagellates utilize polyphosphates or pyrophosphates as energyrich intermediates in their metabolism leading to the synthesis of energy. On the other hand, the more advanced flagellates utilize ATP.

SUBSTRATE UTILIZATION

Among hemoflagellates, such as the trypanosomes, there is no doubt that glucose is by far the most important exogenous substrate, although other hexoses and glycerol can also be utilized by certain species (Table 4.7). *Trypanosoma brucei brucei*, for example, can take up glucose equal to its own dry weight in 1 to 2 hours. Amino acids and short-chain lipids can also be utilized as substrates for the synthesis of energy among most parasitic flagellates. In Table 4.8 are listed the amino acids taken up by some parasitic flagellates.

TRANSPORT INTO CELL

We are just beginning to understand how various molecules are transported into the body of flagellates. However, evidences suggest that mediated systems are involved and that more than one transport site exists. For example, *Trypanosoma equiperdum* includes three sites: two for the transport of glycerol and one for hexoses. In other species, such as *T. brucei gambiense*, there are two sites for the uptake of hexoses. Specifically, it takes up glucose and mannose at one site and fructose at a second. In the case of *T. lewisi*, one site is specific for glucose while the second for fractose and

mannose (see Honigberg, 1967, for review). The uptake of amino acids possibly occurs at the same sites where the hexoses are taken up.

The mechanism of lipid absorption by flagellates remains somewhat obscure. Wotton and Becker (1963) have suggested that a combination of direct diffusion, ion pump mechanisms, endocytosis, direct porosities, and the rearrangement of protein molecules on the outer surface of the surface membrane are involved.

METABOLISM

Trypanosomes living in their vertebrate hosts' blood can be divided into three groups based on the end products of their aerobic glucose metabolism (Table 4.9). One group, consisting of the **brucei** group,[*] produce mainly pyruvate. The second group, consisting of *Trypanosoma cruzi*[†] and *T. lewisi*, produce mainly CO_2 but also some acetate and succinate. The third group, containing *Trypanosoma vivax*,[‡] appears to be

---

[*] In recent years there has been the tendency to consider the species of *Trypanosoma* of the **brucei** group (p. 131) as belonging to the subgenus *Trypanozoon*.

[†] If the nomenclatural system explained above is accepted, then *Trypanosoma cruzi* is considered a member of the subgenus *Schizotrypanum*, hence *Trypanosoma (Schizotrypanum) cruzi*.

[‡] If one accepts *Trypanozoon* and *Schizotrypanum* as valid subgenera, then *Trypanosoma vivax* should be assigned to the subgenus *Duttonella*, hence *Trypanosoma (Duttonella) vivax*, and *Trypanosoma congolense* to the subgenus *Nannomona*, hence *Trypanosoma (Nannomonas) congolense*.

Table 4.7. Carbohydrates and Related Compounds Utilized by Parasitic Flagellates[a,b]

Species	Pentoses — Arabinose	Rhamnose	Xylose	Hexoses — Fructose	Galactose	Glucose	Mannose	Disaccharides — Lactose	Maltose	Saccharose	Trisaccharide — Raffinose	Polysaccharides — Dextrin	Inulin	Soluble Starch	Alcohols — Glycerol	Mannite	Sorbite	Glucoside — Amygdalin
Trypanosoma																		
T. equiperdum			○	●	●	●	●		●									
T. brucei brucei				●	○	●	●		●						●	○		
T. lewisi				●		●	●		●									
T. cruzi						●			●									
Leishmania																		
L. donovani				●	●	●	●		●	●	●							
L. braziliensis				●	●	●			●	●	●							
L. tropica				●	●	●			●	●	●							
Leptomonas ctenocephali				●		●												
Herpetomonas																		
H. culicidarum	●		●	●	●	●	●		●	●	●	●	●		●			●
H. muscidarum	●			●	●	●	●	●	●	●	●	●	●		●			
H. parva				●	●	●	●		●	●	●	●	●		●			
H. media				●	●	●	●			●	●					●		
Tritrichomonas foetus				●	●	●	●	●	●	●	●	●				●	●	
Trichomonas vaginalis				●	●	●			●			●					●	

[a] Reorganized after von Brand, 1973.

[b] ●, used by flagellate; ○, very slightly used.

intermediates, producing pyruvate, other organic acids, and CO_2.

Trypanozoon Metabolism

The metabolism of glucose to pyruvate by the long slender form of bloodstream trypomastigotes of the subgenus *Trypanozoon* involves the glycolytic pathway and is associated with the concomitant production of energy and reduced coenzymes as shown in Fig. 4.53.

Available evidence indicates that the glycolytic enzymes of these trypanosomes are similar to those occurring in mammals. It is noted that some enzymes of the phosphogluconate pathway are present. These probably function only in the synthesis of pentoses and NADPH. Also, most of the enzymes of the Krebs cycle and the respiratory chain are absent, and mitochondria, although present, include very few, poorly developed tubular cristae.

Although oxygen is consumed by members of the subgenus *Trypanozoon*, its utilization is not inhibited by cyanide, indicating that it does not involve the cytochrome system (specifically $a + a_3$). It is now known that the combined activity of two enzymes, L-α-glycerophosphate dehydrogenase and L-α-glycerophosphate oxidase, reoxidizes NADH and produces water (Fig. 4.54). Cyanide has no effect and there is no linked energy production. This system appears to be unique to salivarian trypanosomes.

The L-α-glycerophosphate oxidase of salivarian trypanosomes has no parallel in mammals and hence has been extensively studied because it could be a target for rational chemotherapy (see Cheng, 1977, for a discussion of this approach to chemotherapy). It is now known that L-α-glycerophosphate oxidase is a particulate enzyme that contains iron and copper and

Table 4.8. Amino Acids Absorbed by Some Parasitic Flagellates[a]

Species	Alanine	Arginine	Asparagine	Aspartic Acid	Cysteine	Glutamic Acid	Glutamine	Glycine	Histidine	Isoleucine	Leucine	Lysine	Methionine	Phenylalanine	Proline	Serine	Threonine	Tryptophan	Tyrosine	Valine	Method[b]
Crithidia fasciculata		×							×	×	×	×	×	×				×	×	×	E
Leishmania donovani			×				×														O
Leishmania tarentolae		×				×			×	×	×	×	×	×	×	×	×	×	×	×	E
Strigomonas oncopelti		×		×		×	×														O
Strigomonas oncopelti	×			×																	L
Strigomonas oncopelti													×								E
Tritrichomonas foetus						×	×														O
Tritrichomonas foetus		×							×	×	×	×	×	×	×	×	×			×	E
Trichomonas vaginalis		×			×				×				×								D
Trypanosoma lewisi, B			×	×		×	×														O
Trypanosoma lewisi, B			×	×			×														D
Trypanosoma brucei rhodesiense, C						×	×														O

[a] Data compiled by von Brand, 1973.
[b] E, essential in culture; O, stimulated O_2 consumption; L, incorporated labeled compound; D, deaminated compound; B, bloodstream form; C, culture form.

Table 4.9. End Products of Aerobic Glucose Metabolism in Three Groups of Bloodstream Trypanosomes.

	Carbon dioxide	Glycerol	Ethanol	Lactate	Pyruvate	Acetate	Succinate	Citrate
Trypanosoma brucei brucei	−	+	−	−	+ +	−	−	−
Trypanosoma brucei rhodesiense	±	+	−	−	+ +	−	−	−
Trypanosoma brucei gambiense	−	+	−	−	+ +	−	−	−
Trypanosoma evansi	−	+	−	−	+ +	−	−	−
Trypanosoma equinum	−	+	−	−	+ +	−	−	−
Trypanosoma equiperdum	−	+	−	−	+ +	−	−	−
Trypanosoma cruzi	+ +	−	−	±	−	+	+	−
Trypanosoma lewisi	+ +	−	−	±	−	+	+	−
Trypanosoma vivax	+	+	−	±	+	−	+	−
Trypanosoma congolense	+	+	−	±	−	+	+	−

$$Glucose + 2 ADP + 2P_i + 2NAD^+$$
$$\rightarrow 2 \text{ Pyruvate} + 2 ATP + 2 NADH + 2H^+$$

Fig. 4.53

is localized in the mitochondria. High O_2 tensions are required by the system. It is also known that this enzyme system is inhibited by various hydroxamic acids such as *m*-chlorobenzhydroxamic acid and salicylhydroxamic acid. Tests, however, have revealed that although either of these acids will completely inhibit O_2 utilization, neither is lethal to the trypanosomes. This indicates that the operation of L-α-glycerophosphate oxidase is not essential for the parasites's survival, a conclusion confirmed by the anaerobic survival of bloodstream trypanosomes. This finding casts doubt on whether an inhibitor of the L-α-glycerophosphate oxidase system will alone be an effective antitrypanosomal agent. However, it is of interest to point out that salicylhydroxamic acid and glycerol combined is a potent trypanocidal agent.

In the absence of O_2, ATP production by the bloodstream form of *Trypanosomas brucei* resulting from glucose catabolism is half as much as in the presence of O_2, and glycerol and pyruvate are produced in equimolar amounts. The possible pathways are presented in Figure 4.55.

Schizotrypanum Metabolism

In members of the subgenus *Schizotrypanum* metabolism leading to energy production is quite different from that of members of the subgenus *Trypanozoon*. Evidences gathered on bloodstream trypomastigotes of *Schizotrypanum* indicate that glucose is catabolized via glycolysis to pyruvate. Again, the phosphogluconate pathway probably proceeds only as far as the production of pentose and NADPH, although in some strains at least about 25% of glucose catabolism proceeds via this pathway rather than by glycolysis (Bowman and Flynn, 1976).

Pyruvate can be further metabolized in four ways in *Schizotrypanum*. In three of these, NADH is reoxidized concomitantly. About 55% of pyruvate enters the Krebs cycle and is fully oxidized to CO_2. The

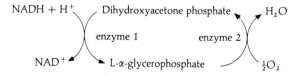

Fig. 4.54. The L-α-glycerophosphate oxidase cycle in blood stream *Trypanosoma brucei*. Enzyme 1 = L-α-glycerophosphate dehydrogenase; enzyme 2 = L-α-glycerophosphate oxidase.

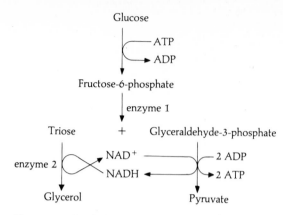

Fig. 4.55. Catabolism of glucose in blood stream *Trypanosoma brucei* under anaerobic conditions. Enzyme 1 = novel aldolase; enzyme 2 = glycerol dehydrogenase.

NADH produced is reoxidized via the cytochrome system, which is also available to reoxidize the NADH generated in glycolysis. The reaction of oxygen with the reduced terminal-electron acceptor is partially cyanide sensitive, indicating the occurrence of cytochrome $a + a_3$, but other terminal oxidases are also present. The presence of mitochondria with well-developed cristae supports the existence of these pathways. A small amount of pyruvate is excreted, and about 6% is reduced to lactate by the enzyme lactate dehydrogenase and is excreted. An additional 12% approximately is metabolized to succinate and about 17% to acetate. The present status of our understanding of glucose catabolism by the bloodstream trypomastigotes of *Trypanosoma* (*Schizotrypanum*) *cruzi* is presented in Fig. 4.56.

Bloodstream *Schizotrypanum* trypomastigotes can also oxidize fatty acids, yielding CO_2 and energy. Probably, acetyl CoA is produced during this process and is oxidized in the Krebs cycle.

From this discussion it is apparent that the bloodstream trypomastigotes of *Schizotrypanum* can obtain energy from several sources, including respiratory chain phosphorylation coupled to electron transport and substrate level phosphorylation during glycolysis and also during the production of succinate and acetate. Metabolism of *Trypanosoma* (*Schizotrypanum*) *lewisi* is probably similar.

The occurrence of well-developed mitochondrial cristae in the intracellular amastigote of *T. cruzi* suggests that its metabolism is similar to that of the bloodstream trypomastigotes. In fact, Rogerson and Gutteridge (1980) have reported that a complete glycolytic pathway and a complete Krebs cycle occur in the culture, bloodstream, and intracellular forms of *T. cruzi*.

Metabolism of African Trypanosomes

There is yet little information on the metabolism of *Trypanosoma* (*Duttonella*) *vivax* and *Trypanosoma* (*Nannomonas*) *congolense*; however, they appear to be inter-

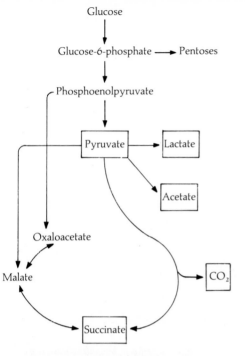

Fig. 4.56. Glucose catabolism in bloodstream form of *Trypanosoma cruzi*. The end products are in boxes.

mediates between members of the subgenera *Trypanozoon* and *Schizotrypanum* (Table 4.9). Both of these African species possess tubular mitochondrial cristae, and some of the Krebs cycle enzymes are present, although it is unlikely that either a fully functional Krebs cycle or cytochrome system exists.

HOST-PARASITE RELATIONSHIPS

What does a parasite derive from and inflict on its host? Besides mechanical damage and morphologic change, the host-parasite relationship is often a physiologic one. Such pathophysiologic effects of flagellates, especially the hemoflagellates, have been reviewed in the volume edited by Kreier (1977) and the symposium proceedings sponsored by the Ciba Foundation (1974). The diseases that such flagellates as *Trypanosoma brucei gambiense* and *Leishmania donovani* cause have been discussed. Quite often, however, the effect of a parasite on its host is subclinical and non-damaging.

Lincicome (1971) has advanced the concept that a reciprocal exchange of chemical substances occurs between parasite and host. In the case of *Trypanosoma lewisi* and its rat host, the enhanced growth of the host can be interpreted as evidence for a substance supplied by the parasite. Similarly, the enhanced growth of mice infected with *T. duttoni* serves as evidence.

Numerous studies have been carried out on the pathophysiologic effects of parasitic flagellates. Some of these are summarized below.

DISTURBANCES OF CARBOHYDRATE METABOLISM

It has been known since 1925 that trypanosomiasis can result in hypoglycemia. Such disturbances are now believed to be mediated by the malfunction of such organs as the adrenals, pancreas, and thyroid rather than due to direct utilization of glucose by the parasites (von Brand, 1973). It is also known that trypanosomiasis interferes with the normal synthesis and deposition of glycogen in the liver and other organs.

DISTURBANCES IN LIPID METABOLISM

The blood levels of some lipids appear to be quite variable in trypanosomiasis and leishmaniasis. Some have reported no change in blood cholesterol levels, others have reported an increase, and still others have reported a decrease (see von Brand, 1973). Similar contradictions exist in respect to phospholipids and fatty acids. There is agreement, however, that the neutral fat content in the liver is increased in cases of *Trypanosoma equiperdum*-caused trypanosomiasis.

DISTURBANCES IN PROTEIN METABOLISM

As expected, there are quantitative alterations in the serum protein fractions of hosts harboring various species of parasitic flagellates. These changes generally reflect increases in globulins and a compensatory decrease in the albumin fraction. Since globulins are constituents of antibodies, it is not surprising that when flagellates are exposed to the circulatory system, as in the case of the trypanosomes, there is an increase in the various globulin fractions.

PHYSICAL AND CHEMICAL FACTORS WITHIN THE HOST

Various physical and chemical factors, such as the oxidation-reduction potential and temperature, play important roles in the survival and normal metabolism of endoparasitic flagellates. From *in vitro* observations, the trichomonads appear to favor specific pH ranges. *Tritrichomonas foetus* thrives best in a pH range of 7.0 to 7.6, *Trichomonas vaginalis* from 5.4 to 6.0, and *Trichomonas gallinae* from 6.5 to 7.5. Most trichomonads multiply best at 37°C; however, strain differences may affect optimum growth temperatures. Nevertheless, almost all species multiply in a temperature range of 32° to 40°C. When maintained at 35° to 39°C, the minimal time required for one division is 5 to 7 hours.

HOST SEX AND AGE

Sex and age play important roles in parasitic infections. For example, male mice are more susceptible to *Trypanosoma cruzi* than are females, and adult humans become hypoglycemic when infected with *Leishmania donovani*, whereas children commonly become hyperglycemic. Sufficient evidences are now available to support the concept that adult male hosts, both vertebrates and invertebrates, are generally more susceptible to parasitism than are adult females.

PARASITE TOXICITY

Parasite toxicity is often suspected in the case of flagellates, although seldom proved. The substantiated cases appear to be limited to the trypanosomes. Concerning *Trypanosoma brucei brucei* and *T. brucei gambiense*, Laveran (1913) has reported that 0.12–0.15 g of dried trypanosomes injected into mice kills them. Schilling and Rondoni (1913) have amplified this experiment and found that the trypanosomes become toxic 1 hour after death and the toxic effect is lost after 18 hours of storage, suggesting that autolytic decomposition processes produce the toxic effect. Coudert and Juttin (1950) have indicated that some material must be secreted by *Trypanosoma cruzi*, because when extracts of this flagellate are injected into tumors in experimental animals, there is regression of growth. The toxicity of substances secreted by *T. cruzi* has also been postulated to account for degeneration of the host's nerve cells, especially in cardiac muscle.

It has also been reported that toxicity may occur when large numbers of trypanosomes are destroyed *in vivo* as the result of immunologic reactions and chemotherapy. Thus, toxic reactions in humans have been attributed to the destruction of both *Trypanosoma brucei rhodesiense* and *T. brucei gambiense* following therapy. Similar reactions in pigs infected with *Trypanosoma simiae* have been attributed to the mass lysis of parasites due to immunologic reactions.

HOST DIETS

Numerous investigators have indicated that the host's diet influences the degree of parasitemia in the case of hemoflagellates. Becker *et al.* (1943) have shown that if vitamin B_1 and B_6 are added in large dosages to the diet of rats infected with *Trypanosoma lewisi*, the number of parasites increases and the infection persists longer. These workers attributed the results to the inhibition of the production of ablastin. Caldwell and György (1943, 1947) obtained similar results when their rats were fed a diet deficient in biotin.

Not only do dietary changes influence the degree of parasitemia, they can actually influence resistance. For example, hosts maintained on vitamin B complex-deficient diets are more resistant to *Trypanosoma equiperdum*, and pigeons, which are normally resistant to *Trypanosoma brucei brucei*, when deprived of vitamin B complex can be infected. In mice, if vitamin E is given in the diet, the multiplication of *T. congolense* is accelerated; if cod liver oil is given, the rate of multiplication is slowed down. The same is true in rats infected with *T. vivax* (Godfrey, 1958).

Ritterson and Stauber (1949) have reported that the host's protein intake influences the course of leishmaniasis in the hamster. Protein-deficient diets lead to earlier emaciation and death, whereas excess dietary protein appears to favor host survival.

PHYSICAL FACTORS IN THE MACROENVIRONMENT

Certain physical factors in the environment are important in host-parasite relationships. It has been demonstrated by Wood (1954) that environmental temperatures of 22° to 23°C retard the appearance of the metacyclic form of *Trypanosoma cruzi* in the feces of *Triatoma protracta*, but temperatures from 28° to 34.5°C increase the number of metacyclic forms.

It is noted that since the invertebrate hosts of flagellates are ectothermic and are readily maintained under a variety of ambient physical conditions, these hosts should be utilized more often in studies of this nature.

It should be obvious that a multitude of factors, some yet unknown, may singly or jointly influence the host-parasite relationship. The future will undoubtedly witness many more investigations of host-parasite relationships, including immunologic and biochemical investigations as well as further studies of ecologic and physiologic aspects of parasitism.

REFERENCES

Abou Akkada, A. R., and Howard, B. H. (1960). The biochemistry of rumen protozoa. III. The carbohydrate metabolism of *Entodinium*. *Biochem. J.* **76**, 445–451.

Actor, P. (1960). Protein and vitamin intake and visceral leishmaniasis in the mouse. *Exp. Parasitol.* **10**, 1–20.

Baker, J. R. (1977). Systematics of parasitic protozoa. *In* "Parasitic Protozoa" (J. P. Kreier, ed.), Vol. 1, pp. 35–56. Academic Press, New York.

Ball, G. H. (1968). Organisms living on and in Protozoa. *In* "Research in Protozoology" (T.-T. Chen, ed.), Vol. 3, pp. 566–718. Pergamon Press, New York.

Becker, C. D. (1977). Flagellate parasites of fish. *In* "Parasitic Protozoa" (J. P. Kreier, ed.), Vol. 1, pp. 357–416. Academic Press, New York.

Becker, E. R., Manresa, M., and Johnson, E. M. (1943). Reduction in the efficiency of ablastin action in *Trypanosoma lewisi* infection by withholding panthothenic acid from the host's diet. *Iowa State Coll. J. Sci.* **17**, 431–441.

Bowman, I. B. R., and Flynn, I. W. (1976). Oxidative metabolism of trypanosomes. *In* "Biology of the Kinetoplastida" (W. H. R. Lumsden and D. A. Evans, eds.), Vol. 1, pp. 435–476. Academic Press, New York.

Bradley, R. E., and Reid, W. M. (1966). *Histomonas meleagridis* and several bacteria as agents of infectious enterohepatitis in gnotobiotic turkeys. *Exp. Parasitol.* **19**, 91–101.

Cailleau, R. (1938). Le cholésterol et l'acide ascorbique, facteurs de croissance pour le flagellé tétramitidé *Trichomonas foetus* Riedmüller. *C. R. Soc. Biol.* (*Paris*) **127**, 861–863.

Caldwell, F. E., and György, P. (1943). Effect of biotin deficiency on duration of infection with *Trypanosoma lewisi* in the rat. *Proc. Soc. Exp. Biol. Med.* **53**, 116–119.

Caldwell, F. E., and György, P. (1947). The influence of biotin deficiency on the course of infection with *Trypanosoma lewisi* in the albino rat. *J. Infec. Dis.* **81**, 197–208.

Camp, R. R., Mattern, C. F. T., and Honigberg, B. M. (1974). Study of *Dientamoeba fragilis* Jepps and Dobell. I. Electronmicroscopic observations of the binucleate stages. II. Taxonomic position and revision of the genus. *J. Protozool.* **21**, 69–82.

Chang, K.-P., and Dwyer, D. M. (1978). *Leishmania donovani* hamster macrophage interactions in vitro: cell entry, intracellular survival, and multiplication of amastigotes. *J. Exp. Med.* **147**, 515–529.

Cheng, T. C. (1970). "Symbiosis: Organisms Living Together." Bobbs Merrill, Indianapolis, Indiana.

Cheng. T. C. (1977). The control of parasites: the role of the parasite. Uptake mechanisms and metabolic interference in parasites as related to chemotherapy. *Proc. Helminth. Soc. Wash.* **44**, 2–17.

Cheng, T. C. (1983). Cutaneous lesions due to nonarthropod parasites. *In* "Cutaneous Infestations of Man and Animal" (L. C. Parrish, W. B. Nutting, and R. M. Schwartzman, eds.), pp. 237–254. Prager, New York.

Ciba Foundation (1974). "Trypanosomiasis and Leishmaniasis with Species Reference to Chagas' Disease." Elsevier, Amsterdam, Holland.

Clark, T. B., Kellen ,W. R., Lindegren, J. E., and Smith, T. A. (1964). The transmission of *Crithidia fasciculata* (Léger, 1902), in *Culiseta incidens* (Thomson). *J. Protozool.* **11**, 400–402.

Cleveland, L. R. (1925a). The effects of oxygenation and starvation on the symbiosis between the termite *Termopsis*, and its intestinal flagellates. *Biol. Bull.* **48**, 309–326.

Cleveland, L. R. (1925b). Toxicity of oxygen for protozoa *in vivo* and *in vitro*; animals defaunated without injury. *Biol. Bull.* **48**, 455–468.

Cleveland, L. R. (1960). Effects of insect hormones on the protozoa of *Cryptocercus* and termites. *In* "Host Influences on Parasite Physiology" (L. A. Stauber, ed.), pp. 5–10. Rutgers University Press, New Brunswick, New Jersey.

Conner, R. L. (1967). Transport phenomena in Protozoa. *In* "Chemical Zoology" (M. Florkin, B. T. Scheer, and G. W. Kidder, eds.), Vol. I, pp. 309–350. Academic Press, New York.

Corliss, J. O. (1967). Systematics of the phylum Protozoa. *In* "Chemical Zoology" (M. Florkin, B. T. Scheer, and G. W. Kidder, eds.), Vol. 1. Protozoa, pp. 1–20. Academic Press, New York.

Cosgrove, W. B. (1963). Carbohydrate utilization by trypanosomids from insects. *Exp. Parasitol.* **13**, 173–177.

Coudert, J., and Juttin, P. (1950). Note sur l'action d'un lysat de *Trypanosoma cruzi* vis-à-vis d'un cancer greffé du rat. *C. R. Soc. Biol.* **144**, 847–849.

Craun, G. F. (1979). Waterborne outbreaks of giardiasis. *In* "Waterborne Transmission of Giardiasis" (W. Jakubowski and J. C. Hoff, eds.), pp. 127–147. U.S. Environmental Protection Agency, Cincinnati, Ohio.

D'Alesandro, P.A. (1970). Nonpathogenic trypanosomes of rodents. *In* "Immunity to Parasitic Animals" (G. J. Jackson, R. Herman, and I. Singer, eds.), pp. 691–738. Appleton-Century-Crofts, New York.

D'Alesandro, P. A. (1972). *Trypanosoma lewisi*: Production of exoantigens during infection in the rat. *Exp. Parasitol.* **32**, 149–164.

DeRaadt, P., and Seed, J. R. (1977). Trypanosomes causing disease in man in Africa. *In* "Parasitic Protozoa" (J. P. Kreier, ed.), Vol. 1, pp. 175–237. Academic Press, New York.

Dobell, C. (1940). Researches on intestinal protozoa of monkeys and man; life history of *Dientamoeba fragilis*: observation, experiments, and speculations. *Parasitology* **32**, 417–461.

Dodge, J. D. (1979). The phytoflagellates: fine structure and phylogeny. *In* "Biochemistry and Physiology of Protozoa" (M. Levandowsky and S. H. Hutner, eds.), Vol. 1, 2nd ed, pp. 7–57. Academic Press, New York.

Doll, J. P., and Franker, C. K. (1963). Experimental histomoniasis in gnotobiotic turkeys. I. Infection and Histopathology of the bacteria-free host. *J. Parasitol.* **49**, 411–414.

Dunmonde, D. C. (1973). Significance of *in vitro* studies of immunity in leishmaniasis. *In* "Leishmaniasis Symposium" pp. 1–23. King's College, Cambridge, England.

Dwyer, D. M. (1973). Amastigote and promastigote stages of *Leishmania donovani*: An immunologic and immunochemical comparison. Prog. Protozool. Proc. 4th Int. Cong. Protozool. p. 129. Univ. Clermont (Clermont-Ferrand), France.

Eisen, H. N. (1974). "Immunology." Harper & Row, New York.

Fairbairn, D. (1958). Trehalose and glucose in helminths and other invertebrates. *Can. J. Zool.* **36**, 787–795.

Felice, R. P. (1952). Studies on the cytology and life history of a *Giardia* from the laboratory rat. *Univ. Calif. Publ. Zool.* **57**, 53–143.

Feng, S. Y., and Stauber, L. A. (1968). Experimental hexamitiasis in the oyster *Crassostrea virginica*. *J. Invert. Pathol.* **10**, 94–110.

Fife, E. H. Jr. (1977). *Trypanosoma* (*Schizotrypanum*) *cruzi*. *In* "Parasitic Protozoa" (J. P. Kreier, ed.), Vol. 1, pp. 135–173. Academic Press, New York.

Franker, C. K., and Doll, J. P. (1964). Experimental histomoniasis in gnotobiotic turkeys. II. Effects of some cecal bacteria on phatogenesis. *J. Parasitol.* **50**, 636–640.

Garnham, P. C. C., and Humphrey, J. H. (1969). Problems in leishmaniasis related to immunology. *Curr. Top. Microbiol. Immunol.* **48**, 20–42.

Gerzeli, G. (1955). Ricerche istochimiche ed istomofologiche sui tripanosomidi (microscopia a contrasto di fase ed interferenziale). *Riv. Parasitol.* **16**, 209–215.

Gibbons, I. R., and Rowe, A. J. (1965). Dynein—a protein with adenosine triphosphatase activity from cilia. *Science* **149**, 424.

Goble, F. C. (1966). Pathogenesis of blood Protozoa. *In* "Biology of Parasites" (E. J. L. Soulsby, ed.), pp. 237–254. Academic Press, New York.

Godfrey, D. G. (1958). Influence of dietary cod liver oil upon *Trypanosoma congolense, T. cruzi, T. vivax* and *T. brucei. Exp. Parasitol.* **7**, 255–268.

Goodwin, T. W. (1979). Isoprenoid distribution and biosynthesis in flagellates. *In* "Biochemistry and Physiology of Protozoa" (M. Levandowsky and S. H. Hutner, eds.), Vol. 1, 2nd ed, pp. 91–120. Academic Press, New York.

Greenblatt, C. L., Jori, L. A., and Cahnmann, H. J. (1969). Chromatographic separation of a rat serum growth factor required by *Trypanosoma lewisi. Exp. Parasitol.* **24**, 228–242.

Grell, K. G. (1973). "Protozoology." Springer-Verlag, New York.

Grewal, M. S. (1957). Pathogenicity of *Trypanosoma rangeli* Tejero, 1920 in the invertebrate host. *Exp. Parasitol.* **6**, 123–130.

Grimstone, A. V. (1959). Cytology, homology and phylogeny—a note on "organic design." *Am. Natur.* **93**, 273–282.

Grimstone, A. V. (1961). Fine structure and morphogenesis in Protozoa. *Biol. Rev.* **36**, 97–150.

Guthrie, R. (1946). Studies of the growth requirements of *Trichomonas foetus* Riedmüller. Ph.D. dissertation, University of Minnesota, Minneapolis. Minnesota.

Gutteridge, W. E., and Coombs, G. H. (1977). "Biochemistry of Parasitic Protozoa." University Park Press, Baltimore.

Hall, R. P. (1965). "Protozoan Nutrition." Blaisdell, New York.

Hatano, S., Ishikawa, H., and Sato, H. (eds.) (1979). "Cell Motility: Molecules and Organization." University Park Press, Baltimore.

Hayashi, S. (1961). How cells move. *Sci. Am.* **205**, 184–204.

Hirschler, J. (1914). Über Plasmastrukturen (Goligi'scher Apparat, Mitochondrien n.a.) in den Tunicaten-, Spangein- und Protozoenzellen. *Anat. Anz.* **47**, 289–311.

Hoare, C. A. (1972). "The Trypanosomes of Mammals. A Zoological Monograph." Blackwell, Oxford, England.

Hollande, A. (1952). Ordre des Bodonides. *In* "Traité de Zoologie" (P. P. Grasse, ed.), Fasc. I. Masson, Paris.

Honigberg, B. M. (1967). Chemistry of parasitism among some protozoa. *In* "Chemical Zoology." (M. Florkin, B. T. Scheer, and G. W. Kidder, eds.), Vol. 1, pp. 695–814. Academic Press, New York.

Honigberg, B. M. (1978). Trichomonads of importance in human medicine. *In* "Parasitic Protozoa" (J. P. Kreier, ed.), Vol. II, pp. 276–454. Academic Press, New York.

Hutt, M. S. R., Koberle, F., and Salfelder, K. (1973). Leishmaniasis and trypanosomiasis. *In* "Tropical Pathology" (H. Spencer, ed.), pp. 351–398. Springer-Verlag, Berlin, Germany.

Hungate, R. E. (1939). Experiments on the nutrition of *Zootermopsis* II. The anaerobic carbohydrate dissimilation by the intestinal protozoa. *Ecology* **20**, 230–245.

Hyman, L. H. (1940). "The Invertebrates: Protozoa through Ctenophora." McGraw Hill, New York.

Inoki, S., Nakanishi, K., and Nakabayashi, T. (1959). Observations on *Trichomonas vaginalis* by electron microscopy. *Bikens J.* **2**, 21–24.

Iwata, S., and Araki, T. (1960). Studies on giardiasis. *Oska Ika Daigaku* **6**, 92–106.

Jakubowski, W., and Hoff, J. C. (eds.) (1979). "Waterborne Transmission of Giardiasis." U.S. Environmental Protection Agency, Cincinnati, Ohio.

Johnson, P. T., and Hertig, M. (1970). Behavior of *Leishmania* in Panamanian phlebotomine sandflies fed on infected animals. *Exp. Parasitol.* **27**, 281–300.

Katzin, L. I., and Kirby, H. (1939). The relative weights of termites and their protozoa. *J. Parasitol.* **5**, 444–445.

Kein, R. L. (1961). Homeostatic mechanisms for cation regulation in *Acanthamoeba* sp. *J. Exp. Biol.* **37**, 407–416.

Kreier, J. P. (ed.) (1977). "Parasitic Protozoa" Vol. I. Academic Press, New York.

Kreier, J. P. (ed.). (1978). "Parasitic Protozoa" Vol. II. Academic Press, New York.

Krough, A. (1941). "The Comparative Physiology of Respiratory Mechanisms." University of Pennsylvania Press, Philadelphia, Pennsylvania.

Kruszynski, J. (1951). A microchemical study of *Plasmodium gallinaceum* by microincineration. *Ann. Trop. Med. Parasitol.* **45**, 85–91.

Kruszynski, J. (1952). A microchemical study of *Plasmodium berghei* by microincineration, with a note on the microscopical demonstration of calcium. *Ann. Trop. Med. Parasitol.* **46**, 117–120.

Kudo, R. R. (1966). "Protozoology," 5th ed. Charles C Thomas, Springfield, Illinois.

Lainson, A. (1983). The American leishmaniases: Some observations on their ecology and epidemiology, *Trans. R. Soc. Trop. Med. Hyg.* **77**, 569–596.

Laird, M. (1956). Aspects of fish parasitology. *Proc. Second Joint Symp. Sci. Soc. Malaya and Malayan Math. Soc.*, pp. 45–54.

Laird, M. (1961). Microecological factors in oyster epizootics. *Can. J. Zool.* **39**, 449–485.

Langridge, D. F. (1966). Flagellated protozoa (Trypanosomidae) in the honey bee, *Apis mellifera*, in Australia. *J. Invert. Pathol.* **8**, 124–125.

Laveran, C. L. A. (1913). Trypanotoxines. Essais d'immunisation contere les trypanosomes. *Bull. Soc. Pathol. Exot.* **6**, 693–698.

Lee, J. J., Hunter, S. H., and Bovee, E. C. (eds.) (1985). "An Illustrated Guide to the Protozoa." Society of Protozoologists, Lawrence, Kansas.

Levine, N. D. (1970). Taxonomy of the Sporozoa. *J. Parasitol.* **34** (Sect. II), 208–209.

Levine, N. D. (1971). Taxonomy of the piroplasms. *Trans. Am. Microsc. Soc.* **90**, 2–33.

Levine, D., and Bradley, C. A. (1940). Further studies on the pathogenicity of *Trichomonas gallinae* for baby chicks. *Poultry Sci.* **19**, 205–209.

Levine, N. D., Corliss, J. O., Cox, F. E. G., Devoux, G., Grain, J. Honigberg, B. M., Leedale, F. F., Loeblich, A. R. III, Lom, J., Lynn, D. Merinfeld, E. G., Page, F. C., Poljansky, G., Sprague, V., Vavra, J., and Wallace, F. G. (1980). A newly revised classification of the Protozoa. *J. Protozool.* **27**, 37–58.

Lewis, D. H. (1975). Ultrastructural study of promastigotes of *Leishmania* from reptiles. *J. Protozool.* **22**, 344–352.

Lincicome, D. R. (1971). The goodness of parasitism: a new hypothesis. *In* "Aspects of the Biology of Symbiosis" (T. C. Cheng, ed.), pp. 139–227. University Park Press, Baltimore, Maryland.

Lindberg, O., and Ernster, L. (1954). "Chemistry and Physiology of Mitochondria and Microsomes." Protoplasmatologia. Springer-Verlag, Vienna.

Loran, M. R., Kerner, M. W., and Anderson, H. H. (1959). Dependence of *Entamoeba histolytica* upon associated streptobacillus for metabolism of glucose. *Exp. Cell Res.* **10**, 241–245.

Lumsden, W. H. R. (1972). Immune responses to hemoprotozoa. I. trypanosomes. *In* "Immunity to Animal Parasites" (E. J. L. Soulsby, ed.), pp. 287–300. Academic Press, New York.

Manners, D. J., and Ryley, J. F. (1955). Studies on the metabolism of the protozoa. 6. The glycogens of the parasitic flagellates *Trichomonas foetus* and *Trichomonas gallinae*. *Biochem. J.* **59**, 369–372.

Mansfield, J. M. (1977). Nonpathogenic trypanosomes of mammals. *In* "Parasitic Protozoa" (J. P. Kreier, ed.), Vol. I, pp. 297–327. Academic Press, New York.

Masson, F. M., and Oxford, A. E. (1951). The action of the ciliates of sheep's rumen upon various water-soluble carbohydrates, including polysaccharides. *J. Gen. Microbiol.* **5**, 664–672.

McDougald, L. R., and Reid, W. M. (1978). *Histomonas meleagridis* and relatives. *In* "Parasitic Protozoa" (J. P. Kreier, ed.), Vol, II, pp. 140–161. Academic Press, New York.

Meyer, E. A., and Radulescu, S. (1979). Giardia and Giardiasis. *Adv. Parasit.* **17**, 1–47.

Min, H. S. (1965). Studies on the transport of carbohydrate in *Crithidia luciliae*. *J. Cell. Comp. Physiol.* **65**, 243–248.

Min, H. S., and Cosgrove, W. B. (1963). Entrance of carbohydrates into the cells of *Crithidia luciliae*. *J. Protozool.* (Suppl.) **10**, 19.

Morrison, D. B., and Jeskey, H. A. (1947). The pigment, lipids and proteins of the malaria parasite (*P. knowlesi*). *Fed. Proc. Fed. Am. Soc. Exp. Biol.* **6**, 279.

Moulder, J. W. (1948). The metabolism of malaria parasites. *Annu. Rev. Microbiol.* **2**, 101–120.

Mukkada, A. J., Meade, J. C., Glaser, T. A., and Bonventre, P. F. (1985). Enhanced metabolism of *Leishmania donovani* amastigotes at acid pH: An adaptation for intracellular growth. *Science* **229**, 1099–1101.

Muller, M. (1967). Digestion. *In* "Chemical Zoology" (M. Florkin, B. T. Scheer, and G. W. Kidder, eds.), Vol. I, pp. 351–380.

Murray, M., Morrison, W. I, and Whitelaw, D. D. (1982). Host susceptibility to African trypanosomiasis: trypanotolerance. *Adv. Parasitol.* **21**, 2–68.

Neal, R. A., Garnham, P. C. C., and Cohen, S. (1969). Immunization against protozoal diseases. *Br. Med. Bull.* **25**, 194–201.

Nilsson, J. R. (1979). Phagotrophy in *Tetrahymena*. *In* "Biochemistry and Physiology of Protozoa" (M. Levandowsky and S. H. Hutner, eds.), Vol. 2, 2nd ed, pp. 339–379. Academic Press, New York.

Paillot, A. (1927). Sur deux protozoaires nonveaux parasites des chenilles de *Pyrausta nubilalis* Hb. *C. R. Acad. Sci.* (*Paris*) **185**, 673–675.

Pettigrew, G. W. (1979). Structural features of protozoan cytochromes. *In* "Biochemistry and Physiology of protozoa" (M. Levandowsky and S. H. Hutner, eds.), Vol. 1,

2nd ed, pp. 59–90. Academic Press, New York.

Ritterson, A. L., and Stauber, L. A. (1949). Protein intake and leishmaniasis in the hamster. *Proc. Soc. Exp. Biol. Med.* **70**, 47–50.

Rogerson, G. W., and Gutteridge, W. E. (1980). Catabolic metabolism in *Trypanosoma cruzi*. *Intl. J. Parasitol.* **10**, 131–135.

Rudzinska, M. A., and Trager, W. (1957). Intracellular phagotrophy by malaria parasites: an electron microscope study of *Plasmodium lophurae*. *J. Protozool.* **4**, 190–199.

Rudzinska, M. A., and Trager, W. (1959). Phagotrophy and two new structures in the malaria parasite *Plasmodium berghei*. *J. Biophys. Biochem. Cytol.* **6**, 103–112.

Scheltema, R. S. (1962). The relationship between the flagellate protozoon *Hexamita* and the oyster *Crassostrea virginica*. *J. Parasitol.* **48**, 137–141.

Schilling, C., and Rodoni, P. (1913). Tossine tripanosomiche e immunita di fronte ai tripanosomi. *Sperimentale Arch. Biol. Norm. Patol.* **67**, 595–613.

Scholtyseck, E. (1979). "Fine Structure of Parasitic Protozoa." Springer-Verlag, Berlin, Germany.

Scott, G. H., and Horning, E. S. (1932). The structure of opalinids, as revealed by the technique of microincineration. *J. Morphol.* **53**, 381–388.

Shrivastava, K. K., and Shrivastava, G. P. (1974). Two cases of *Trypanosoma* (*Herpetosoma*) species infection of man in India. *Trans. Roy. Soc. Trop. Med. Hyg.* **68**, 143–144.

Soltys, M. A., Thompson, S. M. R., and Woo, P. T. K. (1973). Experimental transmission of *Trypanosoma brucei* and *Trypanosoma congolense* in rats and guinea pigs through skin and intact mucous membranes. *Ann. Trop. Med. Parasitol.* **67**, 399–402.

Stauber, L. A. (1963). Immunity to *Leishmania*. *Ann. N. Y. Acad. Sci.* **113**, 409–417.

Stauber, L. A. (1966). Characterization of strains of *Leishmania donovani*. *Exp. Parasitol.* **18**, 1–11.

Stauber, L. A., Ochs, J. Q., and Coy, N. H. (1954). Electrophoretic patterns of the serum proteins of chinchillas and hamsters infected with *Leishmania donovani*. *Exp. Parasitol.* **3**, 325–335.

Taliaferro, W. H. (1924). A reactions product in infections with *Trypanosoma lewisi* which inhibits the reproduction of the trypanosomes. *J. Exp. Med.* **39**, 171–190.

Taliaferro, W. H. (1932). Trypanocidal and reproduction-inhibiting antibodies to *Trypanosoma lewisi* in rats and rabbits. *Am. J. Hyg.* **16**, 32–84.

Taylor, A. E. R., and Baker, J. R. (eds.) (1978). "Methods of Cultivating Parasites *in vitro*." Academic Press, New York.

Trager, W. (1974). Nutrition and biosynthetic capabilities of flagellates: problems of *in vitro* cultivation and differentiation. *In* "Trypanosomiasis and Leishmaniasis with Special Reference to Chagas' Disease." pp. 225–245. Ciba Foundation Symposium 20, Elsevier, Amsterdam, Holland.

Trager, W. (1970). "Symbiosis." van Nostrand Reinhold, New York.

von Brand, T. (1946). "Anaerobiosis in Invertebrates." Biodynamica Monographs, Normandy, Missouri.

von Brand, T. (1952). "Chemical Physiology of Endoparasitic Animals." Academic Press, New York.

von Brand, T. (1975). "Biochemistry of Parasites," 2nd ed. Academic Press, New York.

Watkins, R. (1971). Histology of *Rhodnius prolixus* infected with *Trypanosoma rangeli*. *J. Invert. Pathol.* **17**, 59–66.

Weinman, D. (1977). Trypanosomiases of man and macaques in South Asia. *In* "Parasitic Protozoa." Vol. I. (J. P. Kreier, ed.), pp. 329–355. Academic Press, New York.

Whitfield, P. J. (1979). "The Biology of Parasitism: An Introduction to the Study of Associating Organisms." University Park Press, Baltimore, Maryland.

W.H.O. (1984). "The Leishmaniases: Report of a WHO Expert Committee." Technical Report Series 701. World Health Organization, Geneva, Switzerland.

Willems, R., Massart, L., and Peeters, G. (1942). Uber den kohlehydratstoffwechsel von *Trichomonas hepatica*. *Naturwissenschaffen* **30**, 159–170.

Willey, R., Bowen, W. R., and Durban, E. (1970). Symbiosis between euglena and damsel fly nymphs is seasonal. *Science* **170**, 80–81.

Williams, P., and Vasconcellos Coelho, M. de (1978). Taxonomy and transmission of *Leishmania*. *Adv. Parasit.* **16**, 1–42.

Williamson, J., and Desowitz, R. S. (1961). The chemical composition of trypanosomes. I. Protein, amino acid and sugar analysis. *Exp. Parasitol.* **11**, 161–175.

Wolfe, M. S. (1979). Managing the patient with giardiasis: clinical, diagnostic and therapeutic aspects. *In* "Waterborne Transmission of Giardiasis." (W. Jakubowski and J. C. Hoff, eds.), pp. 39–48. U.S. Environmental Protection Agency, Cincinnati, Ohio.

Woo, P. T. K. (1977). Salivarian trypanosomes producing disease in livestock outside of sub-Saharan Africa. *In* "Parasitic Protozoa." Vol. I, (J. P. Kreier, ed.), pp. 269–296. Academic Press, New York.

Wood, F. W. (1954). Environmental temperature as a factor in development of *Trypanosoma cruzi* in *Triatoma protracta*. *Exp. Parasitol.* **3**, 227–233.

Wotton, R. M., and Becker, D. A. (1963). The ingestion of particulate lipid containing a fluorochrome dye, acridine orange, by *Trypanosoma lewisi*. *Parasitology* **53**, 163–167.

Yasuda, S., and Dusanic, D. G. (1971). Serologic characterization of somatic antigens of forms of *Trypanosoma lewisi* from the bloodstream and cultures. *J. Infect. Dis.* **123**, 544–547.

Zuckerman, A. (1975). Current status of the immunology of blood and tissue protozoa. I. *Leishmania*. *Exp. Parasitol.* **38**, 370–400.

Zuckerman, A., and Lainson, R. (1977). *Leishmania*. *In* "Parasitic Protozoa." (J. P. Kreier, ed.), Vol. I. pp. 57–133, Academic Press, New York.

CLASSIFICATION OF MASTIGOPHORA (THE FLAGELLATES)*

PHYLUM SARCOMASTIGOPHORA

Single type of nucleus; sexuality, when present, essentially syngamy; with flagella, pseudopodia, or both types of locomotor organelles.

Subphylum Mastigophora

One or more flagella typically present in trophozoites; asexual reproduction basically by intrakinetal (symmetrogenic) binary fission; sexual reproduction known in some groups.

CLASS PHYTOMASTIGOPHOREA

Typically with chloroplasts; if chloroplasts lacking, relationship to pigmented forms clearly evident; mostly free living.

ORDER CRYPTOMONADIDA
ORDER DINOFLAGELLIDA

Two flagella, inserted apically or laterally, one ribbon shaped with paraxial rod and single row of fine hairs, the other smooth or with two rows of stiffer hairs; chloroplasts typically golden-brown or green; storage products starch and fat; cells flattened or of complex symmetry with transverse and ventral grooves and often with armor of cellulose plates; nucleus unique among eukaryotes in having chromosomes that consist primarily only of nonprotein-complexed DNA; mitosis intranuclear; flagellates, coccoid unicells, colonies, and simple filaments; sexual reproduction present. (Genera mentioned in text: *Oodinium, Chytriodinium, Duboscquella, Haplozoon, Apodinium, Blastodinium*.)

ORDER EUGLENIDA

Two (rarely more) flagella, one or both emerging from an anterior invagination of the cell; emergent flagella with single row of fine hairs; flagella with paraxial rods; chloroplasts grass-green, absent in many genera; storage products: paramylon, fat, and cyclic metaphosphates; cell with helical symmetry, naked but with complex pellicle of interlocking proteinaceous strips; nonspindle intranuclear mitosis; individual flagellates or colonies. (Mostly free living; few symbiotic species.)

Suborder Eutreptiina
Suborder Euglenina

Two flagella, one emergent from cell invagination, highly mobile, the other short and nonemergent; one genus colonial, several with envelopes.

Suborder Rhabdomonadina
Suborder Sphenomonadina
Suborder Heteronematina
Suborder Euglenamorphina

Three or more emergent flagella; endozoic in digestive tracts of tadpoles and arthropods. (Genus mentioned in text: *Colacium*.)

ORDER CHRYSOMONADIDA
ORDER HETEROCHLORIDA
ORDER CHLOROMONADIDA
ORDER PRYMNESIDA
ORDER VOLVOCIDA
ORDER PRASINOMONADIDA
ORDER SILICOFLAGELLIDA

CLASS ZOOMASTIGOPHOREA

Chloroplasts absent; one to many flagella; amoeboid forms, with or without flagella, in some groups; sexuality known in few groups; a polyphylectic group.

---

*Based on the recommendation of the Committee on Systematics and Evolution of the Society of Protozoologists, consisting of Levine *et al.* (1980). Diagnoses of taxa comprised of free-living species not presented.

ORDER CHOANOFLAGELLIDA
ORDER KINETOPLASTIDA

One of two flagella arising from depression; flagella typically with paraxial rod in addition to axoneme; single mitochondrion (nonfunctional in some forms) extending length of body as a single tube, hoop, or network of branching tubes, usually containing conspicuous Feulgen-positive (DNA-containing) kinetoplast located near flagellar kinetosomes; Golgi apparatus typically in region of flagellar depression, not connected to kinetosomes and flagella; parasitic (majority of species) and free living.

Suborder Bondonina

Typically with two flagella; typical, often large, adbasal kinetoplast or DNA arranged in several discrete bodies (polykinetoplastic condition) or dispersed throughout mitochondrion (pankinetoplastic condition); free living or parasitic. (Genera mentioned in text: *Cryptobia, Proteromonas, Embadomonas, Retortamonas.*)

Suborder Trypanosomatina

Single flagellum either free or attached to body by undulating membrane; kinetoplast relatively small and compact; parasitic. (Genera mentioned in text: *Herpetomonas, Crithidia, Leishmania, Leptomonas, Phytomonas, Trypanosoma.*)

ORDER PROTERMONADIDA

One or two pairs of flagella without paraxial rods; single mitochondrion, distant from kinetosomes, curving around nucleus, not extending length of body, without Feulgen-positive kinetoplast; Golgi apparatus encircling band-shaped rhizoplast passing from kinetosomes near surface of nucleus to mitochondrion; cysts present; parasitic.

ORDER RETORTAMONADIDA

Two to four flagella, one turned posteriorly and associated with ventrally located cytostomal area bordered by fibril; mitochondria and Golgi apparatus absent; intranuclear division spindle; cysts present; parasitic. (Genera mentioned in text; *Chilomastix, Retortamonas.*)

ORDER DIPLOMONADIDA

One or two karyomastigonts; genera with two karyomastigonts with twofold rotational symmetry or, in one genus, primarily mirror symmetry; individual mastigonts with one to four flagella, typically one of them recurrent and associated with cytostome, or, in more advanced genera, with organelles forming cell axis; mitochondria and Golgi apparatus absent; intranuclear division spindle; cysts present; free living or parasitic.

Suborder Enteromonadina

Single karyomastigont containing one to four flagella; in genera with more than single flagellum, one recurrent; frequent transitory forms with two karyomastigonts resulting from delayed cytokinesis; cysts in at least one genus; parasitic.

Suborder Diplomonadina

Two karyomastigonts; body with twofold rotational symmetry, or bilateral symmetry in one genus; each mastigont with four flagella, one of them recurrent; with variety of microtubular bands; cysts present; free living or parasitic. (Genera mentioned in text: *Giardia, Hexamita.*)

ORDER OXYMONADIDA

One or more karyomastigonts, each containing four flagella typically arranged in two pairs in motile stages; one or more flagella may be recurrent, adhering to body surface for greater or lesser distance; kinetosomes of flagellar pairs connected by paracrystalline structure in which are embedded anterior ends of axostylar microtubules; one to many axostyles per organism, contractile in many genera; mitochondria and Golgi apparatus (including parabasal apparatus) absent; division spindle intranuclear; cysts in some species; sexuality in some species; parasitic. (Genera mentioned in text: *Oxymonas, Pyrosonympha, Dinenympha, Notila, Saccinobaculus.*)

ORDER TRICHOMONADIDA

Typically karyomastigonts with four to six flagella, but with only one flagellum in one genus and no flagella in another; karyomastigonts and akaryomastigonts in one family with permanent polymonad-organization; in mastigont(s) of typical genera, one flagellum recurrent, free or with proximal or entire length adherent to body surface; undulating membrane, if present, associated with adherent segment of recurrent flagellum; pelta and noncontractile axostyle in each mastigont, except for one genus; hydrogenosomes present; true cysts infrequent, known in very few species; all or nearly all parasitic. (Genera mentioned in text: *Histomonas, Dientamoeba, Trichomonas, Pentatrichomonas, Tritrichomonas.*)

ORDER HYPERMASTIGIDA

Mastigont system with numerous flagella and multiple parabasal bodies; barren kinetosomes, resembling in arrangement kinetosomes of trichomonads and associated with rootlet filaments characteristic of these flagellates, present in many genera; flagella-bearing kinetosomes distributed in complete or partial circle, in plate or plates, or in longitudinal or spiral rows meeting in a centralized structure; one nucleus per cell; cysts in some species; sexuality in some species; all parasitic.

Suborder Lophomonadina

Extranuclear organelles arranged in one system; typically all old structures resorbed in division and daughter organelles formed *de novo*. (Genera mentioned in text: *Lophomonas, Joenia, Microjoenia.*)

Suborder Trichonymphina

Two or occasionally four mastigont systems; symmetry basically two- or fourfold rotational; typically equal separation of mastigont systems in division, with total or partial retention of old structures when new systems are formed. (Genera mentioned in text: *Barbulanympha, Rhynchonympha, Trichonympha, Urinympha, Pseudotrichonympha, Microspironympha.*)

5

SARCODINA AND OPALINATA: THE AMOEBAE AND OPALINIDS

SUPERCLASS SARCODINA

The Sarcodina includes all the amoebae. These proto-zoans, unlike the flagellates, generally do not possess a form-retaining pellicle; rather, the body surface is covered by an extremely flexible unit membrane that permits the body cytoplasm to flow in any direction, resulting in constant alterations of the body form. Certain free-living amoebae, such as the freshwater *Arcella* sp., do possess rigid tests; however, none of the parasitic forms maintain such surface shells. Because of the flexibility of the body, the amoebae are con-sidered asymmetric; furthermore, there is no constant differentiation between an anterior and a posterior end.

All amoebae are capable of producing **pseudo-podia**, which are the locomotor and food-acquiring organelles. These body extensions result from the outward flow of cytoplasm, pushing the cell mem-brane ahead of the direction of flow. The presence of pseudopodia in the trophozoite stage usually is con-sidered to be the main distinguishing characteristic of the Sarcodina. Some flagellates are capable of pseudo-podial movement at some stage during their life cycles, and some amoebae possess flagella during their developmental stages. In general, however, fla-gellates use flagella as their main mechanisms of loco-motion, and amoebae use pseudopodia.

Generally, the ecto- and endoplasms of parasitic amoebae are much more distinct than those of free-living species. In the formation of certain types of pseudopodia, both areas of the cytoplasm are in-volved, whereas in other types only the ectoplasm contributes. The type of pseudopodium formed by amoebae is quite specific and often is used along with other characteristics, in the identification of species.

Amoebae cannot swim; motility by pseudopodia is dependent on the presence of a substratum on which they can glide. The endoparasitic species are com-monly found in the alimentary tracts of their hosts, where they are intimately associated with the intes-tinal lining. The ectoparasitic species adhere to the body surfaces of their hosts.

Some amoebae are pathogenic to humans and do-mestic animals. Other essentially nonpathogenic species live in close association with these pathogens and, hence, are important in medical and veterinary protozoology from a diagnostic standpoint.

The characteristics of the more commonly encoun-tered species of parasitic amoebae are listed in Table

5.1 to illustrate how these organisms can be morphologically distinguished from one another.

Reproduction among parasitic amoebae usually occurs by mitotic division followed by cytokinesis, and is commonly referred to as binary fission. Some free-living species also are capable of sexual reproduction involving flagellated or amoeboid gametes and the process of meiosis at some point in their life cycles.

CYSTS

Cyst formation is common among parasitic amoebae, enabling these animals to resist unfavorable conditions outside their hosts and to be eventually transmitted to other hosts; hence, cysts are the infective forms of most amoebae. Some species do not encyst (or cysts have not been found), and transmission from host to host occurs via the trophozoite form. The morphology of amoebic cysts is quite consistent and commonly is used as a laboratory diagnostic feature. Some of the cysts commonly encountered in the medical diagnostic laboratory are depicted in Fig. 5.1.

MORPHOLOGY OF PSEUDOPODIA

Pseudopodia are of four types: filopodia, lobopodia, rhizopodia, and axopodia.

Filopodia are filamentous pseudopods composed completely of ectoplasm (Fig. 5.2). This type of amoeboid projection is found almost exclusively among free-living species, such as *Amoeba radiosa* and various testaceans. **Lobopodia** are the most common form found among parasitic amoebae (Fig. 5.2). This type of pseudopodium is blunt and is composed of both ecto- and endoplasm, or of ectoplasm alone. In most species, the formation of lobopodia is gradual; by watching living specimens, one can see the gradual flow of granular cytoplasm into the broad projection. In *Entamoeba histolytica** and *Endamoeba blattae*, the latter a parasite of cockroaches, the lobopodia are shot out suddenly. **Rhizopodia** are filamentous but are even finer than filopodia (Fig. 5.2), often branching and anastomosing, and are found among free-living marine species, primarily the Foraminiferida. **Axopodia** differ from the other three types of pseudopodia in that they are more or less permanent, and within each is an axial rod composed of a bundle of fibrils (microtubules) that are inserted near the center of the body (Fig. 5.2). This type of semipermanent pseudopod is generally long and needlelike and is found among the pelagic freshwater Heliozoea. For a detailed discussion of amoeboid movement, see Jahn and Bovee (1967).

The other body organelles of the Sarcodina are common to all protozoans and have been considered in Chapter 4.

---

*It is noted that González-Robles and Martínez-Palomo (1983) have demonstrated that two pathogenic strains of *Entamoeba histolytica* maintained in culture are capable of forming filopodia in addition to lobopodia.

Table 5.1. Some Parasitic Amoebae of Humans and Domestic Animals

Parasitic Amoeba	Principal Hosts	Habitat	Main Characteristics	Disease
Entamoeba				
E. histolytica	Humans, other mammals	Large intestine	15–60 μm in diameter, distinct ecto- and endoplasms, vacuoles enclose host cells, centric endosome	Amoebiasis
E. coli	Humans, monkeys, pigs	Large intestine	15–50 μm in diameter, sluggish, vacuoles do not enclose host cells, eccentric endosome	Nonpathogenic
E. gingivalis	Humans, dogs, cats, monkeys	Mouth	5–35 μm in diameter, distinct ecto- and endoplasms, vacuoles rarely enclose erythrocytes	Periodontitis?
E. bovis	Cattle	Rumen	Approximately 20 μm in diameter	Nonpathogenic
E. gallinarum	Chickens, turkeys	Caecum	9–25 μm in diameter, centric endosome	Nonpathogenic
E. anatis	Ducks	Intestine	Similar to *E. histolytica*	Enteritis?
E. cuniculi	Rabbits	Intestine	Similar to *E. coli*, 10–20 μm in diameter	Nonpathogenic
E. polecki	Pigs	Intestine	5–15 μm in diameter, resembles precystic stage of *E. histolytica*	Nonpathogenic
Endolimax nana	Humans, monkeys	Large intestine	6–15 μm in diameter, granular and vacuolated cytoplasm, sluggish, vacuoles do not enclose host cells	Nonpathogenic
Iodamoeba butschlii	Humans, pigs	Intestine	8–20 μm in diameter, broad pseudopodia, sluggish, large glycogen mass	Usually nonpathogenic

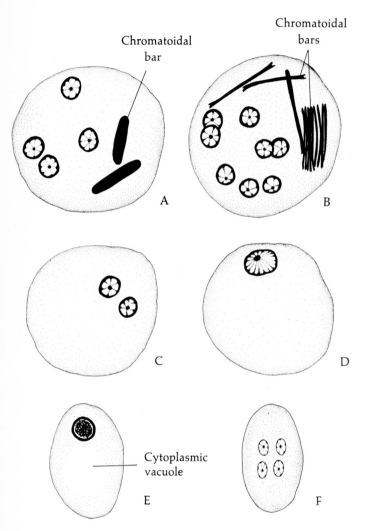

Fig. 5.1. Cysts of some amoebae parasitic in humans and domestic animals. A. Mature cyst of *Entamoeba histolytica* with four characteristic nuclei and chromatoidal bars. **B.** Mature cyst of *Entamoeba coli* with eight characteristic nuclei and jagged chromatoidal bars. **C.** Cyst of *Entamoeba bovis* from cattle with two nuclei. **D.** Cyst of *Entamoeba polecki* from pigs with one characteristic nucleus. **E.** Cyst of *Iodamoeba butschlii* from humans with typical cytoplasmic vacuole. **F.** Cyst of *Endolimax nana* from humans with four characteristic nuclei.

CLASSIFICATION

The subphylum Sarcodina is divided into two superclasses: Rhizopoda and Actinopoda. The majority of the parasitic species belong to the Rhizopoda, although the Actinopoda includes a few species that are parasitic in plants and animals. The major distinguishing

characteristics of these superclasses, as well as the subordinate classes, subclasses, and orders that include parasitic species, are given at the end of this chapter. The monograph by Page (1976) should be consulted by those interested in the taxonomy of amoebae.

SUPERCLASS RHIZOPODA

ORDER AMOEBIDA

The Amoebida includes almost all the species of amoebae parasitic in animals, including humans. With certain exceptions, most of the species are either non-pathogenic or only mildly pathogenic. This order includes the families Amoebidae, Valkampfiidae, Schizopyrenidae, Endamoebidae, and Paramoebidae.

Amoebidae

The family Amoebidae consists mostly of free-living species, including the familiar laboratory animal *Amoeba proteus*. However, a few representatives, such as *Sappinia diploidea*, have been found in the feces of various animals. These are not true parasites but are coprozoic, accidental phoronts, that is, they feed on rich organic materials such as feces. The relatively frequent occurrence of coprozoic amoebae suggests that their affinity for feces and the physicochemical characteristics associated with it may have influenced the initiation of the adaptation to parasitism.

Genus *Hydramoeba*. The genus *Hydramoeba* of the family Amoebidae is represented here by *H. hydroxena*, an ecto- and endoparasite of hydras and freshwater medusae (Fig. 5.3). Its distribution is uncertain, although it has been reported from North Carolina, Virginia, the Soviet Union, and Japan. The trophozoite, measuring 25–200 μm in diameter, commences as an ectoparasite, primarily on the tentacles of hydras, but gradually migrates inward, ingests the host's epidermal cells, and becomes established in the gastrovascular cavity where it actively feeds on gastrodermal cells. Reynolds and Looper (1928) have demonstrated that infected hydras die in 6 days, but I have maintained infected hydras for as many as 10 days. Rice (1960) has reported that this amoeba can destroy the medusa of *Craspedocusta sowerbyi*, a freshwater jellyfish, in 6 days, but apparently does not attack polyps. The host-parasite relationship appears to be an obligatory one, because once removed from its host, the amoeba disintegrates in 4 to 10 days. When the host is killed, the amoeba encysts. The cyst is spherical, measures approximately 28 μm in diameter, and contains a large nucleus, nematocysts from the decreased host, and a conspicuous vacuole. The nucleus of *H. hydroxena* is interesting in that it possesses refractile granules along the inner margin of the nuclear membrane, and there are spokelike radiations from the large endosome.

During the relationship between *Hydramoeba hydroxena* and hydras, the amoeba could be interpreted to be a predator that is gradually devouring its prey. This indicates that at one level a parasite may be difficult to distinguish from a predator, just as at another level a true parasite may be difficult to distinguish from a commensal.

Host Resistance. Stiven (1965) has reported on an interesting aspect of the relationship between the green hydra, *Chlorohydra viridissima*, and *Hydramoeba hydroxena*. Before summarizing his findings, it is important to know that *C. viridissima* is green by virtue of green mutualistic algae (zoochlorellae) present in its tissues. Stiven has found that, when maintained at 25°C, the probability of dying from the amoebic infection is significantly less for hydras with algae than for albinos, i.e., hydras deprived of their mutualistic algae. At 15°C, the proportion of hydras dying was very low in both groups, and hence the suspected

influence of the zoochlorellae appears to be slight (Fig. 5.4). Furthermore, hydras harboring zoochlorellae infected with *H. hydroxena* tend to survive longer than infected albinos in the light, but in darkness their longevities were identical. It has been suggested that hydras with green algae survive attack by the amoeba better than do albinos because of the better nutritional state of the former, since it is known that the hydra is metabolically enhanced by the zoochlorellae. On the other hand, it also has been suggested that the zoochlorellae-bearing hydra has a higher level of resistance because the amoebae acquire food (host cells) more readily from the albino hydra's epidermis. The mutualistic algae, therefore, may be

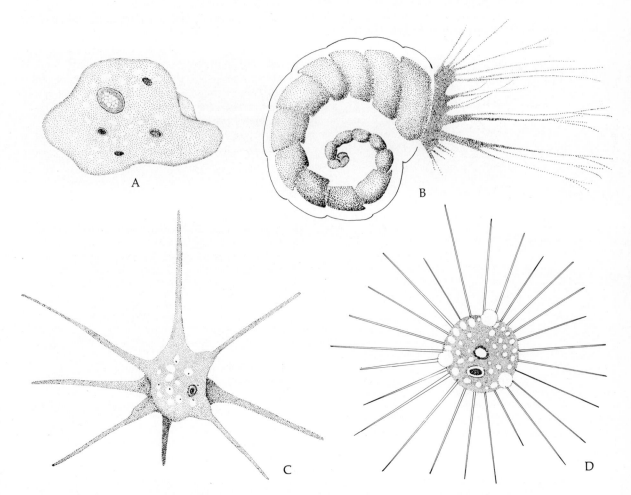

Fig. 5.2. Types of pseudopodia found among amoebae. A. *Endamoeba blattae* from cockroaches with lobopodia. **B.** *Amoeba radiosa*, a free-living species, with filopodia. **C.** *Peneroplis pertusus*, a free-living foraminiferan amoeba, with rhizopodia. **D.** *Actinophrys sol*, a free-living heliozoan amoeba, with axopodia. (All figures redrawn after Kudo, 1966. Protozoology, 5th ed. Charles C Thomas.)

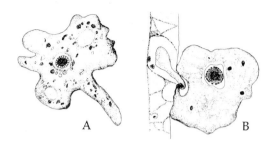

Fig. 5.3. *Hydramoeba hydroxena*, **a parasite of hydra.** **A.** Trophozoite. (Redrawn after Entz, 1912.) **B.** Section of parasitized hydra showing epithelial cell being ingested by *H. hydroxena*. (Redrawn after Hegner and Taliaferro, 1924.)

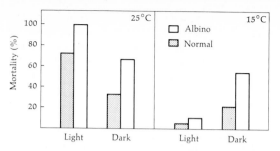

Fig. 5.4. **Proportion of normal and albino** *Chlorohydra viridissima* **dying from infection with** *Hydramoeba hydroxena.* Each group consisted of 18 individuals. (Redrawn after Stiven, 1965.)

instrumental in supplying some substance to the intercellular ground substance of the epidermal cells of the host, which acts to partially inhibit the feeding enzymes of the amoebae.

Vahlkampfiidae, Hartmannellidae, Acanthamoebidae

Genus *Vahlkampfia.* Well-known to marine parasitologists, *Vahlkampfia patuxent* is found in the alimentary canals of oysters. It may reach 140 μm in diameter. This species is undoubtedly an intestine-

dwelling phagotroph that may be only a nonpathogenic facultative parasite. It is characterized by a broad, fan-shaped pseudopodium (Fig. 5.5).

Pathogenic Soil Amoebae. During the past two decades, considerable interest has centered on a group of small, free-living amoebae belonging to the genera *Naegleria* of the family Vahlkampfiidae, *Hartmannella* of the closely related family Hartmannellidae, and *Acanthamoeba* of the family Acanthamoebidae (Fig. 5.6). Not only are these amoebae, which normally occur in fresh water and soil, capable of facultative parasitism in humans and other primates, but certain species and strains of *Acanthamoeba* and *Naegleria* are highly pathogenic.

The facultative parasitic nature of these amoebae was first recognized when *Acanthamoeba* tropho-

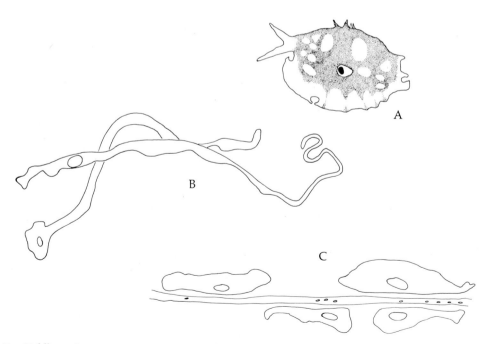

Fig. 5.5. *Vahlkampfia patuxent.* **A.** Trophozoite. **B.** Greatly elongated trophozoites growing on agar slides to which a drop of 0.7% NaCl had been added. **C.** Trophozoites crawling along mould hypha. (Redrawn after Hogue, 1921.)

zoites were found contaminating monkey kidney cell cultures (Jahnes *et al.,* 1957). Since it was determined that the amoebae could not have been introduced during handling of the cultures, they must have been present in the monkey kidneys at excision. Studies by Culbertson *et al.* (1959) have revealed that when experimentally introduced into the nasal passages of mice and monkeys, *Acanthamoeba** can cause meningoencephalitis. This information, coupled with the finding of *Hartmannella* and *Acanthamoeba* in throat swabs from humans (Wang and Feldman, 1961, 1967), aroused considerable interest. In addition, what had been designated "Ryan's virus" in patients suffering from meningoencephalitis was later identified as a species of *Acanthamoeba**. Finally, a number of cases of amoebic meningoencephalitis have been reported in California, New York, Texas, Florida, Virginia, Puerto Rico, southern Austrialia, England, and Czechoslovakia in persons who had been swimming in warm lakes and streams, estuaries, indoor swimming pools and even hot spring spas. The Roman Baths at Bath, England, had to be closed in 1982 as a result of contamination with *Naegleria gruberi*. These cases of meningoencephalitis have been attributed to species of *Acanthamoeba*** and *Naegleria* (especially *N. gruberi*). It is thus apparent that these free-living amoebae are not only capable of facultative parasitism in primates when introduced primarily through the nasal passages, but are also highly pathogenic. Infected individuals portray one of two distinct clinical syndromes (Duma, 1972). (1) The first is an acute, fulminant, rapidly fatal illness that usually affects children and young adults. This is usually associated with *Naegleria* infections. (2) The second, caused by *Acanthamoeba*, often is in the form of insidious neurologic changes in debilitated or immunosuppressed patients. The central nervous systems of such individuals are presumably infected secondarily, and death occurs after a more chronic course.

For a review of these facultatively parasitic small amoebae, see Griffin (1978) and Warhurst (1985).

Facultative parasitism by *Hartmannella* spp. is not restricted to vertebrate hosts. Cheng (1970) has reported that *H. tahitiensis,* normally a soil amoeba, will invade oysters under environmental stress. Also, Richards (1968) has shown that other species of *Hartmannella* will invade freshwater snails. Thus, as in the case of the flagellate *Hexamita nelsoni,* (p. 136), the principle that certain free-living organisms will engage in facultative parasitism when the host is under stress is again exemplified.

*This amoeba was originally identified as *Hartmanella,* but it is now known to be a species of *Acanthamoeba.* Members of Hartmanella are nonpathogenic (Warhurst, 1985).

***Acanthamoeba culbertsoni, A. polyphaga, A. castellanii,* and *A. rhysodes.*

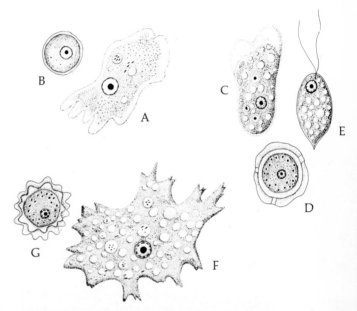

Fig. 5.6.   Some free-living amoebae capable of facultative parasitism. A. Trophozoite of *Hartmannella* sp. with characteristically well-developed ectoplasm and vesicular nucleus with large endosome, and relatively few cytoplasmic vacuoles. **B.** *Hartmannella* cyst with characteristic smooth wall. **C.** Trophozoite of *Naegleria* with typical lobopodia, differentiated ectoplasm, and large numbers of contractile and food vacuoles. **D.** *Naegleria* cyst with wall bearing several apertures. **E.** Biflagellated stage of *Naegleria.* **F.** Trophozoite of *Acanthamoeba* sp. with poorly defined ectoplasm. **G.** *Acanthamoeba* cyst with two membranes, with outer one being highly wrinkled and mammillated.

Endamoebidae

Members of the family Endamoebidae are exclusively parasitic.

Genus *Endamoeba.* The genus *Endamoeba** includes numerous species found in the intestines of invertebrates. The members are distinguished from the closely related species of *Entamoeba* by the morphology of their nuclei. A single nucleus is found in each specimen, and a conspicuous nuclear membrane is present (Fig. 5.7). The nucleoplasm is divided into a peripheral granular zone and a central hyaline zone. The two zones are separated by a ring of endosomes. *Endamoeba blattae* is a common species encountered in the colons of cockroaches. It measures 10–15 μm in

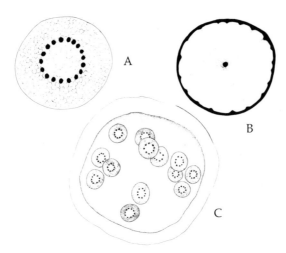

Fig. 5.7. Difference between the nuclei of *Entamoeba* and *Endamoeba*. A. Nucleus of *Endamoeba* showing inner ring of endosomal granules and finer peripheral nucleoplasmic granules. **B.** Nucleus of *Entamoeba* showing centrally located endosome and spokelike nucleoplasmic striations. **C.** Multinucleated cyst of *Endamoeba blattae*.

diameter and possesses broad lobopodia. *Endamoeba simulans*, measuring 50–150 μm in diameter, occurs in the guts of termites.

The various species of *Endamoeba* have multinucleate cysts (Fig. 5.7), which are ingested by the host to establish the infection.

One of the curiosities of parasitology is the occurrence of endamoebae in opalinids (p. 168). Stabler and Chen (1936) and Chen and Stabler (1936) have reported *Endamoeba* spp. in the cytoplasm of opalinids collected from the United States, Panama, Brazil, Chile, and Uruguay in the New World and from Egypt, China, and Sri Lanka. The interesting feature of this relationship is that opalinids are themselves parasites of frogs and toads; hence these endamoebae are **hyperparasites**. These amoebae apparently produce no serious effect on their protozoan hosts. The *Endamoeba* trophozoites measure 5.3–14.3 μm in diameter and appear to feed on the endospherules of the opalinids (Fig. 5.8). Cysts of these endamoebae, found both within and outside of opalinids, measure 9.4 μm in diameter, and are typically uninucleate when occurring within the host protozoan (Fig. 5.8); however, cysts with one, two, and four nuclei have been found outside of hosts. Since the opalinids are undoubtedly ancient organisms, these endamoebae may also be evolutionarily ancient.

Genus *Entamoeba*. The genus *Entamoeba* includes numerous species parasitic in vertebrates and invertebrates. These amoebae possess a vesicular nucleus with a centrically or eccentrically placed endosome and with chromatin granules along the inner surface of the nuclear membrane (Fig. 5.7). Undoubtedly the best known species is the human parasite *Entamoeba histolytica* (Fig. 5.9), the causative agent of amoebic dysentery (or amoebiasis), first discovered in Russia by Lösch in 1875. *E. histolytica* is globally distributed, but its incidence varies in different areas; for example, Santos Zetina (1940) has reported an infection rate as high as 85% in Merida, Yucatan, and Belding (1952) has reported this amoeba in 13.6% (range, 0.8–38%) of 10,867 persons comprising ten surveys conducted in the United States. It is known that incidences are high in Mexico, China, India, and sections of South America. It is important to remember that amoebiasis is not restricted to the tropics and subtropics, since it is found in temperate and even in arctic and antarctic zones. In the United States, the incidence has varied from 1.4% in Tacoma, Washington, to 36.4% in rural Tennessee. The manner of infection—through the ingestion of cysts—causes the incidence to be considerably higher in densely populated institutions. For example, Faust (1931) found a 35.5% incidence in the clinics of New Orleans, and

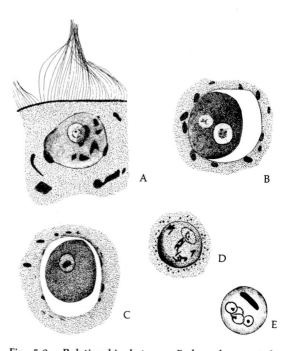

Fig. 5.8. Relationship between *Endamoeba* sp. and the opalinid *Zelleriella*. A. *Endamoeba* trophozoite in peripheral cytoplasm of *Zelleriella* showing ingested endospherules. **B.** *Endamoeba* trophozoite with nucleus enclosing stainable material surrounding endosome. **C.** *Endamoeba* trophozoite showing characteristic features. **D.** Typical uninucleate cyst of *Endamoeba* within *Zelleriella*. **E.** Tetranucleate cyst of *Endamoeba* outside of *Zelleriella* enclosing chromatoidal bar. (Redrawn after Stabler and Chen, 1936.)

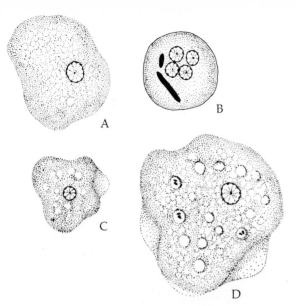

Fig. 5.9. **Stages in the life cycle of** *Entamoeba histolytica.* **A.** Precystic stage. **B.** Mature cyst. **C.** Metacystic stage. **D.** Fully grown trophozoite.

Reardon (1941) found a 40% incidence among the patients in a mental hospital in Georgia.

Life Cycle of *Entamoeba histolytica.* The uninucleate trophozoite of *E. histolytica* (Fig. 5.9) occasionally inhabits the lower portions of the small intestine, but is most common in the colon and rectum of humans and other primates, closely associated with the mucosa. Other mammals, such as dogs and cats, can also be infected. The motile trophozoite measures 15–60 μm in diameter (usually 18–30 μm) and is typically monopodial—that is, it produces one large pseudopodium at a time in an eruptive manner. The cytoplasm, which is indistinctly differentiated into ecto- and endoplasms, contains food vacuoles that enclose the host's erythrocytes, leucocytes, fragments of epithelial cells, and bacteria. Within the host's gut, the trophozoites multiply asexually by binary fission. Electron microscopy has revealed that mitochondria do not occur in the cytoplasm of *E. histolytica*, which is consistent with our knowledge that this amoeba is essentially anaerobic. Smooth endoplasmic reticulum and Golgi complex have been found (El-Hashimi and Pittman, 1970; Griffin and Juniper, 1971; Proctor and Gregory, 1972), and there is one report of the occurrence of rough endoplasmic reticulum (Lowe and Maegraith, 1970). Ribosomes abound in the cytoplasm of *E. histolytica*. For a detailed review of the fine structure of this amoeba, see Albach and Booden (1978).

Not all trophozoites of *E. histolytica* are tissue invaders, but in certain instances they do invade the lining of the victim's large intestine by migrating be-

tween loosened epithelial cells (Takeuchi and Phillips, 1975). Proteolytic enzymes secreted by the amoebae are involved. Within the gut wall, some trophozoites are carried away by the blood circulation to the liver, lungs, brain, and other organs in which these amoebae become lodged and from abcesses. The liver is without question the most frequently infected extraintestinal organ. Within the liver, the trophozoites actively feed on cells, causing severe damage that generally leads to secondary complications (Fig. 5.10). Invasion of tissues other than the intestinal mucosa is known as **secondary amoebiasis**.

From time to time certain intestinal trophozoites assume the precystic form (Fig. 5.9); the body becomes spherical and smaller, food inclusions are extruded, pseudopods are less frequently and sluggishly formed, and chromatoidal bodies (bars) begin to appear. Amoebae at this stage are preparing to encyst. Cysts of *E. histolytica* are spherical and surrounded by a refractile cyst wall (Figs. 5.1, 5.9). The size varies between 3.5 and 20 μm in diameter (depending on whether the cysts are of the large or small strain). When seen in lugol iodine preparations, each cyst contains four prominent vesicular nuclei with more or less centrally situated endosomes. Also characteristic of mature cysts is the presence of elongate chromatoidal bars that, unlike those in the cysts of *Entamoeba coli*, possess rounded rather than jagged ends. These chromatoidal bars represent stored nutrients, glycogen in part, and are gradually utilized as an energy source. These bars also include crystalline aggregates of ribosomes.[*] Such cysts, which are the infective forms, are passed out of the host in feces.

Cysts of *E. histolytica* are highly resistant to desiccation and certain chemicals. Cysts in water can survive up to 1 month, whereas those in feces on dry land survive for more than 12 days. They can survive temperatures up to 50°C, which is their thermal death point. They also can survive up to 4 hours in 1% formaldehyde, up to 7 hours in a 1:100 dilution of phenol, and for 36 to 72 hours in 0.5% chlorine.

When food or water contaminated with *E. histolytica* cysts is ingested by the host, the cysts pass along the alimentary tract to the ileum, where excystation occurs. *In vitro* studies suggest that excystment does not occur immediately. Cysts placed in fresh culture medium at body temperature excyst in 5 or 6 hours.

[*]Siddiqui and Rudzinska (1965) has demonstrated that the chromatoid bodies observed in *Entamoeba invadens* trophozoites cultured axenically include RNA and proteins. These molecules, at the electron microscope level, appear as helical ribonucleoprotein bodies.

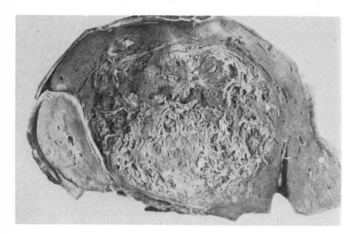

Fig. 5.10. **Abscess in human liver due to *Entamoeba histolytica*.** (After Hunter, Swartzwelder, and Clyde, 1976.)

Upon rupture of the cyst wall, a single multinucleate metacystic amoeba emerges and immediately proceeds to undergo binary fission, giving rise to eight small, uninucleate young trophozoites (known as amoebulae) (Fig. 5.9). Amoebulae pass into the large intestine where they feed, grow, and begin to reproduce by binary fission.

Multiplication of this species thus occurs at two stages during the life cycle—by binary fission in the intestine-dwelling trophozoite stage, and by nuclear division followed by cytokinesis in the cystic and metacystic stages.

Pathogenicity and Nonpathogenicity. A curious aspect of *E. histolytica* infections in humans is the apparent nonpathogenicity in some individuals and the lower incidence of amoebiasis in the temperate zones. Asymptomatic hosts serve as carriers of infective cysts and, hence, are significant from the standpoint of public health. In addition, flies have been incriminated in the spread of this and other amoebae, because cysts can survive for some time in the gut of flies, later to be regurgitated or passed out in feces onto human food. The reason for the apparent nonpathogenicity of certain strains or races of *Entamoeba histolytica* is still not completely understood. This subject has been reviewed critically by Elsdon-Dew (1968).

As stated, the hosts' reactions to this parasite vary greatly. When it occurs in humans living in temperate zones, it is seldom pathogenic, but in the tropics and subtropics the disease known as **amoebiasis** occurs frequently. This discrepancy suggests that two races exist, one predominantly in the tropics and the other in the temperate regions. The tropical race is larger, measuring 15–60 μm in diameter, depending on its metabolic activity, which in turn is influenced by its immediate environmental conditions. The temperate race is smaller, measuring 8–10 μm. To complicate matters, the larger and supposedly pathogenic race is believed to have two forms, the so-called minuta or small cyst form, and the magna or large cyst form. The minuta form is said to be a nonpathogenic lumen dweller, feeding on bacteria. Despite this convenient system of classification based on morphologic-pathogenicity criteria, the situation is far from being clear cut because exceptions do occur. Furthermore, the role of the host in the varied pathogenesis is important but not yet competely defined. Factors such as differential susceptibility of various racial groups (Elsdon-Dew, 1968), diet (for example, a deficiency of vitamin C is known to lower resistance to amoebiasis), stress (Wessenberg, 1974), fatigue, and cell-mediated immunity influence the extent of invasion by *E. histolytica*.

Human amoebiasis also portrays a spectrum of symptoms; even the supposedly highly pathogenic, large, tropical race may not cause symptoms in some individuals.

In humans, *E. histolytica* may invade the intestinal mucosa in varying degrees. They may cause minute submucosal lesions or massive pitlike ones (Fig. 5.11). The host may suffer little intestinal discomfort or rapidly fatal dysentery.

Recently, Lushbaugh *et al.* (1979) have advanced our knowledge of the pathogenicity of *E. histolytica* by finding a heat-labile protein with cytotoxic and enterotoxic activities isolated from cultured amoebae.

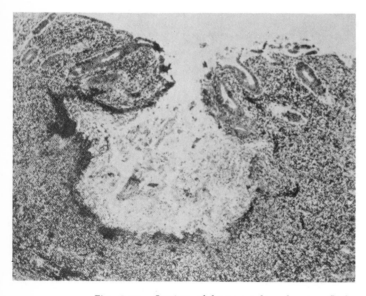

Fig. 5.11. **Section of human colon showing flask-shaped chronic amoebic ulcer involving the mucosa and submucosa.** (From Armed Forces Institute of Pathology.)

They have shown that this toxin has a molecular weight of 25,000–35,000. Bos (1979), studying the same molecule, has reported that the molecular weight is 35,000–45,000. Lushbaugh *et al.* (1980) have also demonstrated that the more virulent strains of *E. histolytica* produce more cytotoxic activity per trophozoite than do the less virulent strains. Furthermore, the cytotoxin is inhibited by specific γG produced by rabbits or infected humans. These findings all support the idea that this toxin is important in the pathogenesis of amoebiasis. Other evidence suggests that this toxin, at least the cytotoxic portion, is a lysosomal enzyme very similar to cathepsin B. For those interested in further details, the review by Pittman (1980) is recommended.

Just how widespread is amoebiasis, and what are its implications on society? It has been stated that this parasite is worldwide in its distribution, with its greatest incidence in the tropics. However, in the United States, the incidence is increasing as a result of entry of migrant workers from countries with high incidences and the sexual practices among homosexuals in our society. Nevertheless, a survey carried out by a special committee of the National Academy of Sciences-National Research Council of the United States (1962) has revealed that the greatest incidence by country occurs in south central and southeast Asia, where 15 to 24 countries reported the presence of *E. histolytica* (see Table 5.2). Numerically, one authority has estimated that 400 million people world-wide harbor this parasite but among these 80% are without acute symptoms. If one assumes that some of those without acute primary or secondary amoebiasis do suffer from some degree of debility, then it is a fair estimate that 100 million people suffer from the disease! Included in this number are those dying from acute dysentery and invasive abscesses as well as those with vague abdominal pains, chronic fatigue, headaches, and other nonspecific symptoms of ill health.

It is extremely difficult to estimate what the 100 million cases of amoebiasis mean in terms of non- or poor productivity or in terms of dollars. Jones (1967) has estimated that if one assumes that the modest sum of $20 (US) is the average required for diagnosis, treatment, and hospitalization in each case, then the phenomenal sum of $2 billion is the medical cost of amoebiasis per year! One would think that in our affluent society sincere attempts would be made to curtail this disease. Unfortunately, even with the wealth of the United States, research and training pertaining to the diagnosis, control, and treatment of amoebiasis have not received what is required, not to mention the less-developed countries, especially those in the tropics, where the incidence of amoebiasis is highest.

Bacteria and *Entamoeba histolytica*. It has been reported that nonpathogenic enteric bacteria may cooperate with *E. histolytica* in producing pathogenicity through tissue invasion. Phillips and co-workers have demonstrated that in germ-free guinea pigs no invasion of the intestinal wall occurs when *E. histolytica* is introduced, but amoebic lesions are produced in hosts harboring a single bacterial symbiont. This type of action, involving the complementary activities of two organisms to produce an effect, is known as **synergism**. The assistance provided by the bacteria is not direct. Rather, it is indirect, having to do with the consumption of oxygen. It is known that *E. histolytica* does not thrive very well in the presence of oxygen. In all probability, the accompanying bacteria reduce the oxygen tension in the area of amoebic infection, permitting the amoeba to proliferate normally and invade tissues. This hypothesis is borne out by experiments in which reducing chemicals, such as cysteine or thioglycolate, are introduced with amoebae. In such instances, despite the lack of bacteria, lesions are formed.

Other *Entamoeba* Species. *Entamoeba hartmanni* is the specific name given to what was formerly considered the small race of *E. histolytica*. It is nonpathogenic. This amoeba measures less than 10 μm in diameter in the trophozoite form and less than 9 μm in the cyst form.

Entamoeba coli (Fig. 5.12), a widely distributed intestinal (colonic), nonpathogenic amoeba of humans and other animals, measures 15–40 μm in diameter

Table 5.2. Distribution of Amoebiasis in 169 Countries from which Data were Available[a]

Caribbean, Central and South America (46 Countries)	Africa (58 Countries)	Southwestern Asia (18 Countries)	South Central and Southeast Asia (24 Countries)	Oceania (23 Countries)	Total (169 Countries)
17	22	6	15	5	65

[a] Data given as number of countries reporting disease. Courtesy of National Academy of Sciences–National Research Council Publication No. 996, 1962.

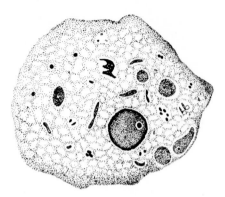

Fig. 5.12. *Entamoeba coli.* Drawing of trophozoite from human intestine. Notice characteristic eccentric endosome in nucleus.

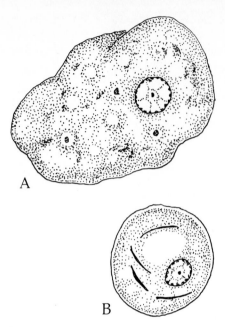

Fig. 5.13. *Entamoeba polecki.* **A.** Trophozoite. **B.** Cyst portraying typical tapering chromatoidal bars and inclusion mass.

(average, 20–35 μm) in the trophozoite form, and 10–30 μm in diameter in the rounded cystic form. The incidence, like that of *E. histolytica*, varies in different areas. For example, Meleney (1930) found that 31.7% of a sampling of Tennesseans harbored *E. coli*; Faust (1930) found that 26.1% of individuals examined in Wise County, Virginia, harbored this amoeba; and Boeck (1923), in an extensive survey of 8029 individuals from all sections of the United States, found that 19.6% were infected.

Entamoeba coli does not ingest or invade host tissues. The food vacuoles present in its heavily granulated cytoplasm enclose bacteria, yeast cells, and other fragments of intestinal debris. For this reason, this amoeba is commonly referred to as a commensal; however, its obligatory dependency on its host suggests that it should be considered a nonpathogenic parasite. *E. coli* infections are generally detected by finding the cysts in feces. The mature cyst characteristically includes eight vesicular nuclei with eccentrically placed endosomes. Younger cysts may contain one, two, or four nuclei. The chromatoidal bodies present in cysts are irregular, with distinctly jagged ends (Fig. 5.1). Although erythrocytes are occasionally found in food vacuoles of *E. coli*, and the cysts are commonly found in diarrheic feces, there is no other evidence to indicate that this species is pathogenic.

The life history of *Entamoeba coli* parallels that of *E. histolytica*—that is, it undergoes the precystic, cyst, metacystic, and trophozoite stages; infection of the host is initiated through the ingestion of cysts.

Entamoeba polecki (Fig. 5.13), measuring 10–25 μm (average, 12 μm) in the trophozoite form and 10–18 μm in the cyst form, is another species commonly found in intestines of pigs, cattle, goats, monkeys, dogs, and has been occasionally reported in humans. This species appears to be an intermediate nonpathogenic form between *E. histolytica* and *E. coli*.

Entamoeba gingivalis (Fig. 5.14), the first parasitic amoeba known from humans (reported by Gros in 1849), is a widely distributed species found in the tartar and debris surrounding teeth. Although food vacuoles within this species enclose host cells, leucocytes, bacteria, and, in culture, erythrocytes, there is little indication that it is pathogenic. Its presence, however, is commonly associated with periodontitis. Cysts are not known for *E. gingivalis*, and transmission is believed to be direct, i.e., via transmission of tropho-

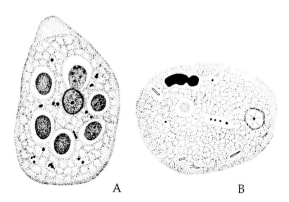

Fig. 5.14. Trophozoites of other *Entamoeba* spp. A. *Entamoeba gingivalis* from human gingival tissue. **B.** *Entamoeba invadens* from the colon of various reptiles, especially snakes. (Redrawn after Geiman and Ratcliffe, 1936.)

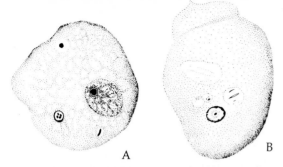

Fig. 5.15. Trophozoites of representatives of *Entamoeba*. A. *Entamoeba citelli* from the caecum and colon of the striped ground squirrel. (Redrawn after Becker, 1926.) **B.** *Entamoeba terrapinae* from colon of turtles. (Redrawn after Sanders and Cleveland, 1930.)

zoites, hence it has been designated as "the kissing amoeba."

Another species, *Entamoeba invadens* (Fig. 5.14), causes amoebiasis in lizards and snakes, but is non-pathogenic in turtles. It measures 9–38 μm in diameter in the trophozoite form and 11–20 μm in the four-nucleate cystic form. It can produce lesions in the stomachs, large and small intestines, and livers of snakes and lizards. It is suspected that infections can be transmitted from turtle to snake with the former serving as the carrier. The life cycle of *E. invadens* parallels that of *E. histolytica*. Lamy's (1948) report that *E. invadens* can be cultured healthily in the presence of host liver tissue, but without bacteria, represents the first successful cultivation of a species of *Entamoeba* in a bacteria-free (axenic) medium. Such cultures grow best at temperatures ranging from 24 to 30°C.

Various other species of *Entamoeba* parasitize domestic and wild animals, but none of these appear to be pathogenic. *Entamoeba bovis* occurs in the rumen of cattle; *E. gallinarum* lives in the cecal contents of chickens and turkeys; *E. gedoelsti* and *E. equi* occur in horses; *E. cuniculi* infects rabbits; *E. muris* occurs in rats and mice; and *E. testudinis* and *E. terrapinae* occur in turtles. Other representative species of *Entamoeba* are depicted in Fig. 5.15.

A few species of *Entamoeba* parasitize insects and probably other arthropods, but none are pathogenic.

Genus *Endolimax*. The genus *Endolimax* of the Endamoebidae includes species found in the colons of humans, other mammals, birds, amphibians, and even cockroaches. The individual species are apparently host specific. *Endolimax nana* is a nonpathogenic parasite in the colon of humans, although periodically it has been suspected of being pathogenic because it is commonly found in diarrheic stools. Some surveys show an incidence of nearly 30%. The trophozoite is monopodial and the lobopodium is broad. The single,

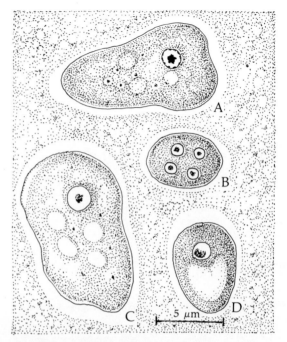

Fig. 5.16. Parasitic amoebae. A. *Endolimax nana* trophozoite with glycogen vacuoles and food vacuoles. **B.** *Endolimax nana* cyst. **C.** *Iodamoeba butschlii* trophozoite. **D.** *Iodamoeba butschlii* cyst with large eccentric karyosome and large glycogen mass.

vesicular nucleus contains a large, irregular endosome. This species is definitely not a tissue invader, since in all known cases the food vacuoles contain bacteria. The life cycle of *E. nana* is typical of cyst-forming endoparasitic amoebae. The trophozoites measure 6–18 μm in diameter (Fig. 5.16) and the characteristic cysts, measuring 5–14 μm at their greatest diameter, are ovoid with one or two nuclei in immature specimens and four in mature ones (Fig. 5.1). Very often a large glycogen body is seen in the cytoplasm, and sometimes a chromatoidal body occurs. Small and large races (strains) have been reported, the only difference being their sizes.

An interesting aspect of the excystation process of *Endolimax nana* is that the tetranucleate encysted form escapes through an extremely minute pore in the cyst wall. Upon escaping from the cyst, the multinucleate amoeba undergoes a series of mitotic divisions during which a portion of the cytoplasm is pinched off with each nucleus so that uninucleate trophozoites are produced. The trophozoites are active feeders and multiply quite rapidly by binary fission.

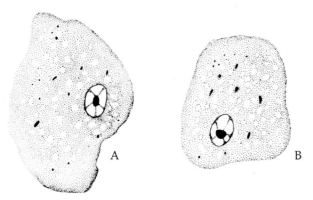

Fig. 5.17. **Endolimax** trophozoites. **A.** *Endolimax ranarum* from intestine of frogs. **B.** *Endolimax blattae* from intestine of cockroaches.

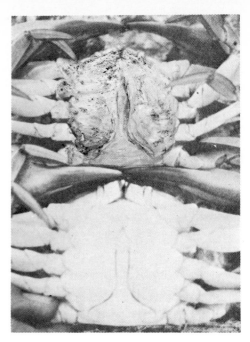

Fig. 5.18. **Gray crab disease.** Infected crab showing gray ventral body surface (above) and uninfected crab (below).

Endolimax gregariniformis, measuring 4–12 μm in diameter, is a species occasionally encountered in chickens and turkeys. The cyst is typically uninucleate. *Endolimax ranarum* is commonly found in the colon of frogs (Fig. 5.17). The cyst of this species is octonucleate and measures 10–25 μm in diameter. *Endolimax blattae* is found in cockroaches (Fig. 5.17); the trophozoites are 4–16 μm in diameter, and the octonucleate cysts measure 6–11 μm at their greatest diameter. Similar amoebae have been reported from a variety of mammals, reptiles, birds, and amphibians.

Genus *Iodamoeba*. The genus *Iodamoeba* includes one species, *I. butschlii* (Fig. 5.16). This intestinal amoeba is very common in primates and pigs. Up to 50% of pigs in some areas harbor this parasite. It may also occur in humans but the incidence is usually less than 5%. The trophozoite found in the host's intestine is 6–25 μm in diameter, with food vacuoles containing bacteria and yeast cells. The most characteristic feature of this genus is the large vesicular nucleus, which includes a large endosome surrounded by rounded granules (spherules). The cyst, measuring 6–16 μm in greatest diameter, varies from rounded to triangular and contains a large nucleus with a large endosome (Fig. 5.1). Within the cytoplasm of the cyst is a large glycogen body, which is often partially wrapped around the nucleus. Because the cysts are uninucleate, the escaping form does not undergo a series of cytoplasmic divisions. The metacystic amoeba leaves through a minute pore in the cyst wall and moves with unusual rapidity. Moisture and warmth are apparently all that it required to induce excystation. *I. butschlii* is generally not considered to be pathogenic, although Derrick (1948) has reported a case of a Japanese prisoner of war who

apparently died from a heavy infection of this amoeba. It is possible that occasional pathogenic strains may occur, or the nonpathogen may become pathogenic under certain conditions such as extreme malnutrition of the host.

Paramoebidae

The members of the family Paramoebidae are of both biologic and economic interest. This family includes *Paramoeba*. In 1966, Sprague and Beckett described a disease syndrome accompanying mortalities of crabs, *Callinectes sapidus*, in eastern Maryland and Virginia known as the "gray crab disease" (Fig. 5.18). Affected crabs portray a grayish discoloration of the ventral body surface, and their blood always contains large numbers of amoebae named *Paramoeba perniciosa* by Sprague *et al.*, (1969). This amoeba, like all other members of the Paramoebidae, includes two nucleus-like bodies that are morphologically different (Fig. 5.19). One is a true nucleus, whereas the nature of the other, designated the nebenkörper, has been elucidated by Perkins and Castagna (1971) through electron microscopy. They have demonstrated that the nebenkörper consists of one or two eukaryotic nuclei and a prokaryoticlike nucleoid with cytoplasm surrounding each of the two regions. The cytoplasm contains ribosome-like particles and phagosomes, the latter containing cytoplasm of the amoeba (Fig. 5.20). It was concluded that the "nebenkörper" is a discrete organism, a protistan hyperparasite, and not an organelle of the amoeba.

Paramoeba perniciosa measures 5–15 μm in diameter, although smaller and larger individuals are occasionally seen. The true nucleus is typically vesicular and in the larger specimens measures about 3 μm in diameter. The hyperparasite is about 3 × 1.5 μm.

PHYSIOLOGY AND BIOCHEMISTRY OF PARASITIC AMOEBAE

Oxygen Requirements

From observing *Entamoeba histolytica* in culture, it is apparent that this amoeba is an anaerobe but can withstand small quantities of oxygen. The elevation of oxygen tension to levels not toxic to free-living amoebae is toxic to this parasite. CO_2 also appears to be necessary. *Entamoeba coli* and *E. hartmanni* behave in much the same manner as *E. histolytica* in their sensitivity to oxygen.

Growth Requirements

Information pertaining to the growth requirements of parasitic amoebae is limited to a few species, such as *Entamoeba histolytica*, which have been successfully cultured *in vitro*. This topic has been critically reviewed by Neal (1968), Fulton (1969), and Taylor and Baker (1978).

Attempts to culture *E. histolytica in vitro* as an approach to ascertaining its growth requirements have only recently produced meaningful information. Beginning with Boeck and Drbohlav in 1925, many investigators have reported success at culturing this amoeba; however, the methods employed involved such undefined ingredients as liver extract, egg, blood, peptone, and serum that they did not lend themselves to specific nutritional studies. Furthermore, until 1961, none of the successful cultures were axenic, although some were monoxenic (with a single species of bacteria present). It should be noted that dead vegetative bacterial cells are sometimes employed as a substitute for live cells. Axenic culturing of *E. histolytica* is now possible (Diamond, 1968).

The volume edited by Taylor and Baker (1978) contains the different culture media currently available for *Entamoeba histolytica* and other amoebae.

GROUPS

Strains of *Entamoeba histolytica*. As a result of attempts to culture *E. histolytica*, at least two groups of this amoeba have been identified. One is the so-called

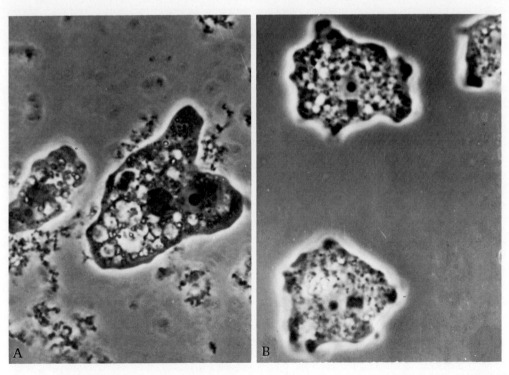

Fig. 5.19. *Paramoeba perniciosa.* **A.** Large trophozoite fixed and stained with dilute methylene blue. **B.** Typical large, living trophozoites. (After Sprague *et al.*, 1969.)

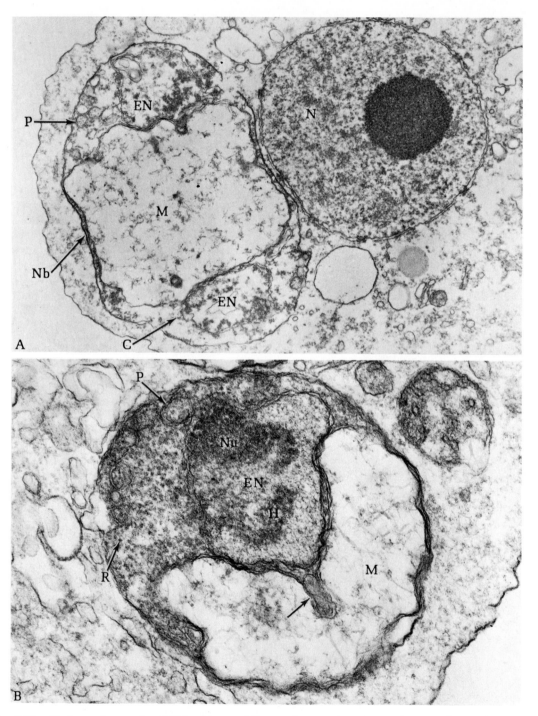

Fig. 5.20.   Electron micrographs of Nebenkörper of *Paramoeba perniciosa*. A. Bipolar or predivision form of the Nebenkörper (Nb). Other structures shown are the host cell nucleus (N), polar caps of Nebenkörper containing eukaryotic nuclei (EN), prokaryoticlike nucleoid or Mittlestück region of Nebenkörper (M), and cytoplasm (C) of Nebenkörper containing phagosomes (P). × 22,400. **B.** Unipolar form of Nebenkörper containing eukaryotic nucleus (EN), prokaryoticlike nucleoid (M), phagosomes (P), and ribosomelike particles (R) in cytoplasm. Also shown are nucleolarlike region (Nu), and invagination of cytoplasm into Mittlestück (arrow). × 62,300. (After Perkins and Castagna, 1971.)

classical **37°C group** (consisting of several strains) and the other is the **Laredo-type group** (also consisting of several strains). The 37°C group grows only at this temperature, whereas the Laredo-type group can be propagated at room temperature (Goldman, 1969). The 37°C strains possess about five times more DNA than the Laredo-type strains. Gelderman *et al.* (1971) have suggested that these represent different species.

Carbohydrate Metabolism

Most of the biochemical studies on parasitic amoebae have been directed toward *Entamoeba histolytica*. Relative to its carbohydrate requirements, *E. histolytica* requires glucose or a polymer that can be converted to glucose. Polymers such as starch are ingested by this amoeba by phagocytosis, whereas glucose is taken up by a specific transport mechanism. This differs from other amoebae in which glucose is taken up by pinocytosis.

Various enzymes of the classical glycolytic pathway have been found in *E. histolytica*—that is, those involved in the initial stages of the degradation of glucose to triose and the subsequent conversions to phosphoenolypyruvate (Reeves, 1972). Reeves *et al.* (1974, 1976), however, have reported the occurrence of an unusual enzyme that mediates the production of fructose-1,6-diphosphate from fructose 6-phosphate using inorganic pyrophosphate as the phosphate donor.

Classically, the conversion of phosphoenolpyruvate to pyruvic acid involves the enzyme pyruvate kinase. In *E. histolytica*, Reeves (1968) has demonstrated that pyruvate kinase does not occur; rather, this metabolic step is mediated by pyruvate phosphate ligase (phosphoenolpyruvic carboxytransphosphorylase). Reeves has purified and characterized this enzyme. It uses AMP and inorganic pyrophosphate to form pyruvic acid as depicted in Figure 5.21.

The dependency on inorganic pyrophosphate for this reaction is absolute. Furthermore, another enzyme, phosphopyruvate carboxylase, is involved. It catalyzes the formation of inorganic pyrophosphate, which, as indicated above, is used as a substrate.

The end products of glucose degradation—that is, CO_2, ethanol, and acetate—have been identified in *E. histolytica* cultured both anaerobically and with some oxygen present. The relative amounts of these products vary, depending on whether the experiments were carried out anaerobically or aerobically. Figure 5.22 depicts the overall reaction for degradation under anaerobic conditions (Montalvo *et al.*, 1971):

Lipid Synthesis. Parasites in general possess only limited ability to synthesize long-chain fatty acids *de*

$$\text{Phosphoenolypyruvate} + \text{AMP} + \text{Pyrophosphate}$$
$$\rightarrow \text{Pyruvate} + \text{ATP} + P_i$$

Fig. 5.21

$$2\ C_6H_{12}O_6 + H_2O$$
$$\rightarrow 4\ CO_2 + CH_3CO_2H + 3\ C_2H_5OH + 2\ H_2$$

Fig. 5.22

novo and therefore rely to a great extent on the host to provide for their essential lipid requirements. From what is known, lipid metabolism in parasitic amoebae is limited to conservative chain lengthening of absorbed fatty acids by the addition of acetate units.

The most abundant sterol in *Entamoeba* is cholesterol. It is suspected, however, that cholesterol is not synthesized by amoebae, but is acquired from their hosts.

PROTEIN SYNTHESIS

Carter *et al.* (1967) have found that the major *in vitro* protein synthesis by *E. histolytica*, as measured by uptake of radioactively labeled amino acids, is associated with ribosome preparations, which are monomers (ca 74S), dimers, and trimers.

HOST-PARASITE RELATIONSHIPS

How does the parasite affect the host? The diseases caused by pathogenic amoebae are the most readily appreciated effects. The ability of *Entamoeba histolytica* to invade tissues has been attributed to a combination of mechanical ingestion and the ground substance hydrolyzing enzyme hyaluronidase. In addition, a glutaminase and several proteolytic enzymes (a pepsin-like enzyme, trypsin, and gelatinase) have been implicated (von Brand, 1973). However, the finding by Lushbaugh *et al.* (1979, 1980) of a cathepsin B-like secretion correlated with the degree of pathogenicity is of primary interest.

EFFECTS OF HOSTS ON AMOEBAE

Various conditions that influence the host also are known to have their effects on parasitic amoebae. For example, if cockroaches infected with *Endolimax blattae* are starved for 8 weeks, the parasites disappear. Maintenance of hosts at high or low temperatures results in less pathogenicity of amoebae; furthermore, maintenance of hosts at high or low temperatures retards the normal rate of trophic multiplication in all species.

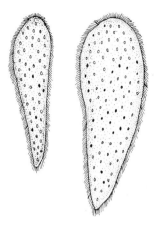

Fig. 5.23. *Opalina hylaxena.* Two individuals from the intestine of the tree frog, *Hyla versicolor.* The larger specimen is about 420 μm long, 125 μm wide, and 28 μm thick. (Redrawn after Metcalf, 1923.)

The host's diet has a great influence on enteric amoebae; high-protein diets are generally unfavorable. Individuals suffering from amoebiasis sometimes lose their symptoms rapidly when maintained on a diet rich in proteins and vitamins. Vitamin deficiencies also inflict profound adverse effects; for example, the lack of vitamin C in guinea pigs, and of niacin in dogs, lowers resistance to amoebiasis.

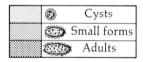

⊙	Cysts
🥚	Small forms
🥚	Adults

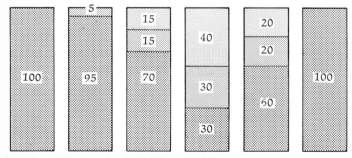

Late April –early Feb.	Early Feb.	Mid-Feb.	Mid-Feb. –early April	Mid-April	Late April –early Feb.
100	5 / 95	15 / 15 / 70	40 / 30 / 30	20 / 20 / 60	100

Pre-oviposition   Oviposition   Post-oviposition

Fig. 5.24. **Relationship between cyst formation by *Opalina ranarum* and the breeding cycle of the frog *Rana temporaria*.** (Redrawn after El Mofty, 1961.)

Another aspect of host-parasite interactions relative to parasitic amoebae has to do with alterations in virulence. For example, Lushbaugh *et al.* (1978, 1980) have demonstrated that the virulence of cultured *Entamoeba histolytica* is enhanced if this amoeba is periodically passed through hamster liver.

The physiology of parasitic amoebae is by no means completely understood. The problems of encystment and amoeboid movement are ones that have challenged biologists for decades. A discussion of the various theories concerning these problems can be obtained from most textbooks in general physiology and cell biology.

SUBPHYLUM OPALINATA

Members of the Opalinata, or opalinids, were originally considered to represent primitive ciliates; however, these protozoans are now considered to be representatives of a subphylum of the Sarcomastigophora.

The opalinids are all parasitic in the colons of amphibians, reptiles, and fish. The taxonomy of this group is not given here in detail but may be found in Levine *et al.* (1980). Asexual reproduction is generally by longitudinal binary fission. However, sexual fusion of gametes has been reported in *Protoopalina intestinalis*, an intestinal parasite of amphibians, and in other species.

The Opalinata includes only one family, the Opalinidae. The best known genera are *Opalina, Protoopalina, Cepedea*, and *Zelleriella.* The review by Wessenberg (1978) of this group of protozoans is recommended.

Genus *Opalina*. The genus *Opalina* includes over 150 species found in the colons of frogs and toads (Fig. 5.23). The parasites' dorsoventrally flattened bodies include many nuclei of the same type and are enveloped by a relatively tough pellicle.

Opalina obtriganoidea, measuring 400–800 μm long by 175–180 μm wide, is found in most common frogs and toads. It is quite representative of the genus, both in size and morphology. Another species, *O. ranarum* (Fig. 5.25) measuring from 500–700 μm long, is also relatively common.

Life Cycle of *Opalina ranarum*. The large, multinucleated trophozoite of *O. ranarum* inhabits the frog's large intestine. El Mofty and Smyth (1959) have reported that, during the nonbreeding, terrestrial phase of the frog *Rana temporaria*, only large trophozoites are found. These divide occasionally by asexual binary fission.

In the spring, just before commencement of egg laying by the host, the multinucleate trophozoite undergoes division (plasmotomy) and gives rise to a number of smaller opalinids, each measuring 30–90 μm long and having fewer nuclei. These smaller forms soon encyst and are passed out of the host. Cysts, which are the infective forms, measure 30–

70 μm in diameter and contain two to twelve nuclei. These are plentiful on the bottoms of ponds when young tadpoles develop. When cysts are ingested by tadpoles, excystation occurs.

The liberated form undergoes a series of divisions culminating in production of gametes. Gametes are of two sizes, the larger female **macrogametes**, and the smaller male **microgametes**. The fusion of a male and a female gamete results in the formation of a diploid **zygote**. The zygote increases in size and undergoes a period of active division, presumably ending only when the opalinid population has reached the limit of the available food supply. Zygotes are also known to undergo encystment, and such cysts are passed to the exterior. When zygote-bearing cysts are ingested by a tadpole, the escaping forms develop into multinucleate adults.

Influence of Host's Hormones. As stated, *Opalina* is capable of both sexual and asexual reproduction. During the nonbreeding period, from late April to early February in Europe, when both male and female frogs are found on dry land, only the "adult" or trophozoite form of *Opalina* is found in their intestines (Fig. 5.25). Trophozoites occasionally reproduce asexually by fission, involving mitosis but not meiosis. Early in February the host's breeding season begins, and dramatic activity on the part of the opalinids occurs. About 14 days before the hosts enter water to copulate, the fission rate of *Opalina* is accelerated, and small precystic forms appear (Fig. 5.25). Each of these encyst and the enclosed cytoplasm includes two to twelve nuclei, with an average of four (Fig. 5.25). These cysts are present in small numbers in frogs copulating in water, appearing slightly earlier in females than in males. After the hosts complete their copulation and ovulation, the number of *Opalina* cysts in their intestines rises sharply but then gradually begins to fall as they are passed out in feces. About 3 months after copulation, cysts cannot usually be found in frogs.

Encystation of *Opalina* trophozoites appears to be influenced by the gonadotropic and/or sex hormones of their hosts. This hypothesis was tested by El Mofty and Smyth (1959) in two ways. (1) The hibernation of potentially sexually mature frogs was enforced by keeping them in a dry habitat. Examination of opalinas from these hosts revealed that none were undergoing encystation. This evidence, although indirect, suggests that the hormonal titers of the hibernating frogs was low and, in turn, the protozoans were not stimulated to encyst. (2) A number of hormones and hormone-containing fluids, such as the urine from pregnant women, were injected into three groups of frogs: normal, hypophysectomized (pituitary removed surgically), and gonadectomized (gonads removed surgically). This was done on prebreeding and postbreeding frogs. As the data in Table 5.3 indicate, encystation was induced among opalinas within 9 to 13 days when pregnancy urine or gonadotropins were injected into prebreeding frogs. Injection of male and female hormones gave the same result, even in gonadectomized and hypophysectomized frogs. This and other experimental results indicate that both sex and gonadotropic hormones in some way stimulate encystation.

It is noted that testosterone proprionate, a male hormone, induces encystation in both prebreeding

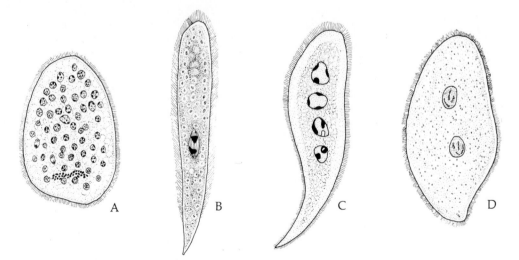

Fig. 5.25. **Some parasitic opalinids. A.** *Opalina ranarum* from large intestine of frogs. **B.** *Protoopalina saturnalis* from large intestine of the marine fish *Box boops*. (Redrawn after Léger and Duboscq, 1904.) **C.** *Cepedea lanceolata* from colon of amphibians. (Redrawn after Bezzenberger, 1903.) **D.** *Zelleriella elliptica* from colon of *Bufo valliceps*. (Redrawn after Chen, 1948.)

Table 5.3. Induction of Encystation of *Opalina*[a]

	Prebreeding Season			Postbreeding Season		
	Normal	*Pituitary Removed*	*Gonads Removed*	*Normal*	*Pituitary Removed*	*Gonads Removed*
Pregnancy urine	+	+	−	−	0	0
Chorionic gonadotropin	+	+	−	−	0	0
Serum gonadotropin	+	+	−	−	0	0
Progesterone	−	−	0	−	0	0
Estrogen	+	+	−	−	0	0
Testosterone proprionate	+	+	+	+	0	0
Epinephrine	+	+	−	+	0	0

[a] The onset of sexual reproduction in *Opalina* is signaled by the appearance of cysts in the host's rectum. +, Cyst present; −, cysts absent; 0, experiment not carried out. (Data from El Mofty, 1959, and El Mofty and Smyth, 1959.)

and postbreeding frogs. Also, the fact that epinephrine, a nonsex hormone, should cause encystation is not surprising since this hormone undoubtedly in some way influences the secretion of either gonadotropic or sex hormones.

In some preliminary experiments, El Mofty (1961) attempted to test the effect of gonadotropic and sex hormones *in vitro*. He was able to maintain *Opalina* in a saline-serum-albumin medium to which was separately added pregnancy urine and gonadotropin. The urine, containing estrogen, caused some precystic forms to appear, followed by encystation, but gonadotropin failed to produce cysts.

But when does *Opalina* undergo sexual reproduction, and how does it enter the frog host? It is recalled that cysts are voided in the host's feces soon after the host's mating process. These, freed in water, are ingested by tadpoles. About 8 hours after ingestion, excystation occurs and a small multinucleate form escapes from each cyst; this is either a **macrogametocyte** or a **microgametocyte**. These migrate posteriorly to the host's cloaca, where as a result of further division, this time meiotic, they give rise to haploid **macrogametes** (ova) and **microgametes** (sperm). A macrogamete and a microgamete fuse to form a diploid zygote, whose fate is at the present uncertain. One hypothesis is that the zygote, through mitosis, gives rise to trophozoites. A second view is that the zygote encysts as a **cystozygote**, which passes out in the host's feces and is ingested by another tadpole. Once ingested, excystment presumably occurs and a uninucleate form escapes which, by repeated division, gives rise to typical trophozoites. In either case, the rectal trophozoites remain with the host through its metamorphosis to the adult stage before the cycle begins again within the onset of the breeding season of the host.

Induction of Sexual Reproduction by Other Agents. In addition to sex hormones, other hormones and compounds have been shown to induce sexual reproduction in opalinids. For example, El-Mofty (1973a,b,c) induced sexual reproduction in *Opalina sudafricana* by injecting its toad host, *Bufo regularis*, with gibberellin A_3, nicotine, or ecdysterone. Also, El-Mofty (1974) has demonstrated that injections of tryptophan metabolites (3-hydroxyanthranilic acid, xanthuremic acid, and DL-kynurene sulfate) into toad hosts induced sexuality of *O. sudafricana*. The same holds true when the urine of patients with bladder cancer, and hence abnormal tryptophan metabolism, was injected into parasitized toads.

Genus *Protoopalina*. Members of *Protoopalina* possess two nuclei and are cylindric or spindle-shaped (Fig. 5.25). All *Protoopalina* species are endosymbionts in amphibians except for *P. saturnalis*, which occurs in the intestine of the marine fish *Box boops*.

Other Opalinid Species. The Opalinidae also includes *Cepedea*, a cylindric or spindle-shaped multinucleated species found in amphibians; and *Zelleriella*, a binucleate flattened form found in amphibians (Fig. 5.26).

Practically no information is yet available about the physiology and biochemistry of opalinids.

REFERENCES

Albach, R. A., and Booden, T. (1978). Amoebae. *In* "Parasitic Protozoa," Vol. II. (J. P. Kreier, ed.), pp. 455–506. Academic Press, New York.

Belding, D. L. (1952). "Textbook of Clinical Parasitology," 2nd ed. Appleton-Century-Crofts, New York.

Boeck, W. C. (1923). Survey of 8,029 persons, in the United States, for intestinal parasites, with special reference for amoebic dysentery among returned soldiers. *Bull. Hyg. Lab. USPHS*, pp. 1–61.

Bos, H. J. (1979). *Entamoeba histolytica*: cytopathogenicity of intact amoebae and cell-free extracts; isolation and characterization of an intracellular toxin. *Exp. Parasitol.* **47**, 369–377.

Carter, W. A., Levy, H. B., and Diamond, L. S. (1967). Protein synthesis by amoebal ribosomes. *Nature (London)* **213**, 722–724.

Chen, T. T., and Stabler, R. M. (1936). Further studies on the Endamoebidae parasitizing opalinid ciliates. *Biol. Bull.* **70**, 72–77.

Cheng, T. C. (1970). *Hartmannella tahitiensis* sp. n., an amoeba associated with mass mortalities of the oyster *Crassostrea commercialis* in Tahiti, French Polynesia. *J. Invert. Pathol.* 15, 405–419.

Culbertson, C. G., Smith, J. W., Cohen, H. K., and Minner, J. R. (1959). Experimental infection of mice and monkeys by *Acanthamoeba*. *Am. J. Pathol.* **35**, 185–197.

Derrick, E. H. (1948). A fatal case of generalized amoebiasis due to a protozoon closely resembling, if not identical with, *Iodamoeba bütschlii*. *Trans. Roy. Soc. Trop. Med. Hyg.* **42**, 191–198.

Diamond, L. S. (1968). Techniques of axenic culture of *Entamoeba histolytica* Schaudinn, 1903 and *E. histolytica*-like amoebae. *J. Parasitol.* **54**, 1047–1056.

Duma, R. J. (1972). Primary amoebic meningoencephalitis. *Crit. Rev. Clin. Lab. Sci.* **3**, 163–192.

El-Hashimi, W., and Pittman. F. (1970). Ultrastructure of *Entamoeba histolytica* trophozoites obtained from the colon and from *in vitro* cultures. *Am. J. Trop. Med. Hyg.* **19**, 215–226.

El Mofty, M. M. (1961). The life cycle of *Opalina ranarum* and an investigation of the factors responsible for controlling its reproductive pattern. Ph.D. Thesis, University of Dublin, Ireland.

El Mofty, M. M. (1973a). Induction of sexual reproduction in *Opalina sudafricana* by injecting its host *Bufo regularis* with gibberellic acid. *Int. J. Parasitol.* **3**, 203–206.

El Mofty, M. M. (1973b). Induction of sexual reproduction in *Opalina sudafricana* parasitic in *Bufo regularis*. *Int. J. Parasitol.* **3**, 265–266.

El Mofty, M. M. (1973c). Ecdysterone induced sexual reproduction in *Opalina sudafricana* parasitic in *Bufo regularis*. *Int. J. Parasitol.* **3**, 863–868.

El Mofty, M. M. (1974). A new biological assay, utilizing parasitic protozoa for screening carcinogenic substances which induce bladder cancer in man and other mammals. *Int. J. Parasitol.* **4**, 47–54.

El Mofty, M. M., and Smyth, J. D. (1959). Endocrine control of sexual reproduction in *Opalina ranrum* parasitic in *Rana temporaria*. *Nature (London)* **186**, 559.

Elsdon-Dew, R. (1968). The epidemiology of amoebiasis. *Adv. Parasitol.* **6**, 1–62.

Faust, E. C. (1930). The *Endamoeba coli* index of *E. histolytica* in a community. *Am J. Trop. Med.* **10**, 137–144.

Faust, E. C. (1931). The incidence and significance of infestation with *Endamoeba histolytica* in New Orleans and the American tropics. *Am. J. Trop. Med.* **11**, 231–237.

Fulton, J. D. (1969). Metabolism and pathogenic mechanisms of parasitic protozoa. *In* "Research in Protozoology" (T. T. Chen, ed.), Vol. 3, pp. 389–504. Pergamon, New York.

Gelderman, A. H., Keister, D. B., Bartgis, I. L., and Diamond,

L. S. (1971). Characterization of the deoxyribonucleic acid of representative strains of *Entamoeba histolytica*, *Entamoeba histolytica*-like amoebae, and *Entamoeba moshkovskii*. *J. Parasitol.* **57**, 906–911.

Goldman, M. (1959). Microfluorimetric evidence of antigenic difference between *Entamoeba histolytica* and *Entamoeba hartmanni*. *Proc. Soc. Exp. Biol. Med.* **102**, 189–191.

González-Robles, A., and Martínez-Palomo, A. (1983). Scanning electron microscopy of attached trophozoites of pathogenic *Entamoeba histolytica*. *J. Protozool.* **30**, 692–700.

Griffin, J. L. (1978). Pathogenic free-living amoebae. *In* "Parasitic Protozoa" Vol. II. (J. P. Kreier, ed.). pp. 507–549. Academic Press, New York.

Griffin, J. L., and Juniper, K. Jr. (1971). Ultrastructure of *Entamoeba histolytica* from human amebic dysentery. *Arch. Pathol.* **21**, 280.

Jahn, T. L., and Bovee, E. C. (1967). Motile behavior of Protozoa. *In* "Research in Protozoology" (T. T. Chen, ed.), Vol. 1, pp. 41–200. Pergamon, New York.

Jahnes, W. G., Fullmer, H. M., and Li, C. P. (1957). Free-living amoeba as contaminants in monkey kidney tissue culture. *Proc. Soc. Exp. Biol. Med.* **96**, 484–488.

Jones, A. W. (1967). "Introduction to Parasitology." Addison Wesley, Reading, Massachusetts.

Lamy, L. (1948). Obtention d'une culture bacteriologiquement pure d'amibes parasites pathogènes (*Entamoeba invadens* Rodhain), ne comportant aucune addition de germes bactériens morts, ai d'aucum extrait microbien. *C. R. Acad. Sci. (Paris)* **226**, 1021–1022. (Also see **226**, 1400–1402).

Levine, N. D., Corliss, J. O., Cox, F. E. G., Deroux, G., Grain, J., Honigberg, B. M., Leedale, G. F., Loeblich, A. R. III, Lom, J., Lynn, D., Merinfeld, E. G., Page, F. C., Poljansky, G., Sprague, V., Vavra, J., and Wallace, F. G. (1980). A newly revised classification of the Protozoa. *J. Protozool.* **27**, 37–58.

Lowe, C. Y., and Maegraith, B. G. (1970). Electron microscopy on axenic strains of *Entamoeba histolytica*. *Am. Trop. Med. Parasitol.* **64**, 293–298.

Lushbaugh, W. B., Kairalla, A. B., Cantey, J. R., Hofbauer, A. F., and Pittman, F. E. (1979). Isolation of a cytotoxin-enterotoxin from *Entamoeba histolytica*. *J. Infect. Dis.* **139**, 9–17.

Lushbaugh, W. B., Kairalla, A. B., Hofbauer, A. F., and Pittman, F. E. (1980). Sequential histopathology of cavitary liver abscess. Formation induced by axenically grown *Entamoeba histolytica*. *Arch. Pathol. Lab. Med.* **104**, 575–579.

Lushbaugh, W. B., Kairalla, A. B., Loadholt, C. B., and Pittman, F. E. (1978). Effect of hamster liver passage on the virulence of axenically cultivated *Entamoeba histolytica*. *Am. J. Trop. Med. Hyg.* **27**, 248–254.

Meleney, H. E. (1930). Community surveys for *Endamoeba histolytica* and other intestinal protozoa in Tennessee; first report. *J. Parasitol.* **16**, 146–153.

Montalvo, F. G., Reeves, R. E., and Warren, L. G. (1971). Aerobic and anaerobic metabolism in *Entamoeba histoly-*

tica. Exp. Parasitol. **30**, 249–256.

Neal, R. A. (1968). The *in vitro* cultivation of *Entamoeba*. Sixth Symp. Br. Soc. Parasitol. (1967), no. 5, pp. 9–26.

Page, F. C. (1976). "An Illustrated Key to Freshwater and Soil Amoebae." Freshwater Biol. Assoc., Sci. Publ. no. 34. Ambleside, Cumbria, England.

Perkins, F. O., and Castagna, M. (1971). Ultrastructure of the *Nebenkörper* or "secondary nucleus" of the parasitic amoeba *Paramoeba perniciosa* (Amoebida, Paramoebidae.) *J. Invert. Pathol.* **17**, 186–193.

Pittman, F. E. (1980). Intestinal amebiasis. *Pract. Gastroenterol.* **4**, 33–38.

Proctor, E. M., and Gregory, M. A. (1972). The ultrastructure of axenically cultivated trophozoites of *Entamoeba histolytica* with particular reference to an observed variation in structural pattern. *Ann. Trop. Med. Parasitol.* **66**, 335–338.

Reardon, L. V. (1941). Incidence of *Endamoeba histolytica* and intestinal nematodes in a Georgia state institution. *J. Parasitol.* **27**, 89–90.

Reynolds, B. D., and Looper, J. B. (1928). Infection experiments with *Hydramoeba hydroxena* nov. gen. *J. Parasitol.* **15**, 23–30.

Rice, N. E. (1960). *Hydramoeba hydroxena* (Entz), a parasite on the fresh-water medusa, *Crapedacusta sowerbi* Lankester, and its pathogenicity for *Hydra cauliculata* Hyman. *J. Protozool.* **7**, 151–156.

Richards, C. S. (1968). Two new species of *Hartmannella* amebae infecting freshwater mollusks. *J. Protozool.* **15**, 651–656.

Reeves, R. E. (1968). A new enzyme with the glycolytic function of pyruvate kinase. *J. Biol. Chem.* **234**, 3202–3204.

Reeves, R. E. (1972). Carbohydrate metabolism in *Entamoeba histolytica*. In "Comparative Biochemistry of Parasites." (H. van den Bossche, ed.), pp. 351–358. Academic Press, New York.

Reeves, R. E., Serrano, R., and South, D. J. (1976). 6-phosphofructokinase (pyrophosphate). Properties of the enzyme from *Entamoeba histolytica* and its reaction mechanism. *J. Biol. Chem.* **251**, 2958–2962.

Santos Zetina, F. (1940). Contribucion al estudio del parasitismo intestinal en Yucatan. *Rev. Med. Yucatan* **20**, 271–277.

Siddiqui, W. A., and Rudzinska, M. A. (1965). The fine structure of axenically-grown trophozoites of *Entamoeba invadens* with special reference to the nucleus and helical ribonucleoprotein bodies. *J. Protozool.* **12**, 448–459.

Sprague, V., and Beckett, R. L. (1966). A disease of blue crabs (*Callinectes sapidus*) in Maryland and Virginia. *J. Invert. Pathol.* **8**, 287–289.

Sprague, V., Beckett, R. L., and Sawyer, T. K. (1969). A new species of *Paramoeba* (Amoebida, Paramoebidae) parasitic in the crab *Callinectes sapidus*. *J. Invert. Pathol.* **14**, 167–174.

Stabler, R. M., and Chen, T. T. (1936). Observations on an *Endamoeba* parasitizing opalinid ciliates. *Biol. Bull.* **70**, 56–71.

Stiven, A. E. (1965). The association of symbiotic algae with the resistance of *Chlorohydra viridissima* (Pallas) in *Hydramoeba hydroxena* (Entz). *J. Invert. Pathol.* **7**, 356–367.

Takeuchi, A., and Phillips, B. P. (1975). Electron microscope studies of experimental *Entamoeba histolytica* infection in the guinea pig. I. Penetration of the intestinal epithelium by trophozoites. *Am J. Trop. Med. Hyg.* **24**, 34–48.

Taylor, A. E. R., and Baker, J. R. (eds.). (1978). "Methods of Cultivating Parasites *in vitro*." Academic Press, London.

von Brand, T. (1973). "Biochemistry of Parasites," 2nd ed. Academic Press, New York.

Wang, S. S., and Feldman, H. A. (1961). Occurrence of acanthamoebae in tissue cultures inoculated with human pharyngeal swabs. *Antimicrob. Ag. Chemother.* **1**, 50–53.

Warhurst, D. C. (1985). Pathogenic free-living amoebae *Parasitol. Today* **1**, 24–28.

Wessenberg, H. S. (1974). The pathogenicity of *Entamoeba histolytica*: is heat stress a factor? *Prospect. Biol. Med.* **18**, 250–266.

Wessenberg, H. S. (1978). Opalinata. In "Parasitic Protozoa" (J. P. Kreier, ed.), Vol. II, pp. 551–581. Academic Press, New York.

CLASSIFICATION OF SARCODINA (THE AMOEBAE)*

PHYLUM SARCOMASTIGOPHORA

Single types of nucleus; sexuality, where present, essentially syngamy; with flagella, pseudopodia, or both types of locomotor organelles.

Subphylum Sarcodina

Pseudopodia, or locomotive protoplasmic flow without discrete pseudopodia; flagella, when present, usually restricted to developmental or other temporary stages; body naked or with external or internal test or skeleton; asexual reproduction by fission; sexuality, if present, associated with flagellate or, more rarely, amoeboid gametes; most species free living.

SUPERCLASS RHIZOPODA

Locomotion by lobopodia, filopodia, or reticulopodia, or by protoplasmic flow without production of discrete pseudopodia.

CLASS LOBOSEA

Pseudopodia lobose or more or less filiform but produced from broader hyaline lobe; usually uninucleate; multinucleate forms not flattened or much-branched plasmodia.

Subclass Gymnamoebia

Without test.

ORDER AMOEBIDA

Typically uninucleated; mitochondria typically present; no flagellate stage.

Suborder Tubulina

Body a branched or unbranched cylinder; no bidirectional flow of cytoplasm; nuclear division mesomitotic. (Genera mentioned in text: *Amoeba, Sappinia, Hydramoeba, Hartmannella, Endamoeba, Entamoeba, Endolimax, Iodamoeba*.)

---

*Based on the recommendation of the Committee on Systematics and Evolution of the Society of Protozoologists consisting of Levine *et al.* (1980). Diagnoses of taxa comprised of free-living species not presented.

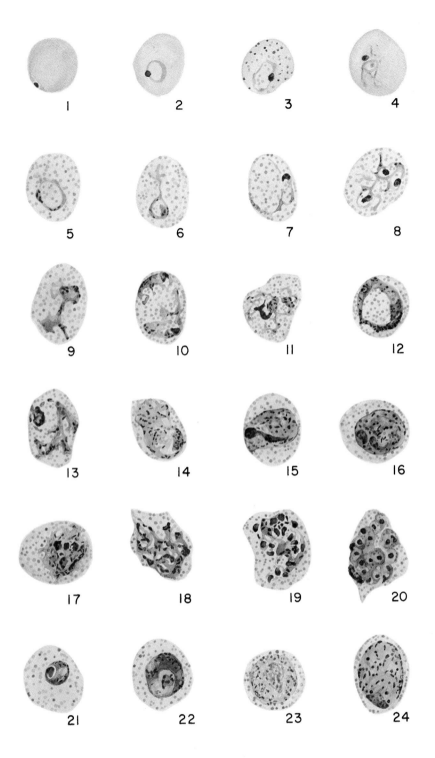

PLATE I *Plasmodium vivax* (Fig. 6.25) **1.** Normal size erythrocyte with marginal ring-form trophozoite. **2.** Young signet-ring form trophozoite in a macrocyte. **3.** Slightly older ring-form trophozoite in erythrocyte showing basophilic stippling. **4.** Polychromatophilic erythrocyte containing young parasite with pseudopodia. **5.** Ring-form trophozoite showing pigment in cytoplasm in an enlarged erythrocyte containing Schüffner's dots. (Schüffner's dots do not occur in all host cells containing growing and older forms of *P. vivax.*) **6, 7.** Medium trophozoite forms. **8.** Three amoeboid trophozoites with fused cytoplasm. **9, 11, 12, 13.** Older amoeboid trophozoites in process of development. **10.** Two amoeboid trophozoites in one host erythrocyte. **14.** Mature trophozoite. **15.** Mature trophozoite with chromatin in process of dividing. **16, 17, 18, 19.** Schizonts showing progressive stages of division. **20.** Mature schizont. **21, 22.** Developing gametocytes. **23.** Mature microgametocyte. **24.** Mature macrogametocyte. (Courtesy of National Institutes of Health.)

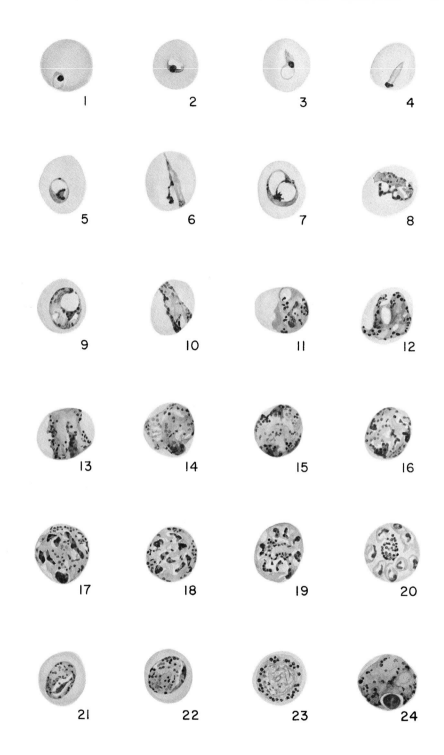

PLATE II *Plasmodium malariae* (Fig. 6.26) **1.** Young ring-form trophozoite. **2, 3, 4.** Young trophozoites in host erythrocytes showing gradual increase of chromatin and cytoplasm. **5.** Developing ring-form trophozoite showing pigment granule. **6.** Early band-form trophozoite (with elongated chromatin and some pigment). **7, 8, 9, 10, 11, 12.** Some variations in the shapes developing tropho-zoites may take. **13, 14.** Mature trophozoites; one in band-form. **15, 16, 17, 18, 19.** Stages in the development of schizont. **20.** Mature schizont. **21.** Immature microgametocyte. **22.** Immature macrogametocyte. **23.** Mature microgametocyte. **24.** Mature macrogametocyte. (Courtesy of National Institutes of Health.)

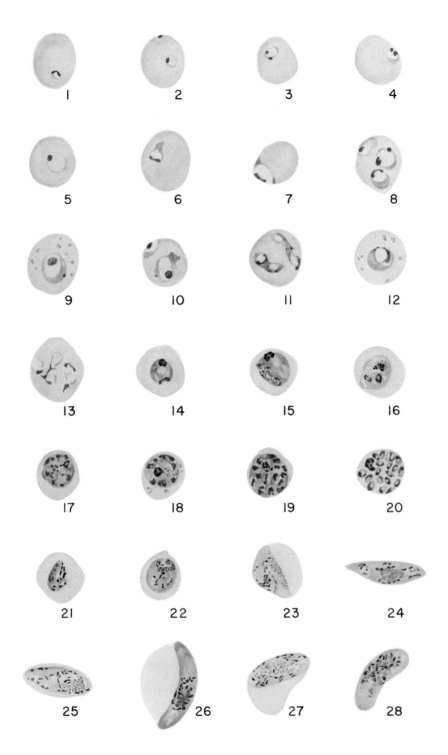

PLATE III *Plasmodium falciparum* (Fig. 6.27) **1.** Very young ring-form trophozoite in host erythrocyte. **2.** Double infection of single cell with young trophozoites; one a "marginal form," the other a "signet-ring" form. **3, 4.** Young trophozoites showing double chromatin dots. **5, 6, 7.** Developing trophozoites. **8.** Three medium trophozoites in one erythrocyte. **9.** Trophozoite showing pigment in erythrocyte containing Maurer's dots. **10, 11.** Two trophozoites in each of two cells showing variations of form. **12.** Nearly mature trophozoite showing haze of pigment throughout cytoplasm; with Maurer's dots in host. **13.** Slender forms. **14.** Mature trophozoite showing clumped pigment. **15.** Initial chromatin division. **16, 17, 18, 19.** Various developmental phases of schizont. **20.** Mature schizont. **21, 22, 23, 24.** Successive forms during development of gametocyte (usually not found in peripheral blood.) **25.** Immature macrogametocyte. **26.** Mature macrogametocyte. **27.** Immature microgametocyte. **28.** Mature microgametocyte. (Courtesy of National Institutes of Health.)

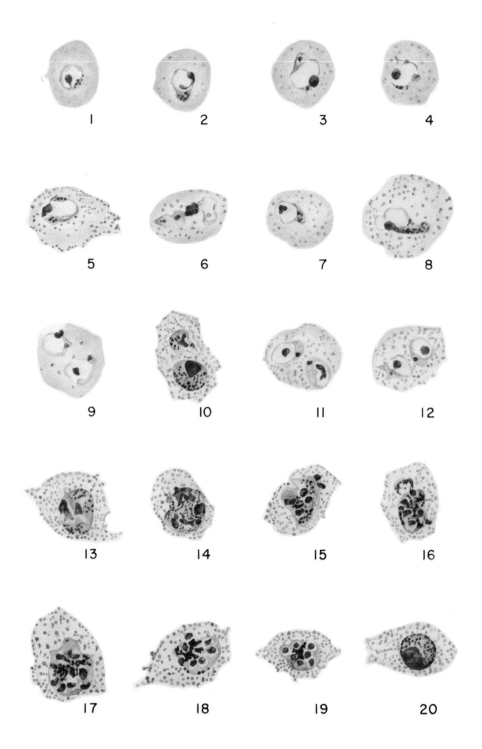

PLATE IV *Plasmodium ovale* (Fig. 6.28) **1.** Young ring-shaped trophozoite. **2, 3, 4, 5.** Older ring-shaped trophozoites. **6, 7, 8.** Older amoeboid trophozoites. **9, 11, 12.** Host erythrocytes doubly infected with trophozoites. **10.** Host erythrocyte doubly infected with young gametocytes. **20.** Mature gametocyte. (Courtesy of National Institutes of Health.)

Suborder Thecina
Suborder Flabellina
Suborder Conopodina
Digitiform or mammilliform, usually blunt, normally unbranched hyaline pseudopodia usually produced from a broad hyaline lobe; not discoid; cysts seldom formed; nuclear division typically mesomitotic. (Genus mentioned in text: *Paramoeba*.)
Suborder Acanthopodina
More or less finely tipped, sometimes filiform, often furcate hyaline pseudopodia produced from a broad hyaline lobe; not regularly discoid; cysts usually formed; nuclear division mesomitotic or metamitotic. (Genus mentioned in text: *Acanthamoeba*.)
ORDER SCHIZOPYRENIDA
Body with shape of monopodial cylinder, usually moving with more or less eruptive, hyaline, hemispheric bulges; typically uninucleate, nuclear division promitotic; temporary flagellate stages in most species. (Genera mentioned in text: *Naegleria, Vahlkampfia*.)
Subclass Testacealobosia
ORDER ARCELLINIDA
ORDER TRICHOSIDA
CLASS ACARPOMYXEA
ORDER LEPTOMYXIDA
ORDER STEREOMYXIDA
CLASS ACRASEA
ORDER ACRASIDA
CLASS EUMYCETOZOEA
Subclass Protosteliia
ORDER PROTOSTELIIDA
Subclass Dictyosteliia
ORDER DICTYOSTELIIDA
Subclass Myxogastria
ORDER ECHINOSTELIIDA
ORDER LICEIDA
ORDER TRICHIIDA
ORDER STEMONITIDA
ORDER PHYSARIDA
CLASS PLASMODIOPHOREA
Obligate intracellular parasites with minute plasmodia; zoospores produced in zoosporangia and bearing anterior pair of unequal, nonmastigonemate flagella; resting spores formed in compact sori or loose clusters within host cells, sexuality reported in some species.
ORDER PLASMODIOPHORIDA
With characters of the class.
CLASS FILOSEA
ORDER ACONCHULINIDA
ORDER GROMIIDA
CLASS GRANULORETICULOSEA
ORDER ATHALAMIDA
ORDER MONOTHALAMIDA
ORDER FORAMINIFERIDA
Suborder Allogromiina
Suborder Textulariina
Suborder Fusulinina
Suborder Miliolina
Suborder Rotaliina

CLASS XENOPHYOPHOREA
ORDER PSAMMINIDA
ORDER STANNOMIDA
SUPERCLASS ACTINOPODA
CLASS ACANTHAREA
ORDER HOLACANTHIDA
ORDER SYMPHYACANTHIDA
ORDER CHAUNACANTHIDA
ORDER ARTHRACANTHIDA
Suborder Sphaenacanthina
Suborder Phyllacanthina
ORDER ACTINELIIDA
CLASS POLYCYSTINEA
ORDER SPUMELLARIDA
Suborder Sphaerocollina
Suborder Sphaerellarina
ORDER NASSELLARIDA
CLASS PHAEODAREA
ORDER PHAEOCYSTIDA
ORDER PHAEOSPHAERIDA
ORDER PHAEOCALPIDA
ORDER PHAEOGROMIDA
ORDER PHAEOCONCHIDA
ORDER PHAEODENDRIDA
CLASS HELIOZOEA
ORDER DESMOTHORACIDA
ORDER ACTINOPHYRIDA
ORDER TAXOPODIDA
ORDER CENTROHELIDA

CLASSIFICATION OF OPALINATA (THE OPALINIDS)*

PHYLUM SARCOMATIGOPHORA
Single type of nucleus; sexuality, when present, essentially syngamy; with flagella, pseudopodia, or both types of locomotor organelles. Opalinata with cilia.
Subphylum Opalinata
Numerous cilia in oblique rows over entire body surface; cytostome absent; nuclear division accentric; binary fission generally interkinetal; known life cycles involve syngamy with anisogamous flagellated gametes; all parasitic.
CLASS OPALINATEA
With characters of the subphylum.
ORDER OPALINIDA
With characters of the class. (Genera mentioned in text: *Opalina, Protoopalina, Cepedea, Zelleriella*.)

---

*Based on the recommendation of the Committee on Systematics and Evolution of the Society of Protozoologists consisting of Levine *et al.* (1980).

6

APICOMPLEXA, MICROSPORA, ASCETOSPORA, AND MYXOZOA

The protozoan phyla Apicomplexa, Microspora, Ascetospora, and Myxozoa were formerly considered to be members of the same subphylum, the Sporozoa; however, recent findings suggest that it is more natural to consider these four groups as distinct phyla (Levine *et al.*, 1980). These groups include parasitic species exclusively. They are either intra- or intercellular parasites of hosts belonging to practically every animal phylum. In vertebrates, these protozoans occur mainly in the blood, the reticuloendothelial system, and the epithelial lining of the intestine; in invertebrates, they are found primarily in the digestive and excretory systems.

The diagnostic characteristics of the members of the Apicomplexa, Microspora, Ascetospora, and Myxozoa, as well as those of the subordinate taxa, are presented at the end of this chapter.

GENERAL LIFE CYCLE PATTERN

Neither the morphology nor the life cycle patterns of the apicomplexans, microsporans, ascetosporans, and myxozoans give much indication of their relationship with the other protozoans. The complexities of their life cycles, which are intimately associated with the physiology of their hosts, suggest that these organisms have long been parasitic. It has been suggested that at least some of these protozoans were in existence by the beginning of the tertiary, about 70 million years ago if not earlier.

Since the life cycle patterns as found among the members of the Apicomplexa, Microspora, Ascetospora, and Myxozoa are both numerous and diverse, it is not possible to give a generalized pattern other than to state that in many life cycles there are usually two or three distinct phases. The three-phase life cycle is by far the more common. The infective form of all of these protozoans is the **sporozoite**. This commonly elongate form is either enveloped or not enveloped within a spore wall. Once introduced into a host, sporozoites, after undergoing certain preparatory stages, reproduce asexually by multiple fission or schizogony. This is the **schizogonic phase**.

The organisms resulting from schizogony are known as **merozoites**. Some of these differentiate into male gametes while others become female gametes. The process by which these gametes develop is known as **gamogony**. When a female gamete and a male gamete fuse, the **zygote** is formed. Gamogony and subsequent fertilization may be considered the **sexual reproductive phase**. The appearance of the zygote marks the end of the reproductive stage and the beginning of the third phase, the **sporogonic phase**. During this phase, the zygote usually develops into one or several **spores**, each containing a distinctive number of **sporozoites**. When sporozoites become established within a host, the schizogonic phase recurs.

Among a few groups, schizogony does not occur. In their life cycles only the sexual and sporogonic phases are present. This is the case among many gregarines (members of the subclass Gregarinia of the Apicomplexa).

With a few exceptions, among species that require two hosts, sexual reproduction takes place in the **definitive** or **primary host** and schizogony occurs in the **intermediate** or **secondary host**.

Spores constitute the transmissive form of the parasite. Spores of species confined to one host usually have a resistant membranous encasement—the **spore membrane**—which enables them to withstand unfavorable conditions while outside the host. Among those species occurring in their hosts' blood, such as the malaria-causing *Plasmodium* spp., the infective forms (sporozoites) are naked—that is, without an encasement.

Many of these protozoans are of medical and veterinary importance because they are responsible for some of our most widespread diseases, such as malaria.

PHYLUM APICOMPLEXA

The phylum Apicomplexa was established to include protozoans that possess a certain combination of structures collectively known as the **apical complex**, which is distinguishable by electron microscopy. The components of this complex, which are described below, include a polar ring(s), micronemes, rhoptries, subpellicular microtubules, micropore(s), and a conoid.

The phylum includes two classes: Perkinsea and Sporozoea. The characteristics of these are presented at the end of this chapter.

Apical Complex

This complex of structures, as stated, includes several features. Specifically, the sporozoites and merozoites possess one or two **polar rings** at the apical terminal (Figs. 6.1, 6.2). These rings are electron-dense structures situated immediately beneath the cell membrane. In members of the suborder Eimeriina (of the class Sporozoea and the subclass Coccidia), a truncated cone of spirally arranged fibrillar structures, known as the **conoid**, is located within the polar rings (Figs. 6.1,

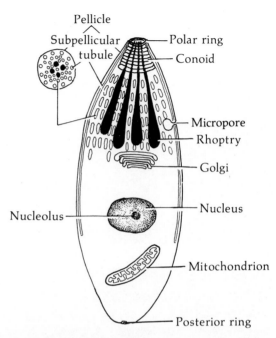

Fig. 6.1. Apical complex. Drawing showing constituents of apical complex of apicomplexan sporozoite.

6.2). **Subpellicular microtubules** radiate from the polar rings and run posteriorly, parallel to the longitudinal body axis (Figs. 6.1, 6.2). These organelles probably serve as support elements and may also be involved in locomotion.

Two or more **rhoptries** extend to the cell membrane within the polar rings (and, if present, the conoid). The rhoptries are electron-dense bodies (Figs. 6.1, 6.2). Smaller, more convoluted elongate bodies, known as **micronemes**, also extend posteriorly from the apical complex (Figs. 6.1, 6.2). The ducts of the micronemes either run anteriorly into the rhoptries or join a common duct system with the rhoptries to lead to the cell surface at the apex. The contents of the micronemes and rhoptries appear to be similar and, when secreted, aid in the penetration of the host cell.

Along the side of the parasite are one or more **micropores** (also known as **cytostomes**) (Fig. 6.1). These function in the intake of nutrients during the intracellular life of the parasite. The edges of the cytostome are marked by two concentric, electron-dense rings, which are situated immediately beneath the cell membrane. As the cytoplasm of the host cell or other nutrients within the **parasitophorous vacuole**, i.e., the host's cytoplasmic vacuole in which the parasite

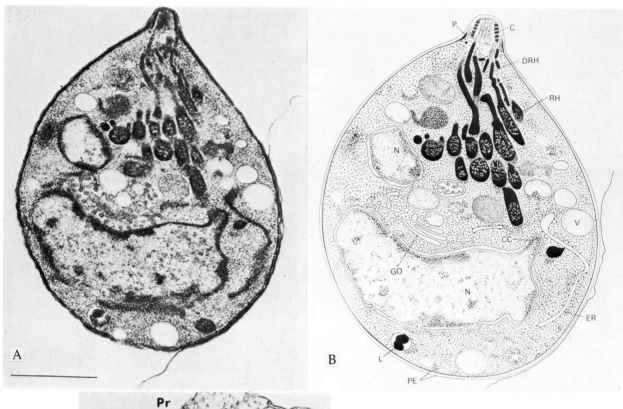

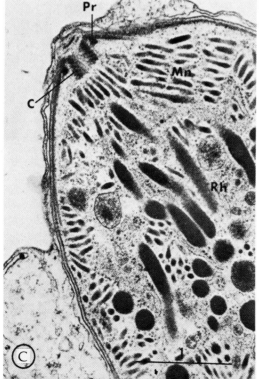

Fig. 6.2. Apical complex. A. Electron micrograph of merozoite of *Toxoplasma gondii* showing certain constituents of apical complex. **B.** Drawing of A. (A and B after Scholtyseck, 1979; with permission of Springer-Verlag.) **C.** Electron micrograph of anterior end of sporozoite of *Eimeria ninakohlyakimovae* showing certain constituents of apical complex. (After Kelley and Hammond, 1972.) C, conoid; CC, centrocone; DRH, rhoptry ductule; ER, endoplasmic reticulum; GO, Golgi; L, lysosome; Mn, micronemes; N, nucleus; P, polar ring; PE, pellicle; Pr, polar ring; Rh, rhoptries; V, vesicle.

resides, is pulled through the rings, the parasite's cell membrane invaginates accordingly and eventually pinches off to form a membrane-bound food vacuole.

With the exception of the micropores, the structures described in this section dedifferentiate and disappear after the sporozoite or merozoite penetrates the host cell and transforms into a trophozoite.

The apical complex is characteristic of sporozoites and merozoites of apicomplexans. It does not occur in microsporans, ascetosporans, or myxozoans.

Probably the most extensively studied member of the Perkinsea is *Perkinsus marinus*. Originally thought to be a fungus and designated *Dermocystidium marinum*, this species was recognized to be economically important in 1949 when it was found to be the causative agent of extensive, warm-weather mortalities of oysters and other shellfish along the Gulf and Florida coasts (Mackin *et al.*, 1950). Most of the information pertaining to this parasite has been reviewed by Ray and Chandler (1955). When seen in histologic sections of infected molluscs, the "spores" occur both intra- and intercellularly. Each spore measures 2–20 μm, occasionally reaching 30 μm in diameter (Fig. 6.3). They are considerably larger in winter than in sum-

mer. The most distinctive feature of *P. marinus* spores is a very large, partially eccentric vacuole, which usually contains one relatively large, polymorphic, refringent inclusion body known as the **vacuoplast**. The usually oval nucleus contains a compact, deep-staining endosome. A reproductive stage, involving multiple fission of the original single nucleus, appears as a spherical body enclosing several daughter cells known as **zoospores**. Each of these is flagellated (Fig. 6.3). The number of zoospores varies from 3 to 50, and these eventually rupture out of the enclosing

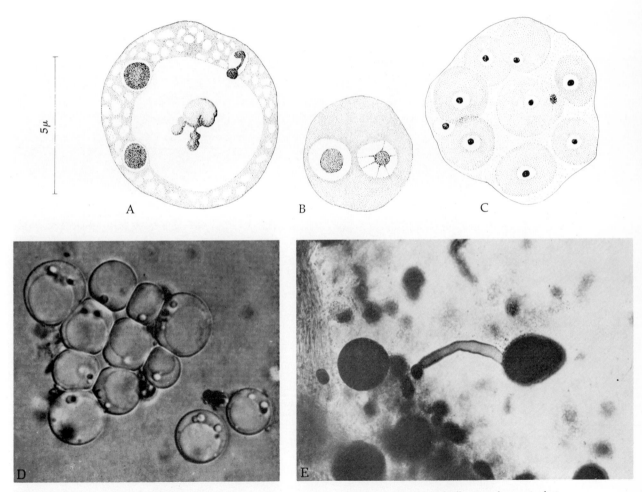

Fig. 6.3. Stages in the development of *Perkinsus marinus*. A. A mature "spore" with markedly irregular vacuoplast, cytoplasmic inclusions, and a very large vacuole. **B.** An immature "spore" with a small vacuole and vesicular nucleus. **C.** Multiple fission resulting in several zoospores. (Redrawn after Mackin *et al.*, 1950.) **D.** Photomicrograph of zoospores from oyster pericardial fluid cultured in fluid thioglycollate medium. **E.** A germinating form of *P. marinus* showing cytoplasmic extension (flagellum) from oyster mantle tissue after incubation for 41 days in seawater containing yeast extract and dextrose. (After Ray and Chandler, 1955.)

spore and infect adjacent tissues or other oysters. Transmission is direct from oyster to oyster.

Perkins (1969) has contributed an electron microscope study of the "spore." Most of the organelles are similar to those of many eukaryote protists; however, funguslike lomasomes occur between the wall and the plasmalemma (Fig. 6.4).

Perkinsus marinus is now known to occur in oysters from the coast of Virginia southward, and along the Gulf coast. It is pathogenic to the molluscan hosts and consequently represent a threat to the oyster industry. No method has yet been devised to control this parasite.

CLASS SPOROZOEA

The class Sporozoea includes three subclasses: Gregarinia, Coccidia, and Piroplasmia. All members of these subclasses are parasitic.

SUBCLASS GREGARINIA

The gregarines are parasites of invertebrates, especially arthropods, molluscs, and annelids. A few have been reported from ascidians, sipunculids, and hemichordates. Since the discovery that molluscs have arisen from a segmented ancestral form, it is now reasonable to postulate that parasitism by gregarines could have been initiated in some common ancestral stock from which segmented animals were derived. Gregarines are most commonly found in the cells lining the coelom and digestive tract and in reproductive organs of the host, although the mature

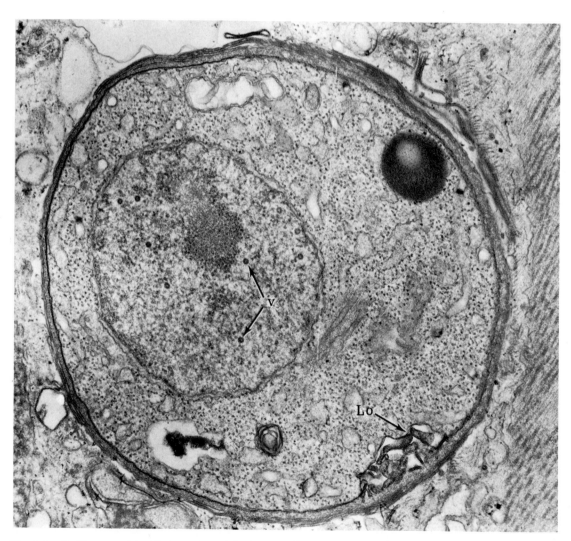

Fig. 6.4. *Perkinsus marinus.* Electron micrograph of "spore" showing viruslike particles in nucleus (v) and lomasome (Lo) associated with wall. × 33,000. (After Perkins, 1969.)

gamonts (i.e., mature trophozoites, also known as **sporadins**) are extracellular. These parasites obtain their nutrition through micropores located on the body surface, and the majority multiply by sexual reproduction.

ORDER EUGREGARINIDA

The order Eugregarinida includes the "true gregarines," which are commonly encountered in insects and other invertebrates. The life cycles of the various species follow an essentially similar pattern.

Generalized Life Cycle

Infection of a new host is established with the ingestion of spores (or **oocysts**) enclosing **sporozoites**. Typically, each spore includes rod- or sickle-shaped sporozoites. Once within the gut of the host, the sporozoites escape from the spore walls and actively penetrate epithelial cells, one sporozoite into each cell. The intracellular sporozoites increase in size and eventually leave the host cell. These larger organisms, now known as **trophozoites** or **gamonts**, are then found in the lumen of the host's gut. After a period of wandering about, each individual, now known as a **gametocyte**, becomes more round and pairs up with a mate. The two then become encysted in a protective envelope known as the **gametocyst**.

Although the members of each pair of gametocytes are indistinguishable, one is a male and the other a female. Once enveloped within a gametocyst, each gametocyte undergoes repeated nuclear divisions, culminating finally in cytoplasmic divisions, with a portion of the parental cytoplasm surrounding each daughter nucleus. The resulting uninucleate bodies are known as **gametes**. The progeny of a male gametocyte are all males, whereas the progeny of a female gametocyte are all females. Because the male and female gametes are morphologically similar, they are known as **isogametes**.

The fusion of a male and a female gamete (known as **isogamy**) results in a **zygote**. Numerous zygotes are formed within a single gametocyst. As soon as the zygotes are formed, a secondary cyst, known as an **oocyst**, or sometimes as a **sporocyst**, is secreted by each zygote around itself. Three rapid divisions occur within the oocyst, resulting in eight sporozoites. At this point, the **gametocyst**, which contains numerous sporozoite-enclosing oocysts, is discharged from the host's gut in feces. It may remain intact and can be ingested by another host, or it may rupture, releasing oocysts, each enclosing eight sporozoites. If the latter occurs, oocysts are ingested by another host.

In the life cycle pattern presented above, sexual reproduction involving isogamy and asexual reproduction involving sporozoite formation (i.e., sporogony) occur but schizogonic reproduction does not occur. As a result, the host is not overwhelmed with parasites, especially tissue-invading parasites, and injury is limited to those host cells actually attacked by the sporozoites.

Suborders of Eugregarinida

The Eugregarinida includes three suborders—Blastogregarinina, Aseptatina, and Septatina. Gamonts of members of Blastogregarinina and Aseptatina possess rather simple bodies that are not compartmentalized, whereas those of members of Septatina are either divided into two chambers (the anterior **protomerite** and the larger, posterior **deutomerite**) or into three parts (an anterior anchoring device known as an **epimerite**, a middle chamber known as the **protomerite**,

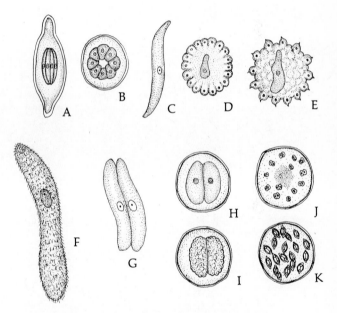

Fig. 6.5. Stages in the life cycle of *Monocystis lumbrici*, a gregarine parasite in the seminal vesicles of earthworms. A. Spore, or oocyst, containing eight sporozoites. **B.** Cross-sectional view of spore enclosing eight sporozoites. **C.** Single liberated sporozoite. **D.** Sporozoite that has entered the multicellular sperm sphere of host (sperm mother cells). **E.** Growth and transformation of sporozoite into trophozoite within the host's sperm sphere. **F.** Mature trophozoite surrounded by thin layer of degenerating sperm-sphere cells, to which the tails of the spermatozoa are attached. **G.** Two trophozoites that are free of degenerated sperm-sphere cells and that have united as gametocytes. **H.** Gametocytes encysted within double-walled gametocyst. **I.** Gametocytes under-going sporulation to form isogametes. **J.** Sexual reproduction during which isogametes unite to form zygotes. **K.** Cyst containing many young spores. The spores have arisen from the division of the zygotes to form sporozoites that secrete the spore capsules around themselves.

and a posterior chamber known as the **deutomerite**). The deutomerite includes the nucleus (see Fig. 6.7 below). For a detailed review of the gregarines, see Manwell (1977).

Suborder Aseptatina. The Aseptatina includes the genus *Monocystis*, the members of which are commonly found in the coelom and in the seminal vesicles of earthworms (Fig. 6.5). In addition, members of the

genera *Enterocystis, Rhabdocystis, Apolocystis,* and *Nematocystis* are all fairly common, and in each case the organism follows the life cycle pattern that does not involve schizogony. Selected members of these and related genera are depicted in Fig. 6.6.

Enterocystis ensis is parasitic in the alimentary tract of the larvae of the ephimerid, *Caenis* sp. (Fig. 6.6). The spores (or oocysts) are elongate ovoid, enclosing eight sporozoites. *Rhabdocystis claviformis* is found in the seminal vesicles of the annelid *Octoplasium complanatum* (Fig. 6.6). The trophozoites of this species measure up to 30 × 300 μm, and the spores are oval with a slight indentation in the middle. *Nematocystis*

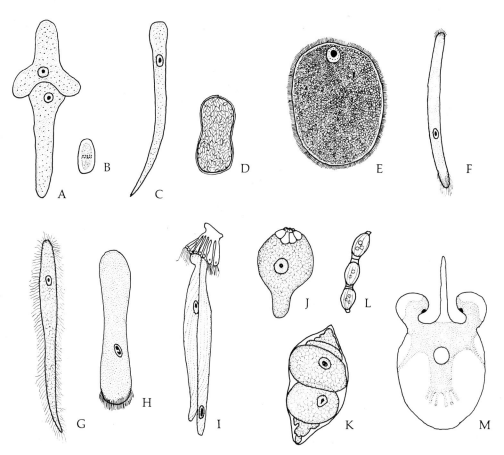

Fig. 6.6.   Some gregarine parasites (Aseptatina). A. two trophozoites or sporadins of *Enterocystis ensis* in typical end-to-end association (syzygy). **B.** Single spore of *E. ensis* from the gut of the larva of the ephemerid *Caenis* sp. (**A** and **B**, redrawn after Zwetkow, 1926.) **C.** Single trophozoite or sporadin of *Rhabdocystis claviformis*, a parasite in the seminal vesicles of the oligochaete *Octolasium complanatum*. **D.** Cyst of *R. claviformis*. (**C** and **D**, modified after Boldt, 1910.) **E.** *Apolocystis gigantea* trophozoite found in seminal vesicles of the oligochaetes *Helodrilus foetidus* and *Lumbricus rubellus*. (Modified after Troisi, 1933.) **F.** *Nematocystis vermicularis* trophozoite from seminal vesicles of oligochaetes. **G.** *Rhynchocystis pilosa* trophozoite from seminal vesicles of oligochaetes. (Redrawn after Hess, 1909.) **H.** *Zygocystis wenrichi* trophozoite from seminal vesicles of *Lumbricus rubellus* and *Helodrilus foetidus*. (Redrawn after Troisi, 1933.) **I.** *Pleurocystis cuenoli* trophozoites in ciliated seminal horn of *Helodrilus longus* and *H. caliginosus*. (Redrawn after Hesse, 1909.) **J.** *Stomatophora coronata* trophozoite with suckerlike anterior epimerite; found in seminal vesicles of oligochaetes of the genus *Pheretima*. **K.** Cyst of *S. coronata*. **L.** Spores of *S. coronata* arranged in typical chain. (**J–L,** redrawn after Hesse, 1909.) **M.** *Choanocystella tentaculata* trophozoite showing typical anterior mobile sucker and a tentacle; found in seminal vesicles of the oligochaete *Pheretima beaufortii* in New Guinea. (Redrawn after Martiis, 1911.)

vermicularis is a parasite in the seminal vesicles of various species of earthworms (Fig. 6.6). The trophozoites measure $1000 \times 100 \ \mu m$ and bear tufts of cilia-like projections at each end. *Apolocystis* includes parasites of the seminal vesicles and coelom of various oligochaetes; the trophozoites are spherical and the spores are biconical (Fig. 6.6).

Members of the genus *Rhynchocystis* have trophozoites whose anterior ends are conical (Fig. 6.6). Many species of *Rhynchocystis* are found in the seminal vesicles of various oligochaetes.

Life Cycle of Monocystis. The life cycles of all of the aseptatinans follow essentially the same pattern. That of *Monocystis lumbrici* (Fig. 6.5) is presented at this point.

Monocystis lumbrici lives in the seminal vesicle of *Lumbricus terrestris* and related earthworms. The annelid becomes infected by ingesting spores each enclosing several sporozoites. Once the spores reach the earthworm's gizzard, they rupture, releasing the sporozoites. These penetrate through the intestinal wall, enter the dorsal blood vessel, and move forward to the hearts. They then exit from the circulatory system and penetrate the host's seminal vesicles, where they enter blastophores (sperm-forming cells) in the vesicle wall. After a brief growth period, during which the sporozoites destroy the developing spermatocytes, the parasites enter the lumen of the vesicle where they mature into **gamonts** (also known as **trophozoites** or **sporadins**). Each gamont measures about 200 μm long by 65 μm wide and is attached to a host cell in the region of the sperm duct. Here the gamonts undergo **syzygy**, i.e., two (sometimes more) gamonts become connected to one another. The anterior organism is termed the **primite**, while the posterior one is known as the **satellite**. After syzygy, the gamonts flatten against each other and secrete a common cyst wall, forming the **gametocyst**. Although two gamonts occur within each cyst, they are morphologically distinct. Each undergoes numerous nuclear divisions and the resulting small nuclei move to the periphery of the cytoplasm. Eventually each nucleus is budded off and becomes surrounded by a piece of the cytoplasm. The resulting cells are gametes. Some of the cytoplasm of each gamont is left behind. These pieces fuse to become the **residual body**.

The male and female gametes of *M. lumbrici* are morphologically distinguishable and hence are known as **anisogametes**. The fusion of two gametes results in the formation of a zygote. Each zygote secretes a **spore** or **oocyst membrane** around itself and three cell divisions ensue. These divisions are known as **sporogony**, and the result is the formation of eight **sporozoites**. It should be apparent that each gametocyst now contains many oocysts. The gametocysts pass out of the host through the sperm duct, and a new host becomes infected by ingesting a gametocyst, or if that is ruptured, by eating an oocyst.

Suborder Septatina. The members of this group are primarily parasites in the alimentary canal of arthropods. Some of the better known genera are described below.

The genus *Gregarina* includes many species parasitic in insects (Fig. 6.7). For example, *G. blattarum* is found in the oriental cockroach, *G. oviceps* in various species of crickets, and *G. polymorpha* in the mealworm, *Tenebrio molitor*.

Harry (1967) has studied the effect of *Gregarina polymorpha* on the larva of *Tenebrio molitor*. He has found that if the hosts are maintained under optimal conditions of temperature, relative humidity, and diet, the parasite does not affect the host's weight or life span; however, if the hosts are grown on a suboptimal diet, the gregarines have a considerable effect on the final pupal weight, which is decreased, and it is difficult for the larva to complete its development. This information serves as an excellent example of the principle that many parasites are nondeleterious when present in healthy hosts, but can be pathogenic when the hosts are unhealthy.

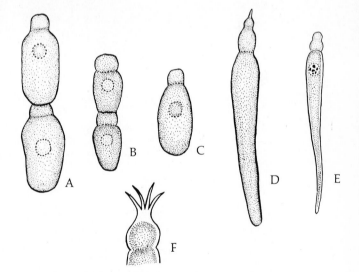

Fig. 6.7. Some representative members of the Septatina. A. *Gregarina blattarum* trophozoites, (or sporadins) from midgut of cockroaches. (Redrawn after Kudo, 1966. "Protozoology" 5th ed. Charles C Thomas.) **B.** *Gregarina oviceps* trophozoites, (or sporadins) from grasshoppers. **C.** *Gregarina locustae* sporadin from the locust *Dissosteria carolina*. (Redrawn after Leidy, 1853.) **D.** *Stylocephalus giganteus* trophozoite, (or sporadin), from various coleoptera. (Modified after Ellis, 1912.) **E.** *Actinocephalus acustispora* trophozoite, (or sporadin) from the gut of the coleopteran *Silpha laevigata*. **F.** Anterior end of *A. acustispora* showing some of the apical processes. (Modified after Léger, 1892.)

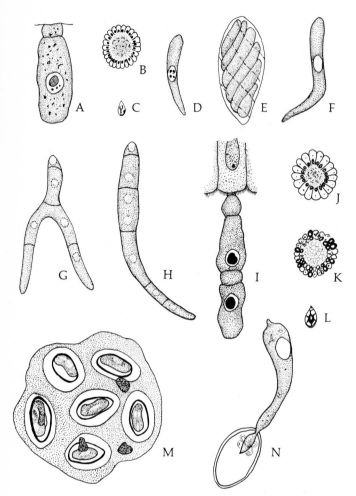

Fig. 6.8. Representatives of *Porospora* and *Nematopsis*. A. *Porospora gigantea* trophozoite attached to the gut of the lobster *Homarus gammarus*. **B.** *P. gigantea* gymnospores in molluscan hosts *Mytilus minimus* and *Trochocochlea mutabilis*. **C and D.** Developing *P. gigantea* sporozoites in molluscan host. **E.** *P. gigantea* sporozoites enveloped by host's phagocyte. **F.** Mature sporozoite of *P. gigantea*. (Modified after Hatt, 1931.) **G. and H.** Trophozoites (or sporadins) of *Nematopsis legeri* in the crustacean *Eriphia spinifrons*. **I.** *N. legeri* trophozoites (or sporadins) attached to gut epithelium of crustacean host. **J.** *N. legeri* gymnospores. **K.** Gymnospores after entering body of molluscan host. **L.** Young sporozoite of *N. legeri*. **M.** *N. legeri* cyst in molluscan host, enclosing six spores. **N.** Germination of spore in gut of crustacean host. (Modified after Hatt, 1931.)

Members of *Gregarina* and related genera are capable of gliding movement in the guts of their arthropod hosts. Mackenzie and Walker (1983) have shown that in the case of *Gregarina garnhami*, parasitic in the midgut of the desert locust, *Schistocerca gregria*, its gliding movement is dependent on contact with a substrate and is accompanied by the formation of a mucus trail. Also, no alteration of body shape occurs during movement.

The genus *Stylocephalus* is characterized by a nipple-shaped epimerite and papillae-covered spores (Fig. 6.7). Several species of this genus have been reported from insects and molluscs.

Members of the genus *Actinocephalus* can be recognized by a sessile epimerite that has eight to ten fingerlike processes at its apex and by their biconical spores (Fig. 6.7). This genus includes *A. parvus* in the gut of the dog flea, *Ctenocephalus canis*.

Among one group of Septatina of considerable interest to marine parasitologists, the members of *Porospora* and *Nematopsis*, two hosts are involved—a crustacean and a mollusc. When naked or well-protected sporozoites enter the stomach or midgut of a specific crustacean host, they develop into typical gamonts with compartmentalized bodies. Two or more of these gamonts undergo syzygy and encyst. Subsequently, their nuclei undergo repeated division, accompanied by cytoplasmic division, so that a large number of daughter cells, known as **gymnospores** (actually gametes), are formed. The gymnospores are located in the host's hindgut and are eventually voided with the feces.

When gymnospores come in contact with the molluscan host, they enter or are engulfed by phagocytes in the mollusc's gills, mantle, or digestive tract. Gymnospores are commonly found in large numbers in their host's gill lacunae. Eventually the gymnospores pair off and fuse, forming zygotes that develop into sporozoites. These sporozoites are either enclosed within membranes or naked, depending on the species.

Sporozoites that are engulfed by phagocytes are ingested by the crustacean host—normally this occurs when young infected molluscs are ingested—the portion of the life cycle that occurs within the crustacean is repeated.

As mentioned, this type of life cycle is characteristic of members of the genera *Porospora* and *Nematopsis*. Members of *Porospora* are further characterized by their sporozoites, which have no protective envelope and are found in molluscan phagocytes (Fig. 6.8). Sporozoites of *Nematopsis*, which are also found in molluscan phagocytes, are enveloped by two membranes (Fig. 6.8).

Porospora and *Nematopsis* are commonly encountered by marine parasitologists since several species occur in economically important marine invertebrates. For example, *Porospora gigantea* alternates between lobster and mussel hosts, with sexual reproduction occurring in the lobster. *Nematopsis legeri*, with gymnospores measuring 7 μm in diameter, *N. ostrearum*, with gymnospores measuring 4 μm in diameter, and *N. prytherchi*, with gymnospores measuring 6 μm in diameter, all utilize crabs and oysters or mussels as hosts. There are high levels of infection of oysters by *N. ostrearum* along the Atlantic and Gulf Coasts of the United States because the gregarious habit of the mud crab provides for recurring contacts. Fortunately, however, there is little evidence that this gregarine is destructive to the oyster under normal conditions. This is probably because of the achievement of a dynamic equilibrium in the oyster between elimination of and reinfection by the parasite (Feng, 1958). Elimination of the parasite is achieved primarily by migration of the sporozoite-laden phagocytes from the oyster.

SUBCLASS COCCIDIA

Coccidians are generally parasites of the epithelia that line the alimentary tracts of vertebrates and invertebrates, but they are also found in the cells of associated glands. On the other hand, some members of the suborder Haemosporina (see p. 193) are blood parasites. Among the Haemosporina are the members of the genus *Plasmodium*, which cause malaria.

An alternation between sexual and asexual reproduction is the general rule among coccidians. Except in a few genera, such as *Aggregata, Lankesterella, Plasmodium, Haemoproteus, Leucocytozoon*, etc., these parasites are associated with only one host. In the case of the exceptions mentioned, a second host, an invertebrate, is required for completion of the life cycles.

As outlined at the end of this chapter, the Coccidia is divided into three orders: Agamococcidiida, Protococcidiida, and Eucoccidiida. The first two include rather obscure species. The emphasis here is being placed on the Eucoccidiida.

ORDER EUCOCCIDIIDA

The Eucoccidiida is divided into three suborders: Adeleina, Eimeriina, and Haemosporina.

Suborder Adeleina

Members of this suborder are characterized by the presence of macro- and microgametocytes which differ in size and are usually associated in syzygy during development.

The genera subordinate to the Adeleina belong to two families. Those genera which parasitize the epithelial lining and associated glands of the gut of invertebrates and occasionally internal organs of vertebrates belong to the family Adeleidae, whereas those which parasitize the blood cells of vertebrates belong to the family Haemogregarinidae.

Family Adeleidae. Members of this family possess sporocysts which are formed in oocysts. From

what is known, probably only one host is involved in the life cycles of these parasites.

Life Cycle of Adelina deronis. *Adelina deronis* (Fig. 6.9) is selected as a representative of the family Adeleidae. The life cycle of this species, which is parasitic in the mesodermal cells lining the coelom of the freshwater oligochaete *Dero limosa* is as follows. Infection of the worm occurs with the ingestion of oocysts (or spores). Each oocyst contains 10 to 14 spherical sporocysts ranging from 6 to 9 μm in diameter. Each sporocyst includes two elongate sporozoites. In the midgut and upper hindgut of the worm, the sporozoites escape upon rupture of the sporocyst and oocyst walls. They then penetrate the gut wall and actively reach and penetrate the peritoneal cells lining the coelom. The sporozoites that fail to penetrate perish. As an intracellular parasite, the sporozoite increases in size and rounds up, often causing a marked vacuolization of the host cell's cytoplasm. No nuclei are visible in the sporozoites until they become schizonts. Merozoites resulting from the division of schizonts are known. Second-generation merozoites, escaping from the host cells, develop into macro- and microgametocytes in the host's coelom.

One gametocyte of each sex enters a new cell in the "pairing process." Instances are known, however, in which a single cell may include one macrogametocyte and two to five microgametocytes. The micro- and macrogametocyte are closely associated while both are

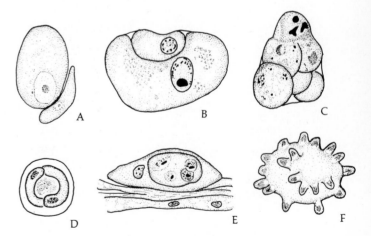

Fig. 6.9. *Adelina deronis*, **a parasite of the oligochaete** *Dero limosa.* **A.** Early pairing of macro- and microgametocyte. **B.** Later stage of pairing, within thin oocyst wall. **C.** Sporoblasts (developing sporocysts) within oocyst; remains of microgametocytes still attached. **D.** Mature sporocyst. **E.** Young schizont in host peritoneal cell. **F.** Schizogony. (Redrawn after Hauschka, 1943.)

undergoing further development. The mature gametes fuse to form the zygote, which secretes a thin oocyst wall. The oocysts mature in the coelom and enclose 10 and 14 sporocysts. It is presumed that oocysts become free and can be ingested by another host only if the host dies and its tissues disintegrate. The life cycle of *A. deronis* is completed in 18 days.

Several genera closely related to *Adelina* are illustrated in Fig. 6.10.

Family Haemogregarinidae. Members of this family require two hosts to complete their life cycles. They inhabit the circulatory system of the vertebrate host and the digestive tract of the invertebrate vector. The genera *Haemogregarina* and *Hepatozoon* are representatives of this family.

Genus Haemogregarina. *Haemogregarina* includes numerous species that utilize various amphibians, reptiles, birds, and a few mammals as the vertebrate host. During their development, the microgametocyte develops into only two or four microgametes, and the sporozoites are not enclosed within outer walls. Lack

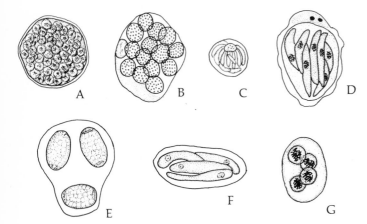

Fig. 6.10. **Some adeleid parasites. A.** Oocyst of *Klossia helicina* parasitic in kidneys of various terrestrial snails; notice typical double envelope and numerous spherical sporocysts. **B.** Renal cell of the mouse, *Mus musculus,* enclosing 14 sporoblasts (developing sporocysts) of *Klossia muris.* **C.** Typical sporocyst of *Klossia muris* enclosing sporozoites. **D.** Oocyst of *Legerella hydropori* parasitic in epithelium of Malpighian tubules of the arthropod *Hydroporus palustris.* **E.** Oocyst of *Chagasella hartmanni* enclosing three developing sporocysts. **F.** Single sporocyst of *Chagasella hartmanni* enclosing four sporozoites; found in gut of the hemipteran *Dysdereus ruficollis.* **G.** Oocyst of *Ithania wenrichi,* a parasite in epithelial cells of the gastric caeca and midgut of the larvae of the crane fly, *Tipula abdominalis.* (**A** and **B,** redrawn after Smith and Johnson, 1902; **D,** redrawn after Vincent, 1927; **E** and **F,** redrawn after Chagas, 1910.)

of any type of encasement around sporozoites is a common evolutionary feature characteristic of many parasites of this group that are introduced into the vertebrate host through the bite of a vector.

Life Cycle of Haemogregarina stepanowi. The life cycle of *H. stepanowi,* a parasite of the turtle *Emys orbicularis,* serves as the pattern to which all haemogregarine life cycles conform (Fig. 6.11).

Infection of the turtle host is established with the introduction of sporozoites into the blood during the feeding process of the leech *Placobdella catenigera.* The elongate sporozoites enter host erythrocytes and grow, becoming V-shaped during the process. Eventually, the two unequal arms of the V fuse, forming an ovoid body. This form is known as a **macroschizont**. The erythrocyte in which it occurs is carried to and becomes lodged in bone marrow. Schizonts of this generation produce 13 to 24 large merozoites that enter other host erythrocytes and transform into **microschizonts**. Each microschizont produces only six smaller merozoites. When these enter host erythrocytes, they become **gametocytes**, ending the schizogonic cycle.

The gametocytes are elongate; the macrogametocyte includes a small nucleus, and the microgametocyte includes a large nucleus and dark-staining transverse bands at the anterior end. No further development occurs in the turtle host.

When a leech ingests infected turtle erythrocytes, the gametocytes are released, and a macrogametocyte joins a microgametocyte in syzygy. A thin oocyst membrane is secreted around the pair. Within, the microgametocyte divides and forms four microgametes. One of these fertilizes the macrogamete, which had differentiated from the macrogametocyte, to form the zygote. Eight sporozoites develop from the zygote, and when these mature, they break out of the thin-walled oocyst into the intestinal lumen. The sporozoites then enter the circulatory system and migrate to the salivary glands. From this site they can be introduced into a new turtle host.

Various invertebrates serve as vectors for haemogregarines. In the case of *Haemogregarina leptodactyli,* a blood parasite of the frog *Leptodactylus ocillatus,* the mite *Acarus* sp. is the vector. Similarly, those hemogregarines which parasitize terrestrial reptiles are transmitted by mites.

Suborder Eimeriina

Coccidians belonging to this suborder are generally intracellular parasites of the epithelial lining of the host's alimentary canal. Both the sexual and schizogonic phases of their life cycles generally occur in one host, although some species exhibit an alternation of hosts. The infective oocysts generally include eight sporozoites. Another characteristic of the members of this suborder is that microgametes and macrogametes develop independently without syzygy.

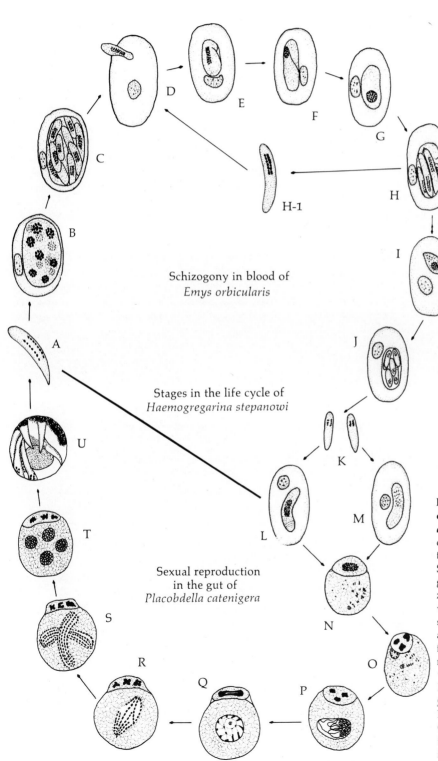

Schizogony in blood of
Emys orbicularis

Stages in the life cycle of
Haemogregarina stepanowi

Sexual reproduction
in the gut of
Placobdella catenigera

Fig. 6.11. Stages in the life cycle of *Haemogregarina stepanowi*, a haemogregarinid parasite of the turtle *Emys orbicularis* and the leech *Placobdella catenigera*. **A.** Single sporozoite. **B–H.** Schizogonic stages. **H–1.** Single merozoite reinfecting an erythrocyte of vertebrate host, thus repeating the schizogonic phase of the cycle. **I** and **J.** Stages during gametocyte formation. **K.** Young macro- and microgametocytes. **L.** Fully developed microgametocyte **M.** Mature macrogametocyte. **N** and **O.** Association of gametocytes. **P.** Fertilization in gut of leech host. **Q–U.** Division of zygotic nucleus to form eight sporozoites. (Modified after Reichenow, 1919.)

The Eimeriina includes over 20 genera, of which only three of the better known are discussed here.

Genus Eimeria. *Eimeria* is a multispecies genus that includes several economically important parasites. The life cycle of *E. tenella*, a species found in the caeca of chickens, is representative of the group (Fig. 6.12). For a detailed review of *E. tenella* and related coccidians of birds, see Ruff and Reid (1977).

Life Cycle of Eimeria tenella. Infection of the fowl occurs when mature oocysts are ingested. Sporozoites escape from the oocyst through a **micropyle** (a minute pore) in the oocyst wall and pass through the surface epithelium of the caecal mucosa into the lamina propria. Here the sporozoites are engulfed by macrophages of the host and transported to cells of the glands of Lieberkühn. As an intracellular parasite, the sporozoite increases in size and schizogony begins. Rapid nuclear division takes place, resulting in a multinucleate schizont. Cytoplasmic divisions follow until a layer of cytoplasm envelops each nucleus. Thus,

within a single host cell, as many as 900 pyriform daughter cells, known as **merozoites**, can be found. A single merozoite measures $2-4$ μm $\times$ $1-1.5$ μm.

When studied with the electron microscope, the merozoite is observed to be enveloped by an outer layer comprised of a unit membrane lined with a dense osmiophilic layer (Fig. 6.13). A micropyle is present. The apical complex is well formed and a few convoluted tubules (sarconemes) are disseminated in the anterior part. Within the cytoplasm are found Golgi complexes, mitochondria, lipid globules, and glucidic grains, in addition to endoplasmic reticulum (Fig. 6.14).

The merozoites escape from the host cell and usually invade another cell in which they increase in size (up to $4-5$ μm in diameter) to become **trophozoites**. These trophozoites repeat the schizogonic cycle. The merozoites resulting from the second schizogonic cycle are larger and more elongate, measuring 15×2 μm. These merozoites may penetrate other cells and repeat the schizogonic cycle, but most of them, upon entering cells, initiate gamogony by becoming male or female gametocytes.

The female, or macrogametocyte, which is slightly larger, develops into a single macrogamete. The male, or microgametocyte, gives rise to numerous biflagellated microgametes. The microgametes escape from the host cell and invade a cell in the vicinity that

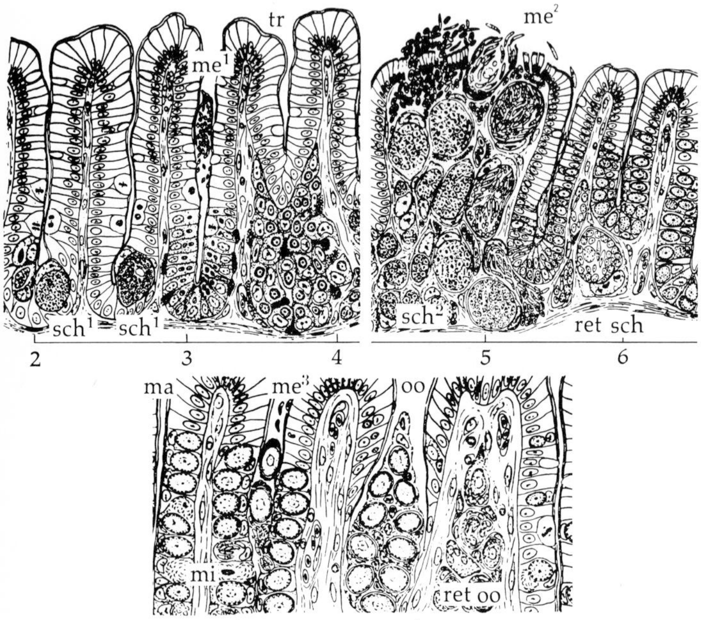

Fig. 6.12. **Diagram illustrating development of *Eimeria tenella* in caecal glands of the chick.** Numbers below line indicate the days of infection. ma, macrogamete; me, merozoite (me^1, me^2, me^3 indicate generation 1, 2, 3 merozoites, respectively); mi, micro- gametocyte; oo, oocyst; ret oo and ret sch, oocysts and schizonts that failed to escape; sch^1 and sch^2, schizonts of generation 1 and 2; tr, young growing trophozoites. (After Tyzzer, 1929.)

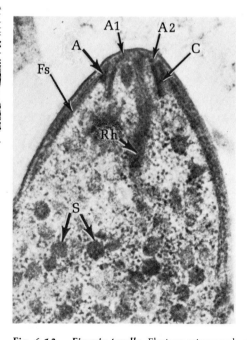

Fig. 6.13. *Eimeria tenella.* Electron micrograph of anterior portion of merozoite showing conoid (C) surmounted by two rings (A1, A2) and surrounded by another (A) from which originate about 26 subpellicular fibrils (Fs). Notice the occurrence of sarconemes (S) and rhoptries (Rh). (After Senaud and Cerná, 1969.)

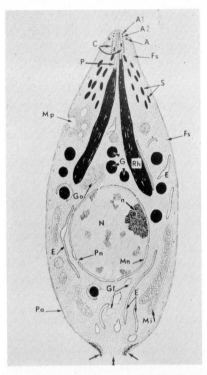

Fig. 6.14. **Schematic drawing showing the ultrastructure of the merozoite of** *Eimeria.* A, Ring from which subpellicular fibrils (Fs) arise; A1 and A2, rings surmounting conoid; C, conoid; E, endoplasmic reticulum; G, lipid globules; Gl, glycogen granules; Go, Golgi; Mi, mitochondria; Mn, nuclear membrane; Mp, micro- pyle; n, nucleolus; N, nucleus; P, peduncle of toxoneme; Pa, unit membrane; Pn, nuclear pore; Rh, rhoptries; S, sarconemes. (After Sénaud and Černá, 1969.)

already harbors a macrogamete. Fertilization takes place within the host cell. The resulting zygote then lays down a cyst wall around itself which is formed from eosinophilic plastic granules of mucoprotein present in the cytoplasm of the macrogamete. These granules pass to the periphery, flatten out, and coalesce, forming the cyst wall after fertilization.

According to Monné and Hönig (1954), the outer layer of the oocyst wall is a quinone-tanned protein, while the inner layer is a lipid lamella closely associated with the protein layer. Completion of the cyst wall marks the transition of the zygote into an oocyst.

The oocysts eventually rupture out of their host cells and enter the intestinal lumen, from whence they are expelled to the exterior in feces. The **prepatent period** (from the time of infection to the appearance of the first oocysts in the feces) is about 7 days. Continuous daily discharge of oocysts in feces is the common occurrence, because the sporozoites initially freed in the host do not all enter host cells simultaneously. Hence, their subsequent development is staggered.

Maturation of oocysts, or **sporogony**, continues after the oocysts are discharged to the exterior. A single sporont is present within each newly passed oocyst. If the required aerobic condition prevails, this sporont undergoes meiotic division, giving rise to four haploid **sporoblasts**, each of which then develops into a **sporocyst**. Two sporozoites develop within each sporocyst. Thus, the mature oocyst, which is the infective form, includes eight sporozoites.

Although oxygen is required for oocysts to mature, the young oocyst that has recently passed out of a host can survive under anaerobic conditions for some time. *Eimeria tenella* infections in chickens, known as coccidiosis, very often are lethal, because hemorrhage and sloughing of the affected tissues are extremely severe. Bloody diarrhea usually occurs. Coccidiosis represents one of the major causes of death on chicken farms. Furthermore, birds that are not killed by the disease become listless, unthrifty, and susceptible to predation and other diseases. The U.S. Department of Agriculture has estimated that in 1965 alone the loss to poultry farmers in the United States was $34,854 million, and this does not include the cost of medicated feeds and added labor.

Oocyst Excystation. Considerable attention has been paid to the excystation of *Eimeria* oocysts. According to Nyberg *et al.* (1968), the stimulus necessary to initiate *in vitro* excystation of *E. tenella* oocysts is provided by exposure to carbon dioxide. This stimulus produces a thinning and indentation at the micropylar region, and the oocysts become permeable to trypsin and bile. Subsequently, the sporozoites become active and escape through the altered micropyle after incubation in the enzyme-bile mixture. These investigators have reported that pretreatment of oocysts with air, nitrogen, oxygen, or helium results in considerably less excystation (Fig. 6.15). Nyberg and Hammond (1964) have shown that in the case of *Eimeria bovis*, a parasite of cows, the stimulus for oocyst excystation is essentially the same.

Other Species of Eimeria. Many other species of *Eimeria* are known. Some of the more common, although not necessarily the most economically important, are *E. stiedae*, *E. magna*, and *E. sciurorum* in rabbits; *E. augusta* and *E. bonasae* in grouse; *E. debliecki* in pigs; *E. arloingi* in sheep and goats; *E. canis* in dogs; *E. smithi*, *E. canadensis*, and *E. bovis* in cattle; *E. cavia* in guinea pigs; *E. nieschulzi* in rats; *E. ranarum* in frogs; *E. pigra* and *E. truttae* in trout; and *E. schubergi* in centipedes. It is thus apparent that *Eimeria* spp. parasitize both warm-blooded and cold-blooded vertebrates as well as invertebrates. For a review of *Eimeria* and related genera of coccidians that para-

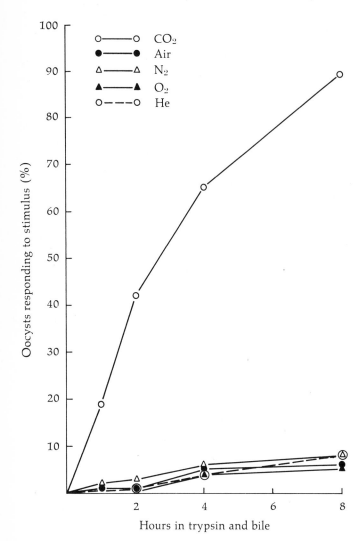

Fig. 6.15. **Effects of CO_2, air, N_2, O_2, or He as stimuli in initiating excystation of *Eimeria tenella* oocysts.** The pretreatment gas phase was for 18 hours at 38°C. Response to the stimulus was determined by examination of oocysts at various time intervals during treatment with 0.5% trypsin and 5.0% bile at 38°C. (After Nyberg *et al.*, 1968.)

sitize mammals other than humans, see Todd and Ernst (1977).

Factors Governing Pathogenicity. The pathogenicity of the various species of *Eimeria* differs. *Eimeria zurnii* and *E. bovis*, for example, are extremely pathogenic to their bovine hosts. Diarrhea is often accompanied by blood, tenesmus, and even death. In one experiment, the feeding of 250,000 to 1 million oocysts of *E. bovis* to calves caused death within 24 to 27 days. On the other hand, other species, such as *E.*

schubergi in centipedes, have little or no effect on their hosts. Not all factors governing pathogenicity in *Eimeria* infections are known. Among the more important factors known are (1) the number of oocysts ingested; (2) the number of merozoite generations that occur and the number of merozoites produced during each schizogonic cycle, since these determine the number of host cells destroyed by each infecting oocyst; (3) the location of the parasites in the host tissues and within host cells; (4) the degree of reinfectivity; (5) the degree of natural or acquired immunity; and (6) possible toxicity of the parasite.

The possible toxicity of *Eimeria* is mentioned since there is some evidence that suggests certain species may release a toxin. Burns and Challey (1959), recognizing that the diverse systemic effects observed in coccidiosis might be due to the presence of a toxic substance associated with the parasite, have sought such a substance in *E. tenella* infections. They prepared extracts from the caecal contents, including oocysts, of infected chicks and injected these intravenously or intraperitoneally into rabbits. The recipients were killed in 16 to 24 hours. Extraction of the toxic substances was effected by precipitation with a 40 to 50% saturation of ammonium sulfate. Further purification was effected in saline solution by adsorption on tricalcium phosphate and by dialysis. The purified toxin, when introduced into rabbits, produced even more severe reactions in the form of marked respiratory distress, prostration, and sometimes convulsions before death within 24 hours. It is notable that unmacerated oocysts did not cause death, suggesting that the toxin occurs within the oocysts. Furthermore, chickens, the natural hosts, are not affected by the toxin as are rabbits. Also, whether the toxin actually plays a role in the coccidiosis syndrome is doubtful since none of the symptoms characteristic of the disease are produced by it. Nevertheless, the toxicity of *E. tenella* extracts to rabbits is real, and the effects do not represent anaphylactic shock because of the delayed nature of the symptoms.

Genus Isospora. Until about 10 years ago, there was considerable confusion as to which species of Eimeriina should be assigned to the genus *Isospora*. It was thought that all coccidians that produce two sporocysts, with each sporocyst containing four sporozoites, belonged to this genus. Since then, the life cycles of several species of this group of coccidians have been elucidated, and it is now clear that some species have only one host (are **monoxenous**), others may have, but do not require, more than one host (are **facultatively heteroxenous**), and still others require more than one host (are **obligatorily heteroxenous**). Dubey (1977) has proposed that only those species that have no intermediate host be considered members of *Isospora*.

As a consequence, the genus *Isospora* as presently defined includes *I. belli*, an uncommon parasite of hu-

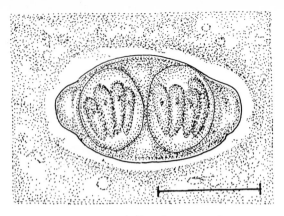

Fig. 6.16. *Isospora belli.* Oocyst enclosing two sporocysts that in turn, enclose four sporozoites each. Bar = 15 μm.

mans (Fig. 6.16). Most of the known cases have been reported in the tropics. *I. belli* oocysts are elongate and ellipsoidal, and measure 20–23 μm × 10–19 μm. Sporulated oocysts each contain two ellipsoidal sporocysts. Each sporocyst measures 9–14 μm × 7–12 μm and contains four crescent-shaped sporozoites and a residual body (leftover cytoplasm of the sporocyst that was not incorporated into sporozoites). Sporulation occurs within 5 days both within the human host and in the external environment (Trier *et al.*, 1974). Thus, both unsporulated and sporulated oocysts may be shed in feces.

Humans become infected with *Isospora belli* by ingesting food contaminated with oocysts. Schizogony and gamogony occur in the upper small intestinal epithelial cells. The disease caused by *I. belli* can be severe. Patients suffer fever, malaise, persistent diarrhea, loss of weight, and even death (Brandberg *et al.*, 1970).

Genus Levineia. Members of *Levineia* have oocysts with two sporocysts, each containing four sporozoites. They are facultatively or obligatorily heteroxenous, and sporogony occurs outside of the mammalian host. Furthermore, they form cysts in tissues other than, or in addition to, the gut of their intermediate hosts.

One of the better known members of *Levineia* is *L. felis*. It is a parasite of cats that utilizes mice and rats (and hamsters in the laboratory) as intermediate hosts. It invades the extraintestinal organs of not only intermediate hosts but also the feline definitive host (Frenkel and Dubey, 1972; Dubey and Frenkel, 1972). Dubey (1977) has summarized what is known about this and related species of *Levineia* from cats and dogs.

Genus Toxoplasma. Our present knowledge of *Toxoplasma gondii*, the only species in this genus, has resulted from intense study during the past 15 years. Originally discovered in 1908 in a desert rodent, the gondi, maintained at the Pasteur Institute in Tunis, it is now known to cause a serious disease in humans

known as toxoplasmosis. This statement requires qualification. Specifically, Krick and Remington (1978) have reported that 50% of the population in the United States harbor *T. gondii*, but most of these individuals are asymptomatic. Clinical toxoplasmosis usually affects only scattered individuals, although small epidemics occur periodically (Maddison *et al.*, 1979). For example, in 1969, 110 university students in São Paulo, Brazil, were diagnosed as suffering from acute toxoplasmosis in a three-month period. Almost all of these individuals admitted to eating uncooked meat. It is now known that inadequately cooked meat, including beef, pork, and lamb, represent a primary source of infection. Freezing meat at −14°C for a few hours, however, will kill all of the infectious oocysts.

Another major source of *Toxoplasma* infection is feral and domestic cats. Any cat, no matter how well cared for, may be passing *Toxoplasma* oocysts, although for only a few days after infection. Congenital toxoplasmosis is a very serious disease. It is passed from pregnant mother to fetus; therefore, a woman who is expecting should not empty a cat's litterbox. Wallace (1971) has found that flies and cockroaches can carry *Toxoplasma* oocysts from cat feces to the dinner table. Also, earthworms can carry oocysts from the graves of cats to the ground surface and thus serve as a source of infection.

Life Cycle of Toxoplasma gondii. The biology, life cycle, pathology, and immunology of *T. gondii* have been authoritatively reviewed by Jacobs (1967) and Frenkel (1973). Readers interested in greater detail should consult these contributions.

Toxoplasma is an intracellular parasite of many types of tissues, especially muscle and intestinal epithelium. In heavy acute infections, the parasite can be found free in the blood and peritoneal exudate. Within the host cell, the parasite usually occurs in the cytoplasm although occasionally it may be found in the nucleus.

The life cycle of this parasite (Fig. 6.17) includes an **intestinal epithelial (enteroepithelial) phase** and an **extraintestinal phase** in domestic cats and other felines but only the extraintestinal phase in other hosts. Sexual reproduction occurs in the cat, but only asexual reproduction occurs in other hosts. Its other hosts include many species of carnivores, insectivores, pigs, rodents, herbivores, primates, and other mammals. It has also been reported from birds. The distribution of *T. gondii* is worldwide.

The extraintestinal phase commences when a cat ingests a sporulated oocyst or sporocyst. Each oocyst (Fig. 6.18) measures 10–13 μm × 9–11 μm. After ingestion, the sporozoites escape from the sporocysts and the oocyst in the host's small intestine. In cats,

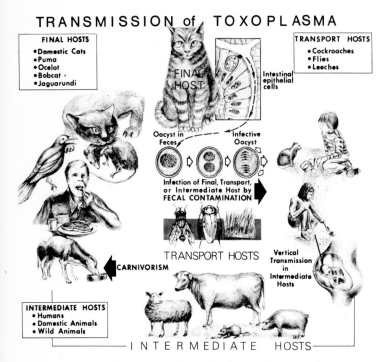

TRANSMISSION of TOXOPLASMA

FINAL HOSTS
• Domestic Cats
• Puma
• Ocelot
• Bobcat
• Jaguarundi

FINAL HOST

TRANSPORT HOSTS
• Cockroaches
• Flies
• Leeches

Intestinal epithelial cells

Oocyst in Feces   Infective Oocyst

Infection of Final, Transport, or Intermediate Host by **FECAL CONTAMINATION**

TRANSPORT HOSTS

Vertical Transmission in Intermediate Hosts

CARNIVORISM

INTERMEDIATE HOSTS
• Humans
• Domestic Animals
• Wild Animals

INTERMEDIATE HOSTS

Fig. 6.17. Life cycle of *Toxoplasma gondii.* Diagram showing methods of transmission. (After Fayer, 1976.)

vacuole, in the host cell before the cell disintegrates, releasing the parasites to invade new cells. Accumulations of tachyzoites in a host cell are known as groups. Tachyzoites are less resistant to stomach juices and therefore are less important sources of infection than are other stages.

As the disease becomes chronic, the zoites that affect brain, heart, and skeletal muscle cells multiply more slowly than during the acute phase. When this occurs, the parasites are known as **bradyzoites** and accumulate in large numbers within a host cell. In time, a tough cyst wall develops around a group of bradyzoites, and the complex is known as a **cyst**. (Fig. 6.19). Such cysts may persist for months or even years, especially in nerve tissue.

The development of cysts coincides with the time at which immunity develops in the host. This immunity, involving both humoral and cellular reactions, is usually permanent. If immunity wanes, released bradyzoites can boost the immunity to its earlier level. Since this immunity to new infections is induced by parasites comprising the first infection, this is an example of premunition.

Usually the cysts are intracellular in host cells; however, they could occur extracellularly as a result of rupturing of the host cell. Bradyzoites are resistant to digestion by trypsin and pepsin, and, when ingested by another host, can establish an infection.

The enteroepithelial phase of toxoplasmosis, which occurs only in cats, is initiated with the ingestion of cysts, bradyzoites, oocysts containing sporozoites, or

some of the sporozoites penetrate intestinal epithelial cells and thus initiate the enteroepithelial phase. Other sporozoites penetrate the mucosa and develop in the underlying lamina propria, mesenteric lymph nodes, and other organs as well as leucocytes.

In hosts other than cats, there is no enteroepithelial development; the sporozoites enter host cells and commence multiplying by **endodyogeny**. These rapidly dividing cells in acute infections are known as **tachyzoites**. Eight to 16 tachyzoites accumulate within a endocytotic vacuole, known as a **parasitophorous**

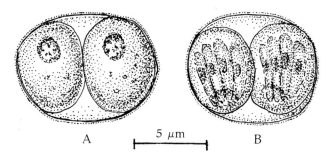

Fig. 6.18. *Toxoplasma gondii.* A. Young oocyst enclosing two sporocysts. **B.** Mature oocyst containing four sporozoites in each sporocyst.

5 μm

A   B

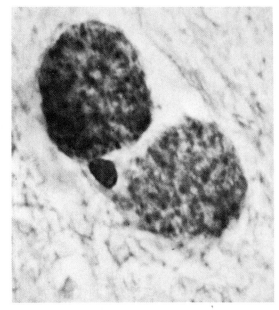

Fig. 6.19. *Toxoplasma gondii.* Cyst in mouse brain. (After Buss in Schmidt and Roberts, 1981; with permission of Mosby.)

occasionally tachyzoites. An additional method by which enteroepithelial toxoplasmosis can be initiated is by the migration of extraintestinal zoites into the intestinal lining of the cat.

Once within a host epithelial cell of the small intestine or colon, the parasite becomes a **trophozoite**, which grows and eventually initiates schizogony.

Frenkel (1973) has reported that at least five different strains of *T. gondii* have been sufficiently studied that characterization of the enteroepithelial stages is possible. The various strains differ primarily in duration of stages and the number and shape of merozoites produced. Basically, from 2 to 40 merozoites are produced by schizogony, endopolyogeny,* or endodyogeny,** and these merozoites initiate subsequent asexual stages. The number of schizogonic cycles varies, but gametocytes are produced within 3 to 15 days after cyst-induced infection. Gametocytes develop in epithelial cells along the length of the small intestine but are more common in the ileum. About 2 to 4% of the gametocytes are male (microgametocytes), and each produces about 12 microgametes.

Oocysts appear in the cat's feces from 3 to 5 days after ingestion of cysts, with peak production from 5 to 8 days. Oxygen is required for the sporulation of oocysts, which requires 1 to 5 days. Extraintestinal development can proceed simultaneously with the enteroepithelial phase in the cat. Dubey (1977) has stated that ingested bradyzoites that penetrate the intestinal wall and multiply as tachyzoites in the lamina propria may become widely disseminated in the cat's extraintestinal tissues within a few hours of infection.

Pathology of Toxoplasmosis. Immunologic surveys have revealed that antibody to *Toxoplasma* is widely prevalent in humans throughout the world; however, clinical toxoplasmosis is rare. Therefore, it is clear that most infections are asymptomatic. Although not all of the influencing factors are yet known, it is certain that the following contribute to the degree of pathology: (1) age of the host, with older hosts being more resistant to the disease; (2) virulence of the strain of *T. gondii* involved; (3) natural susceptibility of the individual host; (4) natural susceptibility of the host species; and (5) the degree of acquired immunity of the host. It is known that white mice are more susceptible than white rats, chickens are more susceptible than carnivores, and pigs are more susceptible than cattle. It is also known that asymptomatic infections can be activated if immunosuppressive drugs, such as corticosteroids, are administered.

Symptomatic toxoplasmosis can be classified as being acute, subacute, chronic, or congenital.

Acute Toxoplasmosis. In most cases of **acute toxoplasmosis**, the intestine is the first site of infection. Cats infected as a result of ingesting oocysts usually reveal little disease beyond loss of some epithelial cells, which are rapidly replaced. In massive infections, however, intestinal lesions can kill kittens in 2 to 3 weeks.

The first extraintestinal sites to be infected in cats and other hosts, including humans, are the mesenteric lymph nodes and the liver parenchyma. The most common symptom of acute toxoplasmosis is painful, swollen lymph glands in the inguinal, cervical, and supraclavicular regions. This symptom may be accompanied by fever, headache, anemia, muscle pain, and sometimes lung complications.

It is noted that tachyzoites proliferate in many tissues and tend to kill host cells faster than the normal turnover rate of such cells. On the other hand, enteroepithelial cells normally live for only a few days, especially at the tips of the intestinal villi. Therefore, the extraepithelial stages of *T. gondii*, especially in such sites as the retina or brain, tend to cause more serious lesions than do those in the intestinal epithelium.

Subacute Toxoplasmosis. If immunity in the host develops slowly, then clinical toxoplasmosis can be prolonged, and the condition is referred to as **subacute toxoplasmosis**. When this occurs, the tachyzoites continue to destroy cells, resulting in the development of extensive lesions in the lung, heart, liver, brain, and eyes. Damage is commonly more extensive in the central nervous system than in non-nerve tissues because of the lower immunocompetence in these tissues.

Chronic Toxoplasmosis. When immunity in the host builds up sufficiently to suppress tachyzoite proliferation, the condition is known as **chronic toxoplasmosis**. This coincides with the formation of cysts. These cysts can remain intact for years, producing no clinical symptoms. A cyst wall may occasionally rupture, releasing bradyzoites. Most of these are killed by the host reactions, although some may form new cysts. Death of bradyzoites stimulates a hypersensitive reaction. In the brain, such sites are gradually replaced by nodules of glial cells. If the number of such nodules is great, the host may develop symptoms of chronic encephalitis accompanied by spastic paralysis in some cases. Repeated infection of retinal cells by tachyzoites can destroy the retina. Furthermore, the occurrence of cysts and their rupturing in the retina and choroid may also lead to blindness. Chronic toxoplasmosis may also cause myocarditis, leading to permanent heart damage and pneumonia.

---

*Endopolyogeny describes the formation of several daughter cells, each with its own surface membrane, while within the mother cell.

**Endodyogeny describes the same phenomenon as endopolyogeny except that only two daughter cells are formed.

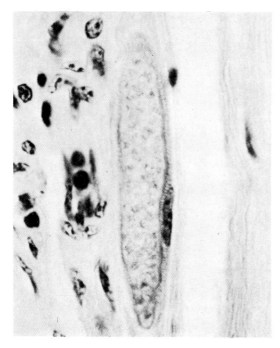

Fig. 6.20. *Sarcocystis.* Sarcocyst in muscle of experimentally infected calf. (Courtesy of Dr. Ronald Fayer.)

Congenital Toxoplasmosis. If an expectant mother contracts acute toxoplasmosis, the parasites often invade the fetus. Fortunately, most neonatal infections are asymptomatic; however, a significant

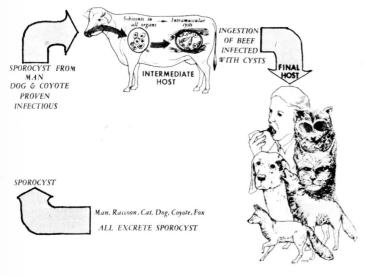

Fig. 6.21. **Life cycle of *Sarcocystis* sp.** Schematic diagram showing how humans can become infected. (Courtesy of Dr. Ronald Fayer.)

number result in death or severe malformation. Such malformations include hydrocephaly (see Fig. 3.21), microcephaly, cerebral calcification, chorioretinitis, and psychomotor disturbances. In children who survive the infection, congenital damage to the brain resulting in mental retardation and epileptic seizures commonly occurs.

Stillborns and spontaneous abortions may result from congenital toxoplasmosis. Although these conditions occur in humans, sheep are particularly susceptible. According to Beverly *et al.* (1971), congenital toxoplasmosis causes half of all abortions in sheep in England and New Zealand.

Genus Sarcocystis. Although protozoan parasites belonging to the genus *Sarcocystis* have been known to occur in the muscle of herbivores for almost 100 years, it was not until Fayer (1972) and Rommel *et al.* (1972) reported their life cycle patterns that it became apparent that they are coccidians of the suborder Eimeriina.

It is now known that the stage occurring in the muscle of herbivores is the **sarcocyst** (Fig. 6.20). Also, it is known that *Sarcocystis* spp. are obligatorily heteroxenous, involving a herbivorous intermediate host and a carnivorous definitive host (Fig. 6.21). Humans serve as definitive host for *Sarcocystis hominis* and *S. suihominis*, but sarcocysts of *S. lindemanni* and several other unidentified species are occasionally found in human muscle (Beaver *et al.*, 1979) (Fig. 6.22). Intermediate hosts of the various species include birds, small rodents, reptiles, and hoofed animals.

Life Cycle of Sarcocystis. Sporocysts, harboring sporozoites, are ingested by the intermediate host. The sporozoites are then released, and they penetrate the intestinal epithelium, become distributed throughout the body, and invade the lining cells of blood vessels in many tissues. There they undergo schizogony (Fig. 6.21), which is usually followed by additional schizogonic generations. Sarcocysts are then formed in skeletal and cardiac muscles (Fig. 6.22) and occasionally in the brain. The sarcocysts of *Sarcocystis* are also referred to as **Miescher's tubules**. These can be seen with the naked eye. The sarcocysts are usually cylindrical or spindle shaped, but they may be irregularly shaped. They occur within a muscle fiber and may reach 1 cm in diameter in some cases; however, they usually are 1–2 mm in diameter and 1 cm or less in length.

The structure of the cyst wall varies with the developmental stage. In some cases the outer wall is smooth; in others it has an outer layer of fibers, known as **cytophaneres**, which radiate out into the muscle. Two distinct regions can be distinguished in the cyst. The peripheral region is occupied by globular **metrocytes**. These resemble typical coccidian merozoites; however, they lack rhoptries and micronemes.

After several divisions, the metrocytes give rise to the more elongate bradyzoites. The latter also resemble typical coccidian merozoites except that they have more micronemes.

When a sarcocyst is ingested by the definitive host, the cyst wall is digested away and the escaping bradyzoites penetrate the lamina propria of the small intestine. At this site they undergo gamogony without an intervening schizogonic generation. The male gametes penetrate the female gametes and the oocyst sporulates in the lamina propria. The oocyst wall is thin and is usually ruptured while passing down the host's intestine; thus, sporocysts (Fig. 6.23), rather than oocysts, are normally passed out in feces. These sporocysts are infective to the intermediate host.

Species of Sarcocystis. Species of *Sarcocystis* are known to utilize humans, canines, felines, primates, and a few other mammals as definitive hosts, and cows, sheep, pigs, and horses as intermediate hosts. The principal species are listed in Table 6.1 along with some of their characteristics. Some are nonpathogenic, whereas others cause such symptoms as loss of appetite, fever, anemia, weight loss, lameness, abortion in pregnant animals, and even death. Dubey (1977) has estimated that 50% of adult swine, sheep, and cattle are infected with *Sarcocystis* spp., some pathogenic, others not.

Other Genera of Eimeriina. Several other genera of coccidians have been assigned to the suborder Eimeriina. These include *Hammondia*, which is very similar to *Toxoplasma* except that it has an obligatory two-host life cycle, and *Frenkelia*, which is very similar to *Sarcocystis* except that its cysts are formed primarily in the host's brain. For a review of these and related genera, see Hammond and Long (1973) and Dubey (1977).

Suborder Haemosporina

Haemosporina is the most extensively studied group of coccidians because it includes members of the genus *Plasmodium*, the malaria-causing parasites. Haemosporinans carry out the schizogonic phase of their life cycles in the blood of a vertebrate host (except in the case of *Haemoproteus*) and both sexual reproduction and sporogony occur in the alimentary canal of an invertebrate. Thus, the sporozoites of haemosporinans are never directly exposed to the macroenvironment, and this is probably the reason why they are not encased within a protective envelope.

The subclass is divided into two families:

Family Plasmodiidae. Cytoplasm includes pigment granules when present in erythrocytes. Presexual phase (schizogony) occurs in blood cells circulating in vertebrate host's peripheral blood.

Family Haemoproteidae. Cytoplasm includes pigment granules when they are present in erythrocytes. Schizogony does not occur in blood cells of peripheral blood. Gametocytes are found as intracellular parasites in peripheral circulation.

Family Plasmodiidae. The Plasmodiidae includes the multispecies genus *Plasmodium*. Among the

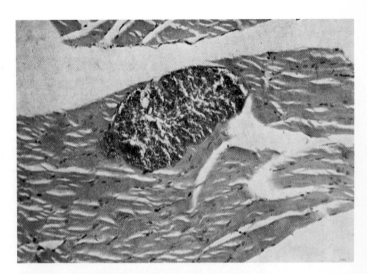

Fig. 6.22. *Sarcocystis.* Photomicrograph of a sarcocyst in human muscle. (After Zaman, 1979; with permission of ADIS Press.)

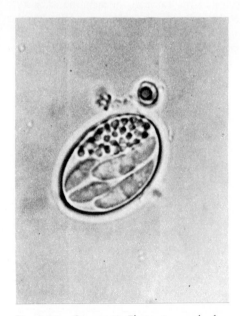

Fig. 6.23. *Sarcocystis.* Photomicrograph of sporocyst enclosing four sporozoites. (Courtesy of Dr. Ronald Fayer.)

species that parasitize vertebrates other than humans are *P. sternoceri* and *P. cnemidophori* in lizards; *P. culesbiana* in frogs; *P. bufonis* in the toad *Bufo americanus*; *P. gallinaceum*, *P. vaughni*, *P. lophurae*, and *P. rouxi* in various birds; *P. kochi*, *P. brasilianum*, *P. knowlesi*, and *P. cynomolgi* in various species of monkeys; and *P.*

berghei in various rodents. Four species—*P. vivax, P. falciparum, P. malariae,* and *P. ovale* (and various strains)—are known to infect humans, causing specific types of malaria. The authoritative volume by Coatney *et al.* (1971) on those species that infect primates is recommended as are the reviews on the plasmodia of reptiles by Ayala (1977), birds by Seed and Manwell (1977), and rodents by Carter and Diggs (1977). Furthermore, the reviews by Collins and Aikawa (1977) and Rieckmann and Silverman (1977) on the plasmodia of nonhuman primates and humans are recommended.

Malaria. Malaria is one of the most vicious diseases of humans. It has played a major role in shaping history and in the decline of civilizations. Human malaria is known to have contributed to the fall of the ancient Greek and Roman empires. Troops in both the Civil War and the Spanish-American War were severely incapacitated by this disease. More than one-quarter of all hospital admissions during these wars were malaria patients. During World War II, malaria epidemics severely threatened both the Japanese and Allied forces in the Far East. In fact, the ultimate success of the United States in Asia may be credited to a large degree to the parasitologists, both civilian and military, who fought and conquered this ruthless enemy. To some degree, the same may be said of the military conflicts in Korea and Vietnam.

How widespread is malaria? It is certainly not limited to the tropics. As recently as 1937 there were at least 1 million cases of malaria each year in United States. In other parts of the world, 52 of 58 countries in Africa, 11 of 18 countries in southwest Asia, all of the 23 political states in the Pacific Basin (Oceania), and 9 of 24 countries in south central and southeast Asia have reported the occurrence of this disease. Figure 6.24 gives some idea of the geographic distribution of malaria before and after control and eradication measures were widely instituted. It has been estimated that in the 1940s there were 350 million cases of malaria throughout the world, which made it the number one disease of humans. About 3 million of these people died annually. Although malaria control programs, primarily mosquito abatement, sponsored by certain cooperating nations and the World Health Organization of the United Nations, have succeeded in stamping out malaria to some extent, reducing the total number to 250 million cases by 1958, it still remains a major health problem in many parts of the world, as American troops in Vietnam can testify. In that part of the world both *Plasmodium vivax* and especially *P. falciparum* occur. Furthermore, periodic outbreaks of malaria in epidemic form still occur in supposedly controlled areas because of relaxation of control and the development of resistance to insecticides by the mosquito vectors, as illustrated by the outbreaks in Sri Lanka and India during 1967–1970. An estimated 300 million people worldwide suffer from malaria (1980) (Warren, 1980). The definitive work on all aspects of malaria research is the three-volume treatise edited by Kreier (1980).

Malaria is usually chronic, debilitating, and periodically disabling, although lethal cases are by no means

Table 6.1. Some Major Species of *Sarcocystis* and Some of Their Characteristics

Species	Definitive Hosts[a]	Natural Intermediate Host	Pathogenicity	Cyst Wall	Sporocysts (μm)
S. cruzi	Dog, coyote, wolf, fox, raccoon	Ox	Pathogenic	0.5 μm (thin)	16 × 11
S. hirsuta	Cat	Ox	Non- or slightly pathogenic	6 μm (thick)	12 × 8
S. hominis	Human, rhesus monkey, baboon	Ox	Non- or slightly pathogenic	5.9 μm (thick) striated	15 × 9
S. ovicanis	Dog	Sheep	Pathogenic	Thick, radially striated	15 × 10
S. tenella	Cat	Sheep	Nonpathogenic	Thin	12 × 8
S. suihominis	Human	Pig	Not known	Not known	13 × 9
S. miescheriana	Dog	Pig	Not known	Not known	13 × 10
S. porcifelis	Cat	Pig	Pathogenic	Not known	13 × 8
S. fayeri	Dog	Horse	Not known	Thin	12 × 8
S. bertrami	Dog	Horse	Not known	Thin	15 × 10
S. muris	Cat	Mouse	Non- or slightly pathogenic	Thick	10 × 8
S. leporum	Cat	Rabbit	Not known	Thick	14 × 9
S. lindemanni	Not known	Human	Nonpathogenic		9 (long)

[a] Infection of definitive host effected by carnivorism.

uncommon. *Plasmodium falciparum* is by far the most lethal of the human malaria-causing species. Mortality as a result of malaria is most pronounced among children younger than 5 or 6 years of age. Individuals beyond that age usually build up antibodies against the parasite. Although these antibodies do not prevent recurrent infections or purge the victim of the parasites, they do permit the body to tolerate the parasite such that it is not consistently lethal. Because of this physical crippling, in areas where malaria is prevalent, the collective number of man-hours lost is a considerable loss to regional or national productivity. It has been estimated that this loss translates into about $2 trillion annually. In India alone, where economic advancement is of the utmost importance, the 100 million cases of malaria means an annual loss of about $500 million. In Mexico the loss approximates $18 million per year. The fascinating history of man's fight against malaria (Harrison, 1978) should be consulted by all students of parasitology.

Eradication of malaria involves therapy for the disease in humans and a continuous battle against the mosquito. The first drug used effectively against malaria was quinine, extracted from the bark of the cinchona tree of South America and other areas in the tropics. This drug, which has been in use since 1640, destroys the erythrocytic stages of *Plasmodium vivax* and of the other species. During World War II when the Japanese occupied the cinchona plantations in Indonesia, it became acutely necessary to find alternative drugs. Consequently, a synthetic drug, atabrine, proved useful in killing the erythrocytic stages of *Plasmodium* and hence suppressed the clinical symptoms. Atabrine, however, is ineffective against the exoerythrocytic stages, and, therefore, malaria patients treated with this drug are susceptible to periodic recurrences when for some physiologic reason, the exoerythrocytic stages still present become active, multiply, and invade red blood cells. The most effective drugs now in use are chloroquine, amodiaquine, chlorquanide, primaquine, and pyrimethamine. Chloroquine and amodiaquine are effective in suppressing clinical symptoms by destroying the erythrocytic stages. These drugs, however, have no effect on the exoerythrocytic stages, and hence complete cure requires treatment with primaquine, a slow-acting drug which will destroy the exoerythrocytic stages, combined with one of the others. For a review of the chemotherapy of malaria, see Peters (1980).

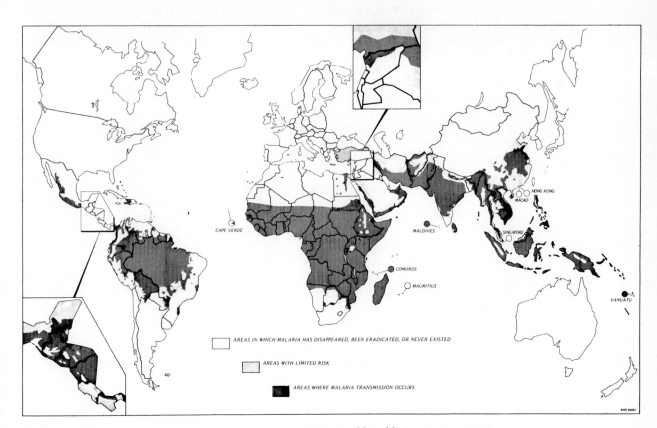

Fig. 6.24.   Epidemiologic assessment of status of malaria, 1982. (World Health Organization, 1984.)

In addition to the chemotherapeutic approach, the development of a protective vaccine against malaria is being aggressively investigated. The reviews by Cochrane *et al.* (1980), Holbrook (1980), Siddiqui (1980), Carter and Gwadz (1980), Diggs (1980), Beaudoin *et al.* (1980), Rieckmann (1980), and Brackett (1980) relative to advances in this area of research are recommended.

Research in the area of malaria chemotherapy has revealed an alarming phenomenon. *Plasmodium* spp., like all organisms, have some genetic instability, and those that because of some mutagenic change become resistant to available drugs, give rise to progeny that inherit this resistance. Thus, in an area where extensive therapy is being carried out, a large percentage of the malaria organisms become drug resistant over time. For this reason, the synthesis of new antimalarial drugs must be a continuous process. Similarly, the use of insecticides to eradicate mosquitoes has revealed the appearance of resistant strains in areas that have been sprayed extensively. In Greece, for example, only a few years after apparently successful control of *Anopheles* with DDT had been initiated, it has been found necessary to use DDT and dieldrin alternately in order to control the malaria-carrying species.

Although, ideally, the search for new antimalarial drugs and new insecticides should be aimed at finding compounds that will block a critical metabolic pathway within the malarial parasite or mosquito, this approach requires a thorough understanding of the biochemical processes occurring within these parasites—a feat that requires highly trained scientists with years of experience in sophisticated techniques. Needless to say, this is an extremely expensive process. Consequently, an empiric trial-and-error approach involving large-scale drug and insecticide screening is usually undertaken.

As a final remark about human malaria, it should be stated that malariologists very often interpret malaria as being a "man-made" disease. This means that with the clearing of forest areas for agriculture, including the building of irrigation canals and the construction of dams, man alters the ecosystem in such a way that areas with previously scant malaria incidence often become highly malarious. It is also noteworthy that in some areas where tractors have replaced buffalo in agriculture, there has been a sudden rise in human malaria. There is some epidemiologic evidence that this is because the anophelene mosquitoes endemic to such areas originally preferred and fed on buffalo, but with the removal of these herds, they feed on humans and thus became transmitters of *Plasmodium* spp.

Genus Plasmodium. Over 50 species of *Plasmodium* are known. Human malaria is generally caused by the four species of the genus *Plasmodium* listed earlier. Closely related to *Plasmodium* are the members of *Leucocytozoon*, *Haemoproteus*, and *Hepatocystis*. The other species of *Plasmodium* are found as parasites of various other vertebrates. The species that cause malaria in birds and *P. berghei* are also interesting in their own right. *Plasmodium berghei* was first discovered in African tree rats and has since been found to be readily transmissible to other rodents, including laboratory rats, mice, and hamsters. Both the avian malaria-causing species and *P. berghei* have become important experimental organisms in research on human malaria. In fact, Sir Ronald Ross, the English Nobel laureate, discovered the transmission of malaria by mosquitoes while studying bird malaria. For a review of the relationship between mosquitoes and avian malaria, see Seed and Manwell (1977).

It should be mentioned briefly that, although *Leucocytozoon* and *Haemoproteus* are found with relative frequency in the erythrocytes of birds, and some species of *Haemoproteus* are encountered in reptiles, these parasites are generally nonpathogenic. However, a few species are pathogenic to birds, including some economically important ones. For example, *Leucocytozoon simondi* is extremely pathogenic to ducks.

Life Cycle of Plasmodium vivax. *Plasmodium vivax* (Fig. 6.25), the most common of the human-infecting species, causes the type known as tertian, benign tertian, or vivax malaria. This type of malaria is characterized by a 48-hour cycle between erythrocytic merozoite production, which is manifested in the host by chills and fever at these intervals.

Infection in humans is established when sporozoites are injected into the blood during the bite of the mosquito vectors, *Anopheles quadrimaculatus* and others (see Table 19.3). It is noted that only female mosquitoes serve as vectors for *Plasmodium* spp., since males feed primarily on plant juices. Once within the victim, the sporozoites almost immediately disappear from the peripheral circulation. Many are destroyed by the host's phagocytes and/or humoral factors; however, some are carried by blood plasma to the liver, where they enter hepatic cells.

Within hepatic cells, exoerythrocytic schizogony occurs, resulting in the formation of numerous **cryptozoites** (a term commonly applied to first-generation exoerythrocytic **merozoites** of *Plasmodium*), which escape and invade other hepatic cells. Cryptozoites undergoing schizogony are known as **exoerythrocytic** or **tissue schizonts**. A second schizogonic phase may occur, giving rise to **metacryptozoites** (second-generation exoerythrocytic merozoites of *Plasmodium*). These escape from liver cells and invade erythrocytes, initiating the erythrocytic phase.

Metacryptozoites invade the host's erythrocytes approximately 6 days after the initial infection in *P. vivax*, 8 days in *P. falciparum*, and 75 hours in the

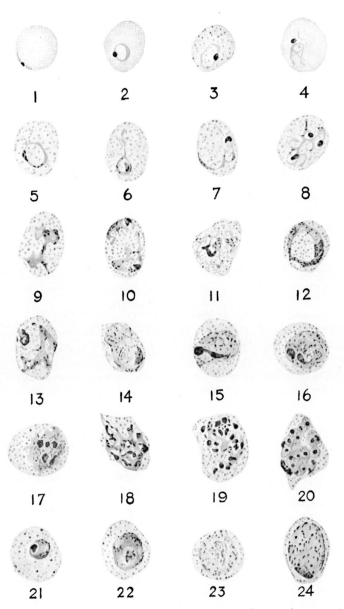

Fig. 6.25. *Plasmodium vivax.* **1.** Normal size erythrocyte with marginal ring-form trophozoite. **2.** Young signet-ring form trophozoite in a macrocyte. **3.** Slightly older ring-form trophozoite in erythrocyte showing basophilic stippling. **4.** Polychromatophilic erythrocyte containing young parasite with pseudopodia. **5.** Ring-form trophozoite showing pigment in cytoplasm in an enlarged erythrocyte containing Schüffner's dots. (Schuffner's dots do not occur in all host cells containing growing and older forms of *P. vivax*.) **6, 7.** Medium trophozoite forms. **8.** Three amoeboid trophozites with fused cytoplasm. **9, 11, 12, 13.** Older amoeboid trophozoites in process of development. **10.** Two amoeboid trophozoites in one host erythrocyte. **14.** Mature trophozoite. **15.** Mature trophozoite with chromatin in process of dividing. **16, 17, 18, 19.** Schizonts showing progressive stages of division. **20.** Mature schizont. **21, 22.** Developing gametocytes. **23.** Mature microgametocyte. **24.** Mature macrogametocyte. (Courtesy of National Institutes of Health.) **See color plates.**

avian *P. gallinaceum.* Thus, the initiation of the erythrocytic stage varies from species to species.

Within the red blood cell occurs the process known as erythrocytic schizogony. Morphologically, the progressive stages of this process are distinguished by the following sequence (Figs. 6.25–6.28):

1. Appearance of the ring stage, which resembles a signet ring, caused by a vacuolated area in the cytoplasm of the parasite. The surrounding cytoplasm is connected to a peripherally located nucleus, which is the "gem" of the ring. When blood smears are stained with Giemsa's or Wright's stain, the ring of cytoplasm appears blue and the nucleus is red.

2. An increase in size readily recognizable by the increase of blue cytoplasm.

3. The appearance of minute red granules in the corpuscular cytoplasm—known as **Schüffner's dots**—is characteristic of *P. vivax.* No such granules appear in *P. malariae* infections, whereas in *P. ovale* infections similar granules are present, but the infected cells are not enlarged and are oval. In *P. falciparum* infections, the parasitized erythrocytes are generally found in the blood in visceral organs. The granules, known as **Maurer's dots**, are fewer and stained deeper with a tinge of red.

4. The beginning of nuclear division. The single nucleus divides; each of the daughter nuclei divides again, resulting in four; and further divisions ensue.

5. The occurrence of segmentation, in which the nuclei are arranged peripherally with clumps of cytoplasm surrounding each nucleus.

6. Merozoite formation, during which each unit, containing one nucleus and surrounding cytoplasm, becomes a merozoite.

These merozoites escape from the erythrocyte and invade fresh cells to repeat the erythrocyte schizogonic cycle. The "ghost cells" left behind after the merozoites escape are destroyed in the spleen. Adrenal dysfunction usually accompanies malaria and brings on the yellowish skin coloration of victims of the disease.

Merozoites that escape from one host erythrocyte must rapidly enter another cell during the erythrocytic phase. Evidence contributed by Trubowitz and Masek (1968) indicates that at least *in vitro,* the host's polymorphonuclear leucocytes can recognize and phagocytose unprotected merozoites, which leads to their destruction. These same host cells will not engulf erythrocytes harboring *Plasmodium.*

It is at the time of the escape of the merozoites that chills and fever beset the host (the condition is known as **paroxysm**). Generations of erythrocytic schizonts occur. In time, some of the merozoites, after invading cells, do not segment but retain a single nucleus and increase in size to become rounded gametocytes (in *P. falciparum* the gametocytes are crescent shaped). The gametocytes are either male or female. The female gametocytes stain a slightly lighter color, but unless viewed by the trained eye, differentiation is difficult. The gametocytes remain intracellular parasites without further developments.

When a mosquito ingests red blood cells containing gametocytes, the sexual cycle is initiated. The male gametocyte (microgametocyte) in the mosquito's gut undergoes a process known as **exflagellation** during which the nucleus divides and a number of "tails" protrude from the cytoplasm. These break away as mature microgametes. The macrogametocyte (female gametocyte) matures into a single macrogamete (the crescent-shaped macrogametocyte of *P. falciparum* becomes rounded). The fusion of gametes results in a zygote that increases in size and elongates to become vermiform and is then known as an **ookinete**.

The ookinete penetrates the stomach lining of the mosquito and in this location increases in size within an enveloping wall that it lays down. This is known as the **oocyst** stage. Within the oocyst, nuclear divisions occur, resulting in a multinucleate condition. Around each nucleus a cytoplasmic aggregation appears. Each of these units is now known as a **sporoblast**.

Within each sporoblast, sporozoites are formed from nuclear and cytoplasmic divisions. The sporoblasts rupture and the sporozoites, over 10,000 strong, fill the cavity of the oocyst.

Within 10 to 24 days after ingestion of the gametocytes by the mosquito, the oocysts burst, releasing the sporozoites into the body cavity. These sporozoites actively migrate to and invade the salivary gland cells of the mosquito host. From their intracellular positions, sporozoites enter the lumen of the gland ducts and are ready to be injected into a human when the mosquito partakes of its blood meal.

For a more detailed review of the life cycle of the malaria-causing parasites, see Wernsdorfer (1980).

Penetration of the Host Cell. Considerable interest has been focused in recent years on how such parasites as the merozoites of *Plasmodium* spp. enter host cells. Ladda *et al.* (1969) have demonstrated by employing electron microscopy that merozoites of *Plasmodium* are engulfed by the host erythrocyte (Figs. 6.29, 6.30). Furthermore, during the engulfment process, the rhoptries and micronemes are believed to secrete surface active molecules that cause the host erythrocyte membrane to expand, resulting in its invagination to form the parasitophorous vacuole. This vacuole envelops the engulfed parasite.

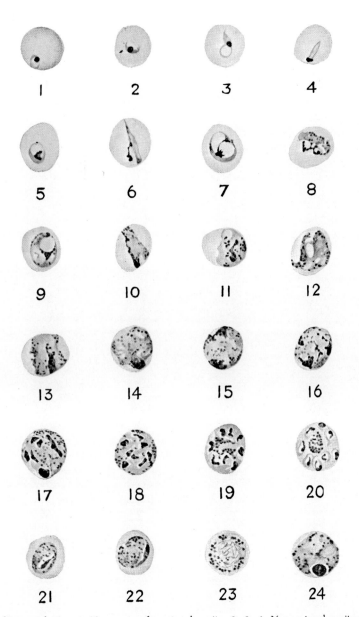

Fig. 6.26. *Plasmodium malariae.* **1.** Young ring-form trophozoite. **2, 3, 4.** Young trophozoites in host erythrocytes showing gradual increase of chromatin and cytoplasm. **5.** Developing ring-form trophozoite showing pigment granule. **6.** Early band-form trophozoite (with elongated chromatin and some pigment.) **7, 8, 9, 10, 11, 12.** Some variations in the shapes developing trophozoites may take. **13, 14.** Mature trophozoites; one in band-form. **15, 16, 17, 18, 19.** Stages in the development of schizont. **20.** Mature schizont. **21.** Immature microgametocyte. **22.** Immature macrogametocyte. **23.** Mature microgametocyte. **24.** Mature macrogametocyte. (Courtesy of National Institutes of Health.) **See color plates.**

The engulfment of merozoites is not totally a passive process. The subpellicular microtubules provide motility during entry. This contention is supported by the disappearance of the microtubular complex when the parasites enter host cells and transform into the nonmotile intracellular form.

In addition to the mechanics of penetration, there is some evidence that certain species of *Plasmodium*, specifically *P. knowlesi* and *P. vivax*, will penetrate only host erythrocytes that possess an antigen known as **Duffy antigen** on their surfaces. This antigen presumably acts as an attachment site recognizable to the

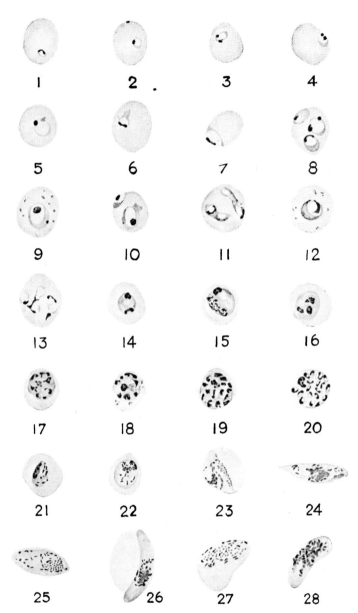

Fig. 6.27. *Plasmodium falciparum.* **1.** Very young ring-form trophozoite in host erythrocyte. **2.** Double infection of single cell with young trophozoites; one a "marginal form," the other a "signet-ring" form. **3, 4.** Young trophozoites showing double chromatin dots. **5, 6, 7.** Developing trophozoites. **8.** Three medium trophozoites in one erythrocyte. **9.** Trophozoite showing pigment in erythrocyte containing Maurer's dots. **10, 11.** Two trophozoites in each of two cells showing variations of form. **12.** Nearly mature trophozoite showing haze of pigment throughout cytoplasm; with Maurer's dots in host. **13.** Slender forms. **14.** Mature trophozoite showing clumped pigment **15.** Initial chromatin division. **16, 17, 18, 19.** Various developmental phases of schizont. **20.** Mature schizont. **21, 22, 23, 24.** Successive forms during development of gametocyte (usually not found in peripheral blood.) **25.** Immature macrogametocyte. **26.** Mature macrogametocyte. **27.** Immature microgametocyte. **28.** Mature microgametocyte. (Courtesy of National Institutes of Health.) **See color plates.**

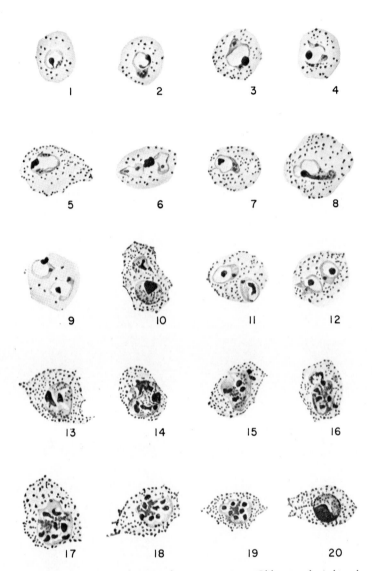

Fig. 6.28. *Plasmodium ovale.* **1.** Young ring-shape trophozoite. **2, 3, 4, 5.** Older ring-shaped trophozoites. **6, 7, 8.** Older amoeboid trophozoites. **9, 11, 12.** Host erythrocytes doubly infected with trophozoites. **10.** Host erythrocyte doubly infected with young gametocytes. **20.** Mature gametocyte. (Courtesy of National Institutes of Health.) **See color plates.**

merozoite (Miller *et al.*, 1975, 1976). At the electron microscope level, there is no junction formed between Duffy-negative erythrocytes and incompatible *Plasmodium* merozoites (Miller *et al.*, 1979). Since blacks commonly possess the genes responsible for the absence of the Duffy antigen, this has been suggested as one reason why they are less susceptible to certain types of malaria.

With the increased application of the tools of cell and molecular biology to resolve parasitologic problems, impressive advances have been made in eluci-

dating the mechanisms involved in the interaction, including penetration of host erythrocytes and malaria parasites. In the case of *Plasmodium falciparum*, the proteins **glycophorins A**, **B**, and **C** are now known to represent receptors in the surface membrane of host erythrocytes for invading merozoites (Butcher *et al.*, 1973; Miller *et al.*, 1975; Pasvol *et al.*, 1984); however, Friedman *et al.* (1984) have suggested that glycophorins A and B are involved in a relatively nonselective, charge-mediated attachment between merozoites and the erythrocyte membrane. Glyco-

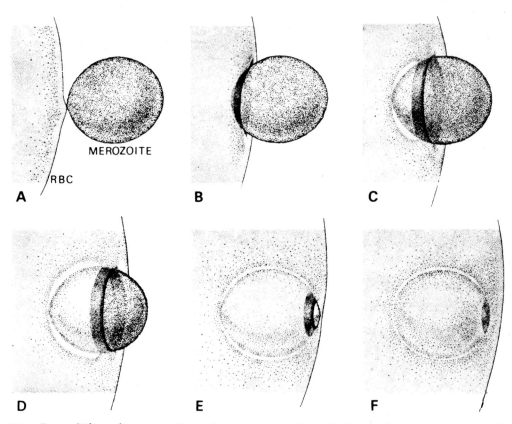

Fig. 6.29. **Entry of *Plasmodium* merozoite into host erythrocyte.** Schematic diagram showing a moving circumferential junction between merozoite and erythrocyte. The moving junction brings the merozoite within an invagination of the erythrocyte membrane. (After Aikawa *et al.*, 1978; with permission of *The Journal of Cell Biology*.)

phorins are recognized by at least two merozoite surface proteins (Perkins, 1984). The gene responsible for a glycophorin-binding protein of *P. falciparum* merozoites has been isolated and successfully employed via genetic engineering to dictate its synthesis in the bacterium *Escherichia coli* (Ravetch *et al.*, 1985).

Additional studies have revealed that **band 3**, a major erythrocyte membrane-spanning protein, mediates specific interaction of malaria parasites and host erythrocytes. In support of this, Miller *et al.* (1983) have demonstrated that monoclonal antibody against rhesus monkey band 3 blocks invasion of monkey erythrocytes by *Plasmodium knowlesi*. Recently, Okoye and Bennett (1985) have demonstrated that purified human erythrocyte band 3 inserted into the surface membrane of human red blood cells in liposomes (artifical phosphololipid vesicles) inhibits the penetration of *Plasmodium falciparum* merozoites.

Furthermore, liposomes containing human band 3 are ten times more effective in inhibiting invasion than those containing pig band 3, suggesting a degree of specificity. These investigators have suggested that liposomes containing band 3 bind to surface sites on the parasites and thereby block their attachment to normally occurring band 3 on the erythrocyte membrane.

Going one step further, Friedman *et al.* (1985) have now demonstrated that the lactosamine chains of erythroglycan, a carbohydrate component of band 3, inhibit invasion of erythrocytes by *P. falciparum*. The relationship between Duffy antigen, glycophorins, and band 3 remains undefined. All three systems may be operative.

Periodicity Among Species. Specific variations occur among the human-infecting species of *Plasmodium*. For example, the time required for completion of the erythrocytic schizogonic cycle within the vertebrate host varies. In *P. malariae* (Fig. 6.26), the causative agent of quartan malaria in both tropical and temperate climes, the period is 72 hours. In *P. ovale* (Fig. 6.28), which causes ovale, or mild, tertian malaria, there is a 49 to 50-hour cycle. There is a 48-hour cycle in *P. falciparum* (Fig. 6.27), the causative agent of subtertian, estivoautumnal, or malignant tertian malaria. Diagnostic differences between the four species of human malaria organisms are given in Table 6.2.

Relapse of Malaria. It has been known for many years that certain malaria patients who had recovered from the disease will undergo relapse. The original explanation given for the cause of relapse is that not all cryptozoites or metacryptozoites escaping from liver cells enter erythrocytes; rather, some remain, protected from antibodies within liver cells. If some physiologic triggering fluctuation occurs within the host, these intracellular parasites become activated and rupture out of liver cells, and this results in another bout of malaria.

More recently, Krotoshi *et al.* (1982) have proposed a second explanation. They maintain that two types of sporozoites exist. One type, when introduced by the mosquito into a human, undergoes the usual exoerythrocytic and erythrocytic phases of development and causes malaria. The second type will also enter a hepatic cell but will remain there unchanged for an indefinite time. This "sleeping" form has been designated as a **hypnozoite**. When some stimulus triggers activation of hypnozoites, a relapse of malaria occurs. The evidences for and against the occurrence of hypnozoites have been summarized by Shortt (1983) and Garnham *et al.* (1983).

Prolonged Prepatent Period. In addition to relapse, it is also known that the prepatent period is unusually long in certain cases of malaria caused by *Plasmodium vivax*. Shute *et al.* (1976) have proposed that two populations of *P. vivax* sporozoites exist. In temperate strains, represented by a North Korean strain, the sporozoites introduced into humans by the mosquito include representatives that require a long prepatent period (LPP) and these occur in great excess over a much smaller proportion of sporozoites that require only a short prepatent period (SPP). On the other hand, in tropical strains of *P. vivax*, the relative

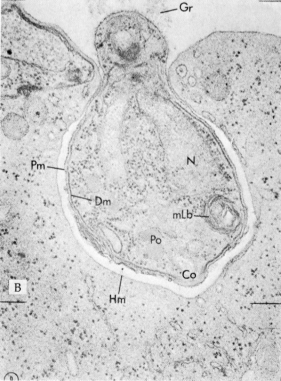

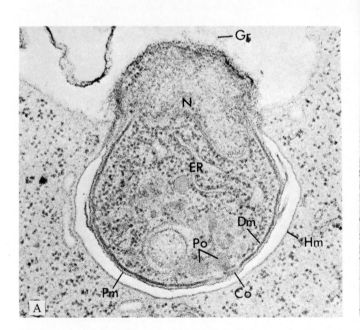

Fig. 6.30. Entry of *Plasmodium* merozoite into host erythrocyte. A. Electron micrograph showing merozoite of *Plasmodium berghei yoeli* with conoid region leading the way into host cell. The erythrocyte membrane enveloping the parasite appears to expand before the advancing conoid pole of the merozoite. × 40,000. **B.** Electron micrograph showing almost complete entry of merozoite of *P. berghei yeoli* into mouse erythrocyte. Notice tight contact between host cell and parasite at initial site of entry. × 45,000. (**A** and **B** after Ladda *et al.*, 1969; with permission of *Journal of Parasitology*.) Co, conoid; Dm, dense inner membrane; ER, endoplasmic reticulum; Gr, adherent granular material; Hm, host cell membrane; mLb, multilaminated membrane body; N, nucleus; Pm, parasite plasmalemma; Po, paired organelles.

proportions of LPP and SPP sporozoites differ, with their numbers being equal, or even with larger numbers of SPP.

It may be that LPP sporozoites are actually hypnozoites. The final answer has yet to come.

Grouping by Life Cycle Patterns. Although the life cycles of the various species of *Plasmodium* are essentially the same, variations in the duration of their exoerythrocytic and erythrocytic phases cause them to be categorized into three types.

1. In the *malariae* group, the exoerythrocytic phase is continuous and periodically erythrocytic merozoites can reinvade hepatic cells and become cryptozoites. For example, in the life cycle of *Plasmodium gallinaceum* of birds, the erythrocytic merozoites can revert to the exoerythrocytic phase. Thus, the exoerythrocytic phase of the cycle in the avian host can last for long periods. The invertebrate hosts in this case are *Aedes* mosquitoes (and others).

2. In the *vivax* group, the exoerythrocytic phase is also long lasting, i.e., involving hypnozoites, accompanied by periodic erythrocytic activity, but it differs from the *malariae* pattern since the erythrocytic merozoites do not reinvade hepatic cells.

3. In the *falciparum* group, the exoerythrocytic phase is short and is followed by the erythrocytic phase. Hence, there are no hypnozoites.

If it is correctly assumed that the exoerythrocytic habitat was that of the prototype, then these three categories suggest an evolutionary progression with the *malariae* group (including *P. gallinaceum* and *P. elongatum* in birds, *P. berghei* in rodents, and *P. mexicanum* in reptiles) being the most primitive. Members of the *falciparum* group then should be considered the most recently evolved.

Plasmodium and the Mosquito Host. As the result of *in vitro* cultivation of the mosquito phase of the avian malaria parasite *Plasmodium relictum*, Ball and Chao (1963) have shown that

1. The mosquito's stomach is not an essential organ for the development of the oocyst; however, it is most easily invaded since it is the organ closest to the developing parasite.

2. Residence in the salivary glands by sporozoites is not required for the latter to become infective.

3. A concentration of oxygen above that found in air is harmful to both the parasite and the mosquito tissues in which it normally develops.

4. The parasite exhibits a critical period extending from the gametocyte stage through the young zygote stage, during which it is much less resistant to low-temperature stress than during the later stages.

Table 6.2. Diagnostic Differences between the Four Species of Human-Infecting *Plasmodium*[a]

	Plasmodium vivax	*Plasmodium malariae*	*Plasmodium ovale*	*Plasmodium falciparum*
Duration of schizogony	48 Hours	72 Hours	49–50 Hours	36–48 Hours
Motility	Active amoeboid until about half grown	Trophozoite slightly amoeboid	Trophozoite slightly amoeboid	Trophozoite active amoeboid
Pigment (hematin)	Yellowish-brown, fine granules and minute rods	Dark brown to black, coarse granules	Dark brown, coarse granules	Dark brown, coarse granules
Stages found in peripheral blood	Trophozoites, schizonts, gametocytes	Trophozoites, schizonts, gametocytes	Trophozoites, schizonts, gametocytes	Trophozoites, gametocytes
Multiple infection in erythrocyte	Very common	Very rare	Rare	Very common
Appearance of infected erythrocyte	Greatly enlarged, pale with red Schüffner's dots	Not enlarged, normal appearance	Slightly enlarged, outline oval to irregular, with Schüffner's dots	Normal size, greenish, basophilic Maurer's clefts and dots
Trophozoites (ring forms)	Amoeboid, small and large rings with vacuole and usually one chromatin dot	Small and large rings with vacuole and usually one chromatin dot, also young band forms	Amoeboid, small and large rings with vacuole	Very small and large rings with vacuole, commonly with two chromatin dots, amoeboid
Segmented schizonts	Fills enlarged RBC, 12–24 merozoites irregularly arranged around mass of pigment	Almost fills normal-sized RBC, 6–12 merozoites regularly arranged around central pigment mass	Fills approx. $\frac{3}{4}$ of RBC, 6–12 merozoites around centric or eccentric pigment mass	Not usually seen in peripheral blood
Gametocytes	Round, fills RBC, chromatin undistributed in cytoplasm	Round, fills RBC, chromatin undistributed in cytoplasm	Round, fills $\frac{3}{4}$ of RBC, chromatin undistributed in cytoplasm	Crescentic- or kidney-shaped, usually free in blood, chromatin undistributed in cytoplasm

[a] Wright's stain is used.

Relative to the last point, Ball and Chao (1964) have demonstrated that when the mosquito *Culex tarsalis* infected with *Plasmodium relictum* is exposed to temperatures above or below 27°C, the optimum temperature for the development of this parasite, dramatic alterations occur. Specifically, at 18°C it requires about twice as long for sporozoites to appear in the mosquito's salivary glands. Also, although sporozoites are immediately infective upon reaching the salivary glands when maintained at 27°C, they do not become infective until 5 days or more following their initial appearance in the glands when maintained at 18°C. At 31°C, the sporozoites appear in the salivary glands in about half the time required at 27°C, but these are also not immediately infective but become so a few days later. Also, maintenance at 31°C for 10 days renders the sporozoites noninfective in subsequent bitings for at least 29 days.

Environmental temperatures are also known to influence the development time of the various species of human-infecting *Plasmodium* in their mosquito hosts. The optimum temperature for development varies with the species. Table 6.3 summarizes the maximum, minimum, and optimum temperatures required by four species of human-infecting *Plasmodium* for development.

Family Haemoproteidae. The family Haemoproteidae includes two genera, *Haemoproteus* and *Leucocytozoon*. Members of *Haemoproteus* are parasitic in erythrocytes and visceral endothelial cells; members of *Leucocytozoon* are parasitic in leucocytes, liver cells, and other cells of their vertebrate hosts. Both an invertebrate and a vertebrate host are necessary for the completion of haemoproteid life cycles. The U-shaped gametocytes of *Haemoproteus* spp. found within the host's red blood cells produce pigment granules. It was while working on the life cycle of *Haemoproteus* that MacCallum, in 1898, discovered

microgametocytes and fertilization, which later led to our understanding of exflagellation (p. 198).

Genus Haemoproteus. Many species of *Haemoproteus* have been described. The vertebrate hosts are usually reptiles and birds. Distribution of the various species is worldwide. Probably the most common and most extensively known species is *H. columbae*, a parasite of the pigeon, *Columba livia* (Fig. 6.31). The life cycle of this species is representative of the group.

Life Cycle of Haemoproteus columbae. When the infected fly vector (*Lynchia* spp., *Pseudolynchia maura*, or *Microlynchia pusilla*) bites a pigeon, the sporozoites are introduced into the bird's blood. These infective forms are carried by blood in plasma to the lungs and other organs, where they invade endothelial cells. Within the endothelial cells, the uninucleate sporozoites, now known as schizonts, undergo rapid nuclear division, resulting in a multinucleate form that fractures into many uninucleate bodies, known as **cytomeres**.

Each cytomere increases in size, and its nucleus divides repeatedly. Infected host cells become hypertrophied and eventually rupture. At the time this occurs, the multinucleated cytomeres also fragment, giving rise to numerous uninucleate merozoites that are expelled when the host cells rupture. The merozoites may reenter endothelial cells and repeat the schizogonic cycle, or enter red blood cells and mature into macro- or microgametocytes. These immature gametes mature within the gut of a fly that ingests infected pigeon blood cells.

The zygote, formed when a micro- and a macrogamete unite, enters the gut endothelium of the fly

Table 6.3. The Influence of Various Temperatures on the Time Required for Development of Four Species of Human-Infecting *Plasmodium* in Their Mosquito Hosts[a,b]

Species	16°C	17°C	18°C	20°C	24°C	30°C	35°C
Plasmodium vivax	—	Halted	—	16 Days	9 Days	7 Days	None develop
Plasmodium falciparum	—	—	Halted	20 Days	11 Days	9 Days	Generally none develop[d]
Plasmodium malariae	Halted	—	—	30 Days	21 Days	15 Days	None develop
Plasmodium ovale			15 Days at 26°C				

[a] Data are based on laboratory observations but approximates what occurs in nature.

[b] In addition to a suitable temperature, a mean relative humidity of over 65% is required for the normal development of the human-infecting plasmodia. (Adapted from Garnham, 1964.)

[c] Although sporogony takes longer at 20°C, it is commonly considered the optimum temperature since more viable sporozoites are produced.

[d] Wenyon (1926) reported that the sporogony of *Plasmodium falciparum* is even further accelerated above 30°C, with sporozoites reaching the salivary glands in less than 5 days.

and forms the oocyst. As in the case of *Plasmodium* spp. sporozoites are formed within oocysts. When the mature oocyst ruptures, the sporozoites are released. These find their way into the hemocoel and eventually into the salivary glands, from whence they can be inoculated into another avian host.

Practically nothing is known about the vectors of most species of *Haemoproteus*. Although most species were thought to be transmitted by hippoboscid flies, a species of *Haemoproteus* parasitic in ducks utilizes biting midges as vectors (Fallis and Wood, 1957). For a definitive review of *Haemoproteus*, see Fallis and Desser (1977).

Genus Leucocytozoon.   Members of *Leucocytozoon* undergo schizogony in cells of visceral organs, often the liver and kidneys, and endothelium of vertebrates. These vertebrates are primarily birds but rarely are other forms (*L. salvelini* occurs in the brook

trout, *Salvelinus fontinalis*). Sexual reproduction of *Leucocytozoon* occurs in blood-sucking insects. The gametocytes are found in leucocytes, or in some species in leucocytes and erythrocytes, of the vertebrate host.

Life Cycle of Leucocytozoon simondi.   O'Roke (1934) has determined the life cycle of *L. simondi*, a pathogen of ducks that utilizes the blackfly *Simulium venustum* as its vector (Fig. 6.32). The developmental pattern is essentially the same as that given for *Haemoproteus columbae*. The schizogonic phase in the duck takes 10 days, whereas the sexual and sporogonic phases in the fly take 2 to 5 days. Many species of *Leucocytozoon* have been reported, including *L. smithi* in domesticated turkeys in southeastern United States and Pennsylvania, *L. bonasae* in the ruffed and spruce grouse, *L. andrewsi* in chickens, and *L. marchouxi* in the mourning dove. Interested readers are referred to the review of *Leucocytozoon* spp. by Fallis *et al.* (1974).

SUBCLASS PIROPLASMIA

The members of this subclass are intracellular parasites of vertebrate erythrocytes or leucocytes and are

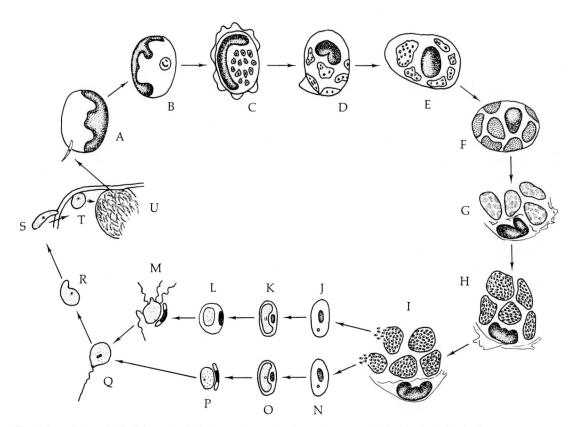

Fig. 6.31.   Stages in the life cycle of *Haemoproteus columbae*, a parasite of the pigeon, *Columba livia*, and flies. A. Sporozoite entering endothelial cell of pigeon. **B.** Growth of schizont within host cell. **C.** Segmentation of multinucleate schizont into uninucleate cytomeres. **D–I.** Development of cytomeres into merozoites. **J–M.** Development of microgametes. **N–P.** Development of macrogametes. **Q.** Joining of micro- and macrogamete in fertilization. **R** and **S.** Ookinetes. **T.** Young oocyst embedded in stomach wall of fly host. **U.** Mature oocyst rupturing and releasing sporozoites. ((Modified after Kudo, 1966. "Protozoology" 5th ed. Charles C Thomas.)

transmitted to their vertebrate hosts by ticks. As indicated in the section devoted to classification at the end of this chapter, there is only one order, Piroplasmida, belonging to this subclass. The principal family is Babesiidae.

Family Babesiidae. The family Babesiidae includes those nonpigment-producing intraerythrocytic parasites of various vertebrates—mostly mammals, amphibians, and reptiles. Various ticks serve as the vectors. Included in this family is the genus *Babesia*.

Life Cycle of Babesia bigemina. *Babesia bigemina* (Fig. 6.33) is found in cattle, causing the commonly lethal hemoglobinuric fever, red-water fever, or Texas fever. It was during studies on the life cycle of this species that the importance of arthropods as vectors for protozoan diseases was demonstrated by Smith and Kilbourne in 1893.

Infection of cattle is initiated with the introduction of **vermicles** into the blood through the bite of the tick *Boophilus annulatus* and related species. There is no exoerythrocytic phase in the life cycle of *Babesia* as there is in *Plasmodium*. After entering host erythrocytes, the parasites, now known as trophozoites, increase in size and usually multiply by binary fission, which may look like budding in the early stages. The intracellular trophozoites, each situated within a parasitophorous vacuole, multiply so rapidly that the resulting merozoites eventually kill the host cell. Subsequently, the escaping merozoites enter other host erythrocytes and divide, and an extremely large population of parasites is built up in a short time.

After the tick has taken a blood meal from an infected cow, the intracellular parasites are released within the tick's gut. Some of these transform into motile, polymorphic isogametes. After fusion of gametes, which usually occurs during the first 24 hours after entering the tick, the cigarshaped zygotes, each measuring 8–10 μm long, penetrate the tick's intestinal epithelium and transform into encysted spheroid sporonts. Each sporont measures up to 16 μm in diameter within 2 days. At this time the sporont nucleus undergoes schizogony, and the resulting vermicles (or merozoites), each measuring 9–13 μm long, migrate

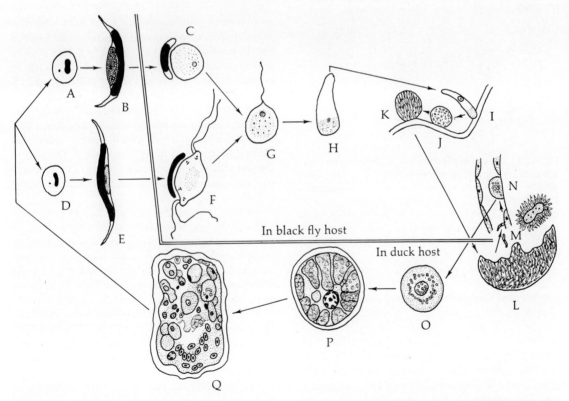

Fig. 6.32. **Stages in the life cycle of *Leucocytozoon simondi*, a parasite of ducks and the blackfly, *Simulium venustum*. A–C.** Development of macrogamete. **D–F.** Development of microgametes. **G.** Fertilization. **H.** Ookinete. **I.** Ookinete penetrating stomach wall of blackfly host. **J.** Young oocyst embedded in stomach wall of blackfly host. **K** and **L.** Development of sporozoites within mature oocyst. **M.** Sporozoites entering endothelial cell of blackfly host. **N–Q.** Schizogony. (Modified after Brumpt, 1949.)

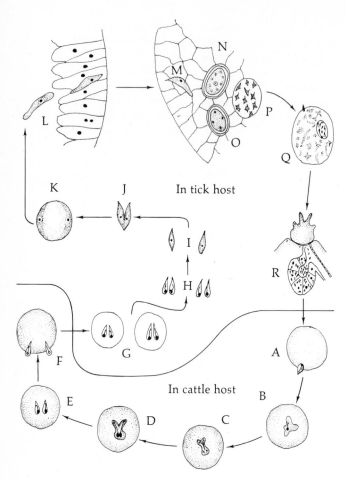

Fig. 6.33. **Stages in the life cycle of *Babesia bigemina*, a parasite of cattle transmitted by ticks. A–F.** Division in cattle erythrocytes. **G** and **H.** Gametocytes. **I.** Isogametes. **J.** Fertilization. **K.** Zygote. **L.** Ookinete penetrating through gut wall. **M.** Ookinete in tick egg. **N–P.** Sporoblast formation. **Q.** Sporokinetes in large tick embryonic cell. **R.** Vermicles (merozoites) in tick salivary gland. (Modified after Dennis, 1932.)

into the host's hemocoel. Here, they enter cells of the Malpighian tubules and, upon becoming established, undergo multiple fission. The resulting merozoites enter the ovaries and the enclosed eggs of the tick. Initially, they are randomly scattered throughout the egg; however, after the infected egg is laid and begins developing, the parasites migrate to and penetrate the intestinal epithelium of the larva. Here, schizogony is repeated, and the resulting merozoites enter the hemocoels of the embryonic ticks, migrate to the salivary glands, and again undergo multiple fission within host cells. The members of this generation of vermicles are piriform, measuring about 2–3 μm long by 1–2 μm wide. They enter the ducts of the salivary glands and are injected into the vertebrate host during feeding by the tick (Rick, 1964).

The passage of *B. bigemina* from a tick to its progeny in the manner described above is known as **transovarian transmission**. The discovery of this mechanism helped to explain how *Boophilis annulatus*, the vector for *B. bigemina* in the Americas, could serve as such an effective transmitter of the parasite from one vertebrate host to another. *B. annulatus* is a one-host tick, i.e., it feeds, matures, and mates on a single cattle host. After feeding and mating, the female tick drops to the ground, lays her eggs, and dies. By this time, via transovarian transmission, the parasites are already in the eggs. The hatching six-legged larval ticks climb onto vegetation and become attached to animals that brush by the plants.

In other parts of the world, two- and three-host ticks serve as the transmitters of *B. bigemina*. In these cases, transovarian transmission is not required and may not occur.

In addition to cattle, *B. bigemina* infects a wide variety of ruminants, including deer, water buffaloes, and zebu. The resulting disease is more severe in adult animals than in calves. Calves less than 1 year old seldom become seriously ill. On the other hand, untreated, adult cattle commonly suffer 50 to 90% mortality. The incubation period is usually 8 to 15 days, but an acutely ill animal may die 4 to 8 days after infection. Infected cattle have high fever (106°–108°F), become dull and listless, and lose appetite. Up to 75% of a host's erythrocytes may be destroyed, and a severe anemia results. Because the body's mechanism for clearance of hemoglobin and its breakdown products is overtaxed, jaundice results, and excess hemoglobin is excreted in urine, which is red. Damage to internal organs also occurs (Levine, 1973). For review of babesiosis in domestic animals, see Mahoney (1977).

Babesia microti. Although human infections with *Babesia* spp. have been known since 1957, in recent years human babesiosis has become sufficiently frequent on Nantucket Island and Martha's Vineyard, both in Massachusetts, on Shelter Island near Long Island, and on eastern Long Island, New York, that it has attracted the attention of medical parasitologists. In each instance the causative agent has been identified as *Babesia microti* (Figs. 6.34, 6.35), a natural parasite of meadow vole and other rodents which is transmitted by ticks, most commonly *Ixodes scapularis* (Spielman, 1976).

Human babesiosis due to *B. microti* mimics relatively mild malaria, although human babesiosis caused by *Babesia divergens*, a cattle parasite in Ireland, can be fatal.

One interesting aspect of human babesiosis is that a large percentage of the recognized cases have oc-

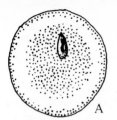
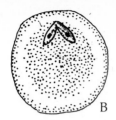

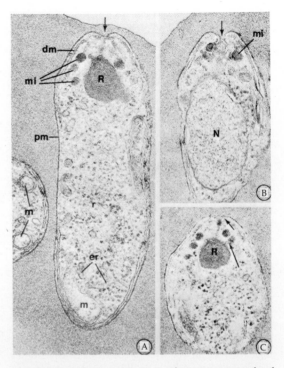

Fig. 6.34. *Babesia microti.* **A.** Trophozoite in erythrocyte of host. **B.** Trophozoite in process of dividing in erythrocyte of host.

curred in individuals who had been splenectomized. Consequently, it was theorized that individuals who have had their spleens removed lack immunocompetence to resist infection. This theory is supported by the finding of Ruebush *et al.* (1981) who have demonstrated that infected *Macaca mulatta* monkeys demonstrated recurrence of severe hemolytic anemia when they were splenectomized. The disease, however, is not restricted to splenectomized humans, as cases on Nantucket Island have revealed. For definitive reviews on various aspects of *Babesia* and babesiosis, see the volume edited by Ristic and Kreier (1981). The reviews by Zwart and Brocklesby (1979) on resistance and immunity to babesiosis and Joyner and Donnelly (1979) on the epidemiology of the disease should also be consulted.

Other Babesia *Species.* Other species of *Babesia* are known to infect horses (*B. equi*), rodents (*B. rodhani*), ground squirrels (*B. wrighti*), cats (*B. felis*), dogs (*B. canis*), hogs (*B. suis* and *B. trautmanni*), goats (*B. motasi*), and other mammals. Ristic and Lewis (1977) have provided a list of species that infect mammals. Very often these infections are acute, leading to death in 7 to 10 days. In chronic infections, jaundice, anemia, and other symptoms occur.

Family Theileridae. Members of the family Theileridae parasitize blood cells of mammals. They are transmitted by hard ticks of the family Ixodidae (p. 582).

Like members of the Babesiidae, the theilerids lack a conoid. Rhoptries, micronemes, subpellicular tubules, and a polar ring occur in the stages found in the tick host.

According to Mehlhorn *et al.* (1979), gamogony occurs in the gut of the nymphal tick, and the resulting kinetes are very similar to the ookinetes of members of Haemosporina. These kinetes grow for a time in the gut cells of the tick, then escape and penetrate the cells of the salivary glands, where sporogony occurs. The resulting cells are the infective sporozoites.

Theileridae includes the genus *Theileria*, several members of which parasitize cattle, sheep, and goats, resulting in a disease known as theileriosis. This disease results in heavy losses of domestic animals in southern Europe, Asia, and Africa.

Fig. 6.35. *Babesia microti.* **A.** Electron micrograph of longitudinal and partial cross-section of merozoites. Notice that at the anterior end the plasma membrane forms an indentation (arrow). × 48,000. **B.** Electron micrograph showing indentation of the plasma membrane at the anterior end (arrow) extending into the cytoplasm. × 50,000. **C.** Electron micrograph showing an extension of the rhoptry reaching close to the indentation of the plasma membrane (short arrow). A connection between the rhoptry and a microneme is apparent (long arrow). × 48,000. (**A, B**, and **C** after Rudzinska and Trager, 1977; with permission of *Canadian Journal of Zoology*.) dm, double membrane situated below plasma membrane; er, rough endoplasmic reticulum; m, mitochondrialike structures; mi, micronemes; N, nucleus; pm, plasma membrane; R, rhoptry.

Theileria parva. This parasite is the causative agent of Coast fever in cattle, zebu, and Cape buffalo in Africa. It is one of the most serious diseases of cattle in eastern and central Africa. Although quite prevalent in southern Africa at one time, it has now been largely eliminated in this area.

When found within the host's erythrocytes, the elongate parasite (Fig. 6.36) measures 1–2 µm in length. When stained with Romanovsky's stain, the intraerythrocytic parasite has blue cytoplasm and a red nucleus at one end. This parasite is transmitted by the ticks *Rhipicephalus evertsi* and *R. appendiculatus*. It

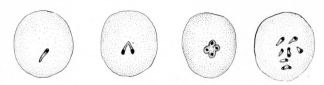

Fig. 6.36. Forms of *Theileria parva* occurring in cattle erythrocytes.

has been successfully maintained in tissue culture by Brocklesby and Hawking (1958).

When a tick feeds on cattle, it injects the sporozoites present in its salivary glands into the mammalian host. These initiate a tissue phase of reproduction. Specifically, they enter lymphocytes within lymphoid tissue, grow, and undergo schizogony. The resulting schizonts, commonly referred to as **Koch's blue bodies**, can be seen in circulating lymphocytes within 3 days after infection. Two types of schizonts have been recognized: **macroschizonts** and **microschizonts**. Multinucleated macroschizonts are those that originally occur in lymph cells. Each of these produces about 90 **macromerozoites**, each measuring 2–2.5 μm in diameter. Some of these enter other lymph cells, especially in fixed tissues, and give rise to subsequent generations of macroschizonts. Others enter lymphocytes and differentiate into microschizonts. Each of these multinucleated bodies produces 80 to 90 **micromerozoites**, each measuring 0.7–1.0 μm wide. If microschizonts rupture while in lymphoid tissues, the escaping micromerozoites enter new lymph cells, maintaining the lymphatic infection. However, if the microschizonts rupture in the circulating blood, the micromerozoites enter erythrocytes and differentiate into trophozoites (also known as **piroplasms**). Piroplasms do not multiply within erythrocytes.

When a tick, no matter at what stage of development, feeds on blood containing piroplasms, it becomes infected. Ingested erythrocytes are digested, and the released piroplasms undergo gamogony and kinete formation. Transovarian transmission does not occur.

In certain species of *Theileria*, for example, *T. parva* and *T. ovis* of sheep and goats, and *T. annulata* of cattle and the water buffalo, sexual reproduction has been reported to occur in the tick host; however, the details are not yet clear. For reviews of what is known about the life cycle of *Theileria*, see Riek (1966) and Barnett (1968, 1977).

Other genera. Also included in the family Theileriidae are *Haematoxenus* in zebu and cattle and *Cytauxzoon* in antelope, both in Africa.

PHYLUM MICROSPORA

The members of this phylum are obligatory intracellular parasites in both vertebrates and invertebrates. As indicated in the classification list at the end of this chapter, these parasites are assigned to one of two classes: Rudimicrosporea and Microsporea.

CLASS RUDIMICROSPOREA

The members of this class are hyperparasites of gregarines (p. 178) in annelids.

CLASS MICROSPOREA

The Microsporea includes comparatively small organisms (3–6 μm) which are parasites of every major group of animals, from other protozoans to humans. They are especially abundant in arthropods, fish, and mammals and are generally pathogenic. Included in this order are several species of some economic importance because they parasitize insects such as silkworms and honey bees. Furthermore, there has been an increased interest in these unicellular parasites because of their possible role as biologic control agents for undesirable invertebrates, especially certain insects and platyhelminths. Canning (1975) has contributed a comprehensive review of microsporeans hyperparasitic in platyhelminths. Furthermore, she has contributed a comprehensive review (1977) of the order Microsporida, which is the largest order of the Microspora.

Life Cycle of Microsporeans

Microsporeans undergo both schizogony and sporogony as intracellular parasites. Very often infected cells portray marked hypertrophy.

The infective form of microsporeans is the spore, which is small (3–6 μm long), spherical, ovoid, or cylindrical (Figs. 6.37, 6.38). The spore wall consists of three layers: an outer dense **exospore**, an electronlucent middle layer known as the **endospore**, and a thin **membrane** enveloping the cytoplasmic contents. The exospore of some species have two to five sublayers. The wall is dense and refractile, and its resistant properties contribute to the survival of the spores.

The most distinctive feature of microsporean spores is the presence of a **polar filament** (Figs. 6.37, 6.38, 6.39). In a resting spore, this hollow filament is tightly coiled. When activated, as explained below, the living component of the spore escapes through this tube into the host cell.

Studies using the electron microscope have revealed the presence of a **polar cap** covering the attached end of the filament and overlying the **polaroplast**. The

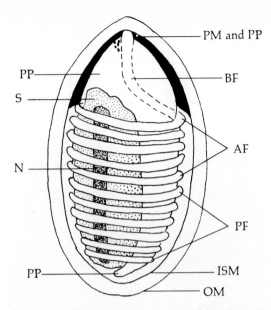

Fig. 6.37. *Thelohania* **spore; a typical microsporean spore.** Schematic drawing showing the structure of the spore of *Thelohania californica*, parasitic in the larvae of the mosquito *Culex tarsalis*, as revealed by electron microscopy. AF, anterior polar filament; BF, basal portion of polar filament; ISM, interior surface membrane; N, nucleus; OM, outer membrane; PF, posterior polar filament; PM, polar mass; PP, polaroplast; S, sporoplasm. (Redrawn after Kudo and Daniels, 1963.)

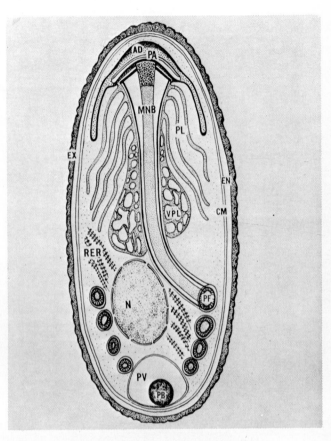

Fig. 6.38. **Microsporidan spore.** Schematic drawing of the fine structure of a spore. AD, anchoring disc of polar tube; CM, cytoplasmic membrane of spore content; EN, endospore; EX, exospore; MNB, manubroid portion of filament; N, nucleus; PA, polar aperture; PB, posterior body; PF, polar tube; PL, lamellae of lamellar polaroplast; PV, posterior vacuole; RER, rough endoplasmic reticulum; VPL, vesicular portion of polaroplast. (After Vávra, 1976; with permission of Plenum Press.)

latter is an anteriorly situated vacuole (Fig. 6.38). The amoeboid **sporoplasm** surrounds the extrusion apparatus. The nucleus and most of the cytoplasm of the sporoplasm lie within the coils of the filament (Fig. 6.38). A **posterior vacuole** occurs at the end opposite the polaroplast (Fig. 6.37). The sporoplasm includes many free ribosomes, some endoplasmic reticula, but no mitochondria. According to Weidner (1972), the membrane and matrix of the polar cap are continuous with the highly pleated membrane comprising the polaroplast. This membrane, in turn, is continuous with the anchoring disc (or base) of the polar filament. When a spore is ingested by a compatible host, a change in the permeability of the polar cap occurs. This permits water to enter the spore, and the filament is expelled explosively by eversion. The stacked membrane in the polaroplast is unfolded as the filament is discharged and contributes to the expelled filament so that the everted filament is considerably longer than when coiled within the spore (Figs. 6.39, 6.40). The extruded filament readily penetrates a host cell, and the sporoplasm flows through the tubular filament and enters the host cell. The end of the filament within the host cell expands to enclose the sporoplasm and becomes the intracellular parasite's new outer membrane.

The stage occurring in the host cell is known as the **trophozoite.** Its nucleus may divide repeatedly, resulting in a multinucleated plasmodium. This is followed by cytoplasmic division, and the process may be repeated. In some species two nuclei are associated intimately, and an organism portraying this condition is referred to as a **diplokaryon** (Fig. 6.41).

The multiple fission of meronts is generally regarded as schizogony, but the process does not appear to be completely analogous to the schizogony found in the Apicomplexa.

Spore formation (sporogenesis) occurs when the nuclear divisions of trophozoites with a single nucleus or two nuclei divide to give rise to nuclei destined

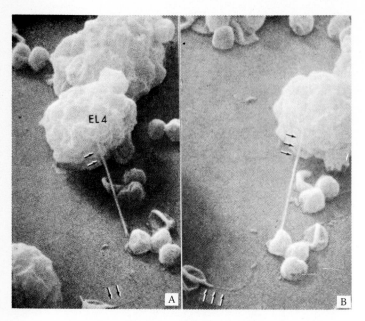

Fig. 6.39. Extruded microsporidan polar filament (discharge tube). A. Scanning electron micrograph of polar filament of spore of *Nosema michaelis*, a parasite of blue crabs, invading EL4 cells. **B.** Scanning electron micrograph of same extruded polar filament penetrating same cell taken from a different angle. × 5000. (After Weidner, 1972; with permission of Springer-Verlag.)

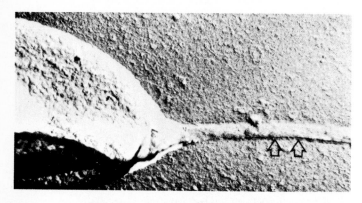

Fig. 6.40. Microsporidan polar filament (discharge tube). Platinum carbon replica of *Nosema michaelis* spore with extruded polar filament (arrows). The filament is 100 nm thick and 50 μm long. × 50,000. (After Weidner, 1976; with permission of *The Journal of Cell Biology*.)

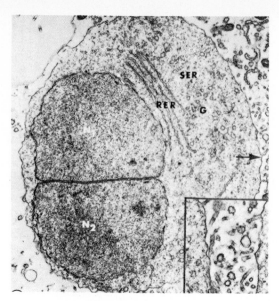

Fig. 6.41. Diplokaryon. Electron micrograph of a portion of a meront of *Pleistophora* sp. containing a diplokaryon (N₁, N₂), rough endoplasmic reticulum (RER), smooth endoplasmic reticulum (SER), and cytoplasmic granules (G). The cell membrane is a unit membrane forming microvillosities (arrow). × 19,600. Insert is an enlarged portion of the cell membrane with microvillosities. × 28,000. (After Vávra, 1976; with permission of Plenum Press.)

to become spore nuclei. According to Hazard *et al.* (1979), the nuclear division preceding sporogony is meiotic, and hence the spores are haploid. This, however, apparently occurs only in the members of certain genera, not including *Nosema*. Haploid spores are not directly infective to new hosts, suggesting that an intermediate host is involved in the life cycle. Restoration of diploid nuclei probably occurs in such a host.

During sporogony, the parasite becomes multinucleate and is known as a **sporogonial plasmodium**. According to Overstreet and Weidner (1974), further differentiation of spores can occur in one of two ways: (1) the cytoplasm surrounding each nucleus becomes segregated around it, forming **sporont-determinate** areas; or (2) an envelope is formed around each nucleus and the surrounding cytoplasm to form a **sporoblast**. Krinsky and Hayes (1978) have demonstrated that a mass of tubules forms in each sporoblast, and these eventually develop into the polar filament and polaroplast. Mitochondria are absent in all of the stages.

For definitive reviews of the structure and development of microsporeans, see Canning (1977) and the volume edited by Bulla and Cheng (1976).

ORDER MINISPORIDA

The members of this small order of the class Microsporea are less complex in structure than the mem-

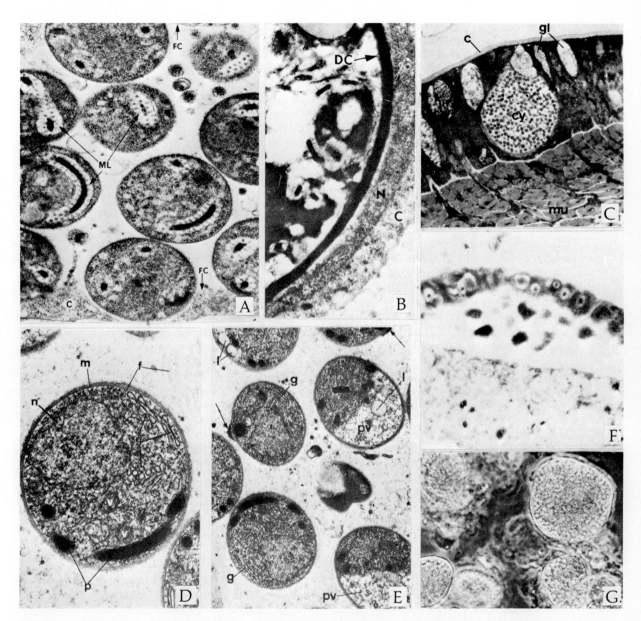

Fig. 6.42. Structure of some microsporidan cysts and spores. A. Electron micrograph of a portion of a cyst of *Chytridiopsis socius*, a parasite of beetles, *Blaps* spp., containing typical spores. The polar filaments have a thick honeycomb middle layer. The cyst is of the so-called fragile type. C, cytoplasm of host cell; FC, fragile type cyst wall; ML, honeycomb middle layer of polar filaments. × 19,000. **B.** Electron micrograph of a portion of the wall of a so-called durable cyst of *C. socius*. C, cytoplasm of host cell; DC, durable cyst wall; N, nucleus of host cell. × 21,000. (**A** and **B** after Sprague, Ormières, and Manier, 1972; with permission of Academic Press.) **C.** Photomicrograph of a portion of the epidermis of the earthworm *Eisenia foetida* showing cyst of *Burkea eisenia* within host cell. × 560. **D.** Electron micrograph of spore of *B. eisenia* containing a double membrane-bound nucleus. The spore is surrounded by a single membrane and covered by a fibrillar coat. × 30,000. **E.** Electron micrograph of several spores of *B. eisenia* showing thickening of the spore surface in region of polar cap (arrows). × 13,600. c, cuticle; cy, cyst; f, fibrillar coat; g, Golgi; gl, gland cells; l, lipid inclusions; m, membrane; mu, body wall musculature; n, double-membrane-bound nucleus; p, polar filament; pv, posterior vacuole. (**C, D, E** after Burke, 1970; with permission of Academic Press.) **F.** Photomicrograph of portion of midgut of the mosquito *Culex territans* showing cysts of *Hessea chapmani* in host cells and free in gut lumen. × 100. **G.** Fresh cysts of *H. chapmani* as observed with phase contrast. × 1000. (**F** and **G** after Clark and Fukuda, 1971; with permission of Academic Press.)

bers of the order Microsporida. Representative genera (Fig. 6.42) include *Burkea*, the species of which are parasitic in the muscles of earthworms; *Chytridiopsis*, parasitic in beetles and myriapods; and *Hessea*, the species of which are parasitic in fish and arthropods.

ORDER MICROSPORIDA

This is by far the largest order of the class Microsporea. The members are more complex in structure than the minisporidans (see diagnosis at end of chapter).

Microsporidan Genera. The Microsporida includes several genera among which the best known are *Nosema, Glugea, Thelohania, Encephalitozoon,* and *Pleistophora*. The members of *Nosema* are characterized by the formation of two spores from each sporont, and those of *Glugea* also are characterized by the development of two spores from each sporont but the presporogonic development is different. In *Thelohania*, each sporont develops into eight sporoblasts and ultimately into eight spores, and in *Pleistophora* each sporont develops into a variable number of sporoblasts, often more than 16, and each of these becomes a spore.

Nosema bombycis (Fig. 6.43) is parasitic in silkworms and causes the fatal pébrine disease which was initially studied by Louis Pasteur in 1870. This parasite is of considerable interest because of its economic importance. It occurs in all tissues of the embryo, larva, pupa, and adult of the silkworm, *Bombyx mori*. The silkworm and other susceptible insects become in-

fected by the ingestion of spores or transovarially. Each oval spore, measuring 3–4 μm × 1.5–2 μm, includes a coiled polar filament that may be as long as 100 μm. The sporoplasm escapes through the extruded and everted polar filament and invades the host's cells. Heavily infected silkworms are characterized by the appearance of numerous minute brownish-black spots scattered over their bodies, and heavily infected larvae cannot spin cocoons and perish without pupating. Transovarial infection occurs when sporonts invade ova and develop into spores. Thus, embryos developing from these ova already include the parasite.

Nosema bryozoides parasitizes germ cells of bryozoans, whereas *N. apis* (Fig. 6.43) affects the digestive tract and indirectly affects the ovaries of worker bees and the queen bee. It causes the so-called *Nosema* disease of bees.

Nosema apis is of economic interest because of its lethality to bees, causing losses to both honey producers and fruit producers (by lower pollination of flowers). In England the disease caused by *N. apis* is often called the Isle of Wight disease. This parasite is encountered most frequently in workers, although drones and queens are also susceptible. The infective spores are oval, measuring 4–6 μm × 2–4 μm, and include a long polar filament. Although *N. apis* is found almost exclusively in the epithelial cells lining the host's gut, the ovary of experimentally infected queen bees undergoes varying degrees of degeneration (Hassanein, 1951), although the eggs are free from infection. Thus, infections, unlike in the case of *N. bombycis*, do not occur transovarially. In addition to bees, *N. apis* will infect other insects. For a review of the biology of this important microsporidan, see Bailey (1959).

Cali (1970), using the electron microscope, has shown that in both *Nosema bombycis* and *N. apis*, a pair of nuclei, collectively known as a diplokaryon, occurs within each parasite in all phases of development. Sometimes multiple diplokarya develop as the result of karyokinesis but not cytokinesis. This double-nucleated condition does not occur in what has been designated *Nosema cuniculi*, and hence it is considered to represent another genus, *Encephalitozoon*. The species *E. cuniculi* is of biomedical interest since it is known to infect small mammals such as rats, mice, hamsters, and rabbits. Furthermore, a species of *Encephalitozoon* has been reported from multiple sclerosis patients in the Soviet Union.

Glugea hertwigi (Fig. 6.43) and *G. mulleri* are parasites of fish. In *G. hertwigi*, the so-called cysts, which are actually units of host cells surrounding sporonts, are quite conspicuous on the host's body surface.

Also included in the order Microsporida are the members of the genus *Amblyospora*. These are virulent pathogens of mosquitoes and blackflies, and hence are prime candidates as biological control agents for

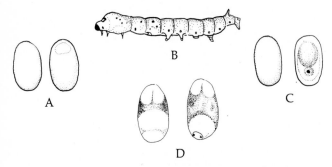

Fig. 6.43. Some microsporidan parasites. A. Fresh spores of *Nosema bombycis*, 3–4 by 1.5–2 μm, a tissue parasite of the silkworm. **B.** Heavily infected silkworm larva showing characteristic pigmentation. (Redrawn after Kudo, 1913.) **C.** Fresh (left) and stained (right) spores of *Nosema apis*, 4–6 by 2–4 μm, a parasite of the gut epithelium and associated tissues of the honeybee. (Redrawn after Kudo, 1921.) **D.** Spores of *Glugea hertwigi*, 4–5.5 by 2–2.5 μm, a parasite in tissues of fish. (Redrawn after Schrader, 1921.)

these noxious insects. It is noted that Sweeney *et al.* (1985) have reported that a species of *Amblyospora* that infects the mosquito *Culex annulirostris* in Australia employs an intermediate host. Specifically, they have reported that this species of *Amblyospora* will infect the copepod *Mesocyclops albicans*, in which it undergoes normal development resulting in the production of spores that are infective to mosquito larvae. This is the first report of an intermediate host involved in the life cycle of a microsporidan.

Other species of Microsporida are depicted in Fig. 6.44. For a complete taxonomic treatment of the Microsporida see the volume edited by Bulla and Cheng (1977).

HYPERPARASITISM BY MICROSPORIDANS

A number of species of *Nosema, Pleistophora, Unikaryon, Microsporidium,* and related genera are known to be hyperparasites in the sporocysts and developing cercariae of trematodes in freshwater gastropods (Canning, 1975). In addition, various species of *Nosema* and *Glugea* are capable of hyperparasitism in larval and adult helminths. For example, both *Nosema legeri* and *N. spelotremae* have been found as hyperparasites in trematode larvae, specifically metacercariae, in marine bivalves; *N. distomi* was described from adult trematodes in *Bufo marinus* in Brazil; *Glugea danilewskyi* and *G. encyclometrae* are hyperparasitic in adult trematodes in snakes; and *Nosema echinostomi* is a hyperparasite in the rediae, cercariae, and metacercariae of echinostomatid trematodes in the snail *Lymnaea limnosa* in France. *N. helminthorum* is hyperparasitic in tapeworms of the genus *Moniezia* in England and Pakistan.

Practically nothing is known about the physiologic aspects and relatively little about the morphologic manifestations of the host-parasite relationship between these hyperparasitic microsporidans and their helminth hosts. However, Dollfus (1946) has reported that *Nosema echinostomi* develops slowly in rediae and cercariae in the snail host. If cercariae are not heavily parasitized, they can develop into metacercariae, but in heavy infections the cercariae are killed. Parasitized rediae lose their characteristic yellowish color and become white. Furthermore, only lightly infected rediae produce cercariae that become parasitized.

Cort *et al.* (1960a) have reported that the development of strigeoid cercariae is distorted if parasitized by *Nosema* (Fig. 6.45). Cercariae emerging from snails do not include microsporidans because infected embryonic cercariae are so injured and distorted they cannot escape from the snail. Strigeoid larvae become heavily infected through experimental feeding of large numbers of *Nosema* spores to snail hosts and their microsporidans develop rapidly, producing spores in about 2 weeks. In nature, however, the number of microsporidans is small and their development is extremely slow, only a few ultimately producing spores. Cort *et al.* (1960b) have suggested that these differences are due to some type of resistance that develops in the natural infections, most of which are probably produced by ingestion of only a few spores. In experimental infections with large number of spores, it has been suggested that the resistance may have been prevented from developing.

It is of interest to note that the use of microsporidans to control schistosome trematodes in their molluscan host has been proposed.

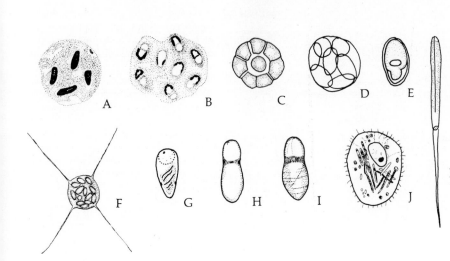

Fig. 6.44. Some microsporidan parasites. A and **B.** Stained sporogonic stages of *Thelohania legeri,* found in fat bodies of larvae of several species of *Anopheles* mosquitoes. **C** and **D.** Mature sporonts of *T. legeri.* **E.** Fresh spore of *T. legeri,* 4–6 by 3–4 μm. (Redrawn after Kudo, 1921.) **F.** Mature sporont of *Trichoduboscquia epeori,* found in fat bodies of mayfly nymphs, 9–10 μm in diameter. **G.** Fresh *T. epeori* spore, 3.5–4 μm long. (Redrawn after Léger, 1926.) **H** and **I.** Stained spores of *Pleistophora longifilis* in testes of *Barbus flaviatilis,* 3–12 by 2–6 μm. (Redrawn after Schuberg, 1910.) **J.** An infected blood cell of *Tubifex tubifex* enclosing *Mrazekia caudata.* (Redrawn after Mrázek, 1897.) **K.** *M. caudata* spore, 16–18 by 1.3–1.4 μm. (Redrawn after Léger and Hesse, 1916.)

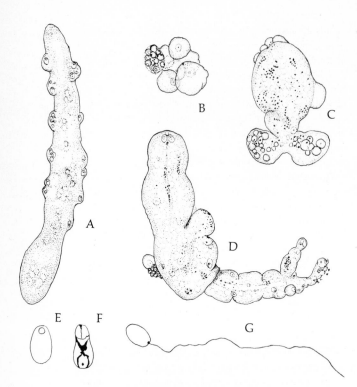

Fig. 6.45. Hyperparasitic *Nosema* in intramolluscan larvae of strigeoid trematodes. A. Young daughter sporocyst of *Posthodiplostomum minimum* showing effect of parasitism. **B.** Mass of embryonic cells from brood chamber of *Apatemon* sp. with one hyperparasitized cell. **C.** Young cercarial embryo of *Apatemon* sp. showing small vesicles and protuberances characteristic of hyperparasitized specimens. **D.** Parasitized cercarial embryo of *Diplostomum flexicaudum* showing vesicles and surface protuberances. **E.** Fresh spore of *Nosema* sp., 3.9–6 by 2.2–3.8 μm. **F.** Stained spore of *Nosema* from section of host tissue. **G.** Fresh spore of *Nosema* with extruded polar filament. (Redrawn after Cort *et al.*, 1960.)

PHYLUM ASCETOSPORA

This relatively small phylum consists exclusively of parasitic species that are characterized by spores lacking polar caps or polar filaments. The ascetosporans are parasitic in the cells, tissues, and body cavities of invertebrates, especially molluscs.

Representative of ascetosporans are members of the genera *Bertramia*, which are coelomic parasites of worms and rotifers; *Anurosporidium*, which are hyperparasitic in the sporocysts of digenetic trematodes; *Urosporidium*, which are found in the coelom of polychaetes; *Coelosporidium*, found in the coelom and Malpighian tubules of cockroaches; and *Haplosporidium*, found in aquatic annelids and molluscs (Fig. 6.46, 6.47). Life cycle data concerning haplosporeans are still sparse.

Life Cycle of Haplosporidium. The genus *Haplosporidium* includes several species found in the gut epithelium and connective tissue of annelids and in various tissues of molluscs.

Although details of the life cycle of the various species remain vague, the pattern consists of a spore, which, when ingested by the host, releases a small amoeboid form, the **amoebula**. The uninucleate or binucleate amoebula may invade a cell or penetrate through the gut wall and continue its development in the host's body cavity. Growth in size is accompanied by nuclear divisions without cytoplasmic divisions, and a multinucleate **plasmodium** is formed. Uninucleate **sporoblasts** appear within the mature plasmodium.

The plasmodium eventually fragments, and each sporoblast later matures into a **spore** which may or may not bear a minute operculum at one end. How mature spores escape from one host and become ingested by another still is undetermined. In some species, the single nucleus within the spore is said to divide, and the daughter nuclei supposedly recombine

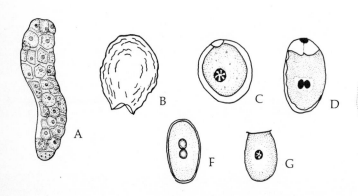

Fig. 6.46. Some ascetosporan parasites. A. Cyst of *Bertramia asperospora* enclosing spores, found in body cavity of rotifers. **B.** Empty cyst of *B. asperospora*. (Redrawn after Minchin, 1912.) **C.** Spore of *Anurosporidium pelseneeri*, a parasite in trematode sporocysts. **D.** Stained spore of *Ichthyosporidium giganteum*, parasitic in various organs of fishes of the genus *Crenilabrus*. **E.** *Coelosporidium periplanetae* trophozoite enclosing spores and chromatoidal bodies; parasitic in Malpighian tubules of cockroaches. **F.** *C. periplanetae* spore, 5.5 by 3 μm or slightly larger. (Redrawn after Sprague, 1940.) **G.** *Haplosporidium limnodrili* spore in gut epithelium of *Limnodrilus* sp. (Redrawn after Granata, 1925.)

as the spore approaches maturity (a process known as **autogamy**). The existence of true gametes has been claimed in some species, but this is questionable.

Haplosporidium nelsoni.   The modern history of the oyster industry in the United States dates back to 1607 when Captain John Smith first sailed in Chesapeake Bay. One of the first benefits reaped by Smith's party was to "... eat some of the oysters, which were very large and delicate in taste." Natural beds of the American oyster, *Crassostrea virginica*, have made possible a major shellfish industry ranging from Texas to Massachusetts. Oystering under sail has served as one of the backbone industries in the Chesapeake and Delaware Bays. At one time in the Maryland waters of the Chesapeake the annual harvest reached 15 million bushels. With the price of the oysters being sometimes as high as $13.00 a bushel, this represented an annual gross income of $195 million. Although these were the peak years, it is not difficult to envision how the oyster industry aided measurably in building up and supporting the many communities along the shores of the Chesapeake and the Delaware until 1959. During that year, a disaster, which is still continuing, hit these bays, especially the Chesapeake.

Actually, the disaster first made its appearance during 1957 in Delaware Bay. Mass mortalities on oyster beds caused great alarm among not only the oystermen but oyster biologists as well. Immediate steps were taken at Rutgers University to investigate the cause.

While examining tissues from moribund oysters, Haskin *et al.* (1966) discovered minute multinucleated bodies scattered throughout their bodies (Fig. 6.47).

There was no doubt that these spheres represented an unknown protozoan parasite which was designated MSX (multinucleate sphere X) and later, after several years of careful study, it was named *Minchinia nelsoni* and is now assigned to the genus *Haplosporidium*.

As *H. nelsoni* spread up the Delaware Bay, it also bridged the oceanic gap to the south and infected oysters not only in the Chesapeake but in Virginia waters as well. It has also spread northward to Long Island Sound. While concerted efforts by several laboratories have failed to stop its spread, much has been learned about its asexual reproduction within oysters, its ecology, and the internal defense mechanisms of oysters. One of the most significant findings was that this parasite cannot survive in waters with salinities less than 15 to 20‰. Thus, oysters growing farther up the bays were not killed. Unfortunately, nature dealt the oyster industry a further blow in the form of severe droughts during 1961–1966. As a consequence, the salinities of waters which were up until then "safe" rose, and *H. nelsoni* invaded farther up the estuaries, killing oysters along its path.

In spite of concerted efforts, it still remains unresolved how oysters become infected, although labo-

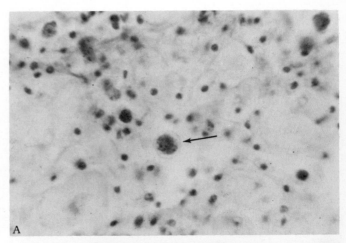

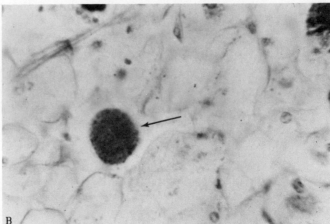

Fig. 6.47. *Haplosporidium nelsoni.* **A.** Multinucleated plasmodium in connective tissue of the American oyster, *Crassostrea virginica.* **B.** Higher magnification of slightly older plasmodium in oyster connective tissue. Infected oysters from New Jersey. Arrow in both photomicrographs points to plasmodium. (After Cheng, 1967.)

ratory and field studies have shown that infections do not occur directly from oyster to oyster. Because of the urgency of the situation, attempts have been made to initiate a program of developing genetic strains of resistant oysters. This has been accomplished with some success by placing uninfected oysters in "infected" waters for a period of time and saving the survivors. The progeny were again exposed to infection and, in time, resistant strains have been developed. However, over the past 12 years *H. nelsoni* and other known diseases and predators of oysters have rendered the industry nearly bankrupt. All one has to do is to visit the once thriving oyster villages along the New Jersey and Maryland coasts and witness the poverty of the proud people and the run-down conditions of their homes and stores and the "MSX plague" becomes evident. Although some oyster fishing still goes on, based on a technologically directed program of harvesting relatively young oysters grown from imported "seed" oysters before they are killed by *H. nelsoni*, the sight of bevies of white-sailed oyster boats gracefully sailing on the Chesapeake is no more, and one more parasite has forced a change in another aspect of the American scene. Yet, if what is left is to be saved (Maryland's take during 1966–1968, the best in many years, was down to 3 million bushels), research on how oysters become infected and how the production of resistant stocks can be accelerated must go on. All this, of course, requires funds that a dying industry cannot afford.

For a review of the biology of *H. nelsoni*, see Cheng (1967), and for detailed accounts of the history, epizootiology, and ecology of this oyster disease, see Ford and Haskin (1982) and Haskin and Ford (1982).

PHYLUM MYXOZOA

The myxozoans are all parasitic. Their diagnostic characteristics are listed in the classification list at the end of this chapter. These protozoans all belong to one class, the Myxosporea.*

Class Myxosporea. Myxosporea includes *Ceratomyxa*, found in the viscera of the rainbow trout, *Salmo gairdnerii*, and in the gallbladder of marine fish; *Trilospora*, *Leptotheca*, and *Myxoproteus*, all found in the gallbladder of various fish; *Mitraspora*, in the kidneys of freshwater fish; *Wardia*, in various tissues of freshwater fish; *Unicapsula*, *Myxidium*, and *Unicauda*, all parasites of fish, amphibians, and reptiles (Figs. 6.48 and 6.49).

Probably the most economically important myxosporean is *Myxosoma cerebralis*. This protozoan infects the nervous system and the auditory organ of such salmonid fish as young trout (*Salmo* spp.), the Coho salmon (*Oncorhynchus kisutch*), chinook salmon (*O. tshawytscha*), and the lake trout (*Salvelinus namaycush*). Infected fish lose their sense of balance and tumble

irradically. For this reason the virulent disease is referred to as the whirling or tumbling disease. In addition, infected fish portray a blackening of the tail up to the anus, and as the disease progresses, skeletal deformities are formed as a result of the penetration of *M. cerebralis* into the cartilage of young fish, thus preventing normal ossification. Such deformities remain even if the fish recovers from the infection although infected fish usually die. It is noted, however, that when *M. cerebralis* infects its original host, the European brown trout (*Salmo trutta*), the symptoms of the whirling disease are not manifested.

M. cerebralis produces spores, each measuring 7–9 μm in diameter and 5 μm thick. They have a protruding rim, and the polar capsules are 4 μm long. These spores are noninfective to other fish, and how the whirling disease is spread has puzzled fish pathologists for years. Recently, Wolf and Markin (1984) have demonstrated that the spores of *M. cerebralis*, when released into the environment, are infective to tubificid oligochaetes. Moreover, they have demonstrated that *Triactinomyxon gyrosalmo*, which had been considered a member of the class Actinosporea, is actually a stage in the life cycle of *Myxosoma cerebralis*. In the tubificid host, the parasite develops into sporocysts, each containing eight spores which are infective to fish. Susceptible fish become infected either by ingesting sporocysts containing spores or infected tubificids. In view of this, Wolf and Markin have suggested that the myxozoan classes Myxosporea and Actinosporea should be considered one class, the Myxosporea. Their suggestion has been adopted.

LIFE CYCLE OF MYXOSPOREANS

Life cycles of members of the various genera of Myxospora are all essentially of the same pattern. When spores are ingested by the host, the sporoplasms escape as amoebulae that burrow through the epithelial lining of the host's gut and migrate to certain visceral organs where they transform into multinucleated plasmodia. This is accomplished by an increase in size accompanied by repeated nuclear divisions.

Within a single plasmodium, groups of nuclei become surrounded by cytoplasm. Each one of these smaller multinucleate units is known as a **sporont**. The sporonts increase in size while their nuclei divide rapidly, resulting in 6 to 18 daughter nuclei within each sporont. If the sporont, depending on the species, develops into a single spore, it is termed a **monosporoblastic** sporont; if it forms two spores, it is known as a **pansporoblastic** or disporoblastic sporont. The development of sporonts into spores is initiated in the center of a plasmodium. The surrounding host tissue becomes hypertrophied, degenerates, and forms a cyst wall around the parasite. Such cysts are visible to the naked eye. If such cysts are located on the host's body surface, they rupture, thus permitting the

---

*Until recently, two classes of myxozoans, Myxosporea and Actinosporea, were recognized; however, the finding by Wolf and Markiw (1984) summarized here supports the recognition of one class, the Myxosporea, only.

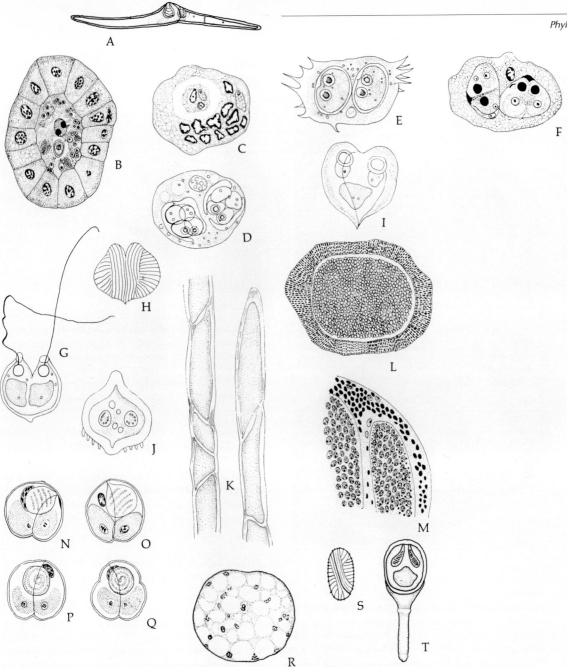

Fig. 6.48.   Some myxosporean parasites. A. *Ceratomyxa mesospora,* parasitic in gallbladder of the marine fish *Cestracion zygaena.* (Redrawn after Davis, 1917.) **B–H.** Stages in the life cycle of *Leptotheca ohlmacheri,* parasitic in nephridial uriniferous tubules of frogs and toads. **B.** Cross section of uriniferous tubule of *Rana pipiens* showing trophozoites and spores. **C.** Trophozoite with bud. **D–F.** Disporous trophozoites. **G.** Spore with extended polar filament. **H.** Surface view of spore. (Redrawn after Kudo, 1922.) **I.** Spore of *Myxoproteus cordiformis,* a parasite in urinary bladder of the marine fish *Chaetodipterus faber.* (Redrawn after Davis, 1917.) **J.** Spore of *Wardia ovinocua,* a parasite in ovary of the freshwater bass *Lepomis humilis,* 10–12 μm wide. (Redrawn after Kudo, 1921.) **K–Q.** Stages in the life cycle of *Unicapsula muscularis,* parasitic in muscles of marine fishes. **K.** Infected muscle fibers of host. **L.** Cross-section of fish infected muscle. **M.** Part of section of infected fish muscle. **N–Q.** Spores of *U. muscularis,* the causative agent of "wormy" halibut. (Redrawn after Davis, 1924.) **R.** Stained young trophozoite of *Myxidium serotinum,* 6.5 by 1.8 mm, with highly vacuolated cytoplasm, parasitic in gallbladder of frogs, toads, and salamanders. **S.** Unstained spore of *M. serotinum* showing characteristic ridges on the membrane, 16–18 by 9 μm. (Redrawn after Kudo, 1943.) **T.** Unstained spore of *Unicauda clavicauda,* parasitic in subdermal connective tissue of the minnow, *Nitropis blennis,* 10.5–11.5 by 8.5–9.5 by 6 μm. (Redrawn after Kudo, 1934.)

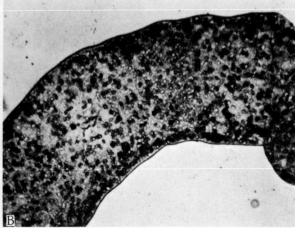

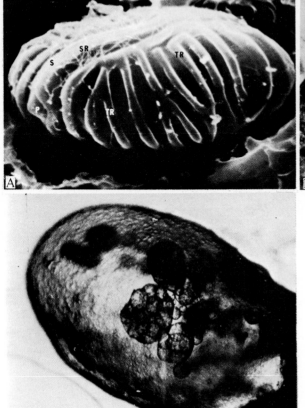

Fig. 6.49. **Stages of *Myxidium serotinum*. A.** Scanning electron micrograph of spore. **B.** Light micrograph of portion of trophozoite enclosing sporoblasts and spores. **C.** Light micrograph of young trophozoites *in situ* in gallbladder of the salamander *Eurycea bislineata*. P, filament eversion pore; S, suture; SR, sutural ridge; TR, traverse ridge. (All micrographs courtesy of J. G. Clark and F. J. Etges.)

fected often develop tumorlike masses (Fig. 6.50). If the internal tissues, especially muscles, are infected, the parasites are so intimately intermingled with the host's tissues that the food value of the fish is destroyed.

PHYSIOLOGY AND BIOCHEMISTRY OF APICOMPLEXA, MICROSPORA, ASCETOSPORA, AND MYXOZOA

OXYGEN REQUIREMENTS

All intraerythrocytic parasites, such as *Haemoproteus* spp. and *Plasmodium* spp., are true aerobes. This is not surprising because blood, especially arterial blood, is rich in oxygen that is incorporated in oxyhemoglobin within erythrocytes. Oxygen tension in the arterial blood of dogs, rabbits, and fish is between 70 and 100 mm Hg, whereas that in the venous blood of humans, horses, and ducks ranges between 37 and 40 mm Hg. The oxygen tension in the blood of these representative vertebrates indicates that in their normal habitats, the blood-dwelling protozoans are in continuous contact with appreciable amounts of oxygen.

Although the intraerythrocytic parasites are aerobic, high concentrations of oxygen are deleterious to

mature spores to escape into water. If the cyst is located internally, escape of the spores must await the death and disintegration of the host. What is being presented here is essentially what occurs in that phase of the life cycle of *Myxosoma cerebralis* in the fish host. Once taken into the invertebrate host, a tubificid annelid in the case of *M. cerebralis*, the spores develop into sporocysts, each containing eight spores (referred to as "actinosporeans" by Wolf and Markin, 1984) in the cells lining the gut of the annelid. Maturation of the eight spores takes 3 to 4 months in the case of *M. cerebralis*.

Some myxosporeans parasitize their host's tissues, living in organs such as the kidneys, muscles, gills, or integument. Others prefer various body cavities, such as the urinary bladder and gallbladder. Most injury caused by myxosporidans results from those species that parasitize tissues. Fish whose integument is in-their growth. For example, Anfinsen *et al.* (1946) have demonstrated that if the environment contains 95% O_2, *Plasmodium knowlesi* cannot be cultured. However, amounts of oxygen ranging from 0.37 to 20% are equally satisfactory for the *in vitro* maintenance of *P. knowlesi*.

The large amount of oxygen utilized by parasitized cells can be employed as an indication of oxygen consumption by intracellular parasites. For example, the oxygen consumption of normal monkey erythrocytes rises from 0.73 mM per hour when 5×10^{12} cells are present, to 51.0 mM per hour when *Plasmodium knowlesi* is present in the cells.

It should be noted, however, that among the intraerythrocytic species, some do not require oxygen during certain phases of their development. For example, during the formation of micro- and macrogametes in *Haemoproteus columbae*, no oxygen is necessary.

Studies on the respiratory rates of *Plasmodium lophurae* and *P. knowlesi* have revealed that respiration is almost completely inhibited by 0.001 molar cyanide, and approximately 64% of respiration is inhibited by carbon dioxide. As in other blood-dwelling protozoans, such as the trypanosomes, the presence of iron-porphyrin respiratory enzymes is indicated. Furthermore, there is some evidence to indicate the presence of flavoprotein systems.

Among the intestine-dwelling species, little oxygen is required. Again, this is not surprising for there is a minimal amount of oxygen in the lumen of the gut of vertebrates and invertebrates. For instance, in the small intestine of such large mammals as horses, cows, and dogs, O_2 tension ranges from 0 to 6 mm Hg.

Little is known about the oxygen requirements of those species that parasitize the lower vertebrates and invertebrates. Undoubtedly the intracellular parasites of lower vertebrates can utilize what oxygen is available within the host's cells.

Hydrogen ion concentration (pH) of the medium in which sporulated oocysts of *Eimeria* are found affects the oxygen uptake to some degree. For example, Smith and Herrick (1944) have demonstrated that when *E. tenella* oocysts are exposed to greatly variable conditions, the oxygen consumption of the oocysts remains unaltered within the remarkably wide pH range of 4.7 to 8.8. However, if the pH goes beyond this range, the rate of oxygen consumption changes.

GROWTH REQUIREMENTS

In order to obtain information about the growth requirements of a parasitic organism it is often necessary to culture the organism *in vitro*. Only after this has been successful can the nutritional requirements be studied. Attempts at culturing certain of these protozoans *in vitro*, especially the malaria-causing members of the genus *Plasmodium*, have met with some

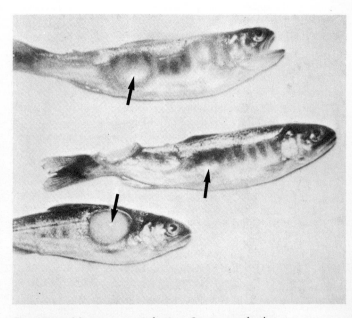

Fig. 6.50. Myxosporean infection. Specimens of coho salmon infected with the myxosporean *Ceratomyxa shasta*. The resulting boil-like disease is fatal. (After Conrad in Hoffman and Meyer, 1974.)

success (see Trager and Jensen, 1980, for review). This is another more applied reason for wanting to develop *in vitro* culture methods for *Plasmodium* spp. Specifically, if a vaccine is to be developed for malaria, then large quantities of the parasite will be required for the preparation of the vaccine. It is impractical to obtain the necessary quantities from infected laboratory animals; therefore, an efficient *in vitro* culture system is essential.

CULTURE OF EXOERYTHROCYTIC STAGES

Two basic types of techniques can be used to culture these stages: (1) placing infected cells or tissues in tissue culture flasks and culturing them by use of standard tissue culture techniques, and (2) maintaining suitable uninfected tissues in tissue culture and attempting to infect these with sporozoites. Of the two methods, considerable success has been attained with the first, but very little success with the second, mainly because of contamination or the failure of sporozoites to penetrate cells.

Species of *Plasmodium* that are normally parasites of birds have been used with considerable success in *in vitro* culture experiments involving tissue culture. In fact, the exoerythrocytic stages of avian plasmodia have been grown routinely in tissue culture for over

30 years. The review by Huff (1964) should be consulted for details of such studies.

It is now known that embryonic cells of birds will sustain plasmodia more efficiently than cells from adult birds (see review by Pipkin and Jensen, 1956). The most successful techniques involve suprachorioallantoic implantation of infected tissue from other embryos. Details of procedures for the long-term culture of avian exoerythrocytic stages have been provided by Beaudoin (1977).

Relative to mammalian plasmodia, Foley *et al.* (1978) successfully cultured rat liver cells infected with the exoerythrocytic stages of *Plasmodium berghei* for up to 44 hours and found the cultures infective when inoculated into recipient rodents.

Culture studies, among other features, have contributed an interesting finding. Specifically, it is now known that host-parasite specificity well known from *in vivo* observations apparently does not hold true *in vitro*. For example, Beaudoin *et al.* (1974) successfully cultivated the bird-infecting *Plasmodium fallax* and *P. lophurae* in embryonic mouse liver cells. Similarly, Strome *et al.* (1979) have successfully cultured the exoerythrocytic stages of *Plasmodium berghei*, a parasite of rodents, in embryonic turkey brain cells.

CULTURE OF ERYTHROCYTIC STAGES

This topic has been reviewed by Trager and Jensen (1980).

Although the erythrocytic stages of *Plasmodium* can be cultured either within their host cells or after they are removed from the cells, the first procedure is the more commonly employed. In either case, it is of critical importance to provide the following conditions: a physical environment with pH, oxygen tension, CO_2 tension, osmotic pressure, and other factors that are comparable to that of blood; suitable nutrients for growth and development; and provision for the

removal of metabolic by-products. By providing these conditions, Geiman *et al.* (1946) have been able to culture *Plasmodium*, using a composite medium that includes inorganic salts, amino acids, vitamins (water-soluble forms), purines, and pyrimidines.

Because erythrocytic stages derive most of their nutrient requirements from host cells, most *in vitro* cultures of these stages require the addition of whole erythrocytes. Trager (1950), however, successfully maintained *Plasmodium lophurae* for 2 to 3 days in media to which were added not whole blood cells but extracts of duck erythrocytes.

Studies on erythrocytic stages of plasmodia cultured *in vitro* have revealed several interesting features relative to nutrition. Trager (1943) has demonstrated the requirement of *P. lophurae* for calcium pantothenate and *P. knowlesi* for *p*-aminobenzoic acid (PABA). The discovery of the requirement for pantothenate led to the demonstration of the antimalarial effect of antimetabolites of this vitamin (Trager, 1971). These compounds, however, were not sufficiently effective to be of practical use. Also, Trager (1977) has shown the requirements for biotin, methionine, and purines by *Plasmodium knowlesi*.

Trager has also demonstrated that this organism requires coenzyme A for optimum growth. From these and similar experiments, it is now known that the addition of methionine, *p*-aminobenzoic acid, glucose, and coenzyme A is important for cultivation of the erythrocytic stages of *Plasmodium* spp. Withdrawal of ascorbic acid from the host's diet affects malarial parasites, but parasites cultured in media that lack ascorbic acid are not appreciably affected, suggesting that this acid acts through the host.

It may be concluded that different species of *Plasmodium* have somewhat different extracellular requirements for survival and pyruvate oxidation. Those of *P. lophurae* are compared with *P. gallinaceum* in Table 6.4.

Although most investigators have incorporated homologous plasma, i.e., plasma from a compatible host, to culture media, it is notable that the origin of the introduced plasma is significant. For example, Glenn and Manwell (1956) have tested the effects of both pigeon and turkey plasma added to cultures of *Plasmodium hexamerium*, a natural parasite in bluebirds and Maryland yellowthroats, under cultivation in the so-called Harvard medium (Ball *et al.*, 1945). It was found that turkey plasma enhanced growth and reproduction, but pigeon plasma did not.

Table 6.4. Comparison of the Extracellular Requirements for Survival and for Pyruvate Oxidation in *Plasmodium lophurae* and *P. gallinaceum*[a]

Plasmodium lophurae	*Plasmodium gallinaceum*
Pyruvate	
Diphosphopyridine nucleotide	Di- and triphosphopyridine nucleotide
Adenosine triphosphate	Adenosine triphosphate
Malate	Malate (or other C_4 dicarboxylic acid)
Coenzyme A	
Leucovorin	
Red blood cell extract	Diphosphothiamin
Gelatin	Manganous chloride

[a] Data compiled by Moulder (1962).

CARBOHYDRATE METABOLISM

Although *Plasmodium* can oxidize various carbohydrates, including fructose, mannose, and glucose, only glucose can satisfy the long-term requirements of growth and reproduction. Furthermore, comparatively little glucose is assimiliated into new parasite proto- plasm. Most of it is oxidized to produce energy. In *Plasmodium gallinaceum*, glucose is quantitatively con- verted to lactic acid under anaerobic conditions. En- zymes of the typical Embden-Meyerhof glycolytic pathway (p. 52) are present, indicating the presence of the phosphorylative glycolytic pathway. The oxi- dative process during carbohydrate metabolism un- doubtedly involves the Krebs or tricarboxylic acid cycle (p. 55).

The malarial organisms are extremely wasteful in their energy-producing mechanisms because they glycolyze much more carbohydrate than they oxidize, resulting in an accumulation of lactic acid, which, if not removed or neutralized in culture media, impairs respiratory activity. It has been estimated that only about 10% of the total required energy is supplied by the anaerobic glycolytic process, whereas the remain- ing 90% comes from the enzymatic transfer to oxy- gen of the electrons liberated during the tricarboxylic acid cycle.

No energy stores have been definitely identified in any of the stages of *Plasmodium* spp. in vertebrates. There is some indication that lipid is stored in the intraerythrocytic stages of *Plasmodium lophurae*. In the case of *Eimeria* spp., mature trophozoites and macro- gametocytes store a polysaccharide very similar to amylopectin, although this molecule occurs in extreme- ly small amounts or not at all in microgametocytes.

Studies designed to demonstrate how simple sugars enter *Plasmodium* have revealed it to be a mediated process. In *P. lophurae* there are two distinct loci with different sugar specificities (Gutteridge and Coombs, 1977).

A major problem an intracellular parasite encoun- ters is obtaining nutrients across two cell membranes, its own and that of the host cell. *Plasmodium* has over- come this obstacle by inducing the host cell membrane partially to lose the ability to regulate the passage of molecules across it. Consequently, the host cell mem- brane for the most part becomes freely permeable to many complex molecules, and therefore these mole- cules can be made available to the intracellular para- site with little or no expenditure of energy.

All *Plasmodium* spp. that have been studied have been found to produce lactate (Table 6.5). No primate-

infecting *Plasmodium* has been shown to include a functional Krebs cycle in most of its life cycle stages, and the mitochondria are without cristae. In these parasites the utilization of oxygen involves a cyanide- sensitive cytochrome oxidase. There is some evidence, however, which suggests the schizont stages utilize oxygen in connection with an active Krebs cycle system.

Plasmodia that naturally infect primates can fix CO_2, a process which probably involves the produc- tion of Krebs cycle intermediates necessary for certain biosynthetic reactions. All *Plasmodium* spp. include a partial phosphogluconate pathway for the production of pentoses. The energy required by the parasite must be satisfied mainly by glycolysis, although additional ATP may be produced in the further metabolism of some intermediates. NADH is reoxidized primarily through the activity of lactate dehydrogenase, but if there is a functional cytochrome system, some NADH may be reoxidized by it. Certainly in those plasmodia which infect rodents a functional cytochrome system occurs in the asexual blood stages. This system in- volves cytochromes and ubiquinone components that are structurally different from those of mammalian cells.

Interestingly, plasmodia that infect birds include mitochondria with well-developed cristae. Further- more, an active Krebs cycle pathway occurs, and me- tabolism of glucose beyond pyruvate exists. The Krebs cycle pathway is utilized primarily to produce amino acids; little CO_2 is evolved. Furthermore, a cyto- chrome system and cytochrome oxidase occur, but their functions in energy production remain unclear. For a review of carbohydrate metabolism in *Plasmo- dium* spp., see Homewood and Neame (1980).

PROTEIN METABOLISM

Although a great deal remains to be learned about protein metabolism in malarial parasites, knowledge in

Table 6.5. End Products of Glucose Catabolism in Selected Malaria-Causing Parasites

	Plasmodium knowlesi (Primate Malaria)	*Plasmodium chabaudi* (Rodent Malaria)	*Plasmodium lophurae* (Avian Malaria)
Percentage (approximate) of glucose carbon excreted as lactate	90	80	40
Other products	Acetate Formate	Unknown	TCA cycle intermediates

this area is advancing rapidly. It is now known that the erythrocytic forms of *Plasmodium* can engulf the cytoplasm of the host erythrocyte, which contains hemoglobin (Fig. 4.14). Malarial organisms contain enzymes that can split hemoglobin into a nonprotein fraction and globin. Also, the protease activity of the erythrocytic stages of *Plasmodium* spp. has been demonstrated to differ from that of the host cell enzymes in optimum pH and sensitivity to inhibitors. The main enzyme in these parasites is an acid protease similar to mammalian cathepsin D. Since the parasites are dependent on these enzymes to break down the host's hemoglobin, it has been suggested that protease inhibitors may be useful in inhibiting the growth by selectively acting on a process that is essential to them but not, at least over a short period, to the host cells.

It has been estimated that approximately 76% of the hemoglobin in an infected erythrocyte is destroyed while harboring a parasite. The nonprotein fraction of the split hemoglobin molecule is hematin, which is not utilized by the parasite but is released as the characteristic malarial pigment, hemozoin, when the parasite ruptures out of the host cell.

Although most of the *in vivo* protein and amino acid requirements of *Plasmodium* are satisfied from host cell proteins, the methionine requirement cannot be totally acquired from this source because very little of this amino acid is present in host cells. As the result, the parasite derives its methionine requirement from the surrounding plasma, indicating that although the protozoan appears to be an intracellular parasite of individual erythrocytes, it is physiologically parasitic on the host cell, the plasma, and other cells as well.

Once taken into the body of the parasite, the bulk of the polypeptides and amino acids are utilized for synthesis and only very small amounts are utilized in oxidative energy- producing processes.

It needs to be noted that *Plasmodium knowlesi*, and probably other species, can synthesize alanine, aspartate, and glutamate from carbohydrates, cysteine from methionine, and methionine from serine and homocysteine. This information indicates that the malaria-causing parasites are not totally dependent on their hosts for all amino acids.

Protein synthesis by cell-free extracts of erythrocytic stages of malaria parasites is inhibited by cycloheximide but not by streptomycin and chloramphenicol. This indicates that the biosynthetic process in *Plasmodium* is probably similar to that in mammalian cells. *Plasmodium knowlesi* includes 80S ribosomes, which are readily dissociated into 60S and 40S subunits, and thus are similar to those of mammalian cells. On the other hand, the rRNA of *Plasmodium* has a

guanine + cytosine content of 37%. This is similar to that of certain free-living protozoans, but quite distinct from the percentage reported for the rRNA of such mammalian cells as reticulocytes.

For a review of the mechanisms involved in the breakdown of host hemoglobin and amino acid metabolism in malarial parasites, see Homewood and Neame (1980).

LIPID METABOLISM

Relatively little is known about lipid metabolism in these protozoans. It is known that the total lipid content in red blood cells infected with *Plasmodium knowlesi* is increased more than 400% over normal. Some 25% of the lipid, mainly cholesterol, is nonsaponifiable. The fatty acid content of infected cells is approximately four to five times that of uninfected cells. Among the fatty acids present in infected cells, there is an unsaturated 18-carbon monocarboxylic fatty acid that possesses lytic properties. It has been suggested that this fatty acid may, in some way, be associated with the rupture of the red blood cell wall when merozoites escape.

In addition to cholesterol, neutral fats, and phospholipids occur in *P. knowlesi*.

Acylglycerols are fatty acid esters of glycerol. The most common acylglycerols contain three fatty acids and are known as triacylglycerols; others contain two fatty acids and are known as diacylglycerols. Erythrocytes infected with *P. knowlesi* contain triacylglycerols, diacylglycerols, and free fatty acids at concentrations greater than those in normal cells. The significance of this phenomenon is still uncertain. These lipids may represent excretions by the parasites or may represent breakdown products of the host cell resulting from parasite activity.

Phosphoglycerides are lipids characteristically present in cell membranes. Erythrocytic stages of *P. knowlesi* contain such phosphoglycerides as phosphatidylcholine, phosphatidylethanolamine, and phosphatidylinositol but no phosphatidylserine. The membranes of the parasite are slightly richer in phosphoglycerides than those of the host erythrocyte. The parasite is capable of hydrolyzing the host cell phosphoglycerides and thus bringing about the greater permeability of the host cell membrane mentioned earlier (p. 223). Also, it results in the greater fragility of infected erythrocytes. Furthermore, it may be one cause of the hemolysis of red cells of malaria victims since the hydrolytic products (lysophosphatidylethanolamine and lysophosphatidylcholine) have hemolytic properties that could damage uninfected erythrocytes.

Malarial parasites utilize exogenous fats. For example, *P. knowlesi* develop in a medium in which the only lipids are stearic acid and cholesterol. How lipids are taken up by these parasites remains unresolved.

Presumably, once taken up, they are hydrolyzed by lipases. Phospholipase A occurs in *P. lophurae*. For a comprehensive review of lipids in protozoans, see Dewey (1967), and for a discussion of lipids in malarial parasites, see Homewood and Neame (1980).

HOST-PARASITE RELATIONSHIPS

EFFECTS OF PARASITE ON HOST

Aside from the obvious clinical symptoms of malaria and other diseases caused by the parasites mentioned in this chapter, an array of other parasite-induced effects are known.

ANEMIAS

In infections with *Babesia* spp. and *Plasmodium* spp., anemias are known to occur. Numerous hypotheses have been postulated to explain the cause of these anemias, especially those resulting from *Plasmodium* infections. (1) It has been suggested that loss of erythrocytes is greater than the rate of replacement during the periods when merozoites escape, hence development of the anemic condition. (2) Parasitized erythrocytes, as stated, are more fragile and therefore are easily ruptured and destroyed. (3) Since the spleen of a malaria victim becomes enlarged, it has been postulated that a lytic substance, lysolecithin, which destroys erythrocytes, is released by the spleen. (4) Parasitized erythrocytes are believed to serve as autoantigenic bodies that bring on production of a specific antibody, hemolysin, in the spleen. Once in the blood, hemolysin destroys red blood cells. A combination of these major hypotheses, all of which are based on experimental evidence, may explain anemia. In humans infected with *Plasmodium falciparum*, the erythrocyte count may drop from 4,600,000–6,200,000/mm³ to 440,000/mm³, and in humans infected with *P. vivax* the count may drop to 560,000/mm³. For a review of erythrocyte destruction mechanisms in malaria, see Seed and Kreier (1980).

ALTERED PROTEIN AND CARBOHYDRATE COMPOSITION OF HOSTS

The carbohydrate and protein compositions of hosts infected with certain protozoans discussed in this chapter are affected. For example, in humans infected with *Plasmodium malariae*, *P. vivax*, and *P. falciparum*, the total serum protein is decreased. An analysis of the protein fractions has revealed that the albumin component is decreased but the globulins are increased. This is not surprising since the globulin fractions, especially gamma globulin, include the antibodies that would be expected to increase upon the introduction of a parasite into the host's blood and tissues. It is noted that Trager (1947) has demonstrated that if *Plasmodium lophurae* within chick erythrocytes is cultured in a medium containing human plasma fractions

rich in albumin and gamma globulin, there is no effect on growth. However, if the medium contains a fraction with alpha globulin and rich in lipoproteins, inhibition of growth occurs. This finding is interesting, as Lambrecht *et al.* (1978) have shown that in individuals infected with *P. vivax*, large molecules of lipoprotein disappear from the plasma.

To the contrary, Ghosh and Sinton (1935) have reported that the blood protein level remains normal in monkeys infected with *P. inui*. This parasite is not pathogenic and has little or no demonstrable effect on its host. That the serum protein is unaltered suggests that the virulence of the parasite is directly correlated with the degree of change in the protein composition of the host's serum.

Waxler (1941), Pratt (1940, 1941), and others have reported that in chickens infected with *Eimeria tenella*, hyperglycemia develops, and there is a decrease in the amount of stored glycogen. This instance serves to demonstrate disruptions in carbohydrate metabolism resulting from protozoan infections. Furthermore, von Brand and Mercado (1956) have reported that in rats infected with *Plasmodium berghei*, less glycogenesis occurs in the liver. Histologic studies revealed that this is due to the apparent nonfunction of certain liver cells. For a review of the known pathologic effects of malaria, see Aikawa *et al.* (1980).

EFFECT ON ENDOCRINE GLANDS

The presence of malarial organisms can affect certain host endocrine glands. For example, in chickens infected with *P. gallinaceum* an increase in the number of adrenal cells occurs. This condition reaches a maximum level at approximately 1 day after the number of parasites reaches its highest level.

EFFECT OF PARASITE SECRETIONS

As in all instances of parasitic infections, parasite secretions may affect the host. This aspect of the host-parasite relationship remains to be completely explored. In one instance, that of *Toxoplasma* infections, the secretion of a toxin is suspected. Weinman and Klatchko (1950) have reported that when secretions of *Toxoplasma* in the peritoneal fluid of infected mice are injected into the veins of parasitized and nonparasitized mice, these animals died immediately. This toxic material is reported to be totally or partially composed of proteins. Confirmation of the existence of this toxin has still to be made.

WEAKENING EFFECTS

Hosts of certain coccidians demonstrate a general weakened condition. For example, in addition to being frequently lethal, *Eimeria tenella,* when present in small numbers, is known to weaken chickens. The gastrocnemius muscle of infected chickens, for example, can perform only 58% of the work done by those of uninfected animals, and muscle fatigue among infected birds sets in before it does in healthy ones (Levine and Herrick, 1954).

CASTRATION

Protozoan parasites have been incriminated in several instances in the castration of their hosts. One such case has been reported by Smith (1905), who found destruction of the testes of the crustacean *Inachus dorsettensis* by *Aggregata eberthi* and feminization of the host's external characteristics.

EFFECTS OF HOST ON PARASITE

SEX HORMONES AND SUSCEPTIBILITY

Evidence indicates that the sex hormones of the host play an important role in governing its susceptibility to some of the protozoans mentioned in this chapter, especially the blood parasites. Chernin (1950) has reported that the degree of parasitemia in ducks infected with *Leucocytozoon simondi* diminishes when female birds commence egg laying. This phenomenon is correlated with female sex hormones. Also, young female chickens (not sexually mature) are more susceptible to *Plasmodium gallinaceum* than males and mature females. In sexually mature, egg-laying ducks there appears to be a higher degree of resistance to *P. lophurae* than in males. Trager and McGhee (1950) have indicated that some substance present in the blood is at least partially responsible for this difference.

AGE RESISTANCE

What has commonly been termed "age resistance" to malaria, especially among avian hosts, may well be partly due to hormonal influences.

EFFECT OF HOST DIET

There is some indication that deficient host diets may result in more severe primary attacks by *Plasmodium* spp. In addition, there is a greater tendency on the part of the host to relapse and become more susceptible to superinfections.

Vitamins are known to play a role in certain coccidian infections. Becker and Dilworth (1941), for example, have reported that in *Eimeria nieschulzi* infections in rats, when vitamin B_1 is present in the host's diet, there is a depressing effect on the development of the parasite as measured by the production of oocysts. On the other hand, vitamin B_6 has a stimulating effect. If both vitamins are given together, there is a strong inhibiting effect.

SICKLE-CELL ANEMIA AND MALARIA

One of the more fascinating discoveries relative to human malaria is that individuals who are heterozygous for the sickle-cell gene are more resistant to the malaria (malignant tertian malaria) caused by *Plasmodium falciparum*. Specifically, individuals are normal as far as hemoglobin synthesis is concerned if they have the genetic constitution of $Hb_1{}^A Hb_1{}^A$ but suffer, often fatally, from sickle-cell anemia if they carry the alleles $Hb_1{}^S Hb_1{}^S$. On the other hand, heterozygous individuals, that is, persons with the genotype $Hb_1{}^A Hb_1{}^S$, are carriers of the sickle-cell trait but are perfectly healthy. Pauling *et al.* (1949) have shown that $Hb_1{}^A Hb_1{}^A$ individuals have normal hemoglobin or hemoglobin A, whereas $Hb_1{}^S Hb_1{}^S$ individuals have hemoglobin S. Hemoglobin A differs from hemoglobin S in that it has glutamic acid residues on each of the identical halves of the hemoglobin molecule. In the S form, one of the negatively charged glutamic acid residues of A is replaced by a valine residue, which is neutral. As a result, hemoglobin S is 50 times less soluble than hemoglobin A, and the former is extremely more viscous than the latter.

In heterozygous $Hb_1{}^A Hb_1{}^S$ individuals both types of hemoglobin occur, the S type constituting from less than one-quarter to nearly one-half of the mixture. According to Allison (1954, 1957), young children between 6 weeks and 6 years old who are heterozygous, and therefore carriers of the sickle-cell trait, are just as susceptible to infection by *Plasmodium falciparum* as normal children, but show lower parasite counts and lower mortality. Thus, the sickle-cell gene, when present in the heterozygous condition, appears to have selective advantage in populations inhabiting areas endemic to this type of malaria. This hypothesis is supported by the fact that the $Hb_1{}^A Hb_1{}^S$ condition is quite prevalent (10–20%) in some areas, such as East Africa, where *P. falciparum* abounds.

Although the cause-and-effect relationship between the lower parasite counts and mortalities in children heterozygous for the sickle-cell trait has not been completely determined, several reasonable postulations have been advanced. Allison (1957) is of the opinion that the trait carriers have sufficient hemoglobin S to limit the multiplication of the parasites, hence, the lower counts. Raper (1959) has found that there is a greater number of gametocyte production in trait carriers, a process that would tend to limit the infection. In addition, Trager (1960) has suggested that merozoites of *P. falciparum* may avoid host erythrocytes enclosing hemoglobin S and Geiman (1964) has suggested that the parasites can invade red blood cells enclosing hemoglobin S but are unable to digest it after phagotrophy. Also, Moulder (1962) has suggested that the lower solubility of hemoglobin S as well as its greater viscosity may hinder the phagotrophy of host material by the intracellular parasites. Finally, Miller *et al.* (1956) have proposed the hypothesis that since parasitized host cells tend to adhere to the capillary walls in peripheral tissues, the loss of oxygen due to parasite consumption and normal loss to surrounding tissues could cause sufficient hypoxia to result in sickling. Such cells would be phagocytosed by leucocytes and the enclosed parasites consequently destroyed.

Friedman (1979) has reported that *Plasmodium falciparum* in sickled cells are killed. Electron microscope studies have revealed that the parasites are disrupted by intrusions of needlelike deoxy-sickle cell hemoglobin aggregates. This is followed by disintegration of cytoplasm and membranes. When present in heterozygous cells, the parasites are not disrupted although extensive vacuolization, a sign of metabolic inhibition, occurs. Friedman has concluded that the resistance of sickle cell gene carriers to malaria is partially due to intracellular destruction of the parasites. It is obvious that the fascinating relationship between *P. falciparum* and the sickle-cell trait requires further study. For a detailed review of the relationship between malaria and the sickle-cell trait, see Honigberg (1967).

PERIODICITY

The synchronized time intervals at which merozoites of specific species of *Plasmodium* escape from host cells represent an interesting phase of host-parasite relationship. Stauber (1939) has shown that the synchronous periodicity of reproduction of three strains of *Plasmodium cathemerium* and one strain of *P. relictum* in birds is affected by high and low temperatures and the amount of light to which the hosts are exposed. Environmental and climatic temperatures are also known to play important roles.

REFERENCES

Aikawa, M., Susuki, M., and Gutierrez, Y. (1980). Pathology of malaria. *In* "Malaria" (J. P. Kreier, ed.), Vol. 2, pp. 47– 102. Academic Press, New York.

Allison, A. C. (1954). Protection afforded by the sickle-cell trait against subtertian malarial infection. *Br. Med. J.* **1**, 290–294.

Allison, A. C. (1957). Malaria in carriers of the sickle-cell trait and in newborn children. *Exp. Parasitol.* **6**, 418–447.

Anfinsen, C. B., Geiman, Q. M., McKee, R. W., Ormsbee, R. A., and Ball, E. G. (1946). Studies on malarial parasites. VIII. Factors affecting the growth of *Plasmodium knowlesi in vitro. J. Exp. Med.* **84**, 607–621.

Ayala, S. C. (1977). Plasmodia of reptiles. *In* "Parasitic Protozoa" (J. P. Kreier, ed.), Vol. III. pp. 267–309. Academic Press, New York.

Bailey, L. (1959). The natural mechanism of suppression of *Nosema apis* Zander in enzootically infected colonies of the honey bee, *Apis mellifera* Linnaeus. *J. Insect Pathol.* **1**, 347–350.

Ball, E. G., *et al.* (1945). *In vitro* growth and multiplication of the malaria parasite, *Plasmodium knowlesi. Science* **101**, 542–544.

Ball, G. H., and Chao, J. (1963). Contributions of *in vitro* culture towards understanding the relationships between avian malaria and the invertebrate hosts. *Ann. N.Y. Acad. Sci.* **113**, 322–331.

Ball, G. H., and Chao, J. (1964). Temperature stresses on the mosquito phase of *Plasmodium relictum. J. Parasitol.* **50**, 748–752.

Barnett, S. F. (1968). Theileriasis. *In* "Infectious Blood Diseases of Man and Animals" (D. Weinman and M. Ristic, eds.), Vol. 2, pp. 269–328. Academic Press, New York.

Barnett, S. F. (1977). Theileria. *In* "Parasitic Protozoa" (J. P. Kreier, ed.), Vol. IV, pp. 77–113. Academic Press, New York.

Beaudoin, R. L. (1977). Should cultivated exoerythrocytic parasites be considered as a source of antigen for a malaria vaccine? *Bull. W.H.O.* **55**, 373–376.

Beaudoin, R. L., Strome, C. P. A., and Clutter, W. G. (1974). Cultivation of avian malaria parasites in mammalian liver cells. *Exp. Parasitol.* **36**, 355–359.

Beaver, P. C., Gadgil, R. K., and Morera, P. (1979). *Sarcocystis* in man: a review and report of five cases. *Am. J. Trop. Med. Hyg.* **28**, 819–844.

Becker, E. R., and Dilworth, R. I. (1941). Nature of *Eimeria nieschulzi* growth-promoting potency of feeding stuffs. II. Vitamins B_1 and B_6. *J. Infect. Dis.* **68**, 285–290.

Beverly, J. K. A., Watson, W. A., and Spence, J. B. (1971). The pathology of the fetus in ovine abortion due to toxoplasmosis. *Vet. Rec.* **88**, 174–178.

Brackett, R. G. (1980). Manufacturing aspects of antiplasmodial vaccine production. *In* "Malaria" (J. P. Kreier, ed.), Vol. 3, pp. 325–329. Academic Press, New York.

Brandberg, L. L., Goldberg, S. B., and Breidenbach, W. C. (1970). Human coccidiosis—a possible cause of malabsorption. The life cycle in small-bowel mucosal biopsies as a diagnostic feature. *N. Engl. J. Med.* **283**, 1306–1313.

Brocklesby, D. W., and Hawking, F. (1958). Growth of *Theileria annulata* and *T. parva* in tissue culture. *Trans. Roy. Soc. Trop. Med. Hyg.* **52**, 414–420.

Bulla, L. A. Jr., and Cheng, T. C. (eds.) (1976). "Comparative Pathobiology Vol. 1. Biology of the Microsporidia." Plenum, New York.

Bulla, L. A. Jr., and Cheng, T. C. (eds.) (1977). "Comparative Pathobiology Vol. 2. Systematics of the Microsporidia." Plenum, New York.

Burns, W. C., and Challey, J. R. (1959). Resistance of birds to challenge with *Eimeria tenella. Exp. Parasitol.* **8**, 515–526.

Butcher, G. A., Mitchell, G. H., and Cohen, S. (1973). Mechanism of host specificity in malarial infection. *Nature* **244**, 40–41.

Cali, A. (1970). Morphogenesis in the genus *Nosema. Proc.*

Fourth Int. Colloq. Insect Pathol., College Park, Maryland, pp. 431–438.

Canning, E. U. (1975). "The Microsporidian Parasites of Platyhelminthes: Their Morphology, Development, Transmission and Pathogenicity." Commonwealth. Inst. Helminthol. Misc. Publ. No. 2. Commonwealth Agric. Bur., Farnham Royal, Bucks, England.

Canning, E. U. (1977). Microsporida. *In* "Parasitic Protozoa" (J. P. Kreier, ed.), Vol. IV, pp. 155–196. Academic Press, New York.

Carter, R., and Diggs, C. L. (1977). Plasmodia of rodents. *In* "Parasitic Protozoa" (J. P. Kreier, ed.), Vol. III, pp. 359–465. Academic Press, New York.

Cheng, T. C. (1967). Marine molluscs as hosts for symbioses: with a review of known parasites of commercially important species. *Adv. Marine Biol.* **5**, 1–424.

Chernin, E. (1950). The relapse phenomenon in *Leucocytozoon simondi* infections in domestic ducks. *J. Parasitol.* **36**, 22–23.

Coatney, G. R., Collins, W. E., Warren, M., and Contacos, P. F. (1971). "The Primate Malarias." U.S. Dept. Health, Education, and Welfare, U.S. Govt. Print. Off., Washington, D. C.

Cochrane, A. H., Nussenzweig, R. S., and Nardin, E. H. (1980). Immunization against sporozoites. *In* "Malaria" (J. P. Kreier, ed.), Vol. 3, pp. 163–202. Academic Press, New York.

Collins, W. E., and Aikawa, M. (1977). Plasmodia of nonhuman primates. *In* "Parasitic Protozoa" (J. P. Kreier, ed.). Vol. III, pp. 467–492. Academic Press, New York.

Cort, W. W., Hussey, K. L., and Ameel, D. J. (1960a). Studies on a microsporidian hyperparasite of strigeoid trematodes. I. Prevalence and effect on the parasitized larval trematodes. *J. Parasitol.* **46**, 317–325.

Cort, W. W., Hussey, K. L., and Ameel, D. J. (1960b). Studies on a microsporidian hyperparasite of strigeoid trematodes. II. Experimental transmission. *J. Parasitol.* **46**, 327–336.

Dewey, V. C. (1967). Lipid composition, nutrition, and metabolism. *In* "Chemical Zoology, Vol. 1. Protozoa" (M. Florkin, B. T. Scheer, and G. W. Kidder, eds.), pp. 161–274. Academic Press, New York.

Dollfus, R. P. (1946). Parasites (animaux et végétaux) des helminthes. Hyperparasites, ennemis et prédateurs des helminthes parasites et des helminthes libres. Encycl. Biol. 37. Lechevalier, Paris.

Dubey, J. P. (1977). *Toxoplasma, Hammondia, Besnoitia, Sarcocystis,* and other tissue cyst-forming coccidia of man and animals. *In* "Parasitic Protozoa" (J. P. Kreier, ed.), Vol. IV, pp. 101–237. Academic Press, New York.

Dubey, J. P., and Frenkel, J. K. (1972). Extra-intestinal stages of *Isospora felis* and *I. rivolta* (Protozoa: Eimeriidae) in cats. *J. Protozool.* **19**, 89–92.

Fallis, A. M., and Desser, S. S. (1977). On species of *Leucocytozoon, Haemoproteus,* and *Hepatocystis. In* "Parasitic Protozoa" (J. P. Kreier, ed.), Vol. III, pp. 239–266. Academic Press, New York.

Fallis, A. M., Desser, S. S., and Khan, R. A. (1974). On species of *Leucocytozoon. Adv. Parasit.* **12**, 1–67.

Fallis, A. M., and Wood, D. M. (1957). Biting midges (Diptera: Ceratopogonidae) as intermediate hosts for *Haemoproteus* of ducks. *Can. J. Zool.* **35**, 425–435.

Fayer, R. (1972). Gametogony of *Sarcocystis* sp. in cell culture. *Science* **175**, 65–67.

Feng, S. Y. (1958). Observations on distribution and elimination of spores of *Nematopsis ostrearum* in oysters. *Proc. Nat. Shellfish. Assoc.* **48**, 162–173.

Foley, D. A., Kennard, J., and Vanderberg, J. P. (1978). *Plasmodium berghei*: Infective exoerythrocytic schizonts in primary monolayer cultures of rat liver cells. *Exp. Parasitol.* **46**, 179–188.

Ford, S. E., and Haskin, H. H. (1982). History and epizootiology of *Haplosporidium nelsoni* (MSX), an oyster pathogen in Delaware Bay, 1957–1980. *J. Invert. Pathol.* **40**, 118–141.

Frenkel, J. K. (1973). Toxoplasmosis: parasite life cycle, pathology and immunology. *In* "The Coccidia" (D. M. Hammond and P. L. Long, eds.), pp. 343–410. University Park Press, Baltimore, Maryland.

Frenkel, J. K., and Dubey, J. P. (1972). Rodents as vectors for feline coccidia, *Isospora felis* and *Isospora rivolta. J. Infect. Dis.* **125**, 69–72.

Friedman, M. J. (1979). Ultrastructural damage to the malaria parasite in the sickled cell. *J. Protozool.* **26**, 195–199.

Friedman, M. J., Blankenberg, T., Sensabaugh, G., and Tenforde, T. S. (1984). Recognition and invasion of human erythrocytes by malarial parasites: Contribution of sialoglycoproteins to attachment and host specificity. *J. Cell Biol.* **98**, 1672–1677.

Friedman, M. J., Fukuda, M., and Laine, R. A. (1985). Evidence for a malarial parasite interaction site on the major transmembrane protein of the human erythrocyte. *Science* **228**, 75–77.

Garnham, P. C. C., Krotoski, W. A., Bray, R. S., Killick-Kendrick, R., and Cogswell, F. (1983). Relapse in primate malaria: a reply. *Trans. Roy. Soc. Trop. Med. Hyg.* **77**, 736–738.

Geiman, Q. M. (1964). Comparative physiology: mutualism, symbiosis, and parasitism. *Ann. Rev. Physiol.* **26**, 75–108.

Geiman, Q. M., Anfinsen, C. B., McKee, R. W., Ormsbee, R. A., and Ball, E. G. (1946). Studies on malarial parasites. VII. Methods and techniques for cultivation. *J. Exp. Med.* **84**, 538–606.

Ghosh, B. N., and Sinton, J. A. (1935). Quantitative changes in the proteins of the blood sera of monkeys infected with malarial plasmodia. *Rec. Malaria Surv. India* **5**, 173–202.

Glenn, S., and Manwell, R. D. (1956). Further studies on the cultivation of the avian malaria parasites. II. The effects of heterologous sera and added metabolities on growth and reproduction *in vitro. Exp. Parasitol.* **5**, 22–33.

Grell, K. G. (1973). "Protozoology." Springer-Verlag, New York.

Hammond, D. M., and Long, P. L. (eds.) (1973). "The Coccidia." University Park Press, Baltimore.

Harrison, G. (1978). "Mosquitoes, Malaria & Man: A History of the Hostilities Since 1880." E. P. Dutton, New York.

Harry, O. G. (1967). The effect of a eugregarine *Gregarina polymorpha* (Hammerschmidt) on the mealworm larva of *Tenebrio molitor* (L.) *J. Protozool.* **14**, 539–547.

Haskin, H. H., and Ford, S. E. (1982). *Haplosporidium nelsoni* (MSX) on Delaware Bay seed oyster beds: a host-parasite relationship along a salinity gradient. *J. Invert. Pathol.* **40**, 388–405.

Haskin, H. H., Stauber, L. A., and Mackin, J. A. (1966). *Minchinia nelsoni* n. sp. (Haplosporida, Haplosporidiidae): causative agent of the Delaware Bay oyster epizootic. *Science* **153**, 1414–1416.

Hassanein, M. H. (1951). Studies on the effect of infection with *Nosema apis* on the physiology of the queen honeybee. *Quart. J. Microsc. Sci.* **92**, 225.

Hazard, E. I., Andreadis, T. G., Joslyn, D. J., and Ellis, E. A. (1979). Meiosis and its implications in the life cycles of *Amblyospora* and *Parathelohania* (Microspora). *J. Parasitol.* **65**, 117–122.

Holbrook, T. W. (1980). Immunization against exoerythrocytic stages of malaria parasites. *In* "Malaria" (J. P. Kreier, ed.), Vol. 3, pp. 203–230. Academic Press, New York.

Homewood, C. A. and Neame, K. D. (1980). Biochemistry of malarial parasites. *In* "Malaria" (J. P. Kreier, ed.), Vol. 1, pp. 346–405. Academic Press, New York.

Honigberg, B. M. (1967). Chemistry of parasitism among some protozoa. *In* "Chemical Zoology Vol. 1. Protozoa" (M. Florkin, B. T. Scheer, and G. W. Kidder, eds.), pp. 695–814. Academic Press, New York.

Huff, C. G. (1964). Cultivation of the exoerythrocytic stages of malarial parasites. *Am. J. Trop. Med. Hyg.* **13**, 171–177.

Jacobs, L. (1967). *Toxoplasma* and toxoplasmosis. *Adv. Parasitol.* **5**, 1–45.

Janiszewska, J. (1956). [Actinomyxidia II. Systematization, sexual cycle, description of new genera and species.] *Wiad. Parazytol.* **2**, 251–252.

Joyner, L. P., and Donnelly, J. (1979). The epidemiology of babesial infections. *Adv. Parasit.* **17**, 115–140.

Kreier, J. P. (1980). "Malaria," Vols. 1, 2, 3. Academic Press, New York.

Krick, J. A., and Remington, J. S. (1978). Toxoplasmosis in the adult. *N. Engl. J. Med.* **298**, 550–553.

Krinsky, W. L., and Hayes, S. F. (1978). Fine structure of the sporogonic stages of *Nosema parkeri. J. Protozool.* **25**, 177–186.

Krotoski, W. A., Collins, W. E., Bray, R. S., Garnham, P. C. C., Cogswell, F. B., Gwadz, F. B., Killick-Kendrick, R., Wolf, R., Sinden, R., Koontz, L. C., and Stanfill, P. S. (1982). Demonstration of hypnozoites in sporozoite-transmitted *Plasmodium vivax* infection, *Am. J. Trop. Med. Hyg.* **31**, 1291–1293.

Ladda, R. L., Aikawa, M., and Sprinz, H. (1969). Penetration of erythrocytes by merozoites of mammalian and avian malarial parasites. *J. Parasitol.* **55**, 633–644.

Lambrecht, A. J., Snoeck, J., and Timmermans, U. (1978). Transient an-alpha-lipoproteinemia in man during infection by *Plasmodium vivax. Lancet* **1**, 1206.

Levine, H. D. (1973). "Protozoan Parasites of Domestic Animals and of Man," 2nd ed. Burgess, Minneapolis, Minnesota.

Levine, L., and Herrick, C. A. (1954). The effects of the protozoan parasite *Eimeria tenella* on the ability of the chicken to do muscular work when its muscles are stimulated directly and indirectly. *J. Parasitol.* **40**, 525–531.

Levine, N. D., Corliss, J. O., Cox, F. E., Deroux, G., Grain, J., Honigberg, B. M., Leedale, G. F., Loeblich, A. R. III,

Dom, J., Lynn, D., Merinfeld, E. G., Page, F. C., Poljansky, G., Sprague, V., Vavra, J., and Wallace, F. G. (1980). A newly revised classification of the Protozoa. *J. Protozool.* **27**, 37–58.

Mackenzie, C., and Walter, M. H. (1983). Substrate contact, mucus, and eugregarine gliding, *J. Protozool.* **30**, 3–8.

Mackin, J. G., Owen, H. M., and Collier, A. (1950). Preliminary note on the occurrence of a new protistan parasite, *Dermocystidium marinum* n. sp. in *Crassostrea virginica* (Gmelin). *Science* **111**, 328–329.

Maddison, S. E., Slemenda, S. B., Teutsch, S. M., Walls, K. W., Kagan, I. G., Mason, W. R., Bell, F., and Smith, J. (1979). Lymphocyte proliferative responsiveness in 31 patients after an outbreak of toxoplasmosis. *Am. J. Trop. Med. Hyg.* **28**, 955–961.

Mahoney, D. F. (1977). *Babesia* of domestic animals. *In* "Parasitic Protozoa" (J. P. Kreier, ed.), Vol. IV, pp. 1–52. Academic Press, New York.

Manwell, R. D. (1977). Gregarines and haemogregarines. *In* "Parasitic Protozoa" (J. P. Kreier, ed.), Vol. III, pp. 1–32. Academic Press, New York.

Mehlhorn, H., Schein, E., and Warnecke, M. (1979). Electron-microscopic studies on *Theileria ovis* Rodhain, 1916: Development of kinetes in the gut of the vector tick, *Rhipicephalus evertsi evertsi* Neumann, 1897, and their transmission within the cells of the salivary glands. *J. Protozool.* **26**, 377–385.

Miller, L. H., Aikawa, M., Johnson, J. G., and Shiroishi, T. C. (1979). Interaction between cytochalasin B-treated malarial parasite and red cells: attachment and junction formation. *J. Exp. Med.* **149**, 172–184.

Miller, L. H., Hudson, D., Renor, J., Taylor, D., Hadley, T. J., and Zilberstein, D. (1983). A monoclonal antibody to rhesus erythrocyte band 3 inhibits invasion by malaria (*Plasmodium knowlesi*) merozoites. *J. Clin. Invest.* **72**, 1357–1364.

Miller, L. H., Mason, S. J., Clyde, D. F., and McGinniss, M. H. (1976). The resistance factor to *Plasmodium vivax* in blacks: the Duffy blood group genotype, Fy Fy. *New Engl. J. Med.* **215**, 302–304.

Miller, L. H., Mason, S. J., Dvorak, J. A., McGinniss, M. H., and Rothman, I. K. (1975). Erythrocyte receptors for (*Plasmodium knowlesi*) malaria: the Duffy blood group determinants. *Science* **189**, 561–563.

Miller, M. J., Neal, J. V., and Livingston, F. B. (1956). Distribution of parasites in the red cells of sickle-cell trait carriers infected with *Plasmodium falciparum. Trans. Roy. Soc. Trop. Med. Hyg.* **50**, 294–296.

Miller, L. H., Shiroishi, T., Dvorak, J. A., Durocher, J. R., and Schrier, B. K. (1975). Enzymatic modification on the erythrocyte membrane surface and its effect on malarial merozoite invasion. *J. Mol. Med.* **1**, 55–63.

Monne, L., and Honig, G. (1954). On the properties of the shell of the coccidean oocyst. *Ark. Zool.* **7**, 251–256.

Moulder, J. W. (1962). "The Biochemistry of Intracellular Parasitism." University of Chicago Press, Chicago, Illinois.

Nyberg, P. A., and Hammond, D. M. (1964). Excystation of

Eimeria bovis and other species of bovine coccidia. *J. Protozool.* **11**, 474–489.

Nyberg, P. A., Bauer, D. N., and Knapp, S. E. (1968). Carbon dioxide as the initial stimulus for excystation of *Eimeria tenella* oocysts. *J. Protozool.* **15**, 144–148.

Okoye, V. C. H. and Bennett, V. (1985). *Plasmodium falciparum* malaria: Band 3 as a possible receptor during invasion of human erythrocytes. *Science* **327**, 169–171.

O'Roke, E. C. (1934). A malaria-like disease of ducks caused by *Leucocytozoon anatis* Wickware. *Bull. Univ. Mich. School Forest. Conserv.*, 44 pp.

Overstreet, R. M., and Weidner, E. (1974). Differentiation of microsporidian spore-tails in *Inodosporus spraguei* gen. et sp. n. *Zeit. Parasitenk.* **44**, 169–186.

Pasval, G., Anstee, D., and Tanner, M. J. A. (1984). Glycophorin C and the invasion of red cells by *Plasmodium falciparum*. *Lancet* **I**, 907–908.

Pauling, L., Itano, H. A., Singer, S. J., and Wells, I. C. (1949). Sickle cell anemia, a molecular disease. *Science* **110**, 543–548.

Perkins, F. O. (1969). Ultrastructure of vegetative stages in *Labyrinthomyxa marina* (=*Dermocystidium marinum*), a commercially significant oyster pathogen. *J. Invert. Pathol.* **13**, 199–222.

Perkins, M. E. (1984). Surface proteins of *Plasmodium falciparum* merozoites binding to the erythrocyte receptor, glyphorin. *J. Exp. Med.* **160**, 788–798.

Peters, W. (1980). Chemotherapy of malaria. *In* "Malaria" (J. P. Kreier, ed.), Vol. 1, pp. 145–283. Academic Press, New York.

Pipkin, A. C., and Jensen, D. V. (1956). Avian embryos and tissue culture in the study of parasitic protozoa. I. Malarial parasites. *Exp. Parasitol.* **7**, 491–530.

Pratt, I. (1940). The effect of *Eimeria tenella* (Coccidia) upon the blood sugar of the chicken. *Trans. Am. Microsc. Soc.* **59**, 31–37.

Pratt, I. (1941). The effect of *Eimeria tenella* (Coccidia) upon the glycogen stores of the chicken. *Am. J. Hyg.* **34**, 54–61.

Raper, A. B. (1959). Further observations on sickling and malaria. *Trans. Roy. Soc. Trop. Med. Hyg.* **53**, 110–117.

Ravetch, J. V., Kochan, J., and Perkins, M. (1985). Isolation of the gene for a glyphorin-binding protein implicated in erythrocyte invasion by a malaria parasite. *Science* **227**, 1593–1597.

Ray, S. M., and Chandler, A. C. (1955). *Dermocystidium marinum*, a parasite of oysters. *Exp. Parasitol.* **4**, 172–200.

Rick, R. F. (1964). The life cycle of *Babesia bigemina* (Smith and Kilbourne, 1893) in the tick vector *Boophilus microplus* (Canastrini). *Aust. J. Agr. Res.* **15**, 802–821.

Rick, R. F. (1966). The development of *Babesia* spp. and *Theileria* spp. in ticks with special reference to those occurring in cattle. *In* "Biology of Parasites" (E. J. L. Soulsby, ed.), pp. 15–32. Academic Press, New York.

Rieckmann, K. H. (1980). Prospects for malaria blood stage vaccine. *In* "Malaria" (J. P. Kreier, ed.), Vol. 3, pp. 321–324. Academic Press, New York.

Rieckmann, K. H., and Silverman, P. H. (1977). Plasmodia

of man. *In* "Parasitic Protozoa" (J. P. Kreier, ed.), Vol. III, pp. 493–527. Academic Press, New York.

Ristic, M., and Kreier, J. P. (eds.) (1981). "Babesiosis." Academic Press, New York.

Ristic, M., and Lewis, G. E. Jr. (1977). *Babesia* in man and wild and laboratory-adapted mammals. *In* "Parasitic Protozoa" (J. P. Kreier, ed.), Vol. IV. pp. 53–76. Academic Press, New York.

Rommel, M., Heydorn, A. O., and Gruber, F. (1972). Beitrage zum Lebenszyklus der Sarkosporidien. III. *Isopora hominis* (Railliet und Lucet, 1981) Wenyon, 1923, eine Dauerform der Sarkosporidien des Rindes und des Schweins. *Berlin Muench. Tieraerztl. Wochenschr.* **85**, 143–145.

Ruebush, T. K. II, Juranek, D. D., Spielman, A., Piesman, J., and Healy, G. R. (1980). Epidemiology of human babesiosis on Nantucket Island. *Am. J. Trop. Med. Hyg.* **30**, 937–954.

Ruff, M. D., and Reid, W. M. (1977). Avian coccidia. *In* "Parasitic Protozoa" (J. P. Kreier, ed.), Vol. III, pp. 34–69. Academic Press, New York.

Seed, T. M., and Kreier, J. P. (1980). Erythrocyte destruction mechanisms in malaria. *In* "Malaria" (J. P. Kreier, ed.), Vol. 2, pp. 1–46. Academic Press, New York.

Seed, T. M., and Manwell, R. D. (1977). Plasmodia of birds. *In* "Parasitic Protozoa" (J. P. Kreier, ed.), Vol. III, pp. 311–357. Academic Press, New York.

Shortt, H. E. (1983). Relapse in primate malaria: its implications for the disease in man. *Trans. Roy. Soc. Trop.* **77**, 734–736.

Shute, P. G., Lupascu, G., Branzei, P., Maryon, M., Constantinescu, P., Bruce-Chwatt, L. J., Draper, G. G., Killick-Kendrick, R., and Garnham, P. C. C. (1976). A strain of *Plasmodium vivax* characterized by prolonged incubation: the effect of numbers of sporozoites on the length of the prepatent period. *Trans. Roy. Soc. Trop. Med. Hyg.* **70**, 474–481.

Siddiqui, W. A. (1980). Immunization against asexual blood-inhabiting stages of plasmodia. *In* "Malaria" (J. P. Kreier, ed.), Vol. 3, pp. 231–262. Academic Press, New York.

Smith, B. F., and Herrick, C. A. (1944). The respiration of the protozoan parasite, *Eimeria tenella*. *J. Parasitol.* **30**, 295–302.

Smith, G. W. (1905). Note on the gregarine (*Aggregata inachi*, n. sp.) which may cause the parasitic castration of its host (*Inachus dorsettensis*). *Mitt. Zool. Sta. Neapel* **17**, 406–410.

Spielman, A. (1976). Human babesiosis on Nantucket Island: Transmission by nymphal *Ixodes* ticks. *Am. J. Trop. Med. Hyg.* **25**, 784–787.

Stauber, L. A. (1939). Factors influencing the asexual periodicity of avian malarias. *J. Parasitol.* **25**, 95–116.

Strome, C. P. A., DeSantis, P., and Beaudoin, R. L. (1979). Cultivation of exoerythrocytic stages of *Plasmodium berghei* from sporozoite to merozoite. *In Vitro* **15**, 531–536.

Sweeney, A. W., Hazard, E. I., and Graham M. F. (1985). Intermediate host for an *Amblyospora* sp. (Microspora) infecting the mosquito, *Culex annulirostris*. *J. Invert. Pathol.* **46**, 98–102.

Todd, K. S. Jr., and Ernst, J. V. (1977). Coccidia of mammals except man. *In* "Parasitic Protozoa" (J. P. Kreier, ed.), Vol. III, pp. 71–99. Academic Press, New York.

Trager, W. (1943). Further studies on the survival and development *in vitro* of a malarial parasite. *J. Exp. Med.* **77**,

411–420.

Trager, W. (1947). The development of the malaria parasite *Plasmodium lophurae* in red blood cell suspension *in vitro*. *J. Parasitol.* **33**, 345–350.

Trager, W. (1950). Studies on the extracellular cultivation of an intracellular parasite (avian malaria). I. Development of the organisms in erythrocyte extracts, and the favoring effect of adenosine triphosphate. *J. Exp. Med.* **92**, 349–365.

Trager, W. (1960). The physiology of intracellular parasites. *Proc. Helminthol. Soc. Wash.* **27**, 221–227.

Trager, W. (1971). Further studies on the effects of antipantothenates on malaria parasites (*Plasmodium coatneyi* and *P. falciparum*) in vitro. *J. Protozool.* **18**, 232–239.

Trager, W. (1977). Cofactors and vitamins in the metabolism of malarial parasites. *Bull. W.H.O.* **55**, 285–290.

Trager, W., and Jensen, J. B. (1980). Cultivation of erythrocytic and exoerythrocytic stages of plasmodia. *In* "Malaria" (J. P. Kreier, ed.), Vol. 2, pp. 271–319. Academic Press, New York.

Trager, W., and McGhee, R. B. (1950). Factors in plasma concerned in natural resistance to an avian malaria parasite (*Plasmodium lophurae*). *J. Exp. Med.* **91**, 365–379.

Trier, J. S., Moxey, P. C., Schimmel, E. M., and Robles, E. (1974). Chronic intestinal coccidiosis in man: intestinal morphology and response to treatment. *Gastroenterology* **66**, 923–935.

Trubowitz, S., and Masek, B. (1968). *Plasmodium falciparum*: phagocytosis by polymorphonuclear leukocytes. *Science* **162**, 273–274.

von Brand, T., and Mercado, T. I. (1956). Quantitative and histochemical studies on glycogenesis in the liver of rats infected with *Plasmodium berghei*. *Exp. Parasitol.* **5**, 34–37.

Wallace, G. D. (1971). Experimental transmission of *Toxoplasma gondii* by filth flies. *Am. J. Trop. Med. Hyg.* **20**, 411–413.

Warren, K. S. (1980). The great neglected diseases of mankind. *In* "Malaria" (J. P. Kreier, ed.), Vol. 3, pp. 335–338. Academic Press, New York.

Waxler, S. H. (1941). Changes occurring in the blood and tissue of chickens during coccidiosis and artificial hemorrhage. *Am. J. Physiol.* **134**, 19–29.

Weidner, E. (1972). Ultrastructural study of microsporidian invasion into cells. *Zeit. Parasitenk.* **40**, 227–242.

Weinman, D., and Klatchko, H. J. (1950). Description of toxin in toxoplasmosis. *Yale J. Biol. Med.* **22**, 323–326.

Wernsdorfer, W. H. (1980). The importance of malaria in the world. *In* "Malaria" (J. P. Kreier, ed.), Vol. 1, pp. 1–93. Academic Press, New York.

Wolf, K., and Markiw, M. E. (1984). Biology contravenes taxonomy in the Myxozoa: new discoveries show alternation of invertebrate and vertebrate hosts. *Science* **225**, 1449–1452.

Zwart, D., and Brocklesby, D. W. (1979). Babesiosis: Nonspecific resistance, immunological factors and pathogenesis. *Adv. Parasit.* **17**, 49–113.

CLASSIFICATION OF APICOMPLEXA, MICROSPORA, ASCETOSPORA, AND MYXOZOA*

PHYLUM APICOMPLEXA

Apical complex (visible with electron microscope), generally consisting of polar ring(s), rhoptries, micronemes, conoid, and subpellicular microtubules present at some stage; micropore(s) generally present at some stage; cilia absent; sexuality by syngamy; all species parasitic.

CLASS PERKINSEA

Conoid forming incomplete cone; zoospores (sporozoites?) flagellated, with anterior vacuole; no sexual reproduction; homoxenous.

ORDER PERKINSIDA

With characters of the class (Genus mentioned in text: *Perkinsus*.)

CLASS SPOROZOEA

Conoid, if present, forming complete cone; reproduction generally both sexual and asexual; oocysts generally containing infective sporozoites which result from sporogony; locomotion of mature organisms by body flexion, gliding, or undulation of longitudinal ridges; flagella present only in microgametes of some groups; pseudopods ordinarily absent, but if present, used for feeding, not locomotion; homoxenous or heteroxenous.

Subclass Gregarinia

Mature gamonts large, extracellular; mucron or epimerite in mature organism; mucron formed from conoid; generally syzygy of gamonts; gametes usually similar (isogamous) or nearly so, with similar numbers of male and female gametes produced by gamonts; zygotes forming oocysts within gametocytes; life cycle characteristically consisting of gametogony and sporogony; in digestive tract or body cavity of invertebrates or lower chordates; generally homoxenous.

ORDER ARCHIGREGARINIDA

Life cycle apparently primitive, characteristically with merogony, gametogony, sporogony; gamonts (trophozoites) aseptate; in annelids, sipunculids, hemichordates, or ascidians.

ORDER EUGREGARINIDA

Merogony absent; gametogony and sporogony present; locomotion progressive, by gliding or undulation of longitudinal ridges, or nonprogressive; typically parasites of annelids and arthropods, but some species in other invertebrates.

Suborder Blastogregarinina

Gametogony by gamonts while still attached to intestine, with anisogamous gametes budding off of gamonts; no syzygy; gametocysts absent; oocysts with 10 to 16 naked sporozoites; gamont composed of single compartment with mucron, without definite protomerite and deutomerite; in marine polychaetes.

Suborder Aseptatina

Gametocysts present; gamont composed of single compartment, without definite protomerite and deutomerite, but with mucron (epimerite?) in some species;

*Based on the recommendation of the Committee on Systematics and Evolution of the Society of Protozoologists consisting of Levine *et al.* (1980). All of the members of these four phyla are parasitic.

syzygy present. (Genera mentioned in text: *Monocystis, Enterocystis, Rhabdocystis, Apolocystis, Nematocystis, Rhynchocystis.*)

Suborder Septatina

Gametocysts present; gamont divided into protomerite and deutomerite by septum; with epimerite; in alimentary canal of invertebrates, especially arthropods. (Genera mentioned in text: *Gregarina, Stylocephalus, Actinocephalus, Porospora, Nematopsis.*)

ORDER NEOGREGARINIDA

Merogony, presumably acquired secondarily; in Malpighian tubules, intestine, hemocoel, or fat tissues of insects.

Subclass Coccidia

Gamonts ordinarily present; mature gamonts small, typically intracellular, without mucron or epimerite; syzygy generally absent, but if present, involves markedly anisogamous gametes; life cycle characteristically consists of merogony, gametogony, and sporogony; most species in vertebrates.

ORDER AGAMOCOCCIDIIDA

Merogony and gametogony absent.

ORDER PROTOCOCCIDIIDA

Merogony absent; in invertebrates.

ORDER EUCOCCIDIIDA

Merogony present; in vertebrates and/or invertebrates.

Suborder Adeleina

Macrogamete and microgamont usually associated in syzygy during development; microgamont producing 1 to 4 microgametes; sporozoites enclosed in envelope; homoxenous or heteroxenous. (Genera mentioned in text: *Adelina, Haemogregarina, Hepatozoon.*)

Suborder Eimeriina

Macrogamete and microgamont developing independently; no syzygy; microgamont typically producing many microgametes; zygote nonmotile; sporozoites typically enclosed in sporocyst within oocyst; homoxenous or heteroxenous. (Genera mentioned in text: *Aggregata, Eimeria, Lankesterella, Isospora, Levineia, Toxoplasma, Sarcocystis, Frenkelia.*)

Suborder Haemosporina

Macrogamete and microgamont developing independently; no syzygy; conoid usually absent; microgamont producing eight flagellated microgametes; zygote motile (ookinete); sporozoites naked, with three-membraned wall; heteroxenous, with merogony in vertebrates and sporogony in invertebrates; transmitted by blood-sucking insects. (Genera mentioned in text: *Plasmodium, Haemoproteus, Leucocytozoon.*)

Subclass Piroplasmia

Piriform, round, rod-shaped, or amoeboid; conoid absent; no oocysts, spores, and pseudocysts; flagella absent; usually without subpellicular microtubules, with polar ring and rhoptries; locomotion by body flexion, gliding or, in sexual stages (in Babesiidae and Theileriidae, at least), by large axopodiumlike organelle; asexual and probably sexual reproduction; parasitic in erythrocytes and some-

times also in other circulating and fixed cells; heteroxenous, with merogony in vertebrates and sporogony in invertebrates; sporozoites with single-membraned wall; vectors are ticks, but vectors of dactylosomatids unknown.

ORDER PIROPLASMIDA

With characters of the subclass. (Genera mentioned in text: *Babesia, Theileria.*)

PHYLUM MICROSPORA

Unicellular spores, each with imperforate wall, containing one uninucleate or dinucleate sporoplasm and simple or complex extrusion apparatus always with polar tube and polar cap; without mitochondria; often, if not usually, dimorphic in sporulation sequence; obligatory intracellular parasites in nearly all major animal groups.

CLASS RUDIMICROSPOREA

Spore with simple extrusion apparatus consisting of polar cap and thick polar tube extending backward from cap, bending laterally and terminating in funnel; polaroplast and posterior vacuole absent; spore spherical or subspherical; sporulation sequence with dimorphism, occurring either in parasitophorous vacuole or in thick-walled cyst; hyperparasites of gregarines in annelids.

ORDER METCHNIKOVELLIDA

With characters of the class.

CLASS MICROSPOREA

Spore with complex extrusion apparatus of Golgi origin, often including polaroplast and posterior vacuole in addition to polar tube and polar caps; polar tube typically filamentous, extending backward from polar cap and coiling around inside of spore wall; spore shape various, depending largely on structure of extrusion apparatus; spore wall with three layers—proteinaceous exospore, chitinous endospore, and membranous inner layer; outer two layers varying considerably in structure; sporocyst present or absent.

ORDER MINISPORIDA

General tendency toward minimum development of accessory spore organelles (components of extrusion apparatus and spore wall) and accompanying tendency toward maximum development of sporocysts; spore without well-developed polaroplast; usually with relatively short polar filament, with little or no endospore; shape spherical or slightly ovoid; merogony present or absent (?); sporulation stages usually separated from host cell cytoplasm by intracellular sporocyst. (Genera mentioned in text: *Burkea, Chytridiopsis, Hessea.*)

ORDER MICROSPORIDA

General tendency toward maximum development and varied specialization of accessory spore organelles (components of extrusion apparatus and spore wall), with accompanying reduction of sporocysts; sporocysts inside host cell present or absent; cysts of other kinds sometimes formed from host cell membrane or other host material; merogony present; spore shape variable.

Suborder Pansporoblastina

Sporulation sequence occurring within more or less persistent intracellular (in host cell) sporocyst; often dimorphic, with another sporulation sequence not involving such membrane; sporoblasts and spores usually uninucleate when membrane present, dinucleate when membrane absent. (Genera mentioned in text: *Thelohania, Pleistophora.*)

Suborder Apansporoblastina

Pansporoblastic membrane usually absent, vestigial when present, never persisting as sporophorous vesicle; sporoblast most often binucleate. (Genera mentioned in text: *Nosema, Encephalitozoon, Glugea, Unikaryon, Amblyospora, Microsporidium* [?].)

PHYLUM ASCETOSPORA

Spore multicellular (or unicellular?); with one or more sporoplasms; without polar capsules or polar filaments; all parasitic.

CLASS STELLATOSPOREA

Haplosporosomes present; spore with one or more sporoplasms.

ORDER OCCLUSOSPORIDA

Spore with more than one sporoplasm; sporulation involving series of endogenous buddings that produce sporoplasm(s) within sporoplasm(s); spore wall entire.

ORDER BALANOSPORIDA

Spore with one sporoplasm; spore wall interrupted anteriorly by orifice; orifice covered externally by operculum or internally by diaphragm. (Genera mentioned in text: *Urosporidium, Anurosporidium, Haplosporidium, Bertramia, Coelosporidium*.)

CLASS PARAMYXEA

Spore bicellular, consisting of parietal cell and one sporoplasm; spore without orifice.

ORDER PARAMYXIDA

With characters of the class.

PHYLUM MYXOZOA

Spores of multicellular origin, with one or more polar capsules and sporoplasms; with one, two, or three (rarely more) valves; all species parasitic.

CLASS MYXOSPOREA

Spore with one or two sporoplasms and one to six (typically two) polar capsules, each capsule with coiled polar; filament; filament function probably anchoring; spore membrane generally with two, occasionally up to six, valves; trophozoite stage well developed, main site of proliferation; coelozoic or histozoic in cold-blooded vertebrates.

ORDER BIVALVULIDA

Spore wall with two valves.

Suborder Biopolarina

Spores in vertebrate host with polar capsules at opposite ends of spore, or with widely divergent polar capsules located in sutural plane or zone. Spores produced in invertebrate host with three polar capsules; membrane with three valves; several to many sporoplasms. (Genera mentioned in text: *Myxidium, Wardia, Myxosoma, Triactinomyxon, Mitraspora, Myxoproteus*.)

Suborder Eurysporina

Spores with two to four polar capsules at one pole in plane perpendicular to sutural plane. (Genera mentioned in text: *Ceratomyxa, Leptotheca, Unicapsula, Unicauda*.)

Suborder Platysporina

Spores with two polar capsules at one pole in sutural plane; spores bilaterally symmetrical, unless with single polar capsule.

ORDER MULTIVALVULIDA

Spore wall with three or more valves. (Genus mentioned in text: *Trilospora*.)

7

CILIOPHORA: THE CILIATES

Members of the phylum Ciliophora, commonly known as ciliates, differ from the other protozoans in that they bear cilia during some stage in their development. These protozoans also possess two types of nuclei—macronucleus and micronucleus—and reproduce asexually by transverse binary fission.

Most ciliophorans are free living, being found in fresh, brackish, and marine waters, but there are several genera of symbiotic forms. Some of these, such as the pathogen *Balantidium coli* in humans, are

true parasites, but by far the majority are epiphoretic and commensalistic species.

Reynolds (1930) has listed 55 species of ciliates known to be attached to the body surfaces of various vertebrates and invertebrates, and Hoffman (1978) has listed and reviewed what is known about those species associated with freshwater fish. These unique species are often categorically classified as parasites, although they are more properly considered epiphoronts, since their relationship with their hosts is generally devoid

Table 7.1. Some Important Parasitic Ciliates of Humans and Economically Important Animals

Organism	Principal Hosts	Habitat	Main Characteristics	Disease
Balantidium coli	Pigs, monkeys, humans	Large intestine	50–100 μm long, large ovoid body, anterior end more pointed, peristome leading into distinct cleft, two contractile vacuoles, kidney-shaped nucleus	Balantidiasis (in humans
Balantidium sp.	Sheep	Intestine	45 μm long, 33 μm wide	Nonpathogenic
Ichthyophthirius multifiliis	Fish	Primarily integument	100–1000 μm long, ovoid, with large cytosome measuring 30–40 μm in diameter	Ichthyophthiriosis
Enchelys parasitica	Trout	Integument	Flask-shaped, cytosome slitlike, about 12 μm long	Enchelysiosis

of any type of physiologic interactions (nutritional, sensory, etc.). The relationship is purely one of mechanical attachment. In a few instances, however, some physiologic bond does exist, for if removed from their hosts the ciliates cannot survive. Such is the case between *Trichodina scorpenae* and its fish host.

At the other extreme, *Ichthyophthirius multifiliis* is a histozoic parasite that inflicts severe damage to its fish host. Another similar parasite, *Enchelys parasitica*, causes a fatal skin disease of the rainbow trout. Some medically and economically important species of ciliates are listed in Table 7.1.

Although there is a lack of fossil evidence, most agree that the ciliates represent an ancient group, although not as ancient as the flagellates and the amoebae. Perhaps for this reason, among others, there are fewer parasitic ciliates than flagellates or amoebae. It is possible that the greater specialization of the ciliates through evolution implies refinements in their physiologic requirements that make adaptation to the parasitic way of life more difficult.

The classification of the Ciliophora is presented at the end of this chapter. For those interested in the taxonomy of this group of protozoans, the monograph by Corliss (1979) is recommended.

MORPHOLOGY

The body shapes and sizes among the various species of ciliates vary greatly, but they all have cilia, which may be uniformly distributed over the entire body surface or grouped in heavier concentrations in certain areas. Furthermore, certain species, such as the group commonly referred to as suctorians, bear cilia only during their developmental stages (swarm cells or swarmers). Once suctorians mature, they develop characteristic tentacles that function in locomotion and nutrient acquisition.

Microscopically, cilia appear as extremely fine,

short, hairlike projections that are usually arranged in definite rows (Figs. 7.1, 7.2). Each cilium is attached at the **basal granule** (also known as the **basal body** or **kinetosome**) situated immediately below the pellicle, which is a definite membrane surrounding the cytoplasm (Fig. 7.1).

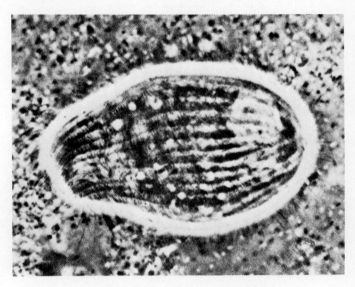

Fig. 7.2. Arrangement of cilia. Photomicrograph of the ciliate *Balantidium coli* from culture showing occurrence of ciliary rows (kineties). Phase contrast. (After Zaman, 1978; with permission of Academic Press.)

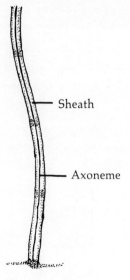

Sheath

Axoneme

Fig. 7.3. Cilia structure. Drawing of the basal portion of a single cilium showing the sheath and axoneme.

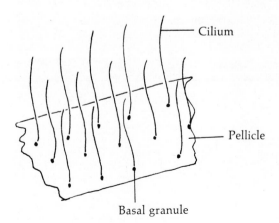

Cilium

Pellicle

Basal granule

Fig. 7.1. Drawing showing cilia arranged in rows.

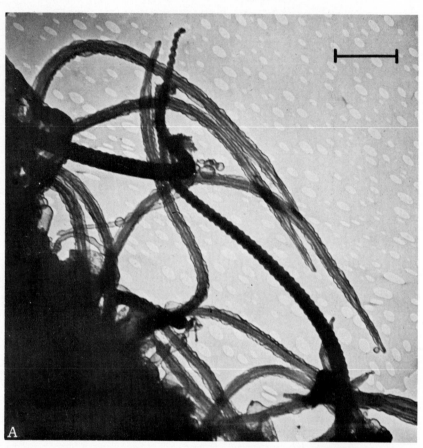

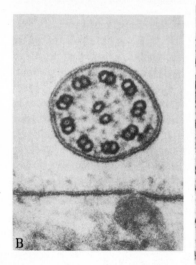

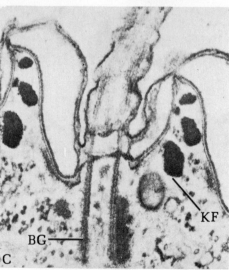

Fig. 7.4. Electron micrographs of cilia. A. Cilia of *Paramecium* showing fibrils within sheath. Bar indicates 1 μm. × 18,000. (Courtesy of T. F. Anderson, Institute for Cancer Research, Philadelphia.) **B.** Cross section of single cilium showing arrangement of microtubules. × 150,000. (After Elliott, 1963.) **C.** Attachment of cilium showing unit membrane nature of pellicular membrane, basal granule (kinetosome) (BG), and kinetodesmal fibrils (KF) in ectoplasm. × 70,000. (After Pitelka, 1970.)

Structurally, the single cilium is similar to a flagellum (p. 111). There is an axoneme surrounded by an elastic cytoplasmic sheath (Fig. 7.3). When studied with the electron microscope, the axoneme is seen to be composed of nine peripheral pairs and one central pair of microtubules (Fig. 7.4), just as in a flagellum.

Several types of subpellicular fibrillar networks have been described in ciliates (Fig. 7.5). These networks have been termed the **silverline**, the **infraciliature**, or the **neuromotor** (neuroneme) system (Hall, 1953). The neuromotor system is thought to be a primitive nervous system that coordinates ciliary motion. This has been postulated because the movement of cilia, unlike the more independent beating of flagella, is well coordinated in a rhythmic or metachronous wavelike pattern (Fig. 7.6). The monograph by Grell (1973), in which are considered the ultrastructure and function of cilia, should be consulted by those interested in these aspects.

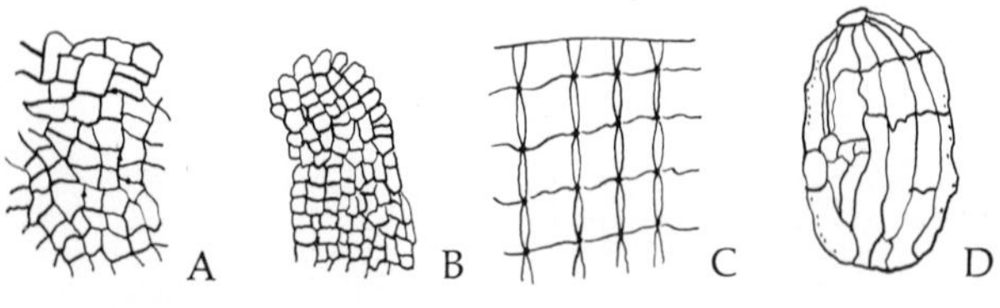

SPECIALIZED CILIARY ORGANELLES

The fusion of adjacent cilia in definite patterns gives rise to various types of specialized organelles. Those commonly found on parasitic species include the **cirri**, composed of a tuft of cilia from two or three adjacent rows that have fused together to give the appearance of long flexible spines; the **membranelle**, composed of two fused layers of cilia (lamellae) forming a triangular platelike structure; and the **paroral membrane** (undulating membrane),* composed of one or more lamellae forming a flap that is generally associated with the oral groove or peristome (Fig. 7.7).

Specialized ciliary organelles are of both functional and diagnostic importance. For example, cirri are commonly found among members of the order Hypotrichida of the subclass Spirotrichia; membranelles are found among the more advanced ciliates except certain members of the class Kinetofragminophora; and undulating membranes are most commonly, although not exclusively, found among the Kinetofragminophora.

Functionally, cilia not only act as mechanisms of locomotion but also serve in the acquisition of nutrients and as sensory organelles. Cirri are associated with specialized movement and are used as stalks or paddles. Undulating membranes and membranelles are generally associated with food acquisition.

Cilia beat in a uniform wavelike fashion (known as **metachronism**), but the mechanisms of movement and coordination have not been totally defined (Fig. 7.6). Probably the most acceptable hypothesis offered

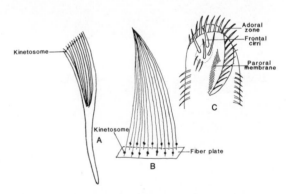

Fig. 7.5. **Silverline systems of ciliates. A.** System of *Prordon teres,* an aquatic ciliate, showing narrow mesh type of organization with some orientation of fibrils. **B.** System showing primitive narrow mesh type of arrangement. **C.** System of *Cinetochilum margaritaceum* showing double striation pattern. **D.** System of *Cyclidium glaucoma* showing striation system. (Redrawn after Klein, 1932.)

Fig. 7.6. **Metachronous movement of cilia in a longitudinal row.**

Fig. 7.7. **Specialized ciliary structures. A.** Anal cirrus (side view) of *Euplotex eurystomus.* **B.** A membranelle of *E. eurystomus.* **C.** Anterior portion of *Stylonychia* (ventral view) showing paroral membrane. (Redrawn after Taylor, 1920.)

thus far to explain ciliary motion is the alternating contraction of one side then the other of the peripheral microtubules (Bradfield, 1955). It has been postulated that the effective stroke is due to the simultaneous contraction of five of the peripheral microtubules along their entire lengths while the remaining four remain idle, but subsequently the latter slowly contract, beginning at the base, producing the recovery stroke. It is also assumed that the triggering mechanism resides in the basal body and that the central

---

*The undulating membrane of ciliates is not to be confused with that of flagellates, which is formed by the undulating distentions of a flagellum attached by desmosomes to the body surface.

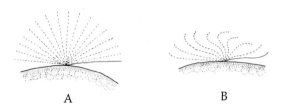

Fig. 7.8. Ciliary movement. A. Pendular movement of single cilium. **B.** Flexural movement of single cilium.

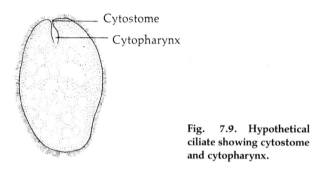

Cytostome

Cytopharynx

Fig. 7.9. Hypothetical ciliate showing cytostome and cytopharynx.

Excretory canal

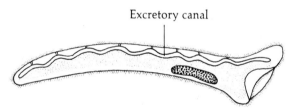

Fig. 7.10. The ciliate *Haptophrya michiganensis* showing longitudinally oriented excretory canal. (Redrawn after MacLannan, 1944.)

microtubules serve as organelles of transmission, which provide rapid conduction up the cilium.

Ciliary motion has been classified into two types: (1) In **pendular** movement, the single cilium sways from one side to the other, flexing only at the base (Fig. 7.8); this type of motion is characteristic of cirri of hypotrichs. (2) In the common **flexural** movement (Fig. 7.8), the cilium first bends at the free end and the bending passes toward the base; during recovery, the cilium straightens out from the base upward.

Sharp (1914) reported that the so-called neuromotor network, which lies beneath the body surface in the ectoplasm, is connected with a "coordinating center" —the motorium. This hypothetical organelle presumably governs the pattern and rhythm of motion. The coordinating function of the motorium also has been suggested in *Chlamydodon* by MacDougall (1928), who reported that if this organelle is destroyed, the coordinated movement of the membranelle is disrupted. The "motorium concept" has been modified in recent years. In brief, the **neuroid** theory postulates nervelike impulses conducted through the protoplasm at the bases of the cilia and the **mechanical** theories that hold that the movement of one cilium mechanically stimulates the next to move (Fawcett, 1961). The first, which is essentially an extension of the motorium concept, is no longer widely accepted since electron microscope studies have failed to reveal any morphologic connections between the fibrillar systems and the basal bodies of cilia; furthermore, a motorium has never been demonstrated. On the other hand, biochemical and physiologic evidences support the mechanical theories, at least in part. For example, glycerin-extracted cilia may be preserved for weeks and then activated by adenosine triphosphate (ATP). However, when such preparations are treated with surface-active agents such as digitonin or saponin, the cilia lose their coordinated beat. It is assumed that the surface-active agents remove a lipid or lipoprotein that is essential for the coordinated beat. This also indicates that elements responsible for contraction (intraciliary excitation) as well as coordination (interciliary conduction) are operative. For a more detailed discussion of the chemical aspects of ciliary and flagellar function as related to protozoans, see Child (1967).

OTHER SPECIALIZED STRUCTURES

In addition to cilia and specialized ciliary organelles, several other types of structures are found among the ciliates. The ingestion of food particles in many species is made possible by a minute "mouth"—the **cytostome**—which leads into the **cytopharynx** (Fig. 7.9). This structure is rarely visible in most ciliates because of the membranelles and/or undulating membranes in the peristomal zone surrounding the cytostome. Cytostomes have been demonstrated in certain parasitic flagellates and amoebae, for example, trophozoites of *Dientamoeba fragilis*, *Trypanosoma cruzi*, and *Entamoeba muris*.

The remaining organelles and cell inclusions in ciliates correspond with those found in other protozoans and have been discussed in Chapter 4.

Interestingly, in the large parasitic ciliate *Haptophrya michiganensis*, found in the gut and gallbladder of salamanders, a specialized contractile canal is oriented along the dorsal side of the elongated body. Located along the wall of this canal are accessory vacuoles that gradually unite with one another and later with the main canal (Fig. 7.10). The contents of the dilated canal empty to the exterior through several short excretory ducts. This unique excretory system rids the body of waste materials, such as carbon dioxide, water, and nitrogenous materials. Because endoparasitic apicomplexans, microsporans, ascetosporans, myxozoans, amoebae, and flagellates live in near-isotonic environments, contractile vacuoles are usually not seen in their bodies. However, vacuoles do exist in parasitic ciliates, including suctorians, but the reason for their occurrence is still uncertain.

CLASSIFICATION

The classification of the phylum Ciliophora has been subjected to considerable revision in recent years. According to Levine *et al.* (1980), this phylum includes three classes: Kinetofragminophorea, Oligohymenophorea, and Polymenophorea. The diagnostic characteristics of these classes and those of the subordinate taxa are presented at the end of the chapter. All three classes include symbiotic species.

CLASS KINETOFRAGMINOPHOREA

The majority of the members of Kinetofragminophorea are free living; however, a number of representatives are symbiotic. For example, members of the suborder Archistomatina (of the subclass Gymnostomatia and order Prostomatida) all occur in the digestive tract of equines. These include members of the genera *Alloiozona*, *Blepharoprosthium*, and *Didesmis* (Fig. 7.11).

Genus *Balantidium*. The best known parasitic representatives of the Kinetofragminophorea are members of the genus *Balantidium*. The numerous species occur in the intestines of vertebrates and invertebrates. The most important of these is *Balantidium coli* (Fig. 7.12), a parasite of pigs and humans. Different strains of *B. coli* exist, for morphologically identical forms have been found in guinea pigs and other vertebrates. Humans most probably acquired the first *Balantidium* infections from pigs when the latter became domesticated and often lived under the same roof. Furthermore, *Balantidium* also has been reported from nonhuman primates, which may serve as reservoirs of human infections.

Although balantidiasis,* the human disease caused by *B. coli*, is most common in the tropics, especially the Philippines, it is by no means limited to warm climates. Globally, balantidiasis occurs in less than 1% of the human population; however, higher incidences occur among institutionalized persons. Also, the incidence is considerably higher among pigs, ranging from 20 to 100%.

The trophozoite of *B. coli*, measuring 30–200 μm (average 50–80 μm) by 20–120 μm, is the largest

---

*The suffix *-iosis* means that a pathologic condition exists; the suffix *-iasis* implies that the disease state cannot be recognized. This "rule," however, is not always observed.

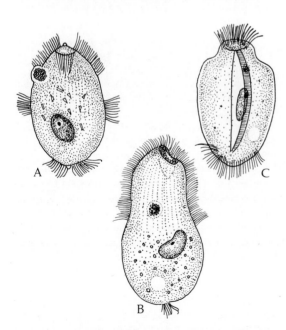

Fig. 7.11. Representative kinetofragminophoreans.
A. *Alloiozona trizona* from caecum and colon of horses. (Redrawn after Hsiung, 1930.) **B.** *Blepharoprosthium pireum* from caecum and colon of horses. (Redrawn after Hsiung, 1930.) **C.** *Didesmis quadrata* from caecum and colon of horses. (Redrawn after Hsiung, 1930.)

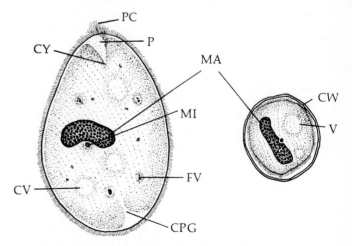

Fig. 7.12. Trophozoite and cyst of *Balantidium coli*, an intestinal parasite of pigs and humans. CPG, cytopyge; CV, contractile vacuole; CW, cyst wall; FV, food vacuole; MA, macronucleus; MI, micronucleus; P, peristome; PC, peristomal cilia; V, vacuole.

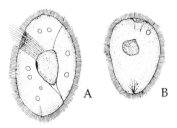

Fig. 7.13. Some trichostomatid ciliates. A. *Isotricha prostoma* from stomach of cattle, 80–195 by 53–85 μm. **B.** *Dasytricha ruminantium* from stomach of cattle, 50–75 by 3–40 μm. (Redrawn after Becker and Talbott, 1927.)

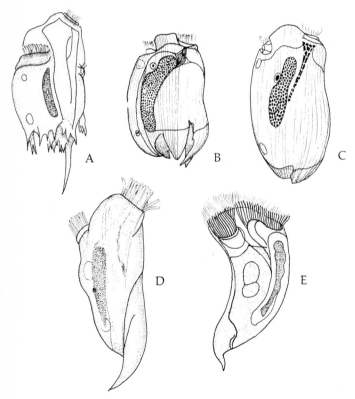

Fig. 7.14. Some entodiniomorphid ciliates. A. *Ophryoscolex caudatus* from stomach of cattle, sheep, and goats, 137–162 by 80–98 μm. (Redrawn after Dogiel, 1927.) **B.** *Diplodinium dentatum* from stomach of cattle, 65–82 by 40–50 μm. (Redrawn after Kofoid and MacLennan, 1932.) **C.** *Eremoplastron bovis* from stomach of cattle and sheep, 52–100 by 34–50 μm. (Redrawn after Kofoid and MacLennan, 1932.) **D.** *Epidinium caudatum* from stomach of cattle, camels, and reindeer, 113–151 by 45–61 μm. (Redrawn after Becker and Talbott, 1927.) **E.** *Cunhaia curvata* from caecum of the guinea pig *Cavia aperea*, 60–80 by 30–40 μm. (Redrawn after Hasselmann, 1926.)

protozoan parasite of humans. The cytostome and cytopharynx are located at the anterior end. There is a distinct peristomal zone lined with coarser cilia. The macronucleus is typically ovoid to reniform; the vesicular micronucleus is spherical (Fig. 7.12).

Most specimens possess two prominent contractile vacuoles, one in the middle and the other near the posterior end of the body. Food vacuoles in the cytoplasm enclose debris, bacteria, starch granules, erythrocytes, and fragments of host epithelium. The usual reproductive method is asexual by transverse fission, with the posterior daughter cell forming a new cytostome after division. Conjugation is also known to occur in this species.

In humans, *B. coli* inhabits the lumen of the colon, from which site it may invade the submucosal tissue, inflicting considerable damage. In a few cases, this ciliate has even been transported by the blood into the spinal fluid. As in *Entamoeba histolytica* infections, diarrhea and secondary complications accompany balantidiasis. Mild and asymptomatic cases in humans are known in which clinical symptoms are slight or cannot be recognized.

Transfer of *Balantidium coli* from host to host occurs via the cystic form. Cysts, measuring 40–60 μm in diameter, are round and the wall consists of two membranes (Fig. 7.12). The macronucleus and contractile vacuoles can be seen within the cyst wall. Cysts are found in the feces of hosts and are generally not considered sites of reproduction, although cysts containing two individuals have been reported.

In addition to humans, other primates, and pigs, *B. coli* has also been reported from rats and guinea pigs. Since their large size makes *B. coli* easy to diagnose, it is difficult to understand why two valuable gorillas died from balantidiasis in the Cincinnati Zoo in 1982. For a detailed review of what is known about this ciliate, see Zaman (1978).

Other Balantidium *Species.* Balantidial infections are common in numerous species of vertebrates, including frogs, toads, fish, tortoises, birds, and cattle. In each of these cases, a different species of *Balantidium* is generally believed to be involved. For example, *B. duodeni* is a fairly common parasite of frogs, especially in warmer climates. Balantidial infections of invertebrates are also fairly common; *B. praenucleatum* is the species found in cockroaches.

RUMEN CILIATES

The rumen of cows and sheep represents a unique ecologic habitat. It is warm (39°C), anaerobic, rich in particulate matter that is more or less resistant to digestion, and usually deficient in soluble nutrients such as amino acids and glucose. Within this part of the digestive tract of ruminants is found a collection of unique ciliates. Hungate (1978) and Coleman (1979) have presented reviews of this mixture of protozoans.

The rumen ciliates have become so adapted to their environment that they are intolerant of all but the slightest alterations in their habitat. These ciliates belong to two orders of the subclass Vestibuliferia of the Kinetofragminophorea: the Trichostomatida (to which *Balantidium* also belongs) and the Entodiniomorphida.

Trichostomatida

The rumen ciliates belonging to this order belong to one of two genera: *Isotricha* and *Dasytricha* (Fig. 7.13). The bodies of these protozoans are uniformly ciliated and have a cytostome characteristically situated in different species.

Entodiniomorphida

The rumen ciliates belonging to this order are characterized by a firm pellicle that is often drawn out posteriorly into spines and by the absence of cilia except on the peristome and in defined bands elsewhere (Fig. 7.14). Representative of the entodiniomorphid ciliates are the genera *Entodinium*, *Diplodinium*, *Ophryoscolex*, and *Cycloposthium* (Fig. 7.14).

SUCTORIANS

As indicated in the classification list at the end of the chapter, one of the subclasses of the Kinetofragminophorea is the Suctoria. The members of this subclass are widespread as ectosymbionts on a variety of marine and freshwater organisms.

The suctorians are an unusual group of protozoans that are now considered to be true ciliates. They possess both macro- and micronuclei and bear cilia during their free-swimming larval stage. As adults, they lose their cilia and develop specialized tentacles that distinguish them from all other protozoans.

The commonly spherical body of adult suctorians is covered with a pellicle and in some species with a more rigid lorica. There is no cytostome; ingestion of food occurs through unique tentacles, each of which is supported by a central axoneme comprised of microtubules. In cross-section the two files of microtubules are seen to be arranged in a daisy-like configuration when studied with the electron microscope (Sundermann and Paulin, 1981). The tentacles may be of two types: the **piercing** type, which has a sharp, pointed terminal and is used to puncture the pellicle of its prey; and the **suctorial** type, which has a rounded or flattened knob at the terminal and is used for attachment and the absorption of nutrients (Fig. 7.15). Suctorians commonly feed on other ciliates, paralyzing their prey with **haptocysts** and drawing in their cytoplasm and inclusions through the tentacles. This food becomes enclosed in vacuoles formed at the proximal end of the tentacles.

Most suctorians are free living, either attached or unattached. A few species, however, have been reported by Reynolds (1930) as epiphoronts on various

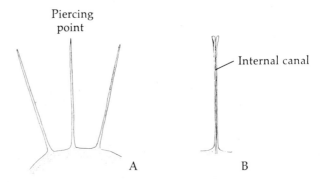

Fig. 7.15. Suctorian tentacles. A. Piercing type. **B.** Suctorial type.

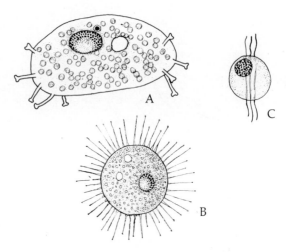

Fig. 7.16. Some parasitic suctorians. A. *Allantosoma intestinalis* attached to various ciliates living in the colon and caecum of horses. (Redrawn after Hsiung, 1928.) **B.** *Sphaerophrya stentoris* parasitic in the ciliate *Stentor*. **C.** *Endosphaera engelmanni* embedded in the cytoplasm of *Opisthonecta henneguyi* and other ciliates. (Redrawn after Lynch and Noble, 1931.)

invertebrates (Fig. 7.16), and others have been reported in other protozoans and in mammals. A few representative endosymbiotic species are considered here.

Suctorian Genera

Allantosoma, a genus with several species found attached to other ciliates in the intestines of horses, possesses an elongate ovoid body with one or more tentacles at each end (Fig. 7.16). The development, life cycles, and general biology of these unique suctorians remain to be resolved.

because while feeding on the cytoplasm, they rapidly destroy their hosts.

The genus *Sphaerophyra* includes mostly free-living species found in fresh water; but one species, *S. stentoris,* is an endoparasite of the ciliate *Stentor* (Fig. 7.16). During the life cycle of *S. stentoris,* two forms are evident; larval swarmer cells, which bear cilia at the posterior end and tentacles at the anterior end and which invade the host's cytoplasm, and adults in the cytoplasm of *Stentor.* The adults are spherical and bear tentacles on their entire body surface.

Endosphaera includes several species that are parasitic in freshwater and marine ciliates (Fig. 7.16).

The various species symbiotic in other ciliates and other protozoans could be considered endopredators

CLASS OLIGOHYMENOPHOREA

Several members of the class Oligohymenophorea are parasitic. Some are facultative, whereas others are obligatory parasites.

From the standpoint of economic importance, probably the most prominent ciliate belonging to this class is *Ichthyophthirius.*

Genus *Ichthyophthirius.* *Ichthyophthirius multifiliis* is a histozoic parasite of fish (Fig. 7.17). This ovoid parasite, which measures 100–1000 μm long, possesses a large cytostome that is 30–40 μm in diameter located at the anterior extremity. The cilia-

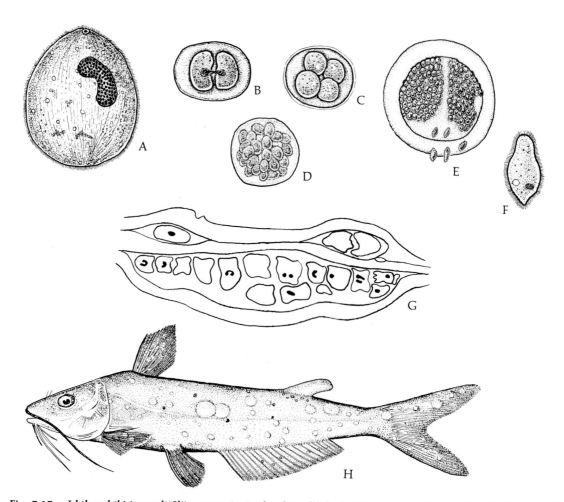

Fig. 7.17. *Ichthyophthirius multifiliis,* a parasite in the skin of fish. A. Mature trophont or trophozoite. **B.** Encysted individual undergoing first division. **C.** Four daughter individuals within cyst wall. **D.** Numerous daughter tomites within cyst wall. **E.** Mature cyst containing numerous tomites. **F.** Young theront that has escaped from cyst. **G.** Section through fin of infected carp showing numerous parasites. **H.** A catfish, *Ameiurus albidus,* heavily infected with *I. multifiliis.* (**A** and **E,** modified after Bütschli, 1882–1889; **B–D,** and **F,** modified after Fouquet, 1876; **G,** redrawn after Kudo, 1946. Protozoology, 3rd ed. Charles C Thomas; and **H,** redrawn after Stiles, 1894.)

tion is uniform. The large macronucleus is horseshoe shaped, with the micronucleus usually situated adjacent to it. The cytoplasm is exceptionally granular and includes numerous fatlike globules and contractile vacuoles. Living in the skin, gills, fins, and eyes of freshwater fish, it causes a skin disease—ichthyophthiriosis (commonly known as "ich")—characterized by the formation of lesions, which often become infected with microbes, resulting in the death of the host. The distribution of *I. multifiliis* is cosmopolitan, being known in freshwater ponds and hatcheries. It is noted that although this ciliate is difficult to find in fish in their natural habitats, it is the most common parasite of fishes in aquaria and hatcheries.

Life Cycle of Ichthyophthirius multifiliis. The life history of this important ciliate has been studied by MacLennan (1935, 1937, 1942). The stages in its life cycle are depicted in Fig. 7.17. The **trophozoites**, which are commonly found in thin-walled cysts underneath the host's epidermis, rotate continuously. They increase in size and divide to form two to four individuals. When the trophozoites (also known as **trophonts**) reach a certain size, they escape from their hosts and drop to the bottom of the aquarium or pond where they secrete a gelatinous cyst around themselves. The cysts are commonly attached to the substratum.

Within the cyst wall, the cytostome is resorbed and the cytoplasm and nucleus divide by transverse division until the cyst contains up to 1000 minute spherical ciliated cells known as **tomites**. Each tomite measures 18–22 μm in diameter. These rounded cells soon elongate, becoming 40–100 μm in length. They eventually rupture out of the cyst wall and are then known as **theronts**. The free-swimming theronts seek out new fish hosts, penetrate the epithelium, and grow to become trophozoites.

Theronts can attack fish during the first 96 hours after excystation but are most effective during the initial 48 hours.

The time required for the complete life cycle depends on the ambient temperature. At the optimum temperature, 24 to 25°C, multiplication within the cyst is about 7 to 8 hours. The free-swimming theronts, which live for about 48 hours if a host is not located, possess only incompletely developed cytostomes, which become completely formed only when the ciliates are firmly established under the host's epidermis. At this site, the trophonts, or trophozoites, actively ingest host cell components. For a detailed review of the biology of *I. multifiliis*, see Hoffman (1978).

Genus *Cryptocaryon.* Closely related to *Ichthyophthirius* is *Cryptocaryon irritans*, a histozoic parasite of marine fish. Like *Ichthyophthirius multifiliis*, this ciliate causes slimy skin on its host as well as lesions on the body and gills. Cryptocaryoniosis may also lead to blindness. It is commonly lethal. For a review of this fish disease, which is commonly referred to as the "white spot" disease, see Sindermann (1970).

Genus *Tetrahymena.* Members of the genus *Tetrahymena* are normally free-living but are commonly found as facultative parasites in various vertebrates and invertebrates (Fig. 7.18). Certain species can be induced to become parasitic by experimentally injecting them into experimental hosts (Thompson, 1958). Corliss (1960) has contributed an essay on the facultative parasitism of *Tetrahymena* spp. and has listed the various animals known to serve as hosts.

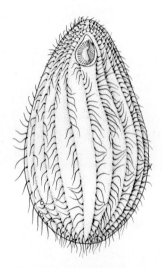

Fig. 7.18. *Tetrahymena pyriformis*, **a ciliated protozoan capable of facultative parasitism.** The body of a living specimen is pliable and measures about 50×30 μm. (Redrawn after Furgason, 1940.)

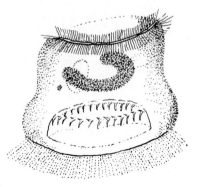

Fig. 7.19. *Trichodina.* An oligohymenophorean ciliate from the skin and gills of frog and toad tadpoles. (Redrawn after Diller, 1928.)

Genus *Trichodina*. Members of the genus *Trichodina* parasitize a wide variety of aquatic invertebrates, fish, and amphibians. Characteristic of these ciliates is the presence of a basal disc containing a crown of hard, pointed teethlike structures that aid these parasites in attaching to their hosts (Fig. 7.19). The number, arrangement, and shapes of these structures are used in specific identification.

When parasitic on fish, *Trichodina* spp. may cause some damage to the gills. Usually, however, they cause little or no injury.

Representative species of *Trichodina* include *T. urinicola* in the urinary bladder of amphibians, *T. californica* on the gills of salmon, and *T. pediculus* on *Hydra*.

CLASS POLYMENOPHOREA

The class Polymenophorea includes primarily large, free-living forms that occur in a variety of habitats; however, several species, especially members of the suborder Clevelandellina (see the end of this chapter) are endoparasitic in the digestive tracts of lower vertebrates, arthropods, and occasionally molluscs and annelids.

Genus *Nyctotherus*. Members of the genus *Nyctotherus* are commonly encountered in the large intestines of amphibians, fish, and invertebrates (Fig. 7.20). The many species have been catalogued by Wichterman (1938). Members of this genus may be recognized by the deep, laterally situated peristome that ends in a cytostome and gullet. The peristome begins at the anterior end and is lined with a row of long cilia. Situated in the cytoplasm is a large reniform macronucleus with a small micronucleus lying in its concavity.

Nyctotherus ovalis in cockroaches and millipedes, measuring 90–185 μm long, and *N. cordiformis* in frogs, measuring 60–200 μm long, are the two most commonly encountered species. These ciliates are nonpathogenic. It is of interest to note that a ciliated cytoproct (primitive "anus") has been reported in *Nyctotherus* (Fig. 7.20).

Life Cycle of Nyctotherus. The trophozoites of *Nyctotherus cordiformis* occur in the rectums of frogs and toads, where they divide occasionally by binary fission. In the spring, when the host's breeding season commences, a change from asexual to sexual reproduction occurs. This change is characterized by the trophozoites dividing more frequently, finally forming mononuclear precystic forms. Once encystation occurs, the cysts pass out into the water with the host's feces. Such cysts are ingested by tadpoles, and the preconjugants excyst. These conjugate and undergo a series of nuclear changes like those that occur in *Paramecium*. After conjugation, the postconjugants separate. These are found almost exclusively in recently metamorphosed frogs, in which the postconjugants undergo a series of binary fission.

As in *Opalina*, the injection of pregnancy urine or male and female hormones into hosts under certain conditions induces the encystation of *N. cordiformis*, suggesting that the process is correlated with the sex hormone levels in the host.

PHYSIOLOGY AND BIOCHEMISTRY OF SYMBIOTIC CILIATES

OXYGEN REQUIREMENTS

The oxygen requirement of endoparasitic ciliates varies among the species and with the location of the organisms within their hosts. It is apparent that both aerobic and anaerobic (perhaps microaerobic) species exist. Daniel (1931) has reported that *Balantidium coli* consumes oxygen and is aerobic. Unlike microaerobes, *B. coli* can be maintained for hours even under high O_2 tensions. On the other hand, Lwoff and Valentini (1948) and others have repeatedly cultured *Nyctotherus* in the absence of O_2, indicating that this endoparasitic species is capable of anaerobic respiration, perhaps as a facultative anaerobe. In the case of ciliates that live in the rumen of cattle, Hungate (1942, 1955) has demonstrated that anaerobic conditions are necessary for *in vitro* culture.

That a functional cytochrome system occurs in *Balantidium* is suggested by Agosin and von Brand (1953), who have demonstrated that a 4.5×10^{-4} M concentration of cyanide will inhibit oxygen consumption. A 10^{-2} concentration of malonate will also inhibit oxygen uptake to some extent.

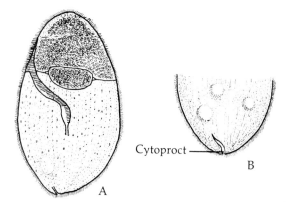

Cytoproct

B

A

Fig. 7.20. *Nyctotherus ovalis*, **a parasite in the colon of frogs. A.** Trophozoite. **B.** Posterior portion of trophozoite showing cytoproct. (**A**, redrawn after Kudo, 1936.)

As stated earlier, unlike other parasitic protozoans, the parasitic ciliates possess cytoplasmic contractile vacuoles, which presumably perform the functions of osmoregulation and excretion of metabolic waste, at least under hyperosmotic conditions.

MacLennan (1933) has demonstrated that in the rumen-dwelling ciliates *Epidinium* and *Ophryoscolex* of cattle, the intervals between contractions of the vacuoles vary from 1 minute to more than 1 hour for the former and from 2 to 4 minutes for the latter. This investigator has pointed out that there is very little absorption or diffusion of water through the cell membrane of these species, for the body walls are relatively impermeable.

Wertheim (1934a, b) has verified MacLennan's results by finding that the intervals between contractions of vacuoles in rumen-dwelling ciliates are lengthy when compared to those of free-living species. He has recorded that in *Entodinium*, *Isotricha*, *Ostrachodinium*, and *Ophryoscolex* the intervals ranged from 1 to 12 minutes. On the other hand, Strelkow (1931a, b) has reported that the intervals are rapid in *Cycloposthium* (25–30 seconds) and *Tripalmaria* (30–45 seconds), two species found in the caecum of horses. Species, and probably even more important, habitat differences, are responsible for the discrepancies in the results. Nevertheless, these observations all indicate that the contractile vacuoles of endoparasitic ciliates are present and functional.

GROWTH REQUIREMENTS

Little is known definitely about the nutrient requirements of symbiotic ciliates. It is known that many of the rumen ciliates can utilize highly complex carbohydrates and cellulose as energy sources and that *Balantidium coli* possesses a starch-splitting enzyme. These species differ from the parasitic flagellates at least in this respect, for the flagellates can utilize only simple sugars in most instances and are dependent on their hosts to degrade the more complex food molecules.

Balantidium coli has been successfully maintained *in vitro* in human serum diluted with saline and kept at 30° to 37°C when accompanying bacteria are present. It is suspected that as in *Entamoeba histolytica* cultures, the bacteria aid in the synthesis of necessary enzymes and coenzymes for the protozoa. Nevertheless, because serum is protein in composition, it is apparent that *B. coli* can utilize proteins as food. This is borne out by the tissue-ingesting habits of the parasite.

Observations on *B. coli* grown in culture have revealed that addition of starch granules is highly favorable. Ciliates can be seen enclosing numerous glycogen granules. In addition, erythrocytes and yeast cells are also avidly devoured. However, if given a choice, *B. coli* preferentially ingests starch granules.

Gutierrez and Davis (1959) have shown that at least two genera of rumen ciliates—*Entodinium* and *Diplodinium*—can utilize bacteria, which are mostly protein, as food.

The utilization of fats has been shown in some instances. In fact, parasitic ciliates usually include lipids in their cytoplasm. It is noted, however, that the so-called lipid bodies in such species as *Balantidium* and *Nyctotherus* probably do not represent pure lipids but a mixture of lipids, proteins, and other substances.

CARBOHYDRATE METABOLISM

All rumen ciliates store amylopectin, which may occur in the cytoplasm as small granules or may be incorporated as larger grains in skeletal plates. This amylopectin is derived from simple sugars or polysaccharides originating in the host's diet. In the case of oligohymenophorans, rapid synthesis of amylopectin from glucose with a consequent increase in cell density has been employed as the basis of a method to separate these protozoans from rumen contents (Oxford, 1951). Up to 70% of the cell dry weight can consist of amylopectin, and if an excessive amount of glucose is available, the protozoans may become overpacked and burst (Sugden and Oxford, 1952).

What exogenous carbohydrates are not converted to stored amylopectin in rumen ciliates are fermented via anaerobic metabolism to a mixture of acids and gas. When starved, these ciliates maintain their fermentative activity at the expense of intracellular amylopectin. Heald and Oxford (1953) and Gutierrez (1955) have demonstrated that the kinetofragminophoreans *Isotricha* and *Dasytricha* produce large quantities of hydrogen and carbon dioxide as well as appreciable amounts of lactic acid, moderate amounts of acetic and butyric acids, and traces of propionic acid as end products of carbohydrate metabolism. On the other hand, Abou Akkada and Howard (1960) and Mah (1963) have shown that two other genera of kinetofragminophoreans, *Entodinium* and *Ophryoscolex*, also produce large quantities of hydrogen and carbon dioxide but only trace amounts of lactic acid via anaerobic carbohydrate metabolism. The main acids produced by these organisms as a result of fermentation are acetic and butyric acids, and small quantities of formic and propionic acids are also produced. For a detailed review of carbohydrate metabolism in protozoans, including ciliates, see Ryley (1967).

PROTEIN METABOLISM

Very little is known about protein metabolism in parasitic ciliates, although considerable information has been accumulated on this aspect of the biochemistry of *Tetrahymena* (Kidder, 1967). In brief, it is now known that most ciliates derive their essential amino acids from their hosts, although limited transamination does occur within their bodies.

LIPID METABOLISM

As is the case with protein metabolism, most of what is known about lipid metabolism in ciliates has been gained through studies on *Tetrahymena* (see Dewey, 1967, for review). Very little is known about this aspect of the biochemistry of parasitic species except some information on rumen ciliates. Viviani *et al.* (1963) have reported the presence of 19 fatty acids, ranging from 12:0 to 21:0, as well as phospholipids in rumen ciliates. Most of the fatty acids are taken up by the parasites from the host (Gutierrez *et al.*, 1962). Furthermore, there is a "pecking order" in which the fatty acids are taken up. For example, in the case of *Isotricha prostoma*, fatty acids are concentrated in the following order: stearate > oleate > palmitate > linoleate, and in the case of *Isotricha intestinalis*, uptake of fatty acids occurs in the following order: palmitate > oleate > linoleate > stearate.

Rumen ciliates are capable of secreting lipolytic enzymes that mediate the breakdown of lipids outside of their bodies prior to uptake (Wright, 1961). However, Coleman and Hall (1969) have found that *Entodinium caudatum* can engulf oil droplets, and Prins (1977) has reported that such droplets are digested within the protozoan's body. For a review of lipid as well as other aspects of metabolism in rumen ciliates, see Coleman (1979).

HOST-PARASITE RELATIONSHIPS

Considerable discussion can be found in the literature as to whether rumen- and reticulum-dwelling ciliates should be considered commensals, parasites, or mutualists. Oxford (1955) and Hungate (1955) have reviewed the arguments relative to this question, and Coleman (1979) has presented a detailed review of this ecologic group of ciliates. It appears that certain species, such as *Diplodinium*, should be considered mutualists because they possess the enzyme cellulase and harbor cellulytic bacteria, and they can aid their hosts in cellulose digestion. Furthermore, the oligohymenophoreans occurring in both artiodactyl and perissodactyl ungulates aid the accompanying bacteria in breaking down carbohydrates, including starch ingested by their hosts. Since these species perform functions their hosts are incapable of performing, they are considered mutualists. However, Oxford (1955) has cautioned against sweeping generalities because some of the so-called mutualists are also injurious to their hosts, causing destruction of B vitamins and production of lactic acid, waste bacterial proteins, and so on.

The stomach-dwelling ciliates represent an interesting group for study, since few general statements can be made that apply to all of them. For example, *Diplodinium maggii* possesses cellulase, but coexisting species of *Isotricha*, *Outschlia*, and *Entodinium* do not. Thus it appears wise to consider each species a distinct physiologic organism despite the close phylogenetic relationship and common habitat of these species.

Several interesting observations have been made about the physiology of the relationship between *Balantidium coli* and its hosts. The clinical symptoms and tissue pathology of *B. coli* infections have been mentioned. Since host cells and cell fragments are known to occur in the food vacuoles of this ciliate, it has been generally assumed that *B. coli* brings about ulceration through mechanical ingestion. However, Tempelis and Lysenko (1957) have demonstrated that hyaluronidase—the enzyme that hydrolyzes hyaluronic acid, a component of the ground substance that binds tissues—is present in *B. coli* and is probably involved in the tissue-invading process.

Although *Balantidium coli* of pigs and humans cannot be conclusively differentiated on morphologic grounds, attempts at infecting man with specimens from pigs have been mostly futile. This then suggests the possible occurrence of at least two physiologically distinct strains, but many authorities are of the opinion that *B. coli* is normally a parasite of pigs and only under rare and ideal conditions does it become established in humans. They believe the so-called failures at infecting humans result from conditions that are less than optimal for the parasite. This concept appears reasonable because natural *B. coli* infections in man are infrequent.

REFERENCES

Abou Akkada, A. R., and Howard, B. H. (1960). The biochemistry of rumen protozoa. 3. The carbohydrate metabolism of *Entodinium*. *Biochem. J.* **76**, 445–451.

Agosin, M., and von Brand, T. (1953). Studies on the respiratory metabolism of *Balantidium coli*. *J. Infect. Dis.* **93**, 101–106.

Brandfield, J. R. G. (1955). Fibre patterns in animal flagella and cilia. *Symp. Soc. Exp. Biol.* **9**, 306–334.

Child, F. M. (1967). The chemistry of protozoan cilia and

flagella. *In* "Chemical Zoology, Vol. 1. Protozoa" (M. Florkin, B. T. Scheer, and G. W. Kidder, eds.), pp. 381–393. Academic Press, New York.

Coleman, G. S. (1979). Rumen ciliate protozoa. *In* "Biochemistry and Physiology of Protozoa" (M. Levandowsky and S. H. Hutner, eds.), Vol. 2, 2nd ed, pp. 381–408. Academic Press, New York.

Coleman, G. S., and Hall, F. J. (1969). Electron microscopy of the rumen ciliate *Entodinium caudatum*, with special reference to the engulfment of bacterial and other particulate matter. *Tissue Cell* **1**, 607–618.

Corliss, J. O. (1960). *Tetrahymena chironomi* sp. nov., a ciliate from midge larvae, and the current status of facultative parasitism in the genus *Tetrahymena*. *Parasitology* **50**, 111–153.

Corliss, J. O. (1979). "The Ciliated Protozoa," 2nd ed. Pergamon, Oxford, England.

Daniel, G. E. (1931). The respiratory quotient of *Balantidium coli*. *Am. J. Hyg.* **14**, 411–420.

Dewey, V. C. (1967). Lipid composition, nutrition, and metabolism. *In* "Chemical Zoology, Vol. 1. Protozoa." (M. Florkin, B. T. Scheer, and G. W. Kiddler, eds.), pp. 161–274. Academic Press, New York.

Fawcett, D. (1961). Cilia and flagella. *In* "The Cell" (J. Brachet and A. E. Mirsky, eds.), Vol. 2, pp. 218–270. Academic Press, New York.

Grell, K. G. (1973). "Protozoology." Springer-Verlag, New York.

Gutierrez, J. (1955). Experiments on the culture and physiology of holotrichs from the bovine rumen. *Biochem. J.* **60**, 516–522.

Gutierrez, J., and Davis, R. E. (1959). Bacterial ingestion by the rumen ciliates *Entodinium* and *Diplodinium*. *J. Protozool.* **6**, 222–226.

Gutierrez, J., Williams, P. P., Davis, R. E., and Warwick, E. J. (1962). Lipid metabolism of rumen ciliates and bacteria. I. Uptake of fatty acids by *Isotricha prostoma* and *Entodinium simplex*. *Appl. Microbiol.* **10**, 548–551.

Hall, R. P. (1953). "Protozoology." Prentice-Hall, Englewood Cliffs, New Jersey.

Heald, P. J., and Oxford, A. E. (1953). Fermentation of soluble sugars by anaerobic holotrich ciliate protozoa of the genera *Isotricha* and *Dasytricha*. *Biochem. J.* **53**, 506–512.

Hoffman, G. L. (1978). Ciliates of freshwater fishes. *In* "Parasitic Protozoa," Vol. II. (J. P. Kreier, ed.), pp. 584–632. Academic Press, New York.

Hungate, R. E. (1942). The culture of *Eudiplodinium neglectum*, with experiments on the digestion of cellulose. *Biol. Bull.* **83**, 303–319.

Hungate, R. E. (1955). The ciliates of the rumen. *In* "Biochemistry and Physiology of Protozoa" (S. H. Hutner and A. Lwoff, eds.), Vol. II, pp. 159–179. Academic Press, New York.

Hungate, R. E. (1978). The rumen protozoa. *In* "Parasitic Protozoa" (J. P. Kreier, ed.), Vol. II, pp. 655–695. Academic Press, New York.

Kidder, G. W. (1967). Nitrogen: distribution, nutrition, and metabolism. *In* "Chemical Zoology. Vol. 1. Protozoa" (M. Florkin, B. T. Scheer, and G. W. Kidder, eds.), pp. 93–159. Academic Press, New York.

Levine, N. D., Corliss, J. O., Cox, F. E. G., Deroux, G., Grain, J., Honigberg, B. M., Leesdale, G. F. Loeblich, A. R. IV, Lom, J. Lynn, D., Merinfeld, E. G., Page, F. C., Poljansky, G., Sprague, V., Varva, J., and Wallace, F. G. (1980). A newly revised classification of the Protozoa. *J. Protozool.* **27**, 37–58.

Lwoff, A., and Valentini, S. (1948). Culture du flagelle opalinide *Cepedea dimidiata*. *Ann. Inst. Pasteur, Paris* **75**, 1–7.

MacDougall, M. S. (1928). Neuromotor system of *Chlamydodon*. *Biol. Bull.* **54**, 471–484.

MacLennan, R. F. (1933). The pulsatory cycle of the contractile vacuoles in the Ophryoscolecidae. Ciliates from the stomach of cattle. *Univ. Calif. Publ. Zool.* **39**, 205–250.

MacLennan, R. F. (1935). Observations on the life history of *Ichthyophthirius*, a ciliate parasitic on fish. *Northwest Sci.* **9**, 12–14.

MacLennan, R. F. (1937). Growth in the ciliate *Ichthyophthirius*. I. Maturity and encystment. *J. Exp. Zool.* **76**, 423–440.

MacLennan, R. F. (1942). Growth in the ciliate *Ichthyophthirius*. II. Volume. *J. Exp. Zool.* **91**, 1–13.

Mah, R. A. (1963). Some physiological studies on the rumen ciliate, *Ophryoscolex purkynei* Stein. *Bacteriol. Proc.*, p. 9.

Oxford, A. E. (1951). The conversion of certain soluble sugars to a glucosan by holotrich ciliates in the rumen of sheep. *J. Gen. Microbiol.* **5**, 83–90.

Oxford, A. E. (1955). The rumen ciliate protozoa: their chemical composition, metabolism, requirements for maintenance and culture, and physiological significance for the host. *Exp. Parasitol.* **6**, 569–605.

Pitelka, D. R. (1961). Fine structure of the silver-line and fibrillar systems of three tetrahymenid ciliates. *J. Protozool.* **8**, 75–89.

Prins, R. A. (1977). Biochemical activities of gut microorganisms. *In* "Microbial Ecology of the Gut" (R. T. J. Clarke and T. Bauchop, eds.), pp. 73–183. Academic Press, New York.

Reynolds, B. D. (1930). Ectoparasitic protozoa. *In* "Problems and Methods of Research in Protozoology" (R. Hegner and J. Andrews, eds.), pp. 11–26. Macmillan, New York.

Ryley, J. R. (1967). Carbohydrates and respiration. *In* "Chemical Zoology. Vol. 1. Protozoa" (M. Florkin, B. T. Scheer, and G. W. Kidder, eds.), pp. 55–92. Academic Press, New York.

Sharp, R. G. (1914). *Diplodinium eucaudatum*, with an account of its neuromotor apparatus. *Univ. Cal. Publ. Zool.* **1**, 43–122.

Sindermann, C. J. (1970). "Principal Diseases of Marine Fish and Shellfish." Academic Press, New York.

Strelkow, A. A. (1931a). Morphologische Studien über oligotriche Infusorien aus dem Darme des Pferdes. II. Cytologische Untersuchungen der Gattung *Cycloposthium* Bundle. *Arch. Protistenk.* **75**, 191–220.

Strelkow, A. A. (1931b). Morphologische Studien über oligotriche Infusorien aus dem Darme des Pferdes. III. Körperbau von *Tripalmaria dogieli* Grassovsky. *Arch. Protistenk.* **76**, 221–254.

Sugden, B., and Oxford, A. E. (1952). Some cultural studies with holotrich ciliate protozoa of the sheep's rumen. *J. Gen. Microbiol.* **7**, 145–153.

Sundermann, C. A., and Paulin, J. J. (1981). Ultrastructural features of *Allantosoma intestinalis*, a suctorian ciliate isolated from the large intestine of the horse. *J. Protozool.* **28**, 400–405.

Tempelis, C. H., and Lysenko, M. G. (1957). The production of hyaluronidase by *Balantidium coli*. *Exp. Parasitol.* **6**, 31–36.

Thompson, J. C., Jr. (1958). Experimental infections of various animals with strains of the genus *Tetrahymena*. *J. Protozool.* **5**, 203–205.

Viviani, R., Gandolfi, M. G., and Borgatti, A. R. (1963). Fatty acid composition of the lipids of bacteria and protozoa of sheep stomach. *Bull. Soc. Ital. Biol. Sper.* **39**, 1831–1835.

Wertheim, P. (1943a). Über die Pulsation der kontraktilen Vakuolen bei den Wiederkäuermageninfusorien. *Zool. Anz.* **106**, 20–24.

Wertheim, P. (1934b). Zweiter Beitrag zur Kenntnis der Vakuolenpulsation bei Wiederkäuerinfusorien nebst einigen biologischen Beobachtungen. *Zool. Anz.* **107**, 77–84.

Wichterman, R. (1938). The present state of knowledge concerning the existence of species of *Nyctotherus* (Ciliata) living in man. *Am. J. Trop. Med.* **18**, 67–75.

Wright, D. E. (1961). Bloat in cattle. XX. Lipase activity of rumen microorganisms. *New Zealand J. Agric. Res.* **4**, 216–223.

Zaman, V. (1978). *Balantidium coli. In* "Parasitic Protozoa" (J. P. Kreier, ed.), Vol. II, pp. 633–653. Academic Press, New York.

CLASSIFICATION OF CILIOPHORA (THE CILIATES)*

PHYLUM CILIOPHORA

Simple cilia or compound ciliary organelles typical in at least one stage of life cycle; with subpellicular infraciliature present even when cilia absent; two types of nuclei, with rare exception; binary fission transverse, but budding and multiple fission also occur; sexuality involving conjugation, autogamy, and cytogamy; nutrition heterotrophic; contractile vacuole typically present; most species free living, but many commensal, some parasitic, and a large number found as phoronts on a variety of hosts.

CLASS KINETOFRAGMINOPHOREA

Oral infraciliature only slightly distinct from somatic infraciliature and differentiated from anterior parts, or other segments, of all or some of somatic kineties; cytostome often apical (or subapical) or midventral, on surface of body or at bottom of atrium or vestibulum; cytopharyngeal apparatus commonly prominent; compound ciliature, oral or somatic, typically absent.

---

*Based on the recommendation of the Committee on Systematics and Evolution of the Society of Protozoologists consisting of Levine *et al.* (1980). Diagnoses of taxa comprised of free-living species not presented.

Subclass Gymnostomatia

Cytostomal area superficial, apical or subapical; circumoral infraciliature without kinetosomal differentiation other than closer packing of kinetosomes; cytopharyngeal apparatus of rhabdos type; toxicysts common; somatic ciliation usually uniform.

ORDER PROSTOMATIDA

Cytostome apical or subapical; circumoral infraciliature involving anterior parts of all somatic kineties; typical polyploid independent macronucleus; body often large; commonly carnivorous.

Suborder Archistomatina

Cytostome apical; simplest type of circumoral infraciliature (closely packed kinetosomes); somatic ciliature mostly in tufts or bands; concrement vacuoles; no toxicysts; all known species principally in equids. (Genera mentioned in text: *Alloiozona*, *Blepharoprosthium*, *Didesmis*.)

Suborder Prostomatina

Cytostome apical, round; circumoral ciliature unspecialized; kineties bipolar, with axial-radial symmetry; no toxicysts.

Suborder Prorodontina

Suborder Haptorina

ORDER PLEUROSTOMATIDA

ORDER PRIMOCILIATIDA

ORDER KARYORELICTIDA

Subclass Vestibuliferia

Apical or near-apical (occasionally at posterior pole) vestibulum commonly present, equipped with cilia derived from anterior parts of somatic kineties and leading to cytostome; cytopharyngeal apparatus resembling rhabdos; free living or parasitic, especially in digestive tract of vertebrates and invertebrates.

ORDER TRICHOSTOMATIDA

No reorganization of somatic kineties at level of vestibulum other than more packed alignment of kinetosomes or addition of supernumerary segments of kineties; many species endosymbiotic in vertebrate hosts.

Suborder Trichostomatina

Somatic ciliature not reduced. (Genera mentioned in text: *Balantidium*, *Isotricha*, *Dasytricha*.)

Suborder Blepharocorythina

Somatic ciliature markedly reduced; all species in herbivorous mammals, especially equids.

ORDER ENTODINIOMORPHIDA

Somatic ciliature in form of unique ciliary tufts or bands, otherwise body naked; oral area sometimes retractable; pellicle generally firm, sometimes drawn out into processes; skeletal plates in many species; symbiotic in mammalian herbivores, including anthropoid apes. (Genera mentioned in text: *Entodinium*, *Ophryoscolex*, *Cycloposthium*, *Diplodinium*, *Ostracodinium*, *Tripalmaria*, *Outschlia*.)

ORDER COLPODIDA

Subclass Hypostomatia

Cytostome nonpolar, on ventral surface; cylindrical or flattened dorsoventrally, often with reduction of somatic ciliature; oral area may be sunk into atrium, with atrial ciliature present; some species astomatous; free living or ecto- or endosymbiotic, principally of invertebrate hosts.

SUPERORDER NASSULIDEA

ORDER SYNHYMENIIDA

ORDER NASSULIDA
Suborder Nassulina
Suborder Microthoracina

SUPERORDER PHYLLOPHARYNGIDEA

With "teethlike" capitula; circumoral ciliature restricted to three short rows of kinetosomes near oral opening; somatic ciliature only on ventral surface, in two asymmetric fields; preoral suture skewed to left; macronucleus commonly heteromerous.

ORDER CRYTOPHORIDA

Three rows of oral ciliature arising from kineties of left field, composed of pairs of kinetosomes with inverted polarity; body dorsoventrally flattened or laterally compressed; ventral ciliature often thigmotactic; many species with "glandular" adhesive organelle near posterior end.

Suborder Chlamydodontina

Ventral ciliature thigmotactic, without specialized glandular organelle; body broad and dorsoventrally flattened, with ventral surface in contact with substrate; macronucleus heteromerous; free living or symbiotic, some species parasitic, harmful to gills of freshwater fish. (Genus mentioned in text: *Chlamydodon*.)

Suborder Dysteriina

Generally reduced ciliature and relatively narrow body; sometimes with very prominent capitula ("teeth"); adhesive organelle well developed, often with protruding mobile appendix (podite); macronucleus heteromerous; species numerous, mainly marine, free living or symbiotic.

Suborder Hypocomatina

Ventral surface densely ciliated; dorsal surface humped; cytopharyngeal tube may protrude from body; adhesive organelle inconspicuous in right ventral pit; macronucleus not heteromerous; ecto- or endosymbionts of marine hosts.

ORDER CHONOTRICHIDA

Variously vase shaped, sessile and sedentary forms; naked, except for ciliature of ventral surface; adhesive organelle active in stalk production; macronucleus heteromerous; reproduction by budding; marine and freshwater species, ectosymbiotic principally on crustaceans.

Suborder Exogemmina

External budding, often with a single bud at a time; body relatively large, long, cylindrical, with well-developed collar, but spines often absent; stalks almost universal; freshwater, brackish, and marine hosts, including algae.

Suborder Cryptogemmina

Internal budding with up to eight tomites in brood pouch; body small, flattened, angular, with spines but reduced collar; stalk may be absent; symbionts of marine crustaceans (including "whale lice").

SUPERORDER RHYNCHODEA

Aberrant, small rostrate forms, with sucking tube and toxicysts; body of mature stage often nearly naked or with somatic ciliature limited to thigmotactic field; buds ("larvae") typically ciliated; commensals and parasites, most commonly on gills of marine bivalves.

ORDER RHYNCHODIDA

With characters of the superorder.

SUPERORDER APOSTOMATIDEA

Cytostome inconspicuous or, in certain stages of polymorphic life cycle, absent; glandular complex (rosette) typically near oral area; in mature forms somatic ciliature spiraled, often widely spaced; commonly with anterior thigmotactic ciliary field; life cycle complex, sometimes involving alteration of hosts; cysts common; most species associated with marine crustaceans.

ORDER APOSTOMATIDA

With characters of the superorder.

Suborder Apostomatina

Cytostome present in both trophont and tomite stages, typically with rosette; hosts mostly marine crustaceans, but a few species associated with polychaetes or freshwater crustaceans.

Suborder Astomatophorina

Cytostome absent; remnants of oral ciliature present; body often elongate, vermiform; marked thigmotactism; strobilation type of budding, producing colonies; life cycles incompletely known; some species in coelomic fluid of amphipods and isopods, others in organs of squid and octopus.

Suborder Pilisuctorina

Trophonts nonciliated, immobile, in cysts; some species impaled on setae of crustacean hosts; migrating tomites, produced by strobilation or budding, flattened and ciliated, but apparently mouthless; on marine crustaceans of various kinds and, perhaps, terrestrial mites.

Subclass Suctoria

Suctorial tentacles, generally multiple, containing haptocysts; adult body sessile and sedentary, seldom with cilia; reproduction by budding; stalk commonly present, noncontractile; conjugation often involving micro- and macroconjugants; migratory larva ciliated (with right field and possibly vestigial left field), without tentacles or stalk; widespread on marine and freshwater organisms, occasionally endosymbiotic.

ORDER SUCTORIDA

With characters of the subclass.

Suborder Exogenina

Budding exogenous, without invagination of parental cortex; some species with both prehensile and suctorial tentacles; larvae of some species long and vermiform, practically devoid of cilia and nonmotile; mostly large, solitary, and marine; free living or ectosymbiotic; some species loricate and colonial. (Genera mentioned in text: *Allantosoma, Sphaerophrya*.)

Suborder Endogenina

Budding endogenous, with larvae free in pouch before emergence; some species stalkless; several with atypically huge ramified body; bundles of tentacles may be branched; migratory larvae often small; diverse habitats, with some species endosymbiotic in various hosts. (Genus mentioned in text: *Endosphaera*.)

Suborder Evanginogenina

Budding involves evagination of entire pouch with bud still attached; large, single larva, flattened, with distinct patterns of ventral ciliature; some adults with branched tentacular bundles; common freshwater or marine ectophoronts, with few endoparasitic species.

CLASS OLIGOHYMENOPHOREA

Oral apparatus, at least partially in buccal cavity, generally well defined, although absent in one group; oral ciliature, clearly distinct from somatic ciliature, consisting of paroral membrane on right side and small number of compound organelles (membranelles, etc.) on left side; cytostome usually ventral and/or near anterior end, situated at bottom of buccal cavity; cysts common; various species loricate; colony formation common in some groups.

Subclass Hymenostomatia

Body ciliation often uniform and heavy; buccal cavity, when present, ventral; sessile forms, stalked or as colony; cyst formation relatively rare; most species in fresh water.

ORDER HYMENOSTOMATIDA

Buccal cavity well defined, containing membranelles with infraciliary bases typically three to four rows of kinetosomes wide; oral area on ventral surface, usually in anterior half of body.

Suborder Tetrahymenina

Uniformly ciliated; three oral membranelles on left and undulating or paroral membrane on right; preoral but no postoral suture; seldom any thigmotactic ciliature, with even caudal cilia uncommon; mucocysts common; mostly free-living, freshwater, microphagous forms, although few species facultative or even obligate endoparasites. (Genus mentioned in text: *Tetrahymena*.)

Suborder Ophryoglenina

Large, primarily freshwater, histophagous forms; life cycle polymorphic, with cyst stage; oral apparatus including three ciliary organelles left; several species causing white spot disease in marine and freshwater fish. (Genera mentioned in text: *Ichthyophthirius, Crytocaryon*.)

Suborder Peniculina

ORDER SCUTICOCILIATIDA

Body uniformly to sparsely ciliated; thigmotactic area common in many species; buccal ciliature often dominated by tripartite (anterior, middle, and posterior segments) paroral membrane on right side; mucocysts and caudal cilia common; mitochondria long, sometimes fused to form gigantic "chondriome"; cysts common.

Suborder Philasterina

Infraciliature of paroral membrane with reduced a and c segments; scutica very transient; mucocysts prominent, rod shaped; numerous species, especially in brackish or marine habitats, including sand; some symbionts of sea urchins, others endosymbionts of molluscs, polychaetes, and other hosts.

Suborder Pleuronematina

Body commonly small or very small; paroral membrane often prominent; cytostome equatorial to subequatorial; infraciliary base of paroral membrane clearly trisegmented, with segment c serving as permanent scutica; caudal cilia often conspicuous; mucocysts prominent; cytoproct typically present; widely distributed, mostly free living, but some symbiotic.

Suborder Thigmotrichina

Prominent thigmotactic region, typically near anterior end, commonly present; cytostome often at or near posterior end; body usually heavily ciliated and laterally compressed; prominent sucker at anterior end in some groups; cytoproct generally absent, all species symbiotic, especially in bivalves and oligochaetes.

ORDER ASTOMATIDA

Body usually large or long, uniformly ciliated; mouth absent; complex intraciliary endoskeleton and often elaborate holdfast organelles (hooks, spines, or sucker) may be present at anterior end; fission may be by budding, with chain formation; cytoproct absent; contractile vacuoles present; all endoparasitic, mostly in oligochaetes (soil, freshwater, marine); few species in other annelids, molluscs, and turbellarian; one major group in caudate amphibians. (Genus mentioned in text: *Haptophrya*.)

Subclass Peritrichia

Oral ciliary field prominent, covering apical end of body and dipping into infundibulum; paroral membrane and adoral membranelles becoming "peniculi" in infundibulum; somatic ciliature reduced to temporary posterior circlet of locomotor cilia; widely distributed species, many stalked and sedentary, others mobile, all with aboral scopula; dispersal by migratory telotroch (larval form); mucocysts and pellicular pores universal; myonemes associated with strong contractility of stalk or parts of body; conjugation total, involving fusion of micro- and macroconjugants.

ORDER PERITRICHIDA

With characters of the subclass.

Suborder Sessilina

Suborder Mobilina

Mobile forms, usually conical or cylindrical (or discoid and orally-aborally flattened) with permanently ciliated trochal band (ciliary girdle); complex thigmotactic apparatus at aboral end, often with highly distinctive denticulate ring; all species ecto- or endoparasites of freshwater or marine vertebrates and invertebrates; forms on gills of fish pathogenic. (Genus mentioned in text: *Trichodina*.)

CLASS POLYMENOPHOREA

Dominated by well-developed, conspicuous adoral zone of numerous buccal or peristomial organelles, often extending out onto body surface; on right side, one or seven lines of paroral ciliature; somatic ciliature complete or reduced, or as cirri; cytostome at bottom of buccal cavity; somatic infraciliature rarely including kinetodesmata; cytoproct often absent; cysts, and especially loricae, very common in some groups; often large and commonly free-living, free-swimming forms in great variety of habitats.

Subclass Spirotrichia

With characters of the class.

ORDER HETEROTRICHIDA

Generally large to very large forms, often highly contractile, sometimes pigmented; body dominated by conspicuous adoral zone, but also commonly bearing heavy ciliation; macronucleus oval or, often, beaded; parasitic and free-living species.

Suborder Heterotrichina

Suborder Clevelandellina

Somatic ciliature well developed, sometimes separated

into distinct areas by well-defined suture lines, several specialized unique fibers associated with kinetosomes; macronuclear karyophore and/or conspicuous dorsoanterior sucker characteristic of many species; endoparasitic in digestive tract of arthropods, especially insects, or lower vertebrates, occasionally in oligochaetes or molluscs. (Genus mentioned in text: *Nyctotherus.*)

Suborder Armophorina
Suborder Coliphorina
Suborder Plagiotomina

Body laterally flattened with conspicuous adoral zone and two diplostichomonad paroral lines on right side; cytostome subequatorial; no macronuclear karyophore; body uniformly ciliated, with groups of cilia highly reminiscent of cirri of primitive hypotrichs; few species, all endosymbiotic in oligochaetes.

Suborder Licnophorina

Hourglasslike shape; prominent oral disc bearing massive wreath of membranelles; conspicuous basal disc at antapical pole, the latter with a particularly complex substructure, serving as attachment organelle; body proper without cilia; several ectosymbiotic species associated with variety of marine invertebrates.

ORDER ODONTOSTOMATIDA
ORDER OLIGOTRICHIDA

Suborder Oligotrichina
Suborder Tintinnina

ORDER HYPOTRICHIDA

Dorsoventrally flattened, highly mobile (yet often thigmotactic), with unique cursorial type of locomotion; body dominated by compound ciliary structures, consisting of prominent adoral zone (of numerous paramembranelles) near anterior end, multiple paroral lines (diplo- or polystichomonads) on right side of peristomial field, and cirri on ventral surface; rows of widely spaced "sensory-bristle" cilia common on dorsal surface; some species loricate, few colony forming; species numerous and very widespread.

Suborder Stichotrichina

Body generally elongate; cirri often small and quite inconspicuous, typically in 3 to 12 longitudinal, sometimes spiraled, rows on ventral surface; predominantly free living, but one species ectosymbiotic on hydras.

Suborder Sporadotrichina

Body often oval to elliptical, cirri, nonaligned, typically heavy and conspicuous, in isolated groups in specific regions of ventral surface; most species free living in widely diverse habitats (freshwater, interstitial, marine, etc.), but few symbiotic with echinoids or ectosymbiotic with several invertebrate groups.

8

INTRODUCTION TO THE SYMBIOTIC FLATWORMS

Turbellaria: The Turbellarians
Aspidobothrea: The Aspidobothrid
Trematodes

The parasitic platyhelminths are among the oldest parasites known. The liver fluke, *Fasciola hepatica*, was reportedly first found by Jehan de Brie in 1379 in the liver of sheep in France. The importance of these worms from the medical and veterinary standpoints cannot be overemphasized. From the zoologic standpoint, these animals are extremely interesting, since they represent an important group of invertebrates the physiology and anatomy of which have been altered to meet the parasitic way of life.

GENERAL MORPHOLOGY

The phylum Platyhelminthes includes various acoelomate, dorsoventrally flattened animals that are commonly known as the flatworms. All members are typically bilaterally symmetrical and possess the flame cell or protonephritic type of excretory (osmoregulatory) system. They usually lack a definitive anus, and also lack skeletal, circulatory, and respiratory systems. The spaces between the body wall and the internal organs are filled with connective tissue fibers and unattached and fixed cells of various types. The intercellular spaces in living specimens are filled with body fluids. The fibers, cells, and spaces between them are referred to collectively as the **parenchyma**.

The platyhelminth body shape varies from a broad, dorsoventrally flattened, leaflike, monozoic animal, such as the liver fluke *Fasciola hepatica*, to a long polyzoic chain of proglottids as among the true tapeworms. Among the cestodes, or tapeworms, the body segments near the anterior or holdfast end of the chain are generally immature and smaller than those near the opposing end. Various intermediate forms exist among both the monozoic and polyzoic flatworms.

The nervous system may be in the form of a primitive epidermal net, as in a few free-living forms. However, most parasitic species possess relatively well-developed nervous systems comprised of a 'brain" or cephalic ganglion consisting of ganglionic cells and fibers that may either form a definite bilobed mass or be arranged in the parenchyma as a ring encircling the anterior end of the alimentary tract.

From the brain numerous nerve fibers are directed anteriorly, laterally, and posteriorly to innervate the various parts of the body. Generally, the number of nerve cords directed posteriorly is consistent in a given group. Transverse commissures usually connect the longitudinally oriented cords (Figs. 8.1, 8.2). The

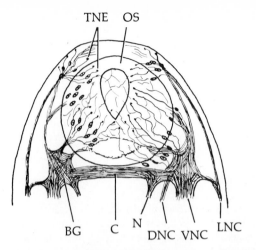

Fig. 8.1. Innervation of the anterior end and oral sucker of a digenetic trematode. BG, brain ganglion; C, connection between two lateral brain ganglia; DNC, dorsal nerve cord; LNC, lateral nerve cord; N, nerve to muscular pharynx; OS, oral sucker; TNE, tactile nerve endings; VNC, ventral nerve cord. (Redrawn after Bettendorf, 1897.)

specific arrangement pattern of the nervous system in each major group of platyhelminths is given in subsequent sections.

Except in a few cases, the parasitic flatworms are monoecious—that is, both male and female reproductive systems are found in the same individual. In sexually mature worms, both the testes and the ovaries often function simultaneously. Internal fertilization is the rule and may be accomplished by self-fertilization, which is extremely rare; by fertilization between proglottids, as in some tapeworms; or by cross-fertilization between two individuals.

Not all platyhelminths are parasitic since a large number of flatworms, members of the class Turbellaria, are free living. A few parasitic turbellarians, however, are known, such as the rhabdocoel *Collastoma pacifica*, found in the gut of the sipunculid worm *Dendrostoma pyrodes* (Kozloff, 1953). In addition, members of the family Fecampiidae of the suborder Lecithophora and those of the suborder Temnocephalida, both of the order Rhabdocoela, are symbiotic.

The more commonly encountered parasitic flatworms belong to the classes Trematoda (the flukes) and Cestoidea (the tapeworms). Adult trematodes and cestodes differ from turbellarians in lacking a ciliated cellular or syncytial epithelial covering on their body surfaces. Instead, the bodies of trematodes and cestodes are covered by an extremely thin layer of syncytially arranged cytoplasm that is connected with underlying cells (Fig. 8.3). This outer cytoplasmic surface is barely visible under the light microscope but can be fully appreciated in electron micrographs. The surface syncytium and the underlying cells are referred to collectively as the **tegument**. It is a living structure consisting of proteins, acid mucopolysaccharides, lipids, lipoproteins, mucoids, alkaline and acid phosphatases, other enzymes, and RNA and is absorptive and possibly secretive in nature. In almost all species the outer delimiting membrane is thrown into folds forming projecting **microvilli**. These presumably enhance absorption of nutrients by extending the body surface. In addition, pinocytotic vesicles have been reported associated with the tegumentary surface in certain species (Burton, 1964), which also serve as uptake mechanisms. For a detailed review of the helminth tegument, see Lee (1966, 1972).

Other details of the platyhelminth body surface are given in the subsequent chapters (pp. 304, 388).

Attachment Mechanisms

The mechanisms of attachment of flatworm parasites on or within their hosts are definitely the result of adaptation to the parasitic way of life. The most common type of attachment organ is the muscular suction cup or sucker,* of which there are various forms. Most

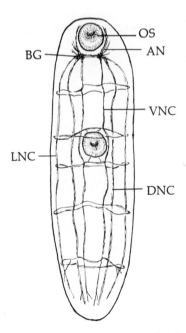

Fig. 8.2. Nervous system of a distomate trematode. AN, anterior nerves; BG, brain ganglion; DNC, dorsal nerve cord; LNC, lateral nerve cord; OS, oral sucker; VNC, ventral nerve cord. (Redrawn after Looss, 1894.)

---

*The term *sucker*, although frequently used, is not completely descriptive. The term *acetabulum* in its broad sense is probably a better one, because these holdfast structures grasp by pinching as well as by sucking.

digenetic trematodes, for example, have two suckers—the **anterior oral sucker** and the **ventral sucker** or **acetabulum.**

Monogenetic trematodes are often characterized by a single large sucker, a number of circularly (or bilaterally) arranged suckers, or a complex muscular, wedge-shaped or spatulate organ known as an **opisthaptor** located at the posterior terminals of their bodies. Tapeworms generally possess true cup-shaped suckers on their scolices or modified sucking grooves known as **bothria.** The taxonomic significance of these sucker shapes is discussed in Chapter 12.

The suckers of flatworms are highly muscular protrusible cups that not only adhere to the surface of the host's tissue, but actually embrace the facing tissue (Fig. 8.4).

In addition to suckers, monogenetic flukes often possess an array of hooks, clamps, and sucker-clamps located on the opisthaptor that aid in attachment by digging into or clamping onto the host's tissue. Some tapeworms, commonly referred to as armed tapeworms, have a spinous rostellum. One or more circular rows of hooks are located at the end of the scolex, and these also become embedded in the host's tissue. These attachment organs are of great importance to flatworms, for in their natural habitats within or on their hosts, the parasites must remain permanently and strongly attached so they cannot be dislodged by intestinal flow of chyme or swept away by

water in the case of aquatic ectoparasitic species (p. 274).

In addition to suckers and other organs of attachment, flatworms also are aided in their adherence to their hosts' tissues by specialized tegumentary microvilli (known as **microtriches** in the case of tapeworms [p. 388]). A further holdfast mechanism found on turbellarians such as *Bdelloura* is surface areas onto which mucous glands secrete. This secretion cements the worms in place.

Alimentary Tracts

Trematodes have incomplete alimentary canals that vary from a single short, blind-sac caecum or two short caeca to two long caeca with side branches (as in the liver fluke *Fasciola hepatica*). In few species the caeca may terminate with anal pores, but the typical condition is a blind gastrovascular cavity that requires that both ingestion of food and egestion of wastes occur through the mouth.

Partially or completely digested nutrients in which the worms are bathed are taken into the caeca, primarily via the mouth and esophagus, and the digested materials are distributed to the body tissues after assimilation by the caecal wall. In the case of incompletely digested materials, further degradation occurs within the caecal lumina by the action of enzymes secreted by certain cells in the caecal wall (Halton, 1967; Bogitsh *et al.*, 1968; Bogitsh and Shannon, 1971). Some nutrients, such as certain amino acids and sugars, can also be absorbed by the tegument. The extreme of parasitic adaptation is exemplified among the cestodes, in which an alimentary tract is com-

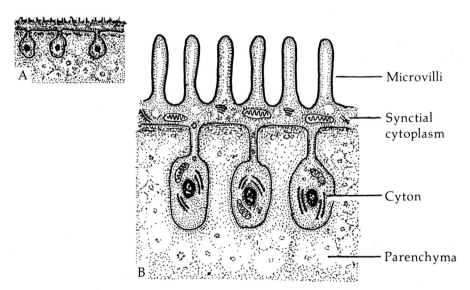

Fig. 8.3. Tegument of parasitic platyhelminths. Semidiagrammatic drawings of a portion of the body surface of parasitic platyhelminths (trematodes and cestodes) as seen in a thin section under the light microscope (**A**) and with the transmission electron microscope (**B**).

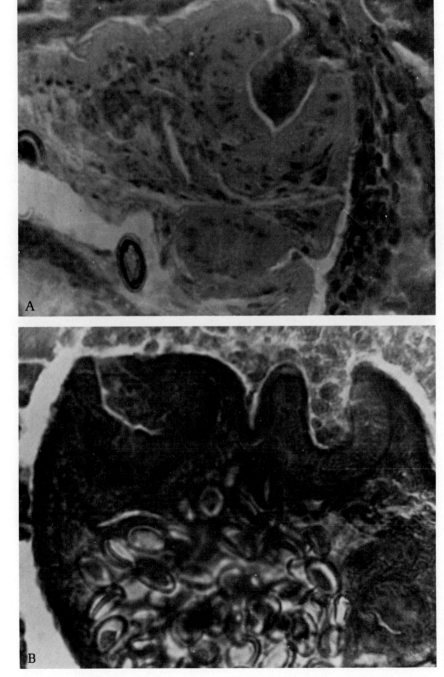

Fig. 8.4. Function of trematode suckers. Photomicrograph showing oral sucker of *Hasstilesia tricolor* holding onto intestinal mucosa of rabbit host.

pletely wanting. In these parasites, nutrients can be taken up only through the tegument.

CHEMICAL COMPOSITION

The chemical composition of endoparasitic platyhelminths has been studied by numerous investigators. Such studies have been carried out for two primary reasons: (1) to determine types of molecules the bodies of these parasites consist of, and (2) to ascertain what immunogenic molecules occur in these parasites.

Proteins

Weinland and von Brand (1926) have reported that 58% of the dry weight in *Fasciola hepatica* is protein, 67% in *Fasciola gigantica*, and 53% in *Paramphistomum explanatum* (Goil, 1958). All three of these parasites are trematodes. The proportion of protein to total dry weight of certain tapeworms has been reported as follows: 60% in *Diphyllobothrium latum* (Smorodintsev and Babesin, 1936), 36% in *Raillietina cesticillus* (Reid, 1942), 33% in *Taeniarhynchus saginatus* (Smorodintsev *et al.*, 1933), 30–36% in *Moniezia expansa* (von Brand, 1933; Wardle, 1937), 21% in *Cittotaenia perplexa* (Campbell, 1960), 32 to 33% in *Hymenolepis diminuta* (Goodchild and Vilar-Alvarez, 1962), and 45% in *Taenia taeniaeformis* (von Brand and Bowman, 1961).

In larval cestodes, Hopkins (1950) has demonstrated that 36% of the dry weight of plerocercoids of *Schistocephalus solidus* is protein, and von Brand and Bowman (1961) have reported that in *Cysticerus fasciolaris* (larva of *Taenia taeniaeformis*) 31% of the dry weight is protein.

The various protein fractions isolated from platyhelminths include albumins, globulins, albumoses, nucleoproteins, keratin, elastin, collagen, and reticulin.

Robinson (1961), by employing chromatographic separation, identified 18 amino acids in acid hydrolyzates of adults of the blood fluke *Schistosoma mansoni*. Cheng (1963), employing similar techniques, identified 15 and 19 bound, and 8 and 14 free amino acids in the sporocysts of the trematodes *Gorgodera amplicava* and *Glypthelmins quieta*, respectively, and 15 and 18 bound and 7 and 14 free amino acids in the free-swimming cercariae of these species. In addition, 17 bound and 13 free amino acids were found in the rediae of another trematode, *Echinoparyphium* sp., and 17 bound and 12 free amino acids in the cercariae of this fluke. A quantitative comparison of these amino acids in hosts and parasites strongly suggests that the parasites obtain their amino acid from their hosts. In the same paper, Cheng has reviewed what was then known about the chemical composition and utilization of amino acids in larval trematodes.

Proteins in platyhelminths are not a part of their stored food, since metazoans do not store proteins per se. Hence, these proteins must be considered functional components of the animal. However, free amino acids do exist in platyhelminths, comprising the amino acid pool, but these are continuously being utilized in the synthesis of proteins and for energy production.

The presence of proteins and their constituents in parasitic flatworms is of particular significance from the viewpoint of immunology (see Chapter 3) since

Table 8.1. Amounts of Glycogen in Some Platyhelminths Expressed as Percentages of the Fresh and Dry Weights of Whole Worms[a]

	Glycogen as Percent of			
	Fresh Weight	Dry Weight	Habitat	Availability of Significant Amount of O_2
Trematodes				
Schistosoma mansoni	—	14–29 ♂	Bloodstream	Yes
		3–5 ♀		
Fasciola hepatica	3.1; 3.7	15; 21	Bile ducts	No
Cestodes				
Diphyllobothrium latum	1.9	20	Intestine	No
Raillietina nodulosus	—	14	Intestine	No
Moniezia expansa	2.7; 3.2	24; 32	Intestine	No
Taenia solium	2.2	25	Intestine	No
Taeniarhynchus saginatus	7.4	60	Intestine	No
Cysticercus fasciolaris	—	28	Liver	?

[a] Data from von Brand, 1973.

nitrogen-containing compounds are among the most active antigens.

Carbohydrates

Glycogen is the major type of carbohydrate stored within the bodies of flatworms. It is primarily stored in the parenchyma and the muscular suckers. In the liver fluke *Fasciola* and related genera, glycogen has been reported in the vitelline glands and in eggs located in the uterus. Similarly, glycogen is present in the vitellaria and sperm of the tapeworm *Hymenolepis diminuta*. On the other hand, no glycogen is present in the vitellaria and in the eggs of the blood fluke *Schistosoma*. Furthermore, Bueding and Koletsky (1950) and Robinson (1956) have reported that in the dioecious *Schistosoma*, less glycogen is present in females than in males.

The amount of glycogen in the bodies of platyhelminths generally does not fluctuate with the season although there are evidences which indicate that the physiologic state of the host frequently has a profound influence on the parasite's glycogen content. For example, starvation of chickens greatly reduces the glycogen content in their intestinal tapeworm *Raillietina cesticillus* (Reid, 1942), and, similarly, starvation of rats reduces the amount of stored glycogen in their intestinal tapeworm *Hymenolepis diminuta* (Read, 1949; Goodchild, 1961). Some known percentages of fresh- and dry-weight glycogen in platyhelminths are given in Table 8.1.

As in the case of proteins, the occurrence of certain polysaccharides, especially polysaccharide-protein complexes, is of immunologic significance since these molecules are antigenic. Such fractions have been repeatedly isolated from various platyhelminths. For example, Maekawa *et al.* (1954) have purified and crystallized an antigenic substance from *Fasciola hepatica* that is largely proteinaceous but also contains carbohydrate.

In addition to glycogen, low-molecular-weight carbohydrates also occur in flatworms. In Table 8.2 are listed the percentages of several sugars found in three species of trematodes as determined by Ueda and Sawada (1968). These sugars, as well as others, including trehalose (Fairbairn, 1958; McDaniel and Dixon, 1967), are presumably employed in energy production via carbohydrate metabolism.

Lipids

Tötterman and Kirk (1939) have reported that 1.6% of the fresh weight of the tapeworm *Diphyllobothrium latum* is lipid, and von Brand (1933) and Smorodintsev and Bebesin (1936) have demonstrated that lipids form 3.4, 1.3, and 1.4% of the fresh weight of the

Table 8.2. Carbohydrates (in Percent of Total Carbohydrates) Identified by Gas Chromatography in Hydrolyzed Tissues of Helminths.[a]

	Carbohydrate						
	Ribose	Xylose	Mannose	Galactose	Glucose	Myoinositol	Unidentified
Clonorchis sinensis	6	0	8	10	73	3	0
Schistosoma japonicum	4	0	12	14	63	7	0
Paragonimus westermani	2	2	4	4	86	0.8	1
Necator americanus	10	7	16	15	26	24	2
Dirofilaria immitis	3	1	8	9	73	4	2

[a] Data after Neda and Swada (1968).

Table 8.3. Major Lipid Fractions of Certain Platyhelminths (in Percent of Total Lipids).[a]

	Phospholipids	Unsaponifiable Matter	Glycerides and Free Fatty Acids
Trematodes			
Schistosoma mansoni	36.6	32.4	31
Gastrothylax crumenifer	16	25	27
Cestodes			
Hymenolepis diminuta	26	9	69
Spirometra mansonoides (adult)	53.4	14.5	25.6
S. mansonoides (larva)	56	18.6	18.2

[a] Data compiled by von Brand (1973).

tapeworms *Moniezia expansa*, *M. denticulata*, and *Taenia solium*, respectively, and 30.1 and 16.2% of the dry weight of *M. expansa* and *M. denticulata*, respectively.

Flury and Leeb (1926) and Weinland and von Brand (1926) have reported that 1.9 to 2.4% of the fresh weight and 12.2 to 13.3% of the dry weight of the liver fluke *Fasciola hepatica* consists of lipids. Fractionation of these lipids has yielded phosphatids, unsaponifiable matter, saturated fatty acids, unsaturated fatty acids, hydroxy fatty acids, and glycerol. Listed in Table 8.3 are the percentages of the major lipid fractions in several species of platyhelminths.

Some studies have been carried out to identify the fats present in platyhelminths. For example, Smith *et al.* (1966) have reported the occurrence of 19 fatty acids in the blood fluke *Schistosoma mansoni* ranging from 12:0 to 24:1 (number of carbons:number of double bonds), and Ginger and Fairbairn (1966) have reported the occurrence of 12 fatty acids in the tapeworm *Hymenolepis diminuta* ranging from 14:0 to 20:2 (or more double bonds). Not all of the lipids present in the parasitic flatworms represent stored nutrients, since some are excretory products.

Those interested in the chemical compositions of platyhelminths should consult the definitive treatise by von Brand (1973).

Inorganic Substances

In addition to proteins, carbohydrates, and lipids, certain inorganic substances are present in flatworms. For example, Weinland and von Brand (1926) have shown that inorganic substances form 1.14% of the fresh weight of *Fasciola hepatica* and 4.9% of its dry weight. Similarly, 0.43% of fresh weight and 4.8% of dry weight of the tapeworm *Diphyllobothrium latum* are composed of inorganic materials (Smorodintsev and Bebesin, 1936). Listed in Table 8.4 are the percentages of inorganic substances in several species of platyhelminths.

From what is known, it is apparent that cestodes possess greater quantities of inorganic substances than trematodes. This can be explained by the accumulation of calcareous corpuscles in the parenchyma of cestodes. These corpuscles are composed of an organic base coupled with inorganic substances. These inorganic materials have been determined to be potassium, sodium, magnesium, calcium phosphate, and sulfate in the body tissues of the larva of the cestode *Taenia taeniaeformis* (=*Cysticercus fasciolaris*) (Salisbury and Anderson, 1939) and CaO, MgO, P_2O_5, and CO_2 in the calcareous corpuscles of *Taenia marginata* (von Brand, 1933). Extremely small quantities of iron have been located in the intestinal epithelia

Table 8.4. Percentage of Inorganic Substances in a Trematode and a Cestode.[a]

| | Inorganic Components in % of Dry Weight | | | | |
	Ca	Mg	K	Na	P
Trematode					
Clonorchis sinensis	0.078	0.019	0.199	0.645	0.532
Cestode					
Hymenolepis diminuta	0.075	0.061	0.836	0.323	

[a] Data compiled by von Brand (1973).

of male *Schistosoma japonicum* and in *Paragonimus ringeri*, both trematodes. It is believed that these traces of iron represent the breakdown product of the host's hemoglobin upon which these worms feed. Dawes (1954) has reported in *Fasciola hepatica* the presence of $NaCl$, KCl, $CaCl_2$, $MgSO_4$, Na_2HPO_4, and $NaHCO_3$. These same inorganic compounds, except Na_2HPO_4, plus NaH_2PO_4 and sodium acetate, apparently occur in *Schistosoma mansoni* (Senft and Senft, 1962).

Potassium, sodium, magnesium, calcium, chlorine, sulfur, iron, phosphorus, and silicon all have been reported from the body fluids of helminths, especially the cystic fluids of the tapeworms *Cysticercus tenuicollis* and *Echinococcus granulosus*.

HOSTS OF PLATYHELMINTHS

The number of hosts required by flatworm parasites to complete their life cycles varies. Among the Digenea, there are always one or more intermediate hosts. Among the Monogenea, no intermediate hosts are required. Among the Cestoidea, the number again varies from one to several. The number of obligate hosts may sometimes indicate phylogenetic relationships among these animals (p. 27).

Parasitic platyhelminths undoubtedly originated from a free-living ancestor, most probably a primitive turbellarian (p. 25). Based on this assumption, it appears that some of the free-living flatworm ancestors became adapted to ectoparasitism and, like the modern Monogenea, did not require an intermediate host. Endoparasites, especially those living in their host's intestine, probably entered their hosts through accidental ingestion, and among those that survived some were able to enter other sites associated with the alimentary tract, such as the bile duct.

Utilization of intermediate hosts, necessary for transport and later for a phase of the parasite's development, may be considered a later development. On the other hand, the intermediate hosts may have been the original definitive hosts. The problem, however, is not so easily reconciled. It is also conceivable that some parasites with more than one intermediate host may have lost their dependency on one or more of their hosts, thus reducing both the number of hosts and the complexity of their life cycles. The two seemingly opposing concepts are discussed on page 27.

The life history patterns of platyhelminths are discussed in following chapters. At this point, however, it should be mentioned that adult flukes and tapeworms usually parasitize vertebrates while larval forms generally are found in invertebrates. A number of exceptions to this rule are known among aspidobothrid trematodes and parasitic turbellarians, the adults of which parasitize various invertebrates. Also, some cestodes have only vertebrate hosts, both intermediate and definitive.

CLASSIFICATION

As indicated at the end of this and the following chapters, the phylum Platyhelminthes includes three classes:* the essentially free-living Turbellaria, the parasitic Trematoda, and the parasitic Cestoidea.

SYMBIOTIC TURBELLARIANS

The turbellarians are primarily free-living predators and scavengers. Most species are found in salt and fresh water, but some inhabit moist terrestrial habitats. There is yet no firm agreement as to how many orders comprise the Turbellaria. According to Scheer and Jones (1968), ten orders may be recognized; however, according to Jennings (1971), an authority on symbiotic turbellarians, the more traditional classification, which recognizes five orders (Acoela, Rhabdocoela, Alloeocoela, Tricladida, and Polycladida) is still the most acceptable (see diagnostic characteristics of these orders and subordinate taxa at the end of this chapter).

A number of species of turbellarians representing at least 27 families and all of the major subdivisions of the class have been reported to be engaged in symbiotic relationships with a variety of hosts. The most common hosts for symbiotic turbellarians are echinoderms, crustaceans, and molluscs. Less common hosts include annelids, sipunculids, arachnids, coelenterates, other turbellarians, and elasmobranch and bony fishes. One striking feature of these associations is their host specificity, i.e., the turbellarians belonging to the same family tend to be associated with one

---

*There have been several schemes proposed for the classification of the Platyhelminthes. At the present there is no firm agreement. It should be mentioned, however, that some are of the opinion that there are four classes in this phylum: Turbellaria, Monogenea, Trematoda, and Cestoidea. Also, Ubelaker (1983) has proposed that Cestoidea should be considered a separate phylum.

Table 8.5. Species of Acoela Associated with Marine Invertebrate Hosts.[a]

Acoel	Host
Avagina glandulifera	*Spatangus purpureus* (Echinoidea)
Avagina incola	*Echinocardium flavescens* (Echinoidea)
	Echinocardium cordatum (Echinoidea)
Aphanastoma sanguineum	*Chirodota laevis* (Holothuroidea)
Aphanastoma pallidum	*Myriothrochus rinkii* (Holothuroidea)
Aechmalotus pyrula	*Eupyrgus scaber* (Holothuroidea)
Otocoelis chirodotae	*Chirodota laevis* (Holothuroidea)
Ectocotyla paguri	Hermit crabs (Crustacea)
Meara stichopi	*Stichopus tremulus* (Holothuroidea)

[a] Data compiled by Jennings (1971).

type of host. For example, almost all of the members of the rhabdocoel family Umagillidae are associated with echinoderms, and the remaining members occur only in sipunculids.

ORDER ACOELA

The acoelous turbellarians are small flatworms that measure from one to several millimeters in length. They are exclusively marine. Their more primitive evolutionary position among turbellarians is reflected by the absence of an excretory system and of a permanent gut lumen. Digestion of phagocytosed food particles occurs in temporary vacuoles within the syncytial lining of the gut (Jennings, 1957).

A few species of Acoela have been reported associated with other marine invertebrates (Table 8.5). One of these, *Ectocotyla paguri*, is an ectosymbiont of hermit crabs along the Atlantic coast of North America. Its symbiotic habit is reflected by the presence of a posteriorly situated adhesive disc, by means of which it is attached to its host.

ORDER RHABDOCOELA

Like the acoels, rhabdocoels are small, but they have a more complex internal anatomy. The gut is saclike, with a permanent lumen and an anterior muscular pharynx that serves a feeding function.

As indicated in the classification list at the end of this chapter, the order Rhabdocoela is divided into four suborders: Notandropora, Lecithophora, Opisthandropora, and Temnocephalida.

Suborder Temnocephalida

Members of the Temnocephalida are all symbionts on freshwater animals, primarily crayfishes, prawns, isopods, and other crustaceans. They are less frequently found on turtles and snails. These turbellarians are attached to the external surfaces or in the branchial chambers of their hosts. The relationship between the host and the symbiont appears to be commensalistic, for rather than deriving nutrient from the host, these helminths feed on small animals and diatoms available in the habitat. For example, *Temnocephala brenesi*, an ectocommensal on the gills of the freshwater prawn *Macrobrachium americanum*, has been shown by Jennings (1968a, b) to feed on protozoans, rotifers, small annelids, crustaceans, and fragments of the host's food. Only one species, *Scutariella didactyla*, found on prawns in Montenegro, apparently feeds on its host's body fluids and, hence, may be considered a parasite.

Although temnocephalids do not normally leave their hosts, attempts to remove them have revealed that some species can live for some time away from the host, whereas others die in a short time. Hence, some physiologic dependence, especially in the latter cases, must exist.

Geographically, these ectosymbionts are limited to tropical, subtropical, and certain temperate areas. Commonly found in South America, Australia, and New Zealand, some species also have been reported from the Balkan Peninsula, India, Central America, and the islands of the South Pacific.

Morphology

Temnocephalid turbellarians are small, measuring between 0.3 and 2 mm in length. They are generally colorless and somewhat translucent. Their body surface is structurally well adapted to symbiosis. For example, *Temnocephala* has 6 prominent anteriorly directed tentacles, *Actinodactylella* has 12 tentacles distributed along the body, and *Caridinicola*, *Monodiscus*, and certain other genera possess two short, stumpy tentacles (Fig. 8.5). Only members of the genus *Didymorchis* lack tentacles. In others, such as *Craspedella* (Fig. 8.6), papillae are present on the dorsal posterior region in addition to tentacles.

In addition to the conspicuous tentacles, almost all temnocephalids possess a posteriorly situated adhesive disc. Members of *Caridinicola* lack an adhesive disc but have in its place a pair of adhesive pits (Fig. 8.5). In some genera an anterior adhesive organ is also present. The adhesive discs of temnocephalids are not as specialized as the suckers of trematodes. By alternately attaching themselves by their tentacles

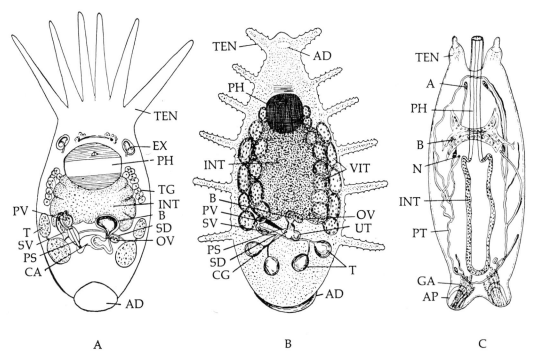

A   B   C

Fig. 8.5. Symbiotic turbellarians. A. *Temnocephala.* (After Haswell, 1893.) **B.** *Actinodactylella.* (After Haswell, 1893.) **C.** *Caridinicola.* (After Plate, 1914.) A, athrocyte; AD, adhesive disc; AP, adhesive pit; B, bursa; CA, common antrum; CG, common gonopore; EX, excretory ampulla; GA, glands of adhesive pit; INT, intestine; N, nephridiopore; OV, ovary; PH, pharynx; PS, penis stylet; PT, protonephridial tubules; PV, prostatic vesicle; SD, sperm duct; SV, seminal vesicle; T, testis; TEN, tentacle; TG, tentacle glands; UT, uterus; VIT, vitellaria.

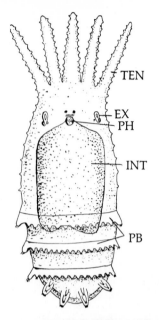

Fig. 8.6. *Craspedella.* Adult specimen showing structures. EX, excretory ampulla; INT, intestine; PB, papillate bands; PH, pharynx; TEN, tentacle. (After Haswell, 1893.)

and adhesive discs, temnocephalids are able to move about on their hosts.

In addition to their locomotor function, the tentacles are also tactile and contribute to capturing prey. Furthermore, these structures apparently can serve as defense mechanisms, since an internal pair of gland cells furnish rhabdites, similar to those of free-living planarians, to the anterior tips of the tentacles.

The syncytial body surface of temnocephalids is different from that of other turbellarians in that there is usually an anucleate cytoplasmic border. Although thinner than the tegument of trematodes and cestodes, this border resembles a tegument.

Internal Structures.

Reproductive System. The temnocephalids are monoecious. There is usually a pair of testes present. Each member of the pair may be divided into two to six parts. Leading from each testis is a **vas efferens**, which unites with its counterpart to form the common **vas deferens** or sperm duct. Sperm within the sperm duct are deposited in the **seminal vesicle**, from which they are conducted to the **cirrus**—an eversible and protrusible structure—or the **penial apparatus**. A group of unicellular glands (**prostate glands**) either empty into the base of the penis or into a prostatic vesicle, which in turn empties into the terminal end of the male reproductive tract.

The female reproductive system consists of a single ovary, from which the **oviduct** arises. Material from the paired yolk glands empties into the female tract. These glands are subdivided in some species and

branched to form an anastomosing network in others. Beyond the point where the ducts from the yolk glands enter, the female tract generally becomes enlarged to form the **seminal bursa**, which is extremely active in digesting and absorbing excess yolk.

In some species, one or more small **seminal receptacles** are present. The **uterus**, which conducts the zygotes away from the seminal bursa, possesses thickened walls of secretory cells that produce the materials that surround the zygote and yolk in forming the capsule. The uterus and the terminal end of the male reproductive system open into a common **genital atrium**. The opening through which the atrium communicates with the exterior is the **gonopore** or genital pore.

Alimentary Tract. The alimentary tract is incomplete. The mouth is midventral, nearer the anterior end of the body, and it leads into the pharynx, which in turn leads into a blind-sac, the intestinal **caecum**.

Jennings (1968b) has presented a comprehensive review of what is known about ingestion and digestion among turbellarians, including the temnocephalids. In brief, while attached to their hosts, most species of *Temnocephala* capture planktonic prey and food fragments from their hosts by employing the anteriorly directed tentacles. The worm then bends ventrally to bring the food to the midventrally situated mouth. During ingestion, the pharynx exerts considerable pressure on the food to macerate it. The gastrodermis lining the intestinal caecum consists of club-shaped gland cells and phagocytic cells. Histochemical investigations have revealed that the gland cells contain an endopeptidase that is discharged into the lumen. By means of this and other hydrolytic enzymes, ingested and partially fragmented food particles are further disintegrated, but digestion is incomplete. Eventually, the partially digested food is phagocytosed, and digestion is completed within the phagocytic cells. Within the food vacuoles of the phagocytes are found an endopeptidase, aminopeptidase (an exopeptidase), alkaline phosphatase, and possibly other enzymes.

Nervous System. The nervous system consists of a cephalic ganglion or "brain" situated near the pharynx (Fig. 8.7). From this nerve center a series of nerve cords arise that form both subepidermal and submuscular nets. The submuscular net contains many sensory nerves which lead from the tentacles to the brain. Also included in the submuscular net are three pairs of longitudinal nerve cords with transverse connectives.

Many species possess a pair of eyes. Each eye is double, with a retinal cell at each end and a pigment

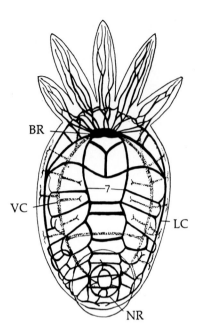

Fig. 8.7. Brain of *Temnocephala*. Drawing showing position of the brain and main nerve cords. BR, brain; LC, lateral nerve cord; NR, nerve ring in adhesive disc; VC, ventral nerve cord. (After Merton, 1913.)

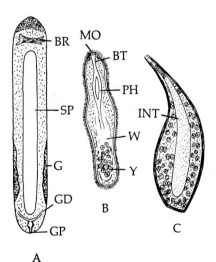

Fig. 8.8. *Fecampia*. A. Adult without mouth or pharynx. **B.** Juvenile stage with digestive system and eyespots. **C.** Adult secreting capsule containing many clusters of eggs and yolk cells. BR, brain; BT, buccal tube; G, hermaphroditic gonad; GD, gonoduct; GP, common gonopore; INT, intestine; MO, mouth; PH, pharynx; SP, space; W, worm inside capsule; Y, yolk in intestine. (After Caullery and Mesnil, 1903.)

cell in between. For a detailed account of the nervous system of turbellarians, see Bullock and Horridge (1965).

Suborder Lecithophora

Another group of symbiotic rhabdocoels, members of the family Fecampiidae of the suborder Lecithophora, are endoparasitic in various marine crustaceans. Adults of the genus *Fecampia* (Fig. 8.8) live in the body cavity (hemocoel) of various marine crustaceans. For example, *F. xanthocephala* is found in the marine isopod *Idotea neglecta*. *Fecampia* is the only known parasitic turbellarian that includes a free-living larval stage that is morphologically distinct from the adult.

Life Cycle of *Fecampia*. The larva of *Fecampia* possesses eyes, a mouth, a long buccal tube, a pharynx, and an intestine. In the posterior end of the body are found the so-called embryonic germinal cells. By some yet unknown means they enter their host's hemocoel and grow to sexual maturity. During maturation, each larva loses its eyes, mouth, and pharynx, while the intestinal caecum remains as a longitudinal slit closed at both ends. Upon completion of these morphogenetic changes, the reproductive structures develop, and the body becomes increasingly more opaque because of an increase in size of the numerous subepidermal cells. The tegument of both the larval and adult stages is ciliated.

Adults escape from the host, presumably by penetrating through the less heavily chitinized areas of the exoskeleton, and drop to the bottom of the sea. Secretions from their subepithelial gland cells form a pear-shaped cocoon, within which they lay their eggs. The eggs are surrounded by capsules, each enveloping two eggs and many yolk globules (Fig. 8.8). Upon completion of oviposition, the adult gradually disintegrates.

Several species of *Fecampia* are known: *F. erythrocephala* occurs in hermit crabs and *F. spiralis* in an Antarctic isopod. Aside from *F. spiralis*, the various species have been found only in European coastal waters.

OTHER SYMBIOTIC TURBELLARIANS

Many other species of turbellarians have been reported to be symbiotic on or in marine animals. In all instances, however, little morphologic difference exists between these symbionts and their free-living relatives. The most striking difference is that rhabdites are absent or present only in restricted parts of the body of the endosymbiotic species. This can be interpreted as an adaptation to endosymbiosis because free-living turbellarians utilize rhabdites as defensive or offensive mechanisms. The endosymbiotic species, being in constant and continuous contact with nourishment and free from predators, have completely or partially lost their rhabdites.

Truly parasitic turbellarians are known to occur in echinoderms, marine molluscs, and the shipworm, *Teredo*. *Paravortex gemellipara* lives in the mussel *Modiolus* off the New England coast (Fig. 8.9). *Paravortex cardii* is found in the ovary of another bivalve, *Cardium edule*; *Oekiocolax plagiostomorum* occurs in the parenchyma of the marine turbellarian *Plagiostomum*, causing parasitic castration; and *Graffilla* occurs in the viscera of marine gastropods and shipworms.

Certain members of the order Alloeocoela are ecto-commensalistic or ectoparasitic on marine invertebrates. For example, certain species of *Hypotrichina* are found on the crustacean *Hebalia* in the Mediterranean. Among the Tricladida, again, a few species are either ectocommensals or ectoparasites. *Syncoelidium pellucidum* and certain species of *Bdelloura*, for example, are found on the horseshoe crab, *Limulus*.

Micropharynx, which is attached to the body surface of rays in northern waters, participates in a true parasitic relationship. This triclad feeds on the host's epithelial cells. Adaptations to parasitism occur in this species; its pharynx is much smaller than that of closely related free-living species, probably reflecting its type of nutrition, and a subventral holdfast organ allows the worm to attach itself to the host.

Many polyclads are found associated with other organisms in a more or less permanent fashion, but the nature of such associations has not been studied in any detail. Although polyclads are found in the shells of marine molluscs, feeding on mucus, detritus, and small organisms, the same species also occur free. Thus, their presence in molluscs is probably accidental and not obligatory. In some cases, a true commensalistic relationship may exist.

Stylochus inimicus, *S. ellipticus*, and *S. frontalis*, the so-called oyster leeches, penetrate through the shells of oysters to destroy the adductor muscle, thus releasing the valves. They then enter through the gaping valves and feed on the oysters' soft parts. The enzyme secreted when the polyclad burrows through the mollusc's shell has not been identified. These polyclads are of economic importance to the oyster industry. It is known that oysters will attempt to defend themselves against *Stylochus* by walling the invader off with a layer of conchiolin secreted by the mantle. Cheng (1967) has presented a review of the biology of these species of polyclad turbellarians.

Certain other species of polyclads associated with marine invertebrates appear to be on the borderline of parasitism. *Euplana takewakii*, for example, is found in the genital bursae of ophiuroids, and *Hoploplana inquilina* is found in the mantle cavities of marine snails.

It is of evolutionary interest to emphasize that certain modern turbellarians do engage in symbiotic relationships. Because present-day trematodes and cestodes are believed to have arisen from some ancient turbellarian ancestors that are also the progenitors of modern turbellarians, it is significant to know that the ability to engage in ecto- and endosymbiosis is still present in the turbellarian stock (see p. 25).

CLASS TREMATODA

All adult trematodes are parasitic, monozoic flatworms. The class is divided into three orders:[*] Aspidobothrea, Monogenea, and Digenea. The ecto- and endoparasitic aspidobothrean trematodes lack an oral sucker but possess an enormous ventral holdfast organ, the **ventral disc**, which is usually subdivided into compartments. No anterior adhesive structures or hooks are present. The single excretory pore is located posteriorly.

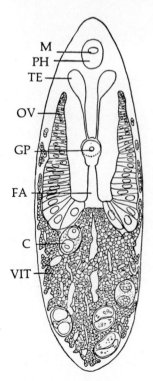

Fig. 8.9. *Paravortex.* Adult of *P. gemellipara* from the mussel *Modiolus.* C, capsule containing two embryos; FA, female atrium (or antrum); GP, common gonopore; M, mouth; OV, ovary; PH, pharynx; TE, testis; VIT, vitellaria. (After Ball, 1916.)

---

[*]There are several classification schemes for the class Trematoda. Another popular one is to divide this class into three subclasses—Digenea, Aspidobothrea, and Didymozoidea—and to consider the Monogenea representative of another class. I prefer the one presented here.

The ecto- or endoparasitic monogenetic trematodes may or may not possess an oral sucker. If one is present, it is weakly developed. A large attachment organ that is armed with hooks, also known as an **opisthaptor**, is present at the posterior end of the body of these trematodes. Generally, a pair of adhesive structures is located anteriorly, as are the paired anterodorsal excretory pores.

The endoparasitic digenetic trematodes all possess a well-developed oral sucker. A second sucker is commonly present on most species. The suckers are not armed with hooks. Generally, a single posterior excretory pore is present. Unlike monogenetic trematodes, which contain only a few large eggs in their uteri, the uterus of digenetic trematodes is usually long and contains numerous eggs.

All trematodes are obligatory parasites as adults. If a free-living phase occurs during their larval development, it lasts for an extremely short period. For example, oncomiracidia of monogenetic trematodes and miracidia and cercariae of certain digenetic trematodes are free-swimming for hours. Since these larvae do not feed, they must quickly enter a suitable host before further development can occur.

ORDER ASPIDOBOTHREA

Aspidobothrea, sometimes referred to as the Aspidogastrea or the Aspidocotylea, includes relatively few species of flatworms that have been considered by various authorities as members of either the Digenea or the Monogenea. However, in recent years most helminthologists consider this group to represent a separate order of the class Trematoda.

Aspidobothrids are primarily endoparasites of ectothermic vertebrates (fish and turtles) (Hendrix and Overstreet, 1977) and invertebrates, both freshwater and marine, but they are most commonly encountered in bivalve and gastropod molluscs. The common species, *Aspidogaster conchicola*, is often found in freshwater clams of the genera *Unio*, *Anodonta*, and *Gonidea*. It is also known to occur in snails of the genera *Viviparus* and *Goniobasis* in North America, usually in the digestive gland. Sometimes as many as 15 to 20 specimens can be found surrounding the heart in the transparent pericardium. When present in the pericardium, it elicits a tissue reaction (Huehner and Etges, 1981). This parasite can also occur in the renal cavity where renal metaplasia is a typical host response (Pauley and Becker, 1968). Ectoparasitic species of aspidobothrids are also known. Rohde (1972) has contributed a definitive review of the morphology and biology of this group of trematodes.

Yamaguti (1963) has presented a complete taxonomic treatise of the group.

Morphology

All aspidobothrids are readily recognized by the large, oval, powerful holdfast organ, the **ventral disc**, which covers the entire ventral surface of the body. This sucker is generally subdivided by septa into two to several longitudinal rows of depressions known as **alveoli**. The number and arrangement of the alveoli are important in identifying these parasites.

The anterior end of the body tapers to a somewhat blunted end where the funnel-shaped mouth is located. Although an oral sucker is said to be absent, some species, such as *Aspidogaster conchicola* and *A. limacoides*, possess a band of muscles that surrounds the mouth, giving the appearance of a rudimentary oral sucker.

The mouth opens into a **prepharynx**, which leads into a muscular **pharynx**, which in turn opens into a long, blind-sac **intestinal caecum** which in most species terminates near the posterior end of the body. The caecum is lined with a layer of columnar epithelium surrounded by two thin muscle layers—an inner longitudinal and an outer circular layer.

In an electron microscopic study of the digestive tract of *Multicotyle purvisi*, Rohde (1971a) has described

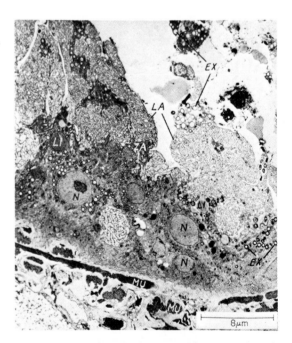

Fig. 8.10. Aspidobothrid caecum. Electron micrograph of a transverse section through caecum of *Multicotyle purvisi.* EX, excretory body; LA, lamellae (of nerves) or lamellar evaginations of surface membrane (of protonephridia); LI, lipoid droplet; MU, muscle cells; N, nuclei; ZA, zona adherens. (After Rhode, 1971; with permission of *Zoologische Jahrbücher.*)

the circular and longitudinal muscle layers of the caecum (Fig. 8.10) and the epithelial lining, the cells of which are separated by lateral, deeply convoluted membranes (Fig. 8.10). The mucosal cells are rich in Golgi complexes, rough endoplasmic reticulum, mitochondria, and invaginations of the basal cell membrane. Also, the free surface of these lining cells is enlarged by numerous lamellae or folds, the presence of which increases the absorptive surface (Fig. 8.10).

Reproductive Systems. Aspidobothrids are monoecious. The reproductive organs resemble those of digenetic trematodes (p. 307), but the male system generally includes one testis, although two are found in some species. The posteriorly located testis leads into the seminal vesicle via the vas deferens. Spermatozoa are expelled from the system through the copulatory organ, the **cirrus**, which may protrude through the genital atrium to the outside by the ventral genital pore at the anterior margin of the opisthaptor. In some genera, such as *Cotylaspis* and *Aspidogaster*, a muscular cirrus sac surrounds the seminal vesicle and the cirrus; in other genera, such as *Lissemysia*, such a sac is absent.

The female system consists of a single ovary located anterior to the testis or testes, near the midlength of the body. The ovary, like that of monogenetic trematodes (p. 281), is commonly folded. Leading from the ovary is the oviduct, which enters the **ootype**. The latter is an enlargement in the oviduct where the eggs are formed.

Three ducts are connected to the ootype or to a portion of the oviduct near the ootype: (1) the **vitelline duct**, which leads into the ootype, originates from the union of the left and right vitelline ducts coming from the vitelline glands; (2) the **Laurer's canal**, which originates at the ootype, is a long tube that is directed posteriorly and either ends blindly in the parenchyma or empties into the excretory vesicle, depending on the species; and (3) the **uterus**, which leads from the ootype to the common **genital atrium**.

The aspidobothrid oviduct is lined with ciliated epithelium (Rohde, 1971b), presumably facilitating the passage of ova into the ootype. It has been postulated by Dawes (1946) that the Laurer's canal is a reservoir for excess shell material, preventing the blockage of the oviduct, vitelline duct, and uterus by such matter. Another interpretation is that this canal is a vestigial vagina. The vitelline glands are distributed as two lateral groups and function identically to those found in the other trematodes and cestodes (p. 308).

Both self- and cross-fertilization between two individuals are possible among the Aspidobothrea. During copulation, the worm inserts the cirrus either into its own uterus or into the uterus of a mate. The spermatozoa pass down the uterus to the ootype, where fertilization takes place. In living specimens, motile sperm can sometimes be seen along the length of the uterus, and for this reason the uterus is considered a uterine seminal receptacle.

Osmoregulatory System. The osmoregulatory (excretory) system of these worms is composed of flame cells (Fig. 8.11) and collecting tubes. The protonephridia have separate excretory vesicles that empty to the outside through either separate pores or through a common pore located on the dorsal surface of the body near the posterior extremity.

Body Tissues. The body surface of the Aspidobothrea, as in all trematodes and cestodes, is covered with a thin tegumentary layer. When examined with

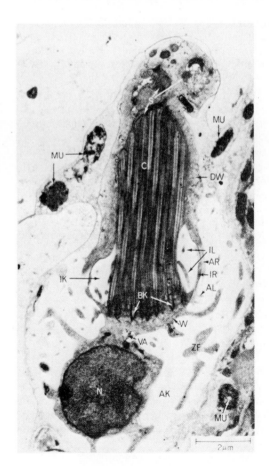

Fig. 8.11. Aspidobothrid flame cell. Electron micrograph of section through a flame cell of *Multicotyle purvisi* adult. AK, external chamber of flame cell; AL, external leptotriches; AR, external rib; BK, basal granule; Cl, cilia; DW, thick-walled upper portion of flame cell; IK, internal chamber of flame cell; IL, internal leptotriches; IR, internal rib; MU, muscle cells; N, nuclei; S, distal processes of ciliary membranes; VA, vacuoles; W, rootlets of cilia; ZF, cytoplasmic processes of flame cell. (After Rhode, 1972; with permission of Academic Press.)

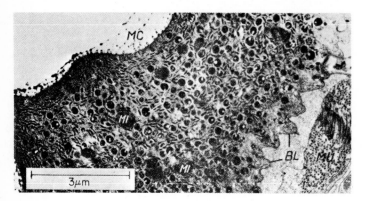

Fig. 8.12. Aspidobothrid tegument. Electron micrograph showing a portion of the tegument of *Multicotyle purvisi* from the dorsal surface in middle of body. BL, basal lamella; MC, mucoid layer; MI, mitochondria; MU, muscle cells. (After Rhode, 1971; with permission of *Zoologische Jahrbücher*.)

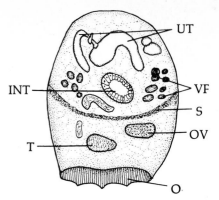

Fig. 8.13. Aspidobothrea structure. Transverse section through region of testis of *Taeniocotyle* showing muscular septum. INT, intestine; O, opisthaptor; OV, ovary; S, septum; T, testis; UT uterus; VF, vitelline follicles.

the electron microscope, it is almost identical to the tegument of digenetic trematodes (Rohde, 1971c), being a syncytium containing large numbers of mitochondria, vacuoles, lamellated bodies, and other organelles. This syncytial, outer stratum is connected by cytoplasmic processes to subtegumental nucleated cells (or cytons) (Fig. 8.12). Immediately underlying the surfacial layer of the tegument is an undulating basal lamina, beneath which are four muscle layers. (1) The outermost layer is composed of circularly arranged muscles. (2) The underlying layer of longitudinal muscles is unique inasmuch as the striated cytoplasm is limited to the periphery of each fiber; hence, they are sometimes described as shallow longitudinal muscles. (3) Next comes a layer of diagonal muscles. (4) The deepest layer is composed of circumferentially arranged myofibers. All these muscles have their origins and insertions in the basal lamina. The spaces between the body wall and the internal organs are filled with loose parenchymal tissue. In the anterior and posterior regions, dorsoventral muscles extend through the parenchyma.

Unique among some aspidobothrids, such as *Taeniocotyle*, is a deep bowl-shaped muscular septum that divides the animal into upper and lower regions (Fig. 8.13). This septum is composed of an upper layer of transverse muscles and a lower layer of longitudinal muscles. The partition is situated such that the upper zone contains the intestinal caecum, the extremities of the reproductive ducts, and the vitelline glands, whereas the lower chamber contains the gonads. The function of this septum is unclear, although it has been suggested that the contraction and extension of the myofibers create a current in the parenchyma that causes the body fluids to circulate.

Biology of Aspidobothreans

The classification of the order Aspidobothrea is presented at the end of this chapter.

General Life Cycle Pattern. The eggs of aspidobothreans are ectolecithal, i.e., most of the yolk, which serves as nutrient for the embryo, is derived from distinct cells (vitelline gland cells) and becomes packaged with the zygote within the eggshell.

The embryos of some species are completely developed within the eggshell when laid by the parent worm. These hatch within a matter of hours. In the case of other species, 3 to 4 weeks of embryonic development is required after the egg is laid. The larva hatching from the egg is known as a **cotylocidium** (Fig. 8.14). It possesses a mouth, pharynx, and simple caecum. A prominent posterioventral sucker is present. This holdfast organ on cotylocidia is not divided into alveoli nor does it bear hooks.

Those species of aspidobothreans that parasitize invertebrate hosts as adults have a direct life cycle, i.e., no intermediate host is involved. On the other hand, those parasitic in vertebrate hosts generally require an intermediate host.

After the cotylocidium enters the host, alveoli begin to develop, tier by tier, in the anterior part of the ventral disc. Then the original disc differentiates into the adult ventral disc through growth and septum formation.

The fine structure of the cotylocidia of *Multicotyle purvisi* and *Cotylogaster occidentalis* has been reported by Rohde (1972) and Frederickson (1978), respectively. When studied with the electron microscope, the tegument of the cotylocidium is essentially identical to that of the adult. The outer syncytial layer

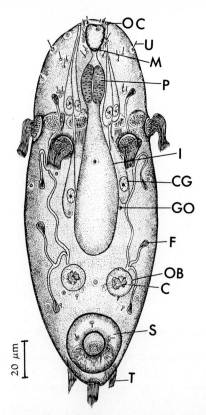

Fig. 8.14. **Aspidobothrid cotylocidium.** Cotylocidium of *Cotylogaster occidentalis*, a parasite of freshwater bivalves. C, concretion; CG, cephalic gland; GO, opening of goblet-like cells; F, flame cell; I, intestine; M, mouth; OB, excretory (osmoregulatory) bladder; OC, opening of cephalic gland; P, pharynx; S, sucker; T, tuft of cilia; U, uniciliate sensory structure. (After Frederickson, 1978; with permission of *Journal of Parasitology*.)

is connected by cytoplasmic bridges with the more deeply situated, nucleated cytons. Rohde (1972) has reported that between the ciliary tufts (Fig. 8.15) and covering most of the body, the tegumental surface of *M. purvisi* bears unique filiform structures termed **microfila** (Fig. 8.16). Each microfilum has one central

Fig. 8.15. Aspidobothrid tegument. Diagrammatic drawing of a portion of the tegument of a cotylocidium of *Multicotyle purvisi* showing ciliary tufts and microfila. BK, basal body; BL, basal lamina; BZ, basal cell membrane; Cl, cilia; EBZ, invaginations of basal cell membrane; FA, fibers; FAD, fascia adherans; MC, mucoid layer; MF, microfila; MI, mitochondria; MO, macula occludens; N, nuclei; O, ovoid bodies; W, rootlets of cilia; ZA, zona adherans. (After Rhode, 1971; with permission of *Zoologische Jahrbücher*.)

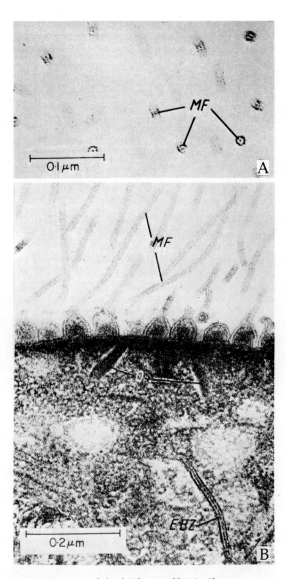

Fig. 8.16. Aspidobothrid microfila. A. Electron micrograph of transverse and oblique sections through microfila of *Multicotyle purvisi* cotylocidium. **B.** Electron micrograph of section through the tegument of *M. purvisi* cotylocidium showing projecting microfila. EBZ, invaginations of basal cell membrane; MF, microfila; O, ovoid bodies. (After Rhode, 1971; with permission of *Zoologische Jahrbücher*.)

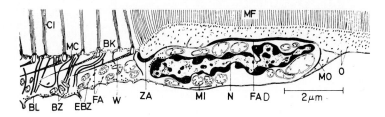

filament and about 9 to 12 peripheral filaments. Microfila differ from microvilli in that there is no cytoplasmic core. Although the function of microfila still remains uncertain, it has been suggested that they provide buoyancy for the cotylocidium to float. Microfila apparently are absent on the cotylocidium of *C. occidentalis*, which bears short microvilli overlayed by an external glycocalyx coat (Fig. 8.17) (Frederickson, 1978).

Life History of Aspidogaster conchicola—*One-Host Cycle.* Only a few life cycles are known among the Aspidobothrea. Huehner and Etges (1977) have demonstrated the complete life cycle of *Aspidogaster conchicola*. Eggs laid by adults in the snail *Viviparus malleatus* are passed out in feces. Embryonated eggs do not hatch until they are ingested by a suitable molluscan host. Huehner and Etges were able to infect both *V. malleatus* and *Goniobasis livescens* with juvenile worms occurring in the intestine and digestive gland of *V. malleatus* after 24 hours. Three growth phases have been recognized. In the initial phase, involving worms less than 0.65 to 9.7 mm long, there is general growth with only slight changes in body proportions, but no development of alveoli on the ventral disc. Most worms 120 days or younger were in this phase.

The second phase, involving worms measuring from 0.65 to 1.65 mm long, is characterized by rapid growth of the ventral disc, decrease in relative size of the oral disc, and development of alveoli on the ventral disc. Furthermore, the stubby tail, characteristic of *A. conchicola* juveniles, disappears shortly after transverse alveolation begins, and the development of the reproductive system is initiated.

The third phase, involving specimens larger than 1.6 to 1.65 mm long, is characterized, among other features, by the complete development of the ovary and testis. Subsequent to this, the vitellaria develop and egg production commences. At 20°C, the development of *A. conchicola* is completed in 270 days in *Viviparus malleatus* and slightly faster in *Goniobasis livescens*. Stages in the life cycle of *A. conchicola* are depicted in Figures 8.18 and 8.19.

It is noted that the juvenile of *Aspidogaster conchicola* hatching from the egg, as well as that of *Aspidogaster indicum* and *Lobatostoma manteri*, are nonciliated. This type of juvenile has been designated as **aspidocidium** by Huehner and Etges. As stated earlier, ciliated aspidobothrean larvae are known as cotylocidia.

Van Cleave and Williams (1943), in an attempt to determine whether turtles can serve as hosts, introduced *A. conchicola* into the stomach of a turtle and found that the parasite clung to the host's stomach wall and lived for 14 days. These results are not particularly surprising since adults of certain other species of *Aspidogaster*, for example, *A. limacoides*, have been found naturally to infect ectothermic vertebrates in the form of freshwater and marine fishes, in

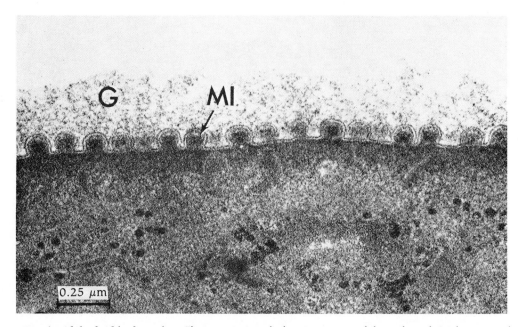

Fig. 8.17. Aspidobothrid body surface. Electron micrograph showing portion of the surface of *Cotylogaster occidentalis* cotylocidium showing blunt microvilli overlayed with glycocalyx. G, glycocalyx; MI, microvillus. (After Federicksen, 1978; with permission of *Journal of Parasitology*.)

addition to molluscs. This suggests that these aspidobothrids may have originally been parasites of molluscs but more recently became adapted to ectothermic vertebrates through ingestion of infected molluscs. It is noted that some species of aspidobothrids show very little host specificity. For example, *C. occidentalis* is able to mature in a fish as well as in a clam, although those in fish are usually larger and produce more eggs. *Aspidogaster limacoides* is found in the intestine of various fishes in the Caspian Sea and the Volga delta.

Life Histories of Lophotaspis *and* Stichocotyle— *Two-Host Cycles.* Wharton (1939) has demonstrated that *Lophotaspis* requires an intermediate host. The larva lives in a marine snail that is ingested by a turtle in which the parasite develops into the adult form. Cunningham (1887), Nickerson (1895), and Odhner (1910) have all concluded that in *Stichocotyle nephropsis*, a parasite of skates, the larva dwells in the intestine of lobsters and develops into the adult only when the intermediate host is ingested by the skate.

It is obvious from the few life cycles presented why some authorities have considered the Aspidobothrea to be related to the Monogenea and others have considered them closely related to the Digenea. Morphologically, these worms are distinct from the members of either order, but their life cycles are either direct—without an intermediate host—as in *Aspidogaster conchicola*, or indirect—with an intermediate host—as in *Lophotaspis* and *Stichocotyle*.

Although certain aspidobothrids require intermediate hosts, the parasite does not undergo asexual multiplication within these hosts, as is the case among digenetic trematodes. The similarities between the aspidobothrids and digenetic and monogenetic trematodes suggest their proximity in phylogenetic relationship to one or the other. All this only serves to strengthen the view that any single characteristic is insufficient in determining the systematic relationships of animals and that all morphologic, developmental, and physiologic evidences must be considered.

Aspidobothrean Physiology and Host-Parasite Relationship

Practically nothing is known about the physiology of the Aspidobothrea. In an attempt to understand the nutritional needs of *Aspidogaster conchicola*, Van Cleave and Williams (1943) maintained specimens in mussel blood at 2°C to 9°C and found that they could be maintained for 75 days. Other media have also been tested with less success. For example, when maintained in 0.75% NaCl, the worms live for only 38 days; in Hedon-Fleig medium, they live for 29 to 38 days.

Several investigators, including myself, have noticed that *Aspidogaster*-infected clams display a less healthy appearance. How this parasite affects the host

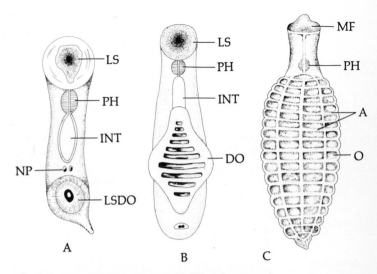

Fig. 8.18. Developmental stages of Aspidogaster. A. Young larva. **B.** Older larva. **C.** Adult of *A. conchicola*. A, alveoli; DO, developing opisthaptor; INT, intestine; LS, larval sucker; LSDO, larval sucker (developing opisthaptor); MF, mouth funnel; NP, nephridiopore; O, opisthaptor; PH, pharynx. (**A** and **B**, redrawn after Voeltzow, 1888. **C**, redrawn after Monticelli, 1892.)

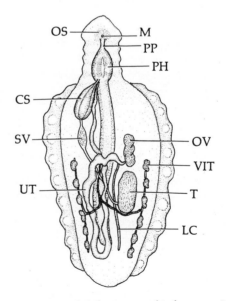

Fig. 8.19. *Aspidogaster conchicola*, a parasite of freshwater clams. CS, cirrus sac; LC, Laurer's canal; M, mouth; OS, oral sucker; OV, ovary; PH, pharynx; PP, prepharynx; SV, seminal vesicle; T, testis; UT, uterus; VIT, vitellaria.

remains to be determined, although it is known that histopathologic changes occur in host tissues surrounding the parasite (Pauley and Becker, 1968; Huehner and Etges, 1981).

Osborn (1905) has reported that the excretory vesicle of the ectoparasitic species *Cotylaspis insignis* pulsates rapidly, excreting water, indicating that the protonephridial system probably serves primarily as an osmoregulatory system. However, most investigators believe that the protonephritic system serves a dual function—osmoregulation as well as excretion.

There is some descriptive information available on the development and hatching of cotylocidia. Such information has been reviewed by Rohde (1972).

REFERENCES

Bogitsh, B. J., and Shannon, A. W. (1971). Cytochemical and biochemical observations on the digestive tracts of digenetic trematodes. VIII. Acid phosphatase activity in *Schistosoma mansoni* and *Schistosomatium douthitti*. *Exp. Parasitol.* **29**, 337–347.

Bogitsh, B. J., Davis, D. A., and Nunnally, D. A. (1968). Cytochemical and biochemical observations on the digestive tracts of digenetic trematodes. I. Ultrastructure of *Haemotoloechus medioplexus* gut. *Exp. Parasitol.* **22**, 96–106.

Bueding, E., and Koletsky, S. (1950). Content and distribution of glycogen in *Schistosoma mansoni*. *Proc. Soc. Exp. Biol. Med.* **73**, 594–596.

Bullock, T. H., and Horridge, G. A. (1965). "Structure and Function in the Nervous Systems of Invertebrates," Vols. I and II. Freeman, San Francisco.

Burton, P. R. (1964). The ultrastructure of the integument of the frog lung-fluke, *Haematoloechus medioplexus* (Trematoda: Plagiochiidae). *J. Morphol.* **115**, 305–317.

Campbell, J. W. (1960). Nitrogen and amino acid composition of three species of anoplocephalid cestodes: *Moniezia expansa*, *Thysanosoma actinoides*, and *Cittotaenia perplexa*. *Exp. Parasitol.* **9**, 1–8.

Cheng, T. C. (1963). Biochemical requirements of larval trematodes. *Ann. N.Y. Acad. Sci.* **113**, 289–321.

Cunningham, J. T. (1887). On *Stichocotyle nephropis*, a new trematode. *Trans. Roy. Soc. Edinburgh* **32**, 273–280.

Dawes, B. (1946). "The Trematoda." Cambridge University Press, London.

Dawes, B. (1954). Maintenance *in vitro* of *Fasciola hepatica*. *Nature (London)* **174**, 654–655.

Fairbairn, D. (1958). Trehalose and glucose in helminths and other invertebrates. *Can. J. Zool.* **36**, 787–795.

Flury, F., and Leeb, F. (1926). Zur Chemie und Toxikologie der Distomen (Leberegel). *Klin. Wochenschr.* **5**, 2054–2055.

Fredericksen, D. W. (1978). The fine structure and phylogenetic position of the cotylocidium larva of *Cotylogaster occidentalis* Nickerson 1902 (Trematoda: Aspidogastridae).

J. Parasitol. **64**, 961–976.

Ginger, C. D., and Fairbairn, D. (1966). Lipid metabolism in helminth parasites. I. The lipids of *Hymenolepis diminuta* (Cestoda). *J. Parasitol.* **52**, 1086–1096.

Goil, M. M. (1958). Protein metabolism in trematode parasites. *J. Helminthol.* **32**, 119–124.

Goodchild, C. G. (1961). Carbohydrate contents of the tapeworm *Hymenolepis diminuta* from normal, bileless, and starved rats. *J. Parasitol.* **47**, 401–405.

Goodchild, C. G., and Vilar-Alvarez, C. M. (1962). *Hymenolepis diminuta* in surgically altered hosts. II. Physical and chemical changes in tapeworms grown in shortened small intestines. *J. Parasitol.* **48**, 379–383.

Halton, D. W. (1967). Observations on the nutrition of digenetic trematodes. *Parasitology* **57**, 639–660.

Hendrix, S. S., and Overstreet, R. M. (1977). Marine aspidogastrids (Trematoda) from fishes in the northern Gulf of Mexico. *J. Parasitol.* **63**, 810–817.

Hopkins, C. A. (1950). Studies on cestode metabolism. I. Glycogen metabolism in *Schistocephalus solidus in vivo*. *J. Parasitol.* **36**, 384–390.

Huehner, M. K., and Etges, F. J. (1977). The life cycle and development of *Aspidogaster conchicola* in the snails, *Viviparus malleatus* and *Goniobasis livescens*. *J. Parasitol.* **63**, 669–674.

Huehner, M. K., and Etges, F. J. (1981). Encapsulation of *Aspidogaster conchicola* (Trematoda: Aspidogastrea) by unionid mussels. *J. Invert. Pathol.* **37**, 123–128.

Jennings, J. B. (1957). Studies on feeding, digestion and food storage in free-living flatworms (Platyhelminthes: Turbellaria). *Biol. Bull.* **112**, 63–80.

Jennings, J. B. (1968a). Feeding, digestion and food storage in two species of temnocephalid flatworms (Turbellaria: Rhabdocoela). *J. Zool.* **156**, 1–8.

Jennings, J. B. (1968b). Nutrition and digestion. *In* "Chemical Zoology" (M. Florkin and B. T. Scheer, eds.), Vol. II. Academic Press, New York.

Jennings, J. B. (1971). Parasitism and commensalism in the Turbellaria. *Adv. Parasitol.* **9**, 1–32.

Kozloff, E. N. (1953). *Collastoma pacifica* sp. nov., a rhabdocoel turbellarian from the gut of *Dendrostoma pyroides* Chamberlin. *J. Parasitol.* **39**, 336–340.

Lee, D. L. (1966). The structure and composition of the helminth cuticle. *Adv. Parasitol.* **4**, 87–254.

Lee, D. L. (1972). The structure of the helminth cuticle. *Adv. Parasitol.* **10**, 347–379.

Maekawa, K., Kitazawa, K., and Jushiba, M. (1954). Purification et cristallisation de l'antigène pour la dermoréaction allergique vis-à-vis de *Fasciola hepatica*. *C. R. Soc. Biol.* **148**, 763–765.

McDaniel, J. S., and Dixon, K. E. (1967). Utilization of exogenous glucose by the rediae of *Parorchis acanthus* (Digenea: Philophthalmidae) and *Cryptocotyle lingua* (Digenea: Heterophyidae). *Biol. Bull.* **133**, 591–599.

Nickerson, W. S. (1895). On *Stichocotyle nephropis* Cunningham, a parasite of the American lobster. *Zool. Jahrb.* **8**, 447–480.

Odhner, T. (1910). *Stichocotyle nephropis* J. T. Cunningham ein aberranter Trematode der Digenenfamilie Aspidogastridae. *Sv. Vet. Acad. Handl.* **45**, 3–16.

Osborn, H. L. (1905). On the habits and structure of *Cotylaspis insignis* Leidy, from Lake Chautauqua, New York, U.S.A. *Zool. Jahrb.* **21**, 201–242.

Pauley, G. B., and Becker, C. D. (1968). *Aspidogaster conchi-*

cola in mollusks of the Columbia River system with comments on the host's pathological response. *J. Parasitol.* **54**, 917–920.

Read, C. P. (1949). Preliminary studies on the intermediary metabolism of the cestode *Hymenolepis diminuta*. *J. Parasitol.* **35** (Suppl.), 26.

Reid, W. M. (1942). Certain nutritional requirements of the fowl cestode *Raillietina cesticillus* (Molin) as demonstrated by short periods of starvation of the host. *J. Parasitol.* **28**, 319–340.

Robinson, D. L. H. (1956). A routine method for the maintenance of *Schistosoma mansoni in vitro*. *J. Helminthol.* **29**, 193–202.

Robinson, D. L. H. (1961). Amino acids of *Schistosoma mansoni*. *Ann. Trop. Med. Parasitol.* **55**, 403–406.

Rohde, K. (1971a). Untersuchungen an *Multicotyle purvisi* Dawes, 1941 (Trematoda: Aspidogastrea). VII. Ultrastruktur des Caecums der freien Larve und der geschlechtsreifen Form. *Zool. Jahrb. Anat.* **88**, 406–420.

Rohde, K. (1971b). Untersuchungen an *Multicotyle purvisi* Dawes, 1941 (Trematoda: Aspidogastrea). I. Entwicklung und Morphologie. *Zool. Jahrb. Anat.* **88**, 138–187.

Rohde, K. (1971c). Untersuchungen an *Multicotyle purvisi* Dawes, 1941 (Trematoda: Aspidogastrea). IV. Ultrastruktur des Integuments der geschlechtsreifen Form und der freien Larva. *Zool. Jahrb. Anat.* **88**, 365–386.

Rohde, K. (1972). The Aspidogastrea, especially *Multicotyle purvisi* Dawes, 1941. *Adv. Parasitol.* **10**, 77–151.

Salisbury, L. F., and Anderson, R. J. (1939). Concerning the chemical composition of *Cysticercus fasciolaris*. *J. Biol. Chem.* **129**, 505–517.

Scheer, B. T., and Jones, E. R. (1968). Introduction to Platyhelminthes. *In* "Chemical Zoology" (M. Florkin and B. T. Scheer, eds.), Vol. II, pp. 287–302. Academic Press, New York.

Senft, A. W., and Senft, D. G. (1962). A chemically defined medium for maintenance of *Schistosoma mansoni*. *J. Parasitol.* **48**, 551–554.

Smith, T. M., Brooks, T. J., and White, H. B. (1966). Thin-layer and gas-liquid chromatographic analysis of lipid from cercariae of *Schistosoma mansoni*. *Am. J. Trop. Med. Hyg.* **15**, 307–313.

Smorodintsev, I. A., and Bebesin, K. V. (1936). Beiträge zur Chemie der Helminthen. Mitt. IV. Die chemische Zusammensetzung des *Diphyllobothrium latum*. *J. Biochem.* **23**, 21–22.

Smorodintsev, I. A., Bebesin, K. V., and Pavlova, P. I. (1933). Beiträge zur Chemie der Helminthen. I. Mitteilung: Die chemische Zusammensetzung von *Taenia saginata*. *Biochem. Z.* **261**, 176–178.

Tötterman, G., and Kirk, E. (1939). Om innehallet av lipoider i *Bothriocephalus latus*. *Nord. Med. Ark.* **3**, 2715–2716.

Ubelaker, J. E. (1983). The morphology, development, and evolution of tapeworm larvae. *In* "Biology of the Eucestoda" (C. Arme and P. W. Pappas, eds.), Vol. 1, pp. 235–296. Academic Press, London.

Ueda, T., and Sawada, T. (1968). Fatty acid and sugar composition of helminths. *Jap. J. Exp. Med.* **38**, 145–147.

Van Cleave, H. J., and Williams, C. O. (1943). Maintenance of a trematode, *Aspidogaster conchicola*, outside the body of its natural host. *J. Parasitol.* **29**, 127–130.

von Brand, T. (1933). Untersuchungen über den Stoffbestand einiger Cestoden und den Stoffwechsel von *Moniezia expansa*. *Z. Vergl. Physiol.* **18**, 562–596.

von Brand, T. (1973). "Biochemistry of Parasites," 2nd ed. Academic Press, New York.

von Brand, T., and Bowman, I. B. R. (1961). Studies on the aerobic and anaerobic metabolism of larval and adult *Taenia taeniaeformis*. *Exp. Parasitol.* **11**, 276–297.

Wardle, R. A. (1937). "Manitoba Essays." 60th Ann. Com. Vol., Univ. Manitoba, Winnipeg. Macmillan, Toronto.

Weinland, E., and von Brand, T. (1926). Beobachtungen an *Fasciola hepatica* (Stoffwechsel und Lebensweise). *Z. Vergl. Physiol.* **4**, 212–285.

Wharton, G. W. (1939). Studies on *Lophotaspis vallei* (Stossich, 1899) (Trematoda: Aspidogastridae). *J. Parasitol.* **25**, 83–86.

Yamaguti, S. (1963). "Systema Helminthum Vol. IV. Monogenea and Aspidocotylea." Interscience, New York.

CLASSIFICATION OF TURBELLARIA (THE TURBELLARIANS)*

CLASS TURBELLARIA

Mostly free-living predators; in terrestrial, freshwater, and marine environments; body surface covered with discrete epithelial cells; no body cavity (coelom) present; bilaterally symmetrical; rhabdites present in most free-living species; osmoregulatory (excretory) system, if present, of the protonephritic type involving flame cells (or bulbs); some species commensalistic, others parasitic on invertebrates, especially echinoderms and molluscs.

ORDER ACOELA

All marine, one to several millimeters long; excretory (osmoregulatory) system absent; pharynx and permanent gut absent; many species without rhabdites; most species free living, feeding on algae, bacteria, protozoans, and other microscopic organisms; temporary gut lined with a syncytium formed when food is ingested; several symbiotic, including parasitic, species. (Genus mentioned in text: *Ectocotyla*.)

ORDER RHABDOCOELA

Small worms; gut saccate, with a permanent lumen; anterior region of digestive tract differentiated into a pharynx (simple, bulbous, or rosulate); free-living species predators of protozoans and small invertebrates; symbiotic, including parasitic, species represented.

Suborder Notandropora (or Catenulida)
Suborder Opisthandropora (or Macrostomida)
Suborder Lecithophora (or Neorhabdocoela)
Freshwater, marine, or terrestrial; with bulbous pharynx; paired nephridia; gonopores ventral; with germovitellaria or separate ovaries and vitellaria (yolk glands); reproduction exclusively sexual; mostly free living, some species symbiotic, including parasitic.

---

*Modified after L. H. Hyman (1951.) "The Invertebrates" Volume II, McGraw-Hill, New York. Diagnoses of taxa comprised of free-living species not presented.

Section Dalyellioida

Without proboscis; mouth at or near anterior tip of body; with single gonopore. (Genera mentioned in text: *Fecampia, Collastoma, Paravortex, Oekiocolex, Graffilla.*)

Section Typhloplanoida

Without proboscis; pharynx usually of rosulate type; mouth ventral, back of anterior tip of body; with one or two gonopores.

Section Kalyptorhyncha

Suborder Temnocephalida

Freshwater; devoid of surface cilia or only slightly ciliated; with 2 to 12 tentacles; posterior end with one or two adhesive discs; with single gonopore. (Genera mentioned in text: *Temnocephala, Scutariella, Actinodactylella, Caridinicola, Monodiscus, Didymorchis, Craspedella.*)

ORDER ALLOEOCOELA

Mostly marine, some in brackish or fresh water, few terrestrial; pharynx simple, bulbous, or plicate; intestine may have short diverticula; flame cells paired, collecting ducts often with more than one main branch; often with more than one nephridiopore; generally with numerous testes; nervous system with three or four pairs of longitudinal trunks.

Suborder Archoophora

Suborder Lecithoepitheliata

Suborder Cumulata (or Holocoela)

Marine or freshwater; with germovitellaria or with separate ovaries and vitelline glands; pharynx bulbous or plicate; intestine mostly without diverticula; male copulatory organ mostly unarmed. (Genus mentioned in text: *Hypotrichina.*)

Suborder Seriata

ORDER TRICLADIDA

Large marine or brackish water elongate worms; with plicate pharynx, usually directed posteriorly; intestine with three diverticulated branches, one oriented forward, two backward; with one pair of ovaries; with two to numerous testes; with vitelline glands; female reproductive system with one (sometimes two) bursa; male reproductive system with a penis; single gonopore; mostly free living, few species symbiotic.

Suborder Maricola (or Retrobursalia)

Marine or in brackish water; with two eyes; bursa, if present, behind penis or else with separate pore; may be paired; sexual reproduction only. (Genera mentioned in text: *Bdelloura, Syncoelidium, Micropharynx.*)

Suborder Paludicola (or Probursalia)

Suborder Terricola

ORDER POLYCLADIDA

Almost exclusively marine; large flattened worms; plicate pharynx opens into a main intestine; intestine with numerous branches radiating to periphery; nervous system with numerous radiating nerve cords; more than two eye spots; numerous, scattered ovaries and testes; without vitelline glands; with one or two gonopores, with male pore, if separated, anterior to female pore.

Suborder Acotelea

Without sucker posterior to female genital pore; eyes not in paired clusters on anterior body margin.

Section Craspedommata

Eyes in a band along entire or part of body margin. (Genus mentioned in text: *Stylochus.*)

Section Schematommata

Without marginal eyes; eyes mostly in four groups positioned well back from anterior body margin. (Genera mentioned in text: *Euplana, Hoploplana.*)

Section Emprosthommata

Suborder Cotylea

With sucker behind female genital pore; with pair of marginal tentacles or, if not, with cluster of eyes in place of each tentacle, or with eyes in a band across anterior body margin; occasionally associated with crustaceans.

CLASSIFICATION OF ASPIDOBOTHREA (THE ASPIDOBOTHRIDS)*

ORDER ASPIDOBOTHREA

With ventral, discoid opisthaptor divided into alveoli and with marginal bodies instead of marginal hooklets; transverse elongate suckers may be arranged on ventral body surface, without marginal organs; mouth with or without liplike processes or lobes; oral sucker present or absent; pharynx well developed; intestine a single median tube; testes single or double, postovarian; cirrus pouch present or absent; genital pore ventral, almost always median, at level of anterior end of opisthaptor; male and female genital pores may be present; vitelline glands follicular or tubular, extending in middle and posterior part of body, usually lateral and posterior to other reproductive organs, exceptionally in middorsal field; excretory pore(s) at posterior terminal of body or dorsoterminal; development direct; endoparasitic, rarely ectoparasitic, in molluscs and ectothermic vertebrates.

Family Aspidogastridae

Body elongate to oval; opisthaptor with numerous shallow loculi; with one or two testes; vitellaria follicular, lateral; parasites of molluscs, fish, and turtles. (Genera mentioned in text: *Aspidogaster, Multicotyle, Cotylaspis, Lissemysia, Cotylogaster, Lophotaspis.*)

Family Stichocotylidae

Body slender and elongate; ventral surface with longitudinal row of separate suckers; with two testes; vitellaria tubular, unpaired; parasites of rays. (Genus mentioned in text: *Stichocotyle.*)

Family Rugogastridae

Body elongate; most of ventral and lateral body surface with transverse rugae; buccal funnel musculature weakly developed; prepharynx, pharynx, and esophagus present; with two caeca; with multiple testes; ovary pretesticular; Laurer's canal present; seminal receptacle absent; vitellaria distributed along caeca; uterus ventral to testes; eggs operculate; parasites of chimaerid fish.

---

*Modified after S. Yamaguti (1963). "Systema Helminthum" Vol. IV. Interscience, New York.

9

MONOGENEA: THE MONOGENETIC TREMATODES

The monogenetic trematodes, commonly referred to as monogeneids or monogeneans, are a group of flukes that as a rule are ectoparasitic, infesting ectothermic vertebrates. A few species, however, have been reported to be attached to aquatic mammals, crustaceans, and cephalopods. At least one species, *Oculotrema hippopotami,* has been reported to parasitize an amphibious mammal; it was discovered in the eye of a hippopotamus in Uganda (Thurston and Laws, 1965). At least three species are known to be true endoparasites; *Dictyocotyle* sp. in the coelom of the ray *Raja lintea, Acolpenteron ureterocetes* in the ureters of freshwater fish, and *Polystoma integerrimum* in the bladder of amphibians. Other species are known to occur in bladders, buccal cavities, and other body spaces that open directly to the exterior.

As ectoparasites, the monogeneids are usually attached to the gills, scales, and fins of fish hosts. In many cases these parasites show a marked preference for a particular gill or gills. For example, *Diclidophora merlangi* attaches most frequently on the first gill of the whiting, and *D. luscae* favors the second and third gills of the pout. All species attached to gills assume the same general position. Llewellyn (1956) has reported that the posterior terminal of the parasite is located closer to the host's gill arch, and the anterior end is closer to the distal end of the host's primary lamellae. In all instances, the mouth of the parasite faces downward. This results in the attached posterior end of the worm lying upstream relative to the current passing over the gill and with the mouth lying downstream.

Serious students of the Monogenea should consult the monographs by Bychowsky (1937, 1957) and Sproston (1946).

MORPHOLOGY

Although the monogenetic trematodes are generally not considered to be serious pathogens of their hosts, they can represent a threat to the health of fish maintained under crowded conditions such as in aquaria and aquaculture ponds. Members of the order Monogenea possess the characteristics of the phylum Platyhelminthes and the class Trematoda, both of which are given in detail in Chapter 8. In addition to these features, monogeneids possess characteristic holdfast organs, paired anterior excretory pores on the dorsal surface of the body, and comparatively short uteri

273

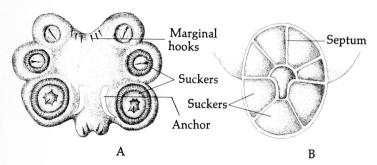

Fig. 9.1. Monogeneid opisthaptors. A. Opisthaptor of *Polystoma.* (Redrawn after Zeller, 1872.) **B.** Opisthaptor of *Tristoma.* (Redrawn after Goto, 1894.)

containing few eggs. Above all, they have direct life cycles that do not include intermediate hosts and reproduce exclusively by sexual means.

Generally, the monogeneids are medium-sized when compared with the other monozoic platyhelminths. These worms seldom exceed 3 cm in length. Marine species are generally larger than freshwater species.

Holdfast Organs

Superficially, the monogeneids are not unlike the digenetic trematodes, except for the presence of a posteriorly situated adhesive structure, the **opisthaptor**, by which the parasite is attached to its host. The opisthaptor is undoubtedly the most specialized of the various types of attachment organs found among platyhelminths. This large, circular, muscular

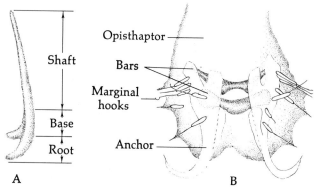

Fig. 9.2. Morphologic features of monogeneid adhesive structures. A. A single anchor showing the three regions. **B.** Posterior terminal of *Dactylogyrus* showing various structures; notice the bars. (Redrawn after Mueller and Van Cleave, 1932.)

disc is generally completely or partially partitioned into smaller suckers by septa (Fig. 9.1).

In the family Gyrodactylidae no secondary divisions are present, but the opisthaptor does bear two to four large hooks known as **anchors**, often supplemented by smaller **haptoral hooks** (Fig. 9.2). In addition to the anchors and haptoral hooks, some species have transversely situated sclerotized **bars**, which may or may not articulate with the anchors (Fig. 9.2).

Anchors, haptoral hooks, and bars are not limited to the Gyrodactylidae, for all monogeneids possess anchors and hooks at some stage during their development.

The hooks of monogeneids, specifically that of *Entobdella diadema*, have been subjected to x-ray diffraction analysis and found to be composed of keratin or a keratinlike protein (Lyons, 1966). Many species possess various numbers of small suckers divided by septa, ranging from 6 to 240, depending on the species. These small suckers are armed with anchors and hooks. In some monogeneids, such as members of the suborder Polyopisthocotylea, there is a single **cotylophore**, which bears multiple suckers or clamps. Figure 9.3 depicts several types of opisthaptors found on monogenetic trematodes.

Llewellyn (1958) has made comparative studies on nine species of the genus *Diclidophora* (Polyopisthocotylea). He has reported that in these species there are both a pair of hinged jaws in each sucker that is operated by intrinsic muscles, and a more powerful extrinsic muscle that acts on a diaphragm, producing suction that is converted into a clamping action.

In addition to the opisthaptor, monogeneids usually possess two anteriorly located auxiliary suckers, the **prohaptors**, one on each side of the mouth. The prohaptors may appear as pits, suckers, or discs. In a few species, an oral sucker occurs.

Alimentary Tract

The internal anatomy of a monogenetic trematode is illustrated in Figure 9.4. The mouth is located at the anterior end of the body, flanked by prohaptors. In some species, such as *Discocotyle sagittata*, a parasite of trout and salmon, the prohaptors are modified as buccal suckers located in the buccal funnel, which is the tube leading from the mouth (Fig. 9.5). Those species with prohaptors not connected with the buccal funnel are considered to be more primitive evolutionarily. The heads of these species are commonly lobed, broadly round, or truncated. In some species the buccal funnel is protrusible, and in such cases, the true mouth lies at the base of the funnel when it is withdrawn. The mouth is seldom surrounded by an oral sucker as it is in digenetic trematodes.

Head glands and **head organs** are commonly present, particularly in the more primitive species. The secretions of the glands are adhesive, reflecting the

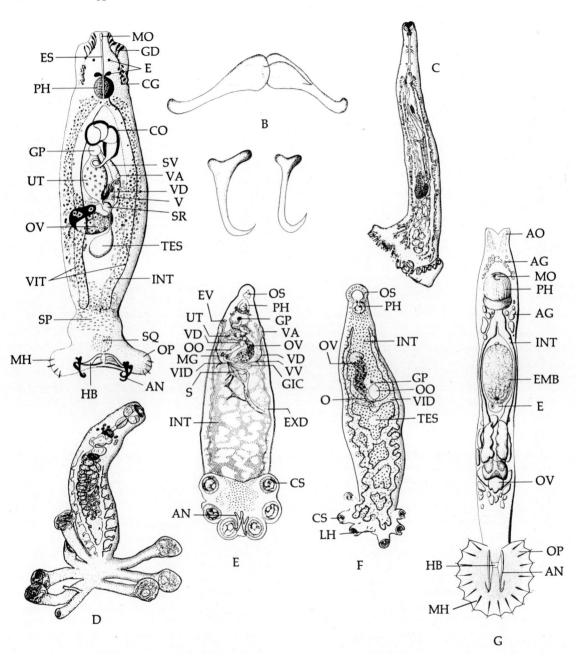

Fig. 9.3. Several species of monogeneids showing various types of opisthaptors. A. *Diplectanum melanesiensis* from the serranid fish *Epinephelus merra.* **B.** Scalelike spines of posterior part of body (above). Dorsal (left) and ventral (right) anchors of *D. melanesiensis.* (**A** and **B**, redrawn after Laird, 1958.) **C.** *Axinoides raphidoma* from Gulf of Mexico fish. (After Hargis, 1956.) **D.** *Choricotyle louisianensis* from gills of the southern whiting. (After Hargis, 1956.) **E.** Bladder generation of *Polystoma integerrimum nearcticum* from urinary bladder of the tree-frog, *Hyla versicolor.* **F.** Gill generation of *P. integerrimum nearcticum.* (**E** and **F**, after Paul, 1938.) **G.** *Gyrodactylus* sp. (Redrawn after Mueller and Van Cleave, 1932.) AG, adhesive glands; AN, anchor; AO, anterior adhesive organ; CG, cephalic glands; CO, copulatory organ; CS, caudal sucker; E, eyespots; EMB, embryo; ES, esophagus; EV, excretory vesicle; EXD, excretory duct; GD, gland duct; GIC, genitointestinal canal; GP, genital pore; HB, bar; INT, intestine; LH, larval hooks; MG, Mehlis' gland; MH, marginal hook; MO, mouth; O, oviduct; OO, ootype; OP, opisthaptor; OS, oral sucker; OV, ovary; PH, pharynx; S, sperm; SP, scalelike spines; SQ, squamodisc; SR, seminal receptacle; SV, seminal vesicle; TES, testis; UT, uterus; VA, vagina; VD, vaginal duct; VID, vitelline duct; VIT, vitellaria; VV, vitellovaginal duct.

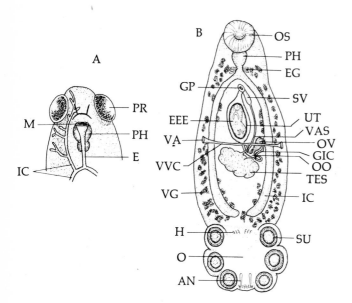

Fig. 9.4. **Anatomy of monogeneids. A.** Anterior terminal of *Tristoma*, parasitic on the gills of swordfish, showing presence of paired prohaptors. **B.** *Polystomoidella oblonga*, parasitic in the urinary bladder of turtles, showing internal structures. AN, anchor; E, esophagus; EEE, egg enclosing embryo; EG, esophageal glands; GIC, genitointestinal canal; GP, genital pore; H, hooks; IC, intestinal caeca; M, mouth; O, opisthaptor; OO, ootype; OS, oral sucker; OV, ovary; PH, pharynx; PR, prohaptor; SU, sucker; SV, seminal vesicle; TES, testis; UT, uterus; VA, vagina; VAS, vas deferens; VG, vitelline glands; VVC, vitellovaginal canal. (**A,** redrawn after Goto, 1894. **B,** redrawn after Cable, 1958.)

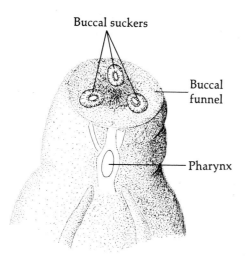

Fig. 9.5. **Anterior end of *Discocotyle sagittata*, parasitic on the gills of sea trout and salmon, showing buccal suckers situated within the buccal funnel.** (Redrawn after Dawes, 1947.)

primary function of these organs—attachment. These areas usually bear dense, long microvilli on the outer surface of the tegument, in contrast to the shorter, less dense microvilli on the rest of the tegument. Near the suckers, certain gland cells, frequently present in the body wall, secrete into the prohaptors and opisthaptors.

The buccal funnel leads into the muscular **pharynx,** which in turn leads into the **esophagus.** The intestinal tract most frequently is in the form of an inverted Y, with the caeca bifurcating from the posterior extremity of the esophagus, which represents the stem of the Y. Modifications of this pattern are known.

In *Dactylogyrus amphibothrium* and related species, the two caeca are not only united posteriorly, but possess lateral **diverticula.** In *Udonella*, the gut is unbranched and resembles a single blind tube with fenestrations in the region of the gonads. In *Polystoma* and *Diclidophora*, the caeca give rise to numerous diverticula, which extend into the disc-shaped opisthaptor. Figure 9.6 shows some of the intestinal tract variations found among the Monogenea.

The intestinal caeca are lined with epithelial cells that are either closely packed or scattered. Rohde (1975) has contributed a review of the fine structure of monogeneids, including the alimentary tract. Using

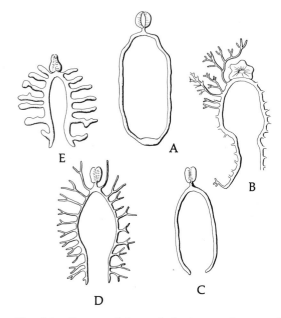

Fig. 9.6. **Some variations of the intestinal caeca of monogeneids. A.** Arrangement of caeca of *Sphyranura*, parasitic on gills of *Necturus*. **B.** Branched caeca of *Benedenia melleni.* (Modified after Jahn and Kuhn, 1932.) **C.** Caeca of *Thaumatocotyle concinna*, attached to the nasal fossae of the stingray. (Modified after Price, 1938.) **D.** Branched caeca of *Microbothrium apiculatum*, attached to the skin of the piked dogfish. (Modified after Price, 1938.) **E.** Branched caeca of *Leptobothrium pristiuri*, attached to the skin of the black-mouthed dogfish. (Modified after Gallien, 1937.)

Fig. 9.7.   Caecal epithelium and secreted cells in caecal lumen. Diagram showing two types of cells lining caecum of *Polystomoides asiaticus* and discharged cells in lumen. BCM, basal cell membrane; BL, basal lamina; G, glycogen; GO, Golgi; L, lamella; LCM, lateral cell membrane; M, mitochondrion; N, nucleus; T, tubular system; V, vacuole; VE, vesicle; Z, zona adherans. (After Rhode, 1973; with permission of *International Journal of Parasitology*.)

the electron microscope, he found that two cell types line the caecal lumen. One type bears long, microvilluslike lamellae along its free border, whereas the second does not (Fig. 9.7). Both types of lining cells include basal nuclei, mitochondria, Golgi complexes, glycogen, and rough endoplasmic reticulum.

The food and feeding habits of the members of the two suborders of the Monogenea—Monopisthocotylea and Polyopisthocotylea—differ. Consequently, it is not surprising that the histologic compositions of the gastrodermis differ, even at the light microscope level. Among members of the Monopisthocotylea, which feed primarily on host epidermis and mucus, this layer is comprised of a continuous layer of cuboidal and columnar cells. On the other hand, among members of the Polyopisthocotylea, which feed primarily on blood and, in some cases, host cells and mucus, the gastrodermis consists of a discontinuous layer of large columnar cells, generally containing varying amounts of hematin granules.

Digestion in monogeneids begins in the caecal

lumen and is completed within cells lining the caeca. A number of enzymes, including a protease, lipase, aminopeptidase, esterase, and alkaline phosphatase, have been reported in the Monogenea (Halton, 1964). For a detailed review of digestion in monogeneids, see Jennings (1972).

Body Wall

The body surface of monogenetic trematodes is covered by a thin tegument. Comprehensive reviews of the fine structure of this tissue have been presented by Lee (1966, 1972), Lyons (1973), and Rohde (1975). All reports indicate that the fine structure is essentially the same as that of the tegument of digenetic trematodes and cestodes (pp. 304, 388). In brief, the tegument is comprised of an anucleate outer syncytial layer connected by cytoplasmic processes to underlying nucleated cytons. The outer level includes mitochondria and extensive smooth membrane systems and is thrown into occasional fingerlike protrusions (Fig. 9.8). In the buccal area, the tegument is covered by small bristlelike structures (Fig. 9.9). Beneath the

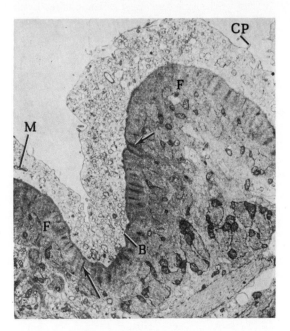

Fig. 9.8.   Electron micrograph of the tegument of the monogeneid *Diclidophora merlangi*, an ectoparasite of the whiting, *Gadus merlangus*, showing major folds and occasional surface protrusions. Arrows indicate projections of the basal lamina into the underlying fibrous layer. ×6000. B, basal lamina; CP, surface protrusion; F, fibrous layer; M, mitochondrion. (After Morris and Halton, 1971.)

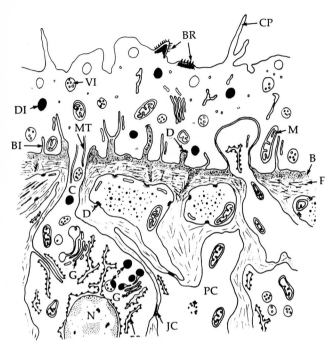

Fig. 9.9. Schematic drawing of structure of tegument of *Diclidophora merlangi* showing cytoplasmic bridge joining outer and inner zones of the tegument. B, basal lamina; BI, basal invagination; BR, surface bristles; C, cytoplasmic bridge; CP, surface protrusion; D, desmosome; DI, dense inclusion; ER, rough endoplasmic reticulum; F, fibrous layer; G, Golgi complex; JC, junctional complex; M, mitochondrion; MT, microtubule; N, nucleus; PC, parenchymal cell; VI, vesicular inclusion. (After Morris and Halton, 1971.)

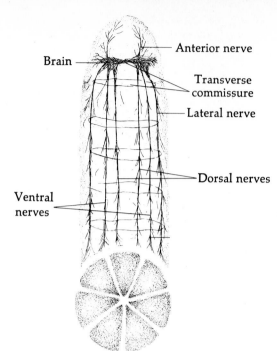

Fig. 9.10. Generalized drawing of a monogenetic trematode, showing arrangement of nervous system.

outer stratum of the tegument is a thin basal lamella comprised of connective tissue fibers. Mediad to this is a thin layer of circumferentially arranged muscles, commonly designated the circular muscles. This layer may be absent in some species.

Beneath the circular muscles is found a thin stratum of diagonally arranged muscles (the diagonal muscles), and under this is a thick layer of well-developed longitudinal muscles. Within the body wall, surrounding the internal organs, are loose parenchymal cells, fibers, and spaces. The cells are generally independent and not syncytially arranged, but in certain species the parenchymal cells situated toward the center (medullary) are discrete, whereas those located on the periphery (cortex) are syncytial. The parenchyma serves as a site of glycogen storage.

Nervous System

The center of the nervous system in the Monogenea generally consists of two large clusters of nerve ganglia, both of which are situated at the anterior end of the body. These two nerve masses are connected by a transverse commissure (Fig. 9.10). From this primitive "brain," nerve fibers arise and extend anteriorly, laterally, and posteriorly. These nerve trunks are scattered throughout the parenchyma but especially in the ventral, lateral, and dorsal regions.

In some species, such as certain members of the families Polystomatidae and Capsalidae, the "brain" forms a circumesophageal ring from which nerve fibers extend anteriorly, laterally, and posteriorly; one pair being dorsal, one pair ventrolateral, and one pair ventral. The ventral nerves, which are the most highly developed, are often connected by a series of transverse commissures. Branches of the nerve fibers innervate the various sucker muscles and other organs of the body.

One or two pairs of eyespots are commonly present in monogeneids, and in most cases these are comparable to those found in certain digenetic trematode cercariae. Each eye is composed of a rounded retinal cell surrounded by rods made up of pigment granules (Fig. 9.11). In a few complex forms, a simple lens is present. Such eyes are most conspicuous in the free-living larval stage, becoming reduced or absent in adults.

Nervous systems among the monogeneids reach an extremely complex condition in members of the genus *Capsala*, in which a rudimentary taste organ is present.

Ciliary Eyes. Lyons (1972) has reported a new type of sense organ in larval *Entobdella soleae*, an ectoparasite of the common sole. These "ciliary eyes" occur on each side of the parasite's cephalic region, situated slightly anteroventral to the four pigmented eyespots. Each round, ciliary eye is intracellular. It does not open to the outside but contains a central cavity lined with a thin rim of cytoplasm (Figs. 9.12 and 9.13) containing the basal bodies of aberrant $9 + 2$ cilia among other organelles. Lyons has attributed a photoreceptive function to these organs. They may produce "off" responses to light and mediate a protective shadow response on the part of the larva. On the other hand, these organs may function in a manner more similar to the pineal eyes of certain vertebrates, i.e., as long-term light receptors setting some diurnal or nocturnal activity. It is noted that Kearn (1973) has demonstrated a diurnal rhythm on the part of *E. soleae* larvae. This rhythm is set by previous exposure to alternating periods of darkness and light, which does not occur if the embryo is maintained in continuous light or darkness. This rhythm is functionally important because it synchronizes hatching of larvae at dawn with the return of nocturnally feeding soles to the sea bed. This and other aspects of the behavior of monogeneids have been reviewed by Llewellyn (1972), and the sense organs of these parasites have been reviewed by Lyons (1972).

Osmoregulatory System

The osmoregulatory (excretory) system among the monogenetic trematodes, like that found in the other platyhelminths, is of the protonephritic type—with flame cells at the end of collecting tubules. Although a meshlike network of tubules has been reported, the general rule is the presence of two main lateral ducts that begin anteriorly and extend posteriorly. Each tube makes a U-curve near the posterior end of the body and then extends anteriorly (Fig. 9.14). Toward the end of each ascending duct there is an enlarged chamber known as the **contractile bladder**. The ducts leaving the bladders empty to the outside through two separate **excretory pores** situated dorsolateral to the mouth. The flame cells are located at the free ends of branches of these main collecting tubes. Rohde (1975) has reviewed the fine structure of the osmoregulatory system of the Monogenea, noting three different arrangements of flame cells: (1) single terminal flame cells, (2) flame-cell complexes consisting of two (or more) flames intimately associated with one region of the collecting tubular system, and (3) lateral (nonterminal) flames found in all but the smallest ducts (Fig. 9.15).

Reproductive Systems

Male System. The male system includes either one testis, as in *Thaumatocotyle*, *Heterocotyle*, and *Leptocotyle*, or two or three testes, as in *Entobdella* and

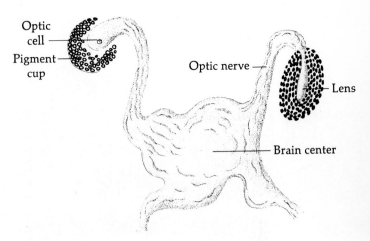

Fig. 9.11. Section showing relation of pigment cup to nerve ending in paired eyespots of parasitic platyhelminth.

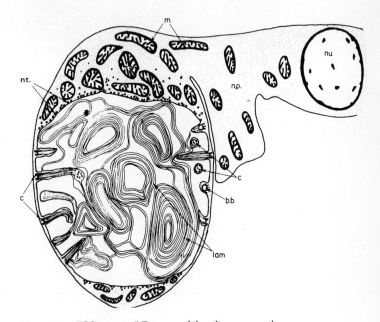

Fig. 9.12. "Ciliary eye." Diagram of the ciliary eye on the oncomiracidium of *Entobdella soleae*. This ciliary organ consists of a single cell with a thick rim of cytoplasm containing mitochondria and nerve vesicles at each pole. The organ is bounded laterally by a thinner cytoplasmic layer containing ciliary basal bodies. The $9 + 2$ cilia project into the lumen of the organ and are associated with and may give rise to lamellate bodies. bb, basal body; c, cilia; lam, lamellate bodies; m, mitochondria; nt, nerve. (After Lyons, 1972; with permission of The Linnean Society of London.)

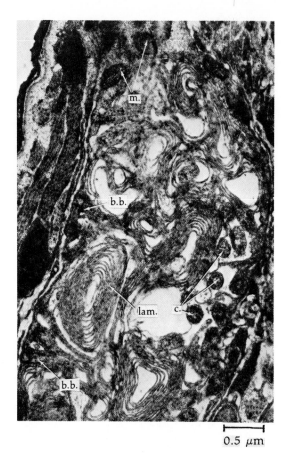

Fig. 9.13. Fine structure of "ciliary eye." Electron micrograph of a section through the ciliary eye of *Entobdella soleae* oncomiracidium. Notice that this organ consists of a single cell with a central cavity into which project 9 + 2 cilia. Lamellate structures composed of thin membrane-bound whorls are present in the lumen of the organ, and mitochondria occur at the poles of the cell. × 14,500. bb, basal bodies of cilia; c, cilia; lam, lamellated bodies; m, mitochondria. (After Lyons, 1972; with permission of The Linnean Society of London.)

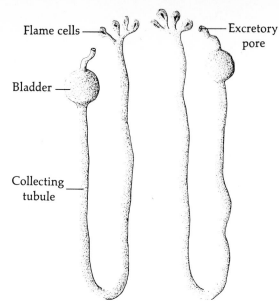

Fig. 9.14. Arrangement of the osmoregulatory (excretory) system as found in most monogenetic trematodes.

Fig. 9.15. Arrangements of flame cells in monogeneid osmoregulatory system. **A.** Single terminal flame cell. **B.** Cluster of terminal flame cells. **C.** Lateral (nonterminal) flame cell.

Monocotyle (Fig. 9.16). Some genera are multitesticular such as *Tritestis* with three testes, *Sphyranura* with 8 to 10, and *Rajonchocotyle* with 200 testes. The primitive number was probably two, as in the Digenea.

A single **vas efferens** arises from each testis. These unite to form the common **vas deferens**, which proceeds to either a **cirrus** (an eversible copulatory organ) or a protrusible **penis**. Prior to entering the cirrus or penis, the vas deferens is enlarged in certain species to form the **seminal vesicle**.

Generally, the cirrus opens to the outside through a

genital atrium located on the ventral body surface behind the caecal bifurcation. In some species the female reproductive system also leads into this chamber, in which case it is a **common atrium**. The aperture of the atrium is the genital pore or **gonopore**.

The degree of development of the copulatory structure varies. In *Merizocotyle*, prostate glands empty into the cirrus. In *Udonella*, a definite cirrus or penis is lacking, self-fertilization being the rule. In *Microcotyle* and *Hexabothrium*, the end of the vas deferens is modified to form a pointed protuberance. In *Diclidophora* and others, the cirrus is armed with hooklets that aid in maintaining it in the atrium of the copulatory mate.

The reproductive systems of monogenetic trematodes are depicted in Figures 9.17 and 9.18.

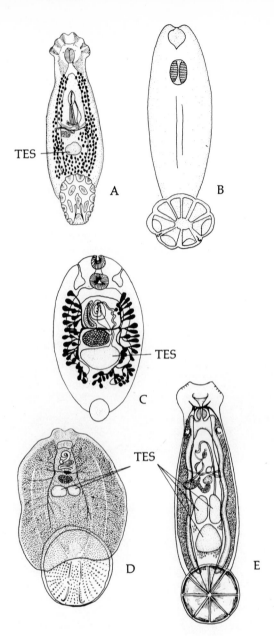

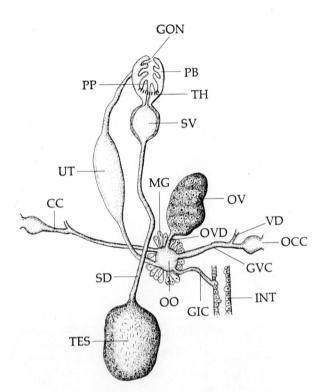

Fig. 9.17.   Reproductive system in monogeneids. Relationship of male and female systems. CC, copulatory canal; GIC, genitointestinal canal; GON, gonopore; GVC, common genitovitelline canal; INT, intestine; MG, Mehlis' gland; OCC, opening of copulatory canal; OO, ootype; OV, ovary; OVD, oviduct; PB, penis bulb; PP, penis papilla; SD, sperm duct; SV, seminal vesicle; TES, testis; TH, thorns (spines); UT, uterus; VD, vitelline duct.

Fig. 9.16.   Some monogenetic trematodes. A. *Thaumatocotyle dasybatis* (Monopisthocotylea) with single testis (TES). (Redrawn after Price, 1938.) **B.** *Heterocotyle pastinacae* (Monopisthocotylea) with single testis that is not shown. (Redrawn after Scott, 1904.) **C.** *Leptocotyle minor* (Monopisthocotylea) with single testis. (Redrawn after Johnstone, 1911.) **D.** *Entobdella hippoglossi* (Monopisthocotylea) with two testes. (Redrawn after Dawes, 1947.) **E.** *Monocotyle* sp. (Monopisthocotylea) with three testes. (Redrawn after Goto, 1894.)

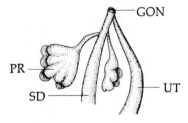

Fig. 9.18.   Type of reproductive system. Terminal of another type of reproductive system as found among monogenetic trematodes. GON, gonopore; PR, prostate glands; SD, sperm duct; UT, uterus.

Female System.   The female reproductive system includes a single ovary, the shape of which may range from globular to elongate and which is usually folded (Fig. 9.17) or lobed. The oviduct arises from the surface of the ovary. The exact positions of the junctions between the **oviduct** and the **seminal receptacle**, the common **vitelline duct**, the **genitointestinal duct**, and the **vagina** vary among the species. Generally, these ducts, which may not all be present, connect with the oviduct before it enlarges to form the ootype.

$$\text{Protein} + o\text{-diphenol} \xrightarrow{\text{polyphenoloxidase}} \text{protein} + \text{quinone}$$

Fig. 9.19

Two types of unicellular glands, collectively known as the **Mehlis' gland**, surround and secrete into the ootype, the so-called **proximal** and **distal** cells (Kohlmann, 1961; Combes, 1968). The secretions of these glands serve primarily as a lubricant that facilitates the passage up the uterus of completely formed eggs from the ootype. One component also contributes to the formation of the egg capsule. According to Kohlmann (1961), the secretion of the thin serous component by the distal cells of the Mehlis' gland in *Polystoma integerrimum*, a bladder parasite of frogs and toads, occurs only during egg laying and has a pH of 4.93. The thicker, mucous secretion is produced continuously by proximal cells. It has a pH of 6.75 to 6.99.

The short uterus may possess a muscular terminal in some species, known as the **metraterm**, which opens into the genital atrium.

The ovary of *P. integerrimum* produces one to three eggs every 10 to 15 minutes, and yolk cells are pressed into the ovovitelloduct (the duct formed from the union of the oviduct and vitelline duct). Each egg contains at least 20 yolk cells, and some spermatozoa. Shell formation is accomplished by quinone tanning (p. 401), which in a simplified form can be expressed as in Figure 9.19.

Mechanical pressure within the ootype liberates the shell grana from the yolk cells. The mucous secretion probably has a lubricating function while the serous secretion contributes to shell formation.

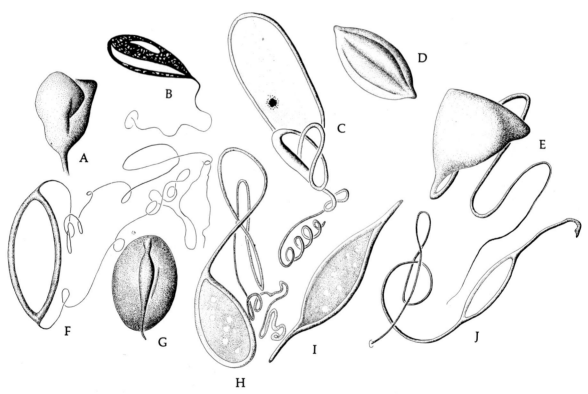

Fig. 9.20. Types of monogenetic trematode eggs. A. Egg of *Thaumatocotyle dasybatis*, parasitic in olfactory organs and gills of the stingray. (Redrawn after Price, 1938.) **B.** Egg of *Udonella caligorum*, parasitic on various fish. (Redrawn after Price, 1938.) **C.** Egg of *Diplozoon paradoxum*, parasitic on gills of various fish. (Redrawn after Dawes, 1946.) **D.** Egg of *Rajonchocotyle alba*, parasitic on gills of the Burton skate. (Redrawn after Fuhrmann, 1928.) **E.** Egg of *Nitzschia monticellii*, parasitic on gills of sturgeons. (Redrawn after Fuhrmann, 1928.) **F.** Egg of *Squalonchocotyle apiculatum*, parasitic on fish. (Redrawn after Fuhrmann, 1928.) **G.** Egg of *Microbothrium apiculatum*, parasitic on piked dogfish and orkneys. (Redrawn after Price, 1938.) **H.** Egg of *Hexabothrium canicula*, parasitic on gills of fish. (Redrawn after Guberlet, 1933.) **I.** Egg of *Erpocotyle laevis*, parasitic on gills of fish. (Redrawn after Guberlet, 1933.) **J.** Egg of *Diclidophora denticulata* parasitic on gills of fish. (Redrawn after Dawes, 1947.)

The seminal receptacle serves as a storage area for spermatozoa received by the female system during copulation.

The **genitointestinal duct**, which occurs only in members of the suborder Polyopisthocotylea, is a unique tube connecting the oviduct with the right intestinal caecum. It is via this structure that excess materials in the ootype are discharged into the caecum and become digested and assimilated.

The vagina of the Monogenea is either single or double, that is, there may be one or a pair of copulation canals, which open to the outside through one or two pores located on the ventral, lateral, or dorsal surfaces of the body. The vaginal pore is independent of the uterine pore (or genital pore) if the uterus enters the common atrium.

In certain species there is an enlarged portion along the length of the vagina known as the **seminal receptacle**. This type of seminal receptacle is a modification of the distinct blind-sac type, which is independent of the vagina. If two vaginae are present, these may unite somewhere along their course.

The cirrus is inserted in the vagina of the female during copulation, and the spermatozoa are introduced down this tubular canal. Some monogeneids do not possess a separate vagina, in which case the uterine pore serves as the site of entrance for spermatozoa. Detailed accounts of copulation in several species are given by Bychowsky (1957), and a review of the functions of the various regions of the reproductive system has been presented by Rohde (1975).

It is noted that the sperm of certain monogeneids are arranged in spermatophores. In *Entobdella diadema*, for example, the spermatophores are produced by a pair of glands lying one on each side of the body near the main nerve cords.

Vitelline glands in monogeneids are generally follicular and lie in the two lateral zones. In *Gyrodactylus*, vitelline glands are absent; instead, the ovary contains cells capable of performing the same function and, hence, is known as an **ovovitellarium**.

The typical shell of a monogeneid egg bears a lidlike cap (the **operculum**) at one end (Fig. 9.20). Fairly commonly, the eggs bear a filament at one or both ends by which they can become attached to the host or to each other in large masses after being laid (Fig. 9.20). The small number of comparatively large, yellowish-brown eggs, generally 1 to 12 in the uterus of an individual worm, is characteristic of monogenetic trematodes.

Monogenetic trematodes produce large numbers of eggs, although relatively few are carried in the uterus at any time. *Polystoma integerrimum*, for example, deposits about 2000 to 2500 eggs during its short laying season. Egg production appears to be influenced by the ambient temperature and oxygen level, increasing with a rise in temperature and/or a fall in oxygen levels. It is noted that the biotic potential of

monogeneids is not as great as that of digenetic trematodes.

GENERAL DEVELOPMENT AND LIFE CYCLE

Eggs lodged in the uterus are soon discharged through the genital pore. These eggs become attached to the host by means of the polar filament or filaments and hatch into ciliated larvae known as **oncomiracidia** (Fig. 9.21). Each free-swimming oncomiracidium seeks out a new host to which it becomes attached and gradually develops into an adult.

Embryonation and hatching of monogeneid eggs increase sharply with increase in temperature (Fig. 9.22) but cease in most species when maintained at 4°C. It is still not known whether a hatching enzyme is present in monogeneid eggs as in the Digenea (p. 311), but Bychowsky (1957) has reported that light initiates the hatching process in some species.

In some species, a host factor is apparently involved in hatching. Euzet and Raibaut (1960) have

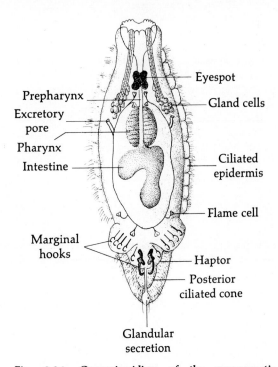

Fig. 9.21. Oncomiracidium of the monogenetic trematode *Gastrocotyle trachuri* parasitic on the gills of the fish *Trachurus trachurus*. (Redrawn after Llewellyn, 1963.)

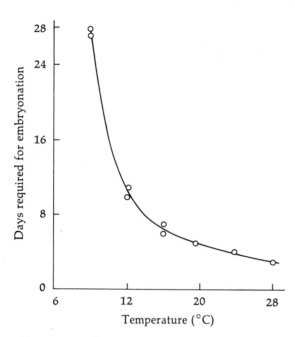

Fig. 9.22. Influence of temperature on embryonation of the eggs of the monogeneid *Dactylogyrus vastator*. (Data from Bychowsky, 1957.)

reported that if eggs of *Erpocotyle torpedinis*, an ecto-parasite of the ray, *Torpedo marmorata*, are placed in water they embryonate but do not hatch. It is only when either the fish host or its excised gills are placed in the water that hatching occurs.

In certain cases, such as *Polystoma integerrimum*, the life cycle is slightly more complex, involving migration of the larvae through the host's alimentary tract. These larvae later become lodged in the bladder, where they reach maturity.

In *Diplozoon*, parasitic on various fish, a ciliated oncomiracidium hatches from the egg. When this free-swimming larva finds and becomes attached to a fish host, it metamorphoses into a **diporpa** larva. When two diporpae make contact, they become intimately associated and develop into adults that remain in permanent copulation (Fig. 9.30; p. 293). This type of relationship is known as **coaptation**.

In another curious group, *Gyrodactylus* spp., commonly found on freshwater fish and amphibians, a form of polyembryony occurs (Kathariner, 1904). Eggs laid by the members of this genus do not contain visible yolk and are fused within the uterus of the parent. During the unusual development, a second embryo is formed within the first, a third within the second, and a fourth within the third. When the first

larva completes its development, it passes out of the parent and immediately becomes attached to the host. This larva soon develops into an adult, but the enclosed larva generations are maintained within.

Ovoviviparous monogeneids are known. Oglesby (1961), for example, has found that in *Polystomoidella oblonga*, a parasite of the musk turtle, *Stenotherus odoratus*, hatching occurs *in utero* and when the larvae are released, they lack cilia but are in an advanced stage of development. These larvae probably develop into adults while attached to the turtle host and, hence, there is no free-swimming stage in the life cycle. Also, unlike digeneans, monogenetic trematodes do not undergo asexual reproduction.

Details of life cycles representative of the Monogenea are given in the following pages. The important fact to remember is that, despite several variations in the development patterns, an intermediate host is never required, and in this respect monogeneids differ from the digenetic trematodes.

A summary of monogeneid life histories is given by Bychowsky (1957).

The Oncomiracidium

The oncomiracidium (Fig. 9.21) is the only larval stage in the life cycle of monogenetic trematodes. This is an elongate, microscopic form which is usually heavily ciliated and bears numerous hooks at its posterior end and is therefore well adapted both for swimming and attaching to a host. The life span of oncomiracidia is short. Unless they find and become attached to a host in about 24 hours, they die. This suggests that there is very little stored nutrient in their bodies or that they have extremely high metabolic rates. Many species, especially those with eyespots, are positively photo-tactic, which may assist them in host finding. In addition to eyespots, monogeneids possess other types of surficial sense organs, including sensory cilia, so-called peg organs, and papillae (see Lyons, 1973, for a review).

Host Finding by Oncomiracidia. During the past two decades there has been considerable interest in host finding by oncomiracidia. Although the results of such studies have been specifically focused, they have also answered some basic questions relative to host specificity; that is, a parasite is associated with a specific host because its infective larva is attracted to that host. By employing the oncomiracidium of *Entobdella soleae* as a model, Kearn (1967) has reported that if offered a choice of isolated scales of its normal host and those of related flatfish, the scales of *Solea solea* (the normal host) are selected almost exclusively. It has been reported that the oncomiracidia almost always settle on the remnant of skin attached to the excised scales, rather than on the bony parts of the scales. Further experiments proved that the mucoid secretions from glands in the fish skin are the primary attractants. Consequently, it was concluded that the oncomiracidia are attached to their natural hosts specifically by chemotactic molecules emitted by glands.

It is noted that the results of Kearn's experiments do not hold true in the case of *Discocotyle* oncomiracidia seeking their trout host (Paling, 1969), nor is chemotaxis apparently responsible for the finding of the bream host by *Diplozoon* larvae (Bovet, 1967). Studies by others (see review by Llewellyn, 1972) have revealed that other types of taxes, especially phototaxis and rheotaxis, are responsible for host finding by oncomiracidia.

BIOLOGY OF MONOGENEIDS

CLASSIFICATION

The current taxonomy of the Monogenea is delineated in a classic monograph by Sproston (1946). The British and European forms have been adequately treated by Dawes (1946). More recently, Yamaguti (1963) has contributed a comprehensive catalog of all known monogeneids. The classic monograph by Bychowsky (1957) is of great value to those who desire a fuller account of these trematodes.

The order Monogenea is divided into two suborders: The **Monopisthocotylea** and the **Polyopisthocotylea**. The diagnostic characteristics of these taxa and those subordinate to them are presented at the end of this chapter.

Suborder Monopisthocotylea

The absence of a genitointestinal canal and of suckerlets or clamps on the single opisthaptor are considered the main distinguishing characteristics of the members of this suborder. Generally, a few adhesive glands are closely associated with the prohaptors. In fact, in certain genera, such as *Dactylogyrus*, prohaptors are lacking, and the adhesive glands empty to the outside through ducts ending in bulbous terminals known as **head organs**.

The sizes and arrangements of the anchors and hooklets on the opisthaptor are important taxonomic characteristics.

Representative genera of the suborder Monopisthocotylea are depicted in Fig. 9.23.

Several complete life cycles are known among the Monopisthocotylea—those of *Gyrodactylus*, *Dactylogyrus*, *Benedenia*, *Ancyrocephalus*, *Acolpenteron*, and others.

Feeding and Food. Prior to presenting the life cycles of some representative monopisthocotyleans, it appears appropriate to present some information pertaining to their food and feeding habits.

Monopisthocotyleans are mainly ectoparasites on their hosts' skin and gills. Consequently, they feed almost exclusively on epidermis and mucus

(Table 9.1). As an adaptation to this type of feeding, their mouths are usually ventrally situated. Kearn (1963a, 1971) has reported in detail the feeding mechanisms of *Entobdella soleae*, an ectoparasite of the sole, *Solea solea*, and *Acanthocotyle* sp., an ectoparasite of another fish, *Raja clavata*. The principal structure associated with feeding is a glandular feeding organ (Fig. 9.24). This organ is capable of protruding and being applied to the surface of the host's epidermis like the circular lip of a cup. The feeding organ is thus held for approximately 5 minutes, during which the approximately 60 gland cells contained therein secrete proteolytic enzymes that digest the circle of epidermis enclosed by the organ (Fig. 9.25). The hydrolyzed host tissues are ingested, but what cannot be digested within the body is eventually egested through the feeding organ.

Since monopisthocotyleans are capable of secreting proteolytic enzymes, it can be speculated that these epidermis feeders could occasionally penetrate deeper and feed on blood from ruptured vessels. Thus, it is not surprising that Kearn (1963b) and Uspenskaya (1962) have reported the occurrence of hemoglobin in the intestinal caeca of several species, including *Amphibdella flarolineata*, *A. torpedinis*, *Dactylogyrus vastator*, *D. solidus*, *Ancylodiscoides parasiluri*, and *Tetraonchus* sp. It is noted, however, that as a rule monopisthocotyleans do not feed on blood as the polyopisthocotyleans do. The review by Jennings (1968) on nutrition and digestion in monogeneans is recommended.

It is of interest to note that Kearn (1963a) is of the opinion that although monopisthocotyleans do cause epidermal lesions on their fish hosts, the resulting damage is negligible under normal circumstances since fish possess high regenerative abilities. For example, the epidermis of the loach, *Misgurnus fossilis*, can regenerate an area 25 to 100 mm^2 within 24–36 hours. What is known about the biology of *Entobdella soleae* has been reviewed by Kearn (1971).

Life Cycle of *Gyrodactylus elegans*. Kathariner (1904) has reported the developmental pattern in *G. elegans*, a common ectoparasite of freshwater and marine fish. The eggs within the parent lack visible yolk. Two such eggs fuse and develop within the parental uterus.

A form of polyembryony takes place in *G. elegans*, although some differential factor is present because not all embryos develop at the same rate. On completion of its development, the first embryo passes to the outside, attaches to the fish host, and develops into an adult. Only then does the second embryo, within the first, mature. Up to four generations may

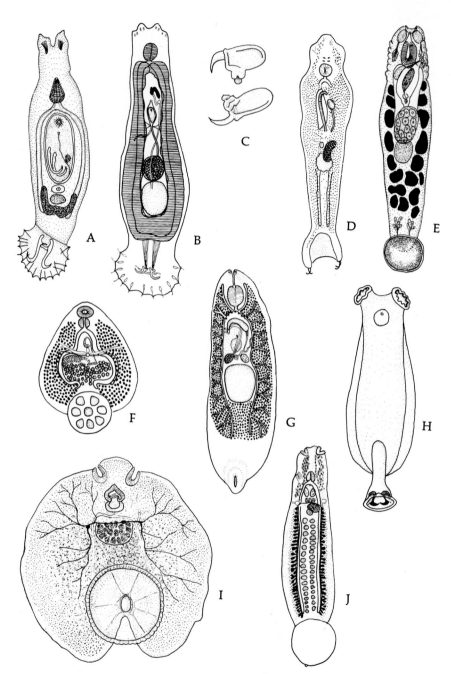

Fig. 9.23. Some monopisthocotyleid monogenetic trematodes. A. *Gyrodactylus* sp. on fish gills; with two head organs. (Redrawn after Bychowsky, 1933.) **B.** *Dactylogyrus* sp. on fish gills; with two pairs of head organs. (Redrawn after Bychowsky, 1933.) **C.** Hooks of *Ancyrocephalus paradoxus*, parasitic on gills of perch. (Redrawn after Lühe, 1909.) **D.** *Diplectanum aequans* on gills of bass and other fish. (Redrawn after MacLaren, 1904.) **E.** *Udonella caligorum* on various fish; prohaptor has a pair of small suckerlike organs. (Redrawn after Price, 1938.) **F.** *Calicotyle kroyeri* in cloaca of various species of rays. (Redrawn after Lebour, 1908.) **G.** *Microbothrium apiculatum* on skin of the piked dogfish, with pair of suckerlike organs in prepharyngeal wall. (Redrawn after Price, 1938.) **H.** *Encotyllabe nordmanni* attached to pharynx of the black sea bream, with thin lateral body margins that are turned ventrally. (Redrawn after Fuhrmann, 1928.) **I.** *Capsala martinieri* on gills of sunfish. (Redrawn after Dawes, 1947.) **J.** *Acanthocotyle* sp. on various fish. (Redrawn after Dawes, 1946.)

develop within the first larva, like a series of Chinese boxes.

Gyrodactylus elegans is sometimes deleterious to its host, depending on the number of parasites present and the physiologic state of the host. Yin and Sproston (1949) have reported an instance in which a number of fantail goldfish were killed by gyrodactyliosis. Death of these fish was attributed to the hypersecretion of mucus on the host's gill surfaces due to the irritation caused by the large number of parasites present. In addition, Van Cleave (1921) has reported the death of bullheads, *Ameiurus melas*, from gyrodactyliosis. With the increase in aquaculture involving the maintenance of fish under crowded conditions, more instances of fish mortality due to gyrodactyliosis are being reported (Sinderman, 1970).

Life Cycle of *Benedenia melleni*. Jahn and Kuhn (1932) have reported the life history of *B. melleni* (Fig. 9.26), a parasite on the Pacific puffer, *Speroides annulatus*, the spadefish, *Chaetodipterus faber*, various species of angelfish, *Angelicthys* spp., and *Pomacanthus* spp.

In *B. melleni*, the egg is tetrahedral and has a single

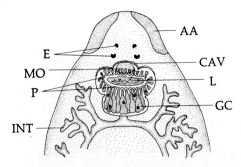

Fig. 9.24. Feeding apparatus of *Entobdella soleae*. Dorsal view of retracted feeding organ showing presence of large number of gland cells. AA, anterior adhesive area; CAV, cavity accommodating feeding organ; E, eyespots; GC, gland cell; INT, intestinal diverticulum; L, lip, tucked inside lumen of feeding organ; MO, mouth; P, papillae. (Redrawn after Kearn, 1963.)

Table 9.1. The Location and Food of Selected Species of Monogenetic Trematodes[a]

Species	Host	Location on Host	Food
Monopisthocotylea			
Entobdella hipoglossi	Halibut	Skin	Epidermis
E. soleae	Sole	Skin	Epidermis
Calicotyle kroyeri	Skate	Cloaca	Epithelium, mucus
Udonella caligorum	Copepod	Skin	Mucus
Leptocotyle minor	Dogfish	Skin	Epidermis[b]
Calceostoma calceostoma	Argyrosomus regium	Skin	Epidermis[b]
Gyrodactylus spp.	Goldfish	Skin, gills	Epithelium[b]
Diplectanum aequans	Bass	Gills	Epithelium[b]
Amphibdelloides maccallumi	Torpedo nobiliana	Gills	Epithelium[b]
Amphibdella torpedinis	T. marmorata	Gills	Blood[c]
A. flavolineata	T. nobiliana	Gills	Blood[c]
Capsala martinieri	Sunfish	Skin	Epidermis[b]
Trochopus sp.	Trigla sp.	Gills	Epidermis[b]
Nitzschia sturionis	Sturgeon	Gills	Epidermis?[d] Blood?
Acanthocotyle spp.	Raia spp.	Skin	Epidermis
Dactylogyrus vastator	Carp?	Gills	Tissue, mucus, blood
Ancylodiscoides parasiluri	?	Gills	Tissue, mucus, blood
Polyopisthocotylea			
Polystoma integerrimum	Frog	Urinary bladder	Blood
Hexabothrium appendiculata	Dogfish	Gills	Blood
Octostoma scombri	Mackerel	Gills	Blood
Discocotyle sagittata	Trout	Gills	Blood
Anthocotyle merluccii	Hake	Gills	Blood
Axine belones	Garfish	Gills	Blood
Diclidophora merlangi	Whiting	Gills	Blood, tissue, mucus
Plectanocotyle gurnardi	Gurnard	Gills	Blood
Diplozoon paradoxum	Minnow	Gills	Blood, tissue
Octodactylus palmata	Ling	Gills	Blood, tissue, mucus

[a] Data assembled by Smyth, 1966.
[b] Based on indirect evidence such as absence of blood pigment.
[c] Other tissues probably also ingested.
[d] Conflicting evidence.

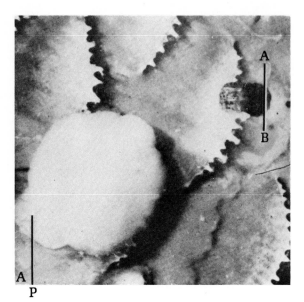

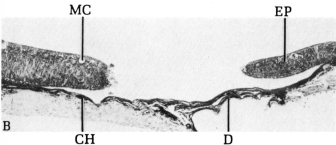

Fig. 9.25. Damage to host caused by *Entobdella*. A. Photograph showing wound (A–B) inflicted by *E. soleae* on the skin of the upper surface of a sole. **B.** Photomicrograph of a section through the wound depicted in **A.** CH, chromatophore; D, dermis; EP, epidermis; MC, mucous cell; P, parasite. (After Kearn, 1963.)

long filament at one end and two shorter filaments at the opposite end that are recurved, giving the appearance of hooks. Unlike *Gyrodactylus elegans*, the parent lays the egg unhatched. On the fourth to sixth day after the egg is laid, the oncomiracidium is visible within the shell.

The fusiform oncomiracidium, which later actively escapes from the eggshell, is approximately 225 μm long by 60 μm wide and bears two bands of cilia at its anterior end, two bands mediolaterally, and two posterolaterally. This larva possesses a folded opisthaptor bearing the definitive anchors that are characteristic of the adult. In addition, a number of marginally situated larval hooks are present.

The oncomiracidium soon loses its ciliated epider-

mis and, concomitantly, the opisthaptor unfolds and the larva becomes attached to the fish host. Only after the larva reaches its host does further development take place (namely, appearance of prohaptors, bifurcation of intestinal caeca, and development of the reproductive organs) and the adult form is attained.

Benedenia melleni injures its host. Because this trematode can attach to the host's epidermis and conjunctiva, the body scales of the fish fall off, exposing large areas of connective and muscle tissue, or the fish becomes blind. Such injuries generally lead to death.

Life Cycle of *Acolpenteron catostomi.* Another example of the direct cycle of the Monopisthocotylea, and all monogeneids for that matter, is that of *A. catostomi*, a species found in the ureters of the hosts—the white sucker, *Catostomus commersonii*, and the Northern hogsucker, *Hypentelium nigricans*. Fischthal and Allison (1942) have reported that single, operculated eggs are found in the parental uterus and also free in the host's ureters. When passed out of the host in urine, each egg contains an uncleaved zygote that develops in water and hatches after 6 to 9 days as an oncomiracidium. This free-swimming larva bears 4 groups of ciliated epithelia, 4 pigmented eyespots, and 14 larval hooklets on a cup-shaped opisthaptor.

The larvae of *A. catostomi* do not require a transfer host since infected fish kept in an isolated tank free from other organisms were infected by second-generation worms in 6 months. The stages in the life cycle of this species are depicted in Fig. 9.27.

Eggs of *Dactylogyrus.* Groben (1940) has reported that *Dactylogyrus vastator*, a gill parasite of carp, possesses two types of eggs the production of which is apparently related to the environmental temperature. One type, laid during the warmer period of the year, develops quite rapidly, whereas the other type, produced in the autumn, develops slowly. Furthermore, the adult worm usually dies after having produced the second type of egg. The second type of egg remains in mud over winter, hatching the following spring. This phenomenon explains increased mortality in carp hatcheries at the end of a cold spell, coinciding with the mass hatching of the "winter eggs." It is possible that similar slow development may exist in other species.

Suborder Polyopisthocotylea

The monogeneids belonging to the suborder Polyopisthocotylea are readily recognized by the presence of multiple posteriorly situated suckers (known as **suckerlets**) or clamps. These suckerlets are borne on a single main disclike "trunk"—the **cotylophore**. A double vagina usually is present along with a genitointestinal canal.

Feeding and Food. The polyopisthocotyleans are mainly gill parasites that feed primarily on blood (Table 9.1), although in some cases they do ingest tissues and mucus. This type of feeding is reflected by the terminal position of their mouths. Furthermore, either there is an oral sucker surrounding the mouth or the mouth is situated in a buccal cavity armed with buccal suckers. A muscular pharynx is always present. These specializations are obviously well adapted for sucking blood.

Ingested blood is rapidly hemolyzed, and partially digested host erythrocytes are commonly found in the intestinal caeca. Halton (1974), for example, has reported the absorption of hemoglobin by the caecal

wall in *Diclidophora merlangi*. The globin moiety is the main nutrient for the parasite, whereas the hematin is not used. The latter is deposited as pigmented granules in the worm's gastrodermal cells, which are eventually eliminated either by direct discharge into the caecal

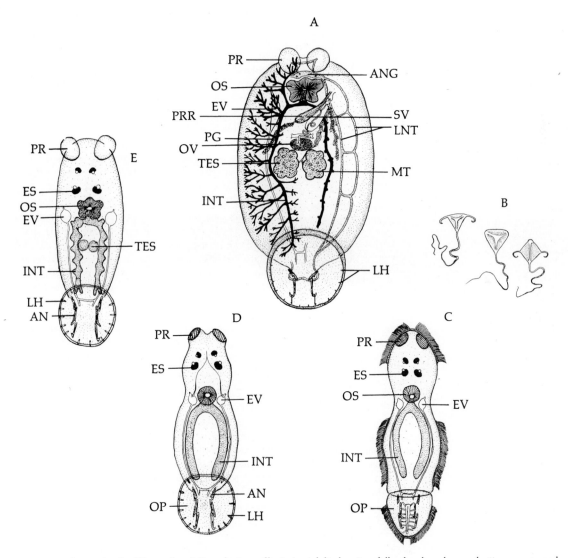

Fig. 9.26. **Stages in the life cycle of *Benedenia melleni*. A.** Adult showing fully developed reproductive organs and accessories, intestinal caeca with secondary diverticula, oral sucker with irregular outline, opisthaptor bearing hooks, and disappearing larval hooks. **B.** Typical eggs, which are discharged from the parent and hatch in water. **C.** Oncomiracidium, drawn from living specimen shortly after hatching, showing six patches of ciliated epidermis, rounded oral sucker, bifurcate intestinal caeca without secondary diverticula, eyespots, and folded opisthaptor. **D.** Oncomiracidium that has shed its ciliated epidermis and has unfolded its opisthaptor. **E.** Young adult from fish host showing beginning of secondary branching of intestinal caeca, beginning of irregular outline of oral sucker, developing testes, and expanded opisthaptor. AN, anchor; ANG, anterior nerve ganglion; ES, eyespot; EV, excretory vesicle; INT, intestine; LH, larval hook; LNT, longitudinal nerve trunks; MT, muscular band through testis; OP, opisthaptor; OS, oral sucker; OV, ovary; PG, prostate gland; PR, prohaptor; PRR, prostatic reservoir; SV, seminal vesicle; TES, testis. (Redrawn after Jahn and Kuhn, 1932.)

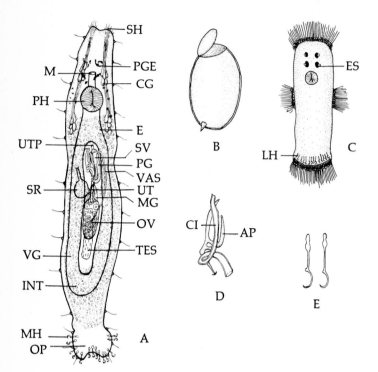

Fig. 9.27. Stages in the life cycle of *Acolpenteron catostomi*. A. Ventral view of adult. **B.** Egg showing operculum and posterior process. **C.** Oncomiracidium showing four zones of ciliated epidermis, four eyespots, and larval hooklets. **D.** Enlarged ventral view of cirrus and accessory piece. **E.** Enlarged drawing of marginal hooks found on opisthaptor of adult. AP, accessory piece; CG, cephalic glands; CI, cirrus; E, esophagus; ES, eyespot; INT, intestine; LH, larval hook; M, mouth; MG, Mehlis' gland; MH, marginal hook; OP, opisthaptor; OV, ovary; PG, prostate gland; PGE, pigment granules of disintegrated eyes; PH, pharynx; SH, sensory hooks; SR, seminal receptacle; SV, seminal vesicle; TES, testis; UT, uterus; UTP, uterine pore; VAS, vas deferens; VG, vitelline glands. (Redrawn after Fischthal and Allison, 1942.)

lumen or by the sloughing of intact epithelial cells. In either case, the granules, independently or within cells, are egested (Llewellyn, 1954; Jennings, 1959).

Life Cycle of *Polystoma integerrimum.* Probably the best known North American species of the Polyopisthocotylea is *Polystoma integerrimum*, a parasite found in the urinary bladders of various frogs and toads. It also occurs in Europe. This unique species has been the subject of numerous investigations and is an interesting but not a typical monogeneid. Stages in the life cycle are depicted in Fig. 9.28.

The monoecious adult inhabits the bladder of amphibians. During the winter months the gonads are nonfunctional, but activity commences with the coming of spring. The worms copulate in the spring, produce from 4 to 122 large eggs per day for 1 week. These eggs are expelled to the exterior.

Embryonic development within the capsule (shell) is affected by temperature. At suitable temperatures above 10°C, the oncomiracidium normally develops in less than 3 weeks. If, however, the temperature drops below 10°C, development may take 6 to 13 weeks.

The correlation between the hatching of *P. integerrimum* eggs and the development and metamorphosis of the frog is one of astounding natural synchronization and suggests a hormonal influence. The barrel-shaped oncomiracidium, which bears 16 arrow-shaped hooklets on its opisthaptor, emerges from the egg and becomes free swimming at the time that the tadpoles lose their external gills and develop internal ones. The larva actively seeks out such a tadpole and enters its gill chamber where it becomes attached to the gill filaments by its armed opisthaptor. In this attached position, development continues for about 8 weeks while the larva subsists on mucus and sloughed host cells.

When the frog undergoes further metamorphosis by losing its gills and developing into a young adult, the worm passes out of the branchial chamber, migrates down the host's alimentary canal, and eventually becomes established in the host's bladder, which by this time has developed. During its migration, the larva loses its ciliated epidermis through atrophy, develops six suckerlets on the cotylophore, loses its larval hooks, and develops adult-type anchors—in other words, the larva matures. In the bladder of the frog, sexual maturity of the parasite is attained within 3 years.

In exceptional situations in which a larva of *P. integerrimum* becomes attached to the external gills of a younger tadpole, an unusual acceleration in larval development takes place. Shortly before the tadpole metamorphoses into an adult, the trematode larva develops into a neotenic form, that is, it becomes sexually mature and produces a single viable egg while remaining somatically immature (Fig. 9.29) (Williams, 1960a, b). Thus, the larval forms can develop in two different ways, depending on the developmental stage of the host.

The correlation between the maturation process of the host and the developmental pattern of the parasite again strongly suggests that the parasite is controlled by the hormonal influence of the host. This has been substantiated by experiments carried out by Miretski (1951), whose studies have revealed that maturation of *Polystoma* is controlled directly or indirectly by the hormonal activity of the frog. Miretski reported that when pituitary extract is injected into an infected immature frog, the polystomes within the frog mature within 4 to 8 days and produce a small number of eggs for approximately 1 week. This period of time corresponds approximately to the time frogs spend in spawning. This synchronized mechanism results in the release of the parasite's eggs only when the frogs enter water to breed. In addition, it also ensures that by the time the oncomiracidia hatch, there are abundant tadpoles available for reinfection.

How the pituitary extract affects the maturation of monogeneids is still uncertain. It is possible that either the increased level of gonadotropins or sex hormones, brought about by the pituitary extract, may be responsible, as is found in the cases of the ciliates *Opalina* and *Nyctotherus* (pp. 169, 244).

Combes (1967) has found that in the case of *Polystoma pelobatis*, a parasite of the frog *Pelobates cultripes*, there are two egg-laying periods (March and October–November) both of which are synchronized with corresponding reproductive periods of the host, which also occurs in two periods.

Recent studies on hormonal action have revealed that in order for a hormone to be effective, the target cells must bear a specific binding (or recognition) protein on their surfaces. It is difficult to imagine that

monogenean cells have evolved specific binding sites on their surfaces for a vertebrate hormone. Consequently, until such have been demonstrated, it appears more reasonable to hypothesize that the parasite is responding to some hormone-induced physiologic change in the host rather than to the hormone directly.

Other representative genera of the Polyopisthocotylea are depicted in Figures 9.30 and 9.31.

In addition to the life cycle of *Polystoma integerrimum*, those of several other polyopisthocotyleans are known. The life cycles of *Sphyranura* and *Diplozoon* are of particular interest because in each case unique variations of the "typical" monogeneid life cycle (such as that of *Benedenia melleni*) are exhibited. In *Sphyranura*, the free-swimming larval stage is absent, whereas in *Diplozoon* the coaptation between two adults is unique.

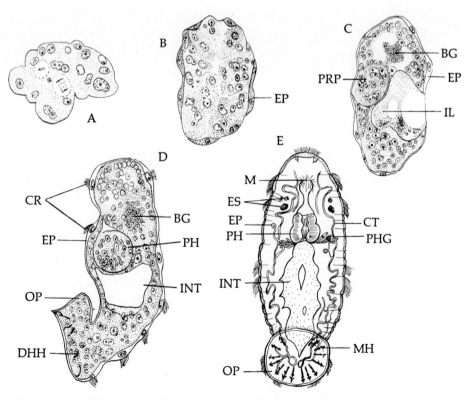

Fig. 9.28. Stages during development of *Polystoma integerrimum*. A. Early stage showing epiboly. **B.** Later stage, epidermis separating from cell mass. **C.** Formation of organs from cell masses, cavity differentiating into digestive lumen. **D.** Later stage. **E.** Larva with patches of ciliated epidermis. BG, brain ganglion; CR, ciliary ridge; CT, collecting tubule (nephridium); DHH, developing haptorial hook; EP, epidermis; ES, eyespots; IL, intestinal lumen; INT, intestine; M, mouth; MH, marginal hook; OP, opisthaptor; PH, pharynx; PHG, pharyngeal glands; PRP, primordium of pharynx. (Redrawn after Halkin, 1901.)

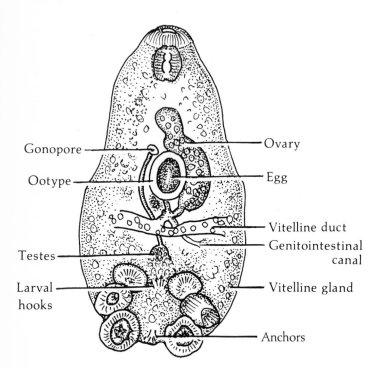

Fig. 9.29. *Polystoma integerrimum.* Neotenic adult from external gills of tadpole. (Redrawn after Williams in Smyth, 1976; with permission of John Wiley & Sons.)

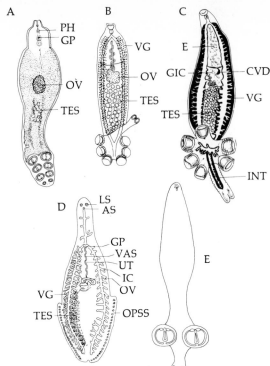

Fig. 9.30. **Some polyopisthocotyleid monogenetic trematodes. A.** *Hexastoma extensicaudum,* on the gills of the tunny, *Thunnus thynnus.* (Redrawn after Dawes, 1940.) **B.** *Erpocotyle* sp., on various marine fish including sharks. (Redrawn after Dawes, 1946.) **C.** *Rajonchocotyloides emarginata* on gills of the thornback ray, *Raja clavata.* (Redrawn after Price, 1940.) **D.** *Microcotyle* sp., on various marine fish. (Redrawn after Hyman, 1951.) **E.** *Anthocotyle merlucii,* on gills of the hake, *Merluccinus merluccius,* and the coalfish, *Gadus virens.* (Redrawn after Cerfontaine, 1896.) AS, anterior sucker; CVD, common vitelline duct; E, eggs; GIC, genitointestinal canal; GP, genital pore; IC, intestinal caecum; INT, intestine; LS, lateral suckers; OPSS, opisthaptor with small suckers; OV, ovary; PH, pharynx; TES, testes; UT, uterus; VAS, vas deferens; VG, vitelline glands.

Life Cycle of *Sphyranura oligorchis.* Alvey (1936) has given an account of the life cycle of *S. oligorchis* (Fig. 9.31), a parasite on the gills of the mudpuppy, *Necturus maculosus.* In this species, the adult releases eggs containing undeveloped zygotes. During the first 4 weeks after the eggs are released into water, embryonic development takes place. The developing larvae, however, are not motile. On approximately the 27th or 29th day, motility is initiated and an active nonciliated larva escapes from each egg. This unique larva appears as a juvenile fluke resembling the adult.

The larva soon sinks to the bottom of the water bed and creeps along the bottom seeking a new host. If another host is not found, the larva soon dies. If it is successful, it becomes attached and within a few days develops into a sexually mature adult. This unique nonciliated larva possesses a well-developed opisthaptor armed with hooks and hooklets and can be thought of as a precocious larva or a juvenile. If the latter interpretation is adopted, then one must surmise that *Sphyranura oligorchis* lacks a larval stage and develops directly from the egg.

Life Cycle of *Diplozoon paradoxum.* The development of *D. paradoxum* is unique. The various stages are depicted in Fig. 9.32. The oncomiracidium hatching from the egg in water is completely covered by a layer of ciliated epithelium. In addition, it possesses two pigmented eye-spots, a well-developed pharynx, and a blind-sac intestinal caecum. The opisthaptor is well developed, bearing a pair of clamps. This ciliated larva is free swimming and, after it searches out a host, becomes attached.

The natural hosts include minnows and other suitable cyprinid fish. While attached to the gills of the fish, each larva metamorphoses, losing its eyespots and developing branched intestinal caeca. The posterior end of the body becomes elongated and wider. A small sucker appears on the midventral surface, and a small conical knob appears at about the same level on the dorsal surface. This larva type is designated a **diporpa**. It does not mature until two diporpae become intimately associated, with the ventral sucker of one grasping the conical knob of the other. By a twist of the two bodies, the free sucker of the one also embraces the conical protuberance of its mate. Not only do the two worms become fused at these points of contact but also at the external openings of their genital ducts. The male pore of the one becomes in apposition with the vaginal pore of the other, and vice versa. Thus the two worms remain in permanent copula. Only after the two worms are thus united do the two additional pairs of clamps take form, the first pair being formed prior to the union. Three pairs of clamps are characteristic of the sexually mature adult.

PHYSIOLOGY AND BIOCHEMISTRY OF MONOGENEIDS

Osmoregulation

Relatively little is known about the physiology of the monogenetic trematodes, and almost nothing is known about the biochemistry of these ectoparasitic worms. Relative to osmoregulation, Wright and Macallum (1887) have observed that the excretory bladder of *Sphyranura osleri* pulsates fairly rapidly during the excretion of its contents. The intervals between pulsations vary from 0.5 to 1.5 minutes. These observations suggest that monogeneids, as

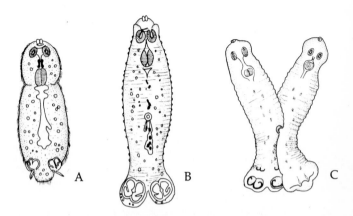

Fig. 9.32. Stages in the life cycle of *Diplozoon paradoxum*. The adults of *D. paradoxum* are depicted in Fig. 9.31 and the egg is depicted in Fig. 9.20. **A.** Free-living oncomiracidium. (Redrawn after Baer, 1952.) **B.** Single diporpa larva. (Redrawn after Baer, 1952.) **C.** Two diporpa larvae in permanent copula. (Redrawn after Zeller, 1892.)

ectoparasites and, hence, primarily surrounded by an aquatic environment, must possess relatively efficient osmoregulatory systems in order to maintain an osmotic equilibrium between the body interior and its surrounding environment. Since the protonephritic

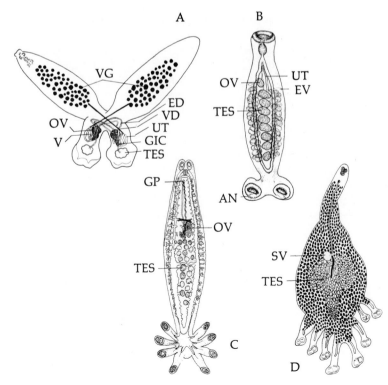

Fig. 9.31. Some polyopisthocotyleid monogenetic trematodes. A. *Diplozoon paradoxum*, on gills of numerous fishes. (Redrawn after Dawes, 1946.) **B.** *Sphyranura* sp., on gills of *Necturus*. **C.** *Octodactylus minor*, on gills of the poutassou, *Gadus poutassou*, and the whiting, *Gadus merlangus*. (Redrawn after Gallien, 1937.) **D.** *Diclidophora merlangi*, on gills of the whiting, *Gadus merlangus*. (Redrawn after Dawes, 1947.) AN, anchor; ED, ejaculatory duct; EV, excretory vesicle; GIC, genitointestinal canal; GP, genital pore; OV, ovary; SV, seminal vesicle; TES, testis; UT, uterus; V, vagina; VD, vitelline duct; VG, vitelline glands.

excretory system is known to remove metabolic wastes from the body as well as to perform the osmoregulatory function, it seems logical to conclude that the rapid pulsations of the excretory vesicle suggest an efficient excretory system.

Oxygen Requirement

Because of the ectoparasitic nature of monogeneids, it is generally assumed that they carry on true aerobic respiration, utilizing the oxygen in the surrounding water. Evidences supporting this contention are behavioral rather than biochemical. In the case of *Entobdella soleae*, a special mechanism has developed to ensure the availability of sufficient oxygen. This trematode is attached to the blind or ventral surface of the sole, *Solea solea*, a fish that spends most of the day partially buried in the bottom of the sea, although it swims freely at night. It is known that unless currents are present, the ocean bottom is usually deficient in oxygen. Consequently, *E. soleae* is exposed to little oxygen in its natural habitat. In order to utilize the amount of oxygen present more efficiently, this parasite undergoes characteristic undulating movements when present in substrates with low oxygen tension. These movements, according to Kearn (1962), consist of wavelike transverse undulations passing along the body from tail to head. As indicated in Fig. 9.33, the undulation rate is inversely proportional to the ambient oxygen tension. In seawater

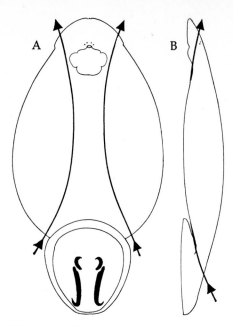

Fig. 9.34. **The main paths of water currents produced by the undulating body movements of *Entobdella soleae*.** **A.** Ventral view. **B.** Lateral view. (After Kearn, 1971.)

containing about 6 ml of O_2 per liter, the undulatory waves average 35 per minute when maintained at 10°C, but the number of waves increases significantly as the amount of O_2 decreases.

In addition to changing the number of these undulations, which serve to agitate the surrounding water and consequently cause more oxygen to pass over the parasite's body surface, *E. solea* also changes the amplitude of the waves and the exposure of its body surface (Fig. 9.34). A similar behavior pattern has been reported in three other monogenean skin parasites, *Pseudocotyle squatinae*, *Acanthocotyle* sp., and *Leptocotyle minor*.

Influence of Temperature

Since monogeneids are almost exclusively ectoparasites, they are exposed to considerably greater variations in ambient temperature than the endoparasitic trematodes. Consequently, the influence of temperature on the biology of these parasites is of some interest. Llewellyn (1957), as the result of studying 13 species of monogeneids removed from marine fish, has reported that when maintained at 3 to 7°C, adult worms survive for 2 to 3 weeks, but few eggs are produced if the ambient temperature drops to below 8°C. Even at 13°C, their normal ambient temperature, egg production usually ceases after 4 days. When maintained at 18°C, the worms do not survive more than 24 hours, but egg production continues for the first 12 hours. When the temperature is elevated to

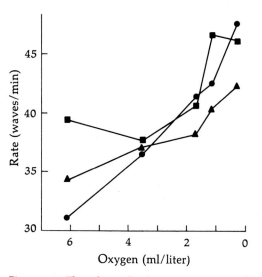

Fig. 9.33. **The relationship between the rate of body undulation and the oxygen content of the medium in three adult individuals of *Entobdella soleae*.** (After Kearn, 1962.)

20°C, the parasites die within 12 hours and generally produce no eggs.

Metabolism

Very few studies have been conducted on the metabolic processes of monogeneids. Because these helminths are ectoparasitic and aerobic, one would expect them to include less stored nutrients when compared to the endoparasitic digenetic trematodes. Indeed, Smyth (1966) has reported that even among monogeneans, the ectoparasitic species store little glycogen, whereas the endoparasitic representatives store about the same amount as the endoparasitic digenetic trematodes (Table 9.2).

Melekh (1963) and Halton (1964) have studied the chemical compositions of several monogeneid species. Their data are presented in Table 9.2. The occurrence of iron in the polyopisthocotyleans listed in Table 9.2 is undoubtedly related to their blood-ingesting habits.

Although it has been claimed that monogeneids, unlike the digenetic trematodes and cestodes, absorb little through their teguments, Halton (1978) has demonstrated that *Diclidophora merlangi*, a gill parasite of fish, can absorb amino acids (alanine and leucine) and glucose from sea water through its tegument. The uptake of amino acid is mediated. It is known that detectable levels of amino acids (0.1–4 μM) occur in sea water.

HOST-PARASITE RELATIONSHIPS

While many cases of infection with monogeneids· cause little distress to their hosts, in some cases the host is injured. It is known that fish can be rendered blind as the result of *Benedenia melleni* becoming attached to their eyes. In addition, *Gyrodactylus*, *Microcotyle*, and *Tetrancistrum* also cause diseases in fish. Gyrodactyliosis is of special concern in aquaria, fish hatcheries, and aquaculture ponds. These parasites can ingest the host's blood from surfacial capillaries. Hypersecretion of mucus by hosts infested with monogeneids is a well-established fact, and this condition is known to clog the gills, thus suffocating the host. *Microcotyle* infestations have caused 90% mortality among angelfish and butterfly fish in the New York Aquarium, brought about through mucus suffocation.

That fish infested with *Benedenia melleni* demonstrate varying degrees of susceptibility and resistance suggests that they may possess a natural immunity or are able to acquire immunity in varying degrees. The black angelfish, for example, is susceptible to *B. melleni* infestations for 1 to 2 weeks as young adults but soon acquire an immunity that lasts until old age, when it once again becomes susceptible. To further test the idea that immunity occurs, Nigrelli (1937) attempted to induce immunity by injecting extracts prepared from living *B. melleni* (from ground, dried worms) and serum from infected fish. His results suggest that the secreted mucus of immunized fish does afford some protection, since worms placed in the mucus died in less than 8 hours, whereas worms placed in the mucus of susceptible fish and sea water survived for 3 days.

It is also known that the minute tears made by monogeneids on the gills and skin of fish are ideal sites for the invasion of secondary bacterial and fungal infections.

Table 9.2. Chemical Composition of Some Monogenetic Trematodes[a]

| | | | | Chemical Composition as % of Dry Weight | | | |
| | | | | | | | |
Species	**Host**	**Location on or in Host**	*Dry Weight as % of Fresh Weight*	*Inorganic Ash*	*Glycogen*	*Protein*	*Iron*
Monopisthocotylea							
Entobdella hippoglossi	Halibut	Skin	16	5.7	0.8	83	—
Calicotyle kroyeri	Skate	Cloaca	12	4.9	1.17	79	—
Polyopisthocotylea							
Polystoma integerrimum	Frog	Urinary bladder	9.5	7.2	8.5	67	0.8
Discocotyle sagittata	Trout	Gills	9.5	6.0	0.9	—	0.4
Diclidophora merlangi	Whiting	Gills	7.5	7.0	1.2	74	0.65
Octodactylus palmata	Ling	Gills	7	6.5	2.5	79	0.7

[a] Data assembled by Halton, 1964.

Host Specificity

Associated with the phenomenon of host-parasite compatibility is that of host specificity. Host specificity has been repeatedly demonstrated among the Monogenea (Jahn and Kuhn, 1932; Bychowsky, 1933; Mizelle et al., 1943; Koratha, 1955; Hargis, 1953, 1957; and others). These workers have pointed out that certain species of monogeneids are highly host specific. Hargis (1957) has suggested that this specificity "may be either physiological and genetic and/or ecological in basis."* The same author has concluded that host specificity among the Monogenea may well be used as a tool in determining the phylogenetic relationships of the hosts parasitized by related parasites, and vice versa.

REFERENCES

Alvey, C. H. (1936). The morphology and development of the monogenetic trematode *Sphyranura oligorchis* (Alvey, 1933) and the description of *Sphyranura polyorchis* n. sp. *Parasitology* **28**, 229–253.

Bovet, J. (1967). Contribution à la morphologie et à la biologie de *Diplozoon paradoxum* V. Nordmann, 1832. *Bull. Soc. Neuchâtel Sci. Natur.*, **90**, 63–159.

Bresciani, J. (1972). The ultrastructure of the integument of the monogenean *Polystoma integerrimum* (Fröhlich 1791). *Roy. Vet. Agr. Univ., Copenhagen, Denmark, Yearbook 1973*, 14–27.

Bychowsky, B. E. (1933). Die Bedeutung der monogenetischen Trematoden für die Erforschung der systematischen Beziehungen der Karpfenfische. Erste Mitteilung. *Zool. Anz.* **102**, 243–251. (Also see *Ibid.*, **105**, 17–38).

Bychowsky, B. E. (1937). Ontogenesis and phylogenetic interrelations of parasitic flatworms. *News Acad. Sci. USSR, Dept. Heath. Nat. Sci.* 1354–1883.

Bychowsky, B. E. (1957). "Monogenetic Trematodes, Their Systematics and Phylogeny" (W. H. Hargis, ed.). AIBS Publ., 1961. Washington, D.C.

Combes, C. (1967). Corrélations entre les cycles sexuels des amphibiens anoures et des Polystomatidae (Monogenea). *C. R. Hebd. Séanc. Acad. Sci. Paris* **264(D)**, 1051–1052.

Combes, C. (1968). Biologie écologie des cycles et biogéographie de Digènes et Monogènes d'amphibiens dans l'est des Pyrénées. *Mem. Mus. Natl. Hist. Nat. A. Zool.* **51**, 1–195.

Dawes, B. (1946). "The Trematoda." Cambridge University Press, London.

Euzet, L., and Raibaut, A. (1960). Le developpement postlarvaire de *Squalonchocotyle torpedinis* (Price, 1942)

(Monogenea, Hexobothriidae). *Bull. Soc. Neuchâtel Sci. Natur.* **83**, 101–108.

Fischthal, J. H., and Allison, L. H. (1942). *Acolpenteron catostomi* n. sp. (Gyrodactyloidea: Calceostomatidae), a monogenetic trematode from the ureters of suckers, with observations on its life history and that of *A. ureteroecetes*. *Trans. Am. Microsc. Soc.* **61**, 53–56.

Groben, G. (1940). Beobachtungen über die Entwicklung verschiedener Arten von Fischschmarotzern aus der Gattung *Dactylogyrus*. *Z. Parasitenk.* **11**, 611–636.

Halton, D. W. (1964). Observations on the nutrition of certain parasitic flatworms (Platyhelminthes: Trematoda). Ph. D. Thesis, University of Leeds, England.

Halton, D. W. (1974). Hemoglobin absorption in the gut of a monogenetic trematode, *Diclidophora merlangi*. *J. Parasitol.* **60**, 59–66.

Halton, D. W. (1978). Trans-tegumental absorption of L-alanine and L-leucine by a monogenean, *Diclidophora merlangi*. *Parasitology* **76**, 29–37.

Hargis, W. J., Jr. (1957). The host specificity of monogenetic trematodes. *Exp. Parasitol.* **6**, 610–625.

Jahn, T. L., and Kuhn, L. R. (1932). The life history of *Epibdella melleni* MacCallum, 1927, a monogenetic trematode parasite on marine fishes. *Biol. Bull.* **62**, 89–111.

Jennings, J. B. (1959). Studies on digestion in the monogenetic trematode *Polystoma integerrimum*. *J. Helminth.* **33**, 197–204.

Jennings, J. B. (1968). Nutrition and digestion. In "Chemical Zoology" (M. Florkin and B. T. Scheer, eds.), Vol. II, pp. 303–326. Academic Press, New York.

Jennings, J. B. (1972). "Feeding, Digestion and Assimilation in Animals," 2nd ed. MacMillan, London.

Kathariner, L. (1904). Ueber die Entwicklung von *Gyrodactylus elegans* v. Nrdm. *Zool. Jahrb.* **70**, 519–550.

Kearn, G. C. (1962). Breathing movements in *Entobdella soleae* (Trematoda, Monogenea) from the skin of the common sole. *J. Mar. Biol. Assoc. U. K.* **42**, 93–104.

Kearn, G. C. (1963a). Feeding in some monogenean skin parasites: *Entobdella soleae* on *Solea solea* and *Acanthocotyle* sp. on *Raja clavata*. *J. Mar. Biol. Assoc. U. K.* **43**, 749–766.

Kearn, G. C. (1963b). The life cycle of the monogenean *Entobdella soleae*, a skin parasite of the common sole. *Parasitology* **53**, 253–263.

Kearn, G. C. (1967). Experiments on host-finding and host-specificity in the monogenean skin parasite *Entobdella soleae*. *Parasitology* **57**, 585–605.

Kearn, G. C. (1971). The physiology and behaviour of the monogenean skin parasite *Entobdella soleae* in relation to its host (*Solea solea*). In "Ecology and Physiology of Parasites" (A. M. Fallis, ed.), pp. 161–187. University of Toronto Press, Canada.

Kearn, G. C. (1973). An endogenous circadian hatching rhythm in the monogenean skin parasite *Entobdella soleae* and the relationship of rhythmical hatching to the activity rhythm of the host (*Solea solea*). *Parasitology* **66**, 101–122.

Kohlmann, F. W. (1961). Untersuchungen zur Biologie, Anatomie, und Histologie von *Polystomum integerrimum* Fröhlich. *Z. Parasitenk.* **20**, 495–524.

Koratha, K. J. (1955). Studies on the monogenetic trematodes of the Texas coast. I. Results of a survey of marine fishes of Port Arkansas, with a review of Monogenea reported from the Gulf of Mexico and notes on euryhalinity, host specificity, and relationships of the remora

---

*For a detailed review of host specificity among monogenetic trematodes and a discussion of infra- and supraspecificity, see Hargis (1957).

and cobia. *Inst. Mar. Sci.* **4**, 234–239.

Lee, D. L. (1966). The structure and composition of the helminth cuticle. *Adv. Parasitol.* **4**, 187–254.

Lee, D. L. (1972). The structure of the helminth cuticle. *Adv. Parasitol.* **10**, 347–379.

Llewellyn, J. (1954). Observations on the food and the gut pigment of the Polyopisthocotylea (Trematoda: Monogenea). *Parasitology* **44**, 428–437.

Llewellyn, J. (1956). The host specificity, microecology, adhesive attitudes and comparative morphology of some trematode gill parasites. *J. Mar. Biol. Assoc. U.K.* **35**, 113–127.

Llewellyn, J. (1957). The larvae of some monogenetic trematode parasites of Plymouth fishes. *J. Mar. Biol. Assoc. U.K.* **36**, 243–259.

Llewellyn, J. (1958). The adhesive mechanisms of monogenetic trematodes; the attachment of species of the Diclidophoridae to the gills of gadoid fishes. *J. Mar. Biol. Assoc. U.K.* **37**, 67–79.

Llewellyn, J. (1972). Behaviour of monogeneans. In "Behavioural Aspects of Parasite Transmission." (E. U. Canning and C. A. Wright, eds.), pp. 19–30. Academic Press, London.

Lyons, K. M. (1966). The chemical nature and evolutionary significance of monogenean attachment sclerites: *Parasitology* **56**, 63–100.

Lyons, K. M. (1972). Sense organs of monogeneans. In "Behavioural Aspects of Parasite Transmission." (E. U. Canning and C. A. Wright, eds.), pp. 181–199. Academic Press, London.

Lyons, K. M. (1973). The epidermis and sense organs of the Monogenea and some related groups. *Adv. Parasitol.* **11**, 193–232.

Melekh, D. A. (1963). Comparative study of glycogen reserves in some ecto- and endoparasites. *Vestn. Leningrad Univ. Biol.* **18**, 5–13.

Miretski, O. Y. (1951). Experiment on controlling the processes of vital activity of the helminth by influencing the condition of the host. *Dokl. Akad, Nauk USSR* **78**, 613–615. (In Russian.)

Mizelle, J. D., LaGrave, D. R., and O'Shaughnessy, R. P. (1943). Studies on monogenetic trematodes. IX. Host specificity of *Pomoxis tetraonchinae*. *Am. Midl. Natur.* **29**, 730–731.

Nigrelli, R. F. (1937). Further studies on the susceptibility and acquired immunity of marine fishes to *Epibdella melleni*, a monogenetic trematode. *Zoologica* **22**, 185–191.

Oglesby, L. C. (1961). Ovoviviparity in the monogenetic trematode *Polystomoidella oblonga*. *J. Parasitol.* **47**, 237–243.

Paling, J. E. (1969). The manner of infection of trout gills by the monogenean parasite *Discocotyle sagittata*. *J. Zool.* **159**, 293–309.

Rohde, K. (1975). Fine structure of the Monogenea, especially *Polystomoides* Ward. *Adv. Parasitol.* **13**, 1–33.

Sinderman, C. J. (1970). "Principal Diseases of Marine Fish and Shellfish." Academic Press, New York.

Smyth, J. D. (1966). "The Physiology of Trematodes" Oliver & Boyd, Edinburgh, Scotland.

Sproston, N. G. (1946). A synopsis of monogenetic trematodes. *Trans. Zool. Soc. London* **25**, 185–600.

Thurston, J. P., and Laws, R. M. (1965). *Oculotrema hippopotami* (Trematoda: Monogenea) in Uganda. *Nature* **205**, 1127.

Uspenskaya, A. V. (1962). Nutrition of monogenetic trematodes. *Dokl. Akad. Nauk USSR* **142**, 1212–1215. (In Russian.)

Van Cleave, H. J. (1921). Notes on two genera of ectoparasitic trematodes from freshwater fishes. *J. Parasitol.* **3**, 33–39.

Williams, J. B. (1960a). The dimorphism of *Polystoma integerrimum* (Frölich) Rudolphi and its bearing on relationships within the Polystomatidae: Part I. *J. Helminthol.* **34**, 151–192.

Williams, J. B. (1960b). The dimorphism of *Polystoma integerrimum* (Frölich) Rudolphi and its bearing on relationships within the Polystomatidae: Part II. *J. Helminthol.* **34**, 323–346.

Wright, R. R., and Macallum, A. B. (1887). *Sphyranura osleri*; a contribution to American helminthology. *J. Morphol.* **1**, 1–48.

Yamaguti, S. (1963). "Systema Helminthum. Vol. 4: Monogenea and Aspidocotylea." Wiley, New York.

Yin, W. Y., and Sproston, N. G. (1949). Studies on the monogenetic trematodes of China; Parts 1–5. *Sinensia* **19**, 57–85.

CLASSIFICATION OF MONOGENEA (THE MONOGENETIC TREMATODES)

Order Monogenea

Body dorsoventrally flattened; hermaphroditic; oviparous or viviparous; with conspicuous posterior adhesive organ (opisthaptor) sometimes divided into loculi; opisthaptor usually with sclerotized anchors, hooks, and/or clamps; commonly with anterior adhesive organ (prohaptor) consisting of one or two suckers, grooves, glands, or expanded ducts from deeper glands; eyes, if present, usually of two pairs; mouth at or near anterior body terminal; pharynx usually present, gut commonly with two simple or branched caeca, after anastomosing posteriorly, rarely a single median tube or sac; with one to many testes; vas deferens commonly convoluted; cirrus present or absent, armed or without spines; male genital pore usually in atrium common with female pore; genital pores ventral or marginal; ovary single, variable in shape; genitointestinal duct present or absent; seminal receptacle present; uterus short; vitelline follicles extensive, usually lateral; vagina absent or present, single or double, with dorsal, ventral, or lateral pore with two lateral osmoregulatory canals, each with expanded vesicle opening dorsally near anterior end; parasites on or in aquatic vertebrates, especially fish, rarely on aquatic invertebrates.

Suborder Monopisthocotylea

Opisthaptor a single unit that may be subdivided into shallow loculi, usually developed directly from larval haptor; one to three pairs of large anchors usually present, commonly with minute marginal hooks; prohaptor glandular or with paired suckers or pseudosuckers; oral sucker absent; eyes often present; genitointestinal canal absent; eggs usually with polar filaments; seminal receptacle is enlargement of vagina.

SUPERFAMILY ACANTHOCOTYLOIDEA

Opisthaptor small, with 14 marginal and two central hooklets, developed separately from larval haptor; haptor muscular, without anchors, but radially arranged spines or septa may be present; parasites of marine fish. (Genus mentioned in text: Family Acanthocotylidae*—*Acanthocotyle.*)

SUPERFAMILY CAPSALOIDEA

Muscular opisthaptor large and circular, often divided by septa into loculi; marginal hooks absent or present; anchors, if present, without connecting bars; prohaptor, if present, with two lateral suckers or one pseudosucker, and with two glandular areas; cirrus without accessory piece; parasites of fishes. (Genera mentioned in text: Family Monocotylidae*—*Dictyocotyle, Thaumatocotyle, Heterocotyle, Monocotyle, Tritestis, Merizocotyle.* Family Capsalidae—*Entobdella, Capsala, Benedenia.* Family Dactylogyridae—*Dactylogyrus, Ancyrocephalus, Tetrancistrum.* Family Microbothriidae—*Leptocotyle, Microcotyle, Leptocotyle.* Family Calceostomatidae—*Acolpenteron.* Family Tetraoncidae—*Tetraoncus*).

SUPERFAMILY GYRODACTYLOIDEA

Rounded or bilobed opisthaptor with one or two pairs of anchors supported by one to three bars; marginal larval hooks present; prohaptor with groups of cephalic glands opening on margin of anterior body terminal; parasites of fish, amphibians, cephalopods, and crustaceans. (Genus mentioned in text: Family Gyrodactylidae*—*Gyrodactylus.*)

SUPERFAMILY UDONELLOIDEA

Muscular opisthaptor lacking septa or hooks; prohaptor poorly developed, with lateral head organs or pseudosuckers; parasites of marine fish or hyperparasitic on copepods on marine fish. (Genus mentioned in text: Family Udonellidae*—*Udonella.*)

Suborder Polyopisthocotylea

Complex opisthaptor commonly subdivided, with suckers, clamps, or anchors; larval haptor absent or reduced to pad supporting terminal anchors; marginal hooklets usually absent; mouth surrounded by sucker, or striated fringe, or with paired suckers inside buccal cavity; prohaptor usually without adhesive glands; eyes usually absent; genitointestinal canal usually present; intestine usually with two caeca, sometimes joined posteriorly; testes usually numerous; eggs usually with polar filaments; seminal receptacle present or absent.

SUPERFAMILY AVIELLOIDEA

Opisthaptor at terminal of long stalk, with six suckers on margin and four large anchors in middle; intestine a simple unbranched sac; prohaptor with two or three

pairs of glandular organs; four eyes present; single testis; eggs without polar filaments; parasites of freshwater teleost fish.

SUPERFAMILY DICLIDOPHOROIDEA

Opisthaptor commonly divided into two lateral rows of four each, with suckers or clamps; prohaptor usually simple; two small suckers in buccal cavity; intestine bifurcate; eyes absent; parasites of fish or hyperparasites on parasitic crustaceans on fish. (Genera mentioned in text: Family Diclidophoridae*—*Diclidophora.* Family Discocotylidae—*Discocotyle.*)

SUPERFAMILY CHIMAERICOLOIDEA

Small opisthaptor near larval haptor; posterior portion of body elongate and slender; opisthaptor with two rows of simple clamps; prohaptor as simple, weak, circumoral sucker; eyes and oval intraoral suckers absent; reproductive organs in anterior portion of body; parasites of chimaerid fish.

SUPERFAMILY DICLYBOTHRIOIDEA

Opisthaptor with three pairs of clamps or suckers, each surrounding large anchor; posterior appendix on opisthaptor with three pairs of large and one pair of very small hooks (in some species a rudimentary pair of suckers present on posterior margin); with two pairs of eyes; intestine bifurcate, branched, anastomosing near posterior end of body; parasites of sturgeons and related fishes. (Genera mentioned in text: Family Hexabothriidae*—*Rajonchocotyle, Hexabothrium, Erpocotyle.*)

SUPERFAMILY MEGALONOCOIDEA

Opisthaptor complex, with three pairs of anchor complexes, each consisting of several sclerotized elements; eyes and marginal hooklets absent; two terminal appendices, each with two pairs of hooks, on posterior margin of opisthaptor of some species; prohaptor consists of two suckers in buccal cavity; parasites of marine teleost fish.

SUPERFAMILY DIPLOZOOIDEA

Adults permanently associated in pairs forming X shape; opisthaptor rectangular or bilobed, with four pairs of, or numerous, clamps; one pair of posterior anchors also present in some species; intestine a single tube with branches; genital pore in posterior half of body; parasites on gills of freshwater fish. (Genus mentioned in text: Family Diplozoidae*—*Diplozoon.*)

SUPERFAMILY MICROCOTYLOIDEA

Opisthaptor symmetrical or asymmetrical, with numerous clamps symmetrically or asymmetrically arranged; anchors may also be present at posterior end; prohaptor of two suckers in buccal cavity; intestine bifurcate, not joined posteriorly; eyes absent; parasites of marine fish.

SUPERFAMILY POLYSTOMATOIDEA

Opisthaptor with two or six well-developed suckers, with or without anchors, with larval hooklets present; oral sucker surrounding mouth; intestine bifurcate, may or may not join posteriorly; eyes usually absent; parasites of fish, amphibians, reptiles, and on eyes of hippopotamus. (Genera mentioned in text: Family Polystomatidae*—*Oculotrema, Polystoma, Polystomoidella.* Family Sphyranuridae—*Sphyranura.*)

---

*For diagnoses of families subordinate to the Monogenea, see Yamaguti (1963).

10

DIGENEA: THE DIGENETIC TREMATODES

The digenetic trematodes, or digeneans, comprise the largest group of monozoic platyhelminths. Within their hosts, these parasites are found in the intestine, gallbladder, urinary bladder, blood, esophagus, and practically every other major organ. Although biologists have been finding new trematodes since the days of Francesco Redi in the 17th century, the wealth of unknown trematode fauna is by no means exhausted. Even today, taxonomists are continuously discovering new species of these parasites in various hosts. The number of known species is well over 40,500. The excellent monograph by Erasmus (1972) on the biology of this group of platyhelminths is recommended.

Morphologically, the digeneans have the characteristics of the Platyhelminthes as given in Chapter 8. Unlike the aspidobothrid and monogenetic trematodes, adult digenetic trematodes usually possess two prominent suckers on their body surface—the **anterior sucker**, often referred to as the oral sucker when it surrounds the mouth, and a ventrally located holdfast sucker called the **acetabulum**. Among certain groups, such as the monostomes, however, only the anterior sucker is present.

From the standpoint of development and life history, the Digenea not only possess by far the most complex developmental cycles among the Platyhelminthes, but also among all members of the Animal Kingdom. During the "typical" life cycle, the worm utilizes two, three, four, or more hosts—one being the definitive host and the remainder being intermediate or paratenic hosts. In each host the parasite takes on one or more different forms. In the intermediate hosts, the parasite assumes various larval forms, each developing asexually from the preceding generation. It is only in the adult that sexual reproduction occurs.

All digenetic trematodes, except the schistosome blood flukes and certain members of the suborder Didymozoida, are monoecious. Sexual reproduction may be brought about by either self-fertilization or cross-fertilization between two individuals. Among the schistosomes, the individuals are either males or females. The blood flukes are of great medical importance because the species that cause schistosomiasis are widespread throughout the tropics and subtropics and are menacing hazards to human health. Numerous other species of trematodes are of medical and veterinary importance because they are the causative agents of human and animal diseases. A list of the more important digeneans is given in Table 10.1.

Table 10.1   Some Digenetic Trematodes Found in Humans and Domestic Animals

Trematode	Principal Hosts	Habitat	Main Characteristics	Disease Caused
Schistosomatoidea				
Schistosoma mansoni	Humans	Blood	Male and female paired, male 6.4–9.9 mm long, female 7.2–14 mm long, conspicuous tuberculations on male, 6–9 testes, eggs 114–175 by 45–68 μm	Schistosomiasis mansoni
S. japonicum	Humans, dogs, cats, rats, mice, cattle, pigs, horses	Blood	Male and female paired, male 12–20 mm long, female 26 mm long, tegument spinous on males, 7 testes, eggs 70–100 by 50–65 μm	Schistosomiasis japonicum
S. haematobium	Humans, monkeys	Blood	Male and female paired, male 10–15 mm long, female 20 mm long, tegumental tuberculations on males, 4–5 testes, eggs 122–170 by 40–70 μm	Schistosomiasis haematobium
Fascioloidea				
Fasciola hepatica	Humans, sheep, cattle	Bile ducts	20–30 mm long, 13 mm wide, cone-shaped process at anterior end, eggs 130–150 by 63–90 μm	Fascioliasis (liver rot)
F. gigantica	Horses, cattle	Bile ducts	Similar to *F. hepatica* only larger	Fascioliasis gigantica
Fascioloides magna	Cattle, horses, sheep	Liver	Larger than *F. hepatica*, often over 200–300 mm long, eggs 109–168 by 75–100 μm	Fascioloidiasis
Fasciolopsis buski	Humans, pigs	Duodenum, jejunum	Broadly ovate, 30–75 mm long, 8–20 mm wide, highly dendritic testes, eggs 130–140 by 80–85 μm	Fasciolopsiasis
Plagiorchioidea				
Dicrocoelium dendriticum	Sheep, cattle, goats, Horses, pigs, rabbits, dogs, humans	Liver, bile ducts	Slender, lancet-shaped, 5–12 mm long, 1 mm wide, extremities pointed, eggs 38–45 by 22–30 μm	Dicrocoeliasis
Opisthorchioidea				
Opisthorchis felineus	Cats, rarely humans	Biliary and pancreatic ducts	Lancet-shaped, rounded posteriorly, 7–12 mm long, 2–3 mm wide, intestinal caeca along entire length of body, eggs 30 by 11 μm	Opisthorchiasis
Clonorchis sinensis	Humans, dogs, cats	Bile ducts	10–25 mm long, 3–5 mm wide, deeply lobed testes, eggs 27.3–35 by 11.7–19.5 μm	Clonorchiasis
Metorchis conjunctus	Dogs, cats, foxes, humans	Gallbladder, bile ducts	1–6.6 mm long, 590 μm to 2.6 mm wide, linguiform, testes slightly lobed, cirrus absent, eggs 22–32 by 11–18 μm	Experimental hosts killed in heavy infections
Parametorchis complexus	Cats	Bile ducts	3–10 mm long, 1.5–2 mm wide, uterus rosette shaped and located in anterior half of body	Cirrhosis of liver (?)
Amphimerus pseudofelineus	Cats, coyotes	Bile ducts	12–22 mm long, 1–2.5 mm wide, uterus with only ascending limb, eggs 25–35 by 12–15 μm	Cirrhosis of liver (?)
Paramonostomum parvum	Ducks	Intestine	Ovoid, 250–500 μm long by 200–350 μm wide	Nonpathogenic
Notocotylus imbricatus	Water fowls	Caecum	2–4 mm long, no acetabulum	Nonpathogenic
Heterophyoidea				
Heterophyes heterophyes	Humans, cats, dogs, foxes	Small intestine	Elongate, pyriform, 1–1.7 mm long, 0.3–0.4 mm wide, small oral sucker	Heterophyiasis
Metagonimus yokogawai	Humans, dogs	Small intestine	Similar to *H. heterophyes* but with acetabulum deflected to right of midline	Metagonimiasis
Apophallus venustus	Cats, dogs, raccoons	Small intestine	950 μm to 1.4 mm long, 250–550 μm wide; no cirrus or cirrus sac; few eggs, 26–32 by 18–22 μm	Nonpathogenic
Cryptocotyle lingua	Usually in fish-eating birds, also found in dogs and cats, very rarely in humans	Small intestine	902 μm to 1.6 mm long, 430–470 μm wide, conspicuous genital pore at anterior margin of acetabulum, anterior portion of body often attenuated, eggs 32–48 by 18–22 μm	Nonpathogenic (?)
Phagicola longa	Dogs, cats, foxes, wolves	Small intestine	500 μm to 1 mm long, 300–400 μm wide, oral sucker surrounded by double row of 16 spines, eggs 18 by 9 μm	Nonpathogenic (?)

Table 10.1. (continued)

Trematode	Principal Hosts	Habitat	Main Characteristics	Disease Caused
Euryhelmis monorchis	Minks	Small intestine	410 μm long, 610 μm wide, only one testis, eggs 29 by 14 μm	Nonpathogenic
Strigeata *Alaria americana*	Dogs	Small intestine	4–5 mm long, posterior portion of body cylindrical, crescentric projection on each side of oral sucker, testes bilobed, eggs 100–134 by 64–80 μm	Nonpathogenic
A. arisaemoides	Dogs, foxes	Small intestine	7–10 mm long, small projections on each side of oral sucker, body constricted, eggs 140 by 90 μm	Nonpathogenic
A. michiganensis	Dogs	Small intestine	1.8–1.91 mm long, right testis bilobed, genital atrium more than twice the size of suckers, eggs 80–140 by 76–80 μm	Nonpathogenic
A. canis	Dogs	Small intestine	2.8–4.2 mm long, small projections on each side of oral sucker, anterior testis lobed, posterior testis horseshoe shaped, eggs 107–133 by 70–99 μm	Nonpathogenic
Strigea falconis	Turkeys	Intestine	Body divided, vitellaria extend into both portions, eggs 110–125 by 75–80 μm	Nonpathogenic
Cotylurus flabelliformis	Ducks, chickens	Small intestine	Body divided, 560–850 μm long, 200 μm wide, eggs 100–112 by 68–76 μm	Nonpathogenic (?)
Postharmostomum gallinum	Chickens	Caecum	Elongate body, 3.5–7.5 mm long, 1–2 mm wide, vitellaria well developed along lateral margins of body	Irritation in heavy infections
Sphaeridiotrema globulus	Ducks, swans	Small intestine	Body subspherical; 500–850 μm long; uterus short, in front of acetabulum, containing 4 to 5 eggs; eggs 90–105 by 60–75 μm	Ulcerative enteritis
Ribeiroia ondatrae	Chickens, fish-eating birds, muskrats	Proventriculus	1.6–3 mm long, testes at posterior end of body, ovary anterior to testes, eggs 82–90 by 45–48 μm	Proventriculitis
Echinostomatoidea *Clinostomum attenuatum*	Chickens	Trachea	5.7 mm long, 1.6 mm wide, dorsal body surface convex, ventral surface concave, oral sucker surrounded by collar	Nonpathogenic
Euparyphium melis	Mink	Stomach, small intestine	Lancet-shaped; 3.86–10.5 mm long; 650 μm to 2.1 mm wide; collar of 27 spines; eggs large, 117–130 by 72–84 μm	Nonpathogenic (?)
Echinoparyphium recurvatum	Chickens, turkeys, usually in water birds	Small intestine	700–4.5 μm long, 500–600 μm wide, collar of 45 spines around oral sucker, four large corner spines, others arranged in two rows	Severe inflammation of small intestine
Echinostoma ilocanum	Humans, rats, dogs	Small intestine	2.5–6.5 mm long, 1–1.35 mm wide, circumoral disc with 49–51 spines, eggs 83–116 by 58–69 μm	Colic, diarrhea
E. revolutum	Chickens, usually in water birds	Caecum, rectum	10–22 mm long, 2–3 mm wide, collar of 37 spines, five grouped together as corner spines, eggs 90–126 by 59–71 μm	Hemorrhagic diarrhea in heavy infections
Himasthla muehlensi[a]	Humans	Intestine	11–17.7 mm long, 0.41–0.67 mm wide, body thin, elongate, collar of 32 spines, two on each side, remaining 28 arranged in horseshoe pattern, eggs 114–149 by 62–85 μm	Gastroenteritis
Hypoderaeum conoideum	Ducks, chickens, pigeons	Small intestine	6–12 mm long, 1.3–2 mm wide, collar of 47–53 spines in two rows, eggs 95–108 by 61–68 μm	Nonpathogenic
Troglotrematoidea *Paragonimus westermani*	Humans, cats	Encapsulated in lungs	Plump, ovoid, 7.5–12 mm long, 4–6 mm wide, scalelike spines, deeply lobed testes, no cirrus pouch or cirrus, ovary lobed, eggs 80–118 by 48–60 μm	Paragonimiasis
P. kellicotti	Humans, dogs, cats, sheep, goats, rats, lions	Encapsulated in lungs	Plump, ovoid, 9–16 mm long, 4–8 mm wide, similar to *P. westermani*, eggs 78–96 by 48–60 μm	Kellicotti paragonimiasis

Table 10.1. (continued)

Trematode	Principal Hosts	Habitat	Main Characteristics	Disease Caused
Nanophyetus salmincola	Dogs, foxes, bobcats, coyotes, cats, raccoons, humans	Intestine	Pyriform, 0.8–1.1 mm long, 0.3–0.5 mm wide, uterus simple with few eggs, vitellaria profuse, eggs 60–80 by 34–35 μm	Salmon poisoning
Sellacotyle mustelae	Mink, foxes	Intestine	355 μm long, 190 μm wide, pyriform, eggs 60 by 54 μm	Slight enteritis
Collyriclum faba	Chickens, turkeys	Encysted in skin	Each cyst with two worms unequal in size 4–5 mm long, 3.5–4.5 mm wide, eggs 19–21 by 9–11 μm	Emaciation and anemia resulting in death
Paramphistomatoidea				
Watsonius watsoni	Humans	Intestine	Pear-shaped, 8–10 mm long, 4–5 mm wide, thick body, acetabulum near posterior end, eggs, 122–130 by 75–80 μm	Severe diarrhea
Gastrodiscoides hominis	Humans	Caecum	Pyriform, 5–10 mm long, 4–6 mm wide, prominent genital cone, large acetabulum covering posterior half of body, eggs 150–152 by 60–72 μm	Mucous diarrhea
Cotylophoron cotylophoron (= *Paramphistomum microbothrioides*)	Cattle	Rumen	3–11 mm long, 1–3 mm wide, conical, convex dorsally, concave ventrally, acetabulum at posterior end, testes large and lobate, eggs 132 by 68 μm	Paramphistomiasis
Paramphistomum cervi	Cattle, moose, deer	Rumen	Similar to *P. microbothrioides*	Nonpathogenic (?)
Zygocotyle lunata	Chickens, usually in water birds	Caecum, small intestine	3–9 mm long, 1.5–3 mm wide, testes lobed, acetabulum at posterior end, ovary behind, posterior testis, eggs 130–150 by 72–90 μm	Nonpathogenic
Cyathocotylioidea				
Mesostephanus appendiculatum	Dogs, cats	Small intestine	900 μm to 1.75 mm long, 400–600 μm wide, large adhesive organ behind acetabulum, genital pore posterior, uterus short with 4–5 eggs, eggs 117 by 63–68 μm	Nonpathogenic (?)
Cyclocoelioidea				
Typhlocoelum cymbium	Ducks, geese	Trachea, bronchi and air sacs	6–12 mm long, 3–6 mm wide, caeca form complete ring, eggs 122–154 by 63–81 μm	Suffocation in heavy infections

[a] This species may be identical to *Himasthla quissetensis*, an intestional parasite of sea gulls, with metacercariae in shellfish.

GENERAL MORPHOLOGY

Since the digenetic trematodes pass through various stages during their life cycles, it is advisable to consider the morphology of each generation separately and to cite the variations commonly encountered. In order to consider the various stages more intelligently, a generalized, hypothetical life cycle pattern is presented below.

GENERALIZED LIFE CYCLE PATTERN

The zygote, resulting from the fusion of the male and female gametes, is encased within an egg shell. The eggs, which are quite numerous, become lodged in the long and often coiled uterus. From their *in utero* position the eggs are released into the host's intestinal lumen through the genital pore and later are passed to the outside in the host's feces. The eggs are deposited in water and eventually hatch, and from each egg a free-swimming **miracidium** emerges.

The miracidium penetrates the integument of a molluscan host (the first intermediate host) and sheds its ciliated epidermis in the process. The naked miracidium develops into a **sporocyst**, which migrates through blood vessels and other spaces and tissues leading to the digestive gland and eventually becomes lodged in that organ. In some species, the eggs enter the molluscan host passively—that is, the eggs are ingested by the host, and the miracidia hatch out within the host's digestive tract and then migrate to the digestive gland.

Although the digestive gland is a common site for further development of the sporocyst, it is not the only one. The gonad, mantle, lymph spaces surrounding the intestine, gill chambers, and other areas may also serve as sites for further development. Once established in a suitable position, the sporocyst increases in size and develops to its mature form.

Within the sporocyst, which is commonly elongate and hollow, germinal cells are present. These cells increase in number by mitotic division and eventually differentiate into **germ balls**. Each germ ball further increases in size and may differentiate into another larval generation called the **redia**.

As in the sporocyst, certain germinal cells in the blood chamber of rediae eventually give rise to tail-bearing **cercariae**, a fourth larval generation. Cercariae escape from their molluscan host and become free swimming. In some species, the cercariae never leave the molluscan host and enter the next host only if the infected mollusc is ingested.

When free-swimming cercariae come in contact with a compatible second intermediate host—often an arthropod or some other invertebrate, or even a vertebrate—they actively penetrate the host's body and encyst. The encysted larva is known as a **metacercaria**. When the second intermediate host is ingested by the vertebrate definitive host, the encysted metacercaria excysts in the host's intestine and gradually matures into the adult.

During the life cycle of various digeneans, it is not uncommon to find mother and daughter generations of sporocysts and rediae. The number of generations of these different larval forms is usually consistent in the members of a specific taxon. Variations of this generalized life history pattern are known. From the foregoing, it is evident that stages in the life cycle of this hypothetical trematode include the adult, egg, miracidium, sporocyst, redia, cercaria, and metacercaria.

THE ADULT

Sexually mature adult digenetic trematodes can be separated into morphologic types.

1. The most frequently encountered morphologic type is the **distome** (Fig. 10.1). In this type, the body is commonly elongate oval. The size, depending on the species, varies from 30 μm to approximately 30 mm in length, as in the liver fluke, *Fasciola hepatica*. In some unusual species, such as those that parasitize scombriid fish, the length may be several inches. Both the anteriorly situated **oral sucker** and the **acetabulum** are present. The mouth is located in the center of the oral sucker, and the intestinal caeca usually are forked, although the caeca of members of the family Cyclocoeliidae are joined posteriorly to form a ring.

2. The **amphistome** type (Fig. 10.1) is characterized by the position of the acetabulum, which is located at the posterior end of the body and is often referred to as the **posterior**

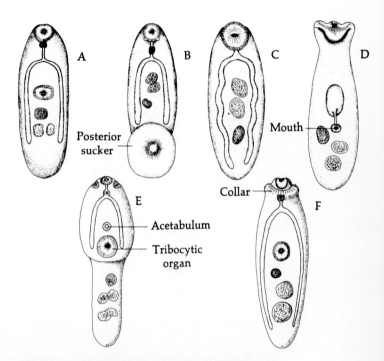

Fig. 10.1. **Major morphologic types of adult digenetic trematodes. A.** Distome. **B.** Amphistome. **C.** Monostome. **D.** Gasterostome. **E.** Holostome. **F.** Echinostome.

sucker. The ovary is located posterior to the testes, an arrangement also found among other types of trematodes, but almost universal among the amphistomes.

3. The **monostome** type is characterized by the presence of only one sucker—the anteriorly located oral sucker (Fig. 10.1).

4. The **gasterostome** type, which is limited to certain species parasitic in fish, is characterized by the mouth being located not in the center of the anterior sucker but in the middle of the ventral sucker (Fig. 10.1). Thus, in this case the ventral sucker is the oral sucker.

5. The **holostome** type, found primarily in the intestine of birds and less frequently in mammals and other vertebrates, is characterized by its elongate body, which is divided into a forebody and a hindbody (Fig. 10.1). The mildly, ventrally concave forebody bears the anterior sucker and the acetabulum. Quite commonly, the anterior sucker is provided with auxiliary suckers flanking it. A special glandular adhesive organ, called the **tribocytic organ**, is located immediately posterior to the acetabulum.

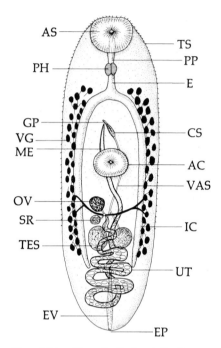

Fig. 10.2. Hypothetical adult distomate digenetic trematode showing major anatomic features. AC, acetabulum; AS, anterior sucker; CS, cirrus sac; E, esophagus; EP, excretory pore; EV, excretory vesicle; GP, genital pore; IC, intestinal caecum; ME, metraterm; OV, ovary; PH, pharynx; PP, prepharynx; SR, seminal receptacle; TES, testis; TS, tegumentary spines; UT, uterus; VAS, vas deferens; VG, vitelline gland.

6. The **echinostome** type is actually a specialized distome, for the positions of the suckers are comparable to those found in distomes (Fig. 10.1). However, it may be considered a distinct type, because there is a collar of large spines surrounding the oral sucker.

These six morphologic types do not necessarily indicate six distinct phylogenetic groups but are merely useful in describing adult trematodes.

Structure of the Adult

Despite superficial differences, the anatomy of the various groups of digenetic trematodes is quite similar. For this reason, the anatomy of a hypothetical distome is being presented to exemplify the various details (Fig. 10.2).

Tegument. The outer surface of the adult digenean is covered by a syncytium known as the tegument. Lee (1966) has reviewed what is known about the structure and composition of this tissue. In brief, under the light microscope the tegument appears as a more or less homogeneous "noncellular" layer about 7–16 μm thick. It shows an affinity for basic stains.

When studied with the electron microscope, the tegument can be seen to consist of two zones. The outer zone, which is separated from the environment by a unit membrane, consists of a cytoplasmic syncytium (Fig. 10.3). Embedded in this layer are mitochondria, endoplasmic reticulum, various types of vacuoles and, in some instances, glycogen granules and other types of inclusions.

The outer surface is usually thrown into folds to form microvilli. These undulations not only serve to increase the absorptive surface, but pinocytotic vesicles are also formed in the crypts between ad-

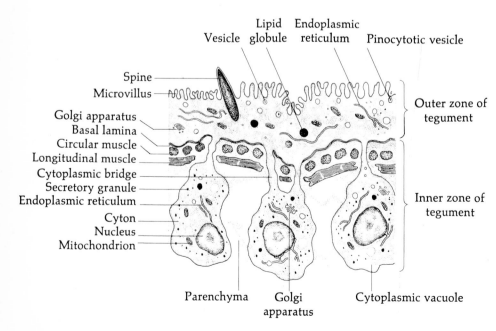

Fig. 10.3. Trematode tegument. Drawing showing fine structure of the tegument of a digenetic trematode (*Fasciola*) composed from numerous electron micrographs.

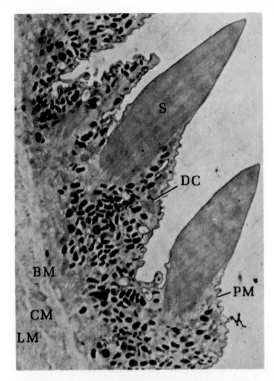

Fig. 10.4.   Tegumental spines. Electron micrograph of a portion of the tegument of the adult of *Haematoloechus medioplexus,* a lung fluke of frogs, showing two embedded spines. BM, basement membrane (= basal lamina); CM, circular muscles; DC, external level of tegument; LM, longitudinal muscles; PM, plasma membrane; S, spine. (After Burton, 1964.)

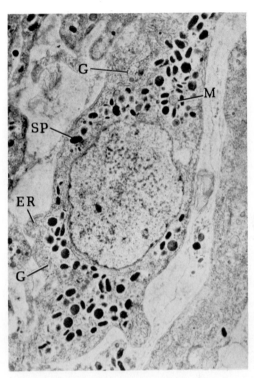

Fig. 10.5.   Cyton. Electron micrograph showing single cyton (perikaryon) of the inner tegumentary zone of adult *Haematoloechus medioplexus,* a lung fluke of frogs. ER, endoplasmic reticulum; G, Golgi apparatus; M, mitochondrion; SP, secretory product. (After Burton, 1964.)

jacent microvilli for the intake of large molecules and particulate materials. Also embedded in the outer zone of some species are tegumentary spines (Fig. 10.4). These are overlayed by the surficial plasma membrane and undoubtedly serve as ancillary holdfast mechanisms *in situ.*

The outer syncytial zone is connected by cytoplasmic bridges to nucleated bodies, known as **cytons** (or **perikarya**), embedded deeper in the parenchyma. The cytons, collectively designated as the inner tegumentary zone, include vacuoles, endoplasmic reticulum, mitochondria, Golgi bodies, glycogen deposits, and various types of vesicles in addition to the nucleus (Figs. 10.3, 10.5).

In the region between the outer and inner tegumentary zones are found several other types of tissues. Specifically, lying immediately mediad to the outer tegumentary zone, and separated from it by a unit membrane, is a thin layer of connective tissue fibers known as the **basal lamella.** Beneath this are a series of circular muscles, under which are the fascicles of longitudinal muscles (Fig. 10.3).

Investigations into the chemical composition of the

adult digenean tegument have revealed the presence of glycogen, nonglycogenic polysaccharides, lipids, acid mucopolysaccharides, and mucoproteins (Pantelouris, 1964; Öhman, 1965; and others). The occurrence of acid mucopolysaccharides is of particular significance, since the molecules are known to be capable of inhibiting various digestive enzymes, and their presence and possible secretion onto the body surface may account for why the intestinal trematodes are not digested by their hosts' enzymes.

In addition to the chemical entities listed, several enzymes have been detected in the trematode tegument. Both acid and alkaline phosphatases have been reported (Yamao, 1952; Lewert and Dusanic, 1961). Esterases have also been detected (Nimmo-Smith and Standen, 1963). In certain specialized cases, other hydrolytic enzymes are associated with certain regions of the tegument. For example, Erasmus and Öhman (1963) have reported the occurrence of aminopeptidase associated with the body surface in the adhesive organ region of *Cyathocotyle bushiensis*, a caecal parasite of ducks. In this case the enzyme is secreted by underlying cells and is active in extracorporeal digestion, that is, food, in this case cells of the host's caecum, is partially predigested outside of the parasite's body prior to ingestion.

Alimentary Tract. Digeneans possess an incomplete digestive system. The mouth is located anteriorly in the center of the oral sucker (except in gasterostomes). The mouth leads into a bulbous muscular **pharynx** via a short **prepharynx**, which may be absent in some species. The pharynx serves primarily as a masticatory organ, from which the ingested food particles pass into the **esophagus**.

The esophagus is lined with a layer of cells. At the posterior terminal of the esophagus the alimentary tract bifurcates, giving rise to two blind-sac **intestinal caeca** that terminate posteriorly in the parenchyma. Both the buccal cavity and esophagus may receive secretions of salivary and esophageal glands, respectively, which contribute to the process of digestion. The lengths of the esophagus and intestinal caeca are commonly employed, in combination with other characteristics, as taxonomic diagnostic features.

Generally, the cells lining the intestinal caeca (gastrodermis) are described categorically as being epithelial; however, they perform more than a lining function. Dawes (1962), who studied the intestinal epithelium of *Fasciola hepatica*, suggested that these cells go through a secretory and an absorptive phase. During the absorptive phase, when nutrient material is present, the cells are short; however, when food materials are absent, they become columnar. During the secretory phase, the cells become filled with secretory materials. When these materials are secreted, the cells collapse, and are later regenerated. The secretions presumably represent digestive enzymes, since others have shown that the caecal cells are rich in a number of enzymes and in ribonucleic acid. The latter suggests that active protein synthesis is occurring. The specific enzymes detected in a number of species of digenetic trematodes are presented in Table 10.2.

The diets of adult digeneans differ, depending on their habitats. For example, in the case of *Fasciola hepatica* parasitic in the bile ducts of sheep and cows, Pearson (1963), by tracing ^{51}Cr-labeled blood, has shown that it ingests blood. In addition, Dawes (1963) and others have reported that this parasite abrades the superficial epithelium of the hypertrophied bile duct in which it lives and feeds on the tissue debris produced. Halton (1964) and others have found that digeneans that are intestinal parasites feed primarily on partially digested intestinal contents along with mucus and mucosal cells of the host's intestinal wall and blood. In the blood-inhabiting forms, such as *Schistosoma* spp., the host's blood is their primary food. Digeneans need not be residents of the vascular system to be exclusively blood ingestors. *Haplometra cylindracea*, a lung parasite of amphibians; *Diplodiscus subclavatus* and *Megalodiscus temperatus*, both rectum parasites of amphibians; and *Haematoloechus medioplexus*, a lung parasite of frogs, also feed on blood (Halton, 1964; Bogitsh, 1972; Morris, 1973).

Lipidlike globules have been reported in the gastrodermis and lumen of a number of digenetic trematodes (see Bogitsh, 1975, and Erasmus, 1977, for reviews). Bogitsh *et al.* (1968), Dike (1969), Bogitsh

Table 10.2. Nature of the Food and Enzymes Present in Some Species of Digenetic Trematodes[a]

Species	Host	Location	Food	Protease	Lipase	Alkaline Phosphatase	Acid Phosphatase	Esterase	Aminopeptidase
Fasciola hepatica	Sheep	Bile duct	Tissue, blood	+	+	+	+	+	−
Schistosoma mansoni	Mouse	Venous system	Blood	+	−	+	+	−	−
Cyathocotyle bushiensis	Duck	Rectum	Tissue	−	−	+	−	+	+
Haplometra cylindracea	Frog	Lungs	Blood	+	−	+	+	+	−
Haematoloechus medioplexus	Frog	Lungs	Blood	+	−	+	+	−	−
Diplodiscus subclavatus	Frog	Rectum	Blood	−	−	+	+	−	−
Opisthioglyphe ranae	Frog	Intestine	Tissue, mucus, blood	−	−	+	+	+	−
Gorgoderina vitelliloba	Frog	Bladder	Tissue, blood	−	−	+	+	+	−

[a] Data compiled by Smith, 1966.

[b] +, present; −, not tested for.

and Shannon (1971), and others have identified a number of acid hydrolases associated with these globules and consequently have identified them as lysosomes. In *Haematoloechus medioplexus*, Davis *et al.* (1968) have reported that the membrane-bound lysosomes are released into the caecal lumen in the confined region between the surface amplifications where extracellular digestion occurs when the lysosomes release their enclosed enzymes. These investigators have designated the released lysosomes as **superficial digestive vacuoles**. Apparently a similar phenomenon occurs among schistosomes (Bogitsh, 1981) and in *Fasciola hepatica* (Robinson and Threadgold, 1975).

In *Schistosoma mansoni*, which feeds almost exclusively on its host's blood, Timms and Bueding (1959) reported the occurrence of a proteolytic enzyme with a pH optimum of 3.9 that is specific for hemoglobin. Dresden and Deelder (1979) have established this hemoglobinase to be very similar to the lysosomal enzyme cathepsin B_1.

Reproductive Systems. The reproductive systems, both male and female, of digenetic trematodes are somewhat similar to those found in the Monogenea and Aspidobothrea.

Male System. The male reproductive system generally includes two testes, although multitesticular species are known; the members of several families (e.g., Spirorchidae, Schistosomatidae) are multitesticular. Digeneans with a single testis also exist. The testes are located in the parenchyma in rather exact positions. In fact, the positions of these gonads are of considerable importance in the identification of species.

Leading from each testis is a **vas efferens**. These ducts generally unite to form the common **vas deferens**, which opens into the **cirrus pouch** (Fig. 10.6). In some species the vasa efferentia enter the cirrus pouch independently. The cirrus pouch (or cirrus sac) is situated at the terminal of the male reproductive system, enclosing the **seminal vesicle**, the **prostate glands**, and the protrusible **cirrus** (Fig. 10.6). In some trematodes a permanent penis is present. In others, the seminal vesicle is located outside the pouch and is known as an **external seminal vesicle** (Fig. 10.6). Not all digenetic trematodes possess a cirrus pouch; in some, the vas deferens empties directly to the outside via the genital pore.

Spermatozoa, formed in the testes, pass down the vasa efferentia and vas deferens to the seminal vesicle and are stored therein. During copulation, sperm are ejected through the eversible cirrus, which is inserted in the **genital atrium** of the female system. The **prostate gland** cells, which surround and empty into the cirrus, are believed to secrete a fluid in which spermatozoa survive. For a review of spermatogenesis in trematodes, see Mohandas (1983).

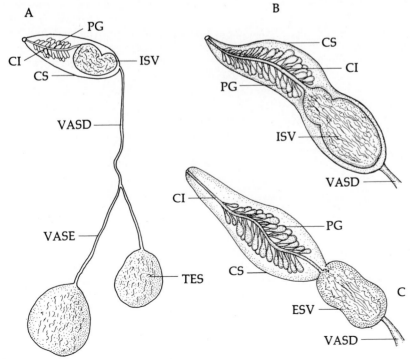

Fig. 10.6. Male reproductive system of digenetic trematode. A. Complete system showing constituent parts. **B.** Cirrus sac of species with internal seminal vesicle. **C.** Terminal portion of species with external seminal vesicle. CI, cirrus; CS, cirrus sac; ESV, external seminal vesicle; ISV, internal seminal vesicle; PG, prostate glands; TES, testis; VASD, vas deferens; VASE, vas efferens.

Female System. The female reproductive system generally includes a single ovary located in the parenchyma, either anterior or posterior to the testes, depending on the species. Ova (actually secondary oocytes) formed within the ovary are released from that organ via a short **oviduct** that opens into a minute chamber, the **ootype**. Three auxiliary organs also empty into the ootype: (1) the **Mehlis' gland**, which is a cluster of unicellular glands surrounding and independently emptying into the ootype; (2) the **common vitelline duct**, which receives the materials from the vitelline glands via the left and right ducts and deposits them in the ootype; and (3) the duct from the **seminal receptacle** (absent in some species), which delivers sperm into the ootype or oviduct.

In some species a fourth structure—the **vitelline reservoir**—is present as a diverticulum of the common vitelline duct. The function of this reservoir is for temporary storage of vitelline materials.

The Mehlis' gland has several functions: (1) These glands secrete a fluid that enhances the hardening or tanning process of newly formed eggs by maintaining the desired pH, redox potential, and so on. (2) The secretion of these cells causes release of the shell globules from the vitelline glands. (3) The secretion forms a thin membrane around the cells forming the egg, and the shell globules then build up from within this membrane. (4) Its secretion lubricates the uterus, facilitating passage of the eggs. (5) Its secretion activates spermatozoa, which are passed down to the ootype.

The **uterus**, which is connected to the ootype, is the long and often convoluted tube in which the formed eggs are transported to the exterior. In some species, the distal segment of the uterus is muscular and by peristaltic movement expels eggs. This muscular portion is referred to as the **metraterm**. The terminal portion of the uterine tract empties into the genital atrium which, in most species, is the common chamber into which the cirrus also opens. The atrium opens to the exterior through the common **genital pore**.

During self- or cross-fertilization (the latter being the more common), the cirrus is evaginated—that is, with the inside surface becoming the outside, much in the same fashion as a sock is pulled inside out—and is inserted in the female aperture. Spermatozoa introduced down the uterus are stored in the seminal receptacle, from which they enter the ootype, where fertilization and egg formation occur.

In some species, a **Laurer's canal** is present. This tube originates from the ootype and passes dorsally through the parenchyma; it may or may not open to the exterior. If the canal opens to the exterior, it may serve as a vagina. This is believed to be the case in the progenitor species.

The **vitelline glands**, collectively known as the vitellaria, are groups of glands (acini) commonly located along the two lateral sides of the body, but in some species they converge along the midline. Arising from each acinus is a **vitelline duct**, which joins other ducts on the same side to form the left and right vitelline ducts, respectively. The left and right ducts either may converge to form the common vitelline duct, which enters the ootype, or may enter the ootype independently.

In addition to contributing yolk material for incorporation within the egg, the vitelline glands also secrete large globules, known as **shell globules**, which envelop the developing eggs and eventually coalesce and become hardened to form the shell. This hardening process involves the tanning of the protein or sclerotin present within the coalesced globules by quinone. Thus, the vitelline glands contribute both yolk and shell materials. Further details of eggshell formation are presented in the next subsection.

Egg Formation. The definitive digenean egg includes a rather complex eggshell. This covering consists of a resistant, tanned protein, which in turn is overlayed by a thin stratum of lipoprotein. Smyth and Clegg (1959) have presented a detailed account of egg formation in digeneans. The following is a brief review.

As ova are released periodically from the ovary, they pass into the ootype. The release of ova is controlled by a sphincter muscle, the **oocapt**, situated at the ovary-oviduct junction. Within the ootype, the ovum is surrounded by vitelline globules coming from the vitelline glands. Under the influence of Mehlis' gland secretions, some of the vitelline globules coalesce to form a semiliquid shell. The remainder of the vitelline globules present within the shell are associated with the fertilized ovum and serve as the food reserve during development. The shape and size of the eggshell is, in part, determined by the shape of the uterus.

Shell precursors from the vitelline glands are in the form of proteins, phenols, and phenolase. These, as stated, coalesce, and as the egg passes up the uterus, the shell becomes tanned and hardened (Fig. 10.7).

In addition to the precursors secreted by the vitelline glands, the Mehlis' gland cells also contribute. Specifically, one type of cell comprising this gland secretes a lipoprotein that envelops the protein shell. A second type of cell secretes mucus, which probably serves to lubricate the uterus for the passage of eggs. In *Fasciola hepatica*, Clegg (1965) has found that the lipoprotein of the Mehlis' gland serves as the template for the deposition of the shell precursors. In other words, in this species, and perhaps in others, there is a lipoprotein layer on the interior as well as the exterior of the protein shell. These lipoproteins contain 61% protein, 38% lipids, and 1% carbohydrate, including hexosamine. Most of the lipids are cholesterol and triglycerides. Only a small amount of phospholipid is present.

Exceptions to the process of egg formation summarized above exist. For example, in *Syncoelium spathulatum*, a gill parasite of the flying fish, *Prognichthys*, Coil and Kuntz (1963) have reported that the protein and phenolic components of the shell are secreted by the uterine wall and deposited on a thin membrane template produced by the Mehlis' gland. It has been suggested also that in the human blood fluke, *Schistosoma mansoni*, the uterine wall may produce a substance necessary for shell formation (Gönnert, 1955).

Excretory System. The excretory or osmoregulatory system of digenetic trematodes is of the protonephritic type. The arrangement of the indivi-

dual flame cells along the collecting tubes is so consistent in a given taxon that their arrangement pattern—known as the **flame cell pattern**—is employed as a tool to indicate phylogenetic relationships. For example, two flukes with similar anatomic features and with identical flame cell patterns could be considered members of the same genus. This concept has been widely adopted despite the fact that exceptions to the rule are known. Flame cell formulae are now employed not only to indicate relationships among adult flukes but to demonstrate affinities between larval (cercarial and metacercarial) and adult forms. The arrangement of flame cells can be expressed in a formula. For example, the pattern found

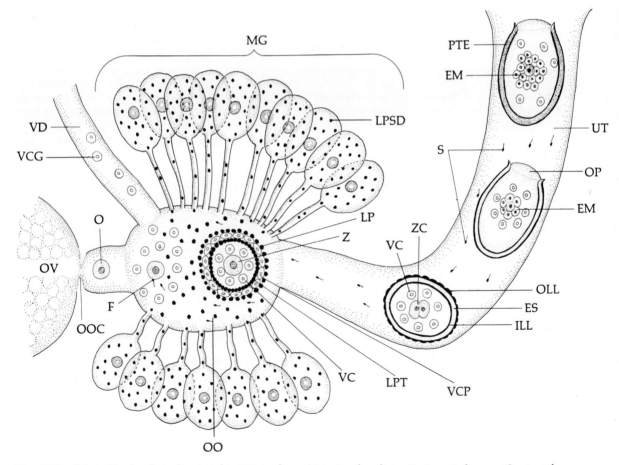

Fig. 10.7. Schematic drawing showing formation of constituents of a digenetic trematode egg. (See text for description.) EM, embryonic miracidium; ES, eggshell; F, fertilization; ILL, inner lipoprotein layer; LP, lipoprotein; LPSD, lipoprotein secretion droplets; LPT, lipoprotein template; MG, Mehlis' gland; O, ovum; OLL, outer lipoprotein layer; OO, ootype; OOC, oocapt; OP, operculum; OV, ovary; PTE, partially tanned eggshell enveloped by lipoprotein layers; S, sperm; UT, uterus; VC, vitelline cell; VCG, vitelline cell (globule); VCP, vitelline cell (proteins, phenols, and phenolase); VD, vitelline duct; Z, zygote; ZC, zygote cleavage.

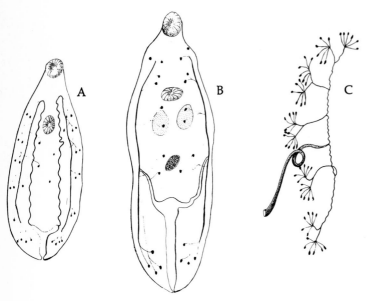

Fig. 10.8. **Flame cell patterns of some digenetic trematodes.** **A.** Pattern of *Opisthorchis pedicellata* (Opisthorchiidae). **B.** Pattern of *Dicrocoelium* sp. (Dicrocoeliidae). **C.** Right half of flame cell system of *Otodistomum* sp. (Azygiidae). (Redrawn after Faust, 1932.)

in *Opisthorchis pellicellata*, a parasite in the gallbladder of fish (Fig. 10.8), can be expressed thus:*

$$2[2 + 2 + 2 + 3 + 3 + 3 + 2] = 34$$

The formula for *Dicrocoelium dendriticum*, a liver fluke of sheep (Fig. 10.8),* can be expressed as:

$$2[2 + 2 + 2 + 2 + 2 + 2] = 24$$

The formula for *Otodistomum veliporus*, parasitic in British skates and rays (Fig. 10.8),* is expressed as:

$$2[(6+6)+(6+6)+(6+6)+(6+6)+(6+6)] = 120$$

Each flame cell leads into a minute **capillary duct** that joins with others to form the anterior and posterior **collecting ducts**. These two ducts on each side of the body empty into an **accessory duct** (or tubule). The two accessory ducts on each side unite to form the left and right **common collecting tubules**. When the common collecting tubules extend to the midregion of the body, they either fuse with the excretory vesicle (**mesostomate** type) or extend to near the anterior end of the body prior to passing posteriorly

---

*The reader must refer to Fig. 10.8 in order to understand how the formula represents arrangement of the flame cells.

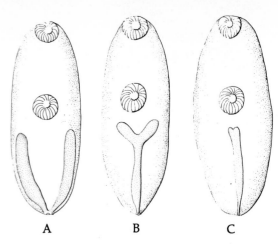

Fig. 10.9. **Shapes of excretory vesicles of digenetic trematodes. A.** V-shaped vesicle. **B.** Y-shaped vesicle. **C.** I-shaped vesicle.

to join the excretory vesicle (**stenostomate** type). It is noted that whether the arrangement of the excretory tubules is of the mesostomate or stenostomate type can be recognized in the cercarial stage.

The excretory vesicle is singular and opens to the exterior through a pore located at or near the posterior extremity of the body. Again, the shape of the vesicle is often used as a taxonomic feature. The shape may be in the form of a V, a Y, or an I (Fig. 10.9).

The flame cells include true cilia with the typical 9 + 2 arrangement of microtubules. Alkaline phosphatase is present in the walls lining the collecting tubules and in the flame cells. The concentration of this enzyme is greatest in the portion of each capillary duct in the immediate proximity of each flame cell. This enzyme functions in the selective absorption of molecules. The primary nitrogenous excretory product of trematodes is ammonia, although uric acid, urea, and even amino acids have been reported. It is still uncertain, however, whether the amino acids given off by schistosomes, for example, are excreted through the osmoregulatory system or through the tegument.

THE EGG

The ovoid digenean egg is typically operculated, i.e., there is a lidlike cap or **operculum** at one end which, when pushed off, permits the miracidium to escape (Fig. 10.10). Gönnert (1962) has presented an account of how the operculum is formed. Eggs of some species possess a minute projection or **boss** at the opposite end. The eggs of schistosomes, however, are not operculated. The shell is split during the hatching process. Within the shell, the fully or partially developed miracidium is surrounded by yolk globules within the vitelline membrane. The latter is formed during the development of the miracidium by the fusion of certain small, surface somatic cells.

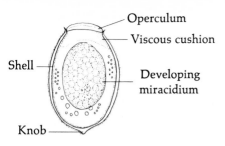

Operculum
Viscous cushion
Shell
Developing miracidium
Knob

Fig. 10.10. A typical digenetic trematode egg. The tanned protein shell is sandwiched between two extremely thin lipoprotein layers. The inner lipoprotein layer may be absent in some species.

Rowan (1956, 1957) has reported that in the egg of *Fasciola hepatica* there is a viscous and granular cushion beneath the operculum. As the hatching process is initiated by the secretion of a proteolytic hatching enzyme by the miracidium, this cushion expands. The hatching enzyme is secreted in response to exposure to light. The enzyme dissolves the cementing material by which the operculum is attached, releasing the operculum. Expansion of the granular cushion, accompanied by exosmosis of salts and other materials from within the egg, pushes off the operculum. The cushion resembles a colloid, is at least partially composed of protein, and changes from a gel to a sol during the expanding process.

Wilson (1968) has presented an alternate interpretation. According to him, the fully developed miracidium is responsible for altering the permeability of the membrane on the inner concave surface of the viscous cushion, which is a fibrillar mucoprotein complex. As a result, the cushion becomes hydrated from its dehydrated or semidehydrated state when fluids from within the egg pass through the now more permeable membrane. The cushion now increases in volume and compresses the miracidium. An internal pressure is gradually built up until the operculum finally flies off. Both the cushion and miracidium are then expelled by the hypertonicity of the egg contents.

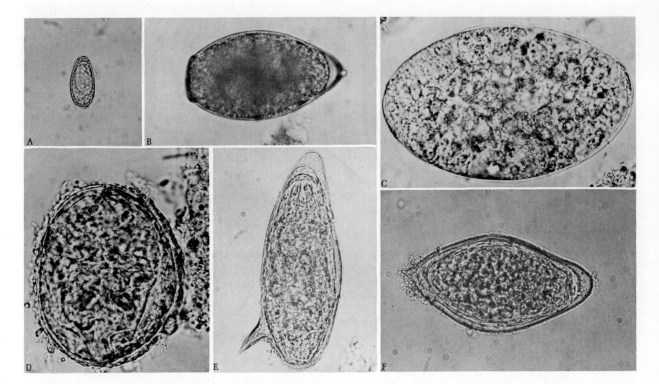

Fig. 10.11. Eggs of some digenetic trematodes parasitic in humans. A. *Clonorchis sinensis* egg, 27–35 by 12–20 μm, flask-shaped with operculum at narrow end. **B.** *Paragonimus westermani* egg, 80–118 by 48–60 μm, oval, with operculum at flattened end. **C.** *Fasciolopsis buski* egg, 130–140 by 80–85 μm, ellipsoidal, with inconspicuous operculum. **D.** *Schistosoma japonicum* egg, 70–100 by 50–65 μm, round to oval, with short lateral spine, and no operculum. **E.** *Schistosoma mansoni* egg, 114–175 by 45–68 μm, elongate oval, with long lateral spine, and no operculum. **F.** *Schistosoma haematobium* egg, 112–170 by 40–70 μm, spindle-shaped, with posterior terminal spine, and no operculum. (All photomicrographs courtesy of Ward's Natural Science Establishment, Inc., Rochester, New York.)

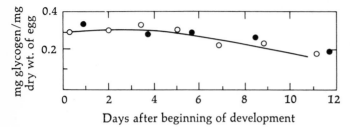

Fig. 10.12. Glycogen in eggs. Graph showing reduction in the glycogen content of *Fasciola hepatica* eggs during embryonation at 25°C. o, Eggs isolated from gallbladder; ●, eggs isolated from uterus of worm. (Data after Hortsmann, 1962.)

As mentioned earlier, the eggs of some species hatch in water, and the free-swimming miracidia penetrate the first intermediate host. In other species, the eggs must be ingested by the molluscan host before they hatch. In the latter case, the miracidium escapes only after the egg comes in contact with the host's digestive fluids. One may also assume that the physical conditions in the host's alimentary tract also influence hatching.

The development of the miracidium within the egg is stimulated by high oxygen tension. Other ambient conditions also influence development. Rowcliffe and Ollerenshaw (1960) have reported that the eggs of *Fasciola hepatica* will not develop outside the pH range of 4.2 to 9. Temperature is another influencing factor. Again, in the case of *F. hepatica* eggs, the miracidium

requires 23 weeks to develop at 10°C; however, it takes only 8 days at 30°C. If the temperature is raised above 30°C, development is slowed, and ceases at 37°C. Similar studies by Bain and Etges (1973) have revealed that hatching of *Schistosoma mansoni* eggs is also influenced by light, temperature, O_2 tension, and redox conditions, although in this case, as stated, the shell is split when the miracidium escapes. Freezing kills trematode eggs.

The characteristics of digenean eggs, especially those of medically important species, are commonly used in diagnosis of infections. The eggs of some of these species are depicted in Figure 10.11. It should be apparent that the digenean egg is not a static stage. In fact, it is rather active metabolically. For example, according to Hortsmann (1962) the glycogen content in *Fasciola hepatica* eggs maintained at 25°C decreases during miracidial development (Fig. 10.12). This indicates that the developing miracidium utilizes both carbohydrate and lipid (yolk) metabolism for energy. Since lipid metabolism is an aerobic process, one would expect to find oxygen consumption. Indeed, this has been shown to be true (Fig. 10.13).

THE MIRACIDIUM

When the egg hatches, a minute, elongate ovoid larva covered with flattened ciliated epidermal plates escapes. This is the miracidium (Fig. 10.14).

Beneath the epidermal plates are found well-developed circular and longitudinal muscles (Fig. 10.15).

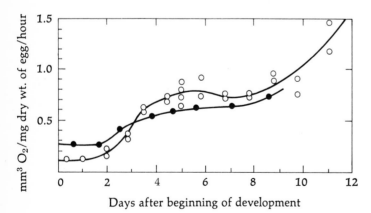

Fig. 10.13. Oxygen consumption of eggs. Graph showing the oxygen consumption of *Fasciola hepatica* eggs during embryonation at 25°C. o, Eggs collected from gallbladder; ●, eggs isolated from uterus of worm. (Data after Horstmann, 1962.)

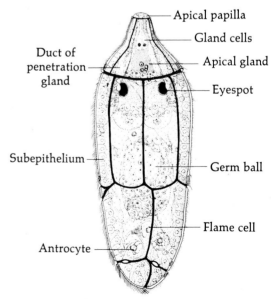

Fig. 10.14. Miracidial morphology. Miracidium of *Heronimus* showing ciliated epidermal plates and certain internal structures. (Modified after Lynch, 1933.)

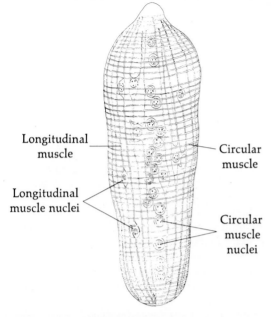

Fig. 10.15. Miracidial musculature. Arrangement of muscles of the miracidium of *Posthodiplostomum cuticola*. Redrawn after Dönges, 1964.)

At the anterior tip of the miracidium is a mobile **apical papilla** or **terebratorium** (Fig. 10.16). On this papilla are four filaments that are connected to nerve cells, which, in turn, are connected with the cerebral ganglion. These filaments are probably chemoreceptors. A multinucleated **apical gland**, located in the parenchyma, empties to the outside near the terebratorium (Fig. 10.17). The secretion of this gland aids in dissolving the host's tissue during the miracidium's penetration process. In addition to the apical gland, additional lateral penetration glands occur. These also deposit lytic enzymes at the base of the papilla (Fig. 10.17). The papilla becomes partially invaginated during penetration and thus functions as a suction cup into which the glandular secretion is secreted. The invagination of the papilla is apparently stimulated by a host factor (Cheng, 1968).

The sensory organs of the miracidium also may include two, sometimes three, eyespots and lateral papillae, one on each side of the body between the

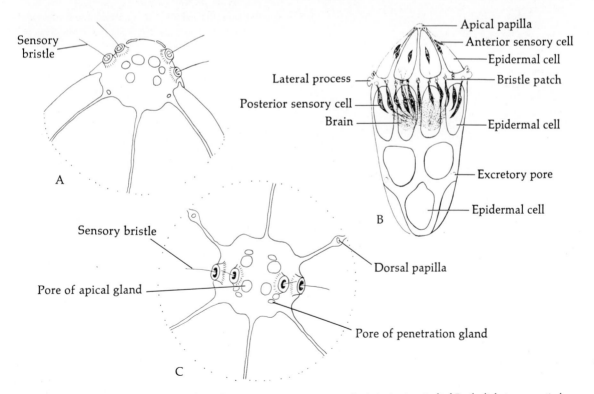

Fig. 10.16. Miracidial sensory bristle and anterior nervous system. A. Anterior terminal of *Posthodiplostomum cuticola* miracidium showing sensory bristles. **B.** Frontal view of anterior terminal of *P. cuticola* miracidium showing sensory bristles and pores. **C.** Nervous system of anterior portion of *Schistosomatium douthitti* miracidium. (**A** and **B**, redrawn after Dönges, 1964. **C**, redrawn after Price, 1931.)

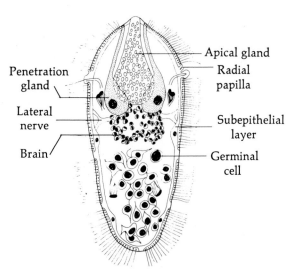

Fig. 10.17. Miracidium of *Schistosomatium douthitti* showing apical and penetration glands. (Redrawn after Price, 1931.)

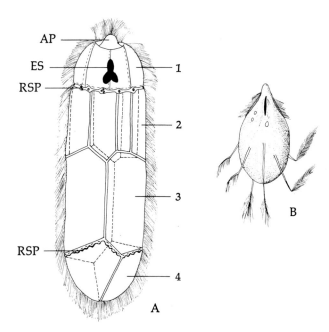

Fig. 10.18. Miracidial morphology. A. Miracidium of *Parorchis acanthus* showing locations of sensory papillae. (Redrawn after Rees, 1940.) **B.** Miracidium of a bucephalid trematode showing stalked tufts of cilia. (Redrawn after Woodhead, 1929.) AP, apical papilla; ES, eyespot; RSP, ring of sensory papillae; 1, first tier of epidermal plates; 2, second tier of epidermal plates; 3, third tier of epidermal plates; 4, fourth tier of epidermal plates.

first and second tiers of ciliated plates. In some species, such as *Parorchis acanthus*, a rectum-dwelling parasite in gulls and flamingos, there are additional sensory papillae that are circularly arranged in the same groove (Fig. 10.18).

The "brain" of the miracidium is a large cephalic ganglion lying in the parenchyma behind the apical region. From this center, nerve fibers innervate various tissues and organs of the body.

The flame cell type of excretory system is present in the miracidium. The body fluids containing wastes are collected by two or three pairs of flame cells and are deposited to the exterior through two laterally situated excretory pores (Fig. 10.19).

During the differentiation of the miracidium within the eggshell, certain germinal cells become trapped in the parenchyma. These develop into germ balls by increasing in size and number and becoming enveloped. These germ balls eventually differentiate into the next larval generation. They are usually located in the posterior portion of the body.

Miracidia of digeneans belonging to the families Bucephalidae and Brachylaemidae are unique in that the body plates are not ciliated. Rather, the cilia are arranged in tufts borne on stalks (Fig. 10.18).

Behavior

The behavior of free-swimming miracidia has been studied to some extent. The fact that most miracidia have eyespots suggests that they are phototactic. Indeed, in a number of species either positive or negative phototaxis occurs, more commonly the former.

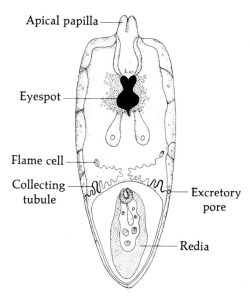

Fig. 10.19. Miracidium of *Parorchis acanthus* showing osmoregulatory (excretory) system. This miracidium is unusual in that it encloses a fully developed redia. (Redrawn after Rees, 1940.)

Furthermore, the phototactic response is influenced by the ambient temperature. For example, Takahashi *et al.* (1961) have reported that at 15°C, *Schistosoma japonicum* miracidia exhibit positive phototaxis at any light intensity; however, if the temperature is elevated to 20°C, they respond only to light intensities of 2000 lux or less. Thus, the relationship of phototactic response and temperature is a linear one. Takahashi *et al.* have proposed that at a constant light intensity, miracidial swimming velocity is a function of temperature. For example, at an intensity of 2500 lux,

$$y = 0.075x - 0.315$$

where y is the swimming velocity (mm/sec) and x is the temperature (°C).

Similarly, it has been shown that miracidial geotactic responses are influenced by temperature. In the case of *S. japonicum*, its innate negative geotaxis is masked by its negative phototaxis and, as the consequence, the miracidia move to the bottom of the aquatic environment.

Similar studies have been carried out on other species of digeneans (Cheng, 1967), and most investigators now agree that these innate behavior patterns aid the miracidium in host localization (p. 48) since they either orient the parasites toward the general location of their hosts or the host behaves similarly and consequently the two end up near each other. The review by Ulmer (1971) on host finding by helminths, including miracidia, is recommended.

Feeding and Metabolism

Miracidia cannot ingest food. They can, however, utilize exogenous glucose. Bryant and Williams (1962) have traced the incorporation of ¹⁴C-glucose and succinate in *Fasciola* miracidia. They have reported that the labeled glucose becomes incorporated into hexose phosphate, phosphoenolpyruvic acid, alanine, and lactic acid. This suggests that the Embden-Meyerhof glycolytic pathway occurs (p. 52). There is no evidence that the complete Krebs' cycle is operative, although two of the enzymes of this cycle, succinic dehydrogenase and fumarase, occur.

THE SPOROCYST

First-generation sporocysts, having differentiated from miracidia, frequently are found between the tubules in the digestive gland of the molluscan host. However, first-generation or mother sporocysts may occur elsewhere in the intermediate host's body. Those of plagiorchioids (members of the superfamily Plagiorchioidea) are attached to the snail's alimentary tract, and those of the schistosomes usually occur in the tentacles or foot, depending on the site of entry of the miracidia. The shape of sporocysts ranges from ovoid buds to elongate, sausage-shaped bodies, to branched structures. The branched forms are found in members of the families Bucephalidae, Heronimidae,

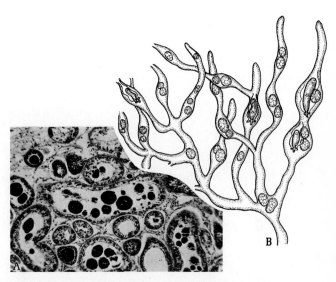

Fig. 10.20.   Branched sporocysts. A. Photomicrograph showing cross section of branches of the sporocyst of *Bucephalus* sp. in the digestive gland of the American oyster, *Crassostrea virginica*. **B.** Drawing of a portion of a branched sporocyst of *Bucephalus* sp. (After Cheng, 1970.)

Brachylaemidae, and to some extent, Dicrocoeliidae. In these, the branching arms ramify throughout the host's tissues (Fig. 10.20).

The sporocyst wall varies in thickness, depending on age and species. The outermost layer, like that of all subsequent stages in the life cycle of digeneans, is comprised of a syncytial tegument observable with the electron microscope. Projecting from the tegument are microvilli (Fig. 10.21), which increase the body surface and hence enhance the absorption of nutrients.

Under the tegument is a thin layer of connective tissue known as the basal lamella. Beneath this layer are found circularly arranged muscles followed by an inner layer of parenchymal cells (Fig. 10.22). The **brood chamber** is a hollow space in the center of the sporocyst within which are found the germ balls that eventually differentiate into the next generation. In living specimens, the brood chamber also contains a fluid. Sporocysts contain no alimentary or reproductive system. A nervous system has been reported in certain species, such as the schistosomes (Théron and Fournier, 1982). Flame cells are generally present. In some, there is an inconspicuous birth pore, through which the fully formed larvae escape. Larvae of those species without a birth pore escape by rupturing the sporocyst wall.

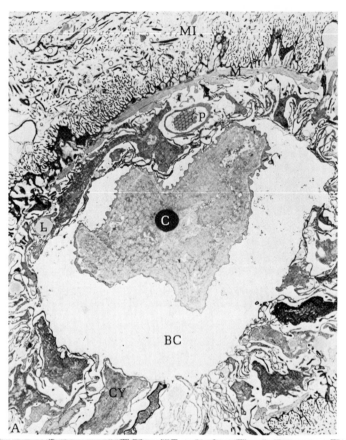

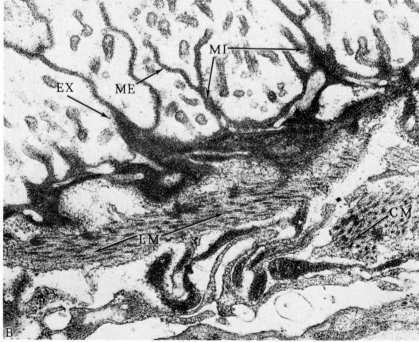

Fig. 10.21. **Daughter sporocyst of *Schistosoma mansoni*. A.** Electron micrograph of cross section of daughter sporocyst embedded in interacinar space of the digestive gland of the snail *Biomphalaria glabrata.* Notice branched microvilli arising from the surface of the sporocyst. **B.** Higher magnification electron micrograph of sporocyst tegument showing branched microvilli arising from the surface. BC, brood chamber; C, section of cercaria; CM, circular muscles; EX, extracellular material adhering to microvillar surface; LM, longitudinal muscles; M, muscle layer; ME, surface membrane; MI, microvilli; P, protonephridium (flame cell).

In some species, mother sporocysts give rise to daughter sporocysts and the daughter sporocysts, rarely, to still further generations of sporocysts. In other species mother sporocysts give rise immediately to the next larval generation—the **redia**—or directly to cercariae (see Fig. 10.23). Although daughter sporocysts for the most part are similar to mother sporocysts, the daughter sporocysts can be distinguished by the inclusion of rediae or cercariae, whichever is the case in the particular species, and by being larger and longer.

Nutrition and Metabolism

Since sporocysts lack mouths, they must derive their nutrients through the body wall. They acquire amino acids, sugars, especially glucose, and certain fatty acids from their molluscan hosts' hemolymph and tissues (Cheng, 1963). The carbohydrates are metabolized in the production of energy, and the nitrogen compounds are utilized for growth and reproduction, although some may also be utilized for energy production. However, since there is very little oxygen in their habitats, it appears unlikely that sporocysts carry on appreciable oxidative metabolism, including the utilization of lipids as an energy source.

THE REDIA

In some species that include a redial stage, these larvae either develop directly from the miracidium or are found within the brood chambers of sporocysts. They eventually escape into the host's tissue, usually in the digestive gland. In echinostomes whose rediae are found in bivalved molluscs, the rediae are commonly found embedded between the inner and outer lamellae of the host's gills.

The redia is elongate and normally possesses two or four budlike projections—the **ambulatory buds** (or **procruscula**)—two located anterolaterally and two posterolaterally (Fig. 10.24). Unlike the sporocyst, the redia possesses a mouth located at the anterior terminal. The mouth leads into a muscular pharynx, which in turn leads into a blind-sac caecum. Rediae are capable of motility through the movement of the ambulatory buds and body contractions, and they ingest host cells.

The rediae of certain trematodes possess a pair of cephalic ganglia on each side of the pharynx. From these ganglia nerve fibers radiate in all directions.

The protonephritic osmoregulatory (excretory) system is present in most rediae. Figure 10.25 depicts the excretory system found in the redia of *Cotylophoron cotylophorum*, an amphistome fluke parasitic in the rumen of cattle. In this species, the collecting ducts on each side empty into separate excretory vesicles, which in turn empty to the exterior through two laterally located excretory pores. The flame cell pattern found in these larvae is of the same general pattern as that

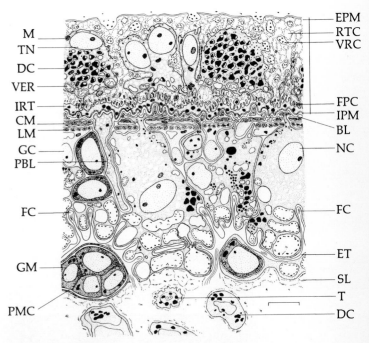

Fig. 10.22. Fine structure of sporocyst. Composite drawing of the body wall of the daughter sporocyst of *Bucephalus haimeanus.* This is believed to be either a young specimen and the nucleated outer stratum (indicated by bar on right) is sloughed during maturation, or James *et al.* have misinterpreted the stratum indicated by the bar as being part of the sporocyst wall when actually it is host tissue. BL, basal lamina; CM, circular muscle; DC, dense cytoplasm; EPM, external plasma membrane of tegument; ET, excretory tubule; FC, flame cell; FPC, fingerlike projections of dense cytoplasm; GC, germinal cell; GM, germinal mass; IPM, internal plasma membrane of tegument; IRT, inner region of tegument; LM, longitudinal membrane; M, mitochondrion; NC, nucleus of cyton of tegument; PBL, projection of basal lamina into subtegument; PMC, plasma membrane of giant somatic cell lining brood chamber; RTC, reticulate tegument cytoplasm; SL, secretion of lining brood chamber; T, tail of developing cercaria; TN, tegument nucleus; VER, vesicular endoplasmic reticulum; VRC, vesicle of reticulate cytoplasm (possibly containing acid mucopolysaccharide). (Redrawn after James *et al.*, 1966.)

found in the adult, but there are fewer individual flame cells.

The histology of the body wall of rediae, including its fine structure, is identical to that of sporocysts. More often than not, a **birth pore** is present, lateral to and near the mouth. Within the brood chamber are found germ balls that either differentiate into another generation of rediae (daughter rediae) or into the next larval generation (cercariae).

Nutrition and Metabolism

Unlike sporocysts, rediae can engulf food in the form of their molluscan hosts' tissues, although absorption of molecules through the tegument also occurs. Because they can ingest, the damage inflicted on molluscs by rediae is usually considerably more severe than that caused by sporocysts. Since most species of rediae live within their hosts' digestive glands and feed on molluscan hemolymph and tissues, especially digestive gland cells, which are very rich in glycogen, both proteins and carbohydrates are available to these larval trematodes.

Cheng and Yee (1968) have demonstrated that in addition to actively ingesting host cells, the hydrolytic enzyme aminopeptidase is present in the redial wall of *Philophthalmus gralli* in the snail *Tarebia granifera*. This enzyme is most probably secreted to the exterior and contributes to the lysis of surrounding

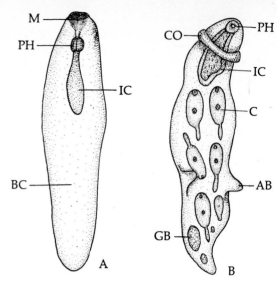

Fig. 10.24. Morphology of rediae. A. Mother redia of *Crepidostomum cornutum.* **B.** Redia of the liver fluke, *Fasciola* sp. AB, ambulatory bud; BC, brood chamber; C, cercaria; CO, collar; GB, germ ball; IC, intestinal caecum; M, mouth; PH, pharynx. (Redrawn after Ross and McKay, 1929.)

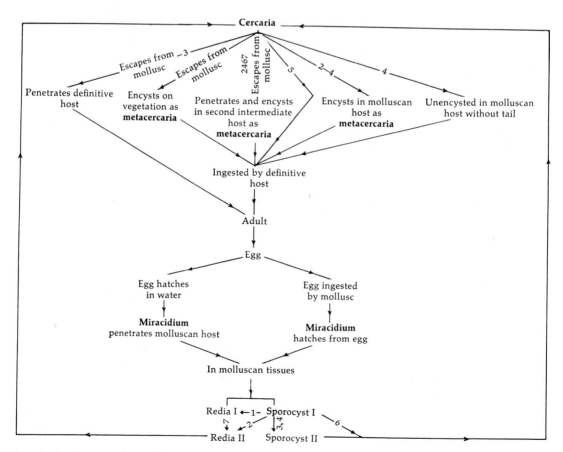

Fig. 10.23. Digenean life cycle patterns. Variations in the life cycle patterns among selected groups of digenetic trematodes. 1, Fascioloidea; 2, some Opisthorchioidea; 3, Schistosomatoidea; 4, some Plagiorchioidea; 5, Azgiidae; 6, Bucephaloidea; 7, Gorgoderidae.

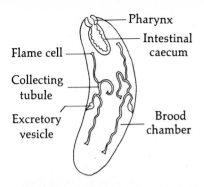

Fig. 10.25. Redia of the amphistome *Cotylophoron* showing osmoregulatory (excretory) system. (Redrawn after Bennett, 1936.)

host cells. Thus, at least this species of redia is capable of partial extracorporeal digestion.

Although there is little oxygen available to rediae in their natural habitats, *in vitro* studies have revealed that they are capable of oxygen utilization. For example, the rediae of *Himasthla quissetensis* regulate their respiration rate even when there is only 0.5% oxygen present (Vernberg, 1963).

As in the case of sporocysts, carbohydrate metabolism is the primary mechanism for energy production in rediae. Exceptions, however, are known. In the case of the rediae of *Himasthla quissetensis*, Vernberg (1963) has reported that the addition of such sugars as glucose, fucose, mannose, and ribose does not increase their respiratory rates (Table 10.3), suggesting that these sugars are not utilized. These rediae, however, have since been shown to utilize glucose. This redia is known to include the pathways for converting

the amino acids proline and glutamic acid to carbohydrates, suggesting that carbohydrate synthesis, at least in part, occurs via these and possibly other amino acids.

Some of the Krebs cycle enzymes have been demonstrated in certain rediae. In *Himasthla quissetensis* rediae, pyruvate, α-ketoglutarate, and succinate occur. Furthermore, malonate, which competitively inhibits succinic dehydrogenase, is known to decrease the uptake of oxygen, suggesting that at least a partial Krebs cycle exists.

Protein and lipid metabolisms have not been examined in rediae.

THE CERCARIA

Cercariae are differentiated from the germ balls in the brood chamber of sporocysts or rediae, which ever may be the pattern in the particular species (Fig. 10.23). After escaping from the brood chamber of the preceding generation, cercariae actively leave the molluscan host and (1) become free swimming in water, actively seeking the next host; (2) encyst on vegetation; (3) occur in the slime trail of terrestrial snails; (4) encyst in the mantle cavity of the mollusc; or (5) develop into a tailless, nonencysted metacercaria within the molluscan host. Thus, except in the first instance, these larvae must await ingestion by the definitive or another intermediate host in order to continue their development.

Cercarial Types

Before describing the internal anatomy, it appears advisable to consider the types of cercariae as based on their external morphology. The separation of cercariae

Table 10.3 Utilization of Substrates by Rediae, Cercariae, and Adults of *Himasthia quissetensis*[a]

Substrate	Percentage Change in O_2 Consumption		
	Rediae	*Cercariae*	*Adult*
Pyruvate	0	—	—
Ketoglutarate	0	—	—
Succinate	+40	0	+27
Malonate	−24	0	−64
Serine	−12	—	—
Proline	+46	0	+16
Glutamic acid	+20	—	+22
Mannose	0	—	+50
Glucose	+15[b]	0	+49
Fucose	0	—	—
Ribose	−11	—	0

[a] Data from Vernberg and Hunter, 1963.

[b] Although Vernberg and Hunter (1963) did not report the utilization of glucose by *H. quissetensis* rediae, we have observed it in our laboratory.

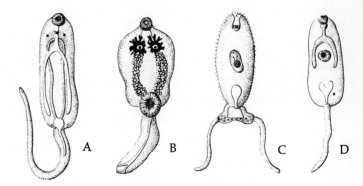

Fig. 10.26. Some morphologic types of cercariae. A. Monostome cercaria (*Notocotylus seineti*). **B.** Amphistome cercaria (*Cercaria frondosa*). **C.** Gasterostome cercaria (*Bucephalopsis gracilescens*). **D.** Distomate cercaria. (**A** and **B,** redrawn after Faust, 1930; **C,** redrawn after Lebour, 1911.)

into the following types, however, by no means indicates that the species belonging to each type are
closely related. These categories are purely descriptive,
although, in many cases, closely related flukes do possess similar cercariae.

1. **Monostome** cercariae possess one sucker
   only—the anteriorly situated oral sucker (Fig.
   10.26). They possess two eyespots and long
   simple tails, usually with pointed, locomotor
   processes directed posteriorly from the posterior
   extremity of the body proper. This type of

cercaria develops in rediae and gives rise to
monostomate adults.

2. **Amphistome** cercariae possess a posterior
   sucker and eyespots (Fig. 10.26). These cercariae
   develop in rediae and give rise to amphistome
   adults (superfamily Paramphistomoidea).

3. **Gasterostome** cercariae possess a mouth located on the ventral surface of the body, and
   they eventually develop into gasterostome
   adults of the family Bucephalidae (Fig. 10.26).

4. **Distome** cercariae possess two suckers—the
   anteriorly located oral sucker and the ventrally
   located acetabulum (Fig. 10.26). This is by far
   the most common type of cercaria.

The four types of cercariae listed above are classified
according to the position and number of body suckers.
The following descriptive types are categorized according to the shape and relative size of their tails.

1. **Pleurolophocercous** cercariae are characterized by a dorsoventral finlike fold along
   the length of the tail (Fig. 10.27). These cercariae commonly possess eyespots and an
   extremely small acetabulum that is readily
   overlooked. Pleurolophocercous cercariae are
   produced in rediae, encyst in ectothermic
   vertebrates, and when ingested by the definitive host develop into adults belonging to the
   superfamily Opisthorchioidea.

2. **Cystocercous** cercariae are characterized by
   a cavity or "cyst" at the base of the tail into
   which the body proper can be withdrawn
   (Fig. 10.27). These are usually parasites of
   amphibians as adults and belong to the family
   Gorgoderidae.

3. **Furcocercous** cercariae are characterized
   by a forked tail (Fig. 10.27). Some species
   may possess eyespots. Those with a pharynx
   are the larvae of holostomes and strigeids,

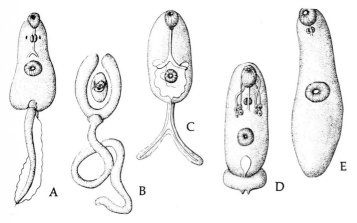

**Fig. 10.27. Some additional morphologic types of
cercariae. A.** Pleurolophocercous cercaria (*Glypthelmins
quieta*). **B.** Cystocercous cercaria (*Gorgoderina* sp.) **C.** Furcocercous cercaria (*Cercaria dichotoma*). **D.** Microcercous
cercaria (*Cercaria brachyura*). **E.** Gymnophallus cercaria.
(**C** and **D,** redrawn after Lebour, 1911.)

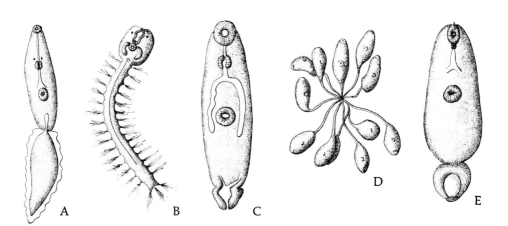

Fig. 10.28. Some additional morphologic types of cercariae. A. Rhopalocercous cercaria (*Cercaria isopori*). **B.** Trichocercous cercaria (*Opechona bacillaris*). **C.** Cercariaeum (*Cercaria politae nitidulae.*) **D.** Rat-king cercariae. **E.** Cotylocercous cercaria.
(**A,** redrawn after Lühe, 1909; **B,** redrawn after Lebour, 1916; **C,** redrawn after Harper, 1929.)

whereas those without a pharynx are schistosomes or schistosome-related species.

4. **Microcercous** cercariae are characterized by a small tail that may be knoblike or conical (Fig. 10.27). These nonswimming larvae do not represent any specific taxonomic group.
5. **Gymnophallus** cercariae are characterized by the complete absence of a tail (Fig. 10.27).
6. **Rhopalocercous** cercariae are characterized by a broad tail that is as wide as or wider than the body proper (Fig. 10.28).
7. **Leptocercous** cercariae are characterized by a straight tail that is slender and narrower than the body proper (Fig. 10.28).
8. **Trichocercous** cercariae are characterized by tails armed with spines or bristles (Fig. 10.28). These are marine forms.
9. **Cercariaea**, like the gymnophallus cercariae, lack a tail but differ from the latter in that they generally do not leave the molluscan host (Fig. 10.28). In the few marine species that do leave their hosts, the cercariae move by inching along in a wormlike crawl.
10. **Rat-king** (rattenkönig) cercariae are all marine and are characterized by their colonial arrangement (Fig. 10.28). The tails of the individuals are joined and the bodies arranged in a radial pattern.
11. **Cotylocercous** cercariae are similar to the microcercous type except that the short tail is shaped like a cup (Fig. 10.28), generally with large gland cells lining the concavity.

Cercariae can also be categorized morphologically by specialized body structures. Some of these are listed below.

1. **Echinostome** cercariae are characterized by the presence of a collar of spines around the anterior sucker (Fig. 10.29). These larvae develop in rediae and mature into adults belonging to the superfamily Echinostomatoidea.
2. **Gymnocephalous** cercariae are typical distomes which, unlike the echinostome cercariae, lack a spinous collar (Fig. 10.29).
3. **Xiphidiocercariae** possess an anterior stylet in the oral sucker (Fig. 10.29). Penetration glands are exceptionally well developed in these cercariae, which develop in sporocysts. Certain xiphidiocercariae possess a bipartite, transparent, fluid-filled sac that overlaps the oral sucker, known as a **virgulate organ**. Such cercariae are known as virgulate xiphidiocercariae. Most, if not all, of these cercariae belong to the family Lecithodendriidae. The material secreted by the virgulate organ aids the cercariae in becoming attached to their hosts while penetrating, and it is also protective.
4. **Ophthalmocercariae** include all the forms that possess eyespots.

Since all of these types of cercariae are distinguished by one major characteristic, combinations of these can be used in describing specific cercariae that possess more than one of these characteristics. For example, the cercaria of *Crepidostomum cornutum*, an intestinal

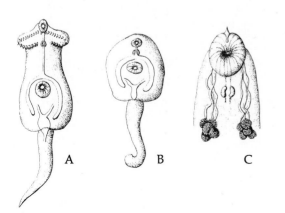

Fig. 10.29. Some additional morphologic types of cercariae. A. Echinostome cercaria (*Echinopharyphium recurvatum*). **B.** Gymnocephalous cercaria (*Fasciola hepatica*). **C.** Anterior portion of xiphidiocercaria showing stylet. (**A** and **B**, redrawn after Harper, 1929.)

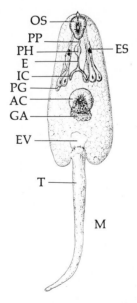

Fig. 10.30. Cercaria of *Crepidostomum cornutum*. AC, acetabulum; E, esophagus; ES, eyespot; EV, excretory vesicle; GA, genital anlagen; IC, intestinal caecum; OS, oral sucker; PG, penetration gland; PH, pharynx; PP, prepharynx; T, tail. (Redrawn after Cheng and James, 1960.)

parasite of fish, can be described as a distomate, lepto-cercous, ophthalmoxiphidiocercaria (Fig. 10.30).

Body Surface

The cercarial body surface is essentially the same as that of the sporocyst, redia, metacercaria, and adult in that it is covered by a tegument (p. 304).

Internal Anatomy

Since the internal anatomy of all cercariae, irrespective of morphologic type, is essentially the same, that of *Crepidostomum cornutum* is used here to illustrate the internal structures (Fig. 10.30).

Alimentary Tract. Except in the gasterostomes, the mouth is situated at the anterior end of the body and is surrounded by the **oral sucker**. Food ingested through the mouth passes down the **prepharynx** and through the **pharynx** and **esophagus** to enter the bifurcate **intestinal caeca**.

Penetration Glands. Cephalic, or **penetration,**

glands are arranged in two general groups, one on each side of the esophagus. All these glands empty to the outside anteriorly via individual ducts.

In some species, the penetration glands are arranged into an anterior, preacetabular group and a posterior, postacetabular group. The functions of the two groups may differ, as in *Schistosoma mansoni* cercariae (Fig. 10.31). In this case, the three pairs of postacetabular glands secrete the lytic substances that aid in penetration, and the two pairs of preacetabular glands secrete a mucoid film believed to be protective in nature (Stirewalt, 1963). The pre- and postacetabular glands of *Schistosoma mansoni* cercariae are chemically distinct as revealed by cytochemistry (Stirewalt and Kruidenier, 1961) (Table 10.4). Studies on the penetration enzymes have revealed a thermolabile, hyaluronidaselike enzyme, a mucopolysaccharidase, and a collagenaselike enzyme (Stirewalt and Evans, 1952; Lee and Lewert, 1957; Lewert and Lee, 1956).

In the cercaria of *Posthodiplostomum minimum*, a parasite of fish, where pre- and postacetabular glands also occur, Bogitsh (1963) has shown that the functions of both sets of glands are identical. Two types of secretions are synthesized in all the glands cells— a glycoprotein and an acid mucopolysaccharide. Bogitsh has interpreted this to mean that the secretions of all glands function in adhesion and probably penetration.

Genital Primordium. Near the acetabulum, embedded in the parenchyma, is found the genital primordium composed of reproductive cells that eventually differentiate into gonads. (Fig. 10.30).

Excretory System. Cercariae have a protonephritic excretory (osmoregulatory) system. However, unlike the protonephritic system in earlier larval

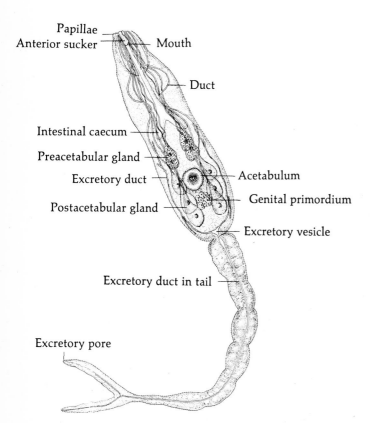

Fig. 10.31. *Schistosoma mansoni* **cercaria showing positions of preacetabular and postacetabular glands.** Bar = 20 μm. (After Cheng and Bier, 1972.)

Table 10.4. Differences in Staining Reactions and Functions between the Pre- and Postacetabular Glands in *Schistosoma mansoni* cercariae[a]

Stains and Functions	Preacetabular Glands	Postacetabular Glands
General description	Macrogranular	Microgranular
General reaction (a)	Acidolphilic	Basophilic
Alizarin (v)	Pink	Not stained
Purpurin (v) (a)	Red	Not stained
Aniline blue (a)	Not stained	Blue
Toluidine blue (a)	Not stained	Blue
Orange G (v) (a)	Not stained	Pale orange
Mallory's triple (a)	Light red	Dark blue
Periodic acid–Schiff	Not stained	Purple
Probable functions	Secretes hyaluronidase and mucopolysaccharidase	Secretes mucus

[a]Data from Stirewalt and Kruidenier, 1961.
[b]a, After fixation; v, vital stain.

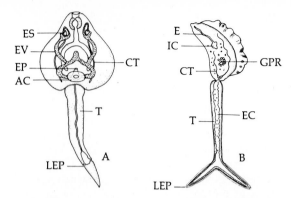

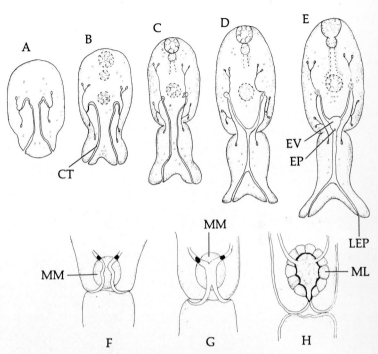

Fig. 10.32. Osmoregulatory (excretory) system of some cercariae. A. *Cotylophoron* cercaria showing excretory canal leading into tail. (Redrawn after Bennett, 1936.) **B.** *Sanguinicola* cercaria showing excretory canal leading to tips of forked tail. (Redrawn after Ejsmont, 1926.) AC, acetabulum; CT, collecting tubule; E, esophagus; EC, excretory canal; EP, excretory pore; ES, eyespot; EV, excretory vesicle; GPR, genital primordium; IC, intestinal caecum; LEP, larval excretory pore; T, tail.

Fig. 10.33. Development of the osmoregulatory (excretory) system in cercariae. A–E. Development of the system in a furcocercous strigeid cercaria. (Redrawn after Komiya, 1939.) Digeneans with this type of excretory vesicle are assigned to the suborder Anepitheliocystidia. **F–H.** Development of the excretory vesicle in *Cercaria dioctorenalis* showing the eventual lining of the vesicle by mesodermal cell. Digeneans with this type of excretory vesicle are assigned to the suborder Epitheliocystidia. (Redrawn after Dobrovolny, 1939.) CT, collecting tubule; EP, excretory pore; EV, excretory vesicle; ML, mesodermal lining; MM, mesodermal mass.

stages—that is, miracidium, sporocyst, and redia—the two lateral collecting ducts empty into a common, posteriorly situated excretory vesicle from which a tube extends posteriorly into the tail. Various numbers and positions of excretory pores are known. For example, in the cercaria of *Cotylophoron cotylophorum* there are two pores, one on each side of the tail (Fig. 10.32). In the furcocercous cercaria of *Sanguinicola* there are two pores, one at the tip of each branch of the fork (Fig. 10.32). The larval excretory pores become nonfunctional and are discarded with the tail when the cercaria develops into the metacercaria.

LaRue (1957) has presented a classification of the Digenea based on the embryonic development of the excretory system and divides the order Digenea into two suborders: Anepitheliocystidia and Epitheliocystidia. Space does not permit detailed descriptions of the known variations of the developmental patterns. However, in the fork-tailed cercariae and others, amphistomes, brachylamids, and fellodistomes, for example, the excretory vesicle is formed through the fusion of the two collecting ducts and does not involve mesodermal cells (Fig. 10.33). In other words, in members of the suborder Anepitheliocystidia (see classification list at the end of this chapter), the embryonic excretory bladder is retained in the adult and is not replaced by mesodermal cells.

In other trematodes, such as plagiorchids, opisthorchids, and allocreadids, the development pattern follows that found in the cercaria of *Cercaria dioctorenalis*, a xiphidiocotylocercous cercaria. During the development of the excretory vesicle in this cercaria, the two ducts fuse in the center of a mesodermal mass, and the resulting excretory vesicle is lined with a

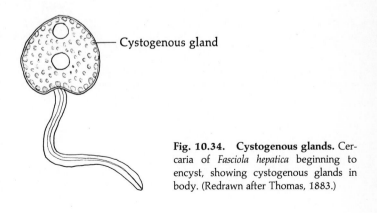

Cystogenous gland

Fig. 10.34. Cystogenous glands. Cercaria of *Fasciola hepatica* beginning to encyst, showing cystogenous glands in body. (Redrawn after Thomas, 1883.)

layer of mesothelium (Fig. 10.33). In other words, in members of the suborder Epitheliocystidia the wall of the embryonic excretory bladder is replaced by mesodermal cells (see classification list at the end of this chapter).

The flame cell pattern found in the cercaria usually coincides with that found in the adult. Thus, if one knows the formula of a particular cercaria, it can be used in associating the larva with the adult.

Cystogenous Glands. In most cercariae conspicuous **cytogenous glands** can be seen in the subtegumental region (Fig. 10.34). The secretion from these glands is responsible for the laying down of the cyst wall when the cercariae are ready to encyst. These glands can be distinguished from the penetration glands by their position and by their affinity for certain stains. For example, they take on a bluish color when stained with dilute thionine or toluidine blue.

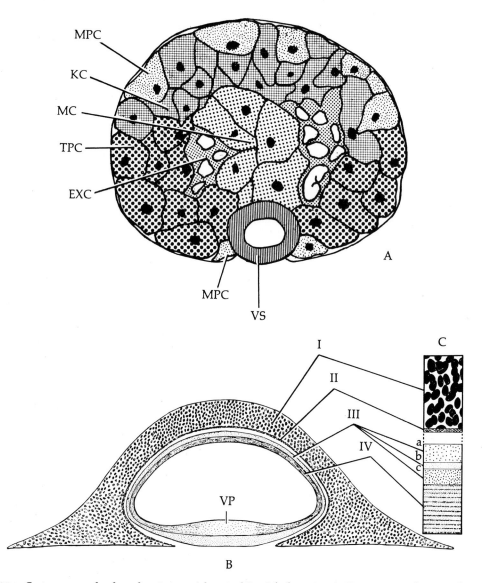

Fig. 10.35. Cystogenous glands and metacercarial cyst of *Fasciola hepatica*. A. Diagrammatic drawing of cercaria in cross section showing distribution of various types of cystogenous glands. **B.** Diagrammatic drawing of metacercarial cyst in section showing layers. **C.** An enlargement of a portion of the dorsal cyst wall showing layers. EXC, excretory canal; KC, keratin cell; MC, mucoprotein cell; MPC, mucopolysaccharide cell; TPC, tanned protein cell; VP, ventral plug; VS, ventral sucker (or acetabulum). (After Dixon, 1965, 1966.)

Considerable attention has been paid to these glands and, as a result, several types of cystogenous gland cells have been described in several species. Only two examples will be presented here.

Dixon (1965) has reported the occurrence of four types of cytogenous gland cells in the cercaria of the liver fluke, *Fasciola hepatica* (Fig. 10.35). Type I cells are ventrolaterally situated and are visible only when the cercaria is still within the snail host since it is at this site that these glands secrete their contents onto the body wall.

Type II and type III cells are dorsally and dorsolaterally situated. They secrete a mucoprotein and an acid and a neutral mucopolysaccharide layer, which together comprise layers II and III of the definitive metacercarial cyst (Fig. 10.35).

Type IV cells are dorsally situated and synthesize and include bacilliform rods that when secreted form the keratin layers of the cyst wall.

In addition to these four cell types, ventral plug cells secrete the ventral plug region of layer IV of the cyst, and ventral cells are situated around the ventral surface. These are believed to play a role in attachment or in formation of the thicker central region of the cyst.

Dixon and Mercer (1964) have studied the fine structure of the cyst of *Fasciola hepatica*. In brief, they have reported that the outer tanned-protein layer is comprised of a meshwork of irregular bodies made up of cigar-shaped particles (Fig. 10.36). The two middle mucopolysaccharide layers are finely fibrous, and the innermost layer is comprised of numerous protein sheets stabilized by disulfide linkages (Fig. 10.36). The innermost layer is formed of tightly wound scrolls synthesized in the cytoplasm of certain cystogenous glands and which, during secretion, are unrolled on the surface of the worm (Fig. 10.37).

Thakur and Cheng (1968) have reported the presence of three types of cystogenous gland cells in the cercaria of *Philophthalmus gralli*, an ocular parasite of birds. When subjected to various histochemical tests, differences in chemical composition have been detected (Table 10.5). They have designated these protein, mucopolysaccharide, and mucoprotein cells (Fig. 10.38). Furthermore, the presence of three walls comprising the metacercarial cyst has been demonstrated, with the outer wall further subdivided into a basal and a peripheral layer. All of the walls include protein but differ in their amino acid compositions. When the chemical compositions of the cyst walls were compared with those of the three types of cystogenous gland cells, it became evident that each layer of the cyst is formed from the secretions of two or more types of cells.

From the examples presented and additional evidences (see Erasmus, 1972), it is becoming apparent that metacercarial cysts are not similar in origin or structure.

Mucoid Glands. Certain glands, known as **mucoid glands**, occur along the midventral surface of

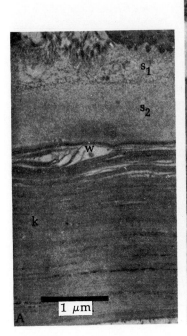
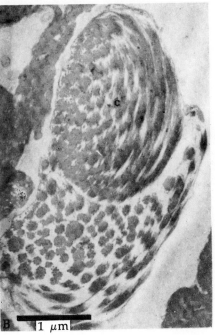

Fig. 10.36. Metacercarial cyst layers of *Fasciola hepatica*. A. Low magnification electron micrograph showing the two middle mucopolysaccharide layers and the inner layer comprised of protein (keratinized) sheets. **B.** Electron micrograph of an aggregate of cigar-shaped bodies occurring in the outer tanned-protein layer. c, cigar-shaped bodies; k, keratinized sheets; s_1, outer mucopolysaccharide layer; c_2, inner mucopolysaccharide layer; w, whorl of keratinized sheets. (**A** and **B**, after Dixon and Mercer, 1964.)

various cercariae (Fig. 10.39). The mucoid secretion of these cells not only protects the cercaria from harmful substances, including its own cytolytic secretions during the penetration process, but also aids movement along the substrate and serves as a lubricant during penetration.

The Metacercaria

The final larval stage of digenetic trematodes is the metacercaria. When the free-swimming cercaria comes to rest on suitable vegetation or penetrates a compatible host, it loses its tail and encysts. Within the cyst wall or walls, the body proper of the cercaria metamorphoses into the metacercaria, which is actually a juvenile replica of the adult. Many cercarial structures, such as the stylet, penetration glands, cystogenous and mucoid glands, and eyespots, soon disappear. The genital primordium differentiates into gonads that are usually nonfunctional; however, in certain progenetic metacercariae, sterile or even fertile eggs are formed. The excretory vesicles in metacercariae are usually greatly distended and include varying amounts of refractile globules.

The histology and fine structure of the various structures found in metacercariae correspond to those of the adult.

Cyst Wall. The composition of the metacercarial cyst wall has been considered earlier in connection with cercarial cystogenous glands. However, it should be mentioned that in species which encyst within an intermediate host, the so-called cyst wall may not be

formed exclusively of materials secreted by the parasite. For example, in the case of *Himasthla quissetensis*, a parasite of gulls that utilizes various species of marine bivalve molluscs as the second intermediate host, the parasite-secreted true wall is commonly enveloped within a tunic comprised of the molluscan host's hemocytes (Cheng *et al.*, 1966).

Metabolism. Very little is known about the metabolism of metacercariae. It is generally assumed that they depend primarily, if not exclusively, on stored nutrients. This is indirectly substantiated by the findings of Vernberg and Vernberg (1971) who reported that the metabolic rate of the metacercariae of *Zoogonus lasius* encysted in the annelid *Leonereis culveri*, as reflected by oxygen utilization, gradually decreases after encystment and as the worm grows, suggesting the gradual reduction of stored nutrients.

Vernberg and Vernberg have also shown that certain metabolic processes in *L. culveri*, the host, are also altered as the result of parasitism. Specifically, uninfected annelids metabolically acclimate to cold temperatures, but infected ones do not. Also, the cytochrome *c* oxidase activity in uninfected *L. culveri* is higher in cold-acclimated specimens than in warm-acclimated ones, whereas in parasitized annelids the pattern is reversed. These data indicate that although morphologically encysted metacercariae appear to be quiescent organisms, in fact, they dynamically affect their hosts.

LIFE CYCLE STUDIES

The complex life history patterns of the Digenea have been studied extensively. Stunkard (1940) has pointed out the importance of this phase of helminthology. Investigations on life cycles are fascinating from the

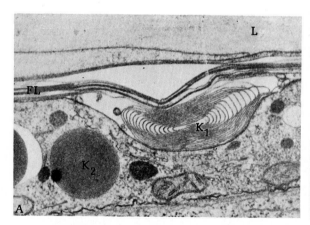

Fig. 10.37.   Keratinized scrolls. A. Electron micrograph of a portion of the edge of a *Fasciola hepatica* metacercaria at a later stage of cyst wall formation showing unrolling of keratinized lamellated body to form the innermost stratum of the cyst. (After Dixon and Mercer, 1964.) **B.** Electron micrograph showing a later stage during the unwinding of a keratinized lamellated body. (After Dixon and Mercer, 1964.) FL, layer of inner cyst wall being formed; K_1, keratinized body in process of being unrolled; K_2, keratinized body; L, formed portion of inner cyst wall.

biologic standpoint because such information can contribute to our understanding of the phylogenetic relationships between species since related trematodes almost always demonstrate similar developmental patterns. Furthermore, a complete understanding of detailed morphology of the various stages, both larval and adult, contributes to our understanding of host-parasite relationships, for the manner in which the parasite is specifically adapted to its host is commonly manifested in its structure. For example, ophthalmoxiphidiocercariae are always free swimming and penetrate their host utilizing their stylets and the secretions from their exceptionally large penetration glands.

From the more practical standpoint, information on the life cycles of medically important species is invaluable because eradication of these species largely depends on knowledge of which snails serve as the intermediate hosts and how the larvae enter humans. If the answers to these questions are known, then various means can be devised to eradicate both the larvae and their intermediate hosts, preventing the completion of their development, that is, entrance into the final host. The human blood fluke, *Schistosoma mansoni*, serves as a good example. If the number of snail hosts for this important human pathogen is reduced, for example through the application of chemicals or biologic control agents, the intramolluscan stages of the parasite die along with the snails. In addition, newly developed miracidia cannot find a suitable host and hence also perish. This type of eradi-

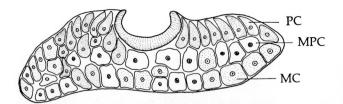

Fig. 10.38. Cystogenous glands of *Philophthalmus gralli*. Diagrammatic drawing of cross section of cercaria through level of the acetabulum, showing distribution of the three types of cystogenous gland cells. MC, mucoprotein cell; MPC, mucopolysaccharide cell; PC, protein cell. (After Thakur and Cheng, 1968.)

cation program, especially chemical control, is now being practiced in many parts of the world while the practicality of employing biologic control agents is being explored.

HOST SPECIFICITY

The Digenea are as a rule not as host specific as the Monogenea. However, the degree of specificity varies

Table 10.5. Cytological and Cytochemical Properties of the Three Types of Cystogenous Gland Cells in *Philophthalmus gralli* Cercariae[a]

Stain Employed	Protein Cells	Mucopolysaccharide Cells	Mucoprotein Cells
		Types of Cystogenous Gland Cells	
Hematoxylin & eosin	Eosinophilic finely granular cytoplasm with vesicular nucleus	Faintly eosinophilic agranular or slightly granular cytoplasm; with vesicular nucleus	Coarsely granular hematoxyphilic cytoplasm; with vesicular nucleus
Mallory's	Red cytoplasmic granules	Light bluish finely granular cytoplasm	Blue cytoplasmic granules
Mercury bromphenol blue (MBM)	Deep blue	Yellowish brown	Deep blue or sometimes yellowish
Periodic acid-Schiff (PAS)	Negative	Negative	Positive
Diastase			Resistant
Light green (counterstain for PAS)	Positive	Positive	Negative
Schiff	Negative	Negative	Negative
Alcian blue	Negative	Positive	Negative
Toluidine blue	Ametachromatic	Gamma metachromatic	Ametachromatic
Millon's (tyrosine)	Positive	Negative	Negative
Coupled tetrazonium reaction	Positive	Negative	Negative
Liebman's	Positive	Negative	Negative
Landing and Hall's (histidine)	Positive	Negative	Negative
Tryptophan			
Indole method	Negative	Negative	Negative
Romieu's method	Negative	Negative	Negative

[a] After Thakur and Cheng, 1968.

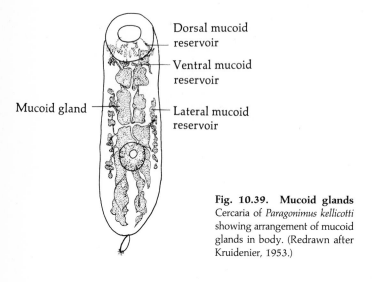

Mucoid gland

Dorsal mucoid
reservoir

Ventral mucoid
reservoir

Lateral mucoid
reservoir

Fig. 10.39. Mucoid glands
Cercaria of *Paragonimus kellicotti*
showing arrangement of mucoid
glands in body. (Redrawn after
Kruidenier, 1953.)

among the various stages. The miracidia, sporocysts, and rediae are usually very host specific and can survive only in one or a limited few closely related hosts, usually molluscs. On the other hand, adults may not be as specific and in many cases will develop in an array of experimental hosts.

The factors that govern host specificity among digenetic trematodes, i.e., the successful establishment of a parasite within either a single species or a group of closely related species of hosts, are not completely understood. Natural resistance and other factors such as incompatible metabolism and environmental stress are undoubtedly responsible for the failure of certain trematodes to infect certain hosts.

GERM CELL CYCLE

Space does not permit a detailed account of the development of each stage in the life history of digenetic trematodes, but it should be apparent that reproduction in the miracidium-sporocyst-redia-cercaria phase is essentially asexual, and the progeny arise through differentiation of germinal cells that are passed on from one generation to the next. This phenomenon, known as the **germ cell cycle**, has been reviewed by Cort *et al.* (1954). The metacercaria and adult are actually stages in the maturation process of the cercaria, the adult being the mature form. It is only in the adult that sexual reproduction occurs. Currently, the most acceptable interpretation of the types of reproduction among digeneans, especially the intramolluscan, asexual phases, is that it is sequential polyembryony, that is, the production of multiple embryos from the

same zygote with no intervening gamete production. It is noted that some evidence shows that certain digeneans can reproduce parthenogenically, both as intramolluscan larvae and as adults (Cable, 1971).

Of particular interest to those interested in gene expression is that all of the stages in the digenean life cycle carry an identical set of genes, yet the morphology or expressions of these genes differs between, for example, a sporocyst and a redia. This implies that certain genes are activated, whereas others are suppressed during the sequence of larval stages. In other words, those genes responsible for the development of an intramolluscan larva into a redia are suppressed, whereas those responsible for development into a sporocyst are activated if the latter is formed. How the activation and/or suppression of genes is brought about remains unresolved.

The life cycle pattern in a given species is almost always the same, and because many researchers have contributed life cycle data for numerous representative trematodes, it is now possible to draw some conclusions relative to developmental patterns in the various taxonomic groups. Figure 10.23 depicts the typical life cycles found in several taxa of the Digenea.

It is of interest to note that a few instances are known in which ambient factors can alter the life cycle pattern of trematodes. One of the most dramatic examples is that reported by Dinnick and Dinnick (1963, 1964). These investigators have reported that during the winter in the highlands of Kenya when the temperature does not rise above 16°C, the life cycle of *Fasciola gigantica* includes only redial generations in the molluscan intermediate host. However, at a constant temperature of 26°C, first-generation rediae alternately produce rediae and cercariae as long as they live. Rediae of the second and perhaps subsequent generations do the same thing until the snail is filled with them, and thereafter, only cercariae are produced. Normally, the life cycle of *F. gigantica* includes a sporocyst and a redial generation, the latter producing cercariae.

BIOLOGY OF DIGENETIC TREMATODES

CLASSIFICATION

The systematics of the order Digenea are being investigated continuously. Other than a few fairly well recognized superfamilies, taxonomic groups above the familial level still are being debated. The British and continental European trematodes have been adequately monographed by Dawes (1946). Yamaguti (1958, 1971) has contributed a taxonomic study of all digenetic trematodes known up to that time, and Schell (1970) has contributed an identification manual to the more common species in North America north of Mexico. In addition, LaRue (1957) has contributed a taxonomic scheme for the higher categories. In addition, the volumes by Erasmus (1972) and Smyth and Halton (1983) are recommended for a detailed review of the biology of this group of platyhelminth parasites.

Digenetic trematodes are separated into three suborders—the Anepitheliocystida, the Epitheliocystida,* and the Didymozoida. As stated earlier (p. 323), in members of the Anepitheliocystida, the embryonic, thin, nonepithelial excretory bladder is retained in the adult rather than replaced by a new epithelial lining derived from mesodermal cells. On the other hand, in members of the Epitheliocystida the wall of the embryonic excretory bladder is replaced by epithelial cells of mesodermal origin. There are other characteristics of both of these two suborders. These are presented in the classification list at the end of this chapter.

ANEPITHELIOCYSTIDIA

The suborder Anepitheliocystidia includes several superfamilies (see classification list at the end of this chapter). The biology of the members of some of these are considered below.

Superfamily Strigeoidea

The strigeoid digeneans are parasites of aquatic reptiles, aquatic birds, and mammals that feed on aquatic

---

*It is noted that some prefer to consider the Digenea as a subclass of the class Trematoda. If that system is employed, then both Anepitheliocystida and Epitheliocystida should be considered superorders subordinate to the subclass Digenea. The Didymozoida is considered by some to be a subclass.

amphibians, fish, and invertebrates. Their distinguishing characteristics are presented in the classification list. The biology of this group of flukes has been extensively reviewed and studied by Dubois (1937, 1938). Dubois's monographs are considered classics on the strigeoid trematodes. The biology of strigeoids parasitic in fish has been summarized by Hoffman (1960, 1967).

The Strigeoidea includes at least six families among which Strigeidae and Diplostomatidae are the best known.

Strigeoid Metacercarial Types. Since the metacercariae of strigeoids are probably the most frequently encountered stage, it should be mentioned that these metacercariae can be separated into three types: (1) The **tetracotyle** type characterized by a pyriform body, with or without a hindbody, but with two well-developed pseudosuckers, two regular suckers, and an auxiliary excretory vesicle (Fig. 10.40). (2) The **neascus** type is characterized by a cup-shaped forebody, a well-developed hindbody, and an auxiliary vesicle but lacks pseudosuckers (Fig. 10.40). (3) The **diplostomulum** type is characterized by a large concave forebody, a small conical hindbody, pseudosuckers, and an auxiliary tubular vesicle (Fig. 10.40). Diplostomula in the host are not surrounded by parasite-elaborated cyst walls.

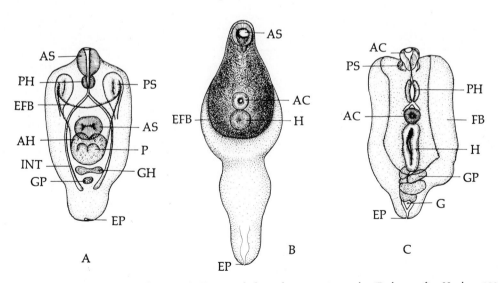

Fig. 10.40. **Strigeoid metacercarial types. A.** Tetracotyle larva from a garter snake. (Redrawn after Hughes, 1929.) **B.** Neascus larva from a bass. (Redrawn after Hughes, 1927.) **C.** Diplostomulum larva. (Redrawn after Bosma, 1934.) AC, acetabulum; AH, anterior lobes of holdfast organ; AS, anterior sucker; EFB, edge of cuplike forebody; EP, excretory pore; GH, glands of holdfast organ; GP, genital primordium; INT, intestine; P, posterior lobes of holdfast organ; PH, pharynx; PS, pseudosucker; H, holdfast organ; FB, forebody; G, gonopore.

Family Strigeidae. The family Strigeidae includes the genera *Strigea* with numerous species parasitic in the intestine of fish-eating scavenger birds, *Ophiosoma* in gulls and bitterns, *Parastrigea* in the domestic duck, and *Cotylurus* in various aquatic birds (ducks, geese, plovers, etc.) (Fig. 10.41). Other representative strigeids are depicted in Figures 10.41 and 10.42.

Life Cycle Pattern. The miracidia of strigeids are free swimming and possess two pairs of flame cells. These ciliated larvae actively penetrate snails, primarily lymnaeid snails, and pass through two sporocyst generations. The daughter sporocysts give rise to furcocercous cercariae each of which possesses a conspicuous pharynx. These free-swimming cercariae penetrate various aquatic animals, for example, snails, fish, leeches, tadpoles, frogs, salamanders, and snakes, and develop into encysted or nonencysted metacercariae. When the definitive host ingests the second intermediate host, the cycle is completed.

Genus Cotylurus. Representative of the strigeid trematodes is *Cotylurus flabelliformis* (Fig. 10.41), a common intestinal parasite of domestic and wild ducks in North America. The adult worm measures 0.55 to 1 mm long and has a cup-shaped forebody. The acetabulum and tribocytic organ (an auxillary holdfast organ) occur in the forebody. The hindbody is short and stout and is curved dorsally.

Life Cycle of Cotylurus flabelliformis. The eggs laid by the adult worm do not contain fully developed miracidia. Rather, development of the miracidium is completed in about 3 weeks after the eggs are passed out in the duck's feces. Upon hatching, the miracidium penetrates freshwater snails belonging to the genera *Lymnaea* and *Stagnicola* and transforms into a mother sporocyst. These first-generation sporocysts give rise to daughter sporocysts, which migrate to the molluscan host's digestive gland and mature. After about 6 weeks they commence to release furcocercous cercariae. Cercariae escaping from the snail are active swimmers, and when they make contact with other snails of the same genera, they penetrate, migrate to the ovotestis, and transform into tetracotyle metacercariae. When a snail harboring tetracotyles is eaten by a duck, the metacercariae mature into adults in about 1 week.

If *C. flabelliformis* cercariae enter a snail other than *Lymnaea* or *Stagnicola*, specifically members of the molluscan families Planobidae or Physidae, they will penetrate sporocysts or rediae of other digenean species already present and develop into tetracotyle metacercariae within them, thus becoming hyperparasites. These tetracotyles will develop into adults if ingested by a duck.

Family Diplostomatidae. The family Diplostomatidae includes the genera *Alaria*, *Uvulifer*, *Diplostomum*, and others. The life cycles of one representative each of *Alaria* and *Uvulifer* are being presented to emphasize several interesting variations as well as to indicate the occurrence of diplostomulum and neascus types of metacercariae.

Genus Alaria. The genus *Alaria* (Fig. 10.42) includes several species, all of which are parasitic in the small intestine of carnivorous mammals. In North

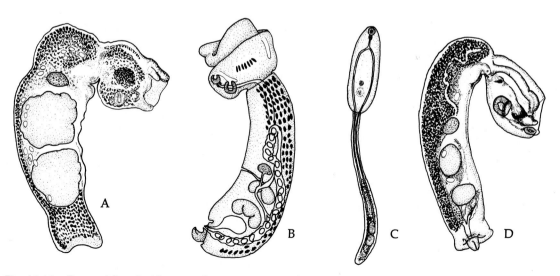

Fig. 10.41. **Some adult strigeid trematodes. A.** *Strigea strigis* from intestine of various fish-eating birds. (Redrawn after Dubois, 1938.) **B.** *Cotylurus erraticus* from intestine of fish-eating birds. (Redrawn after Fuhrmann, 1928.) **C.** *Uvulifer gracilis* from the intestine of the bird *Ceryle lugubris.* (Redrawn after Yamaguti, 1934.) **D.** *Apatemon gracilis* from intestine of fish-eating anseriform birds. (Redrawn after Dubois, 1938.)

America, *Alaria americana* occurs in various species of the canine family. This parasite measures 2.5–4 mm long and has a forebody that is longer than the hindbody. The forebody bears a pair of ventral flaps, and there is a pair of pointed processes on each side of the oral sucker. The elongate tribocytic organ is relatively large and has a ventral depression in its center.

Life Cycle of Alaria. The life cycle of *Alaria* is interesting in that four hosts are commonly involved. The stages are depicted in Figure 10.42. The miracidium is not developed within each egg when the latter is discharged to the exterior in the canine host's feces; however, it is formed and hatches in about 2 weeks. These miracidia are active swimmers and will penetrate several species of freshwater planorbid snails and metamorphose into mother sporocysts. These first-generation sporocysts occur in the snail's renal veins, where they produce daughter sporocysts in about 2 weeks. Subsequently, the second-generation sporocysts migrate to the snail's digestive gland where they reside for approximately 1 year before producing furcocercous cercariae.

The cercariae escape from the first intermediate host during the daylight hours and swim to the water surface, where they hang with their anterior ends pointed downward. If a tadpole swims by, the resulting water currents stimulate each cercaria to swim toward the amphibian second intermediate host. If contact is made, the cercaria penetrates the tadpole, dropping its tail prior to entering. It migrates within the amphibian host's body and in about 2 weeks transforms into a **mesocercaria** (an unencysted stage intermediate between a cercaria and a metacercaria).

If a tadpole or an adult frog harboring mesocercariae is eaten by a canine, the parasites, after being freed from the surrounding host tissues as a result of digestion, penetrate into the coelom, from whence they migrate to the diaphragm and lungs. After about 5 weeks at such a site, each mesocercaria develops into a diplostomulum metacercaria. Such diplostomula migrate up the trachea and down to the small intestine, where they mature in approximately 1 month.

The life cycle of *Alaria* as presented above involves two intermediate and one definitive hosts. However, since tadpoles and frogs are not commonly eaten by canines, usually an additional host is involved. Since such a host is not totally essential for the development of the digenean, it is a paratenic host (p. 4). When this host, commonly but not necessarily a water snake, eats an infected tadpole or frog, the escaping mesocercariae accumulate, sometimes in large numbers as a result of continuous feeding on infected amphibians, falls prey to a canine, the mesocercariae migrate, develop into diplostomula, and eventually reach maturity in the small intestine as described earlier.

Interestingly, Shoop and Corkum (1983) have reported that in the case of *Alaria marcianae*, mesocercariae can be transmitted from mother mice, which are

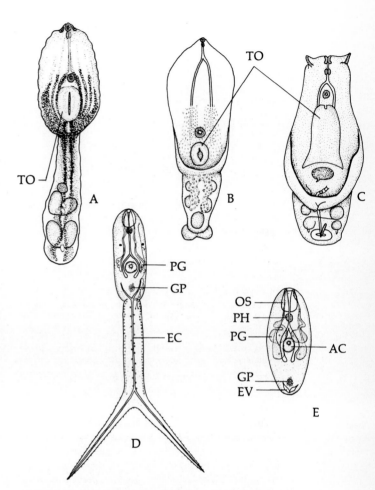

Fig. 10.42. **Some strigeoid trematodes. A.** *Diplostomum spathaceum* from intestine of fish-eating birds. (Redrawn after Dubois, 1938.) **B.** *Neodiplostomum orchilongum* from intestine of hawks. (Redrawn after Noble, 1936.) **C.** *Alaria alata* from intestine of dogs, cats, and foxes. (Redrawn after Baylis, 1929.) **D.** Furcocercous cercaria of *Alaria*. **E.** Mesocercaria of *Alaria* in frogs and tadpoles. (**D** and **E**, redrawn after Bosma, 1934.) AC, acetabulum; EC, excretory canal; EV, excretory vesicle; GP, genital primordium; OS, oral sucker; PG, penetration gland; PH, pharynx; TO, tribocytic organ.

experimental paratenic hosts, to nursing mice by the transmammary route. Transmammary infection appears to be limited to certain diplostomatid trematodes belonging to the genera *Alaria*, *Pharyngostomoides*, and *Procyotrema* and to certain strigeid trematodes belonging to the genus *Strigea* (see Miller, 1981, for review). Also, Shoop and Corkum have demonstrated that nursing raccoons can serve as true definitive hosts for *A. marcianae*, whereas adult raccoons can serve

Fig. 10.43. "**Black spot**" **disease.** A minnow, *Pimephales* sp., infected with strigeoid metacercariae causing "black spot" disease. (Courtesy of Dr. J. S. Mackiewicz.)

only as paratenic hosts, that is, the mesocercariae do not reach sexual maturity. They have coined the term **amphiparatenic host** to designate such hosts.

Alaria infections in the canine host are highly pathogenic, causing severe enteritis and even death. Similarly, the mesocercaria is pathogenic, especially if present in large numbers. Freeman *et al.* (1976) have reported a human death due to the presence of mesocercariae in almost every organ in the body. The individual most likely acquired the infection from eating undercooked frog legs harboring mesocercariae.

Genus Uvulifer. Fishermen are undoubtedly familiar with *Uvulifer* and related strigeoids but do not recognize them as such. *Uvulifer ambloplitis* (and others) is the cause of the so-called black spot disease of fish (Fig. 10.43). Actually, as indicated below, the black spots represent encysted metacercariae surrounded by reaction pigmentation. The adult parasite occurs in the small intestine of the kingfisher, a widely distributed, fish-eating bird in North America. Its ventrally recurved forebody is separated from the longer hindbody by a constriction. Such adult worms measure 1.8–2.5 mm long.

Life Cycle of Uvulifer. The eggs laid by the adult worm do not enclose a fully developed miracidium; however, this ciliated larva is fully formed in about 3 weeks after the eggs are deposited in water in the definitive host's feces. Upon hatching, the miracidia penetrate aquatic snails of the genus *Helisoma* and transform into mother sporocysts. These retain the miracidial eyespots. Daughter sporocysts, formed in mother sporocysts, migrate to the snail host's digestive gland and commence producing cercariae in about 6 weeks. Such cercariae escape from the molluscan host and swim near the water surface. If they come in contact with a fish, they shed their tails and penetrate

the skin. Once inside the dermis, they metamorphose into neascus type metacercariae. Each neascus secretes a hyaline cyst wall around itself, which becomes enveloped by melanin deposited by the host. Kingfishers become infected as a result of ingesting fish harboring neascus metacercariae, which develop into sexually mature adults in about 30 days.

Superfamily Schistosomatoidea

The Schistosomatoidea includes several families of digeneans (see classification list at the end of this chapter) that are unique in that they do not require a second intermediate host, do not have a metacercarial stage, and mature in the blood vascular system of their definitive hosts. Furthermore, most species are dioecious. The schistosomatoids are parasites of fish, turtles, birds, and mammals. Several species are blood parasites of humans, causing the dreaded disease known as schistosomiasis. Those species that occur in fish and turtles belong to the families Aporocotylidae, Sanguinicolidae, and Spirorchidae, whereas those that occur in birds and mammals, including humans, belong to the family Schistosomatidae.

The elongate, slender bodies of the schistosomatoids are well adapted for living in their definitive hosts' vascular systems. They lack a muscular pharynx and may or may not possess an acetabulum. The eggs of schistosomes typically lack an operculum. The life history pattern includes a free-swimming miracidium, two (or more) generations of sporocysts in a molluscan intermediate host, and furcocercous cercariae that penetrate the integument of the definitive host and develop into adults in the blood of the portal system.

Family Schistosomatidae. The family Schistosomatidae includes the genus *Schistosoma*, which, in turn, includes several species parasitic in the blood of humans and other animals. The three major human-infecting species are *S. mansoni*, *S. japonicum*, and *S. haematobium*.

Human Schistosomiasis. In spite of efforts to reduce the incidence of human **schistosomiasis**[*] in the world, the number of cases does not appear to be declining. It is extremely difficult to estimate how many individuals are afflicted with this disease. Conservatively speaking, at least 200 million cases of human schistosomiasis exist in the world. In the Peoples' Republic of China alone, it has been estimated that at least 15 million cases exist, and this disease is regarded as the most serious disease in that country. Another heavily affected nation is Egypt. In lower Egypt, an incidence of 60% has been estimated in most areas, with an incidence as high as 90% in some localities. In all, close to 17 million cases of human schistosomiasis exist in that country. In tropical and subtropical Africa,

---

[*]Known as bilharziasis among European parasitologists. In following the usage described earlier (p. 239), if the disease state is recognizable, the condition should be referred to as schistosomiosis; however, as a result of common usage, schistosomiasis is preferred.

Wright (1972) has estimated that over 91 million human infections exist.

With the advent of jet airplanes, the world is rapidly shrinking. Schistosomiasis is no longer a disease of distant lands. Indeed, human cases of this parasite-caused disease exist in considerable numbers in such metropolitan areas as New York, Chicago, and Philadelphia, primarily among immigrants from endemic areas. Fortunately, the proper snail intermediate hosts cannot survive in temperate climes and consequently the parasite's life cycle has not become established.

Because of worldwide importance of the disease, numerous monographs and multiauthored treatises have been written about it. The volumes by Gelfand (1967) and Jordan and Webbe (1982), and the treatise edited by Ansari (1972) are especially recommended. In addition, Warren (1973a) has contributed a volume devoted to selected abstracts of the literature pertaining to schistosomes and schistosomiasis published between 1852 and 1972, Warren and Newill (1967) have compiled a bibliography of the world's literature for the same span of time, and Warren and Hoffman (1976) have compiled two volumes of abstracts of the complete literature pertaining to these parasites and the diseases they cause between the years 1963 and 1974. Without a doubt, more has been written about the schistosomes and schistosomiasis than about any other group of animals and about any other disease complex.

Schistosomiasis has influenced world history. To cite a recent example, Kierman attributed a decisive role to this disease in the prevention of an assault on Formosa by Communist China early in 1950. In that instance, between 30,000 and 50,000 cases of acute schistosomiasis among soldiers cancelled the attack. During World War II, schistosomiasis among American troops in the Pacific caused great problems. For example, the infection of nearly 2000 troops during the recapturing of Leyte, in 1944, caused the loss of over 300,000 man-days and $3 million for medical care (Fig. 10.44).

Going back in history, there is reason to believe that the curse that Joshua placed on Jericho was the introduction of *Schistosoma haematobium* into the communal well by the invaders. The curse was removed after the abandonment of Jericho, and subsequent droughts have eliminated the snail intermediate host (Hulse, 1971).

Schistosomiasis is known to have occurred in Egypt since ancient times. At least 50 references to the disease can be found in the surviving papyri. Also, calcified eggs of *S. haematobium* have been found in Egyptian mummies dating from 1200 B.C.

Schistosoma mansoni. This parasite is a dioecious species widely distributed in Africa, parts of

Fig. 10.44. Schistosomiasis. Bathing and doing laundry in waters contaminated with schistosome cercariae constitute the major means of human infection in endemic areas. This photograph of a woman washing and children bathing was taken in Lyte, the Philippines. The standing boy's swollen abdomen indicates that he is already infected. (Photograph courtesy of the World Health Organization.)

Brazil, Venezuela, Guyana, Surinam, Puerto Rico, St. Lucia, Martinique, Guadaloupe, and the Dominican Republic (Fig. 10.44). Its occurrence in the New World is not as concentrated or as prevalent as it is in Africa. Its presence in the western hemisphere is believed to be a by-product of the African slave trade.

The females and males occurring in the host's portal blood, usually in the smaller branches of the inferior mesenteric vein in the region of the lower colon, are usually in close association. The male, measuring 6.4–9.9 mm in length, is broader than the female. The two lateral borders of its body are curved to form a groove—the **gynecophoral canal**—in which the female, measuring 7.2–14 mm in length, is snugly held. Minute tegumental tubercles occur on the dorsal body surface of the male. If present on females, these protuberances are limited to the anterior and posterior ends of the body (Figs. 10.45, 10.46). Hockley (1973) has contributed a critical review of the fine structure

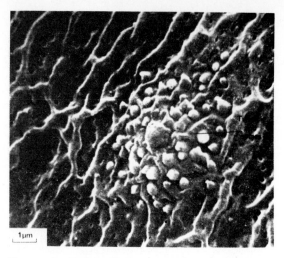

Fig. 10.46. Body surface tubercles and spines. Scanning electron micrograph showing a tubercle surrounded by spines on the body surface of male *Schistosoma mansoni*. (After Hockley, 1973; with permission of Academic Press.)

of the schistosome tegument in all of the life cycle stages. Both suckers are present and are located in the anterior portion of the body, although those of the female are smaller and not as muscular as those of the male. The internal anatomy of this fluke is similar to that of other trematodes except that a pharynx is wanting and the males are multitesticular, with six to nine testes.

Life Cycle of Schistosoma mansoni.   In *S. mansoni*, fertilization occurs when spermatozoa from the male are introduced into the female. The fully formed egg, measuring 114–175 μm by 45–68 μm, is non-operculate and bears a large posterolateral spine (Figs. 10.11, 10.47, 10.48). Such eggs are laid by the female in the smallest blood vessels of the host to which they

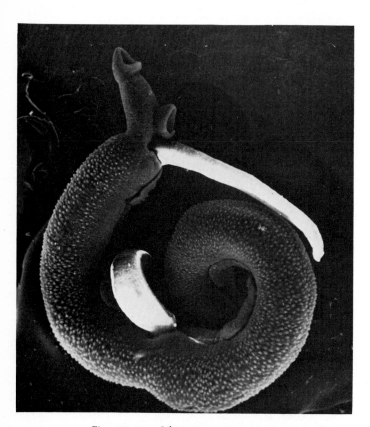

Fig. 10.45. *Schistosoma mansoni.* Scanning electron micrograph of adult female laying in the gynecophoral canal of male. Note presence of tegumental tubercles on surface of male. (Photograph courtesy of Dr. D. A. Erasmus, University College, Cardiff, Wales.)

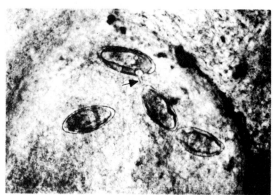

Fig. 10.47. *Schistosoma mansoni.* Photomicrograph of eggs in smear preparation of liver of experimentally infected mouse. Arrow pointing to lateral spine of egg. (Courtesy of Dr. Grover C. Miller, North Carolina State University.)

migrate after copulation. The eggs become free in the mesenteries and are carried by the blood to the proximity of the large intestine, where they gradually penetrate the intestinal wall, become included in the lumen, and are passed out in feces. Commonly, eggs become trapped in various tissues and cause inflammation.

In fact, the most serious pathologic alterations in schistosomiasis patients are caused by eggs entrapped in tissues, especially in the wall of the colon and rectum and in the liver in the case of *S. mansoni*. Eggs lodged at these sites act as foreign bodies and antigens originating with each egg elicit inflammatory reactions involving leucocytic and fibroblastic infiltration. Eventually, a **granuloma** (pseudotubercle) is formed at each egg site (Fig. 10.49). In addition, small abscesses occur and the occlusion of small blood vessels leads to necrosis and ulceration. A detailed review of the pathology of schistosome infections has been contributed by Warren (1973b).

Andrade and Barka (1962), by employing histochemical techniques, have demonstrated that eggs in the liver of experimentally infected mice contain acid phosphatase, nucleotides, ATPase, aminopeptidase, alkaline phosphatase, and a nonspecific esterase. In addition, a complex carbohydrate, which is antigenic, is also present. The reticuloendothelial cells of the host's liver and spleen become sites of high acid phosphatase activity.

When *S. mansoni* eggs come in contact with water, the enclosed miracidia break out of the shells. Although the hatching mechanism is still not completely understood, it is known that the first sign of hatching is activation of the cilia. Subsequently, a line on the side of the egg shell, known as the **stigma**, opens and the miracidium emerges (Figs. 10.48, 10.50). Pan (1980) has contributed an outstanding study of the fine structure of the miracidium of *S. mansoni*.

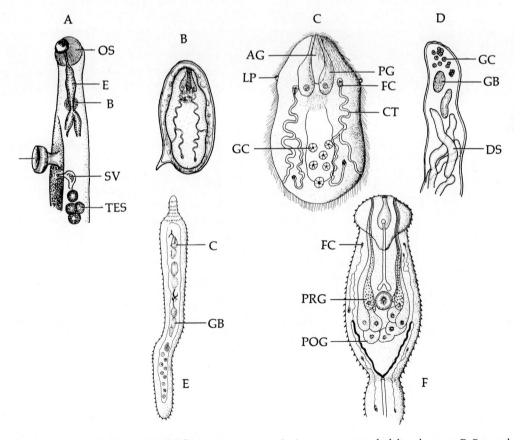

Fig. 10.48. Stages in the life cycle of *Schistosoma mansoni*. A. Anterior portion of adult male worm. **B.** Egg enclosing miracidium. **C.** Miracidium. **D.** Portion of mother sporocyst. **E.** Daughter sporocyst. **F.** Body of cercaria. For a more detailed drawing of the cercaria, see Fig. 10.31. AC, acetabulum; AG, apical gland; B, brain; C, cercaria; CT, collecting tubule; DS, daughter sporocyst; E, esophagus; FC, flame cell; GB, germ ball; GC, germinal cell; LP, lateral papilla; OS, oral sucker; PG, penetration gland; POG, postacetabular gland; PRG, preacetabular gland; SV, seminal vesicle; TES, testis.

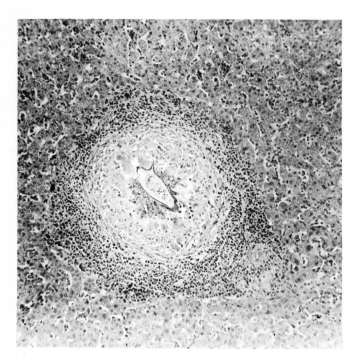

Fig. 10.49. Granuloma (pseudotubercle) surrounding *Schistosoma mansoni* egg. Photograph showing granulomatous cells forming pseudotubercle surrounding *S. mansoni* egg in liver. (Armed Forces Institute of Pathology Negative No. 64–6532.)

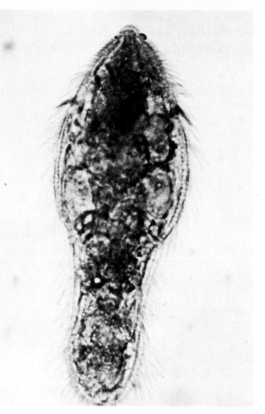

Fig. 10.50. *Schistosoma mansoni.* Photomicrograph of miracidium. (Courtesy of Dr. Grover C. Miller, North Carolina State University.)

The free-swimming miracidia (Fig. 10.50) find and penetrate a suitable pulmonate snail, *Biomphalaria glabrata* or *B. pfeifferi* (Fig. 10.54; Table 10.6), and transform into mother sporocysts (Fig. 10.48), which are located in the region of the host's foot or in the tentacles. Daughter sporocysts (Fig. 10.48), differentiating from germ balls in mother sporocysts, migrate to the host's digestive gland on about the 12th day after miracidial penetration and eventually produce large numbers of furcocercous cercariae (Fig. 10.48). These escape from daughter sporocysts through the birth pore, leave the snail, and become free swimming in water. When cercariae come in contact with human skin in water, they actively penetrate it.

It needs to be mentioned that Hansen *et al.* (1974) have reported the occurrence of granddaughter or second-generation daughter sporocysts in larval *S. mansoni* maintained in culture. Even more interesting is the report by Jourdane and Théron (1980) that if daughter sporocysts of this parasite are surgically transplanted into an uninfected snail, these second-generation sporocysts produce another generation of sporocysts. In fact, these investigators were able to maintain *S. mansoni* for 1 year solely in molluscs, with transplanted sporocysts giving rise to another generation of sporocysts rather than cercariae.

How schistosome cercariae are attracted to mammalian skin has been investigated by Stirewalt (1971). In brief, the following conditions serve as penetration stimuli: a lipid on the skin surface, a warm target membrane in a temperature-differentiated environment, and possibly light.

The penetration process of *Schistosoma* cercariae is understood to some extent (Stirewalt and Hackey, 1956; Stirewalt, 1966). It involves softening of the keratin of the host skin's horny layer by the alkalinity of preacetabular gland secretion (see below), and altering the acellular, epithelial ground substance of the epidermis and dermis coupled with muscular boring movements (Lewert and Lee, 1954; Stirewalt, 1966). In cases of heavy infections, inflammation (cercarial dermatitis) occurs at the sites of penetration.

There are five pairs of readily visible glands in *S. mansoni* cercariae. Two pairs are preacetabular, and three pairs are postacetabular (Cheng and Bier, 1971; Dorsey and Stirewalt, 1971). These secrete to the

Table 10.6. Major Molluscan Hosts of *Schistosoma mansoni*[a]

Species of Snail	Geographic Location
Biomphalaria glabrata	Puerto Rico, Dominican Republic, St. Kitts, Guadeloupe, Martinique, St. Lucia, Guyana, Venezuela, Brazil
B. tenagophilus	Brazil
B. pfeifferi gaudi	Gambia, Guyana, Sierra Leone, Liberia, Mauritania, Senegal, Sudan, Ivory Coast, Ghana, Nigeria
B. sudanica	Cameroon, Congo, Kenya, Uganda, Tanzania, Malawi
B. ruppellii	Congo, Sudan, Eritrea, Ethiopia, Kenya, Uganda, Tanzania
B. pfeifferi	Uganda, Zimbabwe, Tanzania, South Africa, Mozambique, Madagascar
B. choanomphala	Uganda
B. smithi	Uganda
B. stanleyi	Uganda
B. alexandrina	Egypt, Saudi Arabia, Yemen
B. sudanica	Congo
B. bridouxiana	Congo
Tropicorbis centrimetralis	Brazil
T. philippianus	Ecuador[b]
T. chilensis	Chile[b]
T. albicans	Puerto Rico[b]
T. riisei	Puerto Rico[b]

[a] Data modified after Malek and Cheng, 1974.
[b] Experimental infections.

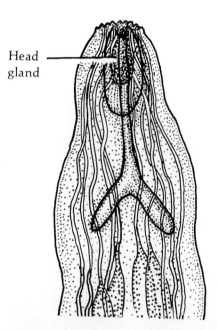

Fig. 10.51. Head gland of *Schistosoma mansoni* cercaria. Drawing of anterior end of cercaria showing position of head gland in oral sucker.

exterior via ducts which open along the anterior margin of the body. In addition, a **head gland** lies within the oral sucker (Fig. 10.51). The postacetabular glands are smaller than the preacetabular ones. They contain finely granular, periodic acid-Schiff staining contents. Some of this material is secreted at each attachment of the oral sucker. These deposits swell in water to provide a sticky mucus for attachment first of the oral sucker, then of the ventral sucker. This secretion may also serve to protect the cercaria and direct and conserve secreted enzymes.

As stated, the preacetabular glands secrete a substance which causes hydrolysis of the ground substance between the skin cells, permitting penetration. In addition to these glands, which permit penetration into and through mammalian skin, there is reason to believe that other gland cells are present that secrete some material which facilitates escape of fully developed cercariae from the molluscan host. These glands, which have not been thoroughly studied, are generally referred to as **escape glands**. Stirewalt (1973) has presented a review of the biology of the various stages in the life cycle of schistosomes, including the cercaria.

Once a cercaria successfully enters the human or some other mammalian host, it is referred to as a **schistosomule**. Rifkin (1971) has studied the penetration of mouse skin by *S. mansoni* schistosomules with the electron microscope. He has found that the

tegument of the schistosomule is essentially the same as those of other parasitic platyhelminths (p. 253). However, tegumentary spines are present, and these tear and destroy the adjacent host epidermal cells during penetration. He has also found a glycocalyx coat on the schistosomular body surface that differs from the mucinlike layer found on cercariae. The nature of this layer remains uncertain. It may be the result of immunologic reaction.

If successful penetration occurs, usually in several minutes, the cercariae lose their tails during the process. Such tailless cercariae, or schistosomules, can be found in the skin for approximately 18 hours after penetration. They then enter the blood circulation and are carried to the host's lungs via the heart. Young adults can be found in the lungs by the second or third day after penetration. Later, by about the 15th day, they can be found accumulated in the liver, where they feed on portal blood and undergo rapid growth.

By approximately the 23rd day postpenetration by cercariae, migration of the young adults to the mesenteric veins takes place, and sexual maturation and mating occur. Egg production generally occurs by the 40th day. If cercariae are taken in with drinking water, these larvae will penetrate through the lining of the

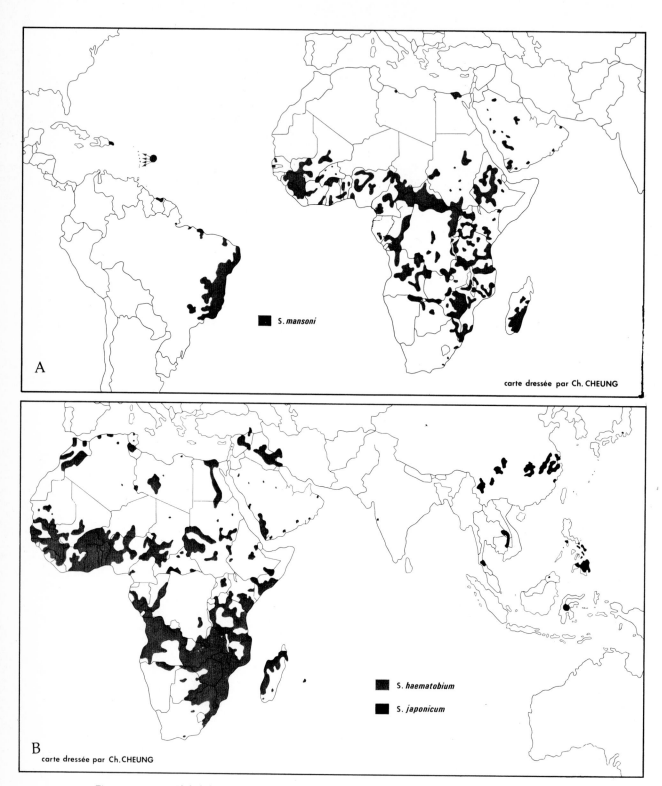

Fig. 10.52. **A.** Global distribution of schistosomiasis due to *Schistosoma mansoni*, 1983. **B.** Global distribution of schistosomiasis due to *Schistosoma haematobium* and *Schistosoma japonicum*, 1983. (World Health Organization, 1984.)

mouth and throat. Rodents can be experimentally infected through the ingestion of infected snails.

The sex of *S. mansoni*, like that of the other dioecious species, is genetically determined at fertilization. A male miracidium eventually gives rise to all male adults and a female miracidium results in female adults.

Schistosoma japonicum. This parasite is geographically confined to the Far East (Fig. 10.52). It is found primarily in Japan, China, the Phillippines, and Sulawesi (Celebes). The life cycle of this species parallels that of *S. mansoni*. The eggs are readily distinguished from those of *S. mansoni* because there is only a very small lateral spine on the shell (Figs. 10.11, 10.53). The entire egg measures 70–100 μm by 50–65 μm. The body surface of the adult male *S. japonicum* bears tegumentary spines in the gynecophoral groove and on the suckers. The body measures about 15 × 0.5 mm and the intestinal caeca unite posteriorly. Females of *S. japonicum* average 20 × 0.3 mm, and thus are longer and narrower than the male.

The snail hosts of *S. japonicum* include amphibious snails of the genus *Oncomelania*. According to Davis (1971), all oncomelanid snails are subspecies of *O. hupensis* (Fig. 10.54). In the Yangtze River drainage system in the People's Republic of China, the intermediate host for *S. japonicum* is *O. hupensis hupensis*; in Japan and parts of southern China, it is *O. hupensis nosophora*; in the Philippines it is *O. hupensis quadrasi*; and in Indonesia it is *O. hupensis lindoensis*. The descriptions and biology of these snails, as well as those that transmit other schistosome species, are presented by Malek and Cheng (1974).

Within the human host, *S. japonicum* adults inhabit the branches of the superior mesenteric vein in the proximity of the small intestine. In addition, the inferior mesenterics and caval system may also be invaded, for the worms tend to migrate away from the liver as they become older. As in the case of *S. mansoni*, the pathogenesis of schistosomiasis japonica is primarily the result of the host's cellular reactions to eggs in tissues. The disease caused by *S. japonicum* is the most severe among those caused by the three major human-infecting schistosomes.

Formosan Strains of S. *japonicum.* There are at least two strains of *S. japonicum* endemic to Formosa

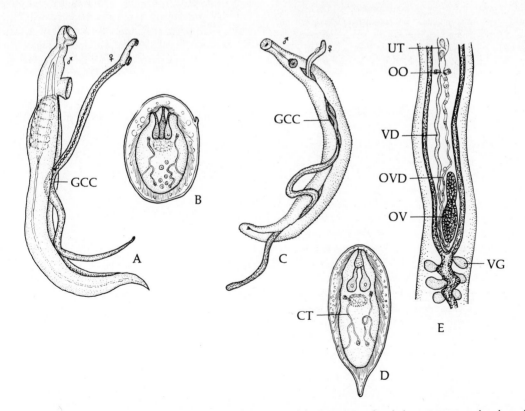

Fig. 10.53. *Schistosoma japonicum* **and** *S. haematobium.* **A.** Adult *S. japonicum* female lying in gynecophoral canal of male. **B.** *S. japonicum* egg enclosing miracidium. **C.** Adult *S. haematobium* female lying in gynecophoral canal of male. **D.** Reproductive system of female *S. haematobium.* **E.** *S. haematobium* egg enclosing miracidium. (**A**, **B**, **D**, and **E**, redrawn after Craig and Faust, 1951.) CT, collecting tubule; GCC, gynecophoral canal; OO, ootype; OV, ovary; OVD, oviduct; UT, uterus; VD, vitelline duct; VG, vitelline gland.

Fig. 10.54. Molluscan intermediate hosts of the three major human-infecting schistosomes. A. *Oncomelania hupensis hupensis* from China. **B.** *Oncomelania hupensis quadrasi* from the Philippines. **C.** *Oncomelania hupensis nosophora* from Japan. **D.** *Biomphalaria glabrata* from Puerto Rico. **E.** *Biomphalaria glabrata* from Brazil. **F.** *Bulinus truncatus* from Egypt. **G.** *Bulinus truncatus rohlfsi* from Senegal. (After Malek and Cheng, 1974; with permission of Academic Press.)

which are noninfectious to humans. These have been designated by Hsü and Hsü (1968) as the Changhua and the Ilan strains. Both strains are essentially noninfective to the snails *Oncomelania hupensis hupensis* from the People's Republic of China and *O. hupensis quadrasi* from the Philippines. Furthermore, both strains are only moderately infective to *O. hupensis nosophora* from Japan. In Formosa, the natural molluscan host is *O. hupensis formosana*; however, it will only infect the Alilao race of this snail and not the Kohsiung race, indicating the existence of race or strain differences among these snails. Those strains of *S. japonicum* which are noninfective to humans are referred to as **zoophilic** strains. In nature, the mammalian hosts are rats, but shrews and bandicoots are also natural hosts, and the Philippine monkey can be infected experimentally.

Schistosoma haematobium. This schistosome is endemic to north Africa (particularly in the valley of the Nile), central and west Africa, and the Near East (Fig. 10.52). It has been estimated that 39.2 million human infections exist; the majority are limited to Africa. Again, the life cycle of *S. haematobium* parallels that of *S. mansoni* except the eggs do not pass into the

lumen of the large intestine but pass into the urinary bladder and are expelled in urine. These eggs measure 112–170 μm by 40–70 μm and are readily identified by a distinct terminal spine (Figs. 10.11, 10.53).

After maturing in the sinusoids of the host's liver, *S. haematobium* adults, as is the case with other human-infecting schistosomes, migrate from the liver. The majority of *S. haematobium* adults reach the vesical, prostatic, and uterine plexuses by way of the inferior haemorrhoidal veins. Adult males may become 15 mm long, whereas females may reach 20 mm in length (Fig. 10.53).

Chimpanzees and other primates can be naturally infected with *S. haematobium*, and in simian hosts adult worms have been found in the urinary bladder, rectum, lungs, liver, appendix, and mesenteric veins.

The snail hosts for *S. haematobium* are listed in Table 10.7. Except for *Ferrissia tenuis* in India and *Planorbarius metidjensis* in Portugal, all of the intermediate hosts belong to the genus *Bulinus* (Fig. 10.54). It is noted that although the transmission of *S. haematobium* occurred in southern Portugal not many years ago, a combination of effective mollusciciding and chemotherapy of schistosomiasis patients has eliminated transmission.

Schistosoma mekongi. In this age, it is relatively rare that one finds a new pathogenic parasite of humans, although it does occur. As a result of American involvement in southeast Asia during the Vietnam War, it is now known that another species of *Schistosoma*, named *S. mekongi* by Voge *et al.* (1978), occurs in Laos and Cambodia. It is a human pathogen most closely related to *S. japonicum*. Its molluscan intermediate host is *Tricula aperta*. Bruce *et al.* (1980) have edited a volume devoted to this parasite.

Table 10.7. The Molluscan Hosts of *Schistosoma haematobium*[a]

Species of Snail	Geographic Location
Bulinus truncatus	Morocco, Tunisia, Algeria, Egypt, Libya, Sudan, Turkey, Syria, Israel, Saudi Arabia, Yemen, Iraq, Iran
B. guernei	Ghana, Cameroon, Gambia, Mauritania
B. truncatus rohlfsi	Ghana, Cameroon, Gambia, Mauritania
B. senegalensis	Senegal, Gambia
B. globosus	Guyana, Gambia, Sierra Leone, Kenya, Liberia, Cameroon, Ghana, Nigeria, Sudan, Angola, Uganda, Tanzania, Zimbabwe, South Africa, Mozambique
B. africanus	South Africa, Zimbabwe, Mozambique
B. jousseaumei	Guyana, Sudan, Gambia
B. nasutus	Kenya
B. coulboisi	Kenya
Ferrissia tenuis	India
Planorbarius metidjensis	Portugal

[a]Data from Malek, E. A., and Cheng, T. C. (1974). Medical and Economic Malacology. Academic Press, New York.

Additional Notes on Pathogenesis. Recall that the major cause of pathologic alterations in the three major types of human schistosomiasis is the eggs which serve as foci for cellular reactions. In addition, the migration of eggs through the urinary bladder wall in schistosomiasis haematobia, and through the colon wall in schistosomiasis mansoni and japonica generally, causes hemorrhage. Consequently, bloody urine (hematuria) or bloody stools usually occur. Also, diarrhea commonly accompanies schistosomiasis mansoni and japonica.

In endemic areas, where reinfections occur, repeated penetration of the intestinal and bladder walls results in extensive scar formation, which prevents migration of the eggs through these structures. As the result, many eggs are transported to the liver and occasionally to other sites, such as the brain and spinal cord. Wherever these aberrant eggs become lodged, local inflammation, fibrosis, and necrosis ensue. Cirrhosis of the liver is not uncommon. Enlargement of the liver and spleen (hepatosplenomegaly) is common (Fig. 10.55). In instances of heavy infections, mature worms may migrate to the brain, lungs, uterus, oviduct, and gonads. As expected, their presence at these sites provoke pathologic changes.

In addition to the types of pathologic manifestations mentioned the antigenicity of the decomposing tissues of dead worms also contribute to the complex pathology of schistosomiasis. For good reviews of various aspects of the pathology of human schistosomiasis, see Mostofi (1967) and Warren (1973b).

Acute cases of schistosomiasis, especially schistosomiasis japonica, which is highly pathogenic, often lead to death.

Immunity. Immunity to human schistosomiasis is not totally understood, although it is known that animals experimentally infected with the human-infecting schistosomes do develop an immunity and at least partial immunity occurs in infected humans. Vogel and co-workers have demonstrated that the monthly exposure of monkeys to moderate numbers of infective cercariae of *S. mansoni* and *S. japonicum*, so as not to elicit clinical schistosomiasis, results in a state of tolerance. Subsequent exposures of these monkeys to cercariae in numbers sufficiently large to kill an unexposed host do not cause symptoms of schistosomiasis, nor does the number of parasites in the body increase. This process of protecting against reinfection as a result of an existing infection is known as **premunition**.

Another approach that has been taken to induce immunization against schistosomiasis is to infect hosts with irradiated cercariae. It is known that if the cercariae of *S. mansoni* and *S. japonicum* are exposed to irradiation greater than 2000 roentgens, the majority of the parasites are prevented from developing into mature, egg-laying adults (Hsü *et al.*, 1963a; Villella and Weinbren, 1965) and hence prevent the develop-

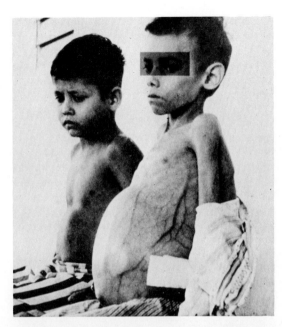

Fig. 10.55. Schistosomiasis. Two Puerto Rican boys infected with *Schistosoma mansoni*. The boy on right is suffering from malnutrition which increases the severity of the disease. Notice his enlarged liver. (From *Phoenix*, University of Michigan.)

ment of egg-induced pathology. At the same time, the host does develop some immunity, although not complete. Hsü *et al.* (1962, 1963b, 1965) have reported encouraging results with irradiated *S. japonicum* cercariae introduced into the rhesus monkey. A detailed review of immunity to schistosomiasis has been presented by Smithers and Terry (1969) and updated by these authors (Smithers and Terry, 1976).

In addition to infecting humans, the three important species of schistosomes can infect various other animals. Natural infections of monkeys and rodents with *S. mansoni* are known. Similarly, monkeys can be naturally infected with *S. haematobium*, and various rodents can be experimentally infected with this parasite. *Schistosoma japonicum* can infect cattle, goats, pigs, dogs, cats, shrews, weasels, various wild rodents, and primates.

Other Schistosomes. Other economically important schistosomes include *Schistosoma bovis* in cattle and sheep. Also, this species can be experimentally induced to mature in guinea pigs. *Schistosoma indicum* is found in the veins of horses, cattle, goats, and buffaloes in India; *S. nasale* in the nostrils of cattle in India; and *S. suis* in the portal veins of pigs and dogs in India.

It is known that Zebu cattle in the White Nile area of the Sudan acquire a high degree of resistance to *Schistosoma bovis* as a result of repeated natural exposures. The adult worms present in these native cattle, however, are not eliminated, although they do show greatly suppressed fecundity. Bushara *et al.* (1983a) have confirmed this fact and also have demonstrated that the resistance is not abrogated when the adult worms are removed by medication. Also, Bushara *et al.* (1983b) have shown that calves can develop partial resistance to *S. bovis* as a result of an initial exposure to 10,000 cercariae, and some resistance can also be stimulated in the absence of the migratory stages of the parasite by transplantation of adult male and female worms, and in the absence of both migratory stages and eggs by transplantation of adult male worms.

Genus Schistosomatium. Among members of *Schistosomatium* found in rodents is *S. douthitti*, a parasite in the hepatic portal system of field mice, albino mice, deer mice, and muskrats. *S. douthitti* is commonly used in the laboratory as an experimental model because it is not infective to humans. This species is dioecious. The body of the male measures from 1.9–6.3 mm in length and is divided into the anterior forebody, which is flattened and occupies two-fifths of the body length, and the hindbody, which is curved to form the gynecophoral groove. There are 14 to 16 testes present in *S. douthitti*, and the body tegument is spinous.

The female, measuring from 1.1–5.4 mm in length, bears spines along the lateral edges of the dorsal and ventral surfaces. These tegumentary spines extend from the anterior portion of the body posteriorly to the level of the ovary. The oral sucker also bears spines. When in copula, the female worm is held within the male's gynecophoral groove with her dorsal body surface against the ventral surface of the male and with only her anterior end protruding.

The eggs of *S. douthitti* are smaller than those of the human-infecting species. They measure 42–80 μm by 50–58 μm and do not possess a lateral spine.

The molluscan hosts for *S. douthitti* include various subspecies of *Lymnaea stagnalis, Physa gyrina elliptica, Lymnaea palustris, Stagnicola exilis,* and *Stagnicola emarginata angulata*. Attempts to infect various mammals with this schistosome have produced varied results, as shown in Table 10.8. Short and Menzel (1959) have reported that some of the eggs of *Schistosomatium douthitti* can develop parthenogenetically, giving rise to haploid females.

Swimmer's Itch. An interesting phase of schistosome biology concerns the so-called cercarial dermatitis or "swimmer's itch." Cercariae of species that normally infect birds and other vertebrates, especially certain mammals, attempt to penetrate the skin of humans and in so doing sensitize the areas of attack, resulting in itchy rashes (Fig. 10.56). Since humans are not suitable definitive hosts for these cercariae, the flukes do not enter the blood and mature, but rather perish after penetrating the skin. In freshwater lakes of North America, cercariae of the genera *Trichobilharzia, Gigantobilharzia,* and *Bilharziella,* which normally infect birds, are the most common dermatitis-producing flukes. Malek (1980) has contributed the most recent comprehensive review of cercarial dermatitis and has listed the various causative cercariae—both freshwater and marine.

One of the most common causative agents of marine swimmer's itch on both the east and west

Table 10.8. Development of *Schistosomatium douthitti* in Various Hosts

Host	Development in Host
Deer mouse	Normal
Field mouse	Normal
Hamster	Normal
Muskrat	Normal
Albino mouse	Normal
Lynx	Normal
Rat	Abnormal
Rabbit	Abnormal
Cat	Abnormal
Monkey	Immunity develops in 3 weeks

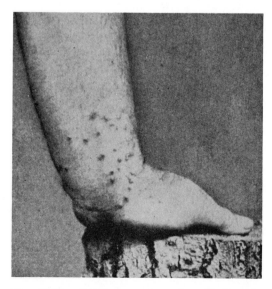

Fig. 10.56. Schistosome dermatitis. Photograph of human wrist showing multiple pustules and some local edema. (Photograph by W. W. Cort.)

coasts of North America is *Microbilharzia variglandis*, a blood parasite of gulls, the cercariae of which develop in the common mudflat snail, *Ilyanassa obsoleta*.

Family Sanguinicolidae. The family Sanguinicolidae includes several monoecious genera primarily parasitic in the blood of fish. These slender trematodes lack suckers and a pharynx, and possess X- or H-shaped intestinal caeca. They possess follicular testes and separate male and female genital pores. Two of the most important genera are *Sanguinicola* and *Cardicola*, found mainly in fish (Fig. 10.57). The members of *Sanguinicola* can be distinguished by several testes arranged in two regular rows, whereas those of *Cardicola* have a single large testis.

Life History of Sanguinicola. Scheuring (1920) reported that eggs of *Sanguinicola inermis* discharged into the blood of fish undergo further development and are eventually located in the capillaries of the host's gills. In this location the eggs hatch and the miracidia escape into the water. The intermediate host is the snail *Lymnaea* sp. The furcocercous cercariae that escape from the gastropod become attached to the gills of carp and actively penetrate into the blood. Some species of *Sanguinicola* do not live in the host's blood vessels but are found in the heart, liver, and other highly vascularized organs.

Cardicola davisi. In North America, a sanguinicolid blood fluke of trout known as *Cardicola davisi* (Fig. 10.57) has caused severe mortality in hatcheries. Adults of this fluke normally reside in the main gill capillaries, lying parallel with the gill cartilages. *C. davisi* differs from most digeneans in that only one egg is produced at a time, and this egg is carried by the blood into the capillaries of the gill filaments, where development of the miracidium is completed and hatching occurs.

The active miracidium works its way through the epithelium to the surface of the gill filament, where a lobule forms. When the lobule ruptures, the miracidium becomes free swimming. After penetrating *Oxytrema* (*Goniobasis*) *circumlineata*, the snail host, the miracidium transforms into a mother sporocyst. Mother sporocysts give rise to daughter sporocysts that in turn give rise to rediae. The furcocercous cercariae are formed within rediae.

These cercariae, like the adults, lack suckers and a pharynx. The caecum of each cercaria terminates in three short bulbous sacs. When a cercaria comes in contact with a fingerling trout, it penetrates through the surface of a fin, actively migrates through veins to the heart and from there to the gill capillaries, where it matures.

Cardicola klamathensis. *C. klamathensis* is closely related to *Cardicola davisi*, but is larger. It is found in the efferent renal veins of trout. The eggs are spherical, and the cercariae bear a longitudinal fin over the dorsal body surface. The gastropod host is *Fluminicola seminalis*.

Family Spirorchidae. The family Spirorchidae includes relatively small distomate and monostomate species found in the blood of reptiles, primarily turtles. The taxonomy of these blood flukes has been reviewed by Stunkard (1923) and Byrd (1939). These trematodes possess an oral sucker but lack a pharynx, and in some species the intestinal caeca are divided, whereas in others there is a single, unbranched caecum. The principal genus is *Spirorchis* (Fig. 10.58) containing several species. Although most commonly found in the blood of turtles, *Spirorchis* can be embedded in the submucosa of the esophagus of turtles. Other representative genera of this family are depicted in Figures 10.57 and 10.58.

Wall (1941) reported the life history of *Spirorchis*. The miracidium is free swimming. There are two generations of sporocysts in the snail host, *Helisoma* spp., and the escaping cercariae are furcocercous. These penetrate the integument of the turtle host.

An interesting aspect of the influence of larval trematodes on their molluscan hosts has been reported by Goodchild and Fried (1963) while studying the small planorbid snail *Menetus dilatatus buchanensis* infected with a species of *Spirorchis*. They reported that infected snails maintained in groups lived longer, began releasing cercariae sooner, and had a greater cercarial output per snail than snails individually maintained. As in many other species of cercariae, the shedding of *Spirorchis* cercariae by *M. dilatatus buchanensis* reflects periodicity, for the cercariae in this case are released at night. Goodchild and Fried were able to reverse the nocturnal shedding by employing a 12-hour alternation of day and night using artificial light. Thus, they have shown that the periodicity is in some way regulated by light.

There are other less well known schistosomelike families, for example, Aporocotylidae, which are parasitic in the blood of fish. The life cycle pattern is similar in all known cases. There is a free-swimming miracidium, mother and daughter sporocysts, a furcocercous free-swimming cercaria, and an adult type of life history pattern.

Superfamily Bucephaloidea

The Bucephaloidea is a relatively small but interesting superfamily. The major family is the Bucephalidae of which *Bucephalus* is representative.

Life Cycle of Bucephalus. The common species *Bucephalus polymorphus* is found in the stomach, intestine, and rarely, in the pyloric caeca of various carnivorous fish (Fig. 10.59). The adult measures 1–6 mm long and possesses the typical ventrally situated

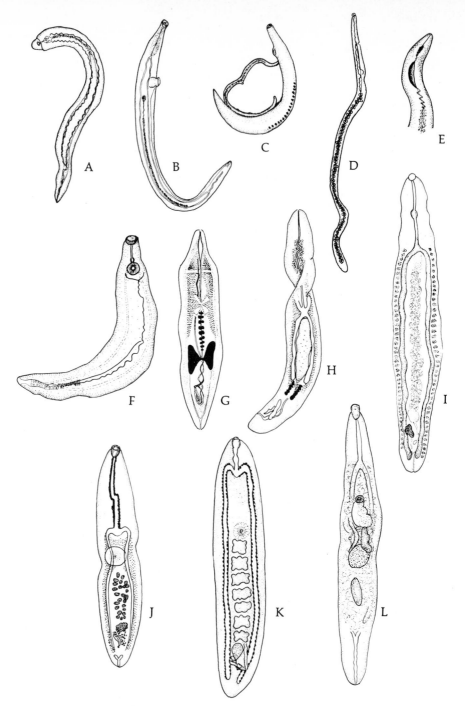

Fig. 10.57. **Some digenetic trematodes parasitic in the circulatory systems of their hosts. A.** Male of *Bivitellobilharzia loxodontae* (Schistosomatidae) in veins of intestinal wall of the elephant, *Loxodonta africana*. (Redrawn after Vogal and Minning, 1940.) **B.** Male of *Ornithobilharzia odhneri* (Schistosomatidae) in portal vein of the bird *Numenius arquatus*. (Redrawn after Faust, 1924.) **C.** Male and female of *Microbilharzia chapini* (Schistosomatidae) in mesentric veins of various birds. (Redrawn after Price, 1929.) **D.** Male of *Trichobilharzia kowalewskii* (Schistosomatidae) in *Anas* spp. (Redrawn after Ejsmont, 1929.) **E.** Anterior end of male *Gigantobilharzia acotylea* (Schistosomatidae) in intestinal veins of gulls and other birds. (Redrawn after Odhner, 1910.) **F.** Male of *Heterobilharzia americana* (Schistosomatidae) in mesentric veins of *Lynx* sp. (Redrawn after Price, 1929.) **G.** *Sanguinicola inermis* (Sanguinicolidae) in heart of the fish *Cyprinus carpio*. (Redrawn after Plehn, 1905.) **H.** *Cardicola cardicola* (Sanguinicolidae) in circulatory system of the marine fish *Calamus bajonado*. (Redrawn after Manter, 1947.) **I.** *Spirorchis innominatus* (Spirorchidae) in blood vessels of various freshwater turtles. (Redrawn after Ward, 1921.) **J.** *Learedius leaaredi* (Spirorchidae) in circulatory system of the marine turtle *Chelone mydas*. (Redrawn after Price, 1934.) **K.** *Plasmiorchis orientalis* (Spirorchidae) in heart of the turtle *Kachuga dhongoka*. (Redrawn after Mehra, 1934.) **L.** *Hapalorhynchus gracile* (Spirorchidae) in artery of the turtle *Chelydra serpentina*. (Redrawn after Stunkard, 1922.)

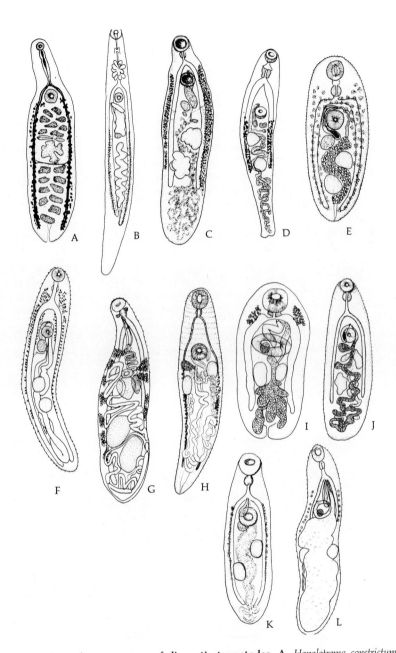

Fig. 10.58. **Representatives of some genera of digenetic trematodes. A.** *Hapalotrema constrictum* (Spirorchidae) in blood vessels and heart of marine turtles. (Redrawn after Looss, 1899.) **B.** *Vasotrema amydae* (Spirorchidae) in circulatory system of freshwater turtles of the genus *Amyda.* (Redrawn after Stunkard, 1926.) **C.** *Astiotrema reniferum* (Plagiorchiidae) in digestive tract of the turtle *Trionyx nilotica;* other species occur in fish. (Redrawn after Loss, 1899.) **D.** *Glossidium pedatum* (Plagiorchiidae) in intestine of freshwater fish of the genus *Bagrus.* (Redrawn after Looss, 1899.) **E.** *Plagiorchis muris* (Plagiorchiidae) in intestine of rats and dogs. (Redrawn after Tanabe, 1922.) **F.** *Haplometra cylindracea* (Plagiorchiidae) in lungs of frogs. (Redrawn after Zeder, 1800.) **G.** *Haematoloechus buttensis* (Plagiorchiidae) in lungs of frogs. (Redrawn after Ingles, 1936.) **H.** *Opisthogonimus philodryadum* (Plagiorchiidae) in digestive tract of snakes. (Redrawn after Lühe, 1900.) **I.** *Stomatrema pusillum* (Plagiorchiidae) in mouth of the water snake *Farancia.* (Redrawn after Guberlet, 1928.) **J.** *Styphlodora serrata* (Plagiorchiidae) in intestine of *Varanus.* (Redrawn after Looss, 1899.) **K.** *Dasymetra conferta* (Plagiorchiidae) in mouth and gastrointestinal tract of snakes. (Redrawn after Nicoll, 1911.) **L.** *Lechriorchis primus* (Plagiorchiidae) in lungs of water snakes. (Redrawn after Stafford, 1905.)

mouth. Eggs laid by the adults are passed out into the water in the host's feces. A unique miracidium with stalked cilia escapes from each egg and is free swimming (Fig. 10.18). When the miracidium comes in contact with freshwater clams of the genera *Unio* and *Anodonta*, it penetrates the first intermediate host and gives rise to highly branched sporocysts that are located in the bivalve's digestive gland (Fig. 10.59).

The sporocysts give rise directly to furcocercous cercariae that escape from the clam and are free swimming (Fig. 10.59). The long forks of the cercarial tail become entangled in the fins of small fish, and thus the cercariae become attached to the second intermediate host. After penetrating, they become encysted under the host's scales and on the fins. Metacercariae, however, are known to encyst in the mouths of some fish. When the small fish is ingested by a larger, carnivorous fish, the metacercariae excyst once they reach the alimentary tract and mature into adults. In certain related species, snails rather than clams serve as the first intermediate host.

Other Bucephalid Species. Probably the best-known bucephalid among marine biologists is *Bucephalus cuculus.* The multibranched sporocyst of this

trematode was first discovered in the American oyster, *Crassostrea virginica*, in Charleston (South Carolina) Harbor. Although the complete life cycle of this parasite remains uncertain, there is little doubt that the adult is an intestinal parasite of a marine or estuarine fish. The eggs passed out in feces probably hatched in water, and the miracidia are swept into oysters by the in-current created by their pumping. Within the molluscan host, the sporocyst, like these of other bucephalids, originates as a subspherical mass that gradually produces rootlike branches. In the case of *B. cuculus* sporocysts in the oyster, the dendritic sporocysts are initially confined in the host's gonad and eventually the branches invade the digestive gland (Fig. 10.20). Invasion of the oyster's gonad results in parasitic castration (p. 14). In addition, the amount of stored glycogen in the hosts is severely reduced, and hence the oysters are watery and lean (Cheng, 1967). On the other hand, the sporocysts of *Bucephalus* are rich in glycogen, and this has led to the misinterpretation that infected oysters are rich in glycogen.

In addition to the numerous species of *Bucephalus*, all of which bear hornlike projections on their cephalic region, the family Bucephalidae also includes *Rhipidocotyle* (Fig. 10.60), *Bucephalopsis* (Fig. 10.60), *Prosorhynchus* (Fig. 10.60), and other genera, all found in various fish. The life cycles of the various bucephalids all follow the same basic pattern as that of *Bucephalus polymorphus.*

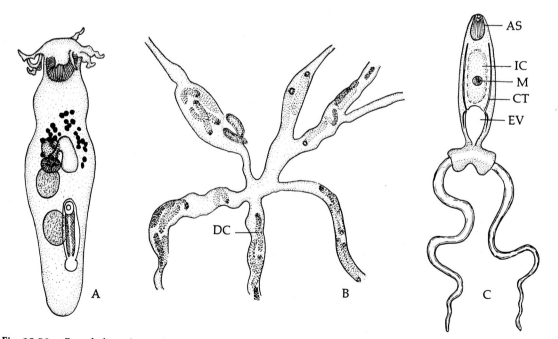

Fig. 10.59. *Bucephalus polymorphus.* **A.** Adult from intestine of various freshwater fishes showing characteristic tentacles extending from head. (Redrawn after Nagaty, 1937.) **B.** Branched sporocyst of *B. polymorphus.* (Redrawn after Palombi, 1934.) **C.** "Oxhead" cercaria showing midventrally situated mouth. (Redrawn after Palombi, 1934.) AS, anterior sucker; CT, collecting tubule; DC, developing cercaria; EV, excretory vesicle; IC, intestinal caecum; M, mouth.

The members of this superfamily are characterized by cercariae and adults that often bear a circumoral collar armed with spines. All are parasites of reptiles, birds, and mammals. The majority of echinostomatoid trematodes belong to the family Echinostomatidae.

Family Echinostomatidae. The family Echinostomatidae includes the genus *Echinostoma*, with over 80 species. *Echinostoma revolutum* is an intestinal parasite of birds, dogs, cats, muskrats, and humans (Fig. 10.61). The biology of this species has been reviewed by Beaver (1937). Echinostomiasis in humans is not very serious, often causing nothing more than diarrhea.

Representatives of other larger genera of the Echinostomatidae are depicted in Figure 10.61.

Life Cycle of Echinostoma revolutum. The life cycle of *E. revolutum* typifies that of most echinostomes. The eggs, which pass out of the host in feces, contain uncleaved zygotes. A miracidium is fully developed in each egg after it has been in water for about two to five weeks. The miracidium hatches from the egg while in water and penetrates snails of the genera *Lymnaea*, *Physa*, and *Bithynia*. Various freshwater pelecypods and gastropods serve as hosts for other echinostomes. A single sporocyst generation and two redial generations occur in the molluscan host. The free-swimming cercariae penetrate and encyst in other molluscs (other members of the family encyst in snails, clams, planaria, tadpoles, and fish), and when snails infected with encysted metacercariae are swallowed by the definitive host, the life cycle is completed.

Almost all echinostome cercariae bear a coronet of spines around the oral sucker. A ringlike collar is usually present, and the horns of the excretory vesicle extend to near the anterior end of the body and include conspicuous globules.

Family Fasciolidae. Another major family of the Echinostomatoidea is the Fasciolidae, the members of which are all large flat distomes parasitic in mammals. The intestinal caeca may be simple or have dendritic lateral branches, as in the common liver fluke *Fasciola hepatica* (Fig. 10.62). The genus *Fasciola* includes *F. hepatica* and *F. gigantica*, both found primarily in the bile ducts of herbivorous mammals.

Human infections with *F. hepatica* are cosmopolitan in distribution and are of increasing importance in Cuba and other Latin American countries, as well as in France and Algeria. Ecologically, sheep- and cattle-raising areas are the primary zones in which human

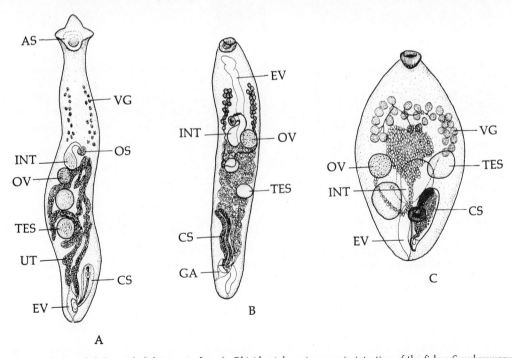

Fig. 10.60. Some adult bucephalid trematodes. A. *Rhipidocotyle pentagonum* in intestine of the fishes *Scomberomorus* and *Thynnus*. (Redrawn after Ozaki, 1934.) **B.** *Bucephalopsis ovata* in intestine of fishes of the genus *Lobotes*. (Redrawn after Yamaguti, 1940.) **C.** *Prosorhynchus magniovatum* in intestine of the marine fish *Conger myriaster*. (Redrawn after Yamaguti, 1938.) AS, anterior sucker; CS, cirrus sac; EV, execretory vesicle; GA, genital atrium; INT, intestine; OS, oral sucker; OV, ovary; TES, testis; UT, uterus; VG, vitelline glands.

infections are found. Fascioliasis is characterized by the destruction of the host's liver tissues, damage to bile ducts, atrophy of the portal vessels, and secondary pathologic conditions. In many cases the disease leads to the death of the host.

Life Cycle of Fasciola hepatica.   *F. hepatica* holds a prominent place in parasitology because its life cycle was the first digenetic trematode cycle to be completely worked out (Thomas, 1883, Leuckart, 1882) and its discovery has stimulated all subsequent life cycle investigations.

The typical eggs are large, 130–150 μm by 63–90 μm, and operculate (Fig. 10.62). These eggs are laid before complete development of the enclosed miracidium and pass into the alimentary tract via the bile ducts. They eventually pass out of the host in feces.

After being in water for 4 to 15 days at 22°C, the miracidium within the shell is completely developed and escapes. This ciliated larva bears eyespots. Within the first 8 hours after hatching, the miracidium must seek out and penetrate a snail (*Lymnaea, Succinea, Fossaria,* or *Practicolella*). During this process, it sheds its ciliated epidermis, and its apical papilla becomes invaginated to form a cuplike structure that is held against the body surface of the snail host. Lytic enzymes secreted into the cavity of the cup facilitate penetration. For a review of the relationship between *F. hepatica* and the molluscan host, see Kendall (1965).

Upon entering the snail, each miracidium metamorphoses into a sporocyst, which in turn gives rise to mother rediae. The mother rediae give rise to daughter rediae (Fig. 10.62). The germ balls in the brood chambers of daughter rediae develop into cercariae, which escape from the snail and become free swimming (Fig. 10.62). When the cercariae reach aquatic vegetation, they lose their tails and encyst as metacercariae (Fig. 10.62). Some cercariae can encyst in water.

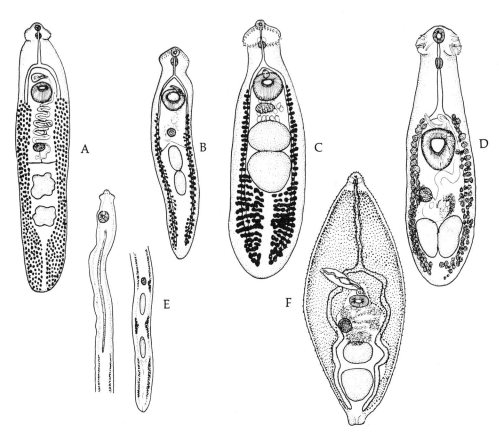

Fig. 10.61.   Some adult echinostomatid trematodes. A. *Echinostoma revolutum* in caeca and rectum of various birds. (Redrawn after Mönnig, 1934.) **B.** *Echinoparyphium elegans* in intestine of the bird *Anas rubripes* and in *Phoenicopterus roseus.* (Redrawn after Looss, 1899.) **C.** *Echinochasmus coaxatus* in intestine of birds of the genus *Colymbus.* (Redrawn after Dietz, 1910.) **D.** *Petasiger skrjabini* in intestine of the bird *Querquedula crecca.* (Redrawn after Bashikirova, 1941.) **E.** *Himasthla rhigedana* in intestine of fishes of the genus *Numenius.* (Redrawn after Dietz, 1909.) **F.** *Pegosomum skrjabini* in the birds *Ardea purpurea* and *Egretta alba.* (Redrawn after Shakhtakhtinskaia, 1949.) Note presence of collar of spines on all species.

Metacercariae swallowed by the definitive host excyst once they reach the duodenum. They then penetrate the intestinal wall and are found in the coelomic cavity. From the body cavity, they penetrate the liver, migrate through the parenchyma, and become established in the bile ducts, where they mature. During migration through the liver, the young adults actively feed on the host's liver cells.

The rate and extent of development of *F. hepatica* within the snail host depend on the degree of infection and on the availability of nutrients within the snail, primarily those stored in the digestive gland.

Fascioliasis in cattle and sheep can be a serious economic problem. For example, Olsen, as the result of an 11-year survey in the United States, has reported that the livers of 1,400,000 cattle and 60,500 calves have been condemned as unfit for human consumption due to fascioliasis. Approximately 1 million pounds of liver were involved. This loss, when transcribed into monetary terms, represented an annual loss of nearly $3 million.

In addition to cattle and sheep, *F. hepatica* has been reported from horses, goats, rabbits, pigs, dogs, and squirrels. Human cases of fascioliasis are known. It is especially common in France, where about 500 persons were infected during 1956–1957. I have seen human fascioliasis in Hawaii which resulted from eating raw or poorly cooked watercress, with attached metacercariae, that had been grown in water downstream from pastures.

It is of interest to note that if raw bovine liver harboring *Fasciola* is eaten by humans, young flukes may become attached to the buccal or pharyngeal

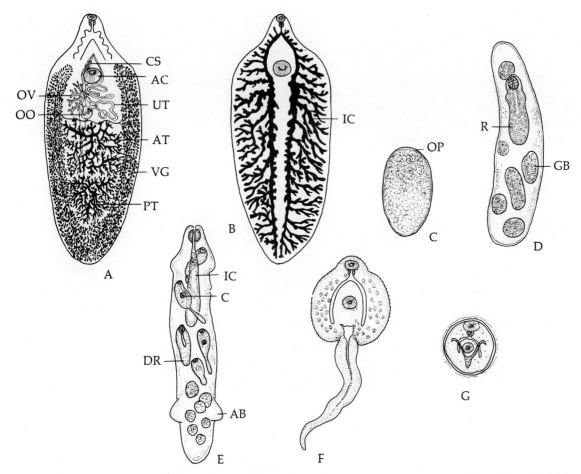

Fig. 10.62.   Stages in the life cycle of *Fasciola hepatica*. A. Adult specimen showing reproductive systems. **B.** Adult specimen showing multibranched intestinal caeca. **C.** Egg showing operculum. **D.** Sporocyst. **E.** Daughter redia. **F.** Cercaria. **G.** Metacercaria encysted on vegetation. AB, ambulatory bud; AC, acetabulum; AT, anterior testis; C, cercaria; CS, cirrus sac; DR, daughter redia; GB, germ ball; IC, intestinal caecum; OO, ootype; OP, operculum; OV, ovary; PT, posterior testis; R, redia; UT, uterus; VG, vitelline glands.

membranes, causing pain, irritation, hoarseness, and coughing. This condition is known as **halzoun**.

Other Fasciolid Trematodes. Other species of *Fasciola* include *F. halli* in cattle and sheep in Louisiana and Texas, and *F. indica* in Indian cattle. In Hawaii, *F. gigantica* is the predominant species that parasitizes cattle. In some areas the incidence of infection may be as high as 80%. The presence of this parasite in Hawaii is not only of concern to cattlemen but is also a public health problem since, like *F. hepatica*, it can infect humans as a result of eating contaminated watercress.

The genus *Fascioloides* includes *F. magna* parasitic in cattle and sheep (Fig. 10.63). Its life history parallels that of *Fasciola hepatica* except that the intermediate snail host can be *Galba*, *Pseudosuccinea*, or *Fossaria*.

In the case of *F. magna* infection in North America, there is reason to believe that domestic cattle have acquired their parasites from deer, exemplifying the principle that domestic animals introduced into a new geographic area often acquire the parasites of wild animals endemic to the region. For a detailed discussion of this phenomenon, see Dogiel (1964).

Other fasciolid trematodes include *Fasciolopsis buski* (Fig. 10.63), which occurs in the intestine of humans and other mammals, especially pigs, and which utilizes the snails *Planorbis coenosus* or *Segmentina largillierti* as intermediate hosts; *Fasciolopsis fuelleborni*, found in humans in India and Egypt; and *Parafasciolopsis fascio-laemorpha* (Fig. 10.63), found in the bile ducts of ungulates in eastern Europe.

Rates of human infections by *Fasciolopsis buski* in Thailand, north of Bangkok, have been known to be as high as 40%. Man becomes infected from eating uncooked nuts of such aquatic plants as *Ipomoea*, *Eichhornia*, and *Pistia* with metacercariae attached.

Superfamily Clinostomatoidea

All members of this superfamily belong to the family Clinostomatidae. All the species are comparatively large and have the anterior sucker and acetabulum closely associated. These flukes also lack a muscular pharynx. Their excretory system is unique in that in addition to the protonephritic system, a secondary network of ramified lacunae is present.

The most common genus of clinostomatoid trematodes is *Clinostomum* (Fig. 10.64) parasitic in the mouth, pharynx, and esophagus of herons, herring gulls, and other fish-eating birds.

Superfamily Brachylaemoidea

The distinguishing characteristics of the Brachylaemoidea are presented in the classification list at the end of this chapter. This superfamily includes, among other smaller families, Brachylaemidae and Leucochloridiidae.

Family Brachylaemidae. Ulmer (1951a, b) has reviewed the systematics of the Brachylaemidae* and

---

*Actually Ulmer (1951a, b) reviewed the taxonomy of the subfamily Brachylaiminae, which has since been elevated to familial rank.

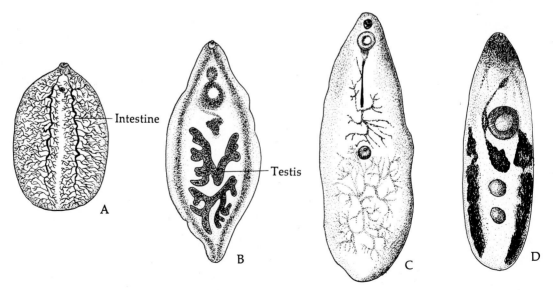

Fig. 10.63. **Some adult fasciolid trematodes. A.** *Fascioloides magna* in liver and lungs of cattle. (Redrawn after Ward and Whipple, 1918.) **B.** *Parafasciolopsis fasciolaemorpha* in bile ducts of the ungulate *Alces alces*. (Redrawn after Ejsmont, 1932.) **C.** *Fasciolopsis buski* in human intestine. (Redrawn after Odhner, 1902.)

Intestine — (label in figure A)

Testis — (label in figure B)

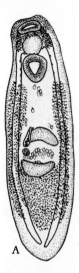

Fig. 10.64. *Clinostomum complanatum.* Adult specimen from buccal cavity and esophagus of various fish-eating birds. (Redrawn after Braun, 1900.)

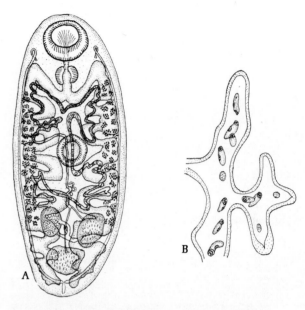

Fig. 10.65. **A brachylaemid trematode. A.** *Postharmostomum helicis* adult from intestine of wild mice, *Peromyscus* spp. **B.** Portion of branched sporocyst of *P. helicis.* (Redrawn after Ulmer, 1951.)

contributed the life history of *Postharmostomum helicis*, a parasite of the wild mice *Peromyscus maniculatus* and *P. leucopus.*

Life Cycle of Postharmostomum helicis. Adults of *P. helicis* (Fig. 10.65) occur in the caecum of their rodent hosts. The eggs include well-developed embryos when they pass out in the host's feces. When such eggs are ingested by the terrestrial gastropod *Anguispira alternata*, miracidia hatch from the eggs and migrate through the intestinal wall into the surrounding connective tissue or into the digestive gland, where they develop into large multibranched mother sporocysts (Fig. 10.65).

The mother sporocysts give rise to daughter sporocysts in 7 to 10 days. The daughter sporocysts are localized within the snail's digestive gland, where they increase in size and become branched, thus being very similar to the mother sporocyst. Short-tailed cercariae are produced within the daughter sporocysts in about 12 weeks during the summer. These cercariae escape into the host's mantle cavity through birth pores situated at the tips of the sporocyst branches. The production of cercariae continues for a year or longer, and possibly for the life span of the snail. The cercariae in the mantle cavity escape to the exterior via the host's respiratory pore and become incorporated in the mollusc's slime trail.

Various species of gastropods (*Polygyra, Anguispira, Stenostoma, Deroceras*), crawling over slime that includes cercariae, become infected when the cercariae enter the pore of their primary ureter. The larval trematodes then migrate through the ureters into the kidney and through the renal canal. They then become lodged in the pericardial chamber. The small tail of each cercaria is lost in approximately ten days, but the

metacercariae remain unencysted. Mice and chipmunks become infected when they ingest gastropods harboring metacercariae.

It is of interest to note that gastropods infected with sporocysts of *P. helicis* do not become infected with cercariae. The reason for this remains undetermined.

Once ingested by the mammalian host, the metacercariae reach the caecum and commence feeding on blood. Sexual maturity is attained in 8 to 20 days. The metacercariae of *P. helicis* can become progenetic—that is, produce eggs before attaining the adult form.

Family Leucochloridiidae. Included in this family is the interesting species *Leucochloridium macrostoma* (Fig. 10.66).

Life Cycle of Leucochloridium macrostoma. *L. macrostoma*, a primarily European species found in the caeca and bursa Fabricii of birds, and occasionally in mammals, is an extremely interesting organism, for certain features of its life cycle suggest that it is well adapted and probably a parasite of long standing. Eggs of *L. macrostoma*, passed out in the feces of the avian host, are dropped onto vegetation, and some are ingested by terrestrial and amphibious snails and slugs. *Succinea* is the major molluscan host, although *Planorbis* and *Helix* are also compatible hosts.

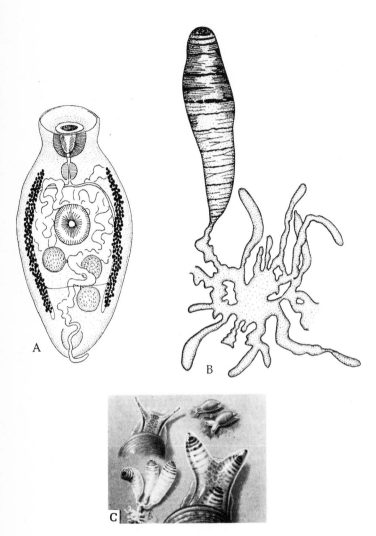

Fig. 10.66. ***Leucochloridium macrostoma.*** **A.** Adult trematode from the caeca and bursa of Fabricii of birds. (Redrawn after Lühe, 1909.) **B.** Branched and pigmented sporocyst teased from tentacle of snail host. (Redrawn after Wesenberg-Lund, 1931.) **C.** *Succinea* snails uninfected and infected with sporocysts of *L. macrostoma;* with latter showing swollen and transparent tentacles enclosing branches. (After Wickler, in Cheng, 1970.)

Miracidia hatching from ingested eggs burrow through the intestinal wall into tissues and transform into mother sporocysts that in turn give rise to daughter sporocysts. As the daughter sporocysts grow, they become highly branched (Fig. 10.66) and migrate toward the host's tentacles. Branches eventually extend into the tentacular cavity.

As a result of the comparatively large diameters of the sporocysts, the host's tentacles become greatly distended and their surfaces become transparent (Fig. 10.66). Because the sporocysts of *Leucochloridium* bear bright green, brown, or orange rings and pulsate continuously, infected snails appear to possess colorful, pulsating, sausage-shaped tentacles, which are noticeable at a distance of several feet. It is believed that the sporocyst-enclosing tentacles attract birds, which peck at them and thus ingest large numbers of encysted tailless cercariae that are located within the daughter sporocysts. Once ingested by a bird, the cercariae escape and migrate to the avian host's cloaca or bursa and mature.

In addition to the method of infection of birds just described, encysted cercariae are also readily released from infected snails, since the swollen snail skin ruptures at the slightest touch. Such cercariae may be individually deposited on vegetation on which infected snails feed, or entire cercariae-enclosing sporocysts may be released and deposited. These sporocysts continue to pulsate and thus presumably attract birds. Young birds are more susceptible to *Leucochloridium* infection than adults. In nature, sporocysts picked up by parent birds are undoubtedly often fed to nestlings.

The unique appearance and behavior pattern of the intramolluscan larvae of *Leucochloridium* suggest that the relationship between this trematode and its molluscan host has resulted from evolutionary changes over a long period. These evolutionary adaptations on the part of the parasite, without doubt, are beneficial to the maintenance of the trematode species, for they enhance the parasite's chances of reaching a definitive host, and thus completing its life cycle, developing to maturity, and producing eggs.

Superfamily Paramphistomoidea

These flukes are characterized by fairly large fleshy bodies that possess a posteriorly situated acetabulum. The best known species include *Watsonius watsoni* and *Gastrodiscoides hominis*, which infect humans (Fig. 10.67), and *Diplodiscus subclavatus* (Fig. 10.67), which live in the rectum of frogs, toads, and newts.

The paramphistomes possess a primitive type of lymphatic system embedded in the parenchyma (Fig. 10.68). Enclosed within the lymphatic canals are primitive blood cells.

Included in the Paramphistomoidea is the family Paramphistomidae. One of the most frequently encountered trematodes belonging to this family in America is *Megalodiscus temperatus*, which occurs in the rectum of frogs. This fluke measures up to 6 mm in length and approximately 2–2.25 mm in thickness.

Life Cycle of Megalodiscus temperatus. Eggs laid by adult flukes in the rectums of frogs and tadpoles pass into water within the hosts' feces. These eggs enclose fully developed miracidia that hatch almost immediately. If young snails (*Helisoma trivolvis, H. campanulatum,* or *H. antrosum*) are present, they are penetrated by miracidia, which then transform into sporocysts.

Three generations of rediae occur in the snail's digestive gland. The last redial generation gives rise to ophthalmocercariae that are positively phototactic (infected snails emit more cercariae during the afternoon and on bright days). If frogs or tadpoles are close by, the cercariae quickly become attached to their skin and encyst. The amphibians become infected when sloughed skin, including encysted metacercariae, is ingested. Excystation occurs in the host's rectum.

When cercariae become attached to the skin of adult frogs, they adhere tenaciously and appear to prefer the pigmented portions of the host's fore- and hindlimbs. Why these positively phototactic cercariae prefer pigmented areas of the host is not understood, but it probably is due to chemotactic response. When cercariae become attached to tadpoles, they adhere lightly and can be readily knocked off. During the metamorphosis of infected tadpoles to adults, adult trematodes that are not expelled from the rectum migrate anteriorly and are commonly found in the host's stomach. On completion of metamorphosis of the host, these adult trematodes once again migrate into the rectum. The physiologic basis for this migratory pattern has not been determined.

Maturation of *M. temperatus* individuals in the frog's rectum is influenced by the number present. Although individuals may mature within 27 days, the normal maturation period is 2 to 3 months. When a large number of flukes are present, maturation is delayed and may take as long as 3 to 4 months.

EPITHELIOCYSTIDA

The suborder Epitheliocystida includes at least five superfamilies (see classification list at the end of this chapter). It bears repeating that this group of digeneans is primarily characterized by the nature of the cells lining the excretory bladder. Specifically, the wall of the embryonic bladder is replaced by epithelial cells of mesodermal origin. The biology of several representative species is considered below.

Superfamily Plagiorchioidea

The Plagiorchioidea is a large superfamily the members of which are parasites of fish, amphibians, reptiles, birds, and mammals. It includes a large number of families among which the more prominent are Plagiorchiidae, Dicrocoeliidae, and Lecithodendriidae. These small-to-medium worms are generally intestinal parasites, but certain species are found in the gallbladder, bile duct, lungs, and pancreatic ducts of their vertebrate hosts. There are life history variations among the different families, a few representatives of which are presented below.

Family Plagiorchiidae. The family Plagiorchiidae includes numerous species parasitic in fish, amphibians, reptiles, birds, and mammals. Those found in fish in-

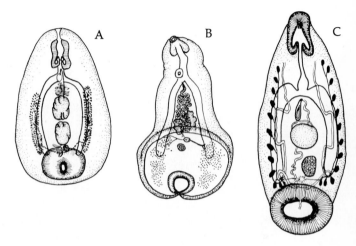

Fig. 10.67. Some paramphistomatoid trematodes. A. *Watsonius watsoni* in small intestine of humans and monkeys. (Redrawn after Stiles and Goldberger, 1910.) **B.** *Gastrodiscoides hominis* in large intestine of humans and other mammals. (Redrawn after Badanin, 1929.) **C.** *Diplodiscus subclavatus* in rectum of amphibians. (Redrawn after Lühe, 1909.)

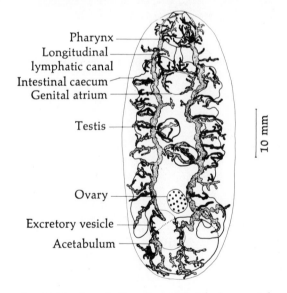

Fig. 10.68. Lymphatic system of *Cotylophoron cotylophorum*; dorsal view. The stippled areas present the longitudinal canals and their branches while the black areas are the branches ventral to organs. (Courtesy of Lin Heng Wan, University of Malaysia.)

clude species of *Astiotrema* (Fig. 10.58) and *Glossidium* (Fig. 10.58). The species of *Glossidium* are limited to freshwater fish. Those found in amphibians include species of *Plagiorchis* (Fig. 10.58) in frogs, toads, and salamanders; *Haplometra* (Fig. 10.58) in the lungs of frogs; and *Haematoloechus* (Fig. 10.58), a large genus with over 40 species, all found in the lungs of various frogs and toads.

Plagiorchiid genera that parasitize reptiles include *Plagiorchis*; *Astiotrema*; *Opisthogonimus* (Fig. 10.58) in the intestine of snakes; *Stomatrema* (Fig. 10.58) in the mouth and esophagus of snakes; *Styphlodora* (Fig. 10.58) with various species in the gallbladders, kidneys, ureters, and intestines of snakes; *Dasymetra* (Fig. 10.58) in the intestine of water snakes; *Ochetosoma* (Fig. 10.69) with numerous species parasitic in the mouth, esophagus, or lungs of snakes; and *Zeugorchis* (Fig. 10.69) in the alimentary canal of various reptiles.

Life Cycle of Ochetosoma aniarum. The life cycles of the reptile-infecting species of Plagiorchiidae are mostly of the same general pattern. Byrd (1935) has reported the life cycle of *Ochetosoma aniarum*, found in the mouth and esophagus of snakes, *Natrix* spp. The eggs laid by the monoecious adults contain fully formed miracidia. These do not hatch until the eggs are ingested by the snail *Physa helei*. It is suspected that the digestive enzymes and the physical parameters (pH, pO_2, pCO_2, etc.) in the gastropods digestive tract play important roles in the hatching process,

activating the miracidia, and perhaps in some way effecting the opening of the operculum. There are two generations of sporocysts in the snail host. The daughter sporocysts give rise to thin and long-tailed xiphidiocercariae, which encyst in tadpoles. When infected tadpoles are ingested by snakes, the cycle is completed.

Life Cycle of Plagiorchis. The second developmental pattern found among the plagiorchids is typified by *Plagiorchis ramianus*, an intestinal parasite of snakes. After the miracidium-enclosing egg is ingested by the snail *Bulinus contortus*, it hatches and the miracidium eventually transforms into a redia instead of a sporocyst. The escaping xiphidiocercariae encyst in dragonfly larvae and hemipteran larvae, and when these are ingested by the definitive host, the cycle is completed.

Plagiorchis *Species.* Numerous species of *Plagiorchis* have been reported from the intestine of birds; some, such as *P. arcuatus* and *P. petrovi*, are found in chickens.

Plagiorchids parasitic in mammals include *Plagiorchis javensis* and *P. philippinensis* in humans, and *P. muris* (Fig. 10.58) in rats, dogs, and also humans and birds.

Life Cycle of Plagiorchis muris. The life history of *Plagiorchis muris* is unusual for the xiphidiocercariae, which are found in sporocysts, encyst in the same sporocysts. The molluscan host is *Lymnaea pervia*. When the snail is ingested by the definitive host, the cycle is completed. In this instance, the snail is the only intermediate host. Yamaguti (1943) did, however, demonstrate that if cercariae of *P. muris* are experimentally removed from the sporocysts, they will encyst in an experimental second intermediate host. This observation strongly suggests that *P. muris* orig-

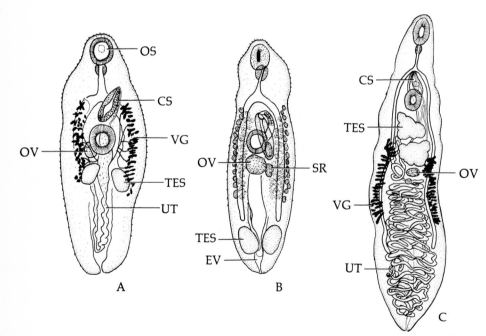

Fig. 10.69. Some adult digenetic trematodes. A. *Ochetosoma ellipticum* (Plagiorchiidae) in mouth, esophagus, and lungs of snakes. (Redrawn after Pratt, 1903.) **B.** *Zeugorchis aequatus* (Plagiorchiidae) in alimentary canal of the snake *Thamnophis sirtalis*. (Redrawn after Stafford, 1905.) **C.** *Dicrocoelium dendriticum* (= *D. lanceatum*) (Dicrocoeliidae) in gallbladder and bile ducts of cattle, sheep, deer, rabbits, humans, and other mammals. (Redrawn after Dawes, 1946.) CS, cirrus sac; EV, excretory vesicle; OS, oral sucker; OV, ovary; SR, seminal receptacle; TES, testis; UT, uterus; VG, vitelline glands.

inally required two intermediate hosts but has recently discarded its requirement for the second host, indicating that hosts can be eliminated during evolution (p. 27).

Family Dicrocoeliidae. The family Dicrocoeliidae includes many genera parasitic in birds, reptiles, and mammals. Undoubtedly the best known of the members is the lancet liver fluke, *Dicrocoelium dendriticum* (Fig. 10.69). This economically important trematode, rarely found in humans, normally lives in the bile ducts and gallbladder of sheep, cattle, deer, woodchucks, rabbits, and other mammals in upper New York State, northern and central Europe, and occasionally elsewhere. *D. dendriticum* possesses an unusual life cycle in that it does not require an aquatic environment at any time. The eggs that pass out from the definitive host are ingested by the land snail *Cionella lubrica*. Although *C. lubrica* is the major molluscan host in the United States and elsewhere, other gastropods can also serve as hosts.

The mother and daughter sporocysts develop in the snail and the xiphidiocercariae are passed out in "slime balls" formed by the snail. These are ingested by *Formica fusca* and other ants. The ant's behavior is modified by the metacercariae encysted near its brain. Infected ants climb up blades of grass and lock on with their mouthparts. When infected ants are eaten along with grass by sheep, the adult parasites develop (Krull and Mapes, 1952; Mapes, 1952).

Family Lecithodendriidae. This family includes numerous species parasitic in the intestine of all classes of vertebrates. Although these parasites are of little economic importance, their biology is extremely interesting. Many species are parasites of chiropterans. The developmental pattern of *Prosthodendrium anaplocami* involves two intermediate hosts—a snail and a mayfly nymph (Etges, 1960).

Superfamily Opisthorchioidea

This superfamily includes several families of digeneans parasitic in fish, amphibians, reptiles, birds, and mammals. Among these, the Opisthorchiidae and Heterophyiidae are being discussed briefly because of their medical importance.

Family Opisthorchiidae. The members of this family are delicate, leaf-shaped flukes with relatively weakly developed suckers. They are parasitic in the biliary system of reptiles, birds, and mammals. At least two species belonging to this family are of medical importance; *Opisthorchis tenuicollis* and *Clonorchis sinensis*. *O. tenuicollis*, also known as *O. felineus* (Fig. 10.70), is parasitic in the bile ducts of fish-eating mammals, including cats, dogs, and humans. It is most common is southern, central, and eastern Europe, Turkey, the southern Soviet Union, Vietnam, India, and Japan. It is present in Puerto Rico and possibly other Caribbean islands. An estimated 1 million or more humans are infected by this parasite and suffer

diarrhea and thickening and eventual erosion of the bile duct wall.

Life Cycle of Opisthorchis tenuicollis. During development of *O. tenuicollis*, the eggs pass out in the host's feces and include fully developed miracidia that do not hatch until they are swallowed by snails of the genus *Bithynia*. In the gastropod, the miracidium hatching from each egg burrows through the snail's intestinal wall and transforms into a sporocyst, which gives rise to one redial generation. The rediae possess neither a birth pore nor ambulatory buds. The pleurolophocercous cercariae (Fig. 10.70) penetrate the skin of cyprinid fish and encyst under the body surface. When infected fish are eaten by the definitive host, the metacercariae develop into adults that live in the bile ducts.

Clonorchis sinensis. C. sinensis, commonly referred to as the Chinese liver fluke, is widely distributed in Korea, Japan, China, Taiwan, and Vietnam, and has been estimated to infect at least 19 million persons. Thickening by scar tissue erodes the epithelial lining of the bile ducts of these patients and eventually occludes them. Furthermore, outpockets of the bile duct wall may form, harboring tightly packed worms. Complete perforation of the bile duct wall may occur. Also, eggs deposited in the liver become encapsulated by connective tissue fibers and blood cells, thereby interfering with liver function. Eggs in the gallbladder often become centers of gallstone formation. In fatal cases of clonorchiasis,* ascites almost always occurs.

Life Cycle of Clonorchis sinensis. The adult of *C. sinensis* is monoecious (Fig. 10.70). It averages 18×4 mm in size and the characteristic light bulb-shaped egg measures 27×16 μm (Fig. 10.11).

The eggs passed out of the mammalian host in feces are readily killed by dessication, but can withstand 0°C for 6 months if kept moist. Such eggs must be ingested by the proper snail host, *Parafossarulus*, *Bulimus*, *Semisulcospira*, *Alocinma*, or *Melanoides*, before they will hatch. The miracidium burrows through the wall of the snail's intestine or rectum and transforms into a sporocyst, which gives rise to rediae that in turn produce pleurolophocercous cercariae. The free-swimming cercariae must penetrate a suitable fish host and encyst as metacercariae. At least 80 species of fish belonging to 10 families, but primarily the Cyprinidae, are suitable second intermediate hosts. Humans and other mammals, as stated, become in-

---

*Technically, the disease should be designated as clonorchiosis; however, as a result of common usage, it is referred to as clonorchiasis.

fected by eating raw or poorly cooked fish. Komiya (1966) has presented a review of *Clonorchis* and clonorchiasis, and Kim and Kuntz (1964) have discussed the epidemiology of this disease in Taiwan.

Reservoir hosts are involved in the transmission of clonorchiasis in endemic areas, including cats, dogs, tigers, foxes, badgers, and mink.

With the increased amount of freshwater fish farming in endemic areas, clonorchiasis has become an increasingly serious problem because oriental aquaculture ponds are commonly fertilized with human manure to enhance the growth of vegetation, which,

in turn, serves as food for fish. In Hong Kong, where fish farming (in the New Territories) is very common, the incidence of human clonorchiasis is about 14%, whereas in more rural endemic areas, the incidence may be as high as 80%.

Family Heterophyidae. The family Heterophyidae includes small distomes and monostomes that live in the intestines of birds and mammals. The genital pore of heterophyids opens into a retractile, suckerlike structure, the **gonotyl**, which is either incorporated in the acetabulum or lies to one side of it. This family includes the genera *Heterophyes*, *Metagonimus*, *Centrocestus*, *Haplorchis*, and *Cryptocotyle*, which all include human-infecting species.

Heterophyes heterophyes is an intestinal parasite of dogs, cats, humans, and other mammals and is commonly found in Asia and Egypt (Fig. 10.71). It has

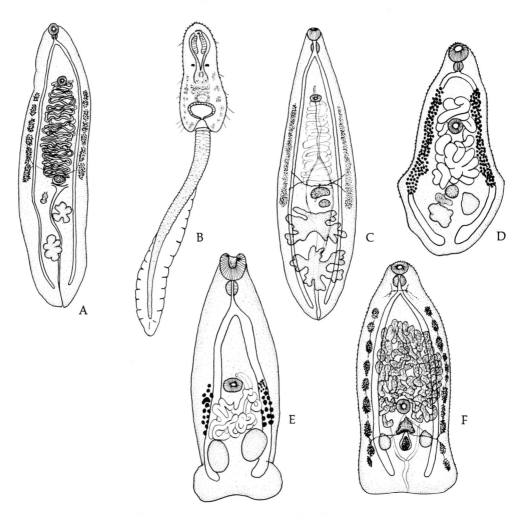

Fig. 10.70. **Some opisthorchiid trematodes. A.** *Opisthorchis tenuicollis* in bile ducts of cats, dogs, and humans. **B.** Cercaria of *O. tenuicollis*. **C.** *Clonorchis sinensis* in bile ducts of humans and other mammals. **D.** *Metorchis albidus* in gallbladder of dogs, wolves, and cats. (Redrawn after Price, 1932.) **E.** *Pseudamphistomum* sp. in bile ducts and gallbladder of mammals, including humans. (Redrawn after Mönnig, 1934.) **F.** *Microtrema truncatum* in digestive tract of cats, dogs, foxes, wolves, and humans. (Redrawn after Looss, 1896.)

been introduced into Hawaii. Its life cycle parallels that of *Opisthorchis tenuicollis*. The molluscan host is usually a member of the family Thiaridae, although *Pirenella* is the common host in the Middle East. In Hawaii, the snail host is *Tarebia granifera*. Individuals suffering from heterophyiasis have inflammatory reaction at sites of contact in the intestine and may also suffer from intestinal pain and mucous diarrhea.

Metagonimus yokogawai is an intestinal parasite of humans, dogs, cats, pigs, and mice in the Far East and countries surrounding the Baltic Sea (Fig. 10.71). Again, infections are caused by the eating of infected raw or poorly cooked fish.

Centrocestus spp. are found in fish as well as in cats, dogs, and humans (Fig. 10.71). The metacercariae of

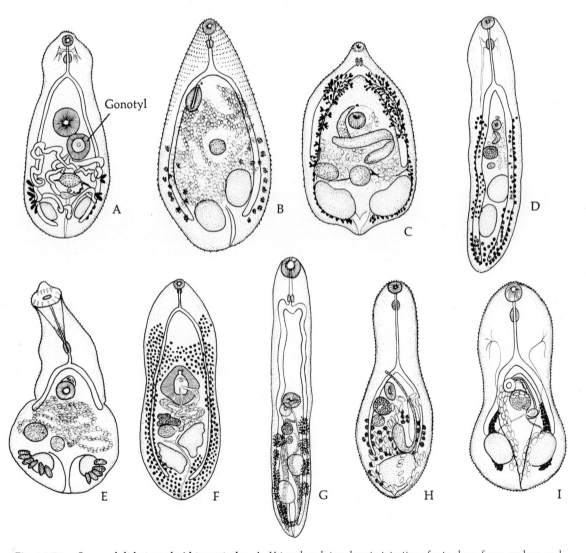

Fig. 10.71. **Some adult heterophyid trematodes. A.** *Heterophyes heterophyes* in intestine of cats, dogs, foxes, wolves, and humans. (Redrawn after Looss, 1896.) **B.** *Metagonimus yokogawai* in intestine of dogs, cats, pigs, mice, and humans. (Redrawn after Leiper, 1913.) **C.** *Centrocestus cuspidatus* in intestine of *Milyus parasiticus*. (Redrawn after Looss, 1896.) **D.** *Apophallus muhlingi* in intestine of various birds, dogs, and cats. (Redrawn after Ciurea, 1924.) **E.** *Phagicola pithecophagicola* in intestine of the mammal *Pithecophaga jefferyi*. (Redrawn after Price, 1935.) **F.** *Cryptocotyle lingua* in various fish-eating birds, dogs, cats, and wolves. (Redrawn after Neveu-Lemaire, 1936.) **G.** *Galactosomum phalacrocoracis* in intestine of the bird *Phalacrocorax pelagicus*. (Redrawn after Yamaguti, 1939.) **H.** *Proceroyum varium* in intestine of the cat. (Redrawn after Onji and Nishio, 1915.) **I.** *Stellantchasmus falcatus* in *Columbus arcticus* and experimentally in the cat and other mammals, including humans. (Redrawn after Onji and Nishio, 1915.)

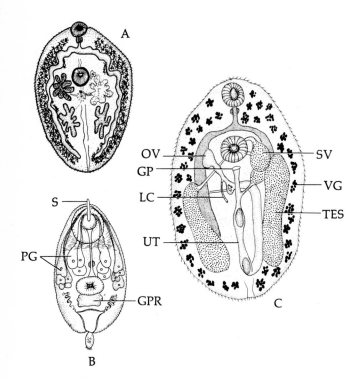

Fig. 10.72. *Paragonimus* and *Nanophyetus*. **A.** Adult of *Paragonimus westermani*. **B.** Cercaria of *P. westermani*. **C.** Adult of *Nanophyetus salmincola*. (Redrawn after Olsen, 1962.) GP, genital pore; GPR, genital primordium; LC, Laurer's canal; OV, ovary; PG, penetration gland; S, stylet; SV, seminal vesicle; TES, testis; UT, uterus; VG, vitelline glands.

these species encyst in cyprinid fish, and when these are ingested by another fish or a mammal, the cycle is completed.

Haplorchis taichui is an intestinal trematode of birds and mammals, including humans. Its life cycle parallels that of *Opisthorchis tenuicollis*.

Another fairly commonly encountered heterophyid in coastal areas is *Cryptocotyle lingua*, an intestinal parasite of fish-eating birds and mammals. This trematode utilized the periwinkle, *Littorina littorea*, as the first intermediate host in which it is represented by a sporocyst and a redial generation. Ophthalmocercariae escaping from rediae are free-swimming in sea water and when they make contact with a shore fish (e.g., cunner, gudgeon, etc.), they penetrate the piscian second intermediate host, and encyst as metacercariae. *C. lingua* cercariae show a predilection for the fin rays and large numbers of metacercariae common occur at such sites surrounded by orange-red or black pigments deposited by the host. When viable metacer-

cariae are ingested by a bird or mammal, including humans, the adult stage is attained.

Numerous other genera of the Heterophyidae are known. Cheng (1973) has presented a summary of these species transmissible to humans via the eating of raw or poorly cooked fish, and Ito (1964) has given an account of heterophyid infections in Japan.

Human cases of heterophyidiasis may involve more than parasitization of the intestinal tract. Eggs of heterophyid trematodes may be carried by the bloodstream to the heart, becoming encapsulated in cardiac muscle. Death of the host due to cardiac arrest may then result. In Hawaii, what is referred to as the "mystery death" among people of Philippino descent has been suspected to be due to aberrant heterophyid eggs. Kean and Brashau (1964) have reported that 14.6% of cardiac failures in the Philippines are due to heterophyid myocarditis.

Host Specificity among Heterophyids. Experimental infection with various species of Heterophyidae indicates that these trematodes are usually not very host specific and can develop in an array of vertebrates. Although these trematodes can develop in most experimental hosts, their behavior pattern in unnatural hosts suggests that such hosts are not completely satisfactory. For example, when heterophyids of night herons are experimentally introduced into mammals, these parasites gradually shift toward the posterior portion of the alimentary tract and are eventually expelled. When heterophyids not normally found in dogs and humans are introduced into these hosts, the parasites become buried deep in the mucous lining of the intestine and their eggs, which are normally passed out in feces, do not escape in this manner. Instead, the eggs are taken up by the blood and lymph and distributed throughout the host's body. Furthermore, many of these worms die after a short period of development. Heterophyids introduced into abnormal hosts generally cause some tissue pathology.

Superfamily Allocreadioidea

The digenean superfamily Allocreadioidea includes many families, whose members are parasites of fish, amphibians, reptiles, and mammals. One of the more important families from the standpoint of public health is the Troglotrematidae.

Family Troglotrematidae. The members of this group of flukes are all small and fleshy. They either are intestine-dwellers or are encysted in the respiratory tract and in connective tissues of birds and mammals. The cercariae are of the microcercous type. Undoubtedly the best known species are *Paragonimus westermani*, *P. kellicotti*, and *Nanophyetus salmincola*.

Life Cycle of Paragonimus westermani. *P. westermani* is an encysted form (encapsulated in host connective tissue) found in the lungs of humans and crab-eating mammals (Fig. 10.72). The adult worms measure 7.5–12 mm long and 4–6 mm wide. This

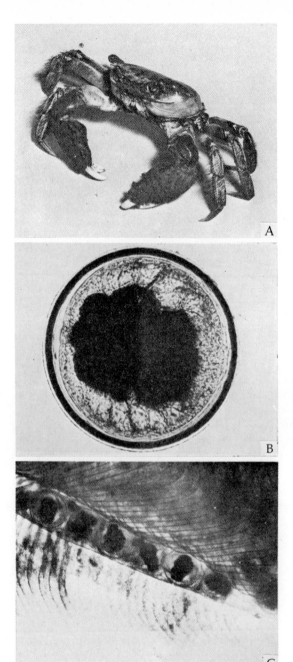

Fig. 10.73. Crab intermediate host of *Paragonimus westermani*. A. *Eriocheir japonicus,* the second intermediate host of *P. westermani* in Taiwan. **B.** A single metacercaria of *P. westermani.* **C.** Several metacercariae encysted in gill filament of crab host. (All photographs courtesy of Dr. Robert E. Kuntz.)

parasite is found primarily in Asia, although it exists in Africa and South and Central America. The eggs are expelled from the definitive host in sputum when the cysts encapsulating the adult worms rupture into the bronchioles and the eggs are coughed up. The miracidia develop in 3 weeks in moist environments. In Asia, eggs and miracidia are found in rice and vegetable paddies, because infected farmers habitually spit while working. This situation is ideal as far as the trematode is concerned, for the snail intermediate hosts are present in the paddies within easy reach.

After penetrating the snail host (*Semisulcospira libertina, S. amurensis, Tarebia granifera,* or *Brotia asperata*), each miracidium gives rise to one sporocyst generation followed by two redial generations. Each sporocyst gives rise to approximately 12 mother rediae, each of which in turn gives rise to 12 daughter rediae.

The cercariae are formed within the brood chambers of the daughter rediae. The escaping microcercous cercariae are 175–240 μm long and possess a spinous tegument and 14 penetration glands (Fig. 10.72). These cercariae begin to escape from the snail host approximately 78 days after infection of the snail. The cercariae crawl rather than swim, and once they make contact with the crustacean second intermediate host, they penetrate the body at various vulnerable sites and then encyst. Noble (1963) has reported that the crustacean host can also become infected from eating infected snails.

In China, Japan, and the Philippines, various species of freshwater crabs serve as suitable second intermediate hosts, and in Korea, a crayfish, *Palaemon nipponensis,* can also serve as the intermediate host. In Taiwan, the crab *Eriocheir japonicus* (Fig. 10.73) is the common second intermediate host.

Although other species of *Paragonimus* tend to encyst in the second intermediate host's cardiac region, *P. westermani* usually form metacercarial cysts in the gills and muscles of the body and legs, and sometimes in the hepatopancreas. Encystment in the heart is also possible. The encysted metacercariae, measuring 0.5 mm or less in diameter, are not doubled over ventrally as are most trematode metacercariae, but lie within the cyst wall in an extended position. When the definitive host, which can be a human, ingests the flesh of infected crustaceans, the parasite excysts in the intestine. However, the fluke does not mature in the intestine; instead it bores through the intestinal wall into the coelom, through the diaphragm, and into the lungs, where it becomes encapsulated.

Sometimes wandering adults do not enter the lungs but become lodged in the spleen, liver, urinary system, intestinal wall, eyes, or muscles, causing tissue damage at these sites. Adult flukes also invade brain tissue, causing a type of eosinophilic meningitis. Encysted worms in the lungs are commonly found in pairs, although a single individual or several specimens may occur within a single cyst.

Within the cysts are found infiltrated host cells and numerous eggs in a reddish semifluid mass. Infected lungs have a peppered appearance because of abscesses. Infected individuals portray such symptoms as coughing, difficulty in breathing, intermittent blood-stained sputum, mild anemia, and slight fever. Fatal cases of paragonimiasis are known, especially if the human host is infected with large numbers of parasites. Discovery of eggs in the sputum is probably the most reliable diagnostic sign. Yokogawa (1965) has contributed a critical review of the genus *Paragonimus* and human paragonimiasis, which should be consulted for further information.

In addition to *P. westermani*, at least two other species, *P. ohirai* and *P. iloktsuenensis*, are known to infect humans in Japan and parts of the People's Republic of China. Brackish-water crabs are the second intermediate hosts for both of these species.

Paragonimus kellicotti is a North American lung fluke of cats, dogs, and pigs. Its life cycle parallels that of *P. westermani*, with crayfish serving as the second intermediate host. The review by Sogandares-Bernal and Seed (1973) on the American species of *Paragonimus*, their biology, and the pathologic changes caused by them is recommended.

Life Cycle of Nanophyetus salmincola. The biology of *Nanophyetus salmincola* (Fig. 10.72) has been reviewed by Millemann and Knapp (1970a). This unique trematode is commonly called the salmon-poisoning fluke because it serves as the vector for the rickettsial organism *Neorickettsia helminthoeca*, which is extremely pathogenic to canine hosts that ingest raw salmon parasitized by *N. salmincola*. Canines that recover from the rickettsial disease either spontaneously or after drug treatment develop a lasting and complete immunity to the rickettsia but not to the trematode. About 90% of dogs naturally infected with the salmon-poisoning disease die.

The eggs of *N. salmincola* do not include fully formed miracidia when passed out of the host. In fact, the miracidium develops extremely slowly. It takes 185 to 200 days before the free-swimming miracidium hatches. The molluscan host is the stream snail *Oxytrema silicula*. Within this gastropod, the miracidia give rise to mother rediae, which in turn simultaneously give rise to daughter rediae and cercariae. The escaping cercariae are microcercous xiphidiocercariae. These penetrate and encyst in the kidney and under the skin of various fish, primarily salmon and trout. The Pacific giant salamander can also serve as a second intermediate host. When infected fish are eaten by dogs, cats, foxes, bears, mink, hogs, or other animals, the metacercariae excyst and develop into adults in the definitive host's small intestine. In addition to the natural definitive hosts listed, a variety of other mammals, primarily canines, felines, and rodents, have been found to be infected (see Millemann and Knapp, 1970a).

Humans can be infected experimentally (Philip, 1958). Nanophyetiasis in humans is of no serious consequence. Although there are some mild pathologic alterations in the intestine of infected dogs, the disease caused by the fluke is unimportant. It is only the rickettsia that is highly lethal to canids. After an incubation period of 6 to 10 days, the body temperature rises to 40 to 42°C, accompanied by edematous swelling of the face and discharge of pus from the eyes. The canid exhibits depression, loss of appetite, increased thirst, vomiting, and diarrhea. Death usually occurs in about 10 to 14 days after onset. As stated, dogs that recover become immune for the rest of their lives.

In Oregon, where *Nanophyetus salmincola* is most extensively distributed in the United States, Gebhardt *et al.* (1966) have reported that during the months of June through October almost 100% of the snail hosts harbor mature cercariae. The range of *N. salmincola* extends beyond western Oregon into southwestern Washington and northwestern California.

Not only is *N. salmincola* important because it is the vector of the salmon-poisoning disease causing rickettsia, the metacercariae are pathogenic to fish (Millemann and Knapp, 1970b). Naturally infected salmonid fish may show such symptoms as exoph-

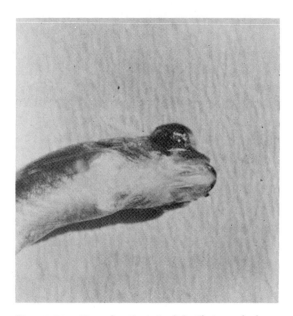

Fig. 10.74. Nanophyetiasis in fish. Photograph showing exophthalmia in an Atlantic salmon, 76 mm long, which had been exposed to *Nanophyetus salmincola* cercariae for 30 minutes approximately 3 weeks before the photograph was taken. (After Millemann and Knapp, 1970.)

thalmia (Fig. 10.74), prolapse of the intestine, damage to fins, gills, eyes, and practically every internal organ. Furthermore, their swimming ability is greatly reduced.

Other Nanophyetus *and Closely Related Species.* Another species of *Nanophyetus*, *N. schikhobalowi*, has been reported from humans in eastern Siberia.

Closely related to *Nanophyetus* spp. is *Sellacotyle mustelae*, a minute species that occurs in the small intestine of mink. The cercariae of this species develop in rediae in the large freshwater viviparous snail *Campeloma rufum*. Metacercariae develop from free-swimming cercariae when they penetrate fish, primarily

the bullhead, *Ameiurus melas*. These metacercariae become infective after 8 days and develop to maturity in the mink in 5 days after ingestion.

Suborder Didymozoida

Members of the suborder Didymozoida represent a small group of trematodes the taxonomic position of

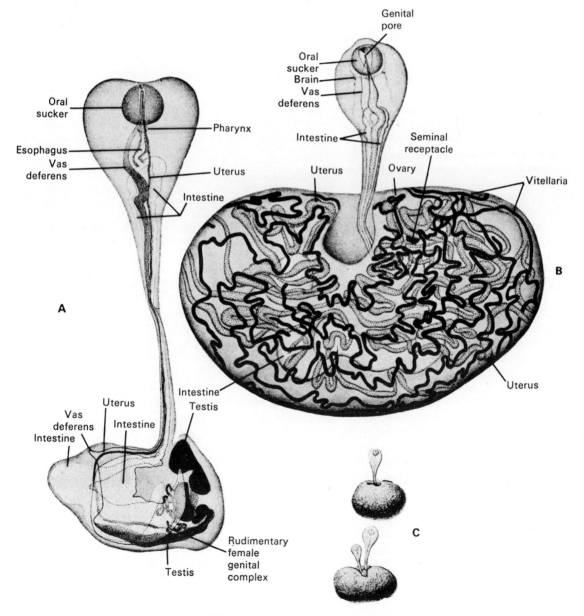

Fig. 10.75. **Sexual dimorphism in *Koellikeria bipartata*, a common parasite of the intestinal mucosa of various tunas. A.** Adult male. **B.** Adult female. **C.** Female (top) and male inside cavity of female (bottom). (After Odhner, 1910.)

which has been debated in recent years. Some are of the opinion that it should be separated from the Digenea and recognized as a distinct order (or subclass) of the class Trematoda, with equal rank as the Digenea. However, following the late S. Yamaguti (1971), the foremost authority on this group of parasitic worms, I still consider it subordinate to the Digenea.

All of the Didymozoida are tissue-dwelling parasites of fish, especially marine species. Most occur encysted within gills, often in pairs, but others are found encysted in the skin, kidney, and ovary. Baer and Joyeux (1961) have presented a review of this group, and Yamaguti (1958) has provided a taxonomic catalogue of all of the species known at that time. In addition, Yamaguti (1970) has contributed a taxonomic study of the species that occur in Hawaiian waters.

The didymozoids are elongate and usually flattened; however, some species are fleshy or lobated. Most species measure only a few millimeters long, but Noble (1975) has described a species of *Nematobothrium* from an oceanic sunfish that is 40 feet long.

Many species of didymozoids are dioecious and portray considerable sexual dimorphism (Fig. 10.75). In the dioecious forms, vestigial reproductive organs of the opposite sex may be present (Self *et al.*, 1963). The anteriorly situated mouth often lacks an oral sucker, although a weakly developed one is present in some species (Fig. 10.76). A ventral sucker is usually absent in adults but may be present in young worms.

One reason for the interpretation that the Didymozoida should be removed from the Digenea is that fragmentary evidences suggest that the life cycle of this group of trematodes may be direct, i.e., they lack an intermediate host. This, however, remains to be proven. Indirect evidence leading to the idea that the life cycle of these trematodes is direct is that the egg contains a nonciliated larva that includes a short gut

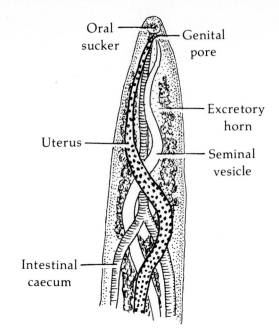

Fig. 10.76. Anterior end of *Nematobibothroides histoidii* from *Mola mola*. Drawing of anterior end of this didymozoid trematode showing small, weakly developed oral sucker. (Redrawn after Noble, 1975; with permission of *Journal of Parasitology*.)

and an oral sucker surrounded by two or more circles of spines. Self *et al.* (1963) have been unsuccessful in infecting snails and crustaceans with the larva of *Ovarionematobothrium texomensis*, a parasite in the ovaries of the buffalo fish in Oklahoma. Since no further attempt to ascertain the life cycle of a didymozoid has been reported, the matter remains uncertain.

Sex determination in the case of dioecious species is interesting. Ishii (1935) has reported that the first worm to occupy a position on its host develops into a female while subsequent arrivals develop into males. The basis for this phenomenon remains unknown.

Table 10.9. Survival Time of Several Species of Digenetic Trematodes under Aerobic and Anaerobic Conditions[a]

Species	Medium	Temperature °C	Survival Time (In Days)	
			Anaerobic	*Aerobic*
Fasciola hepatica	Blood	38–39	1.5	1.5
Fasciola hepatica	Borax-saline and glucose	38	1.5	2
Opisthorchis felineus	Ringer's	37	18	18
Schistosoma mansoni	Serum ultrafiltrate	37	5	12
Sphaerostoma bramae	1% NaCl	Room temperature	5	4
Cryptocotyle lingua (metacercariae)	Modified Ringer's and glucose	Room temperature	4	12

[a]Data after von Brand, 1952.

PHYSIOLOGY AND BIOCHEMISTRY OF DIGENEANS

During the last decade a great deal has been learned about the biochemistry and physiology of digeneans, especially those species of medical importance, such as *Schistosoma* spp. Presented below are brief summaries of what is known about the physiology and biochemistry of this group of parasites. The volumes by Smyth and Halton (1983), Erasmus (1972), and Crompton and Joyner (1980) are recommended to those wishing to supplement this account of the physiology of digeneans, whereas that by Barrett (1981) should be referred to by those wishing a concise review of their biochemistry. In a more restricted article, Coles (1984) has presented a review of what is known about the biochemistry of schistosomes.

OXYGEN REQUIREMENTS

Digenetic trematodes are primarily facultative anaerobes, i.e., they can be maintained under anaerobic conditions but are capable of utilizing oxygen if available. Table 10.9 gives the survival time of several species under both aerobic and anaerobic conditions. In their natural environments, trematodes are subjected to varying amounts of oxygen, depending on the stage of development and the particular host. It is obvious that free-swimming miracidia and cercariae are aerobic, utilizing the oxygen present in the aquatic environment. On the other hand, the intramolluscan larvae, particularly when found in the reproductive systems or in the digestive glands of their hosts, are essentially in anaerobic or microaerobic environments. However, rediae at least utilize oxygen if it is present. Thus intramolluscan larvae depend primarily on carbohydrate metabolism as their energy source.

The differences in the type of metabolism utilized by the various stages in the life cycle of digeneans is reflected at the biochemical level. For example, the adult stages of *Fasciola hepatica* and *Schistosoma mansoni* have probably the highest adenylate cyclase activity in any species of animal (Northup and Mansour, 1978; Higashi *et al.* 1973). As depicted in Fig. 10.77, adenylate cyclase is an enzyme located on the inner surface of cell membranes. When activated (by serotonin in the case of trematodes), it enhances anaerobic glycolysis (p. 52) via a series of intermediate steps (Fig. 10.77). Glycolysis is the main energy-producing mechanism in the adults of *F. hepatica* and *S. mansoni* (Mansour and Mansour, 1962; Mansour; 1962). Recently, Kasschau and Mansour (1982) have studied the levels of adenylate cyclase activity in the cercaria and schistosomule of *S. mansoni*. They have found that the cercaria, a free-living stage in an environment with abundant oxygen, has low adenylate

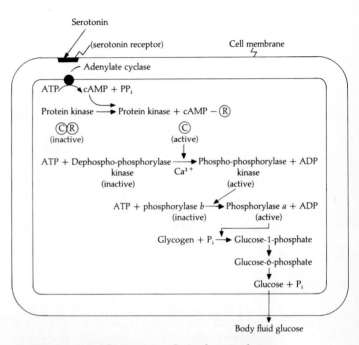

Fig. 10.77. Amplification cascade in the stimulation of glycogenolysis by serotonin in helminth cell to yield body fluid glucose. The amplification produced is about 3-million-fold.

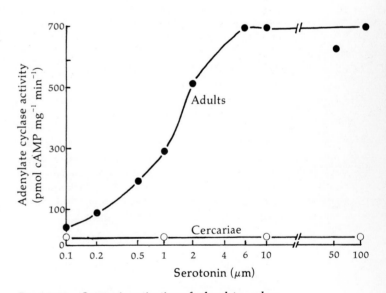

Fig. 10.78. Serotonin activation of adenylate cyclase from adult and cercaria particles of *Schistosoma mansoni*. Representative experiment on adult (•) and cercaria particles (o) in which serotonin was added at the indicated concentrations to adenylate cyclase reaction mixture containing 0.1 mM ATP and 1×10^{-5} M GTP. (After Kasschau and Mansour, 1982; with permission of *Nature*.)

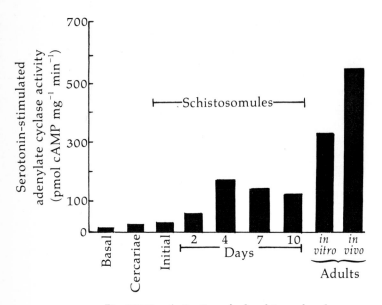

Fig. 10.79. **Activation of adenylate cyclase by sero-tonin and GTP at different stages in the life cycle of *Schistosoma mansoni*.** Early schistosomules (initial) were obtained from culturing cercariae. Adults *in vitro* were obtained from culturing schistosomules for 3 months. Adenylate cyclase activity was measured in a reaction mixture containing 0.1 mM ATP and 100 μm serotonin with 10^{-5} M GTP. Adenylate cyclase activity in particles from schistosomules immediately after preparation from cercariae is designated as "initial." Basal activity for adult worms was 8 pmol cyclic AMP mg^{-1} min^{-1} and for cercariae and schistosomules, <3 pmol cyclic AMP mg^{-1} min^{-1}. (Modified after Kasschau and Mansour, 1982.)

habitats such as mud. This hypothesis is not completely acceptable because a true aerobic ancestor could have gradually inclined toward anaerobism during the process of adapting itself to the endoparasitic way of life. Also, the availability of oxygen in the immediate environment of trematodes need not always indicate that the parasites carry on aerobic metabolism. The most striking example of this is *Schistosoma*, which lives under relatively high oxygen tensions in its adult stage in the blood of its host, yet depends primarily on anaerobic metabolism for energy production. However, in the case of lung flukes (e.g., *Haematoloechus*, *Haplometra*) and esophagus-dwelling flukes (e.g., *Leptophallus*, *Otodistomum*) their ability to survive in the presence of oxygen for relatively long periods *in vitro* suggests that they are true aerobes. The frog lung flukes, *Haematoloechus* sp., actually ingest the host's erythrocytes, which are rich in oxygen.

As the result of quantitative studies, it is now known that most of the factors known to affect the rate of respiration in free-living organisms also affect oxygen consumption in trematodes. Among these, temperature, body mass, and ambient oxygen tension are the most important.

Temperature

The influence of ambient temperature on respiration in ectothermic animals is interesting since there are evidences that the respiratory processes in these organisms are adapted to a particular range of temperatures. If the ambient temperature is raised or lowered from that of the parasite's natural habitat, the respiration rate is dramatically altered, and the parasite dies. For example, in the case of *Gynaecotyla adunca*, a parasite of birds that normally lives at 40°C, its Q_{O_2} is increased if maintained at 41°C; however, the parasite can only survive for a few hours, at a decreased respiratory rate, at 45°C. Similarly, in the case of *Saccocoelium beauforti*, a parasite of fish with a natural maximum ambient temperature of 30°C, there is an increase in Q_{O_2} at temperatures approaching 30°C, but the parasite dies within an hour, during which the Q_{O_2} is depressed, when maintained at 41°C.

The respiration rates of trematodes can also be expressed as temperature coefficients or Q_{10} (i.e., the increase in respiratory rate caused by a 10°C rise in temperature). The Q_{10} values of animals are lowest in the temperature range that approximates that of the natural environment. Hence, the Q_{10} values tend to be low when trematodes are placed at the temperature range normally encountered in their hosts. As examples, the Q_{10} of *Pleurogonius malaclemys*, a turtle parasite (and therefore exposed to a wide temperature range), is high in only the 6 to 12°C range, whereas in *Saccocoelium beauforti* the Q_{10} values are high up to 18°C, and in *Gynaecotyla adunca*, the values are high up to 30°C.

cyclase activity that responds poorly to activation by serotonin (Fig. 10.78). On the other hand, the adenylate cyclase system in schistosomules changes from low serotonin activation, similar to that of cercariae, to one of high activity similar to that found in adults during the initial 4 days (Fig. 10.79), indicating that the schistosomule is the transitional stage from essentially aerobic to anaerobic metabolism. A general review of respiration in trematodes has been contributed by Vernberg (1963). Also, Bryant (1975, 1978) has reviewed the regulation of respiratory metabolism in helminths, including digeneans and Mansour (1984) has reviewed the topic of serotonin receptors in parasitic worms.

Some authors have drawn phylogenetic correlations from the oxygen requirements of parasitic worms. Bunge (1889) assumed that helminths that can survive under anaerobic conditions are descended from ancestral forms that inhabited oxygen-poor

Size

The relationship between body mass (expressed as nitrogen content) and respiration among trematodes has been investigated by Vernberg and Hunter (1959) and others. As a general rule, respiration rate decreases as the worms increase in size. For example, Vernberg and Hunter have shown a significant correlation between body nitrogen content and respiration rate in *Gynaecotyla adunca*; specifically, a fourfold increase in size is correlated with a decrease of QO_2 of approximately 40%.

Oxygen Tension

All animals may be categorized as either **conformers** or **regulators** relative to their response to varying ambient oxygen tensions (p. 38). Among adult digeneans, Vernberg (1963) has shown that *Gynaecotyla adunca* is a conformer at oxygen levels of 1.5 to 5.0%, but below 1.5% it becomes a regulator, maintaining the same level of oxygen consumption from 0.5 to 1.5% oxygen.

The different life cycle stages of digeneans generally utilize oxygen to some extent if it is available, although there is the tendency for the species or stages in the life cycle that are located in regions of low oxygen tension to regulate their respiratory rates. Differences in oxygen consumption by various stages in the life cycle of *Gynaecotyla adunca*, a parasite of birds, are given in Table 10.10. As *G. adunca* adults can be both respiratory conformers and regulators, so can the larvae of certain species. Furthermore, one stage may be a conformer while another is a regulator. This, obviously, has selective advantage for these parasites since they alternate between habitats with different oxygen tensions.

Now that we have seen that trematodes can and do utilize oxygen, the question may be raised as to what significance oxygen consumption has in the metabolism of these parasites? One reason for raising this question is that Bueding (1949), by exposing *Schistosoma mansoni* adults to cyanine dyes that inhibit oxygen uptake, has reported that although aerobic respiration is almost completely inhibited (80%), the glycolytic rate is unaffected, and the production of lactic acid is the same in the presence or absence of the cyanine dyes. These data indicate that although *S. mansoni* adults live in an environment with a high oxygen tension, their metabolism is primarily anaerobic. It has been speculated (Smyth and Halton, 1983) that perhaps the metabolism of *S. mansoni* has become so adapted in response to living in another anaerobic host site, such as the body cavity, occupied during an earlier phase of its evolution, but has retained from its free-living ancestors some ability to use oxygen. The fact that oxygen may still be required in the production of some essential metabolite should also be considered.

Table 10.10. Oxygen Consumption of Various Life Cycle Stages of *Gynaecotyla adunca*[a,b]

	Temperature	Oxygen Consumption in Microliters/Hour/mm³
Free cercariae[c]	30.4	5.35
3–4 Days after penetrating crab[c]	30.4	Not measurable
Immediately prior to encystment[c]	30.4	0.159
Encysted metacercariae	30.4	5.62×10^{-3}/worm
Adults 24 hours after encystment	30.4	0.153
Adults 48 hours after encystment	30.4	0.120
Adults 72 hours after encystment	30.4	0.104

[a] Data after Hunter and Vernberg (1955) and Vernberg and Hunter (1956).
[b] Measurements were made on individual organisms by use of the Cartesian diver respirometer.
[c] Indicates stages later determined not to be those of *G. adunca* but those of *Zoogonus lasius*, a parasite of fish.

As in *S. mansoni*, adults of the liver fluke, *Fasciola*, are also primarily anaerobic; however, in both species the eggshells are hardened by quinone tanning (p. 401), a process involving oxygen, and this may be one reason why oxygen utilization occurs in *Schistosoma* and *Fasciola*.

Respiratory Pigments

Hemoglobin has been found in several species of trematodes. These pigments have absorption spectra comparable to those of mammalian hemoglobin (Table 10.11). Chemical data also support the fact that the pigments in trematodes are hemoglobin. These include (1) deoxygenation under reduced pressure or by the addition of sodium dithionite; (2) the for-

Table 10.11. Absorption Maxima (mμ) of Hemoglobins from Digeneans[a]

	Oxyhemoglobin		Reduced Hemoglobin
Species	α	β	$\beta-\alpha$
Telorchis robustus	575.0		
Alassostoma magnum	575.0		
Fasciola hepatica	580–581	543	
Fasciola gigantica	586	540	
Gastrothylax crumenifer	576	542	550–564
Cotylophoron indicum	585.2	555.8	
Proctoeces subtenuis	579	543	431–558
Fasciolopsis buski	572	538	554–562
Gastrodiscoides hominis	574	538	554–558
Isoparorchis hypselobagri	574	540	550–558
Various mammals	576–578	540–542	430–555

[a] Data compiled by Smyth, 1966, and Haider and Siddiqi, 1976.

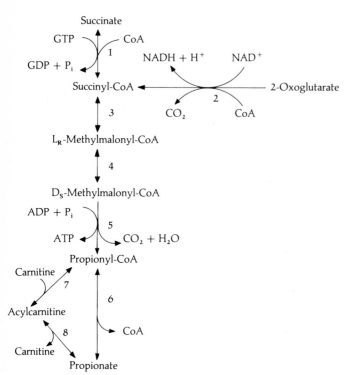

Fig. 10.80. Pathway of propionate formation in *Fasciola hepatica*. The enzymes involved are: **1.** succinate thiokinase or CoA transferase, **2.** 2-oxoglutarate decarboxylase (a minor source of succinyl-CoA), **3.** methylmalonyl-CoA mutase, **4.** methylmalonyl-CoA racemase, **5.** propionyl-CoA carboxylase, **6.** CoA transferase, **7.** carnitine acyltransferase, and **8.** acylcarnitine hydrolase.

mation of pyridine hemochromogen; and (3) the appearance of a bright red color when exposed to carbon monoxide.

Despite the presence of hemoglobin in trematodes, its function remains uncertain. Lee and Smith (1965) point out in their review of hemoglobins in parasitic animals, including trematodes, that this respiratory pigment may serve not in oxygen transport but in some other physiologic function such as storage and the facilitation of diffusion by the Scholander mechanism.

GROWTH REQUIREMENTS

Numerous attempts have been made to culture digenetic trematodes *in vitro*. The rationales underlying most of these have been (1) to understand the nutrient requirements of these parasites, or (2) to find methods to mass produce these worms for vaccine production.

In order to achieve the first goal, the development of chemically defined media is essential. Little success has been achieved to date.

The volume edited by Taylor and Baker (1978) includes the formulary for the culture of parasitic animals, including certain digeneans.

CARBOHYDRATE METABOLISM

Carbohydrates, especially glycogen, constitute the major energy reserve in parasitic helminths, including digeneans. The level of free glucose is extremely low. The outstanding feature of carbohydrate metabolism leading to the synthesis of energy is the production of reduced organic end-products. This persists even under aerobic conditions (Table 2.8).

Acetate and Propionate Producers

In one group of digeneans, represented by *Fasciola hepatica*, acetate and propionate are produced. These acids are produced at a ratio approaching 1:2. Although the pathways leading to the formation of these acids have not been completely studied in most digeneans, they have been examined in the *Fasciola hepatica* (Barrett *et al.*, 1978; Köhler *et al.*, 1978) and the nematode *Ascaris lumbricoides* (p. 519). Assuming that the pathways are similar in all digeneans, if pyruvate and succinate are produced in equal amounts from glucose, this balances relative to oxidations and reductions. The formation of acetate from pyruvate generates an additional mole of the reduced form of nicotinamide adenine dinucleotide (NADH), which could be reoxidized via the reduction of an extra fumarate to succinate. Hence, for every mole of acetate produced, two moles of fumarate would be reduced to succinate, which on decarboxylation yields propionate. This would explain the acetate-to-propionate ratio of 1:2.

In *F. hepatica*, pyruvate is oxidatively decarboxylated via pyruvate dehydrogenase to form acetyl-CoA, which is then cleaved to give acetate. Propionate is produced by succinate being metabolized via succinic thiokinase, methylmalonyl-CoA mutase, methylmalonyl-CoA racemose, propionyl-CoA carboxylase to propionyl-CoA. Propionyl-CoA is then transformed to propionate by mediation of CoA transferase (Fig. 10.80).

Lactate, Acetate, and Propionate Producers

A second group of digeneans, represented by *Dicrocoelium dendriticum*, excrete significant quantities of lactic acid in addition to acetate and propionate or acetate and succinate. It is noted that *Fasciola hepatica*, which is primarily an acetate and propionate producer, will also excrete significant amounts of lactate if glucose is added to the incubation medium. Because *Fasciola* has a limited capacity to fix carbon dioxide, excess carbohydrate is diverted to lactate.

Relative Importance of the Citric Acid Cycle

Such digeneans as *Fasciola hepatica* are carbon dioxide fixers (p. 40). In these, the citric acid cycle does not appear to be involved in carbohydrate metabolism, although certain of the enzymes involved are present. The only function of these enzymes may be the interconversion of carbon skeletons. On the other hand, in other digeneans as *Schistosoma mansoni*, there is evidence for a functional citric acid cycle.

Utilization of Sugars Other Than Glucose

There is some evidence that *Fasciola hepatica* and *Schistosoma mansoni* can utilize hexoses other than glucose. The hexokinase involved in *F. hepatica*, like that of mammals, is relatively nonspecific. Specifically, the enzyme will phosphorylate glucose, fructose, galactose, mannose, and glucosamine. On the other hand, *S. mansoni* possesses a specific kinase for glucose, fructose, mannose, and glucosamine. These kinases phosphorylate the sugars in the 6 position (rather than in the 1 position, as does the fructokinase of mammalian liver).

Pentose Phosphate Pathway

The pentose phosphate pathway (p. 57) is known to occur in *Fasciola*; however, only about 10% of carbohydrate catabolism occurs via this avenue.

For a detailed summary of carbohydrate metabolism in digeneans and other helminths, see Barrett (1981), and for an account of ATP production in *Fasciola hepatica*, see Van Vugt *et al.* (1976). In addition, the general reviews of metabolism in schistosomes and *Fasciola* by Coles (1973, 1975, 1984) are recommended, as is the review by Bryant (1978) on respiratory metabolism in helminths.

PROTEIN METABOLISM

From observations on worms *in vivo*, it is known that the degree and rate of protein synthesis among the Digenea must be considerably higher than in closely related free-living forms. The tremendous output of reproductive cells alone, especially ova, is 10,000 to 100,000 times greater than in free-living relatives, and these reproductive cells are proteinaceous. During the development of larval generations, the multiplication of the amount of protein is manifold. For example, it has been reported that 10,000 cercariae may be derived from a single miracidium of *Schistosoma japonicum*, and in *S. mansoni*, a single miracidium may give rise to as many as 200,000 cercariae. Meyerhof and Rothschild (1940) have reported that in the marine snail *Littorina* infected with *Cryptocotyle lingua*, 1.3 million cercariae are produced in 1 year.

The fecundity of larval trematodes through asexual reproduction suggests that within the body of the

molluscan host, these parasites must have a substantial source of amino acids. Cheng (1963) has investigated this phase of trematode biochemistry. In four species of trematodes it was demonstrated that sporocysts and rediae acquire amino acids from two sources: the free and bound amino acids in the serum of the hosts, and the free amino acids and those amino acids resulting from the degradation of cells in the immediate vicinity of the parasites. Depletion of amino acids has also been demonstrated in the hemolymph of snails infected with *Schistosoma mansoni*.

Although the occurrence of proteases has been investigated in only a few larval trematodes, it is found widely in adults. For example, Rogers (1940) has detected hematin spectroscopically in the intestinal caeca of *Schistosoma mansoni* resulting from the digestion of host blood cells. Also, Timms (1960) has found that schistosomes possess a protease of high specificity that releases tyrosine from globin and hemoglobin. This protease has an optimum pH of 3.9. As another example, Rijavec *et al.* (1962) have reported that *Fasciola hepatica* incubated in radioiodine-labeled blood serum digests the albumins of the blood to amino acids, and the latter are readily absorbed.

Information pertaining to proteolytic enzymes in the intramolluscan stages of digenetic trematodes is still scanty. Cheng and Yee (1968) have found aminopeptidase activity associated with the body surfaces of third-generation rediae and freed cercariae of *Philophthalmus gralli* in the snail *Tarebia granifera*. This exopeptidase is responsible for the lysis of the cytoplasm of surrounding host cells during migration and associated with extracorporeal digestion (i.e., predigestion of host proteins prior to uptake).

As it is now known that the larval and adult teguments of trematodes are absorptive surfaces, at least for certain types of molecules, including amino acids, it should be borne in mind that the absorption of certain nitrogen-containing molecules can occur through the caecal lining as well as the body surface. The passage of amino acids through the tegument can be a two-way street. For example, Kurelec and Ehrlich (1963) have shown that if the mouth of *Fasciola hepatica* is ligated and the worm is incubated in serum, amino acids pass from the worm into serum.

How amino acids and other nitrogen-containing molecules are absorbed either through the gut or the tegument remains incompletely resolved. It is possible that, as in tapeworms (p. 435), the transport mechanism may involve a "carrier." The carrier molecule may be confined to a specific locus in the cell and is coupled with the molecule to be transported, facilitating its passage, or it could be that an enzyme situated

$$\text{L-alanine} + \alpha\text{-ketoglutarate} \underset{\text{transaminase}}{\overset{\text{alanine}}{\rightleftharpoons}} \text{pyruvate} + \text{glutamate}$$

Fig. 10.81. **Transamination.** Formation of pyruvate and glutamate from L-alanine and α-ketoglutarate.

Table 10.12. 2-Oxoglutarate-Linked Transaminases in Selected Species of Digeneans

Species	Donor Amino Acids
Fasciola hepatica	Alanine, arginine, aspartate, leucine, isoleucine, methionine, phenylalanine, proline, tyrosine, valine, ornithine
Schistosoma japonicum	Alanine, arginine, aspartate
Cryptocotyle lingua (rediae)	Alanine, aspartate
Microphallus pygmaeus (Sporocysts)	Alanine, aspartate

in the cell converts the molecule to be transported into a form that is more readily taken in. In mammalian systems, still a third type of carrier is known. This is an expansible carrier in the form of a protein molecule that through folding and unfolding, transfers attached molecules.

Once amino acids enter the body of trematodes, they may be subjected to deamination (p. 55). On the other hand, some evidence indicates the occurrence of transaminases in certain species. By this mechanism, one amino acid can be transformed into another. For example, in *Fasciola hepatica*, Daugherty (1952) and Kurelec and Ehrlich (1963) have shown that alanine and α-ketoglutaric acid are transformed by the action of transaminases to pyruvic acid and glutamic acid (Fig. 10.81). Similarly, Huang *et al.* (1962) have demonstrated that in *Schistosoma japonicum* glutamic-pyruvic and glutamic-oxaloacetic transaminases occur, and these enzymes have an optimum range between pH 7.2 and 7.5.

Among the transaminases that have been examined in digeneans, 2-oxoglutarate linked transaminases appear to be widely distributed. Aspartate: 2-oxoglutarate transaminase and alanine: 2-oxoglutarate transaminase have been found in every helminth investigated. In addition, 2-oxoglutarate transaminases active with a variety of amino acids have been found in several digenean species (Table 10.12).

With the exception of sporocysts, amino acids do not appear to be an important energy source for parasitic helminths, but these nitrogen compounds are utilized for building proteins.

The nitrogenous end-product of metabolism in trematodes appears to be primarily ammonia although small quantities of urea, uric acid, amino acids, and in some cases amines, are also excreted. Some 90% of excreted nitrogen originates from the α-amino nitrogen of amino acids. The remaining 10% originate from the degradation of purines and pyrimidines.

LIPID METABOLISM

Lipase activity has been reported in adult *Fasciola hepatica*, *Schistosoma mansoni*, and other digeneans, suggesting the utilization of fats, although the levels of activity are extremely low. Esterase activity has been demonstrated in other species, especially strigeids (Erasmus and Öhman, 1963). Since most intestinal trematodes live in microaerobic or even essentially anaerobic environments, one would not expect to find high levels of lipid metabolism. On the other hand, lipid metabolism as an energy source is undoubtedly of great importance among free-swimming cercariae (Cheng, 1963). This is supported by the fact that considerable amounts of fatty acids are stored in the bodies of developing cercariae, whereas very little is present in the bodies of adults. This suggests that these fatty acids have been exhausted during the free-swimming, aerobic cercarial stage.

The presence of larval trematodes in the digestive gland of molluscs induces hypersynthesis of neutral fats within the cells of that organ. In time, these neutral fats are broken down to fatty acids, which are readily absorbed through sporocyst walls and are stored, not as neutral fats, but as fatty acids in the bodies of enclosed cercariae.

During starvation the fat contents of trematodes tend to rise. This can be explained by the fact that the higher fatty acids are by-products of carbohydrate metabolism, and since the utilization of carbohydrates is extremely rapid during starvation, increased amounts of higher fatty acids are produced.

β-OXIDATION

In organisms capable of aerobic metabolism, fatty acids are broken down by β-oxidation to give acetyl-CoA, NADH, and reduced flavoprotein (Fig. 10.82). All parasitic helminths studied thus far have revealed some, if not all, of the enzymes of the β-oxidation sequence. Among digeneans, a complete sequence of this group of enzymes has been found in *Fasciola hepatica*. It is noted, however, that although the β-oxidation enzymes are present, the pathway is not active (Barrett, 1981). This has been interpreted to mean that either the citric acid cycle is relatively unimportant in these worms or the nonfunctioning of the pathway is due to the small amount of oxygen present in the habitat. The absence of a classical citric

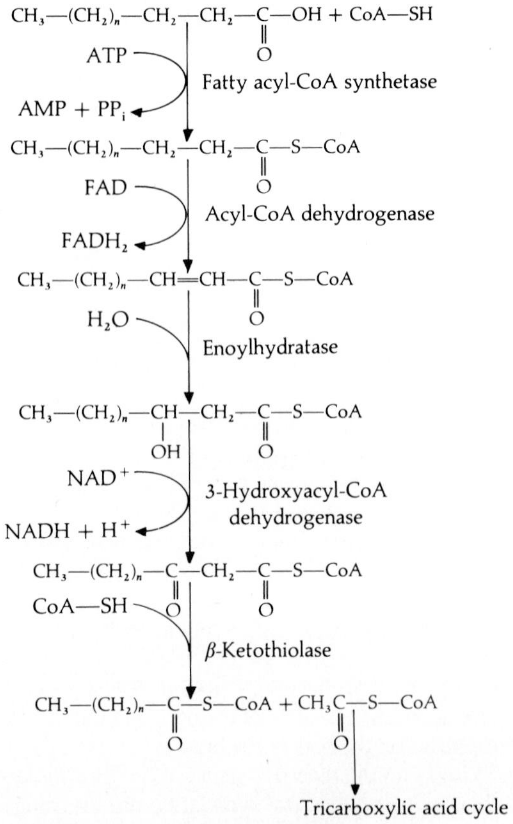

$$CH_3—(CH_2)_n—CH_2—CH_2—C(=O)—OH + CoA—SH$$

Fatty acyl-CoA synthetase
ATP → AMP + PP_i

$$CH_3—(CH_2)_n—CH_2—CH_2—C(=O)—S—CoA$$

Acyl-CoA dehydrogenase
FAD → FADH_2

$$CH_3—(CH_2)_n—CH=CH—C(=O)—S—CoA$$

Enoylhydratase
H_2O

$$CH_3—(CH_2)_n—CH(OH)—CH_2—C(=O)—S—CoA$$

3-Hydroxyacyl-CoA dehydrogenase
NAD^+ → NADH + H^+

$$CH_3—(CH_2)_n—C(=O)—CH_2—C(=O)—S—CoA$$

β-Ketothiolase
CoA—SH

$$CH_3—(CH_2)_n—C(=O)—S—CoA + CH_3C(=O)—S—CoA$$

Tricarboxylic acid cycle

Fig. 10.82. The β-oxidation sequence.

acid cycle would severely limit the further metabolism of the acetyl-CoA produced by β-oxidation. The function of the β-oxidation enzymes in parasitic helminths remains uncertain. They could be associated primarily with the developing eggs within the uterus in preparation for the free-living miracidial stage. Another possible role for these enzymes is involvement in the malonyl-CoA–independent elongation of fatty acids within mitochondria. The second possibility is known to occur in other invertebrates (Oudejans and Van Der Horst, 1974). Finally, these enzymes may be involved in the catabolism of such aliphatic amino acids as valine, leucine, and isoleucine, which involves pathways analogous to those of β-oxidation.

In digeneans, lipids are believed to be the end-products of carbohydrate metabolism. These worms excrete lipid droplets via the excretory system as well as the intestinal caeca (Harris and Cheng, 1973). For example, in *Fasciola hepatica*, lipid excretion can amount to as much as 2% of the total wet weight per 24 hours. Analyses of the excreted materials have revealed that cholesterol, cholesterol esters, triacylglycerols, free fatty acids, and phospholipids are all excreted, suggesting the general loss of lipids, rather than the selective excretion of specialized end products. Since there is no breakdown of fatty acids in adult di-

geneans, body lipids can be turned over only by excretion.

Since lipid catabolism apparently does not occur in adult digeneans, the question can be raised as to why these parasites accumulate lipids. A partial answer lies in the fact that at least a part of the stored lipids is incorporated into eggs. Alternatively, the adult worm may have to take in large amounts of lipids in order to obtain a sufficient amount of a specific fatty acid or fat-soluble vitamin, and the excess lipid is stored and eventually excreted.

Contrary to the endoparasitic stages in the life cycle of digeneans, the free-living stages (i.e., miracidia and cercariae) can carry on β-oxidation and utilize lipids as an energy source. Even then, the amount of stored lipids varies among different species of cercariae, for example, *Schistosoma mansoni* cercariae contain little lipids.

HOST-PARASITE RELATIONSHIPS AMONG THE DIGENEA

Most of the flukes listed in Table 10.1 are detrimental to their hosts. This phase of host-parasite relationship—that is, the production of diseases—is the most obvious effect digeneans have on their hosts. It should be borne in mind, however, that the severity and even the occurrence of a disease is a function of parasite density.

EFFECTS OF ADULTS ON DEFINITIVE HOSTS

Although adult trematodes are not suspected of playing important roles in seriously depriving their hosts of nutrients, it is suspected that they may play an important role in depriving their hosts of vitamins and trace elements, which can result in disrupting host metabolism. Intramolluscan larval trematodes, however, do deplete their host's carbohydrates, lipids, and amino acids, which probably causes some harm to the host. In fact, infected molluscs generally do not survive as long as uninfected ones. However, death is not attributed completely to the loss of stored nutrients but to a large degree to the mechanical and lytic damage caused by the larvae. Also, infected molluscs either cease to produce eggs or their fecundity is greatly reduced (see Malek and Cheng, 1974, for review).

Other forms of metabolic disruption are known. For example, in rabbits infected with *Schistosoma japonicum* or *Clonorchis sinensis*, the blood calcium decreases, and blood potassium and sodium increase. In cattle and sheep infected with *Fasciola hepatica* there is a slight increase of blood chlorides.

Lipid metabolism of the host is disrupted in *Schistosoma japonicum* and *Clonorchis sinensis* infections. Hiromoto (1939) has reported that blood cholesterol of rabbits is raised when *S. japonicum* is present. A parallel condition develops in rabbits infected with *C. sinensis.*

Host protein metabolism is also disrupted. It is known that serum albumin and globulin are decreased and increased, respectively, in humans infected with *Schistosoma haematobium*, and in humans and rabbits infected with *S. japonicum* the total serum protein is decreased but the globulin and nonprotein nitrogen fractions are increased. These findings undoubtedly reflect the fact that antibodies are being produced in mammals parasitized by trematodes, and since antibodies are associated with the globulin fractions, the elevations in globulin content reflect immunologic response.

Although there is considerable information pertaining to disruptions in carbohydrate metabolism in vertebrates due to protozoan parasites, there is considerably less information of this nature relative to helminth parasites, especially trematodes. It is known that in rabbits infected with *Clonorchis sinensis* and in humans infected with *Schistosoma mansoni*, hyper- and hypoglycemia may occur. Similarly, sheep and rabbits infected with *Fasciola hepatica* may reveal hypoglycemia, as do sheep harboring *Paramphistomum microbothrium.*

Most of the work on alteration in carbohydrate metabolism due to trematode infections has been done on schistosomiasis. For example, because the eggs of schistosomes produce extensive necrotic lesions in the host's liver, disturbances in hepatic functions, including glycogen synthesis and detoxification mechanisms, occur in severe cases. Glycosuria and diabetes mellitus have been reported from parasitized humans (Seife and Lisa, 1950). These conditions are due not only to liver damage but also to disruption of the pancreas caused by eggs.

Histopathologically, it is noted that in rabbits infected with *Clonorchis sinensis*, liver polysaccharides are decreased, most pronounced in the peripheral regions of the lobules, and glycogen granules appear in the normally polysaccharide-free epithelium of the bile ducts (Kuwamura, 1958).

TOXINS

Certain endoparasitic trematodes secrete toxic substances. For example, the liver lesions produced by

Fasciola hepatica are only partially due to the traumatic destruction of liver tissue. These lesions are also due to the trematodes' proteolyic, amylolytic, and lipolytic enzymes, to the toxic effect of the excreta of the flukes, and to the absorption of the dying worms' autolytic products. Injections of *F. hepatica* excreta into experimental animals produce toxic reactions, and if the material is injected locally, it produces edema, fever, inflammation, and sometimes anemia. In *S. japonicum*, an ether-soluble hemolysin is present.

EFFECTS OF LARVAE ON MOLLUSCS

Increasingly more investigators are studying the effects of larval trematodes on their molluscan hosts. This topic has been reviewed by Wright (1966), Cheng (1967), and Malek and Cheng (1974). The range of effects varies from none to considerable, the latter being the rule.

The effects of larval trematodes on their molluscan hosts can be categorized in four groups—the effect on the host's digestive gland, the effect on the host's gonads and reproductive structures, the effect on the general physiologic state of the host, and immunologic responses on the part of the host.

Effects on the digestive gland of the host by larval trematodes include (1) accumulation of fatty droplets in the cytoplasm of digestive gland cells; (2) the appearance of vacuoles in the cytoplasm; (3) karyolysis; (4) sloughing of tissues due to mechanical damage; (5) formation of fibromata and granulomata; (6) decrease in the amount of glycogen; (7) rupture of the covering epithelium, resulting in the penetration of foreign bodies; (8) histolytic effects inflicted by the excretory products of the larvae; (9) secretions of granular substances by the host cells; (10) release of cell pigments by host cells; (11) metabolic degeneration of host cells; and (12) displacement and destruction of digestive gland tubules.

Effects on the host's gonads and reproductive structures include (1) direct ingestion of and physiologic damage to the host's gonadal tissue by rediae or mechanical or physiologic destruction of gonads by sporocysts, resulting in parasitic castration; (2) sex reversals; and (3) inhibition of normal gametogenesis through the disruption of normal vascularization, crowding, and toxicity. Effects on the general physiologic state of the host include (1) increased shell growth of the host in some cases,* retardation of growth in others, and no influence on growth in still others, depending on the particular host-parasite association; (2) pigmentation, thinning, and "ballooning"

---

*What has been designated "gigantism" in parasitized molluscs has been questioned by Cheng (1971), who believes that the increased size is due primarily to increased calcium deposition in the shells, at least in most cases.

of shells; (3) a rise in body temperature; (4) blockage and destruction of circulation; (5) increased calcium content in the tissues; (6) decrease in hemoglobin and other hemolymph protein concentrations in the case of *Biomphalaria glabrata* parasitized by *Schistosoma mansoni*; (7) increased heart rate; (8) lowered hemolymph glucose concentration; (9) increased oxygen uptake; and (10) hypersynthesis of lysosomal enzymes within the molluscan host's hemocytes (Cheng *et al.*, 1983).

Immunologic responses in molluscs to larval trematodes have been discussed earlier (p. 68). In brief, these responses are most severe in those host-parasite associations which are of relatively recent origin or are not totally compatible.

EFFECTS OF HOST ON PARASITE

The diet of the host is known to affect trematodes. For example, Krakower *et al.* (1944) have reported that in *Schistosoma mansoni* infection of guinea pigs, although the worms are not affected if the host is maintained on diets deficient in vitamin C, the trematode eggs produced possess weakly formed shells. It is believed that lack of vitamin C affects the shell globule-forming ability of the vitelline glands. These same investigators reported that if the guinea pigs are maintained on diets deficient in vitamin A, more of the initially introduced parasites survive.

In other instances, Rothschild (1939) has found when gulls are deprived of vitamins their *Cryptocotyle lingua* are unaffected, but Beaver (1937) has reported that if pigeons are fed diets deficient in vitamins A and D, the development of their *Echinostoma revolutum* parasites is retarded, or they do not develop at all.

An interesting aspect of the growth pattern of digeneans in their definitive hosts is the effect of parasite density. As an example, Willey (1941) has reported that individuals of the amphistome *Zygocotyle lunata* vary in size within the same host, depending on the number of individuals present. The fewer the worms, the larger the individuals. This phenomenon, known as the **crowding effect**, occurs in other digenean infections also and among cestode infections.

Finally, there is increasing evidence that genetics plays an important role in host-parasite relationships. In brief, certain genetic strains of the same host species may be totally or partially refractory to the same species of parasite. This is vividly demonstrated by the *Biomphalaria glabrata-Schistosoma mansoni* relationship. The Brazilian strain of *B. glabrata* is refractory to the Puerto Rican strain of *S. mansoni*, and the Puerto Rican strain of the snail is refractory to the Brazilian strain of *S. mansoni*. If a Brazilian snail is crossed with a Puerto Rican one, the progeny have a susceptibility index intermediate between the two parental strains (Newton, 1954). There is little doubt that comparable phenomena occur among vertebrate hosts. For a review of the genetic aspects of host-parasite relationships, see the volume edited by Taylor and Muller (1976).

REFERENCES

Andrade, Z. A., and Barka, T. (1962). Histochemical observations on experimental schistosomiasis of mouse. *Am. J. Trop. Med. Hyg.* **11**, 12–16.

Ansari, N. (ed.) (1972). "Epidemiology and Control of Schistosomiasis (Bilharziasis)." University Park Press, Baltimore, Maryland.

Baer, J. G., and Joyeux, C. (1961). Classe des Trématodes (Trématoda Rudolphi). *In* "Traité de Zoologie: Anatomie, Systematique, Biologie. Vol. 4. Part I. Plathelminthes, Mésozoaires, Acanthocéphales, Némertiens." (P. P. Grasse, ed.). Masson & Cie, Paris.

Bain, R. D., and Etges, F. J. (1973). *Schistosoma mansoni*: Factors affecting hatching of eggs. *Exp. Parasitol.* **33**, 155–167.

Barrett, J. (1981). "Biochemistry of Parasitic Helminths." University Park Press, Baltimore, Maryland.

Barrett, J., Coles, G. C., and Simpkin, K. G. (1978). Pathways of acetate and propionate production in adult *Fasciola hepatica. Int. J. Parasitol.* **8**, 117–123.

Beaver, P. C. (1937). Experimental studies on *Echinostoma revolutum* (Froel.), a fluke from birds and mammals. *Ill. Biol. Monogr.* **15**, 1–96.

Bogitsh, B. J. (1963). Histochemical observations on the cercariae of *Posthodiplostomum minimum. Exp. Parasitol.* **14**, 193–202.

Bogitsh, B. J. (1972). Cytochemical and biochemical observations on the digestive tracts of digenetic trematodes. IX. *Megalodiscus temperatus. Exp. Parasitol.* **32**, 244–260.

Bogitsh, B. J. (1975). Cytochemical observations on the gastrodermis of digenetic trematodes. *Trans. Am. Microsc. Soc.* **94**, 524–528.

Bogitsh, B. J. (1982). *Schistosoma mansoni*: cytochemistry and morphology of the gastrodermal Golgi apparatus. *Exp. Parasitol.* **53**, 57–67.

Bogitsh, B. J., and Shannon, A. W. (1971). Cytochemical and biochemical observations on the digestive tracts of digenetic trematodes. VIII. Acid phosphatase activity in *Schistosoma mansoni* and *Schistosomatium douthitti. Exp. Parasitol.* **29**, 337–347.

Bogitsh, B. J., Davis, D. A., and Nunnally, D. A. (1968). Cytochemical and biochemical observations on the digestive tracts of digenetic trematodes II. Ultrastructural localization of acid phosphatase in *Haematoloechus medioplexus. Exp. Parasitol.* **23**, 303–308.

Bruce, J., Sornmani, S., Asch, H. L., and Crawford, K. A. (eds.) (1980). "The Mekong Schistosome." *Malacol. Rev. Suppl.* **2**, 1–282.

Bryant, C. (1975). Carbon dioxide utilization, and the regulation of respiratory metabolic pathways in parasitic helminths. *Adv. Parasitol.* **13**, 35–69.

Bryant, C. (1978). The regulation of respiratory metabolism in parasitic helminths. *Adv. Parasitol.* **16**, 311–331.

Bryant, C., and Williams, J. P. G. (1962). Some aspects of the metabolism of the liver fluke, *Fasciola hepatica* L. *Exp. Parasitol.* **12**, 372–376.

Bueding, E. (1949). Metabolism of parasitic helminths. *Physiol. Rev.* **29**, 195–218.

Bunge, G. (1889). Weitere Untersuchungen über die Athmung der Würmer. *Z. Phys. Chem.* **14**, 318–324.

Bushara, H. O., Majid, B. Y. A., Majîd, A. A., Khitma, I., Gameel, A. A., Karib, E. A. Hussein, M. F., and Taylor, M. G. (1983a). Observations on cattle schistosomiasis in the Sudan, a study in comparative medicine. V. The effect of Praziquantel therapy on naturally acquired resistance to *Schistosoma bovis. Am. J. Trop. Med. Hyg.* **32**, 1370–1374.

Bushara, H. O., Gameel, A. A., Majid, B. Y. A., Khitma, I., Haroun, E. M., Karib, E. A., Hussein, M. F., and Taylor, M. G. (1983b). Observations on cattle schistosomiasis in the Sudan, a study in comparative medicine. VI. Demonstration of resistance to *Schistosoma bovis* challenge after a single exposure to normal cercariae or to transplanted adult worms. *Am. J. Trop. Med. Hyg.* **32**, 1375–1380.

Byrd, E. E. (1935). Life history studies of Reniferinae parasitic in reptilia of the New Orleans area. *Trans. Am. Microsc. Soc.* **54**, 196–225.

Byrd, E. E. (1939). Studies on the blood flukes of the family Spirorchiidae. Part II. Revision of the family and descriptions of new species *J. Tenn. Acad. Sci.* **14**, 116–161. (Also see *ibid.* **13**, 133–136.)

Cable, R. M. (1971). Parthenogenesis in parasitic helminths. *Am. Zool.* **11**, 267–272.

Cheng, T. C. (1963). Biochemical requirements of larval trematodes. *Ann. N.Y. Acad. Sci.* **113**, 289–321.

Cheng, T. C. (1967). Marine molluscs as hosts for symbioses. *Adv. Mar. Biol.* **5**, 1–424.

Cheng, T. C. (1968). The compatibility and incompatibility concept as related to trematodes and molluscs. *Pac. Sci.* **22**, 141–160.

Cheng, T. C. (ed.) (1971). "Aspects of the Biology of Symbiosis." University Park Press, Baltimore, Maryland.

Cheng, T. C. (1973). Human parasites transmissible by seafood—and related problems. *In* "Microbial Safety of Fishery Products" (C. O. Chichester and H. Graham, eds.) pp. 163–189. Academic Press, New York.

Cheng, T. C., and Bier, J. W. (1972). Studies on molluscan schistosomiasis: an analysis of the development of the cercaria of *Schistosoma mansoni. Parasitology* **64**, 129–141.

Cheng, T. C., and Yee, H. W. F. (1968). Histochemical demonstration of aminopeptidase activity associated with the intramolluscan stages of *Philophthalmus gralli* Mathis and Léger. *Parasitology* **58**, 473–480.

Cheng, T. C., Shuster, C. N., Jr., and Anderson, A. H. (1966). A comparative study of the susceptibility and response of eight species of marine pelecypods to the trematode *Himasthla quissetensis. Trans. Am. Microsc. Soc.* **85**, 284–295.

Cheng, T. C., Howland, K. H., Moran, H. J., and Sullivan, J. T. (1983). Studies on parasitic castration: aminopeptidase activity levels and protein concentrations in *Ilyanassa obsoleta* (Mollusca) parasitized by larval trematodes. *J. Invert. Pathol.* **42**, 42–50.

Clegg, J. A. (1965). Secretion of lipoprotein by Mehlis' gland in *Fasciola hepatica. Ann. N.Y. Acad. Sci.* **118**, 969–986.

Coil, W. H., and Kuntz, R. E. (1963). Observations on the histochemistry of *Syncoelium spathulatum* n.sp. *Proc. Helminthol. Soc. Wash.* **30**, 60–65.

Coles, G. C. (1973). The metabolism of schistosomes: a review. *Int. J. Biochem.* **4**, 319–337.

Coles, G. C. (1975). Fluke biochemistry—*Fasciola* and *Schistosoma. Helminthol. Abstr.* **44**, 147–162.

Coles, G. C. (1984). Recent advances in schistosome biochemistry. *Parasitology* **89**, 603–637.

Cort, W. W., Ameel, D. J., and Van der Woude, A. (1954). Germinal development in the sporocysts and rediae of the digenetic trematodes. *Exp. Parasitol.* **3**, 185–225.

Daugherty, J. W. (1952). Intermediary protein metabolism in helminths. I. Transaminase reactions in *Fasciola hepatica. Exp. Parasitol.* **1**, 331–338.

Davis, D. A., Bogitsh, B. J., and Nunnally, D. A. (1968). Cytochemical and biochemical observations on the digestive tracts of digenetic trematodes. I. Ultrastructure of *Haematoloechus medioplexus* gut. *Exp. Parasitol.* **22**, 96–106.

Davis, G. M. (1971). Mass cultivation of *Oncomelania* (Prosobranchia: Hydrobiidae) for studies of *Schistosoma japonicum. In* "Culturing *Biomphalaria* and *Oncomelania* (Gastropoda) for large-scale studies of schistosomiasis." *Bio-Medical Reports,* 406 Med. Lab. No. 19, 85–161.

Dawes, B. (1946). "The Trematoda." Cambridge University Press, London.

Dawes, B. (1962). A histological study of the caecal epithelium of *Fasciola hepatica* L. *Parasitology* **52**, 483–493.

Dawes, B. (1963). Some observations of *Fasciola hepatica* L. during feeding operations in the hepatic parenchyma of the mouse, with notes on the nature of liver damage in this host. *Parasitology* **53**, 135–143.

Dike, S. C. (1969). Acid phosphatase activity and ferritin incorporation in the ceca of digenetic trematodes. *J. Parasitol.* **55**, 111–123.

Dinnick, J. A., and Dinnick, N. N. (1963). Effect of the seasonal variations of temperature on the development of *Fasciola gigantica* in the snail host in the Kenya highlands. *Bull. Epizool. Dis. Africa* **11**, 197–207.

Dinnick, J. A., and Dinnick, N. N. (1964). The influence of temperature on the succession of redial and cercarial generations of *Fasciola gigantica* in a snail host. *Parasitology* **54**, 59–65.

Dixon, K. E. (1965). The structure and histochemistry of the cyst wall of the metacercaria of *Fasciola hepatica* L. *Parasitology* **53**, 215–226.

Dixon, K. E., and Mercer, E. H. (1964). The fine structure of the cyst wall of the metacercaria of *Fasciola hepatica. Quart. J. Microsc. Sci.* **105**, 385–389.

Dogiel, V. A. (1964). "General Parasitology" (Revised and enlarged by Y. I. Polyanski and E. M. Kheisin. English translation by Z. Kabata.) Oliver & Boyd, Edinburgh.

Dorsey, C. H., and Stirewalt, M. A. (1971). *Schistosoma mansoni*: fine structure of cercarial acetabular glands. *Exp. Parasitol.* **30**, 199–214.

Dresden, M. H., and Deelder, A. M. (1979). *Schistosoma mansoni*: Thiol proteinase properties of adult worm "hemoglobinase." *Exp. Parasitol.* **48**: 190–197.

Dubois, G. (1937). Contribution a l'étude des Diplostomes d'oiseaux (Trematoda: Diplostomatidae Poirier, 1886) du Musée de Vienne. *Bull. Soc. Neuchâtel Sci. Natur.* **62**, 99–128.

Dubois, G. (1938). Monographie des Strigeida (Trematoda). *Mem Soc. Neuchâtel Sci. Natur.* **5**, 1–535.

Erasmus, D. A. (1972). "The Biology of Trematodes." Edward Arnold, London.

Erasmus, D. A. (1977). The host-parasite interface of trematodes. *Adv. Parasitol.* **15**, 201–242.

Erasmus, D. A., and Öhman, C. (1963). The structure and function of the adhesive organ in strigeid trematodes. *Ann. N.Y. Acad. Sci.* **113**, 7–35.

Etges, F. J. (1960). On the life history of *Prosthodendrium* (*Acanthatrium*) *anaplocami* n.sp. (Trematoda: Lecithodendriidae). *J. Parasitol.* **46**, 235–240.

Freeman, R., Stuart, P. F., Cullen, J. B., Ritchie, A. C., Mildon, A., Fernandes, B. J., and Bonin, R. (1976). Fatal human infection with mesocercariae of the trematode *Alaria americana. Am. J. Trop. Med. Hyg.* **25**, 803–807.

Gebhardt, G. A., Millemann, R. E., Knapp, S. E., and Nyberg, P. A. (1966). "Salmon poisoning" disease. II. Second intermediate host susceptibility studies. *J. Parasitol.* **52**, 54–59.

Gelfand, M. (1967). "A Clinical Study of Intestinal Bilharziasis (*Schistosoma mansoni*) in Africa." Edward Arnold, London.

Gönnert, R. (1955). Schistosomiasis studies. II. Über die Eibildung bei *Schistosoma mansoni* und das Schicksal der Eier im Wirtsorganismus. *Z. Tropenmed. Parasit.* **6**, 33–52.

Gönnert, R. (1962). Histologische Untersuchungen über den Feinbau der Eibildungstätte (Oogenotop) von *Fasciola hepatica. Z. Parasitenk.* **21**, 475–492.

Goodchild, C. G., and Fried, B. (1963). Experimental infection of the planorbid snail *Menetus dilatatus buchanensis* (Lea) with *Spirorchis* sp. (Trematoda). *J. Parasitol.* **49**, 588–592.

Halton, D. W. (1964). Observations on the nutrition of certain parasitic flatworms (Platyhelminthes: Trematoda). Ph.D. Thesis, University of Leeds, England.

Hansen, E. L., Perez-Mendez, G., Yarwood, E., and Buecher, E. J. (1974). Second-generation daughter sporocysts of *Schistosoma mansoni* in axenic culture. *J. Parasitol.* **60**, 371–372.

Harris, K. R., and Cheng, T. C. (1973). Histochemical demonstration of fats associated with intestinal caeca of *Leucochloridiomorpha constantiae. Trans. Am. Microsc. Soc.* **92**, 496–502.

Higashi, G. I., Kreiner, P. W., Kierns, J. J., and Bitensky, M. W. (1973). Adenosine 3′, 5′-cyclic monophosphate in *Schistosoma mansoni. Life Sci.* **13**, 1211–1220.

Hiromoto, T. (1939). Chemische Untersuchungen des Blutes bei experimenteller Kaninchenschistosomiasis japonica. *Okayama Igakkai Zasshi* **51**, 1633–1637. (English summary).

Hockley, D. J. (1973). Ultrastructure of the tegument of *Schistosoma. Adv. Parasitol.* **11**, 233–305.

Hoffman, G. L. (1960). Synopsis of Strigeoidea (Trematoda) of fishes and their life cycles. *Fish. Bull. U.S. Fish. Wildl. Serv.* **175**, pp. 439–469. U.S. Govt. Printing Office, Washington, D.C.

Hoffman, G. L. (1967). "Parasites of North American Freshwater Fishes." Univ. California Press, Berkeley, California.

Hortsmann, H. J. (1962). Sauerstoffverbranch und Glykogengehalt der Eier von *Fasciola hepatica* während der Entwicklung der Miracidien. *Z. Parasitenk.* **21**, 437–445.

Hsü, H. F., Hsü, S. Y. L., and Osbourne, J. W. (1962). Immunization against *Schistosoma japonicum* in rhesus monkeys produced by irradiated cercariae. *Nature* **194**, 98–99.

Hsü, H. F., Davis, J. R., Hsü, S. Y. L., and Osbourne, J. W. (1963a). Histopathology in albino mice and rhesus monkeys infected with irradiated cercariae of *Schistosoma japonicum. Z. Tropenmed. Parasit.* **14**, 240–261.

Hsü, H. F., Hsü, S. Y. L., and Osbourne, J. W. (1963b). Further studies on rhesus monkeys immunized against *Schistosoma japonicum* by administration of X-irradiated cercariae. *Z. Tropenmed. Parasit.* **14**, 402–412.

Hsü, H. F., Hsü, S. Y. L., and Osbourne, J. W. (1965). Immunization against *Schistosoma japonicum* in rhesus monkeys. *Nature* **206**, 1338–1340.

Hsü, S. Y. L., and Hsü, H. F. (1968). The strain complex of *Schistosoma japonicum* in Taiwan, China. *Z. Tropenmed. Parasitol.* **19**, 43–59.

Huang, T. Y., T'ao, Y. H., and Chu, C. H. (1962). Studies on transaminases of *Schistosoma japonicum. Chinese Med. J. Peking* **81**, 79–85.

Hulse, E. V. (1971). Joshua's curse and the abandonment of ancient Jericho: schistosomiasis as a possible medical explanation. *Med. Hist.* **15**, 376–386.

Ishii, N. (1935). Studies on the family Didymozoidae (Monticelli, 1888). *Jap. J. Zool.* **6**, 279–335.

Ito, J. (1964). *Metagonimus* and other human heterophyid trematodes. *In* "Progress of Medical Parasitology in Japan" (K. Morishita, Y. Komiya, and H. Matsubayashi, eds.), Vol. 1, pp. 317–393. Meguro Parasitological Museum, Tokyo.

Jordan, P., and Webbe, G. (1982). "Schistosomiasis: Epidemiology, Treatment, and Control." Wm. Heineman Medical Books, London.

Jourdane, J., and Théron, A. (1980). *Schistosoma mansoni*: cloning by microsurgical transplantation of sporocysts. *Exp. Parasitol.* **50**, 349–357.

Kasschau, M. R., and Mansour, T. E. (1982). Serotonin-activated adenylate cyclase during early development of *Schistosoma mansoni. Nature* **296**, 66–68.

Kean, B. H., and Breslau, R. C. (1964). "Parasites of the Human Heart." Grune & Stratton, New York.

Kendall, S. B. (1965). Relationships between the species of *Fasciola* and their molluscan hosts. *Adv. Parasitol.* **3**, 59–98.

Kim, D. C., and Kuntz, R. E. (1964). Epidemiology of helminth diseases: *Clonorchis sinensis* (Cobbold, 1875) Looss, 1907 on Taiwan (Formosa). *Chinese Med. J.* **11**, 29–47.

Köhler, P., Bryant, C., and Behm, C. A. (1978). ATP synthesis in a succinate decarboxylase system from *Fasciola hepatica* mitochondria. *Int. J. Parasitol.* **8**, 399–404.

Komiya, Y. (1966). *Clonorchis* and clonorchiasis. *Adv. Parasitol.* **4**, 53–106.

Krakower, C. A., Hoffman, W. A. and Axtmayer, J. H. (1944). Granulación en la cubierta de los huevos de *Esquistosoma mansoni*. *Puerto Rico J. Publ. Health* **19**, 669–677.

Krull, W. H., and Mapes, C. (1952). Studies on the biology of *Dicrocoelium dendriticum* (Rud., 1819) (Trematoda: Dicrocoeliidae), including its relation to the intermediate host, *Cionella lubrica*. *Cornell Vet.* **42**, 252–285. [Also see *ibid.* **42**, 277–285; 339–351; 464–489; 603–605; (1953) *ibid.* **43**, 119–202; 389–401].

Kurelec, B., and Ehrlich, I. (1963). Über die Natur der von *Fasciola hepatica* (L.) *in vitro* ausgeschiedenen Amino- und Ketosäuren. *Exp. Parasitol.* **13**, 113–117.

Kuwamura, T. (1958). [Studies on experimental clonorchiasis, especially on the histochemical change in the liver.] *Shikoku Igaku Zasshi,* **12**, 28–57. (In Japanese.)

LaRue, G. R. (1957). The classification of digenetic Trematoda: a review and a new system. *Exp. Parasitol.* **6**, 306–349.

Lee, D. L. (1966). The structure and composition of the helminth cuticle. *Adv. Parasitol.* **4**, 187–254.

Lee, D. L., and Lewert, R. M. (1957). Studies on the presence of mucopolysaccharidase in penetrating helminth larvae. *J. Infect. Dis.* **101**, 287–294.

Lee, D. L., and Smith, M. H. (1965). Hemoglobins of parasitic animals. *Exp. Parasitol.* **16**, 392–424.

Leuckart, K. G. F. R. (1882). Zur Entwickelungsgeschichte des Leberegels. Zweite Mittheilung. *Zool. Anz.* **5**, 524–528.

Lewert, R. M., and Dusanic, D. G. (1961). Effects of a symmetrical diaminodibenzylalkane on alkaline phosphatase of *Schistosoma mansoni*. *J. Infect. Dis.* **109**, 85–89.

Lewert, R. M., and Lee, C. L. (1954). Studies on the passage of helminth larvae through host tissues. I. Histochemical studies on extracellular changes caused by penetrating larvae. II. Enzymatic activity of larvae *in vitro* and *in vivo*. *J. Infect. Dis.* **95**, 13–51.

Lewert, R. M., and Lee, C. L. (1956). Quantitative studies of the collagenase-like enzymes of cercariae of *Schistosoma mansoni* and the larvae of *Strongyloides ratti*. *J. Infect. Dis.* **99**, 1–14.

Malek, E. A. (1980). "Snail-Transmitted Diseases." Vols. I and II. CRC Press, Boca Raton.

Malek, E. A., and Cheng, T. C. (1974). "Medical and Economic Malacology." Academic Press, New York.

Mansour, T. E. (1962). Effect of serotonin on glycolysis in homogenates from the liver fluke *Fasciola hepatica*. *J. Pharmacol. Exp. Therap.* **135**, 94–101.

Mansour, T. E. (1984). Serotonin receptors in parasitic worms. *Adv. Parasitol.* **23**, 2–36.

Mansour, T. E., and Mansour, J. M. (1962). Effect of serotonin (5-hydroxytryptamine) and adenosine 3′,5′ phosphofructokinase from the liver fluke *Fasciola hepatica*. *J. Biol. Chem.* **237**, 629–634.

Mapes, C. R. (1952). *Cionella lubrica* (Muller), a new intermediate host of *Dicrocoelium dendriticum* (Rud., 1819) Looss, 1899. *J. Parasitol.* **38**, 84.

Meyerhof, E., and Rothschild, M. (1940). A prolific trematode. *Nature* **146**, 367–368.

Millemann, R. E., and Knapp, S. E. (1970a). Biology of *Nanophyetus salmincola* and "salmon poisoning" disease. *Adv. Parasitol.* **8**, 1–41.

Millemann, R. E., and Knapp, S. E. (1970b). Pathogenicity of the "salmon poisoning" trematode, *Nanphyetus salmincola*, to fish. *In* "A Symposium on Diseases of Fishes and Shellfishes" (S. F. Snieszko, ed.), pp. 209–217. Am. Fish. Soc. Spec. Publ., No. 5. Washington, D.C.

Miller, G. C. (1981). Helminths and the transmammary route of infection. *Parasitology* **82**, 335–342.

Mohandas, A. (1983). Platyhelminthes—Trematoda. *In* "Reproductive Biology of Invertebrates. Vol. II. Spermatogenesis and Sperm Function." (K. G. Adiyodi and R. G. Adiyodi, eds.). pp. 105–129. John Wiley, New York.

Morris, G. P. 1973. The fine structure of the cecal epithelium of *Megalodiscus temperatus*. *Canad. J. Zool.* **51**, 457–460.

Mostofi, F. K. (ed.) (1967). "Bilharziasis." Springer-Verlag, New York.

Newton, W. L. (1954). Tissue response to *Schistosoma mansoni* in second generation snails from a cross between two strains of *Australorbis glabratus*. *J. Parasitol.* **40**, 353–355.

Nimmon-Smith, R. H., and Standen, O. D. (1963). Phosphomonoesterases of *Schistosoma mansoni*. *Exp. Parasitol.* **13**, 305–322.

Noble, G. A. (1963). Experimental infection of crabs with *Paragonimus*. *J. Parasitol.* **44**, 352.

Noble, G. A. (1975). Description of *Nematobibothrioides histoidii* (Noble, 1974) (Trematoda: Didymozoidae) and comparisons with other genera. *J. Parasitol.* **61**, 224–227.

Northrup, J. K., and Mansour, T. E. (1978). Adenylate cyclase from *Fasciola hepatica*. 2. Role of guanine nucleotides in coupling adenylate cyclase and serotinin receptors. *Molec. Pharmacol.* **14**, 820–833.

Öhman, C. (1965). The structure and function of the adhesive organ in strigeid trematodes Part II. *Diplostomum spathaceum* Braun, 1893. *Parasitology* **55**, 481–502.

Oudejans, R. C. H. M., and Van Der Horst, D. J. (1974). Aerobic-anaerobic biosynthesis of fatty acids and other lipids from glycolytic intermediates in the pulmonate land snail *Cepaea nemoralis* (L.). *Comp. Biochem. Physiol.* **47B**, 139–147.

Pan, S. C.-T. (1980). The fine structure of the miracidium of *Schistosoma mansoni. J. Invert. Pathol.* **36**, 307–372.

Pantelouris, E. M. (1964). Localization of glycogen in *Fasciola hepatica* L. and an effect of insulin. *J. Helminthol.* **38**, 283–286.

Pearson, I. G. (1963). Use of chromium radioisotope ^{51}Cr to estimate blood loss through ingestion by *Fasciola hepatica*. *Exp. Parasitol.* **13**, 186–193.

Philip, C. B. (1958). A helminth replaces the usual arthropod as vector of a rickettsial-like disease. *Proc. Tenth Int. Congr. Entomol. Montreal*, 1956, Vol. 3, pp. 651–653.

Rifkin, E. (1971). Interaction between *Schistosoma mansoni* schistosomules and penetrated mouse skin at the ultrastructural level. *In* "Aspects of the Biology of Symbiosis" (T. C. Cheng, ed.), pp. 25–43. University Park Press, Baltimore, Maryland.

Rijavec, M., Kurelec, B., and Ehrlich, I. (1962). Ueber den

Verbrauch der Serumalbumine *in vitro* durch *Fasciola hepatica. Biol. Glas.* **15**, 103–107.

Robinson, G., and Threadgold, L. T. (1975). Electron microscope studies of *Fasciola hepatica*. XII. The fine structure of the gastrodermis. *Exp. Parasitol.* **37**, 20–36.

Rogers, W. P. (1940). Haematological studies on the gut contents of certain nematode and trematode parasites. *J. Helminthol.* **18**, 53–62.

Rothschild, M. (1939). A note on the life cycle of *Cryptocotyle lingua* (Creplin) 1825 (Trematoda). *Nov. Zool.* **41**, 178–180.

Rowan, W. B. (1956). The mode of hatching of the egg of *Fasciola hepatica. Exp. Parasitol.* **5**, 118–137.

Rowan, W. B. (1957). The mode of hatching of the egg of *Fasciola hepatica*. II. Colloidal nature of the viscous cushion. *Exp. Parasitol.* **6**, 131–142.

Rowcliffe, S. A., and Ollerenshaw, C. B. (1960). Observations on the bionomics of the egg of *Fasciola hepatica. Ann. Trop. Med. Parasit.* **54**, 172.

Schell, S. C. (1970). "How to Know the Trematodes." Wm. C. Brown, Dubuque, Iowa.

Scheuring, L. (1920). Die Lebensgeschichte eines Karpfenparasiten (*Sanguinicola inermis* Plehn). *Allg. Fishchwirtschaftztg.* **45**, 225–230.

Seife, M., and Lisa, J. R. (1950). Diabetes mellitus and pylephlebitic abscess of the liver resulting from *Schistosoma mansoni* infestation. *Am. J. Trop. Med.* **30**, 769–772.

Self, J. T., Peters, L. E., and Davis, C. E. (1963). The egg, miracidium, and adult of *Nematobothrium texomensis* (Trematoda: Digenea). *J. Parasitol.* **49**, 731–736.

Shoop, W. L., and Corkum, K. C. (1983). Transmammary infection of paratenic and definitive hosts with *Alaria marcianae* (Trematoda) mesocercariae. *J. Parasitol.* **69**, 731–735.

Short, R. B., and Menzel, M. Y. (1959). Chromosomes in parthenogenetic miracidia and embryonic cercariae of *Schistosomatium douthitti. Exp. Parasitol.* **8**, 249–264.

Smithers, S. R., and Terry, R. J. (1969). The immunology of schistosomiasis. *Adv. Parasitol.* **7**, 41–93.

Smithers, S. R., and Terry, R. J. (1976). The immunology of schistosomiasis. *Adv. Parasitol.* **14**, 399–422.

Smyth, J. D., and Clegg, J. A. (1959). Egg-shell formation in trematodes and cestodes. *Exp. Parasitol.* **8**, 286–323.

Smyth, J. D., and Halton, D. W. (1983). "The Physiology of Trematodes." 2nd ed. Cambridge University Press, Cambridge, England.

Stirewalt, M. A. (1963). Chemical biology of secretions of larval helminths. *Ann. N.Y. Acad. Sci.* **113**, 36–53.

Stirewalt, M. A. (1966). Skin penetration mechanisms of helminths. *In* "Biology of Parasites" (E. J. L. Soulsby, ed.), pp. 41–59. Academic Press, New York.

Stirewalt, M. A. (1971). Penetration stimuli for schistosome cercariae. *In* "Aspects of the Biology of Symbiosis" (T. C. Cheng, ed.), pp. 1–23. University Park Press, Baltimore, Maryland.

Stirewalt, M. A. (1973). Important features of the schistosomes. *In* "Epidemiology and Control of Schistosomiasis" (N. Ansari, ed.), pp. 17–31. University Park Press, Baltimore, Maryland.

Stirewalt, M. A., and Evans, A. S. (1952). Demonstration of an enzymatic factor in cercariae of *Schistosoma mansoni* by the streptococcal decapsulation test. *J. Infect. Dis.* **91**, 191–197.

Stirewalt, M. A., and Hackey, J. R. (1956). Penetration of host skin by cercariae of *Schistosoma mansoni*. I. Observed entry into skin of mouse, hamster, rat, monkey and man. *J. Parasitol.* **42**, 565–580.

Stirewalt, M. A., and Kruidenier, F. J. (1961). Activity of the acetabular secretory apparatus of cercariae of *Schistosoma mansoni* under experimental conditions. *Exp. Parasitol.* **11**, 191–211.

Stunkard, H. W. (1940). Life history studies and the development of parasitology. *J. Parasitol.* **26**, 1–15.

Takahashi, T., Mori, K., and Shigeta, Y. (1961). Phototactic, thermotactic and geotactic responses of miracidia of *Schistosoma japonicum. Jap. J. Parasitol.* **10**, 686–691.

Taylor, A. E. R., and Baker, J. R. (eds.). (1978). "Methods of Cultivating Parasites *in vitro*." Academic Press, London.

Taylor, A. E. R., and Muller, R. (eds.). (1976). "Genetic Aspects of Host-Parasite Relationships." Blackwell Sci. Publ., Oxford, England.

Thakur, A. S., and Cheng, T. C. (1968). The formation, structure, and histochemistry of the metacercarial cyst of *Philophthalmus gralli* Mathis and Léger. *Parasitology* **58**, 605–618.

Théron, A., and Fournier, A. (1982). Mise en évidence de structures nerveuses dans le sporocyste-fils du Trématode *Schistosoma mansoni*. C. R. Acad. Sci. Paris, Sér. III. **294**, 365–369.

Thomas, A. F. W. (1883). The life history of the liver fluke (*Fasciola hepatica*). *Quart. J. Microsc. Sci.* **23**, 99–133.

Timms, A. R. (1960). Schistosome enzymes. *In* "Host Influence on Parasite Physiology" (L. A. Stauber, ed.), pp. 41–49. Rutgers University Press, New Brunswick, New Jersey.

Timms, A. R., and Bueding, E. (1959). Studies of a proteolytic enzyme from *Schistosoma mansoni. Br. J. Pharmacol.* **14**, 68–73.

Ulmer, M. J. (1951a). *Postharmostomum helicis* (Leidy, 1847) Robinson, 1949, (Trematoda), its life history and a revision of the subfamily Brachylaeminae. Part I. *Trans. Am. Microsc. Soc.* **70**, 189–238.

Ulmer, M. J. (1951b). *Postharmostomum helicis* (Leidy, 1847) Robinson, 1949, (Trematoda), its life history and a revision of the sub-family Brachylaeminae. Part II. *Trans. Am. Microsc. Soc.* **70**, 319–347.

Ulmer, M. J. (1971). Site-finding behavior in helminths in intermediate and definitive hosts. *In* "Ecology and Physiology of Parasites" (A. M. Fallis, ed.), pp. 123–159. University of Toronto Press, Toronto.

Van Vugt, F., Kalaycioglu, L., and Van den Bergh, S. G. (1976). ATP production in *Fasciola hepatica* mitochondria. *In* "Biochemistry of Parasites and Host-Parasite Relationships" (H. van den Bosche, ed.), pp. 151–166. North-Holland, Amsterdam.

Vernberg, W. B. (1963). Respiration of digenetic trematodes. *Ann. N.Y. Acad. Sci.* **133**, 261–271.

Vernberg, W. B., and Hunter, W. S. (1959). Studies on oxygen consumption in digenetic trematodes. III. The relationship of body nitrogen and oxygen uptake. *Exp. Parasitol.* **8**, 76–82.

Vernberg, W. B., and Vernberg, F. J. (1971). Respiratory

metabolism of a trematode metacercaria and its host. *In* "Aspects of the Biology of Symbiosis" (T. C. Cheng, ed.), pp. 91–102. University Park Press, Baltimore, Maryland.

Villella, J. B., and Weinbren, M. P. (1965). Abnormalities in adult *Schistosoma mansoni* developed from gamma-irradiated cercariae. *J. Parasitol.* **51**, Sect. 2, 42.

Voge, M., Bruckner, D., and Bruce, J. I. (1978). *Schistosoma mekongi* sp.n. from man and animals, compared with four geographic strains of *Schistosoma japonicum. J. Parasitol.* **64**, 577–584.

Wall, L. D. (1941). Life history of *Spirorchis elephantis* (Cort, 1917), a new blood fluke from *Chrysemys picta. Am. Midl. Natur.* **25**, 402–412.

Warren, K. S. (1973a). "Schistosomiasis. The Evolution of a Medical Literature. Selected Abstracts and Citations, 1852–1972." Massachussetts Institute of Technology Press, Cambridge, Massachusetts.

Warren, K. S. (1973b). The pathology of schistosome infections. *Helminth. Abstr.* Ser. A **42**, 591–633.

Warren, K. S., and Hoffman, D. B. Jr. (1976). "Schistosomiasis III. Abstracts of the Complete Literature 1963–1974." Vols. I and II. John Wiley, New York.

Warren, K. S., and Newill, V. A. (1967). "Schistosomiasis. A Bibliography of the World's Literature from 1852 to 1962." Vols. I and II. Press Western Reserve University, Cleveland, Ohio.

Willey, C. H. (1941). The life history and bionomics of the trematode, *Zygocotyle lunata* (Paramphistomidae). *Zoologica* **26**, 65–88.

Wilson, R. A. (1968). The hatching mechanism of the egg of *Fasciola hepatica* L. *Parasitology* **58**, 79–89.

Wright, C. A. (1966). The pathogenesis of helminths in the mollusca. *Helminth. Abstr.* **35**, 207–225.

Wright, W. H. (1972). A consideration of the economic import of schistosomiasis. *Bull. W.H.O.* **197**, 559–566.

Yamaguti, S. (1943). Cercaria of *Plagiorchis muris* (Tanabe, 1922). *Annot. Zool. Jap.* **22**, 1–3.

Yamaguti, S. (1958). "Systema Helminthum," Vol. I (two parts). Interscience, New York.

Yamaguti, S. (1970). "Digenetic Trematodes of Hawaiian Fishes." Keigaku, Tokyo, Japan.

Yamaguti, S. (1971). "Synopsis of Digenetic Trematodes of Vertebrates." Keigaku, Tokyo.

Yamao, Y. (1952). [Histochemical studies on endoparasites. V. Distributions of the glyceromonophosphatases in the tissues of flukes, *Eurytrema coelomaticum, Dicrocoelium lanceatum,* and *Clonorchis sinensis*] *J. Coll. Arts Sci. Chiba Univ.* **1**, 9–13. (In Japanese.)

Yokogawa, M. (1965). *Paragonimus* and paragonimiasis. *Adv. Parasitol.* **3**, 99–158.

CLASSIFICATION OF DIGENEA (THE DIGENETIC TREMATODES)

SUBORDER ANEPITHELIOCYSTIDA

Embryonic excretory vesicle (bladder) retained in adult, not replaced by mesodermal cells; vesicular wall thin, not epithelial; cercaria with forked or simple, straight tail; cercaria without oral stylet.

Superfamily Strigeoidea

Body of adult usually divided into anterior and posterior portions by constriction; pharynx present; accessory suckers and/or penetration glands often present on anterior portion of body; genital pore usually terminal; cercaria furcocerous, with two long rami; parasites of reptiles, birds, and mammals. (Genera mentioned in text: Family Strigeidae*—*Strigea, Ophiosoma, Parastrigea, Cotylurus.* Family Cyathocotylidae—*Cyathocotyle.* Family Diplostomatidae—*Posthodiplostomum, Alaria, Uvulifer.*)

Superfamily Schistosomatoidea

Dioecious adults in vascular system of definitive host; pharynx absent; no second intermediate host in life cycle; cercaria furcocercous with relatively short rami; oral sucker of cercaria replaced by protractile penetration organ; cercarial eyespots either pigmented or not; parasites of fishes, reptiles, birds, and mammals. (Genera mentioned in text: Family Schistosomatidae*—*Schistosoma, Schistosomatium, Trichobilharzia, Bilharziella, Microbilharzia.* Family Sanguinicolidae—*Sanguinicola, Cardicola.* Family Spirorchidae—*Spirorchis.*)

Superfamily Bucephaloidea

Mouth midventral on both cercaria and adult; acetabulum absent; cercarial tail very short, with two very long rami; parasites of fish and amphibians. (Genera mentioned in text: Family Bucephalidae*—*Bucephalus, Rhipidocotyle, Bucephalopsis, Prosorhynchus.*)

Superfamily Echinostomatoidea

Adult and cercaria usually with circumoral collar, commonly armed with spines; parasites of reptiles, birds, and mammals. (Genera mentioned in text: Family Fasciolidae*—*Fasciola, Fascioloides, Parafasciolopsis, Fasciolopsis.* Family Philophthalmidae—*Parorchis, Philophthalmus.* Family Paramphistomatidae—*Cotylophoron, Paramphistomum, Zygocotyle.* Family Echinostomatidae—*Himasthla, Echinostoma*).

Superfamily Clinostomatoidea

Cercaria with short rami on tail; cercarial oral sucker replaced with protractile penetration organ; cercarial ventral sucker rudimentary; pigmented eyespots on cercaria; parasites of reptiles, birds, and mammals. (Genus mentioned in text: Family Clinostomatidae*—*Clinostomum.*)

Superfamily Brachylaemoidea

Genital pore in posterior half of body; cercaria usually develops in terrestrial snails; cercaria with or without tail; parasites of amphibians, birds, and mammals. (Genera mentioned in text: Family Brachylaemidae*—*Postharmostomum.* Family Leucochloridiidae—*Leucochloridium.*)

Superfamily Paramphistomoidea

Adults always amphistomous; no second intermediate host in life cycle; cercaria without penetration glands; cercaria monostome or amphistome; parasites of fish, amphibians, birds, and mammals. (Genera mentioned in text: Family Paramphistomatidae*—*Diplodiscus, Megalodiscus, Watsonius, Gastrodiscoides.*)

Superfamily Azygioidea

Adult often very large; cercaria furcocystocercous; cercarial oral sucker present, but ventral sucker sometimes absent; parasites of fish. (Genus mentioned in text: Family Azygiidae*—*Otodistomum.*)

---

*For diagnoses of families subordinate to the Digenea, see Yamaguti (1958).

Superfamily Cyclocoeloidea

Adult elongate and flat; acetabulum rudimentary or absent; oral sucker present or absent; intestinal caeca united at posterior end of body; cercaria with short, bilobed tail or no tail; parasites of the respiratory system of birds.

Superfamily Fellodistomatoidea

Adult plump; intestinal caeca may be united posteriorly; cercaria furcocercous, develops in marine bivalves; parasites of marine fish.

Superfamily Transversotrematoidea

Adult transversely elongated; intestinal caeca fused at distal ends; cercaria with short rami; base of cercarial tail with two appendages; pharynx absent; parasites in dermis of fish.

Superfamily Notocotyloidea

Adult monostomous; may have tegumental glands on ventral surface; cercaria monostomous; without pharynx; parasites of reptiles, birds, and mammals. (Genus mentioned in text: Family Pronocephalidae*—*Pleurogonius*.)

SUBORDER EPITHELIOCYSTIDIA

Wall of embryonic excretory vesicle (bladder) replaced by epithelial cells of mesodermal origin; cercaria with simple, straight tail; cercarial oral stylet commonly present.

Superfamily Plagiorchioidea

Adults distomate; cercaria distomate, with oral stylet; metacercaria usually in invertebrates; parasites of fish, amphibians, reptiles, birds, and mammals. (Genera mentioned in text: Family Plagiorchiidae*—*Haplometra, Haematoloechus, Astiotrema, Glossidium, Plagiorchis, Opisthogonimus, Stomatrema, Styphlodora, Dasymetra, Ochetosoma, Zeugorchis*. Family Dicrocoeliidae—*Dicrocoelium*. Family Lecithodendriidae—*Prosthodendrium*. Family Brachycoeliidae—*Leptophallus*. Family Microphallidae—*Gynaecotyla*.)

Superfamily Allocreadioidea

Adult one of several morphologic types; with or without eyespots; oral sucker usually simple, but some with appendages; acetabulum in anterior half of body; testes in posterior half of body; ovary pretesticular; cercaria one of several morphologic types; cercaria usually with eyespots; parasites of fish, amphibians, reptiles, and mammals. (Genera mentioned in text: Family Allocreadiidae*—*Crepidostomum*. Family Zoogonidae—*Zoogonus*. Family Troglotrematidae—*Paragonimus, Nanophyetus, Sellacotyle*.)

SUBORDER DIDYMOZOIDA

Adults are tissue-dwelling parasites of fish, especially marine species; often occur in pairs; many species dioecious, with considerable sexual dimorphism; oral sucker commonly absent; ventral sucker absent in most species or present in young worms only. (Genera mentioned in text: Family Didymozoidae*—*Nematobothrium, Ovarionematobothrium*.)

Superfamily Opisthorchioidea

Cercaria with well-developed penetration glands; cercarial oral sucker protractile; cercarial ventral sucker rudimentary; cercarial tail one of several types; parasites of fish, amphibians, reptiles, birds, and mammals. (Genera mentioned in text: Family Opisthorchiidae*—*Opisthorchis, Clonorchis*. Family Heterophyidae—*Heterophyes, Metagonimus, Centrocestus, Haplorchis, Cryptocotyle*.)

Superfamily Hemiuroidea

Tegument usually without spines; testes usually anterior to ovary; vitellaria usually situated behind ovary; tail appendage present on some species; parasites of fish. (Genus mentioned in text: Family Syncoeliidae*—*Syncoelium*.)

Superfamily Isoparorchioidea

Body large, usually plump; testes behind acetabulum; vitellaria posterior; parasitic in swim bladder of fish.

SPECIES OF UNDETERMINED SUPERFAMILY MEMBERSHIP MENTIONED IN TEXT

Family Haploporidae*—*Saccocoelium*.

11

CESTOIDEA: THE TAPEWORMS
CESTODARIA: THE UNSEGMENTED TAPEWORMS

The Cestoidea constitutes another interesting group of parasitic flatworms. Members of this class possess all the characteristics of the phylum Platyhelminthes presented in Chapter 8.* In addition, they lack a mouth and digestive tract, and like the other parasitic platyhelminths, their body surfaces are covered with a tegumental layer. Many of the Cestoidea are true tapeworms of the subclass Eucestoda, whereas a smaller, less well-known group belongs to the subclass Cestodaria. The true tapeworms generally possess segmented bodies with a specialized holdfast organ, the **scolex**, at the anterior end that is embedded in the intestinal mucosa of the host. A few tapeworms, however, do not possess segmented bodies, and still others are segmented internally but lack external body divisions. More detailed accounts of these forms are given in Chapter 12. The cestodarians, on the other hand, do not possess segmented bodies and may superficially resemble the trematodes.

One phase of host-parasite relationship among the

Cestoidea is that concerned with species that parasitize humans and domestic animals, causing varying degrees of pathologic alterations. They are, therefore, of great interest to the medical and veterinary professions. While effects of this type of parasitic infection may be regarded as disease causing, from the biologic standpoint the resulting conditions might best be regarded as manifestations of host response to parasitism. Such is the case of the broadfish tapeworm, *Diphyllobothrium latum*, which becomes parasitic in humans when larvae are ingested in raw or poorly cooked fish. As an intestinal parasite, it may produce no symptoms or may provide severe systemic toxemia and even a type of anemia (p. 438). Another example is the dwarf tapeworm, *Vampirolepis nana*, which may cause eosinophilia and severe systemic toxemia, especially in children.

Not only are the morphology and life cycles of tapeworms of interest to biologists, but also recent studies on the biochemical physiology of these parasites have greatly enhanced our understanding of parasitism. Advances made in understanding the cell biology, biochemistry, and immunology of tapeworms, specifically *Hymenolepis diminuta*, a favored experimental model, are reflected in the volume edited

---

*Based on his interpretation of embryologic evidence, Ubelaker (1983) has proposed that the Cestoidea should be considered a distinct phylum.

by Arai (1980). In addition, the two-volume treatise edited by Arme and Pappas (1983) includes reviews of practically all aspects of the biology and biochemistry of the Eucestoda.

SUBCLASS CESTODARIA

The cestodarians, which are all endoparasites in the intestine and coelomic cavities of various fish and rarely in reptiles, show similarities to both true tapeworms and to trematodes. Nevertheless, since these animals lack a digestive tract and possess fairly well-developed parenchymal muscles, most modern zoologists consider them more closely related to the Eucestoda than to the Trematoda. Their position in the class Cestoidea is suggested not only by the two previously mentioned characteristics, which they share in common with true tapeworms, but also because their larvae resemble those of the Eucestoda. The major difference is that the cestodarian larva—the **lycophore**—characteristically bears ten hooks, whereas the larva of tapeworms have six. On the other hand, the cestodarians are similar to the trematodes in that some species possess suckers similar to those of digenetic flukes. Furthermore, the cestodarians lack scolices, and have only one set of hermaphroditic reproductive organs. The apparent lack of strobilation (proglottization) has caused the cestodarians to be described as monozoic in contrast to the polyzoic condition of the Eucestoda.

Cestodarians are believed to be ancient parasites that were once a flourishing group, which originated in freshwater hosts, in which many modern members of one order, the Amphilinidea, are still found, but which later became marine along with their hosts. These hosts were most probably primitive elasmobranchs, of which the chimaerids are among the only survivors today. Indeed, almost all of the known modern members of another order, the Gyrocotylidea, are intestinal parasites of marine chimaerid fish. If this hypothesis is valid, then these worms must have originated during the early Mesozoic and possibly during the Paleozoic periods. Aside from anatomic characteristics, the fact that cestodarians parasitize only primitive fish suggests their ancient origin.

Wardle and McLeod (1952) recognize three orders subordinate to the subclass Cestodaria: Amphilinidea, Gyrocotylidea, and Biporophyllidea, although the last mentioned is now almost uniformly regarded as the proglottid of a tetraphyllid cestode (p. 426). Members of the two valid subclasses, Amphilinidea and Gyrocotylidea, are quite different from each other, even in certain aspects of their basic histology. These differences make it impractical to discuss a general anatomic pattern. Rather, the patterns are expounded on in the respective sections devoted to each order.

Body Tissues

The tegument of cestodarians is essentially identical to that of the true tapeworms except that electron-dense tips are absent on the projecting microtriches (p. 388) (Cole, 1968; Lyons, 1969). When examined with the light microscope, the tegument is thinner in the amphilinids than in the gyrocotylids. Immediately underneath the tegument is a basal lamina of connective tissue fibers. Internal to this are two layers of muscles, the outer being circularly arranged and the inner being longitudinally arranged. These subtegumental muscles are better developed in gyrocotylids. Certain large parenchymal cells are present in the subtegumental region of amphilinids. At least some of these are cytons (or perikaryons), which are connected by cytoplasmic bridges with the surficial, syncytial layer of the tegument (Cole, 1968). The parenchymal muscles of the gyrocotylids are also generally better developed than those of the amphilinids. However, the long boring muscles, which extend from the proboscis to the so-called **anchor cells** in the posterior region of the body, are very well developed in amphilinid cestodarians. Being acoelomate animals, the spaces between the body wall and the internal organs are filled with a meshlike parenchyma, similar to that found in trematodes and cestodes. Simmons (1974) has presented a detailed account of the parenchyma of *Gyrocotyle*.

BIOLOGY OF CESTODARIANS

The diagnostic characteristics of the two cestodarian orders—Amphilinidea and Gyrocotylidea—are given in the classification list at the end of this chapter.

ORDER AMPHILINIDEA

Adults of the members of this order are parasitic in the coelom of sturgeons, other primitive fish, and tortoises. These ribbonlike monozoic worms measure up to 37 cm in length.

Proboscis and Related Structures

The proboscis, located at the anterior end of the body, is a powerful boring apparatus. It can be retracted into the body proper. A bundle of well-developed muscle fibers extends posteriorly from within the proboscis to certain large cells—the **anchor cells**—located at the posterior end of the body. These muscles, which are twisted anteriorly, cause the proboscis to twist partially during the boring process. In addition, they support the anterior end of the worm during boring and drag the posterior end of the body through the perforation made by the proboscis. True retractor muscles are also present, which are responsible for extension and retraction of the proboscis.

Within the proboscis in some species are found clusters of **frontal glands** that secrete through minute pores on the proboscal surface. The exact chemical nature and function of the secretion have not been studied, but it is assumed to play a lytic role during penetration.

Osmoregulatory System

The osmoregulatory (excretory) system in members of this order is poorly known. It is known, nevertheless, that in *Amphilina* each flame cell includes a group of 18 to 30 flame bulbs (or cells) (Fig. 11.1). Two lateral canals lead into a common excretory vesicle located at the posterior end of the body, which empties to the exterior through a single excretory pore.

Nervous System

The nervous system of amphilinids is similar to that of other platyhelminths. There is a large ganglionic mass located immediately behind the proboscis. From this primitive "brain," two main nerve trunks are directed anteriorly and two posteriorly. The posteriorly directed trunks are joined near the posterior by a commissure.

Reproductive Systems

The cestodarians are monoecious, with a complete set of male and female reproductive organs.

Male System. In amphilinid cestodarians, the male system consists of follicular testes, which are generally limited to the two lateral fields (Fig. 11.1). However, in *Amphilina* the testes are scattered throughout the parenchyma. An individual **vas efferens** arises from each follicle, and these unite to form the common **vas deferens** (or sperm duct), which is directed posteriorly and opens to the exterior through the male genital pore situated at the posterior. Prior to opening to the exterior, the vas deferens is provided with a muscular zone into which prostate glands secrete. In some amphilinids, a **genital papilla** is present at the copulatory terminal.

Female System. The female system consists of a single ovary located near the posterior of the body (Fig. 11.1). An oviduct arises from the surface of the ovary and eventually enlarges to become a winding uterus, which terminates anteriorly at the uterine pore near the proboscis. Branching from the oviduct is the slender **vagina**, which extends posteriorly and opens to the outside in one of three ways, depending on the

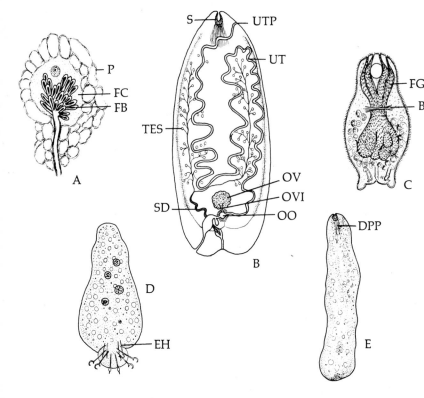

Fig. 11.1. Morphology and stages in the life cycle of *Amphilina*. A. Flame cell with cluster of flame bulbs. (Redrawn after Hein, 1904.) **B.** Adult *Amphilina foliacea* showing male and female reproductive systems. (Redrawn after Benham, 1901.) **C.** Lycophore showing large frontal glands. **D.** Procercoid. (**C** and **D,** redrawn after Janicki, 1928.) **E.** Plerocercoid. B, brain; DPP, developing proboscis protractor muscles; EH, embryonic hook; FB, flame bulbs; FC, flame cell; FG, frontal glands; OO, ootype; OV, ovary; OVI, oviduct; P, parenchyma; S, sucker; SD, sperm duct; TES, testis; UT, uterus; UTP, uterine pore.

species. The vagina can join the vas deferens and open through a common genital pore, open through a separate female genital pore, or bifurcate and open through two separate pores. Vitelline glands are poorly developed and are arranged in two lateral bands, with the common vitelline duct entering the oviduct near the vagino-oviductal junction.

During copulation, spermatozoa are introduced into the vagina through the female or common genital pore, depending on which is present. The spermatozoa ascend the vagina and enter the oviduct, where fertilization takes place. Vitelline secretions are laid down around the fertilized ovum to form part of the egg. The eggs pass up the uterus and, in some ovoviviparous species, hatch before their expulsion from the uterine pore. In oviparous species, development of the enclosed embryo is completed outside the host.

The embryology of *Gephyrolina paragonopora* has been studied along with a portion of its life cycle by Woodland (1923), and the life cycle of *Amphilina foliacea* also has been studied (Salensky, 1874; von

Janicki, 1928). Development in both species is essentially the same in that the zygote is enveloped by a coat of yolk material within a thin, irregularly ovoid shell. Embryonic development occurs while the egg is *in utero*. The escaping ciliated larva of *A. foliacea*—known as a **decacanth larva** or **lycophore** (Fig. 11.1)—is approximately 100 μm long and has 10 hooks embedded in the posterior end. The lycophore of *G. paragonopora* is not ciliated. Both have clusters of large frontal gland cells that open to the outside anteriorly.

Woodland has postulated that the lycophore of *Gephyrolina* escapes through the uterine pore of the parent only after the parent has burrowed through the body wall of the fish host, because he has observed that lycophores that are hatched prematurely within

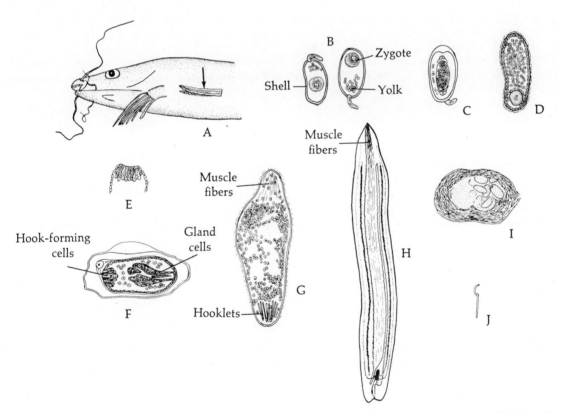

Fig. 11.2. Stages in the life cycle of *Gephyrolina paragonopora*. A. Adult escaping from body cavity of fish host. **B.** Eggs from the proximal segment of uterus enclosing zygote, yolk, and shell. **C.** Egg from medium segment of uterus enclosing morula-stage embryo surrounded by invested membrane. **D.** Longitudinal section of early larva showing one large blastomere. (Shell not shown.) **E.** Section through posterior end of larva showing hooklet-forming cells. **F.** Section of early larva showing gland cells and hooklet-forming cells. **G.** Older larva with fully developed hooklets at posterior end. **H.** Sexually mature adult, dorsal view, showing reproductive organs. **I.** Young cyst developing from mesentery of fish host and enclosing degenerate larva and disintegration products. **J.** Larval hooklet from posterior end of fully developed larva. (Redrawn after Woodland, 1923.)

the host's body cavity become surrounded with histolytic tissue and are eventually destroyed.

Amphilinid Genera

None of the members of the Amphilinidea are common. The type genus, *Amphilina*, is probably the most frequently encountered. *Amphilina foliacea*, a parasite in the coelom of European sturgeons, *Acipenser* spp., was first described by Rudolphi in 1819; he originally thought it was a trematode. Several other species of *Amphilina* have been recorded in various species of sturgeons. In North America, *A. bipunctata* has been reported from a sturgeon in Oregon. The genus *Gephyrolina* includes several species from siluroid fish in India and Brazil (Fig. 11.2). *Gigantolina* includes extremely large forms, measuring up to 380 mm in length, found in the fish *Diagramma crassispinum* in Sri Lanka. A ribbonlike amphilinid, *Austramphilina elongata*, occurs in the body cavity of the tortoise *Chelodina longicollis* in Australia (Fig. 11.3).

Life Cycle of *Amphilina foliacea*. Von Janicki (1928) has contributed the only completely known amphilinid life history, that of *A. foliacea*. The ciliated lycophore, although fully developed, does not escape from the eggshell until the egg is ingested by the intermediate host—an amphipod crustacean. Within the gut of the amphipod, the lycophore hatches and sheds its ciliated epidermis. Secretions of the frontal glands aid the now naked lycophore in penetrating the gut wall of the intermediate host, and the lycophore finally inhabits its hemocoel. In this location, the lycophore undergoes dedifferentiation, that is, it loses most of its body structures, including the frontal

glands, and develops a rounded tail somewhat similar to the cercomer of some cestode larvae. The 10 larval hooks of the lycophore are now enclosed within this tail. This larval form may be referred to as the **procercoid** (Fig. 11.1).

As the procercoid develops, its body elongates, the tail is shed, the adult proboscis is formed along with the frontal glands, and the body muscles begin to take form. This stage is now known as a **plerocercoid** (Fig. 11.1). When the amphipod host is ingested by a sturgeon, the plerocercoid burrows through the fish's gut wall and enters the coelom, where it attains sexual maturity. In *Amphilina* and related genera, such as *Gigantolina* (Fig. 11.3), the 10 embryonic hooks are maintained throughout life in the parenchyma near the posterior end.

ORDER GYROCOTYLIDEA

Adults of the gyrocotylid cestodarians are primarily intestinal parasites of marine chimaerid fish and are quite different from the amphilinids. In fact, upon gross inspection, the two orders bear little resemblance to each other. The gyrocotylids possess a wrinkled bundle of posterior flaps called the **rosette**, which resembles a miniature carnation, at the posterior end of their bodies surrounding a conical depression (Figs. 11.4, 11.5).

These parasites employ the rosette to cling to the coelomic wall of the fish host. A muscular, eversible, cup-shaped proboscis (also known as a sucker or acetabulum) is located at the anterior end (Fig. 11.6). Other characteristics of the gyrocotylids are the ruffled body margins and tegumentary spines at the anterior and posterior ends of the body. Spines located at the posterior end are arranged in front of the rosette. The parenchymal muscles are well developed, as is a

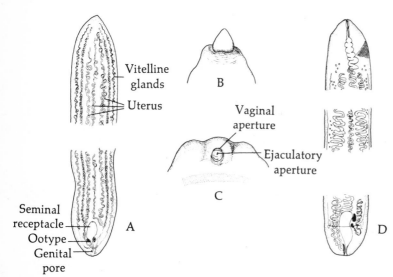

Fig. 11.3. Some cestodarians. A. *Gigantolina magna* from intestine of *Diagramma crassispinum* from Sri Lankan waters. (Redrawn after Southwell, 1928.) **B.** Anterior terminal of *Gephyrolina paragonopora.* **C.** Posterior terminal of *G. paragonopora.* (**B** and **C**, redrawn after Woodland, 1923.) **D.** *Austramphilina elongata* from the body cavity of the tortoise *Chelodina longicollis* in Australia. (Redrawn after Johnston, 1931.)

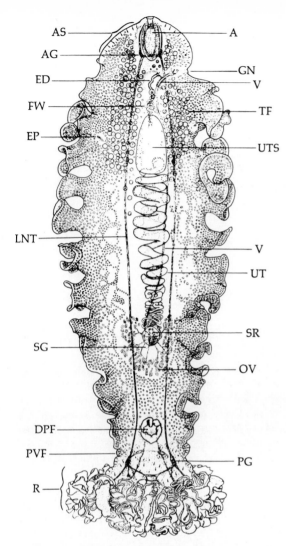

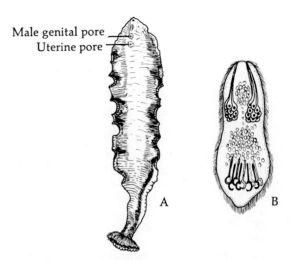

Fig. 11.5. *Gyrocotyle urna.* **A.** Fully extended adult specimen, ventral view. (Redrawn after Wardle, 1932.) **B.** Lycophore larva showing two groups of glands and 10 larval hooks. (Redrawn after Ruszkowski, 1932.)

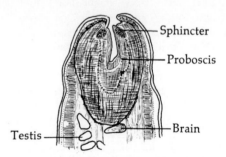

Fig. 11.6. *Gyrocotyle.* Longitudinal section through the proboscis of an adult. (Redrawn after Watson, 1911.)

Fig. 11.4. *Gyrocotyle fimbriata,* **dorsal view.** A, acetabulum; AG, anterior ganglion; AS, acetabular spines; DPF, dorsal pore of funnel; ED, ejaculatory duct; EP, excretory pore; FW, Fovea wageneri; GN, genital notch; LNT, longitudinal nerve trunk; OV, ovary; PG, posterior ganglion; PVE, posterior limit of vitelline follicles; R, rosette; SG, shell gland; SR, seminal receptacle; TF, testicular follicles; UT, uterus; UTS, uterine sac; V, vagina. (After Lynch, 1945.)

Watson (1911) has described two ridges and pits at the anterior tip of the proboscis of *Gyrocotyle,* which are richly innervated and are presumed to be tactile.

Osmoregulatory System

The osmoregulatory system consists of a network of partially ciliated vessels and minute tubules bearing flame cells at their terminals. The main ducts empty to the outside of the body through two nephridiopores (excretory pores) at the anterior end of the body. There is no excretory vesicle (Fig. 11.4).

Reproductive System

The male reproductive system consists of numerous testes scattered anteriorly throughout the parenchyma (Fig. 11.4).

sphincter located at the posterior end of the proboscis, and the retractor muscles, which control the eversion and retraction of the proboscis.

Nervous System

The nervous system of gyrocotylids is unique in that it appears to be better developed posteriorly. Immediately in front of the rosette are located a commissure, two ganglia, and a nerve ring (Fig. 11.4). Anteriorly, a commissure is located behind the proboscis, and two lateral nerve cords extend posteriorly from it. The nerve cords join the posteriorly situated ganglia.

The individual vasa efferentia unite to form a common vas deferens, or sperm duct, which leads to the male genital pore situated on a minute genital papilla projecting from the body surface. This papilla is laterally located on the ventral body surface near the anterior end (Fig. 11.5). Prior to entering the papilla, the vas deferens first forms a **spermiducal vesicle** followed by a muscular area.

The female reproductive system consists of many ovarian follicles in the posterior region of the body (Fig. 11.4). A single **oviductule** arises from each follicle. These oviductules unite to form the **oviduct**, which leads to the **ootype** before enlarging to become the **uterus**. The uterus is coiled and terminates at the female genital pore situated on the middorsal line in the anterior part of the body, posterior to the proboscis. Vitelline glands are scattered throughout the parenchyma, especially in the two lateral regions. The common vitelline duct enters the oviduct, as do the Mehlis' gland cells and the duct of the large **seminal receptacle**, if present. The seminal receptacle is the storage site for sperm, which enter the anteroventrally located **vaginal pore** (female genital pore) during copulation and which swim down to the oviduct which empties into the receptacle.

According to Smyth (1969), egg formation in cestodarians, at least the gyrocotylid *Amphiptyches urna*, is very similar to that in trematodes. Specifically, ova are released periodically from the ovarian follicles and pass down the oviduct to the ootype, where fertilization occurs. Mature vitelline cells pass from the vitelline glands to the ootype, where they release globules of shell precursor which coalesce to form a capsule. Secretion from the Mehlis' gland cells may also play a role in this release. The capsule, which is pliable at first, contains the ovum and remains of the vitelline cells or yolk, which serve as nourishment during subsequent development. As the operculate egg passes along the uterus, it becomes darker and hardens as the result of quinone tanning (p. 401).

Gyrocotylid Genera

The Gyrocotylidea includes the genera *Gyrocotyle* (Figs. 11.4, 11.5) and *Gyrocotyloides* (Fig. 11.7), both parasitic in chimaeroid fish. In *Gyrocotyloides* the surface tegument is smooth (devoid of spines) and the rosette is poorly developed.

Of interest to American parasitologists is *Gyrocotyle fimbriata* (Fig. 11.4), an intestinal parasite of the ratfish, *Hydrolagus colliei*, which occurs along the Pacific coast of the United States. This parasite averages 32 mm in length, but ranges from 12 to 65 mm. Simmons and Laurie (1972) have reported that *G. fimbriata* often

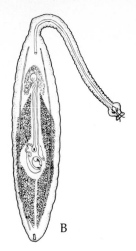

Fig. 11.7. ***Gyrocotyloides.*** *G. nybelini* in the primitive fish *Chimaera monstrosa*. (Redrawn after Fuhrmann, 1931.)

occurs in the same ratfish as a related species, *G. parvispinosa*, in Puget Sound; however, the latter species occupies a more posterior habitat in the host's valvular intestine. The size of both of these species of gyrocotylideans is directly related to the sizes of their hosts, with larger fish harboring larger worms.

Life Cycle of *Gyrocotyle urna*. Complete life history data on gyrocotylids are wanting. Ruszkowski (1932) and Simmons (1974), however, have reported phases of the developmental cycle of *Gyrocotyle* (Fig. 11.5), a parasite of chimaeroid fish. In this species, the eggs secrete a gelatinous capsule a few hours after being ejected through the uterine pore. The *in utero* development of the lycophore is incomplete; hence, further development ensues only after the operculated egg escapes into water. Again, the parent bores through the body wall of the fish host to lay its eggs. Subsequent development is shown in Table 11.1.

Lycophores of *Gyrocotyle urna* (Fig. 11.5) and *G. rugosa* possess a ciliated epidermis, although gyrocotylid lycophores generally lack a ciliated epidermis, possessing instead an embryonic membrane, presumably epidermal, between the larva and the capsule (shell). Ruszkowski's attempts to infect various molluscs (*Cardium*, *Lima*, and *Mytilus*) were unsuccessful, but he did report finding larvae (procercoids) 3–5 mm

Table 11.1.   Development of *Gyrocotyle urna*

Days after Leaving Parent	Stage of Development
5–7	Yolk cells fuse
10–15	Embryonic hooks (10) appear
12–20	First movement observed
25–30	Lycophore fully developed, leaves shell

long in fish hosts on which the rosettes had already formed and the ten embryonic hooks grouped together where the rosettes had formed. This finding suggests that no intermediate host is necessary, but many workers believe that larval development in an unknown intermediate host is required. Von Linstow's (1903) *Gyrocotyle medusarum*, found in a jellyfish, *Phyllorhiza rosacea*, in the Pacific, is thought by those who take the latter view to be a gyrocotylid larva using the jellyfish as its intermediate host. Simmons and Laurie (1972) have suggested that in the case of *Gyrocotyle parvispinosa* and *G. fimbriata* "massive" infections in the ratfish host may result from the ingestion of adult parasites that have escaped from moribund hosts (a common phenomenon among these parasites) and which contain postlarval young worms in their parenchyma. If this hypothesis can be validated experimentally, then these two species of *Gyrocotyle* the life cycle may be direct, not requiring an intermediate host. Adults of *Gyrocotyle* commonly occur in pairs within the fish host (Dienske, 1968; Simmons and Laurie, 1972). This has been interpreted to mean that a pair of worms forms a "functional sexual unit" to ensure cross-fertilization (Halvorsen and Williams, 1968).

Life Cycle of *Gyrocotyle rugosa*. Manter (1951) has reported that the eggs of *Gyrocotyle rugosa*, a parasite of the elephant fish, *Callorhynchus milii*, commonly found in New Zealand waters, hatch almost immediately in seawater. The emerging ciliated lycophores readily penetrate samples of the host's mucus and pieces of tissues cut from the spiral valve of the elephant fish. However, the lycophores demonstrate no affinity for hermit crabs and molluscs placed in their presence. Once within the tissues of the elephant fish, the larvae enter the muscular layers, a few also entering blood vessels. Manter has also found numerous postlarval forms in the mucus from the anterior part of the spiral valve of the host.

Dollfus (1923) and Lynch (1945) have presented extensive bibliographies on *Gyrocotyle*.

PHYSIOLOGY OF CESTODARIANS AND HOST-PARASITE RELATIONSHIPS

Practically nothing is known about the physiology of cestodarians because these monozoic worms are seldom encountered, let alone maintained in the laboratory where they can be studied. It is apparent that their body forms and structures are the results of adaptive evolution correlated with their parasitic way of life. The rosette of gyrocotylids and the powerful proboscis of amphilinids are manifestations of such adaptations that facilitate host-parasite contact and penetration. That cestodarians lack an alimentary tract suggests that they acquire their nutrients by absorption through the body tegument. Indeed, fine structural studies on the tegument by Lyons (1969) and Cole as reported by Simmons (1974) strongly suggest that this is the case. In fact, Simmons *et al.* (1972) have demonstrated that *Gyrocotyle fimbriata* readily accumulates adenine, thymine, and thymidine, although these compounds do not appear to become incorporated into nucleic acids, at least during the initial 24 hours after uptake.

The muscular proboscis of cestodarians is needed by adults to bore through the body wall of their fish host in order to release larvae or eggs to the outside.

Some interest has been directed to the function of the frontal glands of the Amphilinidea. Von Janicki (1928) has reported that these glands in the lycophore secrete a substance that aids in the larval penetration of the intestinal wall of the amphipod intermediate host. On the other hand, Woodland (1923) believes that these structures, as found in adults, are not actually glands but are fibers of the boring muscle.

Interestingly, adult amphilinids are parasitic in the coelomic cavities of their hosts, one of the sites commonly occupied by the larvae of certain cestodes. Some have interpreted this to mean that modern amphilinids are neotenic larvae of ancestral tapeworms whose definitive hosts have become extinct. Such neotenic larvae have become able to attain sexual maturity progenetically and no longer require the original definitive host for the completion of their life cycles. This interpretation seems reasonable and is generally accepted.

REFERENCES

Arai, H. P. (ed.) (1980). "Biology of the Tapeworm *Hymenolepis diminuta.*" Academic Press, New York.

Arme, C., and Pappas, P. W. (eds) (1983). "Biology of the Eucestoda," vols. 1 and 2. Academic Press, London.

Cole, N. C. (1968). "Morphological Studies on Cestodaria." M. A. Thesis, Univ. California, Berkeley.

Dienske, H. (1968). A survey of the metazoan parasites of the rabbitfish, *Chimaera monstrosa* L. (Holocephali). *Netherl. J. Sea Res.* **4**, 32–58.

Dollfus, R. P. (1923). L'orientation morphologique des *Gyrocotyle* et des cestodes en général. *Bull. Soc. Zool. Fr.* **48**, 205–242.

Halvorsen, O., and William, H. H. (1968). Studies on the helminth fauna of Norway. IX. *Gyrocotyle* (Platyhelminthes) in *Chimaera monstrosa* from Oslo Fjord, with emphasis on its mode of attachment and a regulation in the degree of infection. *Nytt. Mag. Zool.* **15**, 130–142.

Lynch, J. E. (1945). Redescription of the species of *Gyrocotyle* from the ratfish, *Hydrolagus colliei* (Lay and Bennett), with notes on the morphology and taxonomy of the genus. *J. Parasitol.* **31**, 418–446.

Lyons, K. M. (1969). The fine structure of the body wall of *Gyrocotyle urna*. *Z. Parasitenk.* **33**, 95–109.

Manter, H. W. (1951). Studies on *Gyrocotyle rugosa* Diesing, 1850, a cestodarian parasite of the elephant fish, *Callorhynchus milii*. *Zool. Publ. Victoria Univ. College, New Zeal.* No. 17.

Ruszkowski, J. S. (1932). Etudes sur le cycle évolutif et sur la structure des cestodes de mer. II. Sur les larves de *Gyrocotyle urna* (Gr. et Wagen.). *Bull. Int. Acad. Pol. Sci. Lett. Cl. Sci. Math. Natur. B*, pp. 629–641.

Salensky, W. (1874). Ueber den Bau und die Entwicklungsgechichte der *Amphilina* (*Monostomum foliaceum* Rud.). *Z. Wiss. Zool.* **24**, 291–342.

Simmons, J. E. (1974). *Gyrocotyle*: a century-old enigma. *In* "Symbiosis in the Sea" (W. B. Vernberg, ed.), pp. 195–218. University of South Carolina Press, Columbia, South Carolina.

Simmons, J. E., and Laurie, J. S. (1972). A study of *Gyrocotyle* in the San Juan Archipelago, Puget Sound, U.S.A., with observations on the host, *Hydrolagus colliei* (Lay and Bennett). *Int. J. Parasitol.* **2**, 59–77.

Simmons, J. E., Buteau, G. H., Jr., MacInnis, A. J., and Kilejian, A. (1972). Characterization and hybridization of DNAs of gyrocotylidean parasites of chimaeroid fishes. *Int. J. Parasitol.* **2**, 273–278.

Smyth, J. D. (1969). "The Physiology of Cestodes." Freeman, San Francisco, California.

Ubelaker, J. E. (1983). The morphology, development and evolution of tapeworm larvae. *In* "Biology of the Eucestoda," vol. 1. (C. Arme and P. W. Pappas, eds.), pp. 235–296. Academic Press, London.

von Janicki, C. (1928). Die Lebensgeschichte von *Amphilia foliacea* G. Wagener, Parasiten des Wolga-Sterlets. Nach Beobachtungun und Experimenten. *Arb. Biol. Wolga-Station* No. 10, Saratov.

von Linstow, O. F. B. (1903). Parasiten, meistens Helminthen, aus Siam. *Arch. Mikrosk. Anat. Entwicklungsmech.* **62**, 108–121.

Wardle, R. A., and McLeod, J. A. (1952). "The Zoology of Tapeworms." University of Minnesota Press, Minneapolis, Minnesota.

Watson, E. E. (1911). The genus *Gyrocotyle* and its significance for problems of cestode structure and phylogeny. *Univ. Calif. Publ. Zool.* **6**, 353–468.

Woodland, W. N. F. (1923). On *Amphilina paragonophora* sp. n. and a hitherto undescribed phase in the life history of the genus. *Quart. J. Microsc. Sci.* **67**, 47–84.

CLASSIFICATION OF CESTODARIA (THE CESTODARIANS)

Subclass Cestodaria

Monozoic flatworm; without mouth or digestive tract; body surface covered by syncytial tegument; with protonephritic osmoregulatory (excretory) system; monoecious, with one set of reproductive organs; with relatively well-developed parenchymal muscles; with or without suckers or rosette; lycophore larva with ten hooks; parasitic in primitive fish and tortoises.

ORDER AMPHILINIDEA

Body elongate and dorsoventrally flattened; with calcareous corpuscles in parenchyma; tegument without armature; protrusible proboscis at anterior end; frontal gland present; without suckers; male and female genital pores separate but closely situated to each other at posterior end of body; testes follicular, usually as two lateral bands, but scattered throughout parenchyma in some species; ovary singular, in posterior half of body; uterus preovarian, with N-shaped coil, opening in proximity of proboscis; vitellaria consisting of two lateral bands; two lateral osmoregulatory canals and branches forming peripheral network; excretory vesicle at posterior end of body; adults parasitic in coelom of sturgeons, other primitive fish (Osteoglossidae, Haemulidae, Siluridae, Acipenseridae), and tortoises. (Genera mentioned in text: *Amphilina*, *Gephyrolina*, *Gigantolina*, *Austramphilina*.)

ORDER GYROCOTYLIDEA

Body with ruffled margins; with tegumentary spines; with rosette at posterior body terminal; with muscular cup-shaped eversible proboscis (also referred to as a sucker or acetabulum) at anterior end; sphincter at posterior end of proboscis, with retractor muscles attached to it; ovary posterior; uterus with extensive lateral loops, terminating at midventral gonopore in anterior half of body; testes anterior; male genital pore near anterior end of body; adults parasitic in intestine of primitive fish (Holocephali). (Genera mentioned in text: *Gyrocotyle*, *Amphiptyches*, *Gyrocotyloides*.)

12

EUCESTODA: THE TRUE TAPEWORMS

The Eucestoda,* or true tapeworms, includes a large and important group of the Platyhelminthes. These are by far the most highly specialized metazoan parasites known. All adult members of this subclass** are endoparasitic in the alimentary tract and associated ducts of various vertebrates. During their life cycles, one or two (or more) intermediate hosts are required, in each of which the tapeworms undergo a phase of their development.

Adult tapeworms are dorsoventrally flattened and often opaquely white; however, some species display coloration, usually yellow or gray. These pigments are presumably due to exogenous materials absorbed by the parasite.

The origin and phylogeny of tapeworms are subjects of much speculation. Most modern biologists agree that the Eucestoda originated from a platyhelminth ancestor that arose from a primitive prototype—

the common stock from which modern trematodes also arose. Wardle and McLeod (1952, Chapter 4) and Stunkard (1983) have presented excellent accounts not only of the origin and evolution of the subclass but of the various subgroups as well.

MORPHOLOGY

The body of the typical cestode is divided into three regions (Fig. 12.1). (1) The **scolex**—the holdfast organ—is the anterior end, and its morphology and dimensions are important in the identification of these worms. (2) The **neck** region, situated immediately posterior to the scolex, is unsegmented, poorly differentiated, and generally narrower than the scolex and the strobila proper, and is the continuously differentiating zone that gives rise to immature proglottids, or body segments, of these worms. (3) The **strobila**, which constitutes the main bulk of the body, is made up of a chain of **proglottids**. The most anteriorly situated proglottids are generally immature; that is, the reproductive organs, although visible, are nonfunctional. Proglottids posterior to the immature proglottids are sexually mature, whereas those located

---

*Members of the subclass Eucestoda are commonly referred to as cestodes.

**As stated in an earlier footnote (p. 378), Ubelaker (1983) is of the opinion that the Cestoidea should be considered a distinct phylum. If his opinion is accepted, then the Eucestoda should be considered a subphylum.

toward the posterior are usually **gravid**, i.e., filled with eggs. In some species, the gravid proglottids drop off from time to time, thus permitting the eggs to escape from the host along with fecal wastes.

Monoecious and dioecious species of cestodes are known. Self-fertilization within a single proglottid and between proglottids, and copulation between two worms, take place among the monoecious species, although the former is extremely rare. Even continuous cross-fertilization between different proglottids of the same tapeworm, known as "selfing," appears to be deleterious. For example, Rogers and Ulmer (1962) have shown that when infections of *Vampirolepis nana*** are established in mice, one tapeworm per mouse, thus assuring self-fertilization, the frequency of abnormalities among cysticercoids (a larval form) is increased. Also, the proportion of eggs developing into cysticercoids when fed to *Tribolium confusum*, the confused flour beetle—a suitable but not an obligatory intermediate host (p. 417)—appears to decrease, and the proportion of cysticercoids developing into adults when fed to mice decreases. In addition, no selfed strain survived beyond the fifth selfed generation, suggesting that cross-fertilization between individuals, resulting in

---

*This species is still referred to as *Hymenolepis nana* by many.

what is genetically termed hybrid vigor, is necessary for the normal perpetuation of the species.

In dioecious species, cross-fertilization is necessary. The occurrence of separate sexes among the Eucestoda is limited to the genus *Dioecocestus*, a parasite of the grebe and ibis (Fig. 12.2); however, anomalies among members of other genera exist.

During development and maturation of reproductive organs, tapeworms demonstrate protandrous hermaphroditism; that is, the male organs become functional before the female organs.

Body Tissues

The basic body tissues—the osmoregulatory structures, the reproductive systems, and the nervous systems—of all tapeworms, except for certain morphologic variations among the different taxa, are essentially the same and are discussed generally rather than separately under each taxonomic grouping.

Body Wall. When studied with the electron microscope, specialized microvilli, known as **microtriches**, can be observed projecting from the outer limiting membrane of the tegument (Fig. 12.3). The dimensions of these projections vary among species and locations. For example, those found on the strobila of *Hymenolepis diminuta* are 750 mm long, whereas those of *Calliobothrium* and *Lacistorhynchus* may reach 2 μm in length (Lumsden, 1966). Microtriches, unlike the tegumental microvilli of trematodes, each include an electron-dense apical tip separated from the more basal region by a multilaminar plate (Figs. 12.4, 12.5). The entire surface of the tegument is covered by a

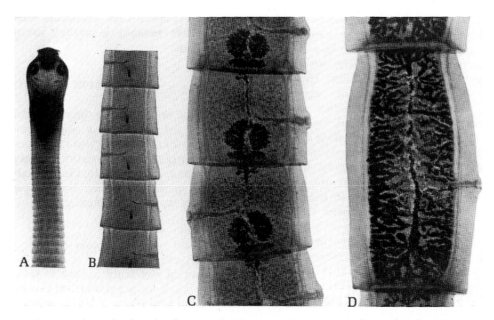

Fig. 12.1.   Regions along the length of a cestode, *Taenia pisiformis.* **A.** Scolex and neck region. **B.** Immature proglottids. **C.** Mature proglottids. **D.** Gravid proglottid. (Courtesy of Ward's Natural Science Establishment, Rochester, New York.)

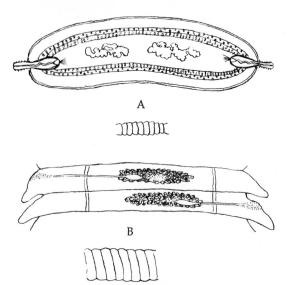

Fig. 12.2. ***Dioecestus acotylus.*** **A.** Transverse section of a proglottid of a male. The irregularly outlined bodies in the center represent testes. **B.** Ventral view of two proglottids of a female specimen. The irregular dark bodies in the middle of the proglottids represent ovaries. (Redrawn after Fuhrmann, 1932.)

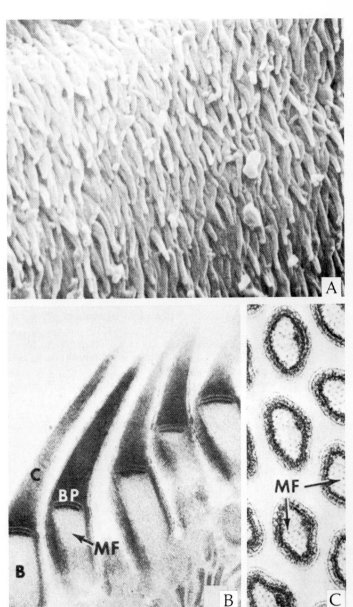

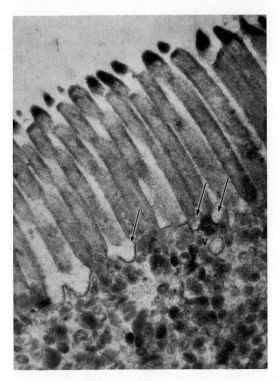

Fig. 12.3. **Cestode microtriches.** Electron micrograph of the body surface of adult *Lacistorhynchus tenuis*, a parasite in the spiral valves of dogfish, *Mustelus canis*, showing microtriches. Note the relationship of certain vesicles in the external level of the tegument with crypts (arrows) situated at the bases of the microvilli, suggesting that the vesicles are endocytotic vesicles. × 24,000. V, vesicle. (After Lumsden, 1966.)

Fig. 12.4. **Microtriches of cestodes. A.** Scanning electron micrograph of posteriorly directed microtriches on the surface of a proglottid of *Hymenolepis diminuta*. (After Ubelaker, Allison, and Specian, 1973; with permission of *Journal of Parasitology*.) **B.** Transmission electron micrograph of sagittal section of tegumental microtriches of *H. diminuta*. **C.** Transmission electron micrograph of cross section through bases of tegumental microtriches of *H. diminuta*, showing orderly array of microfilaments surrounded by accumulation of electron-dense material. B, base of microtrix; BP, multilaminar base plate; C, electron opaque cap; MF, microfilaments. (**B** and **C** after Lumsden and Specian, 1980; with permission of Academic Press.)

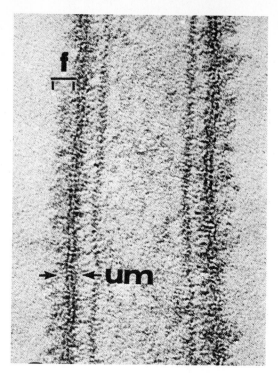

layer of carbohydrate-containing macromolecules known as the **glycocalyx** (Fig. 12.6). Several phenomena known to depend on the adsorption of certain molecules to the glycocalyx include (1) inhibition of such host digestive enzymes as trypsin, chymotrypsin, and pancreatic lipase; (2) absorption of cations; (3) adsorption of bile salts; and (4) enhancement of host amylase activity. In addition, Németh *et al.* (1979) have reported that extracts of certain species of cestodes also inhibit proteases of host origin. Thus, the glycocalyx, in conjunction with other molecules, plays an important role in protection against the host's digestive enzymes, nutrient absorption, and maintenance of the integrity of the parasite's surface membrane.

The distal portion of each microthrix serves as a means of resisting the intestinal current, since the body surface is in intimate contact with the microvilli of the striated border of the cells lining the host's small intestine; and (2) it serves to agitate the microhabitat in the vicinity as the worm moves, stirring up the intestinal fluids so that nutrient materials as well as waste products are in a state of flux. The proximal portion of each microthrix is medullated and could very well serve as sites of absorption.

The cytoplasmic layer from which the microtriches arise is a syncytium. This layer, commonly referred to as the external level of the tegument or as the distal

Fig. 12.6. Cestode glycocalyx. Electron micrograph showing filamentous glycocalyx on the outer surface of the plasmalemma delimiting a tegumental microtrix of *Lacistorhynchus tenuis*, a tapeworm parasite of the dogfish, *Mustelus canis*. The limits of the trilaminate unit membrane covering the microtrix are indicated by arrows (um) while the limits of the glycocalyx are indicated by the bracket at *f*. × 216,000. (After Lumsden, Oaks, and Alworth, 1970; with permission of the *Journal of Parasitology*.)

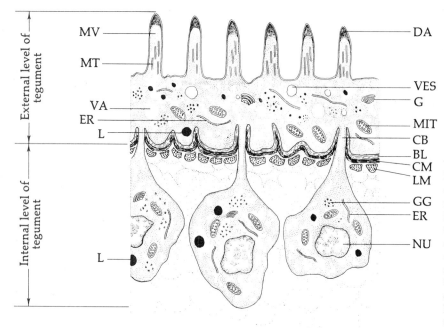

Fig. 12.5. Schematic drawing of fine structure of cestode tegument showing external and internal levels. BL, basal lamella; CB, cytoplasmic bridge; CM, circular muscles; DA, electron-dense apex of microvillus; ER, endoplasmic reticulum; G, Golgi apparatus; GG, glycogen granules; L, lipid droplets; LM, longitudinal muscle; MIT, mitochondrion; MT, microtubule; MV, microvillus; NU, nucleus; VA, vacuole; VES, vesicle.

cytoplasm (Fig. 12.5), contains numerous organelles and structures of varying electron densities; specifically, mitochondria, Golgi complexes, ribosomes, and endoplasmic reticulum have been reported. In addition, a number of vacuoles, vesicles, and rhabdiform organelles usually occur. Glycogen granules also occur in this region in most species.

A number of enzymes have been found to be associated with various organelles embedded in the external layer of the tegument (Table 12.1). Of particular interest are those believed to be involved in "membrane digestion." For example, Taylor and Thomas (1968) have demonstrated that the presence of viable segments of such species as *Hymenolepis diminuta, H. microstoma,* and *Moniezia expansa* increases the rate of hydrolysis of starch by α-amylase *in vitro*. It is believed that certain molecules, including enzymes, secreted from the external level of the cestode tegument are involved in this type of membrane digestion which obviously is advantageous to the parasite.

Some of the vesicles found in the external level of the tegument are believed to be pinocytotic or endocytotic vesicles, i.e., they take up molecules from the exterior (Fig. 12.7). Others may be secretory (or exocytotic), i.e., they carry molecules to the exterior (Fig. 12.7).

The external and internal levels of the tegument are joined by cytoplasmic bridges (Fig. 12.5). Unlike the external level, the internal level consists of discrete cells known as **cytons** or **perikarya**. Each large cell includes a prominent nucleus as well as other structured organelles including mitochondria, Golgi complexes, endoplasmic reticulum, ribosomes, and some vesicles (fewer than in the external level). At least some of the vesicles contribute to the formation of microtriches and hooks when passed into the outer level of the tegument (Mount, 1970; Lumsden *et al.,* 1974). These cytons are rich in glycogen deposits and often include lipid globules (Fig. 12.8).

Immediately beneath the external level of the tegument lies a layer of connective tissue known as the **basal lamina** (Fig. 12.5). Intermingled among the fibers may be vacuoles and granules of varying electron density.

Mediad to the basal lamina is found the body wall musculature, which is arranged in two layers. The outer layer is circularly or transversely arranged, and the inner layer is longitudinally oriented.

A thin layer of neuromuscular cells lies underneath the muscles of the body wall. These are in the form of multipolar bodies that possess cytoplasmic processes, some joining similar processes from neighboring neuromuscular cells and others joining the fine nerve fibers of the nervous system.

Parenchyma. The space enclosed by the body wall, except for that occupied by the reproductive organs, osmoregulatory structures, muscle fibers, and nervous tissue, is filled with a spongy type of tissue

Table 12.1. Some Enzymes in the Tegument of Cestodes as Demonstrated by Histochemistry[a,b]

Species	Alkaline Phosphatase	Acid Phosphatase	Esterase	β-Glucuronidase	Succinic Dehydrogenase	Cytochrome Oxidase	Amylophosphorylase	Transglucosidase
Anoplocephala magna	+	+	·		·			
Anoplocephala perfoliata	+	+	+					
Cysticercus bovis	+	−			·			
Cysticercus fasciolaris	+	+			·			
Cysticercus tenuicollis	+[c]	+[c]						
Davainea proglottina	+		·					
Diphyllobothrium mansoni (plerocercoid)	+	+	·					
Dipylidium caninum	+	+	−					
Echinococcus granulosus	+	−(+)[d]	·		·			
Taenia taeniaeformis	+	+	+	·	+			
Hymenolepis citelli	+	−	+	·	+			
Hymenolepis diminuta (cysticercoid)	+	·	·	·	+			
Hymenolepis diminuta	+	+	+(−)[d]	−	−	+	−	−
Hymenolepis microstoma	+	+	+	·	+	·	·	
Vampirolepis nana	+	+	−	−	+	+	−	−
Vampirolepis nana (cysticercoid)	·	·	·	·	+			
Ligula intestinalis (plerocercoid)	+	+	−		·			
Ligula intestinalis	+	+	−		·			
Moniezia benedeni	+	+	·		·			
Moniezia expansa	+	+	·		·			
Multiceps serialis	+	·			·			
Raillietina cesticillus	+	·			·			
Taenia pisiformis	+	+			·			
Taenia pisiformis (cysticercus)	+[c]	+[c]						
Taeniarhynchus saginatus	+	·			·			

[a] Data compiled by Smyth, 1969.
[b] +, present; −, absent; ·, not tested for.
[c] Demonstrated by biochemical methods only.
[d] Conflicting data.

called the **parenchyma**. Some of the parenchymal cells are syncytially arranged to form a network. In living animals, the spaces present between the cells are filled with fluid. The parenchymal cells serve as sites for the synthesis and storage of glycogen.

Parenchymal Muscles. The presence of parenchymal musculature, as distinguished from the body wall musculature, is characteristic of cestodes. Parenchymal muscles do not exist in trematodes, cestodarians, and turbellarians. One type of these long,

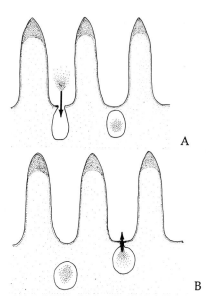

Fig. 12.7. **Diagrammatic drawings of the endocytotic and excretory (or exocytotic) vesicles of the cestode tegument. A.** Endocytotic vesicles. **B.** Excretory (or exocytotic) vesicles.

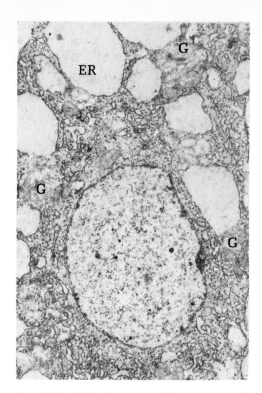

Fig. 12.8. **Cestode tegumental cyton or perikaryon.** Electron micrograph of portion of a cyton in the anterior strobila of *Hymenolepis diminuta.* The perinuclear cytoplasm is filled with profiles of rough endoplasmic reticulum, free ribosomes, and glycogen. Adjacent to the Golgi zones, the endoplasmic reticulum cisternae are commonly dilated and include an amorphous material of moderate electron density. × 14,000. ER, rough endoplasmic reticulum; G, Golgi zones. (After Lumsden, 1966.)

bipolar muscle cells is arranged longitudinally in a circle lying in the parenchyma approximately equidistant between the body wall and an imaginary body axis, dividing the parenchyma into the outer **cortical** and inner **medullary** zones (Fig. 12.9). The muscle cells lie between bundles of fibers that have their origin in the cells. The other type of parenchymal muscles are circularly arranged and line the inner surfaces of the longitudinal muscles, one group running from one lateral margin of the body to the other in the dorsal half of the body, and the other oriented in the same fashion but located ventrally. The two bow-shaped bundles enclose the medullary parenchyma. These circularly arranged muscles are laid down by the longitudinally arranged ones and are inserted terminally in the tegument by fine cytoplasmic processes.

Subtegumentary Muscles. The most complex set of muscles in a tapeworm's body controls the actions of the holdfast structures on the scolex. Basically, these are sets of criss-crossing muscle fibers attached to the inner surfaces of the suckers, permitting their contraction (Fig. 12.10). These are specialized subtegumentary muscles and are not related to the parenchymal musculature.

Calcareous Corpuscles. Many species of cestodes, and some trematodes, contain large numbers of concretions in their parenchyma known as **calcareous corpuscles**. These spherical bodies, which are most

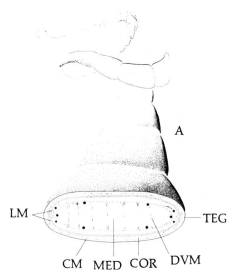

Fig. 12.9. **Tapeworm morphology.** Cross-section of a proglottid showing positions of the cortical and medullary parenchyma. CM, circular muscles; COR, cortical parenchyma; DVM, dorsoventral muscles; LM, longitudinal muscles; MED, medullary parenchyma; TEG, tegument.

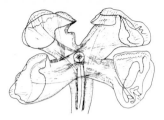

B

Fig. 12.10. **Tapeworm holdfast organ.** Expanded holdfast of *Anthobothrium auriculatum* showing criss-crossing muscle fibers. (Redrawn after Rees, 1943.)

noticeable in larval forms, consist of an organic base bound to inorganic materials. A number of investigators (von Brand *et al.*, 1960, 1965a, b; von Brand and Weinbach, 1965; Scott *et al.*, 1962; and others) have investigated the chemical nature of these noncellular, refractile bodies. The organic base contains DNA, RNA, proteins, glycogen, mucopolysaccharides, and alkaline phosphatase. The inorganic material consists primarily of calcium, magnesium, phosphorus, and carbon dioxide, and traces of metallic elements (Table 12.2).

Electron microscope studies have revealed that the corpuscles are formed within cells and appear as concentric lamellae (von Brand *et al.*, 1960; Scott *et al.*, 1962). The mineral content is amorphous but can be converted to a crystalline form by heating (Fig. 12.11).

The functions of the calcareous corpuscles are not completely known. Because these bodies disappear more rapidly in the absence than in the presence of oxygen, von Brand *et al.*, (1960) have suggested that one function is to buffer anaerobically produced acids. It has also been suggested that the corpuscles may

Fig. 12.11. **Calcareous corpuscle.** Electron micrograph of KOH isolated corpuscle. × 5000. (After Scott *et al.*, 1962.)

serve as a reservoir for inorganic ions. If this is true, tapeworms could derive carbon dioxide from the corpuscles for fixation, obtain magnesium ions that catalyze several steps involved in the Embden-Meyerhof glycolytic pathway (p. 52), and derive phosphates employed in phosphorylative pathways.

Smyth (1967) has reported that the first indication of proglottid formation in *Echinococcus* is a gradual

Table 12.2. Trace Elemental Components of Calcareous Corpuscles of *Echinococcus granulosus* as Determined by Spectroscopic Analysis[a,b]

	Specimens from		
	Chile	*New Zealand*	*Lebanon*
Aluminum	+	+	+
Boron	−	−	−
Cadmium	+	+	+
Copper	−	−	−
Iron	+	+	+
Lead	−	−	−
Manganese	−	+	+
Nickel	+	+	+
Sodium	+	+	+
Strontium	+	+	+
Tin	−	−	−
Titanium	+	+	+

[a] Data after von Brand *et al.*, 1965.
[b] +, Present; −, absent.

Table 12.3. Incorporation of Inorganic Orthophosphate ($^{32}P_i$) into isolated Calcareous Corpuscles of *Taenia taeniaeformis*[a]

	6°C		37°C	
Time (min)	*Radioactivity (cpm/mg dry wt)*	*P_i Uptake (μ moles)*	*Radioactivity (cpm/mg dry wt)*	*P_i Uptake (μ moles)*
0	2.5×10^3	2.6	3.1×10^3	2.3
5	3.6	4.2	5.0	5.0
10	4.6	4.4	5.9	5.2
15	4.5	4.8	5.3	5.1
60	6.3	5.9	6.5	6.5
120	6.2	6.3	8.2	7.6

[a] Data after von Brand and Weinbach, 1965.

reduction in the number of calcareous corpuscles, although they reappear when the third proglottid develops. This may mean that the molecules and ions stored therein are utilized during this stage of development.

Experiments by von Brand and Weinbach (1965) with *Taenia taeniaeformis* have shown that worms exposed to ^{32}P rapidly accumulate radioactive phosphorus in the calcareous corpuscles. Its uptake, which occurs by diffusion, is influenced by sodium ions and carbon dioxide (Table 12.3). For a detailed review of calcium corpuscles in cestodes, see von Brand (1973).

Osmoregulatory System

The osmoregulatory or excretory system in the Eucestoda is essentially the same as the protonephritic type

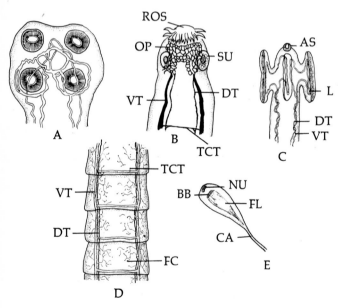

Fig. 12.12. Morphology of the osmoregulatory (excretory) system of cestodes. A. Scolex of *Proteocephalus torulosus* showing single ring type of connection of the osmoregulatory canals. (Redrawn after Wagner, 1917.) **B.** Scolex of *Taenia* showing network type of osmoregulatory plexus. (Redrawn after Riehm, 1881.) **C.** Scolex of *Echeneibothrium* showing lateral loop type of connection of the osmoregulatory canals. (Redrawn after Hyman, 1951.) **D.** Osmoregulatory canals in consecutive proglottids. **E.** Single flame cell. AS, anterior sucker; BB, basal bodies; CA, capillary; DT, dorsal excretory tubules; FC, flame cell; FL, cilia or "flame"; L, loop of collecting tubule within bothridium; NU, nucleus; OP, osmoregulatory plexus; ROS, rostellum; SU, sucker; TCT, transverse collecting tubule; VT, ventral collecting tubule.

found in trematodes (p. 309). The typical cyclophyllidean system is given here to familiarize the reader with the terminology.

Four main **collecting canals** (or ducts) extend the entire length of the strobila. The two ventral canals are ventrolaterally located, and the two dorsal canals are dorsolateral. All these canals are situated in the peripheral zone of the medullary parenchyma. A single transverse canal connects the two ventral canals at the posterior end of each proglottid. The ventral vessels carry the fluid away from the scolex, the dorsal vessels toward it. Within the scolex the four longitudinal canals may be joined by a network of canals or by a single ring vessel, or the dorsal and ventral canals on each side may be joined by a simple connection with no apparent exchange between the two sides (Fig. 12.12).

Along the length of the ventral canals, a series of secondary tubules give rise to tertiary tubules. At the free end of the terminal tubules are flame cells generally arranged in groups of fours. The individual flame cell is a stellate body with granular cytoplasm and a nucleus (Fig. 12.12). The "flame" is actually a group of cilia that arise from a concave basal plate located near the cell nucleus. The cilia, which have the typical $9+2$ arrangement of microtubules, are enveloped in the funnel-shaped enlargement of the free end of the tubule. Fluid collected through the flame cells is passed down the tubules into the main tubes. In certain species, such as *Hymenolepis diminuta*, the collecting ducts are lined with cilia.

In young worms there is an excretory vesicle in the terminal proglottid into which the ventral tubes empty. However, as the neck region produces more proglottids, the older proglottids are pushed further back, and eventually the segment containing the vesicle is broken off. In most older specimens the posterior ends of the tubes open independently to the exterior.

In certain species of tapeworms, the osmoregulatory system is more complex than that described. For example, in the pseudophyllideans *Ligula* and *Schistocephalus*, a complex network of vessels is located in the cortical parenchyma, in addition to the 16 main longitudinal canals found in these species.

Webster and Wilson (1970) have analyzed the fluid within the osmoregulatory system of *Hymenolepis diminuta* and found it contains glucose, soluble proteins, lactic acid, urea, and ammonia.

Nervous System

The nervous system of cestodes is relatively complex considering the lack of marked coordination in these animals. The most striking "coordinated" movement is the contraction of the body, for example, when a worm is placed in tap water.

Because no delimiting sheath is found on the nerve fibers of tapeworms, it was extremely difficult to study

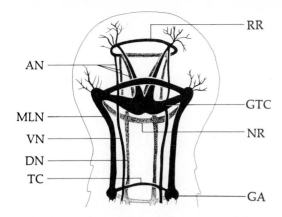

Fig. 12.13. **Cestode brain.** The nervous system in the anterior end of *Moniezia*. AN, anterior nerve; DN, dorsal nerve; GA, ganglion at origin of transverse commissure; GTC, ganglionated transverse commissure; MLN, main lateral nerve; NR, nerve ring; RR, rostellar ring; TC, transverse commissure in proglottid; VN, ventral nerve.

the neuroanatomy of these animals until histochemical techniques became available. Now, by employing cholinesterase as a marker, the nervous system can be readily traced.

The "brain" lies in the scolex and is more or less rectangular (Fig. 12.13). Two large major nerve trunks extend posteriorly along the entire length of the strobila. Two shorter nerve trunks extend anteriorly to innervate the tissues anterior to the "brain." The anterior long side of the rectangle is the **anterior commissure,** the posterior long side is the **posterior commissure.** The rectangle itself is the **cephalic ganglion.** Modifications of this pattern are known.

Nerves extend from the "brain" or cerebral ganglionic mass to the muscles, tegument, and reproductive systems.

A number of investigators, including Lee *et al.* (1963), have demonstrated sensory nerve endings in the tegument of cestodes. Electron microscope studies have revealed that these endings resemble those of other invertebrates morphologically. Each ending is bulbous, with an elongate distal process projecting beyond the tegumental surface (Morseth, 1967) (Fig. 12.14).

Since adult tapeworms maintain almost continuous attachment to their host's intestinal mucosa by their scolices, the muscles associated with scolical suckers and other types of holdfast structures must be in a state of continuous contraction. This is believed to be made possible by the presence of specialized "stretch receptors" (Rees, 1966). The dendrites of these nerve units are deeply embedded in the muscle mass, and stretch deforms the dendritic endings with subsequent changes in sensory frequency. The exact functional mechanisms involved in the stretch receptors of cestodes is still uncertain, but Rees has proposed the

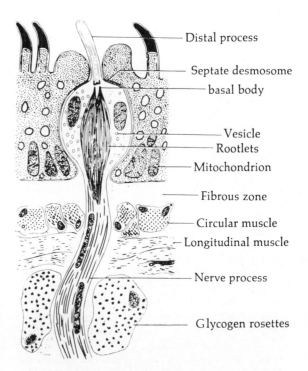

Fig. 12.14. **Cestode nerve ending.** Schematic drawing of a longitudinal section through a sensory ending in the tegument of *Echinococcus granulosus*. (After Morseth, 1967.)

following: (1) Tegumental receptors are stimulated upon surface contact with the host's gut, and the stimuli are transmitted to the central nervous system and from there, via motor fibers, to the muscles of the holdfast organs. (2) The holdfasts contract and become attached. (3) Contraction of the muscle distorts the dendrites of the stretch receptor, initiating an impulse that is transmitted to the nerve. (4) As long as the holdfast remains attached, the stimulus persists and the muscles remain contracted. In addition, the action may be modulated by inhibitory fibers from the nerve cord, which synapse on either the muscles or the dendrites of the receptors.

The chemistry of neurotransmitter substances in cestodes is similar to that of other animals. Cholinesterase has been demonstrated in a number of species (Graff and Read, 1967, and others), and an acetylcholinelike substance is present in cestode extracts (Artemov and Lure, 1941).

A relatively recent development in the neurobiology of cestodes is the discovery of neurosecretory cells (Davey and Breckenridge, 1967), although neurosecretion is well established in almost all of the

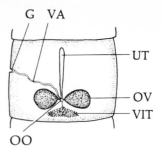

Fig. 12.16. **Female reproductive system of** *Taenia.* G, genital pore; OO, ootype; OV, ovary; UT, uterus; VA, vagina; VIT, vitellaria.

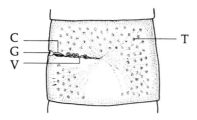

Fig. 12.15. **Male reproductive system of** *Taenia.* The male and female (Fig. 12.16) systems are actually in the same proglottid. C, cirrus sac enclosing cirrus; G, genital pore; T, testis; V, vas deferens.

major free-living invertebrate classes. In *Hymenolepis diminuta,* bipolar neurosecretory cells are clustered in the rostellum. In the cysticercoid, these cells are apparently inactive and do not stain with paraldehyde-fuchsin; however, upon feeding of this larva to a rodent host, granules of fuchsinophilic material appear in the worm's axons on 16 to 18 days postinfection. This supposed neurosecretory activity ceases on about the 40th day postinfection. The exact function(s) of this neurosecretory system remain unexplored, but since its activity commences just prior to strobilization, Smyth (1969) has speculated that the two processes are interrelated.

Reproductive Systems

The reproductive systems of cestodes resembles that of digenetic trematodes except in some (Cyclophyllidea) where the uterus ends blindly, a separate vaginal canal occurs, and the genital pore is lateral in position.

Male System. The male reproductive system consists of one or many testes situated in the medullary parenchyma (Fig. 12.15) except in certain subfamilies of the Proteocephala, in which the testes are arranged in the cortical parenchyma. From each testis, a single vas efferens arises, and if more than one testis is present, the vasa efferentia unite to form a common vas deferens. In certain species there is an enlargement of the vas deferens—the **seminal vesicle**—for storage of spermatozoa. The vas deferens enters the cirrus, which is located within a cirrus pouch. For a detailed review of spermatogenesis in cestodes, see Davis and Roberts (1983b).

In certain species, the seminal vesicle is located within the cirrus pouch and is known as an internal seminal vesicle. In others, the vesicle is located outside the pouch and is hence an external seminal vesicle. In still others, both an internal and an external seminal vesicle are present. Within the cirrus pouch, there are generally glandular cells, collectively known as the

prostate gland, which open into the cirrus through cytoplasmic ducts. The cirrus is eversible or protrusible and usually empties into a common male and female external opening—the **genital pore**—situated at the surface opening of the cup-shaped **atrium.**

Female System. The female reproductive system consists of a single lobed or unlobed ovary from which an oviduct arises (Fig. 12.16). A controlling sphincter, known as the **oocapt**, is situated at the junction of the ovary and oviduct. The oviduct leads to a minute chamber called the **ootype**, where various components of the egg are assembled. Also leading into the ootype are (1) the Mehlis' gland cells, which surround the ootype; (2) the single common vitelline duct is formed from the union of many primary vitelline ducts arising from the vitelline glands (these glands range from a single compact body to numerous individual follicles scattered throughout the parenchyma); and (3) the duct of the seminal receptacle, the latter being an enlarged portion of the vaginal tube, which generally opens into the common genital atrium. For a detailed review of oogenesis, oviposition, and oosorption in cestodes, see Davis and Roberts (1983a).

The tube leaving the ootype is the uterus. In some tapeworms, such as members of the Pseudophyllidea, the uterus opens to the outside of the body through the genital pore. In such cases, the proglottid continually forms eggs, which are expelled. In other species, such as members of the order Cyclophyllidea, the uterus is a blind sac. The proglottids of these species become filled with eggs in distended uteri and eventually break away from the strobila and rupture, permitting the eggs to escape, or the proglottids may remain intact, still containing the eggs when passed out of the host. In other cyclophyllideans, such as *Dipylidium caninum,* the so-called double-pored tapeworm of dogs, the gravid uterus breaks up to form egg capsules, each containing from 1 to approximately 60 eggs. Such eggs are passed out of the host either in sloughed proglottids or in egg capsules.

The vitelline glands secrete material that contributes to the formation of the yolk and shell, whereas the function of the Mehlis' gland is identical with its function in trematodes (p. 308). It secretes a thin membrane around the zygote and associated vitelline cells.

During copulation, the cirrus of one proglottid may be introduced into the vagina—a tube connecting the genital atrium with the seminal receptacle—of another proglottid of the same worm or into a proglottid of another worm. Spermatozoa are stored in the seminal receptacle, from which they enter the ootype for fertilization. As indicated earlier, cross-fertilization between two worms appears to be necessary, at least periodically, to permit hybridization, thus guarding against the deleterious effects of selfing.

Although the ootype is the site of egg formation in most species, in such tapeworms as *Hymenolepis* spp., the envelopes surrounding the zygote or blastomeres (the **inner membrane**, the **embryophore**, and the **capsule** or shell) are laid down after the cleaving zygote enters the uterus, rather than in the ootype.

Details of cestode prelarval embryology are known only for a few pseudophyllideans and cyclophyllideans (Rybicka, 1966; Ubelaker, 1983a, b). Among pseudophyllideans, cleavage is total and equal, one early blastomere becoming separated and later flattening to form the vitelline membrane. Among cyclophyllideans, cleavage is total but unequal.

LIFE CYCLE STUDIES

Generally, cestode life cycles are not as complicated as those of digeneans because they usually do not involve asexually reproductive larval phases. However, most tapeworms also require at least one or two intermediate hosts.

Life cycle patterns among the Eucestoda are of considerable phylogenetic importance, for they often indicate the membership of particular species in specific orders. Unfortunately, the complete representative developmental patterns of all tapeworm orders are as yet not known. The patterns, some yet incomplete, for the Trypanorhyncha, Tetraphyllidea, Proteocephala, Cyclophyllidea, and Pseudophyllidea, however, are known. Some variations in the life cycles of cestodes are given in Table 12.4. For a detailed review of the current status of our knowledge about cestode life cycles, see Voge (1967) and Freeman (1973).

Trypanorhynchan Life Cycle Pattern

No complete life cycle is known among the Trypanorhyncha. However, the following pattern, based on the life cycle of *Lacistorhynchus tenuis* as reported by Riser (1956), gives some indication of how this group develops.

Adult *Lacistorhynchus tenuis* lives in the intestinal spiral valve of sharks. Eggs discharged by the adult tapeworms pass into sea water and a ciliated larva, the **coracidium**, hatches from each egg. The coracidia are

ingested by the splash-pool copepod *Tigriopus fulvus*. The larval stage within the hemocoel of this crustacean is the tailless **procercoid**. When experimentally fed to fish, the procercoids did not undergo further development, and therefore the complete life cycle is not known.

Tetraphyllidean Life Cycle Pattern

As is the case with trypanorhynchs, very little information is available on the life cycles of tetraphyllideans. Reichenbach-Klinke (1956) has reported that in the case of *Acanthobothrium coronatum*, a parasite of elasmobranchs, developing procercoids occur in small crustaceans, and the next larval stage, **plerocercoids**, occurs in sardines. When the latter are fed to sharks, adult cestodes develop from them. From this information it may be concluded that tetraphyllids also have two intermediate hosts, the first crustacean and the second a small bony fish.

Proteocephalan Life Cycle Pattern

In the order Proteocephala, eggs containing larvae known as **oncospheres** leave the host in feces. Generally these eggs are ingested by a copepod, in which the oncospheres (or hexacanth larvae), each bearing three pairs of hooks, escape and actively penetrate the gut wall, reaching the host's hemocoel. In this position, the oncospheres develop into **procercoid** larvae. The procercoids often lack the characteristic **cercomer**—a small caudal appendage commonly found on this type of larva. If a cercomer is present, the characteristic hooks embedded in it are often absent. When the crustacean host is ingested by a definitive host, the procercoids invade such tissues as hepatic, muscular, and intestinal epithelium and develop into **plerocercoid** larvae with invaginated scolices. These larvae migrate back into the lumen of the gut and metamorphose into strobilate adults.

Cyclophyllidean Life Cycle Pattern

In the Cyclophyllidea, the oncosphere bearing six hooks, also known as the **hexacanth larva**, remains passive in the eggshell or the unciliated embryophore (a flexible membrane) until the embryo is ingested by a vertebrate or invertebrate intermediate host.

In species that normally utilize an invertebrate intermediate host, usually an arthropod, the oncosphere, upon hatching, penetrates through the host's gut wall into the hemocoel and develops into a solid **metacestode** (the preadult stage(s) that occurs in an intermediate host) known as a **cysticercoid.** The scolex of the cysticercoid is everted. In those that utilize a vertebrate intermediate host, the oncosphere

Table 12.4. Some Variations in the Life Cycle Patterns of Cestodes[a]

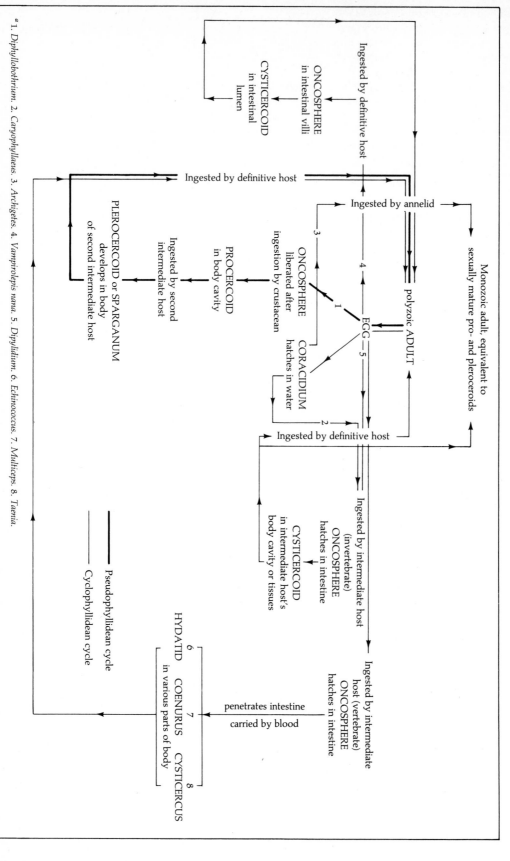

[a] 1. *Diphyllobothrium.* 2. *Caryophyllaeus.* 3. *Archigetes.* 4. *Vampirolepis nana.* 5. *Dipylidium.* 6. *Echinococcus.* 7. *Multiceps.* 8. *Taenia.*

penetrates the host's intestinal lining and enters a venule. It is then carried in blood to some other area of the body where it develops into a metacestode with a scolex inverted in a vesicle (or bladder) and hence is known as a bladderworm or **cysticercus**. When several invaginated scolices develop within a bladderworm, as in *Taenia multiceps*, the metacestode is referred to as a **coenurus**.

Penetration by oncospheres into the intermediate host's intestinal wall is made possible by secretions from glands present in each oncosphere. Such secretions act on the ground substance of the host's mucosa and also have a cytolytic effect.

In *Echinococcus granulosus*, the metacestodes display a third variation of scolex formation. In these, daughter and granddaughter cysts are formed, the daughters originating as invaginations on the wall of the mother and the granddaughters as invaginations on the walls of the daughters. The walls of the second- and third-generation cysts, in turn, give rise to a number of protoscolices, which protrude into the cystic spaces referred to as **brood capsules**. Thus, a large bladder-worm, known as a **hydatid cyst**, is formed enclosing numerous protoscolices, each of which can develop into an adult worm. A hydatid cyst commonly measures 10 mm or more in diameter, and may contain thousands of protoscolices. When the intermediate host harboring hydatid cysts, cysticerci, or coenuri is ingested by the definitive host, each scolex develops into an adult.

Evagination of the scolices of certain tapeworm larvae, once they reach the definitive host's intestine, is sometimes referred to as **excystation** because these larvae may be enveloped by a protective cyst wall or capsule. Rothman (1959) has shown that larvae of such species as *Hymenolepis diminuta*, *H. citelli*, *Vam-pirolepis nana*, and *Oochoristica symmetrica* require the host's bile salts for activation and excystation, whereas the larvae of *Taenia taeniaeformis* do not. Furthermore, the larvae of *T. taeniaeformis*, *H. diminuta*, *H. citelli*, and *Vampirolepis nana* require a proteolytic enzyme to dissolve their cysts. In addition, environmental temperature may influence excystation. The influences of bile salts, pepsin, trysin, and temperature on the excystation of certain cyclophyllidean cestodes are listed in Table 12.5.

Pseudophyllidean Life Cycle Pattern

In the Pseudophyllidea, the oncosphere is covered with a ciliated **embryophore**. The larva hatching from the egg, known as a **coracidium**, is free swimming. Coracidia are ingested by the first intermediate host, usually a copepod, and in the intestine of this host the coracidia shed their ciliated coats while penetrating the gut wall. In the copepod's hemocoel, the larvae develop into elongate oval **procercoids** that retain the six larval hooks. The hooks are situated in a caudal protuberance, the **cercomer**. When the first intermediate host is ingested by a second intermediate host, procercoids develop into solid, wormlike **plerocercoids**, each with an adult scolex. Finally, when the plerocercoid is ingested by the definitive host, the strobilate adult form of the parasite develops.

Pseudophyllidean eggs are similar to those of digenetic trematodes in that there is an operculum at one end. For the coracidium to escape, the operculum must be released. Studies have shown that the eggs of

---

Table 12.5. Some Factors Contributing to the Excystation of Larval Cestodes[a]

Cestode Species	Acid Pepsin (Effect on Cyst Digestion)	Trypsin (Effect on Excystment)	Bile Salts (Effect on Excystment) Bile Salts	Temperature (Effect on Excystment)		Reference
				18–26°C	37°C	
Hymenolepis citelli	Initiates	None[b]	Activation only	None	Excyst	Rothman, 1959
H. diminuta	Initiates	None[b]	Some excystment	None	Excyst	Rothman, 1959
Vampirolepis nana	Initiates	None[b]	Activation only	None	Excyst	Rothman, 1959
Taenia taeniaeformis	Essential	None	Some excystment	Excyst	Excyst	Rothman, 1959
Taenia solium	Essential	None[b]	Excyst		Excyst	Butning, 1927
T. pisiformis	Inessential		Excyst	Excyst	Excyst	DeWaele, 1934; Edgar, 1941
T. tenuicollis			Excyst	Excyst	Excyst	Malkani, 1933
Taeniarhynchus saginatus			Excyst	Excyst	Excyst	Malkani, 1933
Oochoristica symmetrica	Inessential	None	Excyst	Excyst	Excyst	Rothman, 1959
Raillietina kashiwariensis	Inessential	Excyst[c]	Some excystment	None	Excyst[d]	Sawada, 1959

[a] After Read and Simmons, 1963.
[b] Produces excystment if bile salts are present.
[c] Pancreatin active, lipase relatively inactive, amylase not tested.
[d] Temperature at 40 to 42°C.

Table 12.6. Light Requirement for Hatching of Eggs of Some Pseudophyllidean Cestodes[a]

Cestode Species	Light Requirement for Hatching
Diphyllobothrium latum	Essential
D. ursi	Not essential
D. dalliae	Not essential
D. dendriticum	Essential
D. oblongatum	Essential (?)
Schistocephalus solidus	Essential
Spirometra mansonoides	Essential (?)
Triaenophorus lucii	Not essential

[a] After Smyth, 1963.

a number of pseudophyllideans can hatch in the dark, but light is required in some species. Even in species of the same or closely related genera, the eggs of some require light for hatching, whereas the eggs of others do not. Table 12.6 lists some findings relative to the light requirement for the hatching of pseudophyllidean eggs.

Egg Production

Considerable work has been done to elucidate the mechanisms involved in egg formation in cestodes. It is now recognized that there are four general types of egg-forming systems, and these are reviewed briefly at this point. In following Smyth (1969), these are designated as the (1) pseudophyllidean, (2) *Dipylidium*, (3) *Taenia*, and (4) *Stilesia* types.

Pseudophyllidean-Type Egg. The formation of this type of egg is strikingly similar to that of trematode eggs. This is not surprising since the female reproductive system in tapeworms that form this type of egg, i.e., members of the orders Pseudophyllidea, Trypanorhyncha, and Tetraphyllidea, are very similar to that of digenetic trematodes (Fig. 12.17).

Ova periodically released from the ovary pass to the ootype via the oviduct. A sphincter muscle controls the release of ova. Fertilization occurs in the ootype. Mature vitelline cells containing both shell and yolk precursors pass from the vitelline glands to the vitelline reservoir, where they are stored. From the vitelline reservoir, these cells pass into the ootype

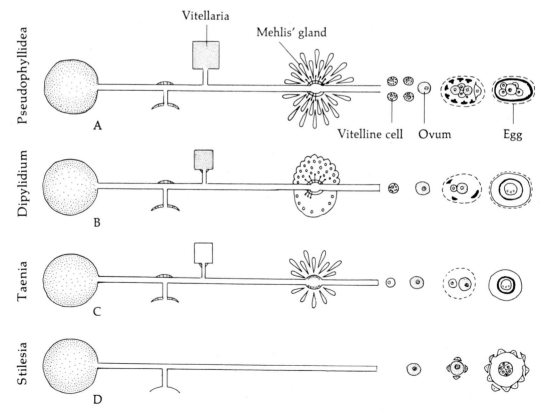

Fig. 12.17. **The four types of egg-forming systems among cestodes. A.** Cestodaria, Pseudopyllidea; Trypanorhyncha, Tetraphyllidea. **B.** Some Cyclophyllidea, e.g., *Dipylidium.* **C.** Taeniidae, e.g., *Taenia.* **D.** Thysanosominae, e.g., *Stilesia.* See text for explanation. (Modified after Löser, 1965.)

where they release the globules of shell precursor which coalesce to form the shell or capsule.

Recall that a group of unicellular glands, collectively known as the Mehlis' gland, surrounds the ootype. According to Löser (1965a, b), there are two types of gland cells—**mucous** and **serous** gland cells. The mucous secretion forms a ground lamella which surrounds the zygote and vitelline cells and on which the shell precursor substance is laid down. The serous gland cells secrete a surface-active agent which assists in coalescing the shell precursor globules.

When the capsule or shell is first formed, it consists of a highly resistant type of protein known as sclerotin. It is unhardened and encloses the zygote and the remains of the vitelline cells, the latter serving as yolk reserves for the developing embryo.

As eggs pass up the uterus, they become visibly darkened due to a process known as tanning (Fig. 12.18). Tanning or sclerotinization occurs when an *o*-quinone is derived enzymatically from an *o*-phenol in the presence of oxygen. The quinone reacts with free NH_2 groups on adjacent protein chains to form strong covalent cross-links (Fig. 12.18), thus forming a stable protein. This process is summarized in Figure 12.19.

The fully developed pseudophyllidean-type egg has a thick sclerotin capsule, commonly with a lidlike operculum at one end.

***Dipylidium*-Type Egg.** This type of egg is characteristic of members of such genera as *Dipylidium*, *Moniezia*, and *Hymenolepis*. Furthermore, with minor variations, it is characteristic of the order Cyclophyllidea, except for the family Taeniidae and the subfamily Thysanosominae.

As in the case of pseudophyllideanlike eggs, fertilization of *Dipylidium*-like eggs occurs in the ootype. As depicted in Fig. 12.17 an outer and inner envelope are formed as well as an embryophore and an oncospheral membrane covering the oncosphere. The entire egg may or may not be covered by a capsule. If one exists, it is generally poorly developed. The capsule, outer envelope, inner envelope, and oncospheral membrane comprise the embryonic envelopes.

phenolase
Phenol→Quinone
Protein→Protein
Tanned protein = Sclerotin

Figure 12.19. Summary of sclerotin formation

In those eggs with a capsule, for example *Dipylidium* eggs, its chemical nature is not clearly understood. On the basis of staining reactions, it has been reported to be composed of sclerotin. Pence (1967), who has examined this layer both with the electron microscope and histochemically, reported that the capsule is homogeneously periodic acid-Schiff-positive, suggesting that it is a polysaccharide or glycoprotein (Fig. 12.20). This, however, has not negated the occurrence of sclerotin. What Pence has designated the "cytoplasmic layer" is believed to represent the outer and inner envelopes, which are known to be cellular. Therefore, it is not surprising that he found lipid droplets, mitochondria, and glycogen embedded therein.

Both serous and mucous gland cells occur in the Mehlis' gland of cestodes producing this type of egg, and these cells presumably play the same roles as homologous cells in pseudophyllideans (Fig. 12.17). In addition, apocrine gland cells occur in the uterine wall of *Dipylidium*-type egg producers, whose secretion forms a globular stratum of acid mucopolysaccharide-rich material on the outside of the capsule. This layer is referred to as the **uterine capsule** (Fig. 12.21). There is also a thin embryophore present, which when examined with the electron microscope appears to be comprised of two layers of rods at right angles to each other (Fig. 12.20).

Dipylidium-type eggs enclose a fully developed hexacanth embryo or oncosphere when laid.

***Taenia*-Type Egg.** This type of egg is characteristic of members of such genera as *Taenia* and *Echino-*

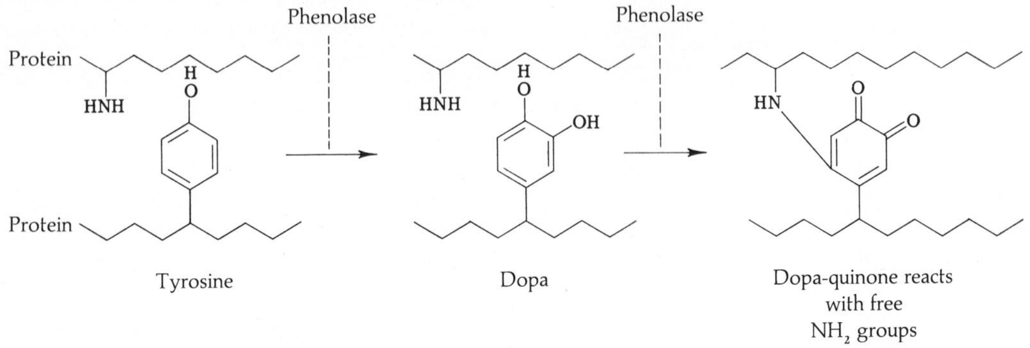

Fig. 12.18. Chemical reactions involved in quinone tanning.

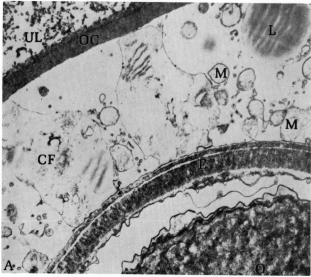

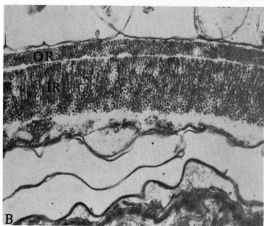

Fig. 12.20. *Dipylidium* egg structures. A. Electron micrograph of section of outer layers surrounding the oncosphere. × 13,800. **B.** Electron micrograph showing outer and inner layers of rods of the embryophore. CF, cellular fragments of outer envelope; E, embryophore; IR, inner zone of rods; L, lipid; M, mitochondria; O, oncosphere; OC, capsule; OR, outer zone of rods. (After Pence, 1967.)

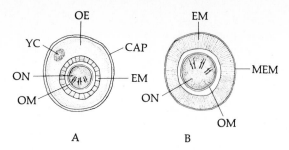

Fig. 12.21. Cestode eggs. A. *Dipylidium* egg. **B.** *Taenia* egg. CAP, capsule; EM, embryophore; MEM, membrane (normally not seen in fecal eggs); OE, outer envelope; OM, oncospheral membrane; ON, oncosphere; YC, yolk cell.

oncosphere is comparatively thick and well developed. It consists of a layer of minute keratin blocks that gives this layer a characteristically radially striated appearance (Fig. 12.21).

The Mehlis' gland complex in tapeworms that form *Taenia*-type eggs includes only mucous gland cells.

***Stilesia*-Type Egg.** Tapeworms forming this type of egg, including members of the genera *Stilesia* and *Avitellina*, lack vitelline glands, and therefore the eggs are formed only from components of the ovum and sperm (Fig. 12.17). However, as the egg passes up the uterine tract, a thick cellular covering is deposited on the surface.

The Mehlis' gland complex is absent, and hence neither mucous nor serous gland cells occur.

BIOLOGY OF EUCESTODA

Several taxonomic treatises are available for identifying the Eucestoda. These include the volumes by Skrjabin (1951), Wardle and McLeod (1952), Wardle *et al.* (1974), Yamaguti (1959), and Schmidt (1985). The smaller volume by Schmidt (1970) is simpler and more useful for beginning students. Species that are parasitic in humans and domestic animals are given in Table 12.7.

As listed in the classification section at the end of this chapter, the subclass Eucestoda is divided into 12 orders. The biology of several of the major orders is considered below.

ORDER CYCLOPHYLLIDEA

The cyclophyllidean cestodes are undoubtedly the most common tapeworms encountered by inland helminthologists because among these tapeworms are found most of the species that commonly parasitize mammals, a large majority of those that parasitize birds, and some that occur in amphibians and reptiles. All members are characterized by four true suckers, or

coccus. Its development is similar to that of the *Dipylidium*-type egg (Fig. 12.17). According to Löser (1965a, b), only one vitelline cell is associated with the zygote. Furthermore, the vitelline cells lack sclerotin precursors and hence no sclerotinized capsule occurs. The egg is enveloped by a very thin membrane that is usually lost by the time the egg is passed from the host. The embryophore surrounding the

Table 12.7. Common Cestodes Parasitic in Humans and Domestic Animals

Cestode	Principal Hosts	Habitat	Main Characteristics	Disease Caused
Pseudophyllidea				
Diphyllobothrium latum	Humans, dogs, cats, minks, bears, other fish-eating mammals	Intestine	Extremely large, 3–10 meters long; consisting of 3000 or more proglottids; scolex 2.5 mm long; 1 mm wide; eggs 70 by 45 μm, operculate	*Diphyllobothriasis* sometimes anemia
Diplogonoporus grandis	Normally a parasite of whales, occasionally in humans	Intestine	1.4–5.9 meters long; two sets of reproductive organs per proglottid; eggs 63–68 by 50 μm, operculate	Diarrhea, constipation, secondary anemia
Spirometra erinaceieuropaei	Dogs, cats, humans	Intestine	85–100 cm long; multiple testes larger than those of *D. latum*; vitellaria numerous; eggs 57–60 by 33–37 μm, operculate	Similar to *D. latum* infections
S. mansonoides	Dogs, cats, usually bobcats, sparaganum occasionally in humans	Small intestine	Rarely over 1 meter long; scolex 200–500 μm wide; bothria shallow; cirrus and vagina open independently on ventral surface; eggs 65 by 37 μm, operculate	Diarrhea and secondary anemia; larvae may cause sparganosis in humans
Ligula intestinalis	Fish-eating birds occasionally in humans	Intestine	Specimens found in humans small, less than 80 cm long	Nonpathogenic
Cyclophyllidea (Anoplocephalidae)				
Bertiella studeri	Monkeys, apes, humans	Small intestine	275–300 mm long; 10 mm wide; scolex subglobose, set off from strobila; eggs irregular in outline, 45–46 by 50 μm, nonoperculate	No apparent symptoms
Anoplocephala magna	Horses	Intestine	350–800 mm long; 20–50 mm wide; 400–500 testes per mature proglottid; eggs 50–60 μm in diameter, nonoperculate	Anoplocephaliasis, secondary anemia
A. perfoliata	Horses	Large and small intestine	10–80 mm long, 10–20 mm wide, scolex 2–3 mm in diameter with lappet behind each sucker	Anoplocephaliasis, secondary anemia
Paranoplocephala mamillana	Horses	Small intestine	6–50 mm long; 4–6 mm wide; suckers slitlike; eggs 50–88 μm in diameter, nonoperculate	Nonpathogenic
Moniezia expansa	Sheep, goats, cattle	Small intestine	Up to 4–5 meters long, 1 cm wide; two sets of reproductive organs per proglottid; proglottid much wider than long; 100–400 testes per segment; eggs 50–60 μm in diameter, nonoperculate	Nonpathogenic
M. benedeni	Sheep, goats, cattle	Small intestine	Up to 4 meters long, larger scolex than *M. expansa*	Nonpathogenic
Thysanosoma actinioides	Sheep, cattle	Small intestine	150–300 mm long; large and prominent suckers; proglottids broader than long with fringe along posterior margin; two sets of reproductive organs per proglottid; eggs expelled in capsules; each egg 19.25–26.95 μm in diameter, nonoperculate	Thysanosomiasis
Aporina delafondi	Pigeons	Small intestine	70–160 mm long, no rostellum, genital pore irregularly alternate, 100 testes per proglottid	Nonpathogenic (?)
Taeniidae				
Taenia solium	Pigs, humans	Small intestine	Up to 2–7 meters long; scolex quadrate with diameter of 1 mm; rostellum armed with double row of hooks; eggs 31–43 μm in diameter, nonoperculate	Taeniasis solium

Table 12.7. (continued)

Cestode	Principal Hosts	Habitat	Main Characteristics	Disease Caused
(*Cysticercus cellulosae*)	Various animals, including humans	Various tissues	Ovoid, whitish, 6–18 mm long	Cysticercosis cellulosae
Taenia taeniaeformis	Cats, dogs, humans	Small intestine	15–60 cm long, 5–6 cm wide, armed rostellum, double row of usually 34 hooks	Taeniasis taeniaeformis
(*Cysticercus crassicollis*)	Rats	Various tissues	—	—
Taenia hydatigena	Dogs	Small intestine	Up to 5 meters long, 4–7 cm wide, scolex armed with double row of 26–44 hooks, 600–700 testes per segment, gravid uterus with 5–10 lateral branches	Taeniasis hydatigena
Taenia ovis	Dogs	Small intestine	Approximately 1 meter long, scolex armed with two rows of 24–36 hooks, 300 testes per proglottid, gravid uterus with 20–25 lateral branches	Taeniasis ovis
T. tenuicollis	Minks	Small intestine	Up to 70 mm long, large suckers, 237–303 μm in diameter, two rows totaling 48 hooks, eggs 17–20 μm in diameter	Taeniasis tenuicollis
T. pisiformis	Dogs, cats, rabbits	Small intestine	500 mm long; 5 mm wide; scolex with double row of 34–48 hooks; genital pores alternate irregularly; gravid uterus with 8–14 branches; eggs 36–40 long, 31–36 μm wide, nonoperculated	Taeniasis pisiformis
Taeniarhynchus saginatus	Cysticercus in cows, adults in humans	Small intestine	10–12 meters long, no rostellum, unarmed, 1000–2000 proglottids, eggs similar to those of *T. solium*	Taeniasis saginatus
Multiceps multiceps	Dogs, foxes, coyotes, humans	Small intestine	Up to 1 meter long, 5 mm wide, scolex armed with double row of 22–32 hooks, approximately 200 testes per proglottid, gravid uterus with 9–26 lateral branches, eggs 31–36 μm in diameter	"Gid" in cysticercus infections; like taeniasis in adult infections
M. serialis	Dogs, occasionally humans	Small intestine	70 cm long, 3–5 cm wide, rostellum with double row of 26–32 hooks, gravid uterus with 20–25 lateral branches	Like taeniasis
Echinococcus granulosus	Hydatid cysts in sheep, horses, deer, pigs, humans	Liver and other organs	Hydatid cysts with thick two-layered wall, filled with fluid (adult morphology given in text)	Hydatid disease
Hymenolepididae *Vampirolepis nana*	Rats, mice, humans	Intestine	25–40 mm long, 1 mm wide, short rostellum with 20–30 hooks in one ring, three testes, eggs 30–47 mm in diameter	Hymenolepiasis nana
Hymenolepis diminuta	Rats, mice, humans, dogs	Intestine	200–600 mm long, 1–4 mm wide, rostellum unarmed, eggs 60–80 by 72–86 μm	Hymenolepiasis diminuta
H. carioca	Chickens, turkeys	Small intestine	300–800 mm long, 500–700 μm wide, rostellum unarmed	Hymenolepiasis carioca
H. cantaniana	Chickens, turkeys, quails	Duodenum	2–12 mm long, rostellum shorter than that of *H. carioca*, otherwise the two species are similar	Hymenolepiasis cantaniana
Fimbriaria fasciolaris Dilepididae	Chickens, ducks	Small intestine	14–85 mm long, with pseudoscolex	Nonpathogenic
Dipylidium caninum	Dogs, cats, foxes, occasionally humans	Small intestine	15–70 cm long, 3 mm wide, conical rostellum, armed with 30–150 hooks, 200 testes per proglottid, eggs in capsules, each egg 35–60 μm in diameter	Chronic enteritis dipylidiasis

Table 12.7. (continued)

Cestode	Principal Hosts	Habitat	Main Characteristics	Disease Caused
Choanotaenia infundibulum	Chickens, turkeys, pheasants	Small intestine	Up to 20 cm long, 1–2 mm wide, posterior proglottids much wider than anterior ones, rostellum with single row of 16–20 hooks	Nonpathogenic
Amoebotaenia sphenoides	Chickens, turkeys	Small intestine	2–4 mm long, entire worm roughly triangular in shape, rostellum with single row of 12–14 hooks, uterus lobed	Nonpathogenic
Metroliasthes lucida	Turkeys	Small intestine	Up to 20 mm long, 1.5 mm wide, scolex unarmed, uterus as two simple round sacs	Nonpathogenic
Davaineidae				
Raillietina cesticillus	Chickens, pheasants	Small intestine	100–130 mm long, 1.5–3 mm wide, scolex broad and flat and about 100 μm in diameter, rostellum armed with double row of 400–500 hooks	Enteritis and hemorrhage
R. tetragona	Chickens, turkeys	Small intestine	Up to 250 mm long, 1–4 mm wide, rostellum with double row of 90–130 hooks, suckers armed with 8–12 rows of hooklets, 6–12 eggs in single capsule	Enteritis and hemorrhage
R. echinobothrida	Chickens, turkeys	Small intestine	Up to 250 mm long, 1–4 mm wide, rostellum with double row of 200–250 hooks, suckers with 8–15 rows of hooklets	Enteritis and hemorrhage
R. salmoni	Rabbits	Small intestine	85 mm long, 3 mm wide, short neck, rostellum with double row of hooks	Nonpathogenic
R. retractilis	Rabbits	Small intestine	35–105 mm long, 3 mm wide, short neck, rostellum with two rows of hooks, genital pore unilateral	Nonpathogenic
Davainea proglottina	Chickens	Small intestine	Up to 4 mm long, usually only 2–5 proglottids, rostellum with 66–100 small hooks, one egg per capsule	General physiologic retardation
D. meleagridis	Turkeys	Small intestine	Up to 5 mm long, composed of 17–22 proglottids, rostellum with double row of 100–150 hooks, suckers armed with 4–6 rows of hooklets	Unknown
Mesocestoididae				
Mesocestoides latus	Cats, skunks, raccoons	Small intestine	12–30 cm long, 2 mm wide, scolex unarmed, vitellaria bilobed in posterior region of proglottid	Nonpathogenic
M. lineatus	Dogs		30 cm to 2 meters long, genital atrium midventral	Nonpathogenic

acetabula, which are symmetrically arranged on the scolex (Figs. 12.22, 12.23, 12.39). Some species also possess an anteriorly projecting rostellum, which may or may not be armed with one or more rows of hooks. The rostellum can be retracted into the rostellar sac within the scolex.

Among the cyclophyllidean tapeworms are the members of the genus *Taenia*. These range from 1 to 2 mm up to 10 meters in length, although the majority are between 1 and 3000 mm.

Considered below are representatives of several cyclophyllidean families.

Family Mesocestoidae

The family Mesocestoidae includes the genus *Mesocestoides*, which includes numerous species commonly found in birds and mammals (Figs. 12.23, 12.24). Unidentified specimens belonging to this genus have been reported from humans in the United States, Denmark, Africa, Japan, China, and Korea.

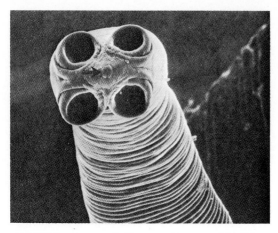

Fig. 12.22. Cyclophyllidean scolex. Scanning electron micrograph of scolex of *Hymenolepis diminuta* showing typical cyclophyllidean suckers. Notice the presence of the so-called apical organ on the rostellum (arrow). (After Ubelaker, Allison, and Specian, 1973; with permission of *Journal of Parasitology.*)

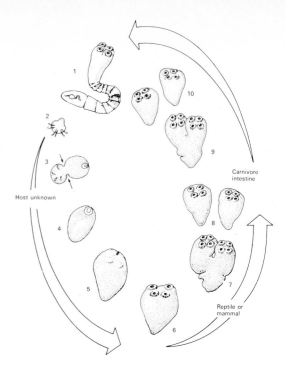

Fig. 12.24. Asexual reproduction of *Mesocestoides corti.* Schematic drawing showing early developmental stages (2–5), tetrathyridium and asexual reproduction in second intermediate host (6–8), and asexual reproduction and subsequent formation of adult worms in intestine of definitive host (9–10). (After Voge, 1969; with permission of University Park Press.)

Voge (1955) has contributed a monograph on the North American species of *Mesocestoides* known at that time.

The metacestode of *Mesocestoides*, known as a **tetrathyridium**, is commonly found in the coelom or peritoneum of dogs, cats, mice, snakes, and other vertebrates. Such a larva resembles a cysticercoid except that its body is long and slender, but not segmented (Fig. 12.23). The anterior end, which includes an invaginated scolex, is comparatively large and bulbous. When the definitive host, including dogs, cats,

racoons, other carnivores, and even birds, ingests the carcass of an intermediate host infected with tetrathyridia, these larvae develop into adults in the host's small intestine. Hexacanths of *Mesocestoides* are not infective to the hosts in which the tetrathyridia are found. Thus, a first intermediate host, yet unknown, most probably is required.

Smyth (1963) has reported the occurrence of penetration gland cells in the oncosphere of *Mesocestoides corti*, a parasite of rodents. During penetration of the gut wall of the first intermediate host, histolytic and cytolytic enzymes secreted by these cells combine with the action of the embryonic hooks in perforating the host's tissues.

An interesting aspect of the reproductive biology of one strain of *Mesocestoides corti* is that these worms are capable of undergoing asexual multiplication in the second intermediate and definitive hosts (Specht and Vogt, 1965). This occurs, however, not by budding, as in hydatids and coenuri, but by longitudinal fission (known as **fissiparity**) of the scolex (Fig. 12.24). An inward projection of the tegument between the suckers on the scolex initiates the fission (Hess, 1980). Note that Conn and Etges (1983) have demonstrated that asexually proliferative *M. corti* tetrathyridia can

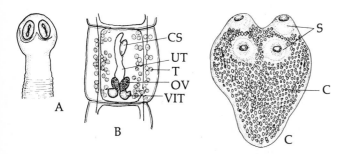

Fig. 12.23. Morphology of *Mesocestoides.* A. Scolex of *Mesocestoides lineatus.* **B.** Mature proglottid of *M. lineatus.* (**A** and **B,** redrawn after Witenberg, 1934.) **C.** Tetrathyridium of *Mesocestoides.* C, calcareous corpuscles; CS, cirrus sac; OV, ovary; S, suckers; T, testis; UT, uterus; VIT, vitelline gland.

be transmitted by the transmammary route from mother mice to offspring.

Family Anoplocephalidae

Members of the family Anoplocephalidae have unusual life cycles. Stunkard, in a series of papers published from 1937 to 1941, has shown that the three representative genera, *Bertiella*, *Moniezia*, and *Cittotaenia*, utilize soil-dwelling oribatid mites as the sole intermediate host. This pattern is true for *Moniezia expansa*, a parasite of lambs (Fig. 12.25). Although oribatid mites are now known to be the intermediate hosts of this economically important tapeworm, eradication programs have been hampered because these microscopic mites are extremely difficult to control.

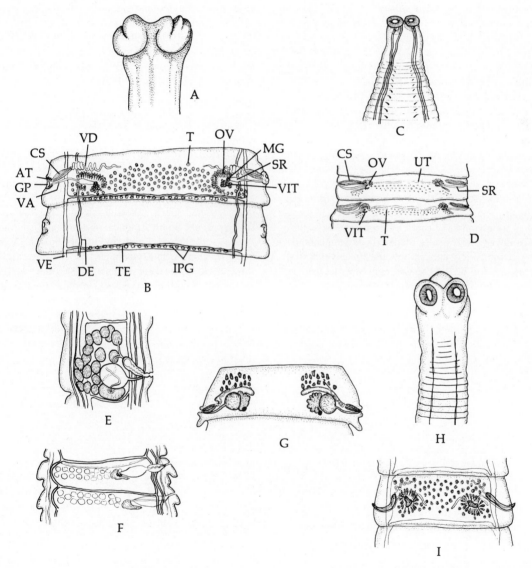

Fig. 12.25. Some anoplocephalid cestodes. A. Scolex of *Moniezia expansa*, the common sheep tapeworm. **B.** Mature proglottids of *M. expansa* showing double set of reproductive organs. (Redrawn after Stiles, 1896.) **C.** Scolex of *Cittotaenia pectinata*. **D.** Mature proglottid of *Cittotaenia*. (**C**, redrawn after Arnold, 1938; **D**, redrawn after Stiles, 1896.) **E.** Mature proglottid of *Andrya primordalis*. (Redrawn after Douthitt, 1915.) **F.** Mature proglottid of *Paranoplocephala mamillana*. (Redrawn after Douthitt, 1915.) **G.** Mature proglottid of *Progamotaenia diaphana*. (Redrawn after Zschokke, 1907.) **H.** Scolex of *Paronia pycnonoti*. **I.** Mature proglottid of *P. pycnonoti*. (**H** and **I**, redrawn after Yamaguti, 1935.) AT, genital atrium; CS, cirrus sac; DE, dorsal osmoregulatory (excretory) tubule; GP, genital pore; IPG, interproglottidal glands; MG, Mehlis' gland; OV, ovary; SR, seminal receptacle; T, testes; TE, transverse excretory tubule; UT, uterus; VA, vagina; VD, vas deferens; VIT, vitelline glands.

To students of wildlife zoology, certain tapeworms of rabbits and hares in North America belonging to the genus *Cittotaenia* are of some interest because of their common occurrence (Fig. 12.25). Other common members of the Anoplocephalidae are depicted in Fig. 12.25.

Life Cycle of *Bertiella studeri*. The life cycle of *Bertiella studeri*, a parasite of Old World primates, has been studied by Stunkard (1940). It is presented here to demonstrate the unusual modification of the cyclophyllidean life cycle pattern as found among the anoplocephalid tapeworms (Fig. 12.26).

Stunkard fed eggs from gravid proglottids of *B. studeri* to 24 species of mites and found that cysticercoids developed in *Scheloribates laevigatus*, *Notaspis coleoptratus*, *Scutovertex minutus*, and *Galumna* spp. These metacestodes are spherical, ovoid, or pyriform, measuring 0.1–0.15 mm in diameter, and possess a cercomer. Like other anoplocephalid cystercoids, this larva possesses a distinctive **pyriform apparatus**, probably employed in host penetration. When infected mites are accidentally eaten by primates, the cysticercoids develop into adults. Stunkard's experimental data do not completely explain how the cycle is completed in nature, for *Bertiella studeri* eggs are not laid and passed out in the host's feces. Instead, gravid proglottids are shed, generally 8 to 16 at a time, and these are passed out of the host. It is suspected, however, that these egg-containing body segments soon disintegrate, releasing the eggs so that they can be ingested by oribatid mites.

B. studeri has been reported many times from humans, especially in southeast Asia, the Philippines, and Indonesia. The scolex on adult worms is unarmed, and the proglottids are considerably wider than they are long (Fig. 12.26). Each egg measures 45–50 μm in diameter.

Other Anoplocephalid Species. In addition to *Bertiella studeri*, two other species belonging to the Anoplocephalidae have been reported to parasitize humans. *Bertiella mucronata*, normally a parasite of New World monkeys, has been reported from children who undoubtedly became infected as a result of close contact with pet or zoo monkeys and the ubiquitous oribatid mites. *Inermicapsifer madagascariensis*, normally parasitic in African rodents, has been reported from humans, especially in South America and Cuba (Baer, 1956).

Family Taeniidae

The family Taeniidae includes many medically and economically important species (See Table 12.7).

Echinococcus granulosus. *Echinococcus granulosus*, the hydatid worm, parasitizes members of the canine family and foxes as adults. In addition, human infections with the hydatid larvae have been reported from Iceland, South Australia, Tasmania, New Zealand, southern South America, northern and southern Africa, southern and eastern Europe, Siberia, Mongolia, northern China, Japan, and the Middle East. In these areas, cattle, sheep, rabbits, horses, and hogs are also infected with hydatid cysts. In the United States, human cases of hydatid disease occur in the regions of the Atlantic and Gulf coasts, the states surrounding the Great Lakes, Missouri, and California. Rausch (1967) has presented a review of the ecology and distribution of *Echinococcus* spp.

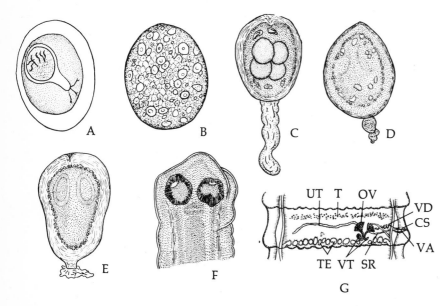

Fig. 12.26. Stages in the life cycle of *Bertiella studeri*. A. Egg removed from gravid proglottid. **B.** Larva from oribatid mite that had been fed eggs. **C.** Immature cysticercoid from oribatid mite. **D.** Cysticercoid from oribatid mite. **E.** Cysticercoid from different genus (*Galumna* sp.) of oribatid mite. **F.** Scolex of mature worm. **G.** Proglottid enclosing reproductive organs that have not completely differentiated. CS, cirrus sac; OV, ovary; SR, seminal receptacle; TE, transverse excretory (osmoregulatory) tubule; T, testes; UT, uterus; VA, vagina; VD, vas deferens; VT, vitelline glands. (All figures redrawn after Stunkard, 1940.)

The adult of *E. granulosus* is typically cyclophyllidean and measures 2–8 mm in length. The strobila generally consists of only three or four proglottids. The rostellum is armed with a double row of 28 to 50 (usually 30 to 36) hooks. In heavy infections, it is not unusual to find hundreds of worms attached to a dog's small intestine. Smyth (1963) has reported the presence of a group of cells lying just beneath the anterior tip of the rostellum. These cells secrete small viscid droplets to the exterior, the function of which is still uncertain. Since the scolex of *E. granulosus* is in close contact with the canine host's intestinal mucosa, this secretion may be immunologically important by acting as an antigen.

Life Cycle of *Echinococcus granulosus*. Stages in the life cycle of *E. granulosus* are depicted in Figures 12.27, 12.28, and 12.29. The life cycle and developmental pattern of this tapeworm is somewhat unusual because hydatid cysts are formed and in which extensive asexual reproduction by budding occurs.

The eggs are nonoperculate, resembling those of other taeniids, and measure approximately 30×38 μm. These are expelled in shed gravid proglottids from which they are later released when the proglottids disintegrate. Eggs gain entrance into the intermediate host either through water or in forage contaminated with feces. Each egg contains a fully developed oncosphere. Usually, human infection develops from intimate association with dogs, but other methods, discussed below, also occur. Infection of children commonly occurs when dogs are permitted to lick children's faces after having groomed themselves. Eggs can also be ingested from contaminated fingers.

Once swallowed, the eggs pass to the duodenum, where they hatch. The escaping oncospheres penetrate the intestinal wall, enter the mesenteric venules, and become lodged in capillary beds of various visceral organs. Among moose, deer, and caribou, hydatid cysts usually develop in the lungs, but among other animals, including humans, these larvae develop in the liver, lungs, and other tissues, the liver being the most common site (Fig. 12.28). Cysts have been known to form in the kidneys, spleen, heart, muscles, brain, and bone marrow. Development of the hydatid from the oncosphere is a slow process, with the host laying down an envelope of connective tissue around the parasite.

In a matter of months, the cyst reaches approximately 10 mm in diameter, whereupon invaginations of its wall produce daughter cysts that project into the lumen of the mother cyst. The cavities are filled with a sterile fluid known as **hydatid fluid**. Within some daughter cysts, tertiary, or granddaughter, cysts form by invagination, each with a cavity known as the **brood chamber**. From the walls of these brood chambers, as well as the walls of mother and daughter cysts, minute worms develop with inverted scolices (Fig. 12.27). Some of these scolices turn right-side-out when they become detached and fall into the brood

chamber. Since each four-suckered scolex lacks an individual bladder, they are called protoscolices, not bladderworms or cysticerci. In older cysts, some containing quarts of hydatid fluid, thousands of minute daughter and granddaughter cysts and free scolices are found as "hydatid sand" in the fluid. Such cysts, known as **unilocular cysts**, generally cause great damage, if not death, to the host, especially if they should rupture and metastasize.

In humans and some domestic animals, the formation of hydatid cysts represents a dead end for the parasite. However, in potential intermediate hosts such as rabbits, foxes, and other wild animals, upon which predators feed, hydatid cysts are ingested. Upon reaching the intestine of the new host, each protoscolex develops into an adult worm. This cycle is interesting biologically because asexual reproduction by endogenous budding takes place during the formation of daughter and granddaughter cysts and scolices. Other than this deviation, the cycle is typically taeniid.

Epidemiologic Notes. Human hydatidosis, as stated, commonly is the result of intimate contact with dogs; however, other unique ethnic customs favor infection. For example, according to Nelson and Rausch (1963), the members of certain primitive tribes in Kenya relish dog intestine roasted on a stick over a open fire. Since cleaning the intestine involves no more than squeezing out its contents and cooking may entail no more than external scorching, viable tapeworm eggs are eaten. Consequently, these people have an incidence of hydatidosis among the highest in the world.

Among leather tanners in Lebanon, the infection rate is also high. Schwabe and Daoud (1961) have reported that the reason is that dog feces is used as an ingredient of the tanning fluid. In handling dog feces including eggs, the tanners contaminate their fingers and accidentally ingest the eggs.

In nature, the carnivore-herbivore relationship, for example, the wolf-moose, wolf-reindeer, and dingo-wallaby relationship, enables *E. granulosus* to complete its life cycle. This is known as **sylvatic echinococcosis.** Humans are seldom involved in this type of cycle. On the other hand, in sheep-raising areas such as Australia, New Zealand, North and South America, Europe, Asia, and parts of Africa, human hydatidosis is a serious problem because of man's close association with sheep dogs.

Stages of Maturation. One of the difficulties in employing zooparasites in developmental biologic studies has been the lack of well-defined stages of development. With the increasing awareness that

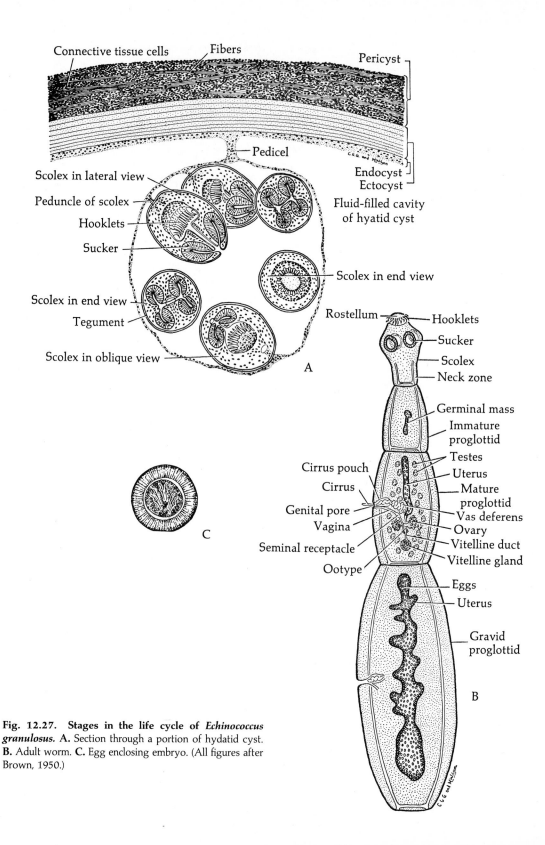

Connective tissue cells
Fibers
Pericyst
Pedicel
Scolex in lateral view
Peduncle of scolex
Hooklets
Sucker
Endocyst
Ectocyst
Fluid-filled cavity
of hyatid cyst
Scolex in end view
Scolex in end view
Tegument
Scolex in oblique view
A

Rostellum
Hooklets
Sucker
Scolex
Neck zone
Germinal mass
Immature
proglottid
Cirrus pouch
Testes
Cirrus
Uterus
Mature
proglottid
Genital pore
Vas deferens
Vagina
Ovary
Seminal receptacle
Vitelline duct
Ootype
Vitelline gland
Eggs
Uterus
Gravid
proglottid
B

C

Fig. 12.27. Stages in the life cycle of *Echinococcus granulosus.* A. Section through a portion of hydatid cyst. **B.** Adult worm. **C.** Egg enclosing embryo. (All figures after Brown, 1950.)

many species of parasitic animals are ideally suited for such studies, a few species have now been examined in sufficient detail so that stages of development have been established. The following stages during the development of the adult of *E. granulosus* have been advocated by Smyth *et al.* (1967). These are depicted in Figure 12.29.

Stage 0. Undeveloped metacestode. This stage is represented by the evaginated but undeveloped protoscolex.

Stage 1. Proglottidization. This stage is represented by the appearance of the first proglottid. The zone where the first intersegmental division is to occur can be recognized as a clear protoplasmic area. The excretory canals also become markedly apparent at about this time.

Stage 2. Appearance of Second Proglottid. This stage is defined by the appearance of the second intersegmental partition.

Stage 3. Early Gametogeny. This stage is characterized by the appearance of testes in the posteriormost (oldest) proglottid.

Stage 4. Genital Pore Formation. This stage is characterized by the appearance of the lateral genital pore.

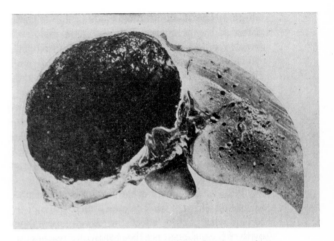

Fig. 12.28. Hydatid cyst. Space formed in human liver resulting from development of an unilocular hydatid cyst. (After Faust, 1937.)

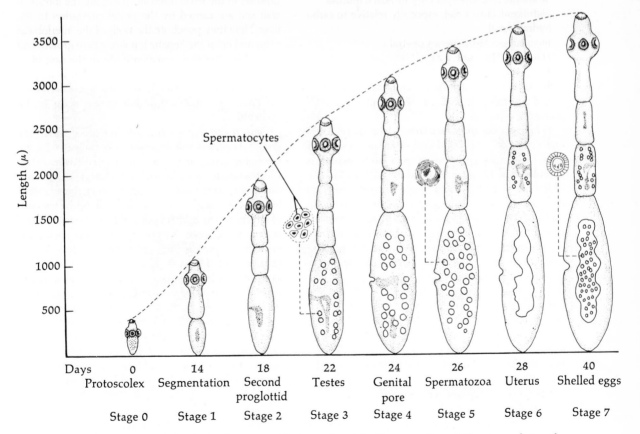

Fig. 12.29. *Echinococcus granulosus* **development.** Criteria employed in the recognition of the stages during the development and maturation of the adult tapeworm. See text for detailed explanation. (Redrawn after Smyth *et al.,* 1967.)

carried by blood or lymph to muscles, viscera, and other organs where they develop into cysticerci.

The cysticerci of *T. solium*, often called *Cysticercus cellulosae*, are ovoid and whitish. These measure 6–18 mm in length. The tiny, dense white spot on one side represents the invaginated scolex. When infected pork is eaten, the scolex evaginates and becomes anchored to the wall of the host's small intestine, and the animal grows to maturity in about 2 to 3 months. Man is the only known natural definitive host, but some development will occur in experimentally infected dogs.

Not only do the adults of *T. solium* parasitize the intestinal tract, but cysticerci also can develop in humans, causing **cysticercosis**. Self-infection with eggs can result either from contaminated fingers or from eggs hatching in the intestine and carried to the stomach by reverse peristalsis. Cysticerci in humans may form in the muscles or subcutaneous tissues, where they do little damage, although tissue response generally occurs. If cysticerci develop in the eye, heart, spinal cord, brain, or some other important organ, the mechanical pressure exerted by these larvae may cause severe neurologic symptoms. Violent headaches, convulsions, local paralysis, vomiting, and optic disturbances are common, sometimes so severe that death results.

When a cysticercus dies, it elicits a severe inflammatory response, which in the brain can cause death. Rarely, a cysticercus may become proliferative, developing branches that infiltrate and destroy the surrounding host tissues. Because of the ability of the cysticerci of *T. solium* to develop in practically every organ in the body, and because of the severity of the resulting pathology, this tapeworm must be considered among the most pathogenic of the human-infecting species.

*Taeniarhynchus saginatus.** This parasite is the most common large tapeworm of humans. It is cosmopolitan in distribution. In parts of the world where raw or rare beef forms a part of the normal diet, as high as 75% of the population may harbor this parasite. The adult is usually 35–60 cm long, although specimens as long as 100–225 cm have been reported. There are approximately 1000 or more proglottids comprising the strobila. Unlike *Taenia solium*, the scolex of *Taeniarhynchus saginatus* is unarmed (Fig. 12.31). The life cycle of this species is similar to that of *T. solium* except that the intermediate host is the cow or some other ungulate. Antelopes, llamas, and

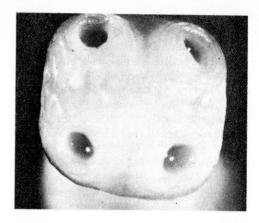

Fig. 12.31. Scolex of *Taeniarhynchus saginatus.* Scanning electron micrograph of frontal view showing absence of rostellum or hooks. (Armed Forces Institute of Pathology Negative No. 65-12073-2.)

giraffes are known to be naturally infected with cysticerci, while lambs and kids can be infected experimentally. Human infections are contracted through the ingestion of cysticerci in beef, particularly the head muscles and the heart. It should be mentioned that the eggs of both *Taenia solium* and *Taeniarhynchus saginatus* can be carried by birds and spread in this fashion.

Saginatus taeniasis** in humans may bring on such symptoms as abdominal pain, excessive appetite, weakness, and loss of weight. These symptoms generally are present in physically weak individuals.

During the life cycle of *T. saginatus*, when eggs are ingested by a suitable intermediate host, they hatch in the duodenum under the influence of gastric and intestinal secretions. The released hexacanths rapidly penetrate the mucosa and are carried to all parts of the body in blood vessels. Commonly, these larvae leave a capillary between muscle cells, enter muscles, and develop into infective cysticerci in about 2 months. Such a larva measures about 10 mm in diameter, is pearly white, and has a single, invaginated scolex.

As in the case of *Taenia solium*, the presence of *T. saginatus* cysticerci, known as *Cysticercus bovis*, generally results in pathologic alternations. For example, when such larvae occur in calves, especially in muscles, they may survive, although conspicuous cellular (and humoral) responses occur. It is noted that cysticerci may survive in one type of host tissue while others in an adjacent tissue become necrotic. This difference can be attributed to nutritional, metabolic, or immunologic differences between tissues. Humans are apparently unsuitable intermediate hosts.

---

* *Taeniarhynchus saginatus* is also known as *Taenia saginata*.

** If clinical manifestations occur, the disease should be referred to as taeniosis; however, as a result of common usage, taeniasis is preferred.

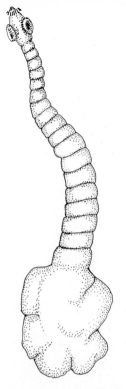

Fig. 12.32. Strobilocercus of *Taenia taeniaeformis.*

Life Cycle of Taenia taeniaeformis. Since *T. taeniaeformis*, also referred to as *Taenia crassicollis* or *Hydatigera taeniaeformis*, is a popular experimental parasite in physiologic and immunologic studies, its developmental cycle will be discussed briefly here. This tapeworm occurs in the small intestine of the cat

and other carnivores, including the lynx and fox. Its scolex is characterized by the presence of prominent suckers and no distinct neck region (Fig. 12.30). Gravid proglottids are shed periodically, and when these disintegrate, the eggs are released and are ingested by various rodents, especially mice and rats. Rabbits, squirrels, and muskrats can also serve as the intermediate host.

Oncospheres liberated from the eggshell in the stomach or small intestine penetrate the mucosa, enter the bloodstream, and are carried to the liver. At this site, an unusual type of larva, known as a **strobilocercus**, develops. The scolex is not invaginated and the bladder is small but there is a conspicuous segmented neck. Such a larva may reach a length of 20 cm (Fig. 12.32). Bullock and Curtis (1920) have reported that strobilocerci in the intermediate host's liver can cause true carcinoma.

In about 30 days, the strobilocerci, known as *Cysticercus fasciolaris*, become infective and if the intermediate host falls prey to a carnivore, the larvae develop into adults in the predator's small intestine. According to Hutchinson (1959), 20–70% of the strobilocercus disintegrates during development to the adult form. During the initial 42 days, the size is doubled during the first 18 days, followed by a period of deceleration. This is followed by the second period of rapid growth. The two periods of rapid growth are separated by the onset of egg production, which occurs between the 16th and 18th days after entering the definitive host.

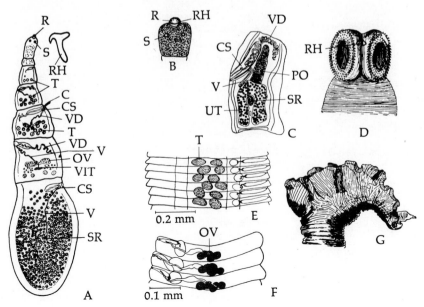

Fig. 12.33. **Some davaineid and hymenolepidid cestodes. A.** *Davainea proglottina* (entire worm), a parasite of fowls. **B.** Scolex of *D. proglottina*. (**A,** after Joyeaux and Baer, 1936; **B,** after Blanchard, 1891.) **C.** Proglottid of *Idiogenes.* (After Clausen, 1915.) **D.** Scolex of *Raillietina echinobothrida* with spiny suckers. (After Blanchard, 1891.) **E.** Successive proglottids of *Diorchis nyrocae* showing male reproductive system. **F.** Successive proglottids of *D. nyrocae* showing female reproductive system. (**E** and **F,** after Long and Wiggins, 1939.) **G.** Scolex and pseudoholdfast of *Fimbriaria fasciolaris,* a parasite of domestic and wild anseriform birds. (After Wolffhügen, 1900.) C, protruded cirrus; CS, cirrus sac; OV, ovary; PO, paruterine organ; R, rostellum; RH, rostellar hooks; S, sucker; SR, seminal receptacle; T, testes; UT, uterus; V, vagina; VD, vas deferens; VIT, vitelline glands.

Family Davaineidae

Members of the family Davaineidae are parasites of birds and mammals. The species are relatively small, measuring 20–80 mm in length. All members possess a single set of reproductive organs per proglottid except *Cotugnia*, which is parasitic in tropical pigeons and parrots. Members of this genus possess two sets of reproductive organs per proglottid.

The more common genera of the Davaineidae are *Davainea*, with several common species in domestic fowls; *Idiogenes* in nondomesticated birds; and *Raillietina* in birds and mammals (Fig. 12.33). *Raillietina* includes *R. madagascariensis*, *R. celebensis*, *R. demerariensis*, *R. asiatica*, and *R. loechesalavezi*, all of which have been reported from humans.

Family Hymenolepididae

Probably the most studied family of the Cyclophyllidea is the Hymenolepididae. Members of this family are small- to medium-sized tapeworms that commonly possess a rostellum armed with a single row of 8 to 30 hooks. However, there are unarmed species, such as *Hymenolepis diminuta*, an intestinal parasite of rodents and rarely of humans.

Common genera include *Hymenolepis*, *Diorchis* (Fig. 12.33) in birds, *Fimbriaria* (Fig. 12.33) in anseriform birds, *Vampirolepis* (Fig. 12.34) and *Pseudohymenolepis* in mammals.

Life Cycle of *Vampirolepis nana**: One-Host Cycle. The life cycle of *V. nana*, a parasite of humans and rodents, is of special biologic interest because it represents a modification of the typical cyclophyllidean life cycle pattern in that the parasite requires only one host in which to complete its development. Stages in the life history of this species are depicted in Figure 12.34.

The adult worm, the smallest of the several species of tapeworms that can infect humans, ranges from 7 to a little over 50 mm in length. The "crowding effect" is demonstrated beautifully in cases of *V. nana* infections, for the size of the worms is inversely proportional to the number of worms present. Adult worms release their oncosphere-containing eggs through their ruptured posterior ends. These eggs pass out, and when ingested by another suitable rodent or human host, they hatch, releasing the oncospheres. These larvae burrow into the villi of the host's intestine, where they develop into cysticercoid

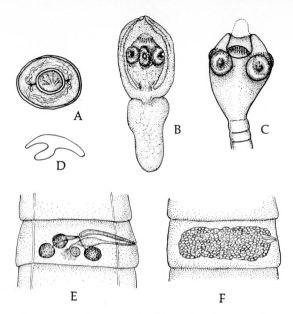

Fig. 12.34. Stages in the life cycle of *Vampirolepis nana*. A. Egg enclosing hexacanth embryo. Note polar filaments attached to each end of embryophore. **B.** Cysticercoid larva with retracted scolex. **C.** Scolex of adult specimen. **D.** Rostellar hook of adult. **E.** Mature proglottid enclosing male and female reproductive organs. **F.** Gravid proglottid showing uterus distended with eggs.

larvae, completing metamorphosis usually by the fourth day. Fully developed cysticercoids escape into the lumen of the gut and, while attached to the gut lining, develop into adults.

This life cycle pattern exemplifies the elimination of an intermediate host by an advanced parasite (p. 27), because experimentally, *V. nana* can be induced to undergo cysticercoid formation in fleas and grain beetles, suggesting that its immediate ancestor utilized an invertebrate intermediate host.

Vampirolepis nana, which is cosmopolitan in distribution, is one of the most common cestode parasites of humans in the world, especially among children. In the southern United States, the incidence is about 1%, whereas in Argentina it is about 9%.

***Hymenolepis diminuta*.** *H. diminuta*, another rodent- and human-infecting species, has the typical two-host cycle, utilizing several grain-ingesting insects as the intermediate host. The cysticercoid larvae are found in the invertebrate host. The stages in its developmental cycle are depicted in Fig. 12.35.

H. diminuta, although very common in rodents, is less frequently found in humans. A single worm may reach a length of 90 cm with a maximum diameter of 3.4–4 mm. The scolex in this species is unarmed and each proglottid is broader than long. The eggs, measuring 60–80 μm in diameter, are usually yellowish brown and spherical. Unlike *V. nana*, the embryophore

---

**Vampirolepis nana* until recently was known as *Hymenolepis nana*. In fact, some still prefer the latter designation.

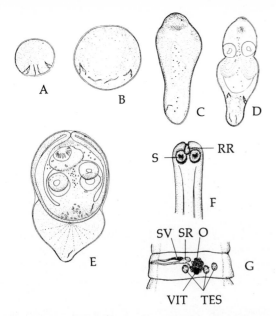

Fig. 12.35. Stages in the life cycle of *Hymenolepis diminuta*. A. Developing embryo as solid sphere, showing paired hooks and external membrane. **B.** Appearance of cavity and dispersal of oncosphere hooks. **C.** Appearance of two body zones, elongation of cavity, and beginning of rostellar and sucker differentiation. **D.** Process of withdrawal, showing separation of neck tissue, which will become the layer immediately enveloping scolex. **E.** Unarmed scolex withdrawn, with large rostellum, and clearly defined suckers—the cysticercoid larva. **F.** Scolex of adult specimen with scolex withdrawn. **G.** Mature proglottid of adult showing male and female reproductive organs. (A–E, after Voge and Heyneman, 1957.) O, ovary; RR, retracted rostellum; S, sucker; SR, seminal receptacle; SV, seminal vesicle; TES, testes; VIT, vitellaria.

in these eggs does not bear conspicuous knobs and filaments at the poles.

As stated, the life cycle of *H. diminuta* includes an intermediate host. Most commonly, adults of the grain beetles, *Tenebrio* and *Tribolium*, and cockroaches serve as the intermediate hosts. Humans become infected when cereals, dried fruits, and other similar foods containing insects that had become infected by ingesting rodent feces, including tapeworm eggs, are eaten.

In cases of hymenolepiasis, severe symptoms of toxicity may occur, especially in children, including abdominal pain, diarrhea, convulsions, and insomnia.

H. diminuta has been a favorite animal for experimental studies for decades. Arai (1980) has edited a volume devoted to various aspects of the morphology, development, physiology, biochemistry, and immunology of this species.

Interactions in the Invertebrate Host. Ubelaker *et al.* (1970) have contributed an electron microscope study of the interaction of *Hymenolepis diminuta* cysticercoids and the beetle host, *Tribolium confusum*. They have found that the surface of the larval tapeworm is

covered with microtriches (microvilli), some branched, as is the case with adult tapeworms (p. 388). Of particular interest is their finding of morphologic suggestions that the microtriches may secrete a lytic substance that causes the lysis of host blood cells that approach the parasite (Fig. 12.36). This suggests that the cysticercoids can protect themselves from the cellular defense mechanisms of the host.

Other *Hymenolepis* Species. Other species of *Hymenolepis* include parasites of ground squirrels (*H. citelli*), rodents (*H. microstoma*), shrews (*H. anthocephalus* and others), muskrats (*H. evaginata* and *H. ondatrae*), various wild ducks (*H. megalops*), anseriform birds (*H. collaris* and *H. tritesticulata*), gulls (*H. ductilis* and others), poultry and game birds (*H. cantaniana*, *H. carioca*, and others), and passeriform birds (*H. corvi*, *H. microcirrosa*, and *H. turdi*).

Family Dilepididae

The family Dilepididae includes, among others, *Dipylidium* in mammals (Fig. 12.37), *Paradilepis* (Fig. 12.37) in various birds, *Liga* (Fig. 12.37) in birds, and *Dilepis* (Fig. 12.37) also in birds. Included also is *Ophiovalipora*, the only genus in this family found in an animal other than a bird or mammal; it is a parasite of snakes.

Life Cycle of *Dipylidium caninum*. The genus *Dipylidium* includes *D. caninum*, a cosmopolitan species found in dogs, cats, and humans, especially children (Moore, 1962). Stages in the life cycle of this species are depicted in Fig. 12.38. *D. caninum* commonly attains a length of 30 cm and possesses a rostellum armed with one to eight (commonly four to six) rows of hooks. This species is easily recognized because each proglottid includes two sets of male and female reproductive systems and a genital pore situated on each side (Fig. 12.38).

The uterus is unusual in that initially it is in the form of a network, but as it becomes filled with eggs, it breaks up into individual uterine balls, each containing 8 to 15 eggs.

Eggs passed out of the vertebrate in packets are ingested by flea larvae of the genera *Pulex* and *Ctenocephalides*, or by the dog louse, *Trichodectes canis*. Oncospheres hatch in the arthropod's gut and burrow through the gut wall, where the parasites develop into cysticercoids when the flea matures.

Infection of the vertebrate host occurs when dogs or cats ingest infected fleas. Infections are established in children when a dog that has just nipped a flea licks the child's mouth or when whole fleas are ingested. Other *Dipylidium* species infect various members of the canine and feline families.

***Dipylidium caninum* and the Flea Host.** Larvae of *D. caninum* are extremely destructive to the larval fleas that serve as intermediate hosts. Pathogenesis in the larval flea occurs in varying degrees. In some instances, mortality as high as 60% has been observed within 24 hours after the cestode eggs are ingested. Deaths occurring during this period represent the initial mortality phase. Just prior to death, the flea larvae become lethargic and translucent white to reddish. The cause of death appears to be primarily the mechanical damage inflicted on the larval flea's gut wall due to penetration by oncospheres migrating to the hemocoel.

A second series of developments, representing the second mortality phase, may cause the death of an additional 20% of the larvae prior to pupation. Instead of spinning their cocoons at the appointed time, some mature larvae wander about or lie dormant for days and eventually die. Other larvae spin their cocoons but delay pupation. During the pupal period, a third series of reactions that represents the third mortality phase causes the death of an additional 10% of the fleas.

Deaths occurring during the second and third mortality phases result from the depletion of reserves in the flea's fat bodies, caused by the rapidly developing cysticercoids and displacement and distortion of organs when large numbers of cysticercoids are present.

Chen (1934) reported that macrocytes, which are normally found only in flea larvae, persist through the pupal and adult stages of infected fleas. These cells represent the defense mechanism of the host, for they

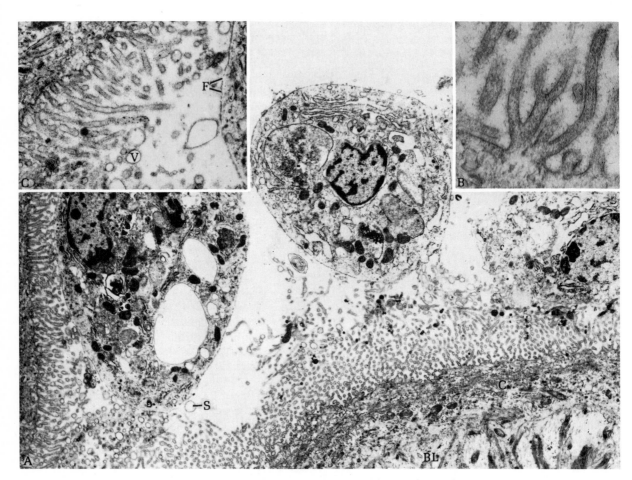

Fig. 12.36. **Electron micrographs of capsule wall of *Hymenolepis diminuta* cysticercoid and associated structures.** **A.** Micrograph showing three hemocytes of insect host in various states of decay, presumably from material released from parasite's microvilli. × 8500. **B.** Structure of parasite's microvilli. × 51,660. **C.** Origin of filamentous coated vesicle from tip of a microvillus. Note the breaks (F) in the hemocyte membrane. × 18,000. (After Ubelaker *et al.*, 1970.) BL, basal lamina of parasite; C, distal cytoplasm of parasite; F, breaks in host hemocyte membrane; S, secreted material (?); V, vesicle.

form capsules enveloping cysticercoids during larval life and may kill the tapeworms. However, if the cysticercoids survive and grow, the capsules of macrocytes become thinner, as no new cells are added. By the time the flea reaches maturity, the capsule surrounding each cysticercoid is no more than a thin web and does not appear to harm the parasite.

ORDER PSEUDOPHYLLIDEA

The pseudophyllidean tapeworms are characterized by true bothria on their scolices. These holdfast organs appear as slitlike depressions on the body surface, although various modifications exist (Fig. 12.39). The scolex is not clearly demarked from the strobila in most species, and a definite neck region is lacking. These worms are comparatively small, ranging from a few millimeters to a few centimeters in length, but certain members of the family Dibothriocephalidae, such as *Polygonoporus*, a parasite of the sperm whale, can reach a length of 30 meters.

Excluding the Cyclophyllidea, the Pseudophyllidea is the largest order of tapeworms. Numerous species have been reported from fish and occasionally mammals. The order is divided into seven families. In Haplobothriidae and Diphyllobothriidae, the genital pore(s) is midventral on the same surface of the dorsoventrally flattened body as the uterine pore. In Bothriocephalidae and Ptychobothriidae, the genital pore(s) is on one surface, whereas the uterine pore opens on the other. In Echinophallidae, Triaenophoridae, and Amphicotylidae, the genital pore(s) opens laterally.

FAMILY HAPLOBOTHRIIDAE

The family Haplobothriidae includes the genus *Haplobothrium*, parasites in freshwater teleost fish (Fig. 12.40). This genus is unique because it appears to be an intermediate form between the Pseudophyllidea and the Trypanorhyncha. The bothria are located at the tips of four tentaclelike projections known as **proboscides**.

During the life cycle of *Haplobothrium globuliforme*, eggs passing out of the definitive host give rise to coracidia, which are ingested by *Cyclops viridis*. After sloughing its ciliated coat, each oncosphere penetrates into the hemocoel of the copepod first intermediate host and develops into a procercoid in about 10 days. When infected *Cyclops* are ingested by a bony fish, each procercoid develops into a plerocercoid bearing proboscides on its scolex. When an infected bony fish is eaten by the freshwater bowfin, *Amia calva*, the plerocercoid matures into the adult also bearing tentaclelike proboscides. Thus, the life cycle of *H. globuliforme* involves three hosts.

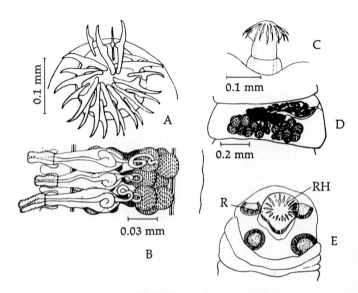

Fig. 12.37. Some cyclophyllidean cestodes. A. Apex of scolex of *Paradilepis brevis* from birds. **B.** Mature proglottid of *P. brevis*. (**A** and **B**, after Long and Wiggins, 1939.) **C.** Scolex of *Liga brasiliensis* from birds with rostellum extended. **D.** Mature proglottid of *L. brasiliensis*. (**C** and **D**, after Ransom, 1909.) **E.** Scolex of *Dilepis*. (After Burt, 1936.) R, rostellum; RH, rostellar hooks.

FAMILY DIPHYLLOBOTHRIIDAE

The Diphyllobothriidae is comprised of numerous genera, among which is *Diphyllobothrium*, including *D. latum*, commonly called the broadfish tapeworm.

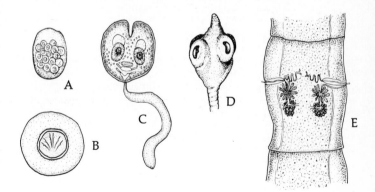

Fig. 12.38. Stages in the life cycle of *Dipylidium caninum*. A. Cluster of eggs in uterine ball. **B.** Egg enclosing three pairs of hooks. **C.** Cysticercoid larva. **D.** Scolex with armed rostellum. **E.** Mature proglottid enclosing two sets of reproductive organs. (**A** and **D**, redrawn after Stiles, 1906.)

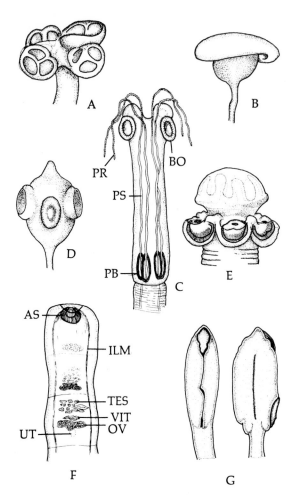

Fig. 12.39. Types of cestode scolices. A. Scolex of *Trilocularia gracilis* (Tetraphyllidea) showing areolae. (Redrawn after Linton, 1924.) **B.** Scolex of *Disculiceps pileatum* (Disculicepitidea.) (Redrawn after Southwell, 1925.) **C.** Scolex of *Gilquinia anteroporus* (Trypanorhyncha). (Redrawn after Hyman, 1951.) **D.** Cyclophyllidean scolex showing prominent acetabula and rostellum. **E.** Scolex of *Nematoparataenia southwelli* (Aporidea). (Redrawn after Fuhrmann, 1933.) **F.** Anterior end of *Nippotaenia chaenogobii* (Nippotaeniidea). (Redrawn after Yamaguti, 1939.) **G.** Scolex of *Adenocephalus pacificus* (Pseudophyllidea), two views. (Redrawn after Nybelin, 1931.) AS, anterior sucker; BO, bothridia; ILM, internal longitudinal muscles; OV, ovary; PB, proboscis bulb; PR, proboscides; PS, proboscis sheath; TES, testes; UT, uterus; VIT, vitellaria.

This species normally is found in various terrestrial and marine fish-eating carnivores, including canines, cats, bears, mongoose, mink, foxes, bears, seals, and sea lions. This tapeworm also parasitizes humans. Human infections have been reported in central Europe,

especially in Finland, Scandinavia, Switzerland, Baltic areas, the Soviet Union, Ireland, Israel, central Africa, Siberia, Japan, Canada, and American states surrounding the Great Lakes.

Life Cycle of *Diphyllobothrium latum*. Stages in the life cycle of *D. latum* are illustrated in Fig. 12.41. Adult worms have been reported as large as 10 meters long and 10–20 mm wide and composed of 3000 or more proglottids. The ovoid eggs do not include completely formed embryos when laid. Such eggs lay dormant in water after passing out of the host and actively consume oxygen. Within the shell, the embryo becomes fully developed in 8 to 12 days. Typical of the Pseudophyllidea, the hexacanth larva is covered by a ciliated **embryophore** and is known as a **coracidium**.

Upon hatching, the coracidium may become free swimming or free crawling but must be ingested by specific copepods, such as *Diaptomus* spp., within 24 hours, or else they perish. Coracidia are quite host specific. If a coracidium successfully enters the appropriate copepod host, it loses its ciliated coat and the naked oncosphere bores through the copepod's intestinal wall into the hemocoel. In 14 to 18 days, the oncosphere transforms into a solid, elongate (500 μm) larva—the **procercoid**. This larva possesses six hooks embedded in its cercomer, which projects from the posterior end.

When an infected copepod is ingested by a fish such as *Esox* spp. in North America and *Salmo* and *Oncorhynchus* in the Far East, the escaping procercoid in the fish host's intestine works its way through the intestinal wall and migrates to the body muscles, where it develops into a long (2–4 cm), solid **plerocercoid**. This larva bears an evaginated scolex at one end.

Unlike most pseudophyllidean plerocercoids, *D. latum* is coiled and can be encapsulated or free in muscle tissue, although the latter is more common. Host reaction in the form of encapsulation of the plerocercoid depends on its location in the fish host. When situated in muscles of the body wall, encapsulation does not usually occur; however, if it should settle in or on the viscera, a capsule is commonly formed by the host around the plerocercoid. Infection of the definitive host, including humans, occurs when plerocercoids in poorly cooked or raw fish are eaten.

Human Diphyllobothriosis. Human infection with *Diphyllobothrium latum* is primarily, although not exclusively, limited to areas where fresh fish are commonly eaten or where the fishing industry, involving handling and cleaning of fish, is carried on. Human infections are contracted when poorly cooked fish, including plerocercoids in the flesh, are ingested or when plerocercoids clinging to the hands of fish cleaners are accidentally ingested.

In Finland, approximately 20% of the human population is infected, whereas in some Baltic communities the prevalence approaches 100%. In North America,

50 to 70% of northern and walleye pike found in some small lakes in the northern United States and Canada harbor plerocercoids of *D. latum*. I have found infections in fish in the Delaware River.

Symptoms of human diphyllobothriosis include abdominal pain, loss of weight, and sometimes, a kind of pernicious anemia (p. 438).

Although the typical pseudophyllidean life cycle includes a free-swimming coracidium, a procercoid in the first intermediate host, a plerocercoid in the second, and the sexually mature adult in the definitive host, certain species, such as *D. latum* and *Spirometra* spp., pass through more than two intermediate hosts. However, the larval stages found in the second and all subsequent intermediate hosts, actually paratenic hosts, remain comparable, merely retaining the plerocercoid form while being transferred through the third and fourth hosts. Plerocercoid larvae of such pseudo-

phyllidean tapeworms are also referred to as **spargana**. In many instances spargana can live in human flesh, causing sparganosis.

Other Diphyllobothriid Genera. Some of the other more prominent genera of the Diphyllobothriidae are *Ligula* in diving and wading birds and *Spirometra* parasitic in various fish-eating birds, dogs, and cats (Fig. 12.40). In these genera, various fish serve as the second or last intermediate host, in which the spargana (plerocercoids) are often encountered encapsulated in the musculature.

Growth Rate. Studies have been carried out to ascertain the growth rates of *Diphyllobothrium* (Smyth,

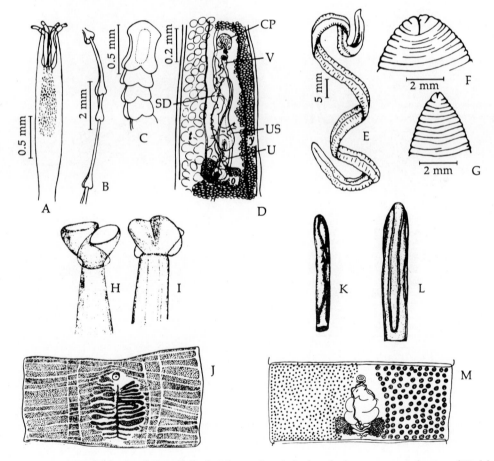

Fig. 12.40. **Some haplobothrid and diphyllobothriid cestodes. A.** Scolex of primary segmented worm of *Haplobothrium globuliforme* from fishes. **B.** 12th, 13th, and 14th proglottids of *H. globuliforme*. **C.** Scolex and first three proglottids of secondary segmented worm of *H. globuliforme*. **D.** Mature proglottid of *H. globuliforme*. (After Cooper, 1914, 1918.) **E.** Entire specimen of *Ligula intestinalis* from intestine of birds. **F.** Anterior end of larval stage of *L. intestinalis*. **G.** Anterior end of adult *L. intestinalis*. (After Lemaire, 1936.) **H** and **I.** Two views of the scolex of *Diphyllobothrium stemmacephalum* from intestine of dolphins. **J.** Gravid proglottid of *D. stemmacephalum*. (After Cohn, 1912.) **K** and **L.** Two views of scolex of *Spirometra mansoni* from intestine of mammals. **M.** Mature proglottid of *S. ranarum*. (**K** and **L**, after Joyeux, 1927; **M**, after Faust *et al.*, 1929.) CP, cirrus pouch; O, ovary; SD, vas deferens; T, testes; U, uterus; US, uterine sac; V, vagina; Y, vitellaria.

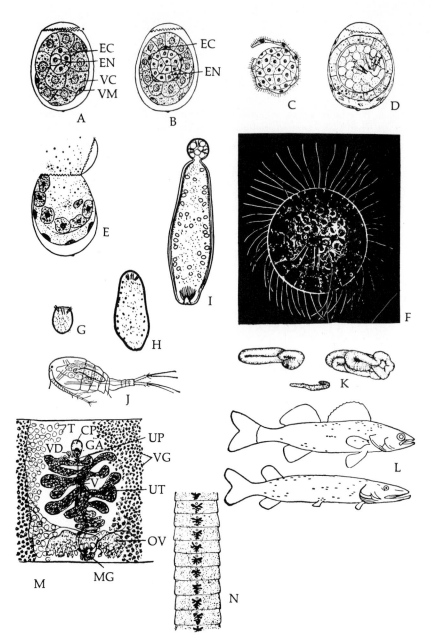

Fig. 12.41. **Stages in the life cycle of *Diphyllobothrium latum*. A** and **B.** Eggs enclosing blastomeres showing origin of ectoderm and entoderm. The embryo has shrunk away from the shell due to fixation. (Modified after Schauinsland, 1886.) **C.** Immature embryo showing development of cilia. (After Schauinsland, 1886.) **D.** Egg a few hours before hatching. **E.** Eggshell immediately after liberation of coracidium. **F.** Coracidium as seen with darkfield microscopy. (After Vergeer, 1928.) **G.** Embryo after shedding ciliated epithelium. **H.** Procercoid growing in *Cyclops*. **I.** Fully developed procercoid from *Cyclops*. **J.** *Cyclops* enclosing fully developed procercoid. (After Brumpt, 1936.) **K.** Plerocercoid larvae removed from fish muscles. **S.** Walleyed pike and pickerel showing distribution of plerocercoids. (After Vergeer, 1928.) **M.** Mature proglottid. **N.** Segment of the strobila. CP, cirrus pouch; EC, ectodermal cell; EN, entodermal cell; GA, genital atrium; MG, Mehlis' gland; OV, ovary; T, testes; UT, uterus; UP, uterine pore; V, vagina; VD, vas deferens; VG, vitelline glands; VM, vitelline membrane; YC, yolk cell.

1963). Growth is defined as an increase in the total mass of protoplasm. In the species that have been studied, an initial lag phase occurs after entry of the parasite into the definitive host. During this phase, there is little growth. The causes for this lag include: (1) the initial establishment in an unfavorable site in the posterior region of the host's intestine, followed by later migration anteriorly to a more favorable site in the duodenum; and (2) an initial period during which intimate contact with the host's intestinal mucosa is being established.

Experimental Sparganosis. Although loss of weight by the host is commonly associated with parasitism, especially among mammals, Mueller (1963) has found that when the spargana of *Spirometra mansonoides* are injected subcutaneously into young female mice, the infected mice show accelerated gain in weight when compared to uninfected control mice (Fig. 12.42) (p. 16).

This parasite-induced weight gain occurs not only in laboratory mice but also in the deer mouse, *Peromyscus*, and in the golden hamster (Mueller, 1965). In the case of the hamster, it was shown that both true growth (as revealed by x-ray examination) and obesity occur. It was reported that the weight gain was additive, increasing with the number of spargana introduced, but reaches a plateau at about 12 worms per mouse weighing 20 g. Beyond this number, there is either no weight gain or increased mortality as the result of parasite invasion of the hosts' pleural cavities or some other vital area. The increased weight, according to Meyer *et al.* (1965), is distributed among the internal organs, liver, skin, and carcass, in that order. Furthermore, of the 61% increase in dry weight, 46% is attributable to lipids, 14% to proteins, and 1% to nucleic acids, glycogen, and ash.

It is noted that Mueller (1965) has reported that the Oriental species, *Spirometra ranarum*, has an even greater stimulatory effect than *S. mansonoides*, while European species stimulate little or no weight gain.

The possible cause of the enhanced growth due to the spargana of *Spirometra* spp. in rodents has been the subject of some study. Mueller (1968) and Mueller and Reed (1968) have reported that the implantation of *S. mansonoides* spargana into hypophysectomized and hypothyroid rats stimulates growth. They have concluded that the secretion of a growth hormonelike substance by the parasites into the altered host is strongly suggested. These results and conclusions have been verified by Steelman *et al.* (1970), who have demonstrated significant increases in the weights of the kidney, thymus, and liver of hypophysectomized rats implanted with spargana. The increase in the weights of these organs, together with a doubling of the width of the epiphyseal cartilage, strongly suggests the action of a growth hormonelike substance. However, Steelman *et al.* have stated that "Growth hormone secretion per se by SMS (*S. mansonoides*

Fig. 12.42. Effect of *Spirometra mansonoides* spargana on growth of mice. Photograph showing experimental mouse (left) and control mouse (right) 69 weeks after start of experiment. The experimental mouse had eight spargana inserted under the dorsal skin. At the time this photograph was taken, the experimental mouse weighed 68.6 grams while the control mouse weighed 38.7 grams. (After Mueller, 1963; with permission of New York Academy of Sciences.)

spargana) seems unlikely but cannot be excluded. Another possible explanation is that a substance is released which causes sulfation factor to be produced. Daughaday and Kipnis ... showed that growth hormone has this action. Alternatively, the growth factor could be sulfation factor like."

In addition to this hormonal action, it appears appropriate to point out that certain aspects of immunology are involved. Specifically, Mueller (1966) has reported that the enhanced growth in rodents is dependent upon the tolerance of the spargana by the host. The difference between the growth rates of parasitized and nonparasitized hosts is correlated with the host's cellular reactions, with the maximum difference occurring when the subdermally situated parasites are not encapsulated. If encapsulation occurs, the weight gain is less apparent. Also, a secondary pyrogenic manifestation also interferes with the effect and usually kills the parasites.

Other Pseudophyllidean Families

The family Bothriocephalidae includes *Bothriocephalus* (Fig. 12.43), which is often encountered in freshwater fish in the United States. The family Ptychobothriidae includes *Ptychobothrium* in elasmobranchs (Fig. 12.43), *Senga* in freshwater labyrinthid and cyprinid fish (Fig. 12.43), *Polyonchobothrium* in African fish (Fig. 12.44), and *Clestobothrium* in fish of the family Merluccidae (Fig. 12.44).

The Echinophallidae, Triaenophoridae, and Amphicotylidae are minor families, although the last mentioned includes *Eubothrium*, with several species fairly common in teleost fish (Fig. 12.44).

ORDER PROTEOCEPHALATA

Members of the order Proteocephalata have been rather comprehensively studied because they are comparatively plentiful. Generally only one family, the Proteocephalidae, is recognized as comprising this order.

Life Cycle of *Ophiotaenia perspicua*. The life cycle of *O. perspicua*, an intestinal parasite of the snake *Natrix rhombifer* (also *N. sipedon* and *Thamnophis sirtalis*), is representative of the pattern of proteocephalid cestodes (Fig. 12.45).

The eggs are voided from the definitive host in feces. Within each eggshell is a completely formed ciliated coracidium which soon hatches in water. When ingested, the freed oncospheres actively penetrate the intestinal wall and enter the hemocoel of copepods, *Cyclops vulgaris*, *C. viridis*, *Mesocylops obsoletus*, or *Microcylops varicans*. Within the crustacean host, the oncospheres develop into procercoids in approximately 14 days. When infected copepods are ingested by tadpoles, the procercoids invade the mesenteries and liver of the amphibian host and develop into encapsulated plerocercoids. The plerocercoids continue to grow, and with the metamorphosis of the tadpole into a frog, the plerocercoids may break out into the host's coelom. When an infected frog is eaten by the definitive host, the plerocercoids develop into adult tapeworms.

Another well-known genus of the Proteocephalata is *Proteocephalus* (Fig. 12.46), the adults of which are intestinal parasites of primarily freshwater fish, although a few species occur in amphibians and reptiles. Aspects of the biology of *Proteocephalus ambloplites* is given below.

Life Cycle of *Proteocephalus ambloplites*. Adults of this tapeworm and related species are readily available almost wherever bass (*Micropterus dolomieui*, *Huro salmoides*), yellow perch (*Perca flavescens*), bowfins (*Amia calva*), and other freshwater fish occur. As the parasite develops and becomes sexually mature, the posterior proglottids become gravid and become detached from the strobila. They pass from the host in feces and upon reaching and absorbing water, they rupture and the enclosed eggs are released, each con-

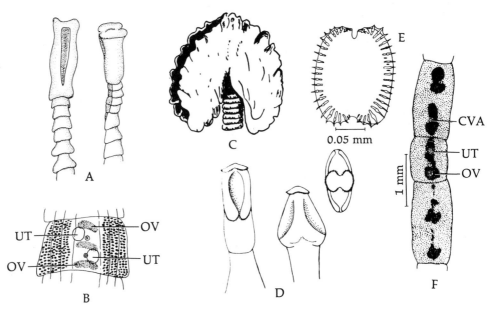

Fig. 12.43. **Some pseudophyllidean cestodes. A.** Two views of the scolex of *Bothriocephalus scorpii*. (Redrawn after Wardle, 1932.) **B.** Gravid proglottid of *B. scorpii*. (Redrawn after Cooper, 1918.) **C.** Scolex of *Ptychobothrium* sp. (After Lönneberg, 1889.) **D.** Three views of the scolex of *Senga besnardi*. **E.** End view of crown of rostellar hook of *S. besnardi*. **F.** Three proglottids of *S. besnardi*. (**D–F**, after Woodland, 1937.) CVA, cirrovaginal atrium; OV, ovary; UT, uterus.

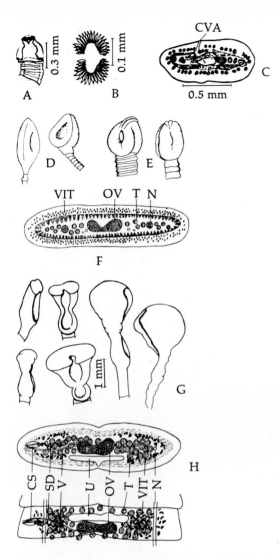

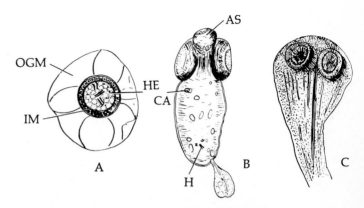

Fig. 12.45. Stages in the life cycle of *Ophiotaenia perspicua*. A. Hexacanth larva enveloped within membrane. **B.** Procercoid. **C.** Anterior end of adult. AS, apical sucker; CA, calcareous bodies; H, hooks; HE, hexacanth; IM, inner membrane; OGM, outer gelatinous membrane. (**A** and **B,** after Thomas, 1931; **C,** after Nybelin, 1917.)

Fig. 12.44. Some pseudophyllidean cestodes. A. Scolex of *Polyonchobothrium gordoni*. **B.** End view of rostellar hooks of *P. gordoni*. **C.** Cross-section through cirrovaginal aperture of *P. gordoni*. (**A–C,** after Woodland, 1937.) **D.** Living scolices of *Clestobothrium crassiceps*. **E.** Fixed scolices of *C. crassiceps*. **F.** Cross section of mature proglottid of *C. crassiceps*. (**D–F,** after Wardle, 1935.) **G.** Scolices of *Eubothrium rugosum*. **H.** Cross-sectional and ventral views of mature proglottid of *E. rugosum*. (**G** and **H,** after Ekbaum, 1933.) CS, cirrus sac; CVA, cirrovaginal atrium; N, nerve cord; OV, ovary; SD, sperm duct (vas deferens); T, testes; U, uterus; V, vagina; VIT, vitelline glands.

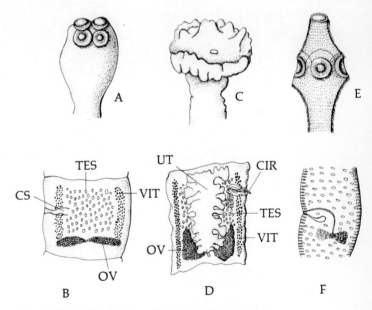

Fig. 12.46. Some proteocephalan cestodes. A. Scolex of *Proteocephalus coregoni*, an intestinal parasite of the fish *Coregonus atikameg*. **B.** Mature proglottid of *P. coregoni*. (**A** and **B,** redrawn after Wardle, 1932.) **C.** Expanded scolex of *Corallobothrium fimbriatum*, an intestinal parasite of siluroid fish. **D.** Mature proglottid of *C. fimbriatum*. (**C** and **D,** redrawn after Essex, 1927.) **E.** Scolex of *Acanthotaenia shipleyi*, an intestinal parasite of reptiles, mainly *Varanus* spp. **F.** Mature proglottid of *A. shipleyi*. (**E** and **F,** redrawn after Southwell, 1930.) CIR, cirrus; CS, cirrus sac; OV, ovary; TES, testes; UT, uterus; VIT, vitellaria.

taining a fully developed unciliated oncosphere. Such eggs are ingested by copepods (*Cyclops, Eucyclops*) and the hatched larva burrows through the intestinal wall into the hemocoel where it develops into a procercoid. Each procercoid develops four suckers like those of the adult. When infected copepods are eaten by young bass or pumpkinseed sunfish, which serve as second intermediate hosts, the procercoids migrate through the intestinal wall into the coelom and metamorphose into plerocercoids. Infection of the definitive host occurs when a second intermediate host harboring plerocercoids is eaten.

Plerocercoids of *P. ambloplites* will develop into adults only if they have completed their development in the second intermediate host. If a bass fry or sunfish harboring immature plerocercoids is eaten by a compatible definitive host, the larval cestodes migrate to the coelom, frequently entering the gonads, where they continue to develop into infective plerocercoids. Infected gonads commonly are destroyed by the parasites, an example of parasitic castration (p. 14). If the fish host is now ingested by another definitive host, the plerocercoids will develop into adults. Sexual maturation of the adult worms is influenced primarily by changes in the definitive host's hormonal pattern,

although ambient temperature is also influential (Hopkins, 1959).

ORDER TETRAPHYLLIDEA

The tetraphyllidean cestodes are readily recognized by four prominent outgrowths on the scolex. These flaplike holdfast organs are properly termed **bothridia** (Figs. 12.39, 12.47), as opposed to the bothria of Pseudophyllidea. Often the single bothridium is subdivided into areolae (or loculi) by septa. Also, frequently there are accessory suckers, hooks, or spines present. An apical, suckerlike organ, known as the **myzorhynchus**, is present on some species.

Life Cycle Pattern

The complete life cycle pattern of the Tetraphyllidea is not known in any of the species. However, larval forms have been described from various hosts and attributed to this order. Baylis (1919), for example, reported a cysticercoid of *Phyllobothrium* encapsulated

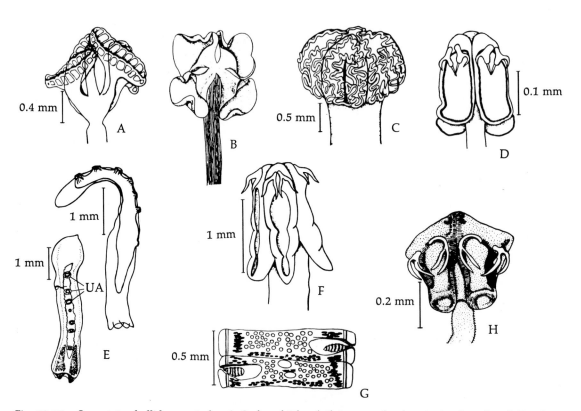

Fig. 12.47. Some tetraphyllidean cestodes. A. Scolex of *Echeneibothrium myzorhynchum*, in the elasmobranch *Raja binoculata.* (After Hart, 1936.) **B.** Scolex of *Anthobothrium cornucopia* in elasmobranchs. (After Southwell, 1925.) **C.** Scolex of *Phyllobothrium lactuca* in elasmobranchs. (After van Beneden, 1850.) **D.** Scolex of *Pedibothrium brevispine* in elasmobranchs. (After Southwell, 1925.) **E.** Proglottids of *Calliobothrium verticillatum* in various elasmobranchs showing uterine apertures (UA). (After Woodland, 1927.) **F.** Scolex of *Acanthobothrium coronatum* in elasmobranchs. **G.** Mature proglottids of *A. coronatum.* (**F** and **G**, after Southwell, 1925.) **H.** Scolex of *Platybothrium hypoprioni* in the elasmobranch *Hypoprion.* (After Potter, 1937.)

in the peritoneum of a dolphin. Similar larvae also have been found on cephalopod molluscs, marine teleostean fish, and marine mammals. *Dinobothrium planum*, another supposedly tetraphyllidean larva, was described by Linton (1922) from *Ommatostrephes illecebrosa*. The scolex of this larva includes two pairs of flattened bothridia arranged back to back. These incomplete pieces of life cycle data suggest that the Tetraphyllidea follow the two-intermediate-host cycle pattern—the two examples cited are in the second intermediate host. The adults are all parasitic in the spiral valve of elasmobranchs, hence the members of this order are exclusively marine.

Several representative genera of the Tetraphyllidea are depicted in Fig. 12.47.

Rees (1967), in her review of the pathogenicity of cestodes, has stated that although those species of Tetraphyllidea which are armed with hooks do tear their hosts' intestinal mucosa, no inflammatory reactions result. For additional reviews on the pathology of infections caused by metacestodes and adult cestodes, see Freeman (1983) and Arme *et al.* (1983), respectively.

ORDER TRYPANORHYNCHA

Members of the Trypanorhyncha are small to medium-sized tapeworms, ranging from 2 to 100 mm in length. They are readily recognized by their four spiny **proboscides**, which can be everted from tubular pockets in the scolex (Figs. 12.39, 12.48). These proboscides are blind tubes and when retracted, their spiny surfaces line the interior of the proboscides. The shapes, sizes, and numbers of spines on the proboscides are taxonomically important. Dollfus (1942, 1946) has contributed taxonomic reviews of this order.

Except in *Hepatoxylon*, in which there is a double set of reproductive organs, all the trypanorhynchid worms possess one set of reproductive organs per proglottid.

All known adult trypanorhynchans are parasitic in the stomach or spiral valves of elasmobranchs, and hence this is also an order of exclusively marine tapeworms.

Life Cycle Data

Complete life cycle data concerning the Trypanorhyncha are lacking. Very few of the species possess operculated eggs, but one species, *Grillotia erinaceus*, is known to have them. The complete life cycle of *G. erinaceus* alone is known because of the work of Ruszkowski (1932), but most parasitologists do not consider the cycle typical of members of this order.

In *G. erinaceus*, the eggs escape from the host, hatch into ciliated coracidia, and are ingested by copepods (*Acartia, Pseudocalanus*, and others). Within the crustacean host, the oncospheres develop into typical procercoids, each with a cercomer. In the second intermediate host, a fish, the procercoids develop into typical plerocercoids. When the second intermediate host is ingested by a selachian host, the adults develop. The plerocercoids of certain species bear a posterior sac, the **blastocyst**, into which the scolex is inverted.

While our knowledge of life cycles in the Trypanorhyncha is incomplete, except for that of *G. erinaceus*, many plerocercoids have been reported in marine snails, bivalves, and cephalopods (Cake, 1976). They

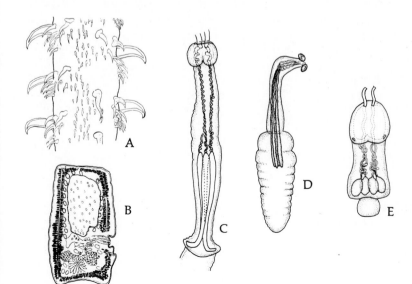

Fig. 12.48. Trypanorhynchan cestodes. A. External surface of metabasal armature of a tentacle of *Grillotia erinaceus*. (Redrawn after Dollfus, 1942.) **B.** Mature proglottid of *G. erinaceus*. (Redrawn after Kükenthal, 1931.) **C.** Scolex of *Dasyrhynchus talismani*. (Redrawn after Dollfus, 1942.) **D.** *Eutetrarhynchus ruficollis* larva. (Redrawn after Eysenhardt, 1829.) **E.** Scolex of *Otobothrium*. (Redrawn after Dollfus, 1942.)

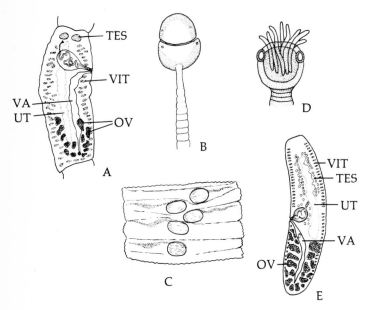

Fig. 12.49.   Some lecanicephalan cestodes. A. Mature proglottid of *Lecanicephalum peltatum.* (Redrawn after Southwell, 1925.) **B.** Scolex of *Tretragonocephalum.* **C.** Immature proglottid of *Tetragonocephalum.* (**B** and **C,** redrawn after Shipley and Hornell, 1906.) **D.** Scolex of *Polypocephalus medusia.* **E.** Mature proglottid of *P. medusia.* (Redrawn after Southwell, 1925.) OV, ovary; TES, testes; UT, uterus; VA, vagina; VIT, vitellaria.

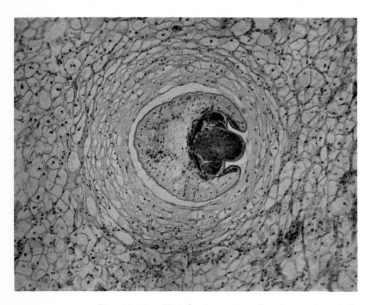

Fig. 12.50.   Histologic section showing metacestode (larva) of *Tetragonocephalum* in connective tissue of the American oyster, *Crassastrea virginica.*

also occur in crustaceans, teleost fish, and, rarely, reptiles. Plerocercoids, such as that of *Nybelinia*, possess a fully developed scolex and may or may not possess a tail. Because these larvae represent the last stage of larval development, ready to be ingested by the definitive host and hence found in a second intermediate host, it is suspected that there is a first intermediate host in which the procercoids are to be found. Several representative genera of the Trypanorhyncha are depicted in Fig. 12.48.

Trypanorhynchan plerocercoid larvae may be so plentiful in the muscles of certain shrimp and fish that their hosts are unsalable; hence these tapeworm larvae are of some economic importance.

ORDER LECANICEPHALIDEA

The Lecanicephalidea includes, among others, *Lecanicephalum* in elasmobranchs, *Tetragonocephalum* in elasmobranchs, and *Polypocephalus* also in elasmobranchs (Fig. 12.49). All known species are intestinal parasites in their elasmobranch hosts, and hence, like the previous two orders discussed, are exclusively marine. The accompanying illustrations give some idea of the variations in the scolical morphology of the various genera (Fig. 12.49).

Life Cycle Data

Complete life cycle data concerning members of this order are also lacking. However, Rifkin and Cheng (1968) and earlier investigators have reported globular larvae, believed to be those of *Tetragonocephalum* (= *Tylocephalum*), occurring in the digestive gland and the zone surrounding the digestive tract of various marine bivalves (Fig. 12.50).

The interaction of *Tetragonocephalum* metacestodes with their molluscan intermediate hosts is an interesting example of molluscan immunology. It is known that the larva of *Tetragonocephalum* is not very host specific and will enter and develop into a metacestode in several species of marine bivalve molluscs; however, the fate of the parasite depends on which species of mollusc it enters. For example, if the coracidium[*] enters the black-lipped pearl oyster, *Margaritifera vulgaris*, there is essentially no cellular reaction on the part of the host and it develops into a metacestode. If it enters the American oyster, *Crassostrea virginica*, the host responds and develops a capsule around the metacestode comprised of fibroblasts, leucocytes, and connective tissue fibers, and the parasite is gradually killed and resorbed (Rifkin and Cheng, 1968). On the

---

[*] Cheng (1966) described what was thought to be the coracidium of *Tylocephalum* (= *Tetragonocephalum*) which does not bear the six hooks characteristic of the first larval stage of cestodes; however, Freeman (1982) has pointed out in a review that this is probably an error.

other hand, if the parasite enters the Japanese littleneck clam, *Tapes semidecussata*, the host reaction is severe and rapid, leading to the almost immediate death and resorption of the metacestode. Host cells responding to the parasitic invasion consist of leucocytes and the so-called brown cells. A comparison of the severity of reactions in various marine bivalves readily demonstrates the principle that immunologic response, especially cellular immunity, is much more severe in less compatible hosts than in compatible hosts of long standing, especially in invertebrate hosts.

OTHER MINOR ORDERS OF TAPEWORMS

As listed in the classification section at the end of the chapter, there are six additional orders of the Eucestoda. These are minor orders which are only being briefly mentioned.

Caryophyllidea

Adults of the order Caryophyllidea are all parasitic in freshwater fish, although several species of *Archigetes* have been reported to mature in oligochaete annelids. These are unsegmented tapeworms (Figs. 12.51, 12.52) that produce operculated, nonembryonated eggs. The scolex is never armed and is usually quite simple. It usually bears shallow loculi, is frilled or smooth (Figs. 12.51, 12.52). Some species lack a scolex and some cause a pocket to form in the wall of the host's intestine in which one or more worms occur.

As shown by Calentine (1964), the life cycle of *Archigetes iowensis* involves catfish, suckers, and min-

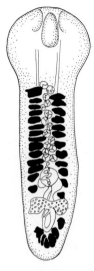

Fig. 12.52. A caryophyllidean cestode. *Penarchigetes oklensis,* an unsegmented monozoic tapeworm. (Redrawn after Schmidt, 1970.)

nows as the most common definitive hosts. Eggs passed out in the feces of the fish host are ingested by oligochaetes. The escaping oncospheres penetrate into the coelom. Each oncosphere develops into a procercoid at this site. Each procercoid includes well-

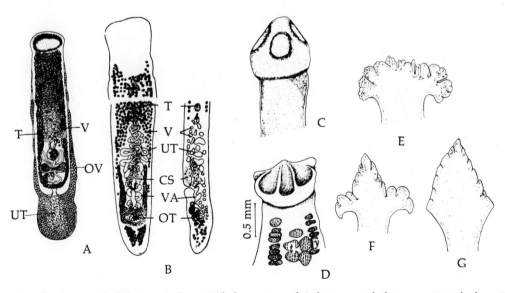

Fig. 12.51. Some caryophyllidean cestodes. A. Whole specimen of *Archigetes cryptobothrius,* parasitic in body cavity of turbificid annelid. (After Wisniewski, 1930.) **B.** Whole specimen and sagittal section of *Caryophyllaeides fennica* from rutilid fish. (After Nybelin, 1922.) **C.** Scolex of *Biacetabulum meridianum* from catostomid fish. (After Hunter, 1927.) **D.** Scolex of *Glaridacris* sp. from catostomid fish. (After Hunter, 1927.) **E–G.** Scolices of *Caryophyllaeus laticeps* from cyprinid and catostomid fish. (After Benham, 1901.) CS, cirrus sac; OT, ootype; OV, ovary; T, testes; UT, uterus; V, vitelline glands; VA, vagina.

Eucestoda: The True Tapeworms

developed reproductive organs. When such procercoids are eaten by a fish, the larva loses its cercomer and becomes an adult.

Wisniewski (1930) has found that some progenetic species of *Archigetes* mature sexually in annelids. She has hypothesized that during the evolutionary history of these species, segmented adults occurred in aquatic reptiles; however, these became extinct with their hosts, but not before a now extinct second larval stage, the plerocercoid, developed the ability to undergo premature sexual maturity within a second intermediate host, i.e., neoteny. If this is true, then the extant species of caryophyllideans may be thought of as neotenic pleocercoids.

For a detailed review of the Caryophyllidea, see Mackiewicz (1972).

Aporidea

This small order of the Eucestoda includes only two known genera: *Gastrotaenia* and *Nematoparataenia*. These are small worms that seldom measure over 3.5 mm in length. All of the species are intestinal parasites of anseriform birds. Like the caryophyllideans, the aporideans do not portray external segmentation. Also, the scolex of *Nematoparataenia* is quite elaborate (Fig. 12.53). Wolffhügel (1938) has suggested that the large cushionlike rostellum is nutrition acquiring in function and that through pulsating movements it attracts a stream of the host's blood and mucus, which are absorbed by glandular cells located within the rostellum. The only two known species occur in the small intestine of swans in Australia, New Zealand, and Sweden.

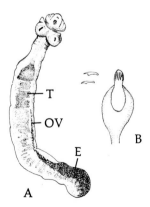

Fig. 12.53. Some aporidean cestodes. A. Entire specimen of *Nematoparataenia southwelli* from swans. (Redrawn after Fuhrmann, 1933.) **B.** Scolex of *Gastrotaenia cygni* from birds. (Redrawn after Wolffhügel, 1938.) E, eggs; OV, ovary; T, testis.

Spathebothriidea

The spathebothriid tapeworms, like members of Aporidea, do not show external segmentation. However, internally there is a linear series of reproductive organs, indicating the presence of incomplete proglottidization. Longitudinal parenchymal muscles are lacking. The adult worms have all been reported from the guts of various fish.

The order includes the genera *Didymobothrium* in marine teleost fish, *Bothrimonus* (Fig. 12.54) in sturgeons, and *Diplocotyle* (Fig. 12.54) in marine teleost fish.

Life Cycle Pattern. In general, eggs passed out of the fish host are ingested by amphipod crustaceans in which the eggs develop into procercoids. When infected amphipods are ingested by a fish, the procercoids develop into sexually functional procercoids (progenetic larvae). When the infected fish host is eaten by the definitive fish host, these procercoids develop into adults.

Diphyllidea

Only two genera of diphyllidean cestodes are known: *Echinobothrium* and *Ditrachybothridium*. The scolex of *Echinobothrium* has a powerful rostellum armed with several rows of large hooks (Fig. 12.55). Two simple bothridia are also present. Each is covered with tiny hairlike spines. The scolex of *Ditrachybothrium* lacks armature.

All of the diphyllideans are parasites of elasmobranchs.

Nippotaeniidea

Hine (1977) has provided a review of this small order of tapeworms. All of the species occur in freshwater gobiid fish in the Soviet Union, Japan, and New Zealand. The scolex bears a simple, apical sucker (Fig. 12.56), and the strobila consists of only a few proglottids.

Litobothridea

This small order includes three species found in thresher sharks in the Pacific. The scolex bears a well-developed apical sucker. Also, the anterior proglottids bear dorsal and ventral spurs which serve to attach the parasite to the host.

PHYSIOLOGY AND BIOCHEMISTRY OF CESTODES

OXYGEN REQUIREMENTS

Although the intestine, the natural habitat of most adult tapeworms, contains a minimum of oxygen, many investigators have found that various tapeworms absorb oxygen in culture media provided that the *in vitro* environment is maintained below 38°C. Furthermore, there is an increased consumption of oxygen following anaerobiosis. This suggests that cestodes are capable of building up an oxygen debt (p. 39).

Investigations of the metabolism of cestodes have revealed that the metabolic by-products of certain species maintained *in vitro* are carbon dioxide, lactic acid, lighter fatty acids, and succinic acid. These metabolites are associated with glycolysis, an anaerobic process, as well as with the citric acid cycle, an aerobic process (Fig. 55). As a result, von Brand (1973) has stated that one cannot generalize about the anaerobic or aerobic needs of tapeworms, and it appears pointless to categorize them one way or the other. Instead, each species must be considered separately relative to its oxygen requirements. Nevertheless, from the information available, cestodes appear to have a predominantly anaerobic metabolism, although aerobic metabolism does occur if oxygen is available.

Aerobic metabolism in cestodes appears to be influenced by the amount of glucose in the medium. In media lacking glucose, the respiratory quotient (RQ) is between 0.5 and 0.9; if glucose is added, the RQ generally is markedly increased. Furthermore, the metabolic rate appears to be influenced by oxygen tension. This is especially true when the tension is at a level of 2–5 vol.%.

Read (1956) has suggested that although the body surface of cestodes is exposed to oxygen from the host's mucosa, the oxygen tension in the medullary

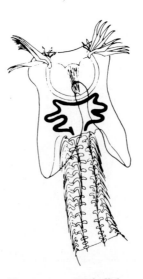

Fig. 12.55.   Diphyllidean scolex. Drawing of scolex showing powerful armature. (After Schmidt, 1970; with permission of William C. Brown Publ.)

parenchyma of a large part of the strobila must be essentially zero. Thus, anaerobic fermentative metabolism most probably occurs here. The end products of such fermentative metabolism in the central regions of each proglottid would diffuse toward the body

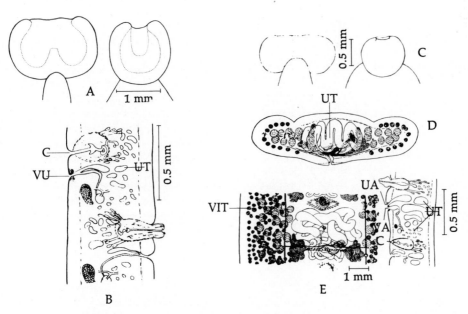

Fig. 12.54.   Some spathebothridean cestodes. A. Scolex of *Bothrimonus fallax* from marine fish. **B.** Sagittal view of two consecutive proglottids. (**A** and **B**, after Nybelin, 1922.) **C.** Scolex of *Diplocotyle olrikii* from fish. **D.** Cross section of mature proglottid. **E.** Ventral view of mature proglottid. (**C–E**, after Nybelin, 1922.) C, cirrus aperture; UA, uterine aperture; UT, uterus; VA, vaginal aperture; VU, vagino-uterine aperture; VIT, vitellaria.

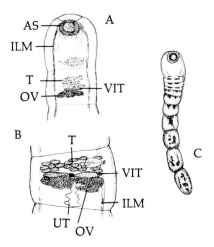

Fig. 12.56. **Some nippotaeniidean cestodes. A.** Anterior end of *Nippotaenia chaenogobii* from the freshwater fish *Chaenogobius*. **B.** Mature proglottid of *N. chaenogobii*. (**A** and **B**, redrawn after Yamaguti, 1939.) **C.** Entire specimen of *Amurotaenia percotti* from the freshwater fish *Percottus*. (Redrawn after Achmerov, 1941.) AS, anterior sucker; ILM, internal longitudinal muscles; OV, ovary; T, testes; UT, uterus; VIT, vitellaria.

surface and become available for further oxidation in the peripheral tissues. Such a mechanism might be considered advantageous in terms of useful energy because high energy-requiring systems, such as the muscular and nervous systems, are for the most part situated in the peripheral regions of the body. This interpretation coincides with the earlier observations

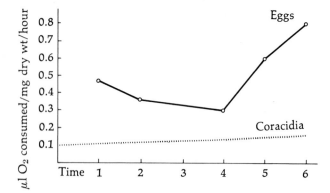

Fig. 12.57. **Oxygen consumption by *Diphyllobothrium latum* eggs and coracidia.** Oxygen consumption at 25°C and 760 mm Hg pressure during embryonation. (Redrawn after Grabiec *et al.*, 1964.)——, Time in days; . . . , time in hours.

by Harnisch who, in a series of papers published between 1932 and 1937, suggested that the apparent absorption of oxygen by tapeworms need not be associated with their respiratory activities *per se*, but is associated with the oxidation of certain products that are produced during anaerobic metabolism and that are not removed from the body.

Cestode eggs also utilize oxygen. For example, when eggs of *Diphyllobothrium latum* are undergoing the first four days of embryonation (at 25°C), oxygen consumption is fairly uniform; however, there is a sharp rise during the subsequent two days (Fig. 12.57). This pattern parallels that found in other invertebrates and is believed to reflect the higher metabolic demands associated with the final stages of morphogenesis of the coracidium.

The respiration of pseudophyllidean eggs is unaffected by the presence or absence of light, but it is completely inhibited by 0.001 M of potassium cyanide, suggesting the presence of a cytochrome system (p. 57).

Some information is available on the respiration of coracidia. Grabiec *et al.* (1964) have reported that the coracidia of *Diphyllobothrium latum* and *Triaenophorus lucii*, the latter a pseudophyllidean parasite of teleost fish, are virtually anaerobic, with only a slight consumption of oxygen. Since polysaccharides and phospholipids occur in the embryophores of these coracidia, and these molecules are quantitatively reduced during the life of the coracidia, it is believed that these stored nutrients are utilized as energy sources via anaerobic metabolism, especially the polysaccharides.

Oxygen utilization by larval and adult cestodes is affected by a number of parameters including the oxygen tension in the immediate environment, the age and nature of the parasite tissue studied, and sometimes the host from which the parasites were taken. For example, it is known that in the case of *Hymenolepis diminuta*, the QO_2 varies with the amount of oxygen available (Read, 1956) (Fig. 12.58). This phenomenon is of considerable importance when one wants to ascertain the true rate of oxygen consumption of cestodes *in vivo*. When in the host's gut, cestodes are exposed to low O_2 tensions in the lumen, higher tensions in the liquid abutting the mucosa, and still higher tensions within the crypts of Lieberkühn.

A gradient exists along the strobila as far as oxygen consumption is concerned. For example, in the case of *Diphyllobothrium* maintained at a tension of 5% O_2, the anterior proglottids take up 91% while the posterior ones take up 66% of what these proglottids utilize when maintained at a tension of 21% O_2. The higher QO_2 of the anterior proglottids reflects the greater metabolic rate associated with the differentiation of the internal organs. Knowing this, it is important to use comparable proglottids maintained at identical oxygen tensions if meaningful comparisons are to be made.

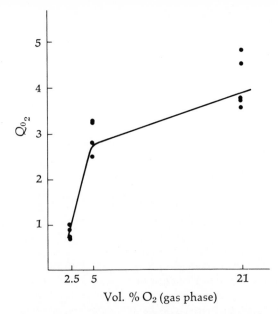

Fig. 12.58. Oxygen consumption by *Hymenolepis diminuta*. Effect of varying oxygen tension on oxygen consumption. (Redrawn after Read, 1956.)

Table 12.9. Utilization of Some Carbohydrates by Cestodes[a]

Species	Glucose	Fructose	Mannose	Galactose	Maltose	Saccharose	Lactose	Trehalose
Calliobothrium verticillatum	×[b]	(×)[b]	(×)	×	—[b]	—	—	—
Cittotaenia sp.	×	—	—	×	×	×	—	—
Hymenolepis citelli	×	—	—	×	—	—	—	—
Hymenolepis diminuta	×	—	(×)	×	—	—	—	—
Vampirolepis nana	×	—	—	×	—	—	—	—
Lacistorhynchus tenuis	×	—	—	×	—	—	—	—
Multiceps serialis, adult	×	×		—		—		—
Multiceps serialis, larva	×	—		—		—		
Taenia taeniaeformis, adult	×	—	(×)	×	—			
Taenia taeniaeformis, larva	×	(×)	(×)	×				

[a] Data compiled by von Brand, 1972.

[b] ×, Compound is utilized; (×), compound is utilized in trace amounts or questionable utilization; —, compound is not utilized.

An example of how the host from which the specimens studied can make a difference has been contributed by von Brand and Bowman (1961). These investigators studied oxygen uptake by the larvae of *Taenia taeniaeformis* from rats and mice and found that the QO_2 of those from rats is higher. They have also shown that larvae from rats respire at a greater rate when incubated in serum than in Tyrode's solution. The oxygen consumption of various forms of several species of cestodes in various media and with or without glucose is given in Table 12.8.

Carbohydrate Metabolism

Carbohydrates are by far the most commonly employed energy source in cestodes. Of the carbohydrates, however, most species can utilize only certain monosaccharides (Read, 1959) (Table 12.9). In *Hymenolepis diminuta*, for example, only glucose and galactose are actively absorbed, whereas fructose, lactose, maltose, and trehalose are not. Upon absorption, the monosaccharides in excess of what is immediately required are converted to and stored as glycogen.

Although the lack of protein in the host's diet has little effect on cestodes, the lack of adequate carbohydrates results in stunting of growth, retardation of egg production, and the development of morphologically abnormal eggs. Interestingly, the growth of worms in hosts fed starch is better than that of worms in hosts fed disaccharides and monosaccharides. This is most probably because the simpler sugars are more

Table 12.8. Rates of Oxygen Consumption of Some Cestodes in the Presence and Absence of Glucose at an Oxygen Tension of Approximately 160 mm Hg[a]

Species	Stage	Temp. (°C)	Oxygen Consumption		RQ	
			Sugar Absent (mm³/mg dry weight/hour)	Sugar Present (mm³/mg dry weight/hour)	Sugar Absent	Sugar Present
Diphyllobothrium latum	Proglottids	37	2.7	15.0		
Diphyllobothrium latum	Plerocercoid	22		0.67		
Echinococcus granulosus	Scolices	37	2.0		0.88	
Hymenolepis diminuta	Adult	37	1.2	3.0	0.51	1.02

[a] Data compiled by von Brand, 1972.

Table 12.10. Glycogen Content in Linear Quarters of the Strobila of *Hymenolepis diminuta*[a]

Worm Number	Anterior Quarter	Second Quarter	Third Quarter	Posterior Quarter
1	21.0	44.2	42.9	40.2
2	23.6	42.7	43.0	26.4
3	22.5	44.8	40.5	27.0

[a] Values are expressed as percentage of the dry weight (Data from Read, 1956).

readily absorbed by the host and hence less sugar is available to the parasites.

Most of the stored nutrients in tapeworms is in the form of glycogen. As is the case with oxygen utilization, a gradient occurs along the strobila relative to glycogen deposition. For example, in *Hymenolepis diminuta* the glycogen content is highest in the region near the scolex, drops in the next 10 cm, rises in the succeeding 10–20 cm, and finally declines in the posteriormost gravid proglottids. If the rat hosts are fasted for 24 hours, there is a marked drop in stored glycogen in the first 20–30 cm sections, but little effect is noticed in the more posterior segments.

If the strobila of *H. diminuta* is divided into four equal sections, differences in the amount of stored glycogen among the quarters are noticeable (Table 12.10).

The nature of the glycogen stored in cestodes has not been studied extensively. In vertebrate muscle, glycogen exists in two forms—that which is soluble and that which is insoluble in trichloroacetic acid. The soluble type is more readily available to meet the metabolic demands of the body and is the first to disappear during exercise. These two types of glycogen exist in *Hymenolepis diminuta*, the soluble type forming approximately 80% of the total body glycogen (Daugherty and Taylor, 1956).

As a rule, larval cestodes include a greater and more consistent amount of glycogen than do adults of the same species. This reflects the more stable environment present in the intermediate host, which is usually the coelomic cavity or tissues.

The catabolism of glycogen in cestodes, as in other helminths, is mediated by phosphorylase to give glucose-1-phosphate and limit dextrins. Complete hydrolysis of the dextrins requires another enzyme, amylo-1,6-glucosidase. The presence of the latter enzyme has been demonstrated in *Hymenolepis diminuta*.

Carbon dioxide fixation in parasites has been discussed earlier (p. 40). It is noted that in such CO_2-fixing cestodes as *Hymenolepis diminuta*, *Ligula intestinalis*, and others, there is consistently considerably more NADP-linked than NAD-linked isocitrate dehydrogenase. This fact is mentioned because, unlike in mammals, the level of isocitrate dehydrogenase (and also that of aconitase) in helminths, including certain cestodes, is frequently low. Sometimes NAD-linked isocitrate dehydrogenase cannot be detected. In mammals, the Krebs cycle flux is thought to proceed entirely via the NAD-linked isocitrate dehydrogenase. Judging from the extremely low levels of this enzyme in CO_2-fixing helminths, this does not appear to be the case. Rather, NADP-linked isocitrate dehydrogenase is apparently the important one relative to the operation of the Krebs cycle.

In addition to glycogen, other types of polysaccharides have been reported in the tissues and fluids of certain cestodes. Kilejian *et al.* (1962), for example, found a polysaccharide with a base of glucosamine, galactose, and/or glucose in the wall of the hydatid cyst, protoscolices, and fluid of *Echinococcus granulosus*. Korc *et al.* (1967) have reported the occurrence of a mucopolysaccharide in the scolices of the same worm. The metabolic function(s) of these compounds remains uncertain, although they could be utilized as energy sources.

In vitro experiments to determine the types of simple sugars taken up by tapeworms have revealed some species-specific differences (see Table 12.9), although almost all species utilize glucose and most also can utilize galactose. The mechanism of uptake has been studied extensively, especially by Read and his associates (see Read, 1961, 1968). It is now apparent that sugars do not enter through the cestode tegument by simple diffusion but by active transport. If simple diffusion occurs, one would expect the rate of entry to be directly proportional to the concentration of the sugar in the incubation medium. This, however, is not the case. In *Hymenolepis diminuta*, for example, Phifer (1960) and Read (1967) have demonstrated that the absorption of glucose is independent of the concentration once beyond the limiting concentration. Specifically, beyond about 1.1×10^{-2} M, increasing the glucose concentration does not increase the absorption rate (Fig. 12.59).

In addition, the uptake of glucose is influenced by the pH. For example, Phifer (1960) has shown that if *H. diminuta* is placed in a glucose concentration of 8.3×10^{-3} M for 30 minutes, there is a major uptake peak at pH 7.5 and a minor one at pH 8.5. These data provide further evidence that the uptake of glucose does not occur by simple diffusion since, if this was the case, pH would not influence the rate of uptake.

It is still uncertain as to what is the nature of the active transport mechanism. It has been suggested by numerous investigators that the enzyme alkaline phosphatase may serve in the phosphorylative uptake

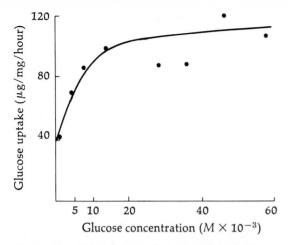

Fig. 12.59.   Glucose uptake by *Hymenolepis diminuta*.
Effect of glucose concentration on its uptake at pH 7.5 in 30 minutes. Optimum absorption occurs at 11×10^{-3} M. (Redrawn after Phifer, 1960.)

of glucose and perhaps other sugars. For a review of membrane transport in helminths, especially cestodes, see Pappas and Read (1975).

As stated earlier, simple sugars taken into the bodies of cestodes are used for energy production and synthesis of other molecules, including glycogen, which is the principal stored nutrient. The fate of carbohydrates relative to metabolism in tapeworms has not been investigated thoroughly except for a few species. Agosin (1957) has shown that in the presence of oxygen, pyruvate is produced by *Echinococcus granulosus*, and there are sufficient data to suggest that the citric acid cycle or some modification of it occurs (Fig. 55). On the other hand, if the worm is exposed to anaerobiosis, there is a high rate of fermentative metabolism, and lactic and succinic acids are usually excreted. In the case of *Taenia taeniaeformis*, von Brand and Bowman (1961) have found that lactic, pyruvic, acetic, and succinic acids are excreted in addition to ethanol. These products suggest the occurrence of both anaerobic and aerobic metabolism.

The reader is referred to the reviews by Read (1968), von Brand (1973), Fioravanti and Saz (1980), Barrett (1981), and Roberts (1983) for greater details about carbohydrate metabolism in cestodes. In brief, it appears that both aerobic and anaerobic metabolic pathways occur and there is considerable difference among species and sometimes even among stages of the same species. Glycolysis appears to be a key pathway for energy metabolism since these parasites usually inhabit regions with reduced oxygen tension. Also, although a complete Krebs cycle is functional in some species, alternate aerobic pathways may also occur. For example, the pentose-phosphate pathway (p. 59) occurs in *Echinococcus granulosus*.

Protein Metabolism

Very little information is available on the protein metabolism of cestodes. Nevertheless, it is known that unlike the protein composition of other animals, the amount of protein comprising the body of tapeworms is less than the sum of the stored glycogen and fats. Studies on the amino acid composition of cestodes have revealed nothing exceptional except that there is a close parallel between the types of amino acids and their concentrational distributions in the worms and in their host's intestinal tissue. Although this suggests that the parasite derives its required amino acids from the host tissues, with which it is in close contact, studies by Chandler and associates (see the review by Read and Simmons, 1963) indicate that at least in the case of *Hymenolepis diminuta* the parasite does not derive its nutrition from direct contact with the host mucosa, but directly from materials passing into the host's intestine. It is emphasized, however, that not all species habitually engage in this type of activity. Such species as *Echinococcus granulosus*, which are in extremely intimate contact with their host's mucosa, may secrete proteolytic enzymes through their teguments, bringing about the hydrolysis of the proteins with which they are in contact, and the resulting amino acids are then absorbed by the tegument (Smyth, 1969). This type of digestive activity has been designated as membrane digestion since the digestive enzymes are situated on or in the membrane where the actual hydrolysis occurs.

Although there is little doubt that amino acids form the bulk of the nitrogen compounds absorbed by cestodes, recent evidences indicate that larger molecules such as dipeptides, polypeptides, or even entire proteins can be taken up by pinocytosis through the tegument (p. 391).

The uptake of amino acids by adult cestodes has been a topic given considerable attention. As is the case with sugars, active transport rather than simple diffusion is operative. In *Calliobothrium verticillatum*, a parasite in the spiral valves of elasmobranchs, for example, Read *et al.* (1960a,b) have shown that there is an accumulation of L-valine against a concentration gradient. The process is very rapid, with the concentration difference of free L-valine inside and outside reaching ratios from 2.6 : 1 to 4.3 : 1 within 40 minutes. Furthermore, the uptake of radioactively labeled L-valine or L-leucine is inhibited by several other amino acids. This is explained by the fact that certain amino acids compete for the same "carrier" or "transport" molecules in the active transport system.

Read and his associates have also demonstrated that in *Hymenolepis diminuta* there are specific loci at

Table 12.11. Amino Acid Synthesis in *Hymenolepis diminuta* from Ammonia[a]

Incubation Media	Amino Nitrogen
Krebs Ringer phosphate	0.13
$+ (NH_4)_2CO_3$	1.4
$+ \alpha$-ketoglutarate	4.8
$+$ pyruvate	6.3
$+$ oxaloacetate	7.4
$+$ glucose	2.8
$+$ succinate	2.4
$+$ glucose $+$ malonate	2.6

[a] Measured as amino nitrogen in milligrams produced by 1 g of worm tissue (dry weight) per hour. Data after Daughterty, 1954.

which amino acids are taken up by active transport. Evidence for this is that some amino acids inhibit the uptake of others, whereas some have no effect. The reason is that in the first instance the amino acids compete for loci and "carriers." The degree of competition may depend on the absolute concentration of amino acids as well as their molar ratios. Further evidence for the occurrence of loci is provided by the fact that certain non-amino acid molecules that have molecular configurations approximating those of amino acids can also compete and, therefore, inhibit the uptake of certain amino acids. Podesta (1980) has contributed a critical review of the membrane biology of *H. diminuta*, including uptake mechanisms.

There is yet no information pertaining to what are the essential amino acids of cestodes, although transamination, i.e., the conversion of one amino acid from another, does occur. In fact, although certain species, such as *Hymenolepis diminuta*, are capable of utilizing the nitrogen of NH_4^+ in the synthesis of amino acids (Table 12.11), transamination appears to be the main mechanism for amino acid synthesis. It is known, for example, that aspartic acid can be converted to glutamic acid, and glutamic acid to alanine by transaminase activity (see Lehninger, 1975, for an explanation of the reactions involved).

There is sufficient evidence to suggest that the usual pathways for the utilization of nitrogen compounds as substrates for energy production occur in cestodes (p. 55), although some of the details of these pathways remain unexplored. The major end products of nitrogen metabolism are urea, uric acid, and ammonia, although a number of other nitrogen-containing compounds have been detected (Table 12.12).

Remember that nitrogen compounds are employed not only in energy production but also in the

Table 12.12. Nitrogen-Containing Excretory Products of Several Species of Tapeworms[a]

Nitrogen-containing Excretory Product	*Hymenolepis diminuta*	*Echinococcus granulosus*[b]	*Taenia tenuicollis*[b] (*Cysticercus tenuicollis*)	*Lachistorhynchus tenuis*	*Taenia taeniaeformis*[b]
Ammonia	+	−	−	+	+
Urea	+	+	+	−	+
Uric acid	−	+	+	−	−
Methylamine	−	−	−	−	+
Ethylamine	−	−	−	−	+
Propylamine	−	−	−	−	−
Butylamine	−	−	−	−	+
Amylamine	−	−	−	−	−
Heptylamine	−	−	−	−	−
Ethylenediamine	−	−	−	−	+
Cadaverine	−	−	−	−	+
Ethanolamine	−	−	−	−	−
1-Amino-2-propanol	−	−	−	−	+
Amino acids	?	+	−	−	+
Creatinine	−	+	+	−	−
Betaine	−	+	−	−	−

[a] Data compiled by Smyth, 1969.
[b] Larvae.

synthesis of soma and reproductive cells, enzymes, and a variety of other types of molecules necessary for life.

Finally, evidence indicates that protein metabolism in larval cestodes is in some way correlated with their infectivity. For example, Hopkins and Hutchison (1958) have demonstrated that the nitrogen composition in the *Taenia taeniaeformis* strobilocercus larvae, all of the same age, is of the same percentage except when heavy infections of over 100 worms exist. During the subsequent growth of these larvae, the nitrogen content drops to the constant level of 4.25 ± 0.25% of the dry weight 67 days after the initial infection of the mouse host. This period corresponds almost exactly with the earliest time (63 days) at which the strobilocercus larvae become infective to the definitive host, the cat. The reason for the correlation between the drop in nitrogen content and infectivity has not been determined. After *T. taeniaeformis* becomes established in the cat, there is a sharp rise in the nitrogen level until it reaches 6.4 ± 0.7% of the dry weight. For a recent review of protein metabolism in cestodes, see Harris (1983).

Lipid Metabolism

The lipid content of cestodes varies greatly. In *Taenia hydatigena*, a parasite of carnivores, lipids form 4.9%

of the dry weight, whereas in *Moniezia expansa*, lipids form 30% of the dry weight. Histochemical examinations have revealed that most of the body lipids are stored in the parenchyma, although almost all organs include some lipids.

In addition to simple lipids, i.e., triglycerol esters of fatty acids (or fats) and esters of fatty acids with complex monohydric alcohols (or waxes), conjugated lipids are also known to occur in certain cestodes such as *Diphyllobothrium latum*, *Hymenolepis diminuta* (Fairbairn *et al.*, 1961), and *H. citelli* (Harrington, 1965). In these, cerebrosides and gangliosides, generally known as glycolipids, have been found not only in the nervous system but in other tissues as well.

Very little is known about lipid metabolism in these animals, and there is no evidence at this time that suggests cestodes utilize fats as stored nutrients. The fatty acids present, especially those found in the excretory system, undoubtedly represent by-products of carbohydrate metabolism. Lipases with low activity are present in *Taenia taeniaeformis* and *T. pisiformis*.

Parasites in general appear to possess only a limited ability to synthesize long chain fatty acids *de novo*. They rely to a great extent on the host to provide their essential requirements. Consequently, it is not surprising that the fatty acid content of cestodes, for example, *Hymenolepis diminuta*, closely resembles that of their hosts. This is verified in the case of *H. diminuta* by altering the host's diet, which, in turn alters the fatty acid and glyceride composition of the tapeworm.

Relative to the uptake and incorporation of fatty acids, studies with labeled compounds have revealed that the surface of *Hymenolepis diminuta* is permeable to acetic, stearic, palmitic, oleic, and linoleic acids (Jacobsen and Fairbairn, 1967; Lumsden and Harrington, 1966).

Mammals can synthesize long chain fatty acids by a reversal of the β-oxidation process, involving the use of acetyl-CoA as the basic two-carbon foundation. The majority of parasites that have been studied appear to be incapable of accomplishing this. The cestodes, like many other parasites, are restricted to conservative chain lengthening of absorbed fatty acids by the addition of acetate units.

There is little information on the mechanisms present in parasites for the introduction of double bonds (desaturation) into long chain saturated fatty acids, nor is much known about the processes involved in triglyceride synthesis. Nevertheless, it is known that *Hymenolepis* will rapidly incorporate glycerol, palmitic acid, and oleic acid into triglycerides and phospholipids. It has been suggested (Chappell, 1980) that this synthesis occurs via steps involving α-glycerophosphate, phosphatidic acid, and diglyceride.

A few tapeworms, for example *Echinococcus granulosus* and *Taenia hydatigena*, are capable of producing energy via lipid metabolism. This is accomplished by employing aerobic β-oxidations. This, however, is the

exception rather than the rule among parasites. In *Hymenolepis* a full complement of β-oxidation enzymes does not occur.

An interesting study relative to lipids in cestodes is that reported by Buteau *et al.* (1971). These investigators have analyzed the fatty acids and sterols of *Lacistorhynchus tenuis*, *Orygmatobothrium musteli*, and *Calliobothrium verticillatum*, three species of tapeworms from elasmobranchs. *L. tenuis* is a trypanorhynchan, whereas *O. musteli* and *C. verticillatum* are tetraphyllideans. They have found that the typically marine fatty acids eicosapentaenoic acid and decosahexaenoic acid together comprise about half of the total fatty acids in each species. It has been postulated that since cestodes are incapable of a significant amount of synthesis, these fatty acids must have been derived from the host. Furthermore, these marine fatty acids are believed to have originated in organisms comprising the elasmobranch hosts' food webs, including dinoflagellates, cryptomonads, and certain fish. This, again, serves as an example of the direct absorption of fatty acids by cestodes from their hosts.

Cholesterol is the major sterol in *L. tenuis*, *O. musteli*, and *C. verticillatum*, although two additional unidentified sterols are also present. For a review of lipid metabolism in cestodes, see Barrett (1983).

Other Nutritional Requirements

As with practically all helminths, the nutritional requirements of cestodes vary with the life cycle stage. For example, the cysticercoid of *Hymenolepis diminuta*, found in the hemocoel of *Tenebrio molitor*, maintained at 25°C, would require considerably less in the way of nutrition than the egg-producing adult in the intestine of the rodent host at 38°C.

Like most metazoans, tapeworms require certain accessory growth factors. The chemical nature of most of these auxiliary factors is still unknown. For example, certain species, such as *H. diminuta*, are dependent on some factor or factors present in brewer's yeast for normal establishment and growth.

Certain vitamins are known to serve as accessory growth factors. *Hymenolepis diminuta* is not dependent on vitamin B_1 in the host's diet for normal establishment, but this vitamin does influence growth. This parasite is not dependent on vitamins A, D, and E for growth, but it is dependent on the fat-soluble vitamins for normal establishment. Although similar information for other species is still scanty, it is strongly suspected that the vitamin requirements for various species differ in both quality and quantity.

One method for studying the growth requirement of cestodes is to employ *in vitro* cultivation methods.

This topic has been reviewed by Evans (1980), who has pointed out that since all presently available culture media contain natural products of unknown chemical composition (serum, liver extract, etc.), this approach has not been highly fruitful in ascertaining the growth requirements of tapeworms. Nevertheless, the *in vitro* culture approach has revealed the beneficial effect of hemin to *Hymenolepis microstoma* (Seidel, 1971, 1975; Khan and DeRycke, 1976) and the need for an external source of vitamin B_6 by the same worm (Roberts and Mong, 1973).

HOST-PARASITE RELATIONSHIPS

The effect of cestodes on their hosts has been little studied aside from the clinical aspects of parasite-caused diseases. *Diphyllobothrium latum* infections in humans cause an anemia similar to pernicious anemia in a small percentage of individuals. Von Bonsdorff (1977) has reviewed the anemia caused by *D. latum*. In Finland this anemia is present in 5 to 10 out of every 10,000 individuals infected with this tapeworm. This anemia develops because *D. latum*, unlike other tapeworms, demonstrates a special affinity for vitamin B_{12}, which plays an important role in hemoglobin formation in the host. This species absorbs 10 to 50 times more vitamin B_{12} than any other species. As the result of depriving the host of this vitamin, hemoglobin synthesis is retarded and the anemia sets in.

In addition to causing this anemia, *D. latum* also disrupts the host's metabolism. For example, Becker (1926) and others have reported that there may be an increase or decrease in the potassium and iron levels in infected humans.

In *Echinococcus granulosus*, Lemaire and Ribère (1935) have reported the presence of proteolytic enzymes in hydatid fluid. The function of these enzymes is uncertain. Apparently, hydatid fluid does not contain toxic substances, and the known cases of deaths resulting from rupture of cysts are due to anaphylactic shock. Only when large quantities of extracted fluid are injected into an experimental animal does death occur; even then, the symptoms are those of shock and not toxicity.

In most cases, adult tapeworms have little visible effect on their hosts except in heavy infections which may result in anemia, loss of weight, and various secondary conditions.

Effect of Host's Diet

The diet and starvation of hosts have considerable effect on their parasites. For example, Reid (1942) has shown that chickens starved for 24 hours expel proglottids of *Raillietina cesticillus*, and when the expelled strobilae are examined, they reveal a marked decrease in stored glycogen. Since glycogen is primarily responsible for muscle energy within the worm, it is suggested that these tapeworms are expelled as the result of their loss of muscular ability to cope with the peristaltic contractions of the host's intestines.

Since tapeworms thrive primarily on carbohydrates in the form of certain simple sugars, hosts maintained on carbohydrate-rich diets consisting of starch enhance the growth of their intestinal helminths. Such is the case in rodents infected with *Hymenolepis diminuta* and in chickens infected with *Raillietina cesticillus*.

Some effects on hosts maintained on vitamin-deficient diets have been recorded relative to the growth of their parasites. In instances where vitamins A, D, and E are withheld, there are no visible effects on the growth of *Hymenolepis diminuta*, but fewer worms become established from the initial infection (Addis and Chandler, 1944, 1946). If vitamin B_1 is withheld, the worms become stunted. Although thiamine is absorbed by *Hymenolepis*, the withholding of this vitamin has no apparent effect on the tapeworm. Apparently it can acquire the minute amount it needs from the traces present in the host. This and similar findings exemplify the principle that, like certain trematodes, some tapeworms are not directly dependent on the vitamins and perhaps other growth factors (e.g., trace elements) found in their host's diets; rather, they acquire what is necessary from host tissues.

In experiments on rats infected with *Hymenolepis* and maintained on various deficient diets Beck (1951) has found that effects on the parasites were particularly noticeable in immature, female and castrated male rats. The same effects were noticeable in normal adult males only if these were maintained on deficient diets over long periods. Beck's results suggest the influence of male sex hormones.

Further evidences of the influence of male hormones are available. For example, the number of *Hymenolepis diminuta* eggs eliminated from the host in 24 hours over a period of 3 months declines in castrated male rats maintained on "complete" diets. The number eliminated is comparable to that eliminated by normal males maintained on "deficient" diets. However, the number of eggs discharged by worms in castrated males maintained on "deficient" diets exhibits a more severe decline, the number being comparable to that eliminated from identically infected normal female rats maintained on "deficient" diets. Administration of testosterone and progesterone (1 mg/day) restores to normal the level of egg output in worms in castrated males. Although progesterone has no effect on worm egg production in female hosts maintained on "deficient" diets, the administration of testosterone, on the other hand, increases the egg count. These results all suggest that the male hormone testosterone does influence the egg-producing ability of *Hymenolepis*.

Effect of Host's Age

Generally, young mice are more susceptible to infections by *Vampirolepis nana*. However, Larsh (1950) has shown that daily feedings of 3.3 mg of thyroid extract to older mice eliminates their resistance to a great degree, increasing their susceptibility to even more than that found in normally raised young hosts.

Host Immunity

The immunology of tapeworm infections has interested parasitologists for some time. This topic has been reviewed by Hopkins (1980) and Rickard (1983). Through the work of H. M. Miller, Jr., it is known that rats can be immunized against new infections of *Taenia taeniaeformis* by the injection of material from adult tapeworms and blood from infected rats. The introduced antigen (in the case of tapeworm extracts) and antibody (in the case of rat blood) did not affect the growth of preexisting worms in the rat's liver. If material from a closely related cestode, *Taenia pisiformis*, is introduced into the host, the host will also develop an immunity to new *T. taeniaeformis* infections, indicating that the antigen-antibody reaction is probably a group reaction directed against related species. Miller has further demonstrated that this induced immunity, which usually lasts for 167 days, can be passed on to offspring, since the young of immunized mothers show some degree of resistance for a few weeks after birth.

In *Vampirolepis nana* infections, the formation of the host's antibody is usually caused by the antigenic substances elaborated by cysticercoids while within the host's villi. Heyneman (1953) has found that immunity developed from an egg infection can affect other worms introduced later in the cysticercoid stage. In this instance, it is not the eggs that elicit antibody formation; rather, it is the parenteral cysticercoids that develop from the eggs that are antigenic.

It is now known that in all cases of invasion by intestinal parasites immunity develops only if the mucosa of the host's intestinal tract is damaged or invaded and antigenic substances are introduced therein. Thus, tapeworms having a parenteral development phase can elicit the synthesis of antibodies. If cysticercoids of *Vampirolepis nana* raised in beetles are fed to rats, no parenteral phase follows, and hence no antibodies are developed. It is not necessary that the parasite include a parenteral phase in its development, because adult trematodes and nematodes that injure their host's enteral mucosa while holding on or feeding, and in so doing inject antigens, can also bring about host antibody synthesis. On the other hand, cestodes and acanthocephalans, both of which lack an alimentary tract, do not cause antibody formation in the host if no parenteral phase exists unless secretory

Table 12.13. Mass Attained by *Hymenolepis diminuta* in Various Hosts[a]

	Host Weight (g)	Worm Volume (ml)
Mouse	27.6	0.28
Hamster	123.0	0.57
Albino rat	377.0	1.11
Hooded rat	452.0	1.33

[a] Data from Read and Voge (1954).

glands are present which introduce antigenic substances, because no antigenic agents are inoculated into the host during their passive feeding process.

Antibodies react against parasites by interfering with their enzymatic activities in connection with nutrition acquisition and consequently with their growth and reproduction, destroy them by combining with their body chemicals, or protect the host by neutralizing toxic products.

Host Species Differences

Not all hosts are equally susceptible to certain cestode infections. For example, not all strains of mice are equally susceptible to *Taenia taeniaeformis*. This natural resistance is apparently genetic in origin, because crossing a resistant male to a less resistant female produces hybrids with less resistance than the father. Natural immunity also may be the reason why some hosts infected by *Diphyllobothrium latum* develop anemia and others do not.

The growth of cestodes also is influenced by differences in host species, as shown by size differences among *Hymenolepis diminuta* adults maintained in four species of mammalian hosts (Table 12.13).

REFERENCES

Addis, C. J., and Chandler, A. C. (1944). Studies on the vitamin requirement of tapeworms. *J. Parasitol.* **30**, 229–236.

Addis, C. J., Jr., and Chandler, A. C. (1946). Further studies on the vitamin requirements of tapeworms. *J. Parasitol.* **32**, 581–584.

Agosin, M. (1957). Studies on the metabolism of *Echinococcus granulosus*. II. Some observations on the carbohydrate metabolism of hydatid cyst scolices. *Exp. Parasitol.* **6**, 586–593.

Arai, H. P. (ed.) (1980). "Biology of the Tapeworm *Hymenolepis diminuta*." Academic Press, New York.

Arme, C., Bridges, J. F., and Hoole, D. (1983). Pathology

of cestode infections in the vertebrate host. *In* "Biology of Eucestoda." Vol. 2. (C. Arme and P. W. Pappas, eds.) pp. 499–538. Academic Press, London.

Artemov, N. M., and Lure, R. N. (1941). On the content of acetylcholine and cholinesterase in the tissues of tapeworms. *Bull. Acad. Sci. URSS. Ser. Biol.* **2**, 278–282. (In Russian.)

Baer, J. G. (1956). The taxonomic position of *Taenia madagascariensis* Davaine, 1870, a tapeworm parasite of man and rodents. *Ann. Trop. Med. Parasitol.* **50**, 152–156.

Barrett, J. (1981). "Biochemistry of Parasitic Helminths." University Park Press, Baltimore, Maryland.

Barrett, J. (1983) Lipid metabolism. *In* "Biology of Eucestoda." Vol. 2. (C. Arme and P. W. Pappas, eds.) pp. 391–419. Academic Press, London.

Baylis, H. A. (1919). A collection of entozoa, chiefly from birds, from the Murman coast. *Ann. Mag. Natur. Hist.* (Ser. 9) **3**, 501–513.

Beck, J. W. (1951). Effect of diet upon singly established *Hymenolepis diminuta. Exp. Parasitol.* **1**, 46–59.

Becker, G. (1926). Der Kaliumgehalt des Serums bei Bothriozephalusträgern. *Acta Soc. Med. Fenn. Duodecin* **7**, 1–8.

Bullock, F. D., and Curtis, M. R. (1920). The experimental production of sarcoma in the liver of rats. *Proc. N.Y. Pathol. Soc.* **30**, 149–175.

Buteau, G. H. Jr., Simmons, J. E., Beach, D. H., Holz, G. G. Jr., and Sherman, I. W. (1971). The lipids of cestodes from Pacific and Atlantic coast triakid sharks. *J. Parasitol.* **57**, 1272–1278.

Cake, E. W. Jr. (1976). A key to larval cestodes of shallow-water, benthic mollusks of the northern Gulf of Mexico. *Proc. Helminth. Soc. Wash.* **43**, 160–171.

Calentine, R. L. (1964). The life cycle of *Archigetes iowensis* (Cestoda: Caryophyllaeidae). *J. Parasitol.* **50**, 454–458.

Chappell, L. H. (1980). "Physiology of Parasites." John Wiley, New York.

Chen, H. T. (1934). Reactions of *Ctenocephaloides felis* to *Dipylidium caninum. Z. Parasitenk.* **6**, 603–637.

Cheng, T. C. (1966). The coracidium of the cestode *Tylocephalum* and the migration and fate of this parasite in the American oyster *Crassostrea virginica. Trans. Am. Microsc. Soc.* **85**, 246–255.

Conn, D. B., and Etges, F. J. (1983). Maternal transmission of asexually proliferating *Mesocestoides corti* (Cestoda) in mice. *J. Parasitol.* **69**, 922–925.

Daugherty, J. W., and Taylor, D. (1956). Regional distribution of glycogen in the rat cestode, *Hymenolepis diminuta. Exp. Parasitol.* **5**, 376–390.

Davey, K. G., and Breckenridge, W. R. (1967). Neurosecretory cells in a cestode, *Hymenolepis diminuta. Science* **158**, 931–932.

Davis, R. E., and Roberts, L. S. (1983a). Platyhelminthes—Eucestoda. *In* "Reproductive Biology of Invertebrates. Vol. I. Oogenesis, oviposition, and oosorption." (K. G. Adiyodi and R. G. Adiyodi, eds.) pp. 109–133. John Wiley, New York.

Davis, R. E., and Roberts, L. S. (1983b). Platyhelminthes—Eucestoda. *In* "Reproductive Biology of Invertebrates.

Vol. II. Spermatogenesis and Sperm Function." (K. G. Adiyodi and R. G. Adiyodi, eds.) pp. 131–149. John Wiley, New York.

Dollfus, R. P. (1942). Etudes critiques sur les Tetrarhynques du Musée de Paris. *Arch. Mus. Nat. Hist. Nat.* (Ser. 6) **19**, 1–466.

Dollfus, R. P. (1946). Notes diverses sur des Tetrarhynques. *Mém. Mus. Nat. Hist. Nat.* N.S. **22**, 179–220.

Evans, W. S. (1980). The cultivation of *Hymenolepis in vitro. In* "Biology of the Tapeworm *Hymenolepis diminuta*." (H. P. Arai, ed.), pp. 425–448. Academic Press, New York.

Fairbairn, D., Wertheim, G., Harpur, R. P., and Schiller, E. L. (1961). Biochemistry of normal and irradiated strains of *Hymenolepis diminuta. Exp. Parasitol.* **11**, 248–263.

Fioravanti, C. F., and Saz, H. J. (1980). Energy metabolism of adult *Hymenolepis diminuta. In* "Biology of the Tapeworm *Hymenolepis diminuta*." (H. P. Arai, ed.), pp. 463–504. Academic Press, New York.

Freeman, R. S. (1973). Ontogeny of cestodes and its bearing on their phylogeny and systematics. *Adv. Parasitol.* **11**, 481–557.

Freeman, R. S. (1982). Do any *Anonchotaenia, Cyathocephalus, Echeneibothrium*, or *Tetragonocephalum* (= *Tylocephalum*) (Eucestoda) have hookless oncospheres or coracidia? *J. Parasitol.* **68**, 737–743.

Freeman, R. S. (1983). Pathology of the invertebrate host metacestode relationship. *In* "Biology of the Eucestoda." Vol. 2. (C. Arme and P. W. Pappas, eds.) pp. 441–497. Academic Press, London.

Gamble, W. G., Segal, M., Schartz, P. M., and Rausch, R. L. (1979). Alveolar hydatid disease in Minnesota. First human case acquired in the contiguous United States. *J. A. M. A.* **241**, 904–907.

Grabiec, S., Guttowa, A., and Michajlow, W. (1964). Investigation on the respiratory metabolism of eggs and coracidia of *Diphyllobothrium latum* (L.) Cestoda. *Bull. Acad. Pol. Sci. Ser. Sci. Biol.* **12**, 29–34.

Graff, D. J., and Read, C. P. (1967). Specific acetylcholinesterase in *Hymenolepis diminuta. J. Parasitol.* **53**, 1030–1031.

Harrington, G. W. (1965). The lipid content of *Hymenolepis diminuta* and *Hymenolepis citelli. Exp. Parasitol.* **17**, 287–295.

Harris, B. G. (1983). Protein metabolism. *In* "Biology of the Eucestoda." Vol. 2. (C. Arme and P. W. Pappas, eds.) pp. 335–341. Academic Press, London.

Hess, E. (1980). Ultrastructural study of the tetrathyridium of *Mesocestoides corti* Hoeppli, 1925: tegument and parenchyma. *Z. Parasitenk.* **49**, 253–261.

Heyneman, D. (1953). Auto-reinfection in white mice resulting from infection by *Hymenolepis nana. J. Parasitol* (Suppl.) **39**, 28.

Hine, P. M. (1977). New species of *Nippotaenia* and *Amurotaenia* (Cestoda: Nippotaeniidae) from New Zealand freshwater fishes. *J. Roy. Soc. N.Z.* **7**, 143–155.

Hopkins, C. A. (1959). Seasonal variations in the incidence and development of the cestode *Proteocephalus filicollis* (Aud. 1810) in *Gasterosteus aculeatus* (L. 1766). *Parasitology* **49**, 529–542.

Hopkins, C. A. (1980). Immunity to *Hymenolepis diminuta. In* "Biology of the Tapeworm *Hymenolepis diminuta*" (H. P. Arai, ed.), pp. 551–614. Academic Press, New York.

Hopkins, C. A., and Hutchison, W. M. (1958). Studies on cestode metabolism. IV. The nitrogen fraction in the large cat tapeworm, *Hydatigera (Taenia) taeniaeformis.*

Exp. Parasitol. **7**, 349–365.

Hutchison, W. M. (1959). Studies on *Hydatigera* (*Taenia*) *taeniaeformis* II. Growth of the adult phase. *Exp. Parasitol.* **8**, 557–567.

Jacobsen, N. S., and Fairbairn, D. (1967). Lipid metabolism in helminth parasites. III. Biosynthesis and interconversion of fatty acids by *Hymenolepis diminuta* (Cestoda). *J. Parasitol.* **53**, 355–361.

Khan, Z. I., and DeRycke, P. H. (1976). Studies on *Hymenolepis microstoma in vitro*. I. Effect of heme compounds on growth and reproduction. *Z. Parasitenk.* **49**, 253–261.

Kilejiian, A., Sauer, K., and Schwabe, C. W. (1962). Host-parasite relationships in echinococcosis. VIII. Infrared spectra and chemical composition of the hydatid cyst. *Exp. Parasitol.* **12**, 377–392.

Korc, I., Hierro, J., Lasalivia, E., Fako, M., and Calcagno, M. (1967). Chemical characterization of the polysaccharide of the hydatid membrane of *Echinococcus granulosus*. *Exp. Parasitol.* **20**, 219–224.

Kumaratilake, L. M., and Thompson, R. C. A. (1982). A review of the taxonomy and speciation of the genus *Echinococcus* Rudolphi 1801. *Zeit. Parasitenk.* **68**, 121–146.

Larsh, J. E., Jr. (1950). The effect of thiouracil and thyroid extract on the natural resistance of mice to *Hymenolepis* infection. *J. Parasitol.* **36**, 473–478.

Lee, D. L., Rothman, A. H., and Senturia, J. B. (1963). Esterases in *Hymenolepis* and in *Hydatigera*. *Exp. Parasitol.* **14**, 285–295.

Lehninger, A. L. (1975). "Biochemistry," 3rd ed. Worth, New York.

Leiby, P. D., Carney, W. P., and Woods, C. E. (1970). Studies on sylvatic echinococcosis. III. Host occurrence and geographic distribution of *Echinococcus multilocularis* in the north central United States. *J. Parasitol.* **56**, 1141–1150.

Lemaire, G., and Ribère, R. (1935). Sur la composition chimique du liquide hydatique. *C.R. Soc. Biol.* **118**, 1578–1579.

Linton, E. (1922). A contribution to the anatomy of *Dinobothrium*, a genus of selachian tapeworms. *Proc. U.S. Nat. Mus.* **60**, 1–16.

Löser, E. (1965a). Der Feinbau des Oogenotop bei Cestoden. *Z. Parasitenk.* **25**, 413–458.

Löser, E. (1965b). Die Eibildung bei Cestoden. *Z. Parasitenk.* **25**, 556–580.

Lumsden, R. D., and Harrington, G. W. (1966). Incorporation of linoleic acid by the cestode *Hymenolepis diminuta* (Rudolphi, 1819). *J. Parasitol.* **52**, 695–700.

Lumsden, R. D., Oaks, J. A., and Mueller, J. F. (1974). Brush border development in the tegument of the tapeworm, *Spirometra mansonoides*. *J. Parasitol.* **60**, 209–226.

Mackiewicz, J. S. (1972). Caryophyllidea (Cestoidea): a review. *Exp. Parasitol.* **31**, 417–512.

Meyer, F., Kimura, S., and Mueller, J. F. (1965). Stimulation of lipogenesis in hamsters by *Spirometra mansonoides*. *J. Parasitol.* **51** (Sect. 2), 57.

Moore, D. V. (1962). A review of human infections with the common dog tapeworm, *Dipylidium caninum*, in the United States. *Southwest. Vet.* **15**, 283–288.

Morseth, D. J. (1967). Observations on the fine structure of the nervous system of *Echinococcus granulosus*. *J. Parasitol.* **53**, 492–500.

Mount, P. M. (1970). Histogenesis of the rostellar hooks of *Taenia crassiceps* (Zeder, 1800) (Cestoda). *J. Parasitol.* **56**, 947–961.

Mueller, J. F. (1963). Parasite-induced weight gain in mice. *Ann. N.Y. Acad. Sci.* **113**, 217–233.

Mueller, J. F. (1965). Further studies on parasitic obesity in mice, deer mice, and hamsters. *J. Parasitol.* **51**, 523–531.

Mueller, J. F. (1966). Host-parasite relationships as illustrated by the cestode *Spirometra mansonoides* In "Host-Parasite Relationships" (J. E. McCauley, ed.), pp. 15–58. Oregon State University Press, Corvallis, Oregon.

Mueller, J. F. (1968). Growth stimulating effect of experimental sparganosis in thyroidectomized and hypophysectomized rats, and comparative activity of different species of *Spirometra*. *J. Parasitol.* **51**, 537–540.

Mueller, J. F., and Reed, P. (1968). Growth stimulation induced by infection with *Spirometra mansonoides* spargana in propylthiouracil-treated rats. *J. Parasitol.* **54**, 51–54.

Nelson, G. S., and Rausch, R. L. (1963). *Echinococcus* infections in man and animals in Kenya. *Ann. Trop. Med. Parasitol.* **57**, 136–149.

Németh, I., Juhász, S., and Baintner, K. (1979). A trypsin and chymotrypsin inhibitor from *Taenia pisiformis*. *Int. J. Parasitol.* **9**, 515–522.

Pappas, P. W., and Read, C P. (1975). Membrane transport in helminth parasites: a review. *Exp. Parasitol.* **37**, 469–530.

Pence, D. B. (1967). The fine structure and histochemistry of infective eggs of *Dipylidium caninum*. *J. Parasitol.* **53**, 1041–1054.

Phifer, K. (1960). Permeation and membrane transport in animal parasites: the absorption of glucose by *Hymenolepis diminuta*. *J. Parasitol.* **46**, 51–62.

Podesta, R. B. (1980). Concepts of membrane biology in *Hymenolepis diminuta*. *In* "Biology of the Tapeworm *Hymenolepis diminuta*" (H. P. Arai, ed.), pp. 505–549. Academic Press, New York.

Rausch, R. L. (1958). *Echinococcus multilocularis* infection. *Proc. Sixth Int. Congr. Trop. Med. Malaria* **2**, 597–610.

Rausch, R. L. (1967). On the ecology and distribution of *Echinococcus* spp. (Cestoda: Taeniidae), and characteristics of their development in the intermediate host. *Ann. Parasitol.* **42**, 19–63.

Read, C. P. (1956). Carbohydrate metabolism in *Hymenolepis diminuta*. *Exp. Parasitol.* **5**, 325–344.

Read, C. P. (1959). The role of carbohydrates in the biology of cestodes. VIII. Some conclusions and hypotheses. *Exp. Parasitol.* **8**, 365–382.

Read, C. P. (1961). The carbohydrate metabolism of worms. *In* "Comparative Physiology of Carbohydrate Metabolism in Heterothermic Animals." University of Washington Press, Seattle, Washington.

Read, C. P. (1967). Carbohydrate metabolism in *Hymenolepis* (Cestoda). *J. Parasitol.* **53**, 1023–1029.

Read, C. P. (1968). Intermediary metabolism of flatworms. *In* "Chemical Zoology" (M. Florkin and B. T. Scheer, eds.), Vol. II, pp. 327–357. Academic Press, New York.

Read, C. P., and Simmons, J. E., Jr. (1963). Biochemistry and physiology of tapeworms. *Physiol. Rev.* **43**, 263–305.

Read, C. P., Simmons, J. E., Jr., Campbell, J. W., and Rothman, A. H. (1960a). Permeation and membrane transport in parasitism: studies on a tapeworm-elasmobranch symbiosis. *Biol. Bull.* **119**, 120–133.

Reid, W. M. (1942). The removal of the fowl tapeworm, *Raillietina cesticillus*, by short periods of starvation. *Poultry Sci.* **21**, 220–229.

Rickard, M. D. (1983). Immunity. *In* "Biology of the Eucestoda." Vol. 2. (C. Arme and P. W. Pappas, eds.) pp. 539–579. Academic Press, London.

Rifkin, E., and Cheng, T. C. (1968). On the formation, structure, and histochemical characterization of the encapsulation cysts in *Crassostrea virginica* parasitized by *Tylocephalum* metacestodes. *J. Invert. Pathol.* **10**, 51–64.

Riser, N. W. (1956). Early larval stages of two cestodes from elasmobranch fishes. *Proc. Helminthol. Soc. Wash.* **23**, 120–124.

Roberts, L. S. (1983). Carbohydrate metabolism. *In* "Biology of the Eucestoda." Vol. 2. (C. Arme and P. W. Pappas, eds.) pp. 343–390. Academic Press, London.

Roberts, L. S., and Mong, F. N. (1973). Developmental physiology of cestodes. XIII. Vitamin B_6 requirement of *Hymenolepis diminuta* during *in vitro* cultivation. *J. Parasitol.* **59**, 101–104.

Rogers, W. A., and Ulmer, M. J. (1962). Effects of continued selfing on *Hymenolepis nana* (Cestoda). *Iowa Acad. Sci.* **69**, 557–571.

Rothman, A. H. (1959). Studies on the excystment of tapeworms. *Exp. Parasitol.* **8**, 336–362.

Ruszkowski, J. S. (1932). Etudes sur le cycle évolutif et sur la structure des cestodes marins. IIIème partie: le cycle évolutif du tetrarhynque *Grillotia erinaceus* (van Beneden, 1858). *Acad. Pol. Sci. Lett., C. R. Mens. Cl. Sc. Math. Natur. Cracovie* **9**, 1–6.

Rybicka, K. (1966). Embryogenesis in cestodes. *Adv. Parasitol.* **4**, 107–186.

Schmidt, G. D. (1970). "How to Know the Tapeworms." Wm. C. Brown, Dubuque, Iowa.

Schmidt, G. D. (1985). "CRC Handbook of Tapeworm Identification." CRC Press, Boca Raton, Florida.

Schwabe, C. W., and Daoud, K. A. (1961). Epidemiology of echinococcosis in the Middle East. I. Human infection in Lebanon, 1949–1959. *Am. J. Trop. Med. Hyg.* **10**, 374–381.

Scott, D. B., Nylen, M. U., von Brand, T., and Pugh, M. H. (1962). The mineralogical composition of the calcareous corpuscles of *Taenia taeniaeformis*. *Exp. Parasitol.* **12**, 445–458.

Seidel, J. S. (1971). Hemin as a requirement in the development *in vitro* of *Hymenolepis microstoma* (Cestoda: Cyclophyllidea). *J. Parasitol.* **57**, 566–570.

Seidel, J. S. (1975). The life cycle *in vitro* of *Hymenolepis microstoma* (Cestoda). *J. Parasitol.* **61**, 677–681.

Skrjabin, K. I. (ed.). (1951). "Essentials of Cestodology," Vols. I and II. Academy of Sciences of the USSR, Moscow. (English translation available from the U.S. Dept. of Commerce, Washington, D. C.).

Smyth, J. D. (1963). Biology of cestode life-cycles. *Commonw. Agr. Bur.* No. 34.

Smyth, J. D. (1967). Studies on tapeworm physiology. XI. *In vitro* cultivation of *Echinococcus granulosus* from the protoscolex to the strobilate stage. *Parasitology* **57**, 111–133.

Smyth, J. D. (1961). "The Physiology of Cestodes." Freeman, San Francisco, California.

Smyth, J. D., Miller, H. J., and Howkins, A. B. (1967). Further analysis of the factors controlling strobilization, differentiation and maturation of *Echinococcus granulosus in vitro*. *Exp. Parasitol.* **21**, 31–41.

Steelman, S. L., Morgan, E. R., Cuccaro, A. J., and Giltzer, M. S. (1970). Growth hormone-like activity in hypophysectomized rats implanted with *Spirometra mansonoides* spargana. *Proc. Soc. Exp. Biol. Med.* **133**, 269–273.

Stunkard, H. W. (1940). Observations on the development of the cestode, *Bertiella studeri*. *Proc. Third Int. Congr. Microbiol. New York*, pp. 461–462.

Stunkard, H. W. (1983). Evolution and systematics. *In* "Biology of the Eucestoda." Vol. 1. (C. Arme and P. W. Pappas, eds.) pp. 1–25. Academic Press, London.

Taylor, E. W., and Thomas, J. H. (1968). Membrane (contact) digestion in the three species of tapeworm *Hymenolepis diminuta*, *Hymenolepis microstoma* and *Moniezia expansa*. *Parasitology* **58**, 535–546.

Ubelaker, J. E. (1983a). The morphology, development. *In* "Biology of the Eucestoda." Vol. 1. (C. Arme and P. W. Pappas, eds.) pp. 139–176. Academic Press, London.

Ubelaker, J. E. (1983b). The morphology, development and evolution of the tapeworm larvae. *In* "Biology of the Eucestoda." Vol. 1. (C. Arme and P. W. Pappas, eds.) pp. 235–296. Academic Press, London.

Ubelaker, J. E., Cooper, N. B., and Allison, V. F. (1970). Possible defensive mechanism of *Hymenolepis diminuta* cysticercoids to hemocytes of the beetle *Tribolium confusum*. *J. Invert. Pathol.* **16**, 310–312.

Voge, M. (1955). North American cestodes of the genus *Mesocestoides*. *Univ. Calif. Publ. Zool.* **59**, 125–156.

Voge, M. (1967). The post-embryonic developmental stages of cestodes. *Adv. Parasitol.* **5**, 247–297.

von Bonsdorff, B. (1977). "Diphyllobothriasis in Man." Academic Press, New York.

von Brand, T. (1973). "Biochemistry of Parasites," 2nd ed. Academic Press, New York.

von Brand, T., and Bowman, I. B. R. (1961). Studies on the aerobic and anaerobic metabolism of larval and adult *Taenia taeniaeformis*. IV. Absorption of glycerol; relations between glycerol absorption and glucose absorption and leakage. *Exp. Parasitol.* **19**, 110–123.

von Brand, T., and Weinback, E. C. (1965). Incorporation of phosphate into the soft tissues and calcareous corpuscles of larval *Taenia taeniaeformis*. *Comp. Biochem. Physiol.* **14**, 11–20.

von Brand, T., Mercado, T. I., Nylen, M. U., and Scott, D. B. (1960). Observations on function, composition and structure of cestode calcareous corpuscles. *Exp. Parasitol.* **9**, 205–214.

von Brand, T., Nylen, M. U., Scott, D. B., and Martin, G. H. (1965a). Observations on calcareous corpuscles in larval *Echinococcus granulosus* of various geographic origins. *Proc. Soc. Exp. Biol. Med.* **120**, 383–385.

von Brand, T., Scott, D. B., Nylen, M. U., and Pugh, M. H. (1965b). Variations in the mineralogical composition of cestode calcareous corpuscles. *Exp. Parasitol.* **16**, 383–391.

Wardle, R. A., and McLeod, J. A. (1952). "The Zoology of

Tapeworms." University of Minnesota Press, Minneapolis, Minnesota.

Wardle, R. A., McLeod, J. A., and Radinovsky, S. (1974). "Advances in the Zoology of Tapeworms, 1950–70." University of Minnesota Press, Minneapolis, Minnesota.

Webster, L. A., and Wilson, R. A. (1970). The chemical composition of protonephridial canal fluid from the cestode *Hymenolepis diminuta. Comp. Biochem. Physiol.* **35**, 201–209.

Wisniewski, L. W. (1930). Das Genus *Archigetes* R. Leuck. Fine studie zur Anatomie, Histogenese, Systematik and Biologie. *Mem. Acad. Si. Lett. Cl. Sci. Math. Nat.*, Ser. B2, 1–160.

Wolffhügel, K. (1938). Nematoparataeniidae. Skolex und Verdaunung. *Z. Infekt. Haustiere* **53**, 9–42.

Yamaguti, S. (1959). "Systema Helminthum. Vol. II: The Cestodes of Vertebrates." Interscience, New York.

Yamashita, J., Ono, Z., Takahashi, H., and Hattori, K. (1955). On the occurrence of *Echinococcus granulosus* (Batsch, 1786) Rudolphi, 1805, in the dog in Rebun Island, and the discussion about the course of infection of the echinococcosis. *Mem. Pac. Agr. Hokkaido Univ.* **2**, 147–150. (In Japanese.)

CLASSIFICATION OF EUCESTODA (THE CESTODES)

Subclass Eucestoda

Polyzoic flatworms (except orders Caryophyllidea and Spathebothriidea); with one or more sets of reproductive organs per proglottid; scolex usually present; shelled embryo with six hooks; parasites of fish, amphibians, reptiles, birds, and mammals.

ORDER CYCLOPHYLLIDEA

Scolex usually with four suckers; rostellum present or absent, armed or not; neck present or absent; strobila usually with distinct segmentation; monoecious (or rarely dioecious); genital pores lateral (ventral in Mesocestoididae); vitelline gland compact, single (double in Mesocestoididae), posterior to ovary (anterior or beneath ovary in Tetrabothriidae); uterine pore absent; parasites of amphibians, reptiles, birds, and mammals. (Genera mentioned in text: Family Hymenolepidae*— *Vampirolepis, Hymenolepis, Diorchis, Fimbriaria, Pseudhymenolepis.* Family Dioecocestidae—*Dioecocestus.* Family Anoplocephalidae—*Moniezia, Oochoristica, Stilesia, Avitellina, Bertiella, Cittotaenia.* Family Taeniidae— *Echinococcus, Hydatigera, Taenia, Taeniarhynchus.* Family Dilepididae—*Dipylidium, Paradilepis, Liga, Dilepis, Ophiovalipora.* Family Mesocestoidae—*Mesocestoides.* Family Davaineidae—*Cotugnia, Davainea, Idiogenes, Raillietina.*)

ORDER PSEUDOPHYLLIDEA

Scolex with two bothria, with or without hooks; neck present or absent; strobila variable; proglottids anapolytic (senile proglottid detached after it has shed enclosed eggs); genital pores lateral, dorsal, or ventral; testes numerous; ovary posterior; vitellaria follicular as in Trypanorhyncha, occasionally in lateral fields but not interrupted by interproglottidal boundaries; uterine pore present, dorsal or ventral; egg usually operculate, containing coracidium; parasites of fish, amphibians, reptiles, birds, and mammals. (Genera mentioned in text: Family Diphyllobothriidae*— *Ligula, Schistocephalus, Polygonoporus, Diphyllobothrium, Spirometra.* Family Haplobothriidae—*Haplobothrium.* Family Bothriocephalidae—*Bothriocephalus.* Family Ptychobothriidae—*Ptychobothrium, Senga, Polyonchobothrium, Clestobothrium.* Family Amphiocotylidae— *Eubothrium.* Family Triaenophoridae—*Triaenophorus.*)

ORDER PROTEOCEPHALATA

Scolex with four suckers, often with prominent apical organ, occasionally with armed rostellum; neck usually present; genital pores lateral; testes numerous; ovary posterior; vitelline glands follicular, usually lateral, either cortical or medullary; uterine pore present or absent; parasites of fish, amphibians, and reptiles. (Genera mentioned in text: Family Proteocephalidae*— *Ophiotaenia, Proteocephalus.*)

ORDER TETRAPHYLLIDEA

Scolex with highly variable bothridia, sometimes also with hooks, spines, or sucker; myzorhynchus present or absent; proglottids commonly hyperapolytic (immature proglottid detached before eggs are formed); hermaphroditic, rarely dioecious; genital pores lateral, rarely posterior; testes numerous; ovary posterior, vitellaria follicular (condensed in *Dioecotaenia* sp.), usually medullary in lateral fields; uterine pore present or not; vagina crosses vas deferens; parasites of elasmobranchs. (Genera mentioned in text: Family Oncobothriidae*—*Calliobothrium.* Family Phyllobothriidae— *Phyllobothrium, Dinobothrium, Orygamtobothrium.*)

ORDER TRYPANORHYNCHA

Scolex elongate, with two or four bothridia and four eversible (rarely atrophied) tentacles armed with hooks; each tentacle invaginates into internal sheath provided with muscular bulb; neck present or absent; strobila apolytic (gravid proglottids disintegrate or become detached), anapolytic, or hyperapolytic; genital pores lateral, rarely ventral; testes numerous; ovary posterior; vitellaria follicular, cortical, and encircling other reproductive organs; uterine pore present or absent; parasites of elasmobranchs. (Genera mentioned in text: Family Lacistorhynchidae*—*Lacistorhynchus, Grillotia.* Family Oncobothriidae—*Acanthobothrium.* Family Hepatoxylidae—*Hepatoxylon.* Family Tentaculariidae— *Nybelinia.*)

ORDER LECANICEPHALIDEA

Scolex divided into anterior and posterior regions by horizontal groove; anterior portion of scolex cushionlike or with unarmed tentacles, capable of being withdrawn into posterior portion, forming a large suckerlike organ; posterior of scolex usually with four suckers; neck present or absent; testes numerous; ovary posterior; vitellaria follicular, lateral or encircling proglottid; uterine pore usually present; para-

---

*For diagnoses of families subordinate to the Eucestoda, see Yamaguti (1959).

sites of elasmobranchs. (Genera mentioned in text: Family Lecanicephalidae*—*Lecanicephalum, Tylocephalum, Polypocephalus.* Family Tetragonocephalidae—*Tetragonocephalum.*)

ORDER CARYOPHYLLIDEA

Scolex unspecialized or with shallow grooves or loculi or shallow bothria. monozoic; genital pores midventral; testes numerous; ovary posterior; vitellaria follicular, scattered or lateral; uterus as a coiled median tube, opening, often together with vagina, near male pore; parasites of teleost fish and aquatic annelids. (Genera mentioned in text: Family Caryophyllaeidae*—*Archigetes, Khawia.*)

ORDER APORIDEA

Scolex with simple suckers or grooves and armed rostellum; proglottids distinguished internally or separate proglottids not evident; genital ducts and pores, cirrus, ootype, and Mehlis' gland absent; hermaphroditic, rarely dioecious; vitelline cells mixed with ovarian cells; parasites of anseriform birds. (Genera mentioned in text: Family Nematoparataeniidae*—*Gastrotaenia, Nematoparataenia.*)

ORDER SPATHEBOTHRIIDEA

Scolex feebly developed, undifferentiated or with funnel-shaped apical organ or one or two hollow, cuplike organs; genital pores and uterine pore ventral or alternating dorsal and ventral; testes in two lateral bands; ovary dendritic; vitellaria follicular, lateral or scattered; uterus coiled; parasites of teleost fish. (Genera mentioned in text: Family Diplocotylidae*—*Didymobothrium, Bothrimonus, Diplocotyle.*)

ORDER DIPHYLLIDEA

Scolex with armed or unarmed peduncle; two spoon-shaped bothridia present, lined with minute spines, sometimes divided by median, longitudinal ridge; apex of scolex with insignificant apical organ or with large rostellum bearing dorsal and ventral groups of T-shaped hooks; genital pores posterior, midventral; testes numerous, anterior; ovary posterior; vitellaria follicular, lateral, or surrounding segment; uterine pore absent; uterus tubular or saccular; parasites of elasmobranchs. (Genera mentioned in text: Family Echinobothriidae*—*Echinobothrium.* Family Ditrachybothriidae—*Ditrachybothrium.*)

ORDER NIPPOTAENIIDEA

Strobila small; scolex with single sucker at apex, otherwise simple; neck short or absent; proglottids each with single set of reproductive organs; genital pores lateral; testes anterior; ovary posterior; vitelline gland compact, single, between testes and ovary; osmoregulatory canals reticular; parasites of teleost fish.

ORDER LITOBOTHRIDEA

Strobila dorsoventrally flattened, with numerous proglottids; scolex a single, well-developed apical sucker; anterior proglottids modified, cruciform in cross section; neck absent; each proglottid with single set of medullary reproductive organs; apolytic or anapolytic; testes numerous, preovarial; genital pores lateral; ovary two or four lobed, posterior; vitellaria follicular, encircling medullary parenchyma; parasites of elasmobranchs.

13

ACANTHOCEPHALA: THE SPINY-HEADED WORMS

The Acanthocephala, or spiny-headed worms, is comprised of a small but interesting and important phylum of helminth parasites. These worms generally measure under 35 mm in length, although *Macracanthorhynchus hirudinaceus*, a common intestinal parasite of pigs, can measure up to 70 cm in length. All species are endoparasitic, living as adults in the alimentary tract of various vertebrates. The number of parasites per host varies from 1 or 2 to as many as 1500.

All the acanthocephalans are elongate, most species being cylindrical and tapering at both ends; however, laterally flattened forms also exist. The body is tubular, with its wall enclosing the pseudocoel (body cavity), in which are suspended the reproductive organs and the proboscis sheath. The main characteristic of the Acanthocephala is the protrusible, armed proboscis at the anterior terminal.

The sexes are separate (dioecious) among these worms, and there is some degree of sexual dimorphism. Female worms, as a rule, are larger than the males, and in species with spines on the body surface, these armatures are better developed in males. Furthermore, the copulatory apparatus of males easily distinguishes them from females.

The biotic potential is great in certain acantho-cephalan species. For example, gravid females of *Macracanthorhynchus hirudinaceus* may contain up to 10 million embryonated eggs at once.

Dorsal and ventral differentiation exists, but is often difficult to discern externally. In naturally curved forms, the convex surface is the dorsal side. In certain genera, such as *Rhadinorhynchus* and *Aspersentis*, the spines on the proboscis are larger ventrally than dorsally (Fig. 13.1).

A digestive tract is completely lacking in members of the Acanthocephala. Like the tapeworms, acanthocephalans, through adaptation to their parasitic way of life, absorb nutrients through the body surface.

MORPHOLOGY

The body of acanthocephalans is divided into two major regions (Fig. 13.2). (1) the **presoma**, which includes the **proboscis** and **neck**, is the anterior region. The proboscis, the anteriorly situated eversible organ, is armed with rows of hooks and spines. The neck is the unarmed region situated immediately posterior to the proboscis. The neck usually equals the proboscis in diameter but various modifications exist among the

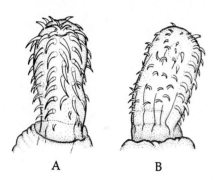

Fig. 13.1. **Proboscis of *Aspersentis*. A.** Ventral view. **B.** Dorsal view. (Redrawn after Van Cleave, 1941.)

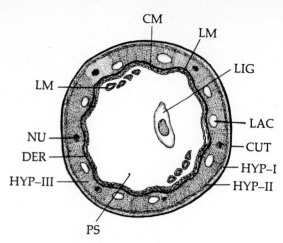

Fig. 13.3. **Cross-section through trunk of an acanthocephalan.** CM, circular muscles; CUT, cuticle; DER, dermis; HYP-I, outermost layer of hypodermis; HYP-II, medial layer of hypodermis; HYP-III, inner layer of hypodermis; LAC, lacuna; LIG, ligament sac; LM, longitudinal muscles; NU, giant nucleus; PS, pseudocoel.

species. (2) The broader body proper, or **trunk**, forms the major part of the animal. It is separated from the presoma by a crease in the tegument, and it may be divided into the smaller anterior portion known as the **foretrunk** and the posterior portion known as the **hindtrunk**.

Acanthocephalan proboscal hooks are defined as the larger recurved projections that are embedded in the hypodermis. These hooks, as those of *Polymorphus*, consist of a core of cytoplasm composed of protein but not carbohydrates and lipoproteins. This

core is surrounded by hard, nonliving material (Crompton and Lee, 1965). Spines are defined as smaller armatures that are not rooted in the proboscis. The larger armatures may be located anteriorly, followed posteriorly by a decreasing gradation of spines, or the larger armatures may be medially situated on the proboscis, with the other spines decreasing in size anteriorly and posteriorly. These hooks and spines are arranged in rows, and their exact sizes and arrangements are used in the identification of species. When the armed proboscis of an acanthocephalan is embedded in the longitudinal muscle layers of the host's gut wall, a fibrous nodule is developed around it.

Immediately posterior and lateral to the neck, the inner layers of the body wall invaginate to form two diverticula that protrude into the pseudocoel. These diverticula are known as **lemnisci**. The sac into which the retracted proboscis fits is known as the **proboscis sheath** or **receptacle**. This structure is muscular and is one cell thick in some species, two in others, and very thick in still others. In one family, the Oligacanthorhynchidae, the wall is very thick dorsally but nonmuscular and thin ventrally.

Body Tissues

The light microscope reveals that the body wall is made up of several recognizable regions (Fig. 13.3). Electron microscopy has revealed that covering the entire body is an extremely thin layer known as the **surface coat** or **glycocalyx**. Immediately underneath the surface coat is the syncytially arranged **hypodermis**, which consists of four distinguishable regions: (1) the outermost stratum, made of radially arranged

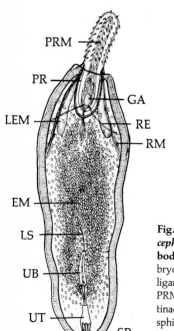

Fig. 13.2. **Gravid female of *Acanthocephalus* (Palaeacanthocephala) showing body structures and organs.** EM, embryos; GA, ganglion; LEM, lemniscus; LS, ligament sac; PR, proboscis receptacle; PRM, proboscis receptacle muscle; RE, retinacula; RM, retractor muscles; SP, sphincter; UB, uterine bell; UT, uterus; VA, vagina. (Redrawn after Yamaguti, 1935.)

striations and referred to as the **striped zone**; (2) the **vesicular zone**, which is the transitional zone between the striped zone and the zone beneath it; (3) the **felt zone**, which is composed of randomly arranged fibers; (4) the next layer, referred to as the **radial zone**, like the outermost zone, consists of radially arranged fibers. This is the thickest of the hypodermal zones and includes a series of channels known as **lacunae**. The lacunae are filled with a fluid that reportedly contains nutrients the animal has absorbed from the environment. Beneath the radial zone is the **basal lamina** which rests on subtegumentary connective tissue.

Although the lacunar channels do not penetrate either the outer layers of the hypodermis or the underlying layers, the nutrient fluid can diffuse throughout the tissues of the body. Thus, the lacunar systems might be thought of as a nutrient-circulating system. The lacunae do not connect with lemnisci, which are considered storage spaces. Miller and Dunagan (1976) have presented a detailed description of the lacunar system of *Macracanthorhynchus hirudinaceus*, including interconnecting channels.

Giant nuclei are found embedded in the radial zone of the hypodermis. These nuclei are so consistent in size, number, and location in certain species that they are commonly used as a taxonomic characteristic. For example, in the family Neoechinorhynchidae, six nuclei are always present; five are located dorsally along the main dorsal lacuna and one ventrally, near the anterior end of the ventral lacuna. In other taxa, the larval stages possess consistent numbers of nuclei, but during the maturation process, these divide (fragment) and give rise to a greater number of smaller nuclei. These unique nuclei of the Acanthocephala may be round, oval, rosette, or even amoeboid in shape.

As depicted in the accompanying illustrations (Figs. 13.4, 13.5), the hypodermis, when studied with the electron microscope, is delineated on its external surface by a three-layer membrane. Medial to this membrane is a homogeneous layer infiltrated with pores leading from the surface to the striped zone.

The striped zone is comprised of a homogeneous material surrounding canals leading from pores on the body surface to the felt zone. Some of these canals are filled with an electron-dense material, giving this zone its striped appearance.

The felt zone contains many fibrous strands extending in various directions, mitochondria with few cristae, and numerous vesicles. Some of the vesicles appear to be connected to the canals of the striped zone, whereas others may be a type of endoplasmic reticulum (Crompton and Lee, 1965).

The radial zone of the hypodermis contains some fibrous strands, although much fewer than the felt zone, and there are large, thin-walled channels present. Mitochondria are quite numerous in this zone

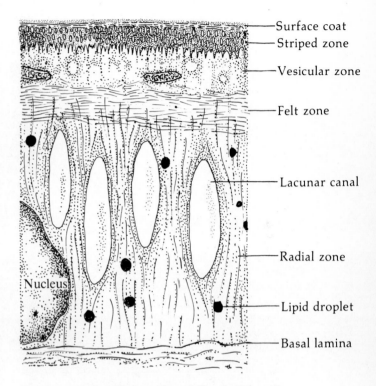

Surface coat
Striped zone

Vesicular zone

Felt zone

Lacunar canal

Radial zone

Nucleus

Lipid droplet

Basal lamina

Fig. 13.4. Body wall of acanthocephalan. Semischematic drawing showing different zones and structures of the body wall of *Moniliformis dubius*. The vesicular zone is a transitional area between the striped zone and the felt zone; it contains vesicles and mitochondria with poorly developed cristae.

and are commonly arranged around the lacunar channels. Lipids also occur in the radial zone as large and small droplets.

The inner boundary of the hypodermis, as seen with the electron microscope, is delineated by a unit membrane. This membrane is usually greatly folded, and the invaginations of the folds often form small vesicles that, according to Crompton and Lee (1965), may be pinocytotic vesicles containing lipid droplets.

Certain enzymes have been demonstrated in the body wall of acanthocephalans using histochemical techniques. Although the surface coat shows no enzyme activity, alkaline phosphatase is present in the hypodermis of some species. Lipase and nonspecific esterase occur in the hypodermis, especially in the proboscis. Aminopeptidase also occurs in the hypodermis of the body.

As has been pointed out by Lee (1966) and Bird and Bird (1969), who have contributed reviews on the body wall of acanthocephalans, the canals opening on

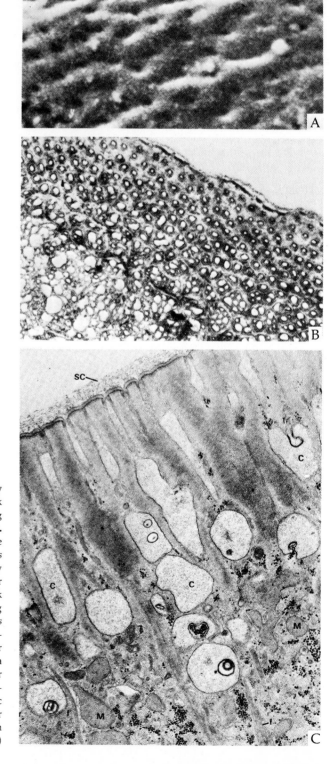

Fig. 13.5. Structure of acanthocephalan body wall. A. Scanning electron micrograph of the trunk surface of *Macracanthorhynchus hirudinaceus* showing "dimples" marking openings of pore canals. **B.** Transmission electron micrograph of an oblique section through the body wall of *Polymorphus minutus* showing sections of pore canals that branch as they descend deeper into the body wall. (**A** and **B**, after Whitfield, 1979; with permission of University Park Press.) **C.** Transmission electron micrograph showing major features of the striped zone of *Moniliformis dubius.* The surface is covered with a finely filamentous surface coat (SC). The surface crypts (C) appear as large scattered vesicular structures. Mitochondria (M), glycogen particles, microtubules, and other cytoplasmic organelles are evident in the inner portion of the striped zone. Bundles of fine cytoplasmic filaments (f) extend between this zone and the deeper cytoplasm of the body wall. × 35,000. (After Byram and Fisher, 1974; with permission of *Tissue and Cell*.)

the body surface of these parasitic worms serve as sites for the uptake of nutrients.

The muscles in the body wall are arranged in two layers. The outer one is circularly arranged, while the inner one is longitudinally oriented. These myofibers are syncytially arranged with scattered nuclei. Since the Acanthocephala are pseudocoelomates, no epithelium lines the body cavity.

In some species, the tegument possesses superficial circular creases suggesting segmentation. However, even in the most severe cases of tegumental folding, there are no indications of internal segmentation except perhaps for the longitudinally arranged muscles that may be attached to the areas where the body wall invaginates. Varying degrees of spination of the tegument exist among the various taxa, ranging from large spines covering the entire trunk to the complete absence of spines.

Very often freshly obtained specimens show pronounced coloration. This is most probably due not to native pigmentation but to the inclusion of colored nutrients absorbed from the host. However, in certain species pigmentation is suspected (von Brand, 1973). It is noted, however, that certain larval acanthocephalans have pigment associated with them. For example, the cystacanth larvae (p. 454) of *Polymorphus minutus* are characteristically surrounded by a bright orange capsule rich in carotenoids, derivatives of cholesterol. Interestingly, it has been suggested by behaviorists that this pigmentation, coupled with the brighter color of infected prawns, renders the larval parasite more obvious to the definitive host, a duck, when it is feeding on freshwater prawns, which are the intermediate hosts.

Proboscis Musculature

The armed proboscis serves as an attachment organ. Within the host's intestine, the proboscis is embedded in the lining of the gut and the hooks anchor the parasite in place. The mechanism behind the extension and retraction of the proboscis is extremely complex in certain species (Fig. 13.6). Generally, a pair of muscles known as **inverter muscles** are attached to the inner surface of the anterior tip of the proboscis. These muscles run posteriorly and are attached at the other end to the base of the proboscis sheath. Some, however, continue through the sheath and split; one group is directed dorsally as the **dorsal inverter**, attaching on that side of the trunk, and the other group is directed ventrally as the **ventral inverter**, attaching on the ventral surface of the trunk. The function of the inverters is to withdraw the proboscis into the receptacle. Another group of muscles, known as the **protrusers**, originate in a circle at the base of the neck and are situated outside the proboscis sheath, attaching along the posterior end of the sheath. These muscles are known as the dorsal, ventral, and lateral

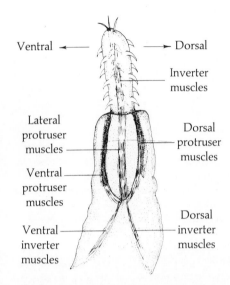

Fig. 13.6.   Proboscal musculature. Anterior end of an acanthocephalan showing major muscles.

protrusers, depending on their location. The function of the protrusers is to evert the proboscis out of the receptacle.

Excretory System

Among acanthocephalans an excretory system exists only in members of the order Archiacanthocephala. No excretory system occurs in the members of the other orders, and waste materials in these diffuse from the body through the body surface, presumably through the surface pores.

In the Archiacanthocephala, the excretory system is of the protonephritic type, consisting of a pair of minute bodies lying on each side of and closely associated with the reproductive organs. Each body is made up of a group of ciliated flame bulbs, which are not individual cells like those found in trematodes and cestodes, for they do not include nuclei. Waste materials in the pseudocoel are brought into the flame bulbs by the ciliary beats. There may be as many as 700 flame bulbs present. From the bulbs the waste is either emptied into a common sac or carried away in a common collecting tubule. In the first system, a common collecting tubule leads from the sac. The two lateral tubules join medially. In male worms the common duct empties into the sperm duct, and in females, the common duct empties into the terminal portion of the uterus, hence they form part of a urogenital system in both sexes.

Nervous System

The "brain" is in the form of a cephalic ganglionic mass located along the ventral border within the proboscis sheath. This structure consists of a mass of nerve fibers enveloped by ganglionic cells. In *Macracanthorhynchus*, there are 86 ganglia; in *Bolbosoma*, there are 73; and in *Hamanniella*, there are 80.

Leading from the cephalic nerve center, certain fibers extend anteriorly, innervating the sensory organs at the tip of the proboscis and on each side of the neck region. Dunagan and Miller (1983) have provided a detailed description of the apical sense organ of *Macracanthorhynchus hirudinaceus*, and Miller and Dunagan (1983) have reported the presence of a multinucleated support cell that serves the apical and lateral sensory organs in the same species. Other fascicles of fibers—generally two lateral bundles—lead posteriorly, giving off branches along their lengths. In males, an auxiliary pair of ganglia, called the **genital ganglia**, is located posteriorly at the base of the cirrus, and these are joined with each other by a ring commissure. Certain branches of the lateral nerve fibers lead into these auxiliary ganglia, and other branches lead from them to innervate the sensory organs of the copulatory organ and bursa. The two main lateral nerve bundles are wrapped within a muscular sheath known as the **retinaculum** (Fig. 13.7).

Reproductive System

The reproductive system in acanthocephalans is unique because the organs are suspended within a ligament sac (Fig. 13.8). This hollow sac of semitransparent connective tissue is attached anteriorly to the posterior extremity of the proboscis sheath or to the nearby body wall. Posteriorly, it is attached to the genital sheath in males and to the uterine bell in females.

In members of the order Archiacanthocephala, the females possess two ligament sacs, one dorsal and one ventral (Fig. 13.8). In these, the posterior attachment of the ventral sac is on the posterior extremity of the body. In the order Eoacanthocephala, there are also two ligament sacs in females. In these two orders, the ovaries are located in the dorsal sac, which communicates anteriorly with the ventral sac through an opening. In the order Palaeacanthocephala, there is only one sac in each sex. Furthermore, in females the sac ruptures when the worms reach sexual maturity, permitting developing germ cells to escape into the pseudocoel.

In addition to the ligament sac(s), a nucleated auxiliary strand—the **ligament strand**—is present. In worms with two sacs, this strand is situated be-

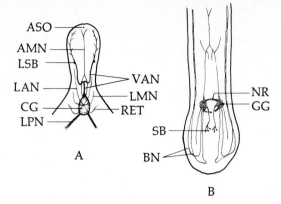

Fig. 13.7. Nervous system of Acanthocephala. A. Nervous system of anterior end of *Macracanthorhynchus hirudinaceus*. **B.** Posterior terminal of male of the same species. (Redrawn after Brandes, 1899.) AMN, anterior median nerve; ASO, apical sensory organ; BN, bursal nerves; CG, cerebral ganglion; GG, genital ganglia; LAN, lateral anterior nerve; LMN, lateral medial nerve; LPN, lateral posterior nerve; LSB, lateral sensory bulb; NR, nerve ring in genital sheath; RET, retinacula; SB, sensory bulb of penis; VAN, ventral anterior nerve.

tween them; in unisaccular forms, the strand runs along the ventral surface of the sac. The function and origin of this strand are hypothetical. It is considered by some to be a vestigial midgut, the remnant of an ancestral alimentary tract.

In male acanthocephalans, there are two rounded or elongated testes that are tandemly arranged within the ligament sac. Arising from the posterior extremity of each testis is an individual sperm duct (vas efferens) that is directed posteriorly. Shortly after passing the level posterior to the posterior testis, one or two minute ducts empty into the sperm ducts. These ducts, known as **cement gland ducts**, arise from the union of many primary ducts that originate in the conspicuous unicellular **cement glands** lying posterior to the caudal testis. In the order Eoacanthocephala, the cement glands form a syncytium with several giant nuclei. A single cement gland duct carries the secretion to a **cement reservoir** from which a pair of ducts extend to join the **sperm duct**. The sperm ducts continue posteriorly within the narrower continuation of the ligament sac known as the **genital sheath**. This sheath containing myofibers from the body wall, surrounds the two sperm ducts that unite to form the single **vas deferens**, which may enlarge to form the **seminal vesicle**. The vas deferens then enters the **cirrus** and terminates in a muscular and eversible cup, the **bursa**.

The ovaries of female acanthocephalans are large, unitary organs in juvenile worms, but are fragmented in adults. The small ovarian bodies of adults are free in the ligament sac and are known as **ovarian balls**. Ova produced by the ovarian balls are fertilized in the pseudocoel by spermatozoa introduced into the female during copulation. Developing eggs are collected by a funnel-shaped **uterine bell**, which although thicker, is continuous with the posterior extremity of the ligament sac. The eggs pass posteriorly through the **uterine tube** into the long muscular **uterus**. A pair of large bell pouches arise as diverticula from the uterine tube and extend anteriorly. The nonmuscular terminal end of the genital tract is known as the **vagina**. For detailed reviews of oogenesis and spermatogenesis in acanthocephalans, see Crompton (1983a, b).

In *Polymorphus minutus*, a parasite of birds, Whitfield (1968, 1970) has studied the histology and function of the uterine bell. This structure consistently includes 17 nuclei, demonstrating organ eutely (i.e., consistent number of cells) characteristic of acanthocephalans. Furthermore, this muscular bell is able to separate mature from immature eggs, permitting only the former to pass into the uterus. This selection is based on the greater size of mature eggs.

During copulation, the bursa of the male everts because of hydrostatic pressure resulting from the injection of fluid into the bursal spaces by the **Saefftigen's pouch** (an extension of the bursa internally into the genital sheath). The bursa is wrapped around the posterior end of the female, the cirrus is introduced into the gonopore, and spermatozoa migrate up the vagina, uterus, and uterine tube. Next, "cement" is secreted from the cement glands and plugs the gonopore and vagina, preventing the escape of spermatozoa. For a detailed review of the reproductive organs and behavior of acanthocephalans, see Crompton (1970).

The acanthocephalan egg possesses three, sometimes four, membranes: (1) The **ovocyte membrane** surrounds the elliptical ovum when it is expelled from the ovarian ball. (2) The **fertilization membrane** is laid down inside the ovocyte membrane after the sperm enters. (3) The **shell membrane** is deposited between the ovocyte and fertilization membranes when development of the embryo is initiated. (4) In some species, an **outer membrane** lies on top of the shell membrane.

The surface texture of the shell membrane varies among the species. In species that utilize a terrestrial intermediate host, the shell is generally hard, whereas in species that utilize an aquatic host, the shell is soft.

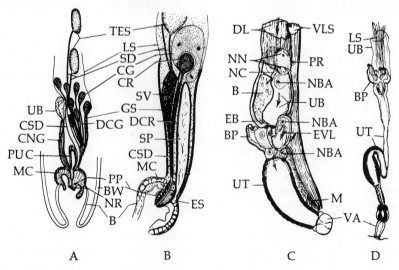

Fig. 13.8. Male and female reproductive systems in acanthocephalans. A. Male reproductive system of *Hamanniella* (Archiacanthocephala), lateral view. **B.** Male reproductive system of *Neoechinorhynchus* (Eoacanthocephala,) lateral view. **C.** Female reproductive system of *Oligacanthorhynchus* (Archiacanthocephala), lateral view. **D.** Female reproductive system of *Bolbosoma* (Palaeacanthocephala.) (**A** and **B**, after Hyman, 1951; **C**, after Meyer, 1931; **D**, after Yamaguti, 1939.) B, bladder or excretory vesicle; BP, bell pockets; BW, body wall; CG, cement glands; CNC, common excretory (nephridial) canal; CR, cement reservoir; CSD, common sperm duct; DCG, ducts of cement glands; DCR, ducts of cement reservoirs; DL, dorsal ligament sac; EG, exit of bladder into bell; ES, entrance of Saefftigen's pouch into bursa cap; EVL, exit into ventral ligament sac; GS, genital sheath; LS, ligament sac; M, membrane, part of ventral ligament sac; MC, muscular cap of bursa; NBA, nuclei of bell apparatus; NC, excretory (nephridial) canal; NN, excretory (nephridial) nuclei; NR, nerve ring; PP, penis papilla; PR, protonephridium; PUC, pouch of urogenital canal; SD, sperm ducts (vas deferens); SP, Saefftigen's pouch; SV, seminal vesicle; TES, testes; UB, uterine bell; UT, uterus; VA, vagina; VLS, ventral ligament sac.

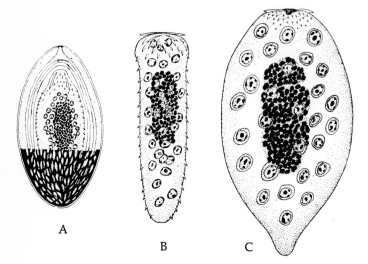

A

B

C

Fig. 13.9.   Egg and acanthor of *Macracanthorhynchus hirudinaceus*. A. Egg with surface appearance shown only in lower portion. **B.** Acanthor, stage I, from lumen of midgut of beetle larva. Note presence of larval rostellar hooks and spinous body. **C.** Acanthor, stage II, after penetrating wall of midgut of beetle larva. (**A,** after Kates, 1943; **B** and **C,** after Van Cleave, 1935.)

Expelled eggs are highly resistant and may remain viable for several months.

The *in utero* embryology of *Macracanthorhynchus hirudinaceus* has been studied by various investigators (Fig. 13.9); however, the reports of Meyer (1928, 1936, 1937, 1938a, b) are considered the classic studies. Zygotic cleavage is of the spiral determinate type, but the positions of the blastomeres are displaced because of the elliptical shape of the eggshell. Before the egg is discharged by the worm, a partially developed acanthor, the first larval stage, is already formed within the thick shell.

CHEMICAL COMPOSITION

The chemical composition of Acanthocephala has been studied in only the more common and larger species. Von Brand (1939) has reported that 70% of the dry weight of *Macracanthorhynchus hirudinaceus* is protein, some of which undoubtedly is represented by scleroproteins of the body wall. In the same species, stored glycogen is found primarily in the body wall (80%), in the ovaries and eggs (12%), and in the body fluid within the lacunar system (8%). Some glycogen is also present in the noncontractile portions of the

muscle cells. Bullock (1949) has stated that the hypodermis is probably the site of glycogenesis.

Von Brand and Saurwein (1942) have reported that, besides glycogen, a small amount of a second type of polysaccharide (galactogen?) has been isolated from *M. hirudinaceus*. In addition, trehalose, glucose, fructose, mannose, and maltose are the main carbohydrate reserves. These are utilized in energy synthesis under aerobic and anaerobic conditions, whereas galactose, which is also present, is utilized only under anaerobic conditions.

Lipids form 1 to 2.1% of the fresh weight of *M. hirudinaceus* (von Brand, 1939, 1940). These lipids have been determined, through fractionation, to be 27% phosphatids, 24% unsaponifiable matter, 2% saturated fatty acids, 32% unsaturated fatty acids, and 2% glycerol. The unsaponifiable material may well be cholesterol. Lipids in this species occur in the body wall, the reproductive organs, and the body fluid. Lipid droplets have also been found in the lacunar system. Bullock (1949) has stated that fat synthesis and hydrolysis, under the control of lipase, also occur in the hypodermis of the trunk and to some extent in the lemnisci.

The body musculature of various acanthocephalans contains different amounts of stored fats. In *Macracanthorhynchus*, there is little fat; but in *Neoechinorhynchus*, *Pomphorhynchus*, and *Echinorhynchus*, there are considerable accumulations. Specific types of lipids are localized in specific areas. Phospholipids, for example, are found primarily in the body wall, accumulating around the lacunae, whereas cholesterol and cholesterol esters are limited to the innermost margins of the body wall.

In addition to proteins, carbohydrates, and lipids, certain inorganic materials are also present. In *M. hirudinaceus*, inorganic materials make up 0.58% of the fresh body weight and 5% of the dry weight. Potassium, sodium, calcium, magnesium, aluminum, iron, manganese, silicon, chlorine, copper, phosphate, sulfate, and carbonate all occur in this species.

LIFE CYCLE STUDIES

An invertebrate intermediate host is required by all known acanthocephalans (Fig. 13.10). Eggs passed out in the feces of the vertebrate host, either on land or in water, contain a partially developed **acanthor**. If the egg is ingested by an invertebrate host (usually an arthropod), development continues, resulting in a fully formed, elongate acanthor that hatches in the gut. For a review of what is known about acanthocephalan eggs, see Crompton (1970).

The acanthor possesses pointed ends, is armed with a rostellum with hooks, and has a spinous body surface. In some species, a definite suture is visible on the shell, which represents the line of rupture upon

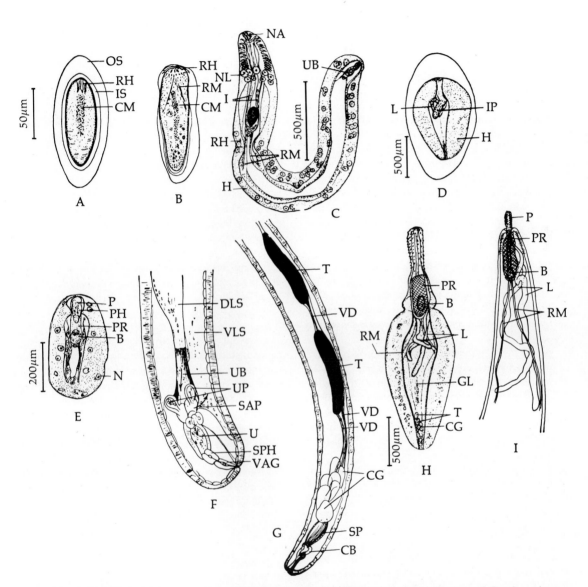

Fig. 13.10. **Stages in the life cycle of *Moniliformis dubius*. A.** Egg. **B.** Acanthor escaping from eggshell and membranes. **C.** Acanthella dissected from enveloping sheath about 40 days after infection. **D.** Cystacanth from body cavity of roach with inverted proboscis, about 50 days after infection. **E.** Median section of acanthor from body cavity of roach, 29 days after infection. **F.** Posterior end of female adult. **G.** Posterior end of male adult. **H.** Cystacanth freed from cyst and with protuded proboscis. **I.** Anterior end of adult worm. (**A–E** and **H**, after Moore, 1946; **F**, **G**, and **I**, after Chandler and Read, 1961.) B, brain; CB, copulatory bursa; CG, cement glands; CM, central nuclear mass; DLS, dorsal ligament sac; GL, genital ligament; H, hypodermis; I, inverter muscle; IP, inverted proboscis; IS, inner shell; L, lemnisci; NA, nuclei of apical ring; NL, nuclei of lemniscal ring; OS, outer shell; P, proboscis; PH, developing proboscis hooks; PR, proboscis receptacle; RH, rostellar hooks; RM, retractor muscles; SAP, sorting apparatus; SP, Saefftigen's pouch; SPH, sphincter; T, testes; U, uterus; UB, uterine bell; UP, uterine pouches; VAG, vagina; VD, vasa differentia; VLS, ventral ligament sac.

hatching. The active acanthor, with the aid of its rostellar hooks and body motions, penetrates the gut wall of the invertebrate host and becomes located in the hemocoel. This penetration process usually takes two to five weeks.

In the hemocoel of the host, the acanthor rounds up and loses its rostellum, hooks, and body spines. It develops a rudimentary proboscis, proboscis sheath, cephalic ganglia, ligament sheath, primitive gonads, etc., and is then known as an **acanthella**. Further differentiation takes place until many of the adult organs become readily recognizable, but the gonads remain nonfunctional. When such a degree of differentiation is reached, accompanied by the elongation of the body, the animal is known as a **juvenile**.

The juvenile is generally enclosed within a sheath and is referred to as a **cystacanth**. It is inactive and its proboscis is retracted. In certain species cystacanths are found encysted in tissues and organs as well as the body cavity of the intermediate host. Complete development within the intermediate host takes 6 to 12 weeks.

When the invertebrate host is ingested by the appropriate vertebrate host, the cystacanth loses its sheath, everts its proboscis, and attaches to the wall of the small intestine where it reaches sexual maturity. In some cases, in addition to the invertebrate intermediate host, a vertebrate or invertebrate paratenic host is involved, in which the cystacanth stage is prolonged.

It is generally agreed that acanthocephalans originated as parasites of fish and the original life cycle involved a single invertebrate intermediate host. More recently acquired paratenic hosts are undoubtedly vertebrates or invertebrates whose internal environments are not conducive to development of the worm to sexual maturity. For example, the normal intermediate host of *Neoechinorhynchus emydis* are ostracods. If infected ostracods are ingested by snails (*Campeloma*, *Pleurocera*, and *Ceriphasia*), the acanthella becomes encysted around the mouth and in the foot of the molluscan paratenic host. That juvenile acanthocephalans are able to survive in paratenic hosts, particularly invertebrates that serve as prey for vertebrates, has made possible the parasitization of such hosts as reptiles, birds, and mammals.

Paratenic hosts are not necessary in the life cycle of acanthocephalans, although in many instances they are advantageous to the parasite because such hosts serve as links in the food chain and carry the immature worm to a suitable definitive host.

Another interesting aspect of acanthocephalan life cycles is the cohabitation of cystacanth and adult worms of the same species within the same fish host. The cystacanths are generally found within the body cavity of the host whereas the adults are within the small intestine. This suggests that the same fish can serve as both the definitive and the intermediate host. Fish may serve as intermediate hosts if the ingested larval acanthocephalan has not completed development in the normal intermediate host. For example, DeGiusti (1939) has shown that when eggs of *Leptorhynchoides thecatus* are fed to amphipods, fully developed infective cystacanths are formed in 32 days. When such larvae are fed to small black bass, they become attached to the fish's intestinal mucosa and develop to maturity. However, when larvae less than 32 days old are fed to black bass, the larvae burrow through the intestinal wall and become reencapsulated as cystacanths in the body cavity. Thus, the black bass can serve as both an intermediate and a definitive host, depending on the degree of maturity of the cystacanth. This unusual phenomenon is advantageous to the parasite because if small black bass harboring reencysted cystacanths are eaten by larger fish, the cystacanths can reach sexual maturity in the larger fish.

Human Infections

Because humans seldom intentionally eat insects, microcrustaceans, toads, or lizards, human infections with acanthocephalans are rare. However, such infections are known (Schmidt, 1971). Specifically, five species have been reported from humans. *Macracanthorhynchus hirudinaceus* and *Moniliformis moniliformis* are the two most commonly reported. In addition, *Acanthocephalus rauschi*, a fish parasite, has been reported from the peritoneum of an Alaskan Eskimo, and *Corynosoma strumosum*, a parasite of seals, has also been reported from humans. The most puzzling report is that of *Acanthocephalus bufonis* in an Indonesian, since this species is normally a parasite of toads. The human must have eaten a paratenic host.

BIOLOGY OF ACANTHOCEPHALANS

As listed in the classification list at the end of this chapter, the phylum Acanthocephala is comprised of three classes: Archiacanthocephala, Palaeacanthocephala, and Eoacanthocephala. The distinguishing characteristics of the members of these classes, as well as those of the subordinate orders, are also presented in the classification list. For those interested in the taxonomy of the Acanthocephala, it is noted that Meyer's (1932, 1933) monograph of the phylum includes all the species described up to 1933, and Ward (1951, 1952) has listed many of the species described since 1933. Yamaguti (1963) has catalogued all the species known up to that date, and Petrochenko (1956, 1958) has assembled the descriptions of all of the species known at the time of writing.

In addition to the general characteristics of all acanthocephalans, members of the Archiacanthocephala exhibit unfragmented, large hypodermal nuclei, and spines are absent from the body trunk. Species of this order are generally parasitic in terrestrial birds and mammals, although a few are found in fish. The archiacanthocephalans employ insects and myriapods as intermediate hosts.

Among the genera represented in this class are *Mediorhynchus*, parasitic in birds; *Gigantorhynchus* in mammals; *Heteracanthorhynchus* in birds; *Oligacanthorhynchus* in fish and birds; *Nephridiorhynchus* in mammals; *Macracanthorhynchus* in mammals, including the common giant hog species *M. hirudinaceus*; *Monili-*

formis in mammals, mostly rodents, including the common species *M. moniliformis* in rats and other small mammals; and *Oncicola* in mammals, including *O. canis*, the species found, sometimes in large numbers, in dogs of North America (Fig. 13.11).

Among members of the Archiacanthocephala, *Macracanthorhynchus hirudinaceus* has been reported in humans, but only rarely. Because of its large size and its easy accessibility from slaughter houses, it has been the subject of numerous physiologic and biochemical studies.

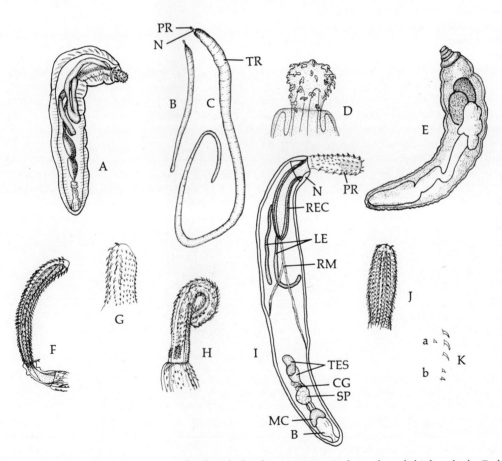

Fig. 13.11.   Some acanthocephalans. A. Male of *Mediorhynchus sipocotensis* (Archiacanthocephala) from birds. (Redrawn after Tubangui, 1935.) **B.** Male of *Macracanthorhynchus hirudinaceus* (Archiacanthocephala). **C.** Female of *M. hirudinaceus.* **D.** Proboscis of *Nephridiorhynchus palawanensis* (Archiacanthocephala) from mammals. (Redrawn after Tubangui and Masilungan, 1938.) **E.** Male of *Oncicola travassosi* (Archiacanthocephala) from mammals. (Modified after Witenberg, 1938.) **F.** Proboscis of *Illiosentis furcatus* (Palaeacanthocephala) from fish. (Redrawn after Van Cleave and Linicome, 1939.) **G.** Proboscis of *Rhadinorhynchus peltohampi* (Palaeacanthocephala) from fish. (Redrawn after Baylis, 1944.) **H.** Proboscis of *Tegorhynchus pectinarius* (Palaeacanthocephala) from fish. (Redrawn after Van Cleave, 1940.) **I.** Male of *Leptorhynchoides thecatus* (Palaeacanthocephala) from fish. (Redrawn after Van Cleave, 1919.) **J.** Proboscis of *Gorgorhynchus lepidus* (Palaeacanthocephala) from fish. (Redrawn after Van Cleave, 1940.) **K.** Proboscis hooks (a) and body spines (b) of *Mehrarhynchus prashadi* (Palaeacanthocephala) from fish. (Redrawn after Datta, 1940.) B, bursa; CG, cement glands; LE, lemnisci; MC, muscular cap of bursa; N, neck; PR, proboscis; REC, receptacle; RM, retractor muscles; SP, Saefftigen's pouch; TES, testes; TR, trunk.

Life Cycle of *Macracanthorhynchus hirudinaceus.* Adults of this species are somewhat flattened and have numerous transverse tegumental wrinkles. Males measure up to 100 mm long, whereas females reach 350 mm or more in length. The proboscis is globular, relatively small, and it bears about five transverse rows of hooks.

Embryonated eggs laid by females pass to the exterior in the host's feces and are ingested by beetles of the genera *Cotinus* and *Phyllophaga*. Once ingested, the eggs hatch within an hour, and the escaping acanthors migrate through the beetle's gut wall into the hemocoel. For the first 5 to 20 days after penetration, the parasite usually remains attached to the outer surface of the beetle's midgut, but as growth continues it becomes detached and is free in the hemocoel. At this site it transforms into an acanthella armed with a proboscis. Finally, the cystacanth stage is reached by the 65th to the 90th day after entering the beetle, and when this stage is ingested by swine, it reaches sexual maturity in two to three months. Female worms produce a maximum of about 260,000 eggs per day. An infected swine may maintain this parasite for a year or more.

Class Palaeacanthocephala

Members of the Palaeacanthocephala include species found almost exclusively in fish, although a few are parasites of aquatic birds, mammals, amphibians, and reptiles. Representatives of the order include *Illiosentis*, *Rhadinorhynchus*, *Tegorhynchus*, and *Pseudorhynchus* (Fig. 13.11). These genera all possess dorsoventral differences in the shapes and sizes of proboscal hooks. *Illiosentis* possesses genital spines in addition to anterior trunk spines.

Other fish-parasitizing genera include *Aspersentis* and *Leptorhynchoides*. The latter includes *L. thecatus*, a common species in the eastern United States (Fig. 13.11). Other representative genera of this order are depicted in Fig. 13.12.

Class Eoacanthocephala

Included in the Eoacanthocephala is the common species *Neoechinorhynchus cylindratus*, which parasitizes freshwater fish in North America. In the life cycle of this species, Ward (1940) demonstrated that ostracods serve as the intermediate host and smaller fish may serve as paratenic hosts. Another species, *Neoechinorhynchus emydis*, is common in turtles in North America. Other members of this class parasitize amphibians.

Other genera comprising the Eoacanthocephala include *Pallisentis*, *Acanthosentis*, *Quadrigyrus*, *Rao-

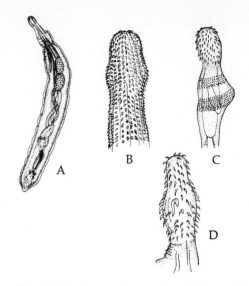

Fig. 13.12. Some palaeacanthocephalans. A. Male of *Polymorphus marilis* from birds. (Redrawn after Van Cleave, 1939.) **B.** Proboscis of *Centrorhynchus conspectus* from birds. (Redrawn after Van Cleave and Pratt, 1940.) **C.** Proboscis of *Bolbosoma thunni* from fish. (Redrawn after Harada, 1935.) **D.** Proboscis of *Corynosoma turbidum* from birds. (Redrawn after Van Cleave, 1937.)

sentis, Octospinifer, Eosentis, and *Tenuisentis*, all fairly common in various piscine hosts (Fig. 13.13).

Life Cycle of *Neoechinorhynchus cylindratus.* The adult of this acanthocephalan occurs in the intestine of a number of species of fish in North America, including *Micropterus dolomieui, Esox niger, E.*

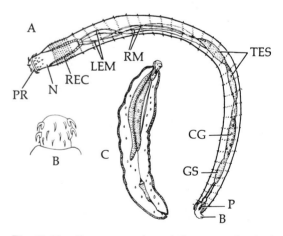

Fig. 13.13. Some eoacanthocephalans. A. Male of *Pallisentis ophiocephali* showing girdles of trunk spines. (Redrawn after Thaper, 1930.) **B.** Proboscis of *Octospinifer*. (Redrawn after Van Cleave, 1919.) **C.** Female of *Eosentis devdevi* in fish. (Redrawn after Datta, 1936.) B, bursa; CG, cement gland; GS, genital sheath; LEM, lemnisci; N, neck; P, penis (cirrus); PR, proboscis; REC, receptacle; RM, retractor muscles; TES, testes.

lucius, *Stizostedion vitreum*, *Amia calva*, *Erimyzon sucetta*, *Anguilla rostrata*, *A. chrysypa*, and *Carpiodes carpio*. Female worms measure 7–11.5 mm long by 0.35– 0.7 mm wide, and male worms measure 4.7–6.3 mm long by 0.36–0.63 mm wide. In females, the globular proboscis measures 0.1–0.14 mm long by 0.16– 0.19 mm broad and bears three rings of six hooks each. One lemniscus is uninucleate, the other is binucleate. The uninucleate one is slightly shorter than the other. The eggs are ellipsoid and measure 50– 60 μm by 17–27 μm. In males, the proboscis is 0.10– 0.14 mm long by 0.15–0.17 mm wide, and the testes are tandemly arranged in front of an elongate, synctial, cement gland with eight nuclei.

Eggs are fully embryonated when layed and are voided in the fish's feces. When the eggs are ingested by ostracods (*Cypria globula*), they hatch and the acanthors burrow through the intestine wall and become located in the hemocoel within 24 hours. If an ostracod ingests large numbers of eggs, the boring action of the acanthors usually kills it. Within the ostracod's hemocoel, growth is slow, with very little change during the first six days. By the 30th day, the parasites usually have reached the definitive size of adult males and females.

A second intermediate host is involved in the life cycle of *N. cylindratus*. This may be blue gill (fish), *Lepomis*, or some other species of centrarchid fish, which ingests infected ostracods. The acanthella is freed in the gut of the fish intermediate host, burrows through the intestine wall, and invades the liver. Bogitsh (1961) has studied the origin and nature of the so-called cyst wall surrounding juvenile *N. cylindratus* in fish liver. He has reported that the cyst is comprised of two layers (Fig. 13.14). The inner one is fibrous and includes diastase-resistant, periodic acid-Schiff-positive material as well as collagenous proteins. The outer layer is composed of fibroblasts. These compressed cells include glycogen as well as proteins and are derived from the host's mesenchymal cells rather than from liver cells, as originally thought. Consequently, the cyst surrounding the parasite is partly formed by a type of host cellular reaction. It is noted that alkaline glycerophosphatase is confined to the outer, fibroblastic layer of the cyst. This enzyme does not occur in the parasite, the inner layer, or in the host's liver cells.

The definitive fish host becomes infected from eating the second intermediate host. When this occurs, the juveniles develop to sexual maturity in about five months.

PHYSIOLOGY AND BIOCHEMISTRY OF ACANTHOCEPHALANS

The physiology, and especially the biochemistry, of acanthocephalans has been little studied when com-

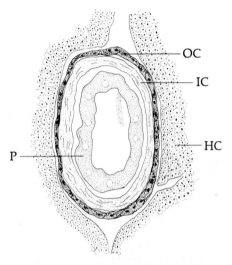

Fig. 13.14. *Neoechinorhynchus cylindratus.* Drawing of a section through an encysted acanthella in the liver of *Lepomis* sp. showing the two layers of the cyst wall. HC, host liver cells; IC, inner layer of cyst; OC, outer fibroblastic layer of cyst; P, parasite. (Redrawn after Bogitsh, 1961.)

pared with trematodes and cestodes. A brief summary is presented below. For a more extensive review, the monograph by Crompton (1970) dealing with the physiologic ecology of these parasites is recommended.

Oxygen Requirements

Von Brand (1940) has suggested that acanthocephalan metabolism is primarily anaerobic *in vivo*, although these worms can survive in the presence of oxygen. Laurie (1957) has demonstrated that *Moniliformis dubius* is capable of anaerobic metabolism and that it ferments glucose, galactose, mannose, fructose, and maltose to acidic catabolites under anaerobic conditions. Even under aerobic conditions, this fermentative type of metabolism is carried on. Thus, as is generally the case among parasitic helminths, the acanthocephalans are best considered facultative anaerobes.

Ability to metabolize under anaerobic and aerobic conditions is exemplified by *Macracanthorhynchus hirudinaceus*, which is capable of absorbing and consuming certain carbohydrates. Under anaerobic conditions, these worms consume 1 g per 100 g wet weight of carbohydrate, whereas under aerobic conditions they consume only 0.8 g. Further evidence that acanthocephalans are more efficient as anaerobes has been contributed by Nicholas and Grigg (1965), who

Table 13.1. Glycogen Content in Organs and Structures of a *Macracanthorhynchus hirudinaceus* Female[a]

Body Wall		Ovaries and Eggs		Body Fluid	
Fresh Substance (%)	*Total Glycogen (%)*	*Fresh Substance (%)*	*Total Glycogen (%)*	*Fresh Substance (%)*	*Total Glycogen (%)*
1.5	80	0.9	12	0.2	8

[a] Data from von Brand (1939).

have found that *Moniliformis dubius** maintained in culture survive best in a gas phase consisting of nitrogen-carbon dioxide rather than oxygen. These data suggest that these worms, as parasites in the intestine, where the oxygen tension is practically nil, are more adapted to anaerobism.

Carbohydrate Metabolism

In analyses of stored nutrients in the body, von Brand (1939) has reported that glycogen occurs in varying quantities in *Macracanthorhynchus hirudinaceus*. His results are presented in Table 13.1. This polysaccharide has since been found in all acanthocephalans that have been examined. Most of the glycogen is stored in the body wall; however, some occurs in the body fluid, which serves as the medium for transporting nutrients. Glycogen found in the ovaries and eggs probably is utilized by the developing acanthors. There is evidence that glycogen is synthesized from glucose, fructose, mannose, and maltose (Laurie, 1959).

In addition to glycogen, the disaccharide trehalose has been demonstrated in *Moniliformis dubius* (Fairbairn, 1958). This sugar has been shown by Fisher (1964), employing radioactively labeled glucose, to be synthesized from glucose.

Acanthocephalans, like other helminths, synthesize pyruvate via the Embden-Meyerhof glycolytic pathway (p. 52); however, Dunagan and Scheifinger (1966) have reported that there is no significant citric acid cycle activity. This conclusion is based on the fact that the enzymes aconitase and isocitric dehydrogenase, both normally involved in the citric acid cycle (p. 55), are present in only insignificant amounts in mitochondrial preparations of *Macracanthorhynchus*

hirudinaceus. This conclusion has been substantiated in other species. For example, *M. dubius* ferments the hexoses that it absorbs, including glucose, fructose, galactose, and mannose, with ethanol and carbon dioxide being the main end products of glycolysis. In addition, small amounts of lactate, succinate, acetate, and butyrate are also produced (Körting and Fairbairn, 1972). An exception to this rule is *Echinorhynchus gadi*, a parasite of cod, which does include an operational citric acid cycle.

Certain acanthocephalans, such as *Moniliformis*, can fix carbon dioxide (p. 40), and the principal enzyme involved is phosphoenolpyruvate carboxykinase (Fig. 13.15).

For a more detailed review of carbohydrate metabolism and related phenomena in acanthocephalans, see Saz (1969).

Protein Metabolism

As expected, proteins comprise a large proportion of the soma of acanthocephalans. For example, von Brand (1939) has reported 70% protein in dry weight analysis of *Macracanthorhynchus hirudinaceus*. Not only are these proteins found in cells, but some also occur in the pseudocoelomic (or perienteric) fluid (Monteoliva Hernández, 1964). In addition, specialized proteins, such as keratins, occur in the egg shell (Monné and Hönig, 1954).

The question of how acanthocephalans take up nitrogen-containing molecules has been of considerable interest. Electron microscope studies have shown that ultramicroscopic pores leading into canals occur in the cuticle and hypodermis of these worms (p. 447). There is little doubt that these serve as sites of entry of certain molecules; however, the entry of amino acids does not appear to occur by simple diffusion. For example, Rothman and Fisher (1964) have shown that in *Macracanthorhynchus hirudinaceus* and *Moniliformis dubius*, L-methionine is taken up against a concentration gradient. Furthermore, the uptake of this amino acid is competitively inhibited by L-isoleucine, L-leucine, L-serine, or L-alanine. These results suggest that, as in the case of cestodes (p. 435), there are specific loci, perhaps the ultramicroscopic pores, where amino acids are absorbed. It is also known that in *M. dubius* the uptake of L-leucine is competitively inhibited by DL-valine, DL-serine, or DL-methionine (Edmonds, 1965).

Relative to the uptake of nitrogen-containing compounds, it is noted that Uglem *et al.* (1973) have demonstrated the occurrence of intrinsic aminopeptidases in the outer tegumental membrane. Due to the oligopeptide-splitting ability of these enzymes, peptides such as leucylleucine and alanylanine exert a competitive inhibition on the active transport of leucine into *M. dubius*. In fact, the peptides cannot interact directly with the leucine-selective neutral amino acid locus, but such loci are situated very close

---

Moniliformis dubius, an intestinal parasite of rats, is most probably identical to *Moniliformis moniliformis*. Actually, *M. moniliformis* is the correct designation; however, *M. dubius* is more frequently employed in the physiologic and biochemical literature.

to the relevant aminopeptidase locations. This relationship means that leucine resulting from the activity of aminopeptidase is readily available for uptake by the transport locus and can compete with exogenous leucine. In fact, Uglem *et al.* have shown that over 90% of the leucine produced by the hydrolysis of leucylleucine at the parasite's body surface does not diffuse into the external medium but is absorbed by the worm.

Although aminopeptidase activity occurs on the surface of adult *M. dubius*, it does not occur on cystacanths removed from the hemocoel of the cockroach intermediate host. However, if the larvae are pretreated with a surface-active agent such as pancreatic lipase or sodium taurocholate, aminopeptidase activity becomes detectable. This is because of some unmasking effect in the surface membrane. It is believed that *in vivo*, agents like bile salts in the intestine of the rat host activate this enzyme.

Very little is known about the pathways by which amino acids are metabolized in acanthocephalans. From the limited information on *M. dubius*, it appears that a modified citric acid cycle may be involved (Graff, 1964; Bryant and Nicholas, 1965).

Lipid Metabolism

Our knowledge of lipid biochemistry as related to the Acanthocephala is limited to the quantitative and some qualitative analyses of lipids. Earlier studies by von Brand (1939, 1940) have revealed 1 to 2.1% lipids in fresh weight analysis, including 27% phosphatids, 2% saturated fatty acids, 32% unsaturated fatty acids, and 2% glycerol. More recent studies by Beames and Fisher (1964) have indicated that the percentages are slightly higher in *Moniliformis dubius*. Specifically, they have reported that in adult females and males, lipids comprise 4.2 and 7.2% of the wet weight, respectively. When subjected to further characterization, 71% of the lipids occur as glycerides, 10% as unsaponifiable lipids, and 19% as phospholipids in females, and 83% as glycerides, 8% as unsaponifiables, and 9% as phospholipids in males. No esterified volatile fatty acids occur even though acetic acid is the major fermentation acid produced.

Lipids apparently are not used as energy sources. Körting and Fairbairn (1972) reported that endogenous lipids are not metabolized during *in vitro* incubation of *M. dubius*. That this is the case is supported by the finding that the enzymes associated with β-oxidation of lipids are present either in very low concentrations or not at all.

An interesting aspect of parasite dependency on the host has been reported by Pflugfelder (1949) on *Acanthocephalus ranae* in frogs. He found that the worm does not produce lipases but absorbs the degradation products of fat digested by host lipases. In this case, the parasite, as far as its fat requirements are

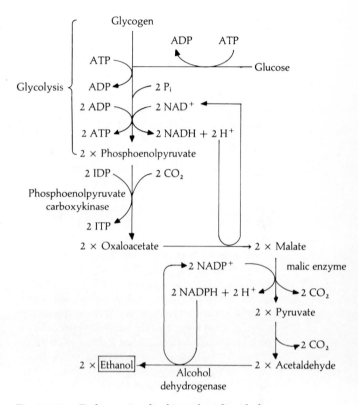

Fig. 13.15. **Pathways involved in carbon dioxide fixation in *Moniliformis dubius*.** Carbon dioxide fixation takes place into phosphoenolpyruvate to give oxaloacetate, which is reduced to malate, thus reoxidizing the glycolytic NADH. The oxidative decarboxylation of malate to pyruvate by the malic enzyme could then generate NADPH for the alcohol dehydrogenase step.

concerned, is dependent on the host for lipid hydrolysis.

Osmoregulation

Van Cleave and Ross (1945) have demonstrated that osmoregulation and tolerance for salinity in the surrounding medium occur among acanthocephalans, specifically in *Neoechinorhynchus emydis*. These worms are normally ventrally flattened *in vivo*, and they maintain this form when placed in a medium containing 0.8 to 0.85% sodium chloride. However, when placed in a 0.7 to 0.75% salt solution, they become turgid and survive less effectively. If placed in tap water, the animals become turgid in 1 hour and usually perish. Similar results can be obtained by injecting salt solutions of given concentrations into the intestines of turtles.

Also experimenting with *N. emydis*, Gettier (1942)

has obtained slightly different results. He has reported that these worms survived best in a solution of 0.5% sodium chloride and 0.22% calcium chloride. Potassium chloride and magnesium chloride were added without benefit. Apparently calcium chloride, even in a more dilute sodium chloride solution, enhances survival. Gettier has confirmed that worms placed in 2% sodium chloride solution died rapidly.

Resistance of Eggs

Manter (1928) has reported that eggs removed from *Macracanthorhynchus hirudinaceus* and *Mediorhynchus* sp. can be induced to hatch by drying and later rewetting them. However, eggs isolated from the host's feces cannot be hatched by this technique.

Eggs of acanthocephalans that utilize a terrestrial intermediate host are extremely resistant to normal environmental conditions. This has been demonstrated by Spindler and Kates (1940), who exposed eggs of *M. hirudinaceus* in soil mixtures, with either the inclusion or exclusion of pig feces, to desiccation by sunlight and found that the exposed eggs remained viable and hatched readily when fed to beetle grubs. Eggs of *M. hirudinaceus* can withstand temperatures from -10 to $45°C$ for as long as 140 days, and can resist desiccation at temperatures up to $39°C$ for as long as 265 days!

HOST-PARASITE RELATIONSHIPS

Adult acanthocephalans demonstrate moderate host specificity. A single species can employ more than one species of host—i.e., the same species may occur in several different families of fish—but a parasite of fish cannot live in birds, mammals, or other vertebrate hosts. In this respect, acanthocephalans resemble many other adult helminth parasites of the intestinal tract.

Effects of Host on Parasite

Burlingame and Chandler (1941) have reported that experimentally starved rats infected with *Moniliformis dubius* lose approximately 30% of the original number of parasites in 5 days; however, this loss is not due to the development of protective host immunity since reinfections are possible. It is now known that complete protective immunity to Acanthocephala generally does not occur. There is instead a "first come, first served" effect, in which worms comprising subsequent infections do not become attached and are passed through the alimentary tract because the favorable positions in the gut are occupied by worms from previous infections. Refractivity of the host to reinfec-

tions is thus a manifestation of parasite competition for nutrients and favorable microhabitats.

Invertebrate intermediate hosts—at least certain ones—respond to the presence of larval acanthocephalans. For example, DeGiusti (1949) has reported that when developing *Leptorhynchoides* is present in the amphipod crustacean *Hyalella*, host epithelial cells proliferate at the site of acanthor penetration in the gut wall. Once the parasite enters the host's hemocoel and encysts, the coelomic side of the developing cyst is covered by motile host blood cells. When the cyst wall ruptures, host cells respond as to a foreign body, and the larval acanthocephalan is enveloped by a layer of amoeboid cells that fuse to form a syncytium. In some cases, the syncytial envelope checks the growth of the parasite and eventually causes its death. If this occurs, the dead parasite is walled off by a brown sclerous membrane.

Similarly, walled larvae of *Macracanthorhynchus hirudinaceus* have been found in Japanese beetle larvae, which are suitable intermediate hosts. Temperature appears to influence the intermediate host's response to acanthocephalan larvae, since they are more often walled off at low than at high temperatures.

It is recalled that one form of immune reaction in arthropods is melanization (p. 69). Brennan and Cheng (1975) have shown that the successful development of the acanthocephalan *Moniliformis dubius* in the cockroach intermediate host, *Periplaneta americana*, depends on the parasite's ability to block the roach's phenoloxidase system, hence inhibiting melanization. Inhibition of the phenoloxidase system is believed to be directly associated with the polyanionic mucins occurring in the envelope surrounding the developing cystacanth. Specifically, these polyanionic mucins complex with the divalent cation Cu^{2+}, which is needed for activation of tyrosinase. As a result of this chemical competition, tyrosinase activity is reduced or completely eliminated, and melanization does not occur.

Effects of Parasite on Host

Exactly how many ways acanthocephalans can injure their definitive hosts remains in doubt. The main type of injury is mechanical, for the proboscal hooks and spines pierce and rupture the lining of the host's intestine, and in the case of *Acanthocephalus anguillae* in fish, even the underlying muscularis may be perforated. In hosts harboring large numbers of worms, or large-sized worms, the amount of nutrients absorbed by the parasites also may be of some consequence.

The number of worms required to kill a host varies. Webster (1943) has suggested that the presence of one or two *Prosthorhynchus formosus* in the intestine is sufficient to kill a robin, and Boyd (1951) has reported that the presence of one to seven acanthocephalans of the same species causes emaciation and blackening of the viscera of starlings. Clark *et al.* (1958) have reported the presence of a large number of *Polymorphus botulus* —up to 610 worms per bird—in dead or dying eider ducks along the New England coast. In contrast to these observations, there are numerous reports of the presence of exceedingly heavy acanthocephalan infections in fish and turtles with much vaguer pathogenicity.

Bullock (1963) has studied the histopathology of the postcaecal intestinal wall and rectum of two species of trout, *Salvelinus fontinalis* and *Salmo gairdneri*, infected with *Acanthocephalus jacksoni*. He reported that there is proliferation of the connective tissue of the lamina propria in the area where the proboscis of the worm is attached. The lamina propria is thickened as a result of the development of collagenous fibrous tissue and increased number of host cells, macrophages, fibroblasts, lymphocytes, polymorphonuclear leucocytes, and granular cells. The cells at the point of attachment are completely destroyed, and the cells in adjacent areas are compressed.

Although no capsule is formed around the parasite, a layer of mucus is interposed between the parasite and the host's epithelium. Bullock is of the opinion that the worm absorbs nutrients from this layer, which contains the end products of the host's digestion as well as secretions from the host's mucosa. On the other hand, Pflugfelder (1956) has reported that a capsule is formed in birds around *Polymorphus* adults, and Prakash and Adams (1960) have reported that a capsule is formed around *Echinorhynchus lageniformis* in the flounder.

Within the fish host, the younger specimens of *Acanthocephalus jacksoni* are more anteriorly attached in the intestine than are older specimens. As the worms grow, they migrate posteriorly, and as a result new lesions are formed each time the worms become reattached. Thus, a number of lesions can result from each worm during its migration. Although wound healing does occur, the lesions, nevertheless, are injurious, since they permit secondary microbial infections. This type of migration is also characteristic of *Acanthocephalus anguillae* infections in fish. Wurmbach (1937) has reported that although the proboscis of *Pomphorhynchus* invariably penetrates both the mucosa and the underlying muscle layer of the fish host, it is not as destructive as *A. anguillae*, for the latter migrates and causes more lesions.

There is little evidence that acanthocephalans secrete toxic substances. However, dogs infected with *Oncicola canis* exhibit symptoms of toxicity that simulate those of rabies. Again, humans infected with *Moniliformis moniliformis* (*M. dubius*) may portray symptoms of toxicity in the form of diarrhea and humming in the ears. The pseudocoelomic fluid of *Macracanthorhynchus hirudinaceus* is capable of slight hemolytic activity in pigs and cattle.

Le Roux (1931) has reported an interesting case of "castration" resulting from acanthocephalan infection. The worm *Polymorphus minutus* does not interfere with the normal functions of the testes of males of the intermediate host, *Gammarus pulex*; but in females, ovarian development is arrested, and the hosts revert to certain juvenile characteristics.

REFERENCES

Beames, C. G., and Fisher, F. M. (1964). A study of the neutral lipids and phospholipids of the acanthocephala *Macracanthorhynchus hirudinaceus* and *Moniliformis dubius*. *Comp. Biochem. Physiol.* **13**, 401–412.

Bird, A. F., and Bird, J. (1969). Skeletal structures and integument of Acanthocephala and Nematoda. *In* "Chemical Zoology" (M. Florkin and B. T. Scheer, eds.), Vol. III, pp. 253–288. Academic Press, New York.

Bogitsh, B. J. (1961). Histological and histochemical observations on the nature of the cyst of *Neoechinorhynchus cylindratus*. *Proc. Helminthol. Soc. Wash.* **28**, 75–81.

Boyd, E. M. (1951). A survey of parasitism of the starling *Sturnus vulgaris* L. in North America. *J. Parasitol.* **37**, 56–84.

Brennan, B. M., and Cheng, T. C. (1975). Resistance of *Moniliformis dubius* to the defense reactions of the American cockroach, *Periplaneta americana*. *J. Invert. Pathol.* **26**, 65–73.

Bryant, C., and Nicholas, W. L. (1965). Intermediary metabolism in *Moniliformis dubius* (Acanthocephala). *Comp. Biochem. Physiol.* **15**, 103–112.

Bullock, W. L. (1949). Histochemical studies on the Acanthocephala. I. The distribution of lipase and phosphatase. *J. Morphol.* **84**, 185–200. (Also see Part II, *ibid.* 201–226).

Bullock, W. L. (1963). Intestinal histology of some salmonid fishes with particular reference to the histopathology of acanthocephalan infections. *J. Morphol.* **112**, 23–44.

Burlingame, P. L., and Chandler, A. C. (1941). Host-parasite relations of *Moniliformis dubius* (Acanthocephala) in albino rats, and the environmental nature of resistance to single and superimposed infections with this parasite. *Am. J. Hyg.* **33**, 1–21.

Clark, G. M., O'Meara, D., and Van Weelden, J. W. (1958). An epizootic among eider ducks involving an acanthocephalid worm. *J. Wildlife Manage.* **22**, 204–205.

Crompton, D. W. T. (1970). "An Ecological Approach to Acanthocephalan Physiology." Cambridge University Press, London.

Crompton, D. W. T. (1983a). Acanthocephala. *In* "Reproductive Biology of Invertebrates. Vol. I. Oogenesis, oviposition, and oosorption." (K. G. Adiyodi and R. G. Adiyodi, eds.) pp. 257–268. John Wiley, New York.

Crompton, D. W. T. (1983b). Acanthocephala. *In* "Reproductive Biology of Invertebrates. Vol. II. Spermatogenesis and Sperm Function." (K. G. Adiyodi and R. G. Adiyodi, eds.) pp. 257–267. John Wiley, New York.

Crompton, D. W. T., and Lee, D. L. (1965). The fine structure of the body wall of *Polymorphus minutus* (Goeze, 1782) (Acanthocephala). *Parasitology* **53**, 357–364.

DeGiusti, D. L. (1939). Further studies on the life cycle of

Leptorhynchoides thecatus. J. Parasitol. **25** (Suppl.), 22.

DeGiusti, D. L. (1949). The life cycle of *Leptorhynchoides thecatus* (Linton), acanthocephalan of fish. *J. Parasitol.* **35**, 437–460.

Dunagan, T. T., and Miller, D. M. (1983). Apical sense organ of *Macracanthorhynchus hirudinaceus* (Acanthocephala). *J. Parasitol.* **69**, 897–902.

Dunagan, T. T., and Scheifinger, C. C. (1966). Studies on glycolytic enzymes of *Macracanthorhynchus hirudinaceus* (Acanthocephala). *J. Parasitol.* **52**, 730–734.

Edmonds, S. J. (1965). Some experiments on the nutrition of *Moniliformis dubius* Meyer (Acanthocephala). *Parasitology* **55**, 337–344.

Fairbairn, D. (1958). Trehalose and glucose in helminths and other invertebrates. *Can. J. Zool.* **36**, 787–795.

Fisher, F. M., Jr. (1964). Synthesis of trehalose in Acanthocephala. *J. Parasitol.* **50**, 803–804.

Gettier, D. A. (1942). Studies on the saline requirements of *Neoechinorhynchus emydis. Proc. Helminthol. Soc. Wash.* **9**, 75–78.

Graff, D. J. (1964). Metabolism of C^{14}-glucose by *Moniliformis dubius* (Acanthocephala). *J. Parasitol.* **50**, 230–234.

Korting, W., and Fairbairn, D. (1972). Anaerobic energy metabolism in *Moniliformis dubius* (Acanthocephala). *J. Parasitol.* **58**, 45–50.

Laurie, J. S. (1957). The *in vitro* fermentation of carbohydrates by two species of cestodes and one species of acanthocephala. *Exp. Parasitol.* **6**, 245–260.

Laurie, J. S. (1959). Aerobic metabolism of *Moniliformis dubius* (Acanthocephala). *Exp. Parasitol.* **8**, 188–197.

Lee, D. L. (1966). The structure and composition of the helminth cuticle. *Adv. Parasitol.* **4**, 187–254.

Le Roux, M. L. (1931). Castration parasitaire et caractères sexuels secondaires chez les gammariens. *C. R. Acad. Sci. (Paris)* **192**, 889–891.

Manter, H. W. (1928). Notes on the eggs and larvae of the thorny-headed worm of hogs. *Trans. Am. Microsc. Soc.* **47**, 342–347.

Meyer, A. (1928). Die Furchung nebst Einbildung, Reifung und Befruchtung des *Gigantorhynchus gigas.* Ein Beitrag zur Morphologie der Acanthocephalan. *Zool. Jahrb.* **50**, 117–218.

Meyer, A. (1932). Acanthocephala. *Bronn's Klass u. Ordnumg. TierReichs* **4**, 1–332.

Meyer, A. (1933). Acanthocephala. *Bronn's Klass u. Ordnumg. TierReichs* **4**, 333–582.

Meyer, A. (1936). Die plasmodiale Entwicklung und Formbildung des Riesenkratzers (*Macracanthorhynchus hirudinaceus*) (Pallas). I. Teil. Mit Forschungsstipendium der Notgemeinnschaft der Deutschen Wissenschaft 1930–1931. *Zool. Jahrb. Abt. Anat.* **62**. 111–172.

Meyer, A. (1937). Die plasmodiale Entwicklung und Formbildung des Riesenkratzers (*Macracanthorhynchus hirudinaceus*). II. Teil. *Zool. Jahrb. Abt. Anat.* **63**, 1–36.

Meyer, A. (1938a). Die plasmodiale Entwicklung und Formbildung des Riesenkratzers (*Macracanthorhynchus hirudinaceus*) (Pallas). III. Teil. *Zool. Jahrb. Abt. Anat.* **64**, 131–197.

Meyer, A. (1938b). Die plasmodiale Entwicklung und Formbildung des Riesenkratzers (*Macracanthorhynchus hirudinaceus*) (Pallas). IV. Allgemeiner Teil. *Zool. Jahrb. Abt. Anat.* **64**, 198–242.

Miller, D. M., and Dunagan, T. T. (1976). Body wall organization of the acanthocephalan, *Macracanthorhynchus hirudinaceus*: A reexamination of the lacunar system. *Proc. Helm. Soc. Wash.* **43**, 99–106.

Miller, D. M., and Dunagan, T. T. (1983). A support cell to the apical and lateral sensory organs in *Macracanthorhynchus hirudinaceus* (Acanthocephala). *J. Parasitol.* **69**, 534–538.

Monné, L., and Honig, G. (1954). On the embryonic envelopes of *Polymorphus botulus* and *P. minutus* (Acanthocephala). *Ark. Zool.* **7**, 257–260.

Monteoliva Hernándex, M. (1964). Estudio electroforetico comparado de tres parasitos intestinales: *Ascaris lumbricoides, Ascaridia galli* y *Macracanthorhynchus hirudinaceus. Rev. Iber. Parasitol.* **24**, 43–49.

Nicholas, W. L., and Grigg, H. (1965). The *in vitro* culture of *Moniliformis dubius* (Acanthocephala). *Exp. Parasitol.* **16**, 332–340.

Petrochenko, V. I. (1956). Acanthocephala of domestic and wild animals. Vol. 1. Akademii Nauk SSSR, Moscow. (English translation: Israel Program for Scientific Translations, 1971).

Petrochenko, V. I. (1958). Acanthocephala of domestic and wild animals. Vol. 2. Akademii Nauk SSSR, Moscow. (English translation: Israel Program for Scientific Translations, 1971).

Pflugfelder, O. (1949). Histologishche Utersuchungen über die Fettresorption darmloser Parasiten: Die Fuktion der Limnisken der Acanthocephalan. *Z. Parasitenk.* **14**, 274–280.

Pflugfelder, O. (1956). Abwehrreaktionen der Wirtstiere von *Polymorphus boschadis* Schr. (Acanthocephala). *Z. Parasitenk.* **17**, 371–382.

Prakash, A., and Adams, J. R. (1960). A histopathological study of the intestinal lesions induced by *Echinorhynchus lageniformis* (Acanthocephala-Echinorhynchidae) in the starry flounder. *Can. J. Zool.* **38**, 895–897.

Rothman, A. M., and Fisher, F. M. (1964). Permeation of amino acids in *Moniliformis* and *Macracanthorhynchus* (Acanthocephala). *J. Parasitol.* **50**, 410–414.

Saz, H. J. (1969). Carbohydrate and energy metabolism of nematodes and acanthocephalans. *In* "Chemical Zoology" (M. Florkin and B. T. Scheer, eds.), Vol. III, pp. 329–360. Academic Press, New York.

Schmidt, G. D. (1971). Acanthocephalan infections of man, with two new records. *J. Parasitol.* **57**, 582–584.

Specht, D., and Voge, M. (1965). Asexual multiplication of *Mesocestoides* tetrathyridia in laboratory animals. *J. Parasitol.* **51**, 268–272.

Spindler, L. A., and Kates, K. C. (1940). Survival on soil of eggs of the swine thornheaded worm, *Macracanthorhynchus hirudinaceus. J. Parasitol.* **26** (Suppl.), 19.

Uglem, G. L., Pappas, P. W., and Read, C. P. (1973). Surface aminopeptidase in *Moniliformis dubius* and its relation to amino acid uptake. *Parasitology* **67**, 185–195.

van Cleave, H. J., and Ross, E. L. (1945). Physiological responses of *Neoechinorhynchus emydis* (Acanthocephala) to various solutions. *J. Parasitol.* **30**, 369–372.

von Brand, T. (1939). Chemical and morphological observations upon the composition of *Macracanthorhynchus hirudinaceus* (Acanthocephala). *J. Parasitol.* **25**, 329–342.

von Brand, T. (1940). Further observations upon the composition of Acanthocephala. *J. Parasitol.* **26**, 301–307.

von Brand, T. (1972). "Biochemistry of Parasites," 2nd ed. Academic Press, New York.

von Brand, T., and Saurwein, J. (1942). Further studies upon the chemistry of *Macracanthorhynchus hirudinaceus. J. Parasitol.* **28**, 315–318.

Ward, H. L. (1940). Studies on the life history of *Neochinorhynchus cylindratus* (Van Cleave, 1913) (Acanthocephala). *Trans. Am. Microsc. Soc.* **59**, 289–291.

Ward, H. L. (1951). The species of Acanthocephala described since 1933. I. *J. Tenn. Acad. Sci.* **26**, 282–311.

Ward, H. L. (1952). The species of Acanthocephala described since 1933. II. *J. Tenn. Acad. Sci.* **27**, 131–149.

Webster, J. D. (1943). Helminths from the robin, with the description of a new nematode, *Porrocaecum brevispiculum. J. Parasitol.* **29**, 161–163.

Whitfield, P. J. (1968). A histological description of the uterine bell of *Polymorphus minutus* (Acanthocephala). *Parasitology* **58**, 671–682.

Whitfield, P. J. (1970). The egg sorting function of the uterine bell of *Polymorphus minutus* (Acanthocephala). *Parasitology* **61**, 111–126.

Wurmbach, H. (1937). Zur Krankheitserregenden Wirkung der Acanthocephalan. Die kratzterergrankung der Barben in der Mosel. *Z. Fisch. Hilfswiss.* **35**, 217–232.

Yamaguti, S. (1963). "Systema Helminthum. Vol. 5: Acanthocephala." Interscience, New York.

CLASSIFICATION OF ACANTHOCEPHALA (THE SPINY-HEADED WORMS)

CLASS ARCHIACANTHOCEPHALA

Main longitudinal lacunar canals dorsal and ventral or dorsal only; few hypodermal nuclei; giant nuclei in lemnisci and cement glands; two ligament sacs persist in females; protonephridia present in one family (Oligacanthorhynchidae); pyriform cement glands separate; eggs oval, usually thick shelled; parasites of birds and mammals; intermediate hosts are insects or myriapods.

ORDER MONILIFORMIDA

Trunk usually pseudosegmented; proboscis cylindrical, with long, approximately straight rows of hooks; sensory papillae present or absent; proboscis receptacle double walled, outer wall with muscle fibers arranged spirally; proboscis inverter muscles pierce posterior end of receptacle or somewhat ventrally; brain near posterior end or near middle of receptacle; protonephridial organs absent. (Genus mentioned in text: Family Moniliformidae*—*Moniliformis*.)

ORDER GIGANTORHYNCHIDA

Trunk occasionally pseudosegmented; proboscis as truncate cone, with approximately longitudinal rows of rooted hooks on the anterior portion, and rootless spines on the basal portion; sensory pits present on apex of proboscis and each side of neck; proboscis receptacle single walled with numerous accessory muscles, complex, thickest dorsally; proboscis inverter muscles pierce ventral wall of receptacle; brain near ventral, middle surface of receptacle; protonephridial organs absent. (Genera mentioned in text: Family

Gigantorhynchidae*—*Mediorhynchus, Gigantorhynchus, Heteracanthorhynchus*; Family Pachysentidae— *Oncicola*.)

ORDER OLIGACANTHORHYNCHIDA

Trunk may be wrinkled but not pseudosegmented; proboscis subspherical, with short, approximately longitudinal rows of few hooks each; sensory papillae present on apex of proboscis and each side of neck; proboscis receptacle single walled, complex, thickest dorsally; proboscis inverter muscle pierces dorsal wall of receptacle; brain near ventral, middle surface of receptacle; protonephridial organs present. (Genera mentioned in text: Family Oligacanthorhychidae*—*Oligacanthorhynchus, Macracanthorhynchus, Hamanniella*).

ORDER APORORHYNCHIDA

Trunk short, conical, may be curved ventrally; proboscis large, globular, with tiny spinelike hooks (which may not pierce the surface of the proboscis) arranged in several spiral rows; proboscis not retractable; neck absent or reduced; protonephridial organs absent.

CLASS PALAEACANTHOCEPHALA

Main longitudinal lacunar canals lateral; hypodermal nuclei fragmented, numerous, occasionally restricted to anterior half of trunk; nuclei of lemnisci and cement glands fragmented; spines present on trunk of some species; single ligament sac of female does not persist throughout life; protonephridia absent; cement glands separate, tubular to spheroid; eggs oval to elongate, sometimes with polar thickenings of second membrane; parasites of fish, amphibians, reptiles, birds, and mammals.

ORDER ECHINORHYNCHIDA

Trunk not pseudosegmented; proboscis cylindrical to spheroid, with longitudinal, regularly alternating rows of hooks; sensory papillae present or absent; proboscis receptacle double walled; proboscis inverter muscles pierce posterior end of receptacle; brain near middle or posterior end of receptacle; parasites of fish and amphibians. (Genera mentioned in text: Family Rhadinorhynchidae*—*Illiosentis, Rhadinorhynchus, Tegorhynchus, Pseudorhynchus, Leptorhynchoides.* Family Gorgorhynchidae—*Aspersentis, Corynosoma.* Family Pomphorhynchidae—*Pomphorhynchus*.)

ORDER POLYMORPHIDA

Proboscis spheroid to cylindrical, armed with numerous hooks in alternating longitudinal rows; proboscis receptacle double walled; with brain near center; parasite of reptiles, birds, and mammals. (Genera mentioned in text: Family Polymorphidae*—*Polymorphus, Bolbosoma, Prosthorhynchus.* Family Echinorhynchidae—*Echinorhynchus, Acanthocephalus*.)

CLASS EOACANTHOCEPHALA

Main longitudinal lacunar canals dorsal and ventral, often no larger in diameter than irregular transverse commis-

---

*For diagnoses of families subordinate to the Acanthocephala, see Yamaguti (1963).

sures; hypodermal nuclei few, large, sometimes amoeboid; proboscis receptacle single walled; proboscis inverter muscle pierces posterior end of receptacle; brain near anterior or middle of receptacle; nuclei of lemnisci few and large; two persistent ligament sacs in female; protonephridia absent; cement gland single, syncytial, with several nuclei, and with cement reservoir appended eggs variously shaped; parasites of fish, amphibians, and reptiles.

ORDER GYRACANTHOCEPHALIDA

Trunk small or medium sized, spined; proboscis small, spheroid, with a few spiral rows of hooks. (Genera mentioned in text: Family Quadrigyridae*— *Quadrigyrus, Raosentis.* Family Pallisentidae— *Pallisentis, Acanthosentis.*)

ORDER NEOECHINORHYNCHIDA

Trunk small to large, unarmed; proboscis spheroid to elongate, with hooks arranged variously. (Genera mentioned in text: Family Neoechinorhynchidae*— *Neoechinorhynchus, Octospinifer, Eosentis.* Family Tenuisentidae—*Tenuisentis.*)

14

NEMATA: THE ROUNDWORMS
ADENOPHOREAN NEMATODES

The relationship of the nematodes or roundworms to other organisms remains uncertain even after over 100 years of debate. Some consider these organisms to constitute an independent phylum, the Nemata (or Nematoda), whereas others include these worms, along with the Rotifera, Gastrotricha, Kinorhyncha, and Nematomorpha, in the phylum Aschelminthes. In another scheme, the nematodes, along with the Nematomorpha, are considered classes of the phylum Nemathelminthes. The most popular view at this time appears to consider the nematodes as representing a distinct phylum, and in following the nomenclature first proposed by the late N. A. Cobb in 1919 and reinstated by B. G. Chitwood in 1958, this phylum is being designated the Nemata.*

Parasitic nematodes are of great interest to biologists because they are plentiful, are frequently encountered as endoparasites, infect a large array of hosts, invertebrates as well as vertebrates, and in some cases are of considerable importance to human and animal health. Furthermore, certain soil-dwelling nematodes are of great interest in agriculture, for they often parasitize economically important plants, inflicting considerable injury and economic loss. Such is the case with the sugar beet nematode, *Heterodera schachtii*. In the majority of plant-parasitizing species, the eggs are deposited either in the roots of the plant host or in the soil. When these eggs hatch, the young worms are found within plant cells or soon become established therein by actively penetrating the roots. These larvae actively ingest plant tissues, often resulting in the formation of galls or "root knots." In some instances, the destruction of plant tissues is so severe that it results in the death of the plant. Horne (1961) has contributed an introductory volume to plant nematodes, and Zuckerman *et al.* (1971) have edited a two-volume treatise devoted to various aspects of the biology of plant parasitic nematodes.

Initially published between 1937 and 1941, the volume by B. G. Chitwood and M. B. Chitwood still remains a classic.* The textbook by A. Maggenti (1981) should also serve as an important reference

---

*Members of the phylum Nemata are commonly referred to as the nematodes although some researchers prefer "nemas" or "nematas."

*I had the privilege of arranging for the reprinting of *Introduction to Nematology* by University Park Press, Baltimore, Maryland, in 1974.

because this author has succeeded in bringing together information on the plant and animal parasitic species as well as the free-living ones.

CHEMICAL COMPOSITION OF NEMATODES

The chemical composition of the parasitic nematodes, especially those that occur in humans and domestic animals, has been investigated to some extent. As is the case with other parasites, the rationales for conducting such studies have centered on two objectives: identifying the molecules responsible for the antigenicity of these parasites, and cataloguing their constituent molecules to gain some insight into their metabolism.

Proteins

In *Ascaris lumbricoides*, a common intestinal parasite of humans and other mammals, proteins represent 48 to 57% of the dry weight; Flury (1912) has reported peptones, albumins, globulins, albumoses, and purine bases as the protein fractions isolated from bodies of this nematode. Similarly, Boudouy (1910) has isolated albumins, albumoses, purine bases, and mucin from *Strongylus equinus*, the double-toothed strongyle nematode found in the caecum of horses. Going one step further, a number of investigators (Pollak and Fairbairn, 1955; Watson and Silvester, 1958; Jaskoski, 1962; and others) have isolated and identified the bound amino acids found in *A. lumbricoides* (Table 14.1). It is suspected that most intestinal nematodes have comparable amino acids comprising their protein fractions.

Not all proteins in nematodes occur as such, since many are conjugated with other categories of molecules, especially carbohydrates. As an example, Kent (1963) has reported the isolation of five protein fractions from *Ascaris lumbricoides* of which only one is carbohydrate free—and it is only weakly antigenic. The four more strongly antigenic, contain from 14 to 76.5% carbohydrates.

The determination of protein concentrations of parasites, including nematodes, is commonly calculated by multiplying their total nitrogen content by 6.25 or one of the other conventional factors. This appears to be a fairly reliable method for nematodes, but not for cestodes in which an abnormally large amount of nonprotein nitrogen occurs.

Not only do proteins occur in the soma of nematodes, but a considerable amount is incorporated in the pseudocoelomic (or perienteric) fluid. This fluid contains albumin and globulins, which are highly anti-

Table 14.1. Amino Acids Identified from Various Tissues of *Ascaris lumbricoides* Adults[a]

Amino Acid	Body	Ovary, Free Acids	Ovary, Protein Acids	Perienteric Fluid	Larvae	Egg, Protein Coat	Egg, Middle Coat	Egg, Vitelline Membrane	Adult, Cuticle
Alanine	×	×	×	×	×				×
Glycine	×	×	×	×	×	×	×		×
Valine	×	×	×	×	×	×			×
Leucine	×	×	×	×	×	×		×	×
Isoleucine	×				×	×		×	×
Proline	×	×	×	×	×	×	×	×	×
Phenylalanine	×	×	×	×	×	×		×	×
Tyrosine	×	×	×	×	×	×	×	×	×
Serine	×			×	×	×	×		
Threonine		×	×	×	×	×			×
Cystine						×	×	×	×	×
Cysteine		×	×						
Methionine				×	×	×			×
Arginine	×			×	×		×	×	×
Lysine	×	×	×	×	×				×
Histidine	×	×	×	×	×				×
Aspartic acid	×	×	×	×	×	×	×	×	×
Glutamic acid	×	×	×	×	×	×			×
Hydroxyproline									×
Tryptophan		×	×	×	×	×		×	
Asparagine		×							
Glutamine		×		×					

[a] Data compiled by von Brand, 1966.

genic. In fact, one of the well-known occupational hazards of individuals working with nematodes is the severe allergic reaction to this fluid. Some parasitologists have had to abandon experiments with nematodes because of the severity of this reaction.

Among proteinaceous molecules, hemoglobin has been reported from some nematodes. In *Ascaris*, for example, two types of hemoglobins have been identified, one in the pseudocoelomic fluid, the other in the body wall. The possible function of these compounds is discussed later (p. 555).

As might be expected, proteins of nematodes can be found in the form of enzymes, neurosecretory hormones, and other metabolic molecules. For a detailed review of the protein composition of nematodes and other parasites, see von Brand (1973).

Carbohydrates

Carbohydrates in nematodes are found both as stored nutrients, primarily in the form of glycogen, and as simple sugars in the body fluids. Fauré-Fremiet (1913) has reported that glucose forms 0.15% of the fresh weight of the body fluid of *Parascaris*, and Rogers (1945) has reported 0.22% in *Ascaris*. The glycogen content has been studied in a number of nematodes. Table 14.2 lists the percentages of glycogen found in several representative species.

The location of glycogen in *Parascaris equorum* has been studied by Toryu (1933) (Table 14.3). He has reported that in male worms, 4.9% of the fresh weight and 96% of the total body glycogen occur in the body wall; in females, 5.8% of the fresh weight and 66% of the total body glycogen are in the body wall. In both males and females, 0.6% of the fresh weight and 9% of the total glycogen are found in the intestinal wall.

The male and female reproductive organs also contain glycogen. In ovarian tissues, glycogen is found primarily in oogonia. There is little or no glycogen in oocytes and little in mature eggs. The total glycogen in the ovaries and eggs of *Parascaris equorum* represents 6.5% of the fresh weight and 23% of the total body glycogen. In testicular tissues, the glycogen amounts to 0.5% of the fresh body weight and 2% of the total glycogen.

As a rule, parasites found in oxygen-poor habitats or in habitats where periodic deficiencies of oxygen exist, for example, the stomach and small intestine, contain more glycogen. This can be explained by the fact that carbohydrates are metabolized much more efficiently under anaerobic conditions by fermentation in energy production.

In addition to the sugars found in body fluids and stored glycogen, other carbohydrates occur in nematodes; for example, the polysaccharides in eggshells, and glucose and trehalose (Table 14.4).

Fats

Fats occur in the body of nematodes. In *Ascaris lumbricoides*, for example, lipids constitute 1.1 to 1.8% of the fresh weight and 10.9% of the dry weight of the body. In the larvae of *Eustrongylides ignotus*, von Brand (1938) has reported that lipids constitute 1.1% of the fresh weight and 4.4% of the dry weight. The fat content of hookworm larvae is considered a good indicator of the physiologic age of the specimens, for the degree of activity parallels the amount of fat content. In fact, hookworm larvae in which the fat content has been completely exhausted are no longer infective.

The most important sites of fat deposition in nematodes are the thickened portions of the body wall musculature. In addition, the subcuticula, intestinal cells, cells of the reproductive system, especially

Table 14.2. Glycogen Content in Nematodes and Its Relation to Habitat and the Availability to Oxygen[a]

Species	Glycogen		Habitat	Availability of Significant Amounts of Oxygen
	Fresh Weight (%)	Dry Weight (%)		
Strongylus vulgaris	3.5		Intestine	?
Ancylostoma caninum	1.6		Intestine	Yes
Ascaridia galli	3.6–4.7		Intestine	?
Parascaris equorum	2.1, 3.8	10, 23	Intestine	No
Ascaris lumbricoides	5.3–8.7	24	Intestine	No
Dirofilaria immitis	1.9	10	Heart	Yes
Litomosoides carinii	0.8	5	Pleural cavity	Yes
Dipetalonema gracilis	0.2		Abdominal cavity	Yes
Trichinella spiralis (larvae)	2.4		Muscle	Moderate
Eustrongylides ignotus (larvae)	6.9	28	Various organs	Yes

[a] Data from various authors.

Table 14.3. Glycogen Content and Distribution in *Parascaris equorum*[a,b]

Sex	Body Wall		Intestine		Uterus		Ovaries and Eggs		Male Reproductive System	
	A	B	A	B	A	B	A	B	A	B
Male	4.9	96	0.6	2	—	—	—	—	0.5	2
Female	5.8	66	0.6	2	1.6	9	0.5	23	—	—

[a] Data from Toryu (1933).
[b] A, percentage of fresh substance; B, percentage of total glycogen.

Table 14.4. Glucose and Trehalose Contents in Some Nematodes[a]

Species	Glucose (%) of Tissue Solids	Trehalose (%) of Tissue Solids
Ascaridia galli	0.78	0.38
Ascaris lumbricoides, perienteric fluid	0.07	4.0
Heterakis gallinae	0.43	0.10
Litomosoides carinii	0.01	0.06
Porrocaecum decipiens, larva	0.16	2.18
Trichinella spiralis, larva	0.04	1.76
Trichuris ovis	0.09	0.48
Uncinaria stenocephala	0.77	0.91

[a] Data from Fairbairn, 1958.

oocytes, oogonia, and ova, and nerve ganglia also contain lipids. Timm (1950) appears to be the first to report that the vitelline membrane of nematode eggs contains myricyl palmitate, which is lipid in nature.

The lipid fractions of nematodes have been determined by Schulz and Becker (1933) to be 6.6% phosphatids, 24.7 to 26% unsaponifiable matter, 30.9% saturated fatty acids, 34.1% unsaturated fatty acids, and 2.4 to 8.8% glycerol.

Remarkable differences exist among species in respect to the length of time fat reserves remain detectable. In embryonated eggs of *Ascaris*, only a slight reduction of fats occurs after 2 years of storage at room temperature, and the fat reserves are exhausted only after 4 years (Münnich, 1965), but in the eggs of *Ascaridia galli*, an intestinal parasite of chickens, most fats are depleted within 10 months (Elliott, 1954). A theory advanced by Engelbrecht and Palm (1964) states that the short-lived larval stages of helminths such as miracidia, cercariae, oncospheres, coracidia, and the larvae of the pinworm, *Enterobius vermicularis*, depend principally on stored glycogen as the energy source, but long-lived ones, such as ascarid larvae, store and utilize lipids primarily.

For a thorough review of the lipid composition of nematodes and other endoparasites, see von Brand (1973).

Inorganic Substances

In addition to proteins, carbohydrates, lipids, and combinations of these, certain inorganic substances are found in nematodes. Weinland (1901) and Flury (1912) have reported that inorganic substances comprise 0.70 to 0.78% of the fresh weight and 4 to 5.1% of the dry weight of *Ascaris lumbricoides*. Rogers (1945) has reported that in the body fluid of *A. lumbricoides* are found traces of potassium, magnesium, sodium, iron, copper, zinc, chlorine, and phosphorus. Among these, sodium is most abundant, followed by chlorine. Von Brand's (1973) monograph should be consulted by those interested in inorganic elements and compounds that occur in endoparasites.

MORPHOLOGY

Nematodes are generally elongate and cylindrical, tapering at both ends. Certain species, such as *Capillaria* spp., possess an almost uniformly cylindrical and extremely thin body. Some sexual dimorphism exists, since the curved posterior end of males has a copulatory organ and other special structures such as alae and papillae (p. 472). Males are usually smaller than females.

The various species of parasitic nematodes differ greatly in size. Some species are microscopic; others measure no more than 1 mm in length; still others, such as the guinea worm, *Dracunculus medinensis*, may reach more than a meter in length. Most species are opaque and whitish when examined in the living state, although some may have absorbed some of the pigments of the surrounding host tissues or fluids.

The structure of nematodes has been studied extensively, especially since the advent of the electron microscope. The volumes by Bird (1971) and Maggenti (1981) are recommended to those desiring comprehensive reviews.

THE CUTICLE

The body surface of parasitic nematodes is covered with an elastic **cuticle**. It is generally smooth; however, various structures such as spines, bristles, warts, punctations, papillae, striations, and ridges may be present (see Chitwood and Chitwood, 1950, and Maggenti, 1981, for reviews of these structures). These specialized cuticular structures may be classified as either ornamentation or sensory organs. Those classed as ornamentation should not be considered nonfunctional. In fact, they generally serve the functions of compensating for exogenous physical stress or aiding in locomotion. Sensory structures enhance the animal's awareness of the ambient environment. In some instances it is difficult to distinguish between ornamentation and sensory structures, since a specific structure often serves dual functions. The arrangements and positions of such structures are also of taxonomic importance. For example, species of *Gnathostoma* possess minute cuticular spines that cover practically the entire body, diminishing in number toward the posterior end; species of *Cooperia* bear longitudinally ribbed depressions and elevations

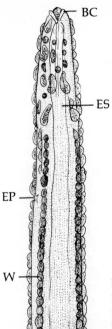

Fig. 14.1. *Gongylonema.* Anterior portion of an adult showing surface structures. BC, buccal capsule; EP, excretory pore; ES, esophagus; W, warts. (Redrawn after Ward, 1916.)

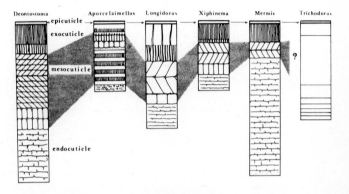

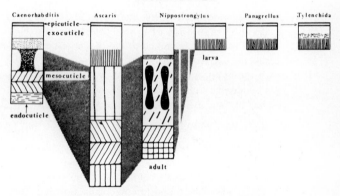

Fig. 14.2. Schematic comparison between nematode cuticular layers. Mesocuticle comparisons are shown by shaded connections. (After Maggenti, 1981; with permission of Springer-Verlag.)

along the length of the body; and species of *Gongylonema* possess wartlike thickenings that cover the body surface (Fig. 14.1).

Cuticular Layers

As the cuticle of nematodes plays an important role in their physiology, considerable attention has been devoted to its fine structure since the advent of the electron microscope. A review of the nematode cuticle has been presented by Lee (1966).

The cuticle covers the entire external surface and also lines the buccal cavity, esophagus (also known as the pharynx), rectum, cloaca, vagina, and excretory pore. When examined with the electron microscope, the occurrence of stratification can be appreciated. Although the four layers of the nematode cuticle described below may not be readily visible in every species, their occurrence is believed to be the rule rather than the exception. These layers are the **epicuticle**, **exocuticle**, **mesocuticle**, and **endocuticle**.

Epicuticle. This layer is a consistent component of all nematode cuticles (Figs. 14.2, 14.3). Typically it is trilaminate, consisting of two electron-dense layers separated by a less electron-dense layer. Chemically, it is similar to keratin in its sulfur content and resembles collagen in its X-ray diffraction pattern. It also contains quinones and polyphenols. In certain species, such as *Trichinella spiralis*, the outer electron-dense stratum of the epicuticle consists of three sublayers, the outermost of which is lipid.

Exocuticle. This layer (Figs. 14.2, 14.3) lies im-

mediately beneath the epicuticle. In most parasitic nematodes, the exocuticle can be divided into the **external exocuticle**, which has no visible substructure, and the radially striated **internal exocuticle**.

Mesocuticle. Among the cuticular layers of nematodes, the mesocuticle shows the greatest variation. It commonly consists of obliquely oriented collagenous, fibrous layers (Fig. 14.2). These vary in number and angular relationship to one another (Fig. 14.4). In *Nippostrongylus*, an intestinal parasite of rodents, the matrix in which the fibers are embedded is filled with fluid. In such genera as *Ascaris* and *Nippostrongylus*, it is the mesocuticle that increases in thickness with age. It is noted that flexibility of the nematode's cuticle is made possible primarily as a result of shifts of the angles of the parallelograms formed by the fibers of the mesocuticle (Fig. 14.4).

Endocuticle. This innermost layer of the cuticle (Fig. 14.2) is also fibrillar; however, the fibers are not well organized into an oblique latticework as in the mesocuticle. Often the pattern is disorganized and appears as overlapping fibers.

Frequent demonstrations of enzymes in the nematode cuticle, for example, the occurrence of esterases and polyphenol oxidase, indicate that the cuticle is not an inert body covering but a metabolically active one.

The cuticle is separated from the underlying tissues by a basal lamina.

The Hypodermis

Under the basal lamina is a thin layer of hypodermis. It may be cellular or syncytial. This layer projects into the pseudocoel along the middorsal, midventral, and lateral lines to form what are known as cords (Figs. 14.5, 14.6). The lateral cords are the largest and contain the excretory canals, if any are present. The nuclei of the hypodermis are found only in the cords. The hypodermal cells contain large amounts of glycogen and fats, and in some species hemoglobin occurs.

Body Wall Musculature

Mediad to the hypodermis is a relatively thick muscular layer. The myofibers associated with the nematode body wall are interesting. All of the muscle cells are spindle-shaped and longitudinally oriented. Each cell is divided into a contractile and a noncontractile portion (Fig. 14.7). The myofibrils are located in the contractile portion, whereas the noncontractile portion contains the nuclei, mitochondria, and stored glycogen granules and lipids. Innervation of these

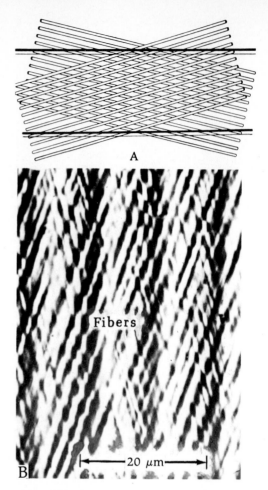

Fig. 14.4. Cuticular fibers. A. Schematic representation of the spiral basketwork of fibers showing the orientation of two of the mesocuticular fiber layers in relation to two of the transverse external annulations in *Ascaris*. (After Harris and Crofton, 1957.) **B.** Photomicrograph of the crossing fibers of the mesocuticle of *Ascaris lumbricoides.* × 1600. (After Bird, 1971.)

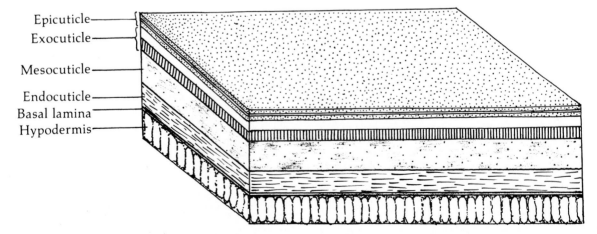

Fig. 14.3. Nematode cuticle. Schematic drawing showing layers of the cuticle.

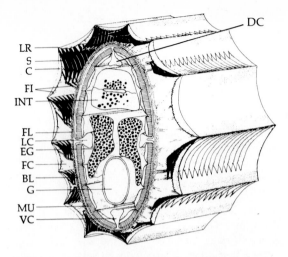

LR
S
C
FI
INT
FL
LC
EG
FC
BL
G
MU
VC
DC

Fig. 14.5. Nematode structure. A stereogram of a section of *Nippostrongylus muris* at midregion of body showing the structure of the cuticle. BL, basal lamella, C, cortex; DC, dorsal cord; EG, "excretory" gland; FC, fluid-filled matrix layer of cuticle; FL, fibrous layer of cuticle; FI, fibrils of collagen; G, gonad; INT, intestine; LC, lateral cord; LR, longitudinal ridge of cuticle; MU, body wall musculature; S, strut (or skeletal rod); VC, ventral cord. (After Lee, 1965.)

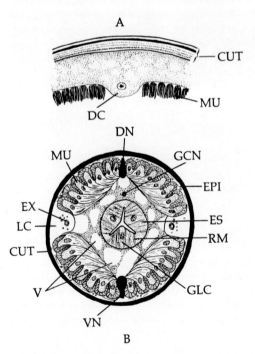

A

CUT
MU
DC

DN
MU
GCN
EPI
EX
LC
ES
RM
CUT
V
GLC
VN

B

Fig. 14.6. Nematode morphology. A. Cross-section through dorsal portion of the body wall of a nematode showing the position of the dorsal cord. **B.** Cross-section through level of the esophagus of *Ascaris* showing positions of cords and other structures. CUT, cuticle; DN, dorsal nerve (associated with dorsal cord); EPI, epidermis (or hypodermis); ES, esophagus; EX, excretory canal; GCN, giant cell nucleus; GLC, gland cell; LC, lateral cord; MU, muscle; RM, radial muscles; V, vacuoles; VN, ventral nerve (associated with ventral cord).

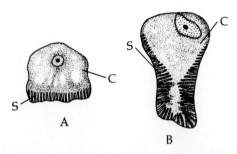

S
C
C
S
A
B

Fig. 14.7. Nematode muscles. A. Platymyarian type of muscle cell. **B.** Coelomyarian type of muscle cell. C, cytoplasmic (nonstriated) portion of cell, which is the noncontractile portion; S, striated (or banded) portion of cell, which is the contractile portion. (Both figures redrawn after Chitwood, 1931.)

muscles involves nerve processes from the noncontractile part of each cell passing to the longitudinal nerves or nerve ring (Fig. 14.8).

The body wall musculature is attached to the cuticle by fibers which originate from the contractile portion of each myofiber, pass through the basal lamina, and are attached in the endocuticle (Fig. 14.9).

Larval Cuticles

The cuticle of nematode larvae is less complex structurally. For example, according to Monné (1955), the cuticle of the larvae of *Dictyocaulus* from the lungs of cattle and *Metastrongylus* from the bronchi of various mammals are structureless, but tanned protein is present. The cuticle of these larvae is chemically distinct from that of *Ascaris* adults since it is not

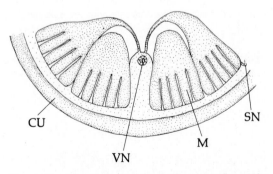

CU
VN
M
SN

Fig. 14.8. Nematode nerve-muscle connections. Schematic drawing showing cytoplasmic processes from body wall muscle cells passing to the ventral nerve situated in the ventral cord. CU, cuticle; M, muscle; SN, subventral nerve; VN, ventral nerve.

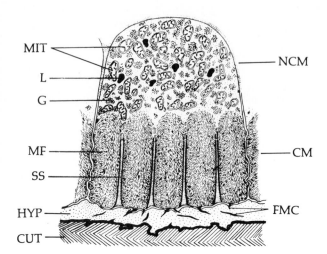

Fig. 14.9. **Attachment of muscle to cuticle.** Semidiagrammatic drawing of a transverse section through a muscle of the body wall of *Nippostrongylus muris* showing fibers attaching muscle to cuticle and other features of the muscle cell as seen with the electron microscope. CM, contractile part of muscle; CUT, cuticle; FMC, fibers attaching muscle to cuticle; G, glycogen; HYP, hypodermis; L, lipid; MF, myofilaments; MIT, mitochondria; NCM, noncontractile part of muscle. (After Lee, 1965.)

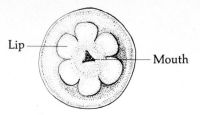

Fig. 14.10. **Nematode structure.** Face on view of a primitive marine nematode showing mouth surrounded by six lips.

affected by the plant enzyme papain, which readily digests the cuticle of *Ascaris* adults.

Some larval cuticles are slightly more complex. Lee (1965) has reported that the cuticle of the third-stage larva of *Nippostrongylus* consists of a thin, outer layer and a thick, inner fluid layer containing widely separated fibers. That of *Trichinella spiralis* larvae consists of two layers separated by a membrane. The inner layer includes an array of very fine fibrils about 40 Å thick. These are arranged circumferentially.

SPECIALIZED STRUCTURES OF THE BODY SURFACE

Lips. In primitive, free-living marine nematodes, the mouth is surrounded by six prominent lips and an array of sensory bristles and papillae (Fig. 14.10). However, in parasitic species, the number of lips usually varies from none to three. It is postulated that in the three-lipped species, the formation of three lips has resulted from fusion of the six lips found in the more primitive forms. If three lips are present, one is dorsal and two are ventrolateral.

Ballonets and Head Bulb. Various specialized structures and sensory organs may be found on certain species surrounding the perioral lips. For example, gnathostomes have four cuticular bulbs known as **bal-**

lonets, located immediately posterior to the lips. These four ballonets are adjacent to one another and form a swollen ring known as the **head bulb** (Fig. 14.11). This structure is armed with spines, the number and arrangement of which are of taxonomic value. In *Gnathostoma*, the head bulb is armed with relatively few circular rows of spines; in *Echinocephalus*, there are numerous rows of minute spines (Fig. 14.11); and in *Tanqua*, transverse striations subdivide the head bulb (Fig. 14.11).

Head Shields. Another type of specialized head organ is the **head shield**. These shields form a cuticular collar located posterior to the lips. The dorsal and ventral portions of the lips are recurved, shielding the lips (Fig. 14.12). Head shields are generally found in the superfamily Spiruroidea (see classification list at the end of Chapter 15).

Cordon. A third type of specialized head organ is the **cordon**. This is a longitudinal cuticular cordlike thickening that varies in number, depending on the particular species (Fig. 14.12).

Amphids. Although not as conspicuous in parasitic nematodes as they are in the free-living species, especially marine species, amphids are present in reduced form on many parasites. These special structures, which are chemoreceptors, appear as a pair of depressions on each side of the cephalic end, lateral or posterolateral to the other specialized head structures. The amphids are richly innervated and are often supplied with a gland (Fig. 14.13).

Phasmids. Another type of sensory organ is the **phasmid**. These minute structures, which function as chemoreceptors, usually occur in pairs, the openings of which are at the terminal minute papillae behind the anus. Again, glands are usually associated with these organs (Fig. 14.13). The presence or absence of phasmids determines whether a particular nematode belongs to the class Secernentea or Adenophorea.

Alae. In addition to the specialized cuticular structures already mentioned, **alae** are present on certain species. These surface structures appear as compressed, longitudinally oriented ridges. Those situated at the anterior margin of the body are known as **cervical alae** (Fig. 14.14), those located toward the posterior terminal, normally in males, are known as caudal alae (Fig. 14.14), and those extending almost the entire length of the body are known as **lateral** or **longitudinal alae**. Alae vary in number from one to four, commonly two, and are primarily cuticular extensions.

Copulatory Bursae. Copulatory bursae are found at the posterior terminal of certain male nematodes, members of the superfamilies Strongyloidea and Dioctophymoidea. These specialized flaplike extensions are commonly supported by riblike muscular rays (Fig. 14.4). The function of the bursa is to enable the male to grasp and envelop the female during copulation. The bursae of strongyloids are strictly cuticular structures, whereas those of dioctophymoids are formed from the flattening of the entire posterior portion of the body.

ALIMENTARY TRACT

The alimentary tract of nematodes is complete. The mouth is located at the anterior tip of the body, and the anus is subterminal.

Buccal Capsule. The mouth opens into the buccal capsule, which is a strongly cuticularized chamber lined, in a number of species, with ridges, rods, and plates, all serving to maintain the shape of the chamber. Buccal capsules are not present in all nematodes. The filarioids (p. 536), for example, possess almost nonexistent cavities. In rhabditoids (p. 501), the buccal capsule is divided into three sections—the anterior vestibule, the **cheilostome**; the middle **protostome**, which is the longest; and the small posterior **telostome**. The buccal capsule is armed with spears, stylets, or teeth in several species. For example, in the Old World hookworm, *Ancylostoma*

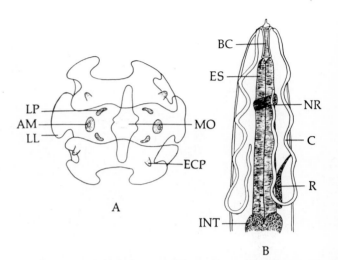

Fig. 14.12. Nematode structures. A. Head-on view of head of *Parabronema*, an intestinal parasite of elephants, showing head shields and cordons. (Redrawn after Baylis, 1921.) **B.** Anterior end of *Dispharynx*, a parasite of birds, showing recurved cordons. (Redrawn after Seurat, 1916.) AM, amphid; BC, ballonet cavity; C, cordon; ECP, external circlet of papillae; INT, intestine; LL, lateral lips; LP, labial papilla; ES, esophagus; MO, mouth; NR, nerve ring; R, renette.

duodenale, there are two laterally situated sets of three teeth, and in the New World hookworm, *Necator americanus*, there are two lateral cutting plates. Because of the presence of these teeth and plates, these worms are known as hookworms.

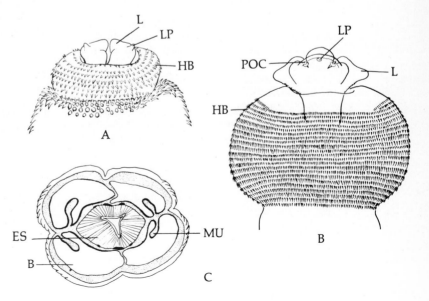

Fig. 14.11. Nematode head bulbs. A. Anterior end of *Gnathostoma* showing head bulb. **B.** Head of *Echinocephalus* showing head bulb. **C.** Cross-section through head bulb of *Tanqua* showing ballonets. B, ballonet cavity; ES, esophagus; HB, head bulb; L, lip; LP, labial papilla; MU, muscle; POC, papilla of outer circlet. (All drawings redrawn after Baylis and Lane, 1920.)

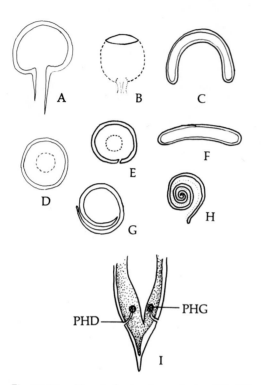

Fig. 14.13. **Nematode structures. A.** Amphid of *Plectus*; **B.** Cyathiform amphid; **C.** Chromadoroidean amphid; **D.** Circular amphid; **E.** Variant of circular amphid; **F.** Chromadoroidean amphid; **G.** Variant of spiral amphid. **H.** Spiral amphid. **I.** Posterior terminal of *Rhabditis* showing phasmids. (**A–H,** redrawn after Stekhoven *et al.,* 1933; **I,** redrawn after Chitwood, 1930.) PHD, duct of phasmid; PHG, gland of phasmid.

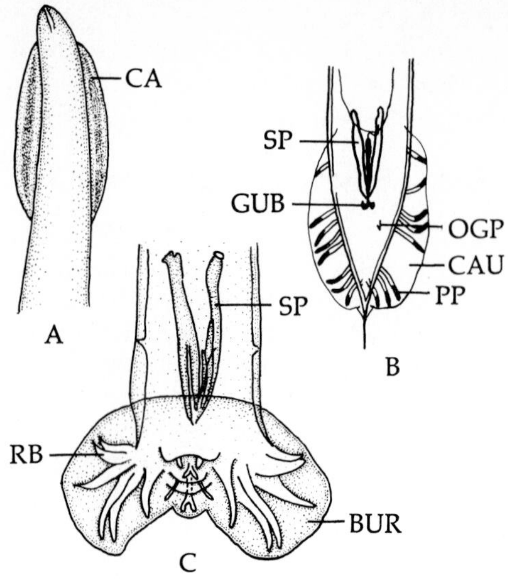

Esophagus.* The buccal capsule leads into the esophagus, which is an elongate structure enveloped by a membranous wall. The lumen of the esophagus is characteristically triradiate in cross section (Fig. 14.15). The esophagus is comprised of syncytially arranged cells.

Within the membranous sheath of the esophagus is a thick muscular layer surrounding the lumen. Three branched esophageal glands are embedded between the muscle fibers—one dorsal and two ventrolateral. Secretions from these glands empty into the esophagus, and pass from there into the buccal capsule.

The esophagus may possess a swollen bulblike

---

*The esophagus of nematodes is sometimes referred to as the pharynx; however, the former term is more commonly used among nematologists.

Fig. 14.14. **Nematode structures. A.** Anterior end of a nematode with cervical alae. **B.** Posterior end of *Rhabditis maupasi* with caudal alae and pedunculated papillae. (Redrawn after Stekhoven and Tennissen, 1938.) **C.** Posterior end of male hookworm showing copulatory bursa. BUR, copulatory bursa; CA, cervical alae; CAU, caudal alae; GUB, gurbernaculum; OGP, ordinary genital papillae; PP, pedunculated papillae; RB, muscular rays of bursa; SP, spicules.

section, which if situated at the posterior terminal of the esophagus is referred to as an **end bulb;*** when located along the midlength of the esophagus, it is known as a **median bulb**. Distinction is made between a true bulb and a pseudobulb. The bulbous chamber of a true bulb is separated from the lumen of the esophagus by sclerotized valves that regulate the aperture of the opening into the bulb. No such valves exist between a pseudobulb and the esophagus proper. In members of the superfamilies Filarioidea and Spiruroidea, the esophagus is muscular only anteriorly, being glandular posteriorly. Among certain ascaroids, such as *Phocanema* (Fig. 14.16), a diverticulum known as the **intestinal caecum** arises from the junction of the esophagus and intestine. Also, in *Contracaecum* and related genera, another diverticulum, known as the **ventricular appendix**, arises from the posterior terminal of the esophagus (Fig. 14.16).

Intestine. The esophagus empties into the intestine through a junction referred to as the **esophago-intestinal valve**. The intestine of nematodes is a straight tube lined with a single layer of cuboidal or columnar epithelium. Electron microscope studies have revealed the occurrence of microvilli projecting from the lining cells. The cells rest on a basal lamina of

---

*The terms **cardiac bulb** and **posterior bulb** also refer to the end bulb.

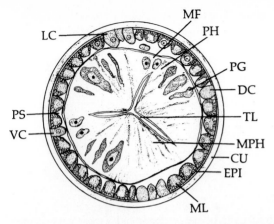

Fig. 14.15. **Nematode esophagus.** Cross-section through level of esophagus showing triradiate lumen and other structures. CU, cuticle; DC, dorsal cord; EPI, epidermis; ES, esophagus; LC, lateral cord; MF, marginal fibers; ML, muscle layer; MPH, muscle fibers of esophagus (pharynx); PG, sections of esophageal glands; PS, pseudocoel; TL, triradiate lumen; VC, ventral cord. (Redrawn after Chitwood, 1931.)

connective tissue fibers, myofibers, or a simple membrane. The epithelial cells are typically uninucleate; however, those found in some strongyloids are multinucleate.

The intestine is divided into the anterior **ventricular region**, the **midregion**, and the posterior **prerectal region**. These regions differ from one another in the shape of the lumen, in the dimensions and contents of the cells, and in their functions. For example, the ventricular region is mainly secretory, whereas the mid- and prerectal regions are absorptive.

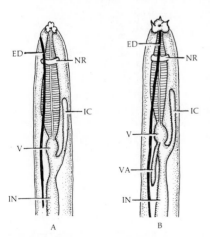

Fig. 14.16. **Anterior ends of some nematode digestive tracts. A.** Anterior end of *Phocanema* sp. showing intestinal caecum. **B.** Anterior end of *Contracaecum* sp. showing intestinal caecum and ventricular appendage. ED, excretory duct; IC, intestinal caecum; IN, intestine; NR, nerve ring; V, ventriculus.

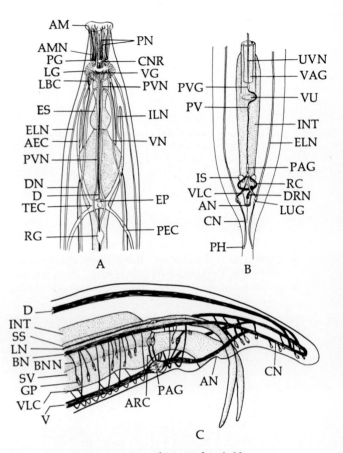

Fig. 14.17. **Nervous system of nematodes. A.** Nervous system in anterior portion of *Cephalobellus*. **B.** Nervous system in posterior portion of same species. (**A** and **B**, redrawn after Chitwood and Chitwood, 1933.) **C.** Nervous system in posterior region of *Ascaris*. (Redrawn after Hesse, 1892.) AEC, anterior excretory canal; AM, amphid; AMN, amphideal nerve; AN, anus; ARC, anorectal connectives; BBN, branch of bursal nerve; BN, bursal nerve; CN, caudal nerve; CNR, circumenteric ring; D, dorsal nerve; DN, dorsal lateral nerve; DRN, dorsorectal nerve; ELN, external lateral nerve; EP, excretory pore; ES, esophagus; GP, genital papilla; ILN, internal lateral nerve; INT, intestine; IS, intestinorectal sphincter; LBC, lateroventral brain connective; LG, lateral ganglion; LN, lateral nerve; LUG, lumbar ganglia; PAG, preanal ganglion; PEC, posterior excretory canal; PG, papillary ganglion; PH, phasmid; PN, papillary nerves; PV, postvulvar ganglia; PVG, prevulvar ganglion; PUN, paired ventral nerves; RC, rectal commissure; RG, retrovesicular ganglion; SS, spicule sheath; SV, seminal vesicle; TEC, transverse excretory canal.

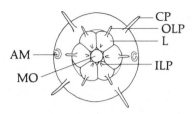

Fig. 14.18. Labial and cephalic papillae. Face on view of a nematode showing the positions of the mouth, lips, amphids, and papillae. AM, amphid; CP, cervical papilla; ILP, inner labial papilla; L, lip; MO, mouth; OLP, outer labial papilla. (Redrawn after Jones, 1959.)

Rectum.   The intestine empties into the rectum, a short, flattened tube joining the intestine and the anus. Most parasitic nematodes possess a number of unicellular **rectal glands**—three in females, six in males—which empty into the rectum. The rectum is lined with a thin layer of cuticle enveloped by a layer of large epithelial cells that is covered by muscle cells.

Anus.   The alimentary tract opens posteriorly through the anus, located subterminally on the ventral body surface. In females of the guinea worm, *Dracunculus medinensis*, and in *Mermis*, the anus is lacking.

NERVOUS SYSTEM

The nervous system consists of a circumenteric ring, composed of nerve fibers and a few ganglia, that surrounds the esophagus. Connected to this ring are a number of ganglia that are also connected to each other by commissures. Six nerves extend anteriorly from the circumenteric ring—two ventrolateral, two lateral, and two dorsolateral. These six fibers and their branches innervate the various organs and tissues in the anterior portion of the body. A middorsal nerve, a midventral nerve, and one to three pairs of lateral nerves are directed posteriorly to innervate the

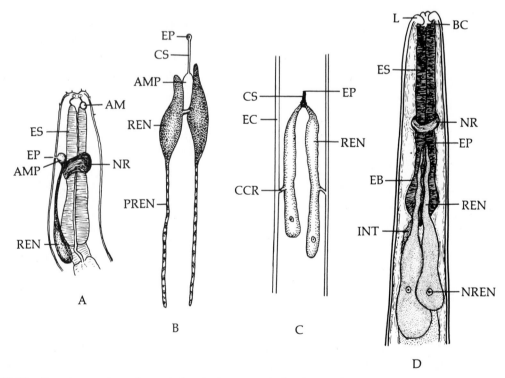

Fig. 14.19.   Excretory system of nematodes. A. Anterior end of *Linhomeus* showing excretory system. (Redrawn after Kreis, 1929.) **B.** Excretory system in larva of *Ancylostoma* with posterior canals developing from renette cells. (Redrawn after Stekhoven, 1927.) **C.** Excretory system of *Oesophagostomum* showing lateral connections of renettes to excretory canals. (Redrawn after Chitwood, 1931.) **D.** Excretory system of *Rhabdias* showing two-celled renette. (Redrawn after Stekhoven, 1927.) AM, amphid; AMP, contractile ampulla; BC, buccal capsule; CCR, connection between renette and excretory canal; CS, common stem; EB, esophageal bulb; EC, excretory canal; EP, excretory pore; ES, esophagus; INT, intestine; L, lips; NR, nerve ring; NREN, nucleus of renette cell; PREN, posterior extension of renette cell; REN, renette cell.

various organs and tissues posterior to the nerve ring. In *Ascaris* and some other genera, a number of commissures join the ventral nerve with the lateral nerves (Fig. 14.17). Many of the branches of the main anteriorly and posteriorly directed nerves terminate as free endings and are sensory in function. For a more detailed discussion of the neuroanatomy of nematodes and their behavior, see Croll and Matthews (1977).

Only recently have attempts been made to study the neurophysiology of nematodes. In *Ascaris*, it is known that the pseudocoelomic fluid contains more sodium (129 mM/liter) than potassium ions (25 mM/liter). Thus, it appears that the nerve fibers function as in other animals, i.e., the action current arises from an influx of external sodium ions, and the resting potential depends on the excess of potassium ions in the axoplasm relative to the external medium.

As is the case in most animals, acetylcholine has been demonstrated in several species. Mellanby (1955) has reported that the anterior portion of *Ascaris*, where the circumenteric ring occurs, contains 15 times more acetylcholinelike activity than strips from the body wall.

Ascaris includes both inhibitory and motor nerves. This is evidenced by the fact that electrical stimulation in the region of the circumenteric ring at frequencies about five per second causes inhibition of both spontaneous and electrically stimulated contractions of the body. Inhibition also occurs in response to shocks shorter than 0.1 msec, indicating that the effect is mediated by nerves. Severing the ventral nerve cord abolishes 80% of the inhibitory effect on body contractions caused by stimuli to the region of the circumenteric ring, and cutting the dorsal nerve cord reduces inhibition even more (Goodwin and Vaughan Williams, 1963).

The effects of a number of drugs on the nervous systems of nematodes have been studied, but nothing unexpected has been observed. The interested reader is referred to Lee (1965) and Jarman (1976) for further information on the neuromuscular physiology of nematodes.

Sense Organs. Nematodes possess mechanoreceptors, chemoreceptors, and photoreceptors. Additional types may also occur. The labial and cephalic papillae situated in the head region (Fig. 14.18) are mechanoreceptors. These are innervated by branches from the papillary nerves, which are known to be cholinergic.

The amphids and phasmids of nematodes are chemoreceptors. Although these organs are receptors, they are also secretory in function. For an authoritative review of nematode sense organs and their secretions, see McLaren (1976). While photoreceptors have been demonstrated only in free-living nematodes, they probably also occur in some parasitic species, especially in the larval, free-living stages.

EXCRETORY SYSTEM

The excretory system of nematodes, when present, is unique. It is not like the protonephritic system of platyhelminths. Each worm is provided with either one or two renettes. A **renette** is a large gland cell that empties to the outside through the excretory pore generally located anteriorly at the level of the circumenteric nerve ring, or slightly anterior or posterior to that level (Fig. 14.19). In some genera, such as *Ancylostoma* and *Oesophagostomum*, there is a tubular extension from the renette. In *Ancylostoma* this extension arises directly from the renette (Fig. 14.19); while in *Oesophagostomum* the canal is lateral to the renette, embedded in the lateral line of the body wall. In the latter case, the canal is joined to the renette by a short duct (Fig. 14.19). Renettes may empty independently to the exterior through the excretory pore, as in *Rhabdias* (Fig. 14.19), or the two renettes may join to form an H. The crossbar of the H is connected with a common contractile ampulla, which in turn expels the excreta by a pulsatory motion through the excretory pore via a common stem. This pattern is found in *Ancylostoma* (Fig. 14.19). The two renettes may also join anteriorly, in which case the excreta is emptied into a common stem leading into the excretory pore. This pattern occurs in *Oesophagostomum* (Fig. 14.19).

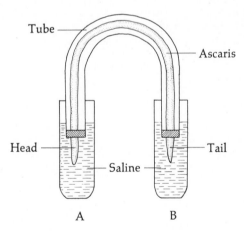

Fig. 14.20. Apparatus for studying excretion in *Ascaris.* The worm is sealed in a U-tube containing moist air. Its head and tail project from the tube and are immersed in saline in containers A and B. Most of the nitrogenous excretory products are detected in saline in container B and little in the saline in container A in which is bathed the cephalic end, including the excretory pore. In this situation *Ascaris* produces more urea and less ammonia than free-swimming individuals. (Redrawn after Savel, 1955.)

The renette and associated ducts serve as absorptive bodies that collect wastes from in the pseudocoel. An excretory system is not present in all parasitic nematodes; for example, members of the superfamilies Trichuroidea and Dioctophymoidea lack an excretory system in their adult stage, but a system similar to that found in *Oesophagostomum* is present in the larvae.

In species without a structured excretory system, metabolic wastes are excreted through the cuticle and through the anus. The major nitrogenous waste product is ammonia. When expelled via the renette system, cuticle, or anus, the ammonia is rapidly diluted to nontoxic levels. Studies on *Ascaris* have shown that when placed in saline, this worm excretes 60% of its total nitrogen as ammonia and only 7% as urea. However, if *Ascaris* is suspended in moist air in a U-tube, with only its head and tail ends in saline (Fig. 14.20), the amount of ammonia excreted falls to 27% of the total nitrogen excreted and urea increases to 52% (Table 14.5). This indicates that when placed under osmotic stress, i.e., when the volume of fluid taken in through the cuticle is lessened, leading to an accumulation of ammonia in the tissues, *Ascaris* switches from ammonia excretion to the production of urea (Weinstein, 1960). This interesting mechanism is of selective advantage to the nematode since, as the result of the switch, there is less danger of the toxic ammonia being accumulated in the vicinity of the worm; rather, the much less toxic urea is excreted.

PSEUDOCOEL

The body cavity of nematodes is known as the pseudocoel because embryologically it does not develop like a true coelom and is not lined with a layer of mesodermal cells. In living nematodes the pseudocoel is filled with a fluid containing mobile phagocytic cells. These cells are derived from certain cells lining the digestive tract.

REPRODUCTIVE SYSTEMS

Nematodes are, as a rule, dioecious animals. The males generally can be distinguished externally from the females by their smaller size, curved posterior ends, and the presence of bursae and other accessory reproductive structures. Although protandric hermaphroditism is found in certain free-living nematodes, true intersexual fertilization is the rule among the parasitic species. However, monoecious species are known that are self-fertilizing hermaphrodites.

Male System. In males, there is generally a single testis; diorchic forms, however, are not uncommon. The testis is tubular and usually convoluted and/or recurved (Fig. 14.21). Two types of testes occur in nematodes—the **telogonic** type, in which the pro-

Table 14.5. Nitrogenous Products Excreted by Some Fasting Nematodes *in Vitro*[a]

Nematode	Total Nitrogen Excreted (mg/g wet wt/24 hr)	Type of Nitrogen Expressed as % of Total					
		Ammonia	*Urea*	*Peptide*	*Amino Acid*	*Amine*	*Uric Acid*
Caenorhabditis (Free-living)	—	+	0	—	+	T	0
Ascaris							
Aerobic	0.4	69	7	21	+	—	0
Anaerobic	0.4	71	6	18	+	—	0
In U-tube	0.4	27	52	19	—	—	0
Larvae						+	—
Nematodirus							
Aerobic	1.36	42	14	35	+	—	3
Anaerobic	1.72	29	4	35	+	—	4
Ascaridia							
Aerobic	0.35	56	12	15	+	—	0
Anaerobic	0.37	59	15	15	+	—	0
Trichinella (larva)							
Aerobic	2.8	33	0	21	28	7	0
Nippostrongylus (3rd-stage larva)							
Aerobic	—	+	0	—	—	+	—

[a] Data after Weinstein, 1955.

[b] +, present; 0, absent; —, no information; T, trace amount.

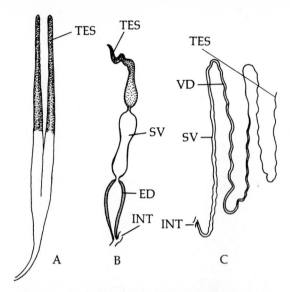

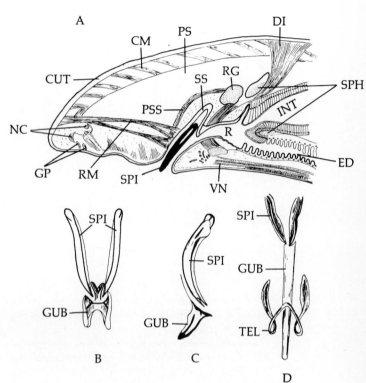

Fig. 14.21. Nematode reproductive system. A. Male reproductive system of *Heterodera marioni* with two parallel testes (diorchic). (Redrawn after Atkinson, 1889.) **B.** Male reproductive system of *Camallanus* with one anterior testis (monorchic). (Redrawn after Törnquist, 1931.) **C.** Male reproductive system of *Ascaris* with single long coiled testis and duct. ED, ejaculatory duct; INT, intestine; SV, seminal vesicle; TES, testis; VD, vas deferens.

Fig. 14.22. Nematode male reproductive system. A. Sagittal section through posterior end of *Ascaris* showing relationship of spicules to rectum. (Redrawn after Voltzenlogel, 1902.) **B.** Spicules and gubernaculum of a chromadoroid nematode. **C.** Side view of same. (**B** and **C**, redrawn after de Man, 1907.) **D.** Male reproductive armature of a strongyloid nematode. (Redrawn after Hall, 1921.) CM, copulatory muscles; CUT, cuticle; DI, dilator of intestine; ED, ejaculatory duct; GP, genital papillae; GUB, gubernaculum; INT, intestine; NC, nerve cells; PS, pseudocoel; PSS, protractor of spicule sheath; R, rectum; RG, rectal gland; RM, retractor muscle; SPH, sphincters; SPI, spicule; SS, spicule sheath; TEL, telamon; VN, ventral nerve.

liferation of germ cells occurs only at the blind end of the elongate testis; and the **hologonic** type, found in members of the superfamilies Dioctophymoidea and Trichuroidea, and in which germ cell proliferation occurs along the entire length of the testis.

Ascaris lumbricoides, the common intestinal nematode of pigs and humans, is a very popular animal in cytologic studies because the various stages of gametogenesis are easily found in sequence along the telogonic testis. The slender **vas deferens** (sperm duct) is continuous with the proximal end of the testis. The vas deferens is directed posteriorly, enlarging terminally to form the muscular **seminal vesicle**. This vesicle leads into the rectum through a muscular **ejaculatory duct** (Fig. 14.21). In certain species, unicellular **prostate glands** occur along the length of the ejaculatory duct, and these glands empty into the duct.

Male nematodes are commonly armed with one or two **copulatory spicules**, the latter being more common. These cuticular structures, which most commonly resemble slightly curved, pointed blades, are inserted within their respective spicule pouches located as side pockets in the rectal (commonly referred to as cloacal) wall. Each spicule is comprised of a cytoplasmic core and is formed by the cells lining the spicule pouch.

In addition to spicules, a sclerotized **gubernaculum** is present in some species. This structure is located along the dorsal wall of the spicule pouch and commonly has incurved margins (Fig. 14.22). The gubernaculum serves as a guide along which the spicules can follow when they are extended. In some members of the superfamily Strongyloidea, another structure, the **telamon**, occurs. It is a partially sclerotized double-bent structure. Each of its two bent arms subtend the lateral walls of the cloaca, and the medial section subtends the ventral wall of the cloaca (Fig. 14.22). The telamon also serves as a guide for the copulatory spicules when they are extended. Movement of the spicules is controlled by antagonistic muscles, i.e., extensors and protractors. The shapes, sizes, and number of copulatory spicules and their auxiliary structures

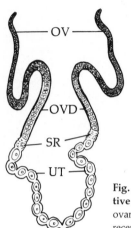

Fig. 14.23. Female reproductive system of *Heterodera*. OV, ovary; OVD, oviduct; SR, seminal receptacle; UT, uterus; VA, vagina. (Redrawn after Nagakura, 1930.)

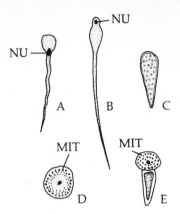

Fig. 14.24. Nematode spermatozoa. A. Sperm of *Passalurus*. (Redrawn after Meves, 1920.) **B.** Sperm of *Trilobus*. (Redrawn after Chitwood, 1931.) **C.** Sperm of *Thoracostoma*. (Redrawn after de Man, 1888.) **D.** Sperm of *Anaplostoma*. (Redrawn after de Man, 1907.) **E.** Sperm of *Parascaris equorum*. (Redrawn after Hyman, 1951.) MIT, mitochondria; NU, nucleus.

are of taxonomic importance. Copulatory spicules are lacking in some nematodes such as the trichina worm, *Trichinella spiralis*.

Female System. Female nematodes usually possess two ovaries (didelphic), one extending anteriorly, the other posteriorly (Fig. 14.23). When only one ovary is present (monodelphic), it extends either anteriorly or posteriorly, depending on the species. The ovaries appear as straight, convoluted, or much folded tubes.

An **oviduct**, lined with columnar epithelium, is continuous with the proximal end of each ovary. At the proximal terminal of each oviduct is a slightly swollen chamber, the **seminal receptacle**, which connects the oviduct to the tubular **uterus**. The two uteri, one associated with each ovary, join in the area of the female **gonopore** to form the **vagina**, which in turn opens to the exterior through the gonopore.

The gonopore is generally located midventrally in the middle third of the body. However, it may be more anteriorly or posteriorly located, depending on the species. In certain members of the superfamilies Strongyloidea and Spiruroidea, the vagina is highly muscular and ejects eggs through muscular contraction; such a vagina is called an **ovijector**.

Fertilization. In parasitic nematodes, copulation between a female and a male is generally necessary for fertilization, except in the genus *Strongyloides* (p. 502). During copulation, the spicules of the male, guided by the gubernaculum and telamon, are inserted into the gonopore of the female. The spermatozoa, which are elongate, conical, or spherical and lack flagellated tail, are capable of amoeboid movement (Fig. 14.24). They migrate up the vagina and uteri to the seminal receptacle, where fertilization occurs. For a detailed review of the gonads, gametogenesis, and fertilization in nematodes, see Foor (1983a,b).

The Egg. The entire sperm enters the ovum or oocyte, and a **fertilization membrane** is secreted by the cell soon after the sperm enters. This membrane gradually increases in thickness to form the chitinous shell. A second membrane, commonly referred to as the **vitelline membrane**, is secreted by the zygote within the shell. This membrane is composed of esterified glycosides with the solubility characteristics of lipids. As the egg passes down the uterus, a third proteinaceous membrane is secreted by the uterine wall and deposited outside the shell. This third membrane is rough in texture (Fig. 14.25) and is absent from the eggs of some species, such as the human whipworm, *Trichuris trichiura* (Fig. 14.26).

Commonly, the eggs of nematodes parasitic in animals are laid by the females and pass from the host in feces. These eggs, depending on the species, may still contain the uncleaved zygote, a group of blastomeres, or even a completely formed larva, as in the human pinworm, *Enterobius vermicularis* (Fig. 14.27).

Hatching. The egg-hatching process among parasitic nematodes occurs at one of two sites: within the host or in the external environment. If the latter is

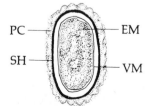

Fig. 14.25. Nematode egg. Fertilized egg of *Ascaris lumbricoides* showing the vitelline membrane, shell, and protein coat. EM, embryo; PC, protein coat; SH, shell; VM, vitelline membrane.

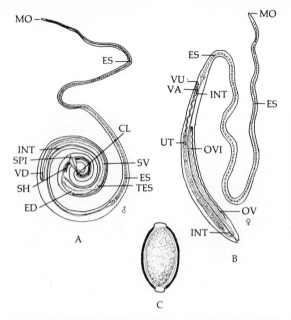

Fig. 14.26. *Trichuris trichiura*. **A.** Adult male. **B.** Adult female. **C.** Egg. CC, cloaca; ED, ejaculatory duct; ES, esophagus; INT, intestine; MO, mouth; OV, ovary; OVI, oviduct; SH, sheath; SPI, spicule; SV, seminal vesicle; TES, testis; UT, uterus; VA, vagina; VD, vas deferens.

Fig. 14.27. **Eggs of *Enterobius vermicularis*, the human pinworm. A.** Photomicrograph with 10 × objective. **B.** Photomicrograph with 43 × objective showing fully developed larvae within eggshells. (After Cheng, 1960.)

the case, a first-stage larva usually escapes. Hatching of eggs in the external environment is partly controlled by such ambient factors as temperature, moisture, and oxygen tension and partly by the larva within the egg. Consequently, an egg will hatch only if the external conditions are suitable, and this ensures that the escaping larva is not exposed to adverse conditions.

Examples of the type of egg described above include those of *Ancylostoma, Trichostrongylus,* and *Nippostrongylus.* The eggs of these genera pass to the exterior in the host's feces, and a free-living, first-stage larva hatches from each. According to Wilson (1958), the larva of *Trichostrongylus retortaeformis,* an intestinal parasite of rabbits, moves freely within the egg and secretes enzymes. The motion, coupled with enzymatic degradation, results in the breakdown of the innermost layer of the eggshell, causing it to become permeable to water. As a result of the uptake of water by the larva, hydrostatic pressure within its pseudocoel increases, and the distended larva exerts pressure on the eggshell, eventually rupturing it.

In the case of eggs of *Nematodirus,* an intestinal parasite of ruminants, they must be exposed to low temperature before they will hatch. This phenomenon has its counterpart in some insects and has obvious survival value for the parasite. Specifically, since *Nematodirus* adults usually do not survive in the vertebrate host from one season to another, eggs deposited in the fall serve to maintain the population. These

eggs in soil not only can survive over winter but are also triggered by the cold weather to hatch in the following spring, and thus continuity of the species is maintained.

As stated, a second category of eggs hatches only after ingestion by a host. In this case the hatching stimulus is provided by the host environment. Rogers (1960) has shown that carbon dioxide is most important in the hatching of *Ascaris* eggs. Although the presence of reducing agents, salts, and variations in pH and temperature all contribute to the hatching process, it is the presence of dissolved gaseous CO_2 and undissociated carbonic acid (H_2CO_3) that induces hatching. Furthermore, CO_2 must be present in the host environment in the proper quantity since high concentrations inhibit hatching. Eggs of *Toxocara,* parasitic in the intestine of carnivores, and *Ascaridia,* in the intestine of fowl, are similarly induced to hatch by conditions in the host's gut. Apparently, exposure to the appropriate CO_2 tension induces the unhatched larva to produce a chitinase, an esterase, and possibly a protease. These enzymes dissolve a portion of the outer layer of the shell, and the larva, still encased by the inner membrane, emerges from a hole, and the inner membrane eventually ruptures, allowing the larva to escape (Fig. 14.28). This mechanism suggests that hydrolytic enzymes secreted by the larva must pass through the inner membrane to act on the outer egg-

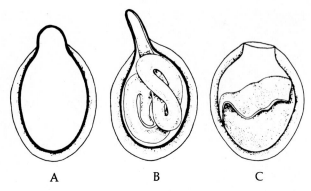

Fig. 14.28. Hatching of *Ascaris* egg. A. The stimulus from the host has produced an internal response from the larva (not shown) resulting in digestion of a hole in the outer layers of the shell through which the flexible inner layer protrudes. **B.** The larva inserts its head into the blister and appears to stretch the inner layer of the shell forcibly. **C.** The inner layer has ruptured and the larva has escaped from the egg, leaving the inner layer of the shell collapsed inside the egg. (After Fairbairn, 1960.)

shell layer, but since the inner membrane is rather impermeable, an alteration in its permeability must occur during the hatching process. This has been demonstrated by Fairbairn (1960) who has found that the sugar trehalose dissolved in the fluid surrounding the larva within the shell does not escape to the exterior unless the larva is stimulated by CO_2.

Molting. All nematodes undergo four molts during their development (Fig. 14.29). The process of molting, also referred to as ecdysis or exsheathing, involves (1) the formation of a new cuticle, (2) the

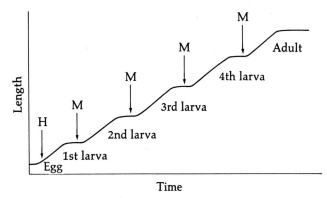

Fig. 14.29. Nematode growth pattern. The life cycle of a hypothetical nematode showing the periods of growth between molts. H, egg hatches; M, molt occurs. (After Lee, 1965.)

loosening of the old cuticle, and (3) the rupturing of the old cuticle and escape of the larva. These phases of molting as applied to arthropods are fairly well understood; however, there are only some preliminary clues in the case of nematodes (Rogers, 1966). Working with *Haemonchus contortus*, the stomach worm of cattle and sheep, and *Trichostrongylus colubriformis*, an intestinal parasite of cattle and sheep, Rogers has shown that nematode larvae secrete an **exsheathing fluid** just prior to the loosening of the old cuticle. The active component in this fluid is extremely labile and will attack the cuticle only from the inside and then only in a special region. This substance has been identified as a highly specific form of the enzyme leucine aminopeptidase.

The secretion of leucine aminopeptidase by the larva is stimulated by carbon dioxide and/or undissociated carbonic acid, but the pH of the medium is also important (Table 14.6). Rogers has postulated that the external stimulant (CO_2) reacts with a "receptor" in the larva, possibly the neurosecretory system, and this causes the release of substances, including leucine aminopeptidase, which affect various target organs and tissues concerned in the general development of the worms, including molting (Fig. 14.30).

Unlike arthropods, nematodes continue to grow between molts, although in many species there is a lag period (**lethargus**) immediately before and after a molt (Fig. 14.29). The occurrence of more or less continuous growth is made possible by the worm's ability to preform new cuticle within the hypodermis before shedding the old one. This process has been described in *Nematospiroides dubius* by Bonner *et al.* (1970).

LIFE CYCLE STUDIES

Detailed representative life cycle patterns of various parasitic nematodes are given under the various taxa listed below. However, it appears appropriate to discuss some of the generalized life history patterns found among the parasites of animals. Table 14.7 depicts some of the life cycle patterns found among superfamilies that include medically and economically important species.

The parasitic habits of nematodes vary from occasional, facultative or accidental parasitism during the larval or adult stages to completely obligatory parasitism in all stages. The following categorization of the degrees of parasitism may be helpful in understanding the progression of host-parasite relationships among the Nemata.

I. Nonparasitic Direct Cycle
  A. Completely Free-Living. Free-living nematodes usually lay their eggs in water or in soil. Developing larvae undergo several molts with little morphologic change and eventually mature into adults. This repre-

sents a direct life cycle without the utilization of hosts, either definite or intermediate.

B. External Feeding. Certain free-living nematodes, while still larvae, pierce and feed on
plant juices and sap. However, the worms
never enter plant tissue. This form of life
might well be considered the twilight zone
between the free-living and ectoparasitic
states.

II. Partial Parasitic Cycle

  A. Larval Form

    1. Host Utilized for Transport. Certain
free-living nematodes require a phase
of parasitic existence during their life
cycle. This parasitic phase occurs during
the larval stages, at which time
the worms penetrate and become ensheathed in insects and depend on the
host to carry them to more favorable
climes. An example of this type of relationship is the larva of *Agamermis decaudata*, parasitic in grasshoppers (Fig.
14.31).

    2. Host Utilized for Food. In another form
of host-parasite relationship, the larval
nematode infects an invertebrate host.
The host does not serve as a mechanism for transport; rather, the nematode larva lives, either ensheathed
or not, within the invertebrate and
remains in a quiescent condition until
the host dies. At that time, the worm
actively feeds on the carcass of the
host.

    3. Host Utilized as Site for Partial Devel-
opment. Among certain free-living
nematodes, the normal cycle is of the
nonparasitic direct type. Occasionally,
however, larvae become parasitic in an

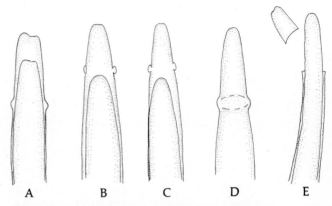

**Fig. 14.30. Stages during exsheathing of the infective
larvae of trichostrongylid nematodes** (e.g., *Trichostrongylus, Haemonchus,* and *Ostertagia*). The sheath, enclosing the
third-stage larva, swells locally near the anterior end (**A, B**)
and separates into two layers (**C**). The inner layers are
digested (**D**), resulting in an area of weakness in the form of
a ring. The tip of the sheath breaks off in the form of a cap
(**E**) and the larva wriggles free. (Redrawn after Lapage,
1935.)

invertebrate host when infective eggs
or ensheathed larvae are ingested.
Within the host, the larvae undergo
development and eventually escape as
adults. Such a nematode generally
undergoes several generations as a
nonparasite before reverting to the
parasitic phase. This type of occasional
parasitism is characteristic of the
Mermithidae.

Table 14.6. The General Effect of Gases, pH, and Reducing Agents on the Exsheathment of Nematodes[a]

| Buffer | Composition on Medium | | Reducing Agent | Exsheathment of Larvae (%) | |
	Gas Phase	pH		*Haemonchus contortus*	*Trichostrongylus axei*
Phosphate	Air	6.0	—	0	0
Phosphate	N_2	6.0	+	0	0
Phosphate	5% CO_2—N_2	±6.0	—	1	4
Phosphate	5% CO_2—N_2	±6.0	+	0	6
Bicarbonate	5% CO_2—N_2	6.0	—	3	5
Bicarbonate	5% CO_2—N_2	6.0	+	14	54
Bicarbonate	5% CO_2—N_2	8.0	—	3	2
Bicarbonate	5% CO_2—N_2	8.0	+	5	21
Bicarbonate	N_2	±8.3	—	0	0
Bicarbonate	N_2	±8.3	+	0	3

[a] The reducing agent was 0.02 M sodium dithionite. (After Rogers, 1960.)

Table 14.7 *Life Cycle Patterns of Some Major Groups of Parasitic Nematodes[a]*

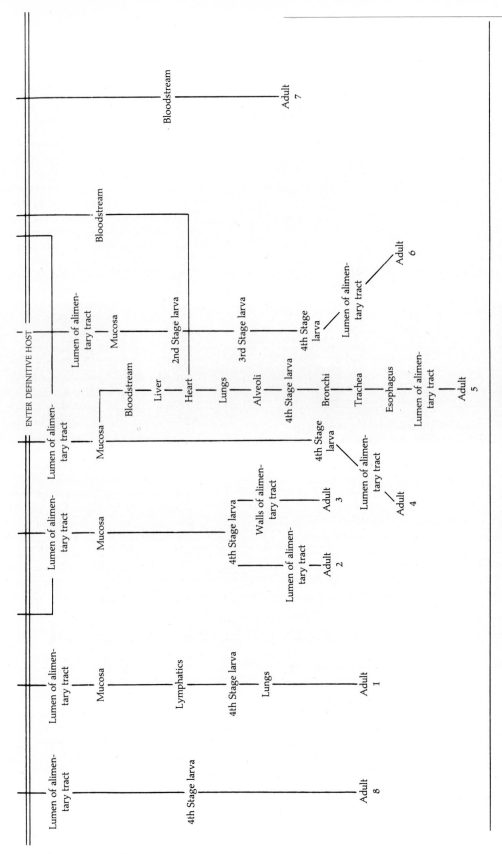

1. Metastrongyloidea; 2. Spiruroidea; 3. Spiruroidea; 4. Many of the Ascaroidea, many of the Strongyloidea, many of the Trichostrongyloidea; 5. Many of the Ascaroidea, many of the Strongyloidea; 6. Trichuroidea; 7. Filarioidea; 8. Oxyuroidea, few Ascaroidea.

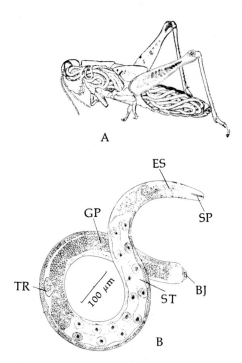

Fig. 14.31. *Agamermis decaudata.* **A.** Larva in grasshopper. **B.** Larva after 6 days in host. (After Christie, 1936.) Bj, breaking joint; ES, esophagus; GP, genital primordium; SP, spear; ST, stichosome; TR, trophosome (intestine).

B. Adult Form (Females Only). Nematodes belonging to this category are free living in soil during their larval stages. On reaching sexual maturity, males and females copulate. The males perish after mating while the females penetrate an invertebrate host in which they produce a new generation of larvae that escape and become free living.

C. Alternately Parasitic and Free-Living Adults. This type of alternating of life cycle is best exemplified by *Strongyloides stercoralis*, an intestinal parasite of humans and other primates (p. 502). The free-living adults are smaller than the parasitic forms. The males and females copulate, and zygotes found inside the resulting eggs often develop so rapidly that they hatch prior to the laying of the eggs. Hence, these nematodes are ovoviviparous. These larvae, known as **rhabditiform** larvae, molt and increase in size to become infective **filariform** larvae. Such larvae enter the host, either through penetration or orally, and develop into

parasitic females.* These females lay parthenogenetic eggs that hatch into rhabditiform larvae, which may pass out of the host and develop into free-living males and females, or repenetrate the host to remain parasitic. Variations in this cycle are discussed on page 503.

III. Total Parasitism
A. One Host
1. Larvae Penetrate Host. In many nematode parasites of plants and animals, the entire life span is spent as an endoparasite. In some endoparasites, the eggs are released or passed out of the host. When these hatch, the young larvae actively bore into the host and become dependent on it throughout the remainder of their lives. Plant parasites belonging to the genus *Heterodera* are included in this category, as are certain parasites of animals.
2. Eggs Ingested by Host. *Enterobius vermicularis*, the common human pinworm, belongs in this category. The eggs are laid outside the host when gravid females expose their gonopores to the exterior to oviposit. Eggs containing larvae are ingested by the host, establishing the infection. The entire life span of the worm, except for the egg, is spent as an endoparasite.
3. Continuous Existence in Host. Very few nematodes are capable of living continuously in a host without spending some phase of the life cycle, even as eggs, outside the host. Most worms belonging to this category are parasites of invertebrates. However, *Probstmayria vivipara*, an oxyurid found in the large intestine of horses, apparently is capable of this type of cycle. *Ollulanus tricuspis*, a stomach-dwelling species in members of the cat family, is also capable of this type of cycle. In these instances, the females are ovoviviparous, giving birth to larvae instead of eggs.
B. Multiple Hosts
1. One Intermediate Host. Nematodes included in this category possess infective larvae that are ingested by or penetrate an invertebrate host in which they continue to develop. However,

*It is now clear that parasitic males of *Strongyloides stercoralis* do not exist and that parasitic females reproduce parthenogenetically (p. 503).

these larvae do not reach sexual maturity until they gain entrance to the definitive host. *Angiostrongylus cantonensis*, the rat lungworm, belongs in this category. During its life cycle, first-stage larvae, hatching from eggs deposited in the rodent host's lungs, migrate up the bronchioles, bronchi, and tracheae and are swallowed. These pass to the exterior in the host's feces. These first-stage larvae enter the invertebrate intermediate host, a mollusc, either by ingestion or by penetration. Within the mollusc, the larvae are found embedded primarily in the foot musculature, where they undergo two molts. When molluscs harboring third-stage larvae are ingested by the rodent host, they migrate to the brain, undergo two more molts, and become young adults. These adults migrate to the host's pulmonary arteries, where they mature (p. 516).

2. Two Intermediate Hosts. Nematodes of this category are exemplified by *Gnathostoma*. During its life history, this worm utilizes *Cyclops* as the first intermediate host. When the arthropod is ingested by a fish, frog, or snake, one of these serves as the second intermediate host. Finally, when the infected fish, frog, or snake is ingested by a

carnivorous mammal, the nematode reaches the definitive host.

It should be apparent that the life cycles of nematodes do not involve asexual reproduction in the larval stages, as found among digenetic trematodes and some cestodes.

Anderson's (1984) essay on the origins of zooparasitic nematodes should be read by those interested in the evolution of parasitism among roundworms.

LARVAL FORMS

As stated, all parasitic nematodes generally undergo four molts (ecdyses) during their life cycles. All of the stages prior to the final molt are generally referred to as first-, second-, third-, or fourth-stage larvae, hereafter designated as L_1, L_2, L_3, and L_4; however, various other designations also are used to describe specific larval forms of nematodes.

Rhabditiform Larva

The L_1 of such parasitic species as *Strongyloides stercoralis* and *Necator americanus* is known as a **rhabditiform larva**. These small larvae each possess an esophagus that is joined to a terminal esophageal bulb by a narrow isthmus (Fig. 14.32).

Filariform Larva

After molting twice, rhabditiform larvae become L_3 or **filariform larvae** (Fig. 14.32). The esophagus in the filariform larva is typically elongate, cylindrical, and without a terminal bulb. Filariform larvae are usually the stage infective to the definitive host.

Microfilaria

The youngest (L_1) larvae of nematodes belonging to the superfamily Filarioidea, for example, those of *Wuchereria bancrofti* and *Loa loa*, are known as **microfilariae**. These larvae (which are discussed in greater detail on pp. 537) are actually embryos rather than fully developed larvae. The body surface is covered by a thin layer of flattened epidermal cells. Within the pseudocoel are found cells or primordia that eventually develop into various adult structures (Fig. 15.41). A conspicuous cord of cytoplasm containing nuclei extends the length of the body and represents the anlagen of the alimentary canal. In addition to this cord, there are: (1) a dash at the anterior tip of the body that is claimed by most to represent the beginning of the mouth; (2) developing nerve ring; (3) a V-shaped invagination that represents the future excretory pore; (4) a renette cell; (5) the so-called G cells,

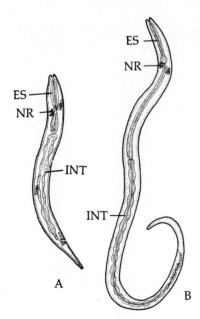

Fig. 14.32. **Nematode larvae. A.** Rhabditiform larva. **B.** Filariform larva. ES, esophagus; INT, intestine; NR, nerve ring.

which eventually develop into a portion of the gut; and (6) a tail spot in the posterior region that represents the developing anus. Microfilariae, generally found in circulating blood and cutaneous tissues, are extremely small, usually measuring between 0.2 and 0.4 mm in length.

BIOLOGY OF THE ADENOPHOREAN NEMATODES

The taxonomy of the Nemata has been in a state of great flux as a result of isolated interests of soil and plant nematologists and of animal parasitologists. However, in recent years there has been a concerted effort to arrive at a more natural classification system based on comparative studies of all species irrespective of their life-styles.

In the description of nematodes, various systems have been devised to express morphologic characteristics, for example, dimensions and ratios, in terms of symbols. The system advocated by de Man is currently the most widely used. According to this method, various body dimensions are given in millimeters (mm) or micrometers (μm) and followed by a series of ratios. The following is a list of symbols designating the various ratios.

a (α) = Body length/greatest body width
b (β) = Body length/distance of esophagus from anterior tip
c (γ) = Body length/distance from anus to tip of tail
V = Distance of vulva from anterior end as percentage of total body length
G_1 = Length of anterior gonad as a percentage of body length
G_2 = Length of posterior gonad as a percentage of body length

Several volumes are available for those interested in the identification of nematodes parasitic in vertebrates. These include those by Yorke and Maplestone (1926) and Yamaguti (1962). Levine (1980) has contributed an outstanding volume dealing with the nematode parasites of humans and domestic animals. For those interested in the higher taxa, the review by M. B. Chitwood (1969) is recommended. Poinar's (1975) treatise on nematodes parasitic in insects is recommended to those interested in this group of parasites. Presented at the end of this and the next chapter is the classification system presented by Maggenti (1981, 1982). In my opinion, it is the most acceptable system today.

As listed, the phylum Nemata is divided into two classes: Adenophorea and Secernentea.

Members of Adenophorea are characterized by an absent or poorly developed excretory system, the absence of phasmids, and well-developed mesenterial tissue. Members of Secernentea are characterized by the occurrence of a well-developed excretory system and the presence of phasmids and weakly developed mesenterial tissue.

CLASS ADENOPHOREA

SUPERFAMILY TRICHUROIDEA

Trichuroidea includes two prominent genera, *Trichuris* and *Trichuroides*. The genus *Trichuris* includes numerous species found in the large intestine and rectum of various mammals, for example, *T. felis* in the caecum and colon of cats, *T. discolor* in cattle, *T. leporis* in rabbits, *T. muris* in rats, *T. suis* in pigs, *T. vulpis* in dogs, *T. ovis* in cattle and sheep, and *T. trichiura* in humans.

All *Trichuris* species possess direct life cycles. For example, *T. trichiura*, the whipworm of humans which is cosmopolitan in its distribution, is generally found in the caecum (Fig. 14.26). The males are slightly smaller than the females, the latter measuring 30–50 mm in length. In both sexes, the long esophagus occupies two-thirds of the body length. The eggs oviposited by females are typically barrel shaped, measuring 50 × 22 μm and containing an uncleaved zygote (Fig. 14.26).

It has been estimated that a single female may lay as many as 1,000 to 46,000 eggs a day. These pass to the exterior in feces and develop slowly in damp soil. The larva within the shell is completely formed in three to six weeks. Developing eggs are fairly resistant to unfavorable conditions. These developing eggs will die, however, in a week's time if exposed to desiccation.

Moisture is necessary for embryonic development. Within the eggshell, the larva does not molt, nor does it hatch while in the soil. Humans are infected when embryonated eggs are ingested through the drinking of contaminated water, in food, or from fingers. The eggshell is destroyed once the egg reaches the host's caecum. The escaping larva burrows into the intervillar spaces, where it remains a few days before reentering the lumen of the caecum, growing there to maturity.

Generally, if a human host harbors fewer than 100 worms, there are no clinical symptoms. However, if the worm burden is greater, trauma caused to the intestinal epithelium and underlying submucosa can result in a chronic hemorrhage that leads to anemia. Other symptoms of trichuriasis include insomnia, nervousness, loss of appetite, vomiting, urticaria, diarrhea, and constipation.

The lesions caused by the anterior ends of worms

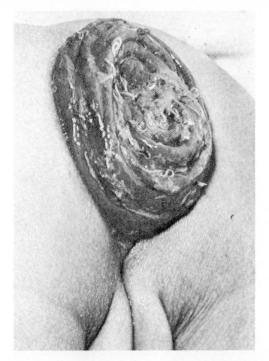

Fig. 14.33. Prolapse of the rectum caused by *Trichuris trichiura*. (After Beck and Davies, 1976; with permission of C. V. Mosby.)

burrowing into the intestinal lining may become secondarily infected by bacteria. This, coupled with allergic responses, may result in colitis, proctitis, and even prolapse of the rectum (Fig. 14.33).

In the canine-infecting *Trichuris vulpis*, Miller (1941) has reported that age resistance is probably nonexistent. Certain dogs, however, are capable of developing an immunity to this parasite. The immunity is manifested by the mass elimination of the nematodes. Nematode parasites that perforate the intestinal lining of their hosts during feeding, and thus introduce antigenic substances, generally elicit the synthesis of rather specific antibodies in their hosts.

Trichuris vulpis adults possess a stylet in the region of the mouth, which, by repeated protrusion and retraction, lacerates host tissues and blood and lymph vessels. This action results in the release of host cell contents and other fluids which, together with the products of histolysis, are ingested by active pumping of the esophagus.

Members of the genus *Trichuroides* are parasites of bats.

SUPERFAMILY TRICHINELLOIDEA

This superfamily includes the species *Trichinella spiralis*. The adult males, measuring 1.4–1.6 mm in length by 0.04–0.06 mm in diameter, are approximately half the length of the females.

During the life cycle of this parasite, one animal serves as both the definitive and the intermediate host, with the larvae and adults occurring in different organs.

Infections are brought about by a host eating meat, commonly poorly cooked pork, containing encapsulated larvae. Once ingested, these larvae (L_1) are released in the duodenum by the action of the host's digestive enzymes. Shortly after being freed, the larvae penetrate the cytoplasm of both absorptive and goblet cells lining the host's mucosa (Wright, 1979). Although apparently little damage is inflicted on the host cells, all four larval stages are intracellular parasites. The larvae, however, do escape and reinvade mucosal cells periodically. Upon reaching sexual maturity, male and female worms copulate in the mucosa. According to Bonner and Etges (1967), sexual attraction is brought about by a chemical signal.

After copulation, which usually occurs about 30 to 35 hours after ingestion of the infective larvae, the adult males either die or pass out of the host while the females burrow deeper into the mucosa and submucosa, sometimes entering and carried in the lymphatic ducts to the mesenteric lymph nodes. During this period, each ovoviviparous female gives birth to living L_1 larvae. It has been estimated that a single female produces about 1500 larvae over a period of 9 to 16 weeks. Eventually, the adult females die. Some of the larvae may escape into the intestinal lumen from whence they are passed out in feces, remaining infective for a brief time (Olsen and Robinson, 1958). Usually, L_1 larvae in the mucosa, each measuring about 0.1 mm long, are carried by the lymphatic and blood vessels into the venous blood, which eventually carries them to the heart.

From the heart the larvae find their way via the hepatic and pulmonary capillaries into the peripheral circulation, which in turn carries them to various tissues of the body. It is only in the striated skeletal and other voluntary muscles, especially those of the diaphragm, jaws, tongue, larynx, and eyes, however, that the larvae develop into the infective stage. According to Shanta and Meerovitch (1967), the larvae commence penetrating muscle cells on about the sixth day after the initial infection. They become established as intracellular parasites within myofibers, usually one larva per cell (Fig. 14.34). On about the 17th day, they commence to coil. At this site, they absorb their nutrients from and through the host cell and reach a length of about 1 mm in approximately 8 weeks, when they become infective. The host cell in which the larva resides has been designated as a **nurse cell** (Purkenson and Despommier, 1974). It

Fig. 14.34. *Trichinella spiralis.* Photomicrograph of larva as intracellular parasite in skeletal myofiber cell. (Courtesy of Nikon Inc.)

undergoes some adaptive changes, and eventually a host-derived double membrane completely surrounds the larva (Despommier, 1975). Eventually, a capsule of host cells and connective tissue fibers is formed around the nurse cell (Fig. 14.35), and calcium becomes deposited in the capsule. Total calcification takes about 10 months.

As the capsule is formed, the enclosed larva enters developmental arrest and can live for many months. Some researchers are of the opinion that they may remain viable for as long as 25 to 30 years. When

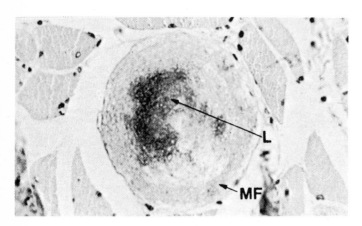

Fig. 14.35. **Calcified** *Trichinella spiralis* **larva.** A very old, calcified larva (L) of *T. spiralis* in a degenerated myofiber (mF). The entire complex is calcified. × 100. (After M. Kenney, 1973; with permission of The Upjohn Co.)

muscle harboring encapsulated larvae is eaten by a compatible host, the freed larvae reinitiate the life cycle.

Trichinella spiralis naturally infects many mammals, including rats, mice, rabbits, cats, dogs, wolves, ferrets, muskrats, badgers, pigs, raccoons, moles, porcupines, bears, and humans. Various birds, including chickens, pigeons, crows, and owls, have been infected experimentally.

If the infective larvae are encapsulated in human myofibers, this represents the end of life of the parasite since, except in rare cases of cannibalism and predation, human flesh is not eaten.

EPIDEMIOLOGY OF TRICHINOSIS

The term **sylvatic trichinosis** has been coined to describe the cycling of the disease between wild carnivores and their prey or carrion. On the other hand, the term **urban trichinosis** has been coined to describe the cycling of the disease among humans, rats, and pigs. In such instances rats and pigs feeding on garbage, including infected pork waste, become infected. Dead or dying infected rats are also eaten by pigs (Fig. 14.36). In turn, raw or poorly cooked pork harboring infective larvae is the vehicle for human infections. Trichinosis in rats alone can be maintained by cannibalism, although rat and mouse feces containing larvae can infect other rodents, pigs, and humans (Levine, 1980). Also, infections can be spread from pig to pig when they nip off one another's tails under crowded conditions (Smith, 1975).

In Alaska, polar bears and black bears are commonly sources of human infection (Maynard and Pauls, 1962). This provides an example of how sylvatic and urban trichinosis overlap.

Human infections are most commonly acquired from eating pork and pork products harboring larvae. I recall serving as an expert witness at a hearing where a coed had become infected as a result of eating inadequately cooked pork sausage. The prevention of trichinosis is still best achieved by eating only thoroughly cooked pork, although freezing at − 15°C for 20 days destroys the infective larvae.

Merkushev (1955) has found that the larvae of certain necrophagous beetles and other insects, which form part of the food of some carnivores, can act as temporary paratenic hosts for *T. spiralis* larvae. Such insect hosts could serve to transfer infections from infected carcasses to carnivores, insectivores, and omnivores. The insects are only temporary hosts, however, since the nematode can survive only six to eight days in insects.

Encysted *Trichinella spiralis* larvae have been found in the flesh of walruses and whales from northern waters. This finding has led to the discovery and confirmation of scavenger-marine mammal relationships. For example, the carcasses of infected walruses and whales dying from natural causes are consumed by a variety of scavengers, ranging from crabs to sea gulls. The *T. spiralis* larvae in the flesh of marine mammals are able to survive in the intestine of the scavengers, especially birds, and are eventually passed out in feces. As the result of the migratory habits of birds, these nematode larvae are transported, as phoronts, over a wide area of sea and shore. When deposited in water, they can be ingested by fish or crustaceans, especially crabs, which also serve as paratenic hosts. When raw or undercooked fish and crabs are ingested by a human or another mammal, the larvae can develop into adults. Both fish and crustaceans can also pick up and transport trichina larvae from eating carcasses of infected marine mammals. As a result of these complicated avenues of infection, the incidence of clinical cases of trichinosis among American and Canadian Eskimos is known to be at least 25%.

Based on some differences in the optimal pH values of certain enzymes and other features, some researchers believe that four subspecies of *T. spiralis* exist: *T. spiralis spiralis* in the temperate zone, *T. spiralis nelsoni* in Africa, *T. spiralis nativa* in the Arctic, and *T. spiralis pseudospiralis* in birds. Whether this designation of subspecies is justified remains to be seen.

Pathology of Trichinosis

The pathogenesis of trichinosis is conveniently considered in three phases: (1) during penetration of adult females into the mucosa; (2) during migration of larvae; and (3) during penetration and encapsulation in muscle cells.

Penetration of Mucosa. When adult female worms penetrate the host's intestinal mucosa, traumatic damage to the intestinal lining occurs (generally between 12 hours and 2 days after infective larvae are eaten). The damage results from the toxicity of the worms' excreta as well as the introduction of bacteria into the wounds. The lesions are characterized by inflammation and pain. Also, the toxic reactions include nausea, vomiting, fever, perfuse perspiration, and diarrhea. Respiratory difficulties may occur, red blotches may erupt on the skin, and the face may swell. This phase of trichinosis usually terminates five to seven days after the appearance of the first symptoms.

Larval Migration. During migration of larvae, damage is inflicted on blood vessels, causing localized edema, especially in the face and hands. Also, wandering larvae (Fig. 14.37) may cause pneumonia, encephalitis, pleurisy, meningitis, nephritis, peritonitis, damage to eyes, and deafness. Deaths due to inflammation of cardiac muscle (myocarditis) have been reported as larvae migrate through that organ.

Intramuscular Phase. About 10 days after the first symptoms appear, the larvae commence penetrat-

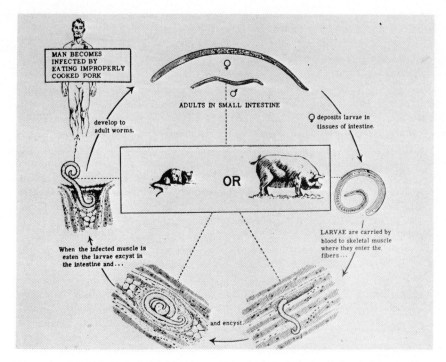

Fig. 14.36. Life cycle of *Trichinella spiralis*. (Courtesy of Naval Medical Center, Bethesda, Maryland.)

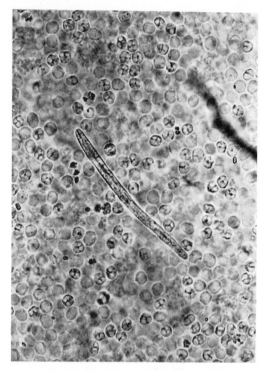

Fig. 14.37.   Migrating *Trichinella spiralis* larva. Photomicrograph of larva migrating in blood of a rat. (After J. Georgi in Schmidt and Roberts, 1981; with permission of C. V. Mosby.)

ing skeletal myofibers. At this time the host experiences intense muscular pain, difficulty in swallowing and breathing, swelling of facial muscles, weakening of blood pressure and pulse, heart damage, and nervous disorders, including hallucinations. Extreme eosinophilia is common. Death is usually caused by heart failure, respiratory complications, toxic reactions, and malfunction of the kidneys. The medical importance of *T. spiralis* is declining in North America and in northern Europe. While Stoll (1947) has estimated 27,800,000 cases of human trichinosis exist worldwide, 21,100,000 in North America, prevalence is considerably less today. For some reason, trichinosis is less common in the tropics but is very common in Mexico, parts of South America, Africa, southern Asia, and the Middle East. Trichinosis remains a threat to human health because of its wide distribution in wild and domestic animals.

Immunity to *Trichinella*

It has been demonstrated that hosts (rats, guinea pigs, pigs) infected with *T. spiralis* develop an immunity,

because upon recovery from the initial infection, they resist reinfection. Although the resistance is never complete, it is transmissible in many cases from immune mothers to their offspring. Young mice can acquire immunity from the mother through her milk (Duckett *et al.*, 1972). *Trichinella* antigen appears within 24 hours when the worms are maintained in serum, and precipitation is known to occur within five days.

Antibodies produced by the host against *Trichinella* are directed against the intestinal stage of the parasite since, if infective larvae are experimentally fed to immune animals, most are eliminated without development.

Levine and Evans (1942) and Gould *et al.* (1955) have demonstrated that partial immunity to *T. spiralis* can be induced by feeding infective larvae irradiated with 3250 and 3750 R of X-rays or 10,000 R of gamma rays from cobalt-60. These dosages sterilize the larvae, but growth is not inhibited. If higher dosages are used, growth of the larvae is inhibited and no immunity is induced. Although host immunity produced by feeding irradiated larvae is incomplete, the number of adults maturing in the intestinal mucosa and the number of larvae that reach skeletal myofibers are greatly reduced (Table 14.8).

Not only are the body tissues *T. spiralis* antigenic, but the excretions and secretions of larvae cultured in nutrient media for two to five days are also antigenic. When excretions and secretions are injected into mice, immunity is manifested by the significantly smaller number of larvae and adults found in the treated hosts. The effect is more readily appreciated among larvae. For a review of immunity to *Trichinella*, see Barriga (1981).

The superfamily Trichinelloidea also includes the genus *Capillaria*, which includes species of economic and medical importance. The members parasitize a wide range of hosts as well as a great variety of locations within the host. Normally, their life cycles are direct, without involving intermediate hosts; however, transport or paratenic hosts are not unusual.

Capillaria includes species parasitic in the crop and esophagus of birds, in mammals, in fish, and in amphibians, especially salamanders. Of the species found in mammals, *C. hepatica* is most frequently encountered. This nematode has been reported in a variety of animals, including humans; however, it most commonly occurs in rodents. The habitat of *C. hepatica* is the host's liver, surrounded by granulomatous reaction cells (Fig. 14.38).

Life Cycle of *Capillaria hepatica*.   Although the life cycle of *C. hepatica* may be interpreted as being direct, it does include a transport host. Few eggs are released in the feces of the host, and these occur only early in the infection. Normally, eggs are released when the infected host is devoured by a predator. The infected liver tissue is digested away, and the eggs are passed through the alimentary tract of the transport host and passed out in feces. Eggs are deposited in soil where they can survive temperatures as low as −15°C, but development requires warmer climates.

Infection of the host is accomplished with ingestion of embryonated eggs. Such eggs hatch in the caecum and the larvae penetrate through the wall, become incorporated in the portal circulation, and are carried to the liver, where they develop to maturity. When experimentally hatched larvae are injected into hosts, these larvae eventually migrate via the bloodstream to the liver.

The additional avenues of escape for eggs of *Capillaria* spp. are displayed by *C. aerophila*, parasitic in the lungs of foxes; *C. annulata*, in the esophagus and crop of turkeys, chickens, and various wild birds; and *C. plica*, found in the kidneys and urinary bladder of dogs and foxes. The eggs of *C. aerophila* are discharged in the lungs and pass up the air passages to the epiglottis. They are then coughed up and ingested and finally pass out in the host's feces. Earthworms of the genera *Lumbricus* and *Allolophora* serve as transport hosts. The eggs of *C. annulata* are oviposited in the esophagus and crop and pass out in feces. *Helodrilus foetidus* and *H. caliginosus*, both earthworms, serve as the paratenic host for this nematode. Eggs expelled from the avian host must be ingested by an earthworm before attaining the infective form. Birds become infected when they ingest infected earthworms. The eggs of *C. plica* are passed out in the urine of the host. A paratenic host is required in this species also. In the case of *C. plica*, *C. annulata*, and *C. aerophila*, the occurrence of a transport host has evolved into a mandatory state.

Many species of *Capillaria* are found in the intestine of their hosts. *Capillaria bursata* and *C. caudinflata* are parasitic in the small intestine of chickens and

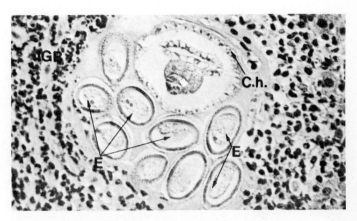

Fig. 14.38. *Capillaria hepatica* in liver. Photomicrograph showing cross-section of an adult *C. hepatica* (C. h.) and its eggs (E) surrounded by intense granulomatous reaction (IGR) in liver. × 250. (After M. Kenney, 1973; with permission of The Upjohn Co.)

other birds, *C. columbae* in the small intestine of pigeons, *C. brevipes* and *C. bovis* in cattle and sheep, and *C. linearis* in the intestine of cats.

As stated, *Capillaria hepatica* can infect humans. Humans, especially children, can become infected by ingesting infective eggs on contaminated fingers. If this occurs, larvae hatching from eggs in the intestine will penetrate through the intestinal wall and some will migrate through various visceral tissues, and hence the condition may be considered a type of

Table 14.8. Effect of Reinfecting Rats with *Trichinella spiralis* Larvae 40 Days after Initial Infection to Show Presence of a Partial Immunity Induced by Using ⁶⁰Co-Irradiated Larvae[a]

Days after Reinfection	Group I (Not Irradiated)		Group II (Irradiated with 10,000 R from ⁶⁰CO)	
	Adult Worms in Intestine	*Larvae Recovered from Muscles*	*Adult Worms in Intestine*	*Larvae Recovered from Muscles*
2	17	181,000	1912	15
	165	32,200		
4	0	166,000	848	11
	8	25,800		
6	0	164,000	42	0
	4	47,240		
8	52	101,000	9	0
	0	640,000		
10	0	156,000	0	22
15	0	190,000	0	0
	0	—		
20	0	96,000	0	496

[a] Data after Gould *et al.* (1955).

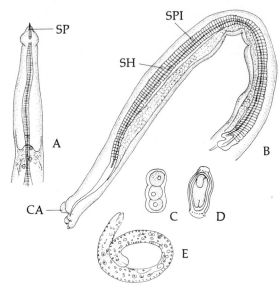

Fig. 14.39. *Capillaria philippinensis* **from a human in the Philippines. A.** Head of adult female showing spear. **B.** Tail of adult male showing spicule, indrawn sheath, and caudal alae. **C.** Egg without shell. **D.** Normal bioperculate egg enclosing developing embryo. **E.** Embryo from adult female. CA, caudal alae; SH, sheath; SP, spear; SPI, spicule. (Redrawn after Chitwood *et al.*, 1968.)

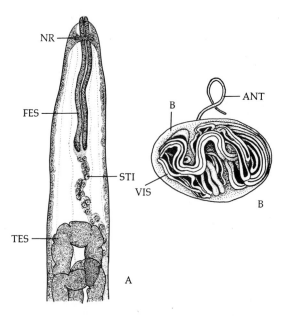

Fig. 14.40. *Cystoopsis acipenseri.* **A.** Anterior portion of male worm. **B.** Female specimen dissected from cyst. ANT, anterior end; B, body; FES, free portion of esophagus; NR, nerve ring; STI, stichosome; TES, testis, VIS, coils of viscera in body. (Redrawn after Janicki and Rasin, 1930.)

"visceral larva migrans" (p. 560). The host's reaction to the condition is manifested by such symptoms as eosinophilia, hepatomegaly, and even pneumonitis if the larvae enter the lungs.

Capillaria philippinensis is of special interest since adults of this genus are usually not found as intestinal parasites of humans, yet *C. philippinensis* was described by Chitwood *et al.* (1968) from the intestine of a man in the Philippines. Furthermore, this species is of considerable public health interest since over 1000 cases of human intestinal capillariasis, with several resulting in death, have been diagnosed in the Phillippines as being caused by it. It is also known to occur in Thailand (Bhaibulaya *et al.*, 1979).

The adult male of *C. philippinensis*, measuring 2.3–3.17 mm long, has, among other characteristics, a tail with ventrolateral expansions which bear at least two pairs of papillae (Fig. 14.39). The female, measuring 2.5–4.3 mm long, is interesting in that it either produces typical bioperculated eggs or eggs without shells (Fig. 14.39). Furthermore, sometimes L_1s develop *in utero*. The complete life cycle of this nematode is not yet known. It has been hypothesized that humans become infected from eating fish (Bhaibulaya *et al.*, 1979). It is undoubtedly a natural parasite of some wild animal which has yet to be identified. It is noted, however, that once established among humans in a village, transmission is readily maintained when soil contaminated with feces including eggs is accidentally ingested.

SUPERFAMILY CYSTOOPSOIDEA

The superfamily Cystoopsoidea includes *Cystoopsis acipenseri* (Fig. 14.40), a parasite in the skin of the Volga sturgeon. The eggs are ingested by amphipods which serve as the intermediate host. In laying their eggs, the female worms rupture out of their encapsulating cysts in the host's skin. Furthermore, the uterus is ruptured while releasing the enclosed eggs. Within the amphipod intermediate host, the eggs hatch and the larvae, each armed with a stylet, penetrate the intestinal wall. After undergoing development in the hemocoel, these larvae encyst in the host's appendicular muscles. When infected amphipods are ingested by a sturgeon, the larvae burrow through the intestinal wall, become encysted in the skin, and grow to maturity.

SUPERFAMILY MERMITHOIDEA

The Mermithoidea includes a group of unique nematodes that are generally free living as adults.

Larval mermithoids are obligate parasites of invertebrates. Although most species parasitize insects (Fig. 14.41, 14.42), others have been reported from spiders, crustaceans, leeches, molluscs, and even other nematodes. Among insect hosts, both terrestrial and aquatic species are parasitized.

Life Cycle of Mermithoids. In most species, the

eggs are deposited in the host's environment and embryonic development commences after oviposition. These eggs, like the adult worms, are susceptible to dessication, and hence successful infection is often dependent on a high moisture content. In all species, the first molt occurs within the egg, and the escaping larva is an L_2. In some cases, a second molt occurs within the egg. If this occurs, the escaping larva is an L_3. This, however, requires further verification. Emergence of the larva is accomplished with the aid of the anteriorly situated stylet and digestive secretions. This stage is referred to as the **infective (or preparasitic) larva**. It is this form that makes contact with the potential host and penetrates it through the cuticle. In some instances, however, for example those species that parasitize simuliids (blackflies), there is reason to believe that they may enter via the alimentary tract.

After entering the host, the parasites usually rest free in the hemocoel, where they undergo rapid growth and development. The second molt occurs at this site, resulting in L_3s. In a few instances, the larval mermithoids occur within their hosts' nervous tissue (brain or ganglia). This behavior is believed to reflect the parasites' avoidance of the insect hosts' lethal internal defense mechanisms. In other instances, larvae occur within membranous envelopes in the host's hemocoel.

After a parasitic existence of from 10 days to several months, depending on the species, the L_3 is ready to leave the host. Upon emergence, the larva is referred to as a **postparasitic larva**. After a period of maturation in the external environment, the final two molts (or molt) occur, and the adult stage is attained. In some species, such as *Hydromermis contorta* and *Capitomermis crassiderma*, both parasites of chironomid insects, the final molt occurs in the host, and it is the adult that emerges.

The development cycle described above is the usual one for mermithoids; however, exceptions are known. For example, in the case of *Mermis nigrescens*, the eggs mature within the body of the adult female worm and are deposited on plants during periodic **ovopositional migrations**, which occur immediately after a rain. The infective larva is completely formed at hatching, and the eggs are thick shelled and resistant to desiccation. Such eggs remain on leaves until eaten by a suitable insect host. They hatch in the host's gut, and the larva enters the hemocoel by penetrating through the intestinal wall. The remainder of the life cycle is similar to that presented earlier.

Effects on Host. In recent years there has developed a great deal of interest in mermithoid nematodes because of their potential use as biologic control agents against undesirable insects, for example, the blackflies that transmit onchocerciasis (p. 542). As a result of this interest, considerable information has evolved about the physiologic and pathologic effects of these nematodes on their hosts. Gordon and

Fig. 14.41. Mermithid nematode. A mermithid in the hemocoel of an adult alfalfa weevil, *Hypera postica*. (After Poinar, 1975; with permission of E. J. Brill.)

Webster (1971, 1972) have provided a review of some of this information, as has Peterson (1985).

It is now known that parasitized insects are rendered sterile. Furthermore, the emergence of mermithoids from their hosts is usually fatal to the latter. Death is

Fig. 14.42. A mermithid nematode emerging from abdomen of an ant. (After H. Lyons in Poinar, 1975; with permission of E. J. Brill.)

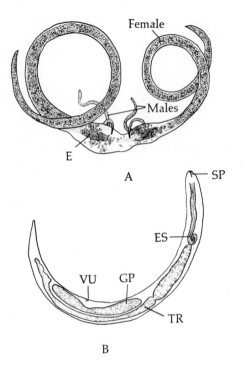

Fig. 14.43.   Mermithid nematodes. A. Female and males of *Tetradonema plicans*. (Redrawn after Hungerford, 1919.) **B.** Young female of *Aproctonema entomophagum*. (Redrawn after Keilen and Robinson, 1933.) E, accumulation of eggs in region of gonopore; ES, esophagus; GP, genital primordium; SP, spear; TR, trophosome; VU, vulva.

generally attributable to depletion of reserve nutrients, a sudden loss of turgor pressure when the parasites emerge, and secondary microbial infections through the emergence site.

Reesimermis nielseni, which is capable of parasitizing at least 18 species of mosquitoes, has become a prime candidate as a biologic control agent against these noxious insects.

For those interested in the taxonomy of mermithoids, the monographs by Nickle (1972) and Poinar (1975) are recommended.

SUPERFAMILY TETRADONEMATOIDEA

The Tetradonematoidea includes *Tetradonema plicans*, *Mermithonema entomophilum*, and *Aproctonema entomophagum* (Fig. 14.43). This group of nematodes is closely related to the Mermithoidea but is generally considered more primitive evolutionarily. Two variations of the mermithoid life cycle pattern are portrayed by *T. plicans* and *A. entomophagum*. In the first, the entire life span is spent as a parasite, whereas in the second only a portion of the cycle is parasitic.

Life Cycle of *Tetradonema plicans*.   The fly *Sciara* is infected in all stages by *Tetradonema plicans*, but especially while in the soil-dwelling larval stage. A single fly may harbor from 2 to 20 worms of both sexes. The fly maggots become infected when they ingest eggs of the parasite. These eggs hatch in the host's intestine, and the larvae burrow through the intestinal wall to become established in the fly's hemocoel. At this site, the nematode feeds on the host's fat bodies and other tissues and attains maturity. Female *T. plicans* measures up to 5 mm in length, whereas the male measures approximately 1 mm. Copulation occurs within the fly, and the eggs are deposited in the host's subcuticular zone near the gonopore. When the fly host dies and disintegrates, the females are released but soon perish, dispersing their eggs in the soil.

Life Cycle of *Aproctonema entomophagum*.   *A. entomophagum* undergoes a life cycle similar to *T. plicans*; however, after copulation within the hemocoel of the host (also *Sciara*), the females escape, lay their eggs, and perish. These eggs hatch in decaying wood, giving rise to free-living larvae that molt after a period of growth. It is the L_2 of the parasite that penetrates the larva of *Sciara*, thus establishing the parasitic phase of its life cycle.

Tetradonematoid nematodes have been described from fungus gnats (Sciaridae), scavenger flies (Sepsidae), midges (Chironomidae), chaoborid flies (Chaoboridae), nitidulid beetles (Nitidulidae), and blackfly larvae (Simuliidae). These parasites generally kill their hosts by depleting their nutrients and permitting microbial infections through the wounds they have caused.

For those interested in the taxonomy of the tetradonematoids, see Poinar (1975).

ORDER MUSPICEIDA

The Muspiceida is a very small order of adenophorean nematodes. The relationship of these worms to the other adenophoreans remains unknown. Representative of this group of parasitic nematodes is *Muspicea borreli*, found in the mouse, *Mus musculus*, by Sambon (1925). Its presence causes a tumorous growth. The characteristics of this order are listed at the end of this chapter.

REFERENCES

Anderson, R. C. (1984). The origins of zooparasitic nematodes. *Canad. J. Zool.* **62**, 317–328.

Barriga, O. O. (1981). "The Immunology of Parasitic Infections." University Park Press, Baltimore, Maryland.

Bhailbulaya, M., Indra-Ngarm, S., and Ananthapruti, M. (1979). Freshwater fishes of Thailand as experimental hosts for *Capillaria philippinensis*. *Int. J. Parasitol.* **9**, 105–108.

Bird, A. F. (1971). "The Structure of Nematodes." Academic Press, New York.

Bonner, T. P., and Etges, F. J. (1967). Chemically mediated sexual attraction in *Trichinella sprialis*. *Exp. Parasitol.* **21**, 53–60.

Bonner, T. P., Menefee, M. G., and Etges, F. G. (1970). Ultrastructure of cuticle formation in a parasitic nematode, *Nematospiroides dubius*. *Zeitschr. Zellforsch. Mikr. Anat.* **104**, 193–204.

Boudouy, T. (1910). Etude chimique de *Sclerostomum equinum*. *Arch. Parasitol.* **14**, 5–39.

Chitwood, B. G., and Chitwood, M. B. (1950). "An Introduction to Nematology" (Reprinted 1974). University Park Press, Baltimore, Maryland.

Chitwood, M. B. (1969). The systematics and biology of some parasitic nematodes. *In* "Chemical Zoology" (M. Florkin and B. T. Scheer, eds.), Vol. III, pp. 223–244. Academic Press, New York.

Chitwood, M. B., Velasquez, C., and Salazar, N. G. (1968). *Capillaria philippinensis* sp. n. (Nematoda: Trichinellida), from man in the Philippines. *J. Parasitol.* **54**, 368–371.

Croll, N. A., and Matthews, B. F. (1977). "Biology of Nematodes." John Wiley, New York.

Despommier, D. (1975). Adaptive changes in muscle fibers infected with *Trichinella spiralis*. *Am. J. Pathol.* **78**, 477–496.

Duckett, M. G., Denham, D. A., and Nelson, G. S. (1972). Immunity to *Trichinella spiralis*. V. Transfer of immunity against the intestinal phase from mother to baby mice. *J. Parasitol.* **50**, 550–554.

Elliott, A. (1954). Relationship of aging, food reserves, and infectivity of larvae of *Ascaridia galli*. *Exp. Parasitol.* **3**, 307–320.

Engelbrecht, H., and Palm, V. (1964). Der endogene Glykogen-und Fettstoffwechsel in seiner Bedeutung für die Differenzierung, Lebens und Infektionsfähigkeit der Entwicklungsstadien parasitischer Würmer. *Zeit. Parasitenk.* **23**, 88–104.

Fairbarn, D. (1960). Physiologic aspects of egg hatching and larval exsheathment in nematodes. *In* "Host Influence of Parasite Physiology" (L. A. Stauber, ed.), pp. 50–64. Rutgers University Press, New Brunswick, New Jersey.

Fauré-Fremiet, E. (1913). Le cycle germinatif chez l'*Ascaris megalocephala*. *Arch. Anat. Microsc.* **15**, 435–578.

Flury, F. (1912). Zur Chemie and Toxokologie der Ascariden. *Arch. Exp. Pathol. Pharmakol.* **67**, 275–392.

Foor, W. E. (1983a). Nematoda. *In* "Reproductive Biology of Invertebrates. Vol. I. Oogenesis, Oviposition, and Oosorption" (K. G. Adiyodi and R. G. Adiyodi, eds.), pp. 223–256. John Wiley, New York.

Foor, W. E. (1983b). Nematoda. *In* "Reproductive Biology of Invertebrates. Vol. II. Spermatogenesis and Sperm Function" (K. G. Adiyodi and R. G. Adiyodi, eds.), pp. 221–256. John Wiley, New York.

Goodwin, L. G., and Vaughan Williams, E. M. (1963). Inhibition and neuromuscular paralysis in *Ascaris lumbricoides*. *J. Physiol.* **168**, 857–871.

Gordon, R., and Webster, J. M. (1971). *Mermis nigrescens*: physiological relationship with its host, the adult locust *Schistocerca gregaria*. *Exp. Parasitol.* **29**, 66–79.

Gordon, R., and Webster, J. M. (1972). Nutritional requirements for protein synthesis during parasitic development of the entomophilic nematode *Mermis nigrescens*. *Parasitology* **64**, 161–172.

Gould, S. E., *et al.* (1955). Studies on *Trichinella spiralis* I-V. *Am. J. Pathol.* **31**, 933–963.

Horne, G. (1961). "Principles of Nematology." McGraw-Hill, New York.

Jarman, M. (1976). Neuromuscular physiology of Nematodes. *In* "The Organization of Nematodes" (N. A. Croll, ed.), pp. 293–312. Academic Press, London.

Jaskoski, B. J. (1962). Paper chromatography of some fractions of *Ascaris suum* eggs. *Exp. Parasitol.* **12**, 19–24.

Kent, N. H. (1963). VI. Current and potential value of immunodiagnostic tests employing soluble antigens. *Am. J. Hyg. Monogr. Ser.* **22**, 30–46.

Lee, D. L. (1965). "The Physiology of Nematodes." Freeman, San Francisco, California.

Lee, D. L. (1966). The structure and composition of the helminth cuticle. *Adv. Parasitol.* **4**, 187–254.

Levine, A. J., and Evans, T. C. (1942). Use of roentgen radiation in locating an origin of host resistance to *Trichinella spiralis* infections. *J. Parasitol.* **28**, 477–483.

Levine, N. D. (1980). "Nematode Parasites of Domestic Animals and of Man," 2nd ed. Burgess Publ., Minneapolis, Minnesota.

Maggenti, A. R. (1981). "General Nematology." Springer-Verlag, New York.

Maggenti, A. R. (1982). Nemata. *In* "Synopsis and Classification of Living Organisms" (S. P. Parker, ed.), pp. 879–929. McGraw-Hill, New York.

Maynard, J. E., and Puals, F. P. (1962). Trichinosis in Alaska. A review and report of two outbreaks due to bear meat, with observations of serodiagnosis and skin testing. *Am. J. Hyg.* **76**, 252–261.

McLaren, D. J. (1976). Sense organs and their secretions. *In* "The Organization of Nematodes" (N. A. Croll, ed.), pp. 139–161. Academic Press, London.

Mellanby, H. (1955). The identification and estimation of acetylcholine in three parasitic nematodes (*Ascaris lumbricoides*, *Litomosoides carinii* and the microfilaria of *Dirofilaria repens*). *Parasitology* **45**, 287–294.

Merkushev, A. V. (1955). Rotation of *Trichinella* infections in natural conditions and their natural foci. *Med. Parazitol. Bolez.* **23**, 125–130.

Miller, M. J. (1941). Quantitative studies on *Trichocephalus vulpis* infections in dogs. *Am. J. Hyg.* **9**, 58–70.

Monné, L. (1955). On the histochemical properties of the egg envelopes and external cuticles of some parasitic nematodes. *Ark. Zool.* **9**, 93–113.

Münnich, H. (1965). Beziehungen wischen dem Fett- und Glykogengehalt von *Ascaris*-Larven und der Dauer ihrer Infektionsfahigkeit. *Zeit. Parasitenk.* **25**, 231–239.

Nickle, W. R. (1972). A contribution to our knowledge of the Mermithidae (Nematoda). *J. Nematol.* **4**, 113–146.

Olsen, O. W., and Robinson, H. A. (1958). Role of rats and mice in transmitting *Trichinella spiralis* through their feces. *J. Parasitol.* **44** (Sect. 2), 35.

Peterson, J. J. (1985). Nematodes as biological control agents: Part 1. Mermithidae. *Adv. Parasitol,* **24**, 307–344.

Poinar, G. O., Jr. (1975). "Entomogenous Nematodes. A Manual and Host List of Insect-Nematode Associations." E. J. Brill, Leiden, The Netherlands.

Pollak, J. K., and Fairbairn, D. (1955). The metabolism of *Ascaris lumbricoides* ovaries. I. Nitrogen distribution. *Can. J. Biochem. Physiol.* **33**, 297–306.

Purkerson, M., and Despommier, D. (1974). Fine structure of the muscle phase of *Trichinella spiralis* in the mouse. In "Trichinosis" (C. Kim, ed.) Intext Publ., New York.

Rogers, W. P. (1945). Studies on the nature and properties of the perienteric fluid of *Ascaris lumbricoides. Parasitology* **35**, 211–218.

Rogers, W. P. (1960). The physiology of infective processes of nematode parasites: the stimulus from the animal host. *Proc. Roy. Soc.* **B152**, 367–386.

Rogers, W. P. (1966). Exsheathment and hatching mechanism in helminths. *In* "Biology of Parasites" (E. J. L. Soulsby, ed.), pp. 33–40. Academic Press, New York.

Sambon, L. W. (1925). Researches on the epidemiology of cancer made in Iceland and Italy. *J. Trop. Med. Hyg.* **28**, 39–71.

Schulz, F. N., and Becker, M. (1933). Über Ascarylalkohol. *Biochem. Z.* **276**, 253–259.

Shanta, C. S., and Meerovitch, E. (1967). The life cycle of *Trichinella spiralis*. II. The muscle phase of development and its possible evolution. *Can. J. Zool.* **45**, 1261–1267.

Smith, H. J. (1975). Trichinae in tail musculature of swine. *Can. J. Comp. Med.* **39**, 362–363.

Stoll, N. R. (1947). This wormy world. *J. Parasitol.* **33**, 1–18.

Timm, R. W. (1950). Chemical composition of the vitelline membrane of *Ascaris lumbricoides* var. *suis. Science* **112**, 167–168.

Toryu, Y. (1933). Contributions to the physiology of *Ascaris*. I. Glycogen content of the ascaris, *Ascaris megalocephala. Cloq. Sci. Rep. Tohoku Imp. Univ.* 4s (Biol.) **8**, 65–74.

von Brand, T. (1938). Physiological observations on a larval *Eustrongyloides* (Nematoda). *J. Parasitol.* **24**, 445–45.

von Brand, T. (1973). "Biochemistry of Parasites," 2nd ed. Academic Press, New York.

Watson, M. R., and Silvester, N. R. (1958). Studies of invertebrate collagen preparations. *Biochem. J.* **71**, 578–584.

Weinstein, P. P. (1960). Excretory mechanisms and excretory products of nematodes: an appraisal. *In* "Host Influence on Parasite Physiology" (L. A. Stauber, ed.), pp. 65–92. Rutgers University Press, New Brunswick, New Jersey.

Weinland, E. (1901). Über Kohlen-hydratzersetzung ohne Sauerstoff-aufnahme bei *Ascaris*, einen tierischen Gärungsprozess. *Z. Biol.* (42 N. F.) **24**, 44–90.

Wilson, P. A. G. (1958). The effect of weak electrolyte solutions on the hatching rate of the eggs of *Trichostrongylus retortaeformis* (Zeder) and its interpretation in terms of a proposed hatching mechanism of strongyloid eggs. *J. Exp. Biol.* **35**, 584–601.

Wright, K. A. (1979). *Trichinella spiralis*: An intracellular

parasite in the intestinal phase. *J. Parasitol.* **65**, 441–445.

Yamaguti, S. (1962). "Systema Helminthum. Vol. III. The Nematodes of Vertebrates" (2 parts). Interscience, New York.

Yorke, W., and Maplestone, B. (1926). "Nematode Parasites of Vertebrates." Blakiston, Philadelphia, Pennsylvania. (Reprinted in 1962 by Hafner Publ. Co., New York).

Zuckerman, B. M., Mai, W. F., and Rohde, R. A. (eds.). (1971). "Plant Parasitic Nematodes," Vols. I and II. Academic Press, New York.

CLASSIFICATION OF ADENOPHOREA (THE ADENOPHOREAN NEMATODES)*

PHYLUM NEMATA

Bilaterally symmetrical unsegmented pseudocoelomates; body generally elongate, cylindrical, covered by cuticle; mouth terminal, surrounded by lips; sexes separate; anterior body characteristically with 16 setiform or papilliform sensory organs and two amphids (chemoreceptors); digestive tract complete, with subterminal anus; excretory system, when present, empties through anterior, ventromedian pore; body musculature limited to longitudinally oriented muscles; no respiratory or circulatory systems; eggs with determinate cleavage, oviparous or ovoviviparous; stages in life cycle are egg, four larval stages, and adult.

Class Adenophorea

Amphids postlabial, variable in shape (porelike, pocketlike, circular, or spiral); cephalic sensory organs setiform to papilloid, postlabial and/or labial; setae and hypodermal glands commonly present; papillae usually present on body; hypodermal cells uninucleate; cuticle usually smooth, but transverse or longitudinal striations may be present; excretory organ, if present, single celled, ventral, and without collecting tubules; caudal glands (three) usually present (absent in most members of Dorylaimida, Mermithida, and Trichocephalida); usually two testes in males, with single ventral series of papilloid or tuboid preanal supplements; male tail rarely with caudal alae; parasitic species associated with invertebrates, vertebrates, and plants; many free living; most marine nematodes belong to this class.

SUBCLASS ENOPLIA

When amphids occur as subcuticular pouches, external openings are transverse (cyathiform); when internal pouch is tubiform, external opening porelike, ellipsoid, or greatly elongated; cephalic sensory organs papilliform or setiform, with or without setae; caudal glands present (in most marine forms) or absent; subventral esophageal glands (five or more) commonly open into buccal cavity through teeth or at anterior esophagus; esophagus cylindrical, conical, or divided into narrow anterior portion and larger, glandular posterior portion; stichosome (a column of rectangular cells, called **stichocytes**, supporting and secreting into esophagus), esophagus (present in

---

*Essentially after Maggenti (1982). Diagnoses of taxa comprised of free-living and/or plant parasitic species not presented. Some of the genera mentioned in this chapter are members of the class Secernentea. Their classification is presented at the end of Chapter 14.

parasitic species); cuticle generally smooth, may bear transverse and/or longitudinal markings; with parasitic representatives.

Order Enoplida
SUBORDER ENOPLINA
Superfamily Enoploidea
Superfamily Oxystominoidea
SUBORDER ONCHOLAIMINA
SUBORDER TRIPYLINA
Superfamily Tripyloidea
Superfamily Ironoidea
Order Isolaimida
Order Mononchida
SUBORDER MONONCHINA
SUBORDER BATHYODONTINA
Superfamily Bathyodontoidea
Superfamily Mononchuloidea
Order Dorylaimida
(With some species being subterranean ectoparasites of plants.)
SUBORDER DORYLAIMINA
(Some members are important parasites of plants that serve as vectors of plant viruses.)
Superfamily Dorylaimoidea
(Members of one family, Longidoridae, include representatives that parasitize plants and act as vectors of plant viruses.)
Superfamily Actinolaimoidea
Superfamily Belondiroidea
Superfamily Encholaimoidea
SUBORDER DIPHTHEROPHORINA
(Some members of this suborder are plant parasites.)
Superfamily Diphtherophoroidea
Superfamily Trichodoroidea
(Important plant parasites that cause "stubby root" and transmit plant viruses.)
SUBORDER NYGOLAIMINA
Order Trichocephalida
With protrusible axial spear in early larval stages; amphids adjacent to lip region; posterior esophageal glands in one or two rows along esophageal lumen, not enclosed by stichosome; stichosome and individual gland openings posterior to nerve ring; males and females with single gonad; germinal zones of male and female gonads extend entire length and form a serial germinal area on one side or around gonoduct; males with one or no spicule; eggs operculate; life cycle either direct (often requiring cannibalism), or indirect (involving arthropod or annelid intermediate host); adults parasitic in vertebrates.
Superfamily Trichuroidea
Stichosome of adults as single row on each side of esophagus (two rows in early larval development); body divided into elongate, narrow anterior end with esophagus with stichosome, and posterior half with reproductive system beginning at esophagointestinal junction; bacillary band (glandular and nonglandular cells of unknown function) occurs laterally; glandular tissue empty to exterior through cuticular pores; males and females with single gonad, reflexed; males with single spicule; eggs operculate—females oviparous, males small and degenerate in some species, in uterus of female; parasitic in humans and other mammals; life cycle direct or indirect. (Genera mentioned in text:

Family Trichuridae*—*Trichuris, Capillaria,* Family Trichosomoididae—*Trichuroides.*)
Superfamily Trichinelloidea
Stichosome as single, short row of stichocytes; body not distinctly divided into two regions; no bacillary band present; female genital pore opening far anterior, in region of stichosome; ovary posterior to stichosome; males with single testis but no spicule; females viviparous. (Genera mentioned in text: Family Trichinellidae*—*Trichinella, Capillaria.*)
Superfamily Cystoopsoidea
Stichosome double in larvae and adults; body not divided into two regions; vulva anterior to stichosome, in region of esophagus; digestive tract ends blindly; males and females with single gonad; males with single spicule; females oviparous; eggs operculate; parasite of sturgeons. (Genus mentioned in text: Family Cystoopsidae*—*Cystoopsis.*)

Order Mermithida
Typically long and slender, over 30 cm long; hypodermis forms eight chords (one dorsal, one ventral, two lateral, and four submedian); anterior chords with nuclei; external amphid porelike or pocketlike; mouth usually terminal, subterminal in some; stichocytes in two rows; intestine degenerate in adults, forms a storage organ (**trophosome**) connected to esophagus by bifurcating or ramifying canal; anterior portion of trophosome overlaps posterior segment of esophagus; gonads in males and females usually paired, occasionally single; males usually with two spicules (some with one); egg modified but nonoperculate; parasites of invertebrates, including insects.
Superfamily Mermithoidea
Body long and slender; cuticle smooth; anterior extremity or lip region with four submedian and two lateral sensory organs; amphids behind lip region; mouth and esophagus nonfunctional in adults; larvae with piercing tooth in mouth region; esophagus of larvae and adult similar; stichosome in two rows of more than four cells; males and females with paired gonads; vulva equatorial, leads to muscular S- or barrel-shaped vagina; males with one or two spicules; numerous papilliform supplements in three rows, post- and preanal; eggs of some species with polar filaments (**byssi**); parasites of invertebrates, especially insects. (Genera mentioned in text: Family Mermithidae*—*Mermis, Hydromermis, Capitomermis, Reesimermis, Agamermis.*)
Superfamily Tetradonematoidea
Esophagus of larvae similar to that of Dorylaimida, narrow anteriorly and swollen posteriorly; stichosome evident in adults, of three or four cells; esophagus reduced anteriorly to simple, hollow tube; larvae with small, axial stylet; males and females with paired

---

*For diagnoses of families subordinate to the Nemata, see Maggenti (1982).

gonads; males with single spicule; preanal supplements may be present; usually parasites of terrestrial and aquatic dipteran insects. (Genera mentioned in text: Family Tetradonematidae*—*Tetradonema, Mermithonema, Aproctonema.*)

Order Muspiceida

Neurosecretory system greatly reduced; amphids and cephalic papillae absent (except in one species); males unknown; digestive tract reduced in females; females didelphic, amphidelphic, and viviparous; parasitic in fish, birds, and mammals, causing tumorous growths. (Genus mentioned in text: Family Muspicidae*—*Muspicea.*)

SUBCLASS CHROMADORIA

Order Araeolaimida

SUBORDER ARAEOLAIMINA
  Superfamily Araeolaimoidea
  Superfamily Axonolaimoidea
  Superfamily Plectoidea
  Superfamily Camacolaimoidea
SUBORDER TRIPYLOIDEA

Order Chromadorida

SUBORDER CHROMADORINA
SUBORDER CYATHOLAIMINA
  Superfamily Cyatholaimoidea
  Superfamily Choanolaimoidea
  Superfamily Comesomatoidea

Order Desmoscolecida

Superfamily Desmoscolecoidea
Superfamily Greeffielloidea

Order Desmodorida

SUBORDER DESMODORINA
  Superfamily Desmodoroidea
  Superfamily Ceramonematoidea
  Superfamily Monoposthioidea
SUBORDER DRACONEMATINA
  Superfamily Draconematoidea
  Superfamily Epsilonematoidea

Order Monhysterida

Superfamily Monhysteroidea
Superfamily Linhomoeoidea
Superfamily Siphonolaimoidea

15

SECERNENTEAN NEMATODES
PHYSIOLOGY AND BIOCHEMISTRY OF NEMATODES

The nematodes discussed in this chapter belong to the class Secernentea. Although the species portray morphologic and life cycle differences, all possess one major characteristic in common, the presence of **phasmids**—the minute, usually paired, chemoreceptors found posterior to the anus. The Secernentea consists of 33 superfamilies (see classification list at the end of this chapter), most of which are parasites of animals.

BIOLOGY OF THE SECERNENTEAN NEMATODES

SUPERFAMILY RHABDITOIDEA

The Rhabditoidea includes an array of species, some free living, others symbiotic in animals, primarily invertebrates, still others symbiotic in or on plants. A few representative genera parasitic in animals are considered below.

Genus *Rhabditis*

Many parasitic species of the Rhabditoidea belong to the genus *Rhabditis*. None of the species, however, are continuously parasitic, i.e., they are parasites during only certain phases of their life cycles. In fact, many of these are only facultative parasites. Many species are found in animal droppings on which they feed, and these are dependent on some coprophagous insect to transport them to new food sites. This dependency on an insect transport host has become so intimate in some species such as *Rhabditis coarctata* that continued development of the nematode is halted if it does not make contact with the insect carrier. In this species, the L_3 (the larval form that has undergone two molts) does not develop into a sexually mature adult until it is carried in the ensheathed form on a dung beetle to fresh dung (Fig. 15.1). If this transport occurs, the larvae escape from their sheaths and undergo two additional molts to become adults.

An intermediate stage of dependency is exemplified by *Rhabditis dubia*. Normally this species can complete its life cycle in cow dung; however, occasionally an L_3 develops that requires being transported by a psychodid fly (p. 645) to fresh dung before it can develop into a sexually mature adult. Earthworms also are known to be suitable transport hosts for some species of *Rhabditis*.

A few species of *Rhabditis* can parasitize mammals, such as *R. strongyloides*, which establishes itself in cutaneous ulcers on dogs. Species that reportedly

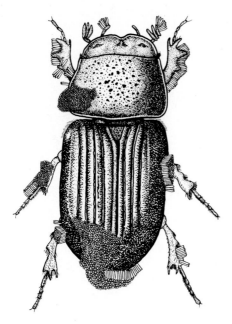

Fig. 15.1. *Rhabditis coarctata.* Numerous ensheathed larvae attached to the dung beetle *Aphodius.* (Redrawn after Triffitt and Oldham, 1927.)

parasitize humans are undoubtedly accidental, temporary parasites.

Other Rhabditoid Genera

Another genus belonging to the Rhabditoidea which is parasitic in animals is *Angiostoma.* This nematode occurs in the intestine of the salamander *Plethodon* and in various land snails.

Members of the genus *Neoaplectana* are facultative parasites of insects (Fig. 15.2). *Neoaplectana glaseri* was introduced into North America and elsewhere in the 1930s as a biologic control agent against the destructive Japanese beetle; however, this experiment has not been totally successful since the nematode commonly loses its infectivity after seven or eight transfers in artificial media.

As in most rhabditoids, *N. glaseri* is normally free living, in this case feeding on insect carcasses. However, occasionally an L_3 develops that requires ingestion or penetration into the larvae or adults of beetles. Within the beetle's gut, these parasitic larvae develop into sexually mature adults, and copulation occurs. The ovoviviparous females each give birth to approximately 15 larvae. During this parasitic phase, the host is killed and the parasitic larvae become free living, feeding on the host's carcass.

Species of *Neoaplectana* can also serve as vectors of

bacterial diseases of insects. A common causative bacterium is *Xenorhabdus nematophilus,* which is actually a mutualist of the nematode, being carried in an expanded pouch in the anterior intestine of infective larvae. When an infective larva enters an insect, a bacterial pellet is ejected from the nematode's alimentary canal into the host's hemolymph. The bacteria then multiply rapidly and cause a lethal septicemia generally in less than 48 hours. Interestingly, further development of the larval nematode vector occurs only when the insect is dead or moribund. Subsequent development of the nematode to the adult stage is rapid, taking no more than 3 to 5 days at 22 to 25°C. Several generations of *Neoaplectana* may occur with the insect carcass, but eventually infective larvae emerge and actively migrate in search of another host.

Genus *Strongyloides*

The best known rhabditoid nematode is the medically important *Strongyloides stercoralis,* an intestinal parasite of humans, other primates, and other mammals. This nematode is found primarily in the tropics and subtropics, but it also occurs in temperate climes. Relative to its incidence, Lopes Pontes (1946) found 24.8% infection among humans in Santa Casa de Misericordia, Rio de Janeiro, and Yoeli *et al.* (1963) found 18% infection among 1437 mental patients in New York City. Stoll (1947) has estimated that 34.9 million human cases of strongyloidiasis exist—21 million in Asia, 900,000 in the Soviet Union, 8.6 million in Africa, 4 million in tropical America, 400,000 in North America, and 100,000 in the Pacific islands.

Two of the sources for human infections are dogs and cats, hence the disease can be zoonotic. In North America, for example, Choquette and Gélinas (1950) have found that 2% of 155 dogs in Montreal, Canada, and Levine and Ivens (1965) have found that 1% of 175 dogs in Illinois were infected with *S. stercoralis.*

Life Cycle of *Strongyloides stercoralis.* The life cycle of *S. stercoralis* can be divided into three phases —the free-living generation, the parasitic generation, and the autoinfection phase.

Free-Living Generation. Free-living *S. stercoralis* occur in moist soil in warm climes. This generation serves as a natural reservoir for infections. Free-living males are fusiform and smaller than females (Fig. 15.3). These males, measuring 0.7 mm in length by 0.04– 0.05 mm in diameter, possess a pointed and ventrally coiled tail. The females, measuring 1 mm in length by 0.05–0.07 mm in diameter, do not possess a coiled tail.

After copulation, the uterus of the female eventually fills with eggs and occupies most of the body space. When these eggs are laid, they contain partially developed larvae. Development is completed in soil within a few hours, and the escaping L_1s are rhabditiform (Fig. 15.3). These larvae feed actively on organic debris, undergo one molt, continue to feed, and gradually develop, after four molts, into sexually mature adults.

The free-living cycle, known as the **heterogonic** cycle, may go on continuously without interruption. However, if the environment becomes unfavorable, the rhabditiform larvae molt and metamorphose into nonfeeding filariform larvae (Fig. 15.3). This larva is the infective form of *S. stercoralis*. When several such free-living generations intervene between parasitic generations, the cycle is said to be indirect.

Parasitic Generation. When filariform larvae come in contact with a human or some other suitable host, they readily penetrate the skin and are carried in the fine dermal blood vessels to larger vessels. Eventually they pass through the right atrium and ventricle of the heart to the lungs via the pulmonary artery. Filariform larvae also can infect dogs and cats through ingestion. In the lungs, the larvae rupture out of the pulmonary capillaries and become lodged in the alveoli.

From here they migrate up the bronchi and trachea to the region of the epiglottis, from where they are coughed up and swallowed. Upon reaching the host's intestine, they mature. Their normal habitat in the host's intestine is in the mucosal lining of the upper small intestine. Here, the adult females produce eggs parthenogenetically. No parasitic male adults exist. It is noted that some female adolescents may penetrate the lining of the bronchi and trachea, mature, and lay their eggs therein, but the majority migrate to the small intestine.

The adult parasitic females are delicate filiform worms that measure up to 2.2 mm long. Its esophagus is long, extending through the anterior third of the body length.

The parthenogenetic eggs hatch into typical rhabditiform larvae, which feed in the intestine but eventually pass out of the host in feces. Larvae in the stools of infected individuals measure 300–800 μm in length. Once established in moist soil, the rhabditiform larvae may develop into free-living adults (**heterogonic development**), or molt and transform into filariform larvae and become infective (**homogonic development**). If homogonic development takes place, the cycle is referred to as direct. Lumen-dwelling rhabditiform larvae also can metamorphose into filariform larvae, penetrate the intestinal wall, migrate to the lungs, and repeat the cycle, resulting in **hyperinfection**.

Autoinfection. As an alternative to hyperinfection, infective filariform larvae, en route to the exterior in feces also can reinfect the host. Specifically, if these larvae come in contact with perianal skin, they can penetrate and be carried to the lungs by the blood. This method of infection is referred to as **autoinfection**. It may be more common than previously thought (Shiroma, 1964).

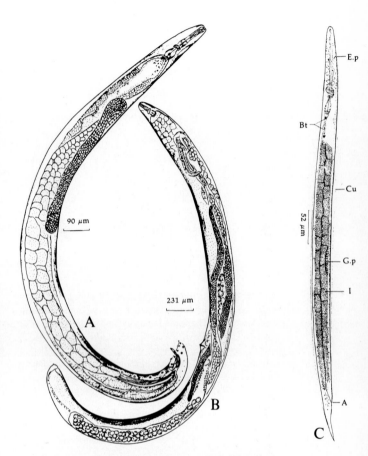

Fig. 15.2. *Neoaplectana carpocapsae.* **A.** Adult male. **B.** Adult female. **C.** Infective larval stage. A, anus; Bt, symbiotic bacteria; Cu, ensheathing second stage cuticle; Ep, excretory pore; Gp, gonad primordium; I, intestine. (All figures after Poinar, 1975; with permission of E. J. Brill.)

Other *Strongyloides* Species

Other *Strongyloides* species of economic importance include *S. papillosus* in sheep. In this species, males are extremely scarce even in the free-living generation. The females usually reproduce parthenogenically. *Strongyloides ransomi* is a fairly common species found in pigs. During its life cycle, unlike *S. stercoralis*, the eggs of the parasitic females do not give rise to rhabditiform larvae in the intestine; rather, the eggs pass to the exterior and then hatch.

Many other species of *Strongyloides* are known. *S. fuelleborni* is a common human parasite in central and eastern Africa (Pampiglioni and Ricciardi, 1971). It has also been reported from other primates. *S. ratti* is a common parasite of rats. Readers interested in

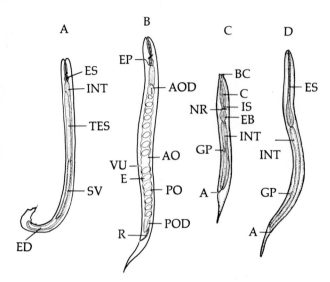

Fig. 15.3. Free-living stages of *Strongyloides ster-coralis*. A. Male from soil. **B.** Female from soil. **C.** Rhabditi-form larva. **D.** Filariform larva. (**A** and **B**, redrawn after Craig and Faust, 1951.) A, anus; AO, anterior ovary; AOD, anterior oviduct; BC, buccal capsule; C, corpus of esophagus; E, eggs *in utero*; EB, end bulb of esophagus; ED, ejaculatory duct; EP, excretory pore; ES, esophagus; GP, genital primordium; INT, intestine; IS, isthmus of esophagus; NR, nerve ring; PO, posterior ovary; POD, posterior oviduct; R, rectum; SV, seminal vesicle; TES, testis; VU, vulva.

species parasitic in domestic animals are referred to Levine (1980).

Immunity to *Strongyloides*

Ample evidence suggests that *Strongyloides* spp. induce at least partial immunity in their hosts. For example, *S. papillosus* is extremely common in lambs, markedly reduced in yearlings, and quite rare in older sheep. Turner (1959) has demonstrated experimentally the development of immunity in parasite-free lambs by exposing them to 10,000, 20,000, or 30,000 larvae every 2 days for 20 days. They were then allowed to graze on a pasture where they were naturally exposed to infective larvae. None of the lambs developed acute signs of strongyloidosis, and they remained strongly resistant to further infection several weeks later when challenged with 300,000 larvae, a dose normally fatal to lambs and kids.

Lambs experimentally immunized against *S. papil-losus* develop at least two antibodies, one associated with the *β*- and the other with the *γ*-globulin fractions of the blood. When larvae of *S. papillosus* are placed in

antisera, oral and anal precipitates appear within 4 hours, and death occurs in 22 hours.

Younger hosts are also more susceptible to other species of *Strongyloides*. This is believed to be due to the lack of protective antibodies that occur in older animals as the result of previous exposure to the parasites.

Differences in infectivity to humans among races of *Strongyloides stercoralis* from different geographic localities have been suggested.

Human *Strongyloidosis*

The pathogenesis of human strongyloidosis can be divided into three phases: cutaneous, pulmonary, and intestinal.

Cutaneous. During invasion of the skin by infective larvae, there is slight hemorrhage and swelling. Also, intense itching occurs. Commonly, the invasion sites are secondarily invaded by infectious microbial agents, resulting in severe inflammation and necrosis.

Pulmonary. During larval migration through the lungs, damage to host tissues may result in massive cellular reactions, which can delay or prevent further larval migration. If this occurs, the parasites may become established in the lungs and commence reproducing as if they had become established in the intestine. When this occurs, the patient suffers from a burning sensation in the chest, coughing, and other symptoms of bronchial pneumonia.

Intestinal. Although young female worms penetrate the intestinal mucosal crypts, where they mature and invade the surrounding tissue, they seldom penetrate deeper than the muscularis mucosa. In some instances, however, this does happen, with the worms randomly migrating through the mucosa and laying several eggs each day. If this occurs, the patient suffers from a localized burning sensation or pain in the abdominal area. Also, destruction of host tissues by adults and larvae usually results in the sloughing of patches of mucosa and the development of fibrotic lesions. If such wounds are invaded by bacteria, death due to septicemia may occur.

As is the case with most parasitic infections, individuals in whom immunologic resistance has been lowered as a result of the other diseases or immunosuppressant drugs portray more severe, sometimes fatal, reactions to strongyloidosis.

SUPERFAMILY STRONGYLOIDEA

The superfamily Strongyloidea includes nematodes found along the entire intestinal tract of reptiles, birds, and mammals. They are also rarely found in fish.

The most frequently encountered members of this superfamily are *Strongylus* in equines, *Ransomus* in rodents, *Cyathostomum* in equines, and *Globocephalus* in primates and sulids.

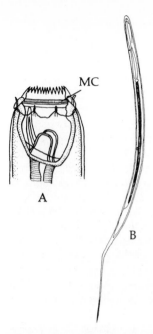

Fig. 15.4. Morphology of some strongyloid nematodes. A. Anterior end of *Strongylus* sp. showing mouth collar. **B.** L₃ of *Strongylus equinus*, the infective stage for horses. MC, mouth collar. (Redrawn after Poluszynski, 1930.)

Genus *Strongylus*

The members of *Strongylus* have mouths that are directed anteriorly and are surrounded by a corona radiata or external leaf-crown arising from the mouth collar (Fig. 15.4).

Strongylus equinus (Fig. 15.4), commonly known as the large palisade worm, occurs in the caecum and, rarely, in the colon of the horse, ass, mule, zebra, and other equines. It is common throughout the world. Both the males and females are dark reddish-gray. The male measures 26–35 mm long by 1.1–1.3 mm wide, and its spicules are slender and 2.6 mm long. The female is 38–55 mm long and 1.8–2.3 mm wide. Its tail is obtuse, and it includes an amphidelphic uterus. A number of other species of *Strongylus* have been recorded from equines (Levine, 1980).

Genus *Cyathostomum*

Cyathostomum coronatum occurs throughout the world in the large intestine of the horse, ass, and zebra. The males are 7–9 mm long, and the females are 9–10 mm long. This and other species of *Cyathostomum* (see Levine, 1980) are found fairly commonly in equines.

Strongyloid Life Cycles. Life cycles among the strongyloids are essentially identical during the initial phases. Eggs passing out of the host contain blastomeres. Development continues in damp soil in the presence of oxygen. The escaping larvae are of the rhabditiform type. These larvae undergo two molts and metamorphose to the strongyliform* L₃, which is ensheathed within the cuticle of the rhabditiform L₂.

The ensheathed larvae are ingested by the horse or whatever the host may be, and once they reach the small intestine, the pattern varies, depending on the species. Some of the known variations are as follows.

1. The strongyliform larvae burrow into the intestinal mucosa, causing inflammation and nodule formation. The larvae dwell in the swellings until after the fourth molt, at which time they reenter the lumen of the intestine. This pattern holds true for *Basicola* in cattle and for *Cyathostomum*.

2. The strongyliform larvae burrow through the intestinal wall and become established in the circulatory system. They are carried through an extensive migratory pathway involving the liver and lungs, and finally become lodged primarily in the anterior mesenteric artery, where they cause aneurysm and thrombosis. This is presumably the pattern in the case of *Strongylus vulgaris* in the horse.

3. The larvae penetrate the intestinal wall, molt, and reenter the lumen of the intestine. This pattern is exemplified by *Strongylus equinus*, also a parasite of horses.

4. The larvae penetrate the intestinal wall but do not migrate. Rather, they come to lie against the intestinal wall under the peritoneum and gradually shift toward the colon and caecum, where they burrow back into the intestinal wall, form nodules, molt, and eventually return to the lumen. This is the case in *Strongylus edentatus* in horses.

SUPERFAMILY DIAPHANOCEPHALOIDEA

This little-known superfamily includes nematodes parasitic in the stomach and intestine primarily of reptiles, although a few occur in amphibians. A representative is *Diaphanocephalus galeatus* (Fig. 15.5) found in the lizard *Tupinambis* in Brazil, Bolivia, Argentina, and Surinam. By far the largest genus belonging to this superfamily is *Kalicephalus*, with numerous species occurring in the alimentary tracts of snakes and lizards all over the world.

---

*The strongyliform larvae of strongyloids are also referred to as filariform larvae.

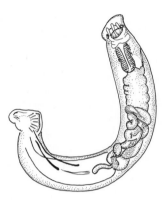

Fig. 15.5. *Diaphanocephalus galeatus,* **a parasite of lizards.** A male specimen. (Redrawn after Ortlepp in York and Maplestone, 1926.)

SUPERFAMILY ANCYLOSTOMATOIDEA

The members of Ancylostomatoidea are commonly referred to as hookworms. Among other characteristics (p. 568), these worms possess heavily sclerotized mouth parts. They are blood feeders occurring in the small intestine of various mammals, reptiles, and amphibians.

Included in this superfamily are the genera *Ancylostoma*, with representatives parasitic in humans, primates, carnivores, and mustelids; *Cyclodontostomum*, with representatives parasitic in rodents; *Galoncus*, with species parasitic in tigers and leopards; and *Gaigeria*, with species parasitic in sheep and goats.

Human Hookworm Disease

Hookworm disease has been, and remains, among the most prevalent and important of the parasitic diseases of humans. Unlike malaria, amoebiasis, or schistosomiasis, hookworm disease may not be clinically spectacular, yet it can affect entire populations by gradually sapping its victims of their strength, vitality, and health. As exemplified in certain parts of the Near East and the Far East, and not too many years ago in the southeastern United States, the victims become lethargic and nonproductive, resulting in economic losses beyond computation. Readers should not be left with the impression that hookworm disease has been controlled or eradicated. It remains a major public health problem in many parts of the world, especially in developing countries.

Two principal species of hookworms infect humans as adults, *Ancylostoma duodenale* and *Necator americanus*. Stoll (1947) has estimated that 72.5 million humans are infected with *A. duodenale*, the majority,

59 million, in Asia. He has also estimated that 384.3 million persons in the world at the time were infected with *N. americanus*, among whom 1.8 million lived in North America. The latter number has been reduced since Stoll's estimate. In North America a more realistic estimate today would be 1 million.

In individuals harboring either *A. duodenale* or *N. americanus*, manifestation of the disease is primarily dependent on two factors: the number of parasites present and the nutritional state of the host. In the case of *N. americanus*, the presence of fewer than 25 worms will not cause recognizable clinical symptoms; the occurrence of 25 to 100 worms usually leads to light symptoms; the presence of 100 to 500 worms produces considerable pathologic alterations and moderate symptoms; the presence of 500 to 1000 worms results in severe pathologic alterations and severe symptoms; and the occurrence of over 1000 parasites leads to grave damage and drastic and often fatal consequences. In the case of *A. duodenale*, because it ingests more blood, fewer worms, about 100, will cause severe symptoms. In all instances, if the host is malnourished, fewer worms will result in graver symptoms.

The pathogenesis of human hookworm disease can be divided into three phases: the invasion (or cutaneous) phase, the migration (or pulmonary) phase, and the intestinal phase.

Invasion Phase. This phase commences when the infective larvae penetrate the skin. Although little damage is inflicted to the superficial skin layers, the larvae, while penetrating blood vessels, stimulate a cellular reaction that may isolate and kill them. This, combined with the stimulatory effects of pyrogenic bacteria introduced into the skin with the invading larvae, results in an urticarial reaction commonly known as ground itch.

Migration Phase. This phase occurs when larvae escape from capillary beds in the lung, enter the alveoli, and progess up the bronchi to the throat. This migration causes hemorrhages that could be serious if a large number of worms are involved. Otherwise, dry coughing and a sore throat are the only symptoms.

Intestinal Phase. This is the most serious phase of hookworm infection. As described later, on reaching the small intestine, young worms become attached to the mucosa by their buccal capsule and teeth and commence to suck blood. When over 25 *Necator americanus* (each of which ingests about 0.03 ml of blood per day) or if 10 *Ancylostoma duodenale* (each of which ingests about 0.15 ml of blood per day) are present, even when about 40% of the iron removed from the host in lost blood is reabsorbed before it is removed from the body in the parasites' egesta, an iron-deficiency anemia develops. This is accompanied by intermittent abdominal pain, loss of appetite, and a craving for soil (geophagy). In heavy infections,

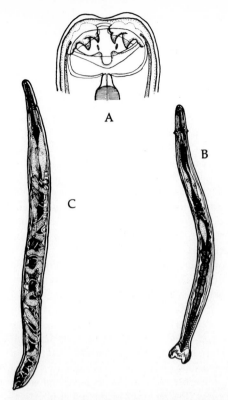

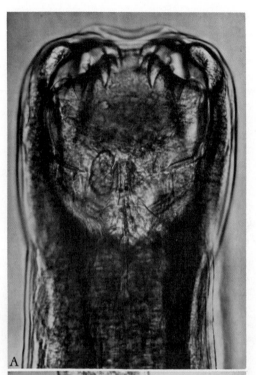

Fig. 15.6. *Ancylostoma duodenale.* **A.** Anterior end showing buccal capsule with two pairs of teeth. (Redrawn after Craig and Faust, 1951.) **B.** Adult male. **C.** Adult female. (**B** and **C**, redrawn after Looss, 1902.)

severe protein deficiency, dry skin and hair, edema, distended abdomen (especially in children), delayed puberty, cardiac failure, mental dullness, and even death may occur.

Larva Migrans

As in schistosome dermatitis, hookworms of animals, for which humans are incompatible hosts, often attempt to penetrate human skin. These normally fail to pass through the stratum germinativum of the skin and migrate for some time at that level, causing a skin eruption known as **cutaneous larva migrans**, or creeping eruption. This aspect of host-parasite relationship is expounded upon more fully in another section (p. 560).

Although skin penetration is the most common avenue of infection, hookworms, especially those of animals, can become established within their hosts through oral ingestion of the strongyliform larvae. Furthermore, Foster (1932) has demonstrated that intrauterine infections are also possible.

Genus *Ancylostoma*

Ancylostoma duodenale (Fig. 15.6), the Old World hookworm, has been reported in southern Europe,

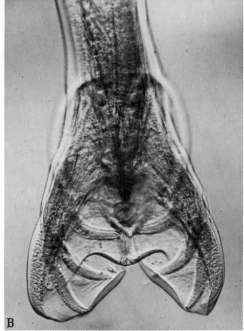

Fig. 15.7. *Ancylostoma caninum,* **the dog hookworm.** **A.** Anterior end of adult showing the three pairs of dorsal teeth. **B.** Copulatory bursa of male. (Courtesy of General Biological Supply House, Inc., Chicago, Illinois.)

North Africa, India, China, Japan, and Southeast Asia. It also has been reported in the New World among Paraguayan Indians, in small areas of the United States, and in the Caribbean. It has been known to occur in coal mines in Belgium and Great Britain, where it was responsible for a type of anemia among miners. This anemia also occurred among tunnel construction workers in Switzerland, Germany, and Italy. At least 65 million humans are infected with this hookworm, the majority, 50 million, in Asia.

The males measure $8-11$ mm $\times 0.5$ mm, whereas the females measure $2-13$ mm $\times 0.6$ mm. The buccal cutting plate is armed with two pairs of teeth (Fig. 15.6). This species is primarily a human parasite, but it can be experimentally induced to parasitize monkeys, dogs, and cats. Its life cycle parallels that of *Necator americanus* described below (p. 509). Note that Schad *et al.* (1984) have reported humans infected with *A. duodenale* from eating raw or poorly cooked meat harboring L_3s.

Ancylostoma braziliense is a parasite of dogs and cats in the tropics and subtropics, including the southeastern United States along the Atlantic and Gulf Coasts. This species has been reported from humans. It is the most common causative agent of cutaneous larva migrans.

Ancylostoma caninum is a relatively large species with females measuring up to 20 mm in length. It is commonly found in dogs, cats, foxes, wolves, and other carnivores in temperate as well as in tropical and subtropical areas (Fig. 15.7).

Considerable research has been conducted on the biology of *Ancylostoma caninum*. Several investigators have reported that dog-adapted strains of this hookworm produce fewer eggs when introduced into cats, while this is not seen in cat-adapted strains in dogs. No explanation for this is yet known. Wells (1931) has reported that a single hookworm may ingest as much as 0.8 ml of blood in 24 hours, although a more conservative amount is 0.1 ml. This alone is severe enough damage, but considerable additional damage is incurred when worms perforate the intestinal mucosa, permitting microbial secondary infections, which usually become necrotic.

Hookworms secrete an anticoagulant during the feeding process. This secretion, originating in the amphidial glands, prevents the clotting of the host's blood during feeding and for a short period afterward. In addition, a proteolytic enzyme is secreted in the esophagus and may play a role in extracorporeal digestion. Age resistance to *A. caninum* occurs as a result of repeated infections. Furthermore, Otto (1941) has demonstrated that the degree of immunity is dependent on the intensity of the initial infection. For example, if a dog is initially challenged with 1000 to 2000 infective larvae, an immunity to a second challenge of 50,000 larvae will result. However, if the initial challenge consists of only 500 larvae, the immunity developed is insufficient to withstand a

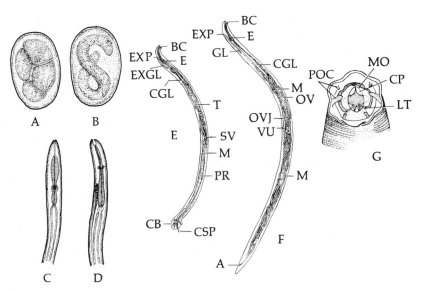

Fig. 15.8. Stages in the life cycle of *Necator americanus*. A. Egg enclosing blastomeres. **B.** Egg enclosing larva. **C.** Rhabditiform larva. **D.** Strongyliform larva. **E.** Adult male. **F.** Adult female. **G.** Anterior end of adult. (**E–G**, after Craig and Faust, 1951.) A, anus; BC, buccal capsule; CB, copulatory bursa; CGL, paired cephalic gland; CP, cutting plates; CSP, copulatory spicules; E, esophagus; EXGL, excretory gland; EXP, excretory pore; L, lips; LT, lateral teeth; M, midgut; MO, mouth; OV, ovary; OVJ, ovijector; POC, papillae of outer circlet; PR, prostate gland; SV, seminal vesicle; T, testis; VU, vulva.

second challenge of 50,000 larvae, and the host will be killed.

Ancylostoma ceylanicum is normally a parasite of carnivores in Sri Lanka, southeast Asia, and the Malaysia Archipelago; however, it has been reported from humans in the Philippines (Velasquez and Cabrera, 1968).

Other Hookworms

Another group of hookworms, including *Uncinaria* in carnivores, *Arthrostoma* in cats, *Monodontus* in rodents and deer, and *Necator* in humans, other primates, mustelids, and rodents, also belongs to the superfamily Ancylostomatoidea. Among these, *Necator* is the principal genus.

Genus *Necator*

Necator americanus (Fig. 15.8), often called the New World, or American, hookworm, is widely distributed in the southern United States, Central and South America, and the Caribbean. It is now known to be indigenous in Africa, India, southeast Asia, China, and southwest Pacific islands. It may have been introduced into the Americas during the slave trade era or even earlier.

The males, measuring 7–9 mm in length by 0.3 mm in diameter, possess a large bursa, whereas females, measuring 9–11 mm in length by 0.4 mm in diameter, are larger and lack a bursa.

Life Cycle of *Necator americanus*. The life history of *N. americanus* is direct, with a free-living larval phase. The adults are parasitic in the small intestine of humans. The worms hold onto the intestinal wall and feed on blood and tissue exudates (Fig. 15.9). Von Brand (1938) has postulated that the erythrocytes ingested by adult hookworms provide much of the oxygen necessary in the worm's metabolism, since these parasites live in a portion of the alimentary tract that contains very little oxygen.

The females of *N. americanus* are prolific egg producers. It has been estimated that a single female can produce 5,000 to 10,000 eggs per day. When the eggs pass out of the host in feces, the embryo is at the four-blastomere stage (Fig. 15.8).

The embryo continues to develop in moist soil that is rich in oxygen. If such an ideal habitat exists, the egg hatches within 24 hours, and the escaping larva is rhabditiform (Fig. 15.8). The rhabditiform larva undergoes two molts and becomes a strongyliform larva, which is characterized by a long cylindrical esophagus with a terminal bulb that is not sharply demarked from the anterior truncate portion of the esophagus (Fig. 15.8). The free-living larvae feed primarily on soil bacteria. A considerable amount of the nutrients, especially lipids, is stored in the body and serves as reserve food until the larva enters the host. When the rhabditiform larva undergoes the second ecdysis, usually on the fifth day, giving rise to the infective

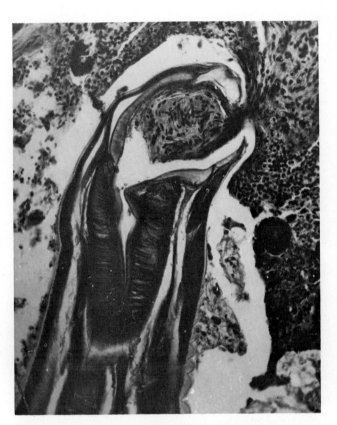

Fig. 15.9. *Necator americanus.* Photomicrograph of adult holding onto host's intestinal mucosa. (Courtesy of Armed Forces Institute of Pathology; Negative No. 33810.)

strongyliform larva, its cuticle is retained as an enclosing sheath. This sheath is usually retained until the infective larva penetrates the skin of the host.

Infection of humans is accomplished when ensheathed strongyliform larvae penetrate the skin, usually of the feet and legs. Although little damage is done during the initial burrowing, the minute perforations on the skin do serve as sites for secondary infections by bacteria and fungi. If large numbers of larvae penetrate a restricted area, dermatitis usually ensues. Once within the dermis, the larvae cause considerable damage while penetrating the minute blood vessels and lymphatics in which they are eventually carried to the right side of the heart. These larvae pass through the heart and are carried to the lungs via the pulmonary artery. Like the larvae of *Ascaris lumbricoides*, the larvae of *N. americanus* rupture out of the capillaries within the lungs and become lodged in the alveoli. From here they migrate up the bronchioles, bronchi, and trachea, and are coughed up and swallowed.

Once the larvae reach the small intestine, they actively burrow into the intervillar spaces, where the third molt occurs on the third to fifth day after entering the host. After the third ecdysis, the larvae become lumen dwellers and increase greatly in size, reaching 3.5 mm in length. A fourth molt follows, after which the worms become adults. The entire life cycle of *N. americanus* takes approximately six weeks.

Genus *Uncinaria*

The genus *Uncinaria* includes *U. stenocephala*, which parasitizes dogs, cats, foxes, and other carnivores. It is commonly found in colder climates such as Canada, whereas *Ancylostoma caninum* is found further south, in the United States.

SUPERFAMILY TRICHOSTRONGYLOIDEA

The Trichostrongyloidea consists of a large group of nematodes that parasitize all groups of vertebrates except fish. The ruminants, however, harbor the great majority of the species. The life cycles are direct, the first three larval stages being saprozoic and the L_3 entering the host ensheathed in the cuticle of the L_2.

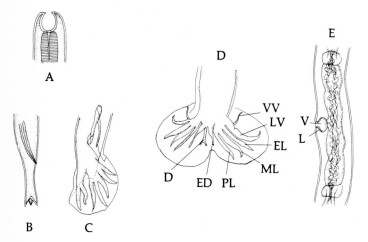

Fig. 15.10. Some trichostrongyloid nematodes. A. Anterior end of *Ollulanus tricuspis*. **B.** Posterior end of adult female *O. tricuspis* showing presence of three cusps. **C.** Posterior end of adult male *O. tricuspis* showing copulatory bursa. (**A–C**, redrawn after Cameron, 1923.) **D.** Copulatory bursa of male *Trichostrongylus axei*. (Redrawn after Ransom, 1911.) **E.** Vulvar region of female *Trichostrongylus colubriformis*. (Redrawn after Kalantaryan, 1928.) D, dorsal ray; ED, externodorsal ray; EL, externolateral ray; L, vulvar lip; LV, lateroventral ray; ML, mediolateral ray; PL, posterolateral ray; V, vulva; VV, ventroventral ray.

Included in this superfamily is *Dictyocaulus* with species parasitic in the bronchi of herbivores, and *Ollulanus* with species parasitic in the stomach and intestine of felines, foxes, pigs, and rodents.

Dictyocaulus arnfieldi is the only lungworm of horses, mules, and asses. Its life cycle is direct. Embryonated eggs are laid by female worms in the bronchioles, some hatching at this site. The unhatched eggs and larvae are coughed up and some leave the host in mucus from the nose or mouth, but the majority are swallowed and passed out in feces. In about a week, the ensheathed L_3 develops. This is the infective form. Horses become infected by ingesting larvae in contaminated food or water. Upon entering the host, the larvae penetrate the intestinal wall and migrate via the lymph channels, blood vessels, and heart to the lungs, where they molt and mature. This parasite is apparently not very pathogenic, although heavy infections do cause coughing and bronchitis.

Ollulanus tricuspis is an interesting parasite found in the stomach of cats. Its life cycle is quite different from those of most parasitic nematodes. The eggs hatch within the body of the female and the L_1 and L_2 develop within the mother. The ensheathed L_3 are found free in the host's stomach, as are the L_4s. All of the larval stages, except the first, possess a tricuspid tail similar to that of the adult female (Fig. 15.10). Since all of the stages in the life cycle of *O. tricuspis* occur in the host's stomach, it was puzzling at first how the infection is transmitted. However, Cameron (1927) has demonstrated that infection could be transmitted by feeding stomach contents or vomitus of infected cats. Thus, it appears that this parasite is normally transmitted when one cat eats the vomitus of another.

Ollulanus tricuspis is not highly pathogenic, although it does cause inflammation of the gastric mucosa and catarrhal gastritis.

Also included in the superfamily Trichostrongyloidea are the genera *Trichostrongylus*, *Cooperia*, *Obeliscoides*, *Ornithostrongylus*, *Haemonchus*, and *Nematodirus*.

Genus *Trichostrongylus*

The genus *Trichostrongylus* includes species parasitic in nonhuman mammals, but *T. extenuatus* and *T. tenuis* have been reported in birds, and *T. orientalis* and *T. probolurus* have been reported infecting humans in Asia and Egypt. These are small, slender worms with a small head and without a buccal cavity or cervical papillae. The copulatory bursa in males has large lateral lobes and a symmetrical dorsal lobe (Fig. 15.10). In females the vulva is a short distance behind the middle of the body and usually has prominent lips (Fig. 15.10).

Trichostrongylus axei. This minute stomach worm of cattle, sheep, and horses is an important trichostrongyloid (Fig. 15.11). Its life cycle is direct. Eggs that pass out of the host include 16 to 32 blastomeres. These hatch in damp soil in 3 to 4 days, and the escaping larvae undergo two molts and are transformed into infective L_3s. These larvae are ensheathed within the cuticle of the L_2s. When ingested by horses, they exsheath and burrow into the gastric mucosa, where they undergo two more molts and attain maturity.

Other *Trichostrongylus* Species

Other familiar species of *Trichostrongylus* include *T. ransomi*, *T. affinis*, and *T. calcaratus*—all found in the small intestine of rabbits; *T. longispicularis* (Fig. 15.11), *T. colubriformis*, and *T. capricola* in the stomach and small intestine of ruminants; and *T. delicatus* in squirrels.

Genus *Cooperia*

All the members of this genus are parasites of ungulates. The cuticle at the anterior end of the body is transversely striated and dilated, giving the head a swollen or bulbous appearance. The cuticle of the remainder of the body has 14 to 16 longitudinal lines, which are transversely striated. The copulatory bursa of males is composed of two large lateral lobes and a small dorsal lobe (Fig. 15.11). The vulva of the female is posterior to the middle of the body and may be covered by a flap.

Cooperia punctata. This species occurs in the small intestine and rarely in the abomasum (the fourth stomach) of cattle, sheep, water buffalo, pronghorn, and several other wild ruminants. Rabbits can be infected experimentally, but the infection rate is low. It is widely distributed and occurs in North America. It is especially important in Hawaii where its presence in large numbers in cattle causes great loss to the beef industry. Its life cycle is direct.

Genus *Obeliscoides*

There are three recognized species in this genus, and all are parasites of lagomorphs. Cervical papillae occur on the species, and the copulatory bursa of males has two large, lateral lobes and a small, well-defined dorsal lobe. The vulva of females is in the posterior quarter of the body.

Obeliscoides cuniculi. This is the stomach worm of rabbits. It is common in several species of cottontail, the marsh rabbit, swamp rabbit, and jackrabbit in the United States, Mexico, and Canada. The adult male is 10–14 mm long with spicules that measure 440–540 μm long, and the female is 15–18.5 mm long.

Genus *Ornithostrongylus*

The members of this genus are attenuated anteriorly. The cuticle is inflated about the head and bears longitudinal lines. The copulatory bursa in males has two well-developed lateral lobes and a very small dorsal lobe. The two spicules are of equal length and each terminates in three points. The tail of females is conical and obtuse, and has a terminal spine. The vulva is situated in the posterior half of the body.

Ornithostrongylus quadriradiatus. This nematode occurs in the small intestine of the domestic

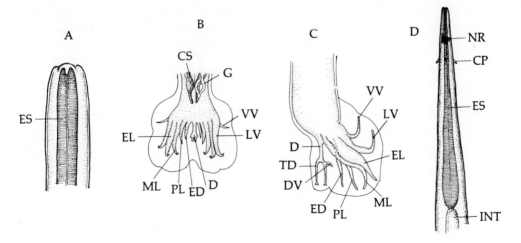

Fig. 15.11. Some trichostrongyloid nematodes. A. Anterior end of *Trichostrongylus axei*. (Redrawn after Yorke and Maplestone, 1926.) **B.** Copulatory bursa of male *Trichostrongylus longispicularis*. (Redrawn after Rose, 1960.) **C.** Side view of copulatory bursa of male *Cooperia punctata*. **D.** Anterior end of *Haemonchus contortus*. (**C** and **D**, redrawn after Ransom, 1911.) CP, cervical papilla; CS, copulatory spicule; D, dorsal ray; DV, ventral branch of dorsal ray; ED, externodorsal ray; EL, externolateral ray; ES, esophagus; G, gubernaculum; INT, intestine; LV, lateroventral ray; ML, mediolateral ray; NR, nerve ring; PL, posterolateral ray; TD, terminal branch of dorsal ray; VV, ventroventral ray.

pigeon, mourning dove, and other species of doves. It occurs in North and South America, South Africa, India, Australia, the Soviet Union, Taiwan, and Sri Lanka. It also occurs in Hawaii. The male is 9–12 mm long, whereas the female is 18–24 mm long. The life cycle is direct.

Genus *Haemonchus*

The members of this genus have a small head region, being less than 50 μm in diameter. The small buccal cavity is surrounded by three inconspicuous lips. A slender tooth (lancet) originates from the dorsal surface of the base of the buccal cavity. The copulatory bursa of males has large lateral lobes and a small, asymmetrical, dorsal lobe. Short spicules are present. The female vulva is situated in the posterior part of the body and is often covered by a prominent flap that projects posteriorly. All of the species are parasites of ruminants.

Haemonchus contortus. This parasite, commonly called the twisted stomach worm, is a blood-sucking nematode occurring in the abomasum of sheep and other ruminants (Fig. 15.11). The males measure 10–20 mm in length, and the females are 18–30 mm long. Freshly acquired specimens are generally reddish due to the host's blood contained within.

Haemonchus contortus is of considerable interest to veterinarians, since it can be lethal in heavy infections. Even in medium-grade infections, the host is weakened physically. It has been estimated that 4000 worms can suck about 60 ml of blood a day, and it is not uncommon to find thousands of worms in a single host.

Life Cycle of Haemonchus contortus. Eggs laid by females pass to the exterior in the host's feces. They hatch under favorable climatic conditions, and the L_1s are free living in soil. Two molts ensue, and hence, three nonparasitic generations occur in soil. The L_3 retains the cuticle after the second molt, so it is unable to feed. It survives at a low metabolic rate on stored nutrients. This is the infective stage of *H. contortus*. It is unable to enter its host by penetration; rather, it must be ingested while the host grazes. On reaching the sheep's abomasum, the larvae undergo two more molts and become adults.

Immunity to *Haemonchus*. Stoll's report in 1928 on the so-called self-cure in sheep to *H. contortus* may be regarded as the beginning of studies on host immunity to helminths. Stoll experimented with two lambs. He infected one with 45 *H. contortus* larvae and left the other uninfected. Both lambs were permitted to graze in the same field. Nineteen days after the first lamb was infected, it began to pass eggs, thus con-

taminating the pasture. The second lamb, which was not experimentally infected, began passing eggs on the 54th day. Approximately 10 weeks after the first egg passing, both lambs reached their peak in egg passing; 10,600 and 13,000 eggs per gram of feces were passed by the originally infected and noninfected lambs, respectively.

A rapid decline in the number of eggs passed followed, until none was passed at the second and third weeks after the peak. Although the lambs continued to ingest up to 14,000 infective larvae per day, no eggs were recovered thereafter. Furthermore, as the lambs continued to be reinfected, the older egg-producing worms were expelled. These findings caused Stoll to describe this phenomenon as "self-cure."

Soulsby (1962) has since demonstrated that, while the L_4s are buried in the mucosa, in contact with blood vessels, there is a rise in the histamine concentration in the blood and a rapid synthesis of antibodies. Since the adults are blood feeders, both histamine and antibodies are ingested along with blood, and this causes the worms to release their hold on the host's mucosa and to be rapidly passed out in feces. Thus, every time an infected sheep becomes reinfected, the older worms are sloughed since the presence of larvae from the second infection in the mucosa causes elevations in both histamine and antibody levels. This interesting mechanism, a sort of parasitologic musical chairs, prevents hyperinfections, i.e., the accumulation of large numbers of worms. A similar phenomenon has been demonstrated in rats parasitized by another nematode, *Nippostrongylus muris*.

Genus *Nematodirus*

Nematodes belonging to this genus are very slender and attenuated anteriorly. The circular mouth is surrounded by a denticulate crown. Behind this crown is an internal circle of six large papillae followed by an external circle of eight small papillae. A dorsal esophageal tooth is present. The anterior end of the body is inflated. The body has about 18 longitudinal striations but no cervical papillae. The copulatory bursa of males consists of two large lateral lobes and a small or indefinite dorsal lobe. The female's vulva is in the posterior part of the body and its tail is usually conical and truncate, with a pointed process at its tip. There are about 28 species known and these occur primarily in ruminants, but also in rodents, lagomorphs, and a few other mammals.

Nematodirus filicollis. *N. filicollis* (Fig. 15.12) is found in the small intestine of the sheep, goat, ox, and various wild ruminants including the bighorn sheep, elk, and species of deer throughout the world. The male measures 10–15 mm long and the female 15–20 mm long.

The life cycle of *Nematodirus* differs from those of

the other trichostrongyloids in that the first two larval molts occur within the eggshell. Eggs passing out of the host in feces contain four to eight blastomeres and subsequent development, although slow, is influenced by the ambient temperature. L_3s may remain within the eggshell for a considerable time before hatching. In fact, they may overwinter within the shell. The definitive host becomes infected when L_3s are ingested while grazing. These larvae molt and remain in the small intestine and develop to adults therein.

Also included in the superfamily Trichostrongyloidea are the genera *Heligmosomum*, *Nematospiroides*, and *Nippostrongylus*. These are briefly discussed below.

Genus *Heligmosomum*

This large genus consisting of at least 17 species includes members that are exclusively parasitic in

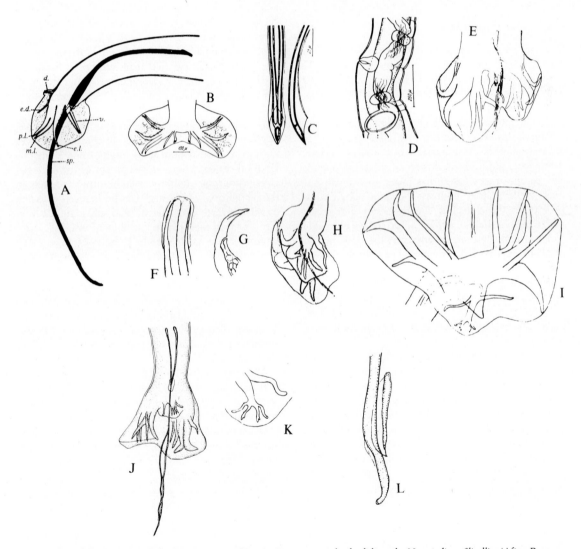

Fig. 15.12. Some trichostrongyloid nematodes. A. Posterior end of adult male *Nematodirus filicollis*. (After Ransom, 1911). **B.** Ventral view of bursa of adult male *Nematodirus filicollis*. **C.** Tips of spicules of adult male *Nematodirus filicollis*. **D.** Vulva of adult female *Nematodirus filicollis*. (**B, C, D** after Raevskaya, 1931.) **E.** Posterior terminal of adult male *Nematospiroides dubius*. **F.** Anterior end of *Nematospiroides dubius*. **G.** Lateral view of posterior end of *Nematospiroides dubius* adult female. **H.** Lateral view of posterior end of *Nematospiroides dubius* adult male. **I.** Bursa of adult male *Nippostrongylus muris*. **J.** Posterior end of adult male *Nippostrongylus muris*. **K.** Dorsal bursal ray of adult male *Nippostrongylus muris*. **L.** Spicule and gubernaculum of adult male *Nippostrongylus muris*. (**E–L** after Travassos, 1937.) d, dorsal ray; e.d., externodorsal ray; e.l. externolateral ray; m.l., mediolateral ray; p.l., posterolateral ray; sp, spicule; v, ventral rays.

rodents. The bodies of these worms are threadlike (filiform), delicate, and generally spirally coiled. Their cuticles are decorated with transverse striations and oblique or longitudinal ridges and are inflated at the anterior end. The mouth is not surrounded by lips. Some of the more common species are *H. costellatum* in *Microtus* spp. in Europe, North America, and the Middle East; *H. halli* in various rodents in Europe; and *H. laeve* in European and North African rodents.

Genus *Nematospiroides*

The members of this genus also have spirally coiled bodies. Their cuticles are decorated with about 30 longitudinal striations, and they are both expanded and annulated at the anterior end. One species, *Nematospiroides dubius*, is a popular experimental animal in immunologic and physiologic studies.

Nematospiroides dubius. *N. dubius* (Fig. 15.12) is parasitic in the small intestine of the house mouse, deermouse, and other wild rodents in North America and Europe. Cross (1960) has found that it will not infect the laboratory rat unless the immunologic competence of the rodent is suppressed or diminished with cortisone, but Liu and Ivey (1961) have been able to infect untreated rats. In any case, the rat is not an ideal host. The partial incompatibility is manifested by the egg production of the parasite. Specifically, Scott *et al.* (1959) have reported that female *N. dubius* produces about 1273 and 1768 eggs per day in mice but only 317 to 436 eggs per day in rats. It is noted that male rats are more susceptible to infection than females.

Life Cycle of Nematospiroides dubius. The life cycle of *N. dubius* is typical of the trichostrongyloids. The males, measuring about 6 mm long and with spicules 540–600 μm long, and females, measuring about 13 mm long, occur in the small intestine of mice. Eggs laid by females measure 75–90 μm × 43–58 μm. These pass out in feces, and if a suitable environment is available, the hatching larvae undergo two molts and the ensheathed L_3s (the infective stage) are attained. Rodents become infected by accidentally ingesting L_3s. Two additional molts occur in the host before the adult stage is attained. The L_4s are usually encysted and embedded in the host's intestinal mucosa, but after the final molt return to the lumen and mature.

Genus *Nippostrongylus*

This genus includes the well-known species *N. muris* (also known as *N. brasiliensis*) (Fig. 15.12). This nematode is a parasite of rats. Originally a parasite of the Norway rat, *Rattus norvegicus*, it has been experimen-

tally maintained in a number of laboratory animals including albino rats, cotton rats, hamsters, and rabbits. Not all these laboratory animals are completely satisfactory hosts since worms grown in some of these animals are stunted or develop abnormally. *N. muris* adults are located in the anterior portion of the small intestine, either in contact with the mucosa or partially embedded in it.

In its natural host, male specimens measure 3–4 mm in length and females measure 4–6 mm in length. In unnatural hosts, such as the cotton rat, males measure only 1.7–3.0 mm in length and females measure 1.7–3.4 mm in length. This nematode is readily recognized by its small head, which bears a cephalic expansion of the cuticle. Furthermore, the body cuticle is transversely striated and has 10 prominent longitudinal ridges. Both the mouth and buccal cavity are small, and two teeth are present, one on each side. Males can be distinguished from females by a prominent copulatory bursa (Fig. 15.12) and by their smaller size.

Life Cycle of Nippostrongylus muris. For normal development, eggs passing out of the host in feces must reach soil abundant in oxygen and moisture. Under these conditions, development of the enclosed larvae progresses until the rhabditiform L_1s hatch. In the laboratory, where eggs can be induced to hatch by mixing them with charcoal and spreading the mixture on moist filter paper placed in a Petri dish, rhabditiform larvae hatch in 18 to 24 hours at 18 to 22°C. The rhabditiform L_1s grow and molt to become rhabditiform L_2s in about 48 hours. Four or five days after the first molt, the L_2s undergo the second ecdysis to become filariform L_3s. Unlike the larvae of *Haemonchus* (p. 512), the filariform larvae of *N. muris* shed their sloughed cuticles (the process is known as **exsheathing**) and actively penetrate the skin of the rat host.

The infective filariform larva of *N. muris* is markedly thermotactic, becoming quite motile if the warmth of an animal's body is made available. This larva is also negatively geotactic. In soil, it actively migrates to the surface and awaits a suitable host. Worms cultured in the laboratory on filter paper migrate to the edges of the paper and extend themselves into the air and wave back and forth.

Infection of rats is generally accomplished by skin penetration, but infections also can be established in the laboratory by oral feeding, although this method of infection is rather ineffective. Furthermore, in the laboratory, rats can be infected by hypodermic injection. Usually the application of approximately 5000 larvae on the skin of rats results in a heavy but nonlethal infection. A much smaller dosage is required if the larvae are injected hypodermically.

Once within the host's blood, the larvae are carried to the lungs via the heart. Here they feed on whole blood and undergo rapid growth and development, culminating in the third ecdysis to become L_4s. These larvae move up the bronchi and trachea and eventually pass down the pharynx to the intestine. The L_4s first appear in the host's intestine about 41 hours after the infection. Approximately 50% of the larvae arrive between 45 and 50 hours after entry. The fourth and final ecdysis occurs in the intestine, and adult males and females develop. Maturation is rapid and eggs appear in the host's feces by the sixth day after infection.

Adults in the rat's intestine feed mainly on blood and host cells. However, some of the intestinal contents are undoubtedly ingested because intestinal flagellates have been reported in the gut of *N. muris.* Only about 60% of the filariform L_3s that penetrate the rat's skin develop to maturity.

SUPERFAMILY METASTRONGYLOIDEA

The superfamily Metastrongyloidea includes the genera *Metastrongylus, Neometastrongylus,* and *Angiostrongylus.* These parasites are filiform and have smooth cuticles. The mouth is surrounded by either three lips (*Neometastrongylus*) or two lateral, trilobed lips (*Metastrongylus*). The buccal cavity is small. They are lung parasites, members of *Metastrongylus* occurring in various mammals, especially pigs; members of *Neometastrongylus* in ungulates only; and members of *Angiostrongylus* in several mammals, especially rodents.

Metastrongylus apri. This nematode occurs in the trachea, bronchi, and bronchioles of the domestic pig, wild boar, and peccary, *Pecari angulatus,* throughout the world. It has also been reported, although rarely in the ox, sheep, dog, and human. It is rather common in the United States, occurring in 69% of pigs in the southern states (Spindler, 1934) and in 31% of pigs in Michigan (Morgan and Hawkins, 1949). The adult male measures 11–26 mm long, and its filiform genital spicules, measuring 3.9–5.5 mm long, terminate in a hook. The adult female measures 28–60 mm long, and its eggs have a corrugated surface. Such eggs are 45–57 μm × 38–42 μm.

Metastrongylus apri is not only important because of its pathogenicity to pigs, causing hemorrhage of the lungs and obstructing breathing, but it also holds a place of special interest in pathobiology since Shope (1941, 1943) found that pig lungworms can serve as vectors for the swine influenza virus. Shope (1958) has also discovered that *M. apri* serves as a carrier of the hog cholera virus. In the case of the swine influenza virus, the viral particles are present in the nematode's eggs and remain viable in the developing larvae in their earthworm intermediate hosts (see below) for as long as 32 months. Sen *et al.* (1961) have confirmed Shope's remarkable finding and have noted that the influenza virus is present in lungworms in a masked

form and a provocative stimulus is required to initiate infection. They have found that a simultaneous infection with another nematode, *Ascaris,* can provide this stimulation.

In the case of the hog cholera virus, the virus is also usually carried in the masked form, and a stimulus also is needed to activate it to bring about pathogenicity. Shope (1958) has found that a concurrent infection with *Ascaris* will also provide the stimulus. Among 282 pigs infected with lungworms (*M. apri* and *M. pudendotectus*), only two developed hog cholera. On the other hand, among 149 pigs infected with *Metastrongylus* and *Ascaris,* 13 (8.7%) came down with hog cholera.

Swine lungworms, besides serving as carriers of the viruses mentioned, are of considerable economic importance because of their own pathogenicity. According to the U.S. Department of Agriculture, these worms cause an annual loss of $3,584,000 in the United States.

Life Cycle of *Metastrongylus apri.* The life cycle of *M. apri* is indirect, i.e., it involves an intermediate host. Embryonated eggs laid by females are coughed up, swallowed, and passed out in feces. They must be ingested by an earthworm if their development is to continue. Several earthworms are known to be suitable intermediate hosts. These include *Lumbricus terrestris, L. rubellus, Eisenia foetida, E. lonnbergi, E. austriaca, Allolobophora caliginosa, Dendrobaena rubida, Bimastus tunuis,* and *Diplocardia* spp. L_1s hatch from eggs in the earthworm's intestine and develop in the walls of the esophagus, crop, gizzard, and intestine. They later enter the circulatory system and accumulate in the heart. This larva molts twice in the earthworm and attains the ensheathed third stage in about 10 to 30 days.

Pigs become infected by ingesting infected earthworms. The L_3s are liberated in the definitive host's intestine. They burrow through the intestinal wall and are carried in lymphatic vessels to the mesenteric lymph nodes. Here they undergo the third molt, and the L_4s are transported to the lungs via the heart by the lymphatic and circulatory systems. The worms undergo the final molt in the lungs and develop to maturity in 24 days or more.

Ewing and Todd (1961) have shown that a type of synergistic relationship exists between *Metastrongylus apri* and the related *M. pudendotectus* in pigs. Specifically, heavier infections can be produced in pigs with smaller numbers of larvae when a mixture of the two species is administered than when either species is introduced alone. For example, if 6000 infective *M. apri* larvae are given to a pig alone, an average of

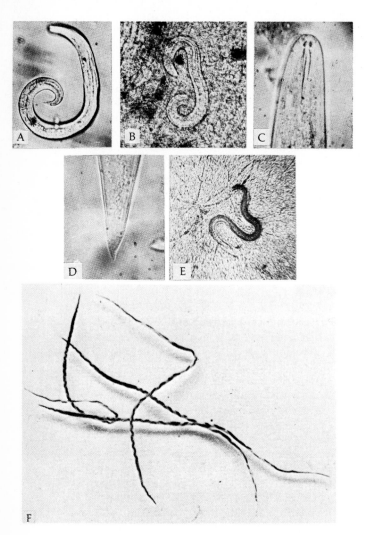

Fig. 15.13. *Angiostrongylus cantonensis.* **A.** L_1 from feces of rat. **B.** Infective L_3 in muscle of a mollusc. **C.** Anterior end of L_3 showing sclerotized stomatorhabdions. **D.** Posterior end of infective L_3. **E.** Developing larva in press preparation of cerebrum of a monkey. (**A–E** after Alicata, 1962; with permission of *Canadian Journal of Zoology*.) **F.** Adult female worms from rat lungs. The "barber pole" appearance is due to the intertwining of the intestine and uterus. (**F** after Zaman, 1979; with permission of Adis Press.)

3.9% is recovered as adults and when 6000 *M. pudendotectus* larvae are given alone, an average of 0.4% is recovered as adults. However, if 3000 larvae of each nematode species are fed to a pig, 60.6% of *M. apri* and 10% of *M. pudendotectus* is recovered as adults. These results indicate that the establishment of both nematode species is enhanced when they occur simultaneously. Furthermore, it has been demonstrated that the maturation of the worms is hastened in double infections.

Genus *Angiostrongylus*

Although considered by some taxonomists to represent a separate superfamily,[*] in following the classification scheme of Maggenti (1981, 1982), this text will include the genus *Angiostrongylus* in the superfamily Metastrongyloidea. Kinsella (1971) has provided a key for the identification of the known species of this genus. All of the species are parasitic in the lungs (pulmonary arteries) and mesenteric arteries of mammals.

Angiostrongylus cantonensis. *A. cantonensis*, the rat lungworm (Fig. 15.13), has come into medical prominence during the past two decades because it has been reported as the causative agent of a type of human meningoencephalitis in Taiwan, Hawaii, Tahiti, the Marshall Islands, New Caledonia, Thailand, New Hebrides, and the Loyalty Islands. Recently it has been found in Cuba.

A. cantonensis is a slender nematode with a simple mouth and no lips or buccal cavity. The adult male measures 15.5–19 mm long, and the adult female measures 21–25 mm long. In females, the intertwining of the intestine and uterus gives the worm a barber-pole appearance. The eggs are thin shelled and include an undeveloped larva when laid.

Life Cycle of Angiostrongylus cantonensis. In infected rats, eggs containing uncleaved zygotes are deposited in the bloodstream, where they become lodged as emboli in the smaller vessels in the lungs. Embryonic development occurs here. The L_1s, rupturing from the thin-shelled eggs, enter the respiratory tract. They then migrate up the trachea and down the alimentary tract, and are discharged in feces. This nematode requires an intermediate host, usually a snail, *Subulina octona*, *Bradybaena similaris*, or *Achatina fulica*; or a slug, *Veronicella leydigi*, *Agriolimax laevis*, or another. Occasionally, larvae of *A. cantonensis* are found in the land planarian *Geoplana septemlineata* and a number of crustaceans. However, neither planarians nor crustaceans can be infected with L_1s and hence are paratenic or transport hosts that acquire L_3s from feeding on naturally infected snails and slugs. The L_1s actively penetrate molluscs, or they may also gain entrance through ingestion. Within the intermediate host, the larvae undergo two molts, and the L_3 is ensheathed. When an infected mollusc is ingested by a rat, the larvae exsheath in the rodent's stomach and burrow through the ileum and become blood borne. The blood-borne larvae in the hepatic portal system are carried to the right chambers of the heart via the hepatic vein and posterior vena cava. The L_3s carried

---

[*] Superfamily Protostrongyloidea.

from the right ventricle to the pulmonary capillaries via the pulmonary artery. From the pulmonary capillaries the larvae are carried to the left chambers of the heart via the pulmonary vein. Still later, the larvae are conducted out of the left ventricle in the aortic arch, common carotid artery, and capillaries of the brain, and eventually they congregate in the central nervous system, especially in the anterior portion of the cerebrum, where they undergo the third and fourth molts. The resulting young adults emerge to the brain surface on the 12th and 14th day, remain in the subarachnoid space for approximately two weeks, then migrate via the venous system through the heart to the pulmonary arteries and tissues of the lungs, where they mature by the 42nd and 45th day in the rat.

In a series of experiments involving the infection of albino rats with known numbers of L_3s, I have shown that if 150 or more infective larvae are fed to each rat, 100 or more of the worms succeed in becoming established in the brain, and the rodent host is killed between the 28th and 36th day postinfection when the young adult worms begin to migrate from the subarachnoid space to the lungs.

Death ensues due to hemorrhage caused by young adult worms as they penetrate the venules on the surface of the brain, interference with the normal flow of blood through the right chambers of the heart, and blockage of the normal circulation through the pulmonary artery. If 100 or fewer infective L_3s are introduced into each rat, fewer than 100 worms—generally between 65 and 80—become established in the brain. In such instances, death does not occur during the period the young adult worms are migrating from the brain. However, if 30 or more worms are present, death generally occurs later during the course of the infection as a result of secondary pulmonary congestion and necrosis of lung tissue.

In addition, young rats infected with *A. cantonensis* do not grow as fast as uninfected ones. Histopathologic examinations of the brains of infected rats have revealed that host cellular response is practically nonexistent. This contrasts with results of similar studies of the brain of infected rabbits, rhesus monkeys, and humans. In these abnormal hosts, the host cellular response is extensive. These data exemplify the principle that parasites in tissues of abnormal hosts generally elicit much more severe cellular response than in normal hosts.

Human Meningoencephalitis. The importance of *A. cantonensis* as a parasite of humans was first reported by Nomura and Lin in 1944 when they recovered six adult worms in the cerebrospinal fluid of a 15-year-old boy in Formosa. The second finding of *A. cantonensis* in a human was by Rosen *et al.* (1962), who reported it in the brain of a mental patient in Hawaii who had died of eosinophilic meningoencephalitis.

Aware of the presence of *A. cantonensis* in Hawaii as a result of its detection in rats by Ash (1962), Alicata postulated that *A. cantonensis* might be of considerable medical importance. After observing a Japanese gardener believed to be suffering from meningoencephalitis after swallowing a slug from an area where slugs were known to harbor L_3s (Horio and Alicata, 1961), Alicata (1962) further postulated that *A. cantonensis* is the causative agent of meningoencephalitis in Hawaii and other areas of the Pacific Basin.

Although *A. cantonensis* was not positively identified in humans, Alicata and Brown (1962), based on field observations, reported that human infections in Tahiti could be acquired from eating raw, freshwater prawns, *Macrobrachium* sp., or native foods that include prawn juice, for they found *A. cantonensis* larvae in prawns. Furthermore, they suggested that human infections could be acquired from eating fresh vegetables and fruits on which small, infected slugs are found. Since *Macrobrachium* is now widely raised in aquaculture ponds throughout the world, including many areas where *A. cantonensis* occurs, the possible transmission of angiostrongylosis must be considered a potential threat.

Continuous surveys have revealed that *A. cantonensis* is regularly found in rats and molluscs on various Pacific islands where human cases of eosinophilic meningoencephalitis occur.

Human cases of meningoencephalitis are characterized by pleocytosis (mostly of eosinophils), headache, stiffness of the neck and back, and paresthesias of various types. Deaths due to meningoencephalitis caused by this parasite are known.

In some cases of angiostrongylosis in experimental animals, adult worms migrate down the spinal cord from the brain and cause paralysis, or migrate to the eyes and cause blindness. In one case of human angiostrongylosis in Thailand, a young adult worm was recovered from the patient's eye. For a review of human angiostrongylosis, see Alicata and Jindrak (1970).

Angiostrongylus costaricensis. A second species of *Angiostrongylus*, *A. costaricensis*, is normally a parasite in the mesenteric arteries of rats in Costa Rica and adjacent states, including Texas, but Morera and Céspedes (1971) have reported more than 70 cases of human infection by this nematode. Unlike *A. cantonensis*, *A. costaricensis* matures in the mesenteric arteries and their branches. At these sites it causes damage to the host's intestinal wall, especially in the caecum and appendix. The arterial wall becomes necrotic and thickened, accompanied by massive eosinophilic infiltration. Blockage of arterioles by larvae and eggs of the parasite contributes to the disease. Infected individuals suffer from high fever and abdominal pain. Infections result from the accidental ingestion of infected molluscan intermediate hosts.

Other Species of *Angiostrongylus*. North American species of *Angiostrongylus* include *A. michiganensis* and *A. gubernaculatus* in mustelid carnivores, and *A. schmidti* in rice rats. None of these have yet been associated with human disease.

Genus *Protostrongylus*

Also included in the superfamily Metastrongyloidea are the members of *Protostrongylus*. These nematodes, along with members of *Pneumostrongylus*, a closely related genus, are characterized by thin filiform bodies, a mouth that is directed anteriorly and either surrounded by lips or not. The buccal capsule is either poorly developed or absent. All of these nematodes are parasitic in either the respiratory or circulatory systems of mammals.

The members of *Protostrongylus* are parasitic in the lungs of herbivores and, rarely, those of carnivores. Most are of considerable importance as pathogens in domesticated and wild ungulates. There are about 27 species, and all have indirect life cycles involving a molluscan intermediate host.

***Protostrongylus rufescens*.** This nematode occurs in the trachea, bronchi, and bronchioles of the sheep, goats, bighorn sheep, and European hare (*Lepus europaeus*), but Joyeux and Gaud (1946) have reported that the form found in rabbits and hares will not infect sheep and goats and have suggested that it be considered a different variety, *P. rufescens* var. *cuniculorum*.

Protostrongylus rufescens is a fairly common parasite in North America, Europe, Israel, Cyprus, Turkey, Iran, India, Africa, and the Soviet Union. For example, one survey revealed that it occurs in 32% of 75 sheep and 50% of 14 goats from New York.

This nematode is commonly called the red lungworm because of its color due to ingested blood. The males measure 16–46 mm long and 120–170 μm wide, and the females measure 25–65 mm long and 150–290 μm wide (Fig. 15.14). This and other members of the genus are pathogenic and have been associated with fatal pneumonia.

Life Cycle of Protostrongylus rufescens. Eggs of *P. rufescens* are deposited in the host's lungs. L_1s hatching from eggs of this nematode migrate up the respiratory tract to the pharynx, where they are swallowed and passed out in feces. These rhabditiform larvae measure 340–400 μm long and about 20 μm wide. They can survive for several months in water

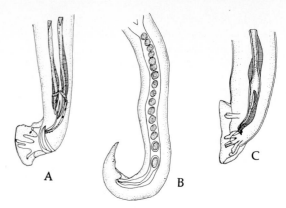

Fig. 15.14. Some metastrongyloid nematodes. A. Lateral view of posterior portion of male *Protostrongylus rufescens*. **B.** Lateral view of posterior portion of female *P. rufescens*. (**A** and **B**, redrawn after Dikmans and Mapes, 1950.) **C.** Lateral view of posterior portion of male *Pneumostrongylus tenuis*. (Redrawn after Dougherty, 1945.)

but will not continue their development until they penetrate an appropriate snail intermediate host. Species of snails known to be compatible include *Helicella ericetorum*, *H. obvia*, and *H. bolli* in Europe; *H. barbesiana*, *H. vestalis*, and *Monacha syriaca* in Israel; and *Helicella rugosiuscula*, *H. gigaxii*, and the slug *Agriolimax kervillei* in the western Mediterranean Basin. Upon making contact with the molluscan host, L_1s penetrate and undergo two molts, thus attaining the ensheathed third stage. The rate of intramolluscan development is dependent on the host species and the ambient temperature. The definitive host becomes infected by accidentally ingesting infected gastropods. From the alimentary tract, the L_3 penetrates into the lymphatic system, undergoes the fourth molt in the mesenteric lymph nodes, and then passes to the lungs (Hobmaier and Hobmaier, 1930).

The other members of the genus *Protostrongylus*, including *P. stilesi*, a common parasite of the bighorn sheep (*Ovis canadensis*) in Canada, have the same life cycle pattern.

Genus *Pneumostrongylus*

Males belonging to this genus have a copulatory bursa that is not divided; the dorsal lobe is as large as the two lateral lobes and is not marked off from them. The only member of this genus is *Pneumostrongylus tenuis*, a parasite of the white-tailed deer and moose in Canada and the eastern United States.

***Pneumostrongylus tenuis*.** This nematode occurs in the cranium of its host, although adults do occur in lungs (Fig. 15.14). It is a fairly common parasite. For example, DeGiusti (1963) found it in 56% of 836 deer in Michigan, Alibasoglu *et al.* (1961) found it in 75% of 80 deer in Pennsylvania, and Smith and Archibald (1967) found it in 63% of 46 white-tailed deer and 80% of 36 moose in the Maritime Provinces in Canada. Anderson (1971) has presented a review of this parasite.

Life Cycle of Pneumostrongylus tenuis.   The life cycle of this nematode has been investigated by Anderson (1963), who has reported that adults in the cranium, spinal cord, and veins of the dura mater of the white-tailed deer produce eggs that may or may not hatch. Eggs and L_1s pass to the lungs via the circulatory system. The larvae break into the alveoli, are coughed up, swallowed, and pass to the exterior in the host's feces. In order to undergo further development, these larvae must be ingested by a suitable gastropod host. A number of snails and slugs have been demonstrated to be compatible intermediate hosts; these include *Zonitoides arboreus*, *Discus cronkhitei*, *Deroceras gracile*, *Stenotrema fraternum*, *Anguispira alternata*, *Triodopsis albolabris*, and *Lymnaea* spp. Within the mollusc, the nematode undergoes two molts and attains the infective L_3 stage in about four days. Anderson (1963) has reported that the L_2 is ensheathed, but not the L_3.

Deer become infected while grazing by eating snails and slugs. Once the intermediate host is digested, the escaping larvae penetrate through the wall of the abomasum and move into the mesentery and omentum. From these sites they invade the spinal cord. The third and last molts occur in the brain or spinal cord, and adults can be found at these sites approximately 50 days postingestion of larvae.

If molluscs harboring L_3s are ingested by sheep, the parasite does not mature.

SUPERFAMILY ASCARIDOIDEA

This superfamily includes parasites of mammals, reptiles, and amphibians which are characterized by the usual presence of three lips, a club-shaped esophagus without a valve, the absence of a ventriculus except in a few genera (*Neoascaris*, *Toxocara*, and *Dujardinascaris*), an intestinal caecum, copulatory spicules in males that are equal or not exactly equal in length, and the occurrence of many papillae on the male tail. The Ascariodidea includes the largest intestinal nematodes.

The ascaridoids are comparatively large and plump worms. The most common and best-known members are *Ascaris lumbricoides*, an intestinal parasite of humans (Fig. 15.15), and *Ascaris suum* of hogs. Although these two species are very similar, they do portray physiologic and morphologic differences. The pig-infecting species normally will not infect humans, and vice versa, and the two species can be distinguished by minor denticle differences.

Genus *Ascaris*

Ascaris lumbricoides.   This species is cosmopolitan in distribution. Stoll (1947) estimated that there were 644 million human infections in the world, of which 3 million occurred in North America, 42 million in tropical America, 59 million in Africa, 488 million Asia, 32 million in Europe, 19.9 million in the Soviet Union, and 500,000 in the Pacific islands. These numbers have not altered significantly. Males, 15–31 cm in length by 2–4 mm in diameter, are smaller than the females, which are 20–40 cm in length by 3–6 mm in diameter (Fig. 15.15). As in most nematodes, the posterior terminal of the males is curved ventrally. The mouth in both sexes is surrounded by a dorsal and two ventrolateral lips (Fig. 15.15).

Life Cycle of Ascaris lumbricoides.   Adults of *A. lumbricoides* live in the small intestine of the host, where they feed on chyme. However, it is suspected that the adults are capable of penetrating the intestinal mucosa to suck blood. Copulation occurs in the host's intestine.

Egg production is prodigious; a single female may contain as many as 27 million eggs. Subspherical eggs laid by fertilized females measure 60–70 μm × 40–50 μm and possess a thin transparent shell covered by an irregular coat of albumin that is usually brownish (Fig. 15.15, 15.16). Unfertilized eggs are often found among fertilized eggs in the host's feces. These eggs are elongated, the albuminous coat is less prominent or absent (Fig. 15.15), and they contain amorphous globules rather than a well-formed zygote.

When fertilized eggs are deposited, the zygote is uncleaved. It is in this state that the eggs pass out in the host's feces. Further development of the zygotes occurs only if the environmental temperature is lower than that of the host's body, i.e., between 15.5 and 35°C. However, development ceases at temperatures below 15.5°C, and the eggs are killed at temperatures above 38°C.

Under natural conditions, eggs develop readily in moist soil in the presence of oxygen. Active larvae are formed within each eggshell in 10 to 14 days, but ingestion of such eggs does not result in an infection. It is only after the L_1 undergoes a molt within the eggshell and matures into the L_2 that the egg becomes infective.

When infective eggs are swallowed, the L_2 hatch in the host's small intestine, and actively burrow through the intestinal lining and are carried to the liver via the mesenteric lymphatics or venules. From the liver, the larvae are transported to the right side of the heart and then to the lungs via the pulmonary artery. The larvae generally remain in the lungs for a few days, where they grow before rupturing out of the pulmonary capillaries and enter the alveoli. From there they move up the bronchioles, bronchi, and trachea to the area of the glottis, from where they are coughed up and swallowed, and pass again down to the small intestine. During this complex migratory process, the individual worms increase from 200 to 300 μm in length to approximately ten times that size. Ecdysis occurs in the small intestine between the 25th and 29th day after ingestion of the eggs. Only larvae that have completed the fourth molt can survive in the intestine and develop to maturity.

The life cycle of most other ascaroids parallels that of *A. lumbricoides*. Sprent (1952) categorizes the migratory behavior of ascaroids in mice into two types—the **tracheal type**, in which the larvae migrate through the viscera and eventually disappear from the tissues, and the **somatic type**, which consists of permanent encystment of living larvae within the tissues. The migratory pattern of *A. lumbricoides* given above represents the tracheal type.

Not all ascaroids possess direct life cycles, that is, cycles not involving an intermediate host. For example, *Ascaris devosi* in the marten and *A. columnaris* in the skunk have rodent intermediate hosts.

Pathology of Ascariasis. Little damage is associated with penetration of the host's intestinal mucosa by newly hatched larvae. However, aberrant larvae migrating in such organs as the spleen, liver, lymph nodes, and brain usually elicit an inflammatory response. Also, when larvae escape from capillaries in the lungs and enter the respiratory system, they cause small hemorrhagic foci. When large numbers of larvae occur, many small blood clots accumulate, leading to Löffler's pneumonia (or *Ascaris* pneumonitis). If large areas of the lungs are affected, death can ensue. Unless large numbers of adult worms are involved, there is little pathology associated with their presence, but symptoms such as abdominal pain, asthma, insomnia, and eye pain may occur. Except for abdominal pain, these symptoms represent allergic responses to metabolic excretions and secretions by the worms.

When large numbers of adults are present, mechanical blockage of the intestinal tract may occur. Further-

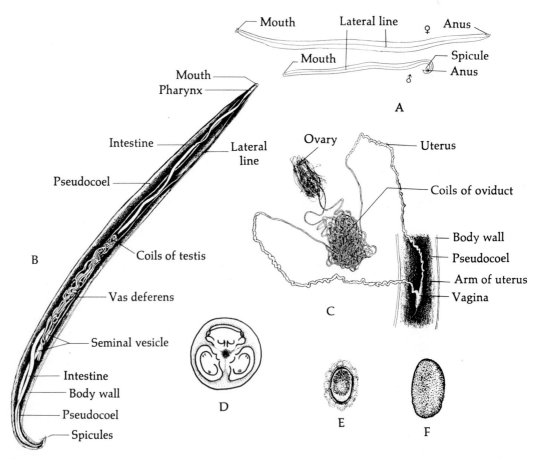

Fig. 15.15. *Ascaris lumbricoides.* **A.** Male and female adults, external view. **B.** Cut-away view of adult male showing internal structures. **C.** Female reproductive system teased out of body. **D.** Face-on view of anterior end. **E.** Fertilized egg. **F.** Unfertilized egg. (**A–C**, after Beck and Braithwaite, 1960.)

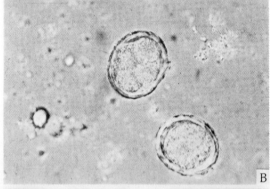

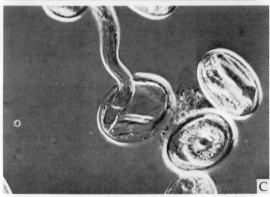

Fig. 15.16. Fertilized eggs of *Ascaris lumbricoides*. A. Freshly passed eggs isolated from feces; these are light to dark brown in color. × 400. **B.** Eggs enclosing embryos undergoing cleavage. × 400. **C.** Eggs containing fully developed larvae; the mammillated layer on the outside of the egg has been chemically removed; the larva escaping from the shell in the center has molted, and its sheath remains in the shell. × 400. (All photographs after Zaman, 1979; with permission of Adis Press.)

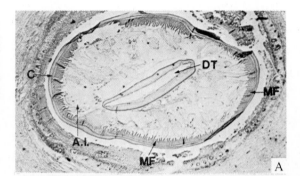

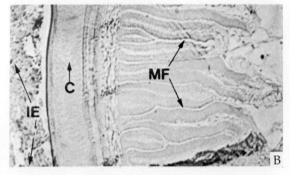

Fig. 15.17. *Ascaris lumbricoides* in appendix. A. Photomicrograph showing cross-section of adult worm in lumen of appendix. × 12. **B.** Photomicrograph showing portion of adult worm in lumen of appendix at higher magnification; note presence of hemorrhagic inflammatory exudate along outer surface of cuticle. × 400. A.l., *Ascaris lumbricoides*; C, cuticle; DT, digestive tract of parasite; IE, inflammatory exudate; MF, muscle fibers of parasite. (**A** and **B** after Kenney, 1973; with permission of The Upjohn Company.)

more, worms may penetrate through the intestinal wall or appendix. If peritonitis develops, death is common. When present in the appendix, *Ascaris lumbricoides* may cause local hemorrhage (Fig. 15.17).

Ascaris suum (= **A. suilla**). This nematode is almost indistinguishable from *A. lumbricoides* except for minor denticle differences. *A. suum* is a very com-

mon parasite of pigs. Its fertilized eggs are extremely durable, remaining viable in damp soil for 4 to 5 years.

Various hypotheses have been advanced to explain why adult nematodes, living in their host's intestine, are not digested by their host's digestive enzymes. The mostly widely accepted theory currently is that *Ascaris* is capable of secreting antienzymes that inhibit the host's digestive enzymes. Indeed, Rhodes *et al.* (1963) have demonstrated that trypsin and chymotrypsin inhibitors are present in the body wall and perienteric (or pseudocoelomic) fluid of *Ascaris suum*.

Genus *Parascaris*

Parascaris equorum. This large intestinal roundworm of horses, also known as *Ascaris megalocephala*, is widely distributed throughout North America (Fig. 15.18). The males measure 150–280 mm in length,

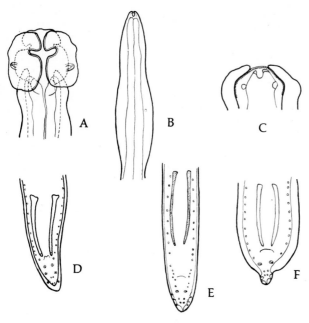

Fig. 15.18. Some nematodes of veterinary importance.
A. Ventral view of an anterior end of *Parascaris equorum*, large roundworm of horses. **B.** Ventral view of posterior end of male *P. equorum*. **C.** Ventral view of anterior end of *Toxascaris leonina*, a parasite of canines and felines. **D.** Ventral view of posterior end of male *T. leonina*. **E.** Anterior end of *Neoascaris vitulorum*, large roundworm of cattle. **F.** Ventral view of posterior end of male *N. vitulorum*. (All figures redrawn after Schiller in Morgan and Hawkins, 1949.)

and the females are 180–350 mm in length. The life cycle of *P. equorum* parallels that of *Ascaris lumbricoides*. This nematode causes severe damage to its host, especially young animals. During migration of the larvae in the lungs, hemorrhages, pneumonia, and fever result. Intestinal obstructions and perforations by the intestinal form have been reported.

Genus *Toxascaris*

Toxascaris leonina.   This parasite is found in the small intestines of dogs, cats, lions, tigers, and other carnivores (Fig. 15.18). During its life cycle, the larvae hatching from the eggs do not migrate through the viscera as in *Ascaris*. Rather, once liberated in the small intestine, the larvae burrow into the crypts of Lieberkühn, the submucosa, and the muscular layer. Within nine to ten days after ingestion of the eggs, the larvae return to the lumen of the small intestine and mature. This pattern is another variation in the life cycle pattern found in the Ascaroidea.

Genus *Neoascaris*

Also included in the Ascariodidea are the genera *Neoascaris* and *Toxocara*. The members of the former are distinguished from those of *Toxocara* by the presence of prominent cervical alae and undivided external projections of the labial palp.

Neoascaris vitulorum.   This parasite is a relatively common large nematode found in the small intestine of cattle, particularly calves (Fig. 15.18). The males measure up to 25 cm in length, while the females are 22–30 cm long. As in some *Ascaris* infections, verminous pneumonia and hemorrhages result from the migration of the larvae through the lungs. These symptoms are more frequently encountered in *N. vitulorum* infections than in *Ascaris* infections. *Neoascaris* infections are probably always prenatally acquired.

Genus *Lagochilascaris*

Included in this genus is *Lagochilascaris minor*. This nematode is believed to be normally an intestinal parasite of cats; however, it has been reported over a dozen times in the tonsils, nose, or neck of humans in northern South America. In human infections, one to nearly 1000 worms occur within an abcess. Furthermore, they can attain maturity in abcesses and produce pitted eggs similar to those of *Toxocara* (Sprent, 1971). How humans become infected remains unknown.

Genus *Toxocara*

The members of this genus possess conspicuous cervical alae and external extensions of the labial palp which are subdivided. There are three lips (Fig. 15.19).

Toxocara canis and T. cati.   The genus *Toxocara* includes, among others, *T. canis* in dogs and *T. cati* in cats and dogs (Fig. 15.19). The life cycles of both species may be similar to that of *Ascaris lumbricoides*—that is, direct and involving the ingestion of embryonated eggs, followed by the complex migration. This is not the only life cycle pattern, however. Sprent (1956) has found that *T. cati* eggs can hatch and develop to L_2s in earthworms, cockroaches, chickens mice, dogs, lambs, and cats. Cats can become infected by ingesting mice harboring larvae in their tissues. In this method of infection, the larvae are mostly confined to the host's digestive tract.

Prenatal infection with *T. canis* is relatively common, but it is not known to occur with *T. cati*.

Occasionally, a syndrome known as **visceral larva migrans** is reported in humans resulting from the ingestion of embryonated eggs of *T. canis* or *T. cati* and the subsequent migration of the larvae in the viscera of the human host (p. 560). The phenomenon is more common with *T. canis*. Although eggs do not develop at temperatures below 12°C, they are not

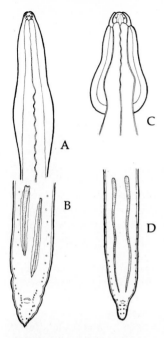

Fig. 15.19. *Toxocara* **spp. A.** Ventral view of anterior end of *Toxocara canis* in dogs. **B.** Ventral view of posterior end of male *T. canis*. **C.** Ventral view of anterior end of *Toxocara cati* in dogs, cats, other felines, and, rarely, humans. **D.** Ventral view of posterior end of male *T. cati*. (All figures redrawn after Schiller in Morgan and Hawkins, 1949.)

killed, since they can survive at temperatures as low as −25°C, provided there is sufficient oxygen.

Life Cycle of Toxocara canis. As a result of research by Sprent (1958) and others, it is now known that two different avenues of development—the **tissue** and the **intestinal phases**—can occur in the case of *T. canis*, depending on the age and the species of the host. In addition, prenatal infection often occurs in dogs, that is, the parasite is passed from mother to offspring via the uterine and umbilical circulation. These findings emphasize that the physiology of the host can be an important factor in influencing the developmental pattern of the parasite, especially among nematodes.

Adult *T. canis* males, measuring up to 10 cm in length, and females, measuring up to 18 cm in length, are found in the small intestine of dogs. The eggs oviposited by females after copulation are unembryonated when passed out in the host's feces. Embryonic development occurs outside the host, and infective eggs enclosing L_2s develop in five to six days, or longer, depending on the ambient temperature. For example, Owen (1930) has reported that during the development of the larva in the egg of *T. canis*, embryogenesis is completed within 35 days when the eggs are exposed to 16.5°C, but development ceases if the eggs are maintained below 12°C, and in as few

as 3.5 days at 30°C. However, death occurs if the eggs are exposed to 37°C.

When infective eggs are ingested by the host, they hatch in the small intestine, and the larvae enter hepatic portal venules and migrate to the lungs via the liver and heart, as in the case of *Ascaris lumbricoides*. From this stage on, the migration route varies, depending on the age and the species of the host.

Tissue Phase. If the host is an older dog, especially one with some immunity acquired from previous infection, some of the larvae migrate to the liver and lungs and remain in the parenchyma of these organs, whereas others continue through the pulmonary veins and heart into the arterial circulation. Larvae in the arterial circulation are carried to the capillaries of the skeletal muscles and kidneys or to the central nervous system, where they remain inactive but alive for a long period. If a female dog becomes pregnant, the dormant larvae are activated by host hormones, re-enter the circulatory system, and are carried to the placenta, where they penetrate to the fetus's circulatory system. They then complete migration through the fetus's lungs and become established in its intestine. Consequently, newborn pups can be infected with *Toxocara* although the mother shows no sign of infection. This represents an example of **paratenesis** since the mother serves as a host to *T. canis* larvae but is not an obligatory host in the parasite's life cycle.

In the case of larvae in rodents, further development is possible if infected rodents are ingested by a dog, especially a young dog. If this occurs, larvae freed in the dog's stomach migrate into the intestine, where they burrow through the intestinal wall and enter the hepatic portal vein. The larvae are carried to the lungs via the liver and heart. Within the lungs, the larvae migrate out of the blood into bronchioles, up the bronchi into the trachea and then are swallowed. Once the larvae reach the small intestine, they develop into adults. The mouse serves as a paratenic host.

Intestinal Phase. If the host is a pup, almost all the larvae reach the lungs, where they escape from the blood vessels and enter alveoli. From these sites, they migrate up the bronchi and trachea and are swallowed. On reaching the small intestine, they develop to maturity. Thus, the migration pattern within the host is essentially the same as that of *Ascaris lumbricoides*.

Anisakid Nematodes

Also included in the superfamily Ascaridoidea are the anisakid nematodes (members of the family Anisakidae). These nematodes have attracted considerable interest in recent years because they are now known to cause gastric and intestinal granulomata and even death in humans (Jackson, 1975). Specifically, members of the genera *Anisakis*, *Belanisakis*, *Phocanema*, *Porrocaecum*, *Paradujardinia*, *Pseudoterranova*, *Cloeoascaris*, *Phocascaris*, and *Contracaecum* are real and potential etiologic agents of human anisakiasis. Cheng (1976) has contributed a review of what is known about the life cycles of the anisakid nematodes and has contributed a taxonomic treatise (Cheng, 1982) on these worms. Most of the anisakids are characterized by the occurrence of an appendix (known as an **intestinal caecum** if it projects anteriorly and as a **ventricular appendix** if it projects posteriorly) projecting from the area of the ventriculus. Also, the position of the excretory pore is an important diagnostic feature.

Life Cycle Pattern of the Anisakids. The life cycles of the members of this group of nematodes involve one or more intermediate hosts. For example, in the life cycle of *Contracaecum*, eggs that pass out in the feces of the definitive host include undeveloped or poorly developed larvae. Larval differentiation continues after the eggs reach water, resulting in the formation of infective larvae armed with a cuticular stylet (also known as a boring tooth).

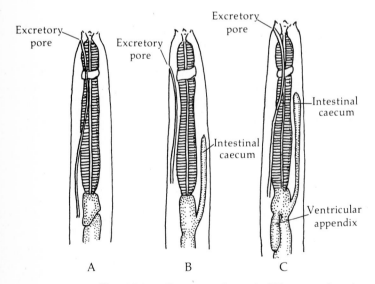

Fig. 15.20. Representative anisakid nematodes. A. Anterior end of *Anisakis* sp. showing excretory pore opening at base of subventral lips. **B.** Anterior end of *Porrocaecum* sp. showing excretory pore opening at level of nerve ring and anteriorly projecting intestinal caecum. **C.** Anterior end of *Contracaecum* sp. showing excretory pore opening at base of subventral lips and presence of anteriorly projecting intestinal caecum and posteriorly projecting ventricular appendix.

The infective larva either hatches or remains in the shell, but in either case it is ingested by the first intermediate host—usually a copepod, an amphipod, a jellyfish, or a fish. Within this host, the larva burrows through the intestinal wall and continues to develop within the body cavity or in the surrounding tissues. Some larvae may encyst. When such an infected host is eaten by the second intermediate host, a fish, the larva again burrows through the intestinal wall and encysts in the host's body cavity or tissues. The liver is a favored site, and as the result of the presence of the worms, the fish host may be deleteriously affected or even die (Margolis, 1970). When a fish containing encysted larvae is eaten by another fish or some other fish-eating animal, the cycle is completed. During maturation, the nematode loses its penetration stylet.

Thomas (1937) has reported that in *Raphidascaris* only one intermediate host is required, a small fish, in which the larvae penetrate the intestinal wall and become encapsulated in the liver and mesenteries. When the small infected fish is eaten by a pike, the cycle is completed. Similarly, a single intermediate host is usually the pattern in the life cycle of *Porrocaecum*.

Genus *Anisakis*. The members of this genus are readily distinguished from the other anisakids by the presence of a bilobed projection on each of the three lips. Furthermore, each of these projections bears a single dentigerous ridge. The excretory pore opens at the base of the subventral lip. All of the species are usually stomach parasites of marine mammals and birds. *Anisakis marina* (Figs. 15.20, 15.24), for example, is a cosmopolitan species occurring in various marine mammals. The infective L_3s occur in various marine fish, and the definitive host becomes infected by eating fish.

The public health importance of *Anisakis* and certain other anisakid nematodes rests with the fact that if humans should eat raw or inadequately cooked fish harboring infective nematode larvae in their flesh, the larvae will commence to burrow through the human stomach or intestinal wall and as a result cause severe eosinophilic granulomas (Fig. 15.21). This human disease, first recognized by Van Thiel *et al.* (1960) in Holland, is now known to occur in Japan (Oshima, 1972) and North America.

This disease occurs most frequently in cultures in which raw or undercooked fish forms part of the normal diet; for example, pickled herring in Scandinavia, and sashimi in Japan. Such dietary habits, of course, are not limited to Scandinavia and the Orient. In the United States, raw fish is commonly eaten in Hawaii and among Eskimoes. Chitwood (1970) has pointed out the danger of this practice and has noted that anisakid nematodes have been found in cod, haddock, and other species of market fish in Connecticut, New York, Maryland, and Canada. Cheng (1976) has reported finding anisakids in 17 species of fish in an extensive survey extending from Georgia to Maine (Figs. 15.22, 15.23).

It is of interest to note that changes in commercial fishing practices have influenced the increased frequency of gastric eosinophilic granuloma formation. The nematode larvae usually occur in the intestinal wall and liver of the fish host, and hence, if commercially caught fish are eviscerated rapidly, most of the worms are removed. In recent years, however, with fishing vessels, especially those from Japan, the Soviet Union, and other European countries, staying out at sea for extended periods of time, the fish are caught and placed on ice or in refrigerators and not eviscerated until the vessels return to port. This delay, for some yet unknown reason, causes the larvae to migrate deeper into the flesh of the fish and consequently remain in the fish when they are cleaned.

Since these nematodes utilize marine mammals as definitive hosts, shifts in the distribution of the mammalian population, especially toward commercial fishing grounds such as the Grand Banks, have led sporadically to greater numbers of anisakids in fish from these areas.

Species of anisakids that are really and potentially pathogenic to humans normally are parasites of marine endothermic vertebrates. Those which utilize fish cannot survive the higher human body temperature (Cheng, 1976).

Genus *Porrocaecum*. *Porrocaecum* spp. are recognized by the occurrence of dentigerous ridges on their three lips and a large intestinal caecum (Fig. 15.20). Species reported from birds include *P. crassum* in *Anas*, *Numida*, *Cairina*, and other birds in Europe; *P. angusticolle* in *Buteo*, *Falco*, *Milvus*, and many other

Fig. 15.22. Anisakid nematodes from marine fish. Specimens of *Contracaecum* sp. removed from intestine and stomach of the cod, *Gadus callarias*, caught off the coast of Massachusetts.

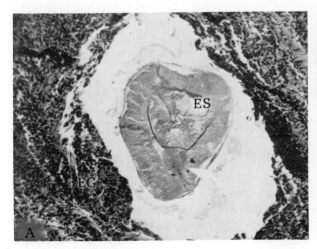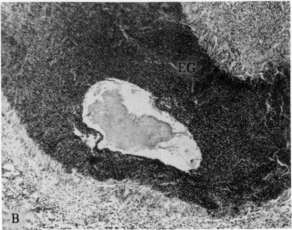

Fig. 15.21. Pathology of anisakiasis. A. Photomicrograph showing cross-section of *Anisakis* larva in human stomach wall, surrounded by eosinophilic granuloma. **B.** Photomicrograph showing oblique section of a degenerate larva encapsulated in a bandlike eosinophilic abscess in the human gastric submucosa. EG, eosinophilic granuloma; ES, esophagus of nematode. (**A** and **B**, after Yokogawa and Yoshimura, 1965.)

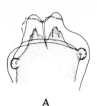

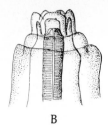

A   B

Fig. 15.24.   Some anisakid nematodes. A. Anterior end of *Anisakis*. (Redrawn after Mosgovoy, 1951.) **B.** Anterior end of *Contracaecum spiculigerum*. (Redrawn after Yamaguti, 1941.)

Fig. 15.23.   Anisakid nematodes in fish. Specimens of *Porrocaecum* sp. migrating from the digestive tract of a summer flounder, *Paralichthys dentatus*, caught off the coast of New Jersey.

avian genera in Europe, Africa, India, Japan, and North America, and *P. wui* in *Corvus*, *Pica*, and *Urocissa* in the People's Republic of China.

In the case of *P. crassum* and others whose life cycles are known, earthworms serve as the intermediate host. Specifically, the nematode eggs are eaten by earthworms, and the larvae, after hatching in the intestine, migrate into blood vessels, undergo two molts, and become infective L_3s. Birds become infected by eating earthworms harboring L_3s.

Species of *Porrocaecum* also occur in mammals, for example, *P. americanum* in various species of shrews and several species in marine mammals. The larvae of those parasitic in marine mammals occurring in fish intermediate hosts can cause human anisakiasis.

Genus *Contracaecum*.   The members of *Contracaecum* are distinguished from the other anisakids mentioned by the absence of dentigerous ridges on the lips. Furthermore, the lips bear anterolateral projections (Figs. 15.20, 15.24). Both an intestinal caecum and ventricular appendix are present. This group of worms are parasitic in birds and fish-eating mammals.

Contracaecum is an extremely large genus with over 50 species. A key to the recognized species is provided by Mosgovoy (1953). Representative of species parasitic in birds is *Contracaecum spiculigerum* (Fig. 15.24). It is cosmopolitan in distribution and occurs in numerous types of birds. During the life cycle of *C.*

spiculigerum, the larvae undergo two molts within the eggshell before hatching. The L_3s are ingested by fish and eventually encapsulated in the fish's musculature. When infected fish are eaten by a bird, for example, the cormorant, the larvae undergo two additional molts and mature within the avian host. Paratenic hosts commonly occur during the life cycle. For example, L_3s can be transferred from infected minnows to carnivorous fish and later passed on to birds.

A large number of species of *Contracaecum* have been reported from mammals. The larvae of some of these, especially those in marine mammals, can cause human anisakiasis.

Pathology of Human Anisakiasis

The clinical symptoms of human anisakiasis include gastric or intestinal pain and vomiting. However, the pain is more generalized and is not as sharp as in acute appendicitis. Moderate leucocytosis generally occurs, although eosinophilia is not apparent. There is no abnormal tension of the abdominal muscles, and no fever results. According to experienced clinicians (Yoshimura, 1966; Ishikura *et al.*, 1969), these signs are important in distinguishing anisakiasis from acute appendicitis and internal obstruction.

The pathology of anisakiasis involves the development of a granuloma surrounding the worm (Fig. 15.21). This can occur in the gastric, small intestinal, or large intestinal walls, the stomach being the most frequent site. According to Kojima *et al.* (1966), the histopathologic lesions can be classified into four types: **phlegmonous**, **abscess**, **abscess-granulomatous**, and **granulomatous**. A fifth type, designated the **foreign body response** type, has been added to this list.

Foreign Body Response Type.   The histopathologic picture associated with this type of response includes infiltration and proliferation of neutrophils associated with a few eosinophils and giant cells. There is little or no edema, but fibrin exudation, hemorrhage, and vascular damage usually occur. Furthermore, granulomas generally develop around the parasite. This type of response is associated with benign clinical symptoms and does not require surgery. According to Japanese investigators (Oyanagi, 1967; Kikuchi *et al.*, 1967; Miyazato *et al.*, 1970), foreign body response results from primary infections, i.e., without presensitization.

Phlegmonous Type. This type of reaction has also been referred to as the Arthus type. There is extensive edematous thickening accompanied by inflitration by lymphocytes, monocytes, neutrophils, and plasma cells. Furthermore, there is an inflammatory response in the associated blood vessels, hemorrhage, and exudation of fibrin. The nematode larva at the center of the reaction complex is usually viable, and a thin layer of tightly packed eosinophils, neutrophils, and histiocytes occurs adjacent to the larva. This type of reaction is frequently associated with acute intestinal anisakiasis within one week of infection.

Abscess Type. This type of reaction is characterized by the accumulation of numerous eosinophils, histiocytes, and lymphocytes surrounding the worm embedded in the submucosa. A distinct granulomatous zone is apparent. Necrosis and hemorrhage with eosinophilic infiltration and fibrin exudation occur in the inner layer of the granuloma. Slight phlegmonous changes are present in the peripheral zone of the granuloma. The cuticle of the nematode larva at the center of the reaction complex is destroyed, and degeneration of the internal structures has commenced. This type of lesion is often associated with chronic cases of gastric and intestinal anisakiasis.

Abscess-Granulomatous Type. This type of reaction is associated with the degraded debris of larval anisakids. The abscess surrounding the residual debris is reduced; however, it is surrounded by a conspicuous tunic of granulomatous cells accompanied by some deposition of collagen. There is less infiltration of eosinophils into the granuloma than in the abscess type. In some cases, lymphocytes, rather than eosinophils, represent the dominant cells. Giant cells are commonly present along the periphery of the decomposed larva. Furthermore, numerous eosinophils have usually invaded the site occupied by the remnants of the parasite. This type of lesion occurs primarily in cases of gastric anisakiasis that are at least 6 months old.

Granulomatous Type. This type of reaction represents an advanced stage of the abscess-granulomatous type. It is characterized by the replacement of the abscess by granulomatous cells coupled with the infiltration by eosinophils. By this stage, the causative parasite is no longer recognizable since it has either totally or almost totally disintegrated. This type of reaction is occasionally found at sites in the gastric and intestinal walls of old infections.

Pathogenesis of Anisakiasis

The mechanisms responsible for the development of the lesions associated with larval anisakids remain uncertain, but there are two hypotheses: the **double hit hypothesis** and the **exacerbation hypothesis**.

Double Hit Hypothesis. According to this hypothesis, which was proposed by Dutch investigators (van Thiel *et al.*, 1960; Kuipers *et al.*, 1963; Arean, 1971), penetration of the gastric or intestinal mucosa by the anisakid larva induces a local, persistent hypersensitivity. If a second larva penetrates near the same site, an eosinophilic phlegmonous inflammation occurs.

This hypothesis has been criticized since it has been demonstrated that a severe reaction occurs in the rabbit stomach as a result of a single infection (Ruitenberg and Loendersloot, 1971). Nevertheless, it is recognized that the reaction to a second infection is more severe than that resulting from the first.

Exacerbation Hypothesis. According to this hypothesis, which is championed primarily by Japanese investigators (Kojima *et al.*, 1966), the anisakid larva, after penetrating into the submucosa of the host's alimentary canal, survives for two or three weeks. During the seventh to tenth days postpenetration, granulation appears around the larva. Subsequently, the living larva sensitizes the newly formed granuloma with its metabolic products. When the worm dies, its cuticle disintegrates and cells and fluid exuding from the dead worm react directly with the sensitized granuloma, resulting in an allergic inflammation around the dead larva. There is some experimental support for this theory (Oyanagi, 1967; Hayasaka *et al.*, 1968). However, although this hypothesis effectively explains the abscess and abscess-granulomatous types of reactions, it does not explain the phlegmonous type of lesions, i.e., a living larva remains at the center of the phlegmon, and severe tissue reaction develops within a few days, but does not last long enough for the sensitization of the surrounding tissues.

Epidemiology of Anisakiasis

There is only one known method by which humans develop anisakiasis. Such infections have resulted from the ingestion of raw or undercooked marine fish harboring infective anisakid larvae.

Very little is presently known about the epidemiology of human anisakiasis. However, since species of larval anisakids that can cause anisakiasis in humans utilize endothermic vertebrates as their natural definitive hosts, in nature, one would expect them to be more abundant in parts of oceans where such animals are more numerous. This need not be the case, however, since fish harboring the infective larvae may migrate, at least periodically, away from areas where endotherms occur. Nevertheless, since the egg, L_1s, and L_2s are essentially sedentary, and L_3s occur in invertebrates incapable of migrating long distances, one would expect to find areas of greater endemicity.

Commercial fishing contributes to the distribution of anisakids. Small fish that fishermen choose not to retain commonly are left lying on deck in the sun for at least three to four hours before they are thrown back into the sea. As the deck is cleared of fish before the next haul, many discarded fish harboring anisakid larvae are thrown back into the sea. Such moribund or dead fish are rapidly eaten by larger fish, and the parasites are passed on to the predators; thus, fishing vessels transport these parasites from one area to another (Cheng, 1976).

The best-known method of preventing human anisakiasis is to avoid raw or undercooked fish. Short of this, the processor must subject the fillet to either heat or extremely low temperatures. Heating is not always a desirable or practical method from the standpoint of the wholesaler or retailer, since 50 to 60°C is required to kill these larvae. On the other hand, normal refrigerator temperatures do not kill anisakid larvae in fillet. Refrigeration at $-20°C$ for up to 52 hours is necessary to kill the parasites (Bier, 1976).

Other methods of preserving fish, such as smoking and marination, are ineffective in killing anisakid larvae. A possible exception is dry salting, provided the salt reaches all parts of the muscles in concentrated form.

Disease in Animals

When adult anisakids occur in their natural definitive hosts, they cause little or no pathology since these are luminal parasites. However, the larvae present in invertebrate and piscine intermediate hosts do cause considerable pathologic alterations. Although not comparable to human anisakiasis, which is also caused by larvae, the pathology associated with anisakiasis in fish, which has been reviewed by Margolis (1970), is different. In fish, the disease is most severe when the larval anisakids are associated with the liver. In fish harboring these parasites in their liver, there is atrophy of that organ and a significant loss in body weight. It remains uncertain whether the damage is due to the parasites per se or to secreted toxins.

SUPERFAMILY OXYUROIDEA

The oxyuroideans are commonly referred to as pinworms. They have simple, direct life cycles, with the egg being the infective stage. Representatives of this

Table 15.1. Some Common Oxyuroid Nematodes

Nematode Species	Host
Blatticola blattae	*Blattella germanica* (German cockroach)
Hammerschmidtiella diesingi	*Periplanata americana* (American cockroach)
Leidynema appendiculata	*Periplanata americana*
Thelastoma bulhoesi	*Periplanata americana*
Thelastoma icemi	*Periplanata americana*
Aspicularis tetraptera	Mouse
Syphacia obvelata	Rat and mouse
Enterobius vermicularis	Human

superfamily are parasites of both vertebrates and invertebrates.

The major diagnostic characteristics of the Oxyuroidea are listed on p. 569.

Some common oxyuroid species that are readily available for laboratory studies are listed in Table 15.1.

Enterobius vermicularis. Undoubtedly the best-known oxyuroid nematode is the human pin- or seatworm, *E. vermicularis* (Fig. 15.25). Children, especially early school-age children, are most frequently infected with this parasite. The geographic distribution of *E. vermicularis* is worldwide, especially in the temperate zones; however, the incidence of infection varies with each locale. For example, Hitchcock (1950) has reported that at least 51% of Alaskan Eskimos are infected; Cates (1953) has reported a 26.85% infection rate among students of five elementary schools in and around Tallahassee, Florida; Zaiman *et al.* (1952) have reported a 58% infection rate among the pupils in a preschool nursery in San Francisco, California; Kessel *et al.* (1954) have reported only a 9% infection rate among five- to nine-year-olds in Tahiti, French Oceania; Ricci (1952) has reported a startling infection rate

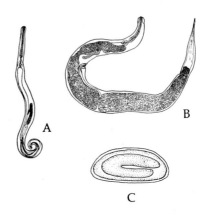

Fig. 15.25. *Enterobius vermicularis.* **A.** Adult male. **B.** Adult female. **C.** Egg enclosing developing larva. (**A** and **B**, after Beaver, 1952.)

of 77.14% among Sicilian children, but in the same population only 6.09% of the adults were infected. Iwanczuk (1953) has reported that 8.6% of 1119 children up to four years of age in Warsaw, Poland, were infected. Cheng (1960), in surveying a group of children from a wide geographic range in the United States, found that 32.93% were infected. This figure probably comes as close as any to portraying the incidence among preschool age children in the United States. At least 500 million persons are infected by this parasite in the world.

Life Cycle of Enterobius vermicularis. The life cycle of *Enterobius* is direct; no intermediate host is required. Within the host's intestine, both the males and females are commonly attached to the epithelial wall. After copulation, gravid females become packed with eggs. These females demonstrate a rhythmic periodicity, migrating posteriorly during the night but seldom during the day, and depositing their eggs, which adhere to the perianal skin folds. Most females migrate back into the large intestine after ovipositing and die; however, some pass to the exterior, while others explode while laying eggs. Males die soon after copulation.

Deposited eggs enclose a motile L_1 larva (Figs. 14.27, 15.25). The transparent eggshell is composed of a hyaline outer albuminous layer, a shell proper composed of two layers of chitin, and an inner embryonic layer.

It has been estimated that a single female deposits from 4,672 to 16,888 (mean, 11,105) eggs. Infections or reinfections become established when these infective eggs are ingested by the host. These eggs are usually picked up from bedclothes or fingernails that become contaminated while the host scratches the itchy perianal zone caused by the migrations of the females. However, the lightweight eggs can be airborne and inhaled. Retroinfections (or retrofections) are possible when larvae hatching from perianally located eggs migrate back up the intestinal tract.

Ingested eggs usually hatch soon after they reach the duodenum. The escaping L_1s migrate posteriorly and undergo three molts in the process and develop into adults by the time they reach the large intestine. The life cycle of the *E. vermicularis* generally takes two months. Eggs are very resistant to drying and remain viable for a week or more under cool, humid conditions. Some claim that they may remain viable for years.

Other Oxyuroid Genera. Other representative members of the superfamily Oxyuroidea from domestic animals are depicted in Figure 15.26.

The life cycle of *Oxyuris equi*, in horses, parallels that of *Enterobius vermicularis*. The life cycles of *Skrjabinema ovis*, the sheep pinworm, and *Passalurus ambiguus*, the rabbit pinworm, are presumably the same.

Numerous other oxyuroid nematodes inhabit the intestines of rabbits, hares, and rodents. These include

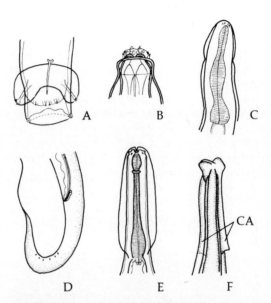

Fig. 15.26. Some oxyuroid nematodes. A. Posterior end of male *Oxyuris equi*, ventral view. **B.** Anterior end of *Skrjabinema ovis*, the sheep pinworm. **C.** Anterior end of *Passalurus ambiguus*, the rabbit pinworm. **D.** Lateral view of posterior end of male *Probstmayria vivipara*, the minute pinworm of horses. **E.** Anterior end of *Dermatoxys veligera*, parasitic in caecum of rabbits. **F.** Anterior end of *Aspiculuris tetraptera*, parasitic in rats and other rodents. (**A–E**, redrawn after Shiller in Morgan and Hawkins, 1949; **F**, redrawn after Hall, 1916.) CA, cervical alae.

Dermatoxys and *Passalurus* in rabbits (Fig. 15.26). Members of both these genera are armed with three buccal teeth. Another commonly encountered oxyuroid nematode is *Probstmayria vivipara*, found in equines the world over. In *P. vivipara*, the horse pinworm, the gravid females give rise to larvae that hatch within the uteri of the parents.

Several genera of Oxyuroidea are parasitic in the intestine of reptiles. These include *Pharyngodon*, *Thelandros*, and *Atractis* in lizards and tortoises (Fig. 15.27), and *Labiduris* in turtles.

Oxyuroids from Invertebrates

Many species of Oxyuroidea have been reported in invertebrate hosts. A few representatives are reviewed at this point.

Most species found in various insects belong to the family Thelastomatidae, members of which are characterized by eight simple papillae that make up the outer circlet surrounding the mouth and by one or no copulatory spicules in males. Representatives found in cockroaches include members of *Leidynema, Hammer-schmidtiella* (Fig. 15.27), and *Blatticola*.

In the life cycle of *Leidynema appendiculata* (Fig. 15.27), the eggs, which pass out in cockroach feces, include partially developed larvae that do not hatch in available culture media or outside the host. However, when the enclosed larvae complete their development, they hatch from the eggshell if the eggs ingested by a roach host reach the host's midgut. Eggs may also be incorporated in the roach's oothecae and serve to infect nymphs when they hatch.

Other thelastomatid genera include *Thelastoma* in cockroaches and scarabaeid beetles (Fig. 15.27) and *Pseudonymus* in water beetles. In *Pseudonymus*, the peculiar eggs are enveloped within two coiled filaments that uncoil when they come in contact with water and become entangled in aquatic plants (Fig. 15.27). Members of *Aorurus* (Fig. 15.27) are found in passalid beetles. The female members of the latter three genera all possess spines in the cervical region.

SUPERFAMILY HETERAKOIDEA

The superfamily Heterakoidea includes several species of economic importance since these are parasites of domestic fowls. For example, *Heterakis* and *Ascaridia* are parasitic in chickens, turkeys, and other fowl.

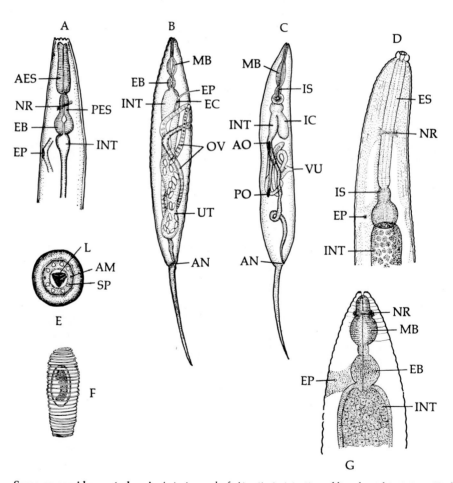

Fig. 15.27. Some oxyuroid nematodes. A. Anterior end of *Atractis*, in intestine of lizards and tortoises. (Redrawn after Thapar, 1925.) **B.** Female *Hammerschmidtiella diesingi* in cockroaches. (Redrawn after Chitwood, 1932.) **C.** Female *Leidynema appendiculata* in cockroaches. (Redrawn after Dobrovolny and Ackert, 1934.) **D.** Anterior end of *Thelastoma*, in cockroaches and scarabaeid beetle larvae. **E.** Face-on view of anterior end of *Thelastoma*. (**D** and **E**, redrawn after Christie, 1931.) **F.** Egg of *Pseudonymus* enclosed in spiral filaments. (Redrawn after Györy, 1856.) **G.** Anterior end of *Aorurus*, in millipedes and beetle larvae. (Redrawn after Christie, 1931.) AES, anterior portion of esophagus; AM, amphid; AN, anus; AO, anterior ovary; EB, end bulb of esophagus; EC, excretory canal; EP, excretory pore; ES, esophagus; IC, intestinal caecum; INT, intestine; IS, isthmus of esophagus; L, lip; MB, median bulb of esophagus; NR, nerve ring; OV, ovary; PES, posterior portion of esophagus; PO, posterior ovary; SP, sensory papilla; UT, uterus; VU, vulva.

Ascaridia galli. The biology of *Ascaridia galli*, the comparatively large intestinal nematode of chickens, has been studied extensively (Fig. 15.28). The males measure 30–80 mm in length, and the females are 60–120 mm long. As in all parasitic nematodes, *A. galli* undergoes four molts, in this case, one while in the eggshell and three within the chicken. Infection of the chicken is established when eggs containing L_2s are ingested. These larvae hatch once the eggs reach the host's small intestine.

For the first nine days, the larvae survive as lumen dwellers or as intervillar space dwellers in the small intestine. On approximately the 10th day, the larvae begin to burrow into the mucosa, causing severe hemorrhage and tissue damage. From the 10th to the 17th day, the larvae are partially lodged in the mucosa, but with their tails protruding into the lumen. Following the 17th day, these nematodes return to the gut lumen dwellers and remain there. The three molts in the host take place at intervals of approximately 6 days.

Resistance to *Ascaridia galli*. Age resistance to *Ascaridia galli* has been demonstrated repeatedly. Chickens older than 3 months and maintained on normal diets are quite resistant; worms that are present are smaller than those in younger hosts. Ackert *et al.* (1939) have attributed resistance in older birds to the increased number of goblet cells present in the intestinal wall because the secretions of these cells inhibit growth and development of the worms. This phenomenon has been demonstrated by Eisenbrandt and Ackert (1941), who reported that duodenal mucosa extracts of dogs and pigs hasten the death of worms maintained *in vitro*. The administration of the sex hormones testosterone and estradiol to male and female birds, respectively, also causes resistance. This resistance is thought to be an indirect rather than a direct effect. Certain genetic strains of chickens are more resistant to *Ascaridia galli* than others.

Heterakis gallinae. This parasite is the common caecal worm of chickens, turkeys, pheasants, and other fowls (Fig. 15.28). Both males and females bear cervical alae, and the males also possess caudal alae and 12 pairs of papillae. The males are 7–13 mm long, and the females are 10–16 mm long. The eggs passed out in the host's feces contain uncleaved zygotes. After embryonated eggs are ingested and hatched in the gizzard or duodenum of the bird, the larvae migrate to the caecum and usually remain in the lumen or

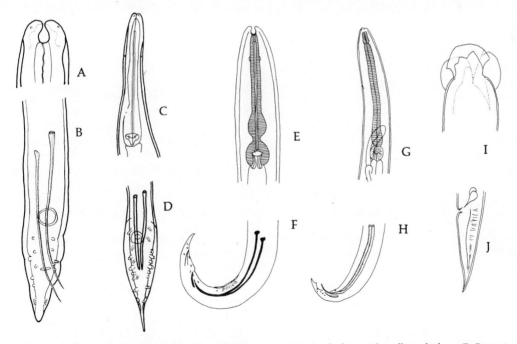

Fig. 15.28.   Some heterakoid and subduroid nematodes. A. Anterior end of *Ascaridia galli*, in chickens. **B.** Posterior end of male *A. galli*. **C.** Anterior end of *Heterakis gallinae*, in fowl. **D.** Posterior end of male *H. gallinae*, ventral view. (**A–D**, redrawn after Schiller in Morgan and Hawkins, 1949.) **E.** Anterior end of *Subulura brumpti*, in fowl. **F.** Lateral view of posterior end of male *S. brumpti*. (Redrawn after Lapage, 1962.) **G.** Anterior end of *Cruzia tentaculata*, in caecum and large intestine of mammals. **H.** Posterior end of male *C. tentaculata*. (Redrawn after Yorke and Maplestone, 1926.) **I.** Dorsal view of anterior end of *Pseudocruzia orientalis*, in pigs. **J.** Posterior end of female *P. orientalis*. (Redrawn after Maplestone, 1930.)

migrate in between the crypts, rarely in the lymph glands.

Heterakis gallinae also undergoes four molts—one within the eggshell while in the soil, the second 4 to 6 days after infection of the host, the third and fourth in 9 to 10 days and 14 days, respectively. Not all the embryonated eggs ingested by the fowl host develop. The percentage of survival is greater when smaller numbers of eggs are ingested by a single host. For example, when 200 eggs are fed to a host, 23% develop; on the other hand, when 1000 eggs are fed to a host, only 0.3% develop.

Although *H. gallinae* is apparently nonpathogenic, it is the vector for *Histomonas meleagridis*, the highly pathogenic "blackhead" disease-causing protozoan (p. 117).

Heterakis spumosa. This nematode is a commonly encountered and widely distributed parasite of wild rats. The sexually mature adults inhabit the caecum and colon. The life cycle is direct.

SUPERFAMILY SUBULUROIDEA

This superfamily includes *Subulura* parasitic in primates and birds, *Cruzia* in turtles and marsupials, and *Pseudocruzia* in swine.

Subulura brumpti. This worm (Fig. 15.28) occurs in the caecum of the chicken, turkey, guinea fowl, bobwhite quail, and other birds. It has been reported in Europe, Asia, Africa, North America, Hawaii, Panama, and Puerto Rico. The male measures 6.9–10 mm long, and the female is 9–13.7 mm long. The anterior end of the body is bent dorsally, and the cervical alae have fine, transverse striations which extend some distance above the length of the body.

An intermediate host is involved in the life cycle of *S. brumpti*. The embryonated eggs passed out in the host's feces must be ingested by an insect host, a beetle or earwig, in which they hatch, and develop to the L$_3$ stage in the hemocoel. Birds become infected from eating infected insects. The third and fourth molts occur in the avian definitive host, and the females start laying eggs about 6 weeks after infection. This nematode is only mildly pathogenic, but when present in large numbers, inflammation of the definitive host's caecum may occur.

Genera *Cruzia* and *Pseudocruzia*

Cruzia spp. (Fig. 15.28) are parasitic in the caecum and large intestine of such mammals as *Didelphys* (opossums) and *Marmota* (marmots). *Pseudocruzia*, represented by *P. orientalis*, occurs in the caecum of pigs in India. It has been suggested that the pig may not be the natural host and that it is usually a parasite of reptiles; however, in Calcutta it is a fairly common parasite, occurring in about 20% of pigs. Morphologically, it is readily identified by its head, which is set off from the rest of the body by a constriction at the base of the lips (Fig. 15.28).

SUPERFAMILY DIOCTOPHYMATOIDEA

The major diagnostic characteristics of the superfamily Dioctophymoidea are listed on p. 570. This superfamily includes the genera *Dioctophyme* and *Soboliphyme*. All the species are parasites in the kidneys, and rarely in the body cavities, of mammals.

Dioctophyme renale. A bright red species, *D. renale*, is generally found in the kidney of carnivores, such as dogs, mink, foxes, bears, and wolves. It is occasionally found in pigs, cattle, horses, and even humans (Figs. 15.29, 15.30). This nematode, which is apparently worldwide in its distribution, is highly destructive, often totally destroying the host's kidney, and may even cause death when both kidneys of the host are involved. It is commonly referred to as the giant kidney worm.

The life span of adult worms ranges from one to three years. With the death of the parasite, the host's kidney, or what is left of it, generally shrivels and disintegrates. *Dioctophyme renale* is one of the largest parasitic nematodes known. The females may measure up to 103 cm in length and have a diameter of 5–12 mm. The males are 35 cm long and 3–4 mm in diameter. The male has a conspicuous, bell-shaped

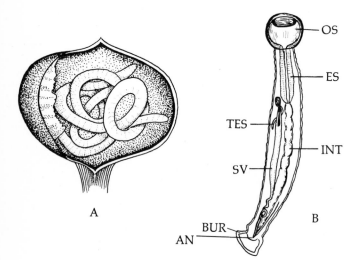

Fig. 15.29. Dioctophymatoid nematodes. A. *Dioctophyme renale* coiled in host's kidney. (Redrawn after Railliet, 1893.) **B.** Male of *Soboliphyme baturini*. (Redrawn after Petrov, 1930.) AN, anus; BUR, bursa; ES, esophagus; INT, intestine; OS, oral sucker; SV, seminal vesicle; TES, testis.

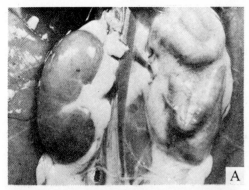

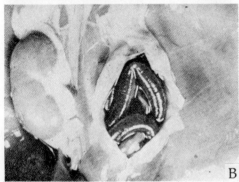

Fig. 15.30. *Dioctophyme renale.* **A.** Kidneys of a ferret; the one on the left is normal while the one on the right is distended by an adult nematode. **B.** Same specimen as **A** but with right kidney opened to show presence of adult worm; the kidney is reduced to a shell. (Photographs by A. E. Woodhead; courtesy of Ann Arbor Biological Center.)

copulatory bursa that lacks supporting rays or papillae.

Life Cycle of Dioctophyme renale.   The life cycle of the giant kidney worm was elucidated by Karmanova (1960), who has also contributed a review of this and related species of nematodes (Karmanova, 1968).

The eggs passed out in the host's urine are thick shelled. They require about six months in water to embryonate. The L_1, which bears an oval spear, hatches after the egg is ingested by an aquatic annelid, *Lumbriculus variegatus*. The L_1s penetrate into the abdominal blood vessels of the annelid, where they eventually develop into L_4s. When an infected annelid is ingested by the definitive host, the larva migrates to a kidney where it matures. A paratenic host may be involved in the transmission of *D. renale*. Specifically, if infected annelids are eaten by any of several species of fish, the escaping nematode larvae can encyst in their muscles or viscera. When the fish paratenic host is eaten by a mammalian definitive host, the L_4s penetrate the duodenal wall and enter a kidney, commonly the right one.

In cases of human infections, the adult worms usually occur in a kidney; however, Beaver and Theis (1979) have reported an L_3 in a subcutaneous nodule.

Loss of kidney function compounded by uremic poisoning results in hosts harboring *D. renale*. Also, the worms may penetrate the greatly reduced renal capsule and wander in the body cavity, inducing host cellular response.

Genus *Soboliphyme*

The genus *Soboliphyme* includes *S. baturini* parasitic in the intestine of cats, sables, and foxes in Kamchatka and Siberia, and in wolverines, ermines, shrews, and martens in North America (Fig. 15.29). This nematode is unique among the members of its superfamily in that an oral sucker surrounds the mouth. The females are oviparous.

SUPERFAMILY SPIRUROIDEA

From what is known, the members of this superfamily include nematodes that require an intermediate host in their life cycles, usually an insect.

Spiruroidea includes such genera as *Spirura* and *Mastophorus*. Members of *Spirura* are characterized by the posterior part of the body being thicker than the anterior and by a more or less spirally twisted posterior portion. A representative of this genus is *S. rytipleurites*, which occurs in the stomach, attached to the mucosa, of the cat, rat, fox, hedgehog, and possibly other mammals in Europe, North Africa, and Madagascar. The males of this species are 18–26 mm long and 600 μm wide, and the females are 28–32 mm long and 800 μm wide.

Members of *Mastophorus* each bear two large, lateral, trilobed lips (pseudolabia). Each of these bears three papillae and is armed with three, five, seven, or nine teeth on its inner surface. A representative of this genus is *M. muris*, which occurs in the stomach of the Norway rat, black rat, house mouse, golden hamster, and other rodents. It has also been reported from the domestic cat and coyote. Its geographic distribution includes North America, Puerto Rico, Israel, North Africa, Madagascar, Europe, the Soviet Union, and Australia. The males measure 12–45 mm long and 0.4–1.2 mm wide, and the females 20–100 mm long and 0.7–2.6 mm wide.

The Spiruroidea also includes *Tetrameres*, the globular stomach worm of poultry. Sexual dimorphism is extreme in this genus. The males are white and threadlike, and rows of spines along the median and lateral lines are usually apparent. The females, on the

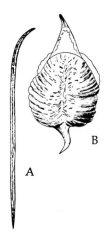

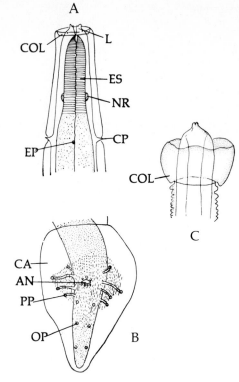

Fig. 15.31. **Tetrameres. A.** Male specimen. **B.** Female specimen. (After Cram, 1931.)

Fig. 15.32. **Physalopteroid nematodes. A.** Anterior end of *Physaloptera*. **B.** Posterior end of male *Physaloptera*. (**A** and **B**, after Ortlepp, 1922.) **C.** Anterior end of *Proleptus* showing cuticular collar. (After Lloyd, 1920.) AN, anus; CA, caudal alus; COL, collar; CP, cervical papilla; EP, excretory pore; ES, esophagus; L, lips; NR, nerve ring; OP, other genital papilla; PP, pedunculated papilla.

other hand, have a red, globular or spindle-shaped body that is divided into four divisions by superficial, longitudinal furrows (Fig. 15.31).

Tetrameres americana occurs in the proventriculus of the chicken, bobwhite quail, and ruffed grouse in North America and Africa. The adult males are 5.0– 5.5 mm long and 116–133 μm wide, whereas the females are 3.5–4.5 mm long and 3 mm wide. When present in small numbers, this parasite does not produce serious symptoms; however, when large numbers occur, the avian host portrays dullness, wasting, emaciation, and even death.

Tetrameres confusa occurs in the proventriculus of the chicken, pigeon, and turkey in South America and the Philippines. The adult males are 4–5 mm long and 150 μm wide, and the females are 3–5 mm long and 2–3 mm wide.

Life Cycle of *Tetrameres*.   During the life cycle of *Tetrameres* spp. an invertebrate intermediate host is involved. This is usually an aquatic crustacean such as an amphipod or a cladoceran; however, grasshoppers, cockroaches, and even earthworms may serve as the intermediate host. The invertebrate host becomes infected as the result of ingesting embryonated eggs passed out in the feces of the definitive host. The L_1s hatching from the eggshell in the arthopod's gut penetrate into the hemocoel and undergo two molts. The resulting L_3s are infective when the intermediate host is eaten by the avian host.

It is noted that L_2s and L_3s in the arthropod host are commonly encapsulated by host leucocytes, and thus give the gross appearance of being encysted.

Spiruroidea also includes *Habronema* with some

species parasitic in equines and others in birds, *Draschia* with species in equines, and *Spiroxys* in turtles.

SUPERFAMILY PHYSALOPTEROIDEA

This superfamily includes a few genera found in the stomachs of amphibians, reptiles, birds, and mammals. The type genus *Physaloptera* includes many species found in a variety of hosts, including amphibians,

Fig. 15.33. ***Abbreviata.*** Head-on view of adult worm. (After Chitwood and Wehr, 1933.)

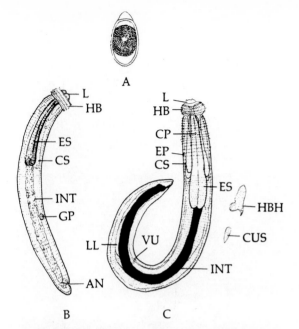

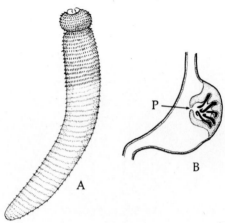

Fig. 15.34. Stages in the life cycle of *Gnathostoma spinigerum*. A. Egg. B. Second-stage larva. C. Third-stage larva. AN, anus; CP, cervical papilla; CS, cervical sac; CUS, cuticular spine of body in proximity of head bulb; EP, excretory pore; ES, esophagus; GP, genital primordium; HB, head bulb; HBH, lateral view of head bulb hook; INT, intestine; L, lip; LL, lateral line; VU, vulva.

Fig. 15.35. *Gnathostoma*. A. Adult worm. B. Cut-away view of stomach of host showing nematode in cyst. Notice presence of pore through which eggs of the parasite escape into lumen of the host's alimentary tract. P, pore. (Redrawn after Miyazaki, 1960.)

reptiles, birds, and mammals (Fig. 15.32). *Proleptus* is found in the stomach and intestine of elasmobranchs (Fig. 15.32). *Skrjabinoptera* is represented by *S. phyrnosoma*, found in horned lizards in the southwestern United States. Another member of the Physalopteroidea is *Abbreviata*, which is found in amphibians, reptiles, and a few mammals (Fig. 15.33). The complete life histories of the physalopteroids mentioned have not been determined, although utilization of two intermediate hosts is suspected.

Of particular medical importance are those physalopteroid nematodes belonging to the genus *Gnathostoma* (Fig. 15.34). The various species have been described from all continents except Europe and Australia. Miyazaki (1960) has contributed a review of the biology and distribution of members of this genus and has listed the various species. Although most of the species have been reported embedded in the stomach wall of mammals, where they are commonly associated with tumorous growths (Figs. 15.35, 15.37), *G. gracile* was reported in the intestine of a fish, *Arapaima gigas*, from Brazil, *G. occipitri* was found subcutaneously in an eagle in Turkestan, and *G. minutum* was found in the connective tissue of a serpent in the Congo. The most important species is *G. spinigerum*, which is the most common causative agent of gnathostomiasis in humans.

Life Cycle of *Gnathostoma spinigerum*. The life cycle and migratory pattern of *G. spinigerum* has been investigated by numerous researchers. The stages in the development of this species are depicted in Fig. 15.34. Usually cats and dogs serve as the definitive host, but humans are fairly commonly infected, especially in southeast Asia. Eggs, when passed from the definitive host, contain one or two blastomeres. When such eggs fall into water of a suitable temperature (27°C is the ideal experimental temperature), a fully developed L_1 is visible in 5 days enveloped in a delicate sheath within the eggshell. This larva hatches in seven days and becomes free swimming. Development continues if the L_1 is ingested by a suitable species of cyclops—i.e., *Mesocylops leuckarti*, *Eucyclops serrulatus*, *Cyclops strenuus*, or *C. vicinus* (Fig. 15.36).

In the cyclops host the larva sloughs off the enveloping sheath and penetrates the gastric wall to become lodged in the host's body cavity. At this site, the larva molts within seven to ten days after ingestion, increases in size, reaching 0.5 mm in length, and becomes an L_2.

If an infected cyclops is ingested by a second intermediate host, usually a fish, reptile, or amphibian, the larva again penetrates the host's gastric wall, migrates to the body musculature, molts, increases to a length of 4 mm, becomes encapsulated within a fibrous membrane, and is then an L_3. Various crustaceans, birds, and mammals, especially rodents, can also serve as the second intermediate host. Infection of the second

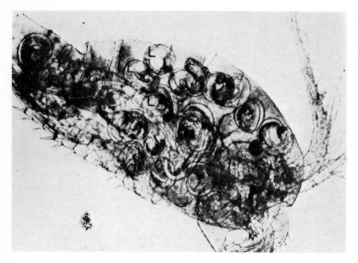

Fig. 15.36. ***Gnathostoma spinigerum* in cyclops host.** Photomicrograph showing L$_3$s of *G. spinigerum* in hemocoel of cyclops intermediate host. (After Daengsvang, 1980; with permission of Southeast Asian Medical Information Center.)

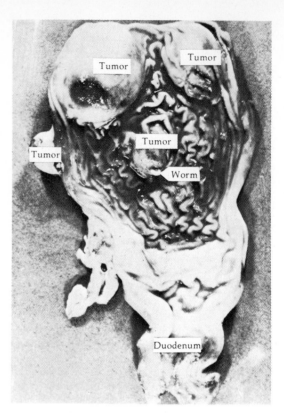

Fig. 15.37. Tumors caused by *Gnathostoma spinigerum*. Photomicrograph showing four tumors protruding from the gastric wall of dog naturally infected with *G. spinigerum.* Notice a worm protruding from one tumor. (After Daengsvang, 1980; with permission of Southeast Asian Medical Information Center.)

intermediate host is known as **primary infection.** Gnathostomes are interesting in that they can pass through several paratenic hosts without further development. Such infections are known as **secondary infections**.

Finally, when hosts harboring L$_3$s are eaten by the definitive host, the larvae penetrate the gastric wall (rarely through the intestinal wall) and enter the liver. From the liver, the larvae migrate into muscles and then into connective tissue, increasing in size and molting twice during the process. When these nematodes attain their maximum size—males 11−25 mm long, females 25−54 mm long—they enter the gastric wall from the exterior and become embedded in typical nodules that possess an opening to the lumen of the stomach (Fig. 15.37). It is through this aperture that eggs are eventually passed out. Such nodules as found in natural hosts are referred to as **gnathostomiasis interna**. In human infections, contracted from eating the raw flesh of second intermediate hosts, the worms never reach maturity and are usually found in subcutaneous tissues. Such a condition is known as **gnathostomiasis externa** (Fig. 15.38). Larva migrans occurs in human infections in which the worm continues moving along under the skin. In either case, humans serve as paratenic hosts for juvenile worms, but are not likely to be involved in transmission. For a comprehensive review of

human gnathostomiasis in Thailand and the causative species, see Daengsvang (1980).

Other Genera

Physalopteroidea also includes the genera *Tanqua*, in the stomachs of lizards and aquatic snakes, and *Echinocephalus*, in the spiral valves of elasmobranchs (Fig. 15.39). In *Echinocephalus*, larvae have been found encapsulated in oysters, which serve as intermediate hosts. *Echinocephalus crassostreai* (Fig. 15.40), for example, was described by Cheng (1975) from the Japanese oyster, *Crassostrea gigas*, from off the coast of the People's Republic of China. The L$_3$ of this nematode will infect mammals, possibly including humans, if eaten in raw oysters.

SUPERFAMILY FILARIOIDEA

This superfamily includes what are generally referred to as the filarial worms. These include some of the most important nematode parasites of humans and domestic animals from the medical and veterinary viewpoints. The long, threadlike adult worms are found in the host's lymphatic glands, tissues, and body cavities, while the L_1s, known as **microfilariae**, live in blood or other tissues (p. 547). Those filarias that parasitize mammals are ovoviviparous, but most species that parasitize birds are oviparous. At the time of larviposition in the ovoviviparous species, the microfilariae uncoil and are released from the female. If the flexible eggshell becomes elongate to form a covering membrane, the larva is termed a **sheathed** microfilaria, but if the shell ruptures and does not form a sheath, the larva is said to be **unsheathed** or naked.

The isolation and identification of microfilariae very often is the only reliable method for determining infections; therefore, the diagnostic characteristics of L_1s of the most prominent species are given below and depicted in Fig. 15.41.

The superfamily Filarioidea includes the genera *Filaria* parasitic in mustelids and *Parafilaria* in cattle. In addition, it includes the genera *Wuchereria, Brugia, Onchocerca, Litomosoides, Loa, Dirofilaria, Mansonella, Setaria,* and others. Representative species of these genera are discussed below.

Genus *Wuchereria*

The members of this genus can be recognized by their enlarged and rounded anterior end, followed by a neck. The mouth is circular and not surrounded by lips. There are two circles of cephalic papillae. The major species is *Wuchereria bancrofti,* a parasite of humans.

Wuchereria bancrofti (Fig. 15.41). This widely distributed parasite is the causative agent of Bancroft's filariasis. It occurs in the Nile Delta, central Africa, Turkey, India, southeast Asia, the Philippines, many Pacific islands, Indonesia, Australia, the Caribbean islands and parts of South America. In brief, this common parasite occurs throughout a broad equatorial belt. It most probably was introduced into the New World by the slave trade. Until the 1920s, it occurred in South Carolina. A Pacific strain occurs throughout the Pacific islands except Hawaii. Millions of human cases of this type of filariasis are in existence.

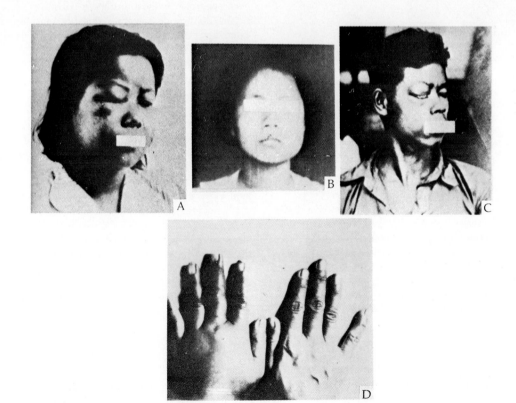

Fig. 15.38. Gnathostomiasis externa. A. Ocular gnathostomiasis causing swelling of eyelids. **B.** Swelling of left cheek due to subcutaneous gnathostomiasis. **C.** Swelling of eyelid due to ocular gnathostomiasis. **D.** Swelling of left hand due to cutaneous gnathostomiasis. (All photographs after Daengsvang, 1980; with permission of Southeast Asian Medical Information Center.)

A

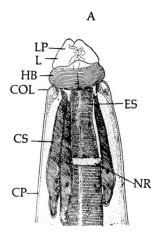

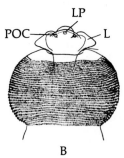

B

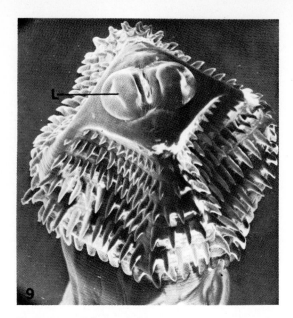

Fig. 15.40. *Echinocephalus crassostreai.* Scanning electron micrograph of cephalic end of L₃ from the Japanese oyster, *Crassostrea gigas.* L, lip. (After Cheng, 1975; with permission of Academic Press, N.Y.)

Fig. 15.39. Some physalopteroid nematodes. A. Anterior end of *Tanqua* showing ballonets and cervical sacs. (After Baylis and Lane, 1920.) **B.** Anterior end of *Echinocephalus.* (After Baylis and Lane, 1920.) COL, collar; CP, cervical papilla; CS, cervical sac; ES, esophagus; HB, head bulb; L, lips; LP, labial papilla; POC, papillae of outer circlet.

Microfilaria of W. bancrofti. The body is 220–300 μm long. The sheath stains red with dilute Giemsa's stain. The tail is not coiled. No nuclei occur in the tail. These L₁s are nocturnal or are only slightly diurnal in their periodicity, depending on the strain (p. 547). They usually live in the blood, but can appear in urine, where they were first discovered by Wucherer in 1866.

Life Cycle of Wuchereria bancrofti. The stages in the life cycle of *W. bancrofti* are depicted in Fig. 15.41. The adult males measure 40 mm in length by 0.1 mm in diameter, and the adult females measure 80–100 mm in length by 0.24–0.3 mm in diameter. These worms are found in the lymph glands and associated ducts of humans. Gravid females give rise to sheathed microfilariae that are 127–320 μm long. Nocturnal periodicity—that is, occurrence of microfilariae in the peripheral circulation at night—has

been reported in most instances. However, certain strains do not demonstrate this phenomenon. For example, in the Pacific strain of *W. bancrofti*, found on various Polynesian islands in the South Pacific, the microfilariae show only slight periodicity, and it is diurnal—that is, they appear in greater numbers in the peripheral blood during the day. The South Pacific strain is said to be **subperiodic**.

Several hypotheses have been contributed to explain nocturnal periodicity. (1) There is chemotactic attraction between the microfilariae and the saliva of mosquito hosts (vectors), which are more plentiful at night. (2) The relaxation of the host during sleep induces the microfilariae to migrate into the peripheral circulation. (3) The migration results from a response to oxygen and carbon dioxide supply. (4) The microfilariae survive for only a short period, and it is during the nocturnal period that they are most abundant and are readily found in the peripheral circulation. None of these hypotheses is completely satisfactory. The phenomenon of microfilarial periodicity is discussed in more detail on p. 547.

In strains that demonstrate nocturnal periodicity, microfilariae are most plentiful in the peripheral circulation between 10:00 P.M. and 2:00 A.M. If these microfilariae are ingested by a mosquito during a blood meal, the mosquito phase of the life cycle commences. Unlike the malaria organisms, filarial worms show little specificity in regard to mosquito hosts. *W. bancrofti* can utilize *Culex* spp., *Aedes* spp., *Mansonia* spp., *Anopheles* spp., and *Psorophora* spp. as vectors equally well.

On reaching the midgut of the mosquito, the microfilariae lose their sheaths within 2 to 6 hours, penetrate the gut wall, and migrate to the thoracic muscles. At this site, the organism becomes shortened into a short, sausage-shaped body measuring 124–250 μm. At this stage of development, the first true molt occurs, after which the tail portion atrophies and the intestinal tract becomes well defined. A second molt follows, and the resulting filiform L_3, measuring 1.4–2 mm in length, migrates anteriorly into the proboscis sheath of the mosquito's mouth parts.

When an infected mosquito feeds again, the larvae can enter the human host through the puncture wound, i.e., they are not inoculated as malarial parasites. The last two molts occur within man. It has been determined that the infection of mosquitoes can only occur if 15 or more microfilariae are present in every 20 mm³ of blood. If there are 100 or more microfilariae in every 20 mm³ of blood, the mosquito is commonly killed. Under experimental conditions, development in the mosquito takes two weeks at 27°C and 90% humidity.

Pathogenicity. *Wuchereria bancrofti* is one of the classic causative agents of elephantiasis in humans. This disease, manifested by enlargement of the limbs, scrotum, and other extremities, is not the initial effect but the result of long-standing infection. As the lymph vessels become blocked by worms, edema occurs, and in time the deposition of connective tissue cells and fibers contribute to elephantiasis. For a detailed review of the disease, see Levine (1980).

Pathogenesis in filariasis is heavily influenced by the immune responses and the degree of inflammation. They are predominantly responses to the adult worms, primarily females. The clinical phases can be divided into the **incubation, acute** (or inflammatory), and **obstructive** stages. The incubation phase, which is largely symptomless, but may include transient lymphatic inflammation accompanied by mild fever and malaise, is initiated at the time of infection and lasts until the first microfilariae appear in the blood. The acute phase is initiated when the females reach maturity and commence to release microfilariae. This phase is characterized by intense lymphatic inflammation, usually in the lower parts of the body. This is accompanied by chills, fever, and toxemia. The symptoms usually subside in a few days but commonly recur at frequent intervals.

The obstructive phase is manifested by a buildup of lymph. This increase in the amount of lymph results in chyluria, or lymph in urine. The obstructive phase eventually leads to elephantiasis (Fig. 15.42).

Genus *Brugia*

Until 1960, several species of filarial worms with similar microfilariae were considered members of the genus *Wuchereria*. In that year, Buckley, as a result of studies of adult worms, established the genus *Brugia*

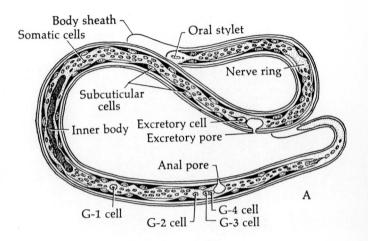

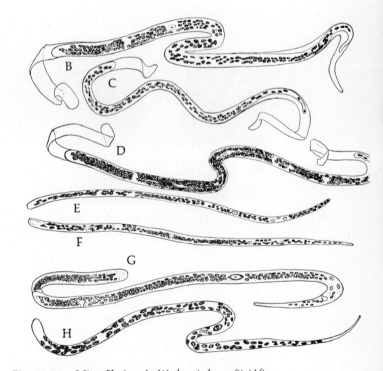

Fig. 15.41.   Microfilariae. A. *Wuchereria bancrofti.* (After Brown, 1950.) **B.** *Wuchereria bancrofti*: sheathed, no nuclei in tip of tail, 270 × 8.5 μm. **C.** *Brugia malayi*; sheathed, two nuclei in tail, 200 × 6 μm. **D.** *Loa loa*: sheathed, nuclei to tip of tail, 275 × 7 μm. **E.** *Acanthocheilonema perstans*: no sheath, tail blunt, with nuclei to tip, 200 × 4.5 μm. **F.** *Mansonella ozzardi*: no sheath, pointed tail without nuclei at tip, 205 × 5 μm. **G.** *Onchocerca volvulus*: no sheath, no nuclei in end of tail, 320 × 7.5 μm. **H.** *Dirofilaria immitis*: no sheath, sharp tail without nuclei at end, 300 × 6 μm. (**B–H**, after Chandler and Read, 1961.)

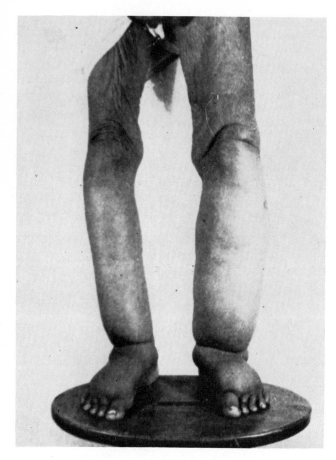

Fig. 15.42. Elephantiasis of human legs. (Courtesy of Armed Forces Institute of Pathology; negative No. A-4430-1.)

to include the "malayi" group consisting of *B. malayi* (Fig. 15.41), which parasitizes humans, other primates, and cats; *B. pahangi*, found in the lymphatics of cats, dogs, tigers, and other wild animals in Malaysia; and *B. patei* found in the lymphatics of various cats and dogs in Kenya. Recently, two additional species of *Brugia*—*B. guyanensis* and *B. beaveri*—have been reported. These are the first *Brugia* species to be reported from the western hemisphere. *Brugia guyanensis* is found in the lymphatic system of the coatimundi, *Nasua nasua vittata*, in Guiana; *B. beaveri* is found in the lymph nodes, skin, and carcass of the raccoon, *Procyon lotor*, in Louisiana.

Brugia malayi causes Malayan filariasis in humans. The range for this species extends from India to China, Japan, Formosa, Malaysia, and Indonesia. A *malayi*-like form has been reported on an island off the

coast of Kenya. Species of the mosquitoes *Mansonia* and *Culex* are the principal vectors, although in Japan *Aedes togoi* appears to be the only vector. It is noted that Gwadz and Chernin (1973) have demonstrated that the infective larvae of *Brugia pahangi* can escape from the mosquito vector and infect mammals via drinking water. This infection route was originally proposed in 1878 by Manson, who first discovered the role of mosquitoes as vectors for *Wuchereria bancrofti* in southern China.

Two strains of *Brugia malayi* are now recognized—the **periodic** and the **semiperiodic** strains. According to Laing *et al.* (1960), the periodic strain is quite host specific and is restricted to humans. Its microfilariae portray nocturnal periodicity. On the other hand, the semiperiodic strain is less host specific, for its hosts include humans, cats, and the macaque monkey, although its normal host is the leaf monkey, *Presbytis*. The periodicity of the microfilariae of the semiperiodic strain is less pronounced.

The microfilariae of the various species of *Brugia* are very similar and extremely difficult to distinguish from one another. *Brugia* is distinguishable from *Wuchereria* by the adults. *Brugia* spp. are smaller, about half the size of *Wuchereria*. The males measure up to 25 mm × 100 μm, and the females measure up to 60 mm × 190 μm. Furthermore, there are typically 11 papillae, known as **anal papillae**, at the posterior end of *Brugia*—four pairs situated ventrolaterally, two pairs situated postanally, and a single large preanal papilla. In *Wuchereria*, there are about 24 anal papillae, 9 to 12 pairs situated ventrolaterally, two pairs postanally, and a single preanal papilla. In addition, there are typically two additional papillae situated between the anal papillae and the tip of the tail in *Brugia* and about four such papillae in *Wuchereria*. The microfilariae of *Brugia* also include nuclei in the tip of the tail while those of *Wuchereria* do not.

Both *W. bancrofti* and *B. malayi* block lymphatic canals when present in large numbers, causing elephantiasis (Fig. 15.42).

Genus *Onchocerca*

The members of this genus are without lips. If cephalic papillae are present, they are extremely small. The cuticle of the adult female is always thick, especially in the middle of the body, whereas that of the male is sometimes thick, with transverse striations, and is reinforced externally by spiral thickenings.

Onchocerca volvulus. This parasite (Fig. 15.41) causes onchocerciasis (Fig. 15.43). It has been estimated that 40 million human cases of this disease exist, of which 30 million are found in tropical Africa and 800,000 in western Guatemala, Colombia, and northeastern Venezuela. It also occurs in Mexico and Arabia. At least 2000 infected persons now have immigrated to London, mostly from Africa (Woodhouse, 1975). Recent estimates show that although

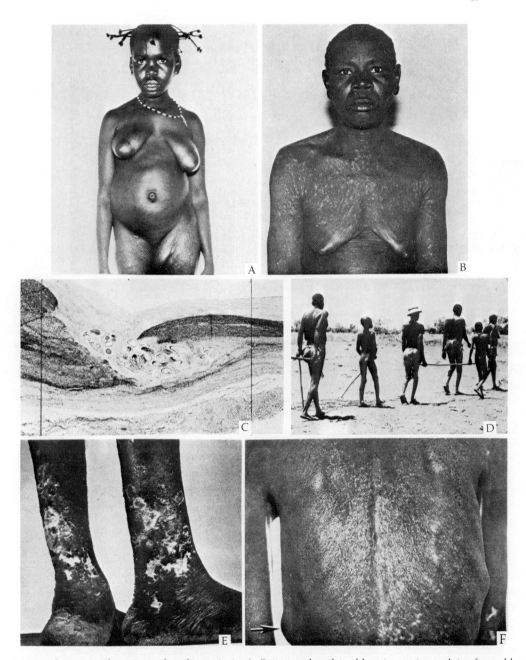

Fig. 15.43.   Some manifestations of onchocerciasis. A. Patient with unilateral hanging groin overlying femoral lymph nodes, an advanced stage of onchocercal lymphadenitis. (Armed Forces Institute of Pathology, Negative No. 68-10066-1.) **B.** A patient from the Ubangi region of Zaire with papular eruptions of the upper trunk, arms, and neck; this is a less common manifestation of onchocercal dermatitis. (Armed Forces Institute of Pathology, Negative No. 68-7835-6.) **C.** Photomicrograph of a section of an aortal wall showing a coiled adult female *O. volvulus.* The worm lies in the media and has disrupted the smooth muscle and elastic fibers; scar tissue has formed around the worm, thickening the overlying intima. (Armed Forces Institute of Pathology, Negative No. 73-6157.) **D.** A now famous World Health Organization photograph showing adults in a west African village blinded by onchocerciasis being led by children. **E.** Male patient with severe onchocercal dermatitis with depigmentation of legs and secondary infection resulting from scratching. (Armed Forces Institute of Pathology, Negative No. 68-8080-1.) **F.** The so-called lizard-skin condition characterized by scaling and altered pigmentation caused by *O. volvulus.* An onchocercoma is present on the left side of the chest (arrow). (Armed Forces Institute of Pathology, Negative No. 69-9769.)

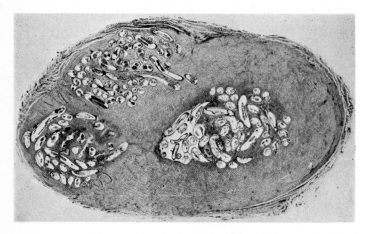

Fig. 15.44. *Onchocerca volvulus.* Photomicrograph of section showing several adult worms coiled within a single nodule. The hyalinized scar tissue surrounding the parasites incarcerates them, preventing migration. (Armed Forces Institute of Pathology, Negative No. 69-3639.)

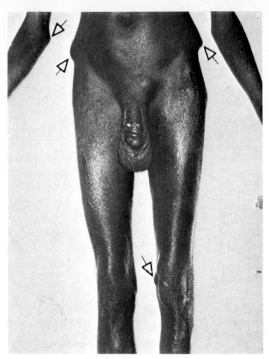

Fig. 15.45. Onchocercomas caused by *Onchocerca volvulus.* A resident of the Ubangi Territory of Zaire with onchocercomas over the elbows, iliac crests, trochanters, and the left knee. (Armed Forces Institute of Pathology, Negative No. 68-7638-3.)

onchocerciasis, or river blindness, is being controlled to some extent in Central and South America, its prevalence is increasing in Africa as a result of the development of irrigation and hydroelectric dam projects.

Microfilaria of Onchocerca volvulus. The body of this microfilaria is 300–350 μm × 5–8 μm (Fig. 15.41). It is unsheathed and the tail is sharply pointed. There are no nuclei in the tail. The larva does not display periodicity. In the host, it is located primarily in the subcutaneous area.

Life Cycle of Onchocerca volvulus. The adult worms dwell in tumors in the subcutaneous connective tissue of humans. In most instances there are two worms, a male and a female, in each nodule. However, numerous worms have been found coiled within a single swelling (Fig. 15.44). The males, measuring 19–42 mm in length, are considerably smaller than the females, which measure 33.5–50 cm in length.

The eggshells rupture upon oviposition and the escaping microfilariae become unsheathed. Such microfilariae are of two sizes, suggesting sexual dimorphism. Instead of being located in the blood circulation, the microfilariae are found in lymphatics, but especially in the connective tissues of skin and also in the corneal conjunctiva. These larvae are positively phototactic. Blackflies, *Simulium* spp., serve as the insect hosts (or vectors) for *O. volvulus*. Wanson (1950) has reported that development in the blackfly is essentially the same as that of *Wuchereria* in mosquitoes.

Human Onchocerciasis. The disease caused by *Onchocerca volvulus* is commonly referred to as river blindness because the microfilariae commonly invade the cornea. Inflammation of the sclera, cornea, iris, and retina is followed by formation of fibrous tissue, leading to impaired vision or total blindness. In addition to this dramatic condition, the presence of microfilariae in the skin often causes a severe dermatitis resulting from an allergic response or toxicity. These areas of the skin commonly become thickened, depigmented, and cracked. These symptoms resemble those of vitamin A deficiency. In fact, it has been suggested that such symptoms reflect the parasite's competition for or interference with vitamin A metabolism.

Adult worms also cause pathologic alterations, although they are less severe. Specifically, adult worms often cause the development of subcutaneous nodules, known as **onchocercomas**, especially over bony prominences (Fig. 15.45). Onchocercomas caused by the Venezuelan and African strains of *O. volvulus* usually occur in the pelvic area, with a few along the spine, chest, and knees. On the other hand, infection with the Central American strain results more commonly in the formation of nodules above the waist, especially on the head and neck. These nodules are relatively benign, consisting of fibrotic host tissues

Fig. 15.46. Elephantiasis caused by *Onchocerca volvulus.* Enlarged scrotum in a patient with severe onchocercal dermatitis. The enlargement had continued steadily over several years, and at the time of its removal, the scrotum weighed 18 kg. (Armed Forces Institute of Pathology, Negative No. 68-8582.)

surrounding adult worms, usually a pair. Note that although subcutaneous onchocercomas are readily excised, the more deeply situated adult worms continue to produce microfilariae, which can migrate to the host's body surface and be transmitted.

Onchocerciasis may result in true elephantiasis (Fig. 15.46). For a detailed discussion of onchocerciasis, see Connor *et al.* (1970), and for a review of the ecology of this parasite, see Duke (1971).

Other *Onchocerca* Species

Other species of *Onchocerca* include *O. gibsoni*, which forms hard nodules in the hide of cattle, destroying its value as leather, and *O. reticulata* in the neck ligaments of horses. Both of these species are found in the United States and are transmitted by *Culicoides* spp. *Onchocerca armillatus* is found in the aorta of cattle in Africa and often causes death through aneurysm. *Onchocerca gutturosa* also occurs in cattle.

Genus *Litomosoides*

Members of this genus have a cuticle that is very finely striated except at the extremities. The head is bluntly rounded, and the mouth is surrounded by neither lips nor papillae.

Litomosoides carinii. This species, a parasite in the pleural cavity of the cotton rat, *Sigmodon hispidus littoralis,* deserves special attention since this filarial worm is commonly used in research laboratories as an experimental animal, especially in the United States and South America, where the rat host is found in open grasslands.

Adults of *L. carinii* are long and thin. The buccal cavity is narrow. The males, which measure 24–26 mm in length, lack alae or papillae at the caudal end. The females, which measure 50–130 mm in length, possess both a bulbous enlargement near the vulva and a long ovijector. Both male and female adults occur in the rat's pleural cavity; occasionally they invade the peritoneal cavity.

Life Cycle of *Litomosoides carinii.* Mature females give birth ovoviviparously to thin, sheathed microfilariae within the host's pleural cavity. These microfilariae migrate to the bloodstream via various routes, the most common being the lungs. The arthropod vector is the tropical rat mite, *Bdellonyssus bacoti.* When mites feed on an infected rat, microfilariae are taken into the mite's intestine, where the larvae burrow through the intestinal wall and become established in the hemocoel.

Approximately 8 days after entering the mite, the larva becomes sausage shaped and undergoes the first ecdysis. The second ecdysis occurs on the 9th or 10th day, and the larva increases in size until it measures 500–950 μm long. By the 14th or 15th day, the L_3 is fully developed and infective. The temperature at which the mite host is maintained influences the time required for development. The periods given above are for worms in mites maintained at 23 to 25°C.

Rats become infected when infective larvae penetrate the skin. In the laboratory, infective larvae removed from mites can be injected subcutaneously to establish an infection. Once within the rat's blood, the larva migrates to the pleural cavity and undergoes the third ecdysis in about a week. Further growth takes place during the next 17 days, the male larva attaining a length of 6.4 ± 0.5 mm and the female attaining a length of 8.7 ± 0.2 mm. External sex characteristics are evident by this time.

The fourth and final molt occurs about 23 or 24 days after the initial infection, followed by growth and sexual maturation. Sexual maturity is reached within 70 to 80 days, and nonperiodic microfilariae can then be found in the host's blood. The peak of microfilarial production occurs between the 17th and

Fig. 15.47. *Loa loa.* Worm migrating across human eye.

20th week after infection, but the adults can live for 60 weeks or more.

Genus *Loa*

All of the species of *Loa* occur in primates. The mouth in members of this genus is simple and without lips, although there are two lateral and four small, submedian cephalic papillae. The cuticle is thick, nonstriated, and ornamented with small bosses (wartlike processes) except at the anterior ends of both sexes and the tail end of males.

Loa loa. This species, commonly known as the eyeworm, dwells in the subcutaneous tissues of humans and baboons in central and western Africa (Fig. 15.41). Stoll (1947) estimated that 13 million cases of human infections existed. The number has not been significantly reduced since then. The threadlike adults are comparatively large. The males measure 30–34 mm in length and 0.35–0.43 mm in diameter; the females measure 50–70 mm in length by 0.5 mm.

Although adult worms are rarely encased in connective tissue capsules, they more often freely migrate in the subcutaneous tissues of the entire body and are sometimes seen passing under the conjunctiva of the eye (Fig. 15.47), hence its common name. Swellings often accompany the migration of this worm, and these may be as large as a chicken's egg. Such swellings result from the host's tissue reacting to toxins secreted by the worms.

Microfilaria of Loa loa. The body is sheathed and 250–300 μm $\times$ 6–8 μm in size (Fig. 15.41). The body stains poorly, or not at all, with Giemsa's stain. The tail is short and recurved and contains nuclei. The body will take on supravital methylene blue stain. This larva occurs in the host's blood.

The sheathed microfilariae are unique among the human-infecting filarial worms in that they demonstrate diurnal periodicity. The insect vector, therefore, must be a daytime biting form, and this is the case, for daylight feeding tabanid flies, *Chrysops dimidiata* and

C. silacea, serve as the vectors. In the fly host, the larvae undergo three larval stages in the thoracic muscles before entering the mouthparts, from which they are introduced into another vertebrate host. As with almost all filarial worms, the last two molts occur within the vertebrate.

Genus *Dirofilaria*

The species of *Dirofilaria* are without lips and essentially have no cephalic papillae. They are parasitic in various mammals.

Dirofilaria immitis. This is the fairly common heartworm of dogs, cats, foxes, and wolves (Fig. 15.41). Surveys have revealed that although this parasite is widely distributed in the United States, it is most frequently encountered in the southern states. The males, measuring 120–250 mm in length by 1 mm in diameter, possess tails that are blunted, armed with caudal alae, and spirally coiled. The females, which measure 250–310 mm $\times$ 1 mm, are larger, and the vulva is situated near the posterior extremity of the esophagus.

The microfilariae (Fig. 15.41), measuring 218–330 μm $\times$ 5–6 μm, are unsheathed and occur in the mammalian host's peripheral circulation. Microfilariae can be found for several months in the blood. A certain degree of periodicity exists, for the maximum number of microfilariae are present between 11:00 P.M. and 3:00 A.M., whereas only half of that number are present between 3:00 A.M. and 7:00 A.M.

Life Cycle of Dirofilaria immitis. The insect vectors for *D. immitis* are mosquitoes. Again, there appears to be little host specificity, for many species of mosquitoes belonging to the genera *Aedes, Culex,* and *Anopheles* can serve as vectors. Other arthropods, such as lice, flies, and ticks, have been suspected as possible vectors. The wide assortment of vectors, however, may be the result of misidentification of filarial larvae in arthropods.

Within 24 to 36 hours after ingestion by the mosquito, the larvae are found primarily in cells of the mosquito's Malpighian tubules. The cell membranes of some of these tubules are ruptured as the result of the penetration. The larvae attain the infective L_3 stage within these cells and reach a length of approximately 900 μm.

The infective larvae escape from the Malpighian tubules and migrate to the hemocoel. Eleven to 12 days after their escape, most of the larvae are found in and among the fat bodies in the lower half of the thorax. From here, they migrate to the mouthparts, where they are in position to be introduced into another vertebrate when the mosquito partakes of another blood meal.

The adults appear in the vertebrate's (usually a dog) heart, particularly the right ventricle and pulmonary arteries, in 8 to 9 months. In addition, *D. immitis* has been found in subcutaneous tissue, in the anterior chamber of the eye, and in pulmonary nodules of the host. Infected animals may suffer from chronic endocarditis, liver enlargement, and inflammation of the kidneys. Ascites tumors associated with this nematode have been reported. Infected dogs become extremely ill due to malfunction of the cardiac valves and often die from cardiac failure. For dog owners, especially in the southern United States, the canine heartworm is a major problem and is responsible for numerous visits to veterinarians.

Other *Dirofilaria* Species

Other species of *Dirofilaria* include *D. repens* found in subcutaneous nodules of the eye, nose, arm, and elsewhere of humans and *D. tenuis* in the subcutaneous tissues of the raccoon in the southern United States, extending from Florida to Texas. It is also known to infect humans. For a review of known cases of humans infected with *Dirofilaria* and other filarial nematodes, see Beaver and Orihel (1965)

Genus *Mansonella*

The members of this genus are characterized by a rounded and swollen anterior end and a smooth cuticle. The posterior end of the male is markedly curved, with a slightly bulbous tip, and with fleshy lappets. The posterior end of the female bears a pair of fleshy lateral appendages.

Mansonella ozzardi. This parasite is also a human-infecting species (Fig. 15.41). Adults are found in body cavities, usually embedded in the visceral adipose tissues, whereas the unsheathed microfilariae occur in blood.

The fly *Culicoides furens* is the vector for *M. ozzardi* in northern South America. *Culicoides paraensis* is probably another suitable vector. Other than local tissue reactions in the form of hydrocoels and lymph swelling, no drastic symptoms are connected with Ozzard's filariasis.

Microfilaria of Mansonella ozzardi. The body is approximately 200 × 5 μm and unsheathed (Fig. 15.41). The tail is pointed and without nuclei. Living specimens are readily stained with methylene blue. It displays no periodicity and is found in the definitive host's blood.

Mansonella perstans. Formerly known as *Dipetalonema perstans* or *Tetrapetalonema perstans*, this nematode (Fig. 15.48) is now assigned to the genus *Mansonella* (Orihel and Eberhard, 1982). It is primarily parasitic in humans, although various primates do serve as natural reservoirs of infection. It occurs in Africa, South America, and the West Indies. Stoll (1947) estimated that 27 million cases of human infections existed, mostly in Africa. The number is probably somewhat less today. Microfilariae of *Mansonella* have been reported in the blood of dogs in the United States; however, these probably represent another species.

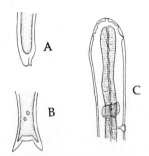

Fig. 15.48. *Mansonella* spp. and *Dipelatonema*. A. Posterior terminal of female *M. perstans*. **B.** Posterior terminal of male *M. perstans* showing bifurcate tail. **C.** Lateral view of anterior portion of *D. reconditum*. (Redrawn after Nelson, 1962.)

Microfilaria of Mansonella perstans. The body is approximately 200 × 4 μm. It is unsheathed and the tail is blunted. The tail contains nuclei. No definite periodicity is evident. In the definitive host, the microfilariae occur in blood.

Life Cycle of Mansonella perstans. The adults are found in the host's body cavities, primarily the peritoneal cavity, although worms reportedly can also be lodged in the pericardial cavity. The males, measuring 45 mm × 60 μm, are smaller than the females, which measure 70–80 mm in length by 120 μm. The tail end of both sexes is curved and is bifurcated to form two nonmuscular flaps (Fig. 15.48).

The fly *Culicoides austeni* is the principal vector. The larvae develop in the fly's thoracic muscles, and the infective form appears eight to nine days after the microfilariae enter the fly.

Mansonella streptocerca. Formerly known as *Dipetalonema streptocerca*, this African parasite is also primarily parasitic in humans, other primates serving as reservoir hosts. *M. streptocerca* is known only in its microfilarial form, which occurs subcutaneously.

Culicoides grahami is the fly vector for *M. streptocerca*. In both *M. perstans* and *M. streptocerca* infections, there appear to be little or no serious consequences. However, local inflammations are associated with the sites at which the nematodes occur in the mammalian host.

Microfilaria of Mansonella streptocerca. The body is approximately 220 × 3 μm. It is unsheathed, and the tail is blunt and hooked. The tail possesses nuclei. The body stains poorly with supravital methylene blue. This larva displays no periodicity and is located in the subcutaneous regions of the host.

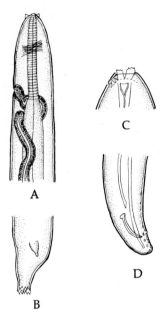

Fig. 15.49. Some additional filarial nematodes. A. Anterior end of *Setaria cervi*, a parasite primarily in the peritoneal cavity of bovines and sheep. **B.** Posterior end of female *S. cervi* (**A** and **B**, redrawn after Yeh, 1959.) **C.** Anterior end of *Stephanofilaria stilesi*, parasitic in skin of abdomen of cattle. **D.** Posterior end of male *S. stilesi*. (**C** and **D**, redrawn after Schiller in Morgan and Hawkins, 1949.)

Genus *Dipetalonema*

There are about 40 species in this genus (Chabaud and Choquet, 1953). The bodies are slender and very thin in the posterior region, and the cuticle is either smooth or very finely striated.

Dipetalonema reconditum. In the United States there is a common species of *Dipetalonema*, probably *D. reconditum*, which is found in dogs (Fig. 15.48). Dog fleas serve as the vector of this species.

Genus *Setaria*

This genus includes approximately 30 species. These are filarial worms primarily of ruminants, equines, and pigs. The species of *Setaria* have mouths surrounded by a protruding cuticular ring that is prolonged anteriorly into four lips. There are four prominent submedian and two small, lateral cephalic papillae. The cuticle is transversely striated.

Setaria cervi. This species (Fig. 15.49) occurs primarily in the peritoneal cavity of the ox, zebu, water buffalo, bison, yak, reindeer, moose, and sheep. It is worldwide in distribution. The adult males measure 40–60 mm long and 380–450 μm wide, whereas the females measure 60–120 mm long and 600–900 μm wide.

When present in the peritoneal cavity of their normal bovine host, *S. cervi* and other species belonging to this genus are apparently nonpathogenic. However, young worms may be found occasionally in the anterior chamber of the eye, where they cause inflammation leading to the development of opacity of the eye and blindness. This condition is most commonly associated with *S. digitata* in the Orient. Dipteran insects (mosquitoes, blackflies, midges, etc.) serve as the vectors for *Setaria* spp.

Genus *Stephanofilaria*

Stephanofilaria is a genus with about five species, all in ruminants. *Stephanofilaria stilesi* (Fig. 15.49) occurs in the skin on the ventral surface of cattle in North America, Hawaii, and the Soviet Union. According to Maddy (1955), it infects 80 to 90% of cattle in the western United States; *S. dedoesi* occurs in the skin of cattle, goats, and water buffaloes in Indonesia. Its incidence in buffaloes is as high as 90% in some areas. A purulent dermatitis is caused by this species as well as by *S. stilesi*. All species of *Stephanofilaria* utilize dipterans, especially flies (hornflies, stableflies, etc.) as intermediate hosts.

Pathogenicity of Filarial Worms in Vectors. In recent years, considerable interest has developed in understanding the relationship between pathogens, including zooparasites, and their invertebrate hosts. In fact, the discipline of invertebrate pathology has come into its own. The rationales for studying these associations stem from (1) understanding these for their own sake; (2) finding possible biologic control agents for invertebrate vectors and other destructive invertebrates, especially in agriculture; and (3) understanding the phylogeny of the disease state, including immune responses. As a result of such studies, some information is available on the relationship between filarial worms and their vectors.

Filarial nematodes cause severe damage to their insect vectors (see Lavoipierre, 1958, and Nelson, 1964, for reviews). In the mosquito *Aedes aegypti* infected with *Dirofilaria*, the pathogenicity is so severe that the mortality is many times that for uninfected vectors. The greatest injury to the mosquito occurs when the filarial worms penetrate the epithelial lining of the midgut. Furthermore, worms in the Malpighian tubules cause degeneration of the cytoplasm of the cells lining the tubules, initially leaving the nuclei intact.

As the result of the destruction of the cytoplasm, cells of walls of the Malpighian tubules resemble distorted, multinucleated cells. By the time the nematodes reach the infective stage, however, the cells lining the tubules are totally disintegrated, such that the worms are enclosed in translucent bags of basement membrane. It has been reported that the pathogenicity of *Dirofilaria* is less severe in unfertilized female mosquitoes than in fertilized ones, although this has not been verified.

Species of insect vectors differ in their ability to survive infections by filarial worms. For example, *Simulium damnosum* cannot survive heavy infections of *Onchocerca volvulus*, that is, when 500 or more worms are present. Similarly, mortality is high when *Simulium callidum* and *S. ochraceum* are infected with large numbers of *O. volvulus*. On the other hand, Dalmat (1955) has found that *Simulium metallicum* is somewhat resistant to heavy infections.

Although the histopathology of blackflies infected with *O. volvulus* has not been studied to any great extent, Lavoipierre (1958) has found that thoracic muscles of *Simulium* in the vicinity of the growing larvae degenerate.

Of particular interest to those concerned with the basic biology of the relationship between larval filarial worms and their arthropod intermediate hosts are the studies by Macdonald (1962a,b) and Townson (1970). According to Macdonald, only mosquitoes, *Aedes aegypti*, homozygous for the gene f^m allow full development of the larvae of *Brugia pahangi*, and according to Townson it is only in mosquitoes with this genotype that there is any marked elevation in mortality from the seventh day on after the infecting blood meal. Townson has also shown that when *A. aegypti* infected with *B. pahangi* and uninfected controls are suspended in a wind tunnel, a large number of the infected mosquitoes will not fly. Part of this can be explained by the fact that, when examined with the electron microscope, separation of the intrasarcosomal cristae occurred in the sarcosomes of some infected fibers in mosquitoes harboring *Brugia*. On the basis of these findings, it has been concluded that in nature a proportion of these mosquitoes that manage to survive the full developmental cycle of *B. pahangi* would be incapable of serving as vectors because of impairment of their ability to fly.

MICROFILARIAL PERIODICITY

As briefly mentioned in connection with certain microfilariae, especially *Wuchereria bancrofti*, a definite periodicity occurs—i.e., large numbers of microfilariae appear in the definitive host's peripheral circulation at certain times during the 24-hour period. This phenomenon is of fundamental interest to biologists because it represents one type of cyclic biologic activity—i.e., the recurrence of a biologic phenomenon.

Microfilarial periodicity, first discovered in 1879 by Patrick Manson in human infections of *Wuchereria bancrofti* in south China, is now known to occur in a number of species of filarial worms (Table 15.2). The

Table 15.2. The Periodicity of Some Species of Microfilariae from Various Hosts[a]

Periodicity	Parasite	Host
Nocturnal	*Wuchereria bancrofti*	Humans
	Brugia malayi	Humans
	B. pahangi	Cat, monkey
	B. patei	Cat
	Dirofilaria immitis	Dog
	D. repens	Dog
	D. corynodes	Monkey
	D. uniformis	Rabbit
	D. tenuis	Raccoon
	Loa loa	Monkey
	Edesonfilaria malayensis	Monkey
	Monnigofilaria setariosa	Mongoose
	Splendidofilaria quiscali	Grackle
	S. columbigallinae	Dove
	S. lissum	Partridge
	Microfilaria sp.	Crows (United States)
	Microfilaria sp.	Canaries (Algiers)
Diurnal	*Loa loa*	Humans
	Wuchereria bancrofti var. *pacifica*[b]	Humans
	Dipetalonema reconditum	Dog
	Ornithofilaria fallisensis	Duck
Absent	*Mansonella ozzardi*	Humans
	Mansonella perstans	Humans
	Dipetalonema magnilarvata	Monkey
	D. witei	Bird
	Litomosoides carinii	Cotton rat
	Setaria cervi	Cattle
	Microfilaria fijiensis	Bat
	Conispiculum guindiense	Lizard
	Foleyella sp.	Frog
	Icosiella sp.	Frog

[a] Data compiled by Hawking, 1962.
[b] Slight periodicity.

timing and extent of such periodicity vary according to the species but are consistent for the species under natural conditions (Figs. 15.50, 15.51).

It bears repeating that the evolution of microfilarial periodicity is of survival value, for it enhances chances for the ingestion of microfilariae by the insect vectors. For example, in the case of the periodic strain of *W.*

bancrofti of humans, the microfilariae are most abundant in the host's peripheral circulation during the night hours when the mosquito vector *Culex fatigans* (and others) are most actively feeding. Similarly, in the case of *Loa loa*, the appearance of microfilariae in large numbers in the host's peripheral circulation during the day coincides with the active feeding period of various *Chrysops* species that serve as vectors.

Interestingly, the so-called Pacific strain (or subperiodic strain) of *Wuchereria bancrofti*, unlike the periodic strain, portrays a certain degree of diurnal periodicity (Fig. 15.50). This opposite rhythmic pattern coincides with the feeding of its mosquito vectors, *Aedes pseudoscutellaris* and others.

In some species, for example, *Litomosoides carinii*, periodicity does not occur. Hawking (1962) has expressed the opinion that possibly these worms have not yet evolved periodic behavior, or perhaps the host's habits are not sufficiently imprinted with the 24-hour rhythm to make the evolution of parasite periodicity feasible. Hawking has also pointed out that (as suggested by the data in Table 15.2) periodicity seems to be well defined among species that parasitize birds and large mammals, especially humans and simians, but it is usually absent among species that parasitize small rodents, amphibians, and reptiles.

Although numerous studies have been carried out in attempts to explain the mechanism(s) responsible for microfilarial periodicity, this aspect of helminth physiology has yet to be completely resolved. It is known, nevertheless, that periodicity is dependent not on the alternation of night and day, but on the 24-hour habits of the host. Therefore, if a host is made to reverse its routine so that it sleeps by day and moves about at night, the periodicity of the microfilariae is reversed. This reversal process takes about a week, indicating that the microfilariae are not affected by sleeping and waking as such but only by the entire 24-hour rhythm. In humans and higher mammals, this rhythm is characterized by decreases in body temperature and oxygen pressure, increases in carbon dioxide tension and body acidity, less excretion of water and chlorides by the kidneys, less adrenal activity, and so forth during sleep. Some or all of these physiologic changes may affect the microfilariae, resulting in the rhythmic cycle.

To understand microfilarial periodicity, it is necessary to know what happens to the microfilariae during the period when they are not present or are only sparsely present in the peripheral circulation. It is now known that when microfilariae are completely or almost completely absent from the peripheral circulation, they accumulate primarily in the capillaries and small blood vessels of the lungs.

Hawking (1962) has suggested that because microfilariae cannot mechanically attach to the blood vessel walls, they may be held there by some kind of electrostatic attraction. When the charge on the microfilariae or on the vessel wall is changed or removed by some

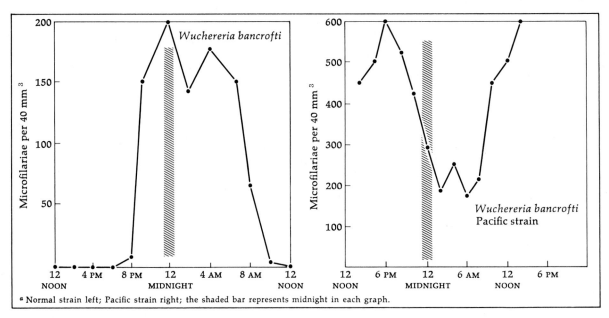

[a] Normal strain left; Pacific strain right; the shaded bar represents midnight in each graph.

Fig. 15.50. Fluctuation in the number of microfilariae in the peripheral circulation of human infected with the normal and pacific strains of *Wuchereria bancrofti*[a]

physiologic alteration, the parasites are again swept into the general circulation. While no direct experimental proof exists for this explanation, the accumulation of microfilariae in the lungs during the host's active phase (when the parasites are not found in great numbers in the peripheral circulation) and the release of microfilariae by some physiologic triggering mechanism into the general circulation during the host's passive (or resting) phase (when the parasites are found evenly distributed throughout the blood and thus appear abundantly in the peripheral circulation) has greatly enhanced our understanding of microfilarial periodicity.

The stimuli controlling alternation between the active and passive phases have been searched for extensively (see Hawking, 1956; McFadzean and Hawking, 1956). The only natural stimuli that may be significant are increased oxygen intake and exercise.

It has been shown that if an individual parasitized by the periodic strain of *Wuchereria bancrofti* inhales air containing increased oxygen tension during the night, there is a marked decrease of microfilariae in the peripheral circulation following the intake of oxygen, and the number of microfilariae in the lungs is increased. Similarly, if the parasitized individual exercises during the night, a marked decrease in the number of microfilariae in the peripheral circulation follows, and their number increases in the lungs.

On the other hand, the intake of oxygen during the day by hosts of the Pacific or subperiodic strain of *W. bancrofti*, which normally displays diurnal periodicity, causes an increase in the number of microfilariae in the peripheral circulation. Similarly, the intake of oxygen by dogs infected with *Dirofilaria immitis* and by monkeys infected with *D. corynodes* causes a rise in the microfilarial count in the peripheral circulation. Thus, it would appear that different species of microfilariae respond differently to such stimuli.

Wang *et al.* (1958) and Fukamachi (1960) have shown that the subcutaneous injection of acetylcholine (12–19 mg/kg of body weight) causes a dramatic increase in the number of *Dirofilaria immitis* microfilariae in the peripheral blood of dogs. Electric shock also produces a rise in the peripheral microfilarial count. Injection of adrenalin, ethylamine, chlorpromazine, atropine, and promethazine, however, has no effect.

Fukamachi has concluded that the periodicity is probably due to some physiologic stimulus to the host's parasympathetic nervous system, which in turn stimulates the worms, since acetylcholine is a parasympathomimetic agent; that is, it mimics the function of the parasympathetic nervous system. However, the intravenous injection of acetylcholine does not cause an increase in the number of microfilariae in the peripheral blood in *Wuchereria bancrofti* infections. Furthermore, there is no evidence for a relationship between the parasympathetic nervous system and the periodicity of human microfilariae. This again indicates that different species and strains of microfilariae respond differently to similar stimuli.

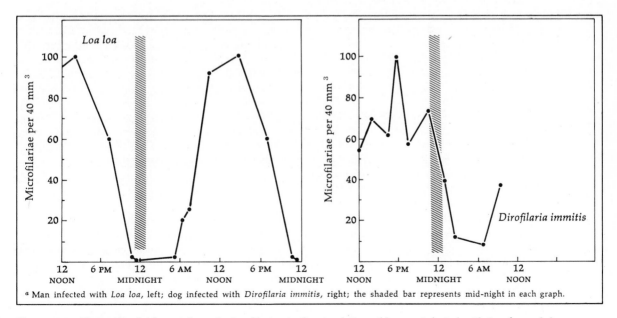

[a] Man infected with *Loa loa*, left; dog infected with *Dirofilaria immitis*, right; the shaded bar represents mid-night in each graph.

Fig. 15.51.   Fluctuation in the number of microfilariae in the circulation of human infected with *Loa loa* and dog infected with *Dirofilaria immitis*[a]

IMMUNITY AND MICROFILAREMIA

Although hosts of filarial nematodes generally maintain their parasites for long periods, sometimes years, it has been repeatedly recognized since Manson's report of 1899 that there is an absence of circulating microfilariae in humans with severe elephantiasis. Such observations suggest that humans develop an immunity to filarial worms that suppresses microfilaremia, i.e., the presence of microfilariae in circulating blood. This phenomenon also occurs in other types of filariasis (Beaver, 1970).

IMMUNITY TO FILARIASIS

Immunity to filariasis, like practically all instances of immunity, can be divided into **natural resistance** and **acquired immunity**. Natural resistance, however, is not immunity per se, but includes a variety of mechanisms other than the host immune reactions which prevent establishment of the parasite. For example, it should be evident that successful invasion by filarial worms is critical for the establishment of these parasites. The infective larvae are deposited on the skin with a drop of fluid, and penetration within a few minutes through the puncture made by the vector while feeding is essential. Hence, it is evident that the microecologic conditions of the host's skin should have considerable influence on the parasite's ability to survive long enough for it to invade. By way of example, Barriga (1981) reports that the consistency of infection of macaques exposed to the infective larvae of *Dirofilaria* spp., which are normally parasites of carnivores, as compared to the rare infection of similarly exposed humans, suggests that skin penetration is an important barrier for human infection. Human skin usually causes rapid dessication and death of the infective larvae.

Another example of natural resistance rests with the correlation between hosts with certain blood types and the incidence of filariasis. The mechanism(s) responsible for this phenomenon remains to be explored, although it is most probably genetic. Also, the sex of the host appears to influence the natural resistance to filariasis. For example, surveys have revealed that infections with *Wuchereria bancrofti*, *Onchocerca volvulus*, and *Mansonella perstans* are more prevalent among males in populations in which women are just as frequently exposed as men. This difference may be due to the influence of sex hormones.

Acquired immunity to filariasis in humans is based on the production of several immunoglobulins, especially IgG and IgE (Ogilvie and Worms, 1976). In nonimmune dogs experimentally infected with *Dirofilaria immitis*, the synthesis of IgM, IgG, and IgE has been detected. These three classes of immunoglobulins have been detected in most animal models employed in the study of filariasis.

In addition to the synthesis of immunoglobulins, cell-mediated immunity, manifested as delayed skin reactions, has been reported but not confirmed. It now appears that only patients with a localized form of onchocercal dermatitis known as *sowda* portray positive delayed skin test. Those infected with the generalized form of filariasis portray the immediate-type reactions only (p. 81).

The protective effect of immunity to filariasis appears to be weak at best, since even in the presence of demonstrable immune responses, filarial infections are chronic. For example, *Wuchereria bancrofti* will survive in the host for over 10 years, *Onchocerca volvulus* for about 16 years, and *Dirofilaria immitis* for about 7 years. Despite the general inefficiency of the immune reactions in filariasis on the current infection, there is some evidence of protection against reinfection. Indirect evidence for this is that adult filarial worms in humans and dogs do not accumulate as would be expected if there is continuous, effective transmission. This phenomenon has been interpreted to mean that older hosts in endemic areas are somewhat resistant to superimposed infections. Among several experiments that has been conducted to duplicate this natural phenomenon is the one by Denham and McGreevy (1977). These investigators have shown that if cats are repeatedly infected with *Brugia pahangi* over a prolonged time period, almost complete resistance to an ulterior challenge develops.

Another aspect of the development of immunity to filarial worms is that referred to as **occult filariasis**. It has been mentioned in the previous subsection. This term describes the condition in which circulating microfilariae are absent, although gravid females occur in the hosts. It has been suggested that the cause of this condition is that prolonged antigenic stimulation or stimulation undisturbed by new infections results in immune responses that suppress the presence of microfilariae in the circulation. For a good review of immunity to filariasis, see Barriga (1981).

SUPERFAMILY DRACUNCULOIDEA

Members of the superfamily Dracunculoidea are slender and large worms. The nematodes are parasitic in the body cavity, its surrounding membranes, and the connective tissues of vertebrates. The best known genus in this superfamily is *Dracunculus*.

Genus *Dracunculus*

The members of this genus are large nematodes with cuticularized thickenings on the head. Sexual dimor-

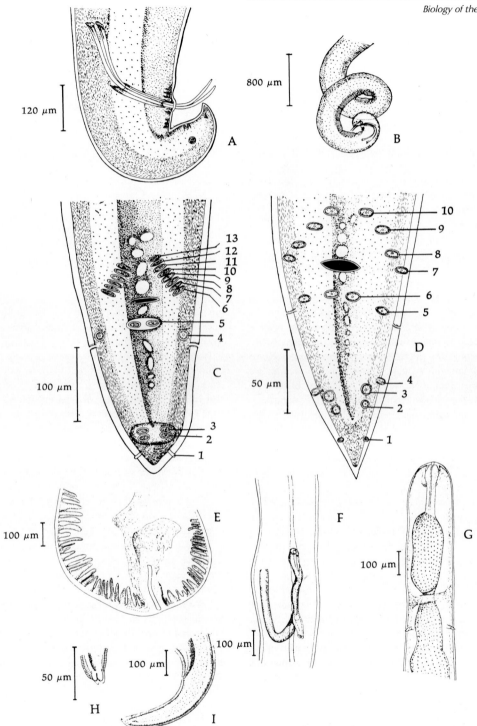

Fig. 15.52. ***Dracunculus.*** **A.** Lateral view of posterior end of *D. medinensis* adult male. **B.** General view of posterior end of *D. medinensis* adult male. **C.** Ventral view of posterior end of *D. dahomensis* adult male. (1, phasmid; 2–5, postanal papillae; 6–13, preanal papillae.) **D.** Ventral view of posterior end of *D. medinensis* adult male. (1–6, postanal papillae; 7–10, preanal papillae.) **E.** Cross section through *D. medinensis* mature female at level of vulva, showing mucoid plug. **F.** Portion of *D. medinensis* female showing vulva, vagina containing mucoid plug, and origin of uteri. **G.** Anterior end of *D. medinensis* adult male showing esophagus. **H.** Tail end of *D. medinensis* immature female showing four caudal mucrones (or cuspids). **I.** Posterior end of immature *D. medinensis* female showing position of anus. (All figures after Moorthy, 1937.)

phism is also extreme in this genus. Undoubtedly the best-known species is *D. medinensis* (Fig. 15.52), the "fiery serpent" of biblical fame, commonly called the guinea worm or Medina worm. It infects humans, dogs, horses, cattle, wolves, various cats, monkeys, and baboons in the Old World, and mink, foxes, raccoons, and dogs in North America. It is widely distributed in Africa, the Near East, the East Indies, and India. Stoll (1947) estimated that there were 48 million cases of human dracunculiasis in the world. The number has been somewhat reduced, but it is probably still over 30 million.

Life Cycle of Dracunculus medinensis. Stages in the life cycle of *D. medinensis* are depicted in Fig. 15.52. The males, measuring 12–40 mm in length and 0.4 mm in diameter, are much smaller than the females, which measure 70–120 cm in length and 0.9–1.7 mm in diameter. No vulva is visible in females. Adult worms usually develop in the viscera, body cavities, and connective tissues of their hosts. On reaching the gravid state, the females migrate to the subcutaneous tissues. By this time, the ovaries are no longer functional, and the uteri are filled with active larvae; hence, this species is ovoviviparous.

In the subcutaneous tissues of the host, the females direct their heads toward the skin, and by secreting an irritant, cause the formation of papules in the host's dermis. Such a papule soon increases in size and resembles a blister from the exterior. When the blister ruptures, it forms a cup-shaped ulcer in the skin. Such eruptions usually occur on the ankles or wrists. When the open ulcer comes in contact with water, a loop of the nematode's uterus prolapses, either through the ruptured anterior end or through the mouth, and ruptures, enabling numerous L_1s to escape into the water.

The rhabditiform L_1s are 500–700 μm long and possess distinctly striated cuticula. They are free living, and their food reserves enable them to live for several days. If ingested by a suitable species of *Cyclops*, they burrow through its midgut and enter the hemocoel (Fig. 15.53). Seldom is there more than one larva in a single *Cyclops*. The presence of more than five or six larvae results in the death of the intermediate host.

Within the hemocoel of *Cyclops*, the single larva, or up to four larvae, continue to develop, undergo two molts, and attain the ensheathed L_3 stage in approximately 20 days. When infected *Cyclops* are swallowed in contaminated water, the larva becomes freed when the intermediate host is digested, and it is stimulated by gastric juices to exsheath in the duodenum of the human host. It burrows through the mucosa, under-

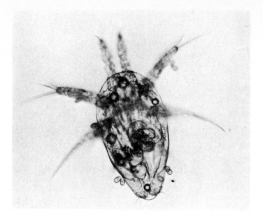

Fig. 15.53. *Dracunculus medinensis* **in** *Cyclops.* Photomicrograph of a nauplius of *Cyclops vernalis* with *D. medinensis* larva in its hemocoel. (After R. Muller in Schmidt and Roberts, 1981; with permission of C. V. Mosby.)

goes two additional molts, and becomes lodged in the liver, body cavity, or subcutaneous tissues, where it matures in 8 to 12 months.

Human hosts of *D. medinensis* usually portray eosinophilia, nausea, vomiting, diarrhea, asthma, and fainting. These symptoms are believed to result from absorption of metabolic wastes produced by the female worm while forming the blister. In addition, the cutaneous ulcer caused by the female worm commonly serves as a site for secondary bacterial and fungal infections. According to Lauckner *et al.* (1961), in parts of Africa such ulcers are the third most common sites of entry of tetanus spores.

Female worms that fail to reach the host's skin often cause complications in deeper tissues in the body. Commonly, they degenerate and during the process release strongly antigenic molecules. This can result in fluid-filled abscesses or cause a type of chronic arthritis when a calcified worm is located near a joint.

Other *Dracunculus* Species. Other species of *Dracunculus* include *D. fuelleborni* in the opossum, *Didelphis aurita*, in Brazil and *D. insignis* in the raccoon, *Procyon lotor*, the dog, and the skunk, *Mephitis mephitis*, in the United States. In both of these species of *Dracunculus*, copepods also serve as the intermediate host.

Other Dracunculoid Genera

Other prominent genera of Drancunculoidea include *Philometra* (Fig. 15.54), found in the body cavities and tissues of fish. Furuyama (1934) has demonstrated that gravid females escape from the fish host and rupture, releasing the larvae, which are then ingested by *Cyclops*.

Seven additional superfamilies subordinate to the class Secernentea should be mentioned briefly at this point. The superfamily Cosmocercoidea includes *Cosmocerca*, with species occurring in the intestine and rarely in the lungs of amphibians. None occur in North America.

The members of the superfamily Seuratoidea are parasites of fish, amphibians, reptiles, birds, and mammals. Among those infecting mammals is *Seuratum*, with species parasitic in rodents and bats.

Members of the superfamily Drilonematoidea are parasites of invertebrates, especially annelids and molluscs. Members of the superfamily Camallanoidea, exemplified by *Camallanus*, are parasites of ectothermic vertebrates that utilize copepods as intermediate hosts. All of the species are viviparous.

Some members of the superfamily Cylindrocorporidea are parasitic, occurring primarily in amphibians and reptiles. A few, such as *Longibucca* spp., are parasitic in bats.

Some members of the superfamilies Aphelenchoidoidea and Criconematoidea are parasites of invertebrates. For a summary of the biology and classification of the aphelenchoidoid nematodes, see Poinar (1975).

Finally, a suborder of secernentean nematodes, the Sphaerulariina, includes parasites of invertebrates, especially insects. Poinar (1975) considers these nematodes to represent an additional superfamily, the Sphaerularioidea. Representative of this group is *Sphaerularia bombi*, one of the most common parasites of bumblebees in Europe and the eastern United States.

PHYSIOLOGY AND BIOCHEMISTRY OF NEMATODES

Oxygen Requirements

One cannot generalize about whether adult intestinal nematodes are completely aerobic or anaerobic, or whether free-living larvae are all completely aerobic. The volume of experimental data now available strongly suggests that each species must be considered relative to its utilization of oxygen. Some survival rates of parasitic nematodes maintained under aerobic and anaerobic conditions are listed in Table 15.3.

To illustrate the differences in oxygen requirement in several intestine-dwelling species, Rogers (1949) has reported that *Nippostrongylus muris* occurs in a region of the small intestine where the oxygen tension is 19 to 30 mm Hg (average, 24 mm Hg), whereas *Nematodirus* spp. are located in a section of the gut where the oxygen tension is lower. These observations strongly suggest that the amount of oxygen present at a particular level of the alimentary tract

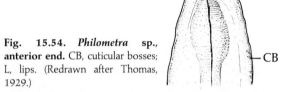

Fig. 15.54. *Philometra* sp., anterior end. CB, cuticular bosses; L, lips. (Redrawn after Thomas, 1929.)

could be at least partially responsible for the location of specific nematode species.

It is known that adult hookworms, which ingest their host's erythrocytes, obtain a portion of their oxygen requirement from these cells. It has been estimated that approximately 0.2 mg of oxygen is available to hookworms from this source. Hookworms and certain other adult nematodes require oxygen in their metabolism. For example, *Nippostrongylus muris* exhibits respiratory quotients (RQs) as high as 6.8. The RQ of most common ascaroids, however, is on the order of 0.4.

On the other hand, horse and pig ascaroids undoubtedly lead an essentially anaerobic existence. Their oxygen uptake, when it occurs, is totally unaffected by cyanide, and cytochrome oxidase has never been demonstrated in their tissues. Although these adult nematodes are essentially anaerobic, they are facultative, rather than obligatory, anaerobes and utilize oxygen if it is available. Ascaroids of pigs and horses utilize oxygen at a relatively high rate after a period of anaerobiosis, indicating the occurrence of an oxygen debt (p. 39). The amount and rate of oxygen uptake is governed by the quantity of end products of anaerobic fermentation accumulated in the tissues.

There is some doubt whether the amount of oxygen required by adults is the same *in vitro* as it is *in vivo*. To illustrate this point, Rogers (1949) has recorded that *Nippostrongylus muris* and *Nematodirus* spp., when in their hosts, utilize only 80 and 40%, respectively, of the oxygen they consume when maintained *in vitro*. In *Haemonchus contortus*, the amount of oxygen required *in vivo* is even lower than that reported for *Nippostrongylus muris* and *Nematodirus* spp. However, worms maintained *in vitro* are considerably more active than those observed *in vivo*. This increased activity is directly related to the amount of oxygen consumed.

The respiratory rate and sometimes the type of respiration (i.e., aerobic or anaerobic) of the egg and larval stages of nematodes generally differ from those of adults (Table 15.4). Eggs of many species, for example, embryonate and hatch in the aerobic environment provided by moist soil. Where investigated, the RQ for eggs varies between 0.6 and 0.9. The consumption of oxygen plus histochemical evidences suggest that the energy required during embryonic development within the egg is provided mainly from the oxidation of fats, although the metabolism of proteins and carbohydrates does contribute some energy. Passay and Fairbairn (1957) have demonstrated that a net decrease in the lipid content in developing eggs corresponds with an increase in carbohydrates. This suggests the synthesis of carbohydrates (trehalose and glycogen) from triglycerides.

The inhibition of respiration in nematode eggs by cyanide, carbon dioxide, or hydrazoic acid indicates dependency on a cytochrome system (Fig. 2.38).

From the standpoint of growth and development of the larva within the egg, the amount of oxygen present is of critical importance. For example, larvae in *Ascaris lumbricoides* eggs can survive a minimum oxygen tension of 30 mm Hg; even then, growth is retarded by 50%. Growth of larvae in eggs of *Parascaris* is barely maintained when the oxygen tension is 5 mm Hg, and is extremely slow at 10 to 80 mm Hg, but is normal at tensions above 80 mm Hg.

Eggs of some nematodes can survive for a period of time in the absence of oxygen, although oxygen is required for complete development and hatching of the larvae. In the egg of *Parascaris*, for example, the embryo can undergo the first cleavage stages under anaerobic conditions but, as stated above, requires oxygen for further development. Again, in *Oxyuris equi*, the embryo can reach the gastrula stage in the

Table 15.4. Respiratory Quotients (RQ) of Various Stages of Species of Nematodes[a,b]

Nematode Species	Respiratory Quotient		
	Eggs	*Larvae*	*Adults*
Haemonchus contortus	0.58–0.60	0.64	—
Ascaris lumbricoides	—	—	3–4
Ascaridia galli	—	—	0.96
Trichinella spiralis	—	1.1	—
Nippostrongylus muris	—	0.66–0.73	0.69
Litomosoides carinii	—	—	0.44
Eustrongylides ignotus	—	1.0	—
Syphacia obvelata	—	—	1.1
Neoaplectana glaseri	—	—	0.59

[a] Oxygen tension is about 160 mm Hg, and sugar is absent. Unless otherwise specified, values are given for 37°–38°C.
[b] Data from von Brand, 1952.

absence of oxygen but requires oxygen from then on. In eggs of *Enterobius vermicularis*, the larva can reach the tadpole stage without oxygen; however, further development requires aerobic conditions.

In the species that have been studied, respiration in hatched larval nematodes is aerobic. The RQ of the infective larvae of *Nippostrongylus* and *Haemonchus* is approximately 0.7. The energy requirements in these larvae are also primarily provided through lipid metabolism. In larvae found in the host's tissues, for example, *Trichinella spiralis* in mammalian muscles and *Eustrongylides ignotus* in fish muscles, the RQ is higher, 1.0 or more, because energy production is provided predominantly through metabolism of the large glycogen supply in the body of the parasites.

Certain larvae can withstand anoxic conditions for some time. For example, Stoll (1940) has reported that the first parasitic molt of *Haemonchus contortus* occurs

Table 15.3. Survival Time of Nematodes *in Vitro* under Anaerobic and Aerobic Conditions[a]

Species	Medium	Temperature (°C)	Survival Time in Days	
			Anaerobic	*Aerobic*
Litomosoides carinii	Ox serum	37	<1	7
Parascaris equorum	Ringer	35–38	2–5	2
Raphidascaris acus	1% NaCl	Room temperature	6	—
Ascaris lumbricoides	1% NaCl	35–38	7–9	6
Trichostrongylus colubriformis	Ringer	37	<1	4–12
T. vitrinus	Ringer	37	<1	4–12
Ostertagia circumcincta	Ringer	37	<1	4–12
Cooperia oncophora	Ringer	37	<2	4–12
C. curticei	Ringer	37	<1	4–12
Nematodirus filicollis	Ringer	37	<2	4–12
Trichinella spiralis (larvae)	Tyrode	37	7 or more	—
Ancylostoma caninum (larvae)	Tap water	17	21	—
Eustrongylides ignotus (larvae)	1% NaCl	37	3	19
	0.85%	20	21	Several months

[a] Data from various authors, mainly after von Brand (1973).

best in cultures with limited oxygen tension. *Trichinella spiralis* can be maintained anaerobically, although oxidative metabolism brings about greater motility.

Uptake of Oxygen

Since nematodes in varying degrees can utilize molecular oxygen, the question may be asked as to how they acquire their oxygen. In the case of the smaller species, oxygen can penetrate the body wall and arrive at the internal organs by diffusion through the pseudocoelomic fluid. Furthermore, the body movements may aid in this process by churning the fluids in the immediate environment. Thus, despite the lack of a structurally recognizable circulatory system or respiratory pigments, small nematodes, because of the greater surface/volume ratio, can derive sufficient oxygen.

In larger nematodes the problem of oxygen uptake becomes more complex. In these, diffusion alone may be insufficient for the distribution of oxygen throughout the body. For example, Lee (1965) has estimated that in *Haemonchus contortus* and *Nematodirus filicollis*, diffusion can provide approximately only 50% of the oxygen requirements of the tissues. It is in these larger species that one would expect to find some oxygen-conducting system and, indeed, in some of these the respiratory pigment, hemoglobin, has been detected.

Hemoglobin in Nematodes

Hemoglobin has been detected in certain nematodes. For example, it is present in *Nippostrongylus muris* at approximately 6 mg per gram of dry weight, and in *Nematodirus* and *Haemonchus contortus* at approximately 0.8 mg per gram of dry weight.

Hemoglobin in many animals occurs in two forms: one in the blood which transports oxygen, and one in tissues, known as **myoglobin**, which stores oxygen. With the exception of *Ascaris*, only one type of hemoglobin usually occurs in nematodes, in the pseudocoelomic fluid. In the case of *Ascaris*, one type of hemoglobin similar to myoglobin occurs in the body wall, and the other is in the pseudocoelomic fluid. The first type functions like myoglobin in that it releases oxygen under anaerobic conditions (Davenport, 1949a). That hemoglobin of nematodes is native and not absorbed from the host has been proven by chemical characterization. In most nematode hemoglobins that have been studied there is one heme moiety per protein molecule, and these hemoglobins are thus similar to vertebrate myoglobin. In the case of *Ascaris*, the host's hemoglobin, according to Smith and Lee (1963), can be broken down and absorbed, and at least the heme moiety can be utilized after slight structural alterations, in the synthesis of nematode hemoglobin.

The survival value of nematode hemoglobin is

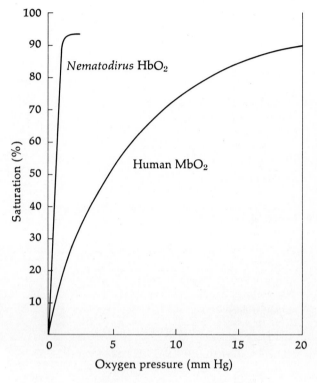

Fig. 15.55. Hemoglobin. Comparison of the affinity of *Nematodirus* hemoglobin and human myoglobin for oxygen. (Data in part after Fairbairn, 1960.)

clearly portrayed by its high affinity for oxygen (Fig. 15.55). Because of this property, nematodes are efficient at extracting oxygen from host tissues and possibly the surrounding environment, but the oxygen tension in the parasite's tissues has to be extremely low before the oxyhemoglobin gives up oxygen. As examples, the oxyhemoglobins of *Heterakis*, *Nippostrongylus*, *Nematodirus*, and *Haemonchus* become deoxygenated only under anaerobic conditions and are reoxygenated in the presence of air (Davenport, 1949b; Rogers, 1949; van Grembergen, 1954).

The hemoglobin in the body wall of *Ascaris* may function not only in the storage of oxygen during anoxia but also in the transfer of oxygen from the cuticle to the pseudocoelomic fluid (Scholander, 1960). *Ascaris* probably does not utilize oxygen in energy production. Rather, it is utilized in other metabolic and biosynthetic processes.

For a detailed review of hemoglobins and other types of pigments in nematodes, see Smith (1969).

Carbohydrate Metabolism

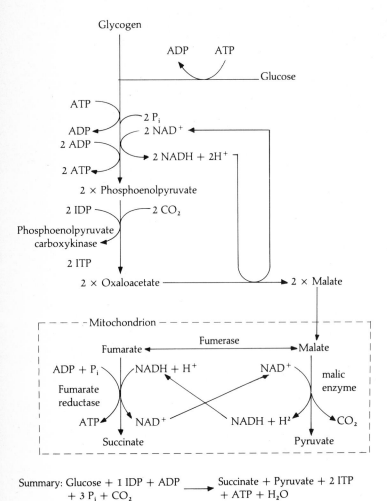

Summary: Glucose + 1 IDP + ADP $\longrightarrow$ Succinate + Pyruvate + 2 ITP
$\qquad$ + 3 P_i + CO_2 $\qquad\qquad\qquad$ + ATP + H_2O

Fig. 15.56. Carbohydrate catabolism in *Ascaris lumbricoides* muscle; including CO_2 fixation.

Culture of Nematodes

Some investigators have been able to maintain parasitic nematodes *in vitro*, others have actually devised media in which the worms grow, and still others have found media in which the worms not only increase in size but also actually molt and develop through the various stages of their normal life cycles. Interested readers are referred to Taylor and Baker (1978), which includes a comprehensive review of known culture methods. It bears recalling, however, that the principal rationales underlying the culture of nematodes are (1) to ascertain their nutritional and growth requirements and (2) to produce large quantities of worms that may be useful for the production of vaccines.

As is the case among most helminth parasites, especially those that live in environments with little or no oxygen, adult nematodes depend primarily on carbohydrate metabolism as the energy source, because proteins and lipids do not lend themselves to the internal oxidation-reduction reactions characteristic of anaerobic processes. Even in nematodes that live in oxygen-rich environments, carbohydrate metabolism predominates. For example, the filarial worm *Litomosoides*, which lives in the tissues or the blood of its host where oxygen is available, produces most of its energy through carbohydrate metabolism. In such cases, however, the nature of carbohydrate catabolism is such that the sugars are not oxidized completely to carbon dioxide and water, indicating that the process is fermentative.

The so-called Pasteur effect has been demonstrated in *Litomosoides carinii* and other species; that is, substrate, especially carbohydrates, is used more rapidly anaerobically than aerobically. This indicates that additional energy is derived per unit of substrate when metabolized in the presence of oxygen.

Glycogen is the main reserve nutrient in the bodies of nematodes. When worms are starved, glycogen is depleted rapidly, resulting in the production of fermentation acids and carbon dioxide. The metabolic end products in all species studied include, for the most part, volatile fatty acids, among which acetic, lactic, butyric, valeric, caproic, and other acids have been identified. Among these, valeric acid or its isomers comprise the major excretory product. For example, in *Trichinella spiralis*, valeric acid represents 80% of the acids produced.

In the case of *Nippostrongylus muris*, Roberts and Fairbairn (1965) have demonstrated that the endogenous carbohydrate reserves are composed of approximately equal parts of glycogen and trehalose, and both are utilized nonpreferentially in the production of energy.

Little is known about glycogensis in nematodes. In *Ascaris*, however, stored glycogen is synthesized from fructose, sorbose, maltose, and sucrose, which are absorbed, but lactose and mannose are not utilized. Furthermore, larger nematodes, such as *Dracunculus* and *Litomosoides*, consume nearly 40 times more carbohydrate per gram of tissue than such smaller species as *Eustrongylides*. The reason for such differences is not known, although the presence of different types of metabolism is suggested. Again, the degree of maturity of the worms studied most probably is responsible for some of these differences. For example, young worms do not require as much carbohydrate as older worms that are producing eggs.

A great deal of progress has been made in recent years in understanding carbohydrate metabolism in nematodes. Most of the studies have involved *Ascaris lumbricoides*, and the results are now considered the prototypes to which the results obtained with other species of helminths are compared. For detailed and authoritative reviews of carbohydrate metabolism in nematodes, see Barrett (1976, 1981). The following is an abbreviated account of the major aspects of carbohydrate metabolism in *Ascaris*.

The process of energy production in adult *Ascaris lumbricoides* appears to be primarily, if not exclusively, anaerobic (Saz and Bueding, 1966; Saz, 1969). The breakdown of carbohydrates results in the accumulation of succinate and a variety of volatile fatty acid end products (Saz and Bueding, 1966). Also, *A. lumbricoides* includes very low levels of cytochrome oxidase activity and therefore is of questionable significance, and the tricarboxylic acid cycle system has been found to be inoperative as an energy-generating mechanism (Kmetec and Bueding, 1961; Ward and Fairbairn, 1970).

In ascarid muscle, Saz and Lescure (1969) have demonstrated that glucose is metabolized via the Embden-Meyerhof pathway, to phosphoenolpyruvate. Since the activity of pyruvate kinase is minimal in ascarid muscle, further utilization of phosphoenolpyruvate requires CO_2 fixation (p. 40). CO_2 fixation is catalyzed by phosphoenolypyruvate carboxykinase, which acts physiologically in the direction of oxalacetate formation. Oxalacetate, in turn, is reduced to malate by the mediation of soluble malate dehydrogenase, with the concomitant regeneration of cytoplasmic NAD^+ (Fig. 15.56).

Cytoplasmic malate then serves as the substrate for mitochondrial metabolism. Specifically, upon entering the mitochondrion, malate undergoes dismutation (oxidation-reduction), half of the malate being oxidized to pyruvate and CO_2, resulting in the generation of intramitochondrial NADH. The remaining malate is dehydrated to form fumarate. Fumarate thus formed, rather than oxygen, then serves as the terminal acceptor for the electron transport mechanism. The dismutation reaction is completed with the reduction of fumarate to succinate via the NADH-coupled fumarate reductase system (Fig. 15.56). Particularly significant is that the reduction of fumarate to succinate results in an anaerobic, electron transport-associated phosphorylation (Saz, 1971).

Succinate and pyruvate, which are the products of malate dismutation, are utilized by *Ascaris* muscle as precursors for the anaerobic formation of the volatile fatty acid end products (Saz and Weil, 1960, 1962; Saz and Lescure, 1969) (Fig. 15.57).

Protein Metabolism

As with most parasitic helminths, the biotic potential of nematodes is great. Although the exact essential amino acids of nematodes have not been determined satisfactorily, the uptake of these must be rapid and must occur in relatively large quantities, especially in

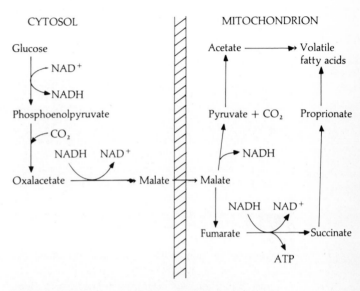

Fig. 15.57. Biochemical pathways leading from cytosol glucose to the formation of mitochondrial volatile fatty acids via succinate and pyruvate in *Ascaris* muscle.

egg-producing individuals. Furthermore, the conversion of proteins and amino acids that are taken into eggs must be highly efficient.

Since nematodes possess an alimentary tract, it is generally assumed that most exogenous proteins are ingested and digested within the alimentary canal. Lee (1962) has shown that secretory activity occurs in the intestine of *Ascaris lumbricoides*. Both holocrinic and merocrinic secretions have been reported. Based on analyses of extracts of the intestines of *Ascaris lumbricoides*, *Strongylus edentatus*, and *Leidynema appendiculata*, the latter a parasite of beetles, Rogers (1941) and Lee (1958) have reported endo- and exopeptidase activity.

Proteins *per se* are not believed to be absorbed by the lining of the digestive tract; however, after luminal digestion, the resulting amino acids, and perhaps peptides, are absorbed via this route. Although the nematode cuticle is permeable to water, some ions, and hydrophobic anthelmintics, it is unlikely that the uptake of nitrogenous compounds via this route is of common occurrence, although in the case of *Ascaridia galli* there is some evidence that absorption via the gut may be supplemented by uptake via the cuticle (Weatherly *et al.*, 1963).

The end products of nitrogen metabolism are known in only a few species, and in these ammonia is given off in relatively large quantities (**ammonotelic metabolism**). For example, Saval (1955) has reported that *Ascaris lumbricoides* excretes 0.39 mg of nitrogen per gram of wet body weight per day, of which 79% is ammonia, 7% urea, and 21% polypeptides. Haskin and Weinstein (1957a,b) have reported that *Trichinella spiralis* larvae excrete 2.8 mg of nitrogen per gram of wet body weight in 24 hours, of which 33% is ammonia, 7% volatile amine nitrogen, 20% peptide nitrogen, and 29% amino acid nitrogen.

Ammonotelic nitrogen metabolism also is found in such free-living aquatic animals as annelids, echinoderms, teleost fish, and reptiles. Among these aquatic organisms, water is readily available in the environment so that ammonia can be diluted and kept below the toxic level. Since ammonia is also highly toxic to nematodes, it is assumed that the intestinal species are capable of deriving water from the host's intestinal contents to dilute the ammonia. An alternative mechanism, i.e., switching from ammonia excretion to urea production, has been presented earlier (p. 478). Since the larvae of certain nematodes pass through a terrestrial phase in which water is not readily available, nitrogen metabolism in these is most probably uricotelic and hence does not involve the production of a large quantity of ammonia.

Evidences supporting the belief that nematodes are capable of synthesizing amino acids have been obtained from three types of experiments: (1) studies on the amino acids required for growth and reproduction in chemically defined media; (2) studies on the incorporation of ^{14}C from substrates in the media into amino acids in the whole organism; and (3) biochemical studies on the nature of reactions occurring in homogenates. For a detailed review of nitrogen metabolism in nematodes, see Barrett (1981).

From what is known, amino acids do not appear to be an important energy source in those species of nematodes parasitic in animals.

Generally, during catabolism, amino acids lose their α-amino nitrogen, and the remaining carbon skeletons are converted to glycolytic or tricarboxylic acid cycle intermediates. These intermediates are then either completely degraded or resynthesized into glucose. Removal of the α-amino nitrogen is accomplished by two primary pathways: transamination and oxidative deamination. In addition, one of a number of specific, nonoxidative deaminases may perform the same function.

Table 15.5. 2-Oxoglutarate-Linked Transaminases in Selected Parasitic Nematodes

Nematode Species	Donor Amino Acids
Ascaris lumbricoides	Alanine, arginine, aspartate, glycine, methionine, phenylalanine, serine, 4-aminobutyrate
Ascaridia galli	Alanine, aspartate
Stephanurus dentatus	Alanine, aspartate

Transamination. By this process, the amino group is transferred from the amino acid to a 2-keto acid, usually 2-oxoglutarate. This is exemplified in Figure 15.58.

This reaction requires pyridoxal phosphate as a cofactor. Most transaminases employ 2-oxoglutarate as one of the amino group acceptors; however, they are less specific toward the other amino donor. 2-oxoglutarate-linked transaminases are widely distributed in parasitic helminths, including nematodes. Listed in Table 15.5 are the donor amino acids in selected species of nematodes associated with 2-oxoglutarate-linked transaminases.

Oxidative Deamination. A common example of oxidative deamination is the catalysis of glutamate by glutamate dehydrogenase. This is depicted in Figure 15.59.

The amino groups, collected into glutamate via transaminase reactions, are released as ammonia. Glutamate dehydrogenase can utilize either NAD or NADP, although NAD is the preferred cofactor.

Glutamate dehydrogenase is widely distributed in helminths, although its regulatory properties have not been studied extensively. That from *Haemonchus contortus* has been reported to be inhibited by AMP, ADP, ATP, aspartate, and thyroxine.

Specific Deaminases. A number of amino acids (arginine, histidine, serine, threonine, glutamine, and aspartate) can be nonoxidatively deaminated by specific deaminases. Some of these enzymes have been found in parasitic nematodes. For example, histidase has been reported in *Ascaridia galli* and *Ascaris lumbricoides*. It catalyzes the reaction shown in Figure 15.60.

Lipid Metabolism

Because lipases and esterases have been detected in the intestine of certain intestinal nematodes, it is generally assumed that digested fats are absorbed through the intestinal wall.

$$\text{L-aspartate} + \text{2-oxoglutarate} \rightleftharpoons$$

$$\text{oxaloacetate} + \text{L-glutarate}$$

Fig. 15.58. Transamination

$$\text{L-glutamate} + H_2O + NAD^+ \ (NADP^+) \rightleftharpoons$$

$$\text{2-oxoglutarate} + NH_4^+ + NADH \ (NADPH)$$

Fig. 15.59. Oxidative deamination

In aerobic nematodes, such as the free-living larvae of human hookworms and *Nippostrongylus*, the amount of lipids in their bodies decreases with age, suggesting that lipid utilization occurs. Similarly, *Trichinella spiralis* larvae maintained *in vitro* utilize fats as an energy source.

In essentially anaerobic adult nematodes, such as *Ascaris*, lipid metabolism is limited. Even after 5 days of starvation, the lipid content is not decreased appreciably. Fairbairn (1957, 1969) and Barrett (1981) have reviewed the studies performed on lipid metabolism in *Ascaris* and other parasitic nematodes. These essays deserve the attention of serious students of helminth physiology and biochemistry.

It has been fairly well documented that carbohydrate and lipid metabolism in nematodes are directly related. An example of this pertains to ascarosides. Flury (1912) and Fauré-Fremiet (1913) independently discovered that *Ascaris lumbricoides* and *Parascaris equorum* contain large quantities of unsaponifiable material that is isolated readily by precipitation from nonpolar solvents, has an empirical formula approximating $C_{32}H_{64}O_4$, melts at 82°C, and forms a diacetate. Although this matter occurs in most tissues of both male and female worms, it was found to be particularly abundant in ovaries. Initially named ascaryl alcohol, the designation was changed to **ascarosides** when it was found that the material is a mixture of closely related glycosides. Since then, Jezyk and Fairbairn (1967) have studied the biosynthesis of ascarosides and have reported that ascarylose (3,6-dideoxy-L-arabinohexose), one of the major constituents, is synthesized in the ovaries of *Ascaris lumbricoides* from either glucose or glucose-1-phosphate by a sequence of reactions that probably involves a nucleotide-diphosphate glucose as an intermediate, and NAD and NADPH as coenzymes. The incorporation of ascarosides into the eggshell renders the shell virtually impermeable to substances other than gases and lipid solvents.

HOST-PARASITE RELATIONSHIPS

The clinical symptoms of nematode-caused diseases in humans and domestic animals are thoroughly discussed in volumes devoted to clinical helminthology. The relationships between animal hosts, including humans, and nematodes is discussed by Rogers (1962), Oliver-González (1969), Bird and Wallace (1969), Hunter *et al.* (1976), and the treatise edited by Binford and Conner (1976). In the following paragraphs, an attempt has been made to summarize some

of the less severe but still important effects nematodes have on their hosts.

Antienzymes

It has been mentioned that *Ascaris* secretes antienzymes the primary function of which is to protect the parasite from being digested by the digestive enzymes of the host. However, the extent of the action of these antienzymes does not stop there. The antitryptic and antipeptic enzymes secreted by *Ascaris* and other nematodes also hinder digestion of a certain amount of the food passing through the host's alimentary tract.

Nutrient Deprivation and Metabolic Changes

Hall (1917) has reported that in one instance a human harbored 11 pounds of *Ascaris*. The amount of food utilized by these worms has been estimated to comprise 8% of the normal 2500 calories of the host. In addition to robbing the host of nutrients, other effects are inflicted. For example, the blood proteins of hosts are altered in most nematode infections. Such changes most probably reflect immune responses on the part of the host (p. 70). Wright and Oliver-González (1943) have pointed out that in dogs and rabbits infected with *Trichinella spiralis*, the plasma albumin fraction decreases while the globulin fractions increase. Weir *et al.* (1948) have pointed out that the total protein in the blood plasma of sheep infected with *Haemonchus contortus* decreases, and Villela and Teixera (1937) reported a decrease in both total protein and albumin in the blood of humans harboring hookworms. Furthermore, in human infections, potassium and sodium levels in blood plasma are increased.

Sugar metabolism in the host is sometimes disrupted by nematodes. In humans, *Ascaris lumbricoides* occasionally causes hypoglycemia. Numerous investigators have shown that in humans, dogs, and rabbits infected with *Trichinella spiralis*, occasional hypoglycemia also occurs.

Nematode Toxins

Toxins have often been suspected as causative agents of nematode-effected metabolic disruptions (see p. 12). The toxicity of nematodes has been reviewed by Schwartz (1921), although a more recent review is needed. Although in some instances incrimination of toxic substances has been unfounded, in other instances it is justified. It is well known that persons working with *Ascaris* and infected animals often develop an allergy. A similar phenomenon is found in some other roundworms (Sprent, 1949). In addition to developing an allergy, infected animals sometimes display other symptoms that disappear when the worms are removed, strongly suggesting that some toxic substance is being secreted.

Various investigators have shown that the pseudo-coelomic fluids of ascaroids are toxic if injected into hosts. Similarly, tissue extracts of such nematodes as *Ascaris*, *Parascaris*, *Strongylus*, and *Dirofilaria* can be lethal when injected into experimental animals.

The nature of nematode toxins has not yet been completely determined. Flury (1912) has suggested that irritation of mucous membranes by ascaroids is due to aldehydes formed during the metabolism of the worms. Bondouy (1908) has reported that the toxic secretion of *Strongylus equinus* is alkaloidlike and is strongly hemolytic. Shimamura and Fujii (1917) have reported that the toxin of *Ascaris* is an albumose-peptone fraction, and Macheboeuf and Mandoul (1939) thought it to be a polypeptide. Deschiens and Poirier (1949) have reported that lesions produced when extracts of *Parascaris equorum* are applied to tissues resemble histamine poisoning.

In addition to damaging tissue by producing lesions and symptoms of toxicity, the secretions of some nematodes actually disrupt the host's metabolism. For example, Mönning (1937) has reported that even when the worms are removed from sheep heavily infected with *Oesophagostomum columbianum*, the hosts die. He interpreted this finding to mean that the defaunated sheep could not overcome the physiologic disarrangements caused by the toxic secretions of the parasites.

Larva Migrans

In connection with nematode host-parasite relationships, the medically important and biologically interesting phenomena collectively known as **larva migrans** should be mentioned. This term describes a nematode's erratic migratory behavior in a vertebrate host in which it cannot mature sexually. In medical parasitology, the human host in which larva migrans occurs is usually a paratenic host, albeit a dead-end host. The vertebrate host need not be a human, but is most commonly a mammal. The physiologic basis for the erratic migration of the larval nematode is poorly understood, but is probably based in the partial infectivity of the parasite and incomplete resistance of the host.

Migration in the skin is commonly referred to as **cutaneous larva migrans** or creeping eruption, whereas migration in internal organs is known as **visceral larva migrans**. These conditions, if they occur in a human, are medically important because they represent parasite-caused pathologic conditions. They

are interesting biologically because they reflect the behavior of parasites in unnatural or abnormal hosts and the responses of such hosts to the parasites.

Although larva migrans in various animals accidentally infected with nematode larvae for which they are not the natural definitive hosts undoubtedly occurs, most research concerning these phenomena has been centered around larva migrans in humans. The experimental or accidental inoculation of larvae of four species of non-human-infecting hookworms—*Ancylostoma braziliense*, *A. caninum*, *Uncinaria stenocephala*, and *Bunostomum phlebotomum*—causes creeping eruption (cutaneous larva migrans) in humans.

Although *A. braziliense* is generally considered a non-human-infecting hookworm, adults of this species have been reported from humans on a number of occasions. Its normal hosts are dogs, cats, and related wild animals. *Ancylostoma caninum* adults, although known in humans, are normally parasites of dogs. Adults of the two other species, *U. stenocephala* and *B. phlebotomum*, are normally parasites of canines and bovines, respectively, and have not been reported from humans.

Cutaneous larva migrans is initiated when the infective larvae of any of these four nematodes penetrate human skin and migrate under the surface, primarily in the dermis, in a tortuous linear fashion. The paths of migration appear as elevations on the skin (Fig. 15.61). These larvae seldom develop to maturity in the abnormal human host. Rarely in the case of *Ancylostoma braziliense*, and even less frequently in the case of *A. caninum*, a few larvae may become blood borne in humans and are carried to the lungs. Eventually, they become established in the intestine, where they develop to maturity, as is normally the case with *A. duodenale* (p. 507).

In addition to the hookworms mentioned, *Strongyloides stercoralis* larvae have been reported to cause cutaneous larva migrans. In these instances, the larvae participate in autoinfection (p. 503)—that is, they penetrate the host's skin in the anal area and produce urticarial swellings that often take on the tortuous linear patterns of creeping eruption.

Visceral larva migrans is commonly, but not exclusively, caused by the adults of ascaroids, which parasitize dogs and cats. Human infections, which are most common in children, are contracted when soil containing embryonated *Toxocara canis* eggs is ingested. Larvae hatching from the eggs penetrate the intestinal wall and soon reach the liver. Although the majority of these larvae remain in the liver, some pass on to the lungs and other areas. In addition, larvae have been known to invade the central nervous system and eyes. Invasion of the eyes leads to formation of a neoplasm that may cause blindness and even death. Hence, removal of the infected eye by surgery becomes necessary.

Although most of the larvae eventually come to

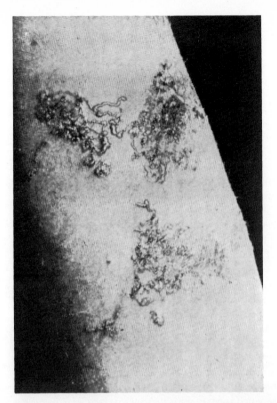

Fig. 15.61. Cutaneous larva migrans in human. Note tortuous canals in skin. (After Weiner, 1960.)

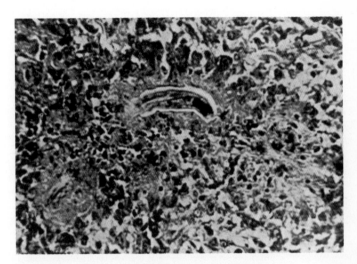

Fig. 15.62. Visceral larva migrans in human. Note section of ascarid larva surrounded by host reaction cells. (After Weiner, 1960.)

rest in one location and become encapsulated by host tissues, for a period of at least several weeks they actively migrate through tissues, leaving long trails of inflammatory and eosinophilic granulomatous reactions (Fig. 15.62). Thus, a few microscopic larvae can produce extensive lesions, especially in the liver and brain. Visceral larva migrans is manifested clinically by eosinophilia, sometimes accompanied by enlargement of the liver, hyperglobulinemia, infiltration of the lungs, and neurologic symptoms.

In addition to *Toxocara canis, T. cati* (p. 522), *Capillaria hepatica* (p. 492), and *Ancylostoma caninum* may possibly be responsible for visceral larva migrans in humans (Karpinski *et al.,* 1956; Weiner, 1960).

It should be emphasized that, although nonhuman hookworms are usually associated with cutaneous larva migrans and *Toxocara canis* is associated with visceral larva migrans in humans, the locations of the lesions and even the resulting symptoms are not completely reliable in the specific identification of the etiologic agent (Beaver, 1956).

During the past few decades, an unusual clinical entity, characterized by acute eosinophilia and asthma-like symptoms and termed **eosinophilic lung** (or tropical eosinophilia), has been identified in humans. The causative agent is believed to be nonhuman-infecting filarial worms because Buckley (1958) was

able to reproduce the disease in a human volunteer by inoculating him with microfilariae obtained from mosquitoes that had fed on a monkey and a cat, respectively. This experiment suggests that eosinophilic lung is closely akin to larva migrans.

Effects of Host on Parasite

The amount of food material within the host governs the maintenance of many nematode parasites. For example, if chickens infected with *Ascaridia galli* are starved for a period of time, the parasites are completely removed in 48 to 96 hours. Analysis of the expelled worms has revealed that their body glycogen is decreased considerably. When host starvation does take place, it usually does not affect larvae, only adults. Again, hosts maintained on complete and normal diets harbor more parasites than otherwise.

Space does not permit elaboration of the role of vitamins in parasite maintenance; however, they do have such a role. For example, in *Ascaridia galli* infections of chickens, deficiency of vitamins A and B complex favors growth of the worms, which attain greater dimensions. Deficiency of vitamin D has no effect. On the other hand, deficiency of vitamin A hinders establishment of *Enterobius vermicularis* infection. These contradictory findings suggest that the influence of vitamins varies among species.

The role of host age and sex in nematode parasitosis is not completely clear. Instances are known in which age resistance or immunity do occur. Todd and

Table 15.6. Some Nematodes Known to Undergo Arrested Development within Their Definitive Hosts during Seasons Unfavorable for Their Free-Living Development[a]

Species	Host	Location	Season	Stage
Ancylostoma duodenale	Human	India	Dry	L_3
Chabertia ovina	Sheep	England	Winter	L_4
Cooperia oncophora	Cattle	Canada	Winter	L_4
Haemonchus contortus	Sheep	Canada, Northern Europe, South Africa, New Zealand, Australia	Winter	L_4
Hyostrongylus rubidus	Swine	England	Winter	L_4
Nematodirus spp.	Sheep	Canada, Scotland	Winter	L_4
Ostertagia ostertagi	Cattle	Canada, Northern Europe, Australia, Israel	Winter Hot, dry	L_4
Oesophagostomum columbianum	Sheep	Australia	Winter	L_4
Dictyocaulus viviparus	Cattle	Europe, Canada	Winter	L_5

[a] Data assembled by Schad, 1977.

Hollingsworth (1952) have reported that more worms develop from a given number of *Ascaridia galli* eggs in male chicks than in females. This suggests sex differences.

Sadun (1948, 1951) has tested the hypothesis that sex hormones may play a role in host-parasite relationships. He injected testosterone benzoate and α-estradiol benzoate into immature male and female chicks, respectively, and found that the growth of *A. galli* in males was temporarily retarded, while that in females was temporarily accelerated. After the initial retardation and acceleration, the parasites were eliminated from the host at a rate greater than normal. Furthermore, the injection of these sex hormones into non-infected hosts increased their resistance to *Ascaridia*.

From such investigations it appears that sex hormones are of considerable importance to the host-parasite relationship. The same holds true for other endocrine secretions. To illustrate this point, Todd (1949) has shown that *A. galli* in chickens suffering from experimentally induced hypothyroidism attain greater lengths.

The effect of mineral-deficient diets on the resistance of fowls to parasitism has been critically reviewed by Gaafer and Ackert (1953).

Arrested Development

In recent years, one aspect of the physiologic ecology of nematodes has attracted some attention. The phenomenon has been designated **arrested development**. This topic has been critically reviewed by Schad (1977).

As a result of following the development of several species of nematodes (Table 15.6), it has been recognized that at least certain species undergo temporary cessation in their development at a precise time during early parasitic development. Such arrested development usually occurs only in a certain proportion of the nematodes within a host while the remainder develop normally. Consequently, a bimodal size distribution is characteristic of the phenomenon and sets it apart from retardation of growth and stunting. Uniform retardation of growth and stunting is characteristic of nematodes in hosts demonstrating innate or acquired resistance.

Arrested development involves only a temporary cessation of development. The generally accepted interpretation of this phenomenon is that it is of survival value. For example, arrested development in some species or in particular geographic variants represents a mechanism which permits the parasites to survive during seasons unfavorable for their external development as free-living larvae. Thus, the phenomenon represents a secondary life history option ensuring survival from year to year and/or transmission to new groups of young susceptible hosts.

Studies aimed at elucidating the mechanisms responsible for arrested development have thus far revealed three major categories of factors: (1) external environmental factors which induce a potential for a diapauselike state and which is transmitted via the host; (2) host factors which determine the host's suitability as an environment for further development or induce arrest; and (3) parasite-related factors, either genetic or density dependent, which, alone or in concert with the first two categories of factors, induce arrest.

As an example of external environmental factors, the ambient temperature is known to induce arrested development. Specifically, Hotson (1967) and Anderson (1972) have reported that in temperate areas where winters are mild but summers are hot and dry, arrested development within the host occurs during summer. The opposite is true in other species, i.e., if the winter temperatures are harsh, the parasites may be arrested during that time of the year.

The age, sex, and species of the host are important examples of host factors. Schad (1977) has cited numerous examples for each of these factors.

Within the realm of parasite-related factors, the genetic constitution of the parasite appears to be of primary importance. Blitz and Gibbs (1972) have reported that a strain of *Haemonchus contortus* from southern Quebec, Canada, undergoes arrest during the autumn. On the other hand, Connan (1975) has reported that this does not occur in a strain from Cambridge, England. This difference is attributable to genetic differences.

Schad (1977) has noted that arrested development does not have a single function. It involves more than synchronizing the life cycle of the parasite with changing external conditions or with events in the life of the host. He has pointed out that storage of larvae in the host in a quiescent form limits large oscillations in parasite biomass, and hence arrest has a regulatory function. Large oscillations place the host and parasite in jeopardy, and arrested development is one mechanism responsible for dampening these oscillations.

Factors Governing Infectiveness

Some of the most important effects hosts have on nematode parasites are those that influence or govern infectiveness. The mere ingestion of eggs or larvae does not ensure infection, since activation of the infective stages also must occur. This activation is usually triggered by physical and chemical components in the host's gut. Among these, dissolved gaseous carbon dioxide or undissolved carbonic acid are the most important. In addition, oxidation-reduction potential, temperature, and pH all influence infectiveness by stimulating secretion of hatching fluids in eggs such as those of ascarids, or secretion of exsheathing fluids in larvae such as those of trichostrongylids. The reviews by Rogers (1962) and Lee (1965) should be consulted for details.

REFERENCES

Ackert, J. E., Edgar, S. A., and Frick, L. P. (1939). Goblet cells and age resistance of animals to parasitism. *Trans. Am. Microsc.* **58**, 81–89.

Alibasoglu, M. D., Kradel, D. C., and Dunne, H. W. (1961). Cerebral nematodiasis in Pennsylvania deer (*Odocoileus virginianus*). *Cornell Vet.* **51**, 431–441.

Alicata, J. E. (1962). *Angiostrongylus cantonensis* (Nematoda: Metastrongylidae) as a causative agent of eosinophilic meningoencephalitis of man in Hawaii and Tahiti. *Can. J. Zool.* **40**, 5–8.

Alicata, J. E., and Brown, R. W. (1962). Observations on the method of human infection with *Angiostrongylus cantonensis* in Tahiti. *Can. J. Zool.* **40**, 755–760.

Alicata, J. E., and Jinkrak, K. (1970). "Angiostrongylosis in the Pacific and Southeast Asia." Charles C Thomas, Springfield, Illinois.

Anderson, N. (1972). Ostertagiasis in beef cattle. *Victoria Vet. Rec.* **30**, 36–38.

Anderson, R. C. (1963). The incidence, development and experimental transmission of *Pneumostrongylus tenuis* Dougherty (Metastrongyloidea: Protostrongylidae) of the meninges of the white-tailed deer (*Odocoileus virginanus borealis*) in Ontario. *Can. J. Zool.* **41**, 775–792.

Anderson, R. C. (1971). Lungworms. *In* "Parasitic Diseases of Wild Mammals" (J. W. Davis and R. C. Anderson, eds.), pp. 81–126. Iowa State Univ. Press, Ames, Iowa.

Arean, V. M. (1971). Anisakiasis. *In* "Pathology of Protozoal and Helminthic Diseases" (S. A. Marcial-Rojas, ed.), pp. 846–851. Williams & Wilkins, Baltimore, Maryland.

Ash, L. R. (1962). The helminth parasites of rats in Hawaii and the description of *Capillaria traverae* sp. n. *J. Parasitol.* **48**, 66–68.

Barrett, J. (1976). Energy metabolism in nematodes. *In* "The Organization of Nematodes" (N. A. Croll, ed.), pp. 11–70. Academic Press, London.

Barrett, J. (1981). "Biochemistry of Parasitic Helminths." University Park Press, Baltimore, Maryland.

Barriga, O. O. (1981). "The Immunology of Parasitic Infections." University Park Press, Baltimore, Maryland.

Beaver, P. C. (1956). Larva migrans, a review. *Exp. Parasitol.* **5**, 587–621.

Beaver, P. C. (1970). Filariasis without microfilaremia. *Am. J. Trop. Med. Hyg.* **19**, 181–189.

Beaver, P. C., and Orihel, T. C. (1965). Human infection with filariae of animals in the United States. *Am. J. Trop. Med. Hyg.* **14**, 1010–1029.

Beaver, P. C., and Theis, J. H. (1979). Dioctophymatid larval nematode in a subcutaneous nodule from man in California. *Am. J. Trop. Med. Hyg.* **28**, 206–212.

Bier, J. W. (1976). Experimental anisakiasis: cultivation and temperature tolerance determinations. *J. Milk Food Technol.* **39**, 132–137.

Binford, C. H., and Connor, D. H. (eds.) (1976). "Pathology of Tropical and Extraordinary Diseases," Vols. 1 and 2. Armed Forces Institute of Pathology, Washington, D.C.

Bird, A. F., and Wallace, H. R. (1969). Chemical ecology of Acanthocephala and Nematoda. *In* "Chemical Zoology" (M. Florkin and B. T. Scheer, eds.), Vol. III, 561–592. Academic Press, New York.

Blitz, N. M., and Gibbs, H. C. (1972). Studies on the arrested development of *Haemonchus contortus* in sheep. I. The induction of arrested development. *Int. J. Parasitol.* **2**, 5–12.

Boudouy, T. (1908). Sur quelques principes constitutifs du *Sclerostomum equinum*. Présence, chez ce parasite, d'un alcaloide cristallisé éminemment hémolytique. *C. R. Acad. Sci. (Paris)* **147**, 928–930.

Buckley, R. (1958). Tropical pulmonary eosinophilia in relation to filarial infections (*Wuchereria* spp.) of animals. *Trans. Roy. Soc. Trop. Med. Hyg.* **52**, 335–336.

Cameron, T. W. M. (1927). Observations on the life history of *Ollulanus tricuspis* Leuck., the stomach worm of the cat. *J. Helminthol.* **5**, 67–80.

Cates, S. S. (1953). The prevalence of pinworm infection among first graders in Tallahassee, Florida, and vicinity. *Quart. J. Fla. Acad. Sci.* **16**, 239–242.

Chabaud, A. G., and Choquet, M. T. (1953). Nouvel essai de classification des filaires superfamilie des Filaroidea. *Ann. Parasitol.* **28**, 172–192.

Cheng, T. C. (1960). A survey of *Enterobius vermicularis* infestation among pre-school age transient children; with the description of a modified Graham swab. *J. Tenn. Acad. Sci.* **35**, 49–53.

Cheng, T. C. (1973). *Echinocephalus crassostreai* sp. nov., a larval nematode from the oyster *Crassostrea gigas* in the Orient. *J. Invert. Pathol.* **26**, 81–90.

Cheng, T. C. (1976). The natural history of anisakiasis in animals. *J. Milk Food Technol.* **39**, 32–46.

Cheng, T. C. (1982). Anisakiasis. *In* "CRC Handbook Series in Zoonoses" (J. H. Steele, ed.), Sect. C., Vol. II,

worm, *Ancylostoma caninum. Am. J. Hyg.* **33**, 39–57.

Owen, W. B. (1930). Factors that influence the development and survival of the ova of an ascarid roundworm, *Toxocara canis* (Werner, 1782) Stiles, 1905, under field conditions. *Minn. Agr. Exp. Sta. Tech. Bull.* **71**, St. Paul, Minnesota.

Oyanagi, T. (1967). Experimental studies on the visceral migrans of gastro-intestinal walls due to *Anisakis* larvae. *Jap. J. Parasitol.* **16**, 470–493. (In Japanese with English summary.)

Pampiglioni, S., and Ricciardi, M. L. (1971). The presence of *Strongyloides fülleborni* von Linstow, 1905, in man in central and east Africa. *Parasitologia* **13**, 257–269.

Passay, R. F., and Fairbairn, D. (1957). The conversion of fat to carbohydrate during embryonation of *Ascaris* eggs. *Can. J. Biochem. Physiol.* **35**, 511–525.

Poinar, G. O. Jr. (1975). "Entomogenous Nematodes. A Manual and Host List of Insect-Nematode Associations." E. J. Brill, Leiden, The Netherlands.

Rhodes, M. B., Marsh, C. L., and Kelley, G. W., Jr. (1963). Trypsin and chymotrypsin inhibitors from *Ascaris suum. Exp. Parasitol.* **13**, 266–272.

Ricci, M. (1952). Sulla diffusione delle parasitosi intestinali in un piccolo cotro siciliano. *Rend. Ist. Super. Sanita* **15**, 57–63.

Roberts, L. S., and Fairbairn, D. (1965). Metabolic studies on adult *Nippostrongylus brasiliensis* (Nematoda: Trichostrongyloidea). *J. Parasitol.* **31**, 129–138.

Rogers, W. P. (1941). Digestion in parasitic nematodes. II. The digestion of fats. III. The digestion of proteins. *J. Heminthol.* **19**, 35–58.

Rogers, W. P. (1949). Aerobic metabolism in nematode parasites of the alimentary tract. *Nature* **163**, 879–880.

Rogers, W. P. (1960). The physiology of infective processes of nematode parasites; the stimulus from the animal host. *Proc. Roy. Soc.* **B152**, 367–386.

Rogers, W. P. (1962). "The Nature of Parasitism." Academic Press, New York.

Rogers, W. P. (1969). Nitrogenous components and their metabolism: Acanthocephala and Nematoda. *In*"Chemical Zoology" (M. Florkin and B. T. Scheer, eds.), Vol. III, pp. 379–428. Academic Press, New York.

Rosen, L., Chappell, R., Wallace, G. L., and Weinstein, P. P. (1962). Eosinophilic meningoencephalitis caused by a metastrongylid lung worm of rats. *J. Am. Med. Ass.* **179**, 620–624.

Ruitenberg, E. J., and Loendersloot, J. (1971). Enzymhistochemisch ouderzoek van *Anisakis* sp. *Tijdschr. Diergeneeskd. Deel.* **96**, 247–260.

Sadun, E. H. (1948). Relation of the gonadal hormones to the natural resistance of chickens and to the growth of the nematode, *Ascaridia galli. J. Parasitol.* **34**, 18.

Sadun, E. H. (1951). Gonadal hormones in experimental *Ascaridia galli* infection in chickens. *Exp. Parasitol.* **1**, 70–82.

Saval, J. (1955). Etudes sur la constitution et le métabolisme protéiques d'*Ascaris lumbricoides* Linné, 1758. I et II. *Rev. Pathol. Comp. Hyg. Gen.* **55**, 52–121, 213–282.

Saz, H. J. (1969). Carbohydrate and energy metabolism of nematodes and acanthocephala. *In* "Chemical Zoology" (M. Florkin and B. T. Scheer, eds.), Vol. III, pp. 329–360. Academic Press, New York.

Saz, H. J. (1971). Anaerobic phosphorylation in *Ascaris* mitochondria and the effects of anthelmintics. *Comp. Biochem. Physiol.* **39B**, 627–637.

Saz, H. J., and Bueding, E. (1966). Relationships between anthelmintic effects and biochemical and physiological mechanisms. *Pharmacol. Rev.* **18**, 971–184.

Saz, H. J., and Lescure, O. L. (1966). Interrelationships between the carbohydrate and lipid metabolism of *Ascaris lumbricoides* egg and adult stages. *Comp. Biochem. Physiol.* **18**, 845–857.

Saz, H. J., and Lescure, O. L. (1969). The functions of phosphoenolpyruvate carboxykinase and malic enzyme in the anaerobic formation of succinate by *Ascaris lumbricoides. Comp. Biochem. Physiol.* **30**, 49–60.

Saz, H. J., and Weil, A. (1960). The mechanism of the formation of α-methylbutyrate from carbohydrate by *Ascaris lumbricoides* muscle. *J. Biol. Chem.* **235**, 914–918.

Schad, G. A. (1977). The role of arrested development in the regulation of nematodes populations. *In* "Regulations of Parasite Populations" (G. W. Esch, ed.). pp. 111–167. Academic Press, New York.

Schad, G. A., Munell, K. D., El Naggar, H. M. S., Page, M. R., Parish, P. K., and Stewart, T. B. (1984). Paratenesis in *Ancylostoma duodenale* suggests possible meat-borne human infection. *Trans. Roy. Soc. Trop. Med. Hyg.* **78**, 203–204.

Scholander, P. F. (1960). Oxygen transport through hemoglobin solutions. *Science* **131**, 585–590.

Schwartz, B. (1921). Hemotoxins from parasitic worms. *J. Agr. Res.* **22**, 379–432.

Scott, J. A., Cross, J. H., Jr., and Dawson, C. (1959). Egg production of *Nematospiroides dubius* in mice and rats. *Texas Rep. Biol. Med.* **17**, 610–617.

Sen H. G., Kelley, G. W., Underdahl, N. R., and Young, G. A. (1961). Transmission of swine influenza virus by lungworm migration. *J. Exp. Med.* **113**, 517–520.

Shimamura, T., and Fujii, H. (1917). Über das Askaron, einen toxischen Bestandteil der Helminthen besonders der Askariden und seine biologische Wirkung (Mitteilung I). *J. Coll. Agr. Imp. Univ. Tokyo* **3**, 189–258.

Shiroma, Y. (1964). Studies on human strongyloidiasis on Okinawa. *Ryuka Kagoshima Med. J.* **34**, 243–266.

Shope, R. E. (1941). The swine lungworm as a reservoir and intermediate host for swine influenza virus. I. The presence of swine influenza virus in healthy and susceptible pigs. II. The transmission of swine influenza virus by the swine lungworm. *J. Exp. Med.* **74**, 41–68.

Shope, R. E. (1943). The swine lungworm as a reservoir and intermediate host for swine influenza virus. III. Factors influencing transmission of the virus and the provocation of influenza. *J. Exp. Med.* **77**, 111–126; 127–138.

Shope, R. E. (1958). The swine lungworm as a reservoir and intermediate host for hog cholera virus. I. The provocation of masked hog cholera virus. *J. Exp. Med.* **107**, 609–622.

Smith, H. J., and Archibald, R. M. (1967). Moose sickness, a neurological disease of moose infected with the common cervine parasite, *Elaphostrongylus tenuis. Can. Vet. J.* **8**, 173–177.

Smith, M. H. (1969). The pigments of Nematoda and Acanthocephala. *In* "Chemical Zoology" (M. Florkin and B. T.

Scheer, eds.), Vol. III, pp. 501–520. Academic Press, New York.

Smith, M. H., and Lee, D. L. (1963). Metabolism of haemoglobin and haematin compounds in *Ascaris lumbricoides. Proc. Roy. Soc.* **B157**, 234–257.

Soulsby, E. J. L. (1962). Antigen-antibody reactions in helminth infections. *Adv. Immunol.* **2**, 265–308.

Spindler, L. A. (1934). The incidence of worm parasites in swine in the southern United States. *Proc. Helminthol. Soc. Wash.* **1**, 40–42.

Sprent, J. F. A. (1949). On the toxic and allergic manifestions produced by the tissues and fluids of *Ascaris*. I. Effect of different tissues. *J. Infec. Dis.* **84**, 221–229.

Sprent, J. F. A. (1952). On the migratory behavior of the larvae of various *Ascaris* species in white mice. I. Distribution of larvae in tissues. *J. Infec. Dis.* **90**, 165–176. (Also see 1953, *Ibid.* **92**, 114–117).

Sprent, J. F. A. (1956). The life history and development of *Toxocara cati* (Schrank, 1788) in the domestic cat. *Parasitology* **46**, 54–78.

Sprent, J. F. A. (1958). Observations on the development of *Toxocara canis* (Werner, 1782) in the dog. *Parasitology* **48**, 184–209.

Sprent, J. F. A. (1971). Speciation and development in the genus *Lagocheilascaris. Parasitology* **62**, 71–112.

Stoll, N. (1940). *In vitro* conditions favoring ecdysis at the end of the first parasitic stages of *Haemonchus contortus* (Nematoda). *Growth* **4**, 383–405.

Stoll, N. (1947). This wormy world. *J. Parasitol.* **33**, 1–18.

Taylor, A. E. R., and Baker, J. R. (eds.) (1978). "Methods of Culturing Parasites *in vitro*." Academic Press, London.

Thomas, L. J. (1937). Life cycle of *Raphidascaris canadensis* Smedley, 1933, a nematode from the pike, *Esox lucius. J. Parasitol.* **23**, 572.

Todd, A. C. (1949). Thyroid condition of chickens and development of parasitic nematodes. *J. Parasitol.* **35**, 255–260.

Todd, A. C., and Hollingsworth, K. P. (1952). Host sex as a factor in development of *Ascaridia galli. Exp. Parasitol.* **1**, 303–304.

Townson, H. (1970). The effect of infection with *Brugia pahangi* on the flight of *Aedes aegypti. Ann. Trop. Med. Parasitol.* **64**, 411–420.

Turner, J. H. (1959). Experimental strongyloidiasis in sheep and goats. II. Multiple infections: development of acquired resistance. *J. Parasitol.* **45**, 76–86.

Velasquez, C., and Cabrera, B. C. (1968). *Ancylostoma ceylanicum* (Looss), in a Filipino woman. *J. Parasitol.* **54**, 430–431.

van Grembergen, G. (1954). Haemoglobin in *Heterakis gallinae. Nature* **174**, 35.

Van Thiel, P. H., Kuipers, F. C., and Roskam, R. T. (1960). A nematode parasitic to herring, causing acute abdominal syndromes in man. *Trop. Geogr. Med.* **12**, 97–113.

Villela, G. G., and Teixeira, J. C. (1937). Blood chemistry in hookworm anemia. *J. Lab. Clin. Med.* **22**, 567–572.

von Brand, T. (1938). The nature of the metabolic activities of intestinal helminths in their natural habitat: aerobiosis or anaerobiosis? *Biodynamica* **42**, 1–13.

Wang, C. F., Lin, C. L., and Chen, W. H. (1958). The mechanism of microfilarial periodicity. *Chinese Med. J.* **77**, 129–135.

Wanson, M. (1950). Contribution a l'étude de l'onchocercose Africaine humain (problèmes de prophylaxis a Léopoldville). *Ann. Soc. Belge Med. Trop.* **30**, 667–863.

Ward, C. W., and Fairbairn, D. (1970). Enzymes of beta-oxidation and the tricaroboxylic acid cycle in adult *Hymenolepis diminuta* (Cestoda) and *Ascaris lumbricoides* (Nematoda). *J. Parasitol.* **56**, 1009–1012.

Weatherly, N. F., Hansen, M. F., and Moser, H. C. (1963). *In vitro* uptake of C^{14}-labeled alanine and glucose by *Ascaridia galli* (Nematoda) of chickens. *Exp. Parasitol.* **14**, 37–48.

Weiner, D. (1960). Larva migrans. *Vet. Med.* **55**, 38–50.

Weir, W. C., Bahler, T. L., Pope, A. L., Phillips, P. H., Herrick, C. A., and Bohstedt, G. (1948). The effect of hemopoietic dietary factors on the resistance of lambs to parasitism with the stomach worm, *Haemonchus contortus. J. Anim. Sci.* **7**, 466–474.

Wells, H. S. (1931). Observations on the blood sucking activities of the hookworm, *Ancylostoma caninum. J. Parasitol.* **17**, 167–182.

Woodhouse, D. F. (1975). Tropical eye diseases in Britain. *Practitioner* **214**, 646–653.

Wright, G. G., and Oliver-González, J. (1943). Electrophoretic studies on antibodies to *Trichinella spiralis* in the rabbit. *J. Infect. Dis.* **72**, 242–245.

Yoeli, M., Most, H., Berman, H. H., and Tesse, B. (1963). I. The problem of strongyloidiasis among the mentally retarded in institutions. *Trans. Roy. Soc. Trop. Med. Hyg.* **57**, 336–345.

Yoshimura, H. (1966). Migration of *Anisakis*-like larvae into the human alimentary tracts, with special reference to its clinical pathology. *Nippon Iji Shinpo*, **2204**, 10–16. (In Japanese.)

Zaiman, H., Leedy, W., and Howard, P. (1952). The incidence of *Enterobius vermicularis* in a metropolitan San Francisco pre-school nursery population. *J. Parasitol.* **38**, 184–185.

CLASSIFICATION OF SECERNENTEA (THE SECERNENTEAN NEMATODES)*

PHYLUM NEMATA (see page 498)
CLASS SECERNENTEA

Amphids usually open to exterior through pores located dorsolaterally on lateral lips or anterior extremity (in some species the amphidial apertures are oval, cleftlike, slitlike, or located postlabially); cephalic sensory organs are situated on lips and are porelike or papilliform, generally 16 in number arranged in two circles (a circumoral circle of 6 and an outer circle of 10), may be reduced in some species; caudal phasmids present; hypodermis uninucleate or multinucleate; cuticle from two to four layers, almost always transversely striated, laterally modified into a "wing" area marked by longitudinal striae or ridges, generally raised slightly above body contour; lateral alae

---

*Essentially after Maggenti (1981). Diagnoses of taxa comprising of free-living and/or plant parasitic species not presented.

may extend out a distance equal to body diameter; esophagus of most species have three esophageal glands, one dorsal (opening in anterior half of body) and two subventral (opening in posterior half of body); excretory system empties ventromedially through cuticularized duct on one or both sides of body; somatic setae or papillae absent on females; caudal papillae may occur on males; male preanal supplements paired and often elaborate; some males with medioventral preanal supplementary papillae; males commonly with caudal alae (known as copulatory bursa).

Subclass Rhabditida

Esophagus of larvae divided into corpus, isthmus, and valved post-corporal bulb; lumen of esophageal bulb expanded into trilobed reservoir lined with cuticle; buccal cavity (stoma) without movable armature and composed of two parts (cheilostome and esophastome), each possibly subdivided into two or more sections; males generally with well developed bursae supported by cuticular rays or papillae.

ORDER RHABDITIA

Number of lips varies from six to none (6, 3, 2, 0); buccal cavity generally tubular but may be separated into five or more sections; esophagus divided into corpus, isthmus, and bulb; terminal excretory duct lined with cuticle and has paired, lateral collecting tubules running posteriorly; females with one or two ovaries; intestinal cells uni-, bi-, or tetranucleate; caudal alae (copulatory bursa), if present, contain papillae rather than supporting rays; parasites of invertebrates and vertebrates.

Suborder Rhabditina

Buccal cavity usually cylindrical, without distinct separation, generally two or more times as long as wide; lips usually distinct, with cephalic sensory papillae and porelike amphids; esophagus divided into corpus (procorpus and metacorpus) and postcorpus (isthmus and valved bulb); females with one or two ovaries; males generally with paired spicules and gubernaculum; caudal alae (copulatory bursa) common (absent in some families); parasites of invertebrates and vertebrates.

Superfamily Rhabditoidea

Well-developed cylindrical buccal cavity (stoma); lips vary from two to six; esophagus, at least in larvae, include muscular posterior bulb with rhabditoid valve; caudal alae of males supported by five to nine papilloid supplements; parasites of invertebrates and vertebrates. (Genera mentioned in text: Family Rhabditidae*—*Rhabditis, Neoaplectana.* Family Angiostomatidae—*Angiostoma,* Family Strongyloididae—*Strongyloides.*)

Superfamily Alloionematoidea
Superfamily Bunonematoidea

Suborder Cephalobina
Superfamily Cephaloboidea
Superfamily Panagrolaimoidea
Superfamily Robertioidea
Superfamily Chambersielloidea
Superfamily Elaphonematoidea

ORDER STRONGYLIDA

Labial region consists of three or six lips or may be replaced by corona radiata; stoma well developed or rudimentary (never collapsed and unobtrusive); esophagus of larvae typically rhabditiform (corpus, isthmus, bulb); esophageal bulb contains typical trilobed rhabdiform valve; esophagus of adults cylindrical to clavate; excretory system includes paired lateral canals and paired subventral glands; females with one or two ovaries and heavily muscular uterus; males with muscular copulatory bursa; with ventroventral, lateroventral, externolateral, mediolateral, posterolateral, externodorsal, dorsoventral, and terminodorsal paired genital papillae; males with paired, equal spicules; adults parasitic in vertebrates; early larval stages bacteria feeders or parasites of annelids and molluscs.

Superfamily Strongyloidea

Stoma well developed and large, variable in shape, hexagonal (in cross section), globular, cylindrical, or infundibuliform, without mandibles; oral opening hexagonal or surrounded by six small lips or corona radiata; no teeth or cutting plates; esophagus of hatched larvae rhabditiform; esophagus of adult cylindrical or clavate; males with well-developed copulatory bursa, rays not fused; adults parasitic in fish (rarely), reptiles, birds, and mammals; early larval stages free living. (Genera mentioned in text: Family Strongylidae*—*Strongylus, Ransomus, Cyathostomum, Globocephalus, Basicola.*)

Superfamily Diaphanocephaloidea

Stoma modified into two massive lateral jaws; corona radiata and lips absent; labial (cephalic) sensory organs (sensilla) of outer circle separate; copulatory bursa of males bell-like or trilobed; adults are intestinal parasites of reptiles and rarely of amphibians; early larval stages free living. (Genera mentioned in text: Family Diaphanocephalidae*—*Diaphanocephalus, Kalicephalus.*)

Superfamily Ancylostomatoidea

Stoma thick-walled, globose, armed or unarmed anteriorly with teeth or cutting plates; without lips or corona radiata; copulatory bursae of males with greatly reduced branches; adults parasitic in intestine of mammals; L_1 and L_2 free living; commonly known as hookworms. (Genera mentioned in text: Family Ancylostomatidae*—*Ancylostoma, Cyclodontostomum, Galoncus, Gaigeria, Bunostomum.* Family Uncinariidae—*Unicinaria, Arthrostoma, Monodontus, Necator.*)

Superfamily Trichostrongyloidea

Oral aperture surrounded by three or six inconspicuous lips (absent in some); corona radiata absent; stoma reduced or collapsed; cuticle of cephalic region commonly inflated; body cuticle thick, often with several longitudinal ridges; esophagus of hatched larva rhabditoid but bulb may be without valve; copulatory bursa of males may be atrophied but lateral rays well developed; adults parasitic in amphibians, reptiles, birds, and mammals; early larval stages free

---

*For diagnoses of families subordinate to the Nemata, see Maggenti (1982).

living. (Genera mentioned in text: Family Trichostrongylidae*—*Trichostrongylus, Cooperia, Obeliscoides, Ornithostrongylus, Haemonchus, Nematodirus.* Family Ollulanidae—*Ollulanus.* Family Dictyocaulidae—*Dictyocaulus.* Family Heligmosomatidae—*Heligmosomum, Nematospiroides, Nippostrongylus.*

Superfamily Metastrongyloidea

Stoma capsule reduced or absent; oral opening may be surrounded by six well-developed or rudimentary lips; submedian papillae of external labial sensory organs (sensilla) not fused; body cuticle not adorned with longitudinal ridges; tail of females asymmetrical, without points or mucrons; rays of copulatory bursa of males reduced and somewhat fused; esophagus of L_1 rhabditoid but without valve in bulb; adults parasitic in mammals. (Genera mentioned in text: Family Metastrongylidae*—*Metastrongylus, Neometastrongylus, Angiostrongylus, Protostrongylus, Pneumostrongylus.*)

ORDER ASCARIDIDA

Oral opening generally surrounded by three or six lips (absent in some species); outer circle of labial papillae usually of eight sensilla (in some species submedians are fused, and hence only four sensilla are observable); paired porelike amphids present; esophagus of some species with short swollen region in stomatal region, followed by cylindrical to club-shaped region, often ending in terminal bulb with typical rhabditoid three-lobed valve; in a few exceptions there are appendices (caeca) extending from posterior region of esophagus; excretory system with lateral collecting tubules, in some species extending posteriorly and anteriorly (H-shaped); males usually with two spicules (none or one in others); with or without gubernaculum; females usually with two ovaries (some have multiple ovaries); adults parasitic in vertebrates.

Superfamily Ascaridoidea

Bodies 1–40 cm long; cuticle thick in larger species, superficially annulated; terminal oral opening usually surrounded by three well-developed lips (two subventral, one dorsal); porelike amphids on subventral lips; stoma poorly developed (collapsed); esophagus cylindrical to clavate; appendage (caecum) may extend from posterior portion of esophagus over anterior portion of intestine; second caecum may be present, extending forward past base of esophagus; females usually with paired ovaries in multiples of three, four, or six; males with two spicules; small gubernaculum present in few species; adults parasitic in vertebrates; life cycle direct or indirect. (Genera mentioned in text: Family Ascarididae*—*Ascaris, Parascaris.* Family Toxocaridae—*Neoascaris, Toxocara, Dujardinascaris, Toxascaris.* Family Anisakidae—*Belanisakis, Phocanema, Porrocaecum, Paradujardinia, Pseudoterranova, Cloeoascaris, Phocascaris, Contracaecum, Raphidascaris, Anisakis.*)

Superfamily Cosmocercoidea

With 3 or 6 lips; ventrolateral papillae present; stoma weakly developed; esophagus subdivided into corpus, isthmus, and posterior bulb; posterior bulb includes valves; esophageal glands uninucleate; no esophageal or intestinal caeca; males may have precloacal sucker; copulatory spicules of males of equal length, oviparous or ovoviviparous; parasites of molluscs, amphibians, reptiles, and mammals. (Genus mentioned in text: Family Cosmocercidae—*Cosmocerca.*)

Superfamily Oxyuroidea

Lips greatly reduced or absent; cephalic sensilla in whorl of eight or four; ventrolateral sensilla absent; stoma vestibular; esophagus variable but posterior bulb always valved; intestinal caeca absent; males may have precloacal suckers; copulatory spicules may be greatly reduced; adults usually parasites of amphibians, reptiles, and mammals. (Genera mentioned in text: Family Oxyuridae*—*Oxyuris, Enterobius, Skrjabinema, Passalurus, Dermatoxys, Probstmayria, Pharyngodon, Thelandros, Atractis, Labiduris.* Family Thelastomatidae—*Leidynema, Hammerschmidtiella, Blatticola, Thelastoma, Pseudonymus, Aorurus.*)

Superfamily Heterakoidea

Lips small and well developed, with eight cephalic sensilla paired into a circlet of four; esophagus of clavate corpus, short isthmus, and valved posterior bulb; subventral esophageal glands binucleate; males with precloacal sucker, surrounded by cuticular ring; paired copulatory spicules present; parasites of amphibians, reptiles, birds, and mammals. (Genera mentioned in text: Family Heterakidae*—*Heterakis.* Family Ascaridiidae—*Ascaridia*).

Superfamily Subuluroidea

Lips small or absent; with external circle of cephalic sensilla consisting of four simple or weakly doubled papillae; usually with three large teeth at base of cheilostome; corpus clavate; isthmus short; posterior bulb usually with valves; subventral esophageal glands binucleate; males with precloacal sucker without cuticular ring; copulatory spicules paired in males; parasites of birds and mammals. (Genera mentioned in text: Family Subuluridae*—*Subulura.* Family Maupasinidae—*Cruzia, Pseudocruzia.*)

Superfamily Seuratoidea

Lip region greatly reduced or absent; ventrolateral cephalic sensilla may be present; submedian sensilla double, rarely single; stoma variable, large in some species, small and weakly developed in others, with teeth in some species; esophagus generally simple, cylindrical or somewhat clavate, without posterior bulb; intestinal caeca present in few species; precloacal sucker may be present on posterior body of males; spicules of equal length and generally accompanied by well-developed gubernaculum; bursa, if present, generally very narrow; parasites of fish, amphibians, reptiles, birds, and mammals. (Genus mentioned in text: Family Seuratidae*—*Seuratum.*)

Suborder Dioctophymatina

Cuticle without endocuticular layer, basally only of oblique fiber layer; external cuticle with annulation but may be ornamented with hooklike spines; esophagus cylindrical but corpus and postcorpus distinguishable internally; postcorpus with three ramifying glands which open to exterior anterior to nerve ring; hypodermis multinucleate; females and males with only one gonad; male tail as expanded and thickened copulatory bursa with bordering papillae; males with single elongate spicule; parasites of amphibians, reptiles, birds, and mammals.

Superfamily Dioctophymatoidea

With characteristics of the suborder. (Genera mentioned in text: Family Diotophymatidae*—*Dioctophyme, Eustrongylides.* Family Soboliphymatidae—*Soboliphyme.*)

Subclass Spiruria

Oral opening surrounded by six lobes (known as pseudolabia) or lips modified into two lateral labia; ventrolateral sensory papillae absent; stoma spacious and globose or long and cylindrical, not distinguishable in certain species (esophagus extends to anterior extremity in these); esophagus divided into two cylindrical parts (with anterior part narrow and posterior part swollen) or esophagus cylindrical and clavate; esophagus not divided into corpus, isthmus, and postcorpus (bulb); postcorpus without valve (even in L_1); excretory system as inverted tuning fork, with collecting tubules in lateral hypodermal chords; vagina often greatly elongated and tortuous; two copulatory spicules of males often unequal in length; copulatory bursa may be present, not supported by muscle-associated papillae.

ORDER SPIRURIDA

Frequently with two lateral lips or pseudolabia (some species with four or more lips, rare species without lips); oral aperture variable in shape, encircled by teeth; amphids laterally situated on anterior extremity; stoma varies from cylindrical and elongate to rudimentary; esophagus generally divided into narrow anterior portion and expanded postcorpus enclosing multinucleate glands; hatched larvae generally provided with cephalic hook and porelike phasmids on tail; parasites of annelids, arthropods, molluscs, and terrestrial and aquatic vertebrates.

Superfamily Spiruroidea

Four lips (pseudolabia) on some species, two in others, lips rarely absent; external circle of cephalic papillae consists of four or eight sensilla (double papillae in some species); cephalic and cervical region ornamented with cordons, collarettes, or rings; oral opening round, hexagonal, or dorsolaterally extended; stoma well developed, may be provided with teeth; vulva generally near middle of body (rarely in posterior region or near esophagus); incomplete longitudinal ridges often present anterior to cloacal aperture on males; parasites of arthropods and vertebrates. (Genera mentioned in text: Family Spiruridae*—*Spirura, Mastophorus, Habronema, Draschia, Spiroxys, Gongylonema.* Family Tetrameridae—*Tetrameres.*)

Superfamily Physalopteroidea

Large paired lateral lips (pseudolabia) usually with teeth; inner whorl of circumoral sensilla reduced or absent, four sensilla of external circle fused; cuticle at anterior body terminal often flexed inward, partially covering pseudolabia; cephalic cuticle of some species bulbous and may bear spines; cordons, collarettes, and rings absent; stoma reduced; caudal papillae on males sometimes pedunculated; caudal alae (or bursa) well developed, often merging ventrally (when caudal papillae not pedunculate, caudal alae absent); female genitalia often number four or more; vulva pre- or postequatorial; parasites of fish, amphibians, reptiles, birds, and mammals. (Genera mentioned in text: Family Physalopteridae*—*Physaloptera, Proleptus, Skjabinoptera, Abbreviata.* Family Gnathostomatidae—*Gnathostoma, Tanqua, Echinocephalus.*)

Superfamily Filarioidea

Oral aperture circular or oval, usually surrounded by eight sensilla of external circle (internal circle absent or consisting of two or four papillae); stoma small and rudimentary esophagus with multincleate glands; corpus and postcorpus not distinct; vulva usually in anterior portion of body; copulatory spicules of males equal or unequal; caudal alae present or absent; no gubernaculum; parasites of amphibians, reptiles, birds, and mammals. (Genera mentioned in text: Family Filariidae*—*Filaria, Parafilaria, Wuchereria, Brugia, Onchocerca, Litomosoides, Loa, Dirofilaria, Mansonella, Dipetalonema.* Family Setariidae—*Setaria, Stephanofilaria.*)

Superfamily Drilonematoidea

Stoma greatly reduced, may be surrounded by rudimentary lips (generally no lips visible) esophagus of adults consists of elongate corpus, slight or no isthmus, and pyriform glandular region, never valved; esophagus short and clavate in some species; females with single elongate ovary; males with paired spicules or none; gubernaculum present or absent; phasmids greatly enlarged, often occupying most of tail width; parasites of annelids and molluscs.

ORDER CAMALLANIDA

Lips absent but cephalic sensilla generally elevated above head contour; stoma varies from massive globe to small vestibule; esophageal glands simple and uninucleate; phasmids large and pocketlike; adults parasitic in vertebrates; utilize copepods as intermediate host.

Superfamily Camallanoidea

Stoma well developed, variable in size, globose or transversely rectangular, supported internally by numerous longitudinal or oblique ridges; cephalic sensilla surrounding stoma, those comprising internal circle minute, external circle comprised of eight partially fused papillae; esophagus short and usually clavate; adults parasitic in fish, amphibians, and reptiles; utilize copepods as intermediate hosts. (Genus mentioned in text: Family Camallanidae—*Camallanus.*)

Superfamily Dracunculoidea

Stoma commonly reduced to small vestibule; full complement of sensilla surrounding oral opening, with internal circle comprised of six well-developed sensilla and external circle of eight separate and well-developed sensilla; vulva in midbody region, atrophied in mature females; posterior intestine atrophied in females; males without well-developed caudal alae, small and postcloacal if present; adults are tissue parasites of fish, reptiles, and mammals. (Genera mentioned in text: Family Dracunculidae*—*Dracunculus.* Family Philometridae—*Philometra.* Family Micropleuridae—*Micropleura.*)

Subclass Diplogasteria

Small to medium-sized worms, seldom over 3 or 4 mm

long; cuticle with annulations which may be crossed by longitudinal striae; labial region may not have well-developed lips but hexaradiate symmetry usually evident; full complement of 16 cephalic sensilla often present, especially on males (in some species inner circle of six may be absent); amphids dorsolateral on lateral lips, generally porelike, although as ovals, clefts, and slits in some species; stoma often armed with large teeth, axial spear, or opposable fossores (some without movable armature), variable in shape; muscular corpus divided into subcylindrical procorpus and muscular (almost always) valved metacorpus followed by isthmus and glandular postcorpus; postcorporal bulb without valve; females with one or two ovaries; males with one testis; paired spicules generally present; gubernaculum may or may not be present; males may have caudal alae; parasitic species associated with few vertebrates but mostly insects and plants.

ORDER DIPLOGASTERIDA
Superfamily Diplogasteroidea
Superfamily Cylindrocorporoidea

Six well-developed lips with cephalic sensilla; small amphids on lateral lips; stoma elongate, one-fourth or more esophageal length; esophagus with corpus enlarged into distinct cylindroid muscular complex; glandular postcorpus slightly swollen at esophagointestinal junction; caudal alae, when present, rudimentary; spicules long and thin; adults intestinal parasites of amphibians, reptiles, and few mammals. (Genus mentioned in text: Family Cylindrocorpidae—*Longibucca*.)

ORDER APHELENCHIDA

Labial cap often set off by constriction; hollow axial spear usually not strongly developed, with or without basal knobs; esophagus usually with large valved metacorporal bulb, often squarish; all esophageal glands empty into metacorpus; females with single anteriorly directed ovary; postuterine sac may or may not be present (serves as spermatheca if present); vulva posteriorly situated; males with or without bursa, when present, genital papillae form rays; males with two or more pairs of caudal papillae; spicules may be slender, slightly arcuate, or thornlike; gubernaculum, if present, forked; parasitic species associated with insects (and a few other invertebrates) and higher plants.

Superfamily Aphelenchoidea
Superfamily Aphelenchoidoidea

Labial cap evident; lip region often expanded and set off from cervical region by incisure; individual lips fused; stylet usually slight, knobs absent or well developed; esophagus immediately posterior to metacorpus; elongate esophageal glands overlap anterior intestine

dorsally; spicules robust, thorn shaped; caudal papillae only slightly raised above body surface; bursa absent; adults parasites of invertebrates, especially insects.

ORDER TYLENCHIDA

Lip region hexaradiate; amphids porelike on lips (oval or cleftlike in some species); internal circle of sensilla may be absent, ten sensilla on outer circle; protrusible spear and three basal knobs usually present (absent in some parasites of insects); esophagus divided into corpus, isthmus, and posterior glandular bulb; corpus divided into procorpus and commonly valved metacorpus; excretory system asymmetrical, with single lateral collecting tube extending the length of the body; females with one or two ovaries; males with one pair of caudal papillae (phasmids); caudal alae present or absent; spicules paired; may be fused at tip; gubernaculum present or absent; adults parasitic in insects and higher plants.

Suborder Tylenchina
(Includes some parasites of plants)
Superfamily Tylenchoidea
(Includes some parasites of plants)
Superfamily Cariconematoidea

Lips in form of disc with four submedian lobes; esophageal corpus divided into procorpus and metacorpus (may be fused); glandular postcorpus cylindrical with slight terminal swelling to pyriform, clearly offset from intestine; females with one anteriorly directed ovary; females from vermiform to obese saccate; adults ectoparasitic on insects.

Suborder Sphaerulariina (or Superfamily Sphaerularioidea)

Lack valved median bulb; with three distinct forms (two free living, one parasitic); free-living forms occur in habitat of young hosts, parasitic form in host's hemocoel; free-living phase of females generally with a stylet and esophagus without valve in metacorpus; orifices of esophageal glands marked by prominent ampullae near ducts; esophageal glands long and overlap anterior intestine; parasitic females grossly enlarged and degenerate to become reproductive sac (or uterus prolapses), and gonadal development occurs outside of body; in such females, stylet is evident but esophagus and intestine degenerate; males all free living; females parasitic in invertebrates, especially insects. (Genus mentioned in text: Family Sphaerulariidae—*Sphaerularia*.)

16

INTRODUCTION TO THE PARASITIC ARTHROPODS

Metastigmata: The Ticks

Members of the phylum Arthropoda represent the largest number of known animals. There are at least 760,000 species of arthropods in existence. According to the most acceptable interpretation at this time, the phylum is divided into four subphyla: Trilobitomorpha, Chelicerata, Crustacea, and Uniramia. The Trilobitomorpha is represented today by fossils only. The Chelicerata includes the class Merostomata (horseshoe crabs), the class Arachnida (scorpions, spiders, harvetsmen, ticks, and mites) (Chapters 16, 17), and the class Pycnogonida (sea spiders) (p. 762). The Crustacea includes the crustaceans (p. 743) and the Uniramia includes the class Insecta (insects) (Chapters 18, 19, 20), the class Chilopoda (centipedes), the class Diplopoda (millipedes), the class Symphyla (symphylans), and the class Pauropoda (pauropodans).

Most arthropods are free living and are found in an array of aquatic (freshwater and marine) and terrestrial habitats. However, some members of the subphylum Crustacea (i.e., Copepoda, Isopoda, Cirripedia, Amphipoda) and the classes Arachnida and Insecta are parasitic. In addition, the larvae of most species of the class Pycnogonida are parasitic.

Although many of these arthropod parasites are of little medical and economic importance, they are of considerable interest to biologists, specifically parasitologists, from the biologic standpoint. On the other hand, some of these parasitic arthropods, such as certain ticks, mites, and insects, are of considerable medical and veterinary importance not only because they cause direct injury to their hosts but also because many serve as vectors for various pathogens, including numerous microorganisms and viruses. As a result of the concentrated interest of certain parasitologists on these disease-transmitting arthropods, a subfield of parasitology known as **medical and veterinary entomology** has become established. Also, with the increasing interest in utilizing pathogenic parasites of insects as biologic control agents and in understanding disease processes in insects, a relatively new subdiscipline known as **insect pathology** has emerged.

Although members of the various classes of arthropods portray a wide range of characteristics, all share a few in common. These triploblastic, bilaterally symmetrical animals all possess segmented bodies bearing jointed appendages. The body surface is always covered by a rigid or semirigid chitinous exoskeleton. These animals possess complete digestive tracts (incomplete in some cirripeds, such as the Rhizocephala) with mouthparts adapted for biting,

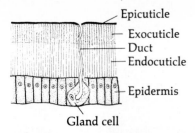

Fig. 16.1. **Arthropod exoskeleton.** Cross-section showing various layers and associated cells and glands.

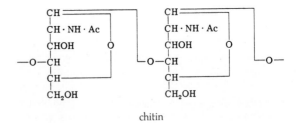

chitin

Fig. 16.2.

chewing, or sucking, depending on the species. An open circulatory system, with or without a dorsally situated heart, pumps the blood (or hemolymph) via arteries into the various organs and body tissues. Blood is returned to the heart through body spaces known as **hemocoels**. In addition, respiratory, excretory, and nervous systems are present.

Arthropods are usually dioecious and exhibit a marked degree of sexual dimorphism. The specific anatomic characteristics of each group considered are given in subsequent sections.

EXOSKELETON

The exoskeleton of arthropods is of considerable interest, not only because it serves as the protective covering but also because the physiology of arthropodan growth is greatly influenced by it. Furthermore, the mobility of arthropods, as well as their reactions to exogenous insults, are influenced by the body surface. Morphologically, the cuticle is comprised of three (sometimes two) layers (Fig. 16.1). These are an outer, extremely thin refractile layer known as the **epicuticle**, which is usually less than 1 μm thick; the middle layer, or **exocuticle**, which is relatively thick and pigmented in most insects; and the innermost **endocuticle**, which is elastic and nonpigmented, and is the thickest of the three cuticular layers. The three cuticular layers are noncellular and are secreted by the cellular **hypodermis** (sometimes referred to as the epidermis) lying directly underneath the cuticle.

The cuticle may be smooth, notched, pitted, ridged, or spinous, depending on the species and the location on the body of the animal. When seen in cross-section, a series of parallel perpendicular lines is found running through the exo- and endocuticles. These lines, known as **pore canals**, have been interpreted to be cytoplasmic processes of the hypodermal cells projecting into the cuticle. Others have suggested that although these canals are of cytoplasmic origin, in time they are transformed into cuticular material. It is possible that both types of pore canals exist. How-

ever, in most cases the canals remain vital cytoplasmic processes.

In addition to the perpendicularly oriented pore canals, a horizontally oriented lamellar system is present in the endocuticle. These lamellae are thicker toward the basal region and originate in the shifting of the hypodermal cells as successive layers of the cuticle are deposited.

The cuticle includes chitin, the chemical composition of which is of a nitrogenous polysaccharidal nature. The empirical formula of chitin is $(C_8H_{13}O_5N)_x$. The molecular configuration of chitin (Fig. 16.2), which is in the form of linked acetylated glucosamine residues, is known from the work of Bergmann and his associates.

Chitin is insoluble in water, dilute acids, bases, alcohol, ether, and other organic solvents. It is soluble, however, in concentrated mineral acids. In addition to chitin, the cuticle of arthropods includes proteins, pigments, salts (for example, $CaCO_3$), and cuticulin; the latter is a complex molecule containing fatty acids and cholesterol. Space does not permit an extensive review of our knowledge of arthropodan integument, including the cuticle, which is a field of specialization in itself. Interested individuals are referred to the review by Locke (1974).

CHEMICAL COMPOSITION

Von Kemnitz (1916) has demonstrated that 43% of the dry weight of the larva of the fly *Gasterophilus intestinalis* is composed of protein. Levenbook (1950), in analyzing the hemolymph of the same larva, found the amino acid components to be glycine, alanine, valine, serine, leucine, isoleucine, phenylalanine, tryosine, glutamic acid, lysine, aspartic acid, arginine, histidine, and proline. There is no doubt that by employing more recent methodology, all of the 20 naturally occurring amino acids will be found present in arthropod hemolymph. Protein accounts for between 39 and 60% of the dry weight of most parasitic insects.

Carbohydrates, primarily glycogen, have been found in various parasitic arthropods. However, there is some doubt whether the glycogen found in these forms is identical to that found in vertebrates because the optical rotation of such glycogens is somewhat different from that of mammalian glycogen. The amount of glycogen present has been determined for various larval flies and ranges between 1.1 and 9.4% of the fresh weight.

Lipids are also present in parasitic arthropods. In *Gasterophilus intestinalis* larvae, lipids constitute 5.2% of the fresh substance and 16.2% of the dry substance. Reinhard and von Brand (1944) have demonstrated that lipids form 26.6% of the dry substance of the rhizocephalan cirriped *Peltogaster paguri*.

Relative to inorganic materials, Levenbook (1950) has demonstrated that the body fluid of *Gasterophilus intestinalis* larvae includes K, Na, Mg, Ca, Zn, Cu, Cl, S, and P. Undoubtedly, similar inorganic materials exist in other parasitic arthropods. Beaumont (1949) and others have found iron in fly larvae. The exact function of the iron is not known, although it is suspected that it serves as an oxygen conductor, as does the iron in the hemoglobin molecule. For a review of what is known about the chemical composition of parasitic arthropods, see von Brand (1973).

THE SYMBIOTIC ARTHROPODS

As stated at the beginning of this chapter, the phylum Arthropoda includes four subphyla. As indicated in the classification lists at the end of Chapters 16 through 21, the subphyla Chelicerata, Crustacea, and Uniramia include parasitic species. Most arthropodan parasites are ectoparasites, but endoparasitic species are known. Discussed in this chapter are the members of the class Arachnida commonly known as ticks. All of these belong to the suborder Metastigmata (or Ixodides).

THE ARACHNIDA

The arthropod class Arachnida includes orders composed of such familiar forms as scorpions, spiders, pseudoscorpions, daddy longlegs, ticks, and mites. The ticks and mites comprise the order Acarina. All these animals are characterized by mouthparts consisting of pedipalps and chelicerae and by the absence of antennae and mandibles. Many are predaceous and

are found in terrestrial and aquatic (both fresh- and saltwater) habitats. However, all ticks and a number of the mites are parasitic, mostly living on animals, both vertebrates and invertebrates. Occasionally they are found within their hosts. A number of species of mites are also parasitic on plants.

The interest of parasitologists in ticks and mites stems from three avenues of investigation—that concerned with the parasitic habits of acarinas, that concerned with their role in the transmission of pathogens (that is, their roles as vectors), and that concerned with their role as intermediate hosts of certain helminths (for example, anoplocephalid cestodes, p. 407). To illustrate the first point, many dog owners are familiar with the skin mites that cause mange, and certain unfortunate individuals may have had first-hand experience with the human-infecting itch mites. The role of ticks as vectors of rickettsiae, spirochaetes, and other pathogenic microorganisms is well known. One of the most feared tick-borne rickettsial diseases in the United States is Rocky Mountain spotted fever, which is transmitted primarily from rabbits and rodents to humans by *Dermacentor andersoni* and related species. A list of tick-borne diseases and their causative agents is given in Table 16.1.

Ticks are characterized by a leathery integument, their larger size (when compared with mites), and a more prominent **capitulum** (also referred to as a **gnathosoma**)—a structure formed from the fusion of the mouthparts and the basis capitulum—and in many species by a piercing hypostome with recurved teeth, lateral teeth on the movable digits of the chelicerae, and other characteristics (see classification list at the end of this chapter).

The body of ticks is segmented, but the segments are not readily visible. The body is more conspicuously divided into two regions: (1) the capitulum, which is the small portion that projects anteriad or anteroventrad and bears the mouthparts (Fig. 16.3); and (2) the body proper. The capitulum is not a true head, although it is commonly referred to as such. It consists of the mouthparts and a basal chitinous segment known as the **basis capitulum**. The ringlike basis capitulum connects the anterior portions of the capitulum to the body proper. The mouthparts, which are located on the capitulum, include three types of structures (Fig. 16.3):

1. The elongate **hypostome** is usually toothed and is medially located, ventrad to the mouth. Its free end projects anteriorly.

2. A pair of **chelicerae** are located on the dorsolateral surfaces of the hypostome, on each side of the mouth. Each chelicera is encased within a sheath that is directly connected with the basis capitulum. Each chelicera is typically divided into three segments, although the divisions are obscure. The free terminal of each chelicera is forked (chelate), giving rise to a

Table 16.1. Major Tick-borne Diseases, Their Etiologic Agents, and Distribution

Disease	Etiologic Agent	Tick Vector	Endemic Area
Texas cattle fever (bovine piroplasmosis, babesiosis, red water fever)	*Babesia bigemina*	*Boophilus annulatus* *B. microplus*	United States southern Florida, Mexico, Central America, South America, Orient, Australia, parts of Africa
		B. decoloratus	Africa
		Rhipicephalus spp.	Africa
		Haemaphysalis punctata	Europe
Equine piroplasmosis	*Babesia caballi*	*Dermacentor* (3 spp.) *Hyalomma* (4 spp.) *Rhipicephalus* (2 spp.)	Africa, USSR, Siberia
	Babesia equi	*Dermacentor* (2 spp.) *Hyalomma* (4 spp.) *Rhipicephalus* (3 spp.)	Eastern USSR, Italy, Africa, India, Brazil
Canine babesiosis	*Babesia canis*	*Rhipicephalus sanguineus*	Many parts of the world
		Hyalomma marginatum	USSR
		Haemaphysalis leachii	South Africa
		Dermacentor reticulatus	Southern Europe
		Ixodes ricinus	Southern Europe
		Dermacentor andersoni	(experimental)
East Coast fever	*Theileria parva*	*Rhipicephalus appendiculatus* *R. capensis* *R. evertsi*	Eastern, central, and southern Africa
Bovine anaplasmosis	*Anaplasma marginale*	*Boophilus annulatus* *B. decoloratus* *B. microplus* *Rhipicephalus simus* *R. bursa* *R. sanguineus* *Ixodes ricinus* *Hyalomma lusitanicum* *Dermacentor variabilis* *D. andersoni* *D. occidentalis*	Worldwide
Rocky Mountain spotted fever	*Rickettsia rickettsi*	*Dermacentor andersoni* *D. variabilis* *D. parumapertus* *D. albipictus* *D. occidentalis*[a] *Amblyomma americanum*[a] *A. cajennense*[a] *Rhipicephalus sanguineus*[a] *Ornithodoros*[a] *Otobius*[a]	United States, parts of Canada, Mexico, parts of South America
Tick-borne typhus	*Rickettsia conorii*	*Rhipicephalus sanguineus* Other spp. of *Rhipicephalus* *Amblyomma* spp. *Haemaphysalis* spp. *Hyalomma*	Area bounding the Mediterranean, other parts of Africa
Queensland tick typhus	*Ricksettsia australis*	*Ixodes holocyclus*	Australia
Q fever (nine-mile fever)	*Coxiella burnetti*	*Dermacentor andersoni* *Amblyomma americanum* *Haemaphysalis humerosa*[a] *Ixodes holocyclus*[a]	Worldword Australia
Colorado tick fever	*Virus*	*Dermacentor andersoni* *D. parumapertus*[a]	Western United States
Tick-borne hemorrhagic fever	*Viruses*	*Hyalomma plumbeum* *H. anatolicum* *Dermacentor pictus* *D. marginatus*	USSR, Asia Minor, southeastern Europe

Table 16.1. (continued)

Disease	Etiologic Agent	Tick Vector	Endemic Area
Tick-borne encephalitides	*Viruses*	*Ixodes ricinus* (western Europe) *I. persulcatus* (eastern Europe) *Dermacentor* spp.[a] *Haemaphysalis* spp.[a]	Europe, Asia, Great Britain
Tularemia	*Pasteurella tularensis*	*Dermacentor andersoni* *D. occidentalis* *D. variabilis* *Rhipicephalus sanguineus* *Amblyomma americanum* *Haemaphysalis leporispalustris* *Ixodes pacificus*[a]	Western United States
Tick-borne relapsing fever	*Borrelia duttoni*[b] *B. venezuelense* *B. hispanica* *B. persica* *B. neotropicales* *B. turicatae* *B. hermsi*	*Ornithodoros moubata* *O. rudis* *O. erraticus* *O. talaje* *O. turicata* *O. hermsi*	Africa central Panama, Colombia, Venezuela, Ecuador Spain, Portugal, North Africa Panama Texas Rocky Mountain and Pacific Coast
	B. parkeri *Borrelia* sp.	*O. parkeri* *O. tholozani*	Rocky Mountain and Pacific Coast Central Asia
Avian spirochaetosis	*Borrelia anserina*	*Argas persicus*	India, Australia, Brazil, Egypt, eastern Asia Minor
Epizootic bovine abortion	*Borrelia*-like spirochete	*O. coriaceus*	Western United States, Mexico

[a] Possible vectors.

[b] The various types of spirochaetes which cause relapsing fever are considered by some to be strains of *Borrelia recurrentis*.

fixed dorsal toothed digit—the **digitus externus**—and a lateral movable **digitus internus**. The teeth are located on the outer surfaces of each chelicera (Fig. 16.3). This pair of appendages function as piercing and tearing structures, by means of which the host's in-

tegument is opened and the entire capitulum, or at least the toothed hypostome, is inserted into the host. These appendages also act as anchors when the parasite is attached.

3. A pair of **palpi** or **pedipalps** arise from the anteroventral margins of the basis capitulum.

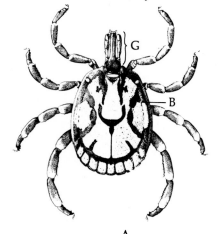

Fig. 16.3. Tick structures. A. Dorsal view of male *Amblyomma* showing two main regions of body. (After Whittick, 1943.) **B.** The chelicerae of *Ixodes reduvius*. Left, dorsal and ventral views of male. Right, dorsal and ventral views of female. (After Vitzthum, 1940.) B, body; C, capitulum or gnathosoma.

In modern ticks, each pedipalp is divided into four segments.* Among the so-called soft ticks (members of the family Argasidae), the pedipalps are flexible and not intimately associated with the hypostome. However, among the hard or wood ticks (members of the family Ixodidae), these palps are rigid and are intimately associated with the hypostome. Furthermore, the fourth segment is reduced and located in a pit on the ventral surface of the third segment in ixodid ticks. The function of the pedipalps is to act as counteranchors while the tick is attached to the host.

When seen from the dorsal aspect, the body proper of argasid ticks is covered with a leathery integument. This surface is ornamented with granulations, tubercles, and, in some, even circular discs. These secondary structures are quite consistent among members of a species and are therefore of taxonomic importance.

There is no **shield** (or **scutum**) on the dorsum of argasid ticks; however, among the ixodid ticks, a shieldlike scutum covers the entire surface of the body in males and only an anteromedial portion in females. The scutum is primarily a protective structure but is also limiting. Male ixodid ticks cannot become bloated with blood while feeding because the non-elastic scutum holds its shape. On the other hand, females, which possess only a small scutum, can and do become greatly distended while engorging blood.

In some ticks, such as *Dermacentor* and *Amblyomma*, the scutum is decorated with silvery streaks. Such ticks are referred to as ornate. The sculptures, furrows, and color patterns found on the dorsal body surface of some ticks are of taxonomic importance. The posterior and posterolateral margins of the dorsum of certain male ticks bear undulations known as **festoons**, which are rarely found on females.

As larvae, all ticks possess three pairs of legs, and a fourth pair appears after the larva molts. The nymphs and adults characteristically bear four pairs of legs. Each leg is subdivided into six segments known as the **coxa**, **trochanter**, **femur**, **genu**, **tibia**, and **tarsus**. The coxa is the most proximal segment. In some species, some of these units are fused; thus, their number is reduced. The legs characteristically terminate in a pair of claws on the tarsi. In some, such as members of the Ixodides, a semitransparent suckerlike **caruncle** (or **pulvillus**) is found on each tarsus.

When the ventral aspect of the tick is examined, the **genital orifice** can be seen located at the angle formed by the two genital grooves, if these are pres-

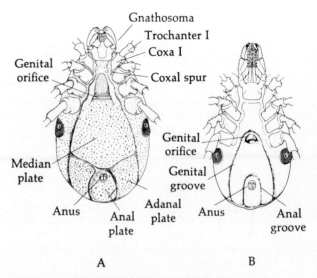

Fig. 16.4. Tick morphology. A. Ventral view of male *Ixodes ricinus*. **B.** Ventral view of female *I. ricinus*. (Both figures after Whittick, 1943.)

ent (Fig. 16.4). If the grooves are absent, the genital orifice is still in the same location, that is, on the midventral line between the first and second pairs of legs. The **anus** is also ventrally located on the midventral lines, approximately equidistant from the level of the fourth pair of legs and the posterior margin of the body. In some species, such as *Dermacentor variabilis*, an **anal groove** is situated posterior to the anus.

The respiratory (breathing) mechanism of adult ticks consists of **tracheae** that branch into **tracheoles**. This array of tubes infiltrates, oxygenates, and removes carbon dioxide from the various areas of the body. The main tracheae open to the exterior through two **spiracles** (also known as stigmata) located laterally near the coxae of the fourth pair of legs in adults. Larval forms may have more than two spiracles.

Body Wall

The cuticle of ticks and mites is subdivided into four layers (Fig. 16.5): (1) The **tectostracum** is the outermost, waxy and unstainable, nonpigmented thin layer. (2) The **epiostracum**, the second layer, is also rather thin. It is composed of polyphenol and cuticulin. (3) The **ectostracum**, or middle layer, is usually pigmented and can be stained with acid dyes. (4) The **hypostracum** is the innermost layer. It is usually pigmented and stains with alkaline dyes. The four layers of the cuticle are not always distinctly recognizable in

---

*According to Baker and Wharton (1952), the pedipalps of modern Acarina are divided into six or fewer segments. For a discussion of the evolution of these appendages, the reader is referred to that volume.

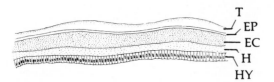

Fig. 16.5. **Tick cuticle.** Cross-section showing stratification. EC, ectostracum; EP, epiostracum; H, hypostracum; HY, hypodermis; T, tectostracum.

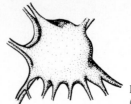

Fig. 16.6. **Brain of a tick.** (Modified after Blauvelt, 1945.)

all Acarina. It should again be emphasized here that the body wall is not an inert covering. It is a dynamic structure intimately correlated with several of the primary physiologic functions of the organism—respiration, sensation, and probably the absorption of certain nutrients and excretion of wastes.

Underneath the cuticle is an epithelial layer known as the **hypodermis**, which is responsible for secretion of the cuticular layers. Various glands, setae, pigment granules, and sense organs are derived from and associated with the hypodermis. Enlarged hypodermal cells in the form of unicellular glands that secrete onto the body surface are quite common among the ticks.

The various systems not previously mentioned in this chapter—the digestive, excretory, circulatory, nervous, and reproductive systems—are essentially the same in ticks and mites and are considered here as representative of both groups.

Digestive System

The alimentary canal of the Acarina can be subdivided into three portions: (1) The **foregut**, of stomadeal (ectodermal) origin, consists of a muscular pharynx leading from the mouth into the tubular esophagus. (2) The **midgut**, or ventriculus, entodermal in origin, is lined with digestive epithelium and has lateral diverticula or caeca (usually a pair) branching from it toward its posterior end. (3) The **hindgut**, of proctodeal (ectodermal) origin, consists of the anterior thin-walled intestine and the posterior muscular rectum, which leads to the exterior through the anus.

Circulatory System

Although hearts are lacking in most acarinas, they occur in certain mites. Circulation is accomplished by colorless hemolymph that infiltrates the various tissues of the body. Blood cells in the form of amoeboid hemocytes are present, although they are not readily demonstrated except during the quiescent stage preceding each molt.

Excretory System

The Acarina all possess at least two mechanisms for excretion. All members possess digestive epithelia lining their ventriculi that can perform the excretory function. The epithelial cells swell during digestion as the result of absorbing excretory products that are later discharged back into the lumen and passed out posteriorly. Acarines also possess one or more of three types of excretory apparatuses: (1) **coxal glands**, which are located adjacent to the coxae of certain legs and empty to the exterior through pores located near the coxae; (2) **entodermal excretory tubules**, which collect the body wastes and empty into the hindgut; and (3) a **modified hindgut** that serves as an excretory vesicle.

Nervous System

The so-called brain of acarines is a large ganglionic mass formed from the fusion of large numbers of ganglia (Fig. 16.6). In larvae, the fusion of these ganglia is often not completed and individual ganglia can be recognized. The ganglionic mass is located around the esophagus, and various nerve cords arise from it. Those innervating the eyes (if present), pedipalps, and chelicerae originate from the portion of the brain dorsal to the esophagus. Those innervating the pharynx, legs, and internal organs originate from the ventral portion of the mass.

Certain sensory structures occur on the body surfaces of ticks and mites. The most common of these are the **setae**, which are hairlike projections from the sclerotized integument. The setae are located on the body proper, as well as on the various appendages. Certain specialized setae, known as **sensillae**, are located on certain mites in the area of the propodosoma in a segmental notch, approximately at the posterior limit of the upper third of the body proper. The notch may or may not be clearly visible. Each sensilla is rooted in a cuticular depression, the **pseudostigma**. The function of such setae is sensory.

Eyes occur on certain mites and a few ticks. Some acarine eyes are rather complex, although they are definitely ocelli rather than compound eyes. The simpler eyes consist of no more than a differentiated portion of the integument, the lens, and a few brown-black pigment granules that are connected to nerve fibers. In many eyeless mites, thin transparent photo-sensitive areas are situated dorsally.

Reproductive System

The acarines are all dioecious. Fertilization is always internal, that is, the spermatozoa are introduced into the female. Tick sperm are transferred in **spermatophores**, each being comprised of an outer bulb (**ectospermatophore**) and an inner capsule (**endospermatophore**) (Feldman-Muhsam, 1967). The mites, as a rule, produce a few eggs at a time, whereas the ticks produce numerous eggs.

The male reproductive system consists of a single testis or a pair of testes. Vasa deferentia conduct the spermatozoa from the testes into the ejaculatory duct, through which the spermatozoa are introduced into the female through the penis. Usually an accessory gland or glands feed into the vasa deferentia.

The female reproductive system includes a single or a pair of ovaries. Leading from each ovary is an oviduct, which in turn opens into the uterus. In most instances the uterus opens to the exterior through the genital pore. During copulation, the spermatozoa are introduced into the uterus; however, in some mites, a vagina is present. A seminal receptacle and accessory glands usually are connected to the uterus. The seminal receptacle not only serves as a storage site for spermatozoa, but in most instances is also the site where male gametes mature (sperm are not mature when they are first introduced into the female). The Acarina are either oviparous or ovoviviparous. The eggs are normally covered by a shell.

Secondary sex characteristics are found in some species. Modified chelicerae on male members of the mite suborder Mesostigmata help transfer sperm into the female, since penes are absent. Many males possess modified legs (one or more pairs) that are used to grasp the female during copulation.

The genital pore of the male and the female is located between the first and second pairs of legs. This orifice is commonly covered by a genital plate. For a detailed account of the fertilization process in ticks, see Feldman-Muhsam (1964, 1979).

PATHOLOGIC ALTERATIONS DUE TO TICKS

The pathologic alterations and diseases of vertebrates associated with tick bites, aside from their role as vectors (p. 575), can be divided into two general categories. (1) **Local damage**—local inflammation and traumatic damage usually occur at the site of attachment. The damage may be a mild inflammation and itching, or it may be more serious, such as the invasion of the auditory canal by the spinose ear tick, *Otobius megnini*, which results in hemorrhage, edema, thickening of the stratum corneum, and partial deafness. Parasitization by *O. megnini* is common among cattle. (2) **Systemic damage**—during the feeding process of ticks, their salivary glands secrete an anticoagulant that prevents clotting so that the host's blood is readily taken in. This anticoagulant may cause a type of sensitization reaction, which is systemic. Tick bites may also cause paralysis. Tick paralysis is common among animals, especially sheep and calves, and occasionally occurs in humans, especially children.

The toxic substance responsible for the paralysis has not been thoroughly studied. From what is known, the substance is usually secreted by salivary glands, although Koch (1967) has reported that the dermal glands of engorged nymphs and females of *Ixodes holocyclus* also secrete toxins. Stone *et al.* (1979), who have contributed a review of tick toxins, suggest that the toxins occur in the hemolymph and are secreted by either the salivary or dermal glands. The highest level of paralyzing toxin occurs toward the end of feeding (Kaufmann, 1976). Chemical studies have revealed that the toxin is associated with a wide range of fractions extracted from salivary glands but especially with proteins in the molecular weight range of 60,000 to 80,000. It is still not known whether these proteins are the toxins or are carrier molecules of toxins of lower molecular weights.

The pathology caused by tick toxins is characterized by ascending flaccid motor paralysis, elevation of body temperature, impairment of respiration, speech, and swallowing, irritation of the eyes, loss of appetite and voice, excessive salivation, and vomiting. Occasionally death occurs as the result of respiratory or cardiac paralysis. If the tick or ticks are removed, the severity of the paralysis diminishes until the condition disappears.

About 12 species of ixodid ticks have been implicated in tick paralysis, among which *Dermacentor andersoni* is the most common in western North America. In the eastern and southern United States, *Dermacentor variabilis*, *Amblyomma americanum*, and *A. maculatum* are known to be responsible. In Australia, *Ixodes holocyclus* is the major causative species. It still remains uncertain why some ticks produce paralysis whereas others of the same species and sex do not. It has been hypothesized that a genetic factor may be involved (James and Harwood, 1969).

SUBORDER METASTIGMATA (IXODIDES)

Members of the suborder Metastigmata, or ticks, are all parasitic and are found on all vertebrates above the fish. As ectoparasites, they feed on the blood and

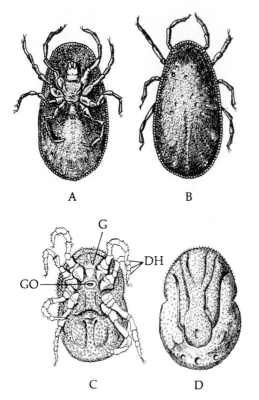

A   B

C   D

Fig. 16.7. Ticks. A. Ventral view of female *Argas persicus.* **B.** Dorsal view of female *A. persicus.* (**A** and **B**, redrawn after Bishopp, 1935.) **C.** Ventral view of female *Ornithodoros moubata.* DH, dorsal humps; G, gnathosoma or capitulum; GO, genital opening. **D.** Dorsal view of female *O. moubata.* (**C** and **D**, redrawn after Nuttall and Warburton, 1908.)

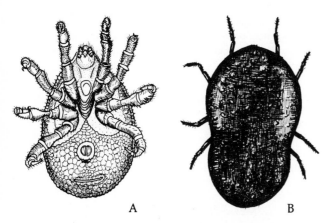

A   B

Fig. 16.8. Ticks. A. Ventral view of female *Antricola coprophilus.* (Redrawn after Cooley and Kohls, 1944.) **B.** Dorsal view of *Otobius megini.* (Redrawn from U.S. Dept. Agr. photograph.)

lymph of their hosts. The published literature pertaining to ticks and tick-borne diseases from about 800 B.C. to 1973 has been compiled by Hoogstraal (1970a,b, 1971, 1972, 1974).

BIOLOGY OF TICKS

The taxonomy of the Metastigmata has been reviewed by Cooley and Kohls (1944) and Baker and Wharton (1952). The volume by Baker and Wharton is especially recommended as an introduction to the subfield of Acarology. Also, the manual by McDaniel (1979) is useful for beginners. The major diagnostic characteristics of this suborder of the Acarina appear in the classification list at the end of this chapter.

The Metastigmata consists of three families: Argasidae (the soft ticks), Ixodidae (the hard ticks), and Nuttalliellidae. The last mentioned is a small family which includes only a single known species, *Nuttalliella namaqua,* which is free living and found under stones in Africa.

FAMILY ARGASIDAE

Members of the Argasidae are distributed throughout the world and are found as ectoparasites on snakes, lizards, turtles, many birds, and some mammals. This family includes the genera *Argas, Ornithodoros, Antricola,* and *Otobius* (Figs. 16.7, 16.8). Several of the species belonging to these genera are of economical and medical importance.

Genus *Ornithodoros*

Ornithodoros, which includes some 90 species, can be distinguished by its more or less dorsoventrally flattened body, which is definitely convex dorsally, and by patterns on the integument formed by the arrangement of mammillae, which extend from the dorsal to the ventral surfaces. All of the species are ectoparasites of mammals, including bats. *Ornithodoros* includes the dreaded *O. moubata* (Fig. 16.7), which is the vector for *Spirochaeta duttoni,* the pathogen that causes African relapsing fever. As the result of crossing and morphologic studies, Walton (1962) has shown that what had been regarded as a single, widespread species of eastern, central, and southern Africa is four distinct species and a subspecies, namely, *O. moubata,* an eyeless species widely distributed in the arid regions of Africa; *O. compactus,* from tortoises in Cape Province and in an area bounded in the north by the Zambesi River; *O. apertus,* a large and rare species from the burrows of porcupines, *Hystrix,* in Kenya, Ghana, and Botswana; and *O. porcinus,* with two subspecies in the burrows and lairs of wild animals in East Africa.

In the Americas, *Ornithodoros talajae, O. venezuelensis, O. turicata, O. hermsi, O. parkeri,* and *O. savignyi* serve as vectors for various spirochaetes that cause relapsing fevers in mammals. Among these, *O. hermsi*, which occurs throughout the Rocky Mountains and Pacific Coast in the United States, is probably the most important vector of *Borrelia recurrentis*, the etiologic agent of one type of relapsing fever. *O. hermsi* is primarily a parasite of rodents. Its life cycle is similar to that of *O. moubata*, presented below. The female lays up to 200 eggs in crevices where the adults hide. The hatching larvae actively seek and feed on the blood of hosts for 12 to 15 minutes. After molting, the two ensuing nymphal instars feed and molt. The adult form is attained after the final molt. In the laboratory it takes about 4 months for eggs to develop to the adult stage.

Onithodoros hermsi, which also serves as a vector for *Borrelia hermsi*, was implicated in an outbreak of tick-borne relapsing fever caused by this spirocheate in Spokane County, Washington, in 1968. This was the largest outbreak of relapsing fever known in the western hemisphere. Among 42 boy scouts, 11 contracted the disease as a result of having slept in old, rodent-infected cabins. Fortunately, all 11 recovered (Thompson *et al.*, 1969).

The most common species of *Ornithodoros* in Africa is *O. moubata*.

Life Cycle of *Ornithodoros moubata*. Adults of *O. moubata* hide in the dust and crevices of huts. This eyeless tick is primarily nocturnal in its feeding habits and takes a relatively short time to engorge itself with the host's blood. The host may be a pig, dog, goat, sheep, rabbit, rodent, or human. Both males and females are bloated and larger than the normal size of 8–11 mm long by 7 mm wide when filled with blood. When alive, *O. moubata* varies from brown to greenish brown in color. Impregnated females deposit their eggs away from the host between blood meals. The eggs are deposited in batches, each batch containing 35 to 340 eggs. A single female can produce a maximum of 1217 eggs in a lifetime.

A larva is completely formed within each egg approximately 7 to 11 days after the eggs are oviposited. The larva is nonmotile within the eggshell, which is split by this time. Within a few hours, the larva undergoes the first molt and transforms into the first nymphal stage. The nymph escapes from the shell and attaches itself to a passing host and remains for approximately 5 to 7 days (or less), feeding on the host's lymph. At the end of the feeding period, the nymph drops from the host and undergoes the second molt, after which it is referred to as the second nymphal instar. It has been reported that six to nine attached feeding periods ensue, each followed by a molt, before the nymph reaches the adult stage. Some populations are parthenogenetic, the progeny being all females (Davis, 1951). Although *O. moubata* does feed on the various animals mentioned, in transmission of the relapsing fever spirochaete, the cycle involves human-to-human feeding rather than animal-to-human feeding.

Variations of the feeding habit exhibited by *Ornithodoros moubata* are found in other members of the genus. For example, the larvae of some species are active and do feed. An interesting physiologic feature of *O. moubata* is that there is no passage between the mid- and hindgut. Consequently, all waste matter remains within the intestinal diverticuli during *O. moubata's* entire life (Enigk and Grittner, 1952).

Genus *Antricola*

Members of *Antricola* can be recognized by their flattened dorsal body wall and by their deeply convex ventral wall (Fig. 16.8). The integument is shiny and smooth except for the presence of tubercles. The body covering is sometimes translucent. Furthermore, the mouthparts of *Antricola* spp. are modified for quick feeding rather than for grasping the host. The species are primarily ectoparasites of birds.

Three species of *Antricola* are known to feed on bats: *A. coprophilus* on bats in Arizona and Mexico, *A. mexicanus* on cave bats in Mexico, and *A. marginatus* on cave bats in Cuba and other parts of the West Indies.

Genus *Argas*

Members of *Argas* are characterized by their flattened bodies and thin lateral margins. The integument is leathery with minute wrinkles interrupted by rounded areas that are pitted at the top and armed with setae located in the pits. The two species found in North America are *A. persicus* and *A. reflexus*, but these are not restricted to North America. In fact, *A. persicus* is a common household pest in Iran and elsewhere (Fig. 16.7).

Argas persicus, commonly known as the fowl tick, not only is irritating to chickens and other birds because of its bite, but also is the vector for *Borrelia gallinarum*, the causative agent of avian spirochaetosis in Brazil, India, Australia, Egypt, and Iran. The symptoms of the disease are diarrhea, loss of appetite, ruffled feathers, convulsions, and eventually death.

Life Cycle of *Argas persicus*. Like *Ornithodoros*, *A. persicus* hides in the nests or near the roosts of their hosts during the day. The nymphs and adults are extremely active at night, climbing onto the birds and engorging themselves with blood. The females lay their reddish brown eggs in clumps of 25 to 100. There are usually several layings between blood meals. It has been estimated that a single female may lay as many as 700 eggs. Such eggs are deposited in the daytime hideouts and hatch in 10 to 28 days,

Fig. 16.9. *Argas persicus.* Larva of the poultry tick.

giving rise to active larvae that bear three pairs of legs (Fig. 16.9). These larvae attach to birds both by day and by night. Once attached, the larvae may remain for as many as 5 days, engorging blood until they resemble reddish balls. During this feeding process, the sites of the bites are severely irritated. After such an extended period of feeding, the larva drops off, hides, and undergoes the first molt in approximately 7 days, developing into the first nymphal instar, which possesses an additional pair of legs, totaling eight legs. The nymphs resemble miniature adults and are nocturnal feeders. In 10 to 12 days, the second molt occurs, and the second nymphal instar is attained. There are three to four such molts, each one sandwiched between feedings. Finally, the adult form is attained. The adult males are approximately 5 mm long by 4.5 mm wide, and the females are 8.5 mm long by 5.5 mm wide. *Argas persicus* inflicts severe damage to birds because of drainage of blood, and the host is often killed. This tick is known to bite humans, causing dermal rashes.

Argas reflexus, the pigeon tick, possesses a narrower anterior end than *A. persicus*. It is ectoparasitic on pigeons and other roosting birds, and frequently attacks humans. Other species of *Argas* include *A. brumpti*, the largest species, measuring from 15–20 mm in length, and *A. vespertilionis*. Both species occur in Africa, and both will bite humans.

Genus *Otobius*

Members of *Otobius* are characterized by a granulated integument and by nymphs that possess integumentary spines. *Otobius* includes the widely distributed *O. megnini*, the spinose ear tick (Fig. 16.8), which is parasitic on the ears of cattle, horses, sheep, mules, dogs, cats, rabbits, deer, and other mammals. It is also known to bite humans. This parasite is widely distributed in the warmer parts of the United States, South America, South Africa, India, and no doubt in other parts of the world.

Life Cycle of *Otobius megnini*.   The larvae of *O. megnini* hatch from comparatively large, dark eggs that are layed on the ground. The hatched larvae climb up the host's legs and body into the inner folds of the ear, where they assume a saccular appearance. After feeding on the host's blood, the larvae molt and transform into the nymphal stage, which is spinose. The nymphs may remain attached to the host's ear for as long as 121 days, after which they drop to the ground and undergo three molts, finally becoming adults.

The adults possess a vestigial hypostome and do not feed. Egg laying begins 12 to 40 days after copulation, which occurs 1 to 3 days after the final molt. The laying period may last as long as 155 days, and there may be as many as 562 eggs produced. If larvae are left unfed at room temperature, they can survive for as long as 63 days, although the average is 44 days. The bites of *O. megnini* and *O. lagophilus*, the latter found attached to the face of rabbits in the western United States and Canada, are vulnerable to secondary infection. Furthermore, *O. megnini* can cause deafness, illness, and even death.

Family Ixodidae

The Ixodidae is the largest of the tick families. The most important genera are *Dermacentor*, *Ixodes*, *Rhipicephalus*, *Haemaphysalis*, *Hyalomma*, *Boophilus*, and *Amblyomma*. Members of these genera can readily be distinguished by the structure of their capitulum (Fig. 16.10).

Genus *Dermacentor*

Ticks belonging to *Dermacentor* are ornate and bear eyes. There are 11 festoons present.

Dermacentor andersoni (Fig. 16.11) is the Rocky Mountain wood tick. It is of great importance because it serves as one of the vectors of *Rickettsia rickettsii*, the causative agent of Rocky Mountain spotted fever. This tick is the principal vector of *R. rickettsii* in the Rocky Mountain states. Copulation occurs on the

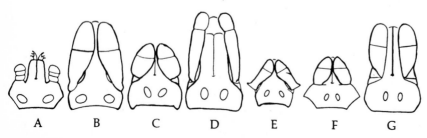

Fig. 16.10. Capitulum of several common genera of ticks. A. *Boophilus.* **B.** *Ixodes.* **C.** *Dermacentor.* **D.** *Amblyomma.* **E.** *Haemaphysalis.* **F.** *Rhipicephalus.* **G.** *Hyalomma.* (After Cooley, 1938, 1946.)

host. In nature this is usually a horse, cow, sheep, deer, bear, or coyote, but this tick will feed on any mammal. It is noted that Burgdorfer and Brinton (1975) have demonstrated that *R. rickettsii* can be transovarially transmitted from female *D. andersoni* to offspring.

Life Cycle of *Dermacentor andersoni*. After copulation, adult ticks again engorge themselves. The females become greatly distended and fall off the host. In approximately 7 days, they begin to oviposit. The egg-laying period may last 3 weeks, during which approximately 6400 eggs are laid.

After 35 days of embryonic development, the hexapod larva, commonly referred to as the **seed tick**, hatches. This larva actively becomes attached to a small mammal, usually a rodent, and feeds on its blood for 3 to 5 days, after which it drops off and, after a period of rest, metamorphoses into an octopod nymph. The nymph hibernates in soil, comes to the surface in the spring and attaches itself to a larger host, on which it engorges for 4 to 9 days. After feeding, the nymph drops off and undergoes the second molt in approximately 14 days. The post-molting form is the adult. These mature individuals, which become attached to a third host (a large animal), are commonly found in endemic areas during the spring months until the end of June.

Dermacentor andersoni, like many other ticks, is capable of surviving for long periods without a blood meal. In some extreme instances, larvae have survived for over 300 days without food. Nymphs can survive for 316 days without food and starved adults can survive up to 413 days.

In addition to serving as a vector for the spotted fever organism, *D. andersoni* serves as transport host for various other pathogenic microorganisms (Table 16.1).

Dermacentor albipictus, the winter or horse tick, differs in habit from other species of *Dermacentor* in that cold weather stimulates the larvae to seek out a host on which they become attached throughout the winter. The larva, nymph, and adult of this species are all found on the same host. Thus, this tick has a one-host life cycle. No dropping off occurs except during the engorged adult stage. Females at this stage lay their eggs on the ground. This species is commonly and widely distributed throughout North America. It causes a general weakening of the host and even death. Death is usually caused by the bacterium *Klebsiella paralytica* which is carried by the tick.

Other *Dermacentor* Species. Other species of *Dermacentor* include *D. variabilis*, the American dog tick, which is the principal vector of the spotted fever organism in the central and eastern United States. It can also serve as the vector for other organisms (Table 16.1). The seed ticks and nymphs of *D. variabilis* are usually found on the same host, a meadow mouse, *Microtus pennsylvanicus*. The adult is a parasite of many large animals, primarily dogs, and hence has a two-host cycle. *Dermacentor occidentalis*, the West Coast tick, is found in California and Oregon. If has been reported on cattle, horses, mules, donkeys, deer, sheep, rabbits, dogs, and humans. The immature

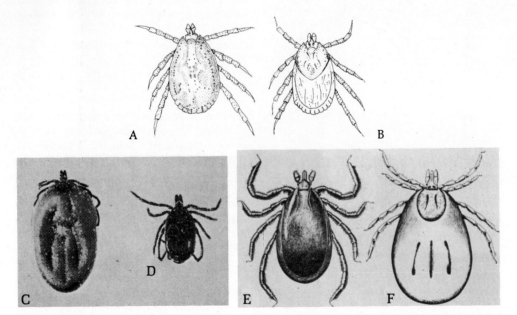

Fig. 16.11.   Ixodid ticks. A. Dorsal view of male *Dermacentor andersoni*. **B.** Dorsal view of unengorged female *D. andersoni*. **C.** Dorsal view of female *Ixodes pacificus*. **D.** Dorsal view of male *I. pacificus*. (**A–D**, after James and Harwood, 1969.) **E.** Dorsal view of male *Ixodes scapularis*. **F.** Dorsal view of female *I. scapularis*. (**E** and **F**, U.S. Dept. Agr. illustrations.)

forms attack rabbits, squirrels, field mice, and skunks.

Genus *Ixodes*

This is a large genus with about 40 species represented in North America. Eyes and festoons are absent, and a characteristic anal groove surrounds the anus in front. For those interested in the identification of the North America genera, the paper by Cooley and Kohls (1945) and the manual by McDaniel (1979) should be consulted. The most frequently encountered species are *I. pacificus* (Fig. 16.11), which occurs on deer, cattle, and dogs in California and other western states and which can bite humans; *I. ricinus*, the European castor-bean tick, which is distributed widely and occurs on a variety of mammals; *I. howelli*, a fairly common bird tick; and *I. scapularis* (Fig. 16.11), the black-legged tick, which is widespread in the southeastern United States and along the East Coast.

Life Cycle of *Ixodes ricinus*. This tick may take as little as 170 days or as long as 3 years to complete its life cycle, depending on environmental conditions.

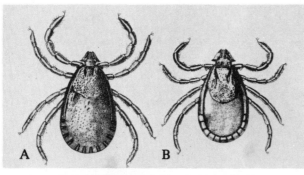

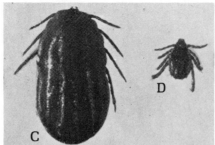

Fig. 16.12.   Ixodid ticks. A. Dorsal view of male *Haemaphysalis leporispalustris.* **B.** Dorsal view of female *H. leporispalustris.* (**A** and **B**, U.S. Dept. Agr. illustrations.) **C.** Dorsal view of engorged female *Boophilus annulatus.* **D.** Dorsal view of male *B. annulatus.* (**C** and **D**, after James and Harwood, 1969.)

The 3-year cycle prevails. Ticks of this species do not feed continuously; rather, they feed for a few days once a year, usually during the spring. During the rest of the 3-year life cycle, the tick is more or less quiescent. Nutrients taken in during the feeding periods are partially utilized by the tick and partially converted to molecules required for egg formation. Immature forms of *I. ricinus* generally feed on smaller hosts than the adults.

Lyme Disease.   In the mid-1970s a new spirochaete-caused disease was recognized. It was named Lyme disease after the town in Connecticut where the disease was first found (Burgdorfer *et al.*, 1982; Steere *et al.*, 1983). The causative agent has been named *Borrelia burgdorfi.* The tick vector is *Ixodes dammini* in the Northeast and Midwest and *Ixodes pacificus* in the western United States. Lyme disease, which begins during the summer, is characterized by a particular skin eruption known as erythema chronicum migrans, which may be followed weeks or months later by migratory polyarthritis. Patients also suffer from carditis and a variety of neurologic disorders. Also, infected persons have sharply rising IgM titers which peak during the third to sixth week of illness, followed by an elevated IgG titer.

Genus *Rhipicephalus*

This genus includes inornate ticks with eyes and festoons. There are about 50 species, of which *R. sanguineus*, the brown dog tick, is cosmopolitan in distribution. McIntosh (1931) has contributed an excellent article about this tick, which serves as the vector for the malignant canine jaundice organism, *Babesia canis*, the highly fatal *Rickettsia canis*, and other microorganisms. The life cycle of this species is similar to that of *Dermacentor andersoni* in that three hosts are involved.

Except for *R. sanguineus*, most species of *Rhipicephalus* are found in Africa, although a few occur in southern Europe and Asia. Most of the species are not very host specific; they can feed on a wide range of mammals. There are a few exceptions, however. For example, *R. distinctus* feeds only on hyraxes. *Rhipicephalus sanguineus* can feed on carnivores and rabbits in addition to dogs, but it ignores ruminants and rarely bites humans.

Genus *Haemaphysalis*

There are 100 or more species of *Haemaphysalis*. These ticks are especially abundant in Asia and the Malagasy Republic, where they primarily parasitize small mammals and birds. The exoskeletons of these comparatively small ticks are not as tough as those found in other members of the Ixodidae. The members are inornate and eyeless.

Haemaphysalis leporispalustris (Fig. 16.12) is a widely distributed species found in North and Central America. It is primarily a parasite of wild rabbits and hence is popularly referred to as the rabbit tick. Nevertheless, it reportedly also parasitizes horses, dogs, cats, and even certain birds. It does bite humans on occasion, but it is not an important vector for disease organisms, although in nature it carries the tularemia and spotted fever organisms among rabbits and, hence, is important from the epidemiologic standpoint.

Other *Haemaphysalis* Species. *Haemaphysalis leachii*, a dog tick found in Africa, Asia, and Australia, is of medical and veterinary importance. In addition to dogs, this tick parasitizes various carnivores, small rodents, and infrequently cattle. It is also a suitable vector for *Babesia canis* in South Africa and tick-borne fevers (Table 16.1).

Haemaphysalis humerosa transmits the so-called Q fever, caused by *Coxiella burnetii*, from bandicoot to bandicoot (*Isodon* spp.) in Australia, but it does not bite humans. Nevertheless, the bandicoot is an important natural reservoir host for the pathogen.

Haemaphysalis concinna is a vector of tick-borne encephalitis in the eastern Soviet Union.

Genus *Hyalomma*

Species of *Hyalomma* are confined to Europe, Africa, and Asia. A key to the known species has been compiled by Hoogstraal (1956).

These are large ticks on which the ornamentation, if present, is confined to pale bands on the legs. The festoons are more or less coalesced. Most of these ticks are ectoparasitic on domestic and wild animals, including horses, sheep, goats, cattle, and occasionally tortoises, lizards, and even birds. Although *Hyalomma* spp. serve as vectors for the protozoan *Theileria* and rickettsiae in domestic animals and rarely in humans, they are equally important in their own right. Their bites are extremely destructive and result in bad wounds that can cause crippling and serve as sites for invasion by screw worms.

Several species of *Hyalomma* serve as vectors for viruses that cause infectious hemorrhagic fevers in the Soviet Union. *Hyalomma aegyptium*, the bont-leg tick, is widely distributed in Europe, Asia, and Africa. Its life cycle generally involves two hosts. The larva and nymph share the same host, commonly rabbits, rodents, and birds. Adults, developing from nymphs that drop off to molt, become attached to larger animals.

Genus *Boophilus*

This genus includes fewer than 10 species, of which *B. annulatus* (Fig. 16.12) is of primary interest. There are no festoons or ornamentations on the species, but eyes are present.

Boophilus annulatus is a vector for *Babesia bigemina*, the Texas cattle fever-causing protozoan (p. 207). It is limited to the southern United States and parts of Mexico. Although primarily a parasite of cattle, *B. annulatus* will attach to horses, mules, sheep, deer, and other animals. Cattle suffering from the Texas fever demonstrate weight loss, reduction in milk production, sterility, and even death, depending on the severity of the infection.

Life Cycle of *Boophilus annulatus*. *Boophilus* spp. are one-host ticks. When females become gravid, they drop off the host and undergo a preoviposition period before laying eggs. A single female lays between 357 and 5105 eggs, averaging 1811 to 4089. The time required for complete development of the larva fluctuates with the temperature—19 days during hot summer months and 180 days during the early autumn months. The larvae (seed ticks) hatching from the eggs are very active. They climb up blades of grass and other vegetation and await a passing cow or some other large herbivore to cling to. *B. annulatus* is parasitic on the same host during its larval, nymphal, and adult stages—a typical one-host cycle. The two molts, one occurring between the larval and nymphal stages and one between the nymphal and adult stages, take place on the host.

Other *Boophilus* Vectors. The life history and developmental pattern of *Babesia bigemina* in the tick and in the bovine host is given on p. 207. The fatality of cattle suffering from this protozoan-caused disease may be as high as 75% in some areas. The disease is endemic in Central and South America, southern Europe, India, parts of Africa, and the Philippines. *Boophilus australis* is the vector in India, the Philippines, and parts of South America; *B. decoloratus* is the vector in Central and South Africa, Australia, the Orient, and southern Florida; and *B. microplus* is the vector in sections of South America and the West Indies. Other species of ticks can also serve as vectors for *Babesia* spp. (Table 16.1).

Genus *Amblyomma*

This genus includes approximately 100 species, most of which have been listed and annotated by Robinson (1926). The species are generally ornate and possess eyes and festoons.

Amblyomma americanum (Fig. 16.13), the Lone Star tick, occurs in Oklahoma, Louisiana, Texas, and other southern and southwestern states. It is also found in Central America and Brazil. This tick occurs on an array of animal hosts, wild and domesticated, ranging from birds to humans. Furthermore, *A. americanum* is of economic importance because it serves as a vector for the spotted fever, Bullis fever, Q fever, and tularemia organisms. The life cycle of *A. americanum* is of the three-host type.

Other *Amblyomma* Species. *Amblyomma cajen-*

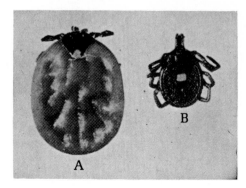

Fig. 16.13. *Amblyomma americanum.* **A.** Dorsal view of engorged female. **B.** Dorsal view of male. (After James and Harwood, 1969.)

nense occurs in the southern United States, Central and South America, and the West Indies. This species parasitizes horses, cattle, sheep, hogs, and humans. It is the main vector for the spotted fever rickettsia in Brazil and Colombia.

Amblyomma also includes *A. hebraeum*, the vector for tick-bite fever in South America, and *A. maculatum*, the Gulf Coast tick, found attached to the ears of cattle. Not only are the sores resulting from the bites of *A. maculatum* irritating and painful, but these lesions are potential sites for screw worm infestations and secondary bacterial and fungal infections. In addition, this species serves as a suitable vector for *Rickettsia* spp. Both *A. hebraeum* and *A. maculatum* have a three-host life cycle. Nymphs of *A. maculatum* have been reported on meadow larks.

PHYSIOLOGY AND BIOCHEMISTRY OF TICKS

Comparatively little is known about the physiology and biochemistry of ticks. The many research papers dealing with the Metastigmata are primarily concerned with taxonomy, morphology, and control aspects of these ectoparasites. This is not surprising because concerted efforts to understand the physiology of ticks are comparatively recent. As in all phases of biology, the form and classification of a group of organisms must be established before one can intelligently evaluate the physiologic (and biochemical) processes associated with them. Furthermore, since the Metastigmata are of medical and veterinary importance, it is not surprising that the control aspects have been investigated extensively.

OXYGEN REQUIREMENTS

That ticks are true aerobes is quite obvious, for they possess well-developed tracheal systems through which gaseous exchange occurs. In addition, it is suspected that the host's erythrocytes ingested by these blood feeders also supply a certain amount of oxygen for metabolism.

COMPOSITION OF BLOOD

The composition of tick blood (hemolymph) remains unknown, although it is suspected to be quite similar to that found in other arachnids, that is, includes either dissolved hemoglobin or hemocyanin. If such is the case, a certain amount of oxygen and carbon dioxide conduction is performed by the blood. The function of the amoeboid hemocytes is not clear, but it is suspected that these are phagocytic and represent one form of internal defense mechanism.

TRANSOVARIAL TRANSMISSION

Transovarial transmission among the Acarina, specifically the Metastigmata, is extremely interesting. Two types of materials are transmitted in this manner.

1. Certain microorganisms for which ticks serve as vectors can be transmitted from one generation to another by infestation of eggs. The protozoans, *Babesia* spp., for example, are passed on in this manner. The discovery of this phenomenon has explained how protozoan parasites can be transmitted from host to host by such species as *Boophilus annulatus*, which almost never migrate from host to host. It is through the passage of the protozoans from mother to offspring, the latter parasitizing other hosts, that the pathogen is spread. Similarly, spirochetes in *Argas* and *Babesia* in *Rhipicephalus sanguineus* are transmitted to other vertebrates via the offspring. It has been mentioned earlier that the spotted fever rickettsia, *Rickettsia rickettsii*, can be transovarially transmitted from female *Dermacentor andersoni* to progeny via eggs. The rickettsiae have been demonstrated in developing oocytes (Fig. 16.14).
2. Hyland and Hammer (1959) have demonstrated that nonliving materials can be transmitted transovarially. These investigators fed adult female *Dermacentor variabilis* glycine tagged with ^{14}C. Later, radioactivity was recorded in the eggs and resulting larvae. The amount of radioactivity, however, diminished by 55% between the eggs and the larvae. This experiment demonstrates that nonliving materials can be transmitted transovarially.

The ability of ticks in all stages to survive for long periods without food has been mentioned. Pavlovsky and Skrynnik (1960) have reported an amazing example of this in *Ornithodoros hermsi*. These investigators found that specimens can live for 3 years without food. Starved ticks are not usually capable of ovipositing because engorgement is a prerequisite to egg laying.

Balashov (1957) has demonstrated that in several species of ticks, such as *Ixodes ricinus, Haemaphysalis punctata, Dermacentor pictus, Rhipicephalus turanicus,* and *Hyalomma plumbeum,* if females are not impregnated on the third or fourth day after the final nymphal ecdysis, their feeding is slowed down or ceases, whereas impregnated females feed healthily. Furthermore, the amount of blood taken in by impregnated females is directly correlated with the quantity and viability of the eggs laid. Maximum weight is attained by engorging females before laying their eggs. If this weight is not attained through the intake of blood, the number of eggs decreases and the percentage of nonviable eggs increases.

THE CUTICLE DURING FEEDING

Lees (1952) has demonstrated that ticks capable of ingesting large volumes of blood are able to secrete new cuticle while attached to the host. He divided the periods of engorgement into two phases. During the first phase there is a gradual increase of body weight and active cuticle synthesis; that is, the endocuticle (ectostracum and hypostracum) increases in thickness. The moderate increase in surface area is due to the molecular growth of the cuticle and not to distention. During the second phase, the cuticle is stretched.

During cuticle synthesis in the first phase, the underlying hypodermis is greatly hypertrophied. The nuclei are enlarged, one or more nucleoli appear, and the cytoplasm becomes heavily concentrated with RNA and alkaline phosphatase. During the stretching process associated with the second phase, cuticular structures become greatly modified, especially the pore canals.

Ixodid ticks undergo cuticle synthesis during all feeding periods, including that of all instars, and hence are slow feeders. On the other hand, among argasid ticks, the larvae commonly undergo cuticle synthesis and therefore feed slowly. However, the nymphs and adults accommodate the smaller blood meals by simply stretching the preformed cuticle and hence are rapid feeders.

RESPONSE TO HUMIDITY

Certain species of ticks exhibit a definite orienting response in a relative humidity gradient. Sonenshine

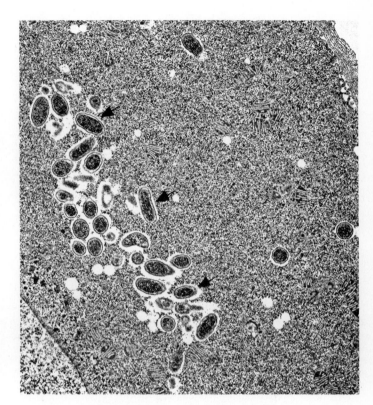

Fig. 16.14. *Rickettsia rickettsii* in *Dermacentor andersoni* oocyte. Electron micrograph showing *R. rickettsii* in perinuclear ooplasm of developing oocyte of *D. andersoni.* × 10,000. (After Burgdorfer and Brinton, 1975; with permission of New York Academy of Sciences.)

(1963), by employing multiple-choice and linear gradient chambers in which the relative humidities (RH) ranged from 23.9–28.4% to 86.6–94.6%, tested the reactions of seven species of ticks. In the case of *Ornithodoros turicata,* 73% of the fasting nymphs and adults placed at the 89% RH position migrated and aggregated in the highest humidity zones. On the other hand, freshly engorged nymphs and adults previously maintained at 80% RH were generally indifferent to the humidity choices, but 63% of the freshly engorged larvae chose the highest humidity zone.

In the case of *Ornithodoros kelleyi* and *Dermacentor andersoni,* a distinct avoiding reaction to high RH zones was exhibited by most specimens. Seventy-one percent of the fasting adults of *O. kelleyi* aggregated in the lowest humidity zones, whereas *D. andersoni* adults preferred zones with low to intermediate RHs, 83% of the ticks aggregating at humidities lower than 56%, including 45% in the lowest humidity zones.

Similar tests with *Ornithodoros parkeri*, *O. savignyi*, *O. delanoei*, and *Argas cooleyi* did not reveal any response.

In order to ascertain the location of the humidity receptors, Sonenshine removed the tarsi of the first pair of legs of *O. kelleyi* females and *D. andersoni* males. As a result, all of these ticks failed to exhibit the avoiding reaction and wandered freely throughout all the humidity zones. Although the exact site and nature of the humidity receptors remain to be studied, Sonenshine's experiments suggest that such receptors may be present on the tarsi.

PROTEIN METABOLISM

Protein biochemistry has been little studied among the ticks. Araman (1979) has reported the following sequence of events related to protein digestion and synthesis in *Rhipicephalus sanguineus* females during development and vitellogenesis.

Ingested host proteins are digested intracellularly by digestive cells in the midgut; limited extracellular digestion also occurs. The breakdown products—smaller peptides and amino acids—are subsequently released into the lumen of the gut and utilized by the tick tissues in the synthesis of enzymes and of structural and vitellogenic proteins. Structural proteins are synthesized primarily during the developmental phase from digestive products of host blood ingested during the growth-feeding phase. On the other hand, vitellogenins and other proteins involved in vitellogenesis are mostly synthesized during the **vitellogenic phase**, utilizing blood meal proteins ingested during the expansion phase of feeding. Most of the digestive products enter the hemocoel where they are utilized by the fat body to produce most of the hemolymph proteins, including vitellogenins. The vitellogenins are then sequestered by the ovary and incorporated into egg yolk proteins. The ovary takes up hemolymph vitellogenins by micropinocytosis. In addition to vitellogenins, the yolk proteins also consist of proteins produced by oocytes in the ovary.

CARBOHYDRATE METABOLISM

It is generally assumed that ticks, being aerobic, undergo carbohydrate metabolism as related to energy production via the conventional pathways presented earlier (p. 52). Glucose is metabolized rapidly (Rodriguez *et al.*, 1979).

LIPID METABOLISM

Again, very little is known about lipid metabolism in ticks. That arachnidans are capable of synthesizing lipids from butanediol is known (Rodriguez *et al.*, 1979).

HOST-PARASITE RELATIONSHIPS

Attraction to Host

As the result of years of attempting to understand the mechanisms involved in the transmission of pathogens by ticks, considerable information is now available on their feeding habits. Those desiring additional information are referred to Hocking (1960) for a comprehensive bibliography, and those interested in the sensory physiology associated with attraction to host and feeding are referred to the reviews by Dethier (1957) and Arthur (1965).

Odor plays an important role in host localization by ticks, although little is known about the identity of the attractive odor components. Phagostimuli that induce and maintain gorging have received attention, especially among the rapid-feeding argasid ticks. For example, *Ornithodoros tholozani* will probe through a membrane of Parafilm (a wax and rubber material) into water, saline isotonic with blood, blood, or a hydrolyzate of washed erythrocytes, if the temperature is maintained at 33°C.

Probing need not lead to actual feeding. For example, *O. tholozani* will engorge only if the system contains blood or laked erythrocytes. However, Galun and Kindler (1965) have demonstrated that isotonic saline is engorged if it contains 10^{-2} M reduced glutathione. Since glutathione occurs in blood at a much lower concentration than 10^{-2} M, other stimulants are also believed to be involved. Glucose at blood concentrations (1 mg/ml) or greater is slightly stimulatory. When present with glucose, several amino acids, especially leucine, stimulate feeding in ticks previously starved for 6 months or longer (Galun and Kindler, 1968). In addition, when combined with 1 mg/ml of glucose, adenosine triphosphate, adenosine diphosphate, inosine or guanine triphosphate, or reduced diphosphopyridine nucleotide induce gorging. Evidently ticks, like blood-sucking insects, have evolved a recognition for vertebrate blood by the use of stimuli from nucleotide phosphates and, in addition, require blood sugar as a complementary stimulation.

Attraction of Tick to Tick

In conjunction with mating among ticks, some work has been performed on the identification of attractants. For example, Berger (1972) has reported that 2,6-dichlorophenol from feeding *Amblyomma americanum* females induces responses in feeding males similar to those observed during mating. Since then, pheromones have been reported to be produced by female ticks of at least four genera of ixodid ticks. It is now known that 2,6-dichlorophenol serves as the female sex attractant in *Dermacentor variabilis, D. andersoni, Rhipicephalus sanguineus,* and *Amblyomma maculatum* (Kellum and Berger, 1977). It needs to be emphasized that this molecule attracts only feeding males.

Other phenols have been reported to act as sex pheromones in hard ticks. For example, Wood *et al.* (1975) have demonstrated that a mixture of phenol and *p*-cresol isolated from female *Rhipicephalus appendiculatus* and *R. puchellus* elicits an investigatory response from sexually active males. Sonenshine *et al.* (1977) have found that these pheromones are produced in a multilobed gland, the **foveal gland**, emptying through pores on the alloscutum of female ticks. Females commence synthesizing and storing sex pheromone soon after molting but do not emit it until stimulated to do so. Sonenshine *et al.* (1979) have proposed that sex pheromones may be employed to control ticks by disrupting reproduction as a result of using pheromone-containing decoys.

In addition to female-produced pheromones, it is also known that males produce attractants (Rechav and Whitehead, 1979). Also, a third type involved in mate location away from the host has been discovered (Leahy *et al.*, 1973, 1975; Treverrow *et al.*, 1977). Consequently, it has been proposed that three categories of tick pheromones be recognized: (1) **female sex pheromones** produced by hard ticks; (2) **assembly pheromones** which are responsible for mate location away from the host (these have been found in both soft and hard ticks); and (3) **male sex pheromones** produced while feeding on the host.

Effects of Parasite on Host

Tick-Host Anemia. Certain ticks, such as *Ixodes dentatus* on wild rabbits and *Dermacentor* spp., cause a noninfectious disease known as tick-host anemia. This anemia, which has been reported in moose, jackrabbits, foxes, and other animals, causes the death of the host, which upon autopsy exhibits pale and watery blood and a definite reduction of hemoglobin. The cause of the disease remains uncertain, although it is suspected that the anemia results from drainage of blood by the parasite, for the anemia is most conspicuous in heavily infected hosts.

Tick-Host Paralysis. As mentioned earlier (p. 579), the bite of certain ticks results in paralysis of the host. The paralysis sets in after the female tick begins feeding. The causative agent is a neurotoxin. Apparently the toxin is not secreted by all females of the same species because the bites of some females do not cause paralysis. Genetics of the tick species may be in part responsible for these differences. It is known that a single female tick can paralyze a 1000-pound bull by the time she is only partially engorged. If the paralysis has not advanced too far, removal of the parasite or parasites results in recovery. Investigations have revealed that age and species resistances exist. Older animals are not as apt to be paralyzed as are young animals, and certain hosts are immune.

Various ticks are known to cause paralysis. In Australia, *Ixodes holocylus* causes paralysis in dogs and humans. *Dermacentor andersoni* bites cause paralysis in humans, cattle, sheep, and bisons in the northwestern United States and southern British Columbia. In the southeastern United States, *D. variabilis* occasionally causes paralysis in humans and dogs. In South Africa and occasionally in Europe, *Ixodes pilosus* has been incriminated. *Amblyomma americanum* bites can cause tick-host paralysis in various mammals. Abbot (1942) and Hughes and Philip (1958) have reviewed this subject.

Mechanical Damage. Bites of certain ticks cause severe mechanical damage to the host's integument and underlying tissues. Several instances of this have been cited previously. Furthermore, the lesions resulting from bites serve as invasion sites for bacterial and fungal infections and as entrance sites for screw worms.

Toxins and Venoms. Among the Ixodidae, *Dermacentor occidentalis, D. variabilis,* and *Ixodes ricinus* cause systemic disturbances suggesting the secretion of some irritant during the bite. Among the Argasidae, the bite of *Ornithodoros moubata* in Africa causes severe swelling and irritation. In parts of California and Mexico, *O. coriaceus* causes a systemic disturbance known as "tlalaja," which is accompanied by severe pain and swelling. The nature of the toxin is not known, although *Ixodes ricinus* secretes an anticoagulant during feeding that prevents clotting of blood at the site of the bite and prevents coagulation of blood within the alimentary tract of the tick. It is suspected that most ticks secrete such a substance, and this may include the toxin.

Transmission of Microorganisms. The role of ticks as vectors for protozoa, bacteria, rickettsiae, and spirochetes has been discussed. Specific vectors for specific microorganisms are listed in Table 16.1. The excellent review by Harwood and James (1979) on the role of ticks as vectors of pathogens is recommended as well as several chapters in the volumes edited by Rodriguez (1979).

It should be stressed that very little is known about metabolism, respiration, food synthesis, and other aspects of tick physiology. This phase of parasitology awaits the contributions of future parasitologists. Arthur (1960) has contributed a monograph on the

Metastigmata that should be consulted as an authoritative source for serious investigators.

REFERENCES

Abbot, K. H. (1942). Tick paralysis: a review. *Proc. Mayo Clin.* **18**, 39–45, 59–64.

Araman, S. F. (1979). Protein digestion and synthesis in ixodid females. *In* "Recent Advances in Acarology" (J. G. Rodriguez, ed.), Vol. I, pp. 385–395. Academic Press, New York.

Arthur, D. R. (1960). "Ticks." Cambridge University Press, London.

Arthur, D. R. (1965). Feeding in ectoparasitic acari with special reference to ticks. *Adv. Parasitol.* **3**, 249–298.

Baker, E. W., and Wharton, G. W. (1952). "An Introduction to Acarology." Macmillan, New York.

Balashov, Y. S. (1957). Gonotropic relationships in ixodid ticks (Acarina, Ixodidae). *Entomol.* **36**, 285–299. (In Russian.)

Beaumont, A. (1949). Contribution à l'étude du métabolisme de l'hemoglobine de *Gasterophilus intestinalis*. *C.R. Soc. Biol.* **142**, 1369–1371.

Berger, R. S. (1972). 2,6-dichlorophenol, sex pheromone of the lone star tick. *Science* **177**, 704–705.

Burgdorfer, W. and Brinton, L. P. (1975). Mechanisms of transovarial infection of spotted fever rickettsiae in ticks. *Ann. N.Y. Acad. Sci.* **226**, 61–72.

Burgdorfer, W., Barbour, A. G., Hayes, S. F., Benach, J. L., Grunwaldt, E., and Davis, J. P. (1982). Lyme disease. A tick-borne spirochetosis? *Science,* **216**, 1317–1319.

Cooley, R. A., and Kohls, G. M. (1944). "The Argasidae of North America, Central America and Cuba." Am. Midl. Natur. Monogr., Notre Dame, Indiana.

Cooley, R. A., and Kohls, G. M. (1945). The genus *Ixodes* in North America. *Nat. Inst. Health Bull. No.* 184.

Davis, G. E. (1942). Species unity or plurality of the relapsing fever spirochaetes. *In* "A Symposium on Relapsing Fever in the Americas," pp. 41–47. AAAS Publ. No. 18.

Davis, G. E. (1951). Parthenogenesis in the argasid tick *Ornithodoros moubata* (Murray, 1977). *J. Parasitol.* **37**, 99–101.

Dethier, V. G. (1957). The sensory physiology of blood-sucking arthropods. *Exp. Parasitol.* **6**, 68–122.

Enigk, K., and Grittner, I. (1952). Die Excretion der Zecken. *Z. Tropenmed. Parasitol.* **4**, 77–94.

Feldman-Muhsam, B. (1964). Some contributions to the understanding of the reproduction of ticks. *Acarologia* (fasc. h.s.), pp. 294–298.

Feldman-Muhsam, B. (1967). Spermatophore formation and sperm transfer in *Ornithodoros* ticks. *Science* **156**, 1252–1253.

Feldman-Muhsam, B. (1979). Copulatory behavior and fecundity of male *Ornithodoros* ticks. *In* "Recent Advances in Acarology" (J. G. Rodriguez, ed.), Vol. II. pp. 159–166. Academic Press, New York.

Galun, R., and Kindler, S. H. (1965). Glutathione as an inducer of feeding in ticks. *Science* **147**, 166–167.

Galun, R., and Kindler, S. H. (1968). Chemical basis of feeding in the tick *Ornithodoros tholozani*. *J. Insect Physiol.* **14**, 1409–1421.

Harwood, R. F., and James, M. T. (1979). "Entomology in Human and Animal Health." 7th ed. Macmillan, New York.

Hocking, B. (1960). Northern biting flies. *Ann. Rev. Entomol.* **5**, 135–152.

Hoogstraal, H. (1956). "African Ixodoidea." Vol. I. Ticks of the Sudan. Research Report NM 005050.29.07, U.S. Navy.

Hoogstraal, H. (1970a). "Bibliography of Ticks and Tick-borne Diseases From Homer (about 800 B.C.) to 31 December 1969." Vol. 1. Spec. Publ. U.S. Naval Med. Res. Unit No. 3, Cairo, Egypt, United Arab Republic.

Hoogstraal, H. (1970b). "Bibliography of Ticks and Tick-borne Diseases From Homer (about 800 B.C.) to 31 December 1969." Vol. 2. Spec. Publ. U.S. Naval Med. Res. Unit No. 3, Cairo, Egypt, United Arab Republic.

Hoogstraal, H. (1971). "Bibliography of Ticks and Tick-borne Disease From Homer (about 800 B.C.) to 31 December 1969." Vol. 3. Spec. Publ. U.S. Naval Med. Res. Unit No. 3, Cairo, Egypt.

Hoogstraal, H. (1972). "Bibliography of Ticks and Tick-borne Diseases From Homer (about 800 B.C.) to 31 December 1969." Vol. 4. Spec. Publ. U.S. Naval Med. Res. Unit No. 3, Cairo, Egypt.

Hoogstraal, H. (1974). "Bibliography of Ticks and Tick-borne Diseases From Homer (about 800 B.C.) to 31 December 1969." Vol. 5. Spec. Publ. U.S. Naval Med. Res. Unit No. 3, Cairo, Egypt.

Hughes, L. E., and Philip, C. B. (1958). Experimental tick paralysis in laboratory animals and native Montana rodents. *Proc. Soc. Exp. Biol. Med.* **99**, 316–319.

Hyland, K. E., Jr., and Hammar, J. L. (1959). Transovarial passage of radioactivity in ticks labeled with C^{14} glycine. *J. Parasitol.* **45**, 24–25.

Kaufmann, W. R. (1976). The influence of various factors on fluid secretions by *in vitro* salivary glands of ixodid ticks. *J. Exp. Biol.* **64**, 727–742.

Kellum, D., and Berger, R. A. (1977). Relationship of the occurrence and function of 2,6-dichlorophenol in two species of *Amblyomma* (Acari: Ixodidae). *J. Med. Entomol.* **13**, 701–705.

Koch, J. H. (1967). Some aspects of tick paralysis in dogs. *New South Wales Vet. Proc.* **3**, 34.

Leahy, M. G., Vandehey, R., and Galun, R. (1973). Assembly pheromone(s) in the soft tick *Argas persicus* (Oken). *Nature* **246**, 515–516.

Leahy, M. G., Karuhize, G., Mango, C., and Galun, R. (1975). An assembly pheromone and its perception in the tick *Ornithodoros moubata* (Murray) (Acari: Argasidae). *J. Med. Entomol.* **12**, 284–287.

Lees, A. D. (1952). The role of the cuticle growth in the feeding process of ticks. *Proc. Zool. Soc. London* **121**, 259–272.

Levenbook, L. (1950). The composition of horse bot fly (*Gasterophilus intestinalis*) larva blood. *Biochem. J.* **47**, 336–346.

Locke, M. (1974). The structure and formation of the integument in insects. *In* "The Physiology of Insecta," 2nd ed. (M. Rockstein, ed.), Vol. VI, pp. 123–213. Academic

Press, New York.

McDaniel, B. (1979). "How to Know the Ticks and Mites." William C. Brown, Dubuque, Iowa.

McIntosh, A. (1931). The brown dog tick. *North Am. Vet.* **12**, 37–41.

Pavlovsky, E. N, and Shrynnik, A. N. (1960). Laboratory observations on the tick *Ornithodoros hermsi* Wheeler, 1935. *Acarologia* **2**, 62–65.

Rechav, Y., and Whitehead, G. B. (1979). Male produced pheromones of Ixodidae. *In* "Recent Advances in Acarology" (J. G. Rodriguez, ed.), Vol. II, pp. 291–296. Academic Press, New York.

Reinhard, E. G., and von Brand, T. (1944). The fat content of *Pagurus* parasitized by *Peltogaster* and its relation to theories of sacculinization. *Physiol. Zool.* **17**, 31–41.

Robinson, L. E. (1926). "Ticks. A Monograph of the Ixodoidea. Part IV: The Genus *Amblyomma*." Cambridge University Press, London.

Rodriguez, J. G. (ed.) (1979). "Recent Advances in Acarology." Vols. I and II. Academic Press, New York.

Rodriguez, J. G., Smith, W. T., Heffron, P., and Oh, S. K. (1979). Metabolism of butanediol in *Tyrophagus putrescentiae*. *In* "Recent Advances in Acarology" (J. G. Rodriguez, ed.), Vol. I, pp. 337–346. Academic Press, New York.

Steere, A. C., Grodzicki, R. L., Kornblatt, A. N., Craft, J. E., Barbour, A. G., Burgdorfer, W., Schmid, G. P., Johnson, E., and Malawista, S. E. (1983). The spirochetal etiology of Lyme disease. *New Engl. J. Med.* **308**, 733–740.

Sonenshine, D. E. (1963). A preliminary report on the humidity behavior of several species of ticks. *In* "Advances in Acarology" (J. A. Naegele, ed.), Vol. I, pp. 431–434. Cornell University Press, Ithaca, New York.

Sonenshine, D. E., Silverstein, R. M., and Homsher, P. J. (1979). Female-produced pheromones of Ixodidae. *In* "Recent Advances in Acarology" (J. G. Rodriguez, ed.), Vol. II, pp. 281–290. Academic Press, New York.

Sonenshine, D. E., Silverstein, R. M., Collins, L. A., Saunders, J., Flynt, C., and Homsher, P. J. (1977). Foveal glands, source of sex pheromone production in the ixodid tick *Dermacentor andersoni* Stiles. *J. Chem. Ecol.* **3**, 695–706.

Stone, B. F., Doube, B. M., and Bennington, K. C. (1979). Toxins of the Australian paralysis tick *Ixodes holocyclus*. *In* "Recent Advances in Acarology" (J. G. Rodriguez, ed.), Vol. I, pp. 347–356. Academic Press, New York.

Thompson, R. S., Burgdorfer, W., Russell, R., and Francis, B. J. (1969). Outbreak of tick-borne relapsing fever in Spokane County, Washington. *J. Am. Med. Assoc.* **210**, 1045–1050.

Treverrow, R. L., Stone, B. F., and Cowie, M. (1977). Aggregation pheromones in 2 Australian hard ticks, *Ixodes holocyclus* and *Aponomma concolor*. *Experientia* **33**, 680–682.

von Brand, T. (1973). "Biochemistry of Parasites," 2nd ed. Academic Press, New York.

von Kemnitz, G. A. H. (1916). Untersuchungen über den Stoffbestand und Stoffwechsel der Larven von *Gasterophilus equi* (Clark), nebst Bemerkungen über den Stoffbestand der Larven von *Chironomus* (spec. ?) L. (Physiologischer Teil). *Z. Biol.* **49**, 129–244.

Walton, G. A. (1962). The *Ornithodoros moubata* super-species problem in relation to human relapsing fever epidemiology. *Symp. Zool. Soc. London No.* **6**, 83–156.

Wood, W. F., Leahy, N. G., Galun, R., Prestwich, G. D., Meinwald, J., Purnell, R. E., and Payne, R. C. (1975). Phenols as pheromones of ixodid ticks: A general phenomenon? *J. Chem. Ecol.* **1**, 501–509.

CLASSIFICATION OF METASTIGMATA (THE TICKS)

ORDER ACARINA

Highly specialized Arachnida with body divided into proterosoma and hysterosoma which are distinguishable as boundary between second and third pairs of legs; segments of mouth and its appendages situated on capitulum (gnathostoma), more or less sharply set off from rest of body; typically with four pairs of legs; usually six podomeres of legs but may vary from two to seven; positions of respiratory and genital openings variable.

Suborder Metastigmata (Ixodides)

Large parasitic acarines known as ticks; mouth with recurved teeth modified for piercing; a tracheal spiracle located behind third or fourth pair of coxae.

Family Ixodidae (Hard Ticks)

Body ovoid; scutum present in all stages; scutum of adult males extends to posterior margin of body; scutum of adult females, like that of larvae and nymphs, restricted to propodosomal zone; capitulum anterior and visible from dorsal view; festoons usually present; eyes, if present, situated dorsally on sides of scutum; segments of pedipalps fused, not movable; porose areas present on base of capitulum in females; stigmatal plates large, posterior to coxa IV; only females distended when engorged with blood; with marked sexual dimorphism. (Genera mentioned in text: *Dermacentor, Amblyomma, Ixodes, Rhipicephalus, Haemaphysalis, Hyalomma, Boophilus.*)

Family Argasidae (Soft Ticks)

Integument leathery in nymphal and adult stages, and wrinkled, granulated, mammillated or having tubercles; scutum absent in all stages; capitulum either subterminal or protruding from anterior margin of body in nymphs and adults, subterminal or terminal in larvae; capitulum lies in distinctly or indistinctly marked depression (camerostome) in all stages; pedipalps freely articulate in all stages; porose area absent in both sexes; eyes usually absent (if present, on supracoxal folds); stigmata near coxa III lack stigmal plates; both sexes distended when engorged with blood; sexual dimorphism slight. (Genera mentioned in text: *Argas, Ornithodoros, Antricola, Otobius.*)

17

THE MITES

The term **mite** has no taxonomic significance. Members of the Acarina that are commonly designated mites belong to six separate suborders (see classification list at the end of this chapter). The phylogeny of these animals is obscure. Most authorities on the group consider them to be polyphyletic in origin.

Extensive monographs of these minute acarines are available. That by Baker and Wharton (1952) is most useful. For those interested in taxonomy, Johnston (1982), Kethley (1982), and O'Conner (1982) should be consulted.

Mites are of interest in parasitology for two main reasons. Some of these minute arthropods are ectoparasitic on animals, both vertebrates and invertebrates, and others on plants; and certain species serve as vectors for microorganisms (Table 17.1). By far, the majority of mites are free living and are found in every imaginable type of habitat, terrestrial as well as aquatic. Mites vary in size, ranging from 0.5 to 2.0 mm in length.

The anatomic terminology for mites is essentially the same as that used for the Metastigmata, considered in the previous chapter. Like those of the ticks, the stages in the life history of the mites include the larva, nymph, and adult forms. However, there are typically a single larval stage, two nymphal stages—the **protonymph** and the **deutonymph**—and the adult. The number of nymphal generations is increased or decreased in some species.

METHODS OF CONTACTING HOSTS

The methods by which parasitic mites make contact with their hosts enable us roughly to divide them into three groups (Camin, 1963). Members of the first group, such as the **feather mites** (Analgesidae, Dermoglyphidae, Proctophyllodidae, etc; p. 608), the **fur mites** (Myobiidae, p. 601), and the **mange mites** (Psoroptidae, p. 607), normally spend their entire life spans on their hosts. Furthermore, they are found on gregarious birds and mammals, and live through many generations on a single host. Transmission from host to host, when it occurs, is direct, resulting from contact between hosts or from contact with materials infested by contact with an infected host.

Members of the second category are nest parasites (**nidicoles**), which generally parasitize nest- and burrow-dwelling vertebrates. The members of the Dermanyssidae (p. 597) and certain members of the

Laelaptidae, which parasitize mammals (p. 596), belong to this group. These mites are hematophagous but generally are found only on their hosts during the feeding periods. They feed less frequently than do members of the first group, but the amount of blood ingested at each feeding is greater. After feeding, these mites drop off their host and are found in or near the host's nest. Thus, they can infest the same hosts when necessary. Transmission of these mites is accomplished in much the same way as in the first group, but there is greater opportunity for these mites to be transferred from one host to another, and they frequently parasitize several hosts of the same species during their life span. For this reason, the nidicolous mites are more important as vectors.

Members of the third category include the larval Trombiculidae, or chiggers (p. 601). Like most of the hard ticks and nidicoles, these mites are attached to their hosts only while feeding. However, these mites feed even less frequently than the nidicoles. A great portion of their life span is therefore spent off the host. Although the chiggers also occur on gregarious vertebrates, as are members of the second group, they are not as host specific and parasitize other types of hosts as well. These mites drop from their hosts practically anywhere and can survive for relatively long periods without a blood meal. When ready to feed again, they climb to some vantage point, such as on the surface of soil or tip of a blade of grass, and await the passing of a suitable host. Thus, the members of the third category are not dependent on the gregarious or nesting habits of their hosts and have the greatest opportunity in each generation to parasitize several hosts of the same species or several species of hosts. Therefore, these acarines are the most important from the standpoint of disease transmission.

BIOLOGY OF MITES

All mites, free living as well as parasitic, are subordinate to one of six suborders: Notostigmata, Tetrastigmata, Mesostigmata, Prostigmata, Astigmata, and Cryptostigmata. Members of the Notostigmata are brightly colored, omnivorous mites that live in leaf litter and beneath stones. Members of the Tetrastigmata (also known as Holothyroidea) are large, predatory mites found in Australia, New Zealand, and the Indo-Pacific. The Mesostigmata includes a large group of free-living as well as parasitic mites. The Prostigmata (also known as Trombidiformes) also includes many parasitic species, as does the Astigmata (also known as Sarcoptiformes). The members of Cryptostigmata are commonly referred to as the oribatid or beetle mites. These are common in leaf mold and soil. Their interest in parasitology stems from their role as intermediate hosts of certain tapeworms (p. 407).

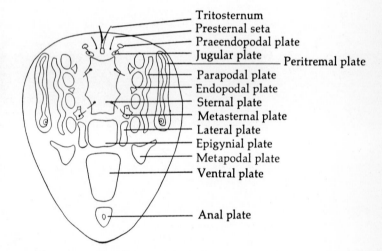

Fig. 17.1. Ventral plates that may be encountered in the Mesostigmata. A median plate, if present, would have the same position as the epigynial plate. (After Baker and Wharton, 1952.)

Table 17.1. Some Representative Etiologic Agents of Diseases Transmitted by Mites

Disease	Etiologic Agent	Mite Vector
Scrub typhus (tsutsugamushi fever)	*Rickettsia tsutsugamushi*	*Trombicula akamushi* (larvae) *T. deliensis* (larvae) *T. pallida*[a] *T. scutellaris*
Rickettsialpox	*R. akari*	*Allodermanyssus sanguineus* *Ornithonyssus (Bdellonyssus) bacoti*[b]
Q. fever	*Coxiella burnetti*	*O. (B.) bacoti*
Meadow mice scrub typhus	—	*Trombicula microti*

[a] Of minor importance.
[b] Potential vector.

Discussed below are the Mesostigmata, Prostigmata, and Astigmata. The diagnostic characteristics of each are listed in the classification section at the end of this chapter. For a practical taxonomic key to the mites as well as ticks, see McDaniel (1979).

SUBORDER MESOSTIGMATA

The Mesostigmata includes free-living as well as symbiotic mites. The heavily sclerotized body plates are quite characteristic (Fig. 17.1). Some 250 species of mesostigmatid mites are parasitic on vertebrates and invertebrates.* Some of the species that parasitize

---

*The true parasitic nature of some of the so-called zooparasites has not been established, and undoubtedly some of these are merely epiphoretic species.

vertebrates are of considerable economic importance and are treated in the following discussion.

The nature of this volume does not permit differentiation in detail of the numerous families of mesostigmatid mites, or the discussion, or even inclusion, of the many genera. Interested readers are referred to the volume by Baker and Wharton (1952), which is most useful in introducing the beginning investigator to the available literature dealing with each taxon.

Parasitic members of this suborder are relegated to seven groups:

Gamasides Group. With forked seta at base of palpal tarsus. Epigynial plate with one pair of genital setae and hinged to ventral plate. Male genital pore in front of sternal plate. Chelicerae of males modified as spermatophores.

Megisthanina Group. Epigynial plate primarily absent in females. Presternal setae flank tritosternum. Sternal plate, which functions as an epigynial plate, covers genital pore. Female genital pore as crescentic fissure between sternal and ventral plates.

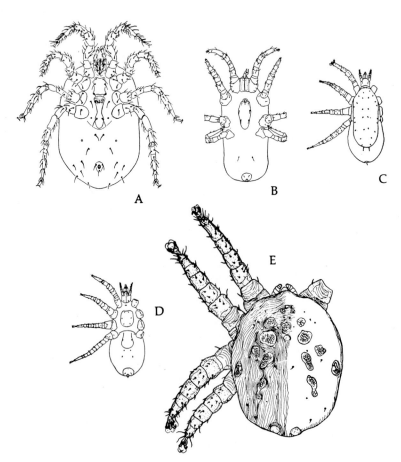

Fig. 17.2.   Some genera of mites. A. Ventral view of female *Raillietia auris*. (After Hirst, 1922.) **B.** Ventral view of male *Pneumonyssus simicola*. (After Vitzthum, 1931.) **C.** Dorsal view of female *Entonyssus glasmacheri*. **D.** Ventral view of female *E. glasmacheri*. (**C** and **D**, after Vitzthum 1935.) **E.** Dorsal view of female *Larinyssus orbicularis*. (After Strandtmann, 1948.)

Liroaspina Group. Lacks epigynial plate. Presternal setae lacking. Female genital pore as large, transverse slit between sternal and ventral plates.

Thinozerconina Group. Females lack lateral plates. Epigynial plate does not articulate at base, is narrow, and has one pair of setae. Male genital pore located in sternal plate.

Diarthrophallina Group. Penis biarticulated, directed posteriorly in groove between coxae of third pair of legs. With prestomal setae. Epigynial plate enclosed in fused sternal and ventral plates, but not separated from ventral plate by suture.

Celaenopsina Group. With a pair of lateral plates developed on ventral surface as anterior elongations of ventral plate, replacing epigynial plate, which is reduced or absent. Median plate absent. Metasternal plates large and uncovered in some, reduced and covered in others, and not attached to sternal plate.

Fedrizziina Group. Females possess median plate (sclerotized portion of vaginal wall) in addition to or in place of epigynial plate. Male genital pore in center of sternal plate, closed by plate attached to margin of pore.

GAMASIDES GROUP

The parasitic representatives of this extremely large group belong to 14 families. Bregetova (1965) has contributed a monograph on the group. Some of the families are discussed below.

Family Raillietidae

This family includes two fairly common ectoparasites of mammals. *Raillietia hopkinsi* is found on the ears of antelopes and *R. auris* on the ears of cattle in North America and Europe (Fig. 17.2). These mites are not true parasites, for they feed on the host's ear wax and sloughed epithelia rather than on living tissues.

Family Halarachnidae

This family includes the genera *Halarachne*, *Pneumonyssus*, and *Orthohalarachne*. Species of *Halarachne* occur in the respiratory passages of seals. Species of *Pneumonyssus* occur in the respiratory tract of mammals, particularly monkeys, apes, and dogs. *Pneumanyssus simicola* is found in the lungs of rhesus monkeys and is capable of causing inflammation (Fig. 17.2). Species of *Orthohalarachne* occur in the air passages, including lungs, of seals. In the United States, *O. zalophi* is found in the nasal passages of the California sea lion. Its presence may produce inflammation, a condition known as **pulmonary** or **nasal acariasis**. It should be apparent that the halarachnid mites are all respiratory tract parasites of mammals. They normally feed on lymph, but they may ingest blood.

Family Entonyssidae

All members of this family are parasitic in the respiratory tract, particularly in the lungs, of snakes. The family includes the genera *Entonyssus* (Fig. 17.2), *Hammertonia*, and *Ophiopneumicola*. The species are ovoviviparous and are widely distributed in America and Africa.

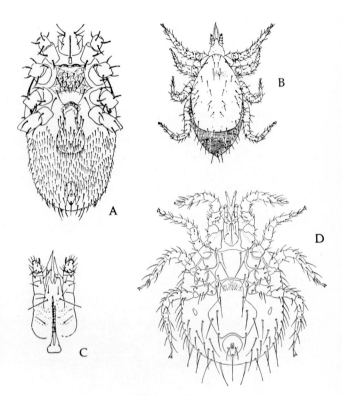

Fig. 17.3. Mites. A. Ventral view of female *Haemogamasus mandschuricus*. (After Vitzthum, 1930.) **B.** Dorsal view of female *Myrmonyssus chapmani*. **C.** Ventral view of capitulum of *M. chapmani*. (**B** and **C**, after Baker and Strandtmann, 1948.) **D.** Ventral view of female *Echinolaelaps echidninus*. (After Hirst, 1922.)

Family Rhinonyssidae

The members of this family are universally distributed and are parasitic in the nasal passages of birds. Among the genera are *Rhinonyssus*, *Larinyssus* (Fig. 17.2), and *Sternostomum*. *Sternostomum* includes *S. rhinoletrum*, which causes catarrhal inflammation in the respiratory passages of fowls.

Family Haemogamasidae

This family includes the genera *Haemogamasus* (Fig. 17.3), *Acanthochela*, and others. Almost all haemogamasids are found on various small mammals in every part of the world. Occasionally, some are found free living and others are found on birds. Those members found on animals are true parasites, feeding on the host's blood, but their attachment to the host is intermittent, and most of the time they are found off the host and in its nest. The life cycle includes the typical larval, protonymphal, deutonymphal, and adult stages. Since there appears to be little, if any, host specificity among these mites, passing from one type of host to another as they do, Baker and Wharton (1952) have suggested that the haemogamasids may be of potential medical importance as vectors for the plague, typhus, and tularemia pathogens.

Family Laelaptidae

This family includes the largest number of parasitic genera. Many laelaptids are found on insects; others are found on mammals. *Myrmonyssus* is a representative genus found on insects (Fig. 17.3). Genera found on mammals include *Echinolaelaps* (Fig. 17.3) and *Haemolaelaps*. *Echinolaelaps* includes *E. echidninus*, which is the vector for the protozoan parasite *Hepatozoon muris* of rats. The mite transmits the protozoan from rat to rat. *Haemolaelaps* includes *H. arcuatus*, which is the vector for *Hepatozoon criceti*, a parasite of hamsters. The larvae and nymphs of parasitic laelaptid mites are mostly lymph feeders, and the adults are hematophagous. The life cycles of these mites, including the egg, larval, protonymphal, deutonymphal, and adult stages are completed between 8 and 28 days.

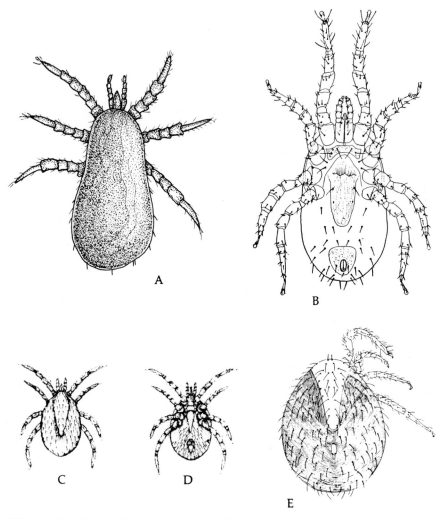

Fig. 17.4.   Mites. A. Dorsal view of female *Dermanyssus gallinae*. **B.** Ventral view of female *D. gallinae*. (**A**, after James and Harwood, 1969; **B**, after Hirst, 1922.) **C.** Dorsal view of *Ornithonyssus bacoti*. **D.** Ventral view of *O. bacoti*. (After Dove and Shelmire, 1932.) **E.** Dorsal view of female *Liponyssoides sanguineus*. (After Baker and Wharton, 1952.)

The gamasid family Dermanyssidae is of great interest in veterinary medicine because it includes several species that inflict considerable injury to domestic animals.

Dermanyssus gallinae, the common red chicken mite, found on chickens throughout the world, also parasitizes turkeys, pigeons, certain wild birds, and, occasionally, humans (Fig. 17.4). This pear-shaped, grayish mite measures 0.6–0.8 mm in length. The males are slightly smaller. Engorged specimens are larger and dark red. Chickens infested with *D. gallinae* often suffer reduced egg-laying capacity; some stop laying completely. The fowls lose weight, become susceptible to other diseases, and even die. Young chicks usually die when attacked by this mite. Thus, large

flocks of chickens can readily be lost when heavily infested. Furthermore, this mite serves as the vector for avian spirochaetes.

Life Cycle of *Dermanyssus gallinae*. This mite is primarily a night feeder, hiding in crevices and under debris during the day. The female begins ovipositing from 12 to 24 hours after feeding and lays up to to seven eggs, which in summer temperatures hatch in 48 to 72 hours. The hexapod larvae are sluggish and do not feed; they transform into protonymphs in approximately 24 to 48 hours. The octopod protonymphs are very active. After taking in a blood

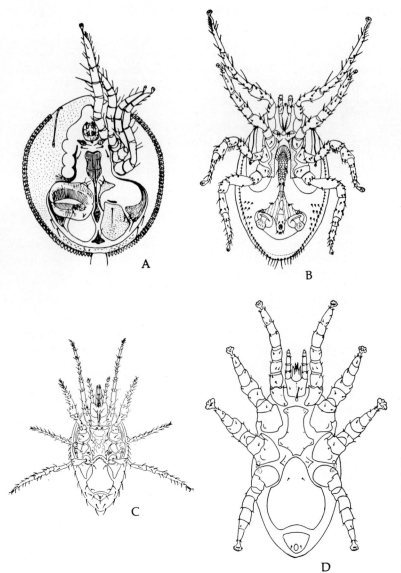

Fig. 17.5. Mites. A. Ventral view of female *Discozercon mirabilis.* (After Berlese, 1910.) **B.** Ventral view of female *Heterozercon oudemansi.* (After Finnegan, 1931.) **C.** Ventral view of female *Pachylaelaps roosevelti.* (After Wharton, 1941.) **D.** Ventral view of female *Jacobsonia tertia.* (After Vitzthum, 1931.)

meal, the protonymphs molt and metamorphose into deutonymphs. Deutonymphs also engorge on the host's blood before molting, giving rise to adults. Third-stage nymphs, although rare, have been reported. The entire life cycle may take as little as 7 days. Thus, if uncontrolled, the mite population in chicken coops multiplies rapidly.

Ornithonyssus bursa, the tropical fowl mite, is widely distributed in Argentina, Brazil, China, India, parts of Africa, Colombia, the Bahamas, and the southern United States. Unlike *Dermanyssus gallinae*, this species is both a day and night feeder on chickens and on the English sparrow, *Passer domesticus*. The adults, measuring approximately 1 mm in length, remain on the host and feed intermittently. The eggs are oviposited either on the host's fluff feathers or in the nest. After 3 days of incubation, the minute six-legged larvae appear. The larva does not feed and molts in about 17 hours, giving rise to the protonymph. The protonymph is an active feeder and, after molting, gives rise to the deutonymph, which in turn feeds, molts, and gives rise to the adult. The entire life cycle takes 8 to 12 days.

Ornithonyssus bacoti, the tropical rat mite, is usually found on the Norway rat, *Rattus norvegicus*, in various parts of the world including the United States (Fig. 17.4). Like *O. bursa*, *O. bacoti* does bite humans, resulting in pain and local inflammation. However, no severe symptoms ensue. This species serves as a vector for the Texas strain of endemic typhus when experimentally tested on infected guinea pigs. It also serves as a vector for *Rickettsia akari*, a pathogen of mice, under experimental conditions. It can also serve as an intermediate host for the cotton rat worm, *Litomosoides carinii* (p. 543). The life history of *O. bacoti* is quite typical—one larval, two nymphal, and an adult stage. Bertram *et al.* (1946) have stated that the unfertilized eggs develop parthenogenically. Mites, by dropping off the host, are capable of traveling relatively long distances; hence, they are widely dispersed.

Liponyssoides sanguineus (Fig. 17.4), the mouse mite, measuring 0.80 × 0.46 mm, is the vector for the rickettsialpox organism, *Rickettsia akari*. The house mouse, *Mus musculus*, serves as the reservoir host, and mites transmit the pathogen from mice to humans. This organism occurs in northern Africa, Asia, Europe, and the United States.

Gamasid Mites of Invertebrates

Several families of the Gamasides group include species parasitic on invertebrates. The family Discozerconidae includes the tropical and subtropical genera *Discozercon*, *Heterozercon*, and *Allozercon*, mostly found ectoparasitic on large millipedes, centipedes, and termites, and sometimes on snakes (Fig. 17.5). A few genera are free living. The family Pachylaelaptidae includes *Pachylaelaps* (Fig. 17.5) and others, mostly found on insects, especially beetles. Some are free living. The family Macrochelidae includes *Macrocheles muscae*, which may or may not be a true parasite. The larvae of this species occur on the eggs of houseflies. Family Iphiopsidae includes *Iphiopsis* and *Jacobsonia* (Fig. 17.5), both found on insects and myriopods, either as true parasites or as epiphoronts. Since true parasites and epiphoretic species are found among closely related species, these mites can be of great importance as experimental animals in studying the transition between the two types of host-symbiont relationships.

OTHER REPRESENTATIVE MESOSTIGMATE

Megisthanina Group. The parasitic members of this group all belong to the genus *Megisthanus*. They are found as ectoparasites on large beetles in moist environments. In the United States, the common species is *M. floridanus*, found on the patent leather bettle, *Popillus disjunctus*.

Liroaspina Group. All members of this group are free living and are found primarily in the tropics. However, Fox (1947) has reported finding *Liroaspis armatus* (Fig. 17.6) on rats in Puerto Rico. There is considerable doubt, however, whether this species is a true parasite. It is suspected that its presence on rats is accidental and temporary.

Thinozerconina Group. Included in this small group is *Dasyponyssus neivai*. This mite was discovered in Brazil on an armadillo.

Diarthrophallina Group. This group includes *Diarthrophallus* and *Passelobia*, both found on various passalid beetles in North, Central, and South America, and in New Guinea. *Diarthrophallus quercus* (Fig. 17.6) is very common on beetles around Durham, North Carolina.

Celaenopsina Group. The genus *Schizogynium* includes species parasitic on beetles in Africa (Fig. 17.6).

Fedrizziina Group. This group includes *Parantennulus* and *Diplopodophilus* (Fig. 17.6). Species of both genera occur on carabid beetles and myriopods.

SUBORDER PROSTIGMATA

Some of the species of the Prostigmata are parasitic as adults; others are parasitic as larvae. Many other species are free-living predators. The parasitic species are found on plants as well as on vertebrates and invertebrates and are assigned to one of the following groups.

Tetrapodili Group. Found on plants. Bodies minute, wormlike, and ringed (annulated). Bodies up to 0.2 mm in length. Only two pairs of legs, which are without claws but possess feather claws. Body with few setae. Capitulum reduced. Chelicerae styletlike. Pedipalps minute, adjacent to rostrum. No respiratory system.

Tarsonemini Group. Parasitic on plants and vertebrates and invertebrates. Minute mites, with or without the usual eight legs. Chelicerae minute and styletlike. Pedipalps tiny and located behind chelicerae. With pair of clublike pseudostigmatic organs between coxae and first and second pairs of legs. Males lack tracheae and stigmata.

Prostigmata Group.* Bodies comparatively

---

*Not to be confused with the suborder Prostigmata to which this group is subordinate.

large, with varying shapes, depending on family. Capitulum usually conspicuous. Usually with eight legs. Stigmata usually present, at base of large chelicerae. Pedipalps larger than those in other groups and not closely adjacent to rostrum.

TARSONEMINI GROUP

The parasitic members of this group include the genera *Podapolipus, Locustacarus* (Fig. 17.7), *Eutarsopolipus, Tarsopolipus,* and *Tetrapolipus.* All species belonging to these genera are ectoparasites of other arthropods. For example, *Podapolipus reconditus* is found under the elytra of European beetles, and *P. bacillus* and *P. grassi* are ectoparasites of grasshoppers.

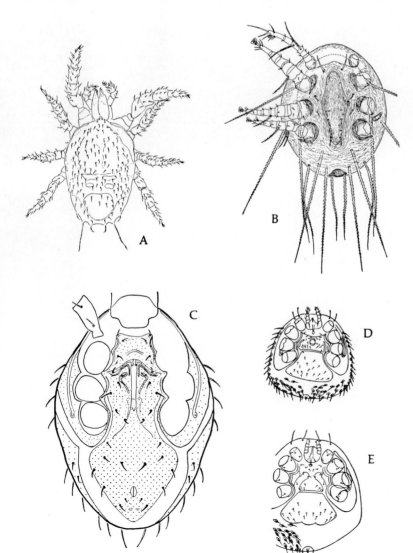

Fig. 17.6. **Mites. A.** Dorsal view of female *Liroaspis armatus.* (After Baker and Wharton, 1952.) **B.** Ventral view of nymph of *Diarthrophallus quercus.* (After Trägårdh, 1946.) **C.** Ventral view of female *Schizogynium intermedium.* (After Trägårdh, 1950.) **D.** Ventral view of male *Diplopodophilus antennophoroides.* **E.** Ventral view of female *D. antennophoroides.* (**D** and **E,** after Willmann, 1941.)

Locustacarus trachealis occurs in the tracheae and air sacs of grasshoppers in the United States and South Africa. It obtains nutrients by piercing the tracheal wall of the host and sucking hemolymph. The genus *Tetrapolipus* includes *T. rhynchophori* found beneath the elytra of the palm weevil, *Rhynchophorus palmarum.*

Life Cycle of *Podapolipus diander.* The life cycle of *P. diander* has been studied by Volkonsky (1940). This mite is a parasite of the grasshopper *Locusta migratoria* in Algeria. The developmental pattern is unusual yet representative of the group.

Young mites are attached to the exoskeleton of the first, second, and third instars of the grasshopper, under the posterior projection of the pronotum. When the host is in its fourth and fifth instar stages, the mite oviposits her eggs under the host's elytra and wings. When the host molts, attaining the adult form, the parasites' eggs have have already hatched, and the young mites become attached to the new exoskeleton of the host, also under the elytra and wings. These immature female mites mature at these sites and lay eggs that hatch into "small" males parthenogenically. These small males copulate with parental females, which then oviposit another type of egg that hatches in five to six days to give rise to females as well as "large" males. These large males are hyperparasitic on the females. The males perforate the abdomen of the females and draw their nutrients, sometimes killing the females.

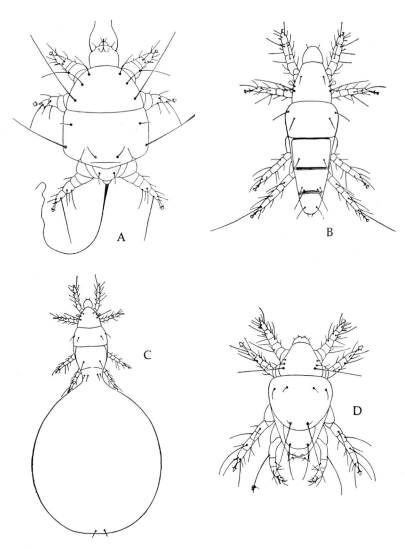

Fig. 17.7. **Mites. A.** Young female of *Locustacarus trachealis.* **B.** Dorsal view of female *Pyemotes ventricosus.* **C.** Gravid female of *P. ventricosus.* **D.** Dorsal view of male of *P. ventricosus.* (All figures after Baker and Wharton, 1952.)

The parasitized females are attached to the intersegmental membranes of the grasshopper's body, where they pierce the integument to draw out nutrient. This position, however, is not permanent, for the female mites gradually migrate posteriorly and congregate at the host's genitalia. The spread of infestations from host to host is accomplished during copulation. The adults are probably neotenic larvae since they possess only six legs. *Podapolipus diander* is thermosensitive, for if the temperature rises above 30°C, the parasite leaves the host and hides either in crevices or on blades of grass, from which it can later attach to another host.

Also included in the Tarsonemini group is *Acarapis woodi*, the Isle of Wight disease mite of bees. This mite is of considerable economic importance in Europe, where commerical beekeeping thrives. The mite is found in the tracheae of bees, where it feeds on the host's body fluid, secretes a toxic substance, and mechanically clogs the air passages, killing the host.

Included in the tarsoneminian genus *Pyemotes* is *P. ventricosus* (Fig. 17.7). This parasite is found on the larvae of various insect pests, such as the grain moth, *Sitotroga cerealella*; the wheat jointworm, *Harmolita grandis*; the peach twig borer, *Anarsia lineatella*; the boll weevil, *Anthonomus grandis*; and the bean and pea weevil, *Mylabris quadrimaculatus*. Because of the fetal parasitic habit of *P. ventricosus* on destructive insects, this species represents one parasite that is economically beneficial. However, this mite bites humans closely associated with the insect hosts, causing a type of acarodermatitis known as "straw itch."

PROSTIGMATA GROUP

Most of the members of this group are free living, either as predators on other mites or on insects or in damp areas such as mossy beds and in humus. Some species, however, are true parasites. The Prostigmata group includes the interesting slug mite, *Riccardoella limacum*, in Europe and the United States. Turk and Phillips (1946), in defining the life cycle of this mite, have reported that it is neotenic. The stages in its life cycle include the typical larva, protonymph, deutonymph, and adult, but it is the deutonymph that contains the fully formed eggs and is either viviparous or ovoviviparous. Eggs found in adult females are not normally developed.

Another prostigmatan species is *Speleognathus sturni* (Fig. 17.8) found in the nasal passages of starlings in the eastern United States. The larva, unlike the other instars, is not parasitic. It is commonly lodged in the mucous secretions of the bird.

Another group of mites belonging to the Prostigmata group, members of the family Myobiidae, possess first pairs of legs that are modified as clasping appendages by which they hold onto the base of the hair of their mammalian hosts or onto the shaft of feathers of avian hosts. Representative genera include

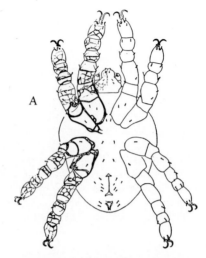

Fig. 17.8. Mite. Ventral view of *Speleognathus sturni* female. (After Boyd, 1948.)

Myobia (Fig. 17.9), *Amorphacarus*, and *Syringophilus* (Fig. 17.9) with various species distributed throughout North America and Europe. These mites are found inside the quills of the avian host's feathers and are believed to feed on the internal cones of the quills. The genus *Harpirhynchus* includes *H. nidulans*, which is found in colonies in the feather follicles of numerous species of birds. Its presence causes tumors and cysts. Members of *Ophioptes* are found in pits in the epidermis of South American snakes. Members of *Psorergates* are of economic significance to sheep farmers in Australia because at least one species, *P. ovis*, living at the base of the hair of sheep, causes chronic irritations.

The 15 or more genera belonging to the family Erythraeidae of the Prostigmata group are found as larvae on insects and arachnids and are represented here by *Leptus* (Fig. 17.9) and *Erythraeus*. The larva of *Leptus atticolus* is attached to the legs of spiders, whereas that of *E. swazianus* is attached by its mouthparts to the undersurface of the wings and on the tympanic membrane of locusts. Adult erythraeids are reddish (hence the familial name) and have legs adapted for running. Adults are predaceous and are commonly found in foliage and humus and on beach sand. The larvae of other species of *Leptus* are found attached to phelangids, pseudoscorpions, and scorpions.

Probably the most irritating mites from the standpoint of injury to humans are the chigger mites. These prostigmatan mites belong to the family Trombiculidae. Their bites commonly result in dermatosis. Also, certain species serve as vectors for pathogens. For those interested in detailed accounts of the role of trombiculid mites as vectors, the papers of Ewing (1944) and Williams (1944) should be consulted. Genera representative of the Trombiculidae include *Euschongastia* (Fig. 17.10), *Trombicula*, and *Doloisia*.

Genus *Trombicula*. This genus includes several frequently encountered species. *Trombicula autumnalis*, the harvest mite, is found throughout Europe. Men working the fields, when bitten, suffer from a severe dermatosis. In the Americas, at least three species are known to attack humans. *Trombicula mansoni* is found along the entire eastern coast of the United States, along the Gulf Coast to Texas, and in isolated locales in Minnesota and Michigan. *Trombicula alfreddugesi* (Fig. 17.10) ranges from New England to Nebraska and south into Florida and Texas, and has also been reported in California, Mexico, and parts of Central and South America. A bright red species, *T. batatas*, is a tropical form sometimes encountered in the southern United States. These three species also cause severe dermatosis in humans. The adults are free living and predaceous, feeding primarily on insect eggs.

Dogs, horses, shrews, and various birds serve as hosts for the parasitic larva of *T. autumnalis*, and turtles, young toads, and snakes are suitable hosts for *T. alfreddugesi*.

Life Cycle of Trombicula batatas. The life cycles of the species of *Trombicula* follow the same basic pattern. The dull orange-colored egg of *T. batatas* is oviposited singly on moist soil. In 4 to 5 days, the eggshell cracks in half, but the larva remains within the cracked shell in the **deutovum** (a chorionic layer secreted by the blastoderm during embryonic development) for another week before the red hexapod larva escapes and actively crawls around seeking a host. Such larvae can survive for approximately 2 weeks without food.

There appears to be little host specificity, since *T. batatas* is known to bite literally up to 100 different vertebrates, including amphibians, reptiles, birds, and mammals. The feeding mechanism is interesting in that the mite does not suck blood; rather, it secretes a lytic substance that dissolves the host's tissue. Such dissolved materials are ingested. After engorging itself on the host's digested tissues for 2 to 10 days, the larva drops off and becomes semidormant, after which it molts and transforms into the nymph.* The nymphal form is maintained for 21 to 52 days, after which the nymph molts and the free-living adult form is attained, surviving up to 45 days.

Trombicula akamushi (Fig. 17.10) and a variety formerly known as *T. deliensis* are vectors of *Rickettsia tsutsugamushi*, the scrub typhus organism. *Trombicula akamushi* occurs in Japan, Formosa, Ceylon, India, the Philippines, Indonesia, and New Guinea. It inhabits partially cultivated land that is inundated by floods during spring and early summer. The mite population usually reaches a peak during July and August. *Trombicula akamushi deliensis* ranges from the southern Soviet Union and Pakistan and India down to New

---

*Some consider the quiescent stage of the larva to be the protonymphal stage.

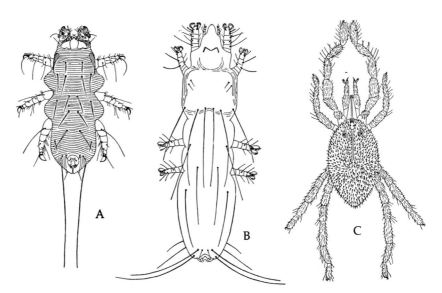

Fig. 17.9. **Mites. A.** Dorsal view of *Myobia musculinus*. (After Baker and Wharton, 1952.) **B.** Dorsal view of female *Syringophilus columbae*. (After Hirst, 1922.) **C.** Dorsal view of female *Leptus hirtipes*. (After Vitzthum, 1926.)

Guinea and Australia. It is characteristically associated with forests.

Scrub typhus (tsutsugamushi fever) is a much dreaded disease, as American combat troops discovered in Asia during World War II. It is not restricted to Asia and Australasia since endemic foci occur in the sourthern Soviet Union. The larval mite serves as the vector in carrying the rickettsiae from the natural reservoirs—voles (*Microtes montebelli*), rats (*Rattus concolor browni, R. flavipectus yunangensis*), and other small mammals—to human hosts. This medically important mite is reddish as an adult and measures 1–2 mm in length. The adults are free living and feed on insect eggs and minute arthropods. Transovarial passage of rickettsiae occurs in nature. The life history of this species parallels that of the other species of *Trombicula* in that there is only one nymphal stage.

Genus *Demodex*. Another group of medically important mites is the demodicid or follicle mites. These belong to the family Demodicidae, which includes a single genus, *Demodex*, with many species. These occur within the hair follicles of their hosts. These minute (0.3–0.4 mm long) mites are wormlike and have four pairs of stubby legs and a striated abdomen.

Demodex folliculorum. *D. folliculorum* (Fig. 17.11), the follicle mite of humans, is found particular-

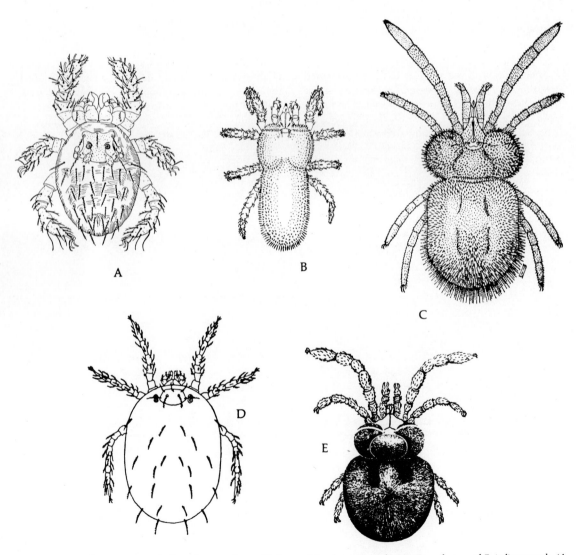

Fig. 17.10.   Chigger mites. A. Dorsal view of parasitic larva of *Euschongastia indica*. **B.** Dorsal view of *E. indica* nymph. (**A** and **B**, after Wharton, 1946.) **C.** *Trombicula alfreddugesi* adult. **D.** *T. alfreddugesi* larva. (**C** and **D**, after Ewing, 1944.) **E.** *Trombicula akamushi* adult. (After Neal and Barnett, 1961.)

Fig. 17.11. Follicle mite. Ventral view of *Demodex folliculorum*. (After James and Harwood, 1969.)

ly around the nose and eyes. It has been estimated that as much as 50% of the world's population harbor this parasite. Their habitat is the hair follicle and associated sebaceous glands. Usually the host is asymptomatic, but these mites may cause inflammation result-

ing in an acnelike condition. The entire life cycle occurs within the follicles. The typical larval, protonymphal, deutonymphal, and adult stages are present.

Other Demodex *Species. Demodex canis* parasitizes dogs around the head. The presence of this mite, along with a staphylococcal bacterium, causes the so-called red mange. *Demodex cati* parasitizes cats, *D. equi* parasitizes horses, and *D. bovis* is found on cattle—all in the head and neck regions. *Demodex bovis* causes depreciation in the value of the skin from these parts for leather. *Demodex phylloides* parasitizes pigs, but unlike the other species is scattered over the entire body, causing white pustules. *Demodex muscardini* is a parasite of the dormouse. For discussions on biology, pathology, and economic losses due to *Demodex* spp., see the volume edited by Parish *et al.* (1983).

Water Mites

Another group of prostigmatan mites that is of interest to parasitologists are the members of the family Hydrachnellidae. These mites form an ecologic rather than a morphologic group. All are found in fresh water. Some are free swimming, and others are nonswimming. Brightly colored species are commonly found in lakes and ponds, along shores, and in streams.

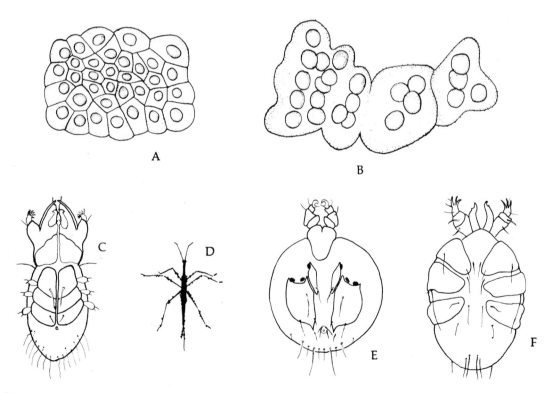

Fig. 17.12. Hydracarina. A. Eggs of *Hydryphantes ruber*. Each egg is surrounded by a gelatinous matrix. **B.** Eggs of *Eylais extendens*. Four eggs are enveloped together within a single capsule. **C.** Ventral view of parasitic larva of *Hydrachna magniscutata*. **D.** Specimen of *Ranatra* sp. heavily parasitized by larvae of *H. magniscutata*. **E.** Ventral view of parasitic larva of *Piona reighardi*. **F.** Ventral view of parasitic larva *of Thyas stolli*. (All figures redrawn after Crowell, 1960.)

Related to the Hydrachnellidae is the family Halacaridae, the members of which are marine mites.

The life cycle of water mites is known only in a general way. There are typically four stages—egg, larva, nymph, and adult. In many species the larvae are parasitic on aquatic insects, whereas the nymphs and adults are free living, predaceous animals, feeding on crustaceans, insect larvae, and other small aquatic organisms. The adults of a few species are parasitic in the gills of freshwater mussels and in the gill chambers of crabs. Water mites can be found at all periods during the year, although adults are most abundant during late summer and fall.

Life Cycle Pattern. The commonly reddish orange eggs of water mites are laid singly, in small clutches, or in masses of up to 400 or more eggs. Each egg may be encapsulated in a gelatinous matrix, or two to four eggs may be arranged in a single capsule (Fig. 17.12). In some species, the entire egg mass, containing 100 or more eggs, may be covered with a gelatinous secretion. The encapsulated eggs may be irregularly arranged, or linearly arranged in short, straight ribbons. Such egg masses are commonly attached to submerged stones, sticks, or aquatic plants. Leaves of *Sphagnum* and *Myriophyllum* are common substrates to which eggs are attached. The eggs of certain species are deposited in cavities made in plant tissues.

The eggs of water mites hatch in less than a week to about 6 months, depending on the species and ambient factors. From each egg capsule a six-legged larva emerges. Some larvae, such as those of *Lebertia*, rupture their capsules by exerting pressure. Others, such as *Thyas*, utilize spines situated on their dorsal body surface to rupture the capsule.

After a short period of free existence in most species, the larvae, which are the parasitic instars, attach themselves to aquatic and semiaquatic insects by means of their well-developed capitulum and derive their nourishment from the host.

After the larval parasitic period, the animal undergoes metamorphosis. Each larva shrinks within its exoskeleton, and within a short period the developing nymph can be seen within. The fully developed nymph eventually escapes from the surrounding larval exoskeleton as a **nymphochrysalis**, either by vigorous activity of the legs or from a transverse slit that develops in the larval skin about two-thirds of the body length from the anterior end (Crowell, 1960). The nymphs, like the adults, possess four pairs of six-segmented legs and are generally free swimming.

After a period of free existence, the nymphs become quiescent (**imagochrysalis stage**) and are attached to plant tissues or in some obscure habitat. From this stage, the adult eventually emerges.

Hosts of Larval Hydracarina. Aquatic and semiaquatic insects are favored hosts of larval water mites.

Very little information is available concerning host–parasite relationships between these mites and their hosts. Marshall and Staley (1929) have reported the presence of unbranched feeding tubes, or **stylostomes**, which connect the larvae of *Lebertia tauinsignata* to their mosquito hosts. Wharton (1954), as a result of histochemical studies, concluded that the stylostome that connects *Trombidium* spp. to their host, the firefly, *Photuris pennsylvanica*, is formed as the result of the precipitation of the mite's salivary secretion by the host's tissues. Solidification occurs at the interface between the insect's blood and the mite's saliva.

Among the hosts of larval water mites found by Crowell (1960) in Ohio, stoneflies (Plecoptera),

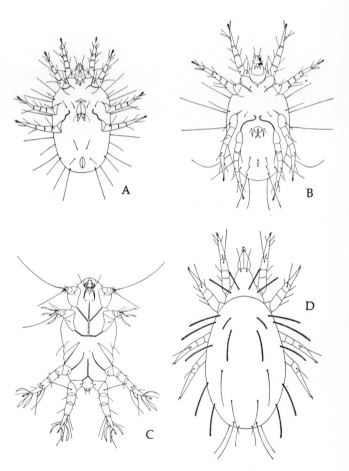

Fig. 17.13. Astigmatan mites. A. Ventral view of female *Ensliniella parasitica*. **B.** Ventral view of male *E. parasitica*. **C.** Ventral view of hypopial nymph of *E. parasitica*. (**A–C**, after Vitzthum, 1925.) **D.** Dorsal view of female *Forcellinia wasmanni*. (After Michael, 1903.)

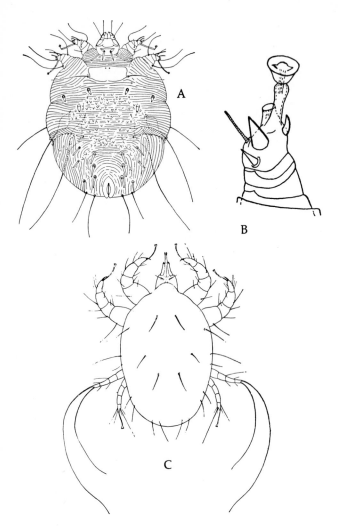

Fig. 17.14. **Itch mites. A.** Dorsal view of female *Sarcoptes scabiei* var. *equi.* (After Hirst, 1922.) **B.** Tarsus I of *Notoedres cati.* (After Grandjean, 1938.) **C.** Dorsal view of *Psoroptes communis.* (After Herms, 1950.)

Uchida and Miyazaki (1935) have reported that in certain areas larval water mites may well serve as biologic control agents. For example, anopheline mosquitoes with five or more larval mites attached to their bodies cannot be induced to bite; hence, the necessary blood meal cannot be obtained.

SUBORDER ASTIGMATA

Members of the Astigmata all belong to one group, the **Acaridiae**. These possess a thin, soft integument, lack stigmata and tracheae, and lack prominent club-shaped pseudostigmatic organs. Caruncles are present on the tarsi, which may or may not bear one claw. Many males possess copulatory suckers on their tarsi and anal region.

ACARIDIAE GROUP

Many members of this group are free living; however, some representatives are parasitic. These parasitic forms can be categorized into three types—those parasitic on invertebrates, those found in the skin and other tissues of warm-blooded vertebrates, and those found on the feathers of birds.

Parasites of Invertebrates

There are many species parasitic on invertebrates. For example, *Ensliniella parasitica* (Fig. 17.13) occurs on the larvae of *Pdynerus delphinalis* (Hymenoptera) in Germany, *Vidia undulata* on *Prosopis conformis* (Hymenoptera) in Italy, and *Forcellinia wasmanni* (Fig. 17.13) in the nest and on ants in Europe, where heavy infestations are lethal.

Parasites in Skin and Other Tissues of Warm-Blooded Vertebrates

Astigmatan mites found in the skin and other tissues of warm-blooded vertebrates include members of the families Sarcoptidae and Psoroptidae. These mites are commonly referred to as the **itch mites**.

Genus Sarcoptes. *Sarcoptes* includes *S. scabiei*, the itch mite of humans (Figs. 17.14, 17.15). This widely distributed mite is one of the earliest known and described parasites. The skin disease caused by *S. scabiei* is commonly referred to as scabies or the "seven-year itch." Females, measuring 0.33–0.45 mm in length by 0.35 mm in width, and males, measuring little more than half that size, both cause scabies; however, the females are by far the most irritating. These mites prefer surfaces of the body where the skin is less thick, such as the breast, penis, shoulder blades, or the area between the fingers. Nevertheless, they do infect other parts of the body. Irritation to the host is caused by the burrowing of the mites, particularly females, and by the secretion and excretion of toxic substances within the host's epidermis. The irritation

various hemipterans of the families Notonectidae, Corixidae, and Nepidae, and dipterans of the families Tendipedidae and Cuclicidae were infected. In addition, dragonflies (Odonata) and caddis flies (Trichoptera) are suitable hosts.

In North America, the larvae of *Hydrachna magniscutata* (Fig. 17.12) is commonly found on such aquatic insects as *Arctocorixa*, the water boatman; *Ranatra*, the water scorpion; and *Notonecta undulata*, the back swimmer. The nymphs and larvae of *Limnesia* and *Piona* parasitize midges, and the larvae of *Thysas* are commonly found on mosquitoes (Fig. 17.12). In fact,

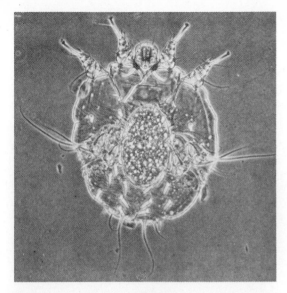

Fig. 17.15. *Sarcoptes scabiei*, the itch mite. (After J. Georgi in Schmidt and Roberts, 1981; with permission of C. V. Mosby.)

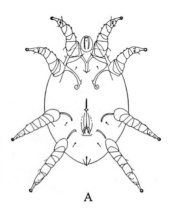

A

Fig. 17.16. *Cytodites.* Ventral view of female *Cytodites nudus.* (After Hirst, 1922.)

is often complicated by scratching and secondary infections.

Life Cycle of Sarcoptes scabiei. Impregnated females of *S. scabiei* lay their rather large eggs (0.15 × 0.1 mm) in tunnels dug during their migration in the host's epidermis. Each female oviposits 10 to 25 eggs at 2- to 3-day intervals for about two months, after which she perishes. The eggs hatch in three to four days, and the larvae actively move about, spreading the infection. The protonymphal and deutonymphal stages are each preceded by a molt, and the adult form is attained in 10 to 12 days. Infections are usually acquired through direct contact.

Varieties of *Sarcoptes scabiei* occur on domestic animals, each causing a skin condition known as mange. For example, *S. scabiei* var. *suis* occurs on pigs, var. *equi* on horses, var. *bovis* on cows, and var. *canis* on dogs. *Sarcoptes scabiei* var. *equi* and var. *canis* are transmissible to humans. For discussions on the biology and pathology of *Sarcoptes* spp., see the volume edited by Parish *et al.* (1983).

Other Sarcoptid Genera. Other genera of the Sarcoptidae include *Notoedres*, with *N. cati*, the feline mange-causing mite (Fig. 17.14), and *Knemidokoptes*, with two important species, *K. mutans*, the scaly-leg mite of chickens, and *K. laevis*, the depluming mite of chickens, pheasants, and geese.

Members of the Psoroptidae are scab-causing mites, primarily on mammals. *Psoroptes* includes *P. communis*, with various varieties that cause scabies (Fig. 17.14). For example, *P. communis* var. *ovis* causes scabies in sheep. Copulation among adults of this species occurs on the host, but the females undergo an

additional molt before ovipositing. The eggs are laid on the wool. *Psoroptes communis* var. *equi* causes the serious scab disease of horses, and *P. communis* var. *bovis* causes psoroptic mange of cattle. On rabbits, *P. communis* var. *cuniculi* causes ear canker.

Other genera of the Psoroptidae include *Caparinia*; *Otodectes*; and *Chorioptes*, with *C. symbiotes*, which causes the foot scab disease of horses, *C. bovis*, the tail mange mite of cattle, and *C. ovis*, the foot mange mite of sheep.

Other Astigmatan Species. *Cytodites nudus* is another astigmatan mite which occurs in the air sacs, respiratory system, and body cavity of chickens, canaries, ruffed grouse, and other birds (Fig. 17.16). Within the body cavity, these mites are attached to the visceral organs such as the liver. If large numbers are present, peritonitis and enteritis ensue. Also, if large numbers are present in the respiratory system, suffocation generally results.

Laminosioptes cysticola is parasitic in various tissues of domestic fowl. This mite is harmful in that it invades subcutaneous tissues in the neck, breast, and flanks. It destroys the connective tissue fibers and myofibers of birds. Whitish calcareous cysts are commonly found in the tissues of birds harboring this mite. Such cysts are not formed around living mites, but represent a host tissue reaction to aggregations of dead mites accumulated within the tissues.

Another group of astigmatan mites generally occurs on the skin of birds. Representative of this group (family Epidermoptidae) are *Epidermoptes bilobatus*, which lives in the skin of chickens and causes a scaly condition known as pityriasis, and *Dermatophagoides scheremetewskyi*, which, in addition to birds, will attack humans and causes a mangelike inflammation (Sasa, 1950; Traver, 1951).

Parasites in Feathers of Birds

Most of the so-called feather mites of birds belong to three families; Analgesidae, Dermoglyphidae, and Proctophyllodidae. Quite representative of the family Analgesidae is *Megninia columbae*, found on pigeons (Fig. 17.17). Dermoglyphidae includes over 35 genera represented here by *Falculifer* and *Pterolichus*, both found on various birds (Fig. 17.17). Proctophyllodidae includes *Trouessartia* (Fig. 17.17).

The cryptostigmatan mites, also referred to as the oribatid mites, are all free living, found primarily in damp places, in grass, mosses, humus, and similar habitats. Their importance in parasitology stems from the role certain members play as intermediate hosts for helminth parasites, specifically cestodes of the family Anoplocephalidae (p. 407). Kates and Runkel (1940) have reviewed the role of these mites as intermediate hosts for tapeworms.

Mites of the suborder Cryptostigmata that serve as intermediate hosts of anoplocephalid cestodes belong to seven families. Some of these are listed in Table 17.2. As stated, the mites are not parasitic. Specimens infected with larval anoplocephalid tapeworms are ingested by foraging definitive hosts of the worms.

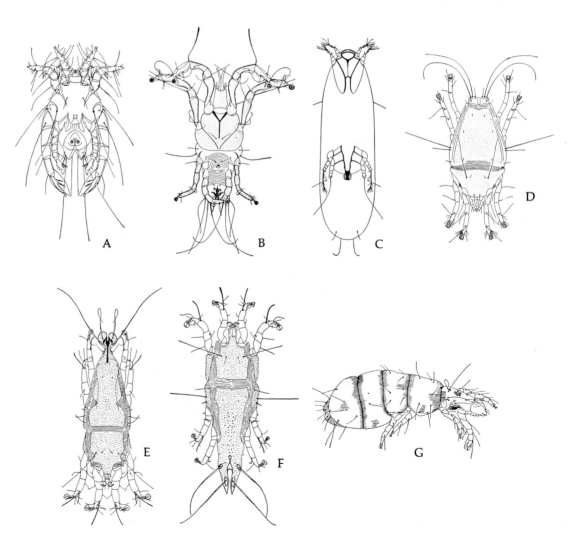

Fig. 17.17.   Feather mites. A. Ventral view of male *Megninia columbae*. **B.** Ventral view of male *Falculifer rostratus*. **C.** Hypopial nymph of *F. rostratus*. (**A–C**, after Hirst, 1922.) **D.** Dorsal view of female *Pterolichus obtusus*. **E.** Dorsal view of male *Trouessartia rosteri*. **F.** Dorsal view of female *T. rosteri*. (**D** and **E**, after Baker and Wharton, 1952.) **G.** Dorsal view of female *Pediculochelus raulti*.

As with the ticks, the voluminous material published on mites is mainly concerned with morphology, taxonomy, and life cycle studies. Relatively little is known about the physiology and host-parasite relationships of these animals.

Oxygen Requirements

The parasitic mites are undoubtedly like their free-living cousins in that they are aerobic. No studies are yet available as to the amount of oxygen required by the various species living in different habitats on their hosts. Efficient tracheal systems in some species suggest their aerobic habits. However, in species that lack such a system, it is generally accepted that gaseous exchange occurs through the thin exoskeleton.

Metabolism

Practically nothing is known about the metabolism of zooparasitic mites. In recent years, however, studies of this nature have been conducted on the two-spotted spider mite, *Tetranychus urticae*, a favored experimental animal of acarine physiologists that feeds on plants. In this species, Ehrhardt and Voss (1961) have found a number of glucosidases that are capable of hydrolyzing various cabohydrates such as maltose, sucrose, trehalose, melibiose, lactose, melezitose, and raffinose to monosaccharides. The presence of α-glucosidase, cellulase, pectinase, and polygalacturonase, on the other hand, could not be demonstrated.

In addition, Mehrotra (1960, 1961, 1963) has shown in *T. urticae* that hexose phosphates can be utilized through the Embden-Meyerhof glycolytic pathway (p. 52), as well as by the hexose monophosphate (pentose phosphate) shunt (p. 59). Mehrotra

Table 17.2. Representative Anoplocephalid Cestodes and Their Mite Intermediate Hosts

Cestode	Vertebrate Host	Mite Intermediate Host	Locality
Bertiella studeri	Primates	*Scutovertex minutus*	Germany
		Notaspis coleoptratus	Germany
		Galumna sp.	Germany
Moniezia expansa	Cattle, lambs, sheep	*Scheloribates laevigatus*	U.S., U.S.S.R.
		Protoschelobates seghettii	U.S.
		Oribatula minuta	U.S.
		Galumna virginiensis	U.S.
		G. obvius	U.S.S.R.
		Galumna sp.	U.S.
M. benedeni	Calves, lambs	*Scheloribates laevigatus*	U.S.S.R.
		Galumna obvius	U.S.S.R.
Moniezia sp.	Ruminants	*Adoristes ovatus*	U.S.S.R.
Anoplocephala perfoliata	Asses, horses	*Carabodes* sp.	U.S.S.R.
		Liacarus sp.	U.S.S.R.
		Scheloribates laevigatus	U.S.S.R.
		S. latipes	U.S.S.R.
		Achipteria sp.	U.S.S.R.
		Galumna nervosus	U.S.S.R.
		G. obvius	U.S.S.R.
A. magna	Assess horses	*Scheloribates laevigatus*	U.S.S.R.
Cittotaenia ctenoides	Common European rabbit	*Scutovertex minutus*	Germany
	(*Oryctolagus cuniculus*)	*Xenillus tegeocranus*	Germany
		Cepheus cepheiformis	Germany
		Liacarus coracinus	Germany
		Scheloribates laevigatus	Germany
		Liebstadia similis	Germany
		Notaspis coleoptratus	Germany
		Galumna nervosus	U.S.
		G. obvius	Germany
C. denticulata	Common European rabbit	*Scutovertex minutus*	Germany
		Xenillus tegeocranus	Germany
		Cepheus cepheiformis	Germany
		Liacarus coracinus	Germany
		Scheloribates laevigatus	Germany
Thysaniezia ovilla	Ruminants	*Scheloribates laevigatus*	U.S.S.R.
		S. latipes	U.S.S.R.
Paranoplocephala mamillana	Horses, rarely tapirs	*Galumna obvius*	U.S.S.R.
		Allogalumna longipluma	U.S.S.R.

has calculated that about 40 to 45% of the glucose is metabolized through the hexose monophosphate pathway. Some of the enzymes of the Krebs cycle have also been shown to be present, suggesting that this metabolic pathway may take place in this mite.

The presence of an active hexose monophosphate pathway in *Tetranychus urticae* may have physiologic significance, since Siperstein and Fagan (1958) have implicated this pathway in the synthesis of lipids. Indeed, McEnroe's (1963) studies indicate that the respiratory quotient (RQ) of the normal summer form of *T. urticae* suggests the synthesis of fats from carbohydrates.

It is not known whether the glucosidases found in *Tetranychus urticae* occur in the zooparasitic species, nor is it known whether both the Embden-Meyerhof and the hexose monophosphate pathways occur in zooparasites.

Because most zooparasitic mites feed on their host's blood, lymph, cells, and cell products, proteolytic enzymes undoubtedly occur.

Relative to the utilization of nutrients by mites, Rodriguez *et al.* (1979) have demonstrated that when *Tyrophagus putrescentiae*, a free-living mite, is reared on a standard casein-wheat germ diet to which is added radioactively labeled 1,3-butanediol, an alcohol, and subsequently analyzed, 62.4% of the recovered label is associated with lipids, 29.6% with proteins, 6.5% with carbohydrates, and 1.5% with nucleic acids. Chromatographic analyses revealed that the major radioactive components of the lipid fraction of mites fed butanediol were triglycerides, which when saponified yielded myristic, palmitic, stearic, and one or more C_{18} unsaturated fatty acids. Thus, butanediol appears to be metabolized in *T. putrescentiae* mainly through lipid pathways.

Osmoregulation and Equilibrium Humidity

Wharton (1960) has pointed out an important aspect of mite physiology, water balance. Inasmuch as these microscopic arthropods are of such minute sizes, they have great difficulty in obtaining and retaining water for use in their metabolism. The mechanism(s) employed by these animals to maintain water balance is not totally understood.

The role of equilibrium humidity in the physiology of mites has been reviewed by Wharton (1963). Since then, Wharton *et al.* (1979) have attempted to define it more precisely. Equilibrium humidity is defined as the lowest relative humidity (RH) at which a living system can achieve equilibrium conditions and still maintain life. From the studies now available, it appears that adult parasitic mites have a high equilibrium

humidity. For example, Lees (1946) has shown that many adult ticks have an equilibrium humidity slightly above 90% RH, and Wharton and Kanungo (1962) have shown that the females of the rodent ectoparasite *Echinolaelaps echidninus* have an equilibrium humidity of about 90% RH.

The significance of equilibrium humidity in animals, especially minute animals such as mites, cannot be overemphasized. In most instances, this is a matter of life or death. If mites are placed in desiccating humidities, they continue to lose water until they succumb to desiccation. On the other hand, if mites are temporarily exposed to an RH above their equilibrium humidity, they may replace their normal water losses by active intake of water from the atmosphere.

In addition to a matter of survival, equilibrium humidity also influences the feeding rate of mites. For example, it is known that the equilibrium humidity of *Echinolaelaps echidninus* is 90% RH below temperature of 32°C. At 32°C and above, this mite cannot maintain its water balance at 90% RH without feeding. Thus, *E. echidninus* will feed more readily at an RH below 90% and at temperatures above 32°C (Wharton and Cross, 1957). Since the damage ectoparasitic mites inflict is almost always associated with feeding, a relative humidity above the the equilibrium humidity of the parasite could possibly reduce the rate of feeding so as to render it relatively harmless.

HOST-PARASITE RELATIONSHIPS

Many ectoparasitic mites are blood ingestors; others feed on epidermal scales and lymph. It is assumed that a certain amount of oxygen is taken into the body via the ingested erythrocytes. Relative to the nocturnal or diurnal feeding habits of certain species, Harrison (1957) has suggested that the feeding pattern of mites could well be influenced by the host's activity. The majority of prostigmatan mites that feed on rats—a nocturnal animal—become detached at night when the host is active. Whether departure from the host is due to a mere physical disruption or the result of physiologic changes within the host remains to be determined. Harrison has also reported that the feeding of prostigmatan mites is influenced by the number of mites attached to a single host. For example, the feeding times of these mites are approximately equal, but if the host is heavily parasitized, the mites show a shorter average feeding time.

Mites secrete an anticoagulant. For example, it is known that in some prostigmatans bunches of grape-like glands secrete a substance into the zone of the mouthparts. Aoki (1957) has demonstrated that the mammal-biting larvae of prostigmatan mites pierce the stratum corneum and stratum lucidum of their host's skin with their cheliceral blades and secrete an enzyme that acts at the point of attachment. The tissues in this area change into a hyaline substance. Interestingly, the process is reversed when the larva detaches itself from the host. The hyaline appearance of the tissues disappears, suggesting that the secreted substance is no longer present.

The nature of this book does not warrant discussion of the histopathology caused by various mites on wild and domestic animals; however, it is of interest to point out that although scalyness and similar conditions are generally present, little host tissue reaction occurs when skin-penetrating mites are present. Yunker and Ishak (1957), for example, have pointed out that when *Knemidokoptes pilae* causes scaly leg or scaly face in the budgerigar, *Melopsithacus undulatus* (lovebird group), an inflammatory reaction takes place in the host's stratum corneum, involving primarily polymorphonuclear leucocytes, some mononuclear leucocytes, and eosinophils, only in the very initial phase. Such tissue response soon disappears, even after the mites become embedded in skin pouches and actively feed on keratin. Again, Pillers (1921) and Griffiths and O'Rouke (1950) have pointed out that inflammatory lesions are absent in the initial stages of scaly-leg infections of fowls caused by *Knemidokoptes mutans*. It is postulated that the exoskeleton of mites serves as a biologically somewhat neutral shield that does not excite host tissue response.

EVOLUTION OF PARASITISM AMONG THE ACARINA

Very little information is available on the evolution of parasitism among the Acarina. From what is known, parasitism among these animals appears to have developed out of their feeding habits—that is, the predators and scavengers gradually became dependent on what are now hosts. Evidence indicates that the parasitic prostigmatan mites originated as parasites of reptiles. From this ancestral group, four others have arisen. The first is represented by a small number of species that parasitize arthropods, particularly those species found in the same habitats as reptiles. The second group became adapted to living in the skins of amphibians. The third group, still parasitic in reptiles, became isolated in the lungs of marine snakes. The fourth group, which is by far the largest, developed as parasites of mammals and became widely distributed. It would also appear that the more primitive prostigmatans are found not only on their hosts but also in the fields where their hosts reside, whereas more recent forms are more intimately associated with their hosts (Audy, 1960).

REFERENCES

Aoki, T. (1957). Histochemical studies on the so-called styostome or hyopharynx in the tissues of the hosts parasitized by the trombiculid mites. *Acta Med. Biol. (Niigata)* **5**, 103–120.

Audy, J. R. (1960). Evolutionary aspects of trombiculid mite parasitization. *In* "Proceedings of the Centenary and Bicentenary Congress of Biology" (R. D. Purchon, ed.), pp. 102–108. University of Malaya Press, Singapore.

Baker, E. W., and Wharton, G. W. (1952). "An Introduction to Acarology." Macmillan, New York.

Bertram, D. S., Unsworth, K., and Gordon, R. M. (1946). The biology and maintenance of *Liponyssus bacoti* Hirst, 1913, and an investigation into its role as a vector of *Litosomoides carinii* to cotton rats, together with some observations on the infection in the white rats. *Ann. Trop. Med. Parasitol.* **40**, 228–252.

Bregetova, N. G. (1956). Key to the families of gamasid mites, Superfamily Gamasoidea. *Opred. Fauny SSR, Zool. Inst. Akad. Nauk SSSR*, **59**, 143–324. (In Russian.)

Camin, J. H. (1963). Relations between host-finding behavior and life histories in ectoparasitic acarina. *In* "Advances in Acarology" (J. A. Naegele, ed.), Vol. 1, pp. 411–424. Cornell University Press, Ithaca, New York.

Crowell, R. M. (1960). The taxonomy, distribution and developmental stages of Ohio water mites. *Ohio Biol. Surv. Bull.* **1**, 1–77.

Ehrhardt, P., and Voss, G. (1961). Die Carbohydrassen der Spinnmilbe *Tetranychus urticae* Koch (Acari Trombidiformes Tetranychidae). *Experientia* **17**, 307.

Ewing, H. E. (1944). The trombiculid mites (chigger mites) and their relation to disease. *J. Parasitol.* **30**, 339–365.

Fox, I. (1947). Seven new mites from rats in Puerto Rico. *Ann. Entomol. Soc. Am.* **40**, 598–603.

Griffiths, R. B., and O'Rouke, F. J. (1950). Observations on the lesions caused by *Cnemidocoptes mutans* and their treatment, with special reference to the use of "gammexane." *Parasitology* **44**, 93–100.

Harrison, J. L. (1957). Additional feeding times of trombiculid larvae. *Stud. Inst. Med. Res. Kuala Lumpur* **28**, 383–393.

Johnston, D. E. (1982). Acari. *In* "Synopsis and Classification of Living Organisms" (S. P. Parker, ed.), pp. 111–116, 145–146.

Kates, K. C., and Runkel, C. E. (1940). Observations on oribatid mite vectors of *Moniezia expansa* on pastures, with a report of several new vectors from the United States. *Proc. Helminthol. Soc. Wash.* **13**, 8–33.

Kethley, J. (1982). Acariformes, *In* "Synopsis and Classification of Living Organisms" (S. P. Parker, ed.), pp. 117–145. McGraw–Hill, New York.

Lees, A. D. (1946). The water balance in *Ixodes ricinus* L. and certain species of ticks. *Parasitology* **37**, 1–20.

Marshall, J., and Staley, J. (1929). A newly observed reaction of certain species of mosquitoes to the bites of larval hydrachnids. *Parasitology* **21**, 158–160.

McDaniel, B. (1979). "How to Know the Ticks and Mites." William C. Brown, Dubuque, Iowa.

McEnroe, W. D. (1963). The role of the digestive system in the water balance of the two-spotted spider mite *Tetrany-*

chus urticae Koch. *In* "Advances in Acarology" (J. E. Naegele, ed.), Vol. I, pp. 225–231. Cornell University Press, Ithaca, New York.

Mehrotra, K. N. (1960). Carbohydrate metabolism in the two-spotted mite (*Tetranychus telarius*). *Bull. Entomol. Soc. Am.* **6**, 151.

Mehrotra, K. N. (1961). Carbohydrate metabolism in the two-spotted mite, *Tetranychus telarius* L. I. Hexose monophosphate cycle. *Comp. Biochem. Physiol.* **3**, 184–198.

Mehrotra, K. N. (1963). Carbohydrate metabolism in the two-spotted spider mite. *In* "Advances in Acarology" (J. A. Naegele, ed.), Vol. I, pp. 232–237. Cornell University Press, Ithaca, New York.

O'Conner, B. M. (1982). Astigmata. *In* "Synopsis and Classification of Living Organisms" (S. P. Parker, ed.), pp. 146–169. McGraw-Hill, New York.

Parish, L. C., Nutting, W. B., and Schwartzman, R. M. (eds.). (1983). "Cutaneous Infestations of Man and Animal." Praeger, New York.

Pillars, A. W. N. (1921). Scaly legs in fowl. *Vet. Rec.* **33** (N. S. I), 827–829.

Rodriguez, J. G., Smith, W. T., Heffron, P., and Oh, S. K. (1979). Metabolism of butanediol in *Tyrophagus putrescentiae*. *In* "Recent Advances in Acarology" (J. G. Rodriguez, ed.), Vol. I, pp. 337–346. Academic Press, New York.

Sasa, M. (1950). Mites of the genus *Dermatophagoides* Bodganoff, 1864, found from three cases of human acariasis. *Jap. J. Exp. Med.* **20**, 519–525.

Siperstein, M. D., and Fagan, V. M. (1958). Studies on the relation between glucose oxidation and intermediary metabolism. I. The infuluence of glycolysis on the synthesis of cholesterol and fatty acids in normal liver. *J. Clin. Invest.* **37**, 1185–1195.

Traver, J. R. (1951). Unusual scalp dermatitis in humans caused by the mite *Dermatophagoides*. *Proc. Entomol. Soc. Wash.* **53**, 1–25.

Turk, F. A., and Phillips, S. M. (1946). A monograph of the slug mite *Riccardoella limacum* (Schrank). *Proc. Zool. Soc. London* **115**, 448–472.

Uchida, T., and Miyazaki, I. (1935). Life-history of watermite parasitic on *Anopheles. Proc. Imp. Acad.* (*Tokyo*) **11**, 73–76.

Volkonsky, M. (1940). *Podapolipus diander*, n. sp. acarien hétérostygmate parasite du criquet migrateue (*Locusta migratoria* L.) *Arch. Inst. Pasteur Alger.* **18**, 321–340.

Wharton, G. W. (1954). Observations on the feeding of prostigmatid larvae (Acarina: Trombidiformes) on arthropods. *J. Wash. Acad. Sci.* **44**, 244–245.

Wharton, G. W. (1960). Water balance in mites. *J. Parasitol.* **46** (Sect. 2), 6.

Wharton, G. W. (1963). Equilibrium humidity. *In* "Advances in Acarology" (J. A. Naegele, ed.), Vol. I, pp. 201–208. Cornell University Press, Ithaca, New York.

Wharton, G. W., and Cross, H. F. (1957). Studies on the feeding habits of laepaptid mites. *J. Parasitol.* **43**, 45–50.

Wharton, G. W., and Kanungo, K. (1962). Some effects of temperature and relative humidity on water balance in females of the spiny rat mite, *Echinolaelaps echindninus*

(Berlese, 1887) (Acarina Laelaptidae). *Ann. Entomol. Soc. Am.* **55**, 483–492.

Wharton, G. W., Duke, K. M., and Epstein, H. M. (1979). Water and the physiology of house dust mites. *In* "Recent Advances in Acarology" (J. G. Rodriguez, ed.), Vol. I, pp. 325–335. Academic Press, New York.

Williams, R. W. (1944). A bibliography pertaining to the mite family Trombidiidae. *Am. Midl. Natur.* **2**, 699–712.

Yunker, C. E., and Ishak, K. G. (1957). Histopathological observations on the sequence of infection in knemidokoptic mange of budgerigars (*Melopsittacus undulatus*). *J. Parasitol.* **43**, 664–672.

CLASSIFICATION OF MITES*

ORDER ACARINA (see p. 591)

Suborder Notostigmata

Suborder Tetrastigmata (Holothyroidea)

Suborder Mesostigmata

With a pair of tracheal spiracles situated beside the coxae of the third pair of legs; body covered with brown plates which vary in number and position; includes both parasitic and free-living species.

GAMASIDES GROUP (see p. 594)

(Genera mentioned in text: Family Raillietidae[†]—*Raillietia*. Family Halarachnidae—*Halarachne, Pneumonyssus, Orthohalarachne*. Family Entonyssidae—*Entonyssus, Hammertonia, Ophiopneumicola*. Family Rhinonyssidae—*Rhinonyssus, Larinyssus, Sternostomum*. Family Haemogamasidae—*Haemogamasus, Acanthochela*. Family Laelaptidae—*Myrmonyssus, Echinolaelaps, Haemolaelaps*. Family Dermanyssidae—*Dermanyssus, Ornithonyssus, Liponyssoides*. Family Discozerconidae—*Discozercon, Heterozercon, Allozercon*. Family Pachylaelaptidae—*Pachylaelaps*. Family Macrochelidae—*Macrocheles*. Family Iphiopsidae—*Iphiopsis, Jacobsonia*.)

MEGISTHANINA GROUP (see p. 594)

(Genus mentioned in text: Family Megisthanidae[†]—*Megisthanus*.)

LIROASPINA GROUP (see p. 595)

(Genus mentioned in text: Family Liroaspidae[†]—*Liroaspis*.)

THINOZERCONINA GROUP (see p. 595)

(Genera mentioned in text: Family Dasyponyssidae[†]—*Dasyponyssus*.)

DIARTHROPHALLINA GROUP (see p. 595)

(Genera mentioned in text: Family Diarthrophallidae—*Diarthrophallus, Passelobia*.)

CELAENOPSINA GROUP (see p. 595)

(Genus mentioned in text: Family Schizogyniidae[†]—*Schizogynium*.)

FEDRIZZIINA GROUP (see p. 595)

(Genera mentioned in text: Family Parantennulidae[†]—*Parantennulus, Diplopodophilus*.)

Suborder Prostigmata (Trombidiformes)

With a pair of spiracles situated anteriorly near the

---

*Diagnoses of groups of mites that are free-living are not presented.

[†] For diagnostic characteristics of the mite families, see Johnston (1982), Kethley (1982), and O'Conner (1982).

mouthparts; includes species parasitic on plants, invertebrates, and vertebrates; also includes free-livng species.

TETRAPODILI GROUP (see p. 599)

TARSONEMINI GROUP (see p. 599)

(Genera mentioned in text: Family Podapolipodidae[†]—*Podapolipus, Locustacarus, Eutarsopolipus, Tarsopolipus, Tetrapolipus*. Family Pyemotidae—*Pyemotes*.)

PROSTIGMATA GROUP (see p. 599)

(Genera mentioned in text: Family Ereynetidae[†]—*Riccardoella*. Family Speleognathidae—*Speleognathus*. Family Myobiidae—*Myobia, Amorphacarus, Syringophilus, Harpirhynchus, Ophioptes, Psorergates*. Family Erythraeidae—*Leptus, Erythraeus*. Family Trombiculidae—*Euschongastia, Trombicula, Doloisia*. Family Demodicidae—*Demodex*. Family Hydrachnellidae—*Lebertia, Thyas, Trombidium, Hydrachna, Limnesia, Piona*. Family Tetranychidae—*Tetranychus*.)

Suborder Astigmata (Sarcoptiformes)

Tracheal system absent (gaseous exchange occurs across weakly sclerotized exoskeleton), includes parasitic and free-living species.

ACARIDIAE GROUP (see p. 606)

(Genera mentioned in text: Family Ensliniellidae[†]—*Ensliniella, Vidia, Forcellinia*. Family Sarcoptidae—*Sarcoptes, Notoedres, Knemidokoptes*. Family Psoroptidae—*Psoroptes, Caparinia, Otodectes, Chorioptes*. Family Cytoditidae—*Cytodites*. Family Laminosioptidae—*Laminosioptes*. Family Epidermoptidae—*Epidermoptes, Dermatophagoides*. Family Analgesidae—*Megninia*. Family Dermoglyphidae—*Falculifer, Pterolichus*. Family Proctophyllodidae—*Trouessartia*.)

Suborder Cryptostigmata

Free living but of interest because of its role as intermediate host of helminths, especially the anoplocephalid tapeworms (p. 407).

18

INTRODUCTION TO THE PARASITIC INSECTS

Siphonaptera: The Fleas
Mallophaga: The Biting Lice
Anoplura: The Sucking Lice

Insects constitute one of the largest known groups of living animals. It has been estimated that over 900,000 species exist. The various species are found in every imaginable type of habitat, be it terrestrial, freshwater, or marine.

Man's interest in insects has formalized into the area of biology known as entomology. Among entomologists, those interested in species that are important as pests of plants and animals have become specialized and have carved out a portion of the field known as economic entomology. Those whose interests are concentrated on species that serve either as disease-causing organisms or as vectors of pathogens have established the field of medical entomology, and those interested in insect diseases and the biologic control of insects are known as insect pathologists. Along similar lines, the broad field of entomology has become subdivided into a number of subdisciplines. Within this and the following two chapters are discussions of those aspects of entomology concerned with the biology of parasitic insects and the role of the hematophagous, or blood-ingesting, species as vectors.

MORPHOLOGY

The morphology of insects varies greatly among species. No other group of animals demonstrates so great a diversity of form as the insects. In addition to the characteristics of the phylum Arthropoda (p. 572), shared with the other classes in the phylum, members of the class Insecta are characterized by three distinct body regions—head, thorax, and abdomen.

The head bears a single pair of antennae and distinct mouthparts adapted for chewing, sucking, or lapping, depending on the feeding habits of the species. The thorax, composed of three somites, bears three pairs of jointed legs and two, one, or no pairs of wings. The abdomen, which is typically composed of 11 or fewer somites, bears the genitalia at its terminal.

EXTERNAL MORPHOLOGY

APPENDAGES OF THE HEAD

Other than the eyes, both compound and simple, which are discussed on p. 620 with the other sense

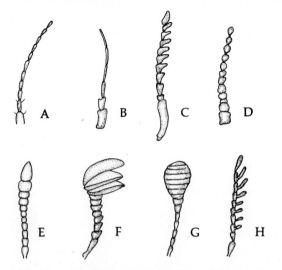

Fig. 18.1.   Types of insect antennae. A. Filiform. **B.** Setaceous. **C.** Serrate. **D.** Moniliform. **E.** Clavate. **F.** Lamellate. **G.** Capitate. **H.** Pectinate. (Redrawn after Comstock, 1949.)

like **lamellate** antennae; the **capitate** antennae, which bear a knob at the distal ends; and the **pectinate** antennae, which are plumose with comblike lateral projections (Fig. 18.1).

Mouthparts

The mouthparts of insects vary greatly, depending on the feeding habits of the species. To become acquainted with the parts, consider those of a familiar insect, the grasshopper, as representative of the Insecta (Fig. 18.2).

The "upper lip" of the grasshopper is the **labrum,** which is not an actual mouthpart. The first true mouthparts are the paired **mandibles,** which lie directly behind the labrum. These are heavily sclerotized and bear jagged teeth along their medial margins. Behind the mandibles are the paired **maxillae.** Each maxilla is jointed, being composed of the basal **cardo,** on which is mounted the **stipes.** The mesially situated **lacinia** and the laterally situated **galea** are both mounted distally on the stipes; the jointed **palpus** (also known as the **maxilliped**) is attached laterally (Fig. 18.3).

The **hypopharynx** is not a true mouthpart; it is an unsegmented outgrowth of the body wall. This tubular structure arises from the ventral membranous floor of the head. The "lower lip" of insects is the **labium.** This heavily sclerotized appendage is attached to the head by the **submentum,** which broadens out distally to form the **mentum.** The distal portion of the mentum is the **ligula.** The **labial palpi** arise from the lateral aspects of the mentum. The homologies and

organs, the head of the "typical" insect bears a pair of antennae and mouthparts.

Antennae

The antennae are a pair of movable, segmented appendages that are usually inserted between the compound eyes. The point of insertion is a socket, commonly surrounded by a sclerotized ring. The morphology of the antennae varies greatly and is important not only in identification but also in the determination of sex in some species.

Antenna types include the thin and threadlike **filiform** antennae; the pointed **setaceous** antennae; the sawlike **serrate** antennae; the beadlike **moniliform** antennae; the club-shaped **clavate** antennae; the leaf-

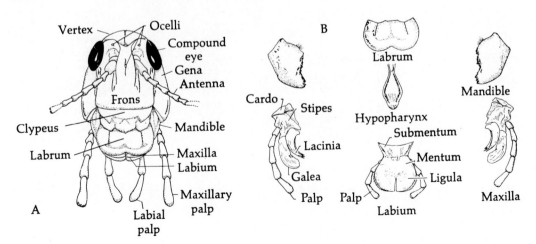

Fig. 18.2.   Mouthparts of the grasshopper. A. Frontal view of head showing anatomic parts. **B.** Mouthparts dissected. (After Storer and Usinger, 1961.)

evolutionary development of the mouthparts of insects are discussed by Ross (1948).

Because the mouthparts of insects are greatly modified in many species, the principal types are listed in the following discussion.

Chewing Type. Mouthparts of this type, as exemplified by those of the grasshopper, are usually found among free-living species, for example, beetles and ants. The mandibles masticate the food, and the maxillae and labium serve to push the particles into the mouth.

Cutting-Sponging Type. The tabanids, or horseflies, of the order Diptera possess mouthparts of this type (Fig. 18.3). The mandibles are in the form of sharp blades, and the maxillae are long and styletlike.

The mandibles and maxillae cut and tear the skin of the host. Blood is collected by a spongelike labium, conveyed to a tube formed by the hypopharynx and the **epipharynx** (another projection of the body wall), and sucked into the esophagus.

Piercing-Sucking Type. The mosquitoes, flies, lice, and bedbugs possess mouthparts of this type (Fig. 18.3). The labrum, mandibles, hypopharynx, and

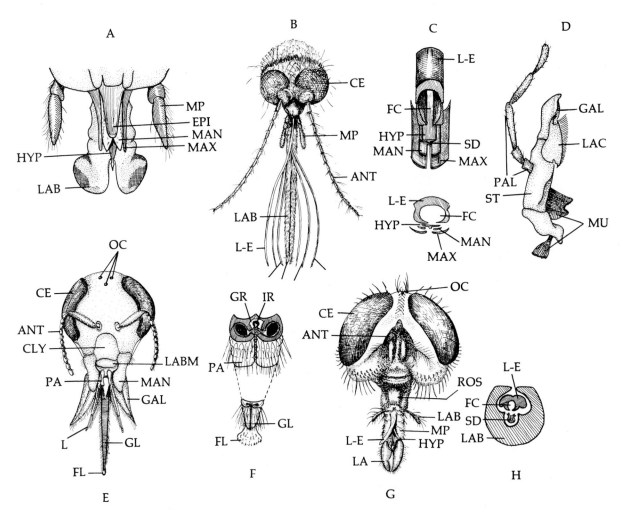

Fig. 18.3. Types of insect mouthparts. A. Cutting-sponging type. (Illinois Natural History Survey illustration.) **B.** Piercing-sucking type. **C.** Parts forming the piercing-sucking tube. (Redrawn after Metcalf and Flint, 1962.) **D.** Maxilla of the cockroach. (Redrawn after Snodgrass, 1935.) **E.** Chewing-lapping type. **F.** Parts forming the tip of the chewing-lapping apparatus. **G.** Sponging type. **H.** Parts forming sponging apparatus. (**F** and **G**, after Metcalf and Flint, 1962.) ANT, antenna; CE, compound eye; CLY, clypeus; EPI, epipharynx; FC, food channel; FL, flabellum; GAL, galeas; GL, glossa; GR, glossal rod; HYP, hypopharynx; IR, inner channel of rod; L, labial palpus; LA, labella; LAB, labium; LABM, labrum; LAC, lacina; L-E, labrum-epipharynx; MAN, mandible; MAX, maxilla; MP, maxillary palpus; MU, muscles; OC, ocelli; PA, paraglossa; PAL, palpus; ROS, rostrum; SD, salivary duct; ST, stipes.

maxillae are long and slender and fit together, forming a hollow tube. The labium is also elongate and wraps around the other parts like a rigid sheath. During feeding, the tube pierces the host's skin like a hypodermic needle, and blood is drawn through it.

Chewing-Lapping Type. Mouthparts of this type are found in bees and wasps (Fig. 18.3). The labrum and mandibles are similar to those found in the grasshopper; however, the maxillae and labium are modified as elongate structures by which the food is drawn up.

Sponging Type. Most nonbiting dipterans possess mouthparts of this type (Fig. 18.3). The parts are similar to those found in the cutting-sponging type, but the mandibles and maxillae are nonfunctional. The remaining parts form a proboscis with a spongelike apex called the **labella**. Liquid foods are conducted to the mouth via minute capillary channels on the labella. Solid food is ingested only after it is dissolved or suspended in deposited saliva.

The cutting-sponging and the piercing-sucking types are the most commonly found among hematophagous insects.

APPENDAGES OF THE THORAX

The thorax of insects is typically divided into three segments—the anterior **prothorax**, the middle **mesothorax**, and the posterior **metathorax**. The dorsal surface is covered by the **notum** (or **tergum**), the lateral aspects by **pleura**, and the ventral surface by the **sternum**.

Legs

The typical legs are found on the thorax, one pair per segment. Each leg consists of six joints known as **coxa**, **trochanter**, **femur**, **tibia**, **tarsus**, and **pretarsus** (sometimes considered a part of the tarsus) (Fig. 18.4).

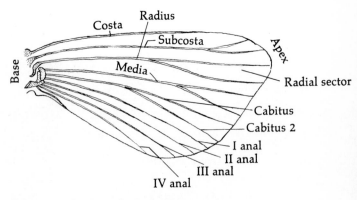

Fig. 18.5. Venation of a "typical" insect wing.

The coxa is the most proximal joint. The shape of the insect leg is modified in many parasitic species.

Wings

Insects are the only group of invertebrates that possess wings. Wings are actually extensions of the body wall. The "typical" insect bears two pairs, namely the **forewings**, which are attached to the mesothorax, and the **hindwings**, which are attached to the metathorax. Each wing exhibits a series of **veins**, which are actually tracheae that extend from the body proper into the wings during development but which harden and become severed at their bases when the wings complete their development and are expanded. The wing **venation**, or vein pattern, is of taxonomic importance. The more typical veins and crossveins are depicted in Figure 18.5.

One or both pairs of wings are greatly modified or reduced, or lacking altogether, in many parasitic insects.

INTERNAL MORPHOLOGY

DIGESTIVE SYSTEM

Insects possess complete alimentary tracts. In addition to the tract, there are various associated glands, for example, salivary glands, gastric caeca, and Malpighian tubules.

The alimentary canal is an asymmetrical tube, oriented lengthwise through the middle of the body (Fig. 18.6). It opens anteriorly through the mouth, which is located in the **preoral cavity**—the space enclosed by the mouthparts. The canal opens to the exterior posteriorly through the anus, which is located on the posteriormost segment of the abdomen.

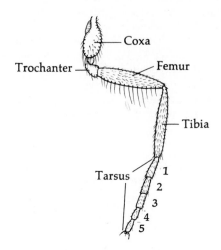

Fig. 18.4. Leg of *Musca domestica*.

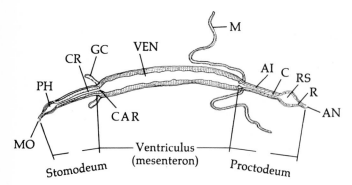

Fig. 18.6. **Insect anatomy.** Constituent parts of the alimentary tract. AI, anterior portion of intestine; AN, anus; C, colon; CAR, cardia, CR, crop; GC, gastric caeca; M, Malpighian tubule; MO, mouth; PH, pharynx; R, rectum; RS, rectal sac; VEN, ventriculus.

Between the mouth and the anus, the alimentary canal is divided into three sections: **stomodeum**, **ventriculus**, and **proctodeum**.

Stomodeum

The anterior portion of the alimentary canal is the stomodeum, which in turn is divided into the anterior, tubular **esophagus**, followed by an enlarged **crop**, which opens into the narrower valvelike **proventriculus**. In free-living insects that ingest solid foods, the proventriculus bears a series of hooks that aid in mastication. If such armatures exist, the proventriculus is called the **gastric mill**.

In most insects, two tubular glands, called the **salivary** or **labial glands**, lie in the body cavity. The ducts of these glands are directed anteriorly. They usually unite anteriorly and conduct glandular secretions into the preoral zone. In hematophagous ectoparasites, the labial glands are quite important because they secrete the anticoagulant that prevents clotting of the host's blood during feeding, and because they are storage sites for protozoan, viral, and other types of microorganisms. These microorganisms are injected into the vertebrate host while the insect vector is feeding.

Ventriculus

The middle section of the alimentary canal is the ventriculus or stomach. This is the site of digestion. In certain hemipterans, this chamber is subdivided into three or four sections. However, in most parasitic species, the ventriculus is without secondary partitions. Associated with the ventriculus are several tubular diverticula, the **gastric caeca**, attached at the

anterior margin of the stomach. These outpocketings function primarily in digestion and to increase the surface area for absorption. For a detailed review of digestion in insects, see House (1974).

Proctodeum

Extending posteriorly from the ventriculus is the proctodeum. The proctodeum is subdivided into the **intestine**, which leads from the stomach, and the **rectum**, which is the enlarged posterior section that communicates with the exterior through the anus. A group of long tubular structures, the **Malpighian tubules**, are found in most insects at the junction of the ventriculus and the proctodeum. These tubes, which vary in number depending on the species, serve an excretory function.

EXCRETORY SYSTEM

The elimination of wastes in insects, that is, excess water, nitrogenous wastes, and salts, is accomplished primarily by the Malpighian tubules. Uric acid, in the form of sodium and potassium salts, is eliminated from the body tissues into the circulating hemolymph in the hemocoel. Eventually, the waste-containing hemolymph circulates near the Malpighian tubules, which absorb the uric acid and discharge it into the intestinal lumen. From the intestinal lumen, uric acid passes into the proctodeum and then to the exterior through the anus. In certain free-living species, such as lepidopterans of the family Pieridae, some body wastes are converted into pigments (uric acid derivatives) and deposited in the cuticle. This mechanism of excretion, however, has not been conclusively demonstrated in parasitic species.

CIRCULATORY SYSTEM

Insects possess an open circulatory system, that is, the hemolymph (or blood) is not confined within vessels as in vertebrates. Instead, the hemolymph is found unrestrained within the coelom, which is referred to as the **hemocoel**. From the hemocoel, the hemolymph seeps into the various tissues of the body. In most insects, however, a "heart" and "aorta" are present, and they aid in the circulation of the hemolymph.

The heart and aorta, which represent areas on the same tube, are dorsally situated, immediately ventral to the dorsal body wall. This circulatory tube, or hemolymph vessel, extends from the posterior extremity to the head, and it is enlarged and muscular posteriorly, constituting the heart (Fig. 18.7). The heart is elongate and chambered, inasmuch as there is a slight swelling of the tube in each of the first nine body somites of the abdomen. Through the pulsatory actions of the heart muscles, hemolymph is forced anteriorly through the aorta into the cavities of the head. From the head, hemolymph flows posteriorly into the hemocoelic zones of the thorax and abdomen.

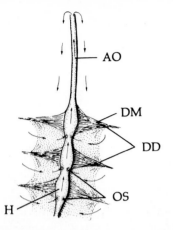

Fig. 18.7. Insect heart. Dorsal view of the three chambers of the heart, aorta, and corresponding part of the dorsal diaphragm. Arrows suggest the course of hemolymph circulation. AO, aorta; DD, dorsal diaphragm; DM, diaphragm muscles; H, heart chamber; OS, ostia. (Redrawn after Snodgrass, 1935.)

Hemolymph is returned into the elongate heart through **ostia**, which are paired openings on each lateral surface of the nine chambers of the heart.

Various auxiliary pulsatile organs, for example, dorsal and ventral diaphragms, have been reported in several free-living species. However, this aspect of insect anatomy has not been critically investigated in the parasitic species. For reviews of the circulatory system, including the heart, and what is known about insect hemocytes, see Jones (1974), Miller (1974), Arnold (1974), Florkin and Jeuniaux (1974), Grégoire (1974), and Rowley and Ratcliffe (1981).

RESPIRATORY SYSTEM

Insects, like some ticks (Metastigmata) and crustaceans, possess a specialized type of respiratory system in the form of a tracheal system. Although the number and arrangement of the tracheal tubes vary greatly, the basic pattern is the same in all species (Fig. 18.8). Opening laterally on the thoracic and abdominal segments is a series of pores known as **spiracles**. The aperture of these "breathing holes" is regulated either by two external movable guard plates, or internally by clamps.

The spiracles lead into individual **tracheae**, which connect at right angles with two longitudinally oriented lateral **tracheal tubes**. Secondary, tertiary, and further branches known as **tracheids** and **trachioles** arise from these lateral tracheal tubes, conducting oxygen to and removing carbon dioxide from the various tissues of the body. In dipteran larvae, a pair of large dorsal tracheal trunks is present on each side of the heart and aorta (Fig. 18.8). These trunks are secondary to the lateral trunks.

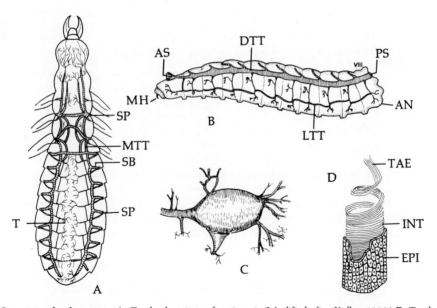

Fig. 18.8. Insect tracheal system. A. Tracheal system of an insect. (Modified after Kolbe, 1889.) **B.** Tracheal system as found in fly larva. **C.** Tracheal air sac. **D.** Structure of tracheal tube. (**B, C, D,** redrawn after Snodgrass, 1935.) AN, anus; anterior spiracle; DTT, dorsal tracheal trunk; EPI, epithelium; INT, intima; LTT, lateral tracheal trunk; MH, mouth hook; MTT, main tracheal trunk; PS, posterior spiracle; SB, spiracular branch; SP, spiracle; T, tracheoles; TAE, taenidium in spiral band of cuticular intima, artificially separated.

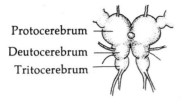

Protocerebrum
Deutocerebrum
Tritocerebrum

Fig. 18.9.   Insect brain. (Modified after Snodgrass, 1935.)

In some species, specialized tracheal air sacs are formed from distended areas of the tracheal trunks (Fig. 18.8). These sacs serve as sites for air storage. Histologically, tracheal tubes consist of a thin outer epithelial covering and an intima of cuticular spirals (Fig. 18.8).

The position and number of spiracles vary greatly among species. For example, on mosquito larvae, only one set of spiracles is present, located on the eighth abdominal segment. On maggots (fly larvae), two pairs of spiracles are present, one pair on the anterior portion of the thorax (known as **prothoracic spiracles**), and one pair on the eighth abdominal segment. When open and functional spiracles are present, the insect is said to possess an open tracheal system. In some species, however, the spiracular openings are vestigial and/or nonfunctional. In these species, gaseous exchange takes place through a thick capillary bed of tracheae located under the integument. These insects are said to possess a closed tracheal system. For a detailed review of the respiration system of insects, see Miller (1974).

NERVOUS SYSTEM

Main Nervous System

The "brain" is in the form of a **supraesophageal ganglion**. This nerve center is composed of three pairs of intimately fused ganglia. The first pair comprises the **protocerebrum**, from which nerves supply the compound eyes and ocelli; the second pair comprises the **deutocerebrum**, which controls the major sympathetic nervous system (Fig. 18.9).

The supraesophageal ganglion is connected to a large **subesophageal ganglion** by two lateral commissures that unite dorsally with the **tritocerebrum**. The tritocerebrum represents the third pair of fused ganglia. Nerve cords arising anteriorly from the subesophageal ganglion innervate the various mouthparts. The main nerves arising from the subesophageal ganglion, however, are the two intimately associated trunks that are directed caudally. Together these two trunks are known as the **ventral nerve cords**.

In each body somite is a fused pair of ganglia, which is joined to the ventral nerve cord by a pair of connectives. Branches arising from the segmental ganglia innervate the various tissues of the body in each segment.

Although primitively a pair of ganglia is present in each body somite, this number is reduced in the Diptera and other forms in which the number of body segments is reduced. In such instances, the ganglia in fused somites also fuse, and hence, the number of discrete ganglia is reduced. For a detailed review of the structure and function of the main nervous system of insects, see Huber (1974).

Stomodeal Nervous System

The preceding discussion pertains to the main nervous system pattern found in insects. In addition, an auxiliary nervous system—the **stomodeal nervous system**—is present in most species. The stomodeal nervous system is so designated because the major bulk of the fibers are located dorsal and lateral to the stomodeum. The center of this system is a **frontal ganglion** located anterior to the supraesophageal ganglion. It is connected with the tritocerebrum by a pair of commissures.

Nerve fibers arising from the frontal ganglion control the involuntary motions of the stomodeum and aorta. In addition to the connectives that join the frontal ganglion to the supraesophageal ganglion, two lateral nerves infiltrate the region of the esophagus, where they are connected with small ganglionic masses. Nerve fibers from these lateral nerves also innervate portions of the stomodeum, the ducts of the salivary glands, the aorta, and certain muscles of the mouthparts.

Sensory Receptors

Sensory receptors of insects are categorically divided into sensory areas and eyes.

Sensory Areas.   Sensory areas are located on various parts of the body. Each is composed of a hairlike seta that projects outward from a minute pore in the cuticle (Fig. 18.10). The base of the seta is connected with a sensory cell, which, in turn, is innervated by a nerve fiber. Such an arrangement represents a tactile sensory area. Response to pressure on the seta occurs via the associated nerve.

Similarly, olfactory and taste areas exist. In these, the sensory mechanism remains almost identical, except that the setae are replaced with a peg, dome, or plate, (Fig. 18.10). The seta, peg, dome, or plate is secreted by specialized epidermal cells known as **trichogen cells**. Another type of cell, known as the **formogen cell**, lays down the socket in which the individual seta is set.

Chemoreception in insects has been reviewed by Hodgson (1974), and mechanoreception has been reviewed by Schwartzkopff (1974).

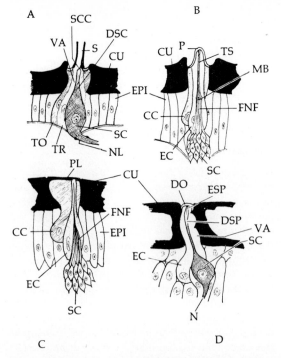

A

B

C

D

Fig. 18.10. **Cuticular structures. A.** Histology of a seta and associated structures. **B.** Histology of a peg and associated structures. **C.** Histology of a plate and associated structures. **D.** Histology of a dome and associated structures. (All figures modified after Snodgrass, 1935.) CC, cap cell; CU, cuticle; DO, dome; DSP or DSC, distal process of sensory cell; EC, enveloping cell; EPI, epithelium; ESP, end of sensory cell process; FNF, fascicle of nerve fibers; MB, minute bodies; N, nerve; NL, neurilemma; P, peg; PL, plate; S, seta; SC, sensory cell; SCC, connection of sensory cell with cuticle; TO, tormogen cell; TR, trichogen cell; TS, terminal strands of fiber connected with cuticle; VA, vacuole.

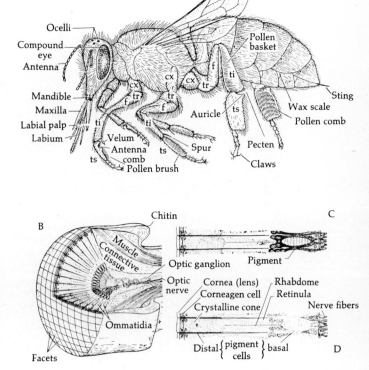

A

B

C

D

E

F

Fig. 18.11. **Insect eyes. A.** Honeybee worker showing positions of compound eyes and ocelli. **B–F.** Arthropod compound eye showing constituent parts. (After Casteel and Imms in Storer, 1951.)

Eyes. Insects commonly possess two types of eyes—the simple eye, or **ocellus**, and the **compound eye**.

The ocellus (Fig. 18.11) differs from the compound eye in that only a single cornea (or lens), a specialized transparent portion of the cuticle secreted by transparent corneagenous cells, is present. Nerve cells form a retina beneath the layer of corneagenous cells. The striated sensitive elements of these cells migrate to form a straight line down one side of each cell, and the lines formed in adjacent cells unite to form the **rhabdom**. A crystalline body lies between the cornea and the nerve cells.

The compound eye is essentially an accumulation of many ocelli. Each unit of a compound eye, known as an **ommatidium**, may be thought of as a single ocellus, because the nerve endings innervating each unit are isolated from the retinae of surrounding ommatidia (Fig. 18.11). Each ommatidium has its own cornea, called a **facet**. Underneath each facet is a

rosette of eight sense cells with a central rhabdom.

Pigment cells surround each lens and each rosette of sense cells. The pigment granules within each pigment cell migrate up and down in synchrony with those in adjacent pigment cells, resulting in unified control of the amount of light that affects the sense cells. The sense cells rest on a basement membrane, which the nerve fibers penetrate.

Insect larvae do not have compound eyes. Development of compound eyes does not commence until the insect reaches the pupal stage. For detailed reviews of the structure and function of ocelli and compound eyes in insects, see Goldsmith and Bernard (1974).

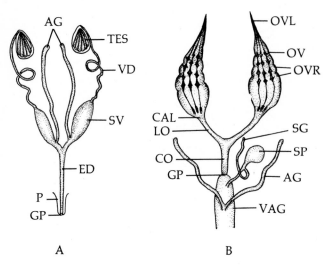

Fig. 18.12.   Insect reproductive systems. A. Male system. **B.** Female system. AG, accessory glands; CAL, calyx; CO, common oviduct; ED, ejaculatory duct; GP, gonopore; LO, lateral oviduct; OV, ovary; OVL, ovarial ligament; OVR, ovariole; P, penis; SG, spermathecal gland; SP, spermatheca; SV, seminal vesicle; TES, testis; VAG, vagina; VD, vas deferens. (Redrawn after Snodgrass, 1935.)

REPRODUCTIVE SYSTEMS

Insects are typically dioecious. Secondary sexual dimorphism is apparent in the external genitalia along with other, less prominent differences.

Male System

In males, the reproductive system consists of two **testes**, which are composed of **tubules** (or **follicles**). The two testes are situated dorsal to the intestine (Fig. 18.12). Arising from each testis is a **vas deferens**, which unites with its mate to form the **ejaculatory duct**. Prior to the union of the vasa deferentia, each enlarges slightly, forming a **seminal vesicle**.

Typically, accessory glands are present and empty into the vasa deferentia prior to their union. The ejaculatory duct opens into the **penis**, which communicates with the exterior through the **genital pore**. Inasmuch as the true penis of insects is membranous, rigidity is effected by the sclerotized **aedeagus**, a part of the external genitalia forming a rigid sheath around the penis.

Female System

In females a series of longitudinally oriented tubular **ovaries** are arranged in two bunches (Fig. 18.12). The posterior terminals of the ovarial units unite to form

the **calyx**, which opens into the **oviduct**. The lateral oviduct from each side unites with the opposite oviduct to form the common oviduct, which opens into the **vagina** (or genital chamber).

Two types of diverticula empty into the vagina—an elongate **spermatheca** (or seminal receptacle), and a pair of **colleterial** (or accessory) **glands**. The vagina opens into the groove of the ovipositor, which is a portion of the external genitalia.

Fertilization

Fertilization among insects is internal. During copulation, spermatozoa are introduced into the vagina and stored in the spermatheca. Ova pass down the oviducts as the result of the peristaltic action of the muscular oviductal walls and pass over the opening of the spermatheca. The spermatozoa fertilize the ova at this time. As eggs are oviposited, the colleterial glands secrete an adhesive material that binds the eggs into masses or glues them to vegetation or to the hair of hosts.

Readers interested in a more detailed discussion of the reproductive systems in insects and the phenomena associated with the reproductive process are referred to the reviews by de Wilde and de Loof (1973a, b).

LIFE CYCLE STUDIES

The nature of this volume does not warrent a detailed account of embryonic development in insects. Interested persons are referred to the review articles by Agrell and Lundquist (1973) and Gilbert and King (1973). Because insect parasitism is not confined to the adult stage—in fact, in many instances it is the larva that is parasitic—the following brief discussion addresses the postembryonic, or larval, development of insects.

The exoskeleton of insects, like that of all arthropods, is a limiting factor. Increase in size or alteration in form cannot be accomplished unless these animals molt. When ecdysis occurs, the sloughed exoskeleton is known as the **exuvia**. The period of time between molts is termed the **stadium**, and the insect form during a stadium is referred to as an **instar**.

Insects vary in the degree of metamorphosis they undergo between the first instar and the sexually mature adult, or **imago**. The degree of metamorphosis can be categorized into three types—the **ametabolous** type, in which metamorphosis is slight and difficult to appreciate; the **hemimetabolous** type, in which metamorphosis is gradual or incomplete; and the **holometabolous** type, in which matamorphosis is complex or complete.

Ametabolous Metamorphosis

Among members of two orders of free-living insects, the Thysanura (bristletails) and Collembola (springtails), newly hatched larvae resemble miniature adults in that they are wingless. During growth, practically no alteration of form, or metamorphosis, occurs. This type of development is known as ametabolous metamorphosis.

Hemimetabolous Metamorphosis

In some insects, newly hatched larvae are quite similar to, and have the same feeding habits as, the imago. There are differences, however. The young are not sexually mature and possess rudimentary reproductive structures. In winged species, the young are wingless. Acquisition of these adult structures occurs during growth. In this type of metamorphosis, the developing instars are referred to as **nymphs**. Hemimetabolous metamorphosis among parasitic insects is exemplified by the biting lice (Mallophaga) and the sucking lice (Anoplura).*

Holometabolous Metamorphosis

This type of metamorphosis is found in the majority of insects. The alteration of form between the postembryonic larva and the imago is quite striking. All the developing instars are wingless except for the preadult instar. The insect undergoes three stages during its development: the feeding grub or **larva**, the nonfeeding **pupa**, and the **imago**.

The larva, which is commonly elongate and wormlike, does not bear visible wings, although imaginal buds (the beginning of wing development) occur internally. The antennae and eyes are often rudimentary. The thoracic legs may or may not be present, and, strangely enough, larvapods may be present on the abdomen of certain species. This is primarily a feeding and growing stage, represented by a few to many instars, depending on the species.

The pupa is a nonfeeding and quiescent stage that possesses many adult characteristics, including appendages and wings, but these are usually held limply against the body. The pupa is the stage during which the body tissues are reorganized for eventual transformation into the adult form. In many insect orders, such as the Lepidoptera (butterflies and moths), the pupa occurs within a cocoon spun by the last larval instar. In others, the pupa is found burrowed in soil. In still others, like some of the Diptera (flies and mosquitoes), the pupa is ensheathed within the sloughed exoskeleton of the last larval instar, which hardens and is known as the **puparium**.

The adult, or **imago**, is the definitive form of an insect. Among holometabolous insects, the adults generally possess mouthparts that are different from those of the larval instars. Among parasitic species, this is exemplified by the flea, which possesses chewing type mouthparts as a scavenging larva, but possesses piercing-sucking mouthparts as an adult.

Hypermetamorphosis

In addition to the previously mentioned three types of postembryonic metamorphosis, an additional type, known as **hypermetamorphosis**, occurs. This type is best considered as a specialized form of holometabolous metamorphosis. In hypermetamorphosis, two or more morphologically distinct types of larvae appear in the life cycle.

Types of Insect Parasitism

Parasitism among insects is not always an adult practice, nor is it always practiced on vertebrate hosts. Insect parasites can be divided into three categories: ectoparasites on vertebrates, for example, lice, flies, and mosquitoes; endoparasites of vertebrates, for example, botflies and warble flies; and parasites of invertebrates. In the first instance, it is the adult that is parasitic, whereas in the last two categories, it is commonly the larvae that are parasitic, although adult parasites of invertebrates also occur. Insects parasitic in other insects are commonly referred to as **parasitoids**. The interesting monograph by Askew (1971) on parasitic insects is recommended, as is the volume by Marshall (1981), which deals with practically all aspects of the ecology of ectoparasitic insects.

BIOLOGY OF PARASITIC INSECTS

There are 26 orders of insects. Of these, eight are known to include parasitic species, which attach to, feed on, or occur within vertebrates and invertebrates. Other orders may include incidental and accidental parasites. The characteristics of the eight major orders that include parasites are presented at the end of this chapter as well at the end of chapters 19 and 20. Considered below are three orders: the Siphonaptera (the fleas), the Mallophaga (the biting lice), and the Anoplura (the sucking lice).

ORDER SIPHONAPTERA

The Siphonaptera includes the fleas. Over 2000 species of these blood-sucking ectoparasites are known. These primarily parasitize mammals, although about 100 species are parasites of birds. The North

---

*Although the Mallophaga and Anoplura are considered to undergo hemimetabolous metamorphosis, these lice do not develop wings as do other hemimetabolous insects. However, the absence of wings on lice is a secondary evolutionary adaptation.

American species have been reviewed by Ewing and Fox (1943), Fox (1940a, b), and Hubbard (1947).

The origin and evolution of fleas remain highly speculative. Most agree, however, that the fleas of mammals evolved from the fleas of birds. Members of *Ceratophyllus*, for example, are clearly derived from forms that parasitize squirrels and other arboreal rodents whose nests provide conditions similar to those of bird nests. It is quite possible that bird fleas, found in nests, have become adapted to tree-climbing mammals, which, in turn, have passed them on to terrestrial mammals.

The more common fleas measure 1.4 to 4 mm in length, the males being slightly smaller than the females. Figure 18.13 depicts the female of *Ctenocephalides felis*, the cat flea. This illustration should provide familiarity with the external anatomy of a flea.

Interest in fleas among parasitologists stems from three avenues of investigation—that concerned with the blood-sucking habits of these pests, that concerned with fleas as vectors for pathogenic microorganisms, and that concerned with the fleas as intermediate hosts of helminth parasites.

Mating Behavior. Humphries (1967a) has studied the mating behavior of the hen flea, *Ceratophyllus gallinae*. This behavior is initiated by apparently fortuitous contact between the abdominal cuticle of the female and the maxillary palps of the male. The female's cuticle possibly carries a pheromone which, when perceived, causes the male to hold its antennae vertically and to push against the female with its head. The female responds by turning away from its mate so that the male can push its head and thorax beneath her abdomen. At this point, the male's erect antennae are used to grasp the ventral part of the female's abdomen, and apparently the stimulation of the inner surfaces of the antennae causes the male to curve its abdomen over its back and thus to bring the genitalia of the mating pair together (Fig. 18.14).

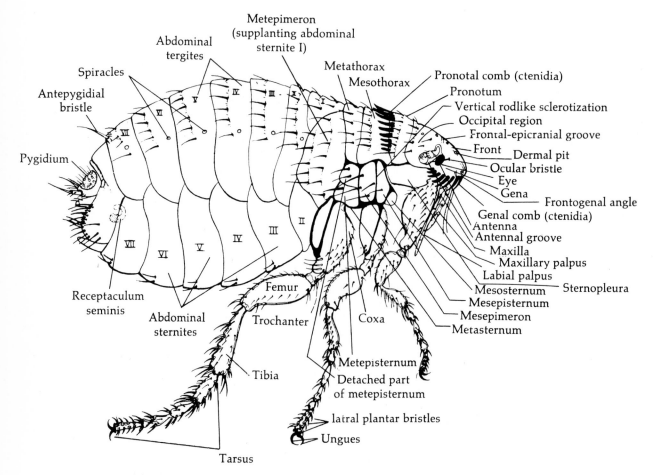

Fig. 18.13. Flea anatomy. Adult female of the cat flea *Ctenocephalides felis*. (Courtesy of the National Center for Communicable Diseases, U.S. Public Health Service, Atlanta, Georgia.)

The inner surfaces of the antennae of many species of male fleas are armed with batteries of suckers (Rothschild and Hinton, 1968), and these serve to hold on to the female. The external genitalia of the male flea is extremely complex, and according to Humphries (1967b) the genitalia are guided into the correct copulatory position by a complex arrangement of lobes and setae. Each step in alignment must be successfully completed before the sequence can proceed. It has been suggested that species-specific differences in male flea genitalia have a species-isolating function, preventing copulation between mates of different species.

Life Cycle Pattern. The life cycles of fleas all follow an essentially similar pattern. The comparatively small, whitish eggs (Fig. 18.15) measure approximately 5 mm in length and are oviposited either on or off the host. If the eggs are oviposited on the host, they are not adhesive and soon are shaken off. They are commonly found in the host's abode. Flea eggs are layed a few at a time, usually 3 to 20. However, during its life span, a female flea can lay many eggs: For example, a single female of the human flea, *Pulex irritans*, can oviposit a total of 448 eggs during a period of 196 days. Because metamorphosis in the Siphonaptera is holometabolic, the hatching form is a larva.

The embryonic development and hatching process of fleas is influenced by the environmental temperature and humidity. High temperatures (35–37°C) inhibit embryonic development. Temperatures ranging from 17 to 23°C bring about hatching in 7 to 9 days, but lower temperatures (11–15°C) lengthen the process to 14 days. Bruce (1948) has reported that growth of *Ctenocephalides felis* is at its optimum when the fleas are maintained at 65 to 90% relative humidity.

The larva possesses a cephalically situated spine that ruptures the egg-shell. The whitish flea larvae are elongate and vermiform (Fig. 18.15), displaying a distinct head but no legs. The body is sparsely covered with bristly hairs, and a pair of tiny hooks, known as **anal struts**, are situated on the last body segment.

The mouthparts of the larva are of the biting type and include mandibles (mandibles are absent in adult fleas). Such larvae are free living, feeding on decaying animal and vegetable matter. *Nosopsyllus fasciatus* larvae subsist totally on the feces of adult fleas. Bird flea larvae utilize the sheaths of feathers and epidermal scales of young birds for food. The larval growth period varies between 9 and 200 days, depending on environmental conditions such as humidity, temperature, and oxygen tension.

Flea larvae usually undergo two molts, increasing in size after each molt. On reaching their maximum size, the third-stage larvae become quiescent, spin a whitish cocoon around themselves, and pupate

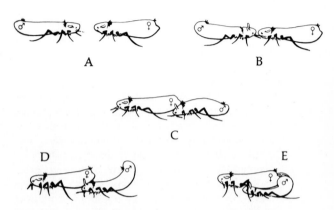

Fig. 18.14.   Mating behavior of *Ceratophyllus gallinae*. **A.** Partners approach each other. **B.** When their maxillary palps make contact, the male's antennae are erected. **C.** Male moves behind female and pushes against her with lowered head, his antennae clasping her abdomen. **D.** Male raises apex of abdomen. **E.** Copulation. (After Humphries, 1967 in Askew, 1971.)

(Fig. 18.15). The pupae remain within the cocoons from 7 days to as long as a year, depending on conditions. The form rupturing out of the cocoon is the adult, which is parasitic.

Emergence of the adult may not be immediate, for fully developed fleas may lie quiescent for many months within the cocoon. When disturbed, even by the slightest vibration, dormant adults are activated and leave the cocoon. Mammals and birds returning to old haunts often become infested with fleas that have remained quiescent for long periods. Similarly, humans entering abandoned abodes can become infested with fleas that become activated because of the disturbance.

Because development of fleas generally occurs off

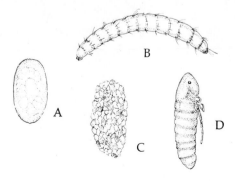

Fig. 18.15.   Stages in the life cycle of a flea. A. Egg. **B.** Larva. **C.** Whitish coccoon. **D.** Pupa.

the host and the parasitic adults must have access to suitable hosts, it is not surprising that fleas are generally associated with hosts that roost or live in dens or nests, and are not usually associated with free-roaming animals.

Host Localization. Fleas are capable of detecting their hosts from a distance not exceeding a few centimeters. From what is known, they are attracted to hosts primarily by their sense of smell, which is sufficiently acute to enable them to distinguish between different species of vertebrates. The odor of an incompatible host may, in some instances, have a repellent effect. For example, horses are not attacked by fleas, and humans working with horses also do not attract fleas.

Although fleas have poorly developed vision since they have no compound eyes, but only a single pair of ocelli, Humphries (1968) has reported that vision is important in the host-finding behavior of *Ceratophyllus gallinae*. Specifically, young fleas emerging from cocoons in the nests of birds in the spring are initially

negatively phototactic and remain in the nests, presumably to mate. However, after a few days the fleas become positively phototactic and negatively geotactic, which results in their climbing vegetation adjacent to nests. From this location they orient themselves toward the light and jump when there is a sudden fall in light intensity as, for example, when a bird passes by.

Despite what appears to be the necessity of close proximity between fleas and hosts, these parasites are highly efficient in locating hosts. For example, Mead-Briggs has reported that when 270 marked fleas, *Spilopsyllus cuniculi*, are released in a meadow with an area of 2000 square yards and three rabbits are liberated in the meadow, within a few days the rabbits are infested with nearly half of the released fleas.

Biology of Fleas. The Siphonaptera includes six families: Tungidae, Pulicidae, Ceratophyllidae, Ischnopsyllidae, Hystinchopsyllidae, and Macropsyllidae.

Family Tungidae

This family includes the chigoe flea, *Tunga penetrans*, and the sticktight flea of poultry, *Echidnophaga gallinacea* (Fig. 18.16). These unique fleas are more or less permanent parasites as adults. They are often referred to as burrowing fleas because the impregnated females become firmly attached to the host's skin and in so doing cause such a severe irritation that the surrounding tissues become swollen, embedding the fleas and giving the appearance that they have burrowed into the skin.

Tunga penetrans (Fig. 18.16), commonly known as the chigoe, chigger, or sand flea, is a native of tropical and subtropical South America and has been introduced into Africa, the West Indies, and elsewhere. There appears to be little host specificity in *T. penetrans*, for in addition to poultry, it attacks humans, dogs, pigs, and other animals. However, it is strictly a parasite of endothermic animals. The embedded females commonly are distended because of the enclosed eggs, which when layed fall to the ground and develop (Fig. 18.16). After laying her eggs, the female is usually expelled from the host by the pressure of the surrounding tissues and dies. The host's skin irritation often opens the way for secondary infection and may become gangrenous. Human infestations by *T. penetrans* are commonly on the foot or between the toes where there is extreme itching, pain, inflammation, and secondary infections (Fig. 18.17).

Echidnophaga gallinacea is usually parasitic on chickens and other birds, but it has been known to attack dogs, humans, rats, and other animals. The adults tend to congregate in masses, usually on the host's head. It is referred to as the sticktight flea because once attached to a host, it seldom migrates. Geographic distribution of this flea is limited to the tropical and subtropical regions of the world.

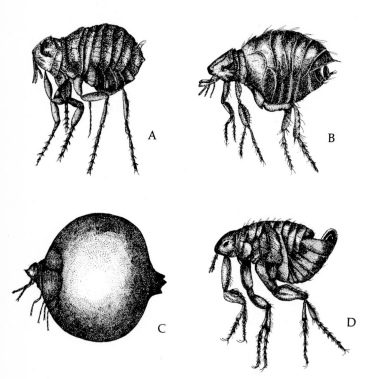

Fig. 18.16.   Fleas. A. *Tunga penetrans.* **B.** Female *Echidnophaga gallinacea.* (Redrawn after Bishopp, 1915.) **C.** Gravid female of the jigger flea *Tunga penetrans.* (Redrawn after Ewing, 1921.) **D.** Male *Pulex irritans,* the human flea. (Redrawn after Bishopp, 1915.)

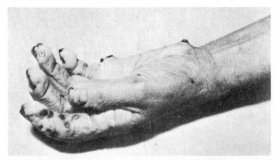

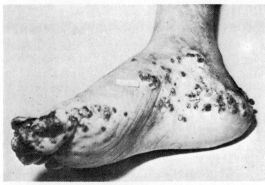

Fig. 18.17. Lesions caused by *Tunga penetrans*. A. Hand showing lesions on fingers and palm. **B.** Foot showing numerous lesions. (After Hunter *et al.*, 1976; with permission of W. B. Saunders.)

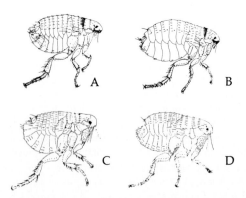

Fig. 18.18. Fleas. A. Male of the cat flea *Ctenocephalides felis*. **B.** Female *C. felis*. **C.** Male of the Oriental rat flea *Xenopsylla cheopis*. **D.** Female *X. cheopis*. (All figures courtesy of National Center for Communicable Diseases, U.S. Public Health Service, Atlanta, Georgia.)

Family Pulicidae

This family includes several species that bite humans. Furthermore, several species are vectors for the plague bacillus, *Yersinia pestis*. *Pulex irritans*, commonly referred to as the human flea, is widely distributed and parasitizes domestic animals as well as humans (Fig. 18.16). This species is capable of transmitting the plague pathogen under laboratory conditions but is not an important vector in nature. The infrequency of the blockage of its esophagus, which results in the regurgitation of alimentary tract contents including plague bacilli, renders it less effective as a vector than other fleas. The role of *P. irritans* as a vector for the filarial worm *Dirofilaria immitis* is discussed on p. 544.

Genus *Ctenocephalides*. *Ctenocephalides* includes approximately 10 species commonly found on carnivores. The most familiar of these are *C. canis*, the dog flea, and *C. felis*, the cat flea (Fig. 18.18). These fleas also are not host specific, for both species will bite both cats and dogs, as well as humans. In addition to causing dermal irritations while feeding, *C. canis* is a suitable intermediate host for the dog tapeworm, *Dipylidium caninum*.

Genus *Xenopsylla*. Over 35 species of *Xenopsylla* are known. *Xenopsylla cheopis*, the Asiatic rat flea, is the most commonly encountered species in this genus (Fig. 18.18). It presumably originated in the

Nile Valley in Africa but has spread over the entire world on rat hosts. A major vector for *Yersinia pestis*, both male and female fleas take in the bacillus from infected rats during feeding.

The stomach of *X. cheopis* has a capacity of 0.5 mm³ and is capable of receiving as many as 5000 bacilli, which continue to multiply inside the flea. The bacteria are limited to the stomach, intestine, and rectum; they are never found in the flea's hemocoel or salivary glands. Consequently, infection is brought about in humans by the rubbing of contaminated flea feces into fresh bite sites. This has since been demonstrated experimentally, although it is not the only method of infection. The other, and probably more common, method of infection from *Xenopsylla cheopis* involves

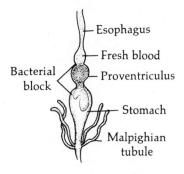

Fig. 18.19. Bacteria in flea proventriculus. Alimentary tract of flea showing proventriculus blocked with bacteria. Notice distention of the esophagus with fresh blood. (Modified after Bacot and Martin, 1914.)

transmission during the bite. This is not accomplished via the salivary glands; instead, a temporary blood clot in the flea's esophagus causes regurgitation of bacteria-including materials up to the level of the clot, which then serves as a culture medium for the bacilli. Such bacteria from time to time pass into the mouthparts and from there into the host (Fig. 18.19).

At least 5 days, and an average of 21 days, occurs between imbibition of bacilli-containing blood and the infective bite. In *Diamanus montanus*, the ground squirrel flea, the average time lapse is 53 days.

Other Xenopsylla *Species.* *Xenopsylla braziliensis*, the African rat flea, is another important species which occurs in Uganda, Kenya, Nigeria, and India. Its specific name arises from the fact that the flea was first described in Brazil, where it had been transported. This flea is a major vector for the plague organism in Africa.

Xenopsylla astia is found in parts of Sri Lanka and India. Epidemiologic evidences indicate that this flea can serve as a vector for *Yersinia pestis*.

Genus ***Hoplopsyllus.*** *Hoplopsyllus anomalus* (Fig. 18.20) is the plague vector of California ground squirrels, *Cittelus beechyi*. This flea also attacks rats.

Family Ceratophyllidae

This flea family includes, among others, the genera *Diamanus*, *Ceratophyllus*, and *Nosopsyllus* (Fig. 18.20).

Diamanus montanus is a commonly encountered ectoparasite of squirrels in the western United States (Fig. 18.20). It is indirectly of medical importance, for it aids in the maintenance of plague infections among squirrels. The egg incubation period in this species is 8 days. The first larval instar lasts for 6 days, the second for 10 days, the third for 12 days, and the pupal stage for 31 days. Under experimental conditions, *D. montanus* can serve as a carrier of *Yersinia pestis*.

Genus ***Ceratophyllus.*** This genus includes *C. gallinae*, the European hen flea. This and *C. niger* are both primarily parasites of hens, but both also attack dogs, cats, and humans. Unlike the sticktight fowl flea, *Echidnophaga gallinacea*, the species of *Ceratophyllus* are not attached to the host's skin, except during feeding. These fleas cause serious injury to hens.

Genus ***Nosopsyllus.*** The European rat flea, *N. fasciatus* (Fig. 18.20), is widely distributed throughout northern Europe and North America. Its chief hosts are rodents; however, it does bite humans and other mammals. Development of this species (which takes 37 days) is optimal at cool temperatures, and the larvae apparently require blood for development. Blood is acquired from the fecal droppings of adult

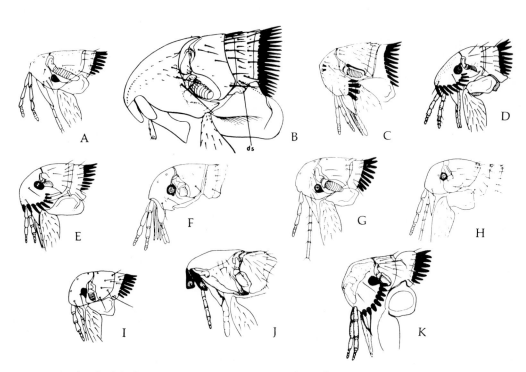

Fig. 18.20. **Heads of adult fleas.** **A.** *Diamanus montanus.* **B.** *Ischnopsyllus elongatus* (ds, dorsal sinus of posternite). **C.** *Leptopsyllus segnis.* **D.** *Ctenocephaloides felis.* **E.** *Ctenocephaloides canis.* **F.** *Pulex irritans.* **G.** *Nosopsyllus fasciatus.* **H.** *Xenopsylla cheopis.* **I.** *Hoplopsyllus anomalis.* **J.** *Myodopsylla insignis.* **K.** *Cediopsylla simplex.* (All figures after Smart, 1943.)

fleas. This species is known to carry *Yersinia pestis* from rodent to rodent, thus maintaining natural reservoirs of the bacilli.

Family Ischnopsyllidae

Ischnopsyllid fleas include members of *Ischnopsyllus* (Fig. 18.20) and other host-specific genera, for all these fleas are parasites of bats and only occasionally are reported from wild mice, *Peromyscus* spp. Little is known about these fleas except that they are not important as vectors for *Yersinia pestis*.

Family Hystrichopsyllidae

The Hystrichopsyllidae includes the common mouse flea, *Leptopsylla segnis*, which is also found on rats (Fig. 18.20). Like all other rodent-infesting siphonapterans, *L. segnis* is distributed all over the world, especially on rodents in seaports and on ships. This species has been found on rats taken from ships in New York and New Orleans, for example.

Family Macropsyllidae

Members of this family are not common. None are known to bite humans or domestic animals, or even to serve as vectors. The only member found in North America is *Rhopalopsyllus*, with two species found on dogs, rats, and opossums in the southern United States, ranging from Florida to Texas.

FLEA-BORNE DISEASES

Plague

The disease commonly referred to as the plague (the "black death" of Europe during the 14th and 17th centuries) is caused by the bacillus *Yersinia pestis*. When this bacterium is introduced into the skin, the lymph glands become inflamed, and the condition is known as **bubonic plague** (Fig. 18.21). When the bacilli become established in the victim's blood, the condition is referred to as **septicemic plague**. If the victim's lungs become involved, it is **pneumonic plague**. Pneumonic plague is a highly contagious pneumonialike disease. Septicemic plague can be distinguished as being either **primary** or **secondary**. Primary septicemic plague is a generalized blood infection with little or no prior lymph node swelling, apparently because the blood is infected so rapidly that the typical inflammation does not develop. Secondary septicemic plague usually develops in pneumonic and sometimes in bubonic plague. Bubonic plague is fatal in about 25 to 50% of untreated cases, and pneumonic and septicemic plague are usually fatal.

The incubation period after transmission by a flea is 2 to 4 days. At that time the victim is afflicted by a chill followed by rapidly rising temperature to 40°C. The lymph nodes draining the infection site swell, becoming hemorrhagic and often necrotic. Initially,

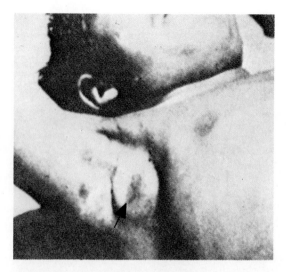

Fig. 18.21. Bubonic plague. Plague bubo in right axilla. (Armed Forces Institute of Pathology, Negative No. ACC 219900-7-B.)

there is mental dullness; however, anxiety followed by delerium, or lethargy and coma ensue. Death occurs within 5 days. If the victim is to recover, the fever commences to drop in 2 to 5 days.

The rapid and serious effects of plague are primarily due to two toxins released by *Yersinia pestis*. These have essentially the same effect, that is, they act on the mitochrondrial membranes of susceptible animals, inhibiting ion uptake and interfering with normal cellular respiration (Kadis *et al.*, 1969). Rats and mice appear to be more sensitive to the toxins than are rabbits, dogs, and primates.

Humans usually acquires infection from rodents via a flea vector because rodents are utilized by *Y. pestis* as reservoir hosts. Plague among rodents is known as **sylvatic plague**.

The literature concerning plague is extensive. Interested individuals are referred to the writings of Hampton (1940), Meyer (1942), and the extensive bibliography by Jellison and Good (1942). In addition, the accounts by Lehane (1969) and Hirst (1953) are recommended. The last plague pandemic began in China during the latter part of the 19th century, reaching Hong Kong and Canton by 1895, and Bombay and Calcutta in 1896, then spread throughout the world. It reached numerous port cities in the United States. Between 1898 and 1908, more than 548,000 persons died per year in India from plague. Between 1900 and 1972, 992 cases were reported in the United States (416 of these in Hawaii) of which

720 were fatal (Pratt and Stark, 1973). The disease still occurs. Between 1958 and 1972, 9 of 51 reported cases in the United States were fatal. The most recent human cases have been rural, resulting from contact with wild rodents or their fleas.

Murine Typhus

In addition to plague, some fleas serve as vectors for the murine (endemic) typhus organism, *Rickettsia typhi* (= *R. mooseri*). This rickettsia is transmitted via con-

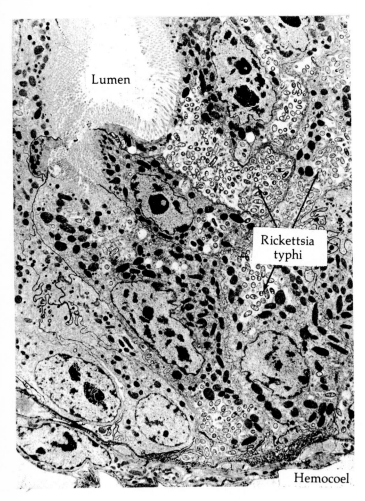

Fig. 18.22. Murine typhus organisms in flea gut. An electron micrograph showing *Rickettsia typhi* as intracellular parasites in cells lining midgut of *Xenopsylla cheopis.* Note that most of the fully differentiated cells contain *R. typhi.* Near the lower left are two cells with basal nuclei; these undifferentiated cells are not infected with rickettsiae. × 2500. (After Ito *et al.,* 1975. Courtesy of New York Academy of Sciences.)

taminated flea feces, which are rubbed into the bite wound. Fleas contract the rickettsiae while feeding either on an infected human or on infected rats and mice, *Rattus norvegicus, R. rattus, R. alexandrinus,* and *Mus musculus.* Furthermore, this rickettsia has been found in the opossum, cottontail rabbit, fox squirrel, Florida skunk, southern weasel, dog, and even the blue jay. Fleas found to be suitable transmitters include *Xenopsylla cheopis, X. astia, Nosopsyllus fasciatus,* and *Leptopsylla segnis.* In addition, the tropical rat mite, *Ornithonyssus bacoti,* and the rat louse, *Polyplax spinulosa,* can serve as vectors. *Echidnophaga gallinacea* has been experimentally determined to be a suitable vector. Once taken into the flea host, *R. typhi* become intracellular parasites within midgut lining cells, where they multiply (Fig. 18.22). The rickettsiae eventually become incorporated in the vector's feces when gut cells enclosing the pathogens are sloughed (Ito *et al.,* 1975).

Distribution of murine typhus includes Mexico, Peru, Columbia, Venezuela, Chile, and northern Argentina. In North America, all the southeastern and southern states in the United States, including Texas, Oklahoma, and parts of southern California are endemic areas, as is Hawaii. In the Old World, *Rickettsia typhi* is found in southern and western Europe, the Soviet Union, north, west, and south Africa, Israel, the South Sea islands, and central and northern China.

In humans, murine typhus is a mild disease that lasts for about 14 days. Infected persons have chills, severe headaches, body pains, and a rash. It is more severe among the elderly. *Xenopsylla cheopis* is the principal vector. In the United States, 5401 cases of murine typhus were reported in 1945. The use of DDT to kill fleas, coupled with rat control programs, have dramatically reduced the number of cases. Between 1969 and 1972, between 18 and 36 cases were reported per year.

Myxomatosis

Rabbits are considered an asset in certain parts of the world and a destructive nuisance in others. For example, in North America and northern Europe, including Britain, the raising of rabbits constitutes a "cottage" industry. On the other hand, the runaway rabbit population in Australia constitutes a threat to vegetable farmers. Myxomatosis, a commonly fatal rabbit disease caused by the myxoma virus, is a threat to rabbit raisers in England. This virus is transmitted by several bloodsucking arthropods, including mosquitoes, fleas, and mites. Among these, the flea *Spilopsyllus cuniculi* is a common vector in Great Britain. The myxoma virus was intentionally introduced into Australia to control the rabbit population, but unforeseen was that in Australia the principal vectors are mosquitoes. Since mosquitoes are most abundant during the warm months when the rabbits have the best chance of surviving myxomatosis and the Australian mosquitoes favor the transmission of an attenuated "field strain" of the myxoma virus which is not as pathogenic, the venture failed. More recently, *Spilopsyllus cuniculi* has been introduced into Australia together with more virulent strains of the virus. This combination holds greater promise for rabbit control (Sobey and Conolley, 1971).

Other Diseases

Other microorganisms associated with fleas include *Pasteurella tularensis*, the tularemia-causing organism in humans, rabbits, and rodents, and *Salmonella enteritidis*, the salmonellosis-causing bacterium in humans.

HELMINTH PARASITES

In addition to serving as vectors for bacteria, viruses, and rickettsiae, certain fleas act as intermediate hosts for helminth parasites (Table 18.1). In addition, *Nosopsyllus fasciatus* and *Xenopsylla cheopis* are principal vectors for the hemoflagellate *Trypanosoma lewisi* in rats (p. 128). Also, *Dipetalonema reconditum*, a filarial nematode of dogs (p. 546), is transmitted by the fleas *Ctenocephalides canis* and *C. felis*.

ORDER MALLOPHAGA

The Mallophaga includes the so-called biting lice, or bird lice. The common name "bird lice" is not preferred because certain members are primarily ectoparasites of mammals. It is estimated there are over 3000 known species of biting lice. These are primarily injurious because of their bites. *Dennyus* on swifts, however, serves as intermediate host for the filarial worm *Filaria cypseli*, and *Trichodectes latus* serves as intermediate host for the dog tapeworm *Dipylidium caninum*. The role of the Mallophaga as transmitters of helminth parasites is certainly minor. Lice, both Mallophaga and Anoplura, may transmit leishmanias (p. 121).

Biting lice range from approximately 1 mm (males of *Goniocotes* found on pigeons) to about 10 mm in length (*Laemobotkrion* found on hawks). Mouthparts of the biting lice are of the chewing type (p. 616), but

they are greatly reduced and are difficult to interpret without critical study.

Mallophagan lice, as well as anopluran lice (p. 635), are presumed to have been derived from primitive forms that lived beneath tree bark and in similar habitats. These progenitors probably first became parasitic on birds, after migrating to their nests. Birds in turn could have passed their lice to arboreal mammals, which later passed the lice to terrestrial mammals. On the other hand, the presence of a very primitive species of lice on a tree shrew, which is a primitive mammal, suggests that the lice infested the shrew without utilizing birds as an intermediate step. Hopkins's (1957) essay on the relationship between lice and mammals should be consulted.

Life Cycle Pattern. The life cycle patterns in the Mallophaga are essentially the same. The development of *Columbicola columbae*, the common pigeon louse, exemplifies the pattern. The entire life cycle of the parasite occurs on the host. As many as 60 whitish opaque eggs of the louse can be found attached to the host's feathers. These eggs undergo a 3- to 5-day incubation if they are maintained in an environmental temperature of 37°C. If the temperature is 33°C, the incubation period is 9 to 14 days; even then, the nymphs that hatch perish in 1 to 6 days. Thus, temperature is of critical importance during the embryonic period.

The young that hatches from the egg is the nymph. There are three nymphal instars, each lasting slightly less than 7 days. The third nymphal instar metamorphoses into the adult. Again, temperature is critical to survival of the adult. *Columbicola columbae* adults live for 30 to 40 days if maintained at 37°C. Lower or higher temperatures shorten the life span.

Matthysse (1946) has studied the life history of *Bovicola bovis*, the red cattle louse, and found that the stages in this species parallel those of *C. columbae*,

Table 18.1. Helminths That May Utilize Fleas as Intermediate Hosts

Helminth Parasite	Definitive Host	Flea Intermediate Host
Dipetalonema reconditum (nematode)	Dogs	*Ctenocephalides felis* *C. canis*
Dipylidium caninum (cestode)	Dogs, cats, certain wild carnivores	*Ctenocephalides felis* *C. canis* *Pulex irritans*
Hymenolepis diminuta (cestode)	Rodents, occasionally humans	*Nosopsyllus fasciatus* *Xenopsylla cheopis*
Vampirolepis nana (cestode)	Rodents, occasionally humans	*Xenopsylla cheopis* *Ctenocephalides canis* *Pulex irritans*

differing only in the lengths of time required to complete each stage. Again, temperature is a critical factor. Transmission of lice from host to host occurs when the hosts' bodies are in contact.

Biology of Mallophagan Lice. Biting lice are divided into three suborders: Amblycera, Ischnocera, and Rhynchophthirina. The last includes only one species, *Haematomyzus elephantis*, parasitic on African and Indian elephants. The Amblycera is further divided into six families among which the more prominent are Gyropidae, Menoponidae, Boopiidae, and Ricinidae. The Ischnocera includes four families among which Trichodectidae and Philopteridae are the principal ones. The diagnostic characteristics of the suborders Amblycera and Ischnocera, along with those of the principal member families, are presented at the end of this chapter.

SUBORDER AMBLYCERA

Amblyceran lice are parasitic on birds and mammals. These insects are readily distinguished from ischnoceran lice by their antennae, which are six-jointed. The antennae of ischnoceran mites are three- or five-jointed.

Family Gyropidae

This family includes two species, *Gyropus ovalis* and *Gliricola percelli*, which are familiar to those who handle laboratory animals, since both are found on guinea pigs. Other members of this family are confined mainly to Central and South America.

Family Menoponidae

The family Menoponidae includes the genera *Menopon, Myrsidea, Colpocephalum, Pseudomenopon, Trinoton*, and *Menacanthus*.

Genus *Menopon*. This genus includes numerous species found on wild and domestic birds.

Menopon gallinae. *M. gallinae*, the most commonly encountered species, is a chicken louse found attached to the shafts of the host's feathers (Fig. 18.23). Since this species is not a blood ingestor but feeds instead on the barbs and scales of feathers, it does little damage to the host by way of irritation. The females, which are about 2 mm long, lay their eggs in clusters on the feathers. The eggs hatch in 2 to 3 weeks.

Menopon gallinae generally does not infest young chicks, presumably because chicks lack well-developed feathers. It does infest ducks, turkeys, and guinea hens, at least when these birds are housed with chickens. The adult lice are extremely hardy and have been kept alive for as long as 9 months.

Other Menopon *Species.* Other species of *Menopon* are parasitic on various birds, for example, *M. aegialitidis* on the killdeer, *Charadrius vociferus*, and *M. leucoxanthum* on the black duck, *Anas rubripes*.

Genus *Menacanthus*. *Menacanthus stramineus*, the common body louse of chickens, is also found on turkeys. It appears to prefer the host's skin to feathers. It is light yellow and approximately 3 mm long. It is destructive in that its bite causes droopiness, weight loss, diarrhea, and reduced egg production. When observed on the host, this louse is extremely active and quick, and migrates over the entire body.

Other Menoponid Genera. *Myrsidea* includes *M. subaequalis*, commonly found on crows and hawks. *Colpocephalum* includes *C. laticeps* on the crow, *Corvus brachyehynchos* and *C. pustulosum* on the pectoral sandpiper, *Calidris melanotos. Pseudomenopon pacificum* is found on coots and grebes.

Trinoton querquedulae is a dark grayish species that measures approximately 4 mm in length (Fig. 18.23). It is a blood ingestor found on ducks.

Family Boopiidae

Boopiidae is a small and little known family with species found on marsupials (kangaroos and wallabies). One species, *Heterodoxus longitarsus*, has been found occasionally on dogs in Australia.

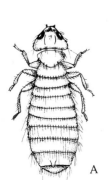

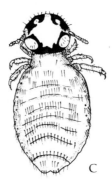

A   B   C   D

Fig. 18.23. **Lice. A.** *Menopon gallinae*, the common shaft louse of poultry. **B.** Anterior end of *Trinoton querquedulae* of ducks. **C.** *Trichodectes canis* of dogs. **D.** *Trichodectes tibialis* of deer.

Family Ricinidae

Ricinidae is another relatively obscure family. The genus *Ricinius* is represented by *R. leptosomus* on the kingbird, *Tyrannus tyrannus*, and *R. lineatus* on the hummingbird, *Archilochus colubris*, in New York State. Another member of this small family is *Trochiloecetes*, represented in New York State by *T. prominens* on the hummingbird.

Amblyceran Host Specificity

Although present data still are scanty, it is becoming more evident that members of the suborder Amblycera demonstrate a certain degree of host specificity. If two avian hosts share a common species of louse, the two hosts are usually closely related. Furthermore, the mammal-parasitizing members of the family Boopiidae are host specific and are restricted to mammals, primarily marsupials.

SUBORDER ISCHNOCERA

As stated, the Ischnoceran lice, which are ectoparasites of birds and mammals, includes two major families—the Trichodectidae, with species parasitic on mammals, and the Philopteridae, with species parasitic on birds.

Family Trichodectidae

Genus *Trichodectes*. Several species of *Trichodectes* are commonly encountered on various domestic animals. Others are found on weasels, badgers, skunks, and other small mammals. *Trichodectes canis* is an irritating louse of dogs throughout the world (Fig. 18.23). This species is broad, short, and approximately 1–2 mm long. Other species encountered in the Americas include *T. breviceps* on llamas in South America and *T. tibialis* on deer in the western United States (Fig. 18.23).

Genus *Damalinia*. Several species of *Damalinia* that infest domestic animals were at one time assigned to *Trichodectes*. However, these species have been transferred to *Damalinia* because their antennae do not demonstrate sexual dimorphism as do those of *Trichodectes*, in which the first segment of the antennae of males is enlarged. *Damalinia bovis* is the most common cattle louse in Britain and is also found in the United States (Fig. 18.24). During winter, this louse is found at the base of the tail, on the shoulders, and along the back of the host. In heavy infestations, the lice are uniformly distributed over the host.

Other species of *Damalinia* include *D. equi* and *D. pilosus* on horses, *D. caprae* on goats, and *D. ovis* and *D. hermsi* on sheep. Although these lice do not produce any serious injury, they do cause skin irritations that can become uncomfortable to the hosts.

Few *Damalinia equi* males have ever been found, and the females reproduce parthenogenically throughout the year. Thus, critical phases in the parthenogenic

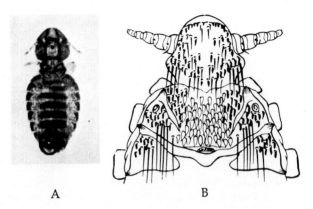

A   B

Fig. 18.24. Lice. A. *Damalinia bovis*, the biting cattle louse. (After James and Harwood, 1969.) **B.** Head and thorax of a seal-infesting sucking louse, *Antarctophthirius*, with typical flattened scales. (Courtesy of U.S. Public Health Service.)

reproduction of *D. equi* are the development of eggs within the female, oviposition, morphogenesis within the shell, and hatching. Murray (1963a) has found that temperature and relative humidity are of critical importance during these phases. If female lice are maintained at 16°C, no egg development within the female takes place, and exposure to 44.5°C for only 1 hour can prevent subsequent oviposition. Fewer eggs are laid at 31°C than at 39°C, and the majority of eggs are oviposited at between 32 and 37°C with 75% relative humidity or less.

Morphogenesis is also influenced by temperature and relative humidity, for only between 31 and 39°C does morphogenesis proceed to completion and hatching occur, provided the relative humidity is less than 90%. A relative humidity of 90% prevents hatching. If eggs are maintained between 27 and 31°C, fewer eggs develop. None reach an advanced stage of development at 42°C. Exposure of eggs at an advanced stage of development for 2 hours at 49°C is lethal, but at least 6 hours of exposure are required to kill eggs at 45°C.

Only during the winter months are temperatures near the horse's skin continuously favorable for egg development in the female louse, for oviposition, and for egg development on the horse. This probably explains the presence of *D. equi* in large numbers on the bodies of horses at the end of winter and their scarcity on the limbs. On the other hand, the temperature within the hair coat on the horse's body during the summer is sufficient to kill the lice, or to reduce the number of eggs laid. Therefore, it is not surprising that the reproduction rate of *D. equi* is greatly reduced during the summer months.

Genus *Felicola*. This genus includes *F. subrostratus*, which parasitizes cats. This species exhibits the generic characteristics of absence of pleural plates and similarity of antennae in both sexes.

Genus *Lepidophthirus*. Members of *Lepidophthirus* and the related genus *Antarctophthirus* are ectoparasites of seals and other marine mammals. They have acquired interesting structural and physiologic adaptive features which are of survival value in their unique habitats. For example, both *Lepidophthirus* and *Antarctophthirus* from seals have scales on their dorsal body surface to retain air when the host dives (Fig. 18.24). In contrast, *Echinophthirus* is found within the seal's nostrils and lacks such scales.

An example of a physiologic adaptation is expressed by *Lepidophthirus macrorhini*, a parasite of the southern elephant seal. Murray and Nicolls (1965) have reported that the host comes ashore only twice a year for periods of 3 to 5 weeks. It is only at this time, when the ambient temperature is 25°C or slightly higher, that the lice can reproduce, and they make the most of the short time available, each female laying up to nine eggs per day. The warmest external parts of the seal's body are the hind flippers, and it is at this site that the lice congregate at sea and on land.

The adaptation of *L. macrorhini* to 25°C or higher, as far as its reproductive habits are concerned, is accentuated when compared to those of *Antarctophthirus ogmorhini*, a parasite of the Weddell seal. This seal inhibits colder water than the elephant seal, and its louse parasite reflects this phenomenon. Murray *et al.* (1965) have found that *A. ogmorhini* is well adapted to cold, being able to survive for 36 hours at −20°C, and reproduces rapidly at temperatures between 5 and 15°C.

Family Philopteridae

Members of this family are either long and slender or broad with rounded abdomens. Philopterid members belonging to the first category include the genera *Lipeurus*, *Ornithobius*, *Columbicola*, and *Esthiopterum*. Those possessing broad bodies and rounded abdomens include the genera *Goniodes*, *Goniocotes*, *Cuclotogaster*, and *Philopterus*.

Again, as among the amblycerid lice, the ischnocerids demonstrate a considerable degree of host specificity.

Genus *Lipeurus*. This genus includes several species encountered on domestic birds. *Lipeurus caponis*, the "variable louse," measuring 2 mm in length, is long and slender and parasitizes chickens; *L. polytrapezius*, also slender and measuring 3–3.5 mm in length, occurs on turkeys; *L. humidianus* occurs on guinea fowl; *L. bidentatus*, a whitish species measuring 1 mm in length occurs on pigeons; *L. squalidus* is a large species that measures 4 mm in length and occurs on ducks and geese; and *L. damicornis*, measuring 2 mm in length, is a broad, brownish species found on pigeons.

The species of *Lipeurus* live among the feathers of both old and young birds. In heavy infestations, areas of the host's skin are made bare by the lice. The greatest amount of injury sustained by the hosts does not result from the feeding habits of these insects, but from irritation of the skin. Infested birds commonly scratch themselves with their claws, further irritating the skin.

When an infested bird is shot, the lice on it die from 2 hours to 2 or 3 days afterward. In extremely rare instances, lice may survive on a dead bird for over a week. Passage of lice from one bird to another most probably occurs while the bodies of the hosts are in contact.

Genus *Ornithobius*. This genus includes *O. cygni*, which measures 1 mm in length. It has a reddish brown head, thorax, and legs, and a white abdomen. This louse is found on swans, whereas the related species, *O. icterodes*, is found on ducks and geese.

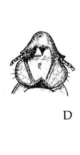

A   B   C

D

Fig. 18.25. Philopterid lice. A. *Gonicotes gigas* of chickens. **B.** Female *Columbicola columbae* of pigeons. **C.** *Esthiopterum crassicorne* of chickens. **D.** Anterior end of *Philopterus dentatus* of ducks.

Genus *Goniodes*. This genus includes *G. dissimilis*, the reddish brown louse found on the feathers of chickens in southern United States. *Goniodes stylifer*, measuring 3 mm in length, is found on turkeys. Other species include *G. minor* on pigeons and *G. meleagridis* on turkeys.

Genus *Gonicotes*. Members of *Gonicotes* can be distinguished from those of *Goniodes* by the absence of prongs on the antennal segments of males. *Gonicotes* includes *G. gigas*, which is the largest of the chicken lice (Fig. 18.25). This species is grayish and 3–4 mm long. *Gonicotes hologaster*, the fluff louse, is closely related but smaller, measuring 0.7–1.3 mm in length. It is also parasitic on chickens; *G. bidentatus* occurs on pigeons.

Other Philopterid Lice. *Columbicola columbae*, a common ectoparasite of pigeons, has been mentioned in connection with the life cycle patterns of biting lice (p. 631) (Fig. 18.25). This very slender species is approximately 2 mm long.

Esthiopterum crassicorne occurs on chicks (Fig. 18.25). The head of this species is elongate in front of the antennae, and the clypeus lacks dorsal spines.

Cuclotogaster heterographus, the chicken head louse, is dark grayish and approximately 2 mm long. It is found on the head and neck of its host and, like *Menopon gallinae*, it lays eggs singly.

Philopterus dentatus, like the other members of the genus, possesses a hornlike projection situated anterior to the antennal insertions (Fig. 18.25). This species parasitizes ducks.

ORDER ANOPLURA

The Anoplura, or sucking lice, are quite similar to the Mallophaga. In fact, in the past these two groups were considered members of the same order. More recent systematists, however, are of the opinion that the drastically different feeding habit of the Anoplura—whose mouthparts are modified for sucking rather than biting—is of sufficient evolutionary importance to merit a distinct order of the Insecta. Members range in length from 2 to 5 mm. All are ectoparasites of mammals. Their diet consists of the host's blood and is sucked through the mouthparts, which are formed as an eversible set of five stylets (p. 616).

About 290 species of sucking lice are known, of which approximately 60 occur in the United States. The species are assigned to one of four families—Echinophthiriidae, Pediculidae, Haematopinidae, and Haematopinoididae. The distinguishing characteristics of each of these are listed in the classification section at the end of this chapter. For a detailed and somewhat different classification scheme of the Anoplura, see Kim and Ludwig (1979).

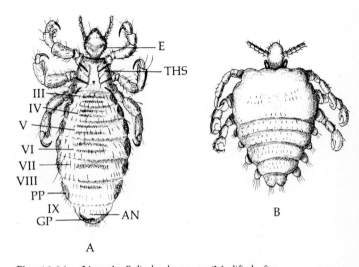

Fig. 18.26. Lice. A. *Pediculus humanus*. (Modified after Keilin and Nuttall, 1930.) **B.** Female *Phthirus pubis*. AN, anus; E, eye; GP, genital pore; PP, pleural plates; THS, thoracic spiracle. III–IX refer to body segments.

Family Echinophthiriidae

The Echinophthiriidae includes a few species found attached to marine mammals such as seals, sea lions, and walruses. For example, *Echinophthirus phocae* is found on seals, and *Antarctophthirus trichechi* is found on the Pacific walrus (Fig. 18.24). Not much is known about the biology of these unusual lice. Investigations on osmoregulation and salt maintenance in these animals should be of considerable interest, for, unlike other lice, they live in the marine environment.

Family Pediculidae

The Pediculidae includes two medically important species that feed on human blood—*Pediculus humanus*, the body louse, and *Phthirus pubis*, the crab louse.

The males of *Pediculus humanus*, a grayish species, are 2–3 mm long, and the females are 3–4 mm long (Fig. 18.26). In the past, most parasitologists recognized two subspecies—*capitis*, found in hair on the head, and *corporis*, found on the body. However, increasing numbers of researchers have come to consider the two subspecies as merely races or forms of the same species. Members of the two races will interbreed. The color of this louse is thought to change, depending on the coloration of the hair of the host, but this strange method of camouflage has not been confirmed.

Life Cycle of *Pediculus humanus*. Development of *P. humanus*, like that of all anopluran and mallo-

Table 18.2. Effect of Temperature on the Incubation of Eggs of *Pediculus humanus*[a]

Temperature (°C)	Incubation Period
22	Will not hatch
24	17–21 days
29	9–11 days
35	5–7 days
38	Will not hatch

[a] Data from Leeson, 1941.

phagan lice, is hemimetabolous (p. 623). The whitish eggs are less than 1 mm long in the *capitis* race, and slightly larger in the *corporis* race. The eggs are attached by a cementlike excretion to hair or clothing. A single female oviposits between 80 and 100 eggs in the *capitis* race and between 200 and 300 in the *corporis* race. Such eggs hatch in 5 to 20 days, and the emerging nymphs are miniature replicas of adults, except that their antennae are composed of three segments (there are five segments in adults).

There are two nymphal instars before the adult form is attained. Development, however, is rapid. The egg-to-egg cycle generally takes 3 weeks. Temperature is critically important during the development process. Leeson (1941) has reported that embryonic development (within the eggshell) ceases if the environmental temperature is below 23°C or above 38°C. Even slight changes in temperature result in marked alterations in developmental time. This phenomenon is demonstrated in Table 18.2.

Upon hatching, the nymphs require almost immediate feeding. If no host is available, they perish within 24 hours. Once reaching maturity, females live for 33 to 40 days, whereas the life span of adult males is slightly shorter.

Male and female *P. humanus* copulate on their host, during which the male crawls underneath the female from behind. When the tips of their abdomens unite, the female rises to a vertical position, lifting the male. The two lice then return to the horizontal position and remain united for 30 minutes or more.

Skin Reactions to *Pediculus humanus*. Not only do the bites of *P. humanus* irritate, but the victim can also become highly sensitive, resulting in tissue reactions that are extremely irritating. Peck *et al.* (1943) have demonstrated that the irritability of louse bites increases with continued recurrence of bites; thus, the dermal reaction is essentially one of hypersensitivity. Human skin subjected to louse bites over long periods often becomes deeply pigmented—a condition known as "vagabond's disease." When such a condition exists, irritability does not occur; instead, the victim becomes immune.

Diseases Transmitted by *Pediculus*. *Pediculus humanus* is a major vector for three important human diseases—relapsing fever, louse-borne typhus, and trench fever.

Relapsing Fever. A cosmopolitan disease, relapsing fever can develop when the spirochaete *Borrelia recurrentis* is inoculated while the victim crushes an infected louse and scratches himself with his contaminated fingernails. This mechanism of infection is quite apparent when one considers the host-parasite relationship between the spirochaete and louse. When the louse takes in spirochaetes during feeding, its gastric juices are detrimental to *B. recurrentis*. In fact, all ingested spirochaetes are killed except for the 1 to 5% that penetrate the alimentary wall. Once these become established in the hemocoel, they begin to multiply. Spirochaetes disappear from the louse's gut 5 to 24 hours after entering but reappear in the hemocoel in 8 to 12 days. These evidences indicate that only through the crushing of the vector's body and the inoculation of contaminated body fluid can infection be initiated. Furthermore, attempts at infection with feces or through bites have proved unsuccessful, although Burgdorfer (1976) has indicated that the spirochaetes can penetrate unbroken skin.

Louse-borne relapsing fever has essentially disappeared from North America, although scattered endemic foci still exist in South America, Europe, Asia, and Africa. There were 4700 cases in Ethiopia in 1971 (Harwood and James, 1979). During the war in Vietnam, an epidemic occurred (Pan American Health Organization, 1973).

Typhus. A much dreaded ancient disease, louse-borne typhus is now limited to Asia, North Africa, and Central and South America. The causative agent is the obligatory intracellular bacterium *Rickettsia prowazeki*. Infection of humans is contracted when contaminated louse feces are rubbed into abraded skin. However, since the rickettsiae remain infective in louse feces if kept dry and at room temperature, it is also possible to contract infections when such fecal particles are inhaled.

Commonly, lice become infected with *R. prowazeki* while feeding on typhus victims during the initial stages of the disease. Once the rickettsiae enter the insect's gut, they penetrate into the epithelial cells and multiply so rapidly that infected cells become greatly distended and usually rupture, releasing the rickettsiae into the lumen. From here, the rickettsiae pass to the exterior in feces. Infected lice usually die within ten days, but if they do not die, their infections are retained for the remainder of their life span.

The symptoms of louse-borne typhus are initiated with high fever (39.5–40°C), which lasts for about 2 weeks. This is accompanied by backache, intense headache, and commonly bronchopneumonia and bronchitis. A rash appears by the fifth or sixth day (Burgdorfer, 1976). After the fever subsides, profuse sweating commences. This is followed by either convalescence or an increased involvement of the central nervous system and death. Those who survive the disease become asymptomatic but are capable of infecting lice for many years.

Trench Fever. This disease is caused by another rickettsia, *Rochalimaea quintana*. Except for a few outbreaks in recent years, trench fever is not very common today. However, it was one of the most frequently encountered diseases during World Wars I and II. As with *R. prowazeki*, *R. quintana* is introduced into victims via contaminated louse feces. Unlike *R. prowazeki*, however, it is an extracellular pathogen in humans. Furthermore, when in the louse *Pediculus humanus*, it survives and multiplies in the lumen of the gut and not in epithelial cells.

Today, small endemic foci of trench fever occur in Egypt, Algeria, Ethiopia, Burundi, Japan, China, Mexico, and Bolivia. An infected person develops headache, body pain, and malaise about 10 to 30 days after entry of *R. quintana*. The body temperature rises rapidly to 39.5 to 40°C. Dizziness and pain in the eyes develop, and a typhuslike rash appears but disappears within 24 hours. The disease is debilitating but nonfatal. Humans are the primary reservoirs for *R. quintana*. It has been recovered from the blood of convalescents as long as 8 years after the initial attack (Burgdorfer, 1976).

Bactericidal Substances in Lice. In addition to the three human pathogens mentioned, human lice can also harbor and transmit *Pasteurella tularensis*, the causative agent of tularemia. Again transmission occurs via louse feces. Except for these, other bacteria are rapidly destroyed when taken into the digestive tract of lice because of bactericidal substances in the alimentary tract. Such substances are present in most insects.

The antibacterial properties of the intestinal contents, however, do not interfere with *P. tularensis*. Furthermore, these properties have no effect on various salmonellas, including the typhoid organism. Salmonellas multiply rapidly in lice and generally kill them in 24 to 48 hours. Even after the lice die, bacteria can survive in the decaying bodies and feces for over a year.

The apparent "immunity" of *Pasteurella tularensis* and salmonellas to the gastric juices of *Pediculus humanus* exemplifies how the compatibility, or for that matter incompatibility, of the vector's chemical composition with the microorganism can influence transmission of microbes by arthropods.

Phthirus pubis. The crab louse is so named because of its crablike ovoid body, which is rather wide (Figs. 18.26, 18.27). Specimens measure 1.5–2 mm in length, the males being slightly smaller and

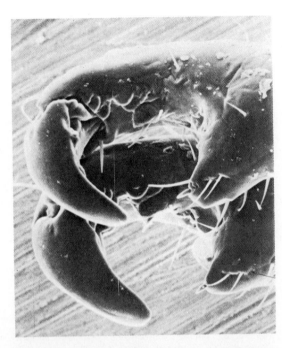

Fig. 18.27. Claws of *Phthirus pubis*. Electron micrograph showing tarsi of second and third legs each with a terminal claw. (Courtesy of Dr. J. Ubelaker.)

grayish. This louse is most frequently found attached to pubic hair, but it also attaches in other hairy regions of the body. The bite of *P. pubis* is extremely irritating, and if infestations occur over long periods, the skin becomes discolored. Transmission of pubic lice occurs through direct contact or from contaminated clothing and toilet seats. A closely related species, *P. gorillae*, is found on gorillas.

Life Cycle of Phthirus pubis. The life history of *Phthirus pubis* parallels that of other lice. Development is of the hemimetabolous type (p. 623). A single female usually oviposits 15 to 25 eggs, which attach to hair (Fig. 18.28). The number of eggs may be greater in some instances. The eggs hatch in 6 or 7 days, and the young undergo three molts and become sexually mature in about 2 to 3 weeks. Adults usually live for a month when on their host. If removed, they die within a day.

Family Haematopinidae

The Haematopinidae includes several important genera, such as *Haematopinus*, *Solenopotes*, and *Linognathus*.

Genus *Haematopinus*. This genus includes several species that parasitize domestic animals.

Fig. 18.28. Nit of *Phthirus pubis* cemented to hair. Notice presence of operculum with pores. (Courtesy of Dr. J. Ubelaker.)

Haematopinus suis, a cosmopolitan species, measuring 5–6 mm in length, is the largest (Fig. 18.29). It is found attached in the skin folds on the neck, ears, abdomen, and legs of pigs, and it will also feed on humans.

Haematopinus tuberculatus, measuring 3.5–5.5 mm in length, is found on cattle in parts of Australia and India. The short-nose cattle louse, *H. eurysternus*,

measures 3.5–4.75 mm in length (Fig. 18.29). *Haematopinus quadripertusus* is the tail louse of cattle.

Haematopinus asini, measuring 2.5–3.5 mm long, attaches at the bases of the mane and tails of horses, mules, and donkeys; *H. ventricosus* is found on rabbits.

All these species, with the exception of *H. tuberculatus*, are found in North America. In fact, *H. asini* is a common parasite of equines in the United States.

Genus *Solenopotes*.   This genus includes several species that infest animals in North America. *Solenopotes capillatus*, the little blue cattle louse, measuring 1.2–1.5 mm long, is widely distributed throughout the world.

Genus *Linognathus*.   This genus includes *L. vituli*, the long-nosed cattle louse (Fig. 18.29); *L. stenopsis* the goat louse; and *L. piliferus*, the common dog louse. One species, *L. pedalis*, the leg louse of sheep, has been reported to cause the host's death if present in large numbers.

Murray (1963b, c), while studying the ecology of sheep lice, found that specific differences exist between the behavioral patterns and survival of *Linognathus pedalis* and the closely related *L. ovillus* at different temperatures. *Linognathus pedalis*, normally found on the legs of sheep, where the temperature near the skin fluctuates greatly, is able to survive prolonged exposure to low temperatures of 2 to 22°C, whereas *L. ovillus*, normally found on the host's face and to a lesser extent on the body, cannot survive exposure to these cold temperatures.

Because the skin temperature of the legs of sheep can drop nearly to that of the environment for considerable periods in cold weather, the ability of *L. pedalis* to survive for several days without feeding at low temperatures appears to be an adaptation to survival in its habitat. Although *L. pedalis* can withstand lower temperatures than *L. ovillus*, the temperatures that are favorable for reproduction of the two species are similar—35°C at 54% relative humidity. Furthermore, engorged females lay approximately 20 times more eggs than unengorged females.

Relative to behavioral patterns, Murray has found that *L. pedalis* is more sedentary and tends to congregate in clusters on the part of the leg covered with hair, whereas *L. ovillus* never forms clusters and its populations are dispersed, being generally more dense in the region where the hair and wool on the sheep's face merge. In both species of lice, only adults are transferred from sheep to sheep.

Family Haematopinoididae

Haematopinoididae includes *Haplopleura aenomydis*, a species commonly found on rats in North America. Various other members of the family are found on gophers in the United States and Canada.

NUTRITION OF LICE

The nutritional physiology and the digestive pro-

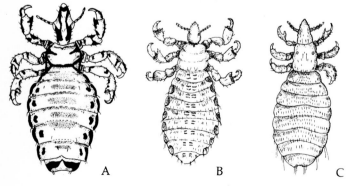

Fig. 18.29.   Lice. A. *Haematopinus suis.* **B.** *Haematopinus eurysternus.* **C.** *Linognathus vituli.* (**B** and **C**, redrawn after Mönnig, 1949.)

cesses of lice, particularly the mallophagan lice, have been reviewed by Waterhouse (1953). Bird lice can feed on the protective sheaths of growing feathers, featherfiber, down, skin-scurf, scabs, blood, their own eggs, and cast skins. They can probably also feed on mucus and sebaceous matter. Some of these materials undoubtedly contain keratin, which the lice can digest with the aid of intracellular mutualistic bacteria in their bodies. Such bacteria pass from louse to louse by way of their eggs.

Some species of mammalian lice can ingest hair, although most species appear to prefer epidermal scales, skin, and wax. Some species feed on blood.

It has been postulated that the reason ant-eating mammals are nearly always free of lice is that the formic acid resulting from digestion of the ants' bodies discourages lice. This hypothesis is in need of verification.

Other aspects of insect physiology are discussed in the last section of Chapter 20.

HOST SPECIFICITY OF LICE

Compared to other arthropod parasites, lice are highly host specific. This is largely because lice seldom leave the body of their host since the entire life cycle can be spent on one animal. Consequently, there is no advantage to the louse habitually leaving its host in

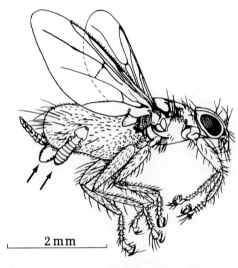

Fig. 18.31. Phoresis. Feather lice, *Sturnidoecus sturni*, being carried by a hippoboscid fly, *Ornithomya fringillina*. (From Askew after Rothschild and Clay, 1952.)

search of new hosts in a hostile environment. Transfer from host to host commonly occurs when these periodically come near each other during copulation or in feeding their young. For example, it is while cuckoos are copulating that lice from one partner is transferred to the other.

Not only are lice comparatively host specific, but they are also habitat specific. Thus, if one harbors several species of lice, the latter do not mix (Fig. 18.30).

It should be mentioned that phoresis plays an important role in the transfer of certain species of lice from one host to another. An example of this has been reported by Rothschild and Clay (1952), who found that lice on starlings can be transported from host to host by the fly *Ornithomya*, which is also a parasite of starlings. The lice commonly become attached to the fly and are transported to another avian host (Fig. 18.31).

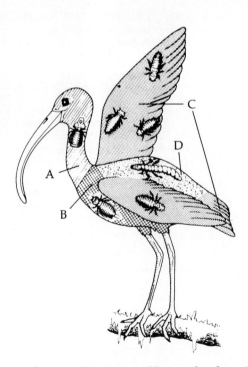

Fig. 18.30. Distribution of lice on the glossy ibis. A. *Ibidoecus bisignatus* on head and neck. **B.** *Menopon plegadis* on breast, abdomen, and flanks. **C.** Species of *Colpocephalum* and *Ferribia* on wings and tail. **D.** *Estiopterum raphidium* on back. (After Dubinin in Dogiel, 1964.)

REFERENCES

Agrell, I. P. S., and Lundquist, A. M. (1973). Physiological and biochemical changes during insect development. *In* "The Physiology of Insecta," 2nd ed. (M. Rockstein, ed.), Vol. I, pp. 159–247. Academic Press, New York.

Arnold, J. W. (1974). The hemocytes of insects. *In* "The Physiology of Insecta," 2nd ed. (M. Rockstein, ed.), Vol. V. pp. 202–254. Academic Press, New York.

Askew, R. R. (1971). "Parasitic Insects." American Elsevier, New York.

Bruce, W. N. (1948). Studies on the biological requirements of the cat flea. *Ann. Entomol. Soc. Am.* **41**, 346–352.

Burgdorfer, W. (1976). Epidemic (louse-borne) typhus. *In* "Tropical Medicine," 5th ed. (G. W. Hunter, J. C. Swartz-walder, and D. F. Clyde, eds.), pp. 102–109. W. B. Saunders, Philadelphia.

de Wilde, J., and de Loof, A. (1973a). Reproduction. *In* "The Physiology of Insecta," 2nd ed. (M. Rockstein, ed.), Vol. I, pp. 12–95. Academic Press, New York.

de Wilde, and de Loof, A. (1973b). Reproduction-endocrine control. *In* "The Physiology of Insecta," 2nd ed. (M. Rockstein, ed.), Vol. I, pp. 97–157. Academic Press, New York.

Ewing, H. E., and Fox, I. (1943). "The Fleas of North America: Classification, Identification, and Geographic Distribution of These Injurious and Disease-Spreading Insects." U.S. Dept. of Agri. Misc. Publ. No. 500. Washington, D.C.

Florkin, M., and Jenniaux, C. (1974). Hemolymph: composition. *In* "The Physiology of Insecta," 2nd ed. (M. Rockstein, ed.), Vol. V, pp. 255–307. Academic Press, New York.

Fox, I. (1940a). "Fleas of Eastern United States." Iowa State College Press, Ames, Iowa.

Fox, I. (1940b). Siphonaptera from western United States. *J. Wash. Acad. Sci.* **30**, 272–276.

Gilbert, L. I., and King, D. S. (1973). Physiology of growth and development: endocrine aspects. *In* "The Physiology of Insecta," 2nd ed. (M. Rockstein, ed.), Vol. I, pp. 249–370. Academic Press, New York.

Goldsmith, T. H., and Bernard, G. D. (1974). The visual system of insects. *In* "The Physiology of Insecta," 2nd ed., (M. Rockstein, ed.), Vol. II, pp. 165–272. Academic Press, New York.

Grégoire, C. (1974). Hemolymph coagulation. *In* "The Physiology of Insecta," 2nd ed. (M. Rockstein, ed.), Vol. V, pp. 309–360. Academic Press, New York.

Hampton, B. G. (1940). Plague in the United States. *Publ. Health Rep.* **55**, 1143–1158.

Harwood, R. F., and James, M. T. (1979). "Entomology in Human and Animal Health," 7th ed. Macmillan, New York.

Hirst, L. F. (1953). "The Conquest of Plague: A Study of the Evolution of Epidemiology." Clarendon Press, Oxford, England.

Hodgson, E. S. (1974). Chemoreception. *In* "The Physiology of Insecta," 2nd ed. (M. Rockstein, ed.), Vol. II, pp. 127–164. Academic Press, New York.

Hopkins, G. H. E. (1957). Host-associations of Siphonaptera. *In* "First Symposium on Host Specificity among Parasites of Vertebrates" (J. Baer, ed.), pp. 64–87. Inst. Zool., Univ. Neuchâtel, Switzerland.

House, H. L. (1974). Digestion. *In* "The Physiology of Insecta," 2nd ed. (M. Rockstein, ed.), Vol. V, pp. 63–117. Academic Press, New York.

Hubbard, C. A. (1947). "Fleas of Western North America." Iowa State College Press, Ames, Iowa.

Huber, F. (1974). Neural integration (central nervous system). *In* "The Physiology of Insecta," 2nd ed. (M. Rockstein, ed.), Vol. IV, pp. 3–100. Academic Press, New York.

Humphries, D. A. (1967a). The mating behaviour of the hen flea, *Ceratophyllus gallinae* (Schrank) (Siphonaptera: Insecta). *Anim. Behav.* **15**, 82–98.

Humphries, D. A. (1967b). The action of male genitalia during the copulation of the hen flea, *Ceratophyllus gallinae* (Schrank). *Proc. Roy. Entomol. Soc. Lond.* **A42**, 101–106.

Humphries, D. A. (1968). The host-finding behavior of the hen flea, *Ceratophyllus gallinae* (Schrank) (Siphonaptera). *Parasitology* **58**, 403–414.

Ito, S., Vinson, J. W., and McGuire, T. J. Jr. (1975). Murine typhus rickettsiae in the oriental rat flea. *Ann. N.Y. Acad. Sci.* **266**, 35–60.

Jellison, W. L., and Good, N. E. (1942). "Index to the Literature of Siphonaptera of North America." U.S. Publ. Health Serv. Nat. Inst. Health Bull. No. 178. Washington, D.C.

Jones, J. C. (1974). Factors affecting heart rates in insects. *In* "The Physiology of Insecta," 2nd ed. (M. Rockstein, ed.), Vol. V, pp. 119–167. Academic Press, New York.

Kadis, S., Montie, T. C., and Ajl, S. J. (1969). Plague toxin. *Sci. Am.* **220**, 93–100.

Kim, K. C., and Ludwig, H. W. (1979). The family classification of the Anoplura. *Syst. Entomol.* **3**, 249–284.

Leeson, H. S. (1941). The effect of temperature upon the hatching of eggs of *Pediculus humanus corporis* DeGeer. *Parasitology* **33**, 243–249.

Lehane, B. (1969). "The Compleat Flea." Viking Press, New York.

Marshall, A. G. (1981). "The Ecology of Ectoparasitic Insects." Academic Press, London and New York.

Matthysse, J. G. (1946). Cattle lice, their biology and control. Agr. Exp. Sta. Bull. No. 832. Cornell University, Ithaca, New York.

Meyer, K. F. (1942). The ecology of plague. *Medicine* **21**, 143–174.

Miller, T. A. (1974). Electrophysiology of the insect heart. *In* "The Physiology of Insecta," 2nd ed. (M. Rockstein, ed.), Vol. V, pp. 169–200. Academic Press, New York.

Murray, M. D. (1963a). Influence of temperature on the reproduction of *Damalinia equi* (Denny). *Austr. J. Zool.* **11**, 183–189.

Murray, M. D. (1963b). The ecology of lice on sheep. III. Differences between the biology of *Linognathus pedalis* (Osborne) and *L. ovillus* (Neumann). *Austr. J. Zool.* **11**, 153–156.

Murray, M. D. (1963c). The ecology of lice on sheep. IV. The establishment and maintenance of populations of *Linognathus ovillus* (Neumann). *Austr. J. Zool.* **11**, 157–172.

Murray, M. D., and Nicholls, D. G. (1965). Studies on the ectoparasites of seals and penguins. I. The ecology of the louse *Lepidophthirus macrorhini* Enderlein on the southern elephant seal, *Mirounga leonia* (L.). *Austr. J. Zool.* **13**, 437–454.

Murray, M. D., Smith, M. S. R., and Soucek, Z. (1965). Studies on ectoparasites of seals and penguins. II. The ecology of the louse *Antarctophthirus ogmorhini* Enderlein on the Weddell seal, *Leptonychotes weddelli. Austr. J. Zool.* **13**, 761–771.

Pan American Health Organization (1973). "Proceedings of

the International Symposium on the Control of Lice and Louse-borne Diseases." PAHO Sci. Publ. No. 263, Washington, D.C.

Peck, S. S., Wright, W. H., and Gant, J. Q. (1943). Cutaneous reactions due to the body louse (*Pediculus humanus*). *J.A.M.A.* **123**, 821–825.

Pratt, H. D., and Stark, H. E. (1973). "Fleas of public health significance and their control." U.S. Dept. H.E.W. Publ. No. (CDC) 75–8267. U.S. Govt. Print. Off., Washington, D.C.

Rothschild, M., and Clay, T. (1952). "Fleas, Flukes and Cuckoos. A Study of Bird Parasites." Collins, London.

Rothschild, M., and Hinton, H. E. (1968). Holding organs on the antennae of male fleas. *Proc. Roy. Entomol. Soc. Lond.* **A43**, 105–107.

Rowley, A. F., and Ratcliffe, N. A. (1981). Insects. *In* "Invertebrate Blood Cells" (N. A. Ratcliffe and A. F. Rowley, eds.), Vol. 2, pp. 421–488. Academic Press, London and New York.

Schwartzkopff, J. (1974). Mechanoreception. *In* "The Physiology of Insecta," 2nd ed. (M. Rockstein, ed.), Vol. II, pp. 273–352. Academic Press, New York.

Sobey, W. R., and Conolley, D. (1971). Myxomatosis: the introduction of the European rabbit flea *Spilopsyllus cuniculi* (Dale) into wild rabbit populations in Australia. *J. Hyg.* **69**, 311–346.

Waterhouse, D. F. (1953). Studies on the digestion of wool by insects. IX. Some features of digestion in chewing lice (Mallophaga) from birds and mammalian hosts. *Austr. J. Biol. Sci.* **6**, 257–275.

CLASSIFICATION OF SIPHONAPTERA (THE FLEAS), MALLOPHAGA (THE BITING LICE), AND ANOPLURA (THE SUCKING LICE), CLASS INSECTA (SEE P. 614)

Order Siphonaptera

Adults small and wingless; ectoparasitic on birds and mammals; bodies laterally compressed; legs long, stout, and spinose; antennae short and clubbed, and fit in depressions alongside of head when not extended; mouthparts of piercing-sucking type; holometabolous metamorphosis.

Family Ceratophyllidae

Three thoracic tergites together longer than first abdominal tergite; head evenly rounded along margin; no vertical suture from dorsal margin of head to bases of antennae; abdominal tergites with at least two rows of setae. (Genera mentioned in text: *Ceratophyllus, Nosopsyllus, Diamanus*.)

Family Pulicidae

Three thoracic tergites together longer than first abdominal tergite; head evenly rounded along margin; no vertical suture from dorsal margin of head to bases of antennae; abdominal tergites with one row of setae. (Genera mentioned in text: *Ctenocephalides, Pulex, Xenopsylla, Hoptosyllus*.)

Order Mallophaga

Ectoparasitic on birds and mammals; bodies small to medium size, usually dorsoventrally flattened and wingless; with chewing mouthparts; antennae short, three- to five-segmented; with reduced compound eyes and no

ocelli; thorax small; legs stout and short; no cerci on abdomen; hemimetabolous metamorphosis.

SUBORDER AMBLYCERA

Antennae six-jointed, short, clavate or capitate, concealed in shallow grooves on underside of head; maxillary palpi four-jointed; mandibles horizontally arranged; parasitic on birds and mammals.

Family Gyropidae

With one claw on tarsi; parasitic on mammals. (Genera mentioned in text: *Gyropus, Gliricola*.)

Family Menoponidae

With two claws on tarsi; parasitic on birds. (Genera mentioned in text: *Menapon, Colpocephalum, Pseudomenopon, Trinoton, Menacanthus, Myrsidea*.)

Family Boopiidae

With two claws on tarsi; parasitic on kangaroos, wallabies, and wombats in Australia. (Genus mentioned in text: *Heterodoxus*.)

Family Ricinidae

With two claws on tarsi; head flattened; parasitic on birds; few species. (Genera mentioned in text: *Ricinus, Trochiloecetes*.)

Family Tungidae

Three thoracic tergites together shorter than first abdominal tergite. (Genera mentioned in text: *Tunga, Echidnophaga*.)

Family Ischnopsyllidae

Three thoracic tergites together longer than first abdominal tergite; with vertical suture extending from dorsal margin of head to bases of antennae; head strongly curved at vertex, with pair of dark anteroventral flaps on each side. (Genus mentioned in text: *Ischnopsyllus*.)

Family Hystrichopsyllidae

Three thoracic tergites together longer than first abdominal tergite; with vertical suture extending from dorsal margin of head to bases of antennae; head strongly curved at vertex, without pair of dark anteroventral flaps on each side; occipital region without dorsal incrassation (thickening). (Genus mentioned in text: *Leptopsylla*.)

Family Macropsyllidae

Three thoracic tergites together longer than first abdominal tergite; head similar to that of Ischnopsyllidae but without pair of dark anteroventral flaps on each side; occipital region with dorsal incrassation. (Genus mentioned in text: *Rhopalopsyllus*.)

SUBORDER ISCHNOCERA

Antennae three- or five-segmented, filiform, not concealed; maxillary palpi absent; mandibles vertically oriented; parasitic on birds and mammals.

Family Trichodectidae

Antennae three-jointed; with one claw on tarsi; parasitic on mammals. (Genera mentioned in text: *Trichodectes, Damalinia, Felicola, Lepidophthirus*.)

Family Philopteridae

Antennae five-jointed; with two claws on tarsi; parasitic on birds. (Genera mentioned in text: *Dennyus,*

Goniocotes, Laemobotkrion, Columbicola, Lipeurus, Ornithobius, Esthiopterum, Goniodes, Cuclotogaster, Philopterus.)

SUBORDER RHYNCHOPHTHIRINA

Parasitic on African and Indian elephants. (Genus mentioned in text: *Haematomyzus.*)

Order Anopura

Ectoparasites of mammals; bodies small to medium size, dorsoventrally flattened, and wingless; mouthparts modified as piercing-sucking organ that is retractile; antennae short, three- to five-segmented; some with legs terminating as hooked claws; thorax fused; abdomen of five to eight distinct segments; hemimetabolous metamorphosis.

Family Echinophthiriidae

Parasitic on marine mammals; bodies robust, covered with short spines and scales; abdomen lacks sclerotized plates except at terminal end and on external genitalia and associated segments; antennae four- or five-jointed; spiracles small; with three pairs of legs, or at least posterior two pairs of legs. (Genera mentioned in text: *Echinophthirus, Antarctophthirus.*)

Family Pediculidae

Parasites of primates, including humans; bodies fairly robust, not covered with dense spines; abdomen armed with pleural plates (paratergites) and with tergal and sternal plates in most species; with well-developed eyes, comprised of pigment granules and a lens; with legs approximately equal in length or with first pair slightly smaller. (Genera mentioned in text: *Pediculus, Phthirus.*)

Family Haematopinidae

Parasites of mammals including pigs, Artiodactyla, and equines; bodies with spines or hairs arranged in rows; scales absent; abdomen with paratergites; tergal and sternal plates present in most species; eyes absent or reduced; antennae five-jointed; all three pairs of legs of similar size; tibiae usually with thumblike process opposing claw. (Genera mentioned in text: *Haematopinus, Solenopotes, Linognathus.*)

Family Haematopinoididae

Parasites on small mammals (shrews, gophers, rats, etc.); bodies relatively small; abdomen with sclerotized plates; eyes present; all three pairs of legs of similar size. (Genus mentioned in text: *Haplopleura.*)

19

DIPTERA: THE FLIES, GNATS, AND MOSQUITOES

The insect order Diptera includes the flies, gnats, and mosquitoes. There are approximately 90,000 known species of dipterans in the world, of which about 17,000 occur in North America. These insects are of great interest to medical and veterinary entomologists because they serve as carriers of many important diseases. Not only are dipterans important as vectors, but many are ectoparasitic bloodsuckers, and still others are endoparasites. For a detailed account of the roles of dipterans in the transmission of diseases, see Greenberg (1971, 1973).

Dipterans can be distinguished from all other insects by the possession of only one pair of membranous wings, the forewings. The hindwings are greatly reduced and appear as a pair of slender, knoblike balancing organs known as **halteres**. The mouthparts of dipterans are modified for piercing and sucking, whereas in others they are modified for rasping and lapping.

Hardy's (1960) monograph of the Nematocera and Brachycera, two dipteran superfamilies, as well as the volume on insects of medical importance edited by Smith (1973), should be consulted by those interested in taxonomy. For the sake of convenience, the following classification is used in this volume.

Suborder Orthorrhapha. Pupa not enclosed in epidermis of larva. Adults emerge from pupal case through a T-shaped anterodorsal split.

Superfamily Nematocera. With long antennae of at least six similar segments. Larvae with well-developed heads.

Superfamily Brachycera. With short antennae consisting of three segments. Third antennal segment may be annulated.

Suborder Cyclorrhapha. Pupa enclosed in epidermis of larva (**puparium**). Adults escape through circular split at one terminal of puparium. Larvae maggotlike.

Superfamily Aschiza. Lacks permanent crescent-shaped mark (the **lunule***) on head. Lacks frontal suture on puparium.

Superfamily Schizophora. Permanent crescent-shaped mark (the lunule) on head. With frontal puparial suture.

---

*The lunule is the scar resulting from the shrinkage of the ptilinium, which is an outgrowth of the head on the pupa used in pushing out the circular opening on the puparial wall during the escaping process of the adult.

SUBORDER ORTHORRHAPHA

The Orthorrhapha includes 32 families, of which 15 are subordinate to the Nematocera and 17 to the Brachycera. Of the orthorrhaphan families, several include parasitic members. Six of these families—Simuliidae, Psychodidae, Ceratopogonidae, Culicidae, Tabanidae, and Rhagionidae—are examined in the following discussion. The first four are subordinate to the Nematocera, and the last two to the Brachycera. Brief descriptions of the dipteran families considered herein are presented at the end of the chapter.

FAMILY SIMULIIDAE (BLACKFLIES)

Over 600 species of simuliid files are known (Smart, 1945; Vargas, *et al.* 1946; Cole, 1969). The bodies of these hematophagous flies range from 1 to 5 mm in length. As with most hematophagous dipterans, it is the female that possesses piercing-sucking mouthparts (p. 616) and feeds on the host. The males possess rudimentary mouthparts and feed on nectar.

Simuliid Life Cycle Pattern

As with all dipterans, the simuliids undergo complete (holometabolous) metamorphosis (p. 623). Adult simuliid flies are generally found in the vicinity of rivers and streams. The eggs are laid either on aquatic plants or on stones beneath the surface of the water. Females of *Simulium maculatum* have been observed to submerge to a depth of 30 cm during oviposition. The eggs are glued to the substratum by a gelatinous coat.

Larvae, hatching from eggs, are somewhat club shaped and swollen posteriorly. They attach to stones and other objects by means of a disclike sucker at the

posterior end of the body. These larvae are invariably aquatic and require swift flowing water for their habitat. They are commonly found congregated in the vicinity of rapids and waterfalls. The larvae pupate in cone-shaped cocoons that are attached in the water. The pupal respiratory organs, composed of long tube-like filaments, protrude from the cocoon and obtain oxygen from the moving water. The form that emerges from the cocoon is the adult. Simuliid adults are hematophagous.

Included in the family Simuliidae are the genera *Cnephia, Simulium,* and *Prosimulium.*

Genus *Cnephia.* *Cnephia* includes several species that feed on the blood of domestic animals and occasionally humans. *Cnephia pecuarum*, the southern buffalo gnat (Fig. 19.1), is an important pest of livestock and humans in the Mississippi Valley during the spring. This species is primarily a diurnal feeder. Its bite causes severe swelling resulting from the material injected during feeding. In heavy attacks, horses, mules, and even cattle have been killed in a matter of hours resulting from the general weakened condition due to blood loss and the effects of the injected substance. In the western and Pacific states, *C. minus* is the common species. This is a smaller and darker species than *C. pecuarum.*

Genus *Simulium.* This genus includes *S. venustum*, a pest of humans in New England, Canada, and the region surrounding the Great Lakes. During June and July, the bites of this blackfly can become quite a hazard to sportsmen.

Simulium vittatum is found in North America and quite frequently in Europe. This species attacks livestock and humans. Similarly, *S. colombaschense*, the golobatz gnat, attacks livestock in middle and southern Europe. Its bite, in instances of heavy infestations, has been reported to kill pigs, sheep, cattle, and horses. In 1923, swarms of *S. colombaschense* spread from the banks of the Danube and descended upon domestic animals in such numbers that about 20,000 horses, cattle, sheep, goats, and pigs are reported to have died from bites. Toxemia and anaphylactic shock, rather than loss of blood, probably accounted for most of these deaths.

The bird-infesting species, *S. occidentale* and *S. meridionale*, commonly called turkey gnats, occur in the southern United States in the late spring. These flies attack the comb and wattles of poultry and by so doing initiate a disease similar to avian cholera.

Two other species, *S. equinum*, a parasite of equines and other animals, and *S. damnosum* (Fig. 19.1), a parasite of humans and cattle in Africa, are also serious hematophagous parasites. What is being referred to as *Simulium damnosum* in this volume actually represents a complex of at least 25 cytotypes, that is, different types distinguishable by differences in their chromosomal numbers and structure (Peterson and Dang, 1981).

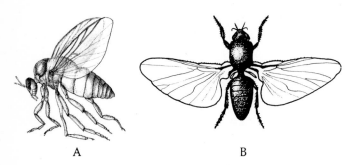

A   B

Fig. 19.1. **Biting dipterans. A.** *Cnephia pecuarum,* the buffalo gnat. (After Garman in James and Harwood, 1969.) **B.** *Simulium damnosum.*

Genus *Prosimulium*. *Prosimulium hirtipes* is a springtime pest of humans and domestic animals in the northwestern United States.

Simuliids as Vectors

In addition to causing irritation through their bites, the simuliid flies serve as vectors for nematodes and protozoan parasites.

Vectors for *Onchocerca volvulus*. *Onchocerca volvulus*, a human-infecting nematode (p. 542), is transmitted by *Simulium damnosum* and *S. neavei* in Africa. *Simulium damnosum* is the major vector in the regions immediately south of the Sahara Desert. This species breeds in large rivers and in small tributaries. *Simulium neavei* is the principal vector in the Congo. It breeds in medium-sized rivers, where the eggs are deposited in clusters on vegetation. The larvae and pupae are found attached (i.e., they are epiphoretic) on the exoskeleton of crabs of the genus *Potamonautes*.

Human onchocercosis also occurs in Mexico, Central America, and parts of South America. In the Americas, *S. ochraceum* is the major vector. This small species measures 1.5–2 mm in length. The thorax is yellowish red, its abdomen is yellow and black, and the legs are black. Dalmat (1955) has studied the development of this fly and has reported that it will breed only in small trickling streams that are relatively free of vegetation. Two other species of *Simulium*, *S. callidum* and *S. metallicum*, are capable of transmitting onchocerosis in the New World.

Vector for *Onchocerca gutterosa*. *Onchocerca gutterosa*, the bovine onchocercosis-causing nematode, is transmitted by *Simulium ornatum* in Europe. Once taken into the ventriculus of the fly, the microfilariae attain the "sausage" form in approximately 10 days. By the 19th or 20th day, the nematodes are found in the fly's thoracic muscles. From here they migrate anteriorly to the mouthparts and are then ready to be introduced into the vertebrate host. This passage of the microfilariae from the fly to the vertebrate host is not a passive one. The worms appear to be attracted by the warmth of the cow's blood and actively migrate from the fly. Not all the nematodes taken into the *Simulium* vector survive; only a small percentage do.

Although the vectors for all other species of *Onchocerca* are not known, it is strongly suspected that in time these vectors will be shown to be simuliid flies.

Vectors for *Leucocytozoon* Species. Species of *Simulium* also serve as vectors for the protozoans *Leucocytozoon* spp. For example, *L. simondi*, a parasite of ducks and geese (p. 206), is transmitted by *Simulium rugglesi* in the northern United States and Canada. Similarly, *L. smithi*, a species injurious to turkeys in the United States, is transmitted by *Simulium jenningsi*, *S. occidentale*, and *S. slossonae*.

Laird (1981) has edited a volume devoted to the systematics, control, physiology, collecting, and rearing of blackflies. Also included in this reference volume are chapters devoted to the predators, parasites, pathogens, and diseases of blackflies.

FAMILY PSYCHODIDAE (MOTH FLIES AND SANDFLIES)

This family includes the genera *Phlebotomus*, *Sergentomyia*, and *Lutzomyia*, the members of which are hematophagous. For a taxonomic treatment of the Psychodidae, see Lewis *et al.* (1977).

Genus *Phlebotomus*. The largest and most important genus of the family Psychodidae is *Phlebotomus*, to which belong a large number of blood-sucking psychodids. These flies, commonly known as sandflies, are seldom more than 4 mm long. As with the simuliid flies, only the females possess piercing-sucking mouthparts and are hemotaphagous. The males are nonparasitic, feeding on moisture. *Phlebotomus* spp. are nocturnal insects, and various species feed on mammals, lizards, snakes, and even caterpillars. The various species, along with those of *Sergentomyia*, are limited to the Old World.

The species of *Phlebotomus* are small, slender specimens with hairy bodies. Their coloration is dull, usually yellowish. The legs are long and lanky and when the flies are at rest, their wings are separated. All of the species are native to the Old World, including the Orient and Africa. In general, these flies are limited to the warmer regions of the world.

Phlebotomus papatasii is the common sandfly in the Old World (Fig. 19.2). It is a serious pest in eastern Europe and the Near East. It is anthropophilic and feeds on blood, biting around the ankles and wrists. This species measures approximately 2.5 mm in length and is yellowish gray with a dull red-brown stripe extending longitudinally down the middorsal line of the thorax and with a reddish brown spot on each lateral surface of the same region. *Phlebotomus argentipes* is widely distributed in areas of India and Burma, where it feeds primarily on cattle blood, although it attacks humans. *Phlebotomus sergenti* is widely distributed in the Near East and northern Africa.

Life Cycle of Phlebotomus *Species.* Life cycle patterns among the various species appear to be similar, differing only in habitats and preferred hosts. The breeding spots of these flies are usually hidden and difficult to find. The elongate eggs are oviposited in small batches under stones, in masonry cracks, between the walls of cesspools, and in other similar out-of-the-way places where the temperature is moderate, the environment dark, and the humidity high. Laying habits of the females vary but can be categorized in one of three patterns. The flies feed and refeed several

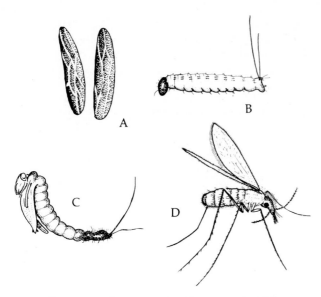

Fig. 19.2. Stages in the life cycle of *Phlebotomus papatasii*. A. Eggs. **B.** Larva. **C.** Pupa. **D.** Adult sandfly. (**B–D**, redrawn after Byam and Archibald, 1921–23.)

times before ovipositing a batch of eggs; oviposition is sandwiched between two blood meals; or the flies feed, oviposit, and die. *Phlebotomus argentipes* demonstrates the third pattern.

The eggs of *P. papatasii* incubate for 9 to 12 days, after which minute whitish larvae with long anal spines and chewing mouthparts emerge (Fig. 19.2). Available information indicates that the incubation period in other species approximates that of *P. papatasii*.

The larvae are free living, feeding on organic debris such as animal excreta. There are four larval instars, the stadia totaling four to six weeks. The fourth larval instar metamorphoses into the pupa, which is not enclosed in a cocoon (Fig. 19.2). The pupal stadium lasts for 10 days. The egg-to-egg cycle requires 7 to 10 weeks. If cold weather sets in prior to completion of the fourth larval instar, this form undergoes diapause, which may last from several weeks to a year.

Genus *Lutzomyia*. Members of the genus *Lutzomyia* are the New World sandflies (Lewis, 1974; Theodor, 1965; Forattini, 1971, 1973). Although species occur as far north as Canada, the medically important ones occur primarily south of Texas. Their life cycles parallel those of *Phlebotomus* spp. These sandflies require a combination of darkness, high humidity, and organic debris on which the larvae feed. Consequently, their breeding sites are commonly

under logs and dead leaves, inside hollow trees, and in animal burrows.

Sandflies as Vectors. Several species of *Phlebotomus* are suitable vectors for the human-infecting species of *Leishmania* (p. 121). *Phlebotomus argentipes* is the vector for *L. donovani* in India; *P. chinensis* and *P. sergenti* are the major vectors in China and Africa, respectively; and *P. major, P. perniciosus,* and *P. longicuspis* are the transmitters in the Mediterranean areas, where dogs serve as natural reservoirs for the flagellate. Generally human infections are acquired when flies are crushed and rubbed into the bites; however, infections can also be acquired through bites.

Phlebotomus papatasii and *P. sergenti* are the principal vectors for *Leishmania tropica*. Again, evidence suggests that infections can be acquired both mechanically and through bites. Not all of the flies that serve as vectors for *Leishmania braziliensis* are clearly defined. It is known, however, that *Lutzomyia* spp. are the principal vectors.

Phlebotomine flies are also known to be the vectors of nonhuman-infecting *Leishmania* spp. The reviews by Lainson and Shaw (1974) and Lewis (1974) should be consulted for details relative to the transmission of *Leishmania* by this group of flies.

Sandfly fever, also known as three-day fever or papatasi fever, is a nonlethal disease endemic to the Mediterranean countries, central Asia, south China, parts of India and Sri Lanka, and sections of South America. Sporadic epidemics have been known to occur. For example, in 1948, three-fourths of the population (1.2 million persons) of Yugoslavia was afflicted (Harwood and James, 1979). The etiologic agent is a virus that is injected into humans during the feeding of the fly, and the clinical symptoms appear after a 3- to 6-day incubation period. The virus can be demonstrated in the victim's blood 24 hours before onset of fever and for 24 hours after the onset. Flies become infective 6 to 8 days after feeding on an infected individual. Transovarial infections in the fly have been suspected. Species of *Phlebotomus* suitable as vectors include *P. papatasii* and *P. sergenti*.

Bartonellosis produces anemia that in humans is sometimes fatal. It is endemic to the areas along the Pacific slope of the Peruvian Andes, including parts of Colombia, Peru, Bolivia, Chile, and Ecuador. The causative bacterium, *Bartonella bacilliformis*, is found in, or occasionally on, erythrocytes in the circulating blood and as intracellular forms in various visceral organs, especially in the endothelial cells of lymph glands. *Bartonella bacilliformis* is transmitted by *Lutzomyia verrucarum* and probably *L. colombiana*, both of which are nocturnal species.

Bartonellosis is also known as Carrion's disease. It occurs in two clinical forms: **Oroya fever** and **verruga peruana**. Oroya fever is the visceral form, accompanied by bone, joint, and muscle pains; anemia; and

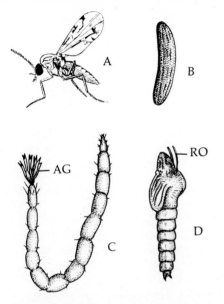

Fig. 19.3. Stages in the life cycle of *Culicoides*. A. Adult. **B.** Eggs. **C.** Larva. **D.** Pupa. (All figures redrawn after Dove *et al.*, 1932.) AG, anal gills; RO, respiratory organ (breathing trumpet.)

jaundice. It is the more virulent form. Verruga peruara is a mild, nonfatal cutaneous disease.

FAMILY CERATOPOGONIDAE (BITING MIDGES, OR NO-SEEUMS)

This family includes more than 50 genera. Not all the ceratopogonid flies are bloodsuckers. Most feed on invertebrates, especially insects. A few feed on ectothermic vertebrates. On the other hand, members of the genera *Culicoides*, *Leptoconops*, *Austroconops*, and *Forcipomyia* include species that are annoying pests that attack mammals, including humans. These midges have been dubbed with such familiar names as punkies, no-see-ums, and black gnats.

Genus *Culicoides*. By far, the most important genus of the Ceratopogonidae is *Culicoides* (Fig. 19.3). Cole (1969) has provided a key to the species found in North America, Barbosa (1947) has contributed a taxonomic study of species found in neotropical regions, Fox (1946) has provided a review of species from the Caribbean, Foot and Pratt (1954) have catalogued species found in the eastern United States, and Campbell and Pelham-Clington (1959) have reviewed the British species of *Culicoides*. Approximately 800 species of this genus have been described.

Familiar species of *Culicoides* include *C. canithorax*, *C. furens*, and *C. melleus*. These species cause a serious economic problem in summer resort areas along the Atlantic Coast of the United States. *Culicoides diabolicus* is a fiercely biting species found in Mexico. *Culicoides peliliouensis* is another fierce biter found along the Carolina coast.

In Britain, *C. impunctatus*, the so-called Scottish midge, occurs in large numbers on mosses and marshy places in Scotland and elsewhere and renders human activity nearly intolerable at certain times.

Not all species of *Culicoides* feed on the blood of mammals, including humans; for example, *C. anophalis* feed on engorged female mosquitoes and hence may be considered hyperparasites.

Life Cycle of Culicoides. The larvae of *Culicoides*, hatching from elongate eggs (Fig. 19.3), are found in mud, sand, and debris at the edges of ponds and other bodies of water. The eggs of certain species are found in rotting organic material far from water. For example, one species breeds in the desert, in rotting saguaro cacti. The segmented vermiform larvae are capable of swimming with an eel-like motion, stopping to rest on floating vegetation.

The free-living larvae are carnivorous or even cannibalistic. As pupation begins, the larvae migrate to the surface of the water, become quiescent, and metamorphose. The pupa possesses a pair of trumpetlike siphons projecting from the thorax (Fig. 19.3). These siphons are specialized spiracles that adhere to the water surface and through which the animal breathes. The postpupal form is the imago.

Temperature is an important factor during development of *Culicoides* spp. Dove *et al.* (1932) have reported that larvae of the salt marsh species survive best at 10°C. In over 200 attempts to maintain larvae between 21 and 32°C, not a single specimen developed to maturity. Temperature also affects adult *Culicoides*. Travis (1949) has reported that among Alaskan species, activity is greatly diminished if the temperature drops to below 13°C, an unusual condition for Alaskan insects, which usually can endure much lower temperatures. Furthermore, Travis has reported that the midges are handicapped in their flight if the wind velocity is greater than 3.5 miles per hour. The correlation between air currents and temperature was studied by Dove *et al.* (1932), who noted that thermotropic responses are marked when there is little air movement. It has been fairly well established that the affinity of *Culicoides* spp. for hosts is one of body heat attraction. This is probably the reason that midges more commonly and abundantly attack humans performing strenuous physical labor.

Several instances have been reported in which species differ in degree of activity, depending on the time of day. For example, *C. canithorax* in Mississippi is more vicious and abundant from 10 A.M. until after dark, whereas *C. mississippiensis* is most active in the early evening.

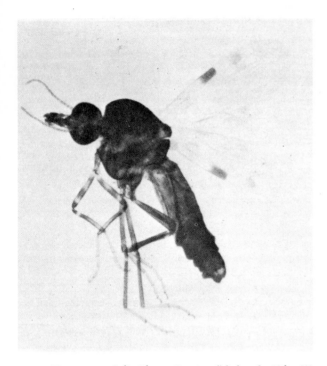

Fig. 19.4. *Culicoides variipenis*, **adult female.** (After W. Buss in Schmidt and Roberts, 1981; with permission of C. V. Mosby.)

Genus *Leptoconops*. *Leptoconops kerteszi*, the Bodega black gnat, and *Leptoconops torrens*, the valley black gnat, are both vicious attackers of humans, domestic animals, and birds in many parts of the United States, particularly along the Pacific Coast, especially in California, and in the southern states. Their bites often result in swellings followed by open exuding lesions. Wirth and Atchley (1973) have contributed a taxonomic review of the North American members of *Leptoconops*.

Ceratopogonid Parasites of Invertebrates. As stated earlier, many species of ceratopogonids are either predators or parasites of invertebrates, especially insects. Among the latter are certain species of *Forcipomyia*. These parasites cling temporarily to the wings of such insects as lacewings and butterflies while they pierce the wing veins and feed on hemolymph. Other species of *Forcipomyia* feed on caterpillars. Similarly, species of *Atrichopogon* feed through the arthrodial membranes of oil beetles and mealy bugs, and *Pterobosca* attack dragonflies and other larger insects.

Culicoides and Related Genera as Vectors. *Forcipomyia utae* and *F. townsendi* are two Peruvian species that may act as vectors for *Leishmania*.

Several species of *Culicoides* are effective vectors of filarial worms that infect humans and domestic animals. *Mansonella perstans*, a human nematode parasite (p. 545), is transmitted by *Culicoides austeni*, a nocturnal species 2 mm long that breeds in banana stumps. This nematode is also transmitted by *C. inornatipennis* and by *C. grahami*, an early morning and evening feeder 1 mm long that also breeds in banana stumps. The human filarial parasite *Mansonella streptocera* (p. 545) is carried by *C. grahami*. *Mansonella ozzardi* (p. 545), another human-infecting filarial worm, is transmitted by *C. furensi*, a species that bears speckles on its mesosternum (Fig. 19.3). The equine filarial worm *Onchocera reticulata* is transmitted by *C. nebeculosus* in England. This species breeds in manure and stagnant water. The bovine filaria *Onchocera gibsoni* is transmitted by a number of species of *Culicoides*.

In addition to serving as intermediate hosts for nematodes, some ceratopogonid midges serve as transmitters of viruses to birds and other animals. For example, *Culicoides variipenis* (Fig. 19.4) transmits *Orbivirus*, the etiologic agent for the so-called blue tongue disease of sheep, cattle, bison, deer, and other ruminants. This disease results in loss of flesh, wool, and poor breeding. It may sometimes lead to death.

For an extensive bibliographic treatment of the literature pertaining to the Ceratopogonidae published between 1758 and 1973, see Atchley *et al.* (1975).

FAMILY CULICIDAE (MOSQUITOES AND GNATS)

The mosquitoes represent the largest group of dipteran pests. There are at least 3000 known species, which are found in every part of the world. These demonstrate little discrimination in sucking blood from humans and animals. Although in the past some investigators have claimed that certain strains of mosquitoes feed only on the blood of one type of host, it is now known that almost all female mosquitoes feed on a variety of hosts, if available.

Many species not only cause great torment through their bites, but also serve as efficient vectors for such dreaded diseases as malaria, yellow fever, dengue, and encephalitis. Indeed, the scourge cast on mankind by these dipterans is beyond description.

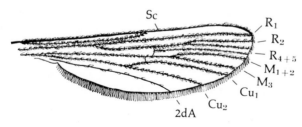

Fig. 19.5. Culicid wing. Wing of mosquito showing venation pattern.

With a few exceptions, most mosquitoes are 2.5–6 mm long. The venation of the single pair of wings is quite characteristic. The veins and posterior margin of each wing are covered with a large number of scales (Fig. 19.5). In my opinion, no better monograph on mosquitoes exists than the extremely complete volume by Horsfall (1955). For those interested in the physiology of mosquitoes, the monograph by Clements (1963) is especially recommended.

Life Cycle of Mosquitoes

Eggs. Most mosquitoes oviposit in water, but certain species, such as *Aedes* spp. and *Psorophora* spp., lay their eggs singly in moist soil. Eggs of these species are quite resistant to desiccation and remain viable in the unhatched state until they are covered with water. In *Psorophora* spp., the spinose protective coat surrounding the egg renders it viable for months or even years.

In species that deposit their eggs in water, some, such as *Anopheles* spp., release their eggs singly in loosely arranged clusters, each one armed with a float of air cells that provide buoyancy (Fig. 19.6). In others, such as *Culex* spp., the eggs are deposited in masses that are vertically arranged as "egg-boats." The arrangements of mosquito eggs, at least among the better-known species, are quite characteristic and are easily recognized. Some of these typical arrangements are depicted in Figure 19.6.

The number of eggs oviposited by a single female varies from species to species, ranging from 40 to several hundred. Such eggs are generally oval, and some bear surface markings.

Larvae. The incubation period varies, ranging from 12 hours to several days. The escaping form is the larva. Although basically all mosquito larvae resemble one another—possessing a breathing siphon (elongate spiracle) on the posterior segment; mandibulate mouthparts; and an elongate, distinctly segmented body bearing setae—species differences do exist, even to the extent that their positions under

the water surface differ. For example, larvae of *Culex* spp. and *Aedes* spp. are attached to the water surface by their siphons, but the bodies are directed downward (Fig. 19.7); among *Anopheles* spp., the larval bodies rest horizontally (Fig. 19.7).

Mosquito larvae are free living, feeding on microorganisms. During the warm months, they undergo four larval stadia of development. The fourth larval instar, after molting, transforms into the pupa. During the four larval stadia, there is a consistent increase in size. The almost microscopic first instar develops into a fourth instar, which measures 8–15 mm in length.

Pupae. The pupa is a nonfeeding but active tissue reorganizational stage (Fig. 19.8). A pair of breathing tubes, located dorsally on the cephalothorax, replaces the caudal siphon of the larva. Pupae are extremely active and are very sensitive to disturbances. They will flutter up and down in the water when disturbed, hence the common designation "tumblers." Although most mosquito pupae acquire their oxygen via breathing tubes that break through the water surface, the pupae of *Mansonia* spp., like their larvae, do not adhere to the water surface. Instead, they obtain the required oxygen by piercing the air channels of the roots of certain aquatic plants with their breathing tubes.

Imagos. Mosquito imagos, or adults, are similar yet are different in minute details, thus accounting for the more than 3000 species that have been described. Figure 19.8 depicts a typical adult and can familiarize the reader with the external anatomic parts. Adults differ not only morphologically, but also ecologically and in their behavior patterns and habits. Space does not permit detailed descriptions of such differences here; however, the biology of the various species is authoritatively described and documented by Horsfall (1955), Christophers (1960), and Clements (1963).

All adult male mosquitoes are vegetarians, feeding on plant juices, but the females are either nectar feeders or bloodsuckers. In parasitology, our interests naturally are concentrated on the hematophagous species.

Biology of the Culicids

The family Culicidae is composed of two subfamilies: Chaoborinae and Culicinae. The Chaoborinae includes a relatively small group of gnats that are not hematophagous and hence are not considered in this volume. The Culicinae includes the mosquitoes. These can be distinguished readily from the gnats by their mouthparts, which are several times longer than the head. Furthermore, the larvae of mosquitoes have laterally inserted antennae on the head.

Numerous taxonomic monographs are available for the identification of mosquitoes from various parts of

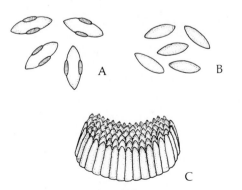

Fig. 19.6. Mosquito eggs. A. Eggs of *Anopheles*. **B.** Eggs of *Aedes aegypti*. **C.** Eggs of *Culex*.

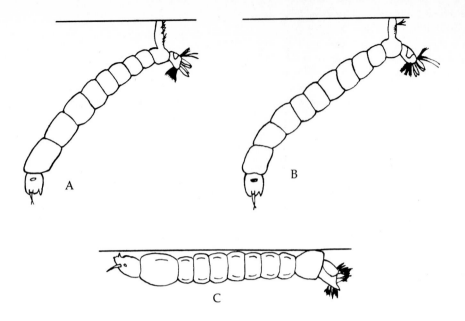

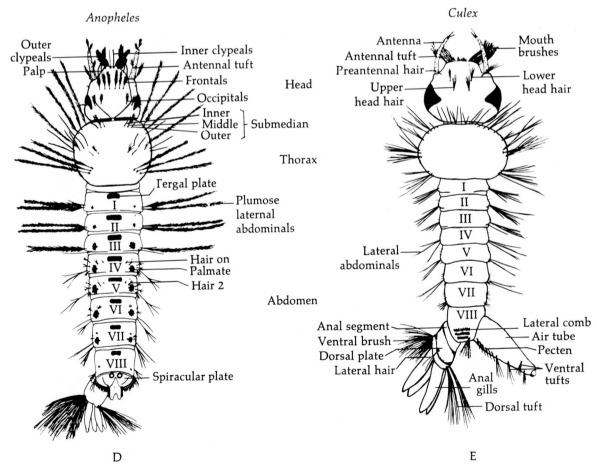

Fig. 19.7. Mosquito larvae. A. Larva of *Culex* adhering to water surface by its breathing siphon. **B.** Larva of *Aedes* attached to water surface. **C.** Larva of *Anopheles* adhering to water surface. **D.** *Anopheles* larva showing external anatomic features. **E.** *Culex* larva showing external anatomic features. (**D** and **E**, National Communicable Disease Center illustrations, U.S. Public Health Service, Atlanta, Georgia.)

the world (Knight and Stone, 1977; Knight, 1978; Carpenter and LaCasse 1955; Carpenter, 1968; Darsie, 1973; Vargas, 1972, 1974; and others). Horsfall (1955) has listed almost all of the species. According to this authority, the Culicinae includes 29 genera. Of these, at least 10 include species found in North America. The largest (in number of species) and most important of these genera from the disease vector standpoint are *Culex, Aedes, Anopheles, Mansonia, Coquillettidia,* and *Culiseta.*

Genus *Culex.* *Culex* includes over 480 species from various parts of the world. Members of the genus can be recognized by the following diagnosis.

Postnotum lacks setae. Scutellum trilobed, one lobe bearing bristles. Palpi short in females, less than one-fifth as long as proboscis. Body humpbacked when at rest. Tip of abdomen blunt in females, covered with broad scales. Postspiracular bristles absent. Winged scales narrow. Larvae with prominent siphon (Fig. 19.7), armed with numerous tufts of hair. Eggs usually deposited in tight, floating masses like rafts on surface water.

Culex pipiens. *C. pipiens,* the common brown house mosquito, is found in many temperate climes. It is a noctural feeder. This domestic species breeds in any body of stagnant water, no matter how small, around the house and in backyards. Although the developmental period is influenced by the prevailing temperature, on the average, eggs require 18 to 24 hours to hatch. The larval instars last 7 days, and the pupal instar lasts approximately 2 days. There appears to be little host specificity on the part of this mosquito, since it feeds on domestic animals, humans, and even birds. However, the type of blood ingested does influence the degree of realization of its biotic potential. For example, Woke (1937) fed 38 specimens of *C. pipiens* on human blood and found that these specimens oviposited 29 egg masses, totaling 2118 eggs, with an average of 73 eggs per mass. When 39 similar mosquitoes were fed on canary blood, they oviposited 22 egg masses, totaling 4473 eggs, with an average of 203.3 eggs per mass. This investigator

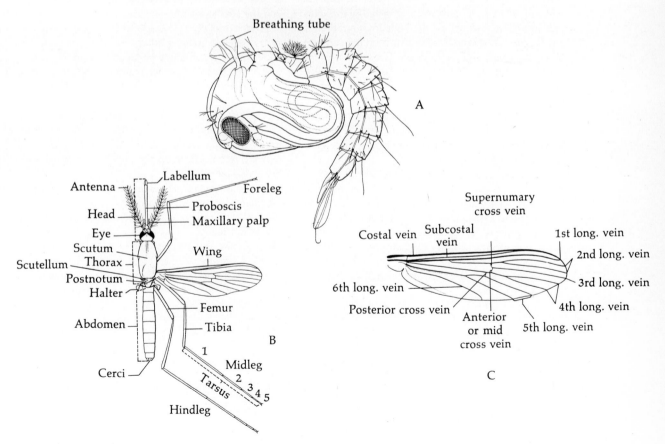

Fig. 19.8. **Mosquito pupa and adult. A.** Pupa of *Anopheles gambiae*. **B.** Generalized drawing of female mosquito. **C.** Mosquito wing. (All figures after Marshall in Smart, 1943.)

computed that *C. pipiens* fed on canary blood produces twice as many eggs per gram of blood ingested.

Actually, what is designated *Culex pipiens* consists of a complex of physiologic species.

As stated earlier (pp. 538, 544), members of the *C. pipiens* complex represent major vectors of the filarial worms *Wuchereria bancrofti* and *Dirofilaria immitis.* They can also transmit avian malaria, avian pox, and encephalitides caused by arboviruses.

Other Culex *Species.* *Culex quinquefasciatus* (= *C. fatigans*) is a species closely related to *C. pipiens,* which is quite widespread in warmer climates, including the southern United States.

Culex tarsalis, a common species in the semiarid western United States, is also found in most of the southern states. It is large and robust and is distinguished by its dark brown to black body and its abdomen, which is ringed by basal segmental bands of yellowish white scales. There is a prominent white band in the middle of the proboscis. This species breeds in water in almost any sunny location. It feeds primarily on birds at night and is attracted to the areas around trees and shrubs by the high concentration of carbon dioxide. It will, however, feed on humans and other mammals, and in so doing transmits western equine encephalitis (WEE) and St. Louis encephalitis. WEE is normally a virus of birds. When introduced into horses, a high degree of mortality commonly occurs. In human infections, it is usually nonfatal; however, it can be severe in children. Infected persons portray fever and drowsiness; hence, it is sometimes referred to as sleeping sickness.

Culex molestus is a common species found in urban areas. This species, unlike the others, can oviposit without first partaking of a blood meal.

Culex tritaeniorhynchus is the most important vector of the Japanese encephalitis virus in the Orient. Numerous species of *Culex,* including some of those mentioned, are suitable vectors for microfilariae, malaria-causing protozoans of birds, and various viruses (Table 19.1).

Genus *Aedes.* This genus includes approximately 800 species, which are well represented all over the world. *Aedes* spp. are the most numerous (in number of species) in North America. Members can be identified by the following diagnosis.

Postnotum lacks setae. Scutellum trilobed. Palpi short in females. Body humpbacked when in resting position. Spiracular bristles present. Abdomen of females pointed and nonmetallic in color with exserted cerci. Claws toothed in females. Larvae with short breathing siphons bearing a single pair of posteroventral tufts of hair. Eggs deposited singly on water surface or on mud.

Aedes vexans. This species is distributed globally. Its coloration varies from brown to gray with uniformly brown wings and tarsi, the latter being banded basally. The females, which are the bloodsuckers, lay their eggs along the edges of rivers. Hatching occurs when water flows over the eggs, permitting the larvae to be totally aquatic. If the eggs are oviposited on the edges of ponds, the same mechanism holds true—water must flood over the eggs as a prerequisite for hatching. This species is a diurnal feeder that occurs in the western United States as well as most of the Holarctic region, South Africa, and the Pacific Basin.

Aedes dorsalis. This species is widely distributed in the United States, Canada, Europe, North Africa, and Taiwan. It can breed in floodwater pools, such as irrigation canals, and in salt marshes. The larvae of *A. dorsalis* have been found in salt pools in Oklahoma and also in pools in Norway where the salt content is 0.07% NaCl. This species will bite humans, horses, cows, sheep, and even birds. Although usually considered a blood feeder, *A. dorsalis,* both males and females, will also feed on plant juices. The females are fierce day biters.

Aedes aegypti. The most notorious of the *Aedes* mosquitoes, this species is the yellow fever mosquito (Fig. 19.9). This species is distributed throughout the warm and humid parts of the world, including Louisiana, Texas, and neighboring states. The adults are characteristically marked with transverse bands of silvery white or yellowish white on the abdomen and with vertical thin stripes on the dorsal surface of the

Table 19.1. Some Viral Diseases Transmitted by Mosquitoes[a]

Viral Disease	Endemic Areas	Major Mosquito Vectors
Yellow fever (YF) (*Flavivirus*)	Humid tropics of Africa; Central and South America	*Aedes aegypti* (urban form) *Haemagogus* spp., *Sabethes* spp., *Aedes* spp. (sylvatic form)
Dengue fever (DEN) (*Flavivirus*)	Primarily Southeast Asia and Carribean	*Aedes aegypti* and other *Aedes* spp.
Chikungunya (CHIK)	Equatorial belt of Africa; south and Southeast Asia	*Aedes aegypti* and other *Aedes* spp., *Mansonia* spp., *Culex* spp.

Table 19.1.   (continued)

Viral Disease	Endemic Areas	Major Mosquito Vectors
Eastern equine encephalitis (EEE) (*Alphavirus*)	Eastern seaboard of Canada south to Argentina, including U.S., Caribbean, Southeast Asia, eastern Europe	*Culiseta melanura* and other bird-biting species (in enzootic cycles) *Aedes taeniorhynchus* *Aedes sollicitans* *Aedes vexans* (in equine and human outbreaks)
Mayaro (MAY)	Bolivia, Brazil, Colombia, Surinam, Trinidad	Primarily *Haemagogus* spp.
Mucambo (MUC) (Considered enzootic strain of Venezuelan equine encephalitis)	Brazil, Trinidad, Surinam, Guiana	*Culex portesi* and other *Culex* spp.
Everglades (EVE) (Considered endemic strain of Venezuelan equine encephalitis)	South Florida	*Culex nigripalpus* and other *Culex* spp.
Ross River (RR) (Rodent reservoir suspected)	Queensland and New South Wales, Australia	*Aedes vigilax* *Culex annulirostris*
Sindbis (SIN)	Africa, India, Philippines, Malaysia, Australia, eastern Europe	*Culex* spp.
O'Nyong Nyong (ONN)	Uganda, Kenya, Congo	*Anopheles funestus* *Anopheles gambiae*
Venezuelan equine encephalitis (VEE) (*Alphavirus*) (Rodent reservoir hosts)	Peru to Texas	*Mansonia* spp. and many other species of *Aedes, Anopheles, Culex, Haemagogus, Psorophora, Sabethes,* and *Wyeomyia*
Western equine encephalitis (WEE) (*Alphavirus*) (Birds and possibly reptiles and amphibians as reservoir hosts)	Western and north central U.S. and adjacent Canada	*Culex tarsalis* and other *Culex* spp., *Culiseta, Psorophora, Anopheles*
Japanese encephalitis virus (JBE) (*Flavivirus*)	Eastern Siberia across Asia to India	*Culex tritaeniorhynchus* and other *Culex* spp., *Aedes albopictus* *Aedes togoi* *Anopheles* spp.
Murray Valley encephalitis (MVE) (Birds suspected as reservoir hosts)	Southeastern Australia, New Guinea	*Culex annulirostris* *Culex bitaeniorhynchus* *Aedes normanensis*
St. Louis encephalitis (SLE) (*Flavivirus*) (Birds as reservoir hosts)	U.S. and Canada	*Culex tarsalis* *Culex pipiens* *Culex quinquefasciatus* *Culex nigripalpus*
West Nile (WN)	Egypt and adjacent African countries, South Africa, India, other Asian localities, France, Cyprus	*Culex* spp. *Anopheles* spp. *Mansonia* spp. (also from ticks)
Banzi (BAN)	Eastern and southern Africa	*Culex rubinotus* and other *Culex* spp.
Bussuquara (BSQ) (Rodent reservoir suspected)	Panama, Colombia, Brazil	*Culex* spp.
Ilheus (ILH)	Panama, eastern South America	*Psorophora* spp.
Spondweni (SPO)	South and west Africa	*Aedes circuluteolus* *Mansonia africana*
Calovo (CVO)	Czechoslovakia, Austria, Yugoslavia	*Anopheles maculipennis*
Guaroa (GRO)	Colombia, Brazil, Panama	*Anopheles* spp.
Tahyna virus (TAH) (Young hares and pigs as reservoir host)	Central Europe	*Aedes* spp.
Rift Valley Fever (RVF)	Kenya, Uganda, Mozambique, Zimbabwe, South Africa	*Aedes caballus* *Culex theileri* *Eretmapodites chrysogaster* *Culex pipiens* and other *Culex* spp.
Avian pox (fowl pox virus) (birds susceptible)		*Culex pipiens* and other *Culex* spp.
Myxomatosis (myxoma virus) (only pathogenic in European rabbit, *Oryctolagus*)	Central and South America, California, Australia	*Culex annulirostris* *Anopheles annulipes* *Aedes* spp. (also simuliids and rabbit flea)

[a] Approximately 90 viruses have been isolated from mosquitoes. These have been catalogued in a volume edited by Berge (1975). Presented here are only some of the more virulent ones.

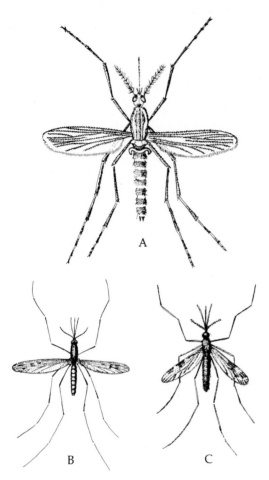

Fig. 19.9. **Adult mosquitoes. A.** *Aedes aegypti*, the yellow fever mosquito. **B.** *Anopheles quadrimaculatus*. **C.** *Anopheles punctipennis*. (All figures redrawn after Matheson, 1950.)

thorax. The legs are banded and the tarsi of the last pair of legs are white.

The eggs of *A. aegypti* are deposited singly on the surface of water. They are dark in color and are surrounded by air cells. It has been estimated that a single female will oviposit approximately 140 eggs if fed on human blood. However, more eggs are produced if they feed on amphibians and reptiles (*e.g.*, frogs and turtles). The eggs are quite sensitive to low temperatures and are usually nonviable if maintained below 10°C. However, they are quite resistant to desiccation. Reed *et al.* (1900) reported that eggs remain viable for 30 days above the water line. Walter Reed and his co-workers, while engaged in their history-making study of yellow fever in Panama, were

able to transport eggs of *A. aegypti* from Havana to Washington, D.C., on filter paper, and 67% of these eggs hatched after 3 months of storage. Similarly, many other investigators have demonstrated the resistance of these eggs to desiccation.

The larval stadia usually last 9 to 12 days; the pupal stadium lasts for only 36 hours.

Aedes aegypti is the principal vector for the yellow fever virus, but other species are also capable of performing this task (Table 19.1).

Other Aedes *Species.*   Other common species of *Aedes* found in North America include *A. sollicitans*, distributed along the Atlantic Coast from Maine to Florida and along the Gulf of Mexico. The breeding sites of this mosquito are salt marshes. The blood-sucking females bite at any time but make no attempt to feed if the temperature drops to 10°C or lower. Temperatures ranging from 20 to 30°C are the most favorable for feeding.

Aedes taeniorhynchus is distributed from New England south to the Guianas along the east coast, and from southern California to Peru in the west. This, too, is a salt-marsh-breeding species and can be recognized by its brown body and white-banded proboscis.

Aedes vexans, *A. dorsalis*, and *A. aegypti* have been cited as freshwater breeders (*A. dorsalis* also breeds in salt marshes); *A. sollicitans* and *A. taeniorhynchus* are cited as salt-marsh breeders. A third category of *Aedes* spp. exists—those that breed in water contained in tree holes. This type includes *A. varipalpis*, a Pacific species; *A. triseriatus*, in the eastern United States; *A. simpsoni* and *A. luteocephalus*, both found in Ethiopia; and *A. seoulensis* in the Far East.

Enhanced transmission of virus.   In an interesting study, Turell *et al.* (1984) have demonstrated that if adult *Aedes taeniorhynchus* females ingest the Rift Valley fever virus, for which they serve as vectors, only 64% of the mosquitoes become infected, and among these, only 5% of refeeding mosquitoes actually transmit the virus to gerbils in the laboratory. On the other hand, ingestion of an identical amount of virus from gerbils concurrently infected with the filarial nematode *Brugia malayi* resulted in 88% of the mosquitoes becoming infected with the virus, and the transmission level by refeeding mosquitoes was increased to 31%. Normally, viral transmission by an arthropod requires that the virus infect cells of the vector's midgut, replicate, then pass into the hemocoel and from there to the salivary glands. Turell and his colleagues have provided some evidence that the enhanced transmission of the Rift Valley fever virus by *A. taeniorhynchus* concurrently infected with *B. malayi* is due to two factors: (1) the presence of microfilariae simplified development of the virus by eliminating the normal replication in the midgut and release into the hemocoel, and (2) by bypassing the replication cycle in the midgut, the time required before the virus can be transmitted to the mammalian host is greatly reduced. Based on this study, it may be concluded that endemic filariasis may promote arbovirus transmission in nature.

Genus *Anopheles.* This genus includes over 400 species which are widely distributed. These can be recognized by the following diagnosis.

Postnotum lacks setae. Scutellum not lobed. Palpi of both sexes usually as long as proboscis. Mandibles and maxillae of females well developed and toothed. Wings usually spotted or mottled. Body not humpbacked when at rest. Larvae rest parallel to water surface. Eggs laid singly with associated floats.

Anopheles quadrimaculatus. This is the common species in eastern North America, being particularly abundant in the southeastern states (Fig. 19.9). This mosquito can be recognized by its dark wings, commonly possessing four black spots formed by aggregations of scales; by its palpi and hind tarsi, which are black; and by the small white "knee spots" at the tips of the femora. In former years, *A. quadrimaculatus* and *A. freeborni* were the major vectors of malaria in North America.

Anopheles punctipennis. This is the most abundant species in North America (Fig. 19.9). It ranges from southern Canada across the United States to Mexico. Environmental conditions are important in its development. For example, Horsfall and Morris (1952) have reported that larvae are four times more plentiful in marshy areas where the summer mean maximum water surface temperature is 18°C than in another area where the temperature is 25°C. Since running water aids in regulating the desirable temperature of the water surface, it is not surprising that the larvae are numerous in running water.

Anopheles pseudopunctipennis. This prominent species is widely spread in South and Central America and Mexico. Its breeding sites are in clear, moving water containing large quantities of filamentous green algae. It is a vicious feeder and attacks humans, cows, sheep, dogs, horses, cats, and even birds. It is an important vector for *Plasmodium* spp.

Other Anopheles *Species.* *Anopheles freeborni* is a uniformly brown species found along the Pacific Coast; *A. albimanus* is an important malaria-transmitting species in the Panama Canal Zone; *A. maculipennis* is the common mosquito in Europe; and *A. sergenti* is a species common in North America and Israel. The breeding sites of the last mentioned species are often under stones and are not easily found.

Genus *Mansonia.* *Mansonia* includes some 50 species, which are widely distributed. These can be recognized by the following diagnosis.

Postnotum lacks setae. Scutellum trilobed. Palpi in females usually one-fourth as long as proboscis or longer. Bodies humpbacked when in resting position. Abdomen of females truncate and nonmetallic. Spiracular bristles absent. Wing scales large and broad. Eggs

deposited as masses that adhere to leaves and other objects, or less commonly as rafts. Siphons of larvae penetrate air cells of plant roots.

Although species of *Mansonia* are not commonly encountered in North America, they are among the most important vectors for microfilariae (*Wuchereria bancrofti, Brugia malayi,* and *Dirofilaria immitis*) (p. 656) and the yellow fever virus in the tropics.

Genus *Coquillettidia.* In recent years, some members of the genus *Mansonia* have been transferred to *Coquillettidia.* The members of both genera are very similar in both morphology and biology. Some species of *Coquillettidia* are capable of transmitting brugian filariasis (p. 540).

Coquillettidia titillans, a tropical species, has invaded southern Florida and Texas. It is found primarily in Central and South America and the Caribbean islands.

The eggs of *C. titillans* are glued to the underside of leaves in water, because the females oviposit while sitting on a leaf with the tip of the abdomen curved beneath the leaf. Larval breathing is accomplished when the siphons puncture the roots of *Pistia.* Although larvae may breathe at the surface of water for several days, without *Pistia* they perish in 5 or 6 days.

Coquillettidia perturbans is widely distributed in North America, ranging from southern Canada to the Gulf of Mexico. It is usually found in timbered areas. It is also found in Great Britain, continental Europe, and Israel. The eggs are deposited on rafts, each raft containing 150 to 308 eggs. Unlike those of *C. titillans,* the larvae of *C. perturbans* can utilize a number of aquatic plants as breathing tubes. Various investigators have reported that *Typha, Limnobium, Pistia, Sagittaria, Nymphea, Pontedaria,* and *Piaropus* serve as the plants for attachment, and discovery of these plants in otherwise suitable sites for development often indicates the presence of the mosquito. In England, *Typha, Glyceria, Acorus,* and *Ranunculus* are the associated plants, but in Norway *C. perturbans* is specifically associated with *Sparganium,* suggesting that this Norwegian form may represent a distinct strain.

Several tropical species of *Mansonia* and *Coquillettidia* are of medical importance as transmitters of microfilariae and viruses. Some of these are listed in Tables 19.1 and 19.2.

Genus *Culiseta.* This genus is represented in North America by approximately eight species or subspecies. Some species feed on mammals, and others feed on birds. *Culiseta inornata* and *C. melanura* are capable of transmitting western and eastern encephalitis. *C. inornata* is more widely distributed than *C. melanura,* being found throughout the northern

Table 19.2. Representative Nematode Parasites and Their Major Mosquito Vectors

Nematode	Mosquito Vector
Wuchereria bancrofti	*Culex* spp. including
	C. quinquefasciatus
	Aedes spp.
	Mansonia spp.
	Anopheles spp.
Brugia malayi	*Culex pipiens quinquefasciatus*
	C. pipiens pallens
	Aedes polynesiensis
	A. togoi
	Anopheles gambiae
	A. funestus
	A. farauti
	A. pyrcanus sinensis
	Coquillettidia spp.
	Mansonia spp.
Dirofilaria immitis	*Culex pipiens*
(in dogs, cats, wild carnivores)	*Aedes aegypti*
	Mansonia spp.

states in the United States and southern Canada. *C. melanura* is restricted to the eastern and central United States.

Pathogens Transmitted by Mosquitoes

Like ticks, mites, and numerous parasitic insects, mosquitoes are not only vicious pests but also serve as vectors for various pathogenic organisms. The major mosquito-transmitted diseases are discussed below.

Malarias. The role of mosquitoes in the transmission of *Plasmodium* spp. has been discussed on p. 194. Relative to the human-infecting malarial organisms, various species of *Anopheles* serve as the only suitable vectors (Table 19.3). However, *Anopheles* spp., *Culex* spp., and *Aedes* spp., primarily the latter two, are the major vectors for avian malarias.

Although specific species of mosquitoes are categorically said to be compatible vectors for specific species of *Plasmodium*, strain differences do exist as far as their susceptibility to the protozoans is concerned. Such variations in the susceptibility of mosquitoes to *Plasmodium* spp. appear to be genetically controlled. For example, Ward (1963) was able to derive a strain of *Aedes aegypti* that was highly resistant to *Plasmodium gallinaceum*, an avian malaria-causing organism, from a susceptible strain by genetic selection. The susceptibility of the resistant strain decreased 98% over 26 generations. This study suggests that differences in the compatibility of mosquito populations for *Plasmodium* spp. may be attributed to variations in gene frequency in different areas. In areas where the mosquito populations are fairly stable in terms of population size and absence of migration and selection, the level of susceptibility should remain constant. On the other hand, considerable variation in susceptibility might be expected in areas where extensive mosquito migration and marked shifts in the size of the population exist.

Yellow Fever. Yellow fever is a viral disease that has plagued mankind for centuries. During the building of the Panama Canal, yellow fever caused so much illness among workers that the project almost came to

Table 19.3. Important Anopheline Vectors of Malaria

Endemic Area and Species of *Anopheles*	Locality	Breeding Sites
United States and Canada		
quadrimaculatus	Eastern, central, and southern U.S. (Gulf to Ontario)	Sunlit, impounded water, marshes, swamps, rice fields
freeborni	Rocky Mountains, New Mexico, Pacific Coast	Sunlit seepage water, irrigation ditches
Mexico, Central America, West Indies		
albimanus	Mexico to Colombia and Venezuela, West Indies	Sunlit brackish and fresh lagoons, swamps, ponds
pseudopunctipennis pseudopunctipennis	Southern U.S. to Argentina	Sunlit streams with green algae, pools
darlingi	Central America to Argentina	Sunlit fresh waters, marshes
aquasalis	Central America to Brazil, West Indies	Brackish marshes, irrigation ditches
punctimacula	Mexico to Brazil, West Indies	Shaded pools, swamps, streams
bellator	West Indies to Brazil	Water at bases of bromeliad leaves
aztecus	Mexico	Sunlit pools with green algae
South America		
cruzii	Brazil	Water at bases of bromeliad leaves
albitarsus	Argentina	Rice fields, marshes, ditches
Europe		
labranchiae labranchiae	Mediterranean Europe and North Africa	Upland streams, rice fields, brackish coastal marshes

Table 19.3. (continued)

Endemic Area and Species of *Anopheles*	Locality	Breeding Sites
labranchiae atroparvus	England, Sweden to Spain and northeastern Italy	Sunlit pools, ponds, marshes
maculipennis messeae	Norway, U.S.S.R., Siberia, Manchuria	Freshwater pools, ponds, marshes
superpictus	Spain, southern Europe, Greece, Asia Minor	Pools in stream beds, irrigation canals, seepages
sacharovi	U.S.S.R., Balkans	Sunlit coastal marshes, fresh and brackish water
North Africa, Middle East		
claviger	Ukraine, Asia Minor, North Africa	Rock pools, wells, cisterns, marshes
pharoensis	North Africa, Israel	Rice fields, swamps
sergentii	North Africa, Israel, Turkey, Syria	Rice fields, irrigation canals, borrow pits, seepage
superpictus		
Central and South Africa		
funestus	Tropical Africa	Ditches, stream margins, swamps, seepage
gambiae	Tropical Africa, Egypt, Arabia	Puddles, pools, sluggish streams
moucheti	Uganda, Congo, Cameroon	Swamps, stream margins
pharoensis	Widely distributed in Africa, Israel	Rice fields, swamps
nili	Widely distributed in central Africa	Shady stream margins
pretoriensis	Widely distributed in central and southern Africa	Sunlit rock pools, stream beds, ditches, hoofprints
Philippine Islands		
minimus flavirostris	Many islands, Indonesia, Celebes	Foothill streams, ditches, wells
mangyanus	Many islands	Stream beds, irrigation ditches
Japan, North China, Korea		
sinensis	Widely distributed	Open clear water, rice fields, swamps, ponds, slow streams
pattoni	North China	Beds of hill streams, rock pools
South and Central China, Burma, Formosa		
minimus	Hilly regions in southern China, Formosa, Burma	Sunlit, slow streams, rice fields, irrigation ditches
sinensis	Plains in central China, Burma	Open clear water, rice fields, swamps, ponds, slow streams
jeyporiensis var. *candidiensis*	Hong Kong area	Rice fields in hill country
culicifacies	Burma	Streams, irrigation ditches
maculatus maculatus	Burma, Vietnam	Stream and river beds, rice fields, pools, lake margins
philippinensis	Burma, Vietnam	Rice fields, pits, swamps, ditches, tanks
India, Sri Lanka		
culicifacies	India, Sri Lanka, Thailand	Streams, irrigation ditches
stephensi	India	Wells, cisterns, roof gutters, water receptacles
maculatus maculatus	India, Sri Lanka	Streams and river beds, rice fields, pools, lake margins
flaviatilis	India, Thailand	Edges of foothill streams, springs, irrigation canals
minimus	Eastern and northern Sri Lanka	Sunlit, slow streams, rice fields, irrigation ditches
India, Sri Lanka	Asia Minor	Stream-bed pools, irrigation canals, seepage, hill district
philippinensis	India	Rice fields, irrigation canals, tanks
Thailand, East India, Malaya		
aconitus	East Indies, Malaya, Vietnam	Rice fields, irrigation ditches, pools in creek beds, reservoirs
nigerrimus	Malaysia, East Indies	Rice fields, impounded water, pits, sluggish streams
maculatus maculatus	Thailand, Malaya, East Indies	Stream and river beds, seepage, lake margins
subpictus subpictus	Malaysia, East Indies	Pits, all sorts of tempoary or permanent collections of water
sundaicus	Thailand, Malaya, East Indies	Seawater lagoons, swamps
umbrosus	Malaysia, Vietnam, East Indies	Shaded jungle pools, mangrove swamps
Australia, Melanesia, Polynesia		
farauti	New Guinea, Solomon Islands, New Hebrides	Fresh or brackish water, all sorts of natural or artificial collections of water, fresh or polluted
punctulatus punctulatus	New Guinea, Solomon Islands	Rain pools, stream margins, hoofprints
bancrofti	New Guinea, northern Australia	Shallow, slow streams

[a] Data from Harwood and James, 1979.

a standstill. It was then that Dr. Walter Reed and his associates, Drs. James Carroll, Jesse W. Lazear, and A. Agramonte, won fame by proving Dr. Carlos Finlay's hypothesis that *Aedes aegypti* serves as transmitter of the pathogen.

Mosquitoes become infected with the virus while feeding on yellow fever victims. A single female may ingest thousands of viruses at one feeding and usually remains infective for the rest of her normal life—200 to 240 days. Under field conditions, the normal incubation period within the vector is 12 hours. However, fluctuations in temperature affect the extrinsic incubation period. If the mosquitoes are exposed to temperatures of 36.8°C, the incubation period is reduced to 4 days, but if they are exposed at 21.2°C, the period is increased to 18 days.

Jungle animals, especially monkeys of the genus *Cebus*, serve as natural reservoirs for the yellow fever virus.

In addition to *Aedes aegypti*, various other species serve as natural and experimental vectors. For example, *Aedes vittatus* is the vector for yellow fever in Egypt. *Haemagogus spegazzinii*, a south American species found in Brazil and Colombia, is a suitable vector, especially in transmitting the virus among monkeys. *Aedes simpsoni* and *A. africanus* have been incriminated as vectors in eastern Africa. Over 30 species of mosquitoes have been shown experimentally to be possible vectors. Some of these are *Aedes fluviatilis, A. scapularis, A. luteocephalus, A. albopictus, Mansonia africana, Haemagogus equinus,* and *H. splendens.*

Dengue Fever. Dengue fever is commonly referred to as breakbone fever. This is another viral disease transmitted by mosquitoes. Various investigators have demonstrated that a number of species are suitable vectors. Among these are *Aedes aegypti, A. albopictus,* and *A. scutellaris. Aedes albopictus* is prevalent in Japan, New Guinea, northern Australia, Malagasy Republic, the Philippines, and Hawaii. It is distinguishable by a silvery stripe on its mesonotum and whitish irregular patches on the lateral aspects of its thorax. *Aedes scutellaris* is a closely related species characterized by whitish wavy lines composed of scales down each side of the thoracic pleura. This mosquito is an important vector of dengue fever in Polynesia.

Equine Encephalitis. Another complex of viral diseases known as encephalitis is mosquito-borne. Although primarily a disease of horses, outbreaks of equine encephalitis among human populations occur periodically. Various strains of the virus exist, most of which are experimentally infective to guinea pigs. Various mosquitoes are suitable transmitters of this

virus; both fresh- and saltwater breeders are represented. Some of the more common species are listed in Table 19.1. It was one of the equine encephalitis viruses (a Venezuelan strain) that caused mass mortalities of horses in Texas during the summer of 1971. A few humans also became infected, and at least one child died. It is noted that there are other viruses transmissible to humans by mosquitoes that are not equine in origin; for example, St. Louis encephalitis (SLE), Japanese B encephalitis (JAP), and Murray Valley encephalitis (MVE) (Table 19.1).

In addition to mosquitoes, the conenose bug *Triatoma sanguisuga*, the tick *Dermacentor andersoni*, and the bird louse *Ornithonyssus sylviarum* are also suitable vectors for the equine encephalitis virus. The louse serves as an important transmitter of the pathogen from bird to bird in the yellow-headed blackbird, *Xanthocephalus xanthocephalus*, which is a natural reservoir.

Various other viruses of humans and domestic animals are transmitted by mosquitoes. Some of these are listed in Table 19.1.

The role of mosquitoes in the transmission of microfilariae has been discussed on page 538. Some of these nematodes and their vectors are listed in Table 19.2.

Behavior of Mosquitoes

Male mosquitoes are capable of detecting and locating females from a range of several centimeters. In the case of *Aedes*, the distance is about 25 cm. The antennae of males are equipped with many whorls of long hairs, and the subbasal segment of each antenna is enlarged to house a battery of sense organs, known as **Johnston's organs**, which is stimulated by the movements of the antenna. The beating of the wings of the female mosquito during flight makes a whining noise, which stimulates the hairs on the male's antennae as well as the antenna shaft to vibrate, and, as a consequence, the Johnston's organs are stimulated. Studies have been made of the frequencies of the sounds emitted by females; for example, the notes produced by *Anopheles maculipennis* buzzes between middle C and E flat. Although male mosquitoes respond to a rather broad range of frequencies, they are suspected to be partially selective in seeking out females of their own species. A male will respond to the "mating call" of a female only if the latter's emission is not disrupted by other sounds (Haskell, 1961). It thus follows that males locate isolated females, and the swarming of mosquitoes has no function in bringing the sexes together.

Female mosquitoes respond to a variety of stimuli in seeking out their hosts. For example, Laarman (1959), working with *Anopheles maculipennis*, has reported that smell is important in locating the host but heat and moisture together also serve as attractants. The thermotaxis is activated by carbon dioxide. Similarly, Wright and Kellogg (1962) have found that *Aedes aegypti* is stimulated to search out a host by an alteration in the ambient carbon dioxide level and is guided to the host's surface partly by color, but mainly by warm, moist convection currents arising from the surface. *Aedes aegypti* is most attracted to dark colors and usually approaches the shaded parts of a host. It still remains uncertain why certain humans do not appear to attract mosquitoes. For example, the entire Eskimo race is said to be refractile.

FAMILY TABANIDAE (HORSEFLIES AND DRAGONFLIES)

The Tabanidae is one of two families of the Brachycera that includes bloodsuckers. The tabanid flies are large and stoutly built and are often beautifully colored. The more common colors are brown, black, orange, or metallic green. They are strong fliers and viciously attack cattle, horses, deer, and other warm-blooded animals, including humans. Only adult females are bloodsuckers; the males feed on plant juices and nectar. The mouthparts of the female are developed for cutting skin and sucking blood that oozes from the wound.

Tabanid Life Cycle Pattern

The life cycles of all tabanid flies follow a basic pattern. The cylindrical eggs, 1–2.5 mm long, are deposited in neatly arranged piles on stems and leaves of aquatic plants or on the leaves of trees, such as willows, that hang over bodies of water (Fig. 19.10). The number of eggs per pile varies from 100 to 1000, and these are cemented together with a gluelike secretion. Some species deposit their eggs in terrestrial habitats such as on logs and mud. These eggs are oviposited during the summer and early fall.

Embryonic development is influenced greatly by environmental conditions, but during the heat of summer the average time ranges from 5 to 7 days. The escaping larvae are vermiform, composed of the head and eleven body somites (Fig. 19.10). Each segment bears a row of warts on which are inserted hairs and setae. Such larvae, upon hatching, either drop into water and migrate to the substratum or burrow into the mud if the eggs are deposited thereon. These larvae are free living, feeding on organic debris while undergoing rapid growth during the summer and fall. They become quiescent with the onset of winter. By the following spring, the larvae will have passed through four to nine instars and are fully grown. The final larval stage migrates to dry ground in preparation for pupation.

Pupae are embedded under the surface of dryer ground (Fig. 19.10). The enclosed pupal stage lasts 2 to 3 weeks in most species; however, the pupae of *Chrysops* metamorphose into adults in less than 2 weeks. The adults rupturing from the cocoon usually hide in foliage, and the females migrate from there as bloodsuckers.

Philip (1947) and Mackerras (1954–1955) have

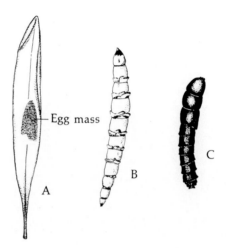

Fig. 19.10. Stages in the life cycle of *Tabanus punctifer.* **A.** Eggs arranged in a mass on a leaf. **B.** Larva. **C.** Pupa.

contributed taxonomic studies on the Tabanidae. This family includes over 2500 species and is divided into two subfamilies—the Pangoniinae, which includes genera that possess apical spurs on their hind tibiae and ocelli in addition to compound eyes; and the Tabaninae, members of which do not possess apical tibial spurs or ocelli.

Subfamily Pangoniinae

The Pangoniinaean tabanids as a rule occur in the warmer parts of the world. They are characterized by a greatly elongated labial proboscis. In males, this proboscis is used to withdraw nectar from deep inside flowers, but in females, the proboscis, along with the mandibles and maxillae, takes in blood.

Genus *Chrysops*. Flies of the genus *Chrysops*, commonly known as deerflies, are worldwide in distribution. At least 80 species are known in North America. These are comparatively small flies with clear wings except for a dark band located vertically down the middle of the wing and a darkly pigmented anterior border (Fig. 19.11). *Chrysops callida*, measuring 7–9 mm in length, is a widely distributed species that is black with lateral pale yellow spots near the base of the abdomen.

Chrysops discalis, measuring 8–10.5 mm long, is a grayish species with black spots on its abdomen (Fig. 19.11). It is fairly common in the western states, including Nevada, Nebraska, and North Dakota. It has also been reported in Manitoba and Saskatchewan. *Chrysops dimidiata* is a southwestern African species that measures 8.5 mm in length and is brownish

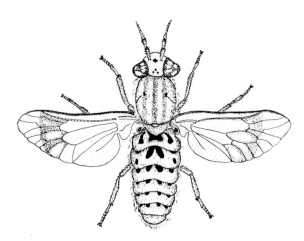

Fig. 19.11. Biting fly. *Chrysops discalis.* (Redrawn after Francis, 1919.)

yellow. In Britain, *C. caecutiens* is the common species.

Genus *Silvius*. Members of *Silvius* are found primarily in Australia, although a few species occur in North America. In *Silvius* spp., the second segment of the antennae is approximately half as long as the first segment.

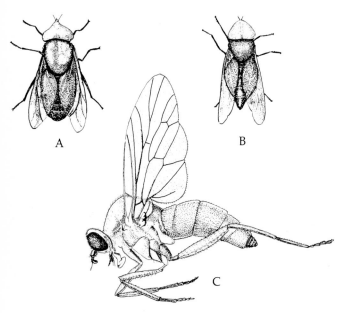

Fig. 19.12. Some biting flies. A. Female *Tabanus atratus,* the black horsefly. **B.** Male *T. atratus.* **C.** *Symphoromyia atripes.*

Subfamily Tabaninae

The tabaninaean tabanids are more common in temperate climes, although exceptions occur, as in the case of the Pangoniinae.

Genus *Tabanus*. The genus *Tabanus*, commonly known as the horseflies, includes some 150 species that are worldwide in distribution. Among the most commonly encountered species in North America is *T. atratus*, the black horsefly, which is 16–30 mm long (Fig. 19.12). This species is found in states east of the Rocky Mountains and in Mexico. Other species include *T. punctifer*, a black and white species found in the western United States, particularly along the Pacific Coast; *T. quinquevittatus*, characterized by its green head; and *T. lineola*, characterized by its brown or black abdomen adorned with three gray stripes. The last two species named are common pests of cattle in the southwestern United States.

Tabanids as Vectors. The loss of blood (100–200 ml per day) by cattle resulting from tabanid flies is a serious problem. Weight loss and decline in milk production commonly result from the bites. In addition to the inflicted injuries, discomforts, and loss of blood, tabanids serve as vectors for bacterial, protozoan, and helminth parasites. Various tabanid flies serve as vectors for *Trypanosoma* spp. (p. 127). *Trypanosoma evansi*, the causative agent of "surra" in horses, cattle, and dogs, is transmitted by *Tabanus striatus*. Similarly, *Trypanosoma equiperdum*, the causative agent for "maldecaderas" in horses and other animals in South America, and *Trypanosoma theileri* of cattle can be transmitted by tabanids. Furthermore, *Trypanosoma vivax*, one of the hemoflagellates that cause sleeping sickness (p. 109), can be transmitted by tabanids where the normal tsetse fly vector is not present, for example, in northern Africa.

Tularemia and anthrax, two bacterial diseases of humans and animals, are transmitted by tabanid flies, among other arthropod vectors. Tularemia is caused by *Francisella tularensis* and anthrax by *Bacillus anthracis*. Tularemia, or deerfly fever, is a disease of humans throughout the United States, Canada, northern Europe, (except Great Britain, Spain, and Portugal), the Soviet Union, Turkey, and Japan. One of the tularemia vectors is *Chrysops discalis*. Rabbits serve as natural reservoirs for the bacterium. Anthrax is a much dreaded disease of cattle, although various other animals and humans are susceptible. *Tabanus striatus* and other tabanids are suitable vectors.

The African eye worm, *Loa loa* (p. 544), is transmitted to humans by *Chrysops dimidiata* and related species. Krinsky (1976) has contributed a review of pathogens known to be transmitted by tabanids.

Behavior of Tabanids. Tabanid flies are attracted to hosts by carbon dioxide and can be captured on sticky traps baited with dry ice. In addition to CO_2, there is little doubt that vision plays an important role in host localization.

Family Rhagionidae (Snipe Flies)

Members of the Rhagionidae, commonly called snipe flies, include both hematophagous and nonhematophagous species. The three important ectoparasitic genera are *Symphoromyia*, *Suragina*, and *Spaniopsis*. These flies are readily recognized by their third antennal segment, which is kidney shaped. Of the approximately 30 species of *Symphoromyia* found in North America, mostly in the mountainous western regions, *S. hirta*, *S. atripes*, *S. kincaidi*, and *S. pachyceras* are the most common. *Symphoromyia hirta*, measuring 7.5 mm in length, is a severe biter of humans and animals; *S. atripes*, a black species with reddish legs and measuring 5.3–8 mm in length, is also a vicious biter (Fig. 19.12); and *S. kincaidi* possesses a black head and thorax but a yellow abdomen.

Suragina longipes is a Mexican species that bites viciously. In Australia, the species of *Spaniopsis* represent the bloodsucking rhagionids. The species of *Spaniopsis* possess an elongated third antennal segment that terminates in a style.

SUBORDER CYCLORRHAPHA

The Cyclorrhapha is divided into two groups—the Aschiza and the Schizophora. Only the family Syrphidae of the Aschiza includes parasitic species. Sixteen families of the Schizophora include parasites, five of which are represented by larvae parasitic in invertebrates, primarily other insects. Although the distinguishing characteristics of the families are briefly presented at the end of this chapter, no attempt is made to give a comprehensive resume of each one. Interested readers are referred to texts and reference volumes dealing with medical and economic entomology for detailed discussions. The treatise by Clausen (1940) should be consulted for detailed information on species that parasitize other insects.

ASCHIZA GROUP

The Aschiza Group includes the family Syrphidae, which, as stated, includes parasitic species. Suster (1960) has contributed a complete review of this family with discussions on species found in Rumania, and Cole (1969) has represented a detailed account of species occurring in western North America.

FAMILY SYRPHIDAE (FLOWERFLIES, OR SYRPHID FLIES)

The colorful adult syrphid flies are free living, usually with dark bodies ornamented with brightly colored spots and bands. They are usually found hovering over flowers, primarily feeding on nectar. Because of their characteristic habit of continuously remaining on the wing near flowers, the common designations "hover flies" and "flowerflies" are often used. A few genera of these flies include parasitic larvae, primarily *Syrphus* and *Tubifera*.

Genus *Syrphus*. Larvae of *Syrphus* spp. are normally predaceous, feeding on plant lice and other plant-infesting insects. A few instances of human gastrointestinal myiasis (infection by dipteran larvae) have been reported, although these reports are in need of confirmation. Human infestations, if such reports are accurate, probably are contracted by ingesting vegetables that contain larvae.

Genus *Tubifera*. This genus includes *T. tenax*, *T. dimidiata*, and *T. arbustorum* (Fig. 19.13). The larvae of these flies cause gastrointestinal myiasis. *Tubifera tenax*, the drone fly, is by far the most common myiasis-causing species. It is worldwide in distribution. The adults are free living and are commonly seen hovering over cesspools, liquid manure, open privies, and bodies of water contaminated with rotting animal matter. The eggs are oviposited in these putrid sites and give rise to larvae that, like those of *Syrphus* spp.,

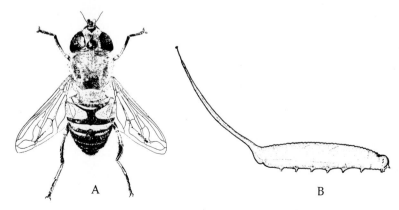

Fig. 19.13. Stages in the life cycle of *Tubifera*. A. Adult drone fly, *T. tenax*. (After James, 1947.) **B.** Rattail larva of *T. dimidiata*.

bear a long anal breathing siphon resembling a tail (Fig. 19.13). Thus, they are commonly called rat-tailed maggots. Animal gastrointestinal myiasis undoubtedly originates through drinking water contaminated with these maggots.

Genus *Helophilus*. This is another genus of the Syrphidae with larvae that cause myiasis. A case of human urinary myiasis is known to be due to penetration up the urinary passage by the maggot. The larvae of *H. pendulus* are similar to those of *Tubifera* spp., but the tracheal tube (or breathing tube) of the former is undulating rather than straight. *Helophilus pendulus* is distributed throughout northern and eastern Europe.

SCHIZOPHORA GROUP

Parasitism by schizophorans can be either external or internal. Ectoparasitism is manifested by the blood-sucking habits of many adult members of this group and by the subcutaneous invasion by larvae of certain members of the Muscidae, Calliphoridae, Sarcophagidae, Gasterophilidae, Oestridae, and Cuterebridae. Endoparasitism is manifested by the larvae of species that cause internal myiasis.

Myiasis

Not only do adults of hematophagous flies bite humans and beasts, but the larvae of various species invade vertebrates. Such infestations are referred to as **myiasis**. If the larvae, or maggots, invade the subdermal regions of their host, the condition is known as **cutaneous myiasis**. If the site of invasion is the host's alimentary tract, the condition is known as **gastrointestinal myiasis**. Similarly, **ocular myiasis**, **urinary myiasis**, and **nasopharyngeal myiasis** are known to occur. With only a few exceptions, the myiasis-causing flies belong to the suborder Cyclorrapha, primarily the Schizophora Group. Myiasis in humans and domestic animals has been authoritatively reviewed by Zumpt (1965) and Harwood and James (1979).

Among the myiasis-causing flies, those which usually invade living tissues are known as **primary invaders**. Eggs of these flies are commonly deposited on fresh wounds—even rather insignificant sores such as those caused by tick bites. The hatching larvae actively feed on living tissues while they burrow into the skin. Adults of these flies are attracted to wounds by the smell of blood. Species that invade decomposing tissues of dead animals are referred to as **secondary invaders**. These maggots, however, attack healthy living tissues when dead tissues are not available.

Among the Muscidae, Calliphoridae, and Sarcophagidae, collectively known as the **Muscoid Group**, myiasis is manifested in four ways: (1) by the sucking of blood; (2) by invasion of wounds and natural cavities, such as the ears, nose, and eyes; (3) by invasion by skin, causing boils in which the larvae dwell; and (4) by living in the stomach or rectum of the host.

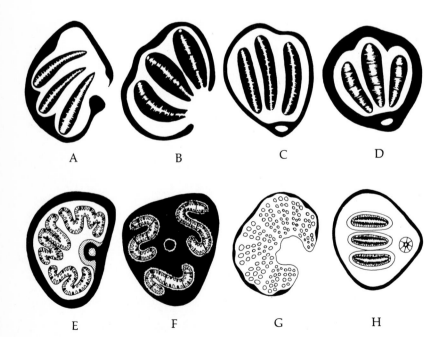

Fig. 19.14. Posterior spiracles of myiasis-causing maggots. (Courtesy of U.S. Naval Medical School, Bethesda, Maryland.) **A.** *Sarcophaga* sp. **B.** *Cochliomyia macellaria*; **C.** *Lucilia sericata*; **D.** *Calliphora erythrocephala*; **E.** *Musca domestica*; **F.** *Stomoxys calcitrans*; **G.** *Hypoderma lineatum*; **H.** *Auchmeromyia luteola*.

Myiasis of small laboratory animals is generally lethal. During primary invasions, highly toxic substances are produced by the growing larvae. These toxins, coupled with tissue destruction, are extremely damaging. Furthermore, the maggots of screwworms are invariably accompanied by a nonpathogenic, although proteolytic, species of bacterium, *Proteus chandleri*, which is practically in pure culture in the wounds a day or so after invasion. Hosts develop an immunity against the toxic secretions of screwworms, but not against the larvae themselves. Although *P. chandleri* is usually present, the toxins of maggots kill most of the bacteria. Thus, the wounds are usually not purulent until the larvae exit, after which purulence ensues, accompanied by production of a copious serosanguineous exudate.

In cases of human myiasis, particularly those caused by the invasion of screwworms up the nose, the nasopharynx, and associated sinuses, death is not uncommon. The development of nervous conditions such as delirium, convulsions, visual disturbances, and loss of speech, accompanied by great pain, generally occurs prior to death if the parasites are not removed.

Myiasis-causing maggots (larvae) are readily recognized by the shape of their posterior spiracles (Fig. 19.14).

The following discussion is a survey of the more commonly encountered genera of the schizophoran families that are of interest to parasitologists.

FAMILY MUSCIDAE (TYPICAL FLIES)

The family Muscidae includes the well-known common housefly, *Musca domestica*, which is worldwide in distribution. Adults are nonparasitic but are household pests. However, they are mechanical transmitters, by their contaminated feet, of such diseases as typhoid fever (etiologic agent, *Salmonella typhi*), cholera (etiologic agent, *Vibrio comma*), and yaws (etiologic agent, *Treponema pertenue*). The volumes by Greenberg (1971, 1973) should be consulted by those interested in the role of flies as transmitters of pathogens.

M. domestica breeds in decaying organic materials, including animal excreta. The females are oviparous, producing 120 to 150 eggs in a batch, with 5 to 20

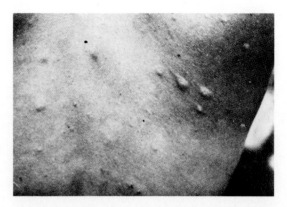

Fig. 19.16. Welts on back from tsetse fly bites. (Armed Forces Institute of Pathology, Negative No. 75-5783.)

batches oviposited during the life span. The larvae have been recovered from human vomitus and feces and passed out in urine (Fig. 19.15). Thus, this fly is capable of causing gastrointestinal and urinary myiasis. Furthermore, dermal myiasis, especially associated with ulcerated wounds, resulting from *M. domestica* invasions are known.

Because the parasitic way of life is not necessary for this larva, these instances of myiasis are referred to as **accidental myiasis**. In gastrointestinal myiasis due to *M. domestica*, the infections are usually traceable to the ingestion of eggs or young larvae.

Genus *Glossina*. This genus includes the tsetse flies, which are important not only as the vectors for certain trypanosomes (p. 130), but also because both the males and females are hematophagous. Their bites result in large welts (Fig. 19.16). Although *Glossina* (Fig. 19.17) was at one time widely distributed, it is

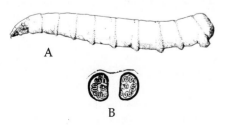

Fig. 19.15. *Musca domestica.* **A.** Larva. **B.** Posterior spiracles of larva. (Both figures after James, 1947.)

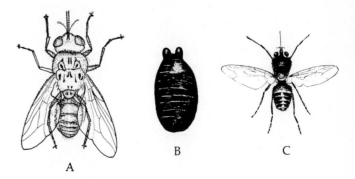

Fig. 19.17. *Glossina*. A. *G. palpalis* in resting position. (Redrawn after Matheson, 1950.) **B.** Pupa of *Glossina* showing two posterior lobes. **C.** *G. morsitans.*

now limited to continental Africa south of the Tropic of Cancer.

Life Cycle of Glossina.   The life cycle of *Glossina* spp. follows the basic pattern of the cyclorrhaphan flies, but a few striking differences exist. The female gives birth to fully developed living larvae. Actually, the larviposited form is the fourth larval instar, for the *in utero* development involves not only embryonic development but also three larval stages. The *in utero* larvae feed on the secretions of specialized glands, commonly referred to as **milk glands**. The fourth-stage larvae are deposited singly at intervals of 10 to 12 days, a total of 8 to 10 larvae being deposited by a single female during her life span. Larvipositing females require a blood meal between extrusions of the larvae, and development of the larvae *in utero* is not completed until blood is ingested by the mother. The extruded larvae are off-white to pale yellow and are not capable of the usual wormlike movements. Instead, they move and burrow by peristalic movement and longitudinal contractions. They are generally larviposited at the base of shrubs and other vegetation, where the soil is damp and loose.

The larvae burrow into the soil and undergo pupation. This usually occurs within 1 hour after birth. The pupa within the puparium is ovoid and brownish black, and bears two characteristic posterior lobes (Fig. 19.17). The usual pupal period lasts from three to four weeks. Soil moisture and temperature are important influencing factors. The form escaping from the puparium is the adult. It is brownish and bears the distinguish-

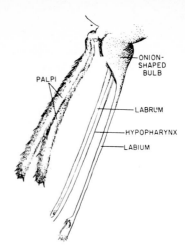

Fig. 19.19.   Mouthparts of *Glossina.* × 17. (After Harwood and James, 1979; with permission of Macmillan.)

ing characteristics depicted in Figures 19.17, 19.18, and 19.19.

Glossina *Species.*   Newstead (1924) has provided descriptions of the 20 species of *Glossina.* The most prominent ones are *G. palpalis,* the major vector of Gambian trypanosomiasis (Fig. 19.17); *G. morsitans,* the major vector of Rhodesian trypanosomiasis (Fig. 19.17); and *G. swynnertoni,* another vector of Rhodesian trypanosomiasis. Although these flies feed on an array of vertebrate hosts, they definitely demonstrate preference for some. For example, *G. palpalis* favors crocodiles, monitor lizards, and other reptiles, and *G. morsitans* prefer pigs and warthogs.

Genus *Stomoxys.*   This genus has been reviewed by Séguy (1935). Members are grayish flies of medium size resembling *Musca domestica* but possessing long, slender horny proboscides that are spear shaped and antennae that are pectinate. The most common species is *S. calcitrans,* the stable fly (Fig. 19.20). This is a vicious bloodsucker that is commonly found in barns and feeds on domestic animals, especially horses. It also invades houses and feeds on human blood. Both sexes are bloodfeeders.

The life history is typically dipteran. The females oviposit from 23 to 100 eggs (average 25–50) at each of four to five layings. The bites of large numbers of flies result in a significant loss of blood and body weight. Instances of cutaneous and intestinal myiasis caused by the larvae are known.

Genus *Haematobia.*   The so-called horn flies belong to the genus *Haematobia.* The most prevalent species in North America is *H. stimulans* (Fig. 19.20) which also occurs in Europe. This fly, a vicious attacker of cattle and other stable animals, is about 4 mm long and resembles *Stomoxys* except that its labium is heavier, its palpi are long, approximating the length of the proboscis, and the antennal aristae are

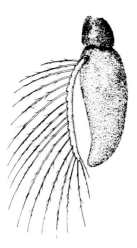

Fig. 19.18.   Antenna of *Glossina* **showing arista with branched hairs.** Much enlarged. (After Harwood and James, 1979; with permission of Macmillan.)

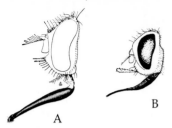

Fig. 19.20. Diptera. A. Lateral view of head of *Stomoxys calcitrans*. (After James, 1947.) **B.** Lateral view of head of *Haematobia stimulans* showing long palpi. (Redrawn after Eldridge in James and Harwood, 1969.)

plumose dorsally. It is called the horn fly because it commonly rests on the horns of cattle, particularly at night. It does leave the host when not feeding. The reddish brown eggs are 1.3–1.5 mm long and, like those of *Stomoxys calcitrans*, are deposited in cow manure, where the larvae are found.

FAMILY CALLIPHORIDAE (BLOWFLIES)

The Calliphoridae includes the so-called blowflies and bluebottle flies. The maggots of many of these flies cause myiasis in domestic animals and humans, whereas still others, such as the cluster fly, *Pollenia rudis*, are parasitic on oligochaetes. The adult flies are not hematophagous. Members of the Calliphoridae are important not only because they cause myiasis, but also as transmitters of microorganisms carried on the hairs of their legs. Hall (1948) has contributed a taxonomic monograph of the blowflies of North America.

Genus *Callitroga*. This genus includes the screwworms. The primary screwworm, *Callitroga hominivorax*, is the most prominent American species (Fig. 19.21). This metallic green fly, with three dark stripes down its thorax and a yellow head and reddish eyes, is distributed from the southern United States to northern Chile. The larva is an obligatory parasite. The females oviposit on fresh wounds on humans and animals. The eggs, ranging from 10 to several hundred in number, are glued in mats along the edge of the open wound. When the larvae hatch, in 1 to 20 hours, they actively burrow into the wound and ingest the host's tissues.

Continued penetration via engorgement results in deep pockets. The posterior spiracles of the larvae are directed toward the opening of the pocket so that breathing is made possible (Fig. 19.21). On reaching maturity in 4 to 8 days, these flesh-eating maggots drop off and pupation takes place in soil. Heavy infestations by screwworms can lead to the death of the host, especially if the larvae invade regions of the nose, eyes, and ears. Human infections by the larvae of *C. hominivorax* are known.

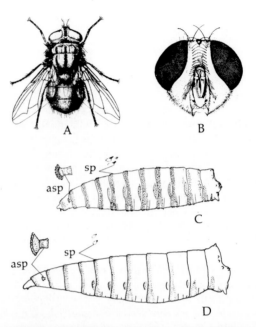

Fig. 19.21. Stages in the life cycle of *Callitroga*. A. Adult female *C. hominivorax*. **B.** Frontal view of head of female *C. hominivorax*. (After Cushman in James, 1947.) **C.** Lateral view of third-stage larva of *C. hominivorax*. **D.** Lateral view of third-stage larva of *C. macellaria*. (Redrawn after Laake *et al.*, 1936.) asp, anterior spiracles; sp, spines.

The secondary screwworm fly, *Callitroga macellaria* (Fig. 19.21), is characterized by its predominantly orange head and by its middorsal longitudinal stripe on the thorax, which does not extend over the scutellum. This species usually does not cause myiasis, being primarily a scavenger in soil or on carrion, but it can become a secondary wound invader. If secondary invasion does occur, the larvae, measuring up to 17 mm in length, do not form pockets as does *C. hominivorax*; instead, they migrate around the wound in the hair.

Genus *Chrysomya*. The larvae of *Chrysomya* spp. may be secondary invaders of sores, or they may cause internal myiasis. These larvae are parasites of mammals, including humans.

Genus *Phormia*. This genus includes the black blowfly, *Phormia regina*, which is characterized by its blackish green or olive-green body, black head, and orange hair on the mesothorax, and by a body length of 7–9 mm. Primarily a saprophagous species, *P. regina* breeds in animal carcasses. However, the larvae do invade healthy tissues and produce myiasis, especially in the southwestern United States.

Other Calliphorid Genera. The larvae of certain species of *Protophormia*, *Cordylobia*, the Tumbu fly (Fig. 19.22), and *Lucilia* can produce cutaneous myiasis.

Calliphora includes two fairly common species, *C. vomitoria* and *C. vicina*. The larvae of both of these calliphorans are flesh ingestors. *Calliphora vomitoria* has black genae clothed with golden-red hairs (Fig. 19.23), whereas *C. vicina* has fulvous genae clothed with black hairs. The eggs of both species are deposited on carrion and rarely on wounds. These hatch in 6 to 48 hours, and the growing larvae feed on flesh for 3 to 9 days. At the end of that period, the larvae are fully grown and drop off. They bury themselves in loose soil or debris and undergo a 2- to 7-day prepupal period. The duration of pupation varies with the environmental temperature, lasting from 10 to 17 days, after which adults emerge. Adults live for approximately 35 days.

Calliphorid Host Specificity. The calliphorid flies as a rule are not attracted to any specific host, nor do the maggots show any preference for any specific sites on their hosts, although the more readily accessible sites are more frequently invaded. The striking exception to the rule appears to be *Lucilia bufonivora*, the larvae of which feed only on living amphibian tissue, especially on toads. In *L. bufonivora*, the female fly deposits her eggs on the surface of the host's skin, usually on the back, and here the eggs hatch within 24 hours. It is believed that hatching is induced by the chemical action of the host's skin glands.

The first-stage larvae migrate along the toad's back until they reach the eyes. When the toad blinks its eyelids, the larvae are carried to the openings of the lacrimal ducts. From these ducts the larvae migrate to the lacrimal glands and from there to the nasal cavity.

Fig. 19.23. *Calliphora vomitoria.* (After James and Harwood, 1969.)

At this site, the second larval stage is reached. The second-stage larvae feed on the host's living tissues and completely destroy the cartilaginous septum of the nasal cavity. The large cavity produced in this way is filled with crawling maggots, which eventually drop out to undergo pupation in soil.

When *L. bufonivora* females are placed among frogs, toads, salamanders, and aquatic newts, they oviposit almost exclusively on toads. This indicates that the fly is attracted by the specific odor of toads and has become adapted both ecologically and physiologically, since the larvae feed on living tissues. The other species of *Lucilia* feed on purulent tissues.

Calliphorid Parasites of Invertebrates. A number of larval calliphorid flies parasitize invertebrates. In these species an interesting form of larval migration occurs. This is strikingly demonstrated by the larva of the cluster fly, *Pollenia rudis.* Adult female flies deposit their eggs in the soil, and the hatching first-stage larvae penetrate the male genital pore of the earthworm *Allolobophora chlorotica.* These maggots enter either the seminal vesicles or the coelomic cavity in somites IX to XII. In this location, the maggots remain motionless for at least eight months and are gradually encapsulated by a wall comprised of the host's connective tissue and blood cells including numerous phagocytes.

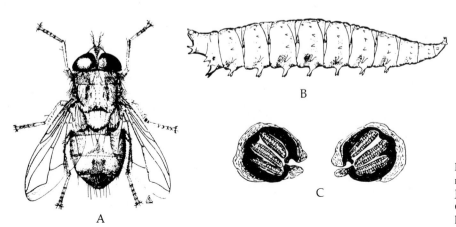

Fig. 19.22. *Cordylobia.* **A.** Adult male of *C. anthropophaga.* (After James, 1947.) **B.** Larva of *C. albiceps.* **C.** Posterior spiracles of *C. albiceps* larva. (**B** and **C**, after Smit, 1931.)

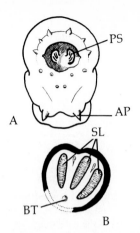

Fig. 19.24.   Posterior spiracles of sarcophagid flies. A. Posterior view of sarcophagid larva. **B.** Single posterior spiracle. (Redrawn after James, 1947.) AP, anal protuberances; BT, buttons; PS, posterior spiracles; SL, slits.

In time, all encapsulated larvae are destroyed except for one. If only one larva enters the earthworm, it is not destroyed. The reason for this selective destruction remains undetermined. The single remaining larva becomes active when the end of its first larval instar period is reached. It breaks through the surrounding tissues and migrates toward the anterior end of the host. This is an unusual sort of migration, for the larva migrates backward, with its posterior end directed forward.

Upon reaching the host's prostomial region, the parasite breaks through the body wall, extrudes its spiracular openings to obtain oxygen from the air, and begins feeding on the surrounding host tissues. As the larva increases in size, the tissues of the host that enclose it gradually disintegrate, exposing the maggot. In time, most of the host's tissues are either ingested or disintegrated, and the larva escapes into the soil and pupates. The third-stage larva of *P. rudis* is entirely saprophagous, feeding on partly destroyed and decomposed host tissue. Larvae at this stage of

development can be maintained on decomposed fragments of earthworms.

Melinda cognata, a calliphorid parasite of the snail *Helicella virgata*, possesses a life cycle similar to *Pollenia rudis* in that the fly larva feeds first on the living tissues of its host and finally devours the dead and decomposed snail. It differs from *P. rudis* in that the adult females deposit their eggs in the snail. It is apparent that both *Pollenia rudis* and *Melinda cognata* are parasites that border on being predators.

FAMILY SARCOPHAGIDAE (FLESH FLIES)

Members of Sarcophagidae, the flesh flies, like the Calliphoridae, are parasitic during their larval stages. Maggots of sarcophagid flies, however, can be distinguished by a girdle of minute spines on each abdominal segment, well-developed, curved mouthparts, and posterior stigmal plates located in a deep cavity. There are three slits on each stigmal plate that are vertically parallel (Fig. 19.24).

The three major genera of the Sarcophagidae are discussed here—*Sarcophaga*, *Wohlfahrtia*, and *Titanogrypha*. Members of these genera can be distinguished by the shape of their antennae. In *Sarcophaga*, the aristae are long and plumose; in *Wohlfahrtia*, they are short and pubescent; and in *Titanogrypha*, they are short and plumose (Fig. 19.25).

The sarcophagid flies are larviparous. The eggs are incubated within a large uterus. Some species deposit their larvae only on dead animals, others on animal wounds, and still others on healthy animals.

Genus *Sarcophaga*.

Sarcophaga haemorrhoidalis.   This species is common throughout North, Central, and South America, Britain, continental Europe, northern China, India, Africa, and Australia (Fig. 19.26). As with all the sarcophagid flies, this species is larviparous (vivipar-

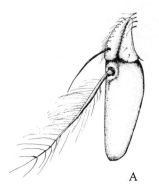
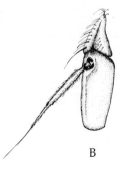
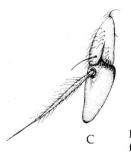

Fig. 19.25.   Antennae of sarcophagid flies. A. *Sarcophaga bullata.* **B.** *Wohlfahrtia magnifica.* **C.** *Titanogrypha alata.* (All figures after James, 1947.)

ous). Larvae are deposited on carrion, feces, rotting meat, and similar breeding spots. These flies are usually scavengers, although they are capable of becoming myiasis-causing organisms by invading wounds or being ingested in contaminated food by the host. The hosts are various mammals, including humans.

Other Sarcophaga *Species.* Other myiasis-causing species of *Sarcophaga* include *S. lambens* in the southern United States and Central and South America; *S. chrysostoma* in Mexico and Central and South America; *S. carnaria* in Europe; and *S. plinthopyga* in the New World (Fig. 19.26). The larva of *S. plinthopyga*, unlike the others, is primarily parasitic in the body of insects. However, cases of human cutaneous myiasis caused by this maggot are known.

The life cycles of these species are typically dipteran.

Genus *Wohlfahrtia*. This genus includes several species whose larvae produce myiasis. *Wohlfahrtia magnifica* occurs in northern Europe, the Soviet Union, the European and African Mediterranean countries, and eastern Africa (Fig. 19.26). As in all sarcophagid species that cause wound myiasis, the adult females deposit larvae in wounds on humans and other mammals, initiating the parasitic condition.

Wohlfahrtia vigil is the most important North American representative (Fig. 19.26). This fly displays a variation of the penetration mechanism employed in initiating cutaneous myiasis in that, unlike *W. magnifica* and the other species, the females are attached to young mammals, including humans. Furthermore, larvae of *W. vigil* are deposited on healthy skin rather than in wounds, and these maggots thus penetrate unruptured skin. Once established in subdermal locations, these larvae cause small abscesslike lesions that measure 6–20 mm in diameter. Even if a larva is

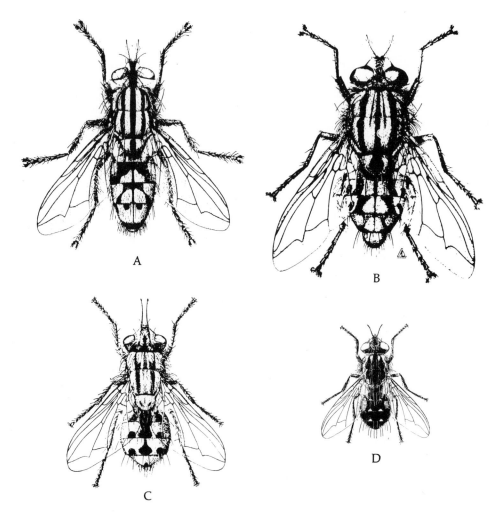

Fig. 19.26. **Sarcophagid flies. A.** Adult male *Sarcophaga haemorrhoidalis.* **B.** Adult male *Sarcophaga plinthopyga.* **C.** Adult female *Wohlfahrtia magnifica.* **D.** Adult female *Wohlfahrtia vigil.* (All figures after James, 1947.)

unsuccessful at penetrating the skin, it causes a marked dermal irritation.

Genus *Titanogrypha.* Members of *Titanogrypha* resemble small *Sarcophaga*. *Titanogrypha alata*, a species found in Florida, southern Texas, and Cuba, is capable of infesting wounds of various mammals, including humans.

Sarcophagid Parasites of Invertebrates. A number of species of larval sarcophagid flies have been reported associated with invertebrate hosts. *Sarcophaga nigriventris*, for example, occurs in land snails, grasshoppers, and beetles; *S. haemorhea* in land snails; *S. scoparia* in gypsy moth pupae; and *S. albicaps* and *S. aratrix* in various Lepidoptera and Coleoptera (van Emden, 1954). When an invertebrate is parasitized, it usually does not live for long. For example, when *Sarcophaga kellyi* deposits larvae on the wings of a locust in flight, the host falls to the ground, and the maggots enter its body through an articulating membrane at the base of the wing. The life span of the locust is limited after that. The larvae of some species, such as *Sarcophaga destructor*, can only enter the bodies of freshly molted locusts.

FAMILY CUTEREBRIDAE (ROBUST BOTFLIES)

The cuterebrids, commonly called robust botflies or warble flies, are another cyclorrhaphan family the members of which are parasitic only during larval stages and thus cause myiasis. All cuterebrids are obligatory parasites as larvae. The major genera of this family are *Cuterebra*, *Dermatobia*, and *Cephenemyia*.

Genus *Cuterebra.* Species of *Cuterebra* are limited to North America. Larvae of most of these species are parasitic on rodents and lagomorphs. The adults are among the largest muscoid flies known, measuring 20 mm or more in length. The bodies are bumblebeelike, and the abdomen is completely, or mostly, black or blue. The thorax is usually covered with dense hair. The larva of *C. buccata* is a common dermal parasite of rabbits.

Cuterebra spp. oviposit near the entrance of the burrows of their hosts, and the hatching larvae attach themselves and proceed to burrow into the host's skin. In all known instances the eggs are oviposited in the abodes of the hosts rather than directly on the hosts. These larvae form cystlike pockets in the subdermal zone that communicate with the exterior via a pore (Fig. 19.27). Human cases of nasal and dermal myiasis caused by these larvae are known.

The infective larvae possess integuments beset with spines (Fig. 19.28). The older specimens are blackish in color. In addition to causing dermal, nasal, and pharyngeal myiasis, these larvae can also invade the cranial cavity.

Other common species of *Cuterebra* include *C. americana* and *C. lepivora* on the western rabbit, *Syvilagus audubonii sanctidiegi*, and *C. emasculator* on the eastern striped chipmunk, *Tamias striatus*. *Cuterebra*

Fig. 19.27. Myiasis. Photograph showing three third-stage larvae of *Cuterebra tenebrosa* in warbles on a bushy-tailed rat, *Neotoma cinerea*. (After C. R. Baird in Harwood and James, 1979; with permission of Macmillan.)

emasculator can destroy its host's testes, causing parasitic castration. Bennett (1955) has reported that *Cuterebra* spp. are almost specific for particular hosts.

Genus *Dermatobia.* This genus includes *D. hominis*, the human botfly (Fig. 19.29). This large

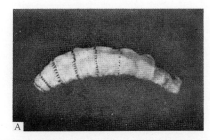

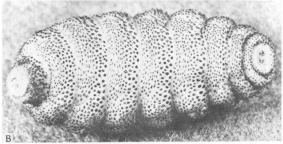

Fig. 19.28. Myiasis-causing dipteran larvae. A. First-stage larva of *Cuterebra* sp. from a case of human myiasis. (After Beachley and Bishopp, 1942.) **B.** Third-stage larva of *Cuterebra jellisoni*, a parasite of cottontail rabbits and jack rabbits. (After A. D. Akre in R. F. Harwood and M. T. James, 1979; with permission of Macmillan.)

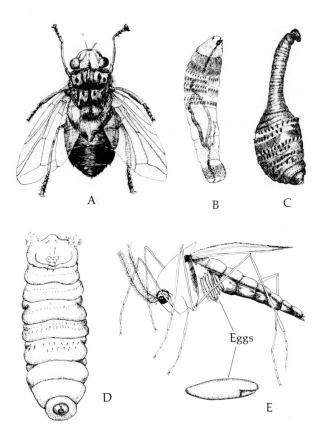

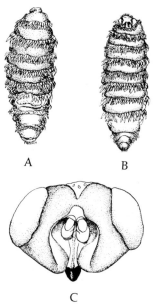

When the carrier feeds on the mammalian host—humans, cattle, dogs, or even rarely birds (*D. hominis* seldom parasitizes members of the Equidae)—the first-stage larvae, which develop within the eggshells (chorion) in 5 to 15 days, escape and enter the host through the bite of the carrier or penetrate the host's skin via a pore or fold.

Each larva penetrates independently and forms a boil-like pouch that communicates with the exterior through a small opening. Larval development within the pocket ensues, and as with other myiasis-causing dipterans, there are three larval instars over a period of five to ten weeks. The mature larva ruptures out of the pocket by enlarging the aperture and falls to the ground to pupate.

When gravid females cannot find a carrier, the eggs are oviposited on vegetation. The larvae hatching from these eggs will survive if they enter a host; however, in most instances they perish. Infestation by *D. hominis* larvae presents a serious problem in tropical and subtropical areas, where cattle and dogs suffering from heavy infestations usually die.

Genus *Cephenemyia*. *Cephenemyia* larvae produce myiasis in deer and reindeer. The maggots most commonly form pockets in the skin of the head and neck (Fig. 19.30). Although *Cephenemyia* was previously suspected of infesting humans, more recent evidence suggests that it does not attack man. Adults of members of this genus are readily identified in that their head bears a narrow epistoma (Fig. 19.30).

Fig. 19.29. Stages in the life cycle of *Dermatobia hominis*. A. Adult female. (Redrawn after James, 1947.) **B.** First-stage larva. (Redrawn after Newstead and Potts, 1925.) **C.** Second-stage larva. (Redrawn after James, 1947.) **D.** Third-stage larva, ventral view. (Redrawn after James, 1947.) **E.** The mosquito, *Psorophora* sp., with *D. hominis* eggs attached. (Redrawn after James, 1947.)

species measures 12 mm in length and possesses a yellow head, dull blue thorax, and metallic blue abdomen. The antennal aristae are pectinate. The three parasitic larval stages are somewhat different, the mature, elongate oval larva being spinose (Fig. 19.29). This larva possesses posterior spiracles, each consisting of three slits located in two cavities, and anterior spiracles that are prominent and flowerlike.

Life Cycle of Dermatobia hominis. The life cycle of this fly is unique in that the gravid females employ a fascinating method of engaging other insects, mainly dipterans, to transport their eggs to the host. This mechanism involves the capturing of other dipterans, usually *Psorophora* spp. (Fig. 19.29) and *Stomoxys* spp., and ovipositing and glueing the *Dermatobia* eggs onto the abdomen of the carrier.

Fig. 19.30. *Cephenemyia*. A. Dorsal view of larva of *Cephenemyia auribarbis*. **B.** Ventral view of larva of *C. auribarbis*. (**A** and **B**, redrawn after Cameron, 1937.) **C.** Frontal view of head of adult female *C. trompe*. (Redrawn after James, 1947.)

Gasterophilidae includes the horse botflies, all of which belong to the genus *Gasterophilus*. The most common species are *G. intestinalis* (Fig. 19.31), *G. haemorrhoidalis*, *G. inermis*, and *G. pecorum*. The first three of these species are cosmopolitan, being found in North America in addition to other parts of the world, whereas *G. percorum* is primarily a European and African species.

Life Cycle of Gasterophilus. Members of the family Equidae are the preferred hosts for the parasitic *Gasterophilus* larvae. Unlike other myiasis-causing larvae, these larvae complete their development in the stomach and intestine of the horse host (thus causing gastrointestinal myiasis) as part of the normal pattern rather than as the result of accidental ingestion (Fig. 19.32). Specific differences occur in the life cycle patterns of the species, but the basic pattern is the same.

The life span of the nonfeeding adults is short. Gravid females deposit their eggs on the hair of the host. These eggs become attached to the hair individually (Fig. 19.33). The sites of egg deposition differ among the species. The light yellow eggs of *G. intestinalis* are deposited all over the host's body but especially on the inner surfaces of the knees; those of *G. haemorrhoidalis* are attached to the fine hairs around the lips; those of *G. nasalis* are attached to hairs under the jaws; those of *G. inermis* to the hairs of the cheeks; and those of *G. pecorum* are deposited not on the host, but on the host's food.

The number of eggs produced by a single female is large. A single female of *G. intestinalis* may deposit as many as 900 eggs in 3 hours. The closer the eggs are laid to the horse's mouth, the smaller their number. For example, *G. haemorrhoidalis* females deposit a mean of 160 to 200 eggs, while females of *G. pecorum* deposit a mean of 2300 to 2500 eggs. This may reflect an adaptation whereby those closest to the mouth have the greatest chance for infecting the host and hence fewer eggs are required to ensure successful establishment.

The incubation period and physical stimuli required for hatching vary among the species. In the case of *Gasterophilus intestinalis*, the eggs hatch in 7 to 14 days in warm summer temperatures, if friction and moisture are provided by the horse's tongue. Similarly, eggs of *G. haemorrhoidalis* require moisture provided by saliva. Hatching under these conditions occurs in only 2 to 4 days. The eggs of *G. nasalis* do not require moisture and hatch in 4 to 5 days.

The modes of entrance of the first-stage larvae of some species differ. When the equine host licks the eggs of *G. intestinalis*, some of the hatched larvae become attached to the lips. These larvae penetrate the mucous membranes of the lips and tongue by utilizing an armature to burrow tunnel-like canals in

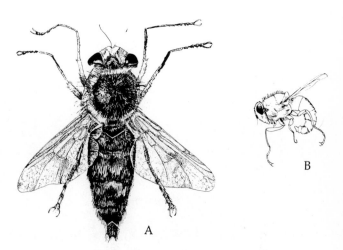

Fig. 19.31. *Gasterophilus intestinalis.* **A.** Dorsal view. **B.** Side view. (Redrawn after James, 1947.)

the subepithelia (Fig. 19.33). From the mouth, the larvae soon migrate into the alimentary canal—the stomach or small intestine—and continue their larval development until the third larval instar is attained. The larvae remain quiescent until the following spring or early summer, when they pass from the host in feces and undergo pupation in soil. The emerging pupae each measures 1.5–2 cm in length (Fig. 19.33).

Fig. 19.32. Gastric myiasis. Photograph showing larvae of *Gasterophilus* attached to lining of stomach of horse. Lesions on the gastric lining indicate sites where larvae have become detached. (After R. D. Akre in Harwood and James, 1979; with permission of Macmillan.)

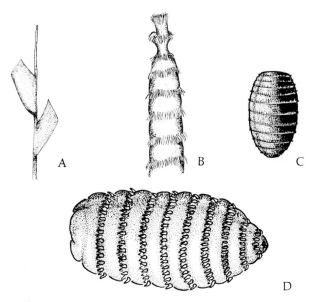

Fig. 19.33. Stages during the development of *Gasterophilus*. A. Eggs attached to host's hair. **B.** Armature at anterior end of *G. intestinalis* larva. **C.** Pupa of *G. intestinalis*. **D.** Mature larva of *G. intestinalis*. (**D**, after Harwood and James, 1979; with permission of Macmillan.)

In *G. haemorrhoidalis*, the path of migration to the stomach parallels that of *G. intestinalis*—that is, burrowing in the subepithelial zone. However, larvae of *G. nasalis* crawl into the mouth and form typical pockets between the molar teeth. Within such pockets

the first stadium is completed, after which the larvae migrate to the stomach or duodenum, where they are commonly found in clusters. Pupation occurs in the early summer when the third larval instars pass from the host in feces.

The path of migration of *G. inermis* is unique. The hatching first larval instars penetrate the host's skin near the cheeks and migrate into the mouth by crawling through the dermis. Once within the buccal cavity, they molt underneath the mucous membrane and then proceed not to the stomach, but to the rectum, where they mature and drop out with the onset of warm weather to pupate in the soil.

The larvae of *G. pecorum* hatch from eggs that are deposited in the host's food. When the larvae come in contact with the horse's lips, they burrow into the mucous membrane and from there to the stomach and rectum.

When humans come in close contact with horses, they can be infected by three species of *Gasterophilus*—*G. intestinalis*, *G. haemorrhoidalis*, and *G. nasalis*. In such cases, the first larval instars penetrate the skin and migrate in a serpentine, random fashion between the layers of the epidermis, causing raised red lines along the paths of migration. The larvae eventually die rather than reach the alimentary tract. This pathology resulting from the parasite's migration is referred to as **myiasis linearis**.

Gastric and intestinal myiasis due to *Gasterophilus* spp. is injurious to the host through (1) blockage of the normal passage of chyme through the alimentary tract, (2) irritation and mechanical injury to the mucous lining of the stomach by the oral hooklets of the maggots, (3) irritation to the intestine, rectum, and anus as the maggots pass posteriorly, and (4) deprivation of some nutrients from the host. In cases of heavy infestation, death commonly occurs.

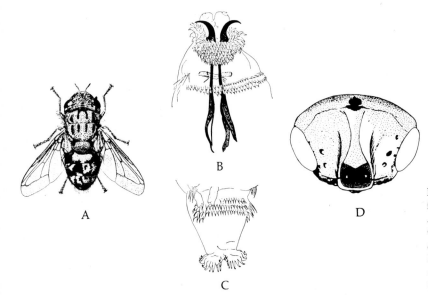

Fig. 19.34. Oestrid flies. A. Adult female *Oestrus ovis*. (After James, 1947.) **B.** Anterior end of first-stage larva of *O. ovis*. **C.** Posterior end of first-stage larva of *O. ovis*. (**B** and **C**, after Galliard, 1934.) **D.** Head of female *Rhinoestrus purpureus* showing rudimentary proboscis and palpi. (After James, 1947.)

Members of this family are commonly referred to as head maggots because the larvae invade and parasitize the nasal cavities and frontal and maxillary sinuses of ruminants, particularly sheep and goats, horses, antelopes, and other hoofed animals (Fig. 19.34). Occasionally, first larval instars parasitize humans. The larvae are robust and only slightly tapered anteriorly. The third larval instar, which is the most commonly encountered parasitic form, possesses a single pair of oral hooks, inconspicuous or non-existent anterior spiracles, and posterior spiracles modified as two sclerotized plates with numerous perforations.

The two prominent genera of the Oestridae are *Oestrus* and *Rhinoestrus*.

Oestrus ovis is globally distributed (Fig. 19.34). The natural host for the parasitic larvae is the sheep; however, human infections are known. The adults are nonfeeders. The females are larviparous and deposit larvae on the nostrils of sheep and goats during the summer and autumn months. A single female may give birth to as many as 60 larvae per hour. These first-stage larvae actively migrate up the host's nostrils and become attached to the mucous membranes lining the nasal and frontal sinuses. In these locations, development continues, passing through two additional larval stadia. During the following spring, the mature larvae, measuring 20–30 mm in length, drop out of the host, usually aided by a sneeze, onto the ground, where they burrow into the soil and pupate. The pupal period lasts for 16 days, after which adults develop.

Rhinoestrus purpureus, the Russian bot, is endemic to France, Spain, Italy, eastern Europe, northern Africa, the Soviet Union, northern China, India, and Central and South Africa (Fig. 19.34). Normal hosts are members of the horse family. The adults, like those of *Oestrus ovis*, are nonfeeders. The females larviposit first larval instars onto the eyes and nostrils of hosts. A single female can larviposit 700 to 800 larvae, in groups of 8 to 40 at a time. Larviposition usually occurs in the fall, and the two additional larval instars occur in the nasal and frontal sinuses of the host. With the onset of spring, these larvae escape, also aided by sneezes, and pupate in soil. If the larvae are deposited in the eye, severe conjunctivitis ensues, often leading to blindness. Human infections, although rare, are known.

FAMILY CHLOROPIDAE (FRIT FLIES, OR EYE FLIES)

Members of the Chloropidae, commonly known as frit flies or eye flies, are not capable of piercing the integument of their mammalian hosts. They are attracted to open wounds, sebaceous materials, and secretions around the eyes. Once attracted to such sites, these minute flies hold onto the edge of the

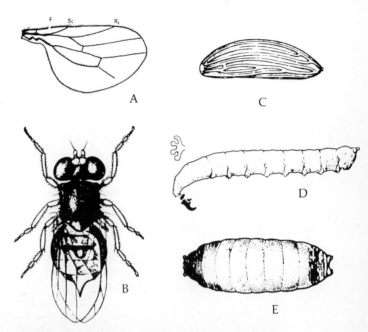

Fig. 19.35.   Chloropidae. A. Wing of *Hippelates*. **B.** Adult *Hippelates pusio*. (Redrawn after Hall, 1932.) **C.** Egg of *H. pusio*. **D.** Larva of *H. pusio* showing cephalopharyngeal skeleton and anterior spiracular process. **E.** Pupa of *H. pusio* (**C–E**, after Herms and Burgess, 1930.)

sores or to the conjunctival epithelium and make minute lesions around the wound or eye with their spinose labella. The two important genera are *Hippelates* and *Siphunculina*.

Hippelates pusio, the eye gnat, is small (about 2 mm long) and bears yellowish legs, eyes, and antennae (Fig. 19.35). This species is found in parts of California, Florida, and other southern states. It is most commonly encountered in areas of sandy or mucky soil undergoing cultivation. The curved and fluted eggs of *H. pusio* are deposited on freshly cultivated soil of decaying organic debris (Fig. 19.35). Such eggs measure about 0.5 mm in length.

After an incubation period of approximately 3 days, the larvae hatch and are free living, ingesting well-aerated organic materials (Fig. 19.35). The larval period usually lasts for 11 days, after which the larvae migrate close to the surface of the incubation medium and pupate (Fig. 19.35). The pupal stadium lasts approximately 5 days.

The direct injury caused by this and related flies is not serious. However, these dipterans are strongly suspected of being able to spread the sore-eye, or pinkeye, disease, a form of contagious conjunctivitis

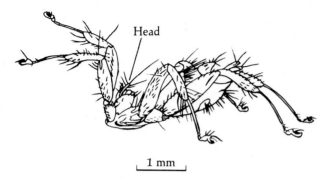

Fig. 19.36.   A nycteribiid. *Stylidia biarticulata.* (After Theodor, 1967.)

of humans and animals. Furthermore, evidence suggests that these flies can be transmitters of *Treponema pertenue*, the yaws-causing spirochaete.

Siphunculina funicola is endemic to India, Sri Lanka, and the East Indies. This fly is attracted to eyes and is strongly suspected of transmitting conjunctivitis. Graham-Smith (1930) has critically reviewed the literature up to that time dealing with Chloropidae and conjunctivitis.

FAMILY NYCTERIBIIDAE (SPIDERLIKE BAT FLIES)

The nycteribiid flies, or bat flies, are parasites of bats in the tropics and subtropics, primarily in the Old World. There are about 250 described species. The development of these flies is unique among insects. The larvae hatch within the body of the mother and grow there to maturity, feeding on the secretions of special milk glands. Mature larvae are deposited in the host's perch, from whence they later crawl onto the host.

There appears to be a certain degree of host specificity among these wingless flies. A single species may infest bats of several genera, but these genera are usually in the same family. Furthermore, the same species of fly is usually not found on both fruit- and insect-eating bats. For example, members of *Cyclopodia* are confined to frugivorous bats of the family Pteropidae. The host-parasite relationship between the bat and the fly has become so intimate that individuals removed from their hosts cannot be induced to feed and rarely live beyond 12 hours. In Britain, *Stylidia biarticulata* is found only on horseshoe bats, *Rhinolophus* spp. (Fig. 19.36).

Nycteribiidae includes such genera as *Eremoctenia* and *Cyclopodia*, all found in lands bordering the Indian Ocean. A few North American species are known. One of the most common North American genera is

Basilia, a taxonomic key to the species of which is given by Peterson (1959).

The progressive atrophy of the wings of nycteribiids and streblids (see below) undoubtedly represents a morphologic adaptation to parasitism, because these flies are almost continuously attached to their hosts, although they are capable of a certain amount of migration on their hosts. The pupae are not found on bats; rather, they are deposited on the walls of caves inhabited by bats. Their adaptation to bats is most probably of secondary origin, for their closest and more primitive relatives are found on other cave-dwelling animals, especially those which share the same caves as bats. Thus, it appears that these flies, or their immediate ancestors, have become adapted to bats only after having been initiated to parasitism on other hosts.

For a taxonomic treatment of the Nycteribiidae, see Theodor (1967).

FAMILY STREBLIDAE (BAT FLIES)

The Streblidae, also known as bat flies, are much better represented in the New World than the Nycteribiidae, although they are also found elsewhere, especially in lands bordering the Indian Ocean. The species, like those of the Nycteribiidae and the Hippoboscidae, include larvae that hatch within the body of the mother. The mature adults, which are attached to bats, are usually but not always wingless. Young adults are winged but commonly lose their wings later in life (Fig. 19.37).

Genus *Ascodipteron.* The members of *Ascodipteron*, found in Asia and Australia, are interesting for the females are embedded in the host (the flying fox bat) rather than attached to its skin. After copulation, the adult female cuts a hole in the host's skin, sheds its wings and legs, and embeds itself in the skin. Once in this position, the body increases in dimension and becomes flask shaped as the result of absorbing nutrients from the host. The head and thorax become invaginated within the bulbous body (Fig. 19.37). No

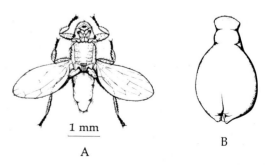

Fig. 19.37.   Bat flies. A. Dorsal view of young adult *Euctenodes mirabilis* still with wings. (After Kessel, 1924.) **B.** A much modified, sessile adult female *Ascodipteron africanum* parasitic on bats. (After Jobling, 1939.)

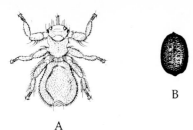

A

B

Fig. 19.38. *Melophagus ovinus.* **A.** Adult. **B.** Pupa (After Smart, 1943.)

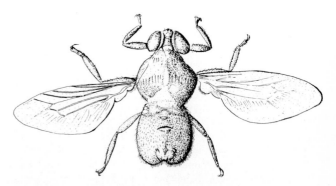

Fig. 19.39. *Hippobosca equina.* A fully winged hippoboscid fly ectoparasitic on horses in Europe. (Redrawn after Askew, 1971.)

reaction on the part of the host has been reported. As the larvae are formed and complete their development within the embedded female, they are expelled through the gonopore at the posterior end of the body which is directed toward the opening of the cavity. These larvae fall to the ground and pupation takes place. This odd behavioral pattern of the females might be considered a type of myiasis, except that the adult, rather than the larva, is the causative agent. Most other streblids deposit their fully developed larvae in bat roosts.

The male of *Ascodipteron* is free living and retains its legs and wings.

FAMILY HIPPOBOSCIDAE (LOUSE FLIES)

The hippoboscid flies, or louse flies, are believed to be closely related to the nycteribiid and streblid flies. The larvae of these flies also occur in their mother. They are not limited as ectoparasites of bats; some are blood-sucking parasites of birds and mammals. Some hippoboscids are winged; others are not; still others possess wings for a time before discarding them. The genera of louse flies that infest mammals include *Melophagus*, *Hippobosca*, and *Lipoptena*.

The 100 species of Hippoboscidae have been treated taxonomically by Maa (1963).

Genus *Melophagus.* The sheep tick or ked, *M. ovinus*, is a common ectoparasite of sheep and goats the world over (Fig. 19.38). This wingless fly is reddish brown and measures 5–7 mm in length.

Life Cycle of Melophagus ovinus. The entire span of *M. ovinus* is spent on the host. It takes approximately 7 days for the fully mature larva to develop within the female. Gravid females give birth to one larva every 7 to 8 days over a period of 4 months, which is the average life span of the adult female. The larva pupates almost immediately after birth.

The pupa, which is contained within the hardened puparial envelope, is glued to the host's wool, commonly on the abdomen, thighs, and shoulders (Fig. 19.38). The environmental temperature influences the pupal period. For example, during the summer months, the pupa metamorphoses into an adult in 19 to 23 days. During the winter, however, if the host is

kept outdoors, the pupal period lasts from 40 to 50 days. It takes 15 to 30 days for the young females to reach sexual maturity. Although a few flies do little damage to sheep, heavy infestations cause severe irritation.

Melophagus ovinus moves about a great deal on the host's body. It does not remain for very long at the surface of the skin, possibly because it is unable to tolerate the higher temperature. The keds are frequently found on the surface of fleece and it is from this position that they are passed from sheep to sheep.

Genus *Hippobosca.* There are about 15 species of *Hippobosca*. Of these, *H. struthionis*, the ostrich louse fly in South Africa, is a parasite of birds; the others are parasites of mammals. *Hippobosca equina* is a species common on equines in Europe (Fig. 19.39); *H. rufipes* parasitizes equines in South Africa; *H. variegata* parasitizes cattle and equines in Europe, Africa, and Asia; *H. camelina* parasitizes camels and dromedaries in Africa; and *H. longipennis* parasitizes dogs in India and certain Mediterranean countries. Functional wings are present on the members of *Hippobosca*.

The paper by Bequaert (1930) concerned with host associations of *Hippobosca* spp. is recommended.

Genus *Lipoptena.* For detailed information on the American species of *Lipoptena*, the paper by Bequaert (1937) should be consulted. These flies possess wings when they emerge from the puparium and are known as **volants**. However, they lose the wings during the maturing process.

Lipoptena depressa. This species is a fairly common parasite of deer in the western United States. Its life cycle, like that of other *Lipoptena* species, is somewhat modified from that of *Hippobosca* in that the pupal stage is not found on the host. Instead, the puparia drop from the host to the ground. The volants, escaping from puparia, fly about among trees. On sensing a deer, they attack it and crawl between the hairs to suck blood. They must find a host within 8 days or they will perish.

Sexual maturity is attained 12 days after feeding, and copulation occurs on the host. A single female larviposits mature larvae singly at 3-day intervals, producing a total of 30 to 35 larvae. Various authors have reported finding feeding adults in "chains"; that is, one fly actually feeds off the host, while the mouthparts of the second fly penetrate the dorsal surface of the abdomen of the first, and the third is attached to the second, and so on.

Other Lipoptena *Species.* Other species of *Lipoptena* include *L. sabulata* on deer in North America; *L. cervi*, the deer ked, on deer in Europe and sections of eastern North America; and *L. mazamae* on deer in Central and South America and the southeastern United States.

Other Hippoboscid Flies. Several species of hippoboscid flies infest birds. Among these are members of the genera *Lynchia*, *Pseudolynchia*, and *Microlynchia*. Representative of *Lynchia* are *L. fusca* and *L. americana*, both parasitic on owls, and *L. hirsuta*, a common parasite of the quail *Lophortyx californica*. *Lynchia hirsuta* can serve as a vector for the blood protozoan *Haemoproteus lophortyx* in quails. Similarly, *L. fusca* can serve as an experimental vector for this protozoan.

Members of *Pseudolynchia* are typified here by the pigeon fly, *P. canariensis*, and *P. brunnea*, which is parasitic on nighthawks. Not only does *P. canariensis* cause much irritation in pigeons, it also serves as a vector for *Haemoproteus columbae* (p. 205).

Microlynchia pusilla is a widely distributed species found on pigeons in the New World.

FAMILY CONOPIDAE (THICKHEADED FLIES)

The conopid flies, commonly referred to as thickheaded flies, are free living as adults, being found on flowers. However, the larvae are parasitic in bumblebees and wasps, and a few species are parasitic in grasshoppers. Adult females deposit eggs on the host's body while in flight. When the eggs hatch, the larvae burrow into the host's abdominal cavity. A representative of this family is *Conops* (Fig. 19.40).

FAMILY PHASIIDAE (PHASIIDS)

The Phasiidae includes a few species that are ectoparasitic on adult beetles (Coleoptera) and a few others species that are ectoparasitic on both nymphs and adults of Hemiptera. Little is known about the parasitic habits of these flies, although it is suspected that

Fig. 19.40.   Conopid and tachinid flies. A. *Conops.* **B.** Adult tachina fly. (Both figures redrawn after Comstock, 1949.)

adults seek out and attach themselves to the hosts from which they draw nutrition. Other phases of their life cycles are spent off the hosts.

FAMILY CORDYLURIDAE (DUNG FLIES)

Adult cordylurid flies are commonly referred to as dung flies, not a very exact name, since these are but one of the many families of flies that occur around dung and refuse. A few cordylurid species possess larvae that are parasitic in caterpillars. These enter their hosts after hatching from eggs oviposited on the host.

FAMILY TACHINIDAE (TACHINA FLIES)

The tachina flies constitute a large family whose members are commonly found among flowers and rotting vegetation (Fig. 19.40). Some 1500 species are listed among the dipteran fauna of North America alone. Parasitologists are interested in these flies because the larvae of all the species are parasitic, primarily in caterpillars. They also parasitize other insects. Cole (1969) has catalogued those species occurring in western North America. The treatise by Clausen (1940) should be consulted for greater detail of the biology of these flies.

The mechanisms employed by the larvae of various species for entering their hosts fall into four categories. (1) Some oviparous species, such as *Tachina larvarum*, deposit their eggs on the integument of the host. When these eggs hatch, the larvae burrow into the caterpillar and grow to maturity. (2) Females of viviparous (larviparous) species, such as *Compsilura concinnata*, deposit their larvae inside the host by puncturing the host's integument, using the sheath of their ovipositors. (3) Some oviparous species, such as *Frontina laeta*, deposit their eggs on foliage that serves as food for the host. When such eggs are ingested, the larvae hatch within the host. (4) In a few rare but remarkable instances, as in the viviparous *Eupeleteria magnicornis*, the females deposit larvae singly in cup-shaped membranous cases. The posterior end of the larva is attached to the inner surface of the cup, and the bottom of the cup (outer surface) is attached to a

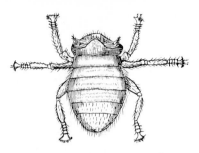

Fig. 19.41. Braulid fly. *Braula caeca.* (Redrawn after Meinert, 1892.)

leaf or twig. In such a position, the larva can move around by stretching its body. When a larva comes in contact with the silk thread on which the caterpillar host moves, it is in position to attach to and penetrate the host when the host migrates past that portion of the silk tract.

FAMILY BRAULIDAE (BRAULID FLIES)

This small family includes *Braula*, commonly known as the bee louse. *Braula caeca* is the common species and is found attached like a louse to honeybees (Fig. 19.41). The females lay eggs that drop into the comb chambers in beehives where they hatch and progress through the developmental stadia. Adults attach themselves to bees and remain attached throughout their life span.

REFERENCES

Atchley, W. R., Wirth, W. W., and Gaskins, C. T. (1975). "A Bibliography and a Keyword-in-Context Index of the Ceratopogonidae (Diptera) from 1758 to 1973." Texas Tech. Press, Lubbock, Texas.

Barbosa, F. A. S. (1947). *Culicoides* (Diptera: Heleidae) da Regiao Neotropica. *An. Soc. Biol. Pernambuco* **7**, 3–30.

Bennett, G. F. (1955). Studies on *Cuterebra emasculator* Fitch, 1856 (Diptera: Cuterebridae) and a discussion of the status of the genus *Cephenemyia* Ltr., 1818. *Can. J. Zool.* **33**, 75–98.

Bequaert, J. (1930). Notes on Hoppoboscidae, 2. The subfamily Hippoboscinae. *Psyche* **37**, 303–326.

Bequaert, J. (1937). Notes on Hoppoboscidae, 5. The American species of *Lipoptena. Bull. Brooklyn Entomol. Soc.* **32**, 91–101.

Berge, T. O. (ed.) (1975). International catalogue of arboviruses including certain other viruses of vertebrates. U.S. Dept. Health, Education, and Welfare. Publ. No. (CDC) 75-8301. Government Printing Office, Washington, D.C.

Campbell, J. A., and Pelham-Clington, E. C. (1959). A taxonomic review of the British species of *Culicoides* Latreille (Diptera, Ceratopogonidae). *Proc. Roy. Soc. Edinburgh* **67**, 181–299.

Carpenter, S. J. (1968). Review of recent literature on mosquitoes of North America. *Calif. Vector Views* **15**, 71–98.

Carpenter, S. J., and LaCasse, W. J. (1955). "Mosquitoes of North America (North of Mexico)." University of California Press, Berkeley, California.

Christophers, S. E. (1960). "*Aedes aegypti* (L.) the Yellow Fever Mosquito: Its Life History, Bionomics and Structure." Cambridge University Press, London.

Clausen, C. P. (1940). "Entomophagous Insects." McGraw-Hill, New York.

Clements, A. N. (1963). "The Physiology of Mosquitoes." Macmillan, New York.

Cole, F. R. (1969). "The Flies of Western North America." University of California Press, Berkeley, California.

Dalmat, H. T. (1955). The black flies (Diptera, Simuliidae) of Guatemala and their role as vectors of onchocerciasis. *Smithson. Misc. Collect.* **125**, 1–425.

Darsie, R. F. (1973). A record of changes in mosquito taxonomy in the United States of America 1955–1972. *Mosq. Syst.* **5**, 187–193.

Dove, W. E., Hall, D. G., and Hull, J. B. (1932). The salt marsh sandfly problem (*Culicoides*). *Ann. Entomol. Soc. Am.* **25**, 505–527.

Forattini, O. P. (1971). [On the classification of the subfamily Phlebotominae in the Americas (Diptera: Psychodidae)]. *Pap. Avul. Zool. S. Saulo* **24**, 93–111. (In Portuguese.)

Forattini, O. P. (1973). "Entomologia Medica. IV. Psychodidae. Phlebotominae Leishmanioses. Bartonelose." Edgard Blucher, São Paulo, Brazil.

Foot, R. H., and Pratt, H. D. (1954). The *Culicoides* of the eastern United States (Diptera, Heleidae). *Publ. Health Monogr. No. 18, U.S. Publ. Health Ser. No. 296.* Washington, D.C.

Fox, I. (1946). A review of the species of biting midges of *Culicoides* from the Caribbean region (Diptera, Ceratopogonidae). *Ann. Entomol. Soc. Am.* **39**, 248–258.

Graham-Smith, G. S. (1930). The Oscinidae (Diptera) as vectors of conjunctivitis, and the anatomy of their mouthparts. *Parasitology* **22**, 457–467.

Greenberg, B. (1971). "Flies and Disease. Vol. I. Ecology, Classification, and Biotic Association." Princeton University Press, Princeton, New Jersey.

Greenberg, B. (1973). "Flies and Disease. Vol. II. Biology and Disease Transmission." Princeton University Press, Princeton, New Jersey.

Hall, D. G. (1948). "The Blowflies of North America." Thomas Say Foundation. Entomol. Soc. Am., College Park, Maryland.

Hardy, D. E. (1960). "Insects of Hawaii. Diptera: Nematocera-Brachycera." Vol. 10. University of Hawaii Press, Honolulu, Hawaii.

Harwood, R. F., and James M. T. (1979). "Entomology in Human and Animal Health," 7th ed. Macmillan, New York.

Haskell, P. T. (1961). "Insect Sounds." Witherby, London.

Horsfall, W. R. (1955). "Mosquitoes. Their Bionomics and Relation to Disease." Ronald Press, New York.

Horsfall, W. R., and Morris, A. P. (1952). Surface conditions limiting larval sites of certain marsh mosquitoes. *Ann. Entomol. Soc. Am.* **45**, 492–498.

Knight, K. L. (1978). "Supplement to a Catologue of the Mosquitoes of the World (Diptera: Culicidae)." Thomas Say Foundation. Entomol. Soc. Am., College Park, Maryland.

Knight, K. L., and Stone, A. (1977). "A Catalog of the Mosquitoes of the World (Diptera: Culicidae)," 2nd ed. Thomas Say Foundation. Entomol. Soc. Am., College Park, Maryland.

Krinsky, W. L. (1976). Animal disease agents transmitted by horse flies and deer flies (Diptera: Tabanidae). *J. Med. Entomol.* **13**, 225–275.

Laarman, J. J. (1959). Host-seeking behaviour of malaria mosquitoes. *Proc. Int. Congr. Zool.* Sect. 8 (paper 24), p. 15.

Lainson, R., and Shaw, J. J. (1974). [The Leishmanias and leishmaniasis of the New World with particular reference to Brazil.] *Bol. Off. Sanit. Panam.* **76**, 93–114. (In Spanish.)

Laird, M. (ed.) (1981). "Blackflies: The Future for Biological Methods in Integrated Control." Academic Press, London.

Lewis, D. J. (1974). The biology of Phlebotomidae in relation to leishmaniasis. *Ann. Rev. Entomol.* **19**, 363–384.

Lewis, D. J., Young, D. G., Fairchild, G. B., and Minter, D. M. (1977). Proposals for a stable classification of the phlebotomine sandflies (Diptera: Psychodidae). *Syst. Entomol.* **2**, 319–332.

Maa, T. C. (1963). Genera and species of Hippoboscidae (Diptera): types, synonymy, habits and natural groupings. *Pacif. Insects* **6**, 1–186.

Mackerras, I. M. (1954–1955). The classification and distribution of Tabanidae. I–III. *Austr. J. Zool.* **2**, 431–554; ibid. **3**, 439–511, 583–633.

Newstead, R. (1924). "Guide to the Study of Tsetse Flies." School of Tropical Medicine, Memoir No. 1, University of Liverpool, England.

Peterson, B. V. (1959). New distribution and host records for bat flies, and a key to the North American species of *Basilia* Ribeiro (Diptera: Nycteribiidae). *Proc. Entomol. Soc. Ont.* **90**, 30–37.

Peterson, B. V., and Dang, P. T. (1981). Morphological means of separating siblings of the *Simulium damnosum* complex (Diptera: Simuliidae). *In* "Blackflies: The Future for Biological Methods in Integrated Control" (M. Laird, ed.), pp. 45–56. Academic Press, London.

Philip, C. B. (1947). A catalog of the blood-sucking fly family Tabanidae of the nearctic region north of Mexico. *Am. Midl. Natur.* **37**, 257–324.

Reed, W., Carroll, J., Agramonte, A., and Lazear, J. W. (1900). The etiology of yellow fever—a preliminary note. *Proc. 28th Annu. Meet. Amer. Publ. Health Ass.* Indianapolis, Indiana.

Séguy, F. (1935). Etude sur les stomoxydines et particulierement des mouches charbonneuses du genre *Stomoxys*. *Encycl. Entomol.* (Ser. B, 2, Diptera) **8**, 15–58.

Smart, J. (1945). The classification of the Simuliidae (Diptera). *Trans. Roy. Entomol. Soc.* **95**, 463–532.

Smith, K. G. V. (ed.) (1973). "Insects and other Arthropods of Medical Significance." Publ. No. 720. British Museum (Natural History), London.

Suster, P. (1960). Diptera Syrphidae. Fauna Republicii Populare Romaine, *Insecta* **11**, 1–286.

Theodor, O. (1965). On the classification of the American Phlebotominae. *J. Med. Entomol.* **2**, 171–197.

Theodor, O. (1967). "An Illustrated Catalogue of the Rothschild Collection of Nycteribiidae." British Museum (Natural History), London.

Travis, B. V. (1949). Studies of mosquito and other biting insect problems in 1949. *J. Econ. Entomol.* **42**, 451–457.

Turell, M. J., Rossignol, P. A., Spielman, A., Rossi, C. A., and Bailey, C. L. (1984) Enhanced arboviral transmission by mosquitoes that concurrently ingested microfilariae. *Science* **225**, 1039–1041.

van Emden, F. I. (1954). Diptera Cyclorrhaga, Calyptrata. *Handb. Ident. Br. Insects* **10(4a)**, 1–133.

Vargas, L. (1972). [Key for the identification of the genera of mosquitoes of the Americas, using female characters.] *Biol. Inform. Direccion Malariol. Saneamiento Ambiental.* **12**, 204–206. (In Spanish.)

Vargas, L. (1974). Bilingual key to the New World genera of mosquitoes (Diptera: Culicidae) based upon fourth stage larvae. *Calif. Vector Views* **21**, 15–18.

Vargas, L., Palacios, A. M., and Najera, A. D. (1946). Simulidos de Mexico. *Rev. Inst. Salub. Enferm. Trop.* **7**, 101–192.

Ward, R. A. (1963). Genetic aspects of the susceptibility of mosquitoes to malarial infection. *Exp. Parasitol.* **13**, 328–341.

Wirth, W. W., and Atchley, W. R. (1973). "A Review of the North American *Leptoconops* (Diptera: Ceratopogonidae)." *Grad. Studies, Texas Tech. Univ.* **5**, 1–57.

Woke, P. A. (1937). Comparative effects of the blood of man and of canary on egg production of *Culex pipiens* Linn. *J. Parasitol.* **23**, 311–313.

Wright, R. H., and Kellogg, F. E. (1962). Response of *Aedes aegypti* to moist convection currents. *Nature (London)* **194**, 402–403.

CLASSIFICATION OF DIPTERA (THE FLIES, GNATS, AND MOSQUITOES)*

Class Insecta (See p. 614)

ORDER DIPTERA

With functional forewings and reduced, knoblike hindwings (**halteres**); mouthparts variable (piercing-sucking or rasping-lapping) as is body form; with complete metamorphosis.

Suborder Orthorrhapha (see p. 643)

SUPERFAMILY NEMATOCERA (see p. 643)

Family Simuliidae

Bodies short and stout. Thorax much arched, giving humpbacked appearance; legs comparatively short; antennae 10- or 11-jointed, slightly longer than the head; ocelli absent; compound eyes on males large and contiguous, those on females widely separated; proboscis not elongate; palpi 4-jointed; wing venation as shown in Fig. 19.42. (Genera mentioned in text: *Cnephia, Simulium, Prosimulium*).

---

*Only those families that include species parasitic on or in animals are presented.

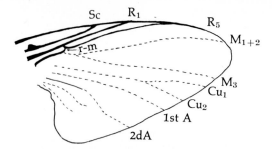

Fig. 19.42. Simuliid wing venation. Wing of *Simulium* showing venation pattern.

Family Psychodidae

Bodies mothlike; antennae slender, clothed with whorls of hair, and long in males; wings with 9 to 11 long, parallel veins, with no cross-veins except at base. (Genera mentioned in text: *Phlebotomus, Sergentomyia, Lutzomyia.*)

Family Ceratopogonidae

Bodies very small and short, 0.6–5.0 mm long, and not very hairy; wings broad, folded over abdomen, often mottled, and with few veins that are not all parallel. (Genera mentioned in text: *Culicoides, Leptoconops, Austroconops, Forcipomyia, Atrichopogon, Pterobosca.*)

Family Culicidae

Bodies slight; abdomen long and slender; wings narrow (Fig. 19.5); antennae with 15 segments, plumose in males; proboscis long and slender; wings with fringe of scalelike setae on margins, compound eyes large, occupying a large portion of surface of head; ocelli lacking. (Genera mentioned in text: *Aedes, Psorophora, Anopheles, Culex, Coquillettidia, Culiceta, Mansonia.*)

SUPERFAMILY BRACHYCERA (see p. 643)

Family Tabanidae

Bodies comparatively large, 7–30 mm long; wings well developed with veins evenly distributed (Fig. 19.43); eyes large and widely separated (dichoptic) in females, contiguous in males (holoptic), antennae usually short and three-jointed, third joint with 4–8 annuli (Fig. 19.44). (Genera mentioned in text: *Chrysops, Silvius, Tabanus.*)

Family Rhagionidae

Bodies trim, of moderate to large size, naked or hairy; males usually holoptic; legs comparatively long, abdomen cone shaped, tapering toward posterior end; three ocelli present; antennae three-jointed; proboscis usually short; wings broad, half open when at rest. (Genera mentioned in text: *Symphoromyia, Suragina, Spaniopsis.*)

Suborder Cyclorrapha

SUPERFAMILY ASCHIZA (OR ASCHIZA GROUP) (see p. 643)

Family Syrphidae

Species vary in form, often called mimic hymenopterous flies, because some resemble bumblebees, honeybees, and wasps; pseudovein (or spurious vein) present between R and M veins (Fig. 19.45);* antennae three-jointed, third segment with dorsal arista or thick-

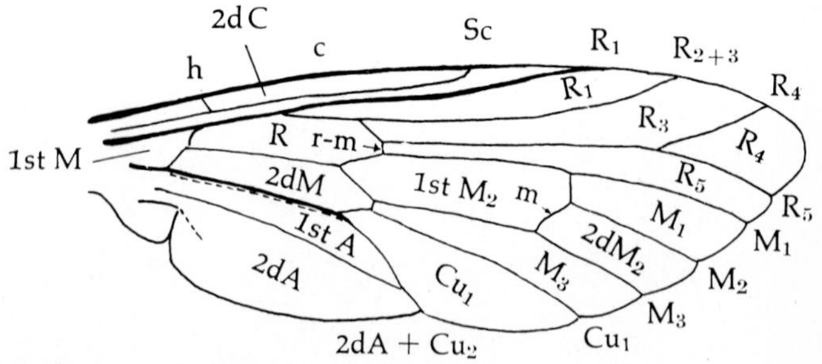

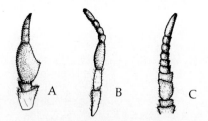

Fig. 19.43. Tabanid wing venation. Wing of *Tabanus* showing venation pattern.

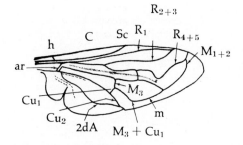

Fig. 19.44. Tabanid antennae. A. Antenna of *Tabanus*. **B.** Antenna of *Chrysops*. **C.** Antenna of *Pangonia*.

Fig. 19.45. Syrphid wing. Wing of *Tubifera* showing venation. (Redrawn after Comstock, 1949.)

ened style; frontal suture on head lacking. (Genera mentioned in text: *Syrphus, Tubifera, Helophilus.*)

SUPERFAMILY SCHIZOPHORA (OR SCHIZOPHORA GROUP) (see p. 643)

Family Muscidae

Typical flies; hypopleural or pteropleural bristles present; basal abdominal bristles reduced; antennae plumose. (Genera mentioned in text: *Musca, Glossina, Stomoxys, Haematobia, Fannia.*)

Family Calliphoridae

Bodies, especially abdomens, are metallic green or blue, less frequently violet or copper colored; large

---

*The spurious vein is absent in some syrphid flies.

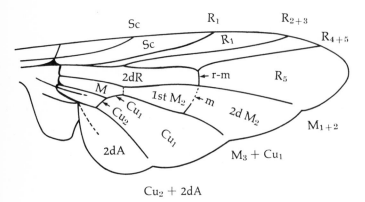

Fig. 19.46. Wing of Gasterophilidae. Wing of *Gasterophilus haemorrhoidalis* showing venation. (After James, 1947.)

flies; antennae are plumose; hypo- and pteropleural bristles present; posteriormost posthumeral bristle present and more ventral than presutural bristle; second ventral abdominal sclerite lies with edges overlying or in contact with ventral edges of corresponding dorsal sclerites. (Genera mentioned in text: *Pollenia, Callitroga, Chrysomya, Phormia, Protophormia, Cordylobia, Lucilia, Melinda.*)

Family Sarcophagidae

Bodies gray or silvery; sides of the head hairy; antennae plumose above and below; compound eyes hairy; proboscis long; palpi rudimentary; Vein M_{1+2} has angular bend and ends considerably before apex of wings. (Genera mentioned in text: *Sarcophaga, Wohlfahrtia, Titanogrypha.*)

Family Gasterophilidae

Bodies beelike; oral opening small; proboscis vestigial;

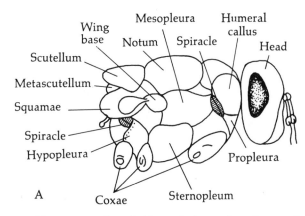

Fig. 19.47. Cuterebrid morphology. Lateral view of anterior portion of cuterebrid showing external parts. (Modified after Jaques, 1947.)

Vein M_{1+2} extends in a nearly straight line toward margin of wing (Fig. 19.46); wings transparent, with dark spots and bands. (Genus mentioned in text: *Gasterophilus.*)

Family Oestridae

Bodies large- or medium-sized and beelike; head large; mouth opening small; mouthparts usually vestigial; vein M_{1+2} bent rather than straight, as in the Gasterophilidae; apical cell of wing closed at margin; postscutellum well developed; squamae large. (Genera mentioned in text: *Oestrus, Rhinoestrus.*)

Family Cuterebridae

Bodies large and beelike; mouth opening small; mouthparts usually vestigial; postscutellum not developed; squamae large; apical cell of wing narrowed at margin. (For external anatomy of anterior portion of body see Fig. 19.47). (Genera mentioned in text: *Cuterebra, Dermatobia, Cephenemyia.*)

Family Chloropidae

Bodies small, rarely exceeding 3 mm in length; with short aristate antennae; squamae greatly reduced or absent; sixth longitudinal and anal veins absent (Fig. 19.35). (Genera mentioned in text: *Hippelates, Siphunculina.*)

Family Nycteribiidae

Parasitic on bats; bodies small, spiderlike, wingless; halteres present; head narrow, folded back in groove on dorsum of thorax when at rest; compound eyes and ocelli vestigial; antennae short, two-jointed; legs long; tarsal claws ordinary. (Genera mentioned in text: *Cyclopodia, Stylidia, Eremoctenia, Basilia.*)

Family Streblidae

Parasitic on bats; bodies small; head of moderate size, with freely movable neck, but not bent back on dorsum; compound eyes vestigial or lacking; ocelli lacking; palpi broad, leaflike; wings vestigial or lacking. (Genus mentioned in text: *Ascodipteron.*)

Family Hippoboscidae

Continuously parasitic on birds and mammals (except pupae of *Lipoptena*); body winged or wingless and compressed; head closely attached to and fitted in notch on thorax; antennae single-jointed, with terminal arista or style, located in depression near mouth, legs stout and short, broadly separated by sternum; tarsal claws strongly armed. (Genera mentioned in text: *Melophagus, Lipoptena, Hippobosca, Lynchia, Pseudolynchia, Microlynchia.*)

Family Conopidae

Head large, broader than thorax; body more or less elongate and naked or sparsely covered with fine hairs; ocelli either present or absent; antennae three-jointed, prominent, and project forward; third segment of antennae either arista or terminal style. (Genus mentioned in text: *Conops.*)

Family Phasiidae

Few species parasitic in adult beetles, others on nymphs and adults of Hemiptera; clypeus more or less protrudes at a low angle like a bridge; vein M_{1+2} bent so that R_5 is narrowed or closed at margin of wing.

Family Cordyluridae

Larvae of some species parasitic in caterpillars; bodies fairly small; subcostal vein separated from vein R_1, nearly half as long as wing, and ends in costa.

Family Tachinidae

Larvae primarily parasitic in caterpillars; bodies usually short, stout, and bristly; both hypopleural and pteropleural bristles present; ventral abdominal sclerite more or less covered by edges of dorsal sclerites. (Genera mentioned in text: *Tachina, Compsilura, Frontina, Eupeleteria.*)

Family Braulidae

Parasitic on honeybees; bodies minute, approximately 1.5 mm long; wingless; halteres lacking; head large; ocelli lacking; compound eyes vestigial; last segment of tarsi with a pair of comblike appendages. (Genus mentioned in text: *Braula.*)

20

HEMIPTERA: THE TRUE BUGS
HYMENOPTERA: THE WASPS
COLEOPTERA: THE BEETLES
STREPSIPTERA: THE TWISTED-WING
INSECTS

ORDER HEMIPTERA

There are some 35,000 hemipteran species in the world, and approximately 13,000 are represented in the Americas. These insects are characterized by four wings. The front pair is thickened at the base, and the thinner extremities overlap at the back. Since the hindwings are completely membranous and are folded beneath the forewings, the forewings are often referred to as wing covers or **hemelytra**. The hindwings are completely lacking among the bedbugs, which are members of the family Cimicidae. Furthermore, the forewings are reduced to two short scalelike pads. The mouthparts of these true bugs are modified for piercing and sucking and are collectively known as the **proboscis**. The proboscis is attached to the front of the head.

There are 34 families of hemipterans. Of these, the Cimicidae, Reduviidae, and Polyctenidae include parasitic species.

Family Cimicidae

The Cimicidae includes the bedbugs and their allies. Approximately 75 species have been described. Their biology, physiology, and taxonomy are discussed in the monograph by Usinger (1966). These bloodsucking insects are characterized by dorsoventrally flattened oval bodies; short, wide heads that are attached to the prothorax; conspicuous four-jointed antennae; proboscides that are three-jointed and lie in grooves beneath the head and thorax; compound eyes without ocelli; and reduced hemelytra.

The cimicids, especially the bedbugs, are still major household pests in many parts of Europe, Asia, Africa, and South America. Periodically, there are still reports of these bugs as bedfellows in North America. Bedbugs are nocturnal, although occasionally they have been reported to suck blood during the day. The reactions of hosts to the bites vary. Generally, the substances secreted by the salivary glands during feeding produce swelling and itching.

Bedbugs are not continuous feeders; instead, they engorge themselves in about five minutes and then drop off. The gregarious habit of bedbugs, which are usually found in large numbers, results in severe multiple attacks on their hosts.

The irritation caused by the piercing of the host's skin is the primary injury inflicted by bedbugs. Although various investigators have reported experimental success in getting bedbugs to serve as vectors for various pathogens, there is little evidence to support the idea that they are important vectors. Crissey (1983) has contributed a brief review of the potential role of bedbugs as vectors.

Life Cycle Pattern. Cimicid life cycle patterns are similar. Metamorphosis is gradual, that is, hemimetabolous (p. 623). In the common bedbug, *Cimex lectularius*, the females oviposit large operculated, yellowish white eggs, each measuring 1.02 × 0.44 mm, in batches in the crevices of bed frames, floors, walls, and similar household sites. Each batch of eggs contains 10 to 50 eggs. A single female deposits 200 to 500 eggs over a period of two to three months, usually during the spring and summer.

The nymphs, which hatch from the eggs after 8 to 10 days of incubation (range 7 to 30 days), resemble adults in possessing greatly reduced hemelytra and lacking hindwings, but the nymphs are white. Several nymphal instars ensue, accompanied by gradual growth, provided the environmental temperature is favorable and the nymphs are fed during each stadium. The nymphal period lasts approximately 6 weeks. This period is prolonged, however, if the environment is not favorable. Bedbug nymphs can undergo extended starvation, sometimes lasting as long as 2 months. Adults of *C. lectularius* are quite sensitive to heat. At about 15.5°C, activity ceases, and most specimens are killed if the temperature reaches 38°C. Adults can also live for long periods without a blood meal, surviving from 17 to 42 days.

The most prominent genera belonging to the Cimicidae are *Cimex* and *Leptocimex*.

Genus *Cimex*. *Cimex lectularius* (Fig. 20.1) is the most common bedbug, being cosmopolitan in its distribution, although it is primarily found in temperate zones. *Cimex hemipterus*, the Indian bedbug, is limited to tropical and subtropical areas. Both species will attack humans but will also feed on other mammals— rats, mice, rabbits, and especially bats. In fact, Askew (1971) has hypothesized that the bedbugs were originally found in caves and have become parasites of cave-dwelling animals, especially bats and certain birds. Humans are believed to have become suitable hosts during their cave-dwelling period. It is noted that of the 22 genera of cimicids, 12 are parasitic on bats. Some modern species of *Climex* will feed on chickens.

Genus *Leptocimex*. This genus includes *Leptocimex boueti*, a tropical species found primarily in west Africa. It will attack humans as well as other mammals, especially bats. It is characterized by very long legs and an elongate, ovoid body.

Family Reduviidae

There are about 2500 known species of reduviid or assassin bugs. These insects feed on the blood of other insects, and some attack humans and other animals. Members of the Reduviidae are commonly

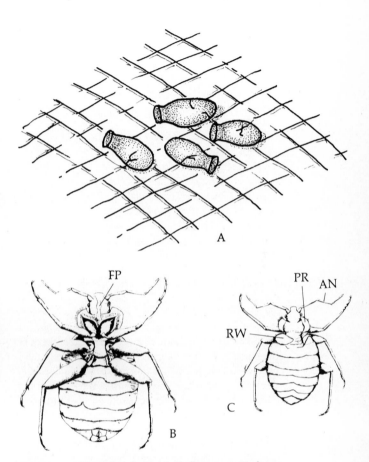

Fig. 20.1. *Cimex lectularius.* **A.** Bedbug eggs. (Redrawn after James and Harwood, 1969.) **B.** Ventral view of adult bedbug. **C.** Dorsal view of adult bedbug. (**B** and **C**, courtesy of Naval Medical School, Bethesda, Maryland.) AN, antenna; FP, flexed proboscis; PR, pronotum; RW, rudimentary wing of adult.

large, measuring 1.5 and 2 cm in length, and some are brightly colored. The proboscis is short, three-jointed, and attached to the tip of the head. When not in use, the proboscis rests within a groove on the prosternum. The antennae are four-jointed. Both compound eyes and ocelli are present, and both pairs of wings are well formed. These bugs are often referred to as assassin bugs, kissing bugs, or cone-nose bugs because of their attacking habits and the shape of their heads.

Life Cycle Pattern. All reduviids demonstrate a similar basic life cycle pattern. The large, smooth, barrel-shaped eggs are deposited on the ground, on trees, or in the dirty corners of houses, depending on the habits of the adults. The eggs are deposited singly or in clusters. A single female may oviposit a few dozen to over 500 eggs, depending on the species. The incubation period varies from species to species and is influenced by the temperature. However, almost all reduviid eggs hatch in 8 to 30 days. The escaping nymph is wingless, and there are generally five nymphal instars.

Both the vicious biting habits of these hemipterans and their roles as transmitters of pathogenic microorganisms have been of great interest to parasitologists.

Usinger (1944) has contributed an annotated taxonomic study on the Reduviidae. Of the many genera, *Triatoma, Rhodnius, Panstrongylus, Melanolestes,* and *Rasahus* are the most important.

Genus *Triatoma*. Several species of *Triatoma* (Fig. 20.2) are naturally infected with the hemoflagellate *Trypanosoma cruzi* (p. 131). Among these, *Triatoma sanguisuga* (the "Mexican bedbug"), found in the United States and Central America, is one of the most common vectors. This bug measures 18–20 mm in length and is characterized by a flattened body that is dark brown and splattered with reddish orange or pinkish areas on the abdomen, on the tips and bases of the hemelytra, and along the lateral and anterior margins of the pronotum. Similarly, *Triatoma protracta* (Fig. 20.2), which is widely distributed along the Pacific Coast of North America, is naturally infected with *Trypanosoma cruzi*. *T. protracta* is also known to be naturally infected in Arizona and Texas. Both of these species are vicious biters of various mammals, including humans. Their bites result in very painful and itchy swellings due to a toxin injected during feeding. In Hawaii, *Triatoma rubrofasciata* produces similar symptoms, and the anaphylactic reactions to the bites result in intense itching welts.

Other species of *Triatoma* known to be naturally infected with *T. cruzi* are *T. infestans* in southern Brazil, Uruguay, Chile, Paraguay, Argentina, and southern Bolivia; *T. dimidiata* in Mexico, Panama, Guatemala, and San Salvador; *T. hegneri, T. rubida, T. barberi,* and *T. gerstaeckeri* in Texas; and *T. rubida* and *T. recurva* in Arizona. For a review of the role of triatomines in the transmission of Chagas' disease caused by *Trypanosoma cruzi*, see Harwood and James (1979). There are more than 10 million infected persons in South and Central America today. A few cases have been reported from Arizona.

Since the *Trypanosoma cruzi* population in the blood of infected humans is small and difficult to detect, the diagnostic technique originated by Brumpt, known as **xenodiagnosis**, is widely used. This technique involves allowing uninfected *Triatoma* to bite the individual and examining the digestive tract of the bug for flagellates after a period of incubation.

Genus *Rhodnius*. *Rhodnius prolixus* (Fig. 20.2) and *R. pallescens* are two representatives of the genus that reportedly are naturally infected with *Trypanosoma cruzi*. The former ranges from Brazil north to Colombia and is found in San Salvador and Mexico, whereas the latter is found in Panama. While working with *R. prolixus* in 1912, Brumpt demonstrated that *T. cruzi* is not transmitted through the bite of the bug. Rather, the infective form of the flagellate passes out in the vector's feces and is mechanically deposited via contaminated hands on the extremely susceptible mucous membranes of the nose, eyes, or mouth, or is rubbed into skin perforations (p. 131). *Rhodnius brumpti* and *R. domesticus* also are found in Brazil.

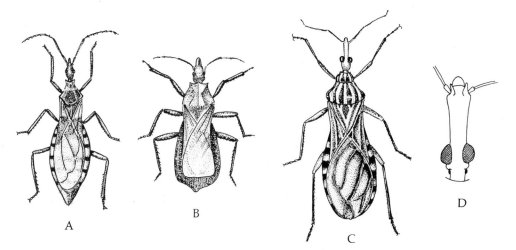

Fig. 20.2. **Some bloodsucking reduviids. A.** Adult *Triatoma sanguisuga*. (Redrawn after Marlatt, 1934.) **B.** Adult *Triatoma protracta*. **C.** Adult *Rhodnius prolixus*. **D.** Head of *R. prolixus* showing insertion of antennae. (**C** and **D**, redrawn after Brumpt, 1910.)

Genus *Panstrongylus.* This genus contains several species, including *P. megistus* and *P. rufotuberculatus*, which are suitable vectors for *Trypanosoma cruzi. Panstrongylus megistus* (Fig. 20.3), which is widely distributed in Brazil, the Guianas, and Paraguay, is a nocturnal hematophagous species, hiding by day in cracks and crevices in houses. The adults, measuring 30–32 mm in length, are black with red markings on the prothorax, wings, and abdomen.

Panstrongylus rubrofasciata, another bloodfeeder, is also widely distributed, being found in the Orient, Ethiopia, Central America, the West Indies, and Florida. The bites of neither *P. megistus* nor *P. rubrofasciata* result in such a severe reaction as those of *Triatoma* spp.

Genus *Melanolestes.* This genus includes *M. picipes* and *M. abdominalis* (Fig. 20.3). *M. picipes* is black and is found under stones, logs, and mosses throughout North America. It bites humans, and the host's reaction to its toxins is severe.

M. abdominalis has habits similar to those of *M. picipes* and is also widely distributed throughout North America. Members of *Melanolestes* are known to serve as vectors for microorganisms.

Genus *Rasahus.* *Rasahus biguttatus* (Fig. 20.4) and *R. thoracicus,* commonly referred to as the corsair bugs, are severe biters and cause damaging inflammations. Neither of these serves as a vector. *Rasahus biguttatus* is found in the southern United States, the West Indies, and South America, whereas *R. thoracicus* is found in the western United States and Mexico.

Family Polyctenidae

The Polyctenidae includes the so-called bat bugs. These little-understood insects are bloodsucking ectoparasites of bats. The familial characteristics include a four-jointed rostrum, shortened three-jointed tarsi, four-jointed antennae, short or vestigial hemelytra, no eyes, and often comblike bands known as **ctenidia**, which are present on the body.

Only a very few polyctenid species are known in North America. *Hesperoctenes longiceps* (Fig. 20.5) occurs on the bat *Eumops perotis californicus* in California. Two other representatives of the genus, *H. hermsi* and *H. eumops,* are known in North America. *Hesperoctenes hermsi* occurs on the bat *Tadarida macrotis* in Texas, and *H. eumops* occurs on *E. perotis californicus* in southern California.

The members of this family were previously assigned to the dipteran family Hippoboscidae, because the bat bugs show an affinity to the hippoboscid flies in being ectoparasites of mammals (bats), and in being larviparous, giving birth to rather advanced larvae. However, the polyctenids are definitely hemipterans rather than dipterans, although their relationship to the other true bugs remains unclear. Their seemingly primitive form, including short legs, reduced wings, and the absence of eyes, undoubtedly reflects their parasitic habit.

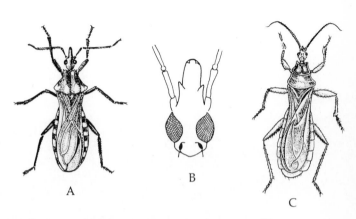

Fig. 20.3. Some bloodsucking reduviids. A. Adult *Panstrongylus megistus.* **B.** Head of *P. megistus* showing insertion of antennae. (Redrawn after Hegner *et al.,* 1929.) **C.** Adult *Melanolestes picipes.* (Redrawn after Matheson, 1950.)

ORDER HYMENOPTERA

The Hymenoptera includes the sawflies, bees, wasps, and ants. These insects are easily recognized by the extremely narrow isthmus that joins the thorax and the abdomen. Other characteristics of this order are presented on p. 705. Approximately 102,000 species of hymenopterans have been described, and some 50% of these, primarily wasps, are parasitic. It is the larvae of hymenopterans that are parasitic, being found in a very great number of insects, spiders, millipedes, and other invertebrates. Many of the parasitic larvae are of economic importance in that they serve as natural pathogens in the control of various destructive insects. These parasitic larval hymenopterans are commonly referred to as **parasitoids**.

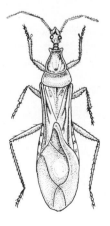

Fig. 20.4. Reduviid. Adult *Rasahus biguttatus.* (Redrawn after Matheson, 1950.)

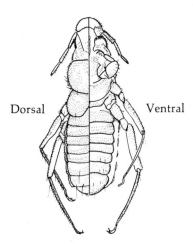

Dorsal　　　Ventral

Fig. 20.5. *Hesperoctenes.* Female *H. longiceps.* (Redrawn after Ferris, 1916.)

Fig. 20.6. **Planidium larva.** Planidium larva of *Perilampus hyalinus.*

Although parasitic hymenopteran larvae begin as parasites, they very often become predators during the end of their larval period and actively eat their hosts.

Only rarely is an adult hymenopteran found to be parasitic. In one such rare instance, a female of *Rielia manticida* was reported by Chopard (1929) to be ectoparasitic at the base of a wing of a praying mantis. At this site, the hymenopteran was observed to gnaw through the chitinous veins of the host's wing and feed on hemolymph.

An unusual series of parasitic adaptations occurs during this relationship. On attaching themselves to the praying mantis, *R. manticida* females cast off their own wings at the bases. When the host oviposits, the eggs are deposited within a foamy mass that later hardens to form the egg capsule. As soon as the host begins ovipositing, the parasitic wasp allows herself to become engulfed within the egg mass and lays her own eggs among those of the host.

If a female becomes attached to a male rather than a female praying mantis, the female wasp will not oviposit, nor will she pass from the female host to a male when the host copulates. The essay on parasitic hymenopterans by Askew (1971) is recommended.

Hyperparasitism. Not only are hymenopterans commonly parasitic, but also certain species are known to be hyperparasitic, that is, a parasite of another parasite. For example, the American species *Perilampus hyalinus* lays its eggs on foliage, and the first larval instar, known as a **planidium** (Fig. 20.6), seeks out and penetrates the body of a caterpillar, for example, that of *Hyphantria*. Once within the caterpillar, the

planidium searches for and enters the body of *Ernestia*, a parasitic tachinid fly. It is only when *P. hyalinus* becomes established as a hyperparasite in *Ernestia* that it continues to complete its development. Other examples of hyperparasitic hymenopterans are given by Askew (1971).

Life Cycle Studies. As a rule, eggs of parasitic wasps are deposited within the host after the integument of the host has been pierced by the ovipositor of the female wasp. Upon hatching, certain of the larvae emerge from and remain on their host as ectoparasites. Most of the larvae, however, remain within their host to pursue their development.

That a given larva may be either an ecto- or an endoparasite, depending on the species of host on which it is found, suggests that the nature of the host is responsible for the type of parasitism practised by the larva. For example, when eggs of the wasp *Dentroster protuberans* are oviposited in the body of the beetle *Myelophilus*, the larvae hatching from the eggs escape to the exterior and become ectoparasitic. If eggs of the same wasp are oviposited in another beetle, *Scotylus*, the larvae remain within the host as endoparasites.

If the wasp larva is endoparasitic, it does not attempt to make contact with the respiratory organs of its host, as endoparasitic dipteran larvae do, although tracheae are present in its body. The exchange of oxygen and carbon dioxide occurs through the entire surface of the hymenopteran's cuticle. This cuticle is so thin that oxygen in the host's hemolymph can be utilized.

The treatise by Clausen (1940), which includes a great deal of information on the parasitic hymenopterans, serves as a standard reference on these parasites.

BIOLOGY OF HYMENOPTERANS

The order Hymenoptera, according to Muesebeck *et al.* (1951), is divided into two suborders—Symphyta and Apocrita. Members of the Symphyta feed on plants except for those belonging to the small, obscure family Orussidae. Members of this family are commonly known as the parasitic wood wasps. These wasps are free living as adults but the larvae are parasitic on the larvae of metallic wood-boring beetles of the family Buprestidae.

Suborder Apocrita includes 59 families,* the majority of which are exclusively parasitic or include parasitic species. These families are divided among 11 superfamilies, some of which are briefly treated in the following discussion.

Superfamily Ichneumonoidea

Members of Ichneumonoidea are wasplike in appearance, but, with a few exceptions, they do not sting. All members are parasitic on or in other insects or other invertebrates. Adult ichneumons vary considerably in size, form, and coloration, but the majority are slender wasps (Fig. 20.7). They differ from the stinging wasps in having longer antennae, which are composed of 16 or more segments (other wasps generally have 12 or 13 segments); in having trochanters composed of two instead of one joint; and in lacking a costal cell in the front wings. In most ichneumons the ovipositor of females is very long, often longer than the body.

Life Cycle of Ichneumons. Ichneumons are parasitic as larvae. The adult female deposits her eggs on or inside the body of the host, which is commonly a lepidopteran larva, although a great variety of immature insects and even spiders can serve as hosts. A certain degree of host specificity is displayed, for most ichneumon species attack only a few types of hosts. If the eggs are laid on the host's epidermis, the newly hatched larvae generally bore into the body. The larvae develop into legless grubs that either become ectoparasitic or remain endoparasitic, depending on the host (p. 686). When pupation begins, the larvae may either remain within the host or emerge and spin their cocoons near the host. The adult is the postpupal form.

Representative of the Ichneumonoidea are *Rhyssella nitida*, the larvae of which parasitize xiphydriid wood wasps; *Phobocampe disparis*, the larvae of which parasitize gypsy moths; and *Tersilochus conotracheli*, the larvae of which parasitize plum curculios (Fig. 20.7). These species are all members of the family Ichneumonoidae.

Another large and beneficial ichneumonoidean family is the Braconidae. Adult braconids are relatively small, rarely measuring over 15 mm in length. Many braconid species have reduced wing venation. The life cycles of braconids are similar to the members of the Ichneumonoidae except that many of them pupate in silken cocoons on the outside of their host's

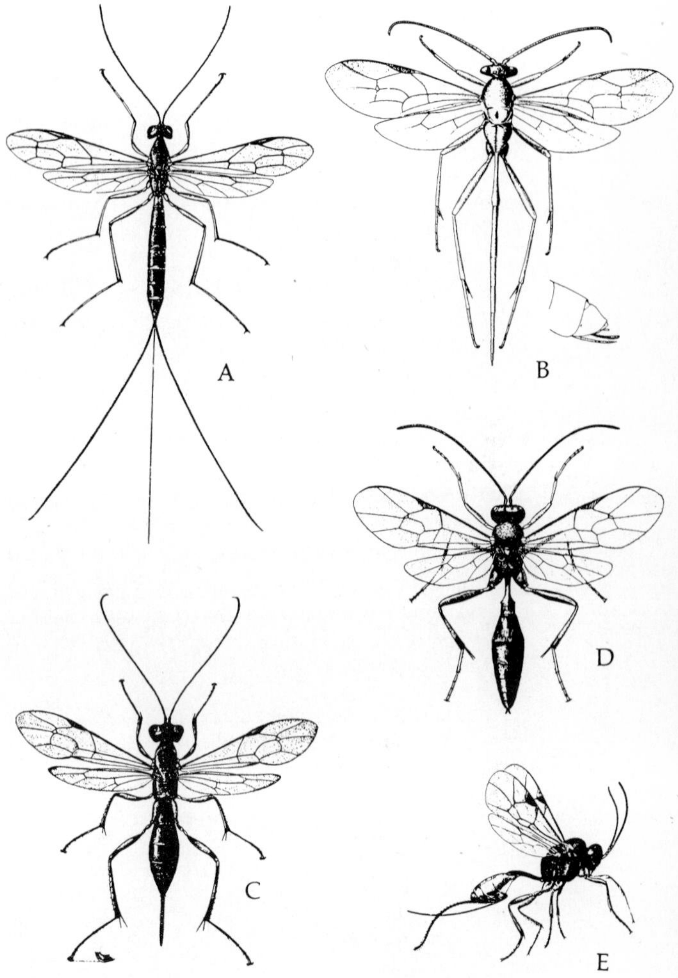

Fig. 20.7. Ichneumonoidea. A. Female *Rhyssella nitida*. **B.** Female *Casinaria texana*. **C.** Female *Phytodietus vulgaris*. **D.** Female *Phobocampe disparis*. **E.** Female *Tersilochus conotracheli*. (**A** and **C**, after Rohwer in Borrer and DeLong, 1954; **B**, after Walley in Borrer and DeLong, 1954; **D** and **E**, courtesy of U.S. Department of Agriculture.)

body, whereas still others spin their cocoons entirely apart from the host.

Genera representative of the Braconidae include *Apanteles*, with *A. melanoscelus*, imported into the United States to control gypsy moth larvae; *Macrocentrus*, including *M. gifuensis*, the larvae of which parasitize larvae of the European corn borer; and *Phanomeris*, including *P. phyllotomae* (Fig. 20.8), the larvae of which parasitize bird leaf-mining sawfly larvae.

---

*The reader is referred to Borrer and DeLong (1954) for a key to and descriptions of these families.

Superfamily Chalcidoidea

Members of the Chalcidoidea, known as chalcid flies, are actually small wasps that measure less than 1–3 mm in length. Their wing venation is reduced, and they usually have elbowed antennae. Chalcidoidean wasps attack the larvae of other parasitic hymenopterans within the body of the primary host and hence are hyperparasites. A few chalcids develop in plant seeds or plant stems.

One of the best known chalcid wasps is *Nasonia vitripennis*,* a member of the family Pteromalidae, because it is sometimes employed as an experimental animal in genetics (Whiting, 1956). This minute black wasp possesses an abdomen that appears more or less triangular in profile.

Life Cycle of *Nasonia*. Pupae of *Sarcophaga* and other sarcophagid flies serve as hosts for *Nasonia vitripennis*, which is parasitic during its larval and adult stages. The adult female seeks a fly puparium and punctures the puparial wall with her ovipositor. A feeding tube (or coagulation tube) is formed from secretions of the ovipositor, connecting the pupa to the tear in the puparial wall. Through this canal, the female ingests the body juices of the fly maggot. This represents the type of parasitism practised by the adult. The adult female not only feeds on the fly pupa, but she oviposits into the puparium by making other holes in the puparial wall and inserting her ovipositor.

The eggs hatch into legless larvae that possess a small head and spiracles arranged in a row on each side of the body (Fig. 20.9). These larvae feed on the fly maggot and, after several molts, transform into pupae that are naked (without puparium or cocoon) (Fig. 20.9). The pupae are at first whitish, but become almost completely black with age. The pupal exuviae are then shed (a process known as **eclosion**), and these immature forms metamorphose into imagos, which gnaw a hole in the puparial wall and emerge (Fig. 20.9).

Not all larvae enclosed with a single fly puparium develop at the same rate. Many undergo larval diapause* (quiescent period). This unique habit of *N. vitripennis* proved at first to be a drawback to geneticists employing this wasp as an experimental animal, because it reduces the number of offspring that can be scored. It is now known that larval diapause can be regulated. If puparia enclosing larvae in diapause are maintained at low temperatures (household refrigerator temperature) for three or more months and then placed in an incubator or at room temperature, all the

---

*Formerly known as *Mormoniella vitripennis*.

*For an excellent account of diapause in arthropods the reader is referred to Lees (1955).

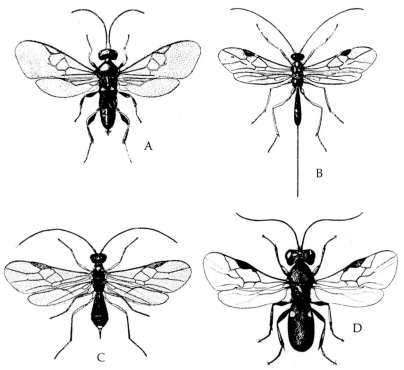

Fig. 20.8. Braconidae. A. *Apanteles diatraeae.* **B.** *Macrocentrus gifuensis.* **C.** *Chelonus texanus.* **D.** *Phanomeris phyllotomae.* (All figures courtesy of U.S. Department of Agriculture.)

larvae pupate. This discovery has greatly enhanced the use of *N. vitripennis* as a genetic tool. In addition, if dark female pupae and freshly enclosed females are maintained below 25°C, optimally at 7°C, for three or more days, the subsequent offspring will undergo larval diapause. This information has also become an asset to geneticists because these parasitic wasps can now be maintained in the diapause state and made to pupate when desired.

Superfamily Proctotrupoidea

All larvae of members of the Proctotrupoidea are parasitic in or on the immature stages of other insects. Adult proctotrupoids are minute black wasps that measure 3–8 mm in length.

Larvae of these wasps parasitize (1) egg capsules of other insects, (2) larvae of other insects, and (3) larvae of parasites of other insects. As an example of the first category of parasitism, larvae of ensign wasps (Fig. 20.10) are parasitic in the egg capsules of roaches. In the second category, larvae of *Pelecinus polyturator* (Fig. 20.10) parasitize larvae of the June beetle. In the third category, some larvae are hyperparasitic in larvae of braconid and chalcid parasites of aphids and scale insects.

Superfamily Scolioidea

All wasps belonging to the Scolioidea are parasitic. During the life cycle of scolioids, the female lays its eggs on its host, usually without injuring the host, and then flies elsewhere to oviposit more eggs on other hosts. Larvae hatching from the eggs are ectoparasitic. They feed on the host and gradually destroy it. Thus, these wasps border between parasitism and predation.

The scolioid wasps may be black, or they may be brightly colored black and yellow. Larvae of the various species are commonly found on the larvae of scarabaeid beetles. One species, *Tiphia popilliavora* (Fig. 20.11), was introduced into the United States to aid in the control of the destructive Japanese beetle. The larvae of *T. popilliavora* feed on the body tissues of the grub of the beetle and thus kill it.

Polyembryony in Parasitic Hymenoptera

The consistent occurrence of **polyembryony**—the production of two or more embryos from the same ovum—is more frequently, but not exclusively, encountered in parasites. Among parasites, this type of embryogenesis is known to occur among the Mesozoa (p. 706), the Trematoda (p. 328), and certain Cestoda (p. 399).* In many ways, it is surprising that polyem-

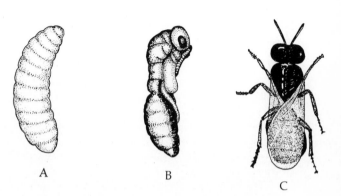

Fig. 20.9. Stages in the life cycle of *Nasonia vitripennis*. A. Larva. **B.** Pupa. **C.** Adult.

bryony is the normal pattern in such advanced animals as certain wasps. It should not be inferred that animals that undergo polyembryony are of common ancestry.

Polyembryony in parasitic hymenopterans, although not a common phenomenon, was first dis-

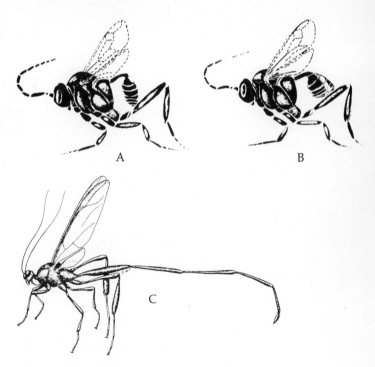

Fig. 20.10. Proctotrupoidea. A. Male ensign wasp, *Prosevania punctata*. **B.** Female *P. punctata*. (**A** and **B**, after Edmunds in Borrer and DeLong, 1964.) **C.** Female *Pelecinus polyturator*. (After Borrer and Delong, 1964.)

---

*The formation of many scolices on a single larval cestode (coenuri and hydatids), each of which later develops into an adult worm, may be considered a form of delayed polyembryony, because the larva, which had developed from a single ovum, gives rise to many individuals.

covered in 1904 by Marchal among certain wasps. His study of this process in *Encyrtus fuscicollis*, a parasite of the caterpillars of *Hyponomeuta* spp., remains the classic one.

In *E. fuscicollis*, the eggs are laid within the caterpillar during July and August, and development commences before the onset of winter. Development comes to a complete halt with cold weather and is resumed the following April. Within the host, the hymenopteran egg is surrounded by a host-elaborated epithelial cyst wall.

From the very beginning, a large nucleus, rich in chromatin, known as the **paranucleus**, is differentiated within the ovum. In addition, a number of smaller nuclei, known as **embryonic nuclei**, are present (Fig. 20.12). The embryonic nuclei are difficult to stain.

The paranucleus undergoes extensive development, becomes lobed, and eventually divides into a number of smaller, amorphous bodies. On the other hand, the embryonic nuclei give rise at an early stage to approximately 100 small aggregates of cells, each one including a germ cell. These aggregates resemble morulae (Fig. 20.12), each eventually becoming an embryo. If a germ cell fails to become incorporated in an embryonic cell aggregate, the resulting larva lacks rudimentary gonads and degenerates without ever going through metamorphosis. Embryogenesis occurs within the egg in the middle of a cytoplasmic mass intermingled with fat-containing paranuclear bodies. The paranucleus thus serves as an amnion and as a trophic layer.

The entire egg is gradually transformed into a long tube in which the larvae are linearly arranged. The host's epithelial cyst persists. It should thus be apparent that the processes involved in the type of polyembryony found in *Encyrtus fuscicollis* are similar to those found in mesozoans, especially the orthonectids (p. 712), and to those found in the formation of the larval stages of digenetic trematodes in molluscs.

Sex determination among the polyembryonic offspring of parasitic hymenopterans still remains puzzling. According to Bugnion, who also studied *Encyrtus fuscicollis*, all offspring from the same egg

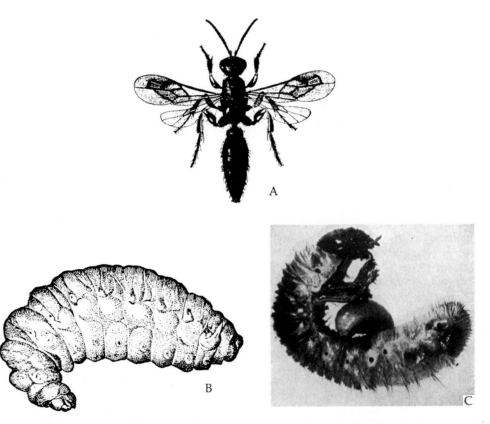

Fig. 20.11. Tiphiid wasps. A. Adult *Tiphia popilliavora*. (Courtesy of U.S. Department of Agriculture.) **B.** *Tiphia* larva. (Courtesy of Illinois State Natural History Survey.) **C.** *Tiphia* larva parasitizing a Japanese beetle grub. (Courtesy of U.S. Department of Agriculture.)

are of the same sex, indicating that sex determination occurs at the beginning of development.

Silvestri, who studied another species, *Litomastix truncatellus*, agrees, but in addition has reported that the offspring from fertilized eggs develop into females, whereas those from parthenogenetic eggs develop into males. The eggs of *L. truncatellus*, a parasite of the caterpillar of *Plusia gamma*, include approximately 100 larvae. Patterson (1917), on the other hand, has reported that among the 177 batches of eggs of *Paracopidosomopis floridanus* studied, 154 included offspring of both sexes. Analysis of his data has revealed that the mixtures of sexes could not be explained by the simultaneous development of several eggs of different sexes in the same host. *P. floridanus* is a parasite of the caterpillar of *Pieris brassicae*, the white cabbage butterfly.

In addition to the species already mentioned, several others undergo polyembryony (Parker, 1931; Paillot, 1937; Clausen, 1940). For example, *Litomastix gelechiae*, a parasite of *Gnorimoschema salinaris*, and *Ageniaspis testaceipes*, a parasite of *Lithocolletis* caterpillars, are known to produce their young by polyembryony. The same holds true for two braconid wasps, *Macrocentrus gifuensis* and *Amicroplus collaris*, the first being parasitic in the caterpillar of *Pyralis* and the second in the caterpillar of *Euxoa segetum*.

The occurrence of polyembryony in certain species of parasitic hymenopterans is undoubtedly the result of evolutionary changes directed by natural selection. Polyembryony could represent a preadapted characteristic carried over from an ancestral form, which benefits the modern species in its propagation, or it could be a relatively recent development. In either case, the modern parasitic hymenopterans in which polyembryony occurs benefit in that a larger number of progeny can thus be produced. In most parasitic animals, this is advantageous to species preservation, for many individuals of each generation die before another suitable host can be found.

The mechanisms responsible for the initiation of polyembryony remain undetermined. Some favor the theory that polyembryony is a characteristic developed in parasitic wasps as a result of the conditions in which the egg of the parasite happens to develop. This theory is primarily based on the findings of Marchal, who believed that polyembryony in *Encyrtus* is associated with the arrest of embryonic development during the winter and the resumption of development in the spring, when the caterpillar host begins to feed. When the host resumes feeding, abrupt osmotic changes occur in the medium in which the eggs are found, and such changes are thought to contribute to the initiation of polyembryony. This theory also appears to hold true in the case of *Polygnotes*, the eggs of which are in the stomach of cecidomyiid larvae, where they undergo very abrupt osmotic changes and, at the same time, are subjected to considerable agitation.

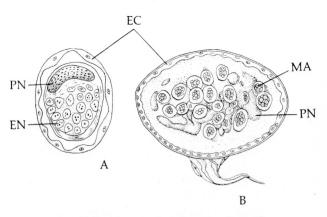

Fig. 20.12. Polyembryony in *Encyrtus fuscicollis*. **A.** Young egg enclosing paranucleus and embryonic nuclei. **B.** Older egg enclosing amorphous paranuclear bodies and embryonic cell aggregates. (Both figures redrawn after Marchal, 1904.) EC, host-elaborated epithelial cyst; EN, embryonic nucleus; MA, morula-like aggregate of embryonic cells; PN, paranucleus.

The hypothesis that osmotic pressure changes may contribute to the initiation of polyembryony is tentative.

ORDER COLEOPTERA

The Coleoptera includes some 250,000 species of beetles and weevils, of which about 27,000 are found in the United States. These insects are readily distinguished by their leathery integument, mandibulate (biting-chewing) mouthparts, and two pairs of wings. The front pair, known as the **elytra**, are horny, heavy, and nonfunctional as organs of flight. The hind pair are membranous and functional, and are folded under the elytra when at rest. Although some coleopterans are wingless, most possess wings. Also characteristic of these insects is the meeting of the elytra along a straight middorsal line when the insect is at rest.

Coleopteran metamorphosis is of the holometabolous type, involving an egg-larva-pupa-imago sequence (p. 623). Beetle larvae, commonly referred to as **grubs**, are characterized by three pairs of well-developed legs. The larvae of weevils, however, are legless.

COLEOPTERANS AS INTERMEDIATE HOSTS

Parasitism is rarely encountered among coleopterans. Most beetles are free living; some, such as various weevils, are extremely destructive to plants of economic importance. Interest in beetles stems from their role as intermediate hosts for helminth parasites and from the few parasitic species. Table 20.1 lists some of the helminth parasites and their coleopteran intermediate hosts.

COLEOPTERANS AS PARASITES

A few members of the families Leptinidae and Platypsyllidae are obligatory parasites. Parasitic members of the Leptinidae include *Leptinillus validis*, an ectoparasite on American beavers; *L. aplodontiae*, on *Apolodontia*, a rodentlike animal found along the Pacific Coast and commonly called the mountain beaver; and *L. testaceus* on mice, voles, and shrews in North America and Europe (Fig. 20.13).

Representative of the Platypsyllidae is *Platypsyllus castoris*, which is a permanent, obligatory parasite of beavers during all stages of its life cycle. This parasite occurs in both Europe and North America.

Occasionally beetles, both larvae and adults, are accidentally ingested by a mammalian host, but they do not remain for long in the host, being rapidly passed out in feces. This type of accidental parasitism is termed **canthariasis**. Larvae of the beetle *Tenebrio molitor*, commonly known as the mealworm, are the most frequently reported accidental "parasites." Undoubtedly this grub gains entrance to the host's alimentary tract by ingestion of contaminated foods. Similarly, *Onthophagus bifasciatus*, *O. unifasciatus*, and

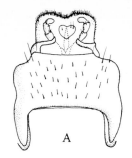

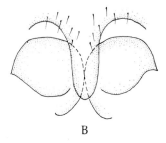

Fig. 20.13. *Leptinillus aplodontiae.* **A.** Labium of beetle. **B.** Presternum of beetle. (**A** and **B**, redrawn after Ferris, 1918.)

Ptinus tectus have been reported as accidental intestinal "parasites."

ORDER STREPSIPTERA

Members of the Strepsiptera, often called twisted-wing insects, are small endoparasites of other insects that include various species of Orthoptera, Hemiptera,

Table 20.1. Representative Helminth Parasites of Vertebrates That Utilize Coleopterans as Intermediate Hosts

Helminth	Vertebrate Host	Coleopteran Host
Nematodes		
Gongylonema pulchrum	Goats, sheep, swine, cattle, occasionally man	Scarabaeidae: *Aphodius, Scarabaeus, Passalurus, Onthophagus;* Tenebrionidae: *Tenebrio molitor*
Ascarops strongylina	Swine	*Copris, Aphodius, Passalurus*
Physocephalus sexalatus	Swine	*Onthophagus, Scarabaeus, Gymnopleurus, Ataenius, Canthon, Phanaeus, Geotrupes*
Spirocerca lupi	Dogs	*Scarabaeus sacer*
Cheilospirura hamulosa	Poultry	*Alphitobius, Gonocephalum, Ammophorus, Anthrenus,* other dermestid beetles, *Alphitophagus,* other fungous beetles
Subulura brumpti	Poultry	*Alphitobius, Gonocephalum, Ammophorus, Dermestes vulpinus*
Acanthocephala		
Macracanthorhynchus hirudinaceus	Swine, rarely man	Scarabaeidae: *Melolontha melolontha, Cetonia aurata, Phyllophaga* spp.
Moniliformis dubius	Rats, occasionally man	*Blaps*
Cestodes		
Raillietina cesticillus	Fowl	Scarabaeidae 2 spp., Tenebrionidae 1 sp., Carabidae 38 spp., including *Amara* and *Pterostichus*
Hymenolepis diminuta	Rats, mice, occasionally man	*Tenebrio molitor* (larva), *Tribolium confusum*
H. carioca	Domestic fowl	*Aphodius, Choeridium, Hister, Anisotarsus* (?)
H. cantaniana	Turkeys	*Ataenius, Choeridium*
Choanotaenia infundibulum	Poultry	Many species of Tenebrionidae

Homoptera, and Hymenoptera. Marked sexual dimorphism exists among these insects. Only the adult males are winged. The forewings are reduced to club-shaped appendages, and the fan-shaped hindwings are large and folded longitudinally when in the resting position (Fig. 20.14). Adult females are legless and larviform. Mouthparts are greatly reduced in these insects (Fig. 20.15). Hypermetamorphosis is the rule (p. 623).

STREPSIPTERAN LIFE CYCLE PATTERN

The strepsipteran life cycle is extremely complex. Adult males are free living, whereas adult females are

endoparasites in the bodies of various insects. The male seeks out and mates with a female.

Strepsipterids are ovoviviparous (larviparous) and extremely prolific. A single female may contain over 2000 minute larvae in her saccular body. The eggs hatch within the female's body, giving rise to first larval instars that are campodeiform and active. These instars are referred to as **triungulins** and possess well-

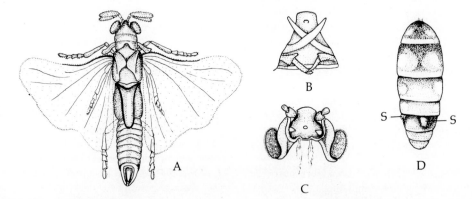

Fig. 20.14. **Strepsiptera. A.** *Ophthalmochlus duryi.* (Redrawn after Pierce, 1909.) **B.** Mouthparts of male *Acroschismus bruesi.* (Redrawn after Pierce, 1909.) **C.** Mouthparts of male *Pentozocera australensis.* (Redrawn after Perkins, 1905.) **D.** Strepsipteran (s) projecting from between abdominal segments of insect host. (Redrawn after Comstock, 1949.)

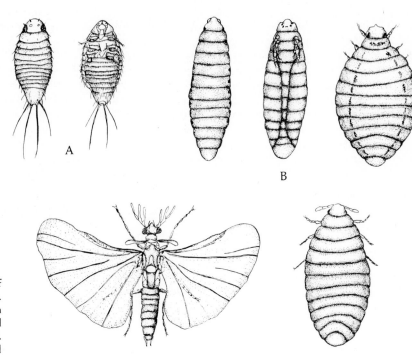

Fig. 20.15. **Stages in the life cycle of** ***Eoxenos laboulbenei.* A.** First-stage larva. **B.** Larvae from body of a thysanuran showing the sides rolled in and expanded after the larva escaped from the host. **C.** Adult male. **D.** Wingless adult female. (All figures redrawn after Silvestri, 1937.)

developed eyes and legs (Fig. 20.14). The triungulins leave the body of the female through unpaired median genital pores located on abdominal segments 2 through 5.

Because the adult female is enveloped within the epidermis of the pupa, the triungulins do not at first escape to the exterior. Instead, they fall into the brood chamber—the space between the female and the puparial envelope. They leave the brood chamber through a slit on the cephalothorax of the puparium, between the head and prothorax, and crawl over the body of the host. This escaping first larval instar represents a temporarily free-living form. It falls to the soil or to vegetation.

If the larva is to survive, it must seek out the larva or nymph of a suitable host and burrow into it. If the hosts are gregarious insects, as are bees or wasps, this is not too great a problem, for it only involves finding a hive. However, if the larva is to parasitize a homopteran, it must crawl over leaves and twigs until such a host is found.

Once established within a compatible host, the larva grows rapidly. After completing the first ecdysis, the second larval instar is legless and becomes increasingly cylindrical (Fig. 20.14). From this stage on, development is different in males and females. There are seven larval instars in all. In males, the cephalothorax is formed during the fifth larval stadium by fusion of the distinct head and thorax, and the seventh instar is enveloped within the epidermis of the sixth. This last larval instar is distinct, for it bears strongly developed appendages. For this reason, it is sometimes referred to as the **prepupa**. At the prepupal stage, the organism first protrudes its anterior end through the intersegmental space on the host's abdomen. The pupa develops within the skin of the seventh larval instar. When the adult metamorphoses from the pupa, it pushes off the operculum present on the puparium and escapes as a winged individual. The free-living males live for only a short while, during which they seek out and copulate with parasitic females.

In females, fusion of the head and thorax also takes place during the fifth larval stadium. The seventh larval instar does not bear appendages, nor is it enveloped in the skin of the sixth. It is also at this stage that the parasite protrudes from the host, but there is no pupal stage. Instead, the larviform adult female develops directly from the seventh larval instar and remains a permanent parasite.

Fertilization of females involves the seminal fluid entering via the route of the triungulin's escape.

Parasitized insects are readily spotted by the trained investigator, since the parasites protrude their anterior ends from between two of the host's abdominal segments (Fig. 20.14). If the protrusion is flat and disc shaped, it is the head of a female; however, if it is rounded and tuberculate, it is the anterior end of the cylindrical body of a male puparium. The most complete account of the life history of a strepsipterid is that given by Nassonow (1892) for *Xenos vesparum*.

EFFECTS OF STREPSIPTERANS ON HOSTS

Larvae of *Xenos vesparum*, lying between the host's organs, absorb the host's hemolymph through their body surfaces. As they increase in size, they can push the surrounding organs out of position, although the mechanical pressure does little damage to the host. The insect host, however, may suffer from malnutrition resulting from the loss of hemolymph.

As many as 31 strepsipteran larvae have been found within the same host. Female hymenopterans appear to be preferred over males as hosts. The heads of parasitized hymenopterans often become smaller, more globular, and more hairy. The pollen-collecting apparatus of female hosts is much reduced, their hind legs are modified to resemble those of males, and the yellow color of the male may even be acquired. In other words, female insect hosts of strepsipterans often become modified to appear as males. Such is the case among parasitized bees, *Andrena chrysosceles* (Smith and Hamm, 1914). Furthermore, the stinger is reduced in size. Parasitic castration occurs in female hosts. Their ovaries become smaller, the oocytes in them degenerate, and such parasitized females are infertile. In males, the copulatory apparatus is reduced and the testes are become smaller although, unlike the ovaries, they remain functional.

BIOLOGY OF STREPSIPTERANS

Numerous species of the Strepsiptera are found in North America. The taxonomic studies of Pierce (1909, 1911) and Bohart (1941) should be consulted by those interested in classification. The volumes by Clausen (1940) and Askew (1971) should also be consulted since these include much information on the biology of strepsipterans.

The order Strepsiptera includes four families; Mengeidae, Stylopidae, Elenchidae, and Halictophagidae.

Family Mengeidae

Members of the Mengeidae are generally considered the least specialized of the strepsipterans. They are mostly free living under stones.

The larvae of one species, *Eoxenos laboulbenei*, are endoparasitic in the body cavity of thysanurans (Fig. 20.15).

Life Cycle of *Eoxenos laboulbenei*. The minute larvae of *E. laboulbenei*, measuring only 200 μm in length, penetrate between the abdominal segments of the body of thysanurans. Here the larvae metamorphose into second-stage larvae that molt after ingesting nourishment and become third-stage larvae. Third-stage larvae are readily recognizable, for their bodies are folded toward the ventral surface and are cigar shaped. Each host generally includes no more than two or three larvae. The host dies when the third-stage larvae emerge from it. The emerged larvae unfold and become flattened. Pupation occurs on the ground.

Adult males are winged, whereas adult females are wingless. Although the female possesses a genital pore, the sperm do not enter this orifice during fertilization. Instead, the male pierces the female's body cuticle with its copulatory apparatus and introduces sperm through the opening. The female's genital pore is apparently used only as a birth pore. *Eoxenos laboulbenei*, like all strepsipterans, is ovoviviparous (larviparous). The larvae emerging from the female seek a new host immediately. Thus, in this mengeid strepsipteran only the larvae are parasitic.

Variations of this life history pattern do occur. Parthenogenetic females exist. They do not emerge from the pupal exuviae, but their vaginal pores do open on the surface. It is through this pore that young larvae emerge. In addition, although thysanurans are generally infected in the autumn and the strepsipteran larvae pass the winter within their hosts, parthenogenetic females, protected by the pupal exuviae, are capable of wintering in soil, and their larvae attack new hosts in the spring.

In addition to *Eoxenos*, Mengeidae includes *Triozocera* (Fig. 20.16).

Family Stylopidae

Stylopidae is the largest family of the Strepsiptera. Most members are parasites of bees, but a few are parasites of wasps. The stylopids are generally considered more specialized than the mengeids. Conversely, the life cycles of stylopids are less complicated, because the number of larval stages is generally reduced. The developmental pattern is essentially that given under "Strepsipteran Life Cycle Pattern" (p. 693).

When *Stylops* (Fig. 20.16) and *Hylecthrus* parasitize solitary bees and *Xenos* parasitizes *Polistes metricus* and other social wasps, there is a synchronization between the life cycle of the parasite and that of the host.

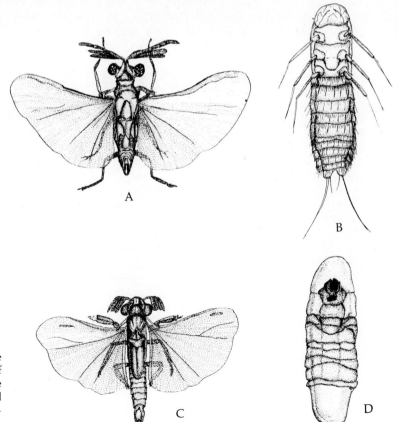

Fig. 20.16. Strepsiptera. A. Male *Triozocera mexicana*. **B.** Ventral view of triunglin of *Stylops californica*. **C.** Male *Halictophagus oncometopiae*. **D.** Ventral view of wingless, adult female *H. oncometopiae*. (All figures after Pierce, 1918.)

Usually the parasite is in the second larval instar when the host larva is about to undergo its nymphal molt. Although never demonstrated, it is possible that an endocrine synchronization takes place during these relationships.

Other Families

Members of the relatively small family Elenchidae are parasites of plant hoppers (Fulgoridae). Members of the Halictophagidae parasitize Hemiptera, Homoptera, and crickets. This large family includes the common genus *Halictophagus* (Fig. 20.16).

INSECT PHYSIOLOGY AND BIOCHEMISTRY AND HOST-PARASITE RELATIONSHIPS

The physiology and biochemistry of insects, whether of free-living or parasitic species, has become a distinct, specialized field of study. With the exception of some of the endoparasitic insects, such as myiasis-causing dipteran larvae, the parasitic larvae of certain hymenopterans, and the strepsipterans, the basic physiologic and biochemical processes with the body of various insects follow essentially comparable pathways. I will not survey the entire field of insect physiology here. Rather, I will attempt to emphasize some aspects of insect physiology that are directly related to host-parasite relationships. Readers interested in the general field of insect physiology should consult the volume edited by Roeder (1953) and the textbook by Wigglesworth (1939), both of which include noteworthy discussions concerning symbiotic insects. The definitive treatise on insect physiology and biochemistry are the volumes edited by Rockstein (1973, 1974a,b,c,d,e).

Oxygen Requirements

Parasitic insects, particularly the hematophagous species, are true aerobes well equipped with complex tracheal systems for the distribution and absorption of oxygen throughout the body tissues and for the elimination of carbon dioxide. However, the amount of oxygen required at each developmental stage generally varies. The sensitivity of the eggs of *Aedes* spp. to oxygen tension is well known. In eggs that are normally laid on moist soil or debris, the reduction of dissolved oxygen from 7 ppm (parts per million) to 3 ppm induces hatching, if the embryonic development is completed. However, a similar reduction of oxygen tension is not necessary to cause hatching of eggs in species of *Aedes* that normally oviposit in water.

The larvae of all mosquitoes require oxygen. For example, if *Culex pipiens* larvae are submerged in water of low oxygen tension, death occurs in approximately one hour. The survival rate of mosquito larvae when thus submerged depends to a great degree on the amount of body surface respiration and hence is directly correlated with size and the permeability of the exoskeleton to molecular oxygen.

In some insects, the tracheal system includes specialized secondary mechanisms that permit more efficient transportation of gases and certain other metabolic end products. Such mechanisms are particularly pronounced among the fleas. Hefford (1938) has reported that tracheal pulsations occur in certain fleas. When the oxygen is absorbed from within the closed tracheal system in these insects, in time the intratracheal pressure falls below the hydrostatic pressure of the surrounding hemocoel, resulting in collapse of the less rigid portions of the tracheal system. When the spiracles are reopened, air rushes in to fill the collapsed portions, inflating them, as well as the air sacs. Thus, the flea possesses a mechanical ventilation that supplements the normal diffusion process. Wigglesworth (1939) has reported that in certain flea and mosquito larvae, liquid is normally found within the tracheoles. This intratracheal fluid is withdrawn into the tissues during muscular activity but is released into the tracheoles again during periods of muscular relaxation. Movement of this fluid is involved in regulation of the passage of oxygen to the tissues and in removal of carbon dioxide and other metabolic end products.

Temperature is a critical factor in insect respiratory activities. In fact, it has been suggested that temperature is the only governing factor in the respiration of *Glossina*. Increases in respiratory rate generally accompany rises in temperature; however, on reaching a critical thermal point, which varies among species, heat inactivation of the animal occurs. Although heat inactivation is markedly noticeable, oxygen consumption does not cease with the "death" of the animal. Instead, in some insects, such as *Calliphora* and *Sarcophaga* larvae, the rates increase concomitantly with the browning of the animals' tissues for 24 hours after heat inactivation is initiated.

In insects that are parasitic in the alimentary tract of their host, respiration is microaerobic (or **oligopneustic**) and possibly nonaerobic (or **apneustic**) (Edwards, 1953).* All or most of the required oxygen is obtained by diffusion from the host's semifluid intestinal environment. The morphology and physiology of endo-

---

*The respiratory activities of arthropods are commonly placed in the following categories: The **apneustic** type, in which no O_2 is utilized from the environment; the **oligopneustic** type, in which only a little O_2 is utilized; the **metapneustic** type, in which considerable O_2 is consumed, although not as much as among free-living forms; and the **amphipneustic** type, in which a large quantity of O_2 is utilized.

parasitic dipterans are greatly influenced by the fact that there are only trace quantities of oxygen within their hosts. These dipterans neither grow rapidly nor feed actively until contact is made with the atmosphere. However, respiration is definitely carried on, often by means of specialized mechanisms.

The oxygen tension in the environment governs the degree of respiratory activity of the larvae. For example, in larvae of *Gasterophilus* spp., first larval instars live in the host's epithelial tissue and are metapneustic, whereas second larval instars are located in the host's pharynx, where there is an abundance of oxygen, and hence are amphipneustic.

In addition to diffusion of gases through the body surface of dipteran larvae, other mechanisms are employed in some instances. For example, in the larva of *Ginglymyia*, an endoparasite of the larva of *Eiophila* (Lepidoptera), the postabdominal spiracles protrude from the body surface and are in direct contact with air bubbles trapped in the web surrounding the host.

In the first larval instar of *Melinda*, which invades the kidney of snails, the posterior spiracles protrude into the pulmonary cavity of the host, thus contacting an oxygen source. In dermal myiasis-causing dipteran larvae, the pore connects the exterior with the sac as an entrance for atmospheric oxygen. Furthermore, the posterior end of the maggot, bearing the spiracles, is as a rule directed toward the aperture.

Since the later larval instars of *Gasterophilus* spp. are spent in the host's stomach or intestine, where there is little oxygen, these maggots are fortified with oxygen stored within the tracheal hemoglobin-containing cells for consumption during this period of oligopneustic or possibly apneustic activity. Keilin and Wang (1946) have reported that the hemoglobin of *Gasterophilus* is composed of two heme-globin units instead of the four units found in nearly all vertebrate hemoglobins.

Bloodsucking insects undoubtedly acquire a certain amount of oxygen from the oxyhemoglobin in the ingested erythrocytes. The hemoglobin ingested by *Rhodnius* is nearly completely digested, for only a very small amount of protohematin can be found in the feces.

Although parasitic hymenopteran larvae possess tracheae, these larvae do not attempt to reach the respiratory organs of their host, where they could obtain oxygen. All gaseous exchanges take place through the extremely thin body cuticle. Oxygen is derived from the host's hemolymph.

Metabolism

Relatively little is known about cellular metabolism in parasitic insects, although there is no reason to believe that it is very different from that of free-living species. Aerobic metabolism appears to predominate in most species. Anaerobic metabolism has been demonstrated in the larvae of *Tenebrio molitor* and undoubtedly occurs in other species. If this larva is maintained at a low oxygen tension (3%) for several hours and then returned to atmospheric oxygen tension, an oxygen debt can be demonstrated, because oxygen uptake is raised above normal for some time. This phenomenon, as in certain other parasites (p. 39), is attributed to the accumulation of intermediary products of anerobic metabolism that become oxidized only after oxygen is restored. Similarly, insects maintained in the complete absence of oxygen demonstrate an oxygen debt when they are returned to an oxygen-containing environment.

During aerobic metabolism, undoubtedly both carbohydrate metabolism, involving utilization of stored glycogen, and lipid metabolism provide most of the required energy. At a given temperature, the metabolic rate of a given species shows a fairly definite relation to size. In general, metabolism per gram of body weight decreases as the insect grows. The metabolic rate declines in such a manner that the respiratory rate is proportional to the mass of the insect multiplied by an exponential factor of two-thirds.

Although certain endoparasitic fly larvae, such as those of *Gasterophilus*, can be maintained for surprisingly long periods in the absence of oxygen, they show almost no accumulation of lactic acid, suggesting an extremely low level or perhaps an absence of anaerobic fermentative metabolism. In these animals, evidence indicates that the glycogen stored in their bodies is coverted to lipids, and the oxygen set free during the process is utilized.

Concerning carbohydrate uptake and metabolism in insects, it is known that the amount of hemolymph sugars varies greatly among species. Analyses of insect hemolymph sugars have revealed that glucose is the predominant sugar in most species, but other sugers may be present as well. Fructose, for example, often occurs in relatively high concentrations, as does the disaccharide trehalose. The presence of trehalose is of particular interest, for it is correlated with the passage of sugar through the intestinal wall in insects. Treherne (1967) has found that glucose is absorbed from the intestine in such a manner that active transport is suggested in addition to diffusion. Glucose need not be the predominant sugar in the hemolymph, for when the glucose concentration in the intestine is low, nearly all sugar in the hemolymph is in the form of trehalose. The synthesis of trehalose from glucose is a part of the active transport mechanism that permits the transfer of glucose from the intestine to the blood. It is now known that trehalose synthesis and breakdown can also occur in the insect's fat body and that trehalose can be converted to glycogen via its initial breakdown to glucose-6-phoshate (Fig. 20.17). For a review of the composition of insect hemolymph, see Florkin and Jenniaux (1974).

Amino acid requirements and utilization by certain insects are listed in Table 20.2.

Although energy metabolism among parasitic insects has not been studied extensively, that of free-living species has been investigated in considerable depth. The volume edited by Downer (1981) should prove to be a valuable reference for those interested in this topic. For a detailed review of protein synthesis in insects in general, see Ilan and Ilan (1974).

Water Absorption

In almost all parasitic insects, especially the ectoparasitic species, an aquatic medium is not required for the absorption of water. In the bedbug *Cimex lectularius*, for example, water can be absorbed from vapor in the air if the air is highly saturated. Flea larvae can also absorb water from vapor in the air, although the relative humidity need be only 50% or slightly higher. It is presumed that absorption of water occurs through the general body cuticle—the mechanism operative in acarines—but experiments demonstrating this among parasitic insects remain sparse. For a review of the permeability of the insect cuticle, see Ebeling (1974).

Growth Requirements

The growth requirements of parasitic insects constitute an extremely complex phase of the physiology of these animals and are by no means completely understood. A review of this topic has been provided by House (1974a). In several instances the fecundity of the adult insect has been shown to be directly correlated with nutrition. In the case of *Stomoxys calcitrans*, for example, if the flies are fed on host's serum or washed blood cells separately, no eggs are produced, but the feeding of a recombination of these two blood fractions results in normal longevity and egg production. This experiment demonstrates that it is not entirely the quantity of blood protein that influences fecundity but the presence of specific proteins

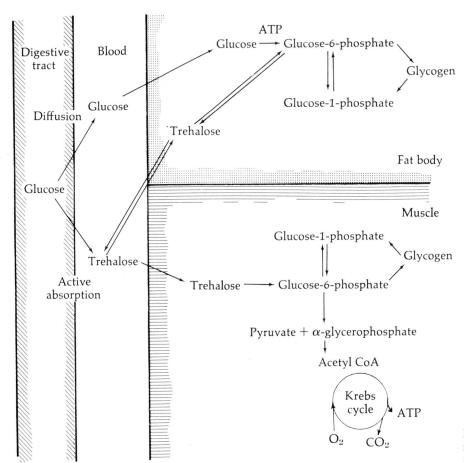

Fig. 20.17. Transport and metabolism of glucose in an insect.

or accessory substances that only the combination of serum and blood cells provides, that is the important factor. Similarly, in *Lucilia sericata*, which normally requires the ingestion of meat juices before oviposition, if the normal diet is replaced by milk, serum, or other protein-rich materials, oviposition does not occur. However, a combination of serum with autolyzed yeast produces oviposition, indicating that this artificial diet includes the substances normally present in meat juices that induce egg production or include sufficiently similar substances to induce egg production (Hobson, 1938).

Various mosquitoes oviposit prior to the intake of a blood meal. This is true in *Aedes scutellaris*, *A. atropalpus*, *A. concolor*, and *Culex pipiens*. However, the number of these **autogenous** eggs depends to a considerable extent on the adequacy of larval diet. Even in these instances, blood meals necessarily precede further period of oviposition, and a minimal quantity of blood is necessary. In *Aedes aegypti*, for example, the females require at least 0.82 mg of blood to induce egg production, although a normal full meal consists of 2 mg. It is now clear that the nutritional requirements for oviposition consist of specific protein as well as quantity.

The nature or quantity of the required proteins inducing egg production varies among types of blood. It has been mentioned earlier that *Culex pipiens* will lay twice as many eggs per milligram of blood ingested if fed on canary blood instead of on human blood. Similarly, *Aedes aegypti* will oviposit more eggs

if fed on guinea pig, rabbit, canary, or frog blood than on human blood. These results strongly suggest degrees of host specificity as measured by egg productivity. Investigations along these lines may well lead to a better understanding of host specificity and other aspects of host-parasite relationships among blood-feeding insects.

Experiments designed to isolate the specific substances required for egg production have revealed that a combination of substances is required. For example, Yoeli and Mer (1938) have shown that if *Anopheles sacharovi* females are fed on raisins, sugar solutions, and hemoglobin solutions, only partially developed ovaries result. If these females are fed serum alone, even less ovarian development takes place. However, if they are fed on serum and washed erythrocytes without hemoglobin, sexual maturity and egg production follow. This suggests that hemoglobin contributes nothing. Greenberg (1951), however, has shown that a diet including hemoglobin is more beneficial to *Aedes aegypti* than blood including erythrocytes with the pigment removed, although the latter diet will bring about egg production of lesser quality. These results indicate that hemoglobin by itself does not elicit production and that blood without hemoglobin is necessary but not ideal. The com-

Table 20.2. A Comparison of Amino Acid Requirements and Utilization by Bacteria, Certain Protozoa, Insects, and Vertebrates[a]

Amino Acid	Bacteria	Protozoa other than Green Flagellates	Insects	Vertebrates
Leucine	ru	ru	R[b]	R
Lysine	ru	ru	R	R
Methionine	ru	ru	R	R
Phenylalanine	ru	ru	R[b]	R
Threonine	ru	ru	R[b]	R
Histidine	ru	ru	R	R
Isoleucine	ru	ru	R	R
Tryptophan	ru	ru	R[b]	R
Valine	ru	ru	R[b]	R
Arginine	ru	ru	R	r
Cystine-cysteine	ru	u	R[b]	N
Glutamic acid	ru	u	r	r
Glycine	ru	ru	r	r
Proline	ru	ru	r	r
Serine	ru	ru	r	N
Alanine	ru	u	r[b]	N
Aspartic acid	ru	u	r[b]	N
Tyrosine	ru	ru	N	N

[a] Data from Scheer, 1963.

[b] Some substitutions are possible; r, required by some; R, required by all tested; u, used by some; N, not required by any. Phenylalanine will substitute for threonine, tryptophan for valine, valine for leucine, tyrosine for phenylalanine, methionine for cystine (except among certain mosquitoes), aspartic acid or glutamic acid for alanine, and glutamic acid for aspartic acid in some species.

bination of the two is a prerequisite for the production of large numbers of eggs. Actually, it is suspected that serum, erythrocyte protoplasm, and hemoglobin, or parts of all three, contribute to normal fecundity.

Among the nutrients required for growth, the amino acids cystine and glycine play important roles in the growth of mosquitoes and probably of most other insects (Table 20.2). If cystine is deprived, even if DL-methionine is present in large quantities, adult mosquitoes found within the puparia will all die with only their anterior ends projecting from the puparia. If glutathione is introduced in place of cystine, normal emergence take place, but the growth rate and survival period of larvae are reduced.

In addition to cystine and glycine, other substances, especially those of the vitamin B complex, make important contributions to growth and metabolism. For example, dietary phenylalanine and tyrosine are important for growth and pigment formation in *Aedes*, but diets deficient in one, but not both, of these affect normal rather than aborted growth.

Trager (1948) and Lichtenstein (1948) have demonstrated in *Aedes aegypti* and *Culex molestus* that thiamine, riboflavin, pyridoxine, niacin, pantothenic acid, biotin, pteroylglutamic acid, and choline are all required accessory growth factors. The role of biotin is of special interest, for if it is partially replaced by oleic acid, lecithin, and related lipids in proper concentrations, larval growth is fairly normal but metamorphosis does not occur. A comparison of the amount of growth factors of the vitamin B complex in various animals is presented in Table 20.3.

In addition to the previously mentioned growth

factors, Subbarow and Trager (1940) have reported that yeast nucleic acid enhances the growth of *Aedes aegypti*. This phenomenon has since been demonstrated in other insects as well.

What are the sources of these growth factors? Some of them undoubtedly are supplied by ingested nutrients, but in some insects, especially the hematophagous species, ingested foods are deficient in some of these requirements. Numerous insects, including many parasitic species, possess mutualists within their gut (extracellular) or in their tissues (intracellular) that supply accessory growth factors, enabling the insects to survive on diets that would otherwise be inadequate (Richards and Brooks, 1958). Yeast and bacteria are the most common mutualists of nonbloodsucking insects that spend their larval stages feeding on diets rich in microorganisms, for example, organic debris.

Blood feeders, such as the louse *Pediculus humanus*, possess minute bacterialike bodies, known as **bacteroids**, found in specialized structures known as stomach discs, or **mycetomes** (mycetocytes) (Fig. 20.18). Aschner (1934), Aschner and Ries (1933), and more recent investigators have demonstrated the dependency of the louse on these mutualists. These investigators have reported that if the hosts are deprived of their intracellular bacteroids while in the egg stage, they die by the fifth or sixth day of larval life. If the bacteroids are removed from the hosts during the third larval stadium, the larvae will develop into adults. In such adults, the males are normal, but the females lay very few eggs and none of these are viable. Brooks (1956, 1960) has summarized most of the information concerning the bacteroid mutualists of cockroaches and has demonstrated that these bacteroids are passed from one generation to the next transovarially. It is generally assumed that in *P. humanus* and other bacteroid-including insects this is also the case. Sacchi *et al.* (1985) have presented a

Table 20.3. Minimal Amounts of Growth Factors of the Vitamin B Complex Required for Normal Growth by Certain Insects and Selected Vertebrates[a,b]

Species	Thiamine	Riboflavin	Pyridoxine	Niacin	Pantothenic Acid	Biotin	Pteroyl-glutamic Acid	Choline
Aedes aegypti	0.1[27]	0.5[27,22]	1.0[23]	1.0[22]	1.0[23]	0.015[20]	+	+
Tribolium confusum	1.0[4]	2.0[4]	1.0[4]	8.0[4]	4.0[4]	0.05[4]	0.25[4]	+
Tenebrio molitor	+	+	+	+	+	−	0.3[5]	−
Blattella germanica	+	+	−	−	−	−	−	2000[15]
Drosophila melanogaster	0.1[27]	0.5[27,22]	1.0[23]	1.0[22]	1.0[23]	0.015[20]	+	+
Chick	0.8[12]	6.0[10]	3.0[11]	18.0[2]	9.0[9]	0.1[8]	0.5[17]	2000[14]
Rat	1.0[12]	3.0[13]	1.5[3]	0[1]	10.0[7]	+	+	200[16]
Dog	+	4.0[6]	+	10.0[31]	+	−	−	+
Monkey	+	0.25[28]	−	−	+	0.1[29]	+	−
Man	1.5[30]	3.0[21]	1.5[30]	20.0[30]	10.0[30]	+	−	−

[a] Quantities given as micrograms/gram of diet or per milliliter of nutrient medium.
[b] Data compiled by Trager, 1953.

detailed description of how bacteroids invade bacteriocytes in developing German cockroaches, *Blattella germanica.*

Various endoparasitic insects, such as certain strepsipteran larvae and the first-stage larvae of some hymenopterans, possess no buccal opening. Furthermore, in some of these the gut is not continuous from mouth to anus. In these endoparasites, nutrients must necessarily be absorbed through the body wall.

Digestion

Parasitic insects that are capable of feeding, unlike the helminths, are able to digest large molecules. This is possible because of enzyme-secreting cells that line the alimentary tract. *Aedes* is a good example in this respect. If starch or other polysaccharides are fed to this mosquito, glycogen is accumulated in the posterior half of the ventriculus before it appears to the fat bodies. This convincingly suggests that the starch and sugar molecules are hydrolyzed in portions of the gut anterior to the posterior half of the ventriculus. Simple sugars are then synthesized as glycogen for storage, suggesting that at least the starch-splitting enzymes must be present. Indeed, this has been shown to be true. In *Anopheles* and *Aedes*, digestion of the complex hemoglobin molecule occurs. Similar processes are found in other bloodsucking insects. Thus, the optimal pH of the gut is of critical importance, for this complements enzymatic action. Table 20.4 compares the gut pH in several insects. For a detailed review of digestion in insects, see House (1974b).

Attraction of Insects to Hosts

The attraction of bloodsucking insects to their hosts is best thought of as manifestations of the sensory physiology of these animals.

Temperature of Host. Temperature plays an important role in attraction. For example, it is known that the bedbug *Cimex* prefers and is attracted to temperatures of approximately 35°C, which is about the temperature of the host's skin. It is also known that fleas are optimally attracted to a temperature of ±30°C. Again, this approximates the average mammalian skin termperature.

If *Pediculus* or *Schistocerca* is placed in an experimental gradient or on a host, it undergoes restless movements until it reaches an area of optimal temperature—26.4 to 29.7°C. It then settles down and begins feeding. Dethier (1957) has summarized experiments performed on two lice, *Pediculus humanus* and *Haematopinus suis*, stating that these bloodsuckers are responsive to light, temperature, humidity, odor, and contact. Normally lice are negatively phototropic, but starved *Pediculus* are positively phototropic.

Humidity. Lice are indifferent when placed in an environment where the relative humidity ranges from 10 to 60%. They choose an area of high humidity

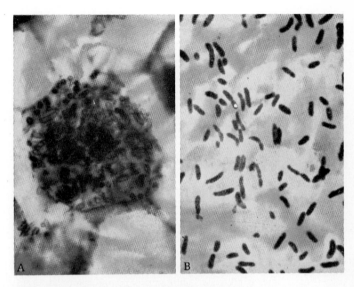

Fig. 20.18. Bacteroids of *Blattella germanica*. A. Single mycetocyte enclosing bacteriods. **B.** Smear of fat body showing bacteriods in various stages. (Courtesy of Dr. M. A. Brooks, University of Minnesota.)

(95%) if the other choices are low (10, 32, and 47%). On the other hand, if the choices include 95, 85, 76, and 60% relative humidity, the lice pick one of the lower zones rather than the 95% zone. It would thus appear that these insects prefer a medium humidity

Table 20.4. Hydrogen Ion Concentration (pH) of Midgut Juices of Some Insects[a]

Species	pH
Thysanura	
Ctenolepisma	4.8–7.0
Orthoptera	
Blattella	6.2
Carausius	6.3
Various grasshoppers	5.8–7.5
Gryllus	7.6
Tettigonia	5.9
Anoplura	
Pediculus	7.2
Odonata	
Anax	6.8–7.2
Hemiptera	
Cimex	6.2
Nezara	7.2
Hymenoptera	
Apis	6.3
Polistes	7.3

[a]Data compiled by Day and Waterhouse in Roeder, 1953.

Fig. 20.19. Mosquito salivary gland. Three-lobed salivary gland of an anopheline mosquito showing inclusion of *Plasmodium* sporozoites. (Redrawn after Wenyon, 1926.)

and avoid any change once they become established to a given humidity.

Environmental Temperature. Activity of biting flies is correlated with environmental temperature. In *Stomoxys calcitrans*, the preferred temperatures range from 22 to 32°C, and the peak of activity occurs at 29°C. In *Fannia canicularis*, imagos of which appear earlier in the year, the preferred temperatures range between 10 and 28°C.

Other Factors. Rahm (1957) has reported that mosquitoes are attracted to dark clothing (optical attraction), humidity, and heat (physical attraction). A combination of these three factors is more attractive than each one singly. Mosquitoes are also attracted to odors (chemical attraction) given off by the hands, although they are not attracted to perspiration.

Receptors. How insects are capable of recognizing chemical and physical signals has not been studied extensively among the parasitic species, although the responsible mechanisms have been investigated extensively among free-living species. Insects possess chemoreceptors as well as mechanoceptors in addition to vision. What is known about chemoreceptors has been reviewed by Hodgson (1974) and what is known about mechanoreceptors has been reviewed by Schwartzkopff (1974). The visual system of insects has been reviewed by Goldsmith and Bernard (1974). In brief, insects, including flies and mosquitoes, possess chemosensory organs made up of many receptor cells. These occur on their bodies, especially on the antennae and mouthparts (Slifer *et al.*, 1959; Slifer and Brescia, 1960). It is generally agreed that insect chemoreceptor cells are modified epithelial cells and that the central axon of the receptor is formed by ingrowth of an extension process toward the central nervous system.

The mechanoreceptors of insects, including mosquitoes, appear as hairs and cuticular sensillae on vari-

ous regions of the body, especially associated with the antennae and appendages. There are also thin areas of the cuticle to which mechanoreceptive functions have been attributed. Each of these receptors is also comprised of a group of innervated cells.

Effects of Parasite on Host

The obvious effects of parasitic insects on their hosts are (1) loss of blood to hematophagous species; (2) introduction of toxins into the host during feeding; (3) infections and general weakening brought about by myiasis-causing species; (4) infection by pathogenic microorganisms carried by these insects; (5) wounds that are subject to secondary infections; and (6) destruction of body tissues, resulting in death when endoparasites are present.

Insect Toxins and Secretions. In *Anopheles*, each salivary gland is made up of three lobes (Fig. 20.19). The lateral lobes include no hemagglutinin and little anticoagulin but do include large amounts of toxic substances. On the other hand, the median lobe contains a strong hemagglutinin and a strong anticoagulin. The toxic secretions of the lateral lobes are responsible for the local inflammation and itchy sensation of mosquito bites.

The toxin of *Aedes* is thermostable. Metcalf (1945) has reported that the anticoagulin is also stable and is active even in a dilution of 1 : 10,000. He also has reported that the hemagglutinin is thermolabile but is active in a dilution of 1 : 1,000,000. The hemagglutinin, the anticoagulin, and the toxin are not found in male mosquitoes, which are not blood feeders.

The toxic effect of bloodsucking insects is not necessarily only local. In some instances the effect is systemic, the toxin being circulated by the bloodstream. Dem'yanchenko (1957) has reported that if bull calves are subjected to mass attacks by the blackfly *Eusimulium pusilla*, there is a rise in blood temperature and leukocyte count; decrease in hemoglobin, erythrocyte, and lymphocyte counts; and disappearance of monocytes. Furthermore, he has found that if an emulsion of the fly's thoracic sections, including the salivary glands, is injected subcutaneously, there is a distinct localized inflammation accompanied by general systemic reactions. These reactions are much more severe if the emulsion is injected intravenously.

The toxicity of myiasis-causing maggots has interested parasitologists for decades. Seyderhelm and Seyderhelm (1914) and Seyderhelm (1918) have reported extraction from *Gasterophilus* larvae of a toxic substance that they termed **oestrin.** They claimed that oestrin is responsible for pernicious anemia of horses, but du Troit (1919) and Marxer (1920) have stated that the Seyderhelms were in error. Zibordi (1920) and du Troit have reported that extracts of *Gasterophilus* larvae may kill horses, or at least effect transitory toxicity. Cameron (1922) has stated that the symptoms of toxicity are mainly due to sensitization from a previous infection. However, Roubaud and Perard (1924) have reported that they were able to produce symptoms of toxicity in small laboratory animals that positively had not previously been exposed to the fly. This question appears to have been at least partially settled by the more modern investigations of Grab (1957) who has reported that in his experiments with 30 horses, symptoms typical of anaphylactic shock, accompanied by changes in the blood chemistry and pathologies of certain internal organs, are effected by single intravenous injections of extract prepared from the larvae of *Gasterophilus*. However, when 62 two- to five-month-old rabbits, guinea pigs, and mice are similarly injected, no reactions occur. Only after the serum from one of the experimentally injected horses is first introduced into the laboratory animal, or upon second challenge with the extract, do the reactions become apparent. Grab's results appear to uphold Cameron's hypothesis that symptoms of toxicity, at least in laboratory animals, result only after a previous challenge. Why the horses reacted to the first injection is not clear, but it is possible that they had been previously infected with *Gasterophilus*—in other words, they had been presensitized—and the reactions actually represented anaphylactic shock as the symptoms suggested.

Relative to cutaneous myiasis-causing maggots, it is known that if the larvae of *Hypoderma* is crushed while in the skin of cattle, fever ensues, suggesting secretion of toxic substances. However, since small laboratory animals are only slightly sensitive to injections of the substance, the reaction in cattle may be anaphylactic.

The topic of immunity to arthropodan parasites has been reviewed by Benjamini and Feingold (1970). The major features have been discussed earlier (p. 91).

REFERENCES

Aschner, M. (1934). Studies on the symbiosis of the body louse. Elimination of the symbionts by centrifugation of the eggs. *Parasitology* **26**, 309–314.

Aschner, M., and Ries, E. (1933). Das Verhalten der Kleiderlaus bei Ausschaltung ihrev Symbionten. Eine experimentelle Symbiosestudie. *Z. Morphol. Okol. Tiere* **26**, 529–590.

Askew, R. R. (1971). "Parasitic Insects." American Elsevier, New York.

Benjamini, E., and Feingold, B. F. (1970). Immunity to arthropods. *In* "Immunity to Parasitic Animals" (G. J. Jackson, R. Herman, and I. Singer, eds.), Vol. 2, pp. 1061–1134. Appleton-Century-Crofts, New York.

Bohart, R. M. (1941). A revision of the Strepsiptera with special reference to the species of North America. *Univ. Calif. Publ. Entomol.* **7**, 91–160.

Borror, D. J., and DeLong, D. M. (1954). "An Introduction to the Study of Insects." Holt, Rinehart and Winston, New York.

Brooks, M. A. (1956). Nature and significance of intracelular bacteroids in cockroaches. *Proc. Tenth Int. Congr. Entomol. (Montreal)*, **2**, 311–314.

Brooks, M. A. (1960). Some dietary factors that affect ovarial transmission of symbiotes. *Proc. Helminthol. Soc. Wash.* **27**, 212–220.

Cameron, A. G. (1922). Bot anaphylaxis. *J. Am. Vet. Med. Ass.* **62**, 332–342.

Chopard, L. (1929). Les parasites de la Mante religieuse, *Riela manticida* Kieff. *Ann. Soc. Entomol. Fr.* **91**, 249–264.

Clausen, C. P. (1940). "Entomophagous Insects." McGraw-Hill, New York.

Crissey, J. T. (1983). The bedbug. *In* "Cutaneous Infestations of Man and Animal." (L. C. Parish, W. B. Nutting, and R. M. Schwartzman, eds.) pp. 296–303. Praeger, New York.

Dem'yanchenko, G. F. (1957). The toxicity of black-fly (Simuliidae) saliva on the organism of farm animals. *Tr. Vses. Nauch-Issled. Inst. Vet. Sanit*, pp. 91–104.

Dethier, V. G. (1957). The sensory physiology of blood-sucking arthropods. *Exp. Parasitol.* **6**, 68–122.

Downer, R. G. (ed.) (1981). "Energy Metabolism in Insects." Plenum, New York.

Du Troit, P. J. (1919). Gastrularven und infektose Anaemie der Pferde. *Monatsh. Prakt. Tierheilko* **30**, 97–118.

Ebeling, W. (1974). Permeability of insect cuticle. *In* "The Physiology of Insecta," 2nd ed. (M. Rockstein, ed.), Vol. VI, pp. 271–343. Academic Press, New York.

Florkin, M., and Jeuniaux, C. (1974). Hemolymph: composition. *In* "The Physiology of Insecta," 2nd ed. (M. Rockstein, ed.), Vol. V, pp. 255–307. Academic Press, New York.

Goldsmith, T. H., and Bernard, G. D. (1974). The visual system of insects. *In* "The Physiology of Insecta," 2nd ed. (M. Rockstein, ed.), Vol. II, pp. 165–272. Academic Press, New York.

Grab, B. G. (1957). The nature of reactive changes in the organism of horses exposed to the parasitism of horse botfly (*Gasterophilus equi*) larvae. *Tr. Kievsk. Vet. Inst.*, pp. 157–172.

Greenberg, J. (1951). Some nutritional requirements of adult mosquitoes (*Aëdes aegypti*) for oviposition. *J. Nutr.* **43**, 27–35.

Harwood, R. F., and James, M. T. (1979). "Entomology in Human and Animal Health," 7th ed. Macmillan, New York.

Hefford, G. M. (1938). Tracheal pulsation in the flea. *Exp. Biol.* **15**, 327–338.

Hobson, R. P. (1938). Sheep blow-fly investigations. Observations on the development of eggs and oviposition in the sheep blow-fly, *Lucilia sericata*. *Mg. Ann. Appl. Biol.* **38**, 383–412.

Hodgson, E. S. (1974). Chemoreception. *In* "The Physiology of Insecta." 2nd ed. (M. Rockstein, ed.), Vol. II, pp. 127–164. Academic Press, New York.

House, H. L. (1974a). Nutrition. *In* "The Physiology of Insecta." 2nd ed. (M. Rockstein, ed.), Vol. V, pp. 1–162. Academic Press, New York.

House, H. L. (1974b). Digestion. *In* "The Physiology of Insecta," 2nd ed. (M. Rockstein, ed.), Vol. V, pp. 63–117. Academic Press, New York.

Ilan, J., and Ilan, J. (1974). Protein synthesis in insects. *In* "The Physiology of Insecta," 2nd ed. (M. Rockstein, ed.), Vol. IV, pp. 355–422. Academic Press, Lew York.

Keilin, D., and Wang, Y. L. (1946). Hemoglobin of *Gasterophilus* larvae. Purification and properties. *Biochem. J.* **40**, 855–866.

Lees, A. D. (1955). "The Physiology of Diapause in Arthropods." Cambridge University Press, London.

Lichtenstein, E. P. (1948). Growth of *Culex molestus* under sterile conditions. *Nature (London)* **162**, 227.

Marxer, A. (1920). Die Beziehungen der *Gasterophilus*-Larven zur infektiösen Anämie. *Z. Immunitaetsforsch. Exp. Ther.* **1** (29), 1–10.

Metcalf, R. L. (1945). The physiology of the salivary glands of *Anopheles quadrimaculatus. Nat. Malaria Soc.* **4**, 271–278.

Muesebeck, C. F. W. *et al.* (1951). Hymenoptera of America North of Mexico. Synoptic catalogue. *U.S. Dept. Agr. Monogr. No.* 2. Washington, D.C.

Nassonow, N. (1892). Contributions a l'histoire des Strepsiptères. *Protok. Obshch. Varshav.* **3**, 1–3 (In Russian.) (Also see 1892. *Congr. Zool*, pp. 174–184.)

Paillot, A. (1937). Le développement embryonnaire d'*Amicroplus collaris* Spin., parasites des chenilles d'*Euzoa segetum* Schiff. *C. R. Acad. Sci. (Paris)* **204**, 810–812.

Parker, H. L. (1931). *Macrocentrus gifuensis* Ashmead, a polyembryonic braconid parasite of the European corn borer. *U.S. Dept. Agr. Tech. Bull.* 230. Washington, D.C.

Patterson, J. (1917). Studies on the biology of *Paracopidosomopis*. I. Data on the sexes. *Biol. Bull.* **32**, 291–305.

Pierce, W. D. (1909). A Monographic Revision of the Twisted Winged Insects Comprising the Order Strepsiptera Kirby. *U.S. Nat. Mus. Bull. No.* 66.

Pierce, W. D. (1911). Strepsiptera. Genera Insect., Fasc. 121.

Rahm, V. (1957). Wichtige Faktoren bei Attraktion von Stechmücken durch den Menschen. *Rev. Suisse Zool.* **64**, 236–246.

Richards, A. G., and Brooks, M. A. (1958). Internal symbiosis in insects. *Ann. Rev. Entomol.* **3**, 37–56.

Rockstein, M. (ed.) (1973). "The Physiology of Insecta," 2nd ed, Vol. I. Academic Press, New York.

Rockstein, M. (ed.) (1974a,b,c,d,e). "The Physiology of Insecta," 2nd ed, Vols. II, III, IV, V, VI. Academic Press, New York.

Roeder, K. D. (ed.) (1953). "Insect Physiology." Wiley, New York.

Roubaud, E., and Perard, C. (1924). Etudes sur l'hypoderme on varron des boeufs; les extraits d'oestres et l'immunisation. *Bull. Soc. Pathol. Exot.* **17**, 259–272.

Sacchi, L., Grigolo, A., Laudani, U., Ricevuti, G., and Dealessi, F. (1985). Behavior of symbionts during oogenesis and early stages of development in the German cockroach, *Blatella germanica* (Blattodea). *J. Invert. Pathol.* **46**, 139–152.

Schwartzkopff, J. (1974). Mechanoreception. *In* "The Phys-

iology of Insecta." 2nd ed. (M. Rockstein, ed.), Vol. II, pp. 273–352. Academic Press, New York.

Seyderhelm, R. (1918). Über die Eigenschaften und Wirkungen des Oestrins und seine Beziehung sur perniziösen Anämie der Pferde. *Arch. Exp. Pathol. Pharmakol.* **82**, 253–326.

Seyderhelm, K. R., and Seyderhelm, R. (1914). Die Ursache der perniziösen Anämie der Pferde; ein Beitrag zum Problem des ultravisiblen Virus. *Arch. Exp. Pathol. Pharmakol.* **76**, 149–201.

Slifer, E. H., and Brescia, V. T. (1960). Permeable sense organs on the antenna of the yellow fever mosquito, *Aedes aegypti* (Linnaeus). *Entomol. News* **71**, 221–225.

Slifer, E. H., Prestage, J. J., and Beams, H. W. (1959). The chemoreceptors and other sense organs of the antennal flagellum of the grasshopper (Orthoptera: Acrididae). *J. Morph.* **105**, 145–191.

Smith, G., and Hamm, A. H. (1914). Studies in the experimental analysis of sex. Part II—on stylops and stylopization. *Quart. J. Microsc. Sci.* **60**, 435–461.

Subbarow, Y., and Trager, W. (1940). The chemical nature of growth factors required by mosquito larvae. II. Pantothenic acid and vitamin B_6. *Gen. Physiol.* **23**, 461–468.

Trager, W. (1948). Biotin and fat-soluble materials with biotin activity in the nutrition of mosquito larvae. *J. Biol. Chem.* **176**, 1211–1223.

Treherne, J. E. (1967). Gut absorption. *Annu. Rev. Entomol.* **12**, 43–58.

Usinger, R. L. (1944). The Triatominae of North and Central America and the West Indies and their public health significance. *Pub. Health Bull.* 288, U.S. Public Health Service, Washington, D.C.

Usinger, R. L. (1966). Monograph of Cimicidae (Hemiptera-Heteroptera). *Thomas Say Foundation* **7**, 1–69; 179–182; 246–585. Ent. Soc. Am., College Park, Maryland.

Whiting, P. W. (1956). *Mormoniella* and the nature of the gene: *Mormoniella vitripennis* (Walker) (Hymenoptera: Pteromalidae). *Proc. Tenth Int. Congr. Entomol.* **2**, 857–865.

Wigglesworth, V. B. (1939). "The Principles of Insect Physiology." Dutton, New York.

Yoeli, M., and Mer, G. G. (1938). The relation of blood feeds on the maturation of ova in *Anopheles elutus. Trans. Roy. Soc. Trop. Med. Hyg.* **31**, 437–444.

Zibordi, D. (1920). Intorno al potere tossico degli estratti di *Gasterophilus equi. Clin. Vet.* **43**, 470–476.

CLASSIFICATION OF HEMIPTERA (THE TRUE BUGS)*

Class Insecta (see p. 614)

ORDER HEMIPTERA

Primarily ectoparasitic; bodies medium size; with piercing-sucking mouthparts forming a beak; antennae six- to ten-segmented; eyes compound and large; with two pairs of wings; **corium** (thickened portion) present at base of outer wing; thinner extremities of outer wing overlap on dorsum when insect is at rest; wing venation reduced; abdomen lacks cerci; holometabolous metamorphosis.

---

*Only families that include species parasitic in or on animals included.

Family Cimicidae

With vestigial hemelytra or wingless; ocelli absent; antennae four-jointed; beak three-jointed; tarsi three-jointed. (Genera mentioned in text: *Cimex, Leptocimex.*)

Family Reduviidae

Beak short, three-jointed, attached to tip of head; distal end of beak rests on prosternum in groove when not in use; ocelli present in winged species (with few exceptions); antennae four-jointed. (Genera mentioned in text: *Triatoma, Rhodnius, Panstrongylus, Melanolestes, Rashus.*)

Family Polyctenidae

Hemelytra vestigial; hind and middle tarsi four-jointed; parasitic on bats. (Genus mentioned in text: *Hesperoctenes.*)

CLASSIFICATION OF HYMENOPTERA (THE WASPS)[*]
Class Insecta
ORDER HYMENOPTERA

Parasitic species small, measuring approximately 0.1 mm in length; integument comparatively heavily sclerotized; mouthparts of chewing type; with two pairs of well-developed wings, if present; wings reduced or absent in some; in winged forms, hindwings smaller than forewings; antennae of 3 to 60 segments; larvae caterpillar-like, with chewing mouthparts; holometabolous metamorphosis or hypermetamorphosis.

Suborder Symphyta (= Chalastogastra)

Abdomen broadly jointed to thorax; winged, with nearly all of the wing veins preserved, although branches of forked veins are modified; ovipositor of females well developed and sheathed.

Family Oryssidae

Adults very active on tree trunks and timber; larvae parasitic on larvae of wood borers (Buprestidae). (Genus mentioned in text: *Oryssus.*)

Suborder Apocrita (= Clistogastra)

First abdominal segment (actually the second) greatly constricted, forming slender waist (**petiole**) between abdomen and wing-bearing region of body; ovipositor and its sheath modified for boring in some, appears as a stinger (with connected poison glands) in others.

SUPERFAMILY ICHNEUMONOIDEA

Wasplike; do not sting (with few exceptions); majority are slender; with long antennae composed of 16 or more segments; trochanters of two joints; without costal cell in forewings; ovipositor of females very long, often longer than body; ecto- or endoparasites of other insects or other invertebrates. (Genera mentioned in text: Family Ichneumonidae[†]—*Rhysella, Phobocampe, Tersilochus.* Family Braconidae—*Apanteles, Macrocentrus, Phanomeris, Amicroplus.*)

SUPERFAMILY CHALCIDOIDEA

Small wasps, less than 1–3 mm long; with reduced wing venation; with elbowed antennae; zooparasitic species attack larvae of other parasitic hymenopterans. (Genera mentioned in text: Family Pteromalidae[†]—*Mormoniella.* Family Chalcididae—*Encyrtus, Perilampus, Paracopidosomopis, Ageniaspis.*)

SUPERFAMILY PROCTOTRUPOIDEA

Adults as minute black wasps, 3–8 mm long; larvae parasitic in or on immature stages of other insects. (Genera mentioned in text: Family Evaniidae[†]—*Prosevania.* Family Pelecinidae—*Pelecinus, Polygnotes.* Family Scelionidae—*Rielia.*)

SUPERFAMILY CYNIPOIDEA

(Most members are gall-forming wasps of plants)

SUPERFAMILY CHRYSIDOIDEA

Small wasps, rarely over 12 mm long; bodies metallic green or blue, coarsely sculptured; ectoparasites of adult wasps or bee larvae.

SUPERFAMILY SCOLIOIDEA

Fair-sized wasps that measure about 1.5 cm or more in length; bodies hairy; black or black and yellow; legs short; larvae commonly on larvae of scarabeid beetles. (Genus mentioned in text: Family Tiphiidae[†]—*Tiphia.*)

CLASSIFICATION OF COLEOPTERA (THE BEETLES)
Class Insecta
ORDER COLEOPTERA

Few species endoparasitic in other insects; with two pairs of wings; forewings (elytra) veinless, hard, covering dorsal aspect of abdomen while in resting position, meeting along middorsal line; mouthparts of chewing type; antennae of 10 to 14 segments; compound eyes conspicuous; legs heavily sclerotized; holometabolous metamorphosis. (Genera mentioned in text: Family Leptinidae[†]—*Leptinillus.* Family Platypsyllidea—*Platypsyllus*; Family Tenebrionidae—*Tenebrio.*)

CLASSIFICATION OF STREPSIPTERA (THE TWISTED-WING INSECTS)
Class Insecta
ORDER STREPSIPTERA

Endoparasitic in other insects; bodies small; only males have wings; forewings reduced to club-shaped appendages; hindwings large, fan shaped, with radiating venation; females legless and larviform; mouthparts vestigial or absent in both sexes; hypermetamorphosis.

Family Mengeidae

Tarsi five-segmented, with two claws. (Genera mentioned in text: *Eoxenos, Triozocerca.*)

Family Stylopidae

Tarsi four-segmented; antennae three- or six-segmented; without claws. (Genera mentioned in text: *Stylops, Hylecthrus, Xenos, Opthalmochlus, Acroschismus, Pentozocera.*)

Family Elenchidae

Tarsi two-segmented; antennae four-segmented; without claws.

Family Halictophagidae

Tarsi three-segmented; antennae seven-segmented, third, fourth, and fifth segments prolonged laterally, seventh segment elongate; without claws. (Genus mentioned in text; *Halictophagus.*)

---

[*]Only taxa that include species parasitic in or on animals included.

[†]For diagnoses of families, see Borror and DeLong (1954).

21

OTHER ZOOPARASITES

As stated in Chapter 1, parasitism is a way of life for which examples can be found in practically every major animal phylum. Although many of the parasitic animals have been considered separately in foregoing chapters, the organisms discussed in this chapter are not by any means less interesting despite the fact that space limitations prevent their being considered in separate chapters. Some of the animals considered in this chapter demonstrate some of the most profound adaptations to parasitism known in the Animal Kingdom. This is particularly true among the mesozoans and molluscs. Students of parasitology should become familiar with the animals presented in the following discussions. Undoubtedly future explorations of these relatively little studied parasites will reveal many exciting adaptations not readily appreciated among the better known categories of animal parasites.

Discussed in this chapter, albeit briefly, are the Mesozoa, a totally parasitic group, as well as parasitic representatives of the phyla Porifera, Coelenterata, Ctenophora, Nematomorpha, Rotifera, Rhynchocoela, Annelida, Mollusca, Tardigrada, Pentastomida, and Echinodermata. Also considered are certain parasitic representatives of the subphylum Crustacea and the class Pycnogonida of the phylum Arthropoda and the subphylum Vertebrata of the phylum Chordata. By far, the majority of these are marine.

PHYLUM MESOZOA

The phylum Mesozoa is a small group of multicellular endoparasites found in the nephridia of squids and octopuses, and in the body spaces and tissues of other marine invertebrates. Although the morphology of both larval and adult forms of the known species has been studied by several workers, the complete life history pattern of one whole group, the Dicyemida, continues to elude biologists.

The relationship of the Mesozoa to the other multicellular invertebrates is uncertain, although some consider the Mesozoa to be highly modified relatives of the digenetic trematodes. Because the mesozoans possess cilia, it has been suggested that they have evolved from ciliates.

Members of this phylum lack digestive, excretory, circulatory, and nervous systems. Their bodies consist of two layers of cells that are not comparable to the epidermis (ectoderm) and endodermis (endoderm) of diploblastic animals (Fig. 21.1). McConnaughey (1968) has presented a review of the Mesozoa.

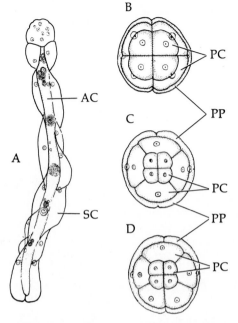

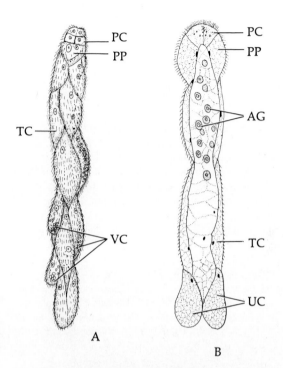

Fig. 21.1. **Morphology of some dicyemids. A.** Rhombogen of *Dicyema briarei*, a parasite of *Octopus briareus*, showing centrally located axial cell (AC) and peripheral somatic cells (SC). (Redrawn after Short, 1961.) **B.** Head-on view of *Dicyema* showing geometric arrangement of polar cap (PC) and parapolar cells (PP). **C.** Head-on view of *Pseudicyema* showing geometric arrangement of polar cap and parapolar cells. **D.** Head-on view of *Dicyemennea* showing geometric arrangement of polar cap and parapolar cells. (**B–D**, redrawn after Nouvel, 1933.)

Fig. 21.2. **Mesozoans. A.** Rhombogen stage of *Dicyema typus* from an octopus showing verruciform cells. **B.** Optical section of a nematogen of *Pseudicyema truncatum* from a squid, *Sepia*. AG, agametes; PC, polar cap; PP, parapolar cell; TC, trunk cell; UC, uropolar cell; VC, verruciform cells. (Both figures redrawn after Whitman, 1882.)

BIOLOGY OF MESOZOANS

Members of the Mesozoa are divided into two classes: Dicyemida and Orthonectida. As indicated in the classification section at the end of this chapter, the Dicyemida are so different from the Orthonectida in many features that it has been suggested that perhaps these two groups should be considered distinct phyla (Dodson, 1956).

CLASS DICYEMIDA

The Dicyemida includes such genera as *Dicyema*, *Pseudicyema*, and *Dicyemennea* (Figs. 21.1, 21.2). Members of these genera are endoparasites in the renal organs of cephalopods. They are distinguished from one another by the number and arrangement of the cells that constitute the anterior end, known as the **polar cap**.

In *Dicyema* there are two tiers of oppositely arranged polar cells (the anterior tier is known as **propolar** and the posterior as **metapolar**), four cells in each tier. In *Pseudicyema* there are also two tiers of four cells each, but the cells composing the tiers are alternately arranged. In *Dicyemennea* there are four

cells in the first tier and five in the second. These tiers are also alternately arranged.

Life Cycle of *Pseudicyema truncatum*

Lameere (1916) and Nouvel (1933) have investigated the life cycle and developmental pattern of *Pseudicyema truncatum*, a species commonly found in renal organs of the European cuttlefish squid, *Sepia officinalis*. The developmental stages are shown in Figure 21.3 and the life cycle is diagrammatically depicted in Figure 21.4. The youngest form, found free in the kidneys of young hosts, is a ciliated larva known as the **larval stem nematogen**. Its body and the body of the **adult stem nematogen** into which it matures are essentially the same as that of the primary nematogen, differing in that they have three axial cells and a larger number of somatic cells arranged in two layers. The outer cilia-bearing layer, known as the **somatoderm** (somatic cells), envelopes the inner **axial cells**.

The somatic cells can be divided into four types. Those comprising the head region or polar cap are termed **propolar** and **metapolar** cells, those making up the trunk proper called **diapolars**, and those comprising the posterior terminal are known as **uropolars**. These four types of cells surround the **axial cells**. In the larval stem nematogen of *P. truncatum* there are four propolars, four metapolars, 17 diapolars and uropolars combined, and three axial cells. The axial cells are the generative (or reproductive) cells of the animal. In the ciliated larval stem nematogen, each axial cell includes one vegetative and one generative nucleus. The generative nucleus soon is surrounded by cytoplasm and becomes an **agamete**.

The ciliated larval stem nematogen develops into the adult stem nematogen made up of the same number of somatodermal cells as the ciliated larva, but during the differentiation from larval to adult stem nematogen, each agamete has divided and is seen as an aggregate of cells. Division of the agametes is strikingly similar to the early cleavage stages in the asexual stages of digenetic trematodes. This similarity, plus the presence of ciliation, as stated, has influenced some investigators to consider the Mesozoa to be closely related to the Digenea (Stunkard, 1954).

The agametes in adult stem nematogens develop into the most common stage, the **primary nematogen**, which is attached to the host's kidney tissue by its polar cap. The primary nematogen of *P. truncatum* consists of four propolars, four metapolars, two parapolars, 10 to 15 diapolars and uropolars combined, and a single axial cell. The development of primary nematogens from agametes within adult stem nematogens is a rather complex process. Each agamete undergoes several divisions, resulting in two types of cells—the smaller **somatic cells** and a larger **axial cell**. The somatic cells then divide repeatedly, forming a group of somatic cells surrounding the axial cell, thus forming a primary nematogen.

Fully developed primary nematogens escape from the parent into the host's kidney fluid. Subsequent generations of identical primary nematogens arise from these, and the number of individuals increases until the renal organ of the cephalopod becomes heavily infected.

A primary nematogen gives rise to others in the following manner. During the morphogenesis of each member of the first primary nematogen generation,

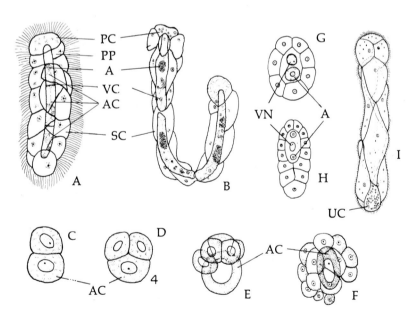

Fig. 21.3. **Stages in the life cycle of *Pseudicyema truncatum*, a parasite of *Sepia*.** **A.** Larval stem nematogen with three axial cells. **B.** Adult stem nematogen, developed directly from larva, with surface cilia omitted. (**A** and **B**, redrawn from Lameere, 1916.) **C.** Two-cell stage during development of a primary nematogen from an agamete. **D.** Three-cell stage during development of a primary nematogen from an agamete. **E.** Six-cell stage during development of a primary nematogen from an agamete. The axial cell is gradually being enveloped by the dividing somatic cells. **F.** Further division of somatic cells during formation of primary nematogen. **G.** Optical section of a developing primary nematogen showing axial cell completely enveloped by smaller somatic cells. The original axial cell nucleus has given rise to an agamete and a vegetative nucleus. **H.** Later stage showing presence of two agametes within the axial cell. **I.** Primary nematogen showing uropolar cell. A, agamete; AC, axial cells; PC, polar cap; PP, parapolar cells; SC, somatic cells; UC, uropolar cell; VC, vegetative cells; VN, vegetative nucleus. (**C–I**, redrawn after Whitman, 1882.)

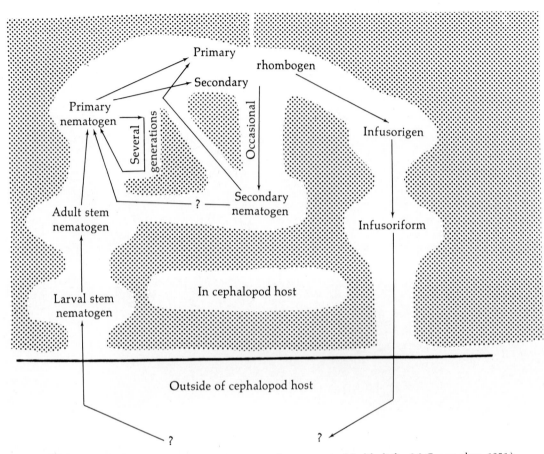

Fig. 21.4. **General life cycle pattern of dicyemid mesozoans.** (Modified after McConnaughey, 1951.)

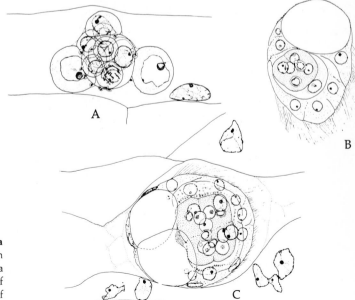

Fig. 21.5. **Stages in the life cycle of *Dicyema briarei*, a parasite of *Octopus*. A.** Infusorigen and paranucleus in axial cell. **B.** Parasagittal optical section of infusoriform larva within axial cell, which is not shown. **C.** Dorsal view of infusoriform larva within axial cell. Note also nuclei of uropolar cells of rhombogen. (**A–C**, after Short, 1961.)

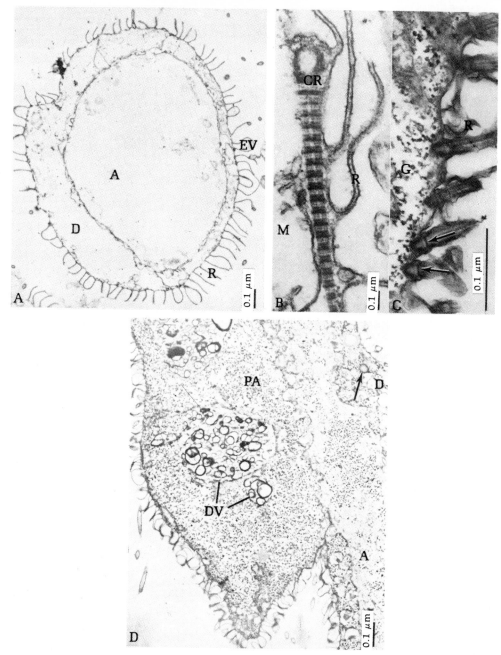

Fig. 21.6. **Electron micrographs of *Dicyema*. A.** Transverse section through diapolar cells and axial cell of *Dicyema aegira* nematogen from *Octopus vulgaris* showing scarcity of organelles, thus imparting hyaline appearance to cells. Notice ruffles on diapolar cells that are fused distally at several locations around the periphery to form endocytic vesicles. × 8500. **B.** Section through periphery of a diapolar cell of *Dicyema aegira* nematogen showing ciliary rootlet cut longitudinally and striated rootlets with a periodicity of about 770 Å. Note presence of dark cross-bands about 485 Å in width, each with five bands within it. Longitudinal filaments can be seen along the length of the rootlet. The surface ruffles are about 340 Å wide and are composed of two 100-Å membranes separated by a cytoplasmic area. × 111,000. **C.** Section through calotte of *Dicyema aegira* nematogen showing basal plates (arrows) within the cilia that are closer together here than on diapolar cells. × 47,500. **D.** Section of *Dicyema typoides* rhombogen from *Octopus vulgaris* showing lamellar inclusions in digestive vacuoles within a parapolar cell. Note storage granule (arrow) in peripheral cells. × 11,500. A, axial cell; CR, ciliary rootlet; D, diapolar cell; DV, digestive vacuole; EV, Endocytic vesicle; G, glycogen; M, mitochondrion; PA, parapolar cell; R, ruffle membrane. (All figures after Ridley, 1968.)

the single nucleus within the single axial cell divides to form two nuclei, one of which remains as the **axial cell nucleus**. The other nucleus is surrounded by cytoplasm and becomes an agamete. This agamete divides rapidly so that by the time the primary nematogen is fully developed, its axial cell includes a large number of agametes. From these agametes, other primary nematogens develop in the same manner as from the agametes in adult stem nematogens.

The production of primary nematogens continues while the host is growing but generally ceases when it attains sexual maturity. At this time the existing primary nematogens either give rise to another generation—the **primary rhombogens**, which develop from agametes within the last generation of primary nematogens—or metamorphose into **secondary rhombogens**. Rhombogens are essentially identical to nematogens except that their somatic cells are filled with lipoproteins (yolklike material) and glycogen and protrude from the body surface of the parasite as **verruciform cells**.

Agametes found in primary rhombogens represent products of the germ cell lineage from agametes in primary nematogens or from agametes carried over from primary nematogens that have metamorphosed into secondary rhombogens. Some of the agametes in rhombogens degenerate, while others divide to form **infusorigens** (Fig. 21.5). The process of infusorigen formation from agametes is comparable to that of the formation of primary nematogens from agametes, except that the cells surrounding the axial cell are not ciliated.

Mesozoan infusorigens are hermaphrodites, for they produce male and female gametes that fuse during sexual reproduction. The female gametes, or ova, result from the rounding up and differentiation of some of the surface cells of the infusorigen, while the agametes in the axial cell undergo meiosis and form sperm that fertilize the ova (McConnaughey, 1951). The resultant zygotes emerge from the surface of the infusorigens and undergo cleavage to form balls of cells that eventually differentiate into ciliated, free-swimming **infusoriform** larvae (Fig. 21.5). This microscopic larva consists of a fixed number of cells: two comparatively large **apical cells** that bear short cilia, and several ciliated cells that cover most of the body surface. In the center of the infusiform larva is the **urn** consisting of four urn cells. The urn is almost completely surrounded by two **capsule cells**. An **urn cavity** occurs anterior to the urn cells. It is bounded anteriorly by two cells the cilia of which extend into the urn cavity.

Infusoriform larvae escape from the parent rhombogen and leave the host. It is suspected that in the marine environment these larvae seek and enter another host in which they undergo further development since they are unable to infect young octopuses directly. The identity of the second host and the

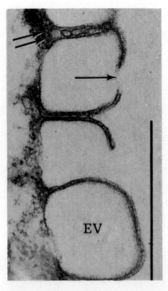

Fig. 21.7. Electron micrograph of section through surface of a diapolar cell of *Dicyema typoides*. The micrograph shows ruffles in process of fusing (arrow) and a large endocytotic vesicle (EV) formed from the fusion of the distal end of two adjacent ruffles. × 41,000 (After Ridley, 1968.)

manner in which young cephalopods become infected remain uncertain.

According to some experts (McConnaughey, 1951), certain rhombogens may cease producing infusorigens and start producing nematogens, known

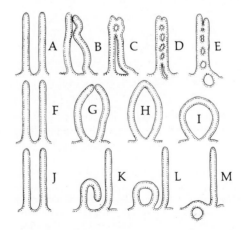

Fig. 21.8. Diagrammatic representation of formation of three types of endocytotic vesicles by *Dicyema* spp. A–E represent the first type, F–I represent the second type, and J–M represent the third type. (After Ridley, 1968.)

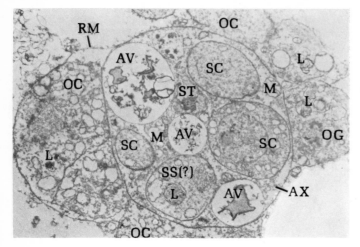

Fig. 21.9. **Electron micrograph of *Dicyema typoides* infusorigen within axial cell of parent rhombogen showing autophagic vacuoles, numerous mitochondria, spermatogenic cells, and a spermatid in axial cell.** AV, autophagic vacuole; AX, axial cell of infusorigen; L, lysosome; M, mitochondrion; OC, oocyte; OG, oogonium; RM, residual membrane; SC, spermatogenic cell; SS, secondary spermatocyte. (After Ridley, 1969.)

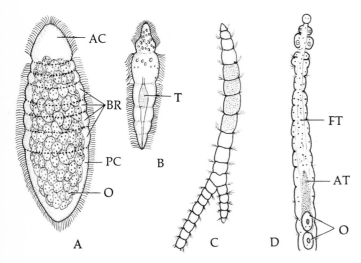

Fig. 21.10. **Some orthonectids.** **A.** Mature female of *Rhopalura granosa*, a parasite of the clam *Heteranomia*. **B.** Mature male of *R. granosa*. **C.** Anterior portion of *Stoecharthrum giardi*, a parasite in the body cavity of the marine annelid *Scoloplos mulleri*. **D.** Anterior portion of *S. giardi* (stained specimen.) AC, anterior cone; AT, anterior testis; BR, body rings; FT, fibrillar axial tissue; O, ovocytes; PC, posterior cone; T, testis. (**A** and **B**, redrawn after Atkins, 1933. **C** and **D**, redrawn after Caullery and Mesnil, 1901.)

as **secondary nematogens**, which in turn produce primary rhombogens (Fig. 21.4).

Rhombogens that remain in the host's renal organ disintegrate in time, and apparently do little damage. They acquire nutrition by absorption and also engulf and digest the host's spermatozoa present in the kidney. Stored glycogen occurs in young nematogens and is gradually used up, but reappears in rhombogens.

Fine Structure

Ridley (1968, 1969) has studied the fine structure of various stages in the life cycles of several species of *Dicyema* occurring in *Octopus vulgaris* in Florida. He has demonstrated that in the nematogen there are very few organelles present, thus imparting a hyaline appearance to the cells (Fig. 21.6). The same is true of rhombogens, although glycogen deposits occur in their cytoplasm, especially in young specimens. The outer surfaces of both nematogens and rhombogens are modified into a system of ruffles, about 1 μm in height, which fuse with each other or with the plasma membrane proper to form endocytotic vesicles (Figs. 21.7, 21.8). These vesicles have been shown experimentally to be capable to taking up ferritin. In addition to these ruffles, which are believed to play a role in the capturing and uptake of nutrients, nematogens and rhombogens also bear typical cilia on their surfaces. It is noted that lysosomelike vesicles have been found within the diapolar cells of nematogens and rhombogens, and these probably are functionally associated with the digestion of ingested and endogenous materials.

Electron microscope studies on the infusorigens and infusoriforms of *Dicyema* spp. have revealed that the hyaline axial cells of infusorigens contain autophagic vacuoles (Fig. 21.9). Glycogen occurs in infusorigen cells, but does not accumulate. On the other hand, glycogen is abundant in the peripheral cells of immature infusoriform larvae, but is less abundant in mature ones. The surfaces of infusoriforms bear typical cilia as well as microvillilike ruffles. Finally, the urn cells include large numbers of secretion granules and dictyosomes. The granules may liquefy and form a cyst.

Pathology

Electron microscope evidences suggest that dicyemids may erode a portion of the brush border of the host's renal appendage and thus may be mildly pathogenic (Ridley, 1968).

CLASS ORTHONECTIDA

The two most common genera in the class Orthonectida are *Rhopalura*, found in brittle stars and other marine invertebrates, and *Stoecharthrum*, a hermaphroditic genus parasitic in marine annelids (Fig. 21.10).

The life cycle pattern of the Orthonectida is better

understood than that of the Dicyemida (Fig. 21.11). While hermaphroditic species exist, the majority of these parasites are dioecious. Almost all of the known species, which are seldom more than 300 μm in length, have been reported from Europe.

Life Cycle of *Rhopalura ophiocomae*

In *R. ophiocomae*, a parasite of the brittle star, *Amphipholis sequamata*, the largest forms are the multi-nucleate **plasmodia** found in the host's gonads (Fig. 21.12), some of which are males and others are females. Within the individual plasmodium, some cytoplasm surrounds each nucleus and through subsequent cytokinetic fragmentation, individual uninucleate **agametes** are formed. These undergo cleavage to form individual solid masses of embryonic cells, with each mass being known as a **morula** (Fig. 21.12).

Differentiation occurs in each morula. The outer surface cells become the **somatoderm** and bear cilia, while the inner cells, 200 to 300 in number, become

sex cells. The body increases in size and becomes elongate (rather long and slender in certain species), with females becoming two to three or four times larger than males. The body surface appears compartmentalized as a result of sutures between the so-called jacket cells. This is most apparent in specimens impregnated with silver nitrate (Fig. 21.13) (Kozloff, 1969). Three distinct regions are recognizable on the body surface of males—an **anterior cone** made up of somatic cells on which the cilia are pointed anteriorly, a **body region** on which the cilia are directed posteriorly, and a **posterior cone** that is marked off from the body. Some of the cells include birefringent bodies that presumably represent stored nutrients (Fig. 21.13).

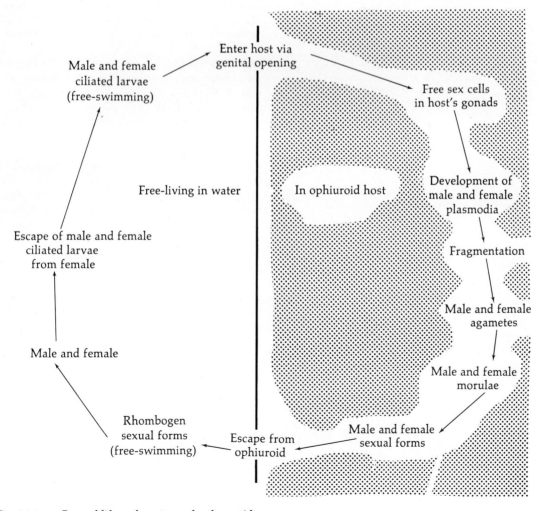

Fig. 21.11.   General life cycle pattern of orthonectid mesozoans.

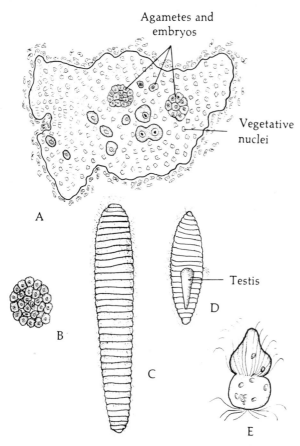

Agametes and
embryos

Vegetative
nuclei

A

B

C

Testis

D

E

Fig. 21.12. Stages in the life cycle of orthonectids. A.
Male plasmodium of *Rhopalura ophiocomae*, a parasite of the
brittle star. **B.** Morula of *R. ophiocomae.* **C.** Free-living
female of *R. ophiocomae.* **D.** Free-living male of *Rhopalura
julini*, a parasite of an annelid. **E.** Ciliated larva of *Rhopalura
granosa*, a parasite of the clam *Heteranomia*. (**A–D,** redrawn
after Caullery and Mesnil, 1901; **E,** redrawn after Atkins,
1933.)

There are two types of females in *R. ophiocomae*:
one elongate, 235–260 μm long by 65–80 μm wide;
the other ovoid, 125–140 μm long by 65–70 μm
wide.

Male and female ciliated sexual forms (commonly
referred to as adults) escape from the host and are
free-swimming in sea water. When a male and female
make contact, the spermatozoa (Fig. 21.13) which are
formed by the sex cells in the male, are introduced
into the sex cell mass of the female through apposed
genital pores (Fig. 21.13). The fertilized ovum undergoes maturation division, during which polar bodies
are extruded on long conical projections. Unequal

cleavage follows the fusion of the pronuclei, resulting
in another morulalike stage with central cells enveloped by smaller peripheral cells. During this phase
of development, chromatin material is eliminated from
some of the cells. Ciliated larvae thus formed mature
in about 24 hours after fertilization, escape from the
genital pore of the female, seek out another host,
invade it via the genital openings, and in the process
lose their outer somatic layer. The inner sex cells
scatter within the host's gonads, primarily in the
gonaducts, and through repeated nuclear division,
male and female plasmodia are formed.

In the hermaphroditic species, both male and female
morulae are formed from the same plasmodium.

Orthonectids that parasitize gonadal tissues of their
hosts destroy germ cells, causing castration. In addition, in the case of *R. ophiocomae*, the plasmodium
stage may invade the brittle star's genitorespiratory
bursae and then spread into the aboral side of the
central disc, around the digestive system, and into the
arms.

It is difficult and speculative to compare the stages
in the life cycle of dicyemid and orthonectid mesozoans. However, the sexual forms of orthonectids resemble the nematogens of dicyemids in that both are
ciliated. On the other hand, they differ in that there
are from several to many more cells in the center of
the sexual forms of orthonectids than there are axial
cells in dicyemids. Furthermore, all the cells comprising the inner cell mass of the sexual forms of
orthonectids are true reproductive cells, being either
ovocytes or spermatogonia, depending upon the sex
of the individual. In dicyemids, only the agametes
enclosed within the axial cells are true reproductive
cells.

The plasmodia of orthonectids probably correspond to the axial cells of dicyemid nematogens,
because these plasmodia represent the central mass of
reproductive cells of ciliated larvae.

Again, there is a difference in that all of the nuclei of
each orthonectid plasmodium eventually become the
nuclei of agametes, while only some of the nuclei in
dicyemid nematogen axial cells become the nuclei of
agametes, the other nuclei becoming the nuclei of
somatic cells.

PHYSIOLOGY OF MESOZOA AND HOST-PARASITE RELATIONSHIPS

Respiration

It is known that cephalopod urine is essentially
anaerobic and, hence, dicyemids living in the cephalopod kidney are believed to be anaerobes that
probably engage in lactic acid fermentation. Nouvel
(1933) has shown that if nematogens and rhombogens
are placed in cephalopod urine or in sea water con-

taining potassium cyanide (which inhibits oxidative respiration but not anaerobic fermentation), the parasites live much longer than do controls not exposed to cyanide. Similarly, vermiforms survive better in media saturated with nitrogen than with oxygen. These observations tend to support the idea that these stages are anaerobic. On the other hand, the swarming infusoriforms which escape from the host and are free swimming in sea water behave differently if exposed to either cyanide or nitrogen. When this occurs, the infusoriforms remain active for only 8 to 12 hours. But if exposed to air, they remain active for several days, thus suggesting that the infusoriforms are aerobic.

Nutrition

According to Nouvel (1933), most, if not all, known species of dicyemids derive their nutrients from their cephalopod host's urine. The main nutrient stored, glycogen, is not demonstrable in germ cells or young embryos, but small deposits first appear when the embryos attain their definitive number of somatic cells and become capable of independent movement. Somatic cells of young vermiform larvae include comparatively large quantities of glycogen. Older nematogens include less glycogen but the amount again increases during the transition to the rhombogen phase. McConnaughey (1968) has presented a good review of the physiology of the Mesozoa.

Host-Parasite Relationships

Although the majority of dicyemids adhere lightly to their hosts' renal tissue by the ciliature of the calotte (designated as the **adhesive tuft**), certain species have more elaborate holdfast mechanisms. For example, in the case of *Dicyemennea brevicephaloides* in Russia, the calotte is expanded and flattened like a mushroom, and is attached to the walls of the host's branchial chambers (Hoffmann, 1965). Similarly, McConnaughey and Overstreet (1962) have reported a species of *Dicyemennea* with a large calotte which is broader than the trunk. Furthermore, a large, slightly branched and heavily ciliated groove traverses the surface of the calotte. This holdfast is employed in a "mouthlike" manner for clinging to tufts of host renal tissue. Even more remarkable, this species is capable of retracting the anterior end of its axial cell, drawing the propolars into a cuplike depression, thus increasing the effectiveness of the mouthlike groove as a gripping device.

PHYLUM PORIFERA

Among the Porifera or sponges, the marine species *Cliona elata* bores into the shells of molluscs. Practically nothing is known about the biology of this sponge, although mass mortalities among oysters

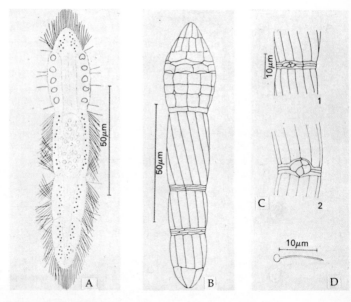

Fig. 21.13. *Rhopalura ophiocomae.* **A.** Drawing of living adult specimen, as seen in optical section, showing distribution of cilia, lipid inclusions, crystal-like inclusion, and testis. **B.** Drawing showing boundaries of jacket cells as revealed by silver nitrate impregnation. **C.** Drawings showing genital pore and adjacent cells in mature (1) and nearly mature (2) specimens as revealed by silver nitrate impregnation. **D.** Living sperm. (After Kozloff, 1969; with permission of *Journal of Parasitology*.)

have been known to occur as the result of the "spice bread disease" caused by *Cliona*.

In addition to *C. elata*, several other species are common along the Atlantic coast of North America. These include *C. vastifica*, *C. lobata*, and *C. truitti*. In each case, the boring sponge inhabits tunnels in the host's shells. These tunnels are bored by the sponges by what is believed to be mechanical action, although the secretion of acids to aid in the fragmentation of the shell may occur (Galtsoff, 1964). These tunnels do not, as a rule, perforate the shell, hence the sponges do not make contact with the oyster's soft tissues. It is only in instances of old and heavy infestations that the tunnels perforate the entire thickness of the shell. Even then, the holes are rapidly mended by the deposition of conchiolin. If this mending process is delayed by adverse conditions, such as temperatures below 7°C, the sponges will make direct contact with the mantle and produce lysis of the epithelium and underlying connective tissue. Grossly, areas on the mantle with sponge infections appear as darkly pigmented pustules.

PHYLUM COELENTERATA

Members of the Coelenterata are all aquatic, primarily marine, and are found attached or free floating. Their diploblastic bodies are either radial or biradial, without a head or other organ systems, and as a rule the alimentary tract is incomplete, that is, there is a mouth and enteron but no anus. The nervous system is diffuse and tactile tentacles may be present surrounding the mouth. All coelenterates possess stinging structures known as nematocysts.

The life cycles of coelenterates are interesting for they undergo metagenesis (or alternatiion of generations), alternating between the asexual **polyp** stage and the sexual **medusa**, or jellyfish stage.

However, as a result of evolutionary changes, the prominence of the medusa and polyp stages may not be equal. For example, in the freshwater jellyfish *Craspedocusta sowerbyi*, the medusa is conspicuous while the minute polyps, resting at the bottom of lakes and ponds, are seldom seen. In some species,

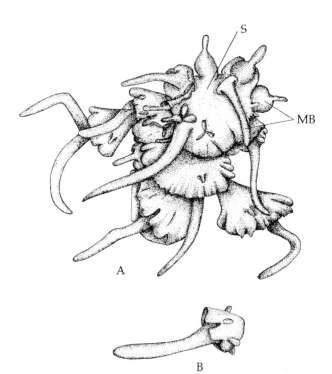

Fig. 21.14. *Cunina peregrina.* **A.** Flattened stolon removed from the medusa *Rhopalonema velatum.* **B.** Medusa of *C. peregrina* produced by budding of parasitic stolon. MB, medusal buds; S, parasitic stolon. (Redrawn after Bigelow, 1909.)

such as the familiar freshwater hydras, the medusa stage is completely absent. For a description of coelenterate morphology, the reader is referred to the volume by Hyman (1940).

Almost all members of the three classes of coelenterates—Hydrozoa, Scyphozoa, and Anthozoa—numbering approximately 10,000 species, are free living. A few species have been reported to be symbiotic, but little is known about the biology of these.

CLASS HYDROZOA

Several species of the class Hydrozoa are known to be symbiotic during some phase of their life cycles. For example, the planula larva of *Cunina proboscidea* is parasitic, attached to another coelenterate, *Geryonia*, and the larva of *C. peregrina* is parasitic on *Rhopalonema velatum*.

Life Cycle of *Cunina*

There is usually no polyp stage in the life cycle of the Narcomedusae, the group to which *Cunina proboscidea* and *C. peregrina* belong. The egg produced by the maternal medusa hatches, and the escaping form is a ciliated **planula larva**, which differentiates into an **actinula larva**, which in turn, differentiates into a medusa.

In the case of *Cunina proboscidea*, Bigelow (1909) has reported that the egg develops in the mesoglea (a jellylike layer between the epidermis and gasterodermis) of the maternal medusa and is enveloped by a nurse cell, or **phorocyte**. The planula larva developing from the egg metamorphoses into a reduced medusa that produces gametes and then degenerates. The planulae resulting from this generation become attached to the host (to *Geryonia* in the case of *C. proboscidea* and to *Rhopalonema* in the case of *C. peregrina*) and develop parasitically. Both *Geryonia* and *Rhopalonema* are also medusae. Each *Cunina* planula becomes flattened out as a stolon that embraces the host (Fig. 21.14) and buds off the definitive medusae (Fig. 21.14).

Nothing is known about the relationship between *Cunina* and *Geryonia* or *Rhopalonema*, although metamorphosis of the *Cunina* planulae into flattened stolons is undoubtedly a modification adaptive to parasitism. Furthermore, this dramatic alteration of body form suggests that the relationships may be of long standing.

Life Cycle of *Polypodium*

The unusual method of medusa production by *Cunina*, that is, by budding from a stolon, also occurs in the parasitic hydrozoan coelenterate *Polypodium*. The polyp stage of this animal is free living. It is found in the Volga River Basin and in the Black and Caspian Seas.

Lipin (1911) has studied aspects of the life cycle of *Polypodium*. The free-living polyp is solitary and

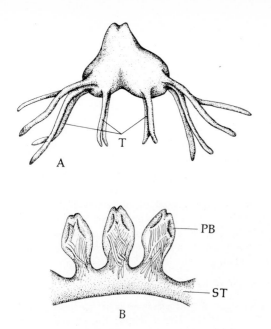

Fig. 21.15. *Polypodium.* **A.** Free-living polyp in walking position. **B.** Portion of parasitic stolon with three polyp buds. (Redrawn after Lipin, 1911.) PB, polyp bud; ST, stolon; T, tentacles.

migrates by employing its eight long and four short aboral tentacles (Fig. 21.15). Such polyps arise as buds from a stolon that is parasitic within eggs inside the ovary of the sturgeon (Fig. 21.15).

When the fish host lays eggs, the stolon escapes and the buds increase in size and mature by an evagination process. After the polyp buds mature and become separated, the stolon disintegrates. Gonads develop in the polyp of *Polypodium*, thus suggesting that a medusa stage does not exist. Nothing is known about the fate of the free-living polyps or how sturgeons become infected. Again, the occurrence of a morphologically modified parasitic stolon in the ovarian eggs of the sturgeon suggests a relationship of long duration. It is noted that Raikova *et al.* (1979) have reported that *Polypodium hydriforme* occurs in eggs of sturgeons in the United States as well as in the Soviet Union.

Zanclea and *Hydrichthys*

In addition to *Cunina* and *Polypodium*, several other hydrozoans are known to be parasitic. Very little is known about the biology of these, although in each case some anatomic modifications exist that can be interpreted as being adaptations to parasitism. For example, the parasitic stages during the life cycles of *Zanclea* and *Hydrichthys* show certain modifications of this type.

The medusae of *Zanclea* are parasitic, being attached to the pharynx of the marine nudibranch *Phyllirhoe.*

Modification of the parasitic medusa's body is apparent, for no tentacles are present. During the relationship between *Zanclea* and *Phyllirhoe*, the latter's tissues are gradually destroyed. Also, as the nudibranch matures, it utilizes the medusa as a pelagic vehicle and later discards it (Martin and Brinckmann, 1963).

Hydrichthys is a colonial hydroid coelenterate that attaches to the body of fish. The hydroids have lost their tentacles and feed on the host's blood and tissues that have been injured by rootlike outgrowths of the parasites.

Hydroids Associated With Marine Molluscs

The most frequently encountered associations involving hydroids are those with molluscan hosts (Rees, 1967). For example, four species of the genus *Eugymnanthea* are known to be symbiotic within the mantle cavities of commercially important pelecypods (Crowell, 1957). *Eugymnanthea inquilina* occurs in the clam *Tapes decussatus* near Naples, Italy; *E. polimantii* occurs in the mussel *Mytilus galloprovincialis* at Taranto, Italy; *E. ostrearum* occurs in the oyster *Crassostrea rhizophorae* in Puerto Rico; and *E. japonica* occurs

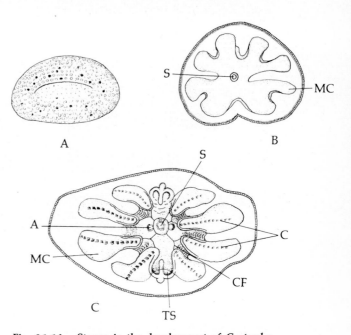

Fig. 21.16. **Stages in the development of *Gastrodes*.** **A.** Bowl-shaped stage parasitic in *Salpa.* **B.** Later stage showing lobed canal system. **C.** Comb-bearing cydippid-type larva. A, anal pore; C, comb rows; CF, ciliated furrow; MC, meridional canals; S, statocyst; TS, tentacle sheath. (Redrawn after Komai, 1922.)

in the oyster *Crassostrea gigas* in Japan. All four species possess a conspicuous, rounded basal disc by which the polyps are attached to their hosts' tissues. For a more detailed account of these coelenterates and citations to the relevant literature, see Cheng (1967).

CLASS ANTHOZOA

Peachia and *Edwardsia* are two sea anemones whose larval stages parasitize other anemones or ctenophores. The coelenterate larvae are found either attached to the body surface of their host or in the gastrovascular cavity. These relationships are more than accidental, for the parasitic anemones each possess a suckerlike ring around the mouth by which they hold on to their hosts.

PHYLUM CTENOPHORA

The ctenophorans, or comb jellies, are almost all free-swimming marine animals. Their unsegmented triploblastic bodies are transparent and biradial, being oriented along an oral-aboral axis. The gastrovascular system consists of a mouth, pharynx, stomach, branched digestive canals, and two anal pores on the aboral surface. The nervous system is diffuse. An aboral sense organ, known as the **statocyst**, is present.

Although superficially similar to certain coelenterates, the ctenophorans lack nematocysts and characteristically bear comblike plates on their bodies at some stage during their development. These animals are hermaphroditic and reproductive cells usually are formed from endodermal cells in the digestive canals. Alternation of generations (metagenesis) does not occur. Most of the species are bioluminescent. The volume by Hyman (1940) and the review by Bouillon (1968) should be consulted for morphological details.

There are approximately 80 species of ctenophorans, among which a few have been reported to be symbiotic. The larvae of *Gastrodes* is a true parasite in the tunicate *Salpa* during its larval stages. The adults are free living.

Life Cycle of *Gastrodes parasiticum*

The life cycle of *Gastrodes parasiticum* has been studied by Komai (1922). The youngest stage found in the tunicate host, *Salpa fusiformis*, is bowl-shaped and is embedded in the mantle (Fig. 21.16). The bowl-shaped form grows into a later larval form possessing four-lobed, armlike canal systems (Fig. 21.16). This larval form metamorphoses into a **cydippid-type larva**, possessing statocyst, tentacles, and eight rows of typically ctenophoran combs. It also possesses a gastrovascular system (Fig. 21.16).

It is at this stage of development that *Gastrodes* leaves its host, settles to the bottom in sea water,

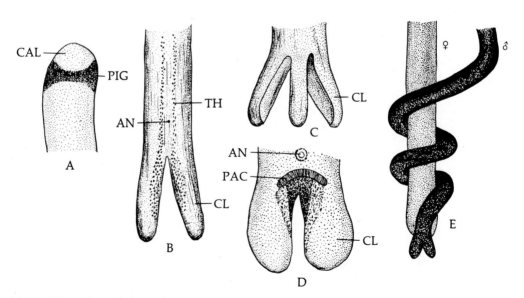

Fig. 21.17. External morphology of nematomorphs. A. Anterior end of *Gordius* showing calotte and pigmented ring. (Redrawn after Heinze, 1937.) **B.** Posterior end of male *Paragordius*. (Redrawn after May, 1919.) **C.** Posterior end of female of *Paragordius*. (Redrawn after Montgomery, 1898.) **D.** Posterior end of male *Gordius*. (Redrawn after Heinze, 1937.) **E.** Male specimen of *Gordius robustus* wrapped around posterior end of female in copula. (Redrawn after May, 1919.) AN, anus; CAL, calotte; CL, caudal lobes; PAC, postanal crescent; PIG, pigment ring; TH, tracts of thorns.

casts off its rows of combs, and flattens out by everting its pharynx. This is the adult form. Eggs are produced in the pharyngeal epithelium of the adult. A typical **planula larva** with solid entoderm develops from each egg. Such a larva bores its way into a new host and again initiates the parasitic phase of its life cycle.

PHYLUM NEMATOMORPHA

Commonly called horse-hair or gordiacean worms, the nematomorphans are long, slender, cylindrical animals. The body surface is covered by a thin layer of cuticle, which is generally rough in texture. The sexes are separate and the individuals lack both lateral cords, which are longitudinal, thickened portions of the body wall, and an excretory system. The adults

are unique because the straight alimentary tract is degenerate at the anterior and posterior ends and hence is not functional.

MORPHOLOGY

Adult nematomorphs commonly reach 0.5−1 m in length and up to 3 mm in diameter. The males are shorter than the females, and the posterior end of the male is usually ventrally coiled. Living specimens vary from yellowish to dark brown.

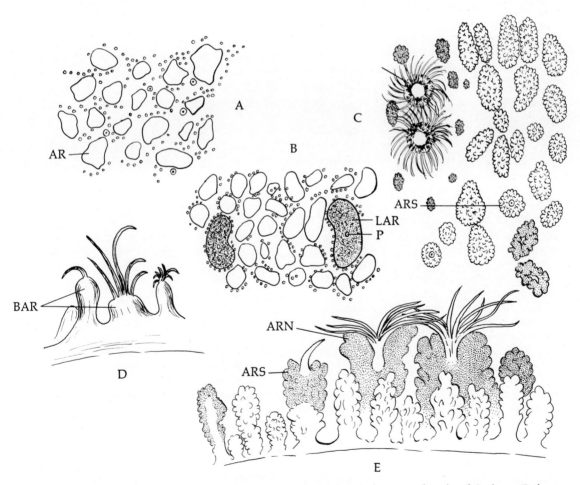

Fig. 21.18. Areoles and other surface structures of nematomorphs. A. Surface view of areoles of *Gordionus*. (Redrawn after Heinze, 1937.) **B.** Surface view of areoles of *Parachordodes*. (Redrawn after Heinze, 1937.) **C.** Surface view of areoles of *Chordodes*. (Redrawn after Camerano, 1897.) **D.** Side view of nematomorph areoles showing some with single and others with several bristles. (Redrawn after Müller, 1927.) **E.** Side view of areoles of *Chordodes*. (Redrawn after Camerano, 1897.) AR, areole; ARN, areole with numerous bristles; ARS, areole with single bristle; BAR, bristle-bearing areoles; LAR, large areole; P, pore on large areole.

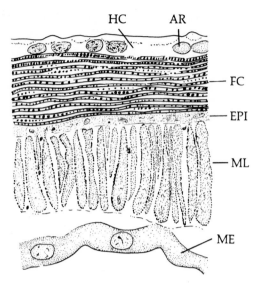

Fig. 21.19. **Section through the body wall of *Para-gordius*.** AR, areole; EPI, epidermis; FC, fibrillar layer of cuticle; HC, homogeneous layer of cuticle; ME, mesenchyme; ML, muscle layer. (Redrawn after Montgomery, 1903.)

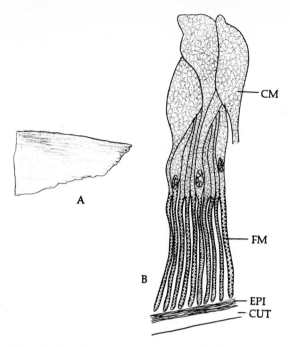

Fig. 21.20. **Nematomorph muscles. A.** Section of a muscle fiber. **B.** Muscle cells of *Nectonema*. CM, cytoplasmic portion of muscle cell; CUT, cuticle; EPI, epidermis; FM, fibrillar layer of cuticle. (Redrawn after Rauther, 1914.)

Unlike nematodes, the bodies of nematomorphs do not taper at both ends; rather, both ends are blunt and rounded. The anteriormost tip of the body—known as the **calotte**—is not pigmented and is set off from the rest of the body by a pigmented collar (Fig. 21.17). The mouth is located on the calotte, either terminally or ventrally. The posterior end of the body is often split into two or three lobes, depending on the sex and genus, and the cloacal aperture is located between the lobes or anteroventrad to these (Fig. 21.17).

The cuticle of many nematomorphs is rough. The roughness is due to rounded or polygonal plates, the **areoles**, arranged on the surface (Fig. 21.18). In some species, the areoles may project as conical papillae bearing bristles (Fig. 21.18) while in others minute pores are present. The interareolar spaces may bear bristles, minute projections, or pores, depending on the species. The aeroles are sensory in function. These represent thickenings of the cuticle.

Body Tissues

When seen in cross section under the light microscope, the body wall consists of three major layers.

1. The outermost cuticular layer is composed of a thin outer homogeneous layer and an inner stratified fibrous layer (Fig. 21.19). The areoles appear as thickenings of the outer homogeneous layer.

2. Beneath the cuticle is a cellular **hypodermis** composed of a single layer of cuboidal to columnar cells. In members of the order Nectonematoidea, a layer of pigment granules is embedded between the cuticle and epidermis. In most nematomorphs, the epidermis is thickened along the midventral body line to form the **ventral cord**. However, in Nectonematoidea, there is a second thickened line, the middorsal **dorsal cord**.

3. Medial to the epidermis is the body wall musculature. These muscles, like those in nematodes, are longitudinally oriented. Except in members of the Nectonematoidea, these muscles are similar to those in acanthocephalans in that the striated portion of contractile fibrils envelops the cytoplasmic portion (Fig. 21.20). In *Nectonema*, the muscles resemble those in nematodes in that there is a slender fibrillar portion and a broadly elongate cytoplasmic portion to each fiber (Fig. 21.20).

Alimentary Tract

Along the middle of the body is found the alimentary tract (Fig. 21.21). The mouth opens on the calotte, and leading from the mouth is a slender **pharynx** that is not hollow but consists of a cord of cells. This non-functional pharynx leads into a mass of cells that

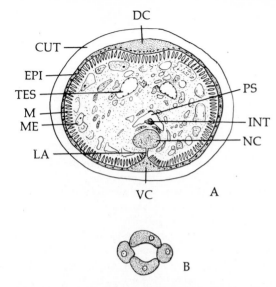

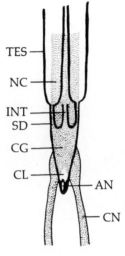

Fig. 21.21. **Nematomorph morphology. A.** Cross-section through *Paragordius*. (Redrawn after Hyman, 1951.) **B.** Cross-section through intestine of *Nectonema*. (Redrawn after Bürger, 1891.) CUT, cuticle; DC, dorsal cord; EPI, epidermis; INT, intestine; LA, lamella connecting nerve cord to ventral cord; M, muscle layer; ME, mesenchyme; NC, nerve cord; PS, pseudocoel surrounding intestine; TES, testis; VC, ventral cord.

Fig. 21.22. **Posterior portion of nematomorph nervous system.** AN, anus; CG, cloacal ganglion; CL, cloaca; CN, caudal nerves extending into caudal lobes; INT, intestine; NC, nerve cord; SD, sperm duct; TES, testis. (Redrawn after Montgomery, 1903.)

presumably represents a pharyngeal bulb. Leading posteriorly from the bulb is the hollow intestine lined with epithelial cells. It has been postulated that since the intestine obviously does not serve a digestive function, it probably is excretory in function.

Posteriorly, the genital ducts empty into the intestine. Immediately after these ducts join the intestine, the tube is enlarged, forming the cloaca, which is lined with cuticle.

The alimentary tract, as described here, is typical of most nematomorphs. In the Nectonematoidea, however, the tract is somewhat different. In these worms, the mouth leads into a hollow cuticularized tube—the **pharynx**—which, in turn, leads into the long intestinal tract. The intestine is composed of two or four large cells that form the wall (Fig. 21.21). These cells may be distinct or syncytially arranged. The intestine of the nectonematoids does not extend to the cloaca; rather, it soon becomes indiscrete and fades out. The cloaca in these worms functions as a portion of the reproductive system.

The space between the body wall and alimentary tract is the **pseudocoel**, which in most nematomorphs is packed with parenchymal cells so that very little empty space is evident. In the Nectonematoidea, however, the pseudocoel is clear and extends from one end of the animal to the other with a septum separating a small chamber anteriorly in the area of the cerebral mass, or brain.

Nervous System

Within the pseudocoel of nematomorphs are found the nervous and reproductive systems. The nervous system consists of a large anteroventral cerebral mass lying within the calotte. Within this mass of ganglia can be found two distinct types of cells: the **giant nerve cells** and the **small nerve cells**. The main nerve of the body is the ventral nerve, which lies within the ventral epidermal cord in the nectonematoid species and dorsal to the ventral cord in the other nematomorphs. In the latter case, the main nerve is connected to the epidermal thickening by a thin nervous lamella.

Toward the anterior terminal of the ventral nerve, the single cord is split into three tracts that enter the cerebral mass, and which presumably terminate as giant nerve cells. The ventral nerve joins a thickened cloacal ganglion posteriorly (Fig. 21.22). Nerve fibers arising from this ganglion innervate the external reproductive structures.

In *Paragordius*, Montgomery (1903) has reported a primitive saclike eye located in the calotte.

A slight modification of the position of the ventral nerve cord is found in the Nectonematoidea. In these worms, the nerve is neither dorsal to nor connected to the ventral cord by the nervous lamella; rather, it is permanently situated in the epidermal thickening.

Reproductive System

All nematomorphs are dioecious. The gonads are located in the pseudocoel surrounded by mesenchyme. In males, the two elongate cylindrical testes

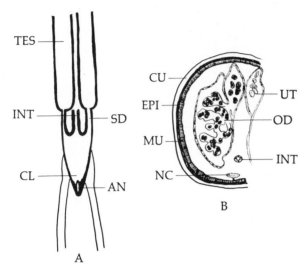

Fig. 21.23. **Reproductive systems of nematomorphs.** **A.** Arrangement of testes in nematomorph. (Modified after Montgomery, 1903.) **B.** Section through female *Paragordius* showing lateral diverticula of ovaries. (Redrawn after Montgomery, 1903.) AN, anus; CL, cloaca; CU, cuticle; EPI, epidermis; INT, intestine; MU, muscle layer; NC, nerve cord; OD, ovarian diverticula; SD, sperm duct; TES, testis; UT, uterus.

series of 3000 to 4000 diverticula extend laterally from each ovary into the pseudocoel (Fig. 21.23). Eggs mature in these side branches before being ejected back into the main tube. The posterior segment of each ovary is known as the **uterus**, which becomes narrower caudally to become the **oviduct**. The two oviducts enter the **antrum** independently.

The antrum is lined with glandular epithelium. It may be considered the anterior portion of the cloaca, but it differs from the cloaca proper because the latter is lined with cuticle. Also arising from the glandular antrum is a slender, anteriorly directed **seminal receptacle**. The cloaca, like that in males, empties to the exterior through the cloacal aperture located in the midst of the caudal lobes.

The reproductive systems described here represent those found in all nematomorphs except for the nectonematoid species. Nectonematoid reproductive systems are not completely understood. Feyel (1936) has reported that in male parasitic juvenile nectonematoids there is a single testis suspended from a dorsal epidermal cord (Fig. 21.24), which leads to the exterior dorsally through a tube. In females, no compact ovary is present; rather, there are individual ovocytes that arise through the differentiation of certain mesenchymal cells. These ovocytes originally are attached to the epidermis but later become free in the pseudocoel. There is a short genital tube at the posterior end of the worm, through which the eggs pass to the exterior. This tube is interpreted to be the vestige of a cloaca.

are situated side by side along the entire length of the body (Fig. 21.23). In some species, the posteriormost segment of each testis is slightly swollen to form the **seminal vesicle**. Each testis empties independently into the cloaca via its own sperm duct.

In young females, the ovaries are grossly indistinguishable from testes, because they are also in the form of two elongate cylindrical tubes longitudinally arranged in the body. However, in older specimens, a

LIFE CYCLE PATTERN

The Nematomorpha are of interest to parasitologists because their larval stages are parasitic in invertebrates. These worms undergo almost all of their development within a freshwater or terrestrial arthropod, at least the members of the class Gordioidea. A few members of the smaller class Nectonematoidea are pelagic in the ocean and their larvae parasitize crabs or hermit crabs. When they emerge, they are identical to the mature adult except that they are not sexually mature, but they soon attain maturity. There appears to be a seasonal preference for emergence, since most species become free living either in late spring or summer. Sexual maturity ensues immediately and copulation occurs.

Copulation is accomplished by the male coiling its posterior end around the posterior end of the female. The spermatozoa are introduced from the cloacal aperture of the male into that of the female (Fig. 21.17). The spermatozoa ascend the cloaca and are stored in the seminal receptacle. Ova, passing through the antrum, are fertilized, and the eggs are passed to the exterior in strips presumably formed from the secretions of the antrum (Fig. 21.25). As many as one

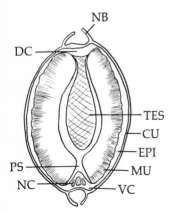

Fig. 21.24. Testis of juvenile nectonematoid. Section through male parasitic juvenile *Nectonema* showing single testis suspended from dorsal epidermal cord. CU, cuticle; DC, dorsal cord; EPI, epidermis; MU, muscle layer; NB, natatory bristles; NC, nerve cord; PS, pseudocoel; TES, testis; VC, ventral cord.

Fig. 21.25. Egg string of nematomorph. (Redrawn after Müller, 1927.)

million or more eggs are laid by each female. This represents one of the highest biotic potentials among animals. Interestingly, adults always emerge near bodies of water, such as lakes, ponds, and streams. Females are generally found along the banks, whereas male are more pelagic.

Larvae enter the arthropod host by direct penetration in some species and through ingestion while in the encysted form in others. The boring ability of certain species involves a specialized boring organ comprised of circularly arranged rows of spines and a proboscis operated by muscles. Zapotosky (1974) has contributed an electron microscope study of the boring organ of the larva of *Paragordius varius* as well as other tissues. Dorier (1925, 1930) has reported that the boring organ is employed by *Gordius aquaticus*, the larvae of which, after hatching from the eggshell, are capable of secreting a mucous cyst wall around themselves. Certain glands located in the anterior portion of the intestine secrete the mucus. In *Chordodes japonensis*, Inoue (1958, 1960) has reported that the larvae emerge from eggs incubated at $23 \pm 3°C$ in approximately 30 days. These larvae are not encysted, nor do they penetrate the body wall of *Culex* and *Chironomus* larvae and *Cloeon* nymphs, all insects, but they are ingested. It is apparent that nematomorph larvae, encysted or nonencysted in water, can reach their arthropod hosts, which include grasshoppers, crickets, beetles, roaches, centipedes, millipedes, and others, only when the host migrates to a location near or is naturally found in the water.

Within the arthropod host, the larvae are situated in the hemocoel, where they gradually develop into juveniles. Development does not involve drastic metamorphosis; rather, the larval structures, such as the body cuticle and muscles, gradually become strengthened. During this developmental period, other larval structures such as the hooks, stylets, and associated muscles degenerate and disappear. In conjunction with the disappearance of these larval structures and the continued growth of other larval structures, certain additional structures, such as the cerebral mass, ventral nerve, and reproductive organs, appear in the juvenile and continue to develop.

In *Chordodes japonensis*, once the larvae are ingested by the insect host, they penetrate through the intes-

tinal wall into the hemocoel and encyst in 2 to 3 days. While it is not clear why these larvae should encyst, it may be that the three species of larval insects tested as experimental hosts were unnatural ones, and the cyst walls served to protect the immature parasites from the defense mechanisms of the host. On the other hand, the cyst walls may represent host encapsulation walling off the invading parasites.

The period of development within the host varies among species. May (1919) has reported that in *Paragordius* and *Gordius* the parasitic phase lasts from several weeks to several months.

BIOLOGY OF NEMATOMORPHS

The phylum Nematomorpha is divided into two orders: Gordioidea and Nectonematoidea. The distinguishing characteristics of these orders are given in the classification section at the end of the chapter.

ORDER GORDIOIDEA

The taxonomy of the order Gordioidea has been reviewed and revised by Müller (1927), Heinze (1934, 1935, 1937), and Carvalho (1942). The order is subdivided into two families: the Chordodidae and Gordiidae.

Family Chordodidae

The family Chordodidae includes, among others, the genera *Neochordodes*, *Pseudochordodes*, *Chordodes*, *Para-*

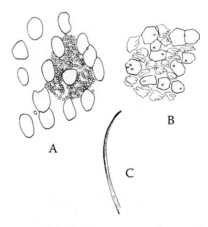

Fig. 21.26. Characteristics of some genera of Nematomorpha. A. Surface areoles of *Neochordodes talensis*. (Redrawn after Carvalho, 1942.) **B.** Surface areoles of *Pseudochordodes pardalis*. (Redrawn after Carvalho, 1942.) **C.** Anterior end of *Chordodes*. (Redrawn after Römer, 1896.)

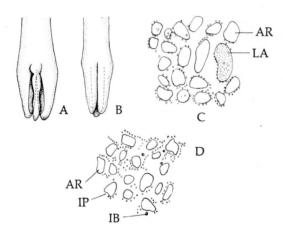

Fig. 21.27. **Characteristics of some genera of Nematomorpha.** **A.** Posterior end of female *Paragordius esavianus*, dorsal view. **B.** Posterior end of male *P. esavianus*, ventral view. (**A** and **B**, redrawn after Carvalho, 1942.) **C.** Areoles of *Parachordodes*. **D.** Areoles of *Gordionus*. (**C** and **D**, redrawn after Heinze, 1937.) AR, areoles; IB, interareolar bristle; IP, interareolar papillae; LA, large areole with central pore.

gordius, *Parachordodes*, and *Gordionus* (Figs. 21.26, 21.27). These genera are distinguished from one another primarily by the number of posterior lobes present and by the types of areoles present.

In *Neochordodes*, which is seldom found in the United States, there are no pore canals in the furrows between the areoles, the posterior end is blunt and without lobes, and only one type of areole is present. In *Pseudochordodes*, also little known in the United States, there is no postcloacal, crescent-shaped fold, the posterior end is bilobed in males, and there are two types of areoles present. In *Chordodes*, the posterior end is not lobed in either sex, and a midventral groove extends the length of the body.

In *Paragordius*, the posterior end is bilobed in males as in *Parachordodes*. However, there is only one type of areole, the posterior lobes are two and one-half to three times as long as they are wide, and the lobes are not armed with hairs or papillae. Several members of *Paragordius* are known in the United States, with *P. varius* being the common species. In *Parachordodes*, not commonly found in the United States, the posterior end of the male is also bilobed. These lobes are shorter and broader than those of *Paragordius*, and they bear rows of bristles. In *Gordionus*, which is represented by several North American species, there is only one type of areole. Furthermore, the two posterior lobes on males are approximately twice as long

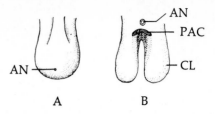

Fig. 21.28. *Gordius.* **A.** Posterior terminal of female specimen. **B.** Posterior terminal of male specimen. AN, anus; CL, caudal lobes; PAC, postanal crescent. (Both redrawn after Heinze, 1937.)

as wide, and they are armed with rows of long papillae on either side of the cloacal aperture.

Family Gordiidae

The family Gordiidae includes *Gordius*. The most common species in the United States is *G. robustus*. The females possess unlobed posterior ends, whereas the males are strongly bilobed posteriorly and are not armed with bristles or tubercles (Fig. 21.28). In addition, there is a distinct crescent-shaped fold in the postcloacal region.

ORDER NECTONEMATOIDEA

The Nectonematoidea includes *Nectonema* (Fig. 21.29). It is characterized by a hollow pseudocoel that extends along the length of the body, a ventral nerve cord located within the ventral epidermal thickening, a dorsal and a ventral epidermal cord, and a single gonad.

The only species known in the United States is *N. agile*, the males of which are 5–20 cm long by 0.3–1 mm in diameter and the females are 3–6 cm long. These pelagic worms are grayish white. They have been reported in sea water near Newport, Rhode Island, and Woods Hole, Massachusetts. Other known species of *Nectonema* have been listed by Feyel (1936).

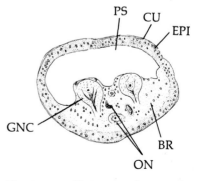

Fig. 21.29. *Nectonema.* **Section through brain.** BR, brain; CU, cuticle; EPI, epidermis; GNC, giant nerve cell: ON, ordinary nerve cells; PS, pseudocoel. (Redrawn after Feyel, 1936.)

The juvenile of *N. agile* parasitizes small crustaceans (*Palaemonetes* sp.).

PHYSIOLOGY OF NEMATOMORPHA

Very little is known about the physiology of nematomorphs. It is known that the adults require an environment rich in oxygen, while the parasitic juveniles live in environments with less oxygen.

Some reports suggest that hosts harboring juvenile nematomorphs actively seek water when the parasites become ready to emerge, while others postulate that juveniles preparing to leave their hosts can sense the presence of water. The coincidence of the host's nearness to water and the emergence of young adults seems to be more than a matter of chance.

Male nematomorphs are more active than females, since males move around on moist banks in a serpentine manner, or swim actively in water. Females, on the other hand, are generally nonmotile, or at most only slightly motile. Unlike many parasites, larval nematomorphs show little host specificity. Some species can develop in an array of beetles and grasshoppers. Occasionally, vertebrates have been reported to serve as hosts for juveniles, especially fish; however, Carvalho (1942) and others have even reported occasional human infestations. In one such instance, a nematomorph was recovered from a patient's urinary passage. Undoubtedly the few known cases of human infestation have originated from drinking water containing larvae or bathing in larvae-containing water.

Since almost all nematomorphs lack a functional alimentary tract, it is presumed that nutrition is absorbed through the body wall as in cestodes and acanthocephalans. May (1919) has reported that juveniles are situated in cavities in the host's viscera. These cavities are presumed to have resulted from digestion of host tissues by digestive enzymes secreted through the parasite's cuticle. Apparently the destruction of host tissues is not sufficient to cause any drastic injury.

PHYLUM ROTIFERA

The rotifers, or wheel animacules, are almost all free living, being found in abundance in fresh water, salt water, leaf axils of mosses, and various other habitats where moisture occurs. The bodies of these animals are more or less cylindrical, and in most species a ciliated disc, the **corona**, is located at the anterior terminal and a forked foot is located at the posterior terminal. The body wall is comprised of a syncytium covered by a thin cuticular layer.

The digestive tract is complete in some (with a mouth and anus) and incomplete in others (with a mouth but no anus). The excretory system of rotifers is similar to that of platyhelminths, since it consists of flame cells and collecting tubules that empty into a vesicle. The nervous system consists of a dorsal ganglic mass, from which nerve fibers arise to innervate the various areas of the body. Some species possess eyespots and sense organs in the form of tuftlike hairs projecting from the body.

Rotifers are dioecious and usually oviparous. Among the free-living species, parthenogenesis has been reported, in addition to sexual reproduction. Development of rotifers is direct; there are no larval stages.

For a more detailed discussion of the Rotifera, see Hyman (1951).

BIOLOGY OF ROTIFERS

The phylum Rotifera is comprised of three classes: Seisonacea, Bdelloidea, and Monogononta. Almost all of the 1400 or more species are free living. However, members of the class Seisonacea, which includes only the genus *Seison*, and *Zelinkiella synaptae* of the class

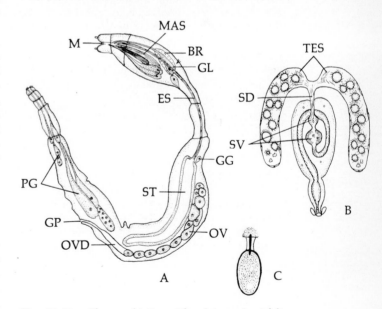

Fig. 21.30. The symbiotic rotifer *Seison*. A. Adult female. (Redrawn after Plate, 1886.) **B.** Reproductive system of male. (Redrawn after Remane, 1929.) **C.** Spermatophore of *Seison*. (Redrawn after Plate, 1886.) BR, brain; ES, esophagus; GG, gastric gland; GL, gland cell; GP, genital pore; M, mouth; MAS, mastax; OV, ovary; OVD, oviduct; PG, pedal glands; SD, sperm duct; ST, stomach; SV, spermatophoral vesicles; TES, testes.

Bdelloidea, and a few species of the Monogononta are symbiotic.

Genus *Seison*

Members of the genus *Seison* comprise a small group of marine rotifers ectosymbiotic on the crustacean *Nebalia* in European waters. The unusual external appearance of the bodies of these rotifers undoubtedly reflects their adaptation to the symbiotic way of life.

The body, measuring up to 3 mm in length, is covered by a segmented cuticle. It is divided into a small oval head, a slender and elongate neck, a thicker fusiform trunk, and a stalklike foot that does not bear typical rotiferan toes but terminates in an adhesive disc for attachment (Fig. 21.30). Typical pedal glands secrete into the disc. The corona is also modified, for instead of numerous cilia, only a few tufts of bristles exist.

Both male and female *Seison* are known. In males, which are slightly smaller and less abundant than females, there is a pair of testes, from which a common sperm duct conducts the spermatozoa to two ciliated chambers, the **spermatophoral vesicle**, in which the sperm are formed into **spermatophores** (Fig. 21.30). These spermatophores are ejected from the male during copulation via the terminal portion of the intestine, which opens on the anterior part of the trunk. In females, there is a pair of ovaries connected with a common oviduct, which in some species opens to the exterior via a genital pore (Fig. 21.30). In others, the oviduct connects with the intestine to form a cloaca, which in turn communicates with the exterior. A female *Seison* produces only one type of egg, and parthenogenesis is not known to occur.

In life, *Seison* moves about the body surface of *Nebalia* in a leechlike manner, alternately attaching its mouth and adhesive disc. It feeds primarily on minute detritus, although it also sucks out the contents of the host's eggs. Because *Seison*, from what is known, does not acquire any metabolic requirements directly from the host, except as a predator of eggs on occasions, it is probably not is a true parasite. However, since it is consistently found on *Nebalia* and there are conspicuous morphologic adaptations, such as the presence of an adhesive disc instead of toes, and reduced ciliature on the corona, it is suspected that the relationship is one of long standing. The relationship may be one of commensalism.

Genus *Zelinkiella*

The class Bdelloidea consists of a large number of free-living rotifers, including some of the most common species. Only the genus *Zelinkiella*—with only one species, *Z. synaptae*—is symbiotic. *Zelinkiella synaptae* is found living in pits in the skin of sea cucumbers (Fig. 21.31). As with *Seison*, the typical rotiferan toes are lacking on *Zelinkiella*; instead, an adhesive disc, on which open the ducts of 12 pedal glands, is present. Nothing is known about the relationship between *Zelinkiella* and its hosts, although the apparent specificity of this rotifer for sea cucumbers and the modification of its toes as a disc suggest that it is more than a recent or temporary relationship.

Symbiotic Monogonontans

Among members of the Monogononta are a few symbiotic species, with some being truly endoparasitic, some in plants and others in animals. For example, *Proales latrunculus* penetrates into the interior of the heliozoan protozoa *Acanthocystis* and feeds on the host's protoplasm, lays about two eggs in the host, then escapes. The eggs, upon developing, also escape from the test of the dead host (Pénard, 1905, 1909). A related species, *Proales gigantea*, is an endoparasite of the eggs of freshwater snails. It enters the egg by boring a hole through the egg membrane and feeds on the contents, causing the death of the snail embryo. While within the host egg, the rotifer lays 7 to 13 eggs that develop very rapidly and the young, after feeding on the remains of the host egg, escape to invade other snail eggs (Stevens, 1912).

There are several species of the Monogononta that are ectosymbiotic. As examples, *Proales gammari* and *Embata parasitica* occur on the legs and gills of amphipods (Plate, 1886); *Pleurotrocha petromyzon* occurs on colonial vorticellids, *Daphnia*, *Cyclops*, insect larvae, *Hydra*, and Budde (1925) has reported it within the valves of dead or dying cladocerans.

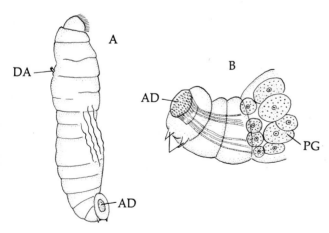

Fig. 21.31.   The symbiotic rotifer *Zelinkiella synaptae*.
A. Adult with corona retracted. **B.** Foot showing adhesive disc and pedal glands. AD, adhesive disc; DA, dorsal antenna; PG, pedal glands. (Both redrawn after Zelinka, 1888.)

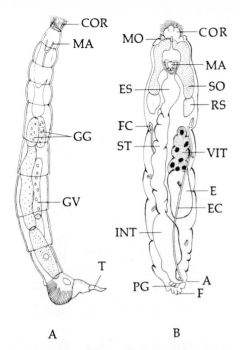

Fig. 21.32. Some parasitic rotifers. A. Female of *Albertia* sp. (Redrawn after Murray, 1905.) **B.** Female of *Albertia vermicularis* (Redrawn after Rees, 1960.) A, anus; COR, corona; E, embryo; EC, excretory canal; ES, esophagus; F, foot; FC, flame cell; GG, gastric glands; GV, germinovitellarium; INT, intestine; MA, mastax; MO, mouth; PG, pedal gland; RS, retrocerebral sac; SO, subcerebral organ; ST, stomach; T, toe; VIT, vitellarium.

More recently, Rees (1960) has reported the occurrence of another endoparasitic rotifer, *Albertia vermicularis* (Fig. 21.32) parasitic in the earthworm *Allolobophora caliginosa*. This species, like the other species of *Albertia*, are wormlike, with a small head and minute rostrum, with neck, eyes (or retrocerebral organ), and with very small toes or none at all.

For a review of the parasitic rotifers, see Hyman (1951) and Rees (1960).

PHYLUM RHYNCHOCOELA

The Rhynchocoela, or Nemertinea, includes the so-called ribbon worms. These slender worms have soft, flat, and unsegmented bodies that are capable of extending and contracting extensively. They are triploblastic and bilaterally symmetrical. An eversible proboscis is present at the anterior end. The digestive tract is straight and complete.

The rhynchocoels may be closely related to the platyhelminths because, like the latter, they are acoelomate, possess a ciliated epidermis like the turbellarians, a protonephridial osmoregulatory system, and a ladderlike nervous system composed of an an-

terior pair of ganglia from which arise two lateral longitudinal nerve cords. Some species possess mid-dorsal and midventral nerve cords in addition to the lateral cords, from which secondary nerve fibers arise.

The Rhynchocoela differs from the Platyhelminthes in that the digestive tract is complete, the reproductive system is simpler, and the sexes are separate. Furthermore, an open circulatory system is consistently present. The treatise by Hyman (1951) should be consulted for details of the morphology of these animals. Gibson's (1972) monograph on the Rhynchocoela includes a review of the morphology, physiology, development, and ecology of this group of worms.

Approximately 500 species of rhynchocoels are known. Almost all are free living and marine, living closely coiled beneath rocks, among algae, or in burrows. Although most species live in the littoral zone, some are found in deep water.

Symbiotic rhynchocoels are rare. *Malacobdella* inhabits the mantle cavity of clams and snails, *Gononemertes* occurs in the atrium of tunicates with its head protruding into the host's pharyngeal cavity, *Nemertopsis actinophila* lives regularly beneath the pedal disc of sea anemones, and *Carcinonemertes* is found in the gills and egg masses of crabs. Most of these rhynchocoels do not appear to be obligate parasites. Rather, they should be considered as commensals, although they may be approaching parasitism for they show adaptive changes characteristic of parasites, such as the loss of eyes and other sense organs, development of adhesive discs for attachment, reduction of the proboscal apparatus, and increased reproductive capacity. *Carcinonemertes* does feed on its host's eggs.

Genus *Malacobdella*

The body of *Malacobdella* is only a few millimeters long and is dorsoventrally flattened (Fig. 21.33). There is a conspicuous adhesive disc at the posterior terminal by which it holds onto its host. Since such a disc is absent in free-living species, its presence could be considered as an adaptation to symbiosis. The adhesive disc is highly innervated, the lateral nerve cords displaying a ganglionic enlargement as they enter the disc. Within the disc, the nerve cords curve around and send secondary nerve fibers to the interior (Fig. 21.33). The multiplicity of gonads reflects an enhanced biotic potential characteristic of parasitic animals.

The circulatory system of *Malacobdella*, like that of other rhynchocoels, consists of blood vessels (or lymphatic canals) with definite walls and with lacunae in the parenchyma that are lined with a delicate membrane (Fig. 21.33). The system is of the open type, that is, the blood (or hemolymph) is not confined within the vessels but also circulates in the parenchymal lacunae. The blood consists of a colorless fluid in which are found amoebocytes (Fig. 21.33).

In the United States, *Malacobdella grossa* is the common species and is widely distributed along the Atlantic coast. It occurs within the mantle cavities of the hard clam, *Mercenaria mercenaria*, and the softshell clam, *Mya arenaria*.

Since adults of *Malacobdella grossa* lack eye-spots, do not become oriented when placed in a stream of water, and are neither attracted nor repelled by potential bivalve hosts, Ropes (1963) has suggested that dispersion of this species most likely occurs during the free-swimming lecithotropic larval phase. This larval phase is apparently of short duration, and the

larvae appear for only a short time in the surface water. From his studies on the distribution of *M. grossa* and the water currents in Nantucket Sound, Massachusetts, Ropes has concluded that surface water current carries the larvae to sites where the molluscan hosts are found and is thus an important factor in the distribution of *M. grossa*.

Three other species of *Malacobdella* are known: *M. japonica* in the clam *Mactra sachalinensis* in Japan, *M. auriculae* in the clam *Chilina dombeiana* in Chile, and *M. minuta* in the clam *Yoldia cooperi* along the Pacific coast of North America.

Genus *Carcinonemertes*

To those interested in marine parasitology, *Carcinonemertes carcinophila* and related species are of particular interest because they are fairly commonly found on the gills and egg masses of various species of crabs (Humes, 1941, 1942). These are small slender worms, varying from less than 1–70 mm in length (Fig. 21.34). A pair of ocelli, or eyespots, is present near the anterior end. The very short proboscis appa-

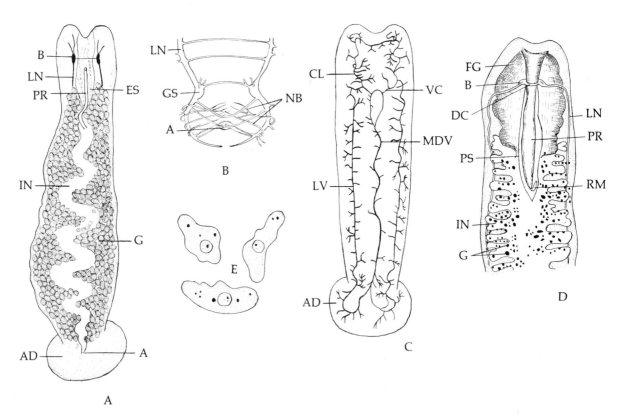

Fig. 21.33. **Symbiotic rhynchocoels. A.** Adult *Malacobdella*. (Redrawn after Guberlet, 1925.) **B.** Posterior end of *Malacobdella* showing nerve supply of adhesive disc. **C.** Circulatory system in adult *Malacobdella*. (**B** and **C**, redrawn after Riepen, 1933.) **D.** Anterior end of adult *Gononemertes* showing proboscal apparatus and numerous glands. (Redrawn after Brinkmann, 1927.) **E.** Hemolymph cells of *Malacobdella*. (Redrawn after Riepen, 1933.) A, anus; AD, adhesive disc; B, brain; CL, cephalic lacuna; DC, dorsal commissure; ES, esophagus; FG, foregut; G, gonads; GS, ganglionic swelling; IN, intestine; LN, lateral nerve cord; LV, lateral blood vessel; MDV, middorsal blood vessel; NB, nerve branches into disc; PR, proboscis; PS, proboscal sheath; RM, retractor muscle; VC, ventral connective.

ratus appears to be functionless. It is scarcely longer than the esophagus and lacks a proboscis sheath and accessory stylets (Fig. 21.34).

Young *Carcinonemertes* live among the host's gills coiled up in mucous sheaths that they secrete. Infestation of mature crabs is most abundant throughout the summer, during the host's spawning season. When female crabs lay their eggs and their egg masses have become attached to the abdominal appendages, the young carcinonemerteans migrate onto the egg masses, on which they feed, grow to sexual maturity, and become ensheathed in mucus.

Carcinonemertean eggs, which include partially developed embryos, are oviposited within the sheaths. After ovipositing, the females move back to the gills of the crab. Pilidiumlike larvae hatch from the eggs and emerge from the sheaths. These larvae are not free swimming but develop into young worms among the egg masses of the host. The young worms either migrate to the gills of the same crab host and secrete individual sheaths or seek other crabs.

If a worm becomes attached to a young crab, its life cycle is cut short when the crab molts, because almost all the rhynchocoels on the gills are discarded with the cast exoskeleton. If the worm becomes attached to an adult host, it develops normally. Although young encapsulated individuals increase in size, it is not clear on what they feed or if they derive their required energy from stored nutrients.

Since a phase of the development of *Carcinonemertes* is dependent on the eggs of the crab host, it is not surprising to find that young individuals that become attached to male crabs are unable to attain sexual maturity. This phenomenon suggests the influence of the host's hormones on the maturation of *Carcinonemertes*. However, this aspect has not yet been studied.

Hopkins (1947), during a survey of *Carcinonemertes* infestation of blue crabs, *Callinectes sapidus*, in Virginia waters of the Chesapeake Bay, found that the appearance of the rhynchocoel could be used with 97% accuracy to determine whether the crab host had spawned or not. Only immature, whitish *Carcinonemertes* were found in the gills of mature crabs that had never spawned. *Carcinonemertes* found in gills of mature crabs that had spawned at least once, or in the gills and egg masses of spawning females, were large and red. Thus, in this instance, the symbiont could be utilized as an indicator of the spawning cycle of the host. The practical importance of this discovery to marine biologists and to the crab industry is obvious.

PHYSIOLOGY OF SYMBIOTIC RHYNCHOCOELS

The physiology of symbiotic rhynchocoels has not been investigated to any extent. Most of these symbionts occupy positions on their hosts where they

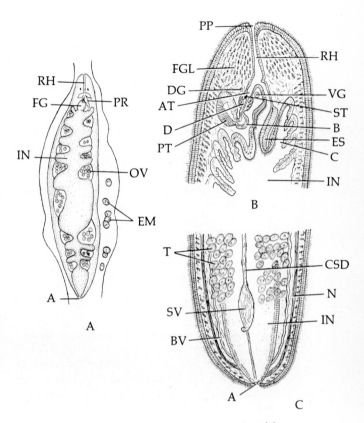

Fig. 21.34. Morphology of *Carcinonemertes*. A. Adult female inside mucous sheath. (Redrawn after Humes, 1942.) B. Sagittal section of anterior end showing frontal glands and reduced proboscis. (Redrawn after Coe, 1902.) C. Frontal section of posterior end of male showing numerous testes and common sperm duct. (Modified after Humes, 1942.) A, anus; AT, anterior tube of proboscis; B, bulb; BV, lateral blood vessel; C, caecum; CSD, common sperm duct; D, diaphragm; DG, dorsal brain ganglion; EM, embryo; ES, esophagus; FG, foregut; FGL, frontal glands; INT, intestine; N, lateral nerve cord; OV, ovaries; PP, proboscis sheath; PR, proboscis; PT, posterior tube of proboscis; RH, rhynchodeum; ST, stylet; SV, seminal vesicle; T, testes; VG, ventral brain ganglion.

can enjoy the ciliary currents that provide both oxygen and food for their host, and they thus can obtain these without exerting themselves. Except for *Carcinonemertes*, which feed on the host's eggs, the other species feed on planktonic fauna carried by their host's ciliary currents.

Eggers (1936) has studied the behavior pattern of *Malacobdella grossa*. This rhynchocoel is sluggish, moving about its host in a leechlike manner by alternately attaching its anterior end and the posterior adhesive disc. When removed from the molluscan host, it only migrates for a short period, after which it becomes attached by its disc and waves its anterior end about in a seeking manner. The dorsal body cilia beat forward while the ventral cilia beat backward. Eggers has shown that unlike other rhynchocoels, the direction of ciliary beat of *M. grossa* cannot be altered nervously. Furthermore, severance of the lateral nerve cords or even the complete removal of the trunk does not cause the adhesive disc to become detached. This suggests that the attachment is controlled primarily by the ganglia present in the disc rather than by the anterior ganglia. When the anterior portion of *Mala-cobdella* is touched, the animal contracts; when the posterior end is touched, it extends. These reflexes are significant in explaining the normal leechlike movements of the animal.

PHYLUM ANNELIDA

The Annelida is composed of segmented worms. Most annelids are free living in aquatic and terrestrial habitats. The phylum Annelida is divided into four classes: Archiannelida, Polychaeta, Oligochaeta, and Hirudinea. None of the Archiannelida is parasitic.

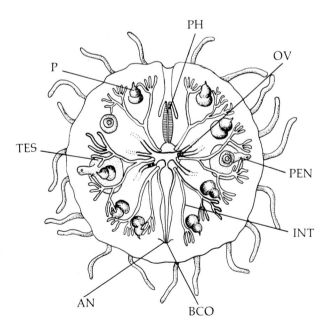

Fig. 21.35. *Myzostomum glabrum.* AN, anus; BCO, opening in body wall through which ova are expelled; INT, intestine; OV, ovarian follicles; P, parapod; PEN, penis; PH, pharynx; TES, testes. (Redrawn after Baer, 1952.)

These small, internally segmented worms are free living in marine waters.

CLASS POLYCHAETA

The class Polychaeta includes the sandworms and tube worms. These annelids are predominantly marine; a few members, such as *Manayunkia speciosa* and *Nereis limnicola*, are freshwater dwellers. In a very few instances parasitism has been reported. For an interesting account of endo- and ectoparasitism among the Polychaeta, the reader is referred to the article by Clark (1956). This author has also included an extensive bibliography on the topic.

Genus *Myzostomum*

This genus includes species that are parasites on echinoderms (Fig. 21.35). Reflective of their adaptation to the parasitic way of life, these annelids have become so altered that they bear little resemblance to other annelids. Most *Myzostomum* species are irregularly discoid and dorsoventrally flattened. Cilia, in the form of cirri, project as tufts around the circumference. Five pairs of papillalike projections are located on the ventral body surface, each bearing one or more setae. Although the body is circular, anteroposterior orientation does exist. The mouth is located anteroventrally. This mouth opens into the muscular pharynx, which is protrusible. The pharynx leads into the saccular stomach, which bears three pairs of lateral, branched caeca (or diverticula). Posteriorly, the stomach opens into the short intestine, which in turn leads into the cloaca. The cloaca opens to the exterior through the anus, which is posteroventral in position (Fig. 21.35). A pair of suckers is found on each lateral margin.

These aberrant annelids are hermaphroditic. Protandrism is normally the condition, that is, the male gonads develop well before the ovaries and discontinue functioning before the ovaries mature. In some species, however, the functional periods of the male and female gonads overlap.

The male system consists of two branched testes, one on each side of the alimentary tract in the coelom. A vas efferens arises from each testis, and the two unite to form the common vas deferens, which opens to the exterior through a pore located adjacent to the third parapodium.

The female system consists of two small ovaries, one on each side of the alimentary tract. These gonads discharge ova into the coelom, which functions as the uterus. A medially located tube, open at one end, collects the fertilized eggs and conducts these to the exterior either through the cloaca or through a gonopore located adjacent to the anus. During copulation, spermatozoa are introduced into the coelom of the female through the same tube.

Some species of *Myzostomum* are found encysted under the integument of echinoderms. In such forms, two individuals, one functioning as a male and the other as a female, occur within the same cyst wall. Other species have been reported endoparasitic in the intestine of echinoderms, but these cases are extremely rare.

Genus *Ichthyotomus*

Another parasitic polychaete is *Ichthyotomus sanguinarius* found on eels. Little is known about the biology of this worm except for its morphology. The dorsoventrally flattened and elongate body, measuring 7–10 mm in length, is quite similar to that of free-living polychaetes except that the modified head lacks the usual appendages. Instead, the worm possesses a pair of crossed stylets that function like a pair of scissors. These structures undoubtedly represent specialized adaptations to parasitism.

This worm attaches to eels near the fins by burrowing a hole in their bodies, using the stylets in the closed position. Once the puncture is made, the stylets are opened and serve as anchors for holding on. The modified perioral zone of the worm consists of a pseudosucker that becomes attached to the wound, and thus, the host's blood is sucked in. An anticoagulant is secreted by the salivary glands during the feeding process. The salivary glands, also known as **hemolytic glands**, are modified poison glands of predaceous polychaetes, and can be considered another type of parasitic adaptation.

Since *Ichthyotomus* attains sexual maturity when it is only 2 mm long and is by then composed of only some 30 of the 70 to 100 somites comprising the body of the fully grown adult, it may be considered a neotenic larva that has become precociously mature, probably as a result of feeding on the blood of its host. Neoteny is always favorable to parasites, because it enables them to produce eggs or larvae within a shorter period.

Although normally a parasite of various species of eels in the mud bottom of the Gulf of Naples in Italy, *I. sanguinarius* will become attached and feed on any fish that remain motionless, such as the electric ray, *Torpedo ocellata*. Thus, it is not as host specific as it would first appear. Apparently the lack of rapid motility on the part of the host is an important factor in host selection.

Genus *Parasitosyllis*

Parasitosyllis is a histozoic parasitic annelid, found on other polychaetes and nemerteans in Tanzania.

CLASS OLIGOCHAETA

Very few parasites are known among the Oligochaeta, but members of the genus *Acanthobdella* are true parasites of salmon in western Siberia and northeastern Europe. The body form of *Acanthobdella* has become so modified that it was considered a leech until

Michaelsen correctly identified it as an oligochaete.*

Aspidodrilus kelsalli is a minute aberrant oligochaete parasitic on earthworms in Sierre Leone. This parasite is conspicuously adapted to parasitism. The entire ventral surface of the posterior half of the body is flattened and slightly concave, thus forming an elongate adhesive disc by which it holds on. In addition, a true sucker in front of the disc also serves as an adhesive organ. *A. kelsalli* utilizes body surface mucus and coelomocytes exuded through the host's dorsal pores as food, in addition to ingesting mud, vegetation, and animal debris. Similarly, *Fridericia parasitica*, found on the body surface of the earthworm *Allolobophora robustra*, feeds on mucus and coelomocytes.

A few additional examples of parasitism among oligochaetes exist. For example, Holt (1963) has reported a branchiobdellid (of the family Branchiobdellidae) oligochaete, known as *Cambarinicola aliena*, to be parasitic in the brood pouches of females of the freshwater isopod, *Asellus bicrenatus* (Fig. 21.36). Although most branchiobdellid oligochaetes are epiphoretic on crayfish (Holt, 1968), *C. aliena* appears to be a true parasite because Holt (1963) has found granular and globular material in the oligochaete's gut that he suspected to be yolk from the host's eggs.

In addition to *Cambarinicola aliena*, *C. branchiophila*, and perhaps other branchiobdellids, also are true parasites. These most probably originated as epiphoronts that entered their hosts adapted to parasitism. *Cambarinicola branchiophila* is confined to the gill chambers of its crayfish host, where it feeds by clipping gill filaments off its host and sucking blood. In both *C. aliena* and *C. branchiophila*, Holt (1963) has reported the presence of weakened body walls and delicate jaws, which he interpreted as adaptations to parasitism.

Other parasitic oligochaetes include certain members of the family Naididae. For example, *Schmardaella lutzi* and *Dero hylae* live in the ureters of tree frogs, *Hyla*, and *Dero floridana* lives in the ureters of the toad *Bufo terrestris* in Florida (Harmon, 1971).

CLASS HIRUDINEA

The Hirudinea includes the leeches. These annelids are commonly found in fresh water, although marine and even a few terrestrial species exist. Leeches differ from other annelids in that their commonly slender, leaf-shaped bodies lack setae. Furthermore, they possess a large posterior sucker and a smaller, anterior sucker,

---

*Some biologists still consider *Acanthobdella* to be a primitive leech.

or pseudosucker, surrounding the mouth. Another characteristic of leeches is that the external body divisions, the **annuli**, do not correspond with the 34 internal true body segments.

Many leeches are predaceous or are scavengers, but some are ectoparasitic on ectothermic and endothermic vertebrates, feeding on their blood. When observed alive, leeches generally exhibit bright coloration, but the pigments do not persist long after capture. Some can swim, while others cannot, but all leeches utilize an inchwormlike crawling motion on a solid substrate. This characteristic creeping movement is facilitated by the holding ability of the anterior and posterior suckers.

Alimentary Tract

The digestive tract of the Hirudinea is complete, leading from mouth to anus. The mouth is located anteriorly in the middle of the oral sucker. It opens internally into the buccal chamber, which in turn leads into the pharynx. The pharynx may be highly muscular, acting as a suction bulb in most bloodsucking species, or it may be only a slender tube, which leads posteriorly into the esophagus. In species with a large pharynx, the esophagus may not be readily visible.

The stomach of leeches generally occupies somites XIII to XIX or XX. It may be pyriform and simple, or it may bear one to 14 pairs of lateral diverticula. These side pouches may be simple or branched. Secondary branches are particularly common, extending from the postcaeca, which are the posteriormost diverticula, in blood-feeding species. The stomach leads into the intestine, which empties into the rectum. The rectum communicates with the exterior through the posteriorly located anus.

Coelom

Unlike the neatly septate coelom found in earthworms, the coelom of leeches is largely obliterated or reduced by the body musculature and parenchymatous tissues. The little space remaining is in the form of longitudinal canals interconnected by small branches. Within these canals are located the nervous, circulatory, excretory, and reproductive systems.

Circulatory System

The closed circulatory system of leeches basically consists of a large dorsal and a large ventral blood vessel, which are connected anteriorly and posteriorly by a number of convoluted smaller vessels. The wall of the anterior segment of the dorsal blood vessel is thickened and contractile, functioning as a heart. In some species of the Rhynchobdellida, the posterior portion of the dorsal vessel is greatly expanded to form a blood sinus enveloping the intestine.

The contractions and expansions of the intestine also aid in forcing the blood anteriorly in the dorsal vessel and posteriorly in the ventral vessel. In members of the family Piscicolidae, lateral pulsating vesicles also aid in forcing the blood to circulate. Numerous minute blood vessels that arise from the two main ones infiltrate the body tissues to form rich beds of capillaries in the areas below the epidermis. Gaseous exchange takes place through the body wall except in a few marine species with lateral gill-like slits.

Excretory System

The excretory system in leeches is comparable to that

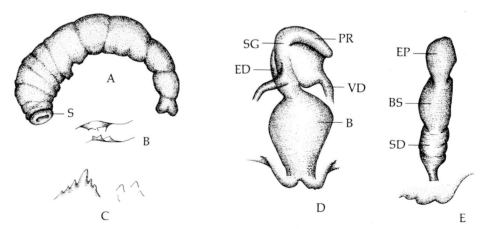

Fig. 21.36. *Cambarinicola aliena.* **A.** Adult removed from brood pouch of female *Asellus bicrenatus.* **B.** Lateral view of jaws. **C.** Dorsal (left) and ventral (right) jaws. **D.** Lateral view of male reproductive system. **E.** spermatheca. B, bursa; BS, spermathecal bulb; ED, ejaculatory duct; EP, ental process; PR, prostate; S, sucker; SD, ectal duct of spermatheca; SG, spermiducal gland; VD, vas deferens. (All figures redrawn after Holt, 1963.)

found in earthworms. There are never more than 17 pairs of nephridia in the Hirudinea. These are segmentally arranged along the middle somites. The collecting terminal consists of a simple ciliated funnel or of a more complex organ. In some leeches, such as *Hirudo* and *Haemopis sanguisuga*, the inner ends of the nephridial tubules have lost their connection with the ciliated funnels and have no excretory function; instead, they are the production sites for coelomocytes (Mann, 1954). The nephridial tubules empty to the exterior through paired nephridiopores located on the body surface in ventral grooves between annuli.

Nervous System

The nervous system consists of an anteriorly situated "brain," located in somites V and VI. The main ventral, paired nerve cord joins the brain at a ganglionic mass located in somites XXV and XXVI. There is a single ganglion in each of somites VII to XXVI along the ventral nerve cord. Nerves arise from these ganglia to innervate the various tissues and organs of the body in their immediate areas.

Sensory Organs

Special sensory organs and structures found on the body surface of leeches include (1) **papillae** (Bayer's organs), which are minute sensory projections plentifully scattered over the body surface; (2) **tubercles**, which are larger retractable protruberances that involve the deeper dermal tissues and muscles and that are covered with papillae; (3) **eyes**, which vary from simple ocelli, consisting of a single visual cell embedded in pigment granules, to rather complex compound eyes found on the anterior end of some members of the family Hirudidae; and (4) **sensillae**, which are comparable to papillae except that a minute hairlike projection is present and that they are confined to certain sensory annuli.

Reproductive System

The reproductive system consists of a male and a female complex. Cross-fertilization between two individuals is the rule.

   Female System.   This consists of two ovaries of varying sizes. A short, narrow oviduct leads from each ovary, and these ducts join to form the vagina. The vagina opens to the exterior through the female gonopore, which is located on the midventral line in somite XI. In some species, such as members of the Hirudidae, a diverticulum, known as the **vaginal caecum**, arises from the vagina and an **albumin gland** is connected with the muscular vagina.

   Male System.   This consists of 4 to 80 pairs of testes located in various somites. Arising from each testis on each side is a **vas efferens**. The vasa efferentia on each side connect with a single **vas deferens**. Each vas deferens gives rise to a **seminal vesicle** at its terminal end. The seminal vesicle is short or long,

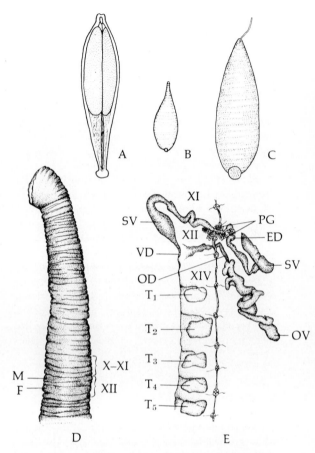

Fig. 21.37.   Leeches. A. Spermatophore of *Glossiphonia* enclosing two oval bundles of spermatozoa. (Modified after Brumpt, 1900.) **B.** Exterior view of *Helobdella.* **C.** External view of *Glossiphonia* with extruded proboscis. **D.** External view of anterior portion of *Illinobdella moorei.* Notice positions of male and female genital pores. **E.** Dissected reproductive organs showing relationship to nerve cord. The Roman numerals in **D** and **E** refer to body somites. (**D** and **E,** redrawn after Moore, 1959.) ED, ejaculatory duct; F, female genital pore; M, male genital pore; OD, oviduct; OV, ovary; PG, prostate glands; SV, seminal vesicle; T₁, 1st testisacs; T₂, 2nd testisacs; T₃, 3rd testisacs; T₄, 4th testisacs; T₅, 5th testisacs; VD, vas deferens.

and straight or convoluted, depending on the species.

   Spermatozoa passing into the seminal vesicles are cemented together and pass into the **ejaculatory duct**. The two ejaculatory ducts, one on each side, unite to form the **glandular atrium**. In all leeches, except in the Hirudidae, the spermatozoa are formed into spermatophores while passing down the ejaculatory duct (Fig. 21.37). In most species, the atrium includes an eversible **bursa** through which the spermatozoa are transferred to the copulatory mate. Members of the Hirudidae possess a penis, which is the copulatory organ.

Fertilization. The mechanisms of sperm penetration and ova fertilization among the leeches vary. In the Rhynchobdellida and Erpobdellidae, the spermatozoa released from the spermatophores onto the body surface of the mate actually penetrate the integument by means of local histolytic activity and reach the ovaries via the coelomic sinuses.

In the Hirudidae, the penis is extruded and the spermatozoa are introduced into the female gonopore. During the breeding season, the inconspicuous **clitellum** secretes rings, much in the same fashion as do earthworms. These rings collect fertilized ova while passing anteriorly over the female gonopore and eventually slip off the anterior end of the body. Such ova-bearing cocoons are deposited in water between May and August. After completing their embryonic development, young leeches escape from the cocoon and become independent. The period required for maturation varies; however, in most cases it takes years.

BIOLOGY OF LEECHES

The class Hirudinea is divided into three orders: Rhynchobdellida, Pharyngobdellida, and Gnathobdellida. The distinguishing characteristics of each of these are presented in the classification section at the end of the chapter. For a monograph of the North American leeches, see Sawyer (1972).

ORDER RHYNCHOBDELLIDA

The feeding habits of members of the Rhynchobdellida are unusual in that a muscular proboscis is present that is actually the anterior end of the pharynx. This structure is protruded and inserted in the host's tissues during feeding.

Rhynchobdellid Genera

The genus *Haementeria* includes *H. officinalis*, the so-called medicinal leech of South and Central America. The species *H.* (= *Placobdella*) *parasitica* is found on snapping turtles.

The genus *Helobdella* (Fig. 21.37), including *H. stagnalis, H. elongata,* and *H. fusca,* has been reported to be a suitable vector for *Trypanosoma inopinatum*, a hemoflagellate of frogs.

The genus *Glossiphonia* (Fig. 21.37) includes *G. complanata, G. heteroclita,* and others found attached to various freshwater snails, and on the larvae of *Chironomus*. Personal observations by the author in Hawaii indicate that *Glossiphonia* spp. attached to the exterior of freshwater snails are capable of causing decalcification of shells, probably by secreting some yet unidentified substance.

The genus *Pontobdella* includes species parasitic on elasmobranch fishes and some species that serve as vectors for *Trypanosoma rajae* of rays. In addition, species of *Pontobdella* may serve as vectors of trypanosomes of other marine fish.

Other members of this order are vectors for trypanosomes and *Cryptobia* of freshwater fish. Meyer (1940, 1946) has reviewed the genera of freshwater rhynchobdellid leeches found on fishes in North America, which include *Piscicola, Piscicolaria,* and *Illinobdella* (Fig. 21.37). These papers should be consulted by those interested in parasitic leeches. The presence of *Piscicola* in fish hatcheries is known to cause serious damage.

In addition to the previously mentioned rhynchobdellid leeches, several others "parasitize" arthropods, especially decapod crustaceans. Little is known concerning the "parasitic" habits of these. It is suspected that in many instances they are merely epiphoretic forms. Meyer and Barden (1955) have catalogued some of the leeches found attached to decapods and have discussed the uncertainty of their parasitic nature.

ORDER PHARYNGOBDELLIDA

The Pharyngobdellida includes *Mooreobdella, Nephelopsis, Erpobdella,* and *Dina*. The last two mentioned genera have been revised in a monograph by Pawlowski (1955).

ORDER GNATHOBDELLIDA

The Gnathobdellida includes *Hirudo medicinalis*, the medicinal leech. The feeding habits of this parasite vary with age. The primary food of young leeches consists of insect hemolymph but the blood of frogs and fish serves this function during the growing period. Adults feed primarily from warm-blooded animals, including humans. In addition to robbing humans of blood, this leech serves as a vector for bacteria and other microorganisms. For example, it can transmit the anthrax bacterium.

Other prominent genera of the Gnathobdellida include *Macrobdella* (Fig. 21.38) and *Limnatis. Limnatis* includes *L. nilotica,* the large horse leech (8–12 cm long), which parasitizes mammals in North Africa, middle Europe, and the Near East. It is commonly used for bloodletting in these areas. Mammals become infested with this leech while drinking contaminated water. Once within the host, it becomes attached to the nasal passage, pharynx, and larynx. If present in large numbers, it can cause asphyxia. Furthermore,

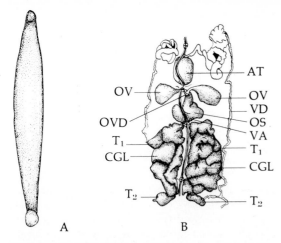

Fig. 21.38. *Macrobdella.* **A.** External view of leech. **B.** Part of reproductive organs of *M. decora.* AT, atrium; CGL, copulatory glands; T_1, 1st testisacs; T_2, 2nd testisacs; OS, ovisac; OV, ovary; OVD, oviduct; VA, vagina; VD, vas deferens. (Redrawn after Moore, 1959.)

heavily parasitized hosts can suffer from anemia because of the loss of blood. In addition, bleeding, nausea, and vomiting, often followed by death due to anemia and suffocation are common.

Another member of *Limnatis, L. africana,* parasitizes humans, monkeys, and dogs in Senegal, Congo, India, and Singapore. Human infestation is most commonly acquired while bathing. The leech can enter the body via the vagina and urethra and cause rather severe bleeding. Again, anemia and even death can result.

Various species of *Dinobdella* occur in India, where these leeches, when swallowed in drinking water, attack the pharynx of cattle.

The genus *Hirudinaria* includes *H. granulosa,* the Indian medicinal leech.

Haemadipsa includes the dreaded land leeches of the Far East. One species, *H. chiliani,* attacks horses and cattle in South America, and heavily parasitized hosts usually perish. Probably the best known species is *H. zeylandica,* which attacks all mammals, including humans, in the tropical jungles of Asia. Adult leeches, measuring 2–3 cm in length, are found on the surfaces of trees and grass and under stones in damp places. They readily attach themselves to the skin of humans and other animals and ingest their blood. This leech is found in India, the Philippines, Australia, and in the Chilian Andes. Other species are found in Japan, China, Chile, and Trinidad.

The genus *Theromyzon* includes species that enter the throats and nostrils of water birds, killing geese and ducks.

The ability of bloodsucking leeches to feed is facilitated by the secretion of anticoagulants into the puncture site.

ECOLOGIC FACTORS

Mann (1955) has contributed a study of some ecologic factors that influence the distribution of leeches. He has shown that the amount of $CaCO_3$ in the environment influences the abundance of certain species. For example, *Erpobdella octoculata* is most abundant in "soft" waters—with 0 to 17 ml of $CaCO_3$ per liter of water—whereas an abundance of *Helobdella stagnalis* is characteristic of "hard" waters—with 60 to 242 ml of $CaCO_3$ per liter of water. These observations suggest that as is the case among most ectoparasites, external environmental factors are just as important to the physiology of leeches as the internal environment of the host is to endoparasites. The work of Bennike (1943) should be consulted because it contains many additional interesting observations on the influence of environmental factors on the distribution of leeches. In addition, the excellent monograph by Mann (1962), which includes much information on the physiology and ecology of leeches, should be consulted.

HOST-PARASITE RELATIONSHIPS

Although commonly referred to as parasites, most leeches are predators, either carnivorous or saprophagous. Only the bloodsucking species can be considered parasites. Most of these are only temporarily attached to their hosts, abandoning them when they become engorged with blood. In a few instances, these parasites have become sedentary and never leave their hosts. This relationship is most prevalent among leeches that are attached to fish, primarily among the rhynchobdellids. From the little information available, these leeches appear to exhibit little or no host selectivity. For example, *Piscicola geometra* is found on many species of freshwater teleost fish, on trout, and even on tadpoles. On the other hand, host specificity is exhibited by *Theromyzon,* which has been found only on aquatic birds, and by *Ostreobdella,* which has been reported only on oysters.

Ozobranchus jantseanus is a leech commonly found attached to the aquatic turtle *Clemmys japonica.* Although the relationship is not an obligatory one, since *O. jantseanus* is also found free living, this leech is particularly adapted to clinging to *C. japonica* because the turtle is in the habit of climbing out of water to sun itself several times a day. During these periods, the leech rapidly dries and shrivels up into a small black disc. However, when the turtle returns to water, the disc swells up and the leech resumes it normal activities. The leech loses four-fifths of its body weight when dried and can remain in a state of anabiosis up to 8 days. The ability of *O. jantseanus* to adapt to anabiosis undoubtedly explains why it can stay attached to its turtle host.

Very seldom does one find external anatomic features among leeches that suggest adaptation to parasitism. In one such instance, *Hemibdella soleae* holds on to the spines on the free edge of the scales of its host, the sole, because its mouth is very narrow and deep. Since only sole scales possess spines of the correct size to fit into the mouth of this parasite, the morphological adaptation is selective in creating a host-parasite relationship. One reason given for the apparent lack of widespread adaptation by parasitic leeches is that they are specialized to begin with and by coincidence their specialization permits the type of temporary parasitism that they practice.

PHYLUM MOLLUSCA

Parasitism is a relatively rare phenomenon among adult molluscs, but in the few reported instances, the degrees of modification in the several parasitic species represent a beautiful series demonstrating progressive stages of adaptation to parasitism. Caullery (1952) has given an excellent account of parasitic molluscs. Parasitism among larval molluscs, known as **protelian** parasitism, is common among freshwater bivalves of the family Unionidae. This form of parasitism is discussed on page 742.

Adult molluscan parasites are limited primarily to members of the class Gastropoda, which includes the snails, slugs, limpets, and related forms. Parasitic gastropods are almost all limited to four families: Capulidae, Eulimidae, Entoconchidae, and Paedophoropodidae.

FAMILY CAPULIDAE

The Capulidae includes prosobranch snails with shells that are slightly coiled or in the form of a simple incurved cone. The only modern genus known is *Thyca*, containing a few species found as ectoparasites of the starfish *Linckia* in the Indian Ocean and in marine waters of the Malay Archipelago. These snails still retain the form of rather typical gastropods but the foot is greatly reduced and the peribuccal region is enlarged to form a larger sucker by which the parasite is attached to its host (Fig. 21.39). A proboscis projects from the center of the sucker that pierces the integument of the host to withdraw fluid nutrients. No radula is present. In species with a long proboscis, the digestive gland and intestine are reduced, but well-developed salivary glands are present, suggesting the comparatively simple nature of its food, which does not require complex digestion. These animals are dioecious and exhibit sexual dimorphism, i.e., the smaller shells of males differ in form from those of females.

Capulids are apparently rather ancient, since fossils of a now extinct genus, *Platyceras*, are known from crinoids and starfishes dating back to the Silurian-Triassic era. The appearance of *Platyceras* is very similar to that of *Thyca*, suggesting that these gastropods have undergone very little evolutionary change in form.

FAMILY EULIMIDAE

Members of Eulimidae possess turreted shells that are

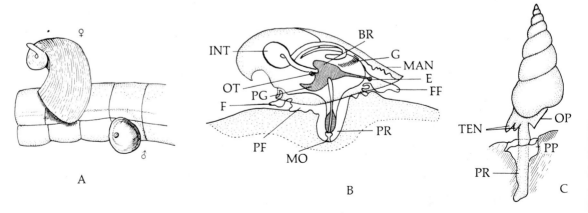

Fig. 21.39. Some parasitic molluscs. A. Male and female specimens of *Thyca stellasteris* showing conspicuous sexual dimorphism. (Redrawn after Koehler and Vaney, 1912.) **B.** Anatomy of *Thyca ectoconcha*. (Redrawn after Sarasin and Sarasin, 1887.) **C.** *Mucronalia palmipedis*. (Redrawn after Koehler and Vaney, 1912.) BR, brain; E, eye; ES, esophagus; F, foot; FF, frontal fold; G, gills (ctenidia); INT, intestine; MAN, mantle; MO, mouth; OP, operculum OT, otocyst; PF, pseudofoot; PG, pedal ganglion; PP, pseudopallium; PR, proboscis; TEN, tentacles.

thin and translucent. This family includes both free-living and parasitic species, the latter being either ecto- or endoparasites of echinoderms. Anatomic adaptations of eulimid snails form a beautiful series showing the different degrees of change resulting from parasitic adaptation. The free-living species generally possess radulae, but these scraping structures do not occur in the parasitic species.

Genus *Pelseneeria*

Members of the hermaphroditic genus *Pelseneeria* are ectoparasitic on holothurians (sea cucumbers). These parasites move over the body surface of the host but never leave it, even to oviposit. *Pelseneeria* possesses a muscular proboscis that penetrates the host's integument and through which nutrients, in the form of body fluids, are drawn out.

In one species, *P. profunda*, there is a fringed collar-like projection—the **pseudopallium**—in the peribuccal zone that partly covers the shell. This structure is definitely an adaptation to parasitism also found in other parasitic gastropods. Although the foot of *Pelseneeria* is better developed than that of *Thyca*, an operculum is also lacking in *Pelseneeria*. This also may be considered an adaptation to parasitism, because an operculum is not found on any other adult parasitic snails except *Peasistilifer* spp.

Genus *Megadenus*

Members of *Megadenus* are parasites of echinoderms in the Bahamas, the Indian Ocean, and in the Yellow Sea. *Megadenus voeltzkowi* and *M. holothuricola* live in the respiratory trees of holothurians, while *M. cysticola* and *M. arrhynchus* are found within tumorlike swellings on the spines of the echinoid *Dorocidaris tiara* and the asteroid *Anthenoides rugulosus*, respectively. These snails are dioecious. In *M. cysticola*, a small male and a large female are found together within the same cystic tumor. *Megadenus holothuricola* and *M. voeltzkowi* possess a proboscis that is thrust through the tracheal wall of the host into the coelom. A proboscis is also present on *M. cysticola*, but not on *M. arrhynchus*. In addition to these modifications, the foot of *Megadenus* has ceased to be a locomotory structure and an operculum is lacking. However, the pedal glands persist. A large pseudopallium is present, the intestine is shortened, and the stomach and digestive gland tend to fuse to form a single organ.

Genus *Peasistilifer*

Several species of *Peasistilifer* attach to ophiuroids, holothurians, sea urchins, and starfishes (Figs. 21.40, 21.41). The shells of these gastropods are well-developed with several apical whorls. An operculum is present and a pseudopallium, although present, is not very well developed. These snails remain attached to the exterior of their hosts, usually at a permanent site. A well-developed, long proboscis penetrates the

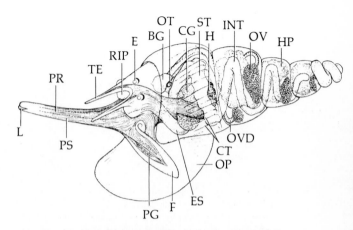

Fig. 21.40. *Peasistilifer nitidula*. Drawing of entire specimen showing positions of internal organs. Reconstructed from serial sections. BG, branchial ganglion; CG, cerebral ganglion; CT, ctenidia; E, eye; ES, esophagus; F, foot; H, heart; HP, hepatopancreas (or digestive gland); INT, intestine; OP, operculum; OT, otocyst; OV, ovary; OVD, oviduct; PG, pedal groove; PR, proboscis; PS, proboscis sheath (=pseudopallium); RIP, rudimentary basal portion of pseudopallium; ST, stomach; TE, tentacle. (After Hoskin and Cheng, 1970.)

host's integument and draws out body fluids. *Peasistilifer* spp. possess no salivary glands, but a well-formed digestive gland is present. One species, *P. variabilis*, is capable of endoparasitism and is sometimes found in the blind-sac alimentary canal of its host, *Synapta soplax*, in the Indian Ocean. These snails are dioecious.

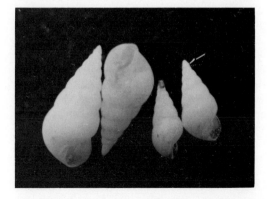

Fig. 21.41. *Peasistilifer nitidula*. Shells from three locations in the Pacific Basin. Left specimen from Tahiti, middle two specimens from Okinawa, right specimen from Fiji. Arrow indicates enlarged mucral third whorl. (Courtesy of Dr. G. P. Hoskin.)

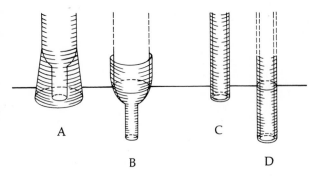

Fig. 21.42. Pseudopallium and proboscis. A. Drawing showing function of pseudopallium prior to penetration of proboscis as postulated by Baer. **B.** Drawing showing folding back of pseudopallium after penetration of proboscis. (**A** and **B**, redrawn after Baer, 1952.) **C.** Diagrammatic drawing showing position of pseudopallium of *Peasistilifer nitidula* prior to penetration of proboscis. **D.** Diagrammatic drawing showing insertion of pseudopallium of *P. nitidula* along with the proboscis after penetration. (**C** and **D**, after Hoskin and Cheng, 1970.)

Hoskin and Cheng (1970) have studied the ecology and microanatomy of *Peasistilifer nitidula* (formerly known as *Mucronalia nitidula*), an ectoparasite of the subtropical sea cucumber *Holothuria atra*. This small, white snail measures 2.0–5.10 mm in length and is negatively phototactic (Figs. 21.40, 21.41). It is attached to its host by a mucoid thread secreted by specialized pedal glands, and if accidentally displaced, it usually becomes reattached by tracing its path by this elastic, mucous thread. *P. nitidula* possesses a pseudopallium, but unlike in most other species of parasitic gastropods, this covering is inserted into the host's body along with the proboscis during feeding (Fig. 21.42). Its nutrient consists of the echinoderm host's body fluids.

Genus *Stylifer*

Members of the genus *Stylifer* possess thin shells and resemble *Peasistilifer*, but they are larger, have no operculum, and the foot is reduced (Fig. 21.43). In some species, the pseudopallium is large and well developed and actually covers the shell (Fig. 21.43). The various species are parasitic on starfish and ophiuroids in Asian and mid-Pacific waters. They possess a long proboscis, which penetrates the host's integument and reaches the coelom to withdraw body fluids. The only species that have been extensively investigated are *S. sibogae* and *S. linckiae*. The former is a hermaphrodite and possess a well-developed male copulatory organ. The female sex products are

flushed out by water drawn into the pseudophallial cavity by the piston action of the proboscis and/or by the contraction and expansion of the pseudophallial wall.

Stylifer linckiae is a common species found associated with the starfish *Linckia multifora* in Hawaiian waters (Fig. 21.43). It is embedded in a gall-like swelling on one or more of the host's arms (Fig. 21.44). A male, which is smaller, and a female are commonly found within the same gall. Eggs laid by the female are passed to the exterior enveloped in a gelatinous matrix through an aperture on the gall. The eggs hatch into veliger larvae (Fig. 21.43) in approximately 10 days and the free-swimming veligers seek out and enter the ambularcral groove of another host and develop to maturity.

Davis (1967) has reported that the presence of *Stylifer linckiae* in an arm of *Linckia multifora* inhibits autotomy (normal loss of an arm). This is probably due to a neurosecretion emitted by the parasite.

There is no doubt that *Stylifer linckiae* derives its nutrients from the host through its long proboscis. Tullis and Cheng (1971) have shown that if the host is fed algae labeled with radioactive carbon, the ^{14}C is eventually taken into the parasitic snails.

Genus *Diacolax*

Diacolax includes only one known species, *D. cucumariae*, an ectoparasite of the holothurian *Cucumaria mendax* in the Falkland Islands. In this greatly modified snail, the pseudopallium covers the entire body, giving the animal the appearance of an oval mass. Attached to the body is a small pointed siphon that leads to the exterior. The anterior end of the animal is embedded in the host. A shell is lacking. Within the ovoid body is the intestine, surrounded by the ovary. The remaining space is filled with developing eggs. The free-swimming larva is a typical veliger.

Genus *Gasterosiphon*

Probably the most specialized of the eulimid parasites are members of the genus *Gasterosiphon* (Fig. 21.45). These snails are endoparasitic in holothurians in the Indian Ocean. Koehler and Vaney (1903) described *G. deimatis* in *Deima blakei*. This gastropod has become so modified in the adult form that it can be recognized as a snail only by the trained eye, and by studying the juvenile stages. This endoparasite communicates with the exterior by means of a siphon connected to a minute pore on the host's body surface. The proximal end of the siphon is connected with the ovoid body which, if cut open, reveals a vestigial foot; an incomplete alimentary tract composed of an esophagus, stomach, and ramifying digestive gland ducts; and a condensed nervous system. Another long slender tube, the proboscis, is attached to the ovoid body at the pole opposite that from which the siphon arises. The proboscis is attached distally to the marginal vessel on the intestine of the host, and blood from this vessel is thus sapped into the parasite.

Gasterosiphon deimatis is a hermaphrodite. The eggs are deposited in the pseudophallial cavity, where they undergo a period of incubation. When sufficiently developed, the eggs are discharged into the sea through the ciliated siphon and hatch, giving rise to free-swimming larvae.

FAMILY ENTOCONCHIDAE

The Entoconchidae includes several genera of molluscs that are extremely modified. None of these possesses a shell, and in many ways they resemble *Gasterosiphon*. Members of Entoconchidae include the genera *Entocolax*, *Entoconcha*, *Enteroxenos*, and *Thyonicola*.

Genus *Entocolax*

Members of the genus *Entocolax* are endoparasitic in the coelom of holothurians (Fig. 21.45). In *E. ludwigii*, a parasite of *Myriotrochus rinkii*, one end of the ovoid body is attached to the host's integument by a short siphon. The main bulk of the parasite is a hollow sac, with the pseudophallium forming the wall. Within this sac are found the ovary and the oviduct. The free end of the animal projects freely in the host's coelom as a tubular proboscis composed of a terminal esophagus and an elongate hepatic intestine.

In *E. schwanzwitschi*, a parasite of *Myriotrochus eurycyclus*, the anatomy is essentially the same as that of *E. ludwigii*. It is also located in the host's coelom, but the siphon is attached to the outer or peritoneal surface of the holothurian's gut, rather than to the integument.

Schwanzwitsch (1917) and Heding (1934) have re-

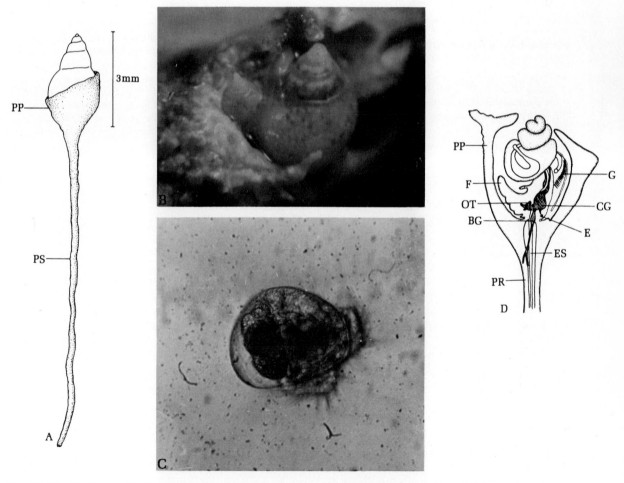

Fig. 21.43. *Stylifer.* **A.** *Stylifer linckiae* removed from its echinoderm host. **B.** Photograph of a large *S. linkiae in situ* with a smaller specimen adjacent to it. **C.** Photomicrograph of a veliger of *S. linkiae.* (**A–C**, courtesy of Dr. R. Tullis.) **D.** *S. linkiae* in cut-away view. (Redrawn after Sarasin and Sarasin, 1887.) BG, buccal ganglion; CG, cerebral ganglion; E, eye; ES, esophagus; F, foot; G, gills; OT, otocyst; PP, pseudopallium; PR, proboscis; PS, proboscis sheath.

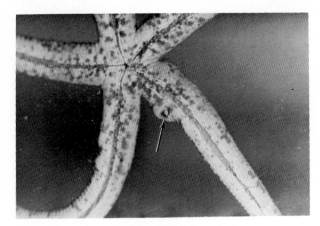

Fig. 21.44. *Stylifer linckiae.* Specimen embedded in gall-like swelling in host's arm (arrow).

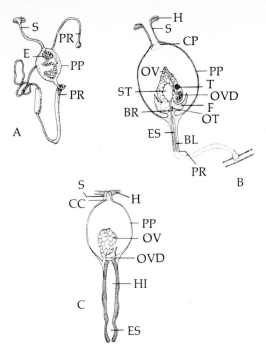

Fig. 21.45. **Parasitic molluscs. A.** Whole specimen of *Gasterosiphon deimatis.* **B.** Cut-away section of *G. deimatis* showing internal anatomy. (**A** and **B**, redrawn after Koehler and Vaney, 1912.) **C.** *Entocolax.* (Redrawn after Vaney, 1914.) BL, blood lacuna; BR, brain; CC, ciliated canal; CP, calcified pseudopallial shell; E, egg mass; ES, esophagus; F, foot; H, host tissue; HI, hepatic intestine; OT, otocyst; OV, ovary; OVD, oviduct; PP, pseudopalium; PR, proboscis; S, siphon; ST, stomach; T, testis.

ported that dwarf males are attached to the inner wall of the female's pseudophallium, where they live permanently. Fertilized eggs in cocoons develop into veliger larvae in the pseudophallial cavity. The larvae are typically gastropodan, with a shell, velum, and foot. These larvae measure no more than 0.5 mm in length.

Infection of the holothurian host occurs when a

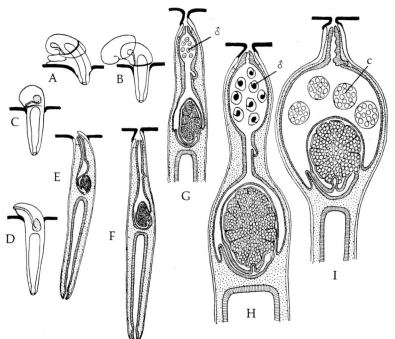

Fig. 21.46. *Entocolax.* **A–F.** Entrance of *Entocolax* into host. **G** and **H.** Hyperparasitic dwarf males in female. **I.** Cocoon filled with ova (c) in female. (After Baer, 1952.)

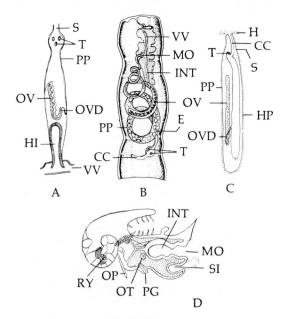

Fig. 21.47. **Parasitic molluscs. A.** *Entoconcha*. (Redrawn after Vaney, 1914.) **B.** *Entoconcha mirabilis* within host, attached to host's ventral blood vessel. (Redrawn after Baur, 1864.) **C.** *Enteroxenos*. (Redrawn after Vaney, 1914.) **D.** Longitudinal section of veliger larva of *Entoconcha*. (Redrawn after Baur, 1864.) CC, ciliated canal; E, egg mass; H, host tissue; HI, hepatic intestine; HP, sheath of host's peritoneum; INT, intestine; MO, mouth; OP, operculum; OT, otocyst; OV, ovary; OVD, oviduct; PG, pedal gland; PP, pseudopallium; RY, residual yolk; S, siphon; SI, saclike invagination; T, testis; VV, ventral blood vessel of host.

young *Entocolax* penetrates the host's surface. *Entocolax ludwigii* penetrates the body wall of the host near the dorsal tentacles, since *M. rinkii*, the host, is normally buried in mud except for its dorsal tentacles and adjacent areas. *Entocolax schwanzwitschi* enters the mouth and penetrates the gut wall of *M. eurycyclus*. The hypothesized method of entrance of this parasite into the host is depicted in Fig. 21.46.

Genus *Entoconcha*

The body of *Entoconcha*, a parasite within the body cavity of holothurians in the Adriatic Sea and in the Philippines, is even more modified than that of *Entocolax*. The mouth, which opens into a hepatic intestine, is attached to the ventral vessel of the host (Fig. 21.47). *Entoconcha mirabilis*, the first parasitic mollusc known, is approximately 8 cm long.

Genus *Enteroxenos*

Adults of *Enteroxenos* are tubular, measuring 100–150 mm in length. They are endoparasites in the body cavity of the holothurian *Stichopus tremulus*. The young specimens are usually found attached to the outer surface of the anterior portion of the host's gut.

Enteroxenos is the most simplified of the parasitic gastropods as a result of evolutionary adaptation. An internal elongate cavity extends the entire length of the body (Fig. 21.47). This tubular chamber opens to the exterior through a minute pore at the anterior end. The gonads are embedded in the body wall with the single elongate ovary lying alongside the tubular cavity and the testis located toward the proximal end. Caullery (1952) has advanced the concept that the entire body of *Enteroxenos* corresponds to a large pseudopallium. Although adults of this parasite are not recognizable as molluscs, the larvae bear molluscan characteristics, including a coiled shell (Fig. 21.47).

Genus *Thyonicola*

Thyonicola mortenseni was described by Mandahl-Barth (1941) within the holothurian *Thyone secreta*, collected at the Cape of Good Hope, South Africa. The adults are long and tubular, measuring a few millimeters to 8 cm in length, and are tangled in knots. The anterior end of the tubular body is attached to the host's intestine. No alimentary tract or other internal organs are present except for the gonads. The testis is situated toward the anterior end of the body, and the ovary is at the posterior end. In gravid individuals the tubular body is filled with thousands of eggs. The veliger larva, unlike the adult, is readily recognized as a gastropod, for it possesses a tiny coiled shell and an operculum.

Life Cycle and Metamorphosis of *Thyonicola mortenseni*. Iwanow (1948) has studied the life cycle and metamorphosis of a parasitic entoconchid, *Parenteroxenos dogieli*, which has since been shown to be *Thyonicola mortenseni*. The eggs are located in cocoons within the parent's tubular body where they undergo extensive development. Cocoons enclosing well-developed larvae are expelled in the feces of the holothurian host. Once the cocoons come in contact with sea water, they burst, releasing larvae that are only about 0.1 mm long. These larvae are unable to survive for any length of time. In order to survive, they must be ingested by a holothurian along with its food.

Once in the host's gut, the larval snail undergoes metamorphic changes. It sheds its thick shell, mantle, and a mass of cells (probably vitelline cells). Furthermore, the primitive stomodeum closes.

The larva now penetrates the host's gut wall with the aid of pedal gland secretions. Once established within the gut wall, its epidermal cilia are lost, and a mass of cells—the **genital primordium**—appears in the mesoderm. As development continues, a cavity in the body of the snail, which has been interpreted to be the stomodeal cavity, increases in size and progresses at one end, eventually reaching and breaking through the epidermis. This cavity communicates with the lumen of the host's gut via a ciliated canal. The large body cavity, known as the **brood pouch**, is believed to have originated from the stomodeal cavity; hence, it is not a pseudopallial cavity.

Concurrent with the development of the elongate brood pouch, the gonads develop from the genital primordium. The ovary, located at the posterior end, increases in size and an oviduct from it opens into the brood pouch. The testicular anlage migrates to the anterior end, becoming located in the mesoderm adjacent to the base of the ciliated canal, and differentiates into the mature testis.

As the juvenile snail increases in size, it expands into the host's coelom, pushing the host's peritoneum with it. Thus, the adult parasite is found projecting into the host's coelomic cavity, with its ciliated canal connected to the host's gut.

The development of *Enteroxenos* is believed to be the same as that of *Thyonicola*, but the eggs of *Enteroxenos* are probably not enclosed in cocoons, and the veliger, unlike that of *Thyonicola*, possesses a heart, statocysts, nephridium, and a better developed nervous system.

FAMILY PAEDOPHOROPODIDAE

The Paedophoropodidae includes *Paedophoropus dicoelobius*, a parasite in the respiratory tract of the holothurian *Eupyrgus pacificus* in the Sea of Japan (Fig. 21.48). This gastropod is dioecious, with male and female intimately associated. There is marked sexual dimorphism, the males being less than half the size of

the females. The largest known female measured 5.5 mm in length. Both sexes possess a long proboscis that penetrates into the host's body cavity and is attached to the host's alimentary tract. Although a shell is absent in adults, this parasitic gastropod, unlike those mentioned earlier, does possess a visceral mass, including a nervous system, a kidney, ovary or testis, and a digestive tube and related structures. There are, however, no gills or heart. Furthermore, a pseudopallium is lacking.

Glands located around the distal portion of the proboscis, near the mouth, are believed to secrete a proteolytic enzyme that aids in the digestion of nutrients. The proboscis leads into a hepatic gut. There are no stomach, terminal gut, or anus.

The larva of *P. dicoelobius* possesses a dextrally coiled shell that is thin and transparent, an operculum, and a large foot. Veliger larvae develop in cocoons.

FAMILY UNIONIDAE

Parasitism occurs among the larvae of freshwater bivalves of the Unionidae. This way of life represents a normal stage in the life cycle of these molluscs. Female unionids incubate their eggs in brood pouches located between the gill lamellae in the water tubes. When the eggs hatch, the larvae, known as **glochidia**, pass out of the parent through the excurrent siphon. Most glochidia are drawn in by the breathing movements of fishes in the vicinity and become attached to the gill or to the body surface of the fish host, where they undergo a period of development, metamorphosing into young clams. If the glochidium is of a species that possesses a pair of hooks, one on each valve (small auxiliary spines may also be present), it generally becomes attached to soft exterior parts of a fish resting at the bottom. If the glochidium does not possess hooks, it clamps onto its host's gill filaments.

Within a few hours after the glochidia become attached to a fish, they become encapsulated by host

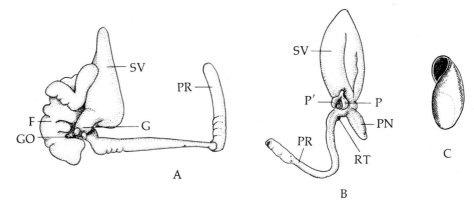

Fig. 21.48. *Paedophoropus dicoelobius.* **A.** Right lateral view of adult female. **B.** Ventral surface of adult male. **C.** Larval shell. F, foot; G, genital regions; GO, region of genital pore; P and P′, left and right lobes of foot; PN, penis; PR, proboscis; RT, rudimentary tentacle; SV, visceral sac. (All figures redrawn after Iwanow, 1937.)

cells. Within the capsules, the glochidia absorb nutrients from the host's epithelium and grow. After developing into young clams, the molluscs open and close their valves until the encapsulating cysts, weakened by then, rupture, and the clams fall to the bottom of the water and gradually mature. In addition to fish, certain glochidia utilize amphibians as hosts (Fig. 21.49).

It is known that in order for glochidia to metamorphose into adult clams, they must make contact with the blood of the fish host. Although the hypothesis has been advanced that the blood of various species of fish may include some specific factor that governs host specificity, Isom and Hudson (1984) have demonstrated that this is not true. They have found that contact of specific glochidia with the blood of many species of fish induces metamorphosis and hence is not a factor in influencing host specificity.

Protelian parasitism occurs among such unionid pelecypods as *Unio, Anodonta, Proptera, Margaritana,* and *Lampsilis.* In Europe, the glochidia of *Anodonta cygnaea, Unio batavus,* and *Margaritana margaritana* are almost always found on cyprinid fishes, with *A. cygnaea* attached to the fins. In the United States, complete larval development of *A. cygnaea* has been shown to occur on axolotls. While host specificity of glochidia has been suggested, there are little data to support this idea.

MOLLUSCAN HOST-PARASITE RELATIONSHIPS

Present knowledge concerning parasitic molluscs is too scanty to permit detailed elaborations on host-parasite relationships, even though these relationships are believed to be ancient. Among the parasitic gastropods, echinoderms appear to be the preferred hosts, which may be attributed to the fact that echi-

noderms are about the only potential hosts sluggish enough to permit snails to climb over them and to establish any type of association.

Another indication of the antiquity of the mollusc-echinoderm association is the fact that when the proboscis of a parasitic mollusc is inserted into the host's tissues, no cellular or humoral reaction occurs on the part of the echinoderm. This lack of response suggests that the host recognizes the parasite as "self" rather than "nonself," or foreign material.

SUBPHYLUM CRUSTACEA

The crustaceans include an array of arthropods such as lobsters, crabs, crayfish, barnacles, and pill bugs. Most members of this subphylum are free living. However, certain members of various classes are parasitic.

The subphylum Crustacea of the phylum Arthropoda (p. 572) includes over 31,300 known species; it is subdivided into eight classes (Barnes, 1980). Among these, four—Copepoda, Branchiura, Cirripedia, and Malacostraca—include symbiotic species. Some of the other classes, such as the Ostracoda, include ectosymbiotic species.

CLASS COPEPODA

The Copepoda includes some 8000 known species of which the majority are free-living aquatic animals. The parasitic members of this class are found as ecto- and endoparasites of vertebrates and invertebrates. The

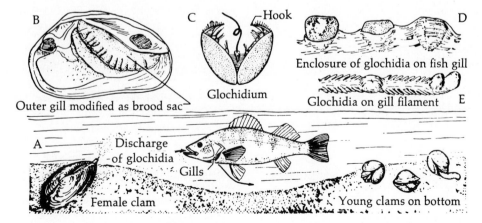

Fig. 21.49. Stages in the life cycle of a unionid mollusc. A. Diagram of the life cycle. **B.** Female's outer gill modified as a brood sac. **C.** Glochidium. **D.** Glochidia enclosed by epithelium on gill of fish host. **E.** Glochidia on a gill filament. (After Storer and Usinger, 1961.)

Other Zooparasites

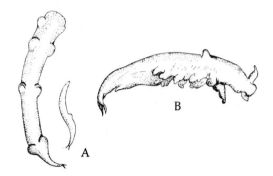

Fig. 21.50. Some parasitic copepods. A. *Ive balanoglossi*; the larger is a female, the smaller is a male. (After Mayer, 1879.) **B.** *Siphonobius gephyreicola* found in *Aspidosiphon brocki*. (After Augener, 1903.)

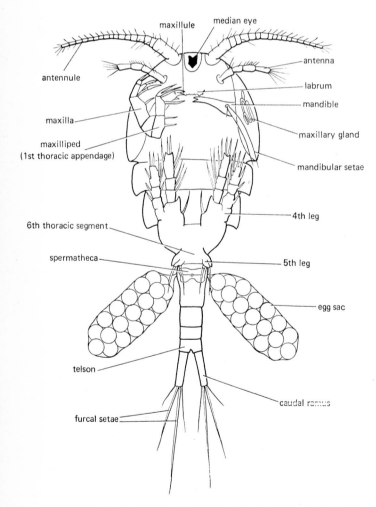

Fig. 21.51. *Cyclops*, a "typical" copepod. Ventral view of adult female. Swimming legs 1, 2, and 3 have been omitted. (After J. Green, 1970.)

ectoparasitic species commonly are found attached to the body surfaces of fish and amphibians, more commonly on fish. On fish hosts, copepods attach to the fins and the gills, and in the mouth. These copepods are true ectoparasites rather than epiphoronts or commensals, for they draw their nourishments from the host's tissues.

Numerous copepods have been reported on or in invertebrates and lower chordates. For example, *Ive balanoglosii* occurs in the coelom of the hemichordate *Glossobalanus minutus* (Fig. 21.50). Other species have been found on the lophophore of the brachiopod *Argyrotheca* and attached to the exterior of chaetognaths. The adult of *Siphonobius gephyreicola* occurs on the rector muscle of the sipunculid *Aspidosiphon brocki* and copepodids (a larval form) within the coelom of this host (Fig. 21.50). In addition, *Octopicola superbus* is fairly commonly found on the octopus, *Octopus vulgaris*, and many species of *Stellicola* occur on sea stars.

Not only are the symbiotic copepods of interest to parasitologists because of their parasitic habits, but the whole group is of interest because many of the free-living species serve as intermediate hosts for helminth parasites, especially cestodes and nematodes.

COPEPOD MORPHOLOGY

The external form of most parasitic copepods, especially members of the order Caligoida, has become so modified as the result of their adaptation to ectoparasitism that they can hardly be recognized as copepods except to the trained eye. By basing the following generalized discussion of the external anatomy of copepods on the most "typical" forms, it is felt the reader can become acquainted with the terminology and interrelationships between parts as applied to these minute crustaceans (Fig. 21.51).

The Body

Basically, the body of a copepod is divided into three regions, **head**, **thorax**, and **abdomen**. These regions are covered by a rigid or semirigid chitinous exoskeleton. The thorax is typically composed of seven segments. The first two segments are fused with the head to form the **cephalothorax**, which is covered dorsally by a large and somwhat flattened chitinous shield, the **carapace**. The seventh, or last, thoracic segment is known as the **genital segment** because it includes the external genitalia. The lines of demarcation between the body segments are commonly fused among the parasitic species. In fact, in some species these sutures are not visible. If the intersegmental zones are not fused and are flexible, they are referred to as the **annular zones**.

The abdomen typically is composed of four segments. However, there may be one or more less in some species and quite often all of the abdominal segments are fused among the parasitic copepods. In members of the Cyclopoida, there is an articulation between fifth and sixth thoracic segments. In these copepods, the portion of the body anterior to the movable joint is known as the **metasome**, and the portion posterior to it is known as the **urosome**.

The Appendages. The appendages of copepods, like those of all arthropods, are jointed. There are typically five pairs of appendages on the head. (1) The first pair is represented by the **first antennae**, which are long and comprised of up to 25 segments. These antennae serve sensory and locomotor functions. Among members of the Cyclopoida, this pair of antennae is usually modified as auxiliary copulatory structures. (2) The second pair are the **second antennae**, which are shorter and may be uniramous (not branched), as is the first pair, or may be biramous (forked). The function of these antennae is primarily sensory, but they also aid in locomotion. (3) The third pair of appendages are the two **mandibles**. (4) The fourth pair are the two **first maxillae**. (5) The fifth pair are the two **second maxillae**.

The last three pairs of head appendages, together with the first pair of thoracic appendages, the **maxillipeds**, comprise the mouthparts and are concerned primarily with the acquisition of food particles. However, on some copepods the second maxillae are modified as holdfast structures.

The thorax bears the ambulatory appendages. The first pair of thoracic appendages, the maxillipeds, are concerned with food acquistion, but thoracic segments 2 to 6 each bear a pair of biramous swimming legs, commonly designated the **metasomal legs**. There may be a vestigial seventh pair of metasomal legs on segment 7. In many parasitic copepods the metasomal legs are absent. If these are present, each is composed of three main joints—the basal **basipodite**, comprised of three segments; the medial **endopodite**, also with three segments; and the laterally located, trisegmented **exopodite**. The endopodite and the exopodite are both based on the distal terminal of the basipodite.

In addition to appendages, other types of smaller outgrowths from the exoskeleton occur. These include the following: (1) **esthetasks**, which are elongate and blunted projections on the antennae that function in a sensory capacity; (2) **hairs**, which are fine, long projections found on various zones of the body but especially on the legs (Fig. 21.52) (these flexible structures serve in sensory, locomotor, food-capturing, and sometimes in balancing roles); (3) **spines**, which are short and stout inflexible rods (Fig. 21.52); (4) **setae**, which are longer and flexible rods (Fig. 21.52).

Internal Anatomy

Below the hypodermis, thick bands of muscles have their origin and insertion on the inner surfaces of the exoskeleton. Parasitic copepods do not have a heart. The reddish, noncellular hemolymph of copepods is

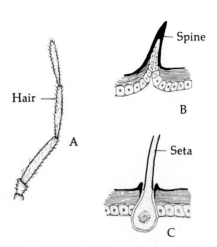

Fig. 21.52. **Body surface structures of arthropods. A.** Hairs on an appendage. **B.** Histologic presentation of a spine. **C.** Histologic presentation of a seta.

circulated through the body cavities (hemocoels) by movements of the body musculature and the alimentary tract. Some members of the order Lernaeopodoida actually possess a closed vessel system. For a review of arthropod hemolymph, see Jeuniaux (1971).

The digestive tract is complete, and the interesting posterior portion is capable of active pulsations made possible by extrinsic muscles. Not only do these pulsations aid in the egestion of wastes, but they also serve as auxiliary mechanisms in respiration, since oxygen enters with incoming water and carbon dioxide is discharged with each contraction. Most of the gaseous exchange, however, occurs through the body surface. For reviews pertaining to digestion in arthropods, see van Weel (1970) and Dadd (1970).

Copepods possess a rather efficient and specialized nervous system in which most of the ganglionic cells are located toward the anterior end of the body (cephalization), and the fibers innervate the various tissues.

COPEPOD LIFE CYCLES

Almost all knowledge of the life cycles of parasitic copepods has been obtained from observations on naturally infected hosts. Few experimental studies are available; hence, certain aspects remain as assumptions. Development and morphogenesis of copepods are interesting as the body form alters drastically during maturation. In fact, the adults of some parasitic species have become so modified in their appearance that they can hardly be recognized as arthropods, let alone copepods.

During development, the egg hatches into a small, active larva known as the **nauplius**. Among free-living species such as the well-known *Cyclops* spp., the nauplius molts five times, and the six nauplius stages are designated nauplius I, nauplius II, nauplius III, etc. Some authors prefer to assign the term **metanauplius** to the stages beyond nauplius I.

Among the parasitic species the number of molts is generally reduced. For example, some members of the order Cyclopoidea have only five nauplius stages. As one nauplius stage molts and metamorphoses into the subsequent stage, the body elongates and the number of appendages increases (Fig. 21.53). Active feeding occurs in between ecdyses.

After feeding, the last nauplius generation, which possesses the full complement of appendages (11 to 12 in free-living forms, number reduced in parasitic species), undergoes another molt and is transformed into the enlarged and elongate **copepodid**. In certain

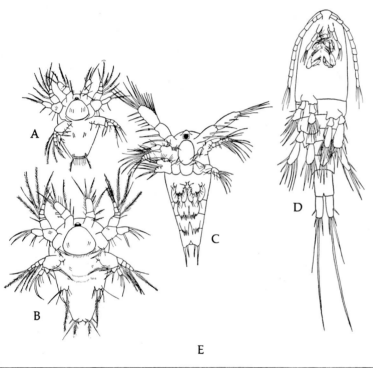

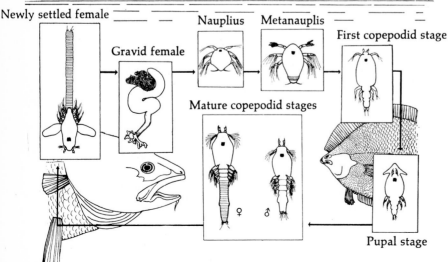

Fig. 21.53. Stages in the life cycle of copepods. A. Nauplius I of *Cyclops*, a free-living species. **B.** Nauplius IV of *Cyclops*. **C.** Nauplius VI of *Diaptomus*, a free-living species. **D.** Copepodid of *Cyclops*. (From Pennak, 1953. "Freshwater Invertebrates of the United States." The Ronald Press, N.Y.) **E.** Stages in the life cycle of *Lernaeocera branchialis*, parasitic on codfish as an adult and on flatfish as a larva. (After Cameron, 1956.)

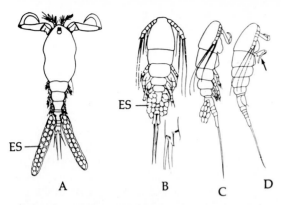

Fig. 21.54. *Ergasilus.* **A.** Dorsal view of female *E. vericolor.* **B.** Dorsal view of female of *E. chautauquaensis.* **C.** Lateral view of female of *E. chautauquaensis.* **D.** Lateral view of male of *E. chautauquaensis.* (**A**, after Wilson, 1911; **B–D**, after Wilson in Edmondson, 1959.) ES, egg sac.

species, such as *Caligus rapax*, commonly found on the pectoral fins of fishes, the nauplius transforms into the **chalimus stage**, which is characterized by a long filament secreted from a frontal head gland. The copepodids possess distinct sutures between the head, thorax, and abdomen. Copepodids feed in between molts, usually four in free-living species and fewer among parasitic members. Finally, the terminal copepodid stage (copepodid V) molts and transforms into the adult (Fig. 21.53).

BIOLOGY OF COPEPODS

There are approximately 8000 known species of copepods, most of which are free living or mere accidental symbionts. These species are divided into seven orders. Most of the parasitic species, however, are limited to the orders Cyclopoida, Caligoida, Lernaeopodoida, Monstrilloida, and Harpacticoida. The diagnostic characteristics of these are presented in the classification section at the end of the chapter. Yamaguti's (1963) monograph on the copepod parasites of fishes should be consulted by those interested in the systematics of this group. For a review of those species parasitic in economically important marine molluscs, see Cheng (1967).

Order Cyclopoida

The order Cyclopoida is comprised of a large number of free-living copepods, including the commonly encountered *Cyclops* spp. A few genera subordinate to this order, however, are parasitic.

Genus *Ergasilus*. The genus *Ergasilus*, including at least 12 freshwater species, is parasitic on the gill filaments of fish (Fig. 21.54). The taxonomy and distribution of members of this genus have been reviewed by Smith (1949). An identification key has been furnished by Roberts (1970).

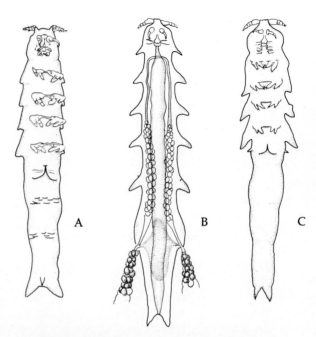

Fig. 21.55. *Mytilicola spp.* **A.** Ventral view of adult male *M. intestinalis.* (Redrawn after Steuer, 1902.) **B.** Ventral view of adult female of *M. orientalis.* **C.** Ventral view of adult male of *M. orientalis.* (**B** and **C**, redrawn after Mori, 1935.)

After nauplii hatch from eggs, they are free swimming, as are the copepodids. However, when copepodids attain sexual maturity, the females become parasitic and attach to their hosts by specialized second antennae, which have become modified as two muscular claws. The sexually mature males are free swimming in most cases, but parasitic males are known. These can be distinguished from the females by the presence of maxillipeds (Fig. 21.54).

All species of *Ergasilus* possess first antennae composed of six segments. The first four pairs of swimmerets are biramous while the fifth pair is reduced and uniramous. During copulation, the female stores sufficient spermatozoa to fertilize all her eggs, which when formed, pass into the egg sacs attached to the genital segment.

Genus *Mytilicola*. Probably the parasitic copepods most familiar to marine biologists are the various species of *Mytilicola* that occur in estuarine and marine molluscs. The most widely distributed of these is *M. intestinalis*, which occurs in the intestines of oysters, clams, mussels, and a number of species of marine snails (Cheng, 1967). The male of the species (Fig. 21.55) measures about 3 mm long, whereas the

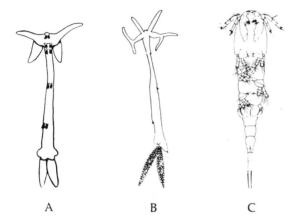

Fig. 21.56. *Lernaea.* **A.** Ventral view of female *Lernaea cruciata* (After LeSeuer, 1824.) **B.** Adult ovigerous female *Lernaea cyprinacea.* **C.** Free-living female of *L. cyprinacea* with legs 2 to 4 shown on one side only. (**B** and **C**, after Edmondson, 1959.)

female is about 8 mm long. A second species, *M. orientalis*, has been introduced into mussels, oysters, clams, and other molluscs in the Pacific Northwest of the United States from imported Japanese oysters. It is pathogenic in that the adults situated in the mollusc's intestine can mechanically injure the host's intestinal epithelium. In fact, Korringa (1950) has reported mass mortalities in commercial mussel beds in Europe due to parasitization by *Mytilicola*. The adult male of *M.*

orientalis is about 4 mm long, whereas the female is 10–12 mm in length (Fig. 21.55).

Life Cycle of Mytilicola. Upon hatching, the nauplii escape into the host's alimentary canal from within the adult female copepod's egg sacs when the sacs rupture at their posterior terminals. These nauplii are enclosed within a membrane but soon emerge. The nauplii and metanauplii are free swimming in sea water. Although in laboratory cultures the first, second, and third copepodid stages can be found free swimming, Grainger (1951) is of the opinion that *Mytilicola* can become parasitic during the first copepodid stage. If this is true, and there is reason to think so (Cheng, 1967), then the development through the second and third copepodid stages to the adult occurs within the molluscan host. Although it is not known how first stage copepodids enter their molluscan hosts, the fact that *Mytilicola* is an intestinal parasite suggests that they are ingested.

Genus *Lernaea.* In *Lernaea* (Fig. 21.56), which includes 13 or more species and formerly assigned to the order Caligoida, it is the vermiform females, bearing hornlike processes on their cephalothorax, that are parasitic. When attached to their piscine and amphibian hosts, the entire anterior end is deeply embedded in the host's flesh and permanently anchored by these hornlike processes. Characteristically, two **ovisacs** (egg sacs) are attached to the posterior end. These females measure 5–23 mm in length.

Haley and Winn (1959) have reported finding *Lernaea cyprinacea* attached to the gills and fins of nine species of fish in Maryland (Fig. 21.57). These parasites, like all ectoparasitic copepods, actually suck and feed on the blood and tissue fluids of their hosts and cause severe damage. Infected foci are commonly hemorrhagic, spongy, and necrotic. Heavy lernaean

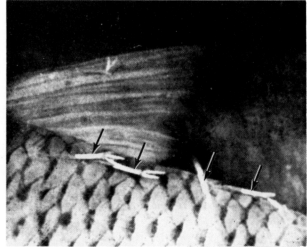

Fig. 21.57. *Lernaea cyprinacea.* Photographs showing copepods parasitic on gill and scales of fish host. (After Haley and Winn, 1959.)

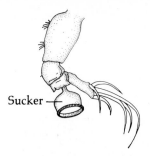

Fig. 21.58. *Myzomolgus stupendus*. Second antenna showing specialized sucker by which the copepod holds on to its host, *Sipunculus nudus*. (After Bocquet and Stock, 1957.)

infections are known to kill the hosts. In this era when fish farming is becoming increasingly more popular, fish kept in crowded ponds are vulnerable to a variety of infectious diseases, including parasitism by *Lernaea* and other copepods.

Life Cycle of Lernaea. During the life cycle of *Lernaea*, the nauplii are free swimming but the copepodids require a temporary host, usually a fish, to which they cling. These parasites copulate while still in the larval form, and after copulation, the males perish and the females leave the temporary host and are free swimming. This free-swimming form (Fig 21.56) is quite different from the parasitic form, to which it metamorphoses once it finds a suitable definitive host and becomes attached to it (Fig. 21.56). Haley and Winn (1959) have found numerous copepodids as temporary parasites on the fish they surveyed and suggested that perhaps these lernaeans can complete their life cycles on either one or two hosts. This hypothesis remains to be explored.

Other Cyclopoids. *Myzomolgus stupendus* and *Catinia plana* have been reported by Bocquet and Stock (1957) attached to sipunculid worms by stalked suckers on the third joint of the second antennae (Fig. 21.58). Several genera of Cyclopoida have been reported on sedentary invertebrates, but it is strongly suspected that these are epiphoronts or commensals rather than true parasites.

Bocquet and Stock, in papers published between 1956 and 1968, have discussed many of the copepods found on invertebrates. Their papers should be consulted by those interested in these symbionts.

One of the more common cyclopoid copepods found on marine fish is *Chondracanthus merlucci* (Fig. 21.59). It occurs within the gill cavity of the fish host. A related species, *Chondracanthus gibbosus*, is found in the same location in the marine angler fish, *Lophius piscatorius*. In both of these species of copepods the appendages are greatly reduced, and the mouth is flanked by sickle-shaped jaws. The female measures about 2.5 cm long, and if it is carefully observed,

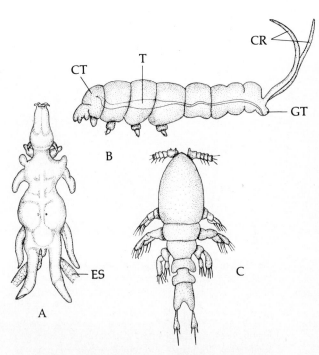

Fig. 21.59. Some parasitic copepods. A. *Chondracanthus merluccii* female. (Redrawn after Markewitsch, 1957.) **B.** Side view of *Cucumaricola notabilis* showing position of left reproductive organ. **C.** First copepodid larva of *C. notabilis*. (**B** and **C**, redrawn after Paterson, 1958.) CR, caudal rami; CT, cephalothorax; ES, egg sac; GT, genital tubercle; T, testis.

minute, maggotlike bodies can be seen attached at the posterior end of the body where the egg sacs are also attached. These are the hyperparasitic males. For authoritative accounts of the taxonomy of chondracanthid copepods from fish, see Ho (1970, 1971).

Another example of a copepod parasitic on an invertebrate host is *Cucumaricola notabilis*, which occurs in the coelom of the holothurian *Cucumaria frauenfeldi*. Paterson (1958) described this greatly modified copepod within amorphous cysts (Fig. 21.59). Enclosed within the cyst are eggs and nauplii, as well as adults. There are two copepodid stages (Fig. 21.59); the first is an active swimming and the second is a quiescent form.

Order Caligoida

The order Caligoida includes parasitic genera exclusively. The most commonly encountered genera are *Lepeophtheirus*, the sea louse (Fig. 21.60); and *Caligus* on marine fish.

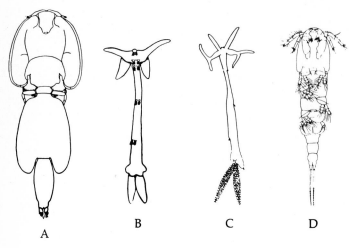

Fig. 21.60. *Lepeophtheirus salmonis*, the "sea louse." Dorsal view of female specimen. (After Kröyer, 1837.)

Genus *Lepeophtheirus*. The genus *Lepeophtheirus* includes several species parasitic on marine fish (Wilson, 1944). Both male and female adults are parasitic, although males are comparatively rare. Unlike the vermiform *Lernaea*, species of *Lepeophtheirus* possess an ovoid or circular dorsoventrally flattened cephalo-

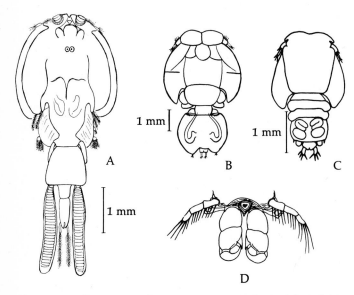

Fig. 21.61. **Some parasitic copepods. A.** *Caligus chelifer* female. (Redrawn after Wilson, 1905.) **B.** Dorsal view of *Dysgamus atlanticus* female. **C.** Dorsal view of *Achtheinus dentatus* male. (**B** and **C**, redrawn after Wilson, 1944.) **D.** Dorsal view of first and second antennae of *Pennella filosa*. (Redrawn after Wilson, 1917.)

thorax covered with a distinguishable carapace. The body segments are fused, and the swimmerets are reduced. One species, *L. salmonis*, is parasitic on salmon, being attached to the body surface near the fish's anus. This species has been reported in fresh water, having been carried to that habitat by its host during its upstream migration. However, the copepod is capable of surviving for only a few days in fresh water, apparently unable to adjust to the change in salinity.

Genus *Caligus*. The species belonging to this genus are primarily ectoparasitic on fish (Fig. 21.61). They are characterized by the presence of two semicircular structures on the frontal margin of the head. Known as **lunules**, these are believed to be sensory organs. In addition, *Caligus* has a suctorial mouth cone. Unlike most other parasitic copepods, *Caligus* adults are capable of swimming and can leave one host to become attached to another.

Other Caligoid Genera. Other genera of the Caligoida include *Dysgamus*, *Achtheinus*, *Teredicola*, *Pennella*, and *Lernaeenicus*, all found on marine fish, including sharks (Fig. 21.61).

Life Cycle of Lernaeenicus. The life cycle of members of this genus is rather intriguing since the form of the animal changes dramatically during the developing copepodid stages. The copepodid I seeks and attaches itself by utilizing its second antennae to the same fish on which the adult females are found. Soon after attachment, the larva cements itself to the host by means of strong frontal filaments.

Subsequent generations of copepodids actually exhibit dedifferentiation in that the swimmerets and mouthparts disappear and the segments of the antennae become invisible. Such copepodids are referred to as **pupae**. At the end of the last pupal copepodid stage, these structures reappear. The young males and females are free swimming, and copulation does not take place on the host. Soon after mating, the males die and the females seek out the same species of host and burrow in. In some instances, *Lernaeenicus* actually embeds its anterior end into the aorta of the host; in other instances, it merely burrows into the tissues of the body, whereupon a tumorous growth results.

Order Lernaeopodoida

The order Lernaepodoida includes the genera *Achtheres*, *Salmincola*, *Brachiella*, *Lernaeosolea*, *Kroyerina*, and *Paeonodes* (Figs. 21.62, 21.63). Of these, only members of the first two genera are found on freshwater fish, whereas the remaining are found on marine fish, primarily on the gills and spiracles of sharks. Identification keys to the species of *Achtheres* are given by Wilson (1915) and Causey (1957). A key to *Salmincola* is also given by Wilson (1915).

The lernaeopodoids are considerably more modified away from the ancestrial copepod form than the caligids. For example, all external signs of segmentation have disappeared in the adult (Figs. 21.62, 21.63). The females of this group of copepods are permanently attached to the exterior of their hosts. The anchoring device, known as the **bulla** (Fig. 21.64), represents solidified secretions from the head and maxillary gland secretions. The maxillae, often very large, are fused to the bulla.

Genus *Achtheres*.   Species *Achtheres* attain an extremely modified form when found attached to their hosts. The females, usually measuring over 3 mm in length, possess soft bellies that are more or less segmented, and lack swimmerets. The second maxillae are greatly modified to form two long, tubular, armlike appendages, the ends of which are attached to a saucerlike bulla. The bulla is embedded in the flesh of

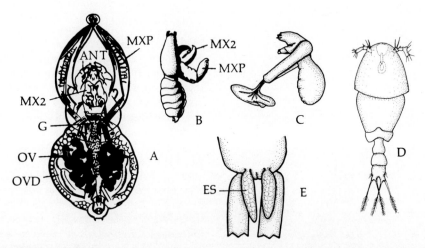

Fig. 21.62.   Parasitic copepods. A. *Achtheres percarum* female. **B.** *A. percarum* male. (After Gerstaecker, 1881.) **C.** Side view of adult *Salmincola beani*. (Redrawn after Wilson, 1915.) **D.** Copepodid larva of *Salmincola edwardsii*, the free-swimming form. **E.** Dorsal view of posterior processes and ovisacs (egg sacs) of *Brachiella squali*. (**D** and **E**, redrawn after Wilson, 1944.) ANT, second antenna; ES, ovisac; G, genital tract; MX2, second maxilla; MXP, maxillipid; OV, ovary; OVD, oviduct.

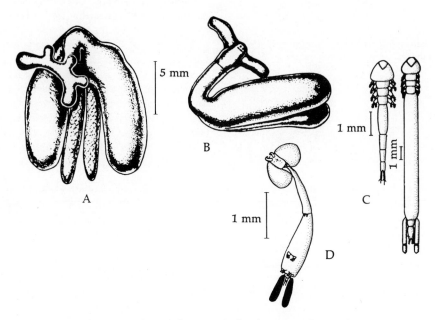

Fig. 21.63.   Parasitic copepods. A. Dorsal view of *Lernaeosolea lycodis*. **B.** Lateral view of same specimen. **C.** Dorsal views of male (left) and female *Kroyerina elongata*. **D.** Ventral view of *Paeonodes exiguus* female with ovisacs (eggs sacs). (All figures after Wilson, 1944.)

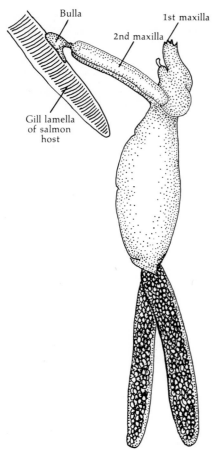

Fig. 21.64. *Salmincola salmonea*, **the so-called "gill maggot" of salmon.** Drawing of live specimen showing bulla at terminal of second maxillae attached to the gill lamella of fish host. (Redrawn after Friend, 1941.)

the host and serves as the absorptive mechanism through which the host's blood is taken in. The males are dwarfed and, when seen, are attached to the host during their early immature stages. They soon become attached to the females before, during, and after copulation, sometimes even permanently after copulation.

Life Cycle of Achtheres. The life cycle of *Achtheres* is somewhat modified from the generalized developmental plan found among copepods. The nauplius never leaves the eggshell; rather, the hatched larva is the copepodid I form, which is free swimming and eventually becomes attached to a host.

Genus *Salmincola*. Species of *Salmincola* are quite similar to those of *Achtheres* except that the body of females is completely unsegmented. The

maxillae of females are also modified as arms, joining at the bulla (Fig. 21.62). The size of the females approximates that of the *Achtheres* species. The males, measuring 1.5 mm or less, lack arms and are attached to the females. Fasten (1919), in reporting the development of *Salmincola edwardsii*, has stated that there is only one free-swimming copepodid stage, and this later becomes attached to a host and undergoes modification of form (Fig. 21.62).

Salmincola spp., as parasites of salmonid fish, are known to cause great damage to hatchery stocks in North America (Kabata, 1970).

Order Monstrilloida

The most familiar genera of the Monstrilloida are *Haemocerca* and *Monstrilla*. The members of this order are free living as adults, but are parasitic as larvae in the blood vessels and hemocoels of annelids or in the body cavities of prosobranch snails.

Genus *Haemocerca*. *Haemocerca danae* is parasitic during its copepodid stages in the body cavities of polychaete worms. The nauplius hatching from the egg is free swimming, and upon coming in contact with the host, burrows through its body wall and transforms into an ovoid mass of cells that migrates to the area of the host's ventral blood vessels. Once in this position, the larva develops two elongate tubular processes through which nutrients are absorbed from the host. The body elongates during this parasitic phase of its life cycle (Fig. 21.65). Once the copepodid reaches the adult form within the enveloping wall, it escapes from the host by rupturing out and becomes free swimming.

Genus *Monstrilla*. *Monstrilla heligolandica* is parasitic on gastropods that are, in turn, parasitic on lamellibranchs.

The nauplius of *Monstrilla* hatching from the egg is without a mouth or gut. It burrows into the host's body, discards its exoskeleton, and its appendages become lost. By the time it reaches the host's body cavity it consists of a naked mass of cells which become enveloped by a thin cuticle. Subsequently, a pair of long flexible appendages develop and it is through these that nutrients are absorbed. In time, the mass of cells differentiates into the organs of the adult. The latter bores its way to the exterior by using rows of hooklike spines surrounding the pointed posterior end of the body. Upon reaching the host's body surface, the thin cuticular envelope bursts and the adult parasite, which resembles *Cyclops*, is freed.

Order Harpacticoidea

The order Harpacticoidea includes *Sacodiscus ovalis*. This parasite occurs on the chela of lobsters, *Homarus americanus*, in Placentia Bay, Newfoundland, and crawls about freely, especially when disturbed. Both the females and males of this copepod are parasitic. In addition, this order also includes *Balaenophilus* on the baleen of whales, and *Neoscutellidium* on the gills of Antarctic fish.

COPEPOD PHYSIOLOGY

Very little is known about the physiology of parasitic copepods. The physiology of Crustacea in general, including information on free-living species, has been compiled in volumes edited by Florkin and Scheer (1970, 1971).

Oxygen Requirements

It is known that parasitic copepods are true aerobic animals that derive their oxygen from the surrounding aquatic media.

Hemoglobin, similar to, although not identical with, mammalian hemoglobin, has been detected in several species. Fox (1953, 1957) has reported hemoglobin in *Dolops ranarum*, a species found on siluroid fish. The presence of hemoglobin is responsible for the red color of copepod blood. Not all copepods, however, possess red blood.

The types of hemoglobin differ among copepods. For example, Fox (1945) has reported that the type found in *Lernaeocera branchialis* is different from that found in *Daphnia* and other free-living species. Presumably hemoglobin facilitates oxygen transport throughout the body tissues, but it is not a necessity, for some species do not possess hemoglobin.

There are no special breathing mechanisms in copepods. The exchange of gases occurs through the exoskeleton and through the posterior terminal of the alimentary tract.

Osmoregulation

Panikkar and Sproston (1941) have demonstrated the ability of *Lernaeocera branchialis*, an ectoparasite on the teleost fish *Pollachius pollachius*, to osmoregulate. These investigators have found that the salt concentration in the body of the copepod is 57 to 82% that of the salt concentration of the surrounding sea water as opposed to a concentration of only 43% within the host itself. If these copepods are removed from their host and isolated in sea water, they become isosmotic. This suggests that while the copepod is in the attached state, the salt concentration within its body is diluted by fluids (blood, etc.) obtained from the host. Not all parasitic copepods are as adaptable as this species; for example, *Lepeoptheirus salmonis* dies in a few days if brought into fresh water by its host.

COPEPOD HOST-PARASITE RELATIONSHIPS

Little is known about host-parasite relationships among copepods. The sometimes drastic body modifications of parasitic copepods while in the adult, attached state undoubtedly result from evolutionary changes influenced by adaptation to the parasitic way of life. The feeding habits of parasitic copepods—that is, ingestion the host's blood and body fluids—cause

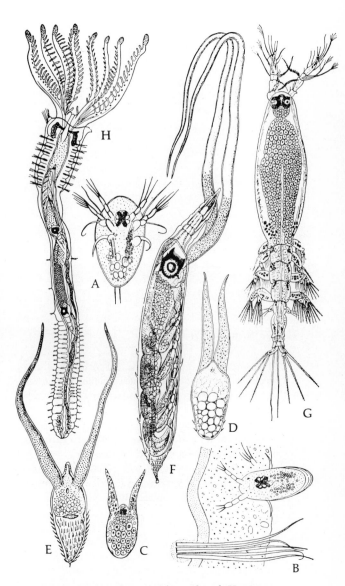

Fig. 21.65. Stages in the life cycle of *Haemocera danae*. A. Nauplius. **B.** Nauplius penetrating host. **C–E.** Successive larval stages showing development of appendages and spinous sheath enclosing larva. **F.** Fully developed copepodid. **G.** Adult female copepod devoid of mouth. **H.** Annelid host with two copepodid larvae within its coelomic cavity. (After Baer, 1952.)

injuries. In nature, the number of parasites is sufficiently sparse so that often no serious effects are noticeable. However, in small isolated areas, such as ponds and hatcheries, copepod parasites, especially lernaeans, cause serious injuries. Furthermore, since

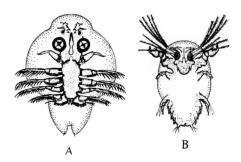

Fig. 21.66. *Argulus.* **A.** Ventral view of female. **B.** Newly hatched larva. (After Cameron, 1956.)

the exoskeleton of these animals is fairly resistant to chemicals, weak acids, and alkalis, infected fish in hatcheries are difficult to treat and frequently must be disposed of.

CLASS BRANCHIURA

Another class of the Crustacea that includes parasitic species is the Branchiura. These minute crustaceans are similar to the copepods; in fact, at one time they were considered to represent an order of the Copepoda, known as the Arguloida. However, in recent years almost all authorities have come to recognize them as being distinct from the copepods.

The branchiurans possess ovoid dorsoventrally flattened bodies (Fig. 21.66). They are parasitic on the body surface or in the branchial chambers of fish.

Genus *Argulus*

The most common members of the Branchiura belong to the genus *Argulus* and are commonly referred to as "fish lice." Many of the species are parasitic on marine fish, and some 15 are found on freshwater fish. At least one species has been reported on an amphibian. The ovoid body is dorsoventrally compressed (Fig. 21.66). The cephalothorax is covered dorsally by the **carapace**, which is recurved medially onto the ventral surface of the body. Specialized respiratory areas are located on the two ventrolateral margins of the carapace. When the animal is viewed from the ventral aspect, the abdominal segments are seen to be fused. There are only four pairs of legs; the fifth and sixth pairs are lacking.

The mouthparts of *Argulus* are greatly reduced, and the most striking feature is the modification of the second maxillae into two suction cups by which the parasite holds onto its host. *Argulus* also possesses a preoral sting by which the animal pierces its host in order to obtain the required blood meal. When seen from the dorsal aspect, two prominent movable compound eyes are visible in the head region.

Both male and female *Argulus* are parasitic. The female, measuring 5–25 mm in length, is larger than the male. Although usually attached to their hosts, these parasites leave their hosts periodically and are free swimming. This habit is particularly true of gravid females, which attach their eggs to sticks and stones at the bottom of the aquatic environment.

The breeding periods of *Argulus* normally occur three times per year. The eggs are laid several hundred at a time and enveloped in a jellylike mass. As each egg is laid, it is fertilized by a sperm stored in the female at the time of copulation. The eggs hatch in approximately 3 weeks. The escaping larva is a copepodid, quite similar to the adult except for modifications of certain appendages. For example, the second antennae are plumose (in some species) and function as locomotor structures, and the second maxillae are modified as clasping structures armed with hooks. In a few days, the copepodid molts; after several such molts, each of which is accompanied by structural modifications, the adult parasitic form is attained.

CLASS CIRRIPEDIA

The cirriped crustaceans, including barnacles, are sessile as adults, commonly attached by antennules located on their heads. The carapace consists of a fleshy mantle, which is armed or not armed with calcareous plates. Typically, the body is indistinctly segmented with six pairs of biramous thoracic appendages known as **cirri**. The abdomen is vestigial. Compound eyes and antennae are absent in the adult, and a heart is lacking. The cirripeds are usually hermaphroditic. The female genital pore is located on the first thoracic segment, and the male pore is located at the posterior end of the body.

CIRRIPED LIFE CYCLE PATTERN

The life cycles of the various species of cirripeds are essentially identical.

Cirripeds are oviparous, and the eggs, upon hatching, give rise to free-swimming nauplii that are distinguishable from nauplii of other crustaceans by the presence of frontal horns on the carapace (Fig. 21.67). The nauplius may undergo several molts. The terminal nauplius gives rise to the **cypris** (Fig. 21.67). The cypris larva is readily distinguishable by its bivalvular shell, its pair of compound eyes, six pairs of two-jointed appendages, which are adapted for swimming, and the single pair of anterior antennules, each of which is armed with a sucking disc at its terminal and has a cement gland at its base. Among free-living species, such as the common barnacles, the cypris

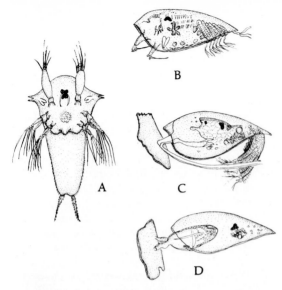

Fig. 21.67. Nauplius and cypris larvae of *Sacculina carcini*. A. Nauplius. **B.** Cypris larva. **C.** Cypris attached to host by its antennae and shedding its locomotory appendages. **D.** Cypris transformed into kentrogen larva and with degenerate structures of cypris still visible. (Redrawn after Delage, 1884.)

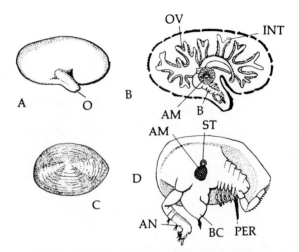

Fig. 21.68. *Baccalaureus japonicus*. A. External view of female. **B.** Female, right valve of carapace removed. **C.** External view of male. **D.** Male, left valve of carapace removed. AM, adductor muscles; AN, antenna; B, body; BC, buccal cone; INT, intestine; PER, pereiopods; O, orifice of carapace; OV, ovary; ST, stomach. (All figures redrawn after Yosii, 1931.)

larva becomes attached to a rock, a plank, or some other solid surface, casts off its shell, loses its eyes, and metamorphoses into the adult, which possesses swimming legs that elongate to become typical cirri.

BIOLOGY OF CIRRIPEDS

The Cirripedia, which includes some 900 known species, is divided into five orders with the free-living barnacles belonging to the Thoracica and other free-living forms belonging to the Acrothoracica. Members of Acrothoracica bore holes in the shells in which they dwell. The parasitic cirripeds belong to one of the three remaining orders: Ascothoracica, Apoda, and Rhizocephala.

Order Ascothoracica

The ascothoracicans are parasites of coelenterates and echinoderms. These animals somewhat resemble ostracods in that the adults do not shed the bivalved cypris larval shell. The alimentary tract and gonads send rootlike branches into the mantle cavity. Nutrition is supplied by the host's body fluids. In the endoparasites, the host's body fluids are absorbed directly into the dendritic alimentary canal. In the ectoparasites, the mouthparts are adapted for piercing the host's integument, and the body fluids are drawn out.

Although the adult body is comparatively small, the shells of certain species, especially those found on coelenterates, are well developed, and both a female

and a small male are found within each shell. Eggs enclosed within the branched ovaries hatch into nauplii that in turn give rise to free-swimming cypris larvae. It is the cypris larva that becomes parasitic in or on the host and develops into adults. *Baccalaureus japonicus* is quite representative of this order (Fig. 21.68).

Order Rhizocephala

The rhizocephalans include the genera *Sacculina* and *Peltogaster*. For those interested in the identification of *Sacculina* and other rhizocephalans, the papers by Smith (1906) and Boschma (1937) should be consulted. Most of our knowledge concerning the parasitic cirripeds has been derived from studies on the two genera mentioned.

Genus *Sacculina*. Day (1935) originally reported the life history of *Sacculina* and his observations have been elaborated on by Faxon (1940). The eggs expelled from the body of the parent hatch into nauplii measuring 0.25 mm in length. These larvae lack an alimentary tract. On molting, the nauplius metamorphoses into a cypris larva that seeks a host, usually a crab. The larva, after becoming attached to the host's hair by its hooked antennae, is referred to as a **kentrogen larva** (Fig. 21.69). The cypris larva is undifferentiated sexually.

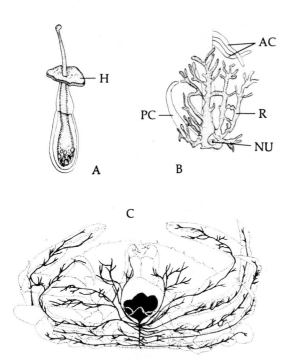

deposition of new exoskeletal materials. When the host molts, a hole is left in the area. In approximately 8 weeks, the parasite emerges as a saclike protrusion enclosing a brood chamber and with rootlike extensions that invade every region of the host's body. The presence of the mature parasite prevents further molting on the part of the crab.

Although rhizocephalans have been considered hermaphroditic, Ichikawa and Yanagimachi (1958), while studying *Peltogasterella socialis*, cast some doubt on this interpretation. At times a crypis larva does not become attached to a crab host. Rather, it settles on an immature rhizocephalan adult. This "parasitized" individual then develops into a "male." The cypris injects a mass of cells into the brood chamber of the adult, and this mass then migrates to the "testes" of the young adult rhizocephalan. The individual cells of the mass differentiate into sperm when the individual reaches sexual maturity in approximately 6 weeks. Ichikawa and Yanagimachi have demonstrated that the "testes" are actually seminal receptacles, and the sperm within are produced by males cells of the cypris. This discovery has revealed that adult rhizocephalans are true females carrying hyperparasitic larval males

Genus *Peltogaster*. Members of *Peltogaster* resemble *Sacculina* morphologically and in their pattern of metamorphosis. They parasitize hermit crabs.

Genus *Thompsonia*. Members of *Thompsonia*, which also parasitize decapods, differ from *Sacculina* in that numerous tumors, rather than a single one, are formed on the alimentary tract of the host. Furthermore, *Thompsonia* does not interfere with the host's molting, and usually several saccular bodies protrude from the body of the host, each enclosing eggs that give rise to cyprii instead of nauplii. When the host molts, these saclike portrusions are cast off, but terminal swellings on the root system within the host give rise to new sacs.

Rhizocephalan Host-Parasite Relationships

Parasitic Castration and Sex Reversal. Rhizocephalan parasites such as *Sacculina* and *Peltogaster* effect some severe changes in their hosts. Among these, parasitic castration is by far the most remarkable. The gonads of the host are affected in two ways. They are retarded in their development, that is, the multiplication and differentiation of the gametocytes are hindered, and actual destruction of the sex cells takes place, resulting in complete atrophy of the gonads. Smith (1906) attributed true disintegration to cell autodigestion, whereas Perez (1933) attributed it to phagocytosis by the follicular cells. The recovery of partially castrated crabs is extremely interesting. Several investigators have reported that when the *Sacculina* parasite is removed, the female host generally regenerates new ovarian tissue, but the male commonly develops complete or partial ovaries instead of testes. Hence, sex reversal has taken place. Reinhard (1956) has pointed

Fig. 21.69. Later stage of development of *Sacculina*.
A. Kentrogon larva. (Redrawn after Delage, 1884.) **B.** Rootlike body of developing *Sacculina* situated along the gut of the crab host between the levels of the anterior and posterior caeca. (Redrawn after Smith, 1910–1911.) **C.** Rootlike extensions of *Sacculina* (black lines) infiltrating various areas of the body of the crab host. (Redrawn after Noble and Noble, 1971.) AC, anterior caeca; H, host tissue; PC, posterior caecum; NU, nucleus; R, rootlike processes of the parasite.

If the host is a freshwater crab, the life cycle is somewhat altered in that the nauplius stage is suppressed and the eggs give rise to cyprii. Once attached to a host, the protoplasm of the kentrogen larva dedifferentiates and becomes a ball of embryonic cells, which is injected into the host via a style in the antennae, which penetrates the host's exoskeleton. This mass of cells is carried in the host's blood to the region of the alimentary canal in the thoracic zone. Once in this location, the mass of cells forms a tumor with rootlike processes wrapped around the alimentary canal (Fig. 21.69).

Further growth is directed posteriorly with the body increasing in size and the dendritic outgrowths increasing. Growth is halted when the junction between the thorax and abdomen is reached. The parasite then exerts pressure against the host's exoskeleton in that region, resulting in prevention of the out, however, that such instances of sex reversal may not be directly attributed to the parasite, because similar occurrences are known among nonparasitized crabs.

Recently, Rubiliani *et al.* (1980) and Rubiliani (1985) have reported that when aqueous extracts of two rhizocephalans, *Sacculina carcini* and *Loxothylacus panopei*, are injected into crabs, *Carcinus maenas* and *Rhithropanopeus harrisii* and *Panopeus herbstii*, respectively, there is degeneration of the testes; specifically, there is degeneration of the germinal epithelium and spermatogonial cells, and absence of secretions in the sinus glands of the central nervous system. These pathologic manifestations were most severe in *Panopeus herbstii* injected with the extract of *Loxothylacus panopei*. Since *Panopeus herbstii* is not a natural host of *Loxothylacus panopei* off the North Carolina coast, it has been concluded that the pathologic effects of rhizocephalans on crab hosts are most severe in unnatural hosts.

Sacculina and related genera often also induce modifications of external secondary sex characteristics in crab hosts. Such modifications include the broadening of the abdomen and feminization of certain appendages in males, and the narrowing of the abdomen and degeneration of pleopods in females (p. 13). These alterations undoubtedly result from damage to gonads and represent manifestations of sex reversal in males.

Similarly, parasitism by *Peltogaster*, *Thompsonia*, and related genera also effects changes in secondary sex characteristics. Reinhard (1956) has given an account of sex alterations resulting from rhizocephalan infections. Various theories have been advanced to explain the alteration of sex in parasitized hosts. The more reasonable ones are based on physiologic deviations from the normal resulting from the presence of the parasite, for example, nutritional deprivation in gonadal tissue and lowering of fat content in the host.

In considering alteration of sex, one might examine the hypothesis offered by Goldschmidt (1923). He assumed that in sex determination opposing genes are present in each individual; hence, the sex of the host depends on the quantitative balance of male and female sex genes. In sex alterations due to rhizocephalans, according to Goldschmidt's postulation, the normal balance is physiologically disrupted by the presence of the parasite, and the male, or more commonly the female, genes become dominant.

Physiologic Alterations. In addition to inducing morphologically appreciable alterations, rhizocephalans also cause physiologic changes. For example, there is an increase of lipochromes in the blood of parasitized *Inachus*; the calcium and magnesium levels in the hemolymph of *Carcinus* are almost doubled when parasitized by *Sacculina*; and in sacculinized *Carcinus*, the fat content in the hepatopancreas (digestive gland) is either increased or decreased and the glyco-

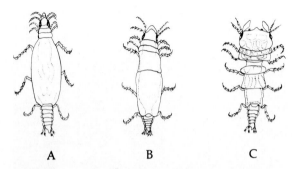

Fig. 21.70. *Gnathia maxillaris.* **A.** Larva. **B.** Adult male. **C.** Adult male. (Redrawn after Sars, 1885.)

gen reserves in the digestive gland and surface epithelium become depleted. Various other similar studies all indicate that the host's metabolism is altered by these parasitic cirripeds.

CLASS MALACOSTRACA

The Malacostraca includes an array of free-living crustaceans such as lobsters, crayfishes, crabs, and shrimps. However, a small number of malacostracans belonging to the orders Isopoda, Amphipoda, and Decapoda, particularly the first, are parasitic.

ORDER ISOPODA

The parasitic isopods belong to one of three suborders—Gnathiidea, Flabellifera, and Epicaridea. The gnathiid isopods are parasites as larvae, the flabelliferans as adults, and the epicarids are parasites both as larvae and as adults. Most of the species are marine.

Suborder Gnathiidea

The gnathiids include the genera *Anceus*, *Praniza*, and *Gnathia* (Fig. 21.70). These animals as larvae are blood-feeding parasites of fish and are attached to the fins of their hosts. These larvae engorge themselves with blood and their bodies become so distended that the last three thoracic segments become completely obscured and remain indistinguishable until the engorged blood is digested. A molt occurs after each of two feeding periods; hence, there are three larval stages. The third larval instar undergoes a final molt and transforms into the adult.

The mouthparts of both male and female adults are completely atrophied except for a pair of powerful mandibles retained by the males. The adults not only do not possess mouthparts, but they also have no

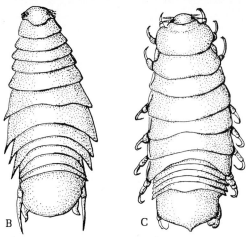

Fig. 21.71. Parasitic isopods. A. An adult female specimen of *Nerocila orbignyi* attached to the head of *Clupea*, a marine sardine. **B.** *Nerocila orbignyi*, a marine form from the mouth of La Plata River. **C.** *Livoneca symmetrica* from fresh water in British Guiana. (All figures are redrawn after Szidat, 1955.)

mouth or gut. The blood taken in by the larvae maintains the adults, which are buried in mud at the bottom of bodies of water. The females lay eggs that hatch into active larvae after incubation. Adult females die soon after incubating their eggs.

Suborder Flabellifera

Some species of adult flabelliferan isopods are ectoparasitic on fish and occasionally on cephalopods. They are found either attached to the gills or to the inside of the mouth. Members of *Ichthyoxenos*, however, are found in pairs, a male and a female, embedded in

minute cavities in the host's skin. These parasitic isopods are not permanently attached, even as adults, for individuals are frequently found in plankton. Host specificity is not demonstrated by flabelliferans. Sexual dimorphism occurs. The males are more similar to the larvae, whereas the females are often asymmetrical. Protandrous hermaphrodites are known in *Nerocila*, *Anicora*, *Cymothoa*, and *Ichthyoxenos*. The functional males of the these protandrous hermaphrodites possess a rudimentary ovary that becomes functional only after the testis ceases to function. In this way, the males become females.

Certain flabelliferans, such as *Cymothoa* and *Nerocila*, are armed with strong hooklike claws (Figs 21.71) and it is by the use of these that they cling to the gills or skin of their fish hosts. *Cymothoa* and *Livoneca* (Fig. 21.71) occur in the gill chambers of fish or mantle cavities of cephalopods, whereas *Nerocila* and *Anilocra* occur on the skin and fins of fish.

According to Menzies *et al.* (1966), *Livoneca convexa* begins its life as a free-living, planktonic organism; however, males, upon reaching adulthood, enter the gill chamber of the marine fish *Chloroscombrus orqueta* whereas the females enter the mouth, possibly via the gill chamber. Pathologic changes in the host are only associated with the male.

Also representative of the Flabellifera are the members of the genus *Aega*. All of the species, except for *A. spongiophila*, are parasitic on fish. *Aega spongiophila* is parasitic on a sponge. All of the species have piercing-sucking mouthparts, and in the fish-parasitizing species this adaptation permits them to suck their host's blood. In addition, the anterior pair of legs bear hooked claws which permit the parasites to cling to their hosts while feeding. After partaking of a blood meal, *Aega* becomes greatly distended and resembles

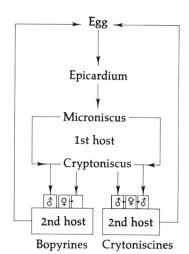

Fig. 21.72. Developmental pattern of epicarid isopods. Two types of life cycles.

a bag of blood. *Aega* will leave its host and swim about. According to Icelandic folklore, a distended *Aega*, known as "Peter's stone," is supposed to possess magical and medicinal properties.

Suborder Epicaridea

The epicarid isopods require two hosts to complete their life cycles (Fig. 21.72). The larva, which hatches from the egg, is known as an **epicaridium**. It resembles a very small isopod but is armed with piercing-sucking mouthparts. Furthermore, epicaridia possess clawlike appendages by which they attach themselves to free-swimming copepods. The larvae undergo six successive molts while attached and feed from the copepod host. There are two distinct larval stages on the copepod. The epicaridium metamorphoses into the **microniscus**, which in turn becomes the **cryptoniscus** (Fig. 21.73). The cryptoniscus leaves the host and becomes attached to the second host, a decapod crustacean living at the bottom of the sea.

The cryptoniscus of bopyrids (a family of the Epicaridea) enters either the branchial chamber or the brood pouch of the crustacean host, undergoes ecdysis, and transforms into a **bopyridium**, which lacks pleopods but retains pereiopods. Sex differentiation occurs when the cryptoniscus larva becomes attached to the second host. Giard and Bonnier (1887, 1893) and Reinhard (1949) have reported that the first larva to become attached to the crab host differentiates into a female and all subsequent larvae become males. If more than one larva enters the crab simultaneously, these all develop into females, but only one eventually reaches sexual maturity while the others disappear. Several males can be present with a single female. If this occurs, the female undergoes further development—the pleopods become flattened and function as breathing organs, a huge brood pouch is formed, and the size increases. The males remain as bopyridia and are considered to be neotenic larvae.

The ambipotent characteristic of the larvae has been demonstrated by Reverberi and Pitotti (1942) and Reverberi (1947), who have found that if young males of *Ione thoracica*, a parasite of *Callianassa*, are removed from the females and transplanted directly to the host, these males alter their sex, becoming females. Furthermore, if juvenile females are purposely placed on adult females, they transform into males. From this and other similar experiments, it appears that development of copepodids into males is brought about by their affixation to females. The genetic basis for this interesting phenomenon remains to be explored.

Among the cryptoniscines (a family within the suborder), the developmental pattern within the second host is different. The cryptonisci entering the crab develop into protandrous hermaphrodites. After attaining the female stage, the isopods degenerate, not

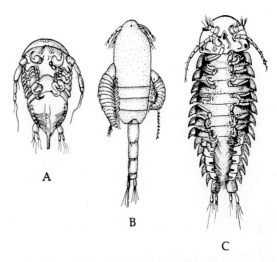

Fig. 21.73. Developmental stages of epicarids. A. Microniscus larva of *Cepon elegans*. (Redrawn after Giard and Bonnier, 1893.) **B.** Microniscus larvae attached to *Calanus elongatus*. (Redrawn after Sars, 1885.) **C.** Cryptoniscus larva of *Portunion kossmanni*. (Redrawn after Giard and Bonnier, 1893.)

as the result of parasitism, but from an excessive production of eggs.

Some representative epicaridean parasites are depicted in Fig. 21.74.

ORDER AMPHIPODA

Certain amphipods are frequently associated with various marine invertebrates. For example, *Phronima sedentaria* is often found in colonies burrowed in tunicates; and the whale louse, *Cyamus*, is found in the epidermis of whales (Fig. 21.75). In *P. sedentaria*, predation, rather than parasitism, is suspected. In *Cyamus*, although nothing except its morphology is known (Lutken, 1873), the animal could well be an ectoparasite, for it appears to be specific for whales. This parasite has a wide geographic range.

In addition to *Cyamus*, four other genera of amphipods have been reported on various marine mammals. All have dorsoventrally flattened bodies, reduced abdomens, and legs armed with claws. The fact that their mouthparts are not adapted for sucking blood has caused some to consider them as "semiparasites" that feed on mucus, bacteria, algae, and other materials on the hosts' skin.

These parasitic amphipods are unique among crustaceans in being unable to swim at any period during their life cycles. The young settle down near the parents so that amphipods of all sizes can be found clustered together on their hosts.

Also representative of this group of parasites is *Syncyamus*, which occurs in the blow hole and angles of the jaw of dolphins. Other genera include *Platycyamus*, *Paracyamus*, and *Isocyamus*. For further details, see Bowman (1955), Margolis (1955), and Leung (1967).

ORDER DECAPODA

Among the decapod crustaceans, certain pea crabs belonging to the genus *Pinnotheres* are parasitic in the mantle cavities of oysters and related molluscs (Stauber, 1945; Christensen and McDermott, 1958; Haven, 1959). Cheng (1967) has presented a critical review of those species associated with commercially important marine molluscs.

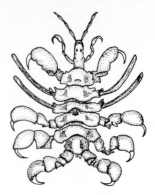

Fig. 21.75. ***Cyamus ceti*, the whale louse.** (Redrawn after Bate and Westwood, 1868.)

Representative of *Pinnotheres* is *P. ostreum* (Fig. 21.76), the so-called oyster crab. This species is commonly found within the mantle cavity of the American oyster, *Crassostrea virginica*, along the Atlantic coast of North America. It is usually the so-called

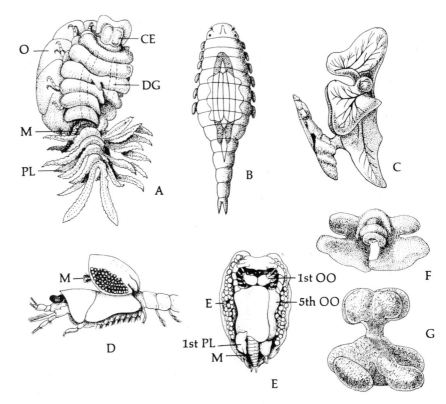

Fig. 21.74. **Some epicardean crustaceans. A.** Dorsal view of gravid female *Cepon elegans*, a parasite of *Pilumnus hirtellus*, with dwarf male attached. (Redrawn after Giard and Bonnier, 1893.) **B.** Male *Portunion maenadis*, greatly enlarged. **C.** Gravid female *P. maenadis*. (**B** and **C**, redrawn after Giard and Bonnier, 1893.) **D.** *Aspidophryxus sarsi* on *Erythrops microphthalma*, with male attached. **E.** Ventral view of *Dajus mysidis*. (Redrawn after Giard and Bonnier, 1893.) **F.** Subadult female *Ancyroniscus bonnieri* before egg deposition. The two pairs of lower lobes belonging to the abdomen are lodged in the visceral cavity of the host. **G.** Adult female *A. bonnieri* after laying eggs. It is reduced to a lobed and closed sac filled with embryos. (**F** and **G**, redrawn after Caullery and Mesnil, 1920.) CE, cephalogaster; DG, digestive gland; E, eggs; M, dwarf male; O, oostegites; 1st OO, first oostegite; 5th OO, fifth oostegite; PL, pleopods.

crab stages (stages I–V) that occur in the oyster (Table 21.1).

Life Cycle of *Pinnotheres ostreum*

The preparasitic larval stages of *P. ostreum* occur in the plankton. These include four **zoeae** followed by one **megalops**. The time required from hatching to molting of the megalops to give rise the first crab stage is 25 days.

The parasitic stages have been studied by Stauber (1945) and Christensen and McDermott (1958). The so-called invasive stage is the first crab stage (Fig. 21.76). It is this form that invades oysters. Six additional developmental stages ensue in the case of females, including the mature adult stage. Males do not develop beyond the **hard stage** (Table 21.1). In both sexes it is the hard stage that is the specialized stage at which copulation occurs. The males leave their hosts at this stage to search for females in another host. It is only during this stage that the crab is free swimming. After copulating with one or more females, which usually takes place in June or July, the males disappear. Thus, males do not survive longer than 1 year. Females become ovigerous (egg bearing)

in their first summer but do not attain maximum size before the second summer. Some do not attain this until their third summer. Fully grown ovigerous females measure 9.4–10.8 mm in width. These carry from 7957 to 9456 eggs. The exact period for which the eggs are carried is not known but it is believed to be from 3 to 5 weeks. Females only produce one batch of eggs during the first year, but may produce two batches during the second and third years.

It is known that in Delaware Bay in the eastern United States, and presumably also in other temperate regions, invasion of oysters by *P. ostreum* is seasonal. Few invasions occur before the beginning of August. It should be noted that the peak of oyster setting occurs during July in Delaware Bay; thus, by the time the first crab stage commences to invade in large numbers, the oyster spat has grown to sufficient size to harbor one or two crabs. A single young oyster measuring 4.2 mm long may harbor two crabs while

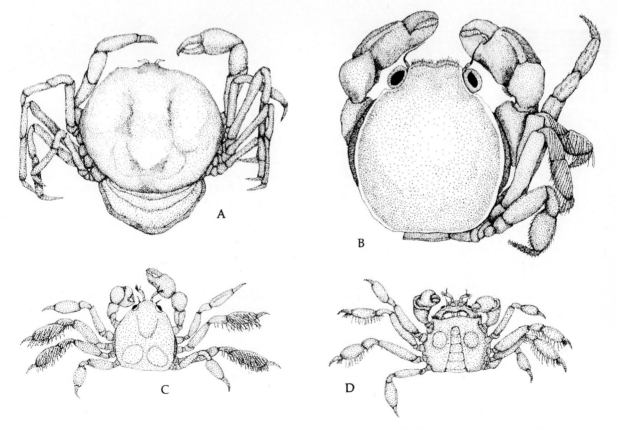

Fig. 21.76. *Pinnotheres ostreum.* **A.** Dorsal view of adult female. **B.** Dorsal view of adult male. **C.** Dorsal view of female stage I (hard stage) crab. **D.** Ventral view of female stage I crab. (**A** and **B**, redrawn after Williams, 1965; **C** and **D**, redrawn after Stauber, 1945.)

larger ones may harbor up to seven. The crab appears to prefer spats (76.7% infestation) over yearlings (54.6% infestation) and older oysters (21.5% infestation), but the survival rate of crabs is better in yearlings and older oysters.

Although the first crab stage (invasive stage) of *P. ostreum* can be found in oysters all winter, growth and development stop in Delaware Bay at the beginning of November when the ambient water temperature begins to drop below 15°C.

Effect on Sex Ratio of Host. Christensen and McDermott (1958) have reported that *Pinnotheres ostreum* may influence the sex ratio among oysters during their second spawning season if they retain their crab parasites from the first year. Their postulation is based on the earlier finding by Awati and Rai (1931) that among 794 uninfested oysters, *Ostrea cucullata*, 41.7% were males, 56.4% were females, and 2.9% were hermaphrodites. On the other hand, among 86 oysters harboring *Pinnotheres* sp., 82.6% were males, 10.4% were females, and 7.0% hermaphrodites. Since

females can be induced to change sex in the laboratory by simple starvation, Awati and Rai concluded that the crab probably interferes sufficiently with food intake of the oyster so that it produces sperm instead of the more "expensive" eggs. That the change in sex is due to reduction in food intake and is not chemically stimulated by some crab-secreted substance appears to be supported by the findings of Amemiya (1935) and Egami (1953), who have shown that the experimental removal of a part of the gill in *Crassostrea gigas* will cause the number of males to far exceed that of females during the breeding season, provided the operation is performed no later than the previous October.

CLASS PYCNOGONIDA

The Pycnogonida (or Pantopoda) is often considered a subphylum of the Arthropoda; however, in following Barnes (1980), it is considered a class of the arthropodan subphylum Chelicerata. The approximately 500 members, commonly known as "sea spiders," are all marine, inhabiting areas extending from the littoral zone to more than 12,000 feet in depth. Their bodies are generally small and superficially spiderlike; hence,

Table 21.1. Postplanktonic Developmental Stages of *Pinnotheres ostreum*[a]

Stage of Development	Range in Carapace Width (mm)	Most Important External Morphologic Characteristics	Biologic Factors
Invasive stage (first crab stage)	0.59–0.73	Flattened carapace and pereiopods. Posterior margins of pereiopods thickened, 3rd and 4th pairs have plumose swimming hairs. Two small, white spots on carapace and on sternum. Carapace hard around these spots	Free swimming until invasion of host. After invasion it is found in all parts of water-conducting system of the host
Pre-hard stages	Male 0.75[b]–2.7[b] Female 0.75[b]–2.7[b]	Rounded carapace. Thin, flexible exoskeleton. Slender pereiopods. No swimming hairs. Large females practically indistinguishable from 2nd stage crabs	Found in all parts of the water-conducting system of the host
Hard stage (Stage I of Stauber, 1945)	Male 1.4–4.6 Female 1.3–2.7	Carapace flattened and very hard. Flattened pereiopods with posterior margins thickened and with plumose swimming hairs on 3rd and 4th pair. Two large, white spots on carapace and on sternum. Males larger on the average than females	Found free swimming and in all parts of water-conducting system of the host. Copulatory stage. Males die in this stage
Stage II	1.3[b]–3.1	Rounded carapace. Thin flexible exoskeleton. Slender pereiopods. No swimming hairs. Abdomen wholly contained in sternal groove. No hairs on pleopods	Never free swimming. Predominantly, possibly always, found only on the gills of the host
Stage III	2.6–4.4	Edges of abdomen extend beyond depression in sternum. First two pairs of pleopods clearly segmented and supplied with a few hairs	Only found on the gills of the host
Stage IV	3.6–8.9	Relative width of abdomen larger than in preceding stage, just reaching coxae of pereiopods in most cases. Pleopods almost fully developed and well supplied with hairs	As in 3rd stage
Stage V (mature female)	4.4–15.1	Abdominal edges covers coxae of pereiopods. Pleopods fully developed. The orange gonads may be seen through the thin carapace	As in 3rd stage

[a] Data compiled by Cheng, 1967.
[b] Approximate measurements.

the common name, although some abyssal species may reach a length of 30 cm. Their thin and short bodies are typically divided into two conspicuous regions, the head and the thorax. The abdomen is vestigial (Fig. 21.77). A long, slender, tripartite feeding apparatus, known as the **proboscis**, is situated at the anterior end of the head segment. The suctorial mouth is located at the anterior tip of the proboscis.

There are usually three or four pairs of jointed appendages on the head segment—the pincerlike **chelifers**, **palpi**, **ovigers**, and the first pair of **walking legs**. In addition, a dorsal tubercle containing four simple eyes is located on the head segment. In certain species either the chelifers and/or palpi are lacking.

The thorax is composed of three or four segments, which may or may not be fused, depending on the species. Generally, there are three pairs of long, jointed, walking legs attached to the thorax. In some species an additional one or two pairs of thoracic walking legs are present. The walking legs are either eight- or nine-jointed.

The pycnogonids are dioecious and oviparous. The eggs are released in a ball-like mass, and the adult males, using their 10-jointed ovigers, carry the eggs. Ovigers are absent on females of certain genera. Development is either direct or indirect. In indirect development, a four-legged larval stage, known as **protonymphon**, occurs (Fig. 21.78).

Pycnogonids are found in all seas except the inner Baltic and Caspian, and they are especially common in the polar seas. The majority of the adults are free living, preying upon various coelenterates and other soft-bodied animals. Some pycnogonids are ectoparasitic on sea anemones.

For detailed accounts of these animals, the monograph of Helfer and Schlottke (1935) and Hedgpeth's (1954) essay on the phylogeny of the Pycnogonida are recommended.

Interest in pycnogonids on the part of marine parasitologists stems from the fact that the protonymphon larvae and juveniles of almost all the species are parasitic in or on hydroids, anemones, nudibranch and bivalve molluscs, ectoprocts, annelids, poriferans, brachiopods, echinoderms, octocorals, and ascidians (Fig. 21.78).

Physiologic aspects of the relationships between larval and juvenile pycnogonids and their hosts are essentially unexplored. The protonymphons, however, possess a number of specialized glands that produce materials for attaching to, or for invading, their hosts (Fig. 21.78). There is no doubt that the relation-

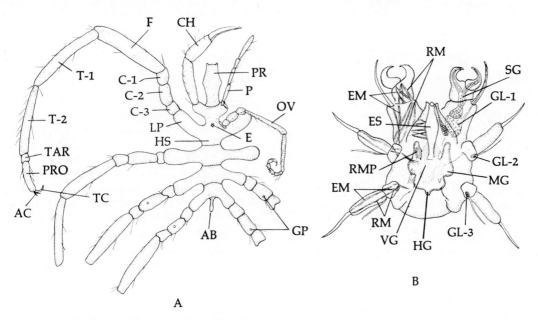

Fig. 21.77. Adult and larval pycnogonids. A. Adult *Nymphon* (Nymphonidae) showing external features. **B.** Protonymphon of *Ammothea echinata* (Ammotheidae) showing muscles (left), digestive and nervous systems (center), and glands (right). AB, abdomen; AC, auxiliary claws; CH, chelifer; E, eggs; EM, extensor muscles; ES, esophagus; F, femur; GL-1, gland of appendage 1; GL-2, gland of appendage 2; GL-3, gland of appendage 3; GP, gonopores; HG, hindgut; MG, midgut; OV, oviger; P, palpus; PR, proboscis; PRO, propodus; RM, retractor muscles; RMP, retractor muscle of proboscis; SG, silk gland of chelifer; T-1, tibial (patella); T-2, tibia 2; TAR, tarsus; TC, terminal claw; VG, ventral ganglion. (Modified after Helfer and Schlottke, 1935.)

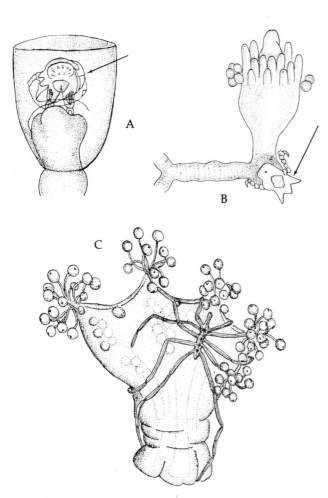

Fig. 21.78. Some parasitic pycnogonids. A. Proto-nymphon larva of *Anoplodactylus pygmaeus* (Phoxichilidi-idae) attached to hydranth of *Obelia*. **B.** Protonymphon larva of *Pycnogonum littorale* (Pycnogonidae) attached to base of the hydrozoan coelenterate *Clava multicornis*, probably in process of penetrating. **C.** Ectoparasitic young adult of *Phoxichilidium femoratum* (Phoxichilidiidae) on the coelenterate *Lucernaria* (All figures redrawn after Helfer and Schlottke, 1935.)

ships are obligatory and parasitic. For a listing of those species of pycnogonids that have been reported from marine invertebrate hosts, see Cheng (1973).

PHYLUM TARDIGRADA

The tardigrades, or "water bears," are almost all free living, and the majority are found in moist terrestrial and freshwater habitats. Some, however, are marine.

The cuticle of these animals lacks true chitin and their bodies are unsegmented: for these reasons, they have been removed from the Arthropoda and considered as representative of a distinct phylum. The bodies are generally more or less cylindrical, with rounded ends, and they measure up to 1 mm in length. There are four pairs of stumpy, unjoined legs, with two or more claws located at the terminal of each leg. The last pair of legs is located at the posterior end of the body.

At the anterior end of the body is found a retractile **snout**. The mouth is situated at the anterior terminal of the snout and is usually armed with a pair of retractable stylets. Circulatory and breathing organs are absent. Materials diffuse and circulate about easily in the body fluid, and the exchange of gases occurs through the body surface. Excretion is the function of the **rectal glands**, which are outpockets of the rectum.

Marcus's (1929) monograph of the Tardigrada and the section in Pennak's (1978) treatise on freshwater invertebrates that deals with the tardigrades are recommended to those desiring details on the morphology and taxonomy of this group.

The tardigrades are dioecious, and a single large saclike gonad is found in both sexes. All known species are oviparous. Most species undergo four to six molts during their development, and sexual maturity is usually attained sometime after the second or third ecdysis.

Tardigrades are most frequently found in lichen, damp moss, and liverwort beds, and less commonly among mosses and algae in fresh water. A few species of tardigrades are symbiotic.

Marine Symbiotic Tardigrades

Less than 10% of the known species are marine. Again, the majority of these are free living, being found among seaweed. However, a few marine species are believed to be symbiotic. For example, Green (1950) found adults of *Echiniscoides sigismundi* in the mantle cavity of the common mussel collected at Whitstable, England. Since Green did not find this tardigrade in the area where the mussels were obtained, despite a careful search of seaweed and other material sieved in a plankton net, he suggested that "some relationship other than a casual one exists between the tardigrades and the mussels." However, if a definite relationship exists between this tardigrade and the mussel, it is a facultative one, for *E. sigismundi* is also known to be free living in seaweed, particularly among the green alga *Enteromopha*, in the Mediterranean, the North Sea, the Caribbean, and even off the coast of China.

Green's discovery of a marine tardigrade associated with a host was not the first. Marcus (1936) has listed several marine species known to be associated with particular animals. For example, *Halechiniscus guiteli*

(Fig. 21.79) is found in and on oyster shells. *Actinarctus doryphorus* is a facultative symbiont on the echinoid *Echinocyamus pusillus*; and *Tetrakentron synaptae* (Fig. 21.79) appears to be a true parasite on the tentacles of the holothurian *Leptosynapta galliennei*. *T. synaptae* possesses a dorsoventrally flattened body that may or may not be the result of its adaptation to parasitism.

Since the symbiotic marine tardigrades, be they obligatory or facultative, have only been found on a few occasions, nothing beyond what has been stated is known about their relationships with their hosts.

Terrestrial Symbiotic Tardigrade

Fox and Garcia-Moll (1962) have reported the only known account of a symbiotic relationship between a tardigrade and a terrestrial snail. Specifically, they have found the larval, preadult, and adult stages of *Echiniscus molluscorum* (Fig. 21.80) in the feces of the land snail *Bulimulus exilis* in Puerto Rico. That these

tardigrades occur in freshly passed feces suggests their being endosymbionts. Furthermore, since *E. molluscorum* was found in snail feces during four different years argues against it being an occasional, or perhaps, accidental symbiont.

PHYLUM PENTASTOMIDA

Members of the phylum Pentastomida, commonly referred to as the tongue worms, possess elongate bodies that are cylindrical in some species and flattened in others. Externally the bodies are annulated, but these rings are not true segments. Adults lack legs, but four or six rudimentary legs are present on

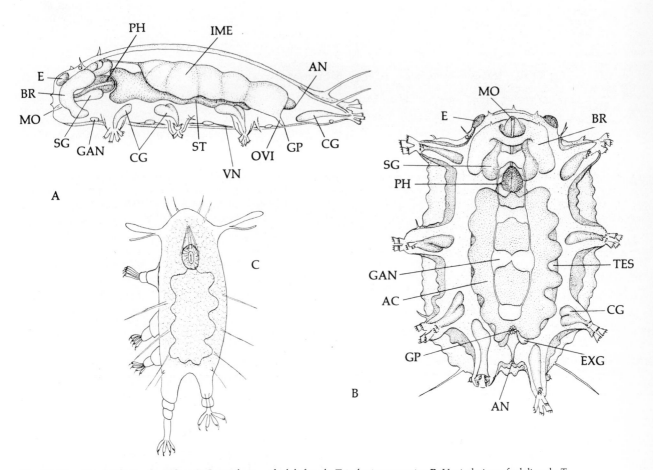

Fig. 21.79. Symbiotic tardigrades. A. Sagittal view of adult female *Tetrakentron synaptae*. **B.** Ventral view of adult male *T. synaptae*. **C.** Dorsal view of adult male *Halechiniscus guiteli*. AC, alimentary canal; AN, anus; BR, brain; CG, claw glands; E, eyespot; EXG, excretory gland; GAN, ventral nerve ganglion; GP, gonopore; IME, immature egg in saclike ovary; MO, mouth; OVI, oviduct; PH, sucking pharynx; SG, salivary gland; ST, stomach; TES, testis; VN, ventral nerve cord. (All figures redrawn after Marcus, 1929.)

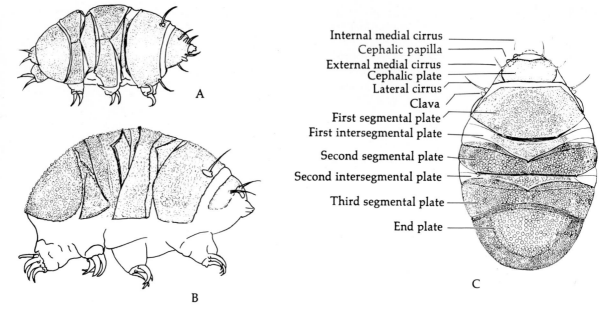

Fig. 21.80. *Echiniscus molluscorum.* **A.** Larva. **B.** Lateral view of adult. **C.** Dorsal view of adult. (After Fox and Garcia-Moll, 1962.)

Table 21.2. Relationships of Pentastomid Life Cycle Stages to Classes of Hosts[a]

Insecta	Pisces	Amphibia	Reptilia	Aves	Mammalia
Cephalobaenida					
Larvae			Larvae		
			Adults	Adults	
Porocephalida					
Porocephaloidea					
Sebekiidae—Subtriquitridae					
	Larvae		Adults		
Other Porocephaloidea					
		Larvae	Larvae		Larvae
			Adults		
Linguatuloidea					
					Larvae
					Adults

[a] After Self, 1969.

larvae. Because of their vermiform bodies, these animals are often erroneously referred to as worms. There are no distinct demarcations between the head, thorax, and abdomen, but the anteriormost portion is often referred to as the cephalothorax and the annulated body is referred to as the abdomen. The greatly modified form of the adults as well as of the larvae undoubtedly reflects their parasitic way of life. There are about 60 species of known pentastomids.

Adults pentastomids are usually parasitic in the respiratory tract and lungs of vertebrates (Table 21.2), although at least one species, *Linguatula serrata*, is known to accidentally migrate into the brain of the definitive host. The larvae are found in the viscera (liver and mesenteric nodes) of the intermediate host, which is usually a mammal or another vertebrate. Thus, these animals are totally parasitic both as larvae and as adults. Hunter and Higgins (1960) have reported the finding of an unencapsulated third-stage larva of *L. serrata* floating free in the anterior chamber of a boy's eye. How the larva arrived there is unknown.

Several monographs have been published on the Pentastomida (Sambon, 1922; Heymons, 1935; Fain, 1961; Nicoli, 1963). In addition, the review by Self

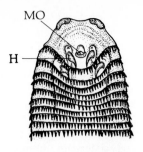

Fig. 21.81. Pentastomid morphology. Anterior end of third-stage larva of *Linguatula serrata* showing mouth with a hook on each side. H, hook; MO, mouth. (Redrawn after Self and Kuntz, 1957.)

(1969), including an extensive bibliography, is available.

PENTASTOMID STRUCTURE

DIGESTIVE SYSTEM

The mouth is located anteriorly and is subterminal. Around the lip is a chitinous ring or **cadre** that serves as a supporting mechanism (Fig. 21.81). The mouth leads into a buccal cavity, which opens into a narrow prepharynx and the muscular pharynx. The pharynx leads into a **cardiac valve** ("valvule cardiale") which is lined with microvilli and which acts passively from pressures in the intestine to inhibit the reverse flow of gut contents (Doucet, 1965). This valve opens posteriorly into the esophagus, which leads into a distended stomach-intestine extending almost the entire length of the body to the anus on the last body segment. The intestinal wall also bears microvilli.

Quite characteristic of adult pentastomids are the two pairs of hollow fanglike hooks, one pair on each side of the mouth, which can be retracted into grooves (Fig. 21.81). In some species, such as *Linguatula serrata*, these hooks are situated on fingerlike projections, the **parapodia**. It has been postulated that these hooks are vestigial appendages. **Frontal glands** (also known as head glands), located at the base of each hook, secrete a lytic substance that liquefies the host's blood or dissolves tissues.

NERVOUS SYSTEM

The most comprehensive study of the nervous system of pentastomatids is that by Doucet (1965). This system is comprised of a circumesophageal ring that is connected with ventrally located ganglia (Fig. 21.82). The ganglia comprising this "brain" are fused in the more advanced species and occur as several interconnected distinct ganglia in the more primitive ones. Either a double or, rarely, a single ventral nerve extends the length of the body, and lateral branches arise

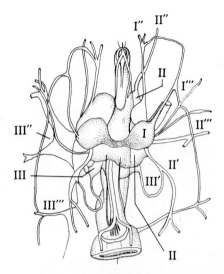

Fig. 21.82. Pentastomid brain. Brain and associated nerves of *Raillietiella boulengeri*. I, II, and III, first, second, and third pairs of ganglia (comprising the brain); I″, II″, III″, common trunks of nerves to dorsolateral, parabuccal-dorsal, parabuccal-ventral, and the anterointernal region of hooks; I‴, II‴, III‴, mixed nerves to lateral sense organs and parietal muscles. (Redrawn after Doucet, 1965.)

from it to innervate the various organs and tissues. Lateral commissures connect the two cords along their lengths like rungs of a ladder. Pentastomids possess no specialized sense organs except for the integument, which is comparatively highly innervated.

REPRODUCTIVE SYSTEM

Pentastomids are dioecious and exhibit a certain degree of sexual dimorphism inasmuch as the males are usually smaller than the females. Males possess a single tubular testis that occupies one-half to one-third of the body cavity. From the testis a pair of **testiducts** arise to conduct the spermatozoa to a short, pear-shaped **vas deferens**. The vas deferens opens to the exterior through a muscular **cirrus**. The cirrus is fitted in a groove of a **dilator organ**, which serves as a guide for the cirrus during copulation. The testiducts serve not only as tubes through which the spermatozoa reach the vas deferens, but also as storage sites. Thus, they are sometimes referred to as **seminal vesicles** (Fig. 21.83). The male genital pore is on the ventral side of the anterior abdominal segment.

In females, two oviducts arise from a single elongate saccular ovary (Fig. 21.83). The **oviducts** conduct the ova to the **uterus**, which is short and simple

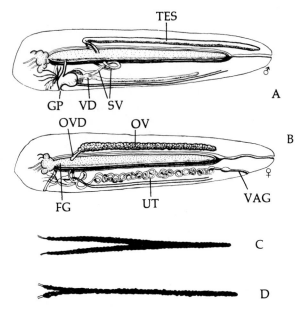

Fig. 21.83. Pentastomid reproductive systems. A. Male system of *Waddycephalus teretiusculus*. **B.** Female system of *W. teretiusculus*. (**A** and **B**, redrawn after Spencer in Heymons, 1935.) **C** and **D.** Representative double female reproductive systems. (Redrawn after Fain, 1961.) FG, frontal gland; GP, genital pore; OV, ovary; OVD, oviduct; SV, seminal vesicle; TES, testis; UT, uterus; VAG, vagina; VD, vas deferens.

in some species and long and convoluted in others. The uterus leads into a short muscular **vagina**, which terminates at the female genital pore. This pore is located ventrally, near the anterior end of the abdomen in some species (members of Cephalobaenida) and near the posterior end in others (members of Porocephalida). Fertilization is internal. That two testiducts in males lead from a single testis and two oviducts lead from a single ovary in females suggests that each sex of the ancestral form possessed two gonads. In fact, in some species of pentastomids the ovary is bipartite at one terminal (Fig. 21.83). These parasites are oviparous.

BODY TISSUES

The body surface of pentastomids is covered with a layer of chitin (Trainer *et al.*, 1975). The body surface of some species bears circular rows of spines. Usually there are transverse rows of cuticular glands with conspicuous pores. According to Banaja *et al.* (1977), these function in the regulation of the hydromineral balance in the hemolymph. Underneath the chitinous

exoskeleton is found the **hypodermis** composed of discrete cells. Intermingled among the cells are unicellular glands, which secrete a substance onto the outer surface of the exoskeleton. The body musculature is mediad to the hypodermis. The striated muscles are metamerically arranged.

The buccal cavity and prepharynx are lined with chitin and are of stomodeal (ectodermal) origin. The buccal cavity and the prepharynx are collectively known as the foregut. The stomach-intestine, or midgut, is lined with a layer of columnar cells that possess microvilli-lined borders. The nuclei of these cells are basally located and are of the vesicular type. These gastrodermal cells rest on a well-defined basement membrane, beneath which is found a thin muscular layer. This layer of myofibers is continuous with the mesenteries, which suspend the gut from the body wall. The pentastomids are true coelomate animals; hence, a peritoneum lines the body cavity.

Rao and Jennings (1959) have reported that two types of cells make up the lining of the midgut. These cells can be distinguished by their characteristic reactions to the periodic acid-Schiff (PAS) reaction and Turnbull's blue reaction for iron. The first type of cell, designated as PAS-positive cells, include large amounts of PAS-positive, diastasefast, Turnbull-negative granules. Such granules collect medially within the cells (toward the lumen). These cells rupture periodically, emptying the granules into the lumen.

The second type of cell, designated here as PAS-negative cells, contains varying amounts of PAS-

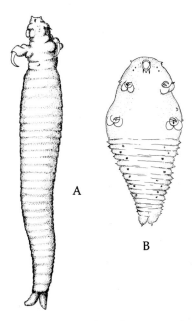

Fig. 21.84. *Raillietiella*. A. Entire specimen of adult *R. mabuiae*. (Redrawn after Baer, 1951.) **B.** Larva of *Raillietiella*, possibly *R. boulengeri*. (Redrawn after Fain, 1961.)

negative granules, which are strongly positive to Turnbull's blue reaction for iron. These iron-containing granules also concentrate at the medial pole of the cells, and eventually the entire mass of cells are extruded from the gastrodermis and are passed out intact in feces.

The hindgut or rectum is also lined with chitin and is proctodeal (ectodermal) in origin.

BIOLOGY OF PENTASTOMIDS

Taxonomic reviews and monographs have been contributed by Sambon (1922), Stiles and Hassall (1927), Heymons (1935), Hill (1948), Fain (1961), and Nicoli (1963). These should be consulted by those interested in identification. The Pentastomida is subdivided into two classes: Cephalobaenida and Porocephalida.

CLASS CEPHALOBAENIDA

Members of the Cephalobaenida are divided into two families. Those parasitic in the lungs of snakes, toads, and lizards are assigned to the Cephalobaenidae; those found in air sacs of birds are assigned to the Reighardiidae.

Family Cephalobaenidae

This family includes the genus *Raillietiella* in ophidians, lacertilians, and birds (Fig. 21.84). Fain (1961) has furnished convincing evidence that *Raillietiella* may have a direct life cycle since primary and older larvae (Fig. 21.84) have been recovered from the tissues of the definitive hosts. Furthermore, Self (1969) has reported that when eggs of *Raillietiella gehyrae* are fed to several species of lizards, migrating, primary larvae can be recovered from the tissues. These data suggest that *Raillietiella* spp. have a direct life cycle. On the other hand, Lavoipierre and Lavoipierre (1966) have demonstrated both natural and experimental larval development of *Raillietiella* in the cockroach *Periplaneta americana*; hence, while allowing for the possibility of a direct life cycle, an intermediate host could be involved. Circumstantial evidence for the occurrence of an intermediate host also exists in the case of *Raillietiella boulengeri*, a parasite of serpents. In this case, lizards are believed to be the intermediate hosts. A schematic representation of stages of cephalobaenid pentastomids as found in different categories of hosts is presented in Table 21.2.

Family Reighardiidae

The family Reighardiidae includes the genus *Reighardia* with species found in the air sacs of gulls and terns. The intermediate host is probably a fish.

CLASS POROCEPHALIDA

Members of the Porocephalida are divided into four families. Those subordinate to the Porocephalidae possess cylindrical bodies and are found in the lungs of reptiles; those subordinate to the Linguatulidae possess flattened bodies and are generally found in the nasal passages of members of the feline and canine families; those belonging to the Sebekidae have small, straight, and completely segmented bodies and occur in crocodilia; and those belonging to the Sambonidae have cylindrical bodies with somewhat dilated posterior ends and are parasites of lacertilians. A key to all the pentastomid genera is provided by Self (1969).

Family Porocephalidae

The Porocephalidae includes the genus *Armillifer* (Fig. 21.85). Adult members of this genus are found in the lungs, trachea, and nasal passages of snakes in eastern Africa and the Orient. The life cycle includes two vertebrate hosts. This is not surprising, for the definitive host is a carnivorous lizard or snake that feeds on small mammals in which the larvae are found.

Eggs oviposited by females in the definitive host are passed to the exterior in the nasal mucus and adhere to vegetation. When such eggs are ingested by the intermediate host, accidentally or intentionally, the larvae, which are already fully formed within the eggshells, hatch and actively burrow through the intestinal wall to become lodged in one of the visceral organs, such as lungs, liver, and kidneys. In such an organ, the larvae undergo further development, molting, and increasing in size. When the infective stage is reached, the larvae drop out of the organ and encyst in the abdominal or pleural cavity.

The intermediate hosts for *Armillifer* spp. include an array of mammals, such as various monkeys, antelopes, hedgehogs, rats, cats, giraffes, and even humans on some occasions. When an infected intermediate host is ingested by the definite host, the infective larvae migrate to the lungs via the throat and trachea.

Human infections by *Armillifer* occur in Africa and the Far East. In Africa, *A. armillatus* larvae have been found in the liver, spleen, and lungs of humans. Such infections could have been contracted in one of three ways, by drinking water contaminated with eggs, by eating vegetables on which the eggs are found, or by handling or eating snakes that serve as the definitive host, and thus become infected with the eggs. In China, the Philippines, and other Asian countries, larvae of *A. moniliformis* have been reported in humans.

In the United States, *Porocephalus crotali* is found in rattlesnakes and water moccasins, and *Kiricephalus coarctatus* (Fig. 21.85) occurs in colubrid snakes. Penn (1942) has reported that *P. crotali* utilizes muskrats and other mammals as the intermediate host.

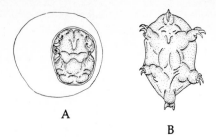

Fig. 21.86. *Porocephalus crotali.* **A.** Egg enclosing larva. **B.** Larva. (Redrawn after Penn, 1942.)

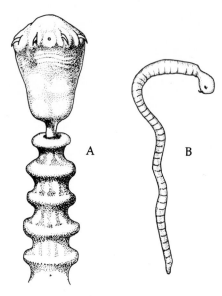

Fig. 21.85. Some pentastomids. A. Anterior end of adult *Armillifer annulatus.* (Redrawn after Baird, 1853.) **B.** Adult *Kiricephalus coarctatus.* (Redrawn after Sambon, 1922.)

Life Cycle of *Porocephalus crotali.* Esslinger (1962a,b,c) has made a detailed study of the life cycle of *P. crotali* and the response of the intermediate host to this parasite. He has reported that the eggs obtained from adults in the lungs of the water moccasin, *Agkistrodon piscivorus,* in Louisiana possess shells composed of four distinct membranes. These eggs are highly resistant to desiccation and enclose fully formed larvae. The larvae bear four legs (Fig. 21.86). From observations on the structure and function of these legs, it was concluded that their operation is based on both muscular and hydraulic mechanisms.

When pentastomid eggs are fed to albino rats and representative infected rats are killed and examined weekly over a period of six months, seven immature stages can be identified within the intermediate host. The form hatching from the egg in the rat's intestine is the **primary larva**. It bears two pairs of legs. It actively migrates through the wall of the host's small intestine and comes to rest in the viscera and associated mesenteries. While migrating through the intestine, the larvae leaves a trail of the host's neutrophils, which undoubtedly have migrated toward the parasite as a manifestation of the host's response to the parasite. Within 7 days after the infection, primary larvae are found in the liver, where their migratory path is marked by mononuclear cells and neutrophils.

Once situated in the liver or some other visceral organ, the parasite undergoes six molts, each one

followed by a period of growth and differentiation. The sixth stage **nymph** shows marked sexual differentiation. While the parasite is undergoing ecdysis, growth, and differentiation, a granulomatous lesion is formed around it in the host's liver. During the first 3 weeks, there is macrophage proliferation as well as production of epithelioid and giant cells in the area of the lesion, accompanied by an accumulation of eosinophils. During the second and third months, the lesion becomes chronic. There are increasing numbers of fibroblasts, fibrous tissue, plasma cells, and lymphocytes. Finally, by the fourth month, the active inflammation subsides. By the sixth month, the sixth stage nymph is surrounded by a dense, hyaline fibrous capsule.

The sixth stage nymph is well developed by the third month and is infective to the reptilian definitive host from then on. Infection is effected by ingestion of the infected rodent host.

Other genera belonging to the Porocephalidae are *Cuberia, Gigliolella, Ligamifer,* and *Kiricephalus* (Fig. 21.85).

Family Linguatulidae

This family includes the genus *Linguatula*, of which the most common species is *L. serrata* (Fig. 21.87) found in nasal passages and frontal sinuses of canines and felines. The adult pentastomid is colorless. The females are 100–130 mm long by 10 mm wide. Within the nasal passages, these parasites dig into the lining and suck blood, causing bleeding, catarrh, and mechanical obstruction to normal breathing. The distribution of *L. serrata* appears to be worldwide and has been reported from Europe, Africa, South and North America, and Asia.

Life Cycle of *Linguatula serrata.* Hobmaier and Hobmaier (1940) have resolved the complete life history of this species. These investigators have reported that the eggs, measuring 90–70 μm, include fully formed larvae with four rudimentary legs. The females produce enormous numbers of eggs; some have reported that a single female can produce several million eggs. The eggshell is resistant and wrinkled. These eggs pass out of the definitive host in nasal discharges and are deposited in water or on vegetation.

When contaminated water or vegetation is swallowed by an intermediate host, the eggs hatch, liberating acariform or primary larvae (Fig. 21.87) that bore through the intestinal wall of the intermediate host (cattle, sheep, rabbits, humans) and become lodged in the viscera, especially in the liver and lungs, or in the mesenteric nodes, where they are covered by host-elaborated capsules. Here the larvae undergo two molts, resulting in a pupalike stage that is devoid of mouthparts, hooks, or body annulations. Such larva measures from 0.35–0.5 mm in length (Fig. 21.87). Further growth and seven additional molts ensue.* The resultant **nymph** (sometimes referred to as the infective larva) (Fig. 21.87) is 4–6 mm long and bears two pairs of hooks and 80 to 90 annular rings. Minute spines project from the posterior margin of each ring. This is the infective form of *L. serrata*. This infective nymph migrates from the visceral organ in which it has developed and encysts in the abdominal or pleural cavity of the intermediate host.

When the intermediate host, especially if a sheep or rabbit, is eaten by the definitive host, the nymphs either escape rapidly from the capsule once they enter the definitive host and cling to the mucous membrane in the host's mouth, or more likely, migrate anteriorly when the surrounding intermediate host's tissues are digested and cling to the lining in the host's mouth. From this position they migrate to the nasal cavities of the host, where they develop rapidly into adults. After copulation, adult females begin ovipositing in about 6 months. Adults are capable of surviving for up to 2 years.

If infected intermediate hosts are ingested by an unnatural definitive host, the nymphs do not migrate to the nasal passages as is normally the case. They do, however, penetrate the intestinal wall and become encapsulated in the body cavity.

Linguatula serrata will invade the nasopharyngeal spaces of humans, causing a syndrome referred to as halzoun, marrara, or nasopharyngeal pentastomiosis. This disease has been reported from India, Turkey, Greece, Morocco, the Sudan, and Lebanon. In Lebanon, it has been linked with the eating of raw or undercooked sheep or goat liver or lymph nodes. In the Sudan, it has been linked with the ingestion of various raw visceral organs of sheep, goats, cattle, or camels. A few minutes to half a hour or more after eating, the patient suffers from a prickling sensation and pain deep in the throat. The pain may later extend to the ears. Edematous congestion of the throat region, the eustachian tubes, nasal passages, conjunc-

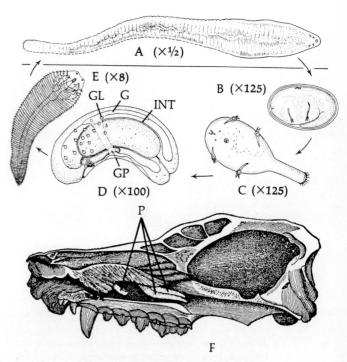

Fig. 21.87. **Stages in the life cycle of *Linguatula serrata*. A.** Adult female from nasal passage of a dog. **B.** Egg containing embryo. **C.** First-stage larva from viscera of sheep, humans, etc. **D.** Third-stage larva, 9th week. **E.** Nymph from sheep liver. **F.** Sagittal view of canine head showing three specimens of *L. serrata*. G, gonad; GL, globule; GP, genital pore; INT, intestine; P, pentastomids. (All figures after Brumpt, 1949; Leuckart, 1860; and Railliet, 1884.)

tiva, and lips follows. Furthermore, nasal and ocular discharges occur, accompanied by sneezing, coughing, and frontal headache. Complications may include abscesses in the auditory canal, facial swelling or paralysis, and sometimes asphyxiation and death (Schacher *et al.*, 1969).

In addition to causing nasopharyngeal pentastomiosis, *Linguatula serrata*, as well as other species of pentastomes, such as *Armillifer armillatus*, *Armillifer mouiliformis*, *Pentastoma najae*, and *Porocephalus* spp., can cause **visceral pentastomiosis**. This condition results from the ingestion of eggs. Developing nymphs encyst in such internal organs as the liver, spleen, lungs, and mesenteries (Fig. 21.88). Visceral pentastomiosis has been reported from Malaysia, the Philippines, Java, China, and Africa (Dönges, 1966; Self *et al.*, 1972, 1975). Commonly, visceral pentastomiosis goes undetected, although some histopatho-

---

*Although a total of nine molts have been reported in *Linguatula serrata*, the more recent work on *Porocephalus crotali* suggest that only six molts occur. It is possible, though unlikely, that these two closely related pentastomids undergo a different number of molts.

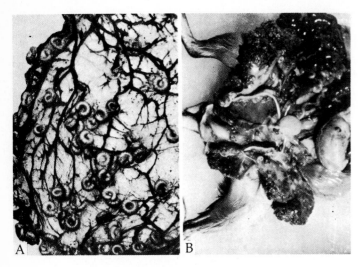

Fig. 21.88. **Visceral pentastomiosis. A.** Photograph of nymphs of *Porocephalus* in the mesentary of the monkey *Cercopithicus aethiops.* (After R. E. Kuntz in Self, 1972; with permission of the American Microscopical Society.) **B.** Massive infection with nymphs of *Porocephalus clavatus* in the rodent *Proechimys guyannensis* produced by feeding the host a female pentastome gravid with eggs. (After C.O.R. Everard in Self, 1972; with permission of the American Microscopical Society.)

logic response does occur. On the other hand, repeated infections can sensitize the patient, resulting in an allergy (Khalil and Schacher, 1965; Self, 1972).

Linguatula rhinaria is a common parasite of livestock. It can become a facultative parasite in humans in both the adult and nymphal stages.

Families Sebekidae and Sambonidae

The family Sebekidae includes the genera *Sebekia, Diesingia, Alofia,* and *Leiperia.* Their hosts are listed in Table 21.3.

The family Sambonidae includes the genera *Sambonia* and *Elenia,* the hosts of which are also listed in Table 21.3.

PHYSIOLOGY OF PENTASTOMIDS

OXYGEN REQUIREMENTS

Although pentastomids possess no respiratory system, it is quite apparent that these parasites are aerobic because they are found in oxygen-rich locations within the host, for example, nasal passages and lungs. Gaseous exchange occurs through the body wall or, as some claim in certain species, through the minute stigmata located on the thin exoskeleton.

The eggs of *Linguatula serrata* and *Porocephalus armillatus* are extremely resistant to adverse environmental conditions (Watson, 1960) and can survive a long time outside the host.

NUTRIENT REQUIREMENTS

Rao and Jennings (1959) have reported that in *Raillietiella agcoi,* a parasite in the lungs of the water snake *Natrix piscator* and other species of snakes, the food consists of host's blood drawn from the blood vessels in the lungs. The hemolytic activity of the frontal glands have been confirmed by these researchers. Only rarely are intact erythrocytes seen in the midgut; hence, hemolysis must have occurred soon after the cells are ingested.

Digestion is mainly intraluminar. However, the intraluminar digestion is supplemented by intracellular digestion within the periodic acid-Schiff-negative cells of the gastrodermis. This hypothesis is verified by the presence of the iron in these cells. Intracellular digestion, however, is not complete. Further breakdown occurs after the PAS-negative cells are expelled into the lumen.

HOST-PARASITE RELATIONSHIPS

Host-parasite relationships among the Pentastomida have not been studied experimentally; however, clinical observations of the effects these parasites have on their hosts have been recorded in some instances. It is known that the developmental stages of *Linguatula serrata* can cause considerable trauma in intermediate hosts, resulting from their perforation of the intestine and migration into the visceral organs. If damage to the liver is severe, hepatitis ensues.

Adult pentastomids have been known to accidentally migrate into the brain of definitive hosts. If this happens, meningitis ensues. Similarly, the presence of these arthropods in the lungs can cause pneumonitis. It is not uncommon for these arthropods to cause death.

Baer (1951) has noted that the location of pentastomids in the air sacs of birds and in the respiratory sinuses of mammals is of evolutionary significance. He has postulated that these animals were originally, and most species still are, parasites of ectothermic animals. During the transition from cold- to warm-blooded hosts, they have become adapted to the coldest and most aerated localities within the warm-blooded hosts. The hypothesis that pentastomids originated as parasites in ectothermic hosts is substantiated by the large number of modern species parasitic in reptiles (Table 21.3). Approximately 43% of the genera and 42% of the species parasitize snakes; 22% of the genera and 28% of the species parasitize crocodiles; and

14% of the genera and 18% of the species parasitize lizards. Only 7% of the known genera parasitize tortoises, birds, and mammals, and a significant percentage of these are parasites of tortoises, which are ectotherms.

Another aspect of pentastomid-host relationship deserves brief mention. Riley *et al.* (1979) have reported that when adult pentastomids feed on their host's tissue fluids and blood cells, they stimulate a strong immune response. The parasites, however, are capable of evading their host's antibodies by secreting a substance from their frontal and subparietal glands. This secreted lamellar material covers the entire body surface and apparently protects the pentastomids from antibody action.

PHYLUM ECHINODERMATA

The echinoderms are all marine. The only known symbiotic echinoderm is the holothurian *Rynkatorpa pawsonii* (Fig. 21.89) reported by Martin (1969) as a commensal attached to the posterior end of an angler fish, *Gigantactis macronema*, caught off the coast of southern California. This sea cucumber is subcylindrical and measures 5.18—7.14 mm long and 1.75—2.52 mm in maximum width. Although this echinoderm is firmly at-

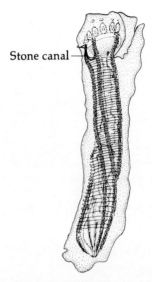

Stone canal—

Fig. 21.89. Symbiotic holothurian. Lateral view of entire specimen of *Rhynkatorpa pawsoni*. (Redrawn after Martin, 1969.)

Table 21.3. Host Distribution by Class of the Major Pentastomid Groups[a]

Ophidia	Lacertilia	Crocodylia	Chelonia	Aves	Mammalia
Order Cephalobaenida	Order Cephalobaenida	Order Porocephalida	Order Porocephalida	Order Porocephalida	Order Porocephalida
Raillietiella	Raillietiella	Sebekia	Diesingia	Reighardia	Linguatula
R. boulengeri	R. geckonis	S. oxycephala	D. ruchgensis	R. sternae	L. serrata
R. furcocerca	R. affinis	S. cesarisi	Butantinella		L. dingophila
Cephalobaena	R. gehyrae	S. divestei	B. megastoma		L. nuttali
C. tetrapoda	R. mabuiae	S. samboni			L. recurvata
Order Porocephalida	R. kochi	S. wedli			
Kiricephalus	R. shipleyi	S. jubini			
K. pattoni	R. giglioli	Leiperia			
K. coarctatus	R. chautedeni	L. cincinnalis			
Porocephalus	R. chamaeleonis	L. gracilis			
P. crotali	Order Porocephalida	Subtriquetra			
P. clavatus	Sambonia	S. subtriquetra			
P. stylesi	S. lohrmanni	S. shipleyi			
P. subulifer	S. solomonensis	Alofia			
Armillifer	S. varani	A. genae			
A. grandis	S. parapodum	A. platycephala			
A. armillatus	Elenia	A. merki			
A. agkistrodontis	E. australis	A. adriatika			
Waddycephalus		A. indica			
W. teretiusculus					
Cubirea					
C. annulata					
C. pomeroyi					
Gigliolella					
G. brumpti					
Ligamifer					
L. mazzai					

[a] After Self, 1969.

tached to the fish host's rough skin and small dermal denticles, no invasion of host tissues occurs. Martin is of the opinion that *R. pawsonii* is a commensal and benefits from the relationship by being transported and being introduced to new feeding areas.

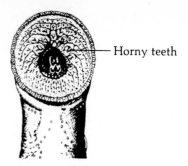

Fig. 21.91. **Sea lamprey.** Frontal view of mouth showing circular rows of chitinous teeth. (After Gage, 1893.)

SUBPHYLUM VERTEBRATA

Although vertebrates are not plentiful or important in active roles as parasites, the biology of these animals is intimately allied with that of invertebrate parasites in their roles as intermediate and definitive hosts, as natural reservoirs, as transport hosts, and as experimental hosts in laboratory investigations.

True parasites, in the biological sense, are rare among the vertebrates. A few animals, such as lampreys and vampire bats, have been considered by some to be parasites. However, they can just as easily be considered predators. The lampreys, vampire bats, candirus, and a few pseudoparasitic vertebrates are briefly mentioned here.

Lampreys

The lampreys belong to the class Cyclostomata, which also includes the hagfish and slime eels. Lampreys, however, are members of the order Petromyzontia, and hagfishes and slime eels belong to the order Myxinoidia. The habitats of lampreys vary, depending upon the species. In North American waters, *Petromyzon marinus*, the sea lamprey, is found in the Atlantic Ocean (Fig. 21.90). It migrates up streams during the spring or early summer to spawn. *Entosphenus tridentatus* is the species found along the Pacific coast, ranging from California to Alaska. It also migrates to fresh water to spawn.

In inland lakes, such as the Great Lakes, *Petromyzon marinus unicolor* is common, and its presence represents a hazard to the freshwater fishing industry. This species is believed to have arisen from marine ancestors that became landlocked during the glacial period and

have acclimated to freshwater living. It, too, migrates up streams to spawn. An anatomy and natural history of the lake lampreys have been contributed by Gage (1893). Other observations on growth and life cycles have been given by Hubbs (1924). The two-volume treatise edited by Hardisty and Potter (1971, 1972) is the most complete treatment available on the evolution, ecology, natural history, cytology, physiology, and biochemistry of lampreys.

The mouth of the lamprey is situated anteroventrally and is suctorial in function. Fleshy papillae or tentacles are located along the circumference of the circular mouth. When these ectothemic parasites attack fish, they become attached by digging into the host's flesh with their buccal teeth (Fig. 21.91). A hole is ripped in the integument and flesh of the fish host by sharp lingual teeth and an anticoagulent known as **lamphredin**, secreted by buccal glands, is injected into the lesion to prevent the host's blood from clotting during feeding. Lamphredin not only prevents coagulation of the fish's blood but also exerts a hemolytic influence on the erythrocytes and induces a lytic action in the torn flesh (Fig. 21.92) (Lennon, 1954). Fish continuously parasitized by lampreys usually die. Hence, if these cyclostomes are considered parasites, they cannot be regarded as very well adapted or efficient ones, for they kill the host on which they depend.

Not all lampreys are bloodsuckers. The so-called brook lampreys, *Lampetra wilderi* and *Entosphenus latottenii*, are not.

Fig. 21.90. **The sea lamprey.** *Petromyzon marinus* attached to whitefish. (Courtesy of U.S. Department of Commerce.)

Myxine limosa, the Atlantic hagfish, is found in North America, ranging from Cape Cod northward. *Myxine glutinosa* is the European species and measures up to three feet in length. *Bdellostoma spp.* occur off the coast of Chile in South America. The Pacific boring hagfish, *Polistotrema stouti*, is found from the coast of lower California to Alaska (Fig. 21.93). Unlike lampreys, slime eels and hagfishes actually burrow into the bodies of their piscine hosts and consume flesh, leaving only bones wrapped in integument. Without doubt, the Myxinoidia must be considered predators rather than true parasites.

The parasitic lampreys, slime eels, and hagfish are all of considerable economic importance because they are extremely destructive to fish. Much of the work done concerning these animals has been for the purpose of control or eradication.

Pseudoparasitism in eels. Pseudoparasitism is encountered among certain eels. Goude and Bean (1895) have reported *Pisododonophis cruentifer* in fish; Deraniyagala (1932) has identified *Ophichthus apicalis* from the body cavities of various percoid fish; Breder and Nigrelli (1943) have reported a large *Myricthys acuminatus* from the coelom of the jewfish, *Promicrops itaiara*; and Breder (1953) has reported *Pisododonophis cruentifer* in a sea bass.

Walters (1955) has cited most of the known instances in which eels of the family Ophichthidae have been found in the coelom of fish. Breder and Nigrelli postulate that such instances of "parasitism" actually represent pseudoparasitism. The engulfed eel tries to escape from the fish and in the process plunges its sharp tail through the gut wall of the host and backs into the coelom. This hypothesis is supported by the fact that ophichthid eels found in the coelom of fish are usually dead and encapsulated within connective tissue cysts, and the anatomy of these eels does not support the idea that they are burrowers that have arrived within their hosts by penetrating from the exterior.

Candirus

Probably the only true endoparasites among vertebrates are the so-called candiru fish found in the upper Amazon in South America. There is some doubt if these are obligatory parasites. One of these, *Vandellia cirrhosa*, a small, slender, naked catfish of the family Trichonycteridae, has been found in the gill chambers, attached to the gills of larger catfish, where they supposedly feed on the gill epithelium and blood of the host fish. *Vandellia cirrhosa* measures no more than 6.5 cm in length and has been reported to migrate up the urethra, anus, or vagina of bathers, causing severe pain when they expand their spine-bearing gill covers (opercula). It is a much feared animal among Amazon Indians, among whom much folklore about this fish

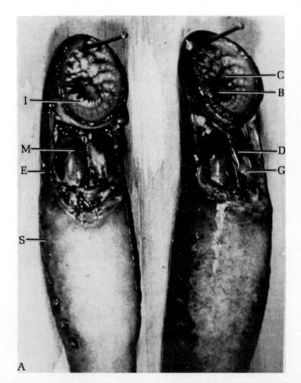

Fig. 21.92. Sea lamprey. A. Ventral view of sucking discs and buccal glands of two sexually mature sea lampreys. **B.** Lytic reaction produced in longnose suckers by subcutaneous injections of lamphredin. B, buccal funnel; C, circumoral cusps; D, duct of buccal gland; E, eye; G, buccal gland; I, infraoral lamina; M, musculus basilaris; S, gill sac. (**A** and **B**, after Lennon, 1954.)

has developed. Undoubtedly, some of the tales relative to attacks on humans by *V. cirrhosa* are merely folklore.

Myers (1944) has reported that another small catfish, *Cetopsis candiru* of the family Cetopsidae, will

Fig. 21.93. *Polistotrema stouti.* The California hagfish showing 12 pairs of gill slits. (After Wolcott, 1946.)

rasp the skin of humans with its teeth when caught and suck blood. This catfish measures 8 inches or less in length.

Vampire Bats

Another group of vertebrates that could be considered parasitic is the vampire bats. These flying mammals belong to the suborder Microchiroptera of the order Chiroptera. Not all bats belonging to this suborder are bloodsuckers. For example, the brown bats *Myotis* and *Eptesicus* are insectivorous. However, the true vampire bats, *Desmodus* spp., are blood ingestors. They are found in the American tropics, where they attack (or parasitize) wild and domestic warm-blooded animals. When attacking, they dig their large canine teeth into the flesh of their victim, usually at the back of the neck or on the body proper, rupture a large blood vessel, and lap up blood. *Desmodus* has been known to bite fowls, particularly chickens, and also dogs, cats, cows, horses, and even humans. Contrary to popular belief, the bite of the vampire bat is not fatal, nor is the loss of the relatively small quantity of blood of serious consequence. The wounds, however, are subject to secondary infection. Furthermore, a form of rabies is carried by these bats, and certain equine trypanosomes utilize *Desmodus* as a vector. These microorganisms are pathogenic and often lethal.

REFERENCES

MESOZOA

Dodson, E. O. (1956). A note on the systematic position of the Mesozoa. *Syst. Zool.* **5**, 37–40.

Hoffman, W. (1965). Mesozoa of the sepiolid, *Rossia pacifica* (Berry). *J. Parasitol.* **51**, 313–320.

Kozloff, E. N. (1969). Morphology of the orthonectid *Rhopalura ophiocomae. J. Parasitol.* **55**, 171–145.

Lameere, A. (1916). Contributions à la conaissance des dicyémides. Première partie. *Bull. Sci. Fr. Belg.* **50**, 1–35. (Also see Deuxième partie. *Ibid* **51**, 347–390).

McConnaughey, B. H. (1951). The life cycle of the dicyemid Mesozoa. *Univ. Calif. Publ. Zool.* **55**, 295–336.

McConnaughey, B. H. (1968). The Mesozoa. *In* "Chemical Zoology" (M. Florkin and B. T. Scheer, eds.), Vol. II, pp. 557–570. Academic Press. New York.

McConnaughey, B. H., and Overstreet, R. (1962). Cited in McConnaughey, B. H. (1968).

Nouvel, H. (1933). Recherches sur la cytologie, la physiologie et la biologie des dicyémides. *Ann. Inst. Oceanogr. (Paris)* **13**, 163–255.

Ridley, R. K. (1968). Electron microscopic studies on dicyemid mesozoa. I. Vermiform stages. *J. Parasitol.* **54**, 975–998.

Ridley, R. K. (1969). Electron microscopic studies on dicyemid mesozoa. II. Infusorigen and infusoriform stages. *J. Parasitol.* **55**, 779–795.

Stunkard, H. W. (1954). The life history and systematic relations of the Mesozoa. *Quart. Rev. Biol.* **29**, 230–244.

PORIFERA

Galtsoff, P. S. (1964). The American oyster, *Crassostrea virginica* Gmelin. *Fish. Bull. Fish. Wildl. Serv. U.S.* **64**, 1–480.

COELENTERATA

Bigelow, H. B. (1909). "The Medusae." *Mus. Comp. Zool., Harvard Univ. Mem.* **37**.

Cheng, T. C. (1967). Marine molluscs as hosts for symbioses. *Adv. Mar. Biol.* **5**, 1–424.

Crowell, S. (1957). *Engymnanthea*, a commensal hydroid living in pelecypods. *Pubbl. Staz. Zool. Napoli* **30**, 162–167.

Hyman, L. H. (1940). "The Invertebrates: Protozoa through Ctenophora," Vol. I. McGraw-Hill, New York.

Lipin, A. (1911). Die Morphologie und Biologie von *Polypodium hydriforme. Zool. Jahrb.* **31**, 317–426.

Martin, W. E., and Brinkmann, A. (1963). Zum Brutparasitismus von *Phyllirrhoe bucephala* Per. and Les. (Gastropoda, Nudibranchia) auf der Meduse *Zanclea costata* Gegenb. (Hydrozoa, Anthomedusae). *Staz. Zool. Napoli Publ.* **33**, 206–223.

Raikova, E., Suppes, C., and Hoffman, G. (1979). The parasitic coelenterate, *Polypodium hydriforme* Ussov, from the eggs of the American acipenseriform *Polyodon spathula. J. Parasitol.* **65**, 804–810.

Rees, B. (1967). A brief survey of the symbiotic associations of Cnidaria and Mollusca. *Proc. Malac. Soc. London* **37**, 213–231.

CTENOPHORA

Bouillon, J. (1968). Introduction to coelenterates. *In* "Chemical Zoology" (M. Florkin and B. T. Scheer, eds.), Vol. II, pp. 81–147. Academic Press, New York.

Hyman, L. H. (1940). "The Invertebrates: Protozoa through Ctenophora," Vol. I. McGraw-Hill, New York.

Kamai, T. (1922). Studies on two aberrant ctenophores— *Coeloplana* and *Gastrodes*. Published by the author. Kyoto, Japan.

Carvalho, J. C. M. (1942). Studies on some Gordiacea of North and South America. *J. Parasitol.* **28**, 213–222.

Dorier, A. (1925). Sur la faculté d'enkystement dans l'eau de la larve du *Gordius aquaticus* L. *C.R. Acad. Sci.* (Paris) **181**, 1098–1099.

Dorier, A. (1930). Recherches biologiques et systematiques sur les gordiaces. *Trav. Lab. Hydrobiol. Piscicult. Univ. Grenoble, France.*

Feyel, T. (1936). Recherches histologiques sur *Nectonema agile* Ver. Etude de la forme parasite. *Arch. Anat. Microsc.* **32**, 197–234.

Heinze, K. (1934). Zur Systematik der Gordiiden. *Zool. Anz.* **106**, 189–192.

Heinze, K. (1935). Über das Genus *Parachordodes* Camerano, 1897 nebst allgemeinen Angaben über die Familie Chordidae. *Z. Parasitenk.* **7**, 657–678.

Heinze, K. (1937). Die Saitenwürmer (Gordioidea) Deutschlands. Eine systematisch faunische Studie über Insectenparasiten aus der Gruppe der Nematomorpha. *Z. Parasitenk.* **9**, 263–344.

Inoue, I. (1958). Studies on the life history of *Chordodes japanensis*, a species of Gordiacea. I. The development and structure of the larva. *Jap. J. Zool.* **12**, 203–218.

Inoue, I. (1960). Studies on the life history of *Chordodes japanensis*, a species of Gordiacea. II. On the manner of entry into the aquatic insect-larvae of *Chordodes* larvae. *Annot. Zool. Jap.* **33**, 132–141.

May, H. G. (1919). Contribution to the life histories of *Gordius robustus* Leidy and *Paragordius varius* (Leidy). *Ill. Biol. Monogr.* **5**, 1–118.

Montgomery, T. H. (1903). The adult organization of *Paragordius varius* (Leidy). *Zool. Jahrb. Abt. Anat.* **18**, 387–474.

Müller, G. W. (1927). Über Gordiaceen. *Z. Morphol. Okol. Tiere* **7**, 134–218.

Zapotosky, J. E. (1974). Fine structure of the larval stage of *Paragordius varius* (Leidy, 1851) (Gordiodea: Paragordidae). I. The preseptum. *Proc. Helminth. Soc. Wash.* **41**, 209–221.

ROTIFERA

Budde, E. (1925). Die parasitischen Rädertiere. *Z. Morphol. Okol. Tiere* **3**, 706–784.

Hyman, L. H. (1951). "The Invertebrates: Acanthocephala, Aschelminthes, and Entoprocta. The Pseudocoelomata Bilateria," Vol. III. McGraw-Hill, New York.

Pénard, E. (1905). Sur un rotifère du genre *Proales*. *Arch. Sci. Phys. Natur.* **20**, 459.

Pénard, E. (1909). Über ein bei *Acanthocystis* parasitisches Rotatorium. *Mikrokosmos* **2**, 135–143.

Plate, L. H. (1886). Untersuchungen einiger aus den Kiemenblättern des *Gammarus pulex* lebenden Ektoparasiten. *Z. Wissensch, Zool.* **43**, 175–241.

Rees, B. (1960). *Albertia vermicularis* (Rotifera) parasitic in the earthworm *Allolobophora caliginosa*. *Parasitology* **50**, 61–66.

Stevens, J. (1912). *Proales gigantea*, a rotifer parasitic in the egg of the water snail. *J. Quekett Microsc. Club* **11** (Ser. 2), 481–486.

RHYNCHOCOELA

Eggers, F. (1936). Zur Bewegungsphysiologie von *Malacobdella grossa*. *Z. Wiss. Zool.* **147**, 101–131.

Gibson, R. (1972). "Nemerteans." Hutchinson University Library, London, England.

Hopkins, S. H. (1947). The nemertean *Carcinonemertes* an indicator of the spawning history of the host, *Callinectes sapidus*. *J. Parasitol.* **33**, 146–150.

Humes, A. G. (1941). The male reproductive system in the nemertean genus *Carcinonemertes*. *J. Morphol.* **69**, 443–454.

Humes, A. G. (1942). The morphology, taxonomy, and bionomics of the nemertean genus *Carcinonemertes*. *Ill. Biol. Monogr.* **18**, 1–105.

Hyman, L. H. (1951). "The Invertebrates: Platyhelminthes and Rhynchocoela," Vol. III. McGraw-Hill, New York.

ANNELIDA

Bennike, S. A. B. (1943). Contributions to the ecology and biology of the Danish fresh-water leeches (Hirudinea). *Folia Limnobiol. Scand.* No. 2.

Clark, R. B. (1956). *Capitella capitata* as a commensal, with a bibliography of parasitism and commensalism in the polychaetes. *Ann. Mag. Natur. Hist.* **9**, 433–448.

Harmon, W. J. (1971). A review of the subgenus *Allodero* (Oligochaeta: Naididae: *Dero*) with a description of *D. (A.) floridana* n.sp. from *Bufo terrestris*. *Trans. Am. Microsc. Soc.* **90**, 225–228.

Holt, P. C. (1963). A new branchiobdellid (Branchiobdellidae: Cambarincola). *J. Tenn. Acad. Sci.* **38**, 97–100.

Holt, P. C. (1968). The Branchiobdellida: epizootic annelids. *Biologist* **50**, 79–94.

Mann, K. H. (1954). The anatomy of the horse leech, *Haemopis sanguisuga* (L.) with particular reference to the excretory system. *Proc. Zool. Soc. London* **124**, 69–88.

Mann, K. H. (1955). Some factors influencing the distribution of fresh-water leeches in Britain. *Proc. Int. Assoc. Theoret. Appl. Limnol.* **12**, 582–587.

Mann, K. H. (1962). "Leeches (Hirudinea). Their Structure, Physiology, Ecology, and Embryology." Pergamon Press, New York.

Meyer, M. C. (1940). A revision of the leeches (Piscicolidae) living on fresh-water fishes of North America. *Trans. Am. Microsc. Soc.* **59**, 354–376.

Meyer, M. C. (1946). A new leech *Piscicola salmositica*. *J. Parasitol.* **32**, 467–476.

Meyer, M. C., and Barden, A. A. Jr. (1955). Leeches symbiotic on Arthropoda, especially decapod Crustacea. *Wasmann J. Biol.* **13**, 297–311.

Powlowski, L. K. (1955). Revision des genres *Erpobdella* et *Dina*. *Bull. Soc. Sci. Let. Lodz* C1.3 **6**, 1–15.

Sawyer, R. T. (1972). "North American Freshwater Leeches, Exclusive of the Piscicolidae, with a Key to All Species." *Ill. Biol. Monogr.* No. 46. University of Illinois Press, Urbana, Illinois.

MOLLUSCA

Arey, L. B. (1924). Observations on an acquired immunity to a metazoan parasite. *J. Exp. Zool.* **38**, 377–381.

Arey, L. B. (1932). A microscopical study of glochidial immunity. *J. Morphol.* **53**, 367–379.

Baer, J. G. (1952). "Ecology of Animal Parasites." University of Illinois Press, Urbana, Illinois.

Caullery, M. (1952). "Parasitism and Symbiosis." Sidgwick & Jackson, London, England.

Davis, L. V. (1967). The suppression of autotomy in *Linckia multifora* (Lamarck) by a parasitic gastropod, *Stylifer linckiae* Sarasin. *Veliger* **9**, 343–346.

Heding, S. G. (1934). *Entocolax trochodotae* n.sp., a new parasitic gastropod. *Vidensk. Medd. Dansk. Naturh. For. Kbh.* **98**, 207–214.

Hoskin, G. P., and Cheng, T. C. (1970). On the ecology and microanatomy of the parasitic marine prosobranch *Mucronalia nitidula* (Pease, 1860). *Proc. Symp. Mollusca*, Pt. III, pp. 780–798. Marine Biology Association India, Bangalor.

Isom, B. G., and Hudson, R. G. (1984). Freshwater mussels and their fish hosts; physiological aspects. *J. Parasitol.* **70**, 318–319.

Iwanow, A. W. (1948). [Metamorphosis of the parasitic snail *Parenteroxenos dogieli* Iwanow]. *Rep. Acad. Sci. U.S.S.R.* **G1**, 765–768. (In Russian.)

Koehler, R., and Vaney, C. (1903). *Entosiphon deimatis*, nouveau mollusque parasite d'une holothurie abyssale. *Rev. Susse Zool.* **11**, 23–41.

Mandahl-Barth, G. (1941). *Tnyonicola mortensi* n. gen. n.sp., eine neue parasitische Schnecke. *Vidensk. Medd. Dansk. Naturh. For. Khh.* **104**, 341–351.

Schwanwitsch, B. N. (1917). Observations sur la femelle et le male rudimentaire d'*Entocolax ludwigii* Voigt. *J. Russe Zool.* **2**, 1–147.

Tullis, R. E., and Cheng, T. C. (1971). The uptake of ^{14}C by *Stylifer linckiae* (Mollusca: Prosobranchia) from its echinoderm host, *Linckia multifora*. *Comp. Biochem. Physiol.* **40B**, 109–112.

CRUSTACEA

Amemiya, I. (1935). Effect of gill excision upon the sexual differentiation of the oyster *Ostrea gigas* Thunberg. *Rep. Japan. Assoc. Adv. Sci.* **10**, 1025–1028.

Awati, P. R., and Rai, H. S. (1931). *Ostrea cucullata*. *Indian Zool. Mem.* **3**, 1–107.

Barnes, R. D. (1980). "Invertebrate Zoology," 4th ed. W. B. Saunders, Philadelphia, Pennsylvania.

Boschma, H. (1937). The species of the genus *Sacculina* (Crustacea Rizocephala). *Zool. Mededel. Rijksmus. Natur. Hist. Leiden* **19**, 187–328.

Bocquet, C., and Stock, J. H. (1957). Copépodes parasites d'invertébrés des côtes de France. I. Sur deux genres de la famille des Clausidiidae, commensaux de mollusques: *Hersilliodes* Canu et *Conchyliuris* nov. gen. *Proc. Kon. Ned.*

Akad. Wetensch. **60**, 212–222.

Bowman, T. E. (1955). A new genus and species of whale-louse (Amphipoda: Cyamidae) from the false killer whale. *Bull. Mar. Sci. Gulf Carib.* **5**, 315–320.

Causey, D. 1957. Parasitic copepoda from Louisiana fresh water fish. *Amer. Midl. Nat.* **58**, 373–383.

Cheng, T. C. (1967). Marine molluscs as hosts for symbiosis. *Adv. Marine Biol.* **5**, 1–424.

Christensen, A. M., and McDermott, J. J. (1958). Life-history and biology of the oyster crab, *Pinnotheres ostreum* Say. *Biol. Bull.* **114**, 146–179.

Dadd, R. H. (1970). Arthropod nutrition. *In* "Chemical Zoology" (M. Florkin and B. T. Scheer, eds.), Vol. V, pp. 35–95. Academic Press, New York.

Day, J. H. (1935). The life-history of *Sacculina*. *Quart. J. Microsc. Sci.* **77**, 549–583.

Egami, N. (1953). Studies on the sexuality in the Japanese oyster, *Ostrea gigas*. VII. Effects of gill removal on growth and sexuality. *Annot. Zool. Jap.* **26**, 145–150.

Fasten, N. (1919). Morphology and attached stages of first copepodid larva of *Salmincola edwardsii* (Olsson) Wilson. *Publ. Puget Sound Biol. Sta. Univ. Wash.* **2**, 153–181.

Faxon, G. E. H. (1940). Notes on the life history of *Sacculina carcini* Thompson. *J. Mar. Biol. Ass. U.K.* **24**, 253–264.

Florkin, M., and Scheer, B. T. (eds.). (1970). "Chemical Zoology," Vol. V, Arthropoda, Part A. Academic Press, New York.

Florkin, M., and Scheer, B. T. (eds.). (1971). "Chemical Zoology," Vol. VI, Arthropoda, Part B. Academic Press, New York.

Fox, H. M. (1945). Haemoglobin in blood sucking parasites. *Nature* **156**, 475.

Fox, H. M. (1953). Crustacea. *Nature* **171**, 162. (Also see "Animal Biochromes and Structural Colours." Cambridge University Press, London.)

Giard, A., and Bonnier, J. (1887). Contributions a l'étude des Bopyriens. *Trav. Inst. Zool. Lille* **5**, 1–151.

Giard, A., and Bonnier, J. (1893). Contributions a l'étude des Epicarides. *Bull. Sci. Fr. Belg.* **25**, 415–493.

Goldschmidt, R. B. (1923). "The Mechanism and Physiology of Sex Determination." Methuen, London. (Trans. W. J. Dakin.)

Grainger, J. N. R. (1951). Notes on the biology of the copepod *Mytilicola intestinalis* Steuer. *Parasitology* **41**, 135–142.

Haley, A. J., and Winn, H. E. (1959). Observations on a lernean parasite of freshwater fishes. *Trans. Am. Fish. Soc.* **88**, 128–129.

Haven, D. (1959). Effect of pea crabs *Pinnotheres ostreum* on oysters *Crassostrea virginica*. *Proc. Nat. Shellfish. Assoc.* **49**, 77–86.

Ho, J.-S. (1970). Revision of the genera of the Chondracanthidae, a copepod family parasitic on marine fishes. *Beaufortia* **17**, 105–218.

Ho, J.-S. (1971). Parasitic copepods of the family Chondracanthidae from fishes of eastern North America. *Smithsonian Contrib. Zool. No. 87*, 1–38.

Ichikawa, A., and Yanagimachi, R. (1958). Studies on the sexual organization of the Rhizocephala. I. The nature of the "testes" of *Peltogaster socialis* Kruger. *Annot. Zool. Jap.* **31**, 82–96.

Jeuniaux, C. (1971). Hemolymph-Arthropoda. *In* "Chemical Zoology" (M. Florkin and B. T. Scheer, eds.), Vol. VI, pp. 64–118. Academic Press, New York.

Kabata, Z. (1970). Crustacea as enemies of fishes. *In* "Diseases of Fishes." Book 1. (S. F. Snieszko and H. R. Axelfod, eds.), pp. 1–171. T. F. H. Publishers, Jersey City, New Jersey.

Korringa, P. (1950). *Vrsserij-Nieuws*, No. 7.

Keung, Y. (1967). An illustrated key to the species of whalelice (Amphipoda, Cyamidae), ectoparasites of Cetacea, with a guide to the literature. *Crustaceana* **12**, 279–290.

Lutken, C. C. (1873). Bidrag til Kundskab om Asterne of Slaegten Latr. elle Hvallusene. *Dan. Vidensk. Selsk.* **10**, 231–284.

Margolis, L. (1955). Notes on the morphology, taxonomy, and synonymy of several species of whale-lice (Cyamidae: Amphipoda). *J. Fish. Res. Bd. Can.* **12**, 121–133.

Menzies, R. J., Bowman, T. E., and Alverson, F. G. (1966). Studies on the biology of the fish parasite *Livoneca convexa* Richardson (Crustacea, Isopoda, Cymothoidae). *Wasmann J. Biol.* **13**, 277–295.

Panikkar, N. K., and Sproston, N. G. (1941). Osmotic relations of some metazoan parasites. *Parasitology* **33**, 214–223.

Paterson, N. F. (1958). External feature and life cycle of *Cucumaricola notabilis* nov. gen. et sp., a copepod parasite of the holothurian, *Cucumaria. Parasitology* **48**, 269–290.

Perez, C. (1933). Processus de résorption phagocytaire des oocytes dans l'ovaire chez les Macropodia succulinées. *C. R. Soc. Biol.* **112**, 1049–1051.

Reinhard, E. G. (1949). Experiments on the determination and differentiation of sex in the bopyrid *Stegophryxus hyptius* Thompson. *Biol. Bull.* **96**, 17–31.

Reinhard, E. G. (1956). Parasitic castration of Crustacea. *Exp. Parasitol.* **5**, 79–107.

Reverberi, G. (1947). Ancora sulle transformazione sperimentale del sesso nei Bopiridi. La transformazione delle femmine giovanili in maschi. *Publ. Staz. Zool. Napoli* **21**, 81–91.

Reverberi, G., and Pitotti, M. (1942). Il ciclo biologica e la determinazione fenotopica del sesso di *Ione thoracia* Montagu, Bopirid e parassito de *Callianassa laticauda* Otto. *Publ. Staz. Zool. Napoli* **19**, 111–184.

Roberts, L. S. (1970). *Ergasilus* (Copepoda: Cyclopoida): revision and key to species in North America. *Trans. Am. Microsc. Soc.* **89**, 134–161.

Rubiliani, C. (1985). Response by two crab species to an extract of a rhizocephalan parasites. *J. Invert. Pathol.* **45**, 304–310.

Rubiliani, C., Payan, G. G., and Rubiliani-Durozoi, M. (1980). Action d'implants et d'homogénats de racines du Rhizocéphale *Sacculina carcini* Thompson chez le crabe mâle *Carcinus maenas* (L.). *C. R. Acad. Sci., Ser. D*, **290**, 355–358.

Smith, G. W. (1906). Rhizocephala. *In* "Fauna and Flora des Golfes von Neapel." Monographie, Vol. 21, pp. 1–123.

Smith, R. F. (1949). Notes on *Ergasilus* parasites from the New Brunswick, New Jersey, area, with a check list of all species and hosts east of the Mississippi River. *Zoologica* **34**, 127–182.

Stauber, L. A. (1945). *Pinnotheres ostreum*, parasitic on the American oyster, *Ostrea* (*Gryphaea*) *virginica. Biol. Bull.* **88**, 269–291.

van Weel, P. B. (1970). Digestion in Crustacea. *In* "Chemical Zoology" (M. Florkin and B. T. Scheer, eds.), Vol. V, pp. 97–115. Academic Press, New York.

Wilson, C. B. (1915). North American parasitic copepods belonging to the Lernaeopodidae, with a revision of the entire family. *Proc. U.S. Nat. Mus.* **47**, 565–729.

Wilson, C. B. (1944). Parasitic copepods in the United States national museum. *Proc. U.S. Nat. Mus.* **94**, 529–582.

Yamaguti, S. (1963). "Parasitic Copepoda and Brachiura of Fishes." Wiley, New York.

PYCNOGONIDA

Cheng, T. C. (1973). "General Parasitology," 1st ed. Academic Press, New York.

Hedgpeth, J. W. (1954). On the phylogeny of Pycnogonida. *Acta Zool.* **35**, 193–213.

Helfer, H., and Schlottke, E. (1935). Pantopoda. *In* "Klassen und Ordnungen des Tierreiches," Vol. 5, Part 4, Book 2. Akademische Verlagsgesellschaft, Leipzig, Germany.

TARDIGRADA

Fox, I., and García-Moll, I. (1962). *Echiniscus molluscorum*, new tardigrade from the feces of the land snail, *Bulimulus exilis* (Gmelin) in Puerto Rico (Tardigrada: Scutechiniscidae). *J. Parasitol.* **48**, 177–181.

Green, J. (1950). Habits of the marine tardigrade, *Echiniscoides skjismundi. Nature* **166**, 153–154.

Marcus, E. (1929). Tardigrada. *In* "Bronns' Klassen und Ordnungen des Tierreichs," Vol. 5, Part 4, Book 3, Akademische Verlagsgesellschaft, Leipzig, Germany.

Marcus, E. (1936). Tardigrada. *In* "Das Tierreich." (Preuss. Akademie Wissensch.), 66, XVI. Berlin and Leipzig, Germany.

Pennak, W. W. (1978). "Fresh-water Invertebrates of the United States," 2nd ed. John Wiley & Sons, New York.

PENTASTOMIDA

Baer, J. G. (1951). "Ecology of Animal Parasites." University of Illinois Press, Urbana, Illinois.

Banaja, A. A., James, J. L., and Riley, J. (1977). Observations on the osmoregulatory system of pentastomids: the tegumental chloride cells. *Int. J. Parasitol.* **7**, 27–40.

Dönges, J. (1966). Parasitäre abdominalcysten bei Nigeriaren. *Z. Trop. Parasitol.* **17**, 252–256.

Doucet, J. (1965). Contribution a l'étude anatomique, histologique et histochimique des pentastomes (Pentastomida). *Mem Off. Rech. Sci. Tech. Outre-Mer.* **14**, 1–150.

Esslinger, J. H. (1962a). Morphology of the egg and larva of *Porocephalus crotali* (Pentastomida). *J. Parasitol.* **48**, 451–462.

Esslinger, J. H. (1962b). Development of *Porocephalus crotali* (Humboldt, 1808) (Pentastomida) in experimental intermediate hosts. *J. Parasitol.* **48**, 452–456.

Esslinger, J. H. (1962c). Hepatic lesions in rats experimentally infected with *Porocephalus crotali* (Pentastomida). *J. Parasitol.* **48**, 631–638.

Fain, A. (1961). Les pentastomides de l'Afrique centrale. *Mus. Roy. Afr. Cent. Terv. Belg. Ann. Ser. 8°, Sci. Zool*, **92**, 1–115.

Heymons, R. (1935). Pentastomida. In "Bronns' Klassen und Ordnungen des Tierreichs," Vol. 5, Sect. 4, Book 1. Akademische. Verlagsgesellschaft, Leipzig, Germany.

Hill, H. R. (1948). Annotated bibliography of the linguatulids. *Bull. S. Calif. Acad. Sci.* **47**, 56–73.

Hobmaier, A., and Hobmaier, M. (1940). On the life cycle of *Linguatula rhinaria*. *Am. J. Trop. Med.* **20**, 199–210.

Hunter, W. S., and Higgins, R. P. (1960). An unusual case of human porocephaliasis. *J. Parasitol.* **46**, 88.

Khalil, G. M., and Schacher, J. F. (1965). *Linguatula serrata* in relations to halzoun and the marrara syndrome. *Am. J. Trop. Med. Hyg.* **14**, 736–746.

Lavoipierre, M. M. J., and Lavoipierre, M. (1966). An arthropod intermediate host of a pentastomid. *Nature* **210**, 845–846.

Nicoli, R. M. (1963). Phylogérèse et systématique le phylum des Pentastomida. *Ann. Parasitol. Hum. Comp.* **38**, 483–516.

Penn, G. H., Jr. (1942). The life history of *Porocephalus crotali*, a parasite of the Louisiana muskrat. *J. Parasitol.* **28**, 277–283.

Rao, H., and Jennings, J. B. (1959). The alimentary system of a pentastomid from the Indian water-snake *Natrix piscator* Schneider. *J. Parasitol.* **45**, 299–300.

Riley, J., James, J. L., and Banaja, A. A. (1979). The possible role of the frontal and sub-parietal gland systems of the pentastomid *Reighardia sternae* (Diesing, 1864) in the evasion of the host immune response. *Parasitology* **78**, 53–66.

Sambon, L. W. (1922). A synopsis of the family Linguatulidae. *J. Trop. Med. Hyg.* **25**, 188–206.

Schacher, J. F., Saab, S., Germanos, R., and Boustany, N. (1969). The aetiology of halzoun in Lebanon: recovery of *Linguatula serrata* nymphs from two patients. *Trans. Roy. Soc. Trop. Med. Hyg.* **63**, 854–858.

Self, J. T. (1969). Biological relationships of the Pentastomida; a bibliography on the Pentastomida. *Exp. Parasitol.* **24**, 63–119.

Self, J. T. (1972). Pentastomiasis: host responses to larval and nymphal infections. *Trans. Am. Microsc. Soc.* **91**, 2–9.

Self, J. T., Hopps, H. C., and Williams, A. O. (1972). Porocephaliasis in man and experimental mice. *Exp. Parasitol.* **32**, 117–126.

Self, J. T., Hopps, H. C., and Williams, A. O. (1975). Pentastomiasis in Africans, a review. *Trop. Geogr. Pathol.* **27**, 1–13.

Stiles, C. W., and Hassall, A. (1927). Key-catalogue of the crustacea and arachnoids of importance in public health. *Hyg. Lab. U.S. Publ. Health Serv. Bull. No.* 148, pp. 197–289.

Trainer, J. E. Jr., Self, J. T., and Richter, K. H. (1975). Ultrastructure of *Porocephalus crotali* (Pentastomida) cuticle with phylogenetic implications. *J. Parasitol.* **61**, 753–758.

Watson, J. M. (1960). "Medical Helminthology." Bailliere, Tindell & Cox, London.

VERTEBRATA

Breder, C. M. Jr. (1953). An ophichthid eel in the coelom of a sea bass. *Zoologica* **38**, 201–202.

Breder, C. M. Jr., and Nigrelli, R. F. (1943). The penetration of a grouper's digestive tract by a sharp-tailed eel. *Copeia* **4**, 162–164.

Deraniyagala, P. E. P. (1932). A curious association between *Ophichthus apicalis* and percoid fishes. *Spolia Zeylan.* **B16**, 355–356.

Gage, S. H. (1893). "The Lake and Brook Lampreys of New York." Wilder Quarterly-Century Book, Ithaca, New York.

Goude, G. B., and Bean, T. H. (1895). Oceanic icthyology, a treatise on the deep-sea and pelagic fishes of the world. U.S. Nat. Mus. Spec. Bull. **2**.

Hardisty, M. W., and Potter, I. C. (eds.) (1971). "The Biology of Lampreys," Vol. 1. Academic Press, New York.

Hardisty, M. W., and Potter, I. C. (eds.) (1972). "The Biology of Lampreys," Vol. 2. Academic Press, New York.

Hubbs, C. L. (1924). The life-cycle and growth of lampreys. *Papers Mich. Acad. Sci.* **4**, 587–603.

Lennon, R. E. (1954). Feeding mechanism of the sea lamprey and its effect on fish hosts. *Fisher. Bull.* **98**, 56, 247–293.

Meyers, G. S. (1944). Two extraordinary new blind nematognath fishes from the Rio Negro, representing a new subfamily of Pygidiidae, with a rearrangement of the genera of the family, and illustrations of some previously described genera and species from Venezuala and Brazil. *Proc. Calif. Acad. Sci.* **23**, 591–602.

Walters, V. (1955). Snake-eels as pseudoparasites of fishes. *Copeia* **2**, 146–147.

CLASSIFICATION OF MESOZOA (THE MESOZOANS

PHYLUM MESOZOA

Body small and ciliated, composed of two layers of cells; without digestive, circulatory, nervous, and excretory systems; endoparasites of marine invertebrates; with complex life cycles.

CLASS DICYEMIDA

Body composed of 20 to 30 cells; with long central axial cell surrounded by single layer of ciliated cells; anterior cells modified for attachment; parasitic in nephridial cavity or attached to renal appendages of vena cava of squids, cuttlefish, and octopods. (Genera mentioned in text; Family Dicyemidae—*Dicyema, Pseudicyema, Dicyemnenea.*)

CLASS ORTHONECTIDA

Dioecious adult unattached within turbellarians, rhynchocoels, polychaetes, bivalve molluscs, brittle stars, annelids, and other marine invertebrates; wormlike body consists of outer layer of ciliated cells enveloping internal mass of gametes. (Genera mentioned in text: Family Rhopaluridae—*Rhopalura, Stoecharthrum.*)

CLASSIFICATION OF PORIFERA (THE SPONGES)*
PHYLUM PORIFERA

Mostly sessile animals with no anterior end; primitively radial, but most species irregular; without mouth and digestive cavity; mostly marine, few in fresh water.

CLASS CALCAREA (= CALCISPONGIATE)
CLASS HEXACTINELLIDA (= HYALOSPONGIAE)
CLASS DEMOSPONGIAE

Skeleton of siliceous spicules or spongin fibers, or combination of both; leuconoid body architecture. (Genus mentioned in text: Family Clionidae—*Cliona*.)

CLASS SCEROSPONGIAE

CLASSIFICATION OF COELENTERATA (THE JELLYFISH AND RELATIVES)**
PHYLUM COELENTERATA

Free swimming or sessile; solitary or colonial; with tentacles surrounding mouth; with specialized cells bearing stinging nematocysts; mostly marine, few in fresh water.

CLASS HYDROZOA

Most species with both polyp and medusa stages in life cycles; mesoglea noncellular; gastrovascular cavity simple, without partitions; gonads derived from epidermis; velum of medusae extends inward from margin of bell, obscuring portion of subumbrellar surface.

ORDER TRACHYLINA
ORDER HYDROIDA

With well-developed polyp generation; medusa present or absent. (Genera mentioned in text: *Craspedocusta, Cunina, Polypodium, Zanclea, Hydrichthys, Eugymnanthea*.)

ORDER ACTINULIDA
ORDER SIPHONOPHORA
ORDER HYDROCORALLINA
CLASS SCYPHOZOA
CLASS ANTHOZOA

Solitary or colonial polypoid coelenterates; medusa absent; mouth leads into tubular pharynx that extends more than half way into gastrovascular cavity; gastrovascular cavity divided by longitudinal mesenteries, or septa, into radiating compartments; gonads gastrodermal; fibrous mesoglea contains cells; nematocysts without operculum (lid).

Subclass Octogorallia (= Alcyonaria)

ORDER STOLONIFERA
ORDER TELESTACEA
ORDER ALCYONACEA
ORDER COENOTHECALIA
ORDER GORGONACEA
ORDER PENNATULACEA

Subclass Zoantharia

Polyps with more than eight tentacles; solitary or colonial.

ORDER ZOANTHIDEA
ORDER ACTINIARIA

Solitary anthozoans with no skeleton; mesenteries in hexamerous cycles, usually with two siphonoglyphs. (Genera mentioned in text: *Edwardsia, Peachia*.)

ORDER SCLERACTINIA (= MADREPORARIA)
ORDER RUGOSA (= TETRACORALLA)
ORDER CORALLIMORPHARIA
ORDER CERIANTHARIA
ORDER ANTIPATHARIA

Subclass Tabulata

CLASSIFICATION OF CTENOPHORA (THE COMB JELLIES)*

PHYLUM CTENOPHORA

Body biradiate; with two tentacles (absent in some) and eight longitudinal rows of ciliary combs; all marine.

CLASS TENTACULATA

With tentacles

ORDER CYDIPPIDA

Body round or oval; tentacles usually branched and retractable into sheath (or pouch). (Genus mentioned in text: *Gastrodes*.)

ORDER LOBATA
ORDER CESTIDA
ORDER PLATYCTENEA

CLASS NUDA

Without tentacles

ORDER BEROIDA

CLASSIFICATION OF NEMATOMORPHA (THE HORSEHAIR WORMS)

PHYLUM NEMATOMORPHA

Body extremely long and threadlike; adults free living in damp soil, fresh water, few marine.

CLASS GORDIOIDEA

With ventral epidermal cord only; ventral nerve distinct from ventral cord; ventral nerve connected to ventral cord by nervous lamella; pseudocoel filled with mesenchymal cells; gonads paired; lateral ovarian diverticula in sexually mature females.

Family Chordodidae

With conspicuous cuticular areoles. (Genera mentioned in text: *Neochordodes, Pseudochordodes, Chordodes, Paragordius, Parachordodes, Gordionus*.)

Family Gordiidae

With inconspicuous cuticular areoles or without areoles. (Genus mentioned in text: *Gordius*.)

CLASS NECTONEMATOIDEA

With hollow pseudocoel that extends along length of body; ventral nerve cord situated within ventral epidermal thickening (ventral cord); dorsal cord also present; gonad single.

Family Nectonematidae

With characteristics of class. (Genus mentioned in text: *Nectonema*.)

---

*Only classes including symbiotic species are characterized.

**Only taxa including parasitic species are characterized.

---

*Only groups including parasitic species are characterized.

CLASSIFICATION OF ROTIFERA (THE ROTIFERS)*

PHYLUM ROTIFERA
Anterior end with ciliated crown (**corona**; in free-living species); posterior end tapering to foot (or disc); pharynx containing movable cuticular pieces (**mastax**); mostly in fresh water, some marine, others in mosses.

CLASS DIGONONTA
With two ovaries; mastax adapted for grinding; one pair of trophi large and flattened, others greatly reduced.

ORDER SEISONIDEA
Elongate body with reduced corona. (Genus mentioned in text: *Seison*.)

ORDER BDELLOIDEA
Anterior end retractile, usually with two trochal discs; cylindrical body telescopic (Genus mentioned in text: *Zelinkiella*.)

CLASS MONOGONTA
With one ovary.

ORDER FLOSCULARIACEA
ORDER COLLOTHECACEA
ORDER PLOIMA
Body with or without lorica; often short, sometimes saclike. (Genera mentioned in text: *Proales*, *Embata*, *Pleurotrocha*.)

CLASSIFICATION OF RHYNCHOCOELA (THE RIBBON WORMS)

PHYLUM RHYNCHOCOELA (=NEMERTINA)
Body long and dorsoventrally flattened; with complex proboscis apparatus; digestive tract complete; body surface covered with ciliated epithelium; mostly marine, some terrestrial and in fresh water.

CLASS ANOPLA
ORDER PALEONEMERTEA
ORDER HETERONEMERTEA
CLASS ENOPLA
Mouth situated anterior to brain; nerve cord within body wall musculature.

ORDER HOPLONEMERTEA
Armed with piercing structure (absent or reduced in some); intestine with lateral diverticula; with dorsal blood vessel. (Genera mentioned in text: *Gononemertes*, *Nemertopsis*, *Carcinonemertes*.)

ORDER BDELLONEMERTEA
Proboscis and esophagus empty into common chamber; proboscis unarmed. (Genus mentioned in text: *Malacobdella*.)

CLASSIFICATION OF ANNELIDA (THE SEGMENTED WORMS)*

PHYLUM ANNELIDA
Body metameric (segmented); with true coelom; with large longitudinal ventral nerve; marine, freshwater, and terrestrial.

---

*Only taxa including symbiotic species are characterized.

CLASS ARCHIANNELIDA**
CLASS OLIGOCHAETA
Freshwater and terrestrial species; metamerism well developed; parapodia absent; prostomium as small rounded lobe or small cone without sensory appendages.

ORDER LUMBRICULIDA
ORDER TUBIFICIDA
Setal bundles usually with two or more setae (rarely absent); setae often hairlike; one pair of testes followed by one pair of ovaries in adjacent segments; male genital pores in segments immediately in front or behind segments containing testes; clitellum one cell thick.

Family Tubificidae
Family Naididae
Very small, predominantly freshwater species; some with elongate proboscis; asexual reproduction common. (Genera mentioned in text: *Schmardaella*, *Dero*.)
Family Branchrobellidae
Small, highly modified worms with sucker at anterior end; with toothed jaw, with weak body wall; epiphoretic or endoparasitic in invertebrates, especially crustaceans. (Genus mentioned in text: *Cambarinicola*.)
Family Phreodrilidae
Family Opistocystidae
Family Dorydrilidae
Family Enchytraeidae
ORDER HAPLOTAXIDA
Family Glossoscolecidae
Family Lumbricidae
Family Ocnerodrilidae
Family Megascolecidae
Family Eudrilidae
CLASS POLYCHAETA
Marine annelids; well-developed metamerism; with each segment bearing a pair of lateral parapodia; prostomium usually with eyes, antennae, and pair of palps; terminal nonsegmented region (**pygidium**) includes anus.

Subclass Errantia
Family Aphroditidae
Family Polynoidae
Family Sigalionidae
Family Phyllodocidae
Family Amphinomidae
Family Pisionidae
Family Alciopidae
Family Tomopteridae
Family Hesionidae
Family Syllidae
Family Nereidae (Genus mentioned in text: *Nereis*.)
Family Nephtyidae
Family Clyceridae
Family Histriobdellidae
Body elongate; proboscis armed with at least two-piece stylet; ectoparasitic.
Family Myzostomidae
Body greatly flattened; commensals and parasites of echinoderms (Genus mentioned in text: *Myzostomum*.)

** The annelid class Archiannelida is now recognized to include an assortment of enigmatic species. It now includes five small families none of which include parasitic species.

Family Icthyotomidae

Body dorsoventrally flattened, elongate; head lacks usually appendages; with two-piece crossed stylet; perioral zone modified as pseudosucker. (Genera mentioned in text: *Ichthyotomus, Parasitosyllis.*)

Subclass Sedentaria

With most of following characteristics: body commonly with regional differentiation; parapodia reduced; without compound setae; prostomium without sensory appendages; head commonly with palps, tentacles, and feeding structures; without teeth or jaws.

> *Family Orbiniidae*
> *Family Spionidae*
> *Family Magelonidae*
> *Family Chaetopteridae*
> *Family Cirratulidae*
> *Family Flabelligeridae*
> *Family Opheliidae*
> *Family Capitellidae*
> *Family Arenicolidae*
> *Family Maldanidae*
> *Family Oweniidae*
> *Family Sabellariidae*
> *Family Pectinariidae*
> *Family Ampharetidae*
> *Family Terebellidae*
> *Family Sabellidae*

Form noncalcareous tubes. (Genus mentioned in text: *Manayunkia.*)

> *Family Serpulidae*

CLASS HIRUDINEA

Commonly in fresh water, some marine and terrestrial; body commonly slender and leaf-shaped; without setae (some exceptions); with large sucker at posterior end and small sucker (or pseudosucker) at anterior end; external body divisions (**annuli**) do not correspond with 34 internal true segments.

ORDER ACANTHOBDELLIDA

Primitive order containing single northern European species parasitic on salmonid fish; with setae on anterior segments; compartmented coelom in five anterior segments.

FAMILY ACANTHOBDELLIDAE

(Same characteristics as order.) (Genus mentioned in text: *Acanthobdella.*)

ORDER RHYNCHOBDELLIDA

Mouth as small pore on anterior sucker through which pharyngeal proboscis can be protruded; without jaws or denticles; blood colorless; circulatory system separate from coelomic sinuses; aquatic.

Family Glossiphonidae

Body flattened, typically with three annuli per segment in midregion of body; ectoparasites of invertebrates and vertebrates. (Genera mentioned in text: *Glossiphonia, Placobdella, Theromyzon, Helobdella, Haementeria, Ostreobdella, Erpobdella.*)

Family Piscicolidae

Body subcylindrical, often with lateral gills; usually with more than three annuli per segment; parasites of marine and freshwater fish, rarely on crustaceans. (Genera mentioned in text: *Piscicola, Piscicolaria, Illinobdella, Pontobdella, Ozobranchus, Hemibdella.*)

ORDER GNATHOBDELLIDA

Mouth large, opening from behind into entire sucker cavity; with noneversible pharynx that acts as suction bulb; with three pairs of jaws; with five annuli per segment; five parts of eyes arranged on arch in somites II to VI; testes large, arranged in metameric pairs; aquatic or terrestrial.

Family Hirudinidae

Primarily amphibious or aquatic bloodsuckers. (Genera mentioned in text: *Hirudo, Haemopis, Macrobdella, Hirudinaria, Limnatis, Dinobdella.*)

Family Haemadipsidae

Terrestrial tropical leeches of Australasian region; attacks primarily warm-blooded vertebrates. (Genus mentioned in text: *Haemadipsa*).

ORDER PHARYNGOBDELLA

Mouth large, opening from behind into entire sucker cavity; pharynx not protrusible, acts as suction pump, extending to somite XIII; three or four pairs of eyes and three muscular pharyngeal ridges; without true jaws or denticles; testes in grapelike bunches; blood red; primarily aquatic, some semiterrestrial.

Family Erpobdellidae

Nearly predacious; with characteristics of order. (Genera mentioned in text: *Erpobdella, Dina, Mooreobdella, Nephelopsis.*)

CLASSIFICATION OF MOLLUSCA (THE SNAILS AND RELATIVES)*

PHYLUM MOLLUSCA

Ventral surface of most species modified as muscular foot of various shapes; dorsal and lateral body surfaces modified as shell-secreting mantle, shell reduced or absent in some species; marine, freshwater, and terrestrial.

CLASS MONOPLACOPHORA
CLASS POLYPLACOPHORA
CLASS APLACOPHORA
CLASS GASTROPODA

Subclass Prosobranchia

Mantle cavity and enclosed organs located anteriorly; aquatic species with one or two gills in mantle cavity; shell usually present; operculum usually present; mostly dioecious; marine, freshwater, and terrestrial.

ORDER ARCHEOGASTROPODA (= DIOTOCARDIA)

Usually with two bipectinate gills, two auricles, and two nephridia; right gill may be reduced or absent.

Family Capulidae

Shell slightly coiled or as simple incurved cone; foot greatly reduced; peribuccal region enlarged as large sucker; proboscis projects from sucker; radula lacking; digestive gland and intestine reduced; with salivary glands; dioecious. (Genera mentioned in text: *Thyca, Platyceras.*)

ORDER MESOGASTROPODA (= TAENIOGLOSSA)

With single monopectinate gill, one auricle, and one nephridium; reproductive system complex, usually

---

*Only taxa including parasitic species are characterized.

with a penis; radula (in free-living species only) composed of seven teeth in transverse row; mostly marine but also some freshwater and terrestrial species.

Family Eulimidae

With thin and translucent turreted shell; radula absent in parasitic species; parasitic species as ecto- and endoparasites of echinoderms. (Genera mentioned in text: *Pelseneeria, Megadenus, Peasistilifer, Stylifer*[†], *Diacolax, Gasterosiphon.*)

Family Paedophorodidae

Highly modified snails; without a shell; pseudopallium absent; with visceral sac containing nervous system, kidney, and ovary or testis; both sexes with long proboscis. (Genus mentioned in text: *Paedophoropus.*)

ORDER NEOGASTROPODA (= STENOGLOSSA)

Subclass Opisthobranchia

With one gill, one auricle, and one nephridium but display detorsion; shell and mantle cavity commonly reduced or absent; secondarily bilaterally symmetrical; hermaphroditic; mostly marine.

ORDER PYRAMIDELLACEA

With shell and operculum; radula replaced by stylet.

Family Pyramidellidae

Shell and operculum present; proboscis with stylet instead of radula; ectoparasites of bivalve molluscs and polychaetes.

ORDER CEPHALASPIDEA
ORDER ACOCHLIDIACEA
ORDER PHILINOGLOSSACEA
ORDER ANASPIDEA (= APLYSIACEA)
ORDER NOTASPIDEA
ORDER SACOGLOSSA
ORDER THECOSOMATA
ORDER GYMNOSOMATA
ORDER NUDIBRANCHIA
ORDER PARASITA

Wormlike body; without shell; endoparasites of echinoderms, especially holothurians.

Family Entoconchidae

Highly modified snails; without a shell; pseudopallium forms body wall; males as hyperparasites of females in some. (Genera mentioned in text: *Entocolax, Entoconcha, Enteroxenos, Thyonicola.*)

Subclass Pulmonata

SUPERORDER SYSTELLOMMATOPHORA
SUPERORDER BASOMMATOPHORA
SUPERORDER STYLOMMATOPHORA
CLASS BIVALVIA (= PELECYPODA)
Subclass Palaeotaxodonta
ORDER NUCULOIDA
Subclass Cryptodonta
ORDER SOLEMYOIDA
Subclass Pteriomorphia
ORDER ARCOIDA

ORDER MYTILOIDA

Subclass Palaeoheterodonta

Valves equal, with few hinged teeth; elongate lateral teeth, if present, not separated from large cardial teeth. cardial teeth.

ORDER UNIONOIDA

Freshwater bivalves.

Family Unionidae

Freshwater bivalves with more or less smooth shells; larvae parasitic. (Genera mentioned in text; *Unio, Anodonta, Proptera, Margaritana, Lampsilis.*)

ORDER TRIGONIOIDA

Subclass Heterodonta

ORDER VENEROIDA
ORDER MYOIDA
ORDER HIPPURITOIDA

Subclass Anomalodesmata

ORDER PHOLADOMYOIDA
CLASS CEPHALOPODA

Subclass Nautiloidea
Subclass Ammonoidea
Subclass Coleoidea

CLASSIFICATION OF CRUSTACEA (THE CRUSTACEANS)*

PHYLUM ARTHROPODA (See p. 572)
Subphylum Crustacea

With two pairs of antennae; with one pair of mandibles; trunk variable.

CLASS CEPHALOCARIDA
CLASS BRANCHIOPODA
CLASS OSTRACODA
CLASS MYSTACOCARIDA
CLASS COPEPODA

ORDER CALANOIDA
ORDER HARPACTICOIDA

Most free living but with occasional species parasitic on crustaceans, whales, and marine fish; both adult males and females parasitic in these species; females with only one egg sac; metasome-urosome articulation between fourth and fifth thoracic segments; antennules very short and second antennae biramous; abdomen almost as wide as thorax; body often linear in shape; approximately 20 parasitic species known. (Genera mentioned in text: *Sacodiscus, Balaenophilus, Neoscutellidium, Stellicola.*)

ORDER CYCLOPOIDA

Parasitic on fish and in a few invertebrates; usually only adult females parasitic, but parasitic males known, for example, *Bomolochus*; body less than 3 mm long and only slightly flattened dorsoventrally; abdominal segments not fused; articulation present between fifth and sixth thoracic segments; fifth metasomal legs and often sixth metasomal legs present; metasome wider than urosome; first antennae with 17 segments or less; approximately 1200 species known. (Genera mentioned in text: *Cyclops, Ergasilus, Mytilicola, Lernaea, Myzomolgus, Catinia, Chondracanthus.*)

ORDER NOTODELPHYOIDA

Symbiotic in tunicates; body not highly modified; prosome-urosome articulation between fourth and fifth

---

[†]Considered by some to represent a separate family, Styliferidae.

* Only taxa including parasitic species are characterized.

postcephalic segments in males, between first and second abdominal segments in females. (Genera mentioned in text: *Ive, Siphonobius*.)

ORDER MONSTRILLOIDA

Mostly parasitic in or on invertebrates; parasitic during larval stages only; free swimming as adults; mouthparts and second antennae lacking in adults that are nonfeeders; approximately 40 species known. (Genera mentioned in text: *Monstrilla, Haemocera*.)

ORDER CALIGOIDA

Parasitic on gills and body surfaces of fish, on aquatic mammals, and rarely on cephalopod molluscs; adults elongate, wormlike; usually more females than males are found as permanent parasites; body 7–15 mm in length, fused, not capable of articulation; body segmentation not visible in some species; approximately 450 species known. (Genera *Lepeophtheirus, Caligus, Dysgamus, Achtheinus, Teredicola, Pennella, Lernaeenicus*.)

ORDER LERNAEOPODOIDA

Parasitic on gills and fins of fish; body 3–8 mm long, fused, not capable of articulation, and not wormlike; cephalothorax not broad and flat; legs lacking in freshwater species; males rarely found, seen as minute individuals clinging to females; approximately 360 species known. (Genera mentioned in text: *Achtheres, Salmincola, Brachiella, Lernaeosolea, Kroyerina, Paeonodes*.)

CLASS BRANCHIURA

With pair of sessile compound eyes; with large shieldlike carapace covering head and thorax; bloodsucking ectoparasites on skin or gill cavity of freshwater and marine fish, and on some amphibians. (Genus mentioned in text: *Argulus*.)

CLASS CIRRIPEDIA

Free living, commonly known as barnacles; all sessile (except parasitic species); marine; symbiotic species (especially parasitic species) highly modified (see text.)

ORDER THORACICA

Free living and commensal barnacles; with six pairs of well-developed cirri; mantle usually covered with calcareous plates.

ORDER ACROTHORACICA

ORDER ASCOTHORACICA

Marine; ecto- or endoparasites of coelenterates or echinoderms; not attached at preoral region; mantle lacks calcareous plates; body bivalved or saclike; cirri numbering six or fewer, usually rudimentary; abdomen present; mouthparts modified for piercing-sucking; antennules often modified for grasping; cement glands lacking; digestive and reproductive systems possess diverticula extending into mantle; usually dioecious, males smaller than females; nauplii lack frontal horns; approximately 15 species known. (Genus mentioned in text: *Baccalaureus*.)

ORDER RHIZOCEPHALA

Primarily parasites of other crustaceans, principally decapods (shrimps, crabs, etc.); adult body greatly modified as thin-walled sac enclosing visceral mass, primarily of gonads; hermaphroditic except Seylonidae; body segmentation not apparent; lacks cirri, sense organs, or alimentary tract; rootlike system extends into interior of host in all directions through which nourishment from host fluids are absorbed; fertilized eggs develop in brood chamber; nauplii escape through aperture at summit of sac, which also permits entrance of oxygenated water; attached to host by short stalk at

base of saccular body; approximately 200 species known. (Genera mentioned in text: *Sacculina, Peltogaster, Peltogasterella, Loxothylacus, Thompsonia*.)

ORDER APODA

Attached by preoral region; mantle lacking; body segmented; abdomen present; cirri lacking; mouth suctorial; alimentary canal incomplete and lacking mid- and hindguts; antennules and cement glands present; hermaphroditic; metamorphosis not known; only one known species.

CLASS MALACOSTRACA

Trunk typically composed of 14 segments and telson (first 8 form thorax, last six form abdomen); all segments bear appendages; first antennae often biramous; mandible usually bears a palp; triturating surface of mandible divided into grinding molar process and cutting incisor; female genital pores on sixth thoracic segment, male genital pores on eighth.

ORDER LEPTOSTRACA
ORDER ANASPIDACEA
ORDER BATHYNELLACEA
ORDER STOMATOPODA
ORDER EUPHAUSIACEA
ORDER DECAPODA

First three pairs of thoracic appendages modified as maxillipeds; remaining five pairs of thoracic appendages are legs (first pair heavier, known as **chelipeds**); head and thoracic segments fused dorsally; sides of overhanging carapace enclose gills in branchial chambers. (Genus mentioned in text: *Pinnotheres*.)

ORDER MYSIDACEA
ORDER CUMACEA
ORDER TANAIDACEA
ORDER ISOPODA

Most species marine, although many fresh water and terrestrial; body dorsoventrally flattened; head usually shield shaped; terga of thoracic and abdominal segments tend to project laterally; without carapace; first one or two thoracic segments fused with head.

Suborder Gnathiidea

First and seventh thoracic segments reduced so that only five large segments visible dorsally; eighth thoracic appendages absent; abdomen small and much narrower than thorax; postlarval manca stage ectoparasitic on marine fish. (Genera mentioned in text: *Anceus, Pranizia, Gnathia*.)

Suborder Anthuridea

Suborder Flabellifera

Body more or less flattened; some abdominal segments may be fused together; last abdominal segment fused with telson; uropods fanshaped, fused with telson to form tail fan; coxae expanded to form coxal plates (may be fused with body); mostly marine, few in fresh water. (Genera mentioned in text: Family Cymothoidae— *Ichthyoxenos, Nerocila, Anicora, Cymothoa, Livoneca, Anilocra, Aega*.)

Suborder Valvifera

Suborder Asellota

Suborder Phreatoicidea

Suborder Epicaridea

With suctorial mouthparts and piercing mandibles; both pairs of maxillae vestigial or absent; females commonly greatly modified, some without segmentation or appendages; parasitic on crustaceans. (Genera mentioned in text: *Cepon, Portunion, Ione, Aspidophryxus, Dajus, Ancyroniscus.*)

Suborder Oniscoidea

ORDER AMPHIPODA

Eyes sessile; without carapace; first (sometimes second) thoracic segment fused with head; without exopodites on thoracic legs; body tends to be laterally compressed; anterior three pairs of abdominal limbs are pleopods, posterior three pairs uropods; gills and heart in thoracic region; abdomen absent in whale lice (Cyamidae); most species marine, but many freshwater and some terrestrial.

Suborder Gammaroidea

Suborder Hyperiidea

Head not fused with second thoracic segment; maxillipeds lack palp; thoracic coxae often small or fused with body; abdomen and pleopods powerfully developed; last two abdominal segments fused; eyes large; body generally transparent. (Genus mentioned in text: *Phronima.*)

Suborder Caprellidea

Head partially fused with second thoracic segment; maxillipeds with palp; thoracic coxae vestigial or absent; abdominal segments much reduced, with vestigial appendages; eyes small; body elongate, cylindrical, or short and flattened; all marine. (Genera mentioned in text: Family Cyamidae—*Cyamus, Synayamus, Platycyamus, Paracyamus, Isocyamus.*)

CLASSIFICATION OF PYCNOGONIDA (THE SEA SPIDERS)

PHYLUM ARTHOPODA (See p. 572)

Subphylum Chelicerata

Without antennae; with one pair of chelicerae; body divided into cephalothorax and abdomen.

CLASS PYCNOGONIDA

Spiderlike; with long legs; with multiple gonopores; legs ovigerous; trunk segmented; with additional pair of walking legs in many species (four pairs); all marine.

Family Nymphonidae

Chelifers present; palpi five-jointed; ten-jointed ovigers present on both males and females; family includes one polymerous (i.e., with many thoracic segments) genus: *Pentanymphon.* (Genus mentioned in text: *Nymphon.*)

Family Callipallenidae

Chelifers present; palpi lacking or reduced; ten-jointed ovigers present on both males and females.

Family Phoxichilidiidae

Chelifers present; palpi lacking; five- to nine-jointed ovigers present on males only. (Genera mentioned in text: *Anoplodactylus, Phoxichilidium.*)

Family Endeidae

Chelifers and palpi absent; seven-jointed ovigers present on males only; includes only one genus: *Endeis.*

Family Ammotheidae

Chelifers usually small and achelate; palpi well developed and comprised of four to ten joints; nine- to ten-jointed ovigers present on both males and females. (Genus mentioned in text: *Ammothea.*)

Family Austrodecidae

Chelifers absent; palpi composed of five to six joints; ovigers of four to seven joints present on both males and females; includes only one genus: *Austrodeus.*

Family Colossendeidae

Chelifers absent except on a few polymerous species; palpi long and composed of nine to ten joints; ovigers ten-jointed and present on both males and females; includes mostly deepwater forms with large proboscides included in four genera: *Pentacolossendeis, Colossendeis, Decolopoda,* and *Dodecolopoda.*

CLASSIFICATION OF TARDIGRADA (THE WATER BEARS)

PHYLUM TARDIGRADA

Body microscopic and segmented; short cylindrical body with four pairs of stubby legs terminating in four to eight claws; body covered by either smooth or ornate cuticle; mostly in damp terrestrial habitats, although freshwater and marine species are known. (Genera mentioned in text: *Echiniscoides, Halechiniscus, Actinarctus, Tetrakentron, Echiniscus.*)

CLASSIFICATION OF PENTASTOMIDA (THE TONGUE WORMS)

PHYLUM PENTASTOMIDA (=LINGUATULIDA)

Body with five anterior protruberances, four being leglike (with claws), two on each side; fifth projection is an anterior, median, snoutlike process bearing mouth; legs in some species reduced to claws; body covered with chitinous cuticle (molts occur during development); digestive tract relatively simple straight tube, with anterior end modified to pump blood from host.

CLASS CEPHALOBAENIDA

Hooks located on parapodia; genital pore in both sexes located at anteroventral margin of abdomen; larvae with six legs.

Family Cephalobaenidae

Parasitic in lungs of snakes, toads, and lizards. (Genus mentioned in text: *Raillietiella.*)

Family Reighardiidae

Parasitic in air sacs of birds. (Genus mentioned in text: *Reighardia.*)

CLASS POROCEPHALIDA

Hooks not located on parapodia but arranged in straight or curved line; genital pore in female located at posteroventral margin of abdomen; larvae with four legs.

Family Porocephalidae

Body cylindrical; parasitic in lungs of reptiles. (Genera mentioned in text: *Armillifer, Cuberia, Gigliolella, Ligamifer, Kiricephalus, Porocephalus.*)

Family Linguatulidae

Body flattened; usually parasitic in nasal passages of canines and felines. (Genus mentioned in text: *Linguatula.*)

Family Sebekidae

Body small, straight, and completely segmented; parasitic in crocodilia. (Genera mentioned in text: *Sebekia, Diesingia, Alofia, Leiperia.*)

Family Sambonidae

Body cylindrical; somewhat dilated posterior end; parasites of lacertilians. (Genera mentioned in text: *Sambonia, Elenia.*)

CLASSIFICATION OF ECHINODERMATA (THE STARFISH AND RELATIVES)

PHYLUM ECHINODERMATA

Body secondarily pentamerous and radial; body wall contains calcareous ossicles usually bearing projecting spines; portion of coelom modified into water-vascular system; all marine.

CLASS CRINOIDEA
CLASS STELLEROIDEA
CLASS ECHINOIDEA
CLASS HOLOTHUROIDEA

Body not drawn out into arms, cucumber shaped; mouth and arms at opposite poles; ambulacral and interambulacral areas arranged meridianally around greatly lengthened polar axis; skeleton reduced to microscopic ossicles; buccal podia modified as ring of tentacles around mouth (in most species). (Genus mentioned in text: *Rhynkatorpa.*)

CLASSIFICATION OF VERTEBRATA (THE VERTEBRATES)

It is beyond the scope of this volume to present even an abbreviated classification of the subphylum Vertebrata of the phylum Chordata. All of the vertebrates that can be considered as parasites belong to the following classes:

CLASS CYCLOSTOMATA (lampreys, hagfish, and eels). The genera mentioned in the text are: *Petromyzon, Entosphenus, Myxine, Bdellostoma, Polistotrema, Pisododonophis, Ophichthus,* and *Myrichthys.*

CLASS OSTEICHTHYES (bony fish). The genera menoned in the text are: *Vandellia* and *Cetopsis.*

ORDER CHIROPTERA (bats) (of the class Mammalia). The genus mentioned in the text is *Desmodus.*

Index

Page numbers in boldface indicate illustrations. Page numbers followed by the letter "t" indicate tabular information.

1893   T. Smith and F. L. Kolbourne discovered that the cattle tick *Boophilus annulatus* is the vector for *Babesia bigemina*, the causative agent of Texas cattle fever. This finding revealed the important role of arthropods as transmitters of pathogens.

1895   D. Bruce discovered that the tsetse fly served as the vector for *Trypanosoma brucei*.

1897   Ronald Ross, an English army doctor in India, found the zygotes of the malaria parasite in anopheline mosquitoes. He was awarded the Nobel Prize in 1902.

1898   P. L. Simond succeeded in transmitting plague from rat to rat via a flea vector.

1900   W. Reed, J. Carroll, J. W. Lazear, and A. Agramonte, while in Cuba, demonstrated that the mosquito *Aedes aegypti* is the vector for yellow fever. Carlos Finlay deserves equal credit for this finding since it was his hypothesis that led to the experimental demonstration by Reed and his co-workers.

1901   R. M. Forde, an Irishman working in West Africa, discovered *Trypanosoma brucei gambiense*, the etiologic agent of Gambian sleeping sickness.

1901   E. Weinland, a German, demonstrated that the metabolism of intestinal worms involves the fermentation of carbohydrates, thus pioneering in the area of helminth biochemistry.

1902   H. Graham, an American pathologist, described dengue in Beirut and demonstrated that the mosquito *Aedes aegypti* is the vector.

1903   E. Marchoux and A. T. Salimben, at the Pasteur Institute in Paris, proved that the fowl spirochetosis-causing organism, *Spirochaeta gallinarum*, is transmitted by the fowl tick, *Argas persicus*.

1903   D. Bruce and D. N. Nabarro demonstrated that the tsetse fly, *Glossina palpalis*, is the vector of *Trypanosoma brucei gambiense*.

1904   A. Looss accidentally infected himself with the larvae of the hookworm *Ancylostoma duodenale* while in Cairo, thus establishing the mode of infection of hookworms.

1906   H. T. Ricketts, an American physician, showed that a tick, *Dermacentor andersoni*, is the transmitter of the etiologic agent for Rocky Mountain spotted fever.

1909   C. Chagas proved that the cone-nosed bug, *Triatoma megista*, is the vector for *Trypanosoma cruzi*, the Chagas' disease-causing flagellate.

1909   C. Nicolle, C. Comte, and E. Conseil in Tunis, and H. T. Ricketts and R. M. Wilder in Mexico proved that the louse
−1910 *Pediculus humanus* is the vector for the typhus-causing rickettsia.

1910   H. B. Fantham described *Trypanosoma brucei rhodesiense*, the causative agent of Rhodesian sleeping sickness, and Kinghorn and York (1912) proved that *Glossina morsitans* is the vector.

1914   K. Miyairi and M. Suzuki in Japan worked out the life cycle of *Schistosoma japonicum*.

1924   (1932), W. H. Taliaferro demonstrated the presence of ablastin and trypanocidal antibodies in *Trypanosoma lewisi* infections in rats, and created great interest in parasite immunology.

1925   D. Keilin at Cambridge, England, elucidated the electron cascade (cytochrome) system by employing parasitic insects and worms.

1928   N. Stoll reported "self-cure" in sheep infected with the nematode *Haemonchus contortus*. This finding stimulated much interest in immunity to metazoan parasites.

1934   C. J. Wesenberg-Lund, a Danish zoologist, reported that echinostome rediae will prey on other rediae and sporocysts within the same snail host. This observation served as the basis for the possible biological control of human-infecting schistosomes in Southeast Asia during the 1960s and 1970s.

1950   E. Bueding elucidated the pathways involved in carbohydrate metabolism of *Schistosoma mansoni*. This report paved the way for the development of a rational approach to chemotherapy.

1953   **J. D. Watson and F. H. C. Crick reported the double helical nature of DNA.**

1956   B. von Bonsdorff, and later W. Nyberg, demonstrated that the pernicious anemia sometimes associated with *Diphyllobothrium latum* infection was due to the high affinity this cestode has for vitamin B_{12}, which it robs from the host. This finding is a classical example of chemical pathology.

1958   W. P. Rogers in Australia elucidated much of the mechanism responsible for the hatching of nematode eggs, that of *Ascaris lumbricoides*.

1958   A. H. Rothman elucidated the role of host bile salts on the development and metabolism of helminths.